McGraw-Hill Series in Water Resources and Environmental Engineering

Consulting Editors

Paul H. King
Rolf Eliassen, Emeritus

Also Available from McGraw-Hill

Schaum's Outline Series in Civil Engineering

Most outlines include basic theory, definitions and hundreds of example problems solved in step-by-step detail, and supplementary problems with answers.

Related titles on the current list include:

Descriptive Geometry
Dynamic Structural Analysis
Engineering Economics
Engineering Mechanics
Fluid Dynamics
Fluid Mechanics & Hydraulics
Introductory Surveying
Mathematical Handbook of Formulas & Tables
Mechanical Vibrations
Reinforced Concrete Design
Statics & Mechanics of Materials
Strength of Materials
Structural Analysis
Structural Steel Design, (LRFD Method)
Theoretical Mechanics

Schaum's Solved Problems Books

Each title in this series is a complete and expert source of solved problems with solutions worked out in step-by-step detail.

Related titles on the current list include:

3000 Solved Problems in Calculus
2500 Solved Problems in Differential Equations
2500 Solved Problems in Fluid Mechanics & Hydraulics
3000 Solved Problems in Linear Algebra
2000 Solved Problems in Numerical Analysis
700 Solved Problems in Vector Mechanics for Engineers: Dynamics
800 Solved Problems in Vector Mechanics for Engineers: Statics

Available at most college bookstores, or for a complete list of titles and prices, write to:
Schaum Division
McGraw-Hill, Inc.
Princeton Road, S–1
Hightstown, NJ 08520

HAZARDOUS WASTE MANAGEMENT

Michael D. LaGrega

Bucknell University
The Environmental Resources Management Group

Phillip L. Buckingham

The Environmental Resources Management Group

Jeffrey C. Evans

Bucknell University

McGraw-Hill, Inc.

New York   St. Louis   San Francisco   Auckland   Bogotá   Caracas
Lisbon   London   Madrid   Mexico City   Milan   Montreal
New Delhi   San Juan   Singapore   Sydney   Tokyo   Toronto

HAZARDOUS WASTE MANAGEMENT
International Editions 1994

Exclusive rights by McGraw-Hill Book Co. - Singapore for manufacture and export. This book cannot be re-exported from the country to which it is consigned by McGraw-Hill.

Information contained in this work has been obtained by McGraw-Hill, Inc., and ERM, from sources believed to be reliable. However, neither McGraw-Hill, ERM, nor their authors guarantee the accuracy or completeness of any information published herein and neither McGraw-Hill, ERM, nor their authors shall be responsible for any errors, omissions, or damages arising out of use of this information. This work is published with the understanding that McGraw-Hill, ERM, and its authors are supplying information but are not attempting to render engineering, legal, accounting, or other professional services. If such services are required, the assistance of an appropriate professional should be sought.

For more information about the Environmental Resources Management Group, call 1-800-544-3117. For more information about other McGraw-Hill materials, call 1-800-2-MCGRAW in the United States. In other countries, call your nearest McGraw-Hill office.

3 4 5 6 7 8 9 0 CMO FC 9 8 7 6

Library of Congress Cataloging-in-Publication Data

LaGrega, Michael D.
    Hazardous waste management / Michael D. LaGrega, Phillip L.
Buckingham, Jeffrey C. Evans.
      p.   cm.
    Includes index.
    ISBN 0-07-019552-8
    1. Hazardous wastes - Management.   I. Buckingham, Phillip L.
II. Evans, Jeffrey C.   III. Title.
TD1030.L34   1994
363.72'876'0973-dc20   94-40044

This book was set in Times Roman by Electronic Technical Publishing Services.
The editors were B. J. Clark and John M. Morriss;
the production supervisor was Paula Keller.
The technical editors at ERM were Karen L. Stein and Paul Nordquist.
The cover was designed by Thomas Hall and Jeffrey Gorham, ERM.
Graphics were provided by the ERM Graphics Department.
Project supervision was done by Eletcronic Technical Publishing Services.

When ordering this title, use ISBN 0-07-113454-9

Printed in Singapore   1000945821

ABOUT THE AUTHORS

Michael D. LaGrega received a Bachelor's degree in Civil Engineering from Manhattan College and a Master's degree and Ph.D. in Environmental Engineering from Syracuse University. Dr. LaGrega is a licensed professional engineer in six states and is board certified in the specialty of hazardous waste management by the American Academy of Environmental Engineers. The author of more than seventy-five publications covering a wide variety of industrial waste topics, Dr. LaGrega's research interests are currently focused in the area of stabilization. He served as Director of Hazardous Waste Planning for the Pennsylvania Department of Environmental Resources in 1983-84. Dr. LaGrega is currently a Principal for the firm of Environmental Resources Management, Inc., and Professor of Civil Engineering at Bucknell University.

Phillip L. Buckingham received a Bachelor's degree in Civil Engineering from North Carolina State University and a Master's degree in Sanitary Engineering from Cornell University. He is a licensed professional engineer in Pennsylvania and Ohio. Mr. Buckingham is a founding Principal of Environmental Resources Management, Inc., and has directed numerous hazardous waste projects including state hazardous waste facilities plans and site remediation and risk assessment studies. He is the author of more than 20 technical publications dealing with a wide range of environmental engineering topics.

Jeffrey C. Evans received a Bachelor's degree in Civil and Environmental Engineering from Clarkson University, a Master's degree in Civil Engineering from Purdue University, and a Ph.D. in Civil Engineering from Lehigh University. Dr. Evans is a licensed professional engineer in Michigan and, prior to joining the academic community, was affiliated with Woodward-Clyde Consultants for more than 10 years. His research interests in the areas of waste containment and environmental geotechnology are reflected in his publications, numbering more than 50. Dr. Evans is currently an Associate Professor of Civil Engineering at Bucknell University and serves as a consultant on a variety of hazardous waste remediation projects.

ABOUT THE ERM GROUP

The **Environmental Resources Management Group (ERM)** was founded by Paul H. Woodruff in 1977. The goals of the founding company continue to be the cornerstone of the entire Group: to provide environmental services of the highest quality on challenging projects; to improve the quality of life for everyone; to operate ERM on a sound business basis; and to have fun doing it.

ERM's management structure is based on one premise–people will perform best when they can control their own destinies and when they are working for their own benefit. It has become almost a cliché to expound the virtues of a management structure that requires creativity and ambition in its employees; however, since its inception, ERM has built its foundation around its people who continue to perform as the company's most valuable asset. An entrepreneurial spirit, combined with in-depth technical expertise, permeates the organization. To really make this idea work, ERM has developed a unique network of member companies whose officers run individual operations and enjoy the benefits of belonging to a large, global organization.

Ownership of ERM is now shared among more than 100 individuals, and that number continues to grow. ERM now employs 2,300 worldwide. In the United States ERM has more than 50 locations with at least one office in each EPA region. Internationally, ERM has locations in Australia, Belgium, Canada, Ireland, Germany, Great Britain, Hong Kong, India, Italy, Mexico, the Netherlands, Portugal, Poland, Spain, Taiwan, and Thailand. Personnel can move from one company to another or from one location to another as their career plans or geographical preferences change over time.

ERM is a technically oriented environmental consulting organization with significant management consulting capabilities. ERM has an experienced staff, supported by state-of-the-art environmental analytical laboratories. The staff includes professionals in the following disciplines: engineering (environmental, civil, chemical, mechanical, geotechnical, construction, and electrical) geology, geophysics, soil science, economics, geography, political science, public administration, hydrology, hydrogeology, biology, toxicology, chemistry, industrial hygiene, regulatory and policy analysis, conflict resolution, business management, computer science, statistics, and environmental planning.

ERM offers services in all environmental areas and technical disciplines: site remediation; management consulting; solid and hazardous waste management; hydrogeology; ground water resources; health, safety, and toxicology; risk assessment, environmental impact assessment, policy and economic analyses, biological investigations; water and wastewater treatment; air quality engineering and management; major hazard assessment; analytical laboratory services; construction and information management services; and drilling services.

ERM's logo is based on the equiangular spiral—a universal form of nature found in a seashell, the curve of the inner ear, and the break of an ocean wave. The spiral backed by a grid represents the combination of science and nature.

To Our Children

CONTENTS

Part II   Current Management Practices

Part III   Treatment and Disposal Methods

Part IV Site Remediation

FOREWORD

As a student and user of this textbook, you are undertaking the mastery of a very important subject. Hazardous waste is one of the most deleterious by-products of the Industrial Revolution.

While there is no doubt that humankind's increasing mastery of natural law has brought us increased life expectancy and a higher quality of life for more people than ever known before, it has also brought the risk of global calamity and impairment of human health. Many of the environmental problems we are dealing with today are an accumulation of 200-plus years of managing the residuals of our industrial output in ways that we now know to be inappropriate. Past practices have resulted in tens of thousands of contaminated sites in the United States and, in proportion to industrial output, elsewhere in the world. The global cost of restoring these sites to minimally acceptable levels will eventually run into the hundreds of billions of dollars. In addition, we are facing the cost of dealing with current and future production and use of products in such a way that they do not impair the environment or human health now or in the future.

Hazardous waste management is truly a multidisciplinary field. You will be called upon to utilize your previous training while at the same time gaining an expanded appreciation for other scientific and nontechnical disciplines. Hazardous waste management is also a multimedia issue because practitioners in this field must recognize and deal with water pollution, air pollution, solid waste, and ground water aspects that affect the environment and human health.

Because the number of problem sites is high, the cost enormous, and the public's patience short, determining what gets cleaned up, to what extent, and who pays for it and how are not solely technical issues. They can and should involve risk assessment as well as technical and economic feasibility. Clearly, education is crucial in providing technical tools for participants in this field—owners, operators, regulators, consultants, contractors, and suppliers. However, in the final analysis, it is the political system, as much as science and technology, that will provide answers to these questions of what, who, and how.

Therefore, education is also crucial for the general public if technology and science are to be appropriately interwoven with the threads of public concern and the economic system in society.

You are encouraged to utilize technology and concepts gained in other fields. We need to continue to make progress in finding more cost-effective tools, both technical and nontechnical, to deal with the management of hazardous waste. This requires the adaptation of technology and concepts gained in other fields. This problem is, after all, a form of societal overhead. This is particularly true with respect to remediating the results of past practices. In our society we have many other serious concerns related to education, housing, health care, and the rebuilding of our infrastructure, to name a few. Each of these areas is suffering from insufficient financial and human resources. Thus, the challenge is to find ways of producing more benefit for less cost.

The very real problem that hazardous waste has created and is creating is compounded by the commingling of technological obstacles; limited financial resources; a general public with an inadequate understanding of basic scientific concepts, including risk assessment; and an extremely complex time- and resource-consuming legal and regulatory process.

My hope is that you will find it worthwhile to study this subject. The future holds substantial challenge and reward for technical and managerial innovation. Finding better ways to do it will help realize competitive advantage in manufacturing and distribution by utilizing technology and management systems that result in higher net yield at a lower environmental cost. Inadequate knowledge and societal habits and attitudes dictated our past hazardous waste management practices. Through education and enlightened self-interest you can help change these attitudes and practices.

Paul H. Woodruff, P.E.
Chairman
The Environmental Resources Management Group

PREFACE

OVERVIEW

The management of hazardous wastes has changed dramatically over the past two decades and will continue to evolve as our knowledge of both the hazards and management methods grows.

This book provides a comprehensive introduction to a complex interdisciplinary field. The authors' objective is to provide a sufficient background to students so that they begin to think about how to approach hazardous waste problems rather than provide simple "cookbook" solutions. To achieve this objective, the book contains a number of unique elements:

- The Landmark Episodes section provides students with an historical understanding of the field of hazardous waste management (Section 1-3).
- Section 1-7 provides an introduction to the important topic of environmental ethics.
- Section 2-1, Environmental Law, puts our complex regulatory system in context and explains how the process works rather than merely providing tables of federal regulations.
- Chapter 4, Fate and Transport, provides a comprehensive treatment of how toxic compounds move in the environment, including ground water and air modeling.
- A discussion of carcinogenesis details cancer-causing chemicals and the uncertainties involved (Section 5-6).
- Section 7-3, Life Cycle Analysis, offers an introduction to the new methods that corporate management uses to minimize hazardous waste.
- Chapter 10 covers the emerging field of bioremediation in depth.
- Detailed case studies give students the necessary equations and design examples to evaluate the effectiveness of different treatment and containment technologies in addressing today's hazardous waste situations.

ORGANIZATION

This book is divided into four major parts:

- Part I provides background material for a complete understanding of hazardous wastes. Graduate students may have encountered much of the material in the first five chapters from other courses. Therefore, this section introduces basics, not comprehensive coverage, on water generation, law, how wastes move through the environment, and their effects on people and the environment.

- Part II examines the methods currently used by management in industry to understand the magnitude of hazardous waste problems and how to avoid many of the problems of the past.

- Part III contains a selection of treatment and disposal methods. Because literally hundreds of such methods exist, the authors chose some of those methods in contemporary practice and a few emerging technologies to permit an in-depth presentation for the processes selected.

- Part IV covers site remediation. This includes the characterization of a site, the assessment of risks that it poses, and the development and selection of remedies.

TO THE INSTRUCTOR

The information in this first edition is sufficient to support a two-semester or three-quarter course. The text is intended for engineering students; however, where possible, we have included background and basic equations to make the text usable by science students as well.

The level of the text is senior undergraduate or first-year graduate students. Draft copies of this text have been used at a number of universities including Bucknell, Penn State (Harrisburg), and Villanova. Students and instructors have been very generous with their comments, and all chapters were rewritten to include requested information and/or topics.

As noted, this book cannot be adequately covered in a single semester so the authors have attempted to provide instructors with a great deal of flexibility in establishing an appropriate syllabus. Chapters have been written to be reasonably self-contained, and numerous references to other sections have been incorporated.

The following outline is for a suggested introductory one-semester course.

Topic	Heading
Fundamentals	
Introduction	Chapter 1
The Regulatory Process	Sections 2-1 through 2-4
Fate and Transport	Sections 4-1, 4-2
Toxicology	Sections 5-1, 5-2, and 5-5
Management Practices	
Environmental Audits	Chapter 6
Waste Minimization	Chapter 7

Treatment Methods
Physico-Chemical Processes	Sections 9-1 through 9-3
Bioremediation	Sections 10-1 through 10-3, Section 10-6
Stabilization	Sections 11-1 through 11-3
Incineration	Sections 12-1 through 12-5
Land Disposal	Sections 13-1, 13-3, and 13-4

Site Remediation
Risk Assessment	Sections 14-1 through 14-4
Remedial Investigations	Sections 15-1 through 15-5
Alternatives Analysis	Chapter 17

This text has gone through numerous reviews and edits. The authors would greatly appreciate any criticisms or suggestions for improvement.

ACKNOWLEDGMENTS

First and foremost, the authors wish to thank Paul Woodruff and the Environmental Resources Management Group for their support and counsel during this undertaking. At Bucknell University, assistance was provided by Lois Engle and Beverly Spatzer. Many individuals assisted in the review of numerous drafts of individual chapters, and each chapter concludes with a separate acknowledgments section.

This book would not have been completed without the exceptional assistance of Jeffrey Gorham, ERM's publications coordinator, and Karen L. Stein, ERM's technical editor. Marie Malmberg, Amaly MacIntyre, Paul Nordquist, Cathy Putz, Joan Plitt, and Sally Wilson provided word processing assistance at ERM. Bucknell University also provided extensive support to this effort and the authors would like to thank specifically Jai B. Kim, Chair, Civil Engineering Department; Thomas Rich, Dean of Engineering; and Larry Shinn, Vice-President for Academic Affairs.

All the figures are original art created by ERM's Graphics Department under the direction of Patricia Ackerman. The graphic artists were Angie Della Camera, Tom Hall, Erin McAllister, and Cindy Pomante.

The editors for McGraw-Hill were B.J. Clark and John M. Morriss, who assisted in many ways throughout the project.

The authors also wish to acknowledge the numerous professors and students who used drafts of the text in classroom testing and provided a great number of suggestions for improvements.

Finally, we thank the following reviewers who provided valuable critiques on the original manuscript: Paul King, Northeastern University; Ralph Kummler, Wayne State University; Raymond Letterman, Syracuse University; Howard Mettee, Youngstown State University; D. William Tedder; and Margaret von Braun, University of Idaho.

Michael D. LaGrega
Phillip L. Buckingham
Jeffrey C. Evans

ACRONYMS

2,4-D	Dichlorophenoxyacetic acid
2,4,5-T	2,4,5-Trichlorophenoxyacetic Acid
A/W	Air to Water ratio (dimensionless)
AA	Atomic Adsorption
ABET	Accreditation Board for Engineering and Technology
ABS	Absorption into Bloodstream (decimal fraction)
acfm	Actual cubic feet per minute
ACS	American Chemical Society
ADI	Acceptable Daily Intakes
ADP	Adenosine diphosphate
AIC	Acceptable Intake Chronic exposures
AICHE	American Institute of Chemical Engineers
AIDS	Acquired Immune Deficiency Syndrome
ALA-D	Aminolevulinc Acid Dehydrase
ANS	American Nuclear Society
ANSI	American National Standards Institute
APC	Air Pollution Control
APCA	Air Pollution Control Association (Now AWMA)
API	American Petroleum Institute
AQCR	Air Quality Control Region
ARARs	Applicable or Relevant and Appropriate Requirements
ASCE	American Society of Civil Engineers
ASTM	American Society for Testing and Materials
AT	Averaging Time
atm	Atmospheres
ATP	Adenosine triphosphate
ATSDR	Agency for Toxic Substances and Disease Registry (U.S.)

AWFSO	Automatic Waste Feed Shutoff
AWMA	Air & Waste Management Association
AWQC	Ambient Water Quality Criteria
AWWA	American Water Works Association
BACT	Best Available Control Technology
BANANA	Build Almost Nothing Anywhere Near Anybody
BAT	Best Available Technology
BCF	Bioconcentration Factor
BDC	Business Development Corporation
BDST	Bed Depth Service Time
BEHP	Bis-2-ethyl hexyl phthalate
BIF	Boilers and Industrial Furnaces
BNA	Base, Neutral, and Acid compound
BOD	Biochemical Oxygen Demand
BOD_5	Five-day Biochemical Oxygen Demand
BTA	Benzyltriethyl Ammonium (clay)
BTU	British Thermal Unit(s)
CAA	Clean Air Act
CAAA	Clean Air Act Amendments
CAD	Computer Aided Design
CAG	Carcinogen Assessment Group (EPA)
Cal WET	California Wet Extraction Test
CASRN	Chemical Abstract Service Registry Number
CATNIP	Cheapest available technology not entailing prosecution
CDI	Chronic Daily Intake
CEC	Cation-exchange Capacity
CEM	Continuous Emission Monitors
CEO	Chief Executive Officer
CERCLA	Comprehensive Environmental Response Compensation and Liability Act (Superfund)
CERCLIS	CERCLA Information Service
CFC	Chlorofluorocarbons
CFM	Cubic feet per minute
CFR	*Code of Federal Regulations*
CFSTR	Continuous Flow Stirred Tank Reactor
CGL	Comprehensive General Liability (insurance)
CH	High plasticity (clay)
CL	Low plasticity (clay)
CMA	Chemical Manufacturers Association
CMS	Corrective Measures Study
CMT	Carcinogenicity, Mutagenicity, and/or Teratogenicity

COD	Chemical Oxygen Demand
CPF	Carcinogen Potency Factor
CSF	Carcinogenic Slope Factor
CSPE	Chlorosulfonated Polyethylene
CWA	Clean Water Act
DCA	Di-chloroacetic acid
DCE	1,1 Dichloroethylene; 1,1 dichloroethane
DDD	1,1-Bis(4-Chlorophenyl)-2,2 Dichloroethane
DDE	1,1-dichloro-2,2-bis(4-chlorophenyl)ethene
DDT	Dichlorodiphenyltrichloroethane
DE	Destruction Efficiency
DEHPA	di-2-ethylhexyl phosphate
DEHP	(Bis)diethyl hexyl phthalate
DI	De-Ionized
DMT	Dilatometer test
DNA	Deoxyribonucleic Acid
DNAPL	Dense Non-Aqueous Phase Liquids
DOD	Department of Defense (U.S.)
DOE	Department of Energy (U.S.)
DOJ	Department of Justice (U.S.)
DOT	Department of Transportation (U.S.)
DRE	Destruction and Removal Efficiency
DVB	Divinylbenzene
dscf	Dry standard cubic feet
dscfm	Dry standard cubic feet per minute
dscm	Dry standard cubic meters
dscmm	Dry standard cubic meters per minute
EA	Endangerment Assessment
ECAO	Environmental Criteria and Assessment Office
ED	Exposure Duration
EDB	Ethylene dibromide
EDC	Ethylene dichloride
EDTA	Ethylene diamine tetra acetic acid
EHS	Environmental Health and Safety
ENFLEX®	Regulatory Database
ENR	Engineering News Record
EP	Extraction Procedure (toxicity test)
EPA	Environmental Protection Agency (U.S.)
EPRI	Electric Power Research Institute
EPM	Equivalent Porous Media
EPTOX	Extraction Procedure Toxicity test

ERM	The Environmental Resources Management Group
ESP	Electrostatic Precipitator
FAST®	Field Analytical Services Technology
FAO	Food and Agriculture Organization (U.N.)
FDA	Food and Drug Administration (U.S.)
FID	Flame-ionization Detector
FIFRA	Federal Insecticide, Fungicide, and Rodenticide Act
FML	Flexible-membrane Liner
FR	*Federal Register*
FS	Feasibility Study
FTIR	Fourier Transform Infra-red (Analysis)
GAC	Granular Activated Carbon
GAO	Government Accounting Office (U.S.)
GC	Gas Chromatograph
GC/MS	Gas Chromatograph Mass Spectrometry
GIS	Geographic Information Systems
GLP	Good Laboratory Practices
GPM	Gallons per minute
GPR	Ground penetrating RADAR
HAP	Hazardous Air Pollutant
HDPE	High-density Polyethylene
HEL	Higher Explosive Limit
HELP	Hydrologie Evaluation of Landfill Performance (Model)
HETP	Height Equivalent of a Theoretical Plate
HHV	Higher Heating Value
HI	Hazard Index
HMCRI	Hazardous Materials Control Research Institute
HP	Horsepower
HPLC	High-performance Liquid Chromatography
HRS	Hazard Ranking System
HSL	Hazardous Substances List
HSWA	Hazardous and Solid Waste Amendments
HTU	Height of a Transfer Unit in a packed column
HxCDDs	Hexachlorodibenzodioxins
IARC	International Agency for Research on Cancer (WHO)
ICAP	Inductively Coupled Argon-plasma (emission spectroscopy)
IPCS	International Programme on Chemical Safety
IRIS	Integrated Risk Information System (EPA database)
IS	Indicator Score

ISC	Industrial Source Complex (model)
ISV	In Situ Vitrification
ISWA	International Solid Waste Association
Kcal	Kilocalorie
kD	Kuderna-Danish (concentration gas chromatography)
kg	Kilograms
Kg-cal	Kilogram-calories
kmol	Kilomoles
kW	Kilowatts
kWh	Kilowatt-hours
LAER	Lowest achievable emmision rate
LC	Liquid Chromatography
LCA	Life cycle analysis
LDPE	Low density polyethylene
LEL	Lower Explosion limit
LEXIS®	Legal database
LF	Landfill
LFG	Landfill gas
LHV	Lower Heating Value
ln	Natural logarithm
LNAPL	Light non-aqueous phase liquids
LOAEL	Lowest observed adverse effect level
LOEL	Lowest observed effect level
LPG	Liquid Propane Gas
MACT	Maximum achievable control technology
MCL	Maximum Contaminant Levels
MCLG	Maximum Contaminant Level Goals
MEI	Maximum Exposed Individual
MEK	Methyl Ethyl Ketone
Me	Metal (cation)
MFO	Mixed function oxidases
MGD	Million Gallons per Day
MGLC	Maximum Ground Level Concentration
MIBK	Methyl Isobutyl Ketone
MIC	Methyl-isocyanate
MLSS	Mixed Liquid Suspended Solids
MLVSS	Mixed Liquor Volatile Suspended Solids
MMD	Methane monooxygenase
MSDS	Material Safety Data Sheets
MSW	Municipal Solid Waste
MT	Metric Tons

MTD	Maximum Tolerated Dose
MWC	Molecular Weight Cutoff
meq	Milliequivalents
mg	Milligram(s)
mgd	Million gallons per day
min	Minutes
mL	Milliliters
mm	Millimeters
mo	Months
mol	Moles
ms	Milliseconds
NA	Not Applicable
NAAQS	National Ambient Air Quality Standards
NAPL	Non-aqueous phase liquids
NC	Non-Carcinogen
NCP	National Contingency Plan
NFPA	National Fire Protection Agency
NIMBY	Not In My Back Yard
NIOSH	National Institute for Occupational Safety and Health
NMP	N-methyl-2-pyrrolidone
NOAA	National Oceanographic and Atmospheric Administration (U.S.)
NOAEL	No-Observable Adverse Effect Level
NOEL	Non-observable effect level
NOS	Not otherwise specified
NPDES	National Pollutant Discharge Elimination System (permit)
NPL	National Priorities List
NRTL	Non-Random Two Liquid (Model)
NSPE	National Society of Professional Engineers
NSPS	New Source Performance Standard
NSR	New Source Review
NTIS	National Technical Information Service
NTU	Number of Transfer Units
NYCRR	New York Code Rules and Regulations
OAU	Organization of African Unity
OCPSF	Organic Chemicals, Plastics and Synthetic Fibers
OECD	Organization of Economic Cooperation and Development
OERR	Office of Emergency and Remedial Response (EPA)
ORP	Oxidation Reduction Potential
OSC	On-scene Coordinator

OSHA	Occupational Safety and Health (Administration or Act)
OSWER	Office of Solid Waste and Emergency Response (EPA)
OTA	Office of Technology Assessment (U.S.)
OTS	Office of Toxic Substances (U.S.)
OVA	Organic Vapor Analyzer
PAC	Powdered Activation Carbon
PACT	Powdered Activated Carbon Treatment
PAH	Polynuclear Aromatic Hydrocarbons
Pa	Pascal = 1 Newton/meter^2
PBB	Polybrominated biphenyls
PC	Personal Computer
PCBs	Polychlorinated Biphenyls
PCC	Toxaphene
PCDD	Polychlorinated dibenzodioxins
PCDF	Polychlorinated dibenzofurans
PCE	Perchloroethylene
PCP	Pentachlorophenol
PFR	Plug Flow Reactor
PICs	Products of Incomplete Combustion
PID	Photoionization Detector
PL	Public Law
PM	Preventive Maintenance
PMN	Premanufacture Notification
PMT	Pressure Meter test
PNAs	Polynuclear Aromatic (hydrocarbons)
POHCs	Principal Organic Hazardous Constituents
POTW	Publicly Owned Treatment Works
PP	Priority Pollutants
PPE	Personal Protective Equipment
PRPs	Potentially Responsible Parties
PSD	Prevent of Significant Deterioration
PVC	Polyvinyl Chloride
PVD	Pore Volume Displacement
PW	Present Worth
PWF	Present Worth Factor
pH	Hydrogen-ion activity
ppb	Part(s) per billion
ppbv	Parts per billion volume
ppm	Part(s) per million

ppmv	Parts per million volume
psi	Pounds per square inch
psia	Pounds per square inch (absolute)
psig	Pounds per square inch (gage)
QA/QC	Quality Assurance/Quality Control
QSAR	Quantitative structural-activity relationship
R&D	Research and Development
RCRA	Resource Conservation and Recovery Act
RD/RA	Remedial Design and Remedial Action
RDF	Refuse-derived Fuel
RFA	RCRA Facility Assessment
RfC	Reference Concentration
RfD	Reference dose
RFI	RCRA Facility Investigation
RI	Remedial Investigation
RI/FS	Remedial Investigation and Feasibility Study
RITZ	Ground water model
RO	Reverse Osmosis
ROD	Record of Decision
RPM	Remedial Project Manager
RQD	Rock Quality Designator
SARA	Superfund Amendments and Reauthorization Act
SB	Soil Bentonite
SBR	Sequencing Batch Reactor
SCBA	Self-Contained Breathing Apparatus
SCC	Secondary Combustion Chamber
SCF	Supercritical Fluid
SCS	Soil Conservation Service (U.S.)
SCWO	Supercritical Waste Oxidation
SDWA	Safe Drinking Water Act
sec	seconds
SEM	Strategic Environmental Management
SF	Slope Factor
SI	International System of Units
SIC	Standard Industrial Code
SM	Percent of Soil Matrix
SNARL	Suggest-No-Adverse-Response Level
SPCC	Spill Prevention, Control and Countermeasures
SPHEM	Superfund Public Health Evaluation Manual (EPA)
SRT	Solids Retention Time
STP	Standard Temperature and Pressure

SVE	Soil Vapor Extractor
SVT	Solvent Vapor Transmissability
SWMU	Solid Waste Management Unit
TCDDs	Tetrachlorodibenzodioxins
TCE	Trichloroethylene
TCLP	Toxicity Characteristics Leaching Procedure
TDS	Total Dissolved Solids (mg/L)
TECE	Tetrachloroethylene
TEF	Toxic Equivalency Factors
THC	Total Hydrocarbons
THM	Trihalomethane
TLV	Threshold Limit Value
TMA	Tetramethyl ammonium (clay)
TNT	Trinitrotoluene
TOC	Total Organic Carbon
TOD	Total Oxygen Demand
TOSCA	Toxic Substances Control Act
TOX	Total Organic Halides
TPH	Total Petroleum Hydrocarbons
TRV	Thermal Relief Valve
TSCA	Toxic Substances Control Act
TSD	Treatment, Storage and Disposal
TSDF	Treatment, Storage, and Disposal Facility
TSDR	Treatment, Storage, Disposal, and Recycling
TSS	Total Suspended Solids
U.S. EPA	U.S. Environmental Protection Agency
U.S.C.	*United States Code*
UF	Ultrafiltration; urea-formaldehyde resin
U.S.G.S.	United States Geological Survey
UN	United Nations
UNEP	United Nations Environment Program
US	United States
USA	United States Army
USAF	United States Air Force
USC	United States Code
UST	Underground Storage Tank
UV	Ultraviolet
VES	Vapor Extraction System
VOC	Volatile Organic Compound
WAS	Waste Activated Sludge
WEF	Water Environment Federation

WES	Waterways Experiment Station (U.S.)
WESTLAW®	Legal database
WHO	World Health Organization
WMA	Waste Management Association
WPCF	Water Pollution Control Federation (Now WEF)
ZOI	Zone of Influence

Symbols

∇	Del operator
∇	Water level
∂	Partial Derivative
[ ]	Molar concentration (moles/Liter)

Metric Prefixes

E	exa	10^{18}
P	peta	10^{15}
T	tera	10^{12}
G	giga	10^9
M	mega	10^6
k	kilo	10^3
h	hecto	10^2
da	deka	10^1
d	deci	10^{-1}
c	centi	10^{-2}
m	milli	10^{-3}
μ	micro	10^{-6}
n	nano	10^{-9}
p	pico	10^{-12}
f	femto	10^{-15}
a	atto	10^{-18}

CHAPTER

1

HAZARDOUS WASTE

If we are going to live so intimately with these chemicals—eating and drinking them, taking them into the very marrow of our bones—we had better know something about their nature and their power.
Rachel Carson, *Silent Spring*

What is **hazardous waste**? This chapter provides a working definition of hazardous waste and discusses its sources of generation and its classification. A review of the historical roots of hazardous waste issues explains why they became the dominant environmental program of the 1980s and will continue as a major program into the 21st Century. The two basic areas of hazardous waste management are presented: 1) management of currently generated waste and 2) remediation of sites contaminated by past practices. The chapter closes with a section on ethics, recognizing that technical decisions made in an arena of imperfect laws can affect the public.

1-1 WORKING DEFINITION

The term **hazardous waste** gained acceptance starting about 1970 with the first national study of the issue, and it became vogue in the mid-1970s with the development of legislative initiatives to regulate it. Long before then the wastes that we now know as "hazardous" were referred to by such terms as special industrial waste or chemical waste. This is still often the case in Europe. The term hazardous waste by itself is ambiguous. A feature of any regulatory program is to provide a legal definition to determine what is and what is not a hazardous waste. Developing a legal definition can take considerable effort with much disagreement. The U.S. Environmental Protection Agency (EPA) took nearly four years from the passage of the nation's first hazardous waste law in 1976 before promulgating regulations that defined hazardous waste. And even then the definition used rather broad terms, included a myriad of exceptions, and has shown the periodic need for refinement. Some exceptions are based less on the waste's inherent hazard than on its generator's political influence.

Other nations have had similar experiences. Each has developed its own administrative definition for identifying and classifying hazardous waste to the particular level of detail necessary to support its legal procedures. These definitions lack scientific rigor. Each reflects the environmental, social, and political policies of a nation's government, and hence approaches, like governments, differ from each other. A practitioner must understand the current legal definitions that apply to a waste; these are introduced in Sec. 2-2. However, this textbook avoids dedicating a number of pages to explain the complexities of a legal definition because a legal definition would periodically change, and today varies from nation to nation. Instead, this chapter presents a practical, working definition.

Developing a working definition first requires defining the term **waste**. A terse definition[1] of waste was prepared in a joint international study as follows: "A waste is a moveable object which has no direct use and is discarded permanently." This definition implies solid waste. In fact U.S. federal hazardous waste laws tend to have stemmed from solid waste programs, and the EPA regulates hazardous waste as a subset of solid waste.

The use of the **solid waste** nomenclature would suggest that the definition covers only solids, sludges, tars, and the like. However, many liquids are considered hazardous waste, typically because of their high strength or because they are a mixture of a hazardous waste with water. Further, many liquid wastes are containerized upon generation and transported as such, thereby making them solid-like. Thus defined, hazardous waste can include solids, sludges, liquids, and containerized gases. It is important to note that hazardous waste excludes that which is discharged directly into the air or water; these wastes are regulated under air and water laws that long predated hazardous waste laws.

It can be seen that the form of a waste is not important when defining whether it is hazardous. Instead, the most critical part of any definition is to include those terms describing the characteristics that cause a waste to be considered "hazardous" (i.e., posing a substantial present or potential danger to human health or the environment). The potential for toxicity, particularly carcinogenesis, has resulted in the greatest public concern and heads any list of characteristics. However, a waste can be considered hazardous if it exhibits any of a variety of other characteristics such as being ignitable, flammable, reactive, explosive, corrosive, radioactive, infectious, irritating, sensitizing, or bioaccumulative. This textbook covers all such waste with the exception of radioactive and infectious waste because these two types require management and technical approaches typically different from those required for other types of hazardous waste.

Therefore, the following working definition of hazardous waste, virtually as prepared under the United Nations Environment Programme auspices in December 1985, serves as a basis for this book:

Hazardous wastes mean wastes [solids, sludges, liquids, and containerized gases] other than radioactive [and infectious] wastes which, by reason of their chemical activity or toxic, explosive, corrosive, or other characteristics, cause danger or likely will cause danger to health or the environment, whether alone or when coming into contact with other waste. . . . [2]

1-2   HISTORICAL ROOTS

Around the beginning of the 1980s, hazardous waste became the leading environmental issue of our society. Now, in the 1990s, while scientific information shows some potentially devastating problems with global ecosystems, hazardous waste still continues to attract far more attention when measured by the amount of federal money spent for environmental programs. As another measure, hazardous waste accounted for about 50% of the $8.2 billion environmental consulting market in 1991 (see Fig. 1-1).

Why has it commanded such importance?

As discussed earlier, a waste may be hazardous for any of a number of reasons. Of these reasons, the potential for some wastes to cause a toxic reaction in humans is preeminent among public concern. It is this concern (if not fear) compounded by misperception, neglect, mistrust, and politics that explains why hazardous waste dominates other environmental issues.

The problem did not emerge overnight. Environmental contamination by toxic substances from waste or other sources has a long history. It seems that many wealthy Romans suffered from lead poisoning two millennia ago; possibly the decline of the Roman Empire was due, at least in part, to lead-induced psychoses among the emperors. However, it is the rapid pace of our technological developments, beginning with the Industrial Revolution, that forms the real roots of the problem as we know it today.

The advent of the Industrial Revolution spurred progress on many fronts. Advances in medical science and public health reduced the death rate, thereby dramatically increasing human population. Personal consumption also grew rapidly as expanding industrial production, resource extraction, and intensive agriculture supplied more goods. With these goods came toxic substances, sometimes as part of the goods themselves, that became residuals after use. Sometimes toxic substances were in the wastes generated when making the goods.

Until recently, government policies did not require appropriate precautions for waste, and few were taken. Simply getting rid of waste was the standard. Relying on

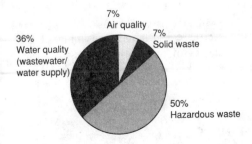

7%
Air quality

7%
Solid waste

36%
Water quality
(wastewater/
water supply)

50%
Hazardous waste

FIGURE 1-1
Hazardous waste dominated the $8.2 billion environmental consulting market in 1991.[3]

the seeming inexhaustibility of the environment was the method. Gradually, exposure to hazardous substances occurred outside the workplace via various environmental pathways. It started first with inorganic compounds such as lead and mercury, and expanded later, upon their introduction in the twentieth century, with synthetic organic compounds. Although technological progress has given us new products that raised our standard of living remarkably, it is the nature of technological development to outpace our knowledge of all that it brings and what that means. The failure in this instance is caused more by ignorance than negligence. The long-term toxic nature of hazardous substances and the inability of the environment to assimilate them completely have done the rest.

The environmental and human health consequences of the residuals and wastes of our technological society were not understood or even recognized initially; it took years or decades for chronic effects to manifest themselves, and those cases were obscured by the fact that everyone is exposed to a wide number of chemicals. This changed when developments in the sciences of epidemiology, toxicology, and analytical chemistry enabled researchers to begin to recognize some noteworthy associations that had been overlooked earlier in our longer-term relationship with toxic chemicals. The landmark cases start not with a hazardous waste but with the effect of DDT residuals on bird populations. Soon afterwards, with mercury poisoning of human populations in Japan and some episodes involving PCBs and dioxin, it became clear that humans were also at risk. Each episode received considerable press in a new and increasingly popular media form that came to be called environmental journalism. Each episode advanced public awareness, heightened public concern, and pushed ahead the environmental movement, eventually prompting the hazardous waste legislation that drives how waste is managed today.

1-3   LANDMARK EPISODES[†]

DDT

Pesticides have continued to be potentially sinister agents in the public eye ever since Rachel Carson's publication of *Silent Spring* in 1962.[4] Ms. Carson riveted world attention to the interconnected web of all life by recounting how DDT residues could be found in deep-sea squid, Antarctic penguins, and the fatty tissues of *Homo sapiens*. In aquatic birds, high levels of DDT were associated with lack of fertility. Later it would be found that DDT inhibits calcium deposition in avian ovaries, leading to egg shells that are too thin to withstand adult weight. In *Homo sapiens* it was not egg shell thinning that was worrisome, but an insidious disease—in laboratory animals DDT exposure was associated with an increased frequency of cancer, one of the first cases associating cancer with exposure to a toxic chemical.

[†]Section 1-3 was prepared by Bruce Molholt, Ph.D., ERM, Inc.

Mercury

Mercury has dramatically different toxicologic properties depending upon its chemical state. As a liquid metal, it was once used to cure constipation, apparently with few adverse side effects. On the other hand, mercury salts, which were used to form felt in the Dutch hat industry, led to the neurologic disorder renowned as being "mad as a hatter." Organic forms of mercury, such as methyl mercury, have proven to be even more pernicious, having caused hundreds of cases of paralysis and sensory loss along Minamata Bay[5] in Japan. Inorganic mercury from a chemical plant became methylated in sediments and then bioaccumulated in shellfish. Because shellfish are the major protein source for much of the local population, this situation was an epidemic waiting to happen. Similar poisoning epidemics have occurred elsewhere (e.g., in Iraq and other countries where persons unknowingly ingested seed grain laced with an organomercury fungicide). However, it was the Minamata disaster and its graphic depiction through Katagiri's remarkable photographs in the late 1960s that heightened global awareness of industrial pollution.

PCBs

Like DDT, polychlorinated biphenyls (PCBs) were produced at about 100 million pounds per year during the 1960s and 1970s. However, unlike DDT, which is useful only as an insecticide, PCBs had multiple uses such as transformer coolant, plasticizer, and in the manufacturing of carbonless paper. Two cases of accidental contamination of rice cooking oil in Japan and Taiwan in the late 1960s and mid-1970s exposed thousands of Asians to high concentrations of PCBs. Miscarriages and birth defects erupted within the exposed populations. Although later it would be shown that these prenatal health problems resulted not from PCBs themselves but from polychlorinated dibenzofurans (PCDFs) that formed when heating the contaminated rice cooking oil, the riveting environmental journalism made the world population vividly aware of the adverse human effects of "PCB" exposure. About the same time in the United States, the contamination of Michigan cattle feed by polybrominated biphenyls (PBBs) not only caused widespread human exposures via milk and other dairy products, but also followed more complex pathways to reach humans.[6] Afflicted cattle were rendered and then used to prepare chicken feed; thousands of human consumers were exposed to PBBs through eggs and egg products. One incident involved 24,000 crates of PBB-contaminated shortcakes which were confiscated in Alabama. PBBs were found in some Michigan mothers' breast milk.

Love Canal

For many, Love Canal will always be the symbol of environmental contamination by hazardous waste. It was this pivotal event that eventually resulted in the passage of the Superfund Act (Comprehensive Environmental Response, Compensation, and Liability Act) in 1980 by the U.S. Congress. Not useful as a canal, this channel was sealed off at the ends and used in the 1940s and 1950s by Hooker Chemical Co. and others as a hazardous waste disposal site. Later, Love Canal would be filled in and covered, and then sold to the Niagara Falls, New York, School District for $1.00. The company informed the school district that the site was used for the disposal of chemical wastes, warning against any excavation and underground construction. This information was added to the deed and copied into subsequent deeds when the school district sold parcels of the property.

The Niagara School District built an elementary school on the site that became quickly surrounded by hundreds of homes, many with young children. By the late 1970s a chemical odor was frequently discernible, especially in the basements of the homes, which, according to the local style, often contained the children's bedrooms. People began experiencing chemically induced problems. Chemical analysis of samples showed various chemical substances at the disposal site, including dioxin.

A local reporter, Michael Brown of the *Niagara Gazette*, decided to follow up on the anecdotal reports of a few diseases among children that seemed to be linked to indoor fumes. Merely by walking around the neighborhood, he soon discovered more than 100 highly believable examples of chemically induced illness. Furthermore, in many basements he himself could smell the chemical fumes, sometimes to the point of nausea and tearing of his eyes. Brown's Pulitzer Prize-winning exposé catalyzed local politics and also reached the outside world.[7] Soon, the local community, led by Lois Gibbs, a housewife with two afflicted children, was demanding federal relief. In one pivotal meeting, the Love Canal activists, discontented with perceived inaction on the part of EPA, locked up two officials overnight until they received assurances from Washington that action would be taken. Ultimately, New York State and the U.S. government (Federal Emergency Management Agency) purchased those homes in the vicinity of the canal. By this time drums were being discovered in many back yards and most homeowners were more than eager to move out (see Fig. 1-2).

EPA later declared Love Canal a Superfund site. Today the site is being remediated. In contrast with general perceptions, the contamination at the site was never proven to have posed the level of risk initially estimated.

FIGURE 1-2
Boarded homes adjacent to Love Canal.

Times Beach

Perhaps the next most publicized episode in America's awakening concern with hazardous waste occurred in Times Beach, Missouri. As with Love Canal, dioxin would become the major toxic chemical of concern. In the late 1960s and early 1970s, wastes were taken from chemical plants near St. Louis and diluted into used crankcase oil (a legal practice at the time) and spread on dirt roads and horse farms for dust control (again, a legal practice at the time). In the early 1970s dioxins were unknown. However, when waste oil was sprayed at a horse farm in May 1971, many animals died. Even after 6 inches of topsoil was removed from contaminated areas, animals continued to die. Chemical tests showed dioxin contamination up to 100 ppm at the village of Times Beach. With the Seveso, Italy chemical factory explosion in 1976, and the widespread toxic effects caused by the dioxin released by this explosion, dioxin (tetrachlorinated dibenzo(p)dioxin or TCDD) would arguably become recognized as the most toxic synthetic chemical yet discovered. The EPA eventually dealt with the Times Beach case by buying all properties in the community and permanently evacuating the residents.

1-4   REGULATORY INITIATIVES

As these and numerous other episodes evolved and received publicity, the public quickly became more aware of the dangers posed to health and the environment by hazardous substances when not managed properly. In response to acute public perceptions, possibly spurred by environmental journalism, the U.S. Congress along with state and local governments, as well as other national governments, enacted sweeping legislation on two fronts:

1. Management of currently generated hazardous waste, and
2. Remediation of contaminated sites.

These laws covered hazardous waste from a "cradle-to-grave" perspective. Prior to these legislative initiatives, economic considerations compelled most generators to dispose of their waste at a low, short-term cost with nominal regard given to the long-term impact. Although the actual risk did not equal the public's perceived risk, the regulatory pendulum had swung dramatically with this legislation.

Currently Generated Waste

In 1970, the U.S. Congress first addressed hazardous waste by including with a solid waste law of minor significance a requirement that the Department of Health, Education, and Welfare (a predecessor to the EPA) conduct a comprehensive investigation of the storage and disposal of hazardous waste. No comprehensive program for regulating hazardous waste existed then. The programs that regulated wastewater effluents and air emissions addressed hazardous waste only in a peripheral manner; the existing solid waste program was small and only addressed municipal waste.

The EPA completed its report to Congress in 1973 and found that ". . . the magnitude of the hazardous waste problem was larger than originally anticipated, and that current disposal practices are generally inadequate."[8] The report cited key examples where indiscriminate disposal of hazardous waste had caused adverse public health impacts, including the hospitalization of several people in 1972 in Minnesota after drinking well water that had been contaminated with arsenic wastes. With the impetus of the findings of this report and subsequent hearings, Congress included in the Resource Conservation and Recovery Act (RCRA) of 1976 Subtitle C empowering federal regulation of hazardous waste for the first time. The law focused on the recovery and recycling of solid waste; however, Subtitle C instituted a program to define which wastes are considered hazardous. Further, it required generators of hazardous waste to track the transportation of waste from the point of generation to its final disposition, to operate treatment and disposal facilities in accordance with established standards, and to otherwise manage hazardous waste properly. Thus, in the United States, hazardous waste regulatory initiatives started as just one small part of a small solid waste program. In a matter of a few years it overshadowed EPA's air and water programs.

The United States was not alone. Other developed nations took initiatives to address hazardous waste at about the same time. For example, the United Kingdom passed its Poisonous Waste Act in 1972 after a highly publicized cyanide episode, and the Federal Republic of Germany amended its solid waste laws in 1976 to regulate industrial waste.[9] As a sidepoint, Germany avoided the use of the term "hazardous" because of the emotional controversy spurred by the term.

Contaminated Sites

In the late 1960s and early 1970s, analytical chemistry techniques advanced to a level of high resolution. When applied to the environmental field, these techniques allowed routine detection and measurement of contaminants at concentrations two or more orders of magnitude smaller than possible just a few years earlier. This new ability to detect compounds at the part per billion level or lower allowed the identification of the widespread presence of potentially toxic compounds in the environment.

Toxic contaminants, deriving from numerous human activities, were detected in the air, water, and land of practically any community. Toxic contamination was no longer considered just limited to isolated cases where the adverse consequences of exposure had manifested themselves. Unfortunately, the scientific knowledge of the toxic effects of exposure to multiple compounds at low concentrations over a long period of time was not well understood and had not kept pace with the ability to detect and measure their concentration. Without this knowledge, but with the media coverage of Love Canal and other early-discovered dump sites, the public's awareness of toxic contamination turned into alarm.

Although the toxic contamination of the environment has numerous sources, of which hazardous waste is just one, sites such as Love Canal where hazardous waste had been mismanaged prompted severe public reaction. Some of the sites contaminated by past practices were indiscriminate dumping grounds; others were bankrupt

FIGURE 1-3
Site contaminated by past practices.

and abandoned manufacturing plants (see Fig. 1-3). Many were former hazardous waste storage, treatment, or disposal facilities, such as Love Canal, that had been subsequently closed, if not abandoned, with minimal plans for further care. It was becoming clear in the late 1970s that potentially thousands of inactive hazardous waste sites existed throughout the nation. Whether these had caused an adverse health effect or had even resulted in exposure was not known; at best the public perceived them as "ticking time bombs."

Neither RCRA nor any other law provided the mechanisms to address inactive sites contaminated by past practices. The U.S. Congress changed that in 1980 with the passage of the Comprehensive Environmental Response, Compensation and Liability Act (CERCLA) or more commonly known as "Superfund." This program seeks to identify sites involving past releases of hazardous substances to the environment and to implement the remedial actions deemed necessary at each site to protect health and the environment.

Foundation of a Regulatory Structure

The two hazardous waste laws (RCRA and CERCLA) were based on the legal concept that generators are liable for the long-term impact of their waste management practices, including their past practices. This contrasted remarkably with the traditional policy that made short-term costs virtually the only economic factor. The enactment and implementation of laws with this foundation have greatly changed the impetus for waste management. They have also fostered a futuristic perspective: most generators endeavor to minimize waste, and many manufacturers even consider the life cycle of their products (from research through manufacturing to use by the consumer and eventually ultimate disposition). The regulatory programs born from RCRA and CERCLA have gained much experience in the years since their enactment, giving considerable insight into particular problem areas that the laws were intended to address. The programs have not slackened—even under considerable scrutiny that claims that they are overreactive, ponderous, costly, and time-consuming. Instead, the programs now carry even greater authority. The public's interest and concern with hazardous waste remain strong.

It should be noted that the RCRA and CERCLA regulatory programs evolved and operated separately. There are perhaps as many uncommon characteristics as common ones between the two programs. Indeed, there are scientific differences: currently generated wastes (i.e., RCRA waste) are usually a residue of an industrial operation and tend to be more concentrated than when found in soils and ground water contaminated by past practices (i.e., CERCLA waste). Even so, the substances in the waste in either case remain the same, as do the inherent hazard they pose to health and the environment. This book approaches the two areas (RCRA and CERCLA) together. The same basic scientific and engineering concepts apply whether the substance is in a waste generated today or was released to the environment years ago in a spill, though the regulatory requirements frequently differ.

Public Perception of Risk

While hazardous waste does pose a real risk to human health and the environment, there is a debate about the magnitude of the risk, especially the risk compared with other sources of environmental contamination. Scientists assert that the public has exaggerated the real risks. In 1987, the EPA conducted a technical ranking of relative risks posed by 31 environmental threats and placed the health risk of hazardous waste sites below several other environmental problems.[10] The relative rankings were based on the comparative evaluations and judgments of a technical panel of 75 experts. In contrast, a concurrent survey conducted by the EPA showed that the public perceived hazardous waste to represent a greater relative risk than most other environmental problems.[11] This is shown in Table 1-1 comparing the EPA's experts' rankings with public perceptions of selected environmental problem areas.

Why do public perceptions of the risk posed by hazardous waste not correlate with expert assessments? The public's perceptions are influenced by a number of non-

TABLE 1-1
Comparison of expert and public opinion regarding environmental problems[12]

Problem	Relative risk	
	EPA expert panel	Public opinion
Hazardous waste sites (active and inactive)	Low-medium	High
Pesticide residue on foods	High	Medium
Indoor air pollutants	High	Low
Consumer exposure to chemicals	High	Low

technical factors such that a real risk is either perceived at a magnified level or simply is deemed unacceptable. Four of these factors are particularly applicable to hazardous waste generation or to a contaminated site, as follows:

- Involuntarily encountered—Because few people actually choose to live or work with hazardous waste, it is perceived as having been imposed upon a community or person. In contrast, smoking is known to be unhealthy, but it is a personal choice—it is voluntary—and thus those who smoke perceive the risk as acceptable.
- Having little apparent benefit—Few see any benefit in hazardous waste, ignoring that it results from some economic activity (e.g., manufacturing). In contrast, riding in an automobile poses greater risk but provides great direct benefit by getting one to where one wants or needs to go.
- Uncontrollable or controlled by others—The landmark contamination episodes such as Love Canal convinced many that hazardous waste is not controllable or at least those in control could not be trusted to manage it properly. In contrast, driving an automobile is controlled by the driver, and "Eighty-five percent of all drivers consider themselves better than average."[13]
- Having unknown but substantial consequence—It can be asserted that science has not developed the information to predict accurately the risk posed by hazardous waste. There is a virtual consensus, however, that birth defects and death are possibilities, however remote, from exposure to toxic substances in hazardous waste.

In aggregate, these four factors illustrate the great perceptual difference. The lack of agreement between public opinion and expert assessments does not suggest that either is wrong or even distorted. It is correctly pointed out that ". . . these are not distortions of risk; they are part of what we mean by the term."[14] In the United States, regulatory programs are more closely aligned with public perception than with scientific assessments; that is, the political system bends to its constituency. Hence, the two federal hazardous waste regulatory programs (RCRA and CERCLA) continue to receive much of the nation's budget for environmental regulation, and together represent a massive and complex set of laws.

1-5 CLASSIFICATION

The basis for determining if a waste is hazardous typically occurs in either of two ways. First, laboratory tests may indicate that it exhibits one or more of the characteristics deemed to make a waste hazardous. Second, it may be on a list of specific wastes compiled by the government because it is known or suspected of having the potential to exhibit hazardous characteristics.

Testing

In the United States, any of the following four characteristics will make a waste hazardous:

- Corrosivity (waste that is highly acidic or alkaline);
- Ignitability (waste easily ignited and thus posing a fire hazard during routine management);
- Reactivity (waste capable of potentially harmful, sudden reactions such as explosions); and
- Toxicity (waste capable of releasing specified substances to water in significant concentrations).

The EPA has defined explicit laboratory procedures to analyze waste for these characteristics. The procedures include specific standards to determine whether the tested waste is hazardous (e.g., a corrosive waste is one with a pH ≤ 2 or ≥ 12.5).

Lists

The cornerstone of most regulatory programs is to itemize specific hazardous wastes into lists. Inclusion in such a list means that the waste is regulated as a hazardous waste. Some programs have prepared "exclusive" lists describing those wastes that are not hazardous, meaning anything not on the list is hazardous. This compares with the U.S. Food and Drug Administration's list of acceptable food additives. Such an approach to regulation was attempted in the United Kingdom but later abandoned.[15]

Most regulatory programs use an inclusive list. Federal regulations in the United States specify four inclusive lists, with some states adding others.[16] Of the four, one lists hazardous waste from **specific sources.** These are wastes generated in specific processes unique to specific industrial groups. Wastes from these sources were tested nationally and found generally to exhibit at least one of the four hazardous characteristics. There are about 100 such wastes, referred to as "K"-wastes, and most made the list because of toxicity. Examples include vacuum stripper discharges from chlordane production (pesticides industry), distillation bottoms from aniline production (organic chemicals industry), and spent pickle liquor from steel finishing operations (iron and steel industry). Many of the listed wastes from specific sources are sludges from air pollution control and wastewater treatment processes (e.g., sludge from the treatment of wastewater from wood preserving processes that use creosote and/or pentachlorophenol [wood preservation industry], emission control dust/sludge from certain steel production processes [iron and steel industry], and API separator sludge [petroleum refining industry]).

A variety of industries may use essentially the same standard operation as part of their particular manufacturing processes. If the waste generated by this standard operation typically is hazardous regardless of the industry, the waste is included on a second U.S. list of hazardous waste (i.e., **non-specific sources**), referred to as "F"-wastes. As examples, the manufacture of a great number of common products typically involves degreasing and electroplating operations. Certain spent halogenated solvents used in degreasing operations and spent cyanide-plating solutions from electroplating operations are F-wastes. Another example is the distillation residue and reactor cleanout waste from production of chlorinated aliphatic hydrocarbons. The wastes from non-specific sources ("F" list) tend to include many different chemical species.

The EPA has two other lists, referred to as "P"- and "U"-wastes, for particular commercial chemical products. These products are regulated as hazardous waste when discarded in an essentially pure form. This includes off-specification products, and container and spill residues.

Some nations have an inclusive listing of chemical substances, either specific or by classes, that indicate a hazard to human health or the environment.[17] Such an example would be PCBs (polychlorinated biphenyls). These nations typically place a concentration limit on such constituents that, when exceeded, automatically classifies the waste as hazardous. The critical concentration limit may rely on leaching or volatility testing. Considering that there are a large number, possibly thousands, of chemical constituents that could render a waste hazardous, the implementation of such regulatory programs can be cumbersome.

Other Classification Systems

Broad definitions and inclusive lists are very important in the regulation of hazardous waste, particularly to ensure that a hazardous waste does not escape the umbrella of regulatory coverage. However, a regulatory classification system has a number of limitations. It tends to ignore relative differences in hazards among waste. All wastes, once deemed hazardous, are accorded the same regulatory coverage. A regulatory classification system does not facilitate definition of hazard nor is it appropriate for evaluating treatment methods.

Consequently, some scientists and officials have proposed alternatives for defining hazardous waste based upon a **degree of hazard** concept. This concept considers individual constituents, their concentration, and their mobility, together with a number of site-specific factors such as potential pathways for migration, exposure routes, dosages, and mitigation. Although this alternative has been proposed, its implementation is impractical.

Another limitation is that a regulatory classification system, because it is driven foremost by the hazard posed by a waste, is limited to identifying a waste's hazardous characteristics and perhaps its constituents; the classification system does not describe adequately (in most cases) the concentration of the constituents nor all important physical and chemical characteristics. Without this information it is impossible to assess the treatability of the waste, its potential for recycling, or its likely fate if released to the environment. An example is the waste coded as D006 in the EPA system. (A waste determined to be hazardous based on testing is given a code, starting with the letter D, to indicate which specific characteristics caused it to be hazardous. In this case, D006 is a waste having a characteristic of toxicity due to the presence of hexavalent chromium.) Such wastes are often generated as sludges or liquids during electroplating operations; however, they are also generated during dust collection or deburring in the fabricated metals industry. These two examples of chromium waste vary considerably in their physical and chemical characteristics as do their risks if released to the environment.

Example 1-1   Waste classification. A waste coded under the EPA classification system as D001 means the waste has been determined to have the characteristic of ignitability. In some states, this is the largest type of waste shipped off-site. What are some examples of wastes labelled as such?

Solution. A waste coded as D001 could be a spent but readily recoverable solvent, a combustible liquid suitable for energy recovery, a low-Btu paint sludge or paint filter suitable for incineration, or even a non-combustible sludge such as tank bottoms removed with the aid of gasoline. The potential treatability for each of these is significantly different.

An alternative to the regulatory classification system is to classify a waste according to the following hierarchy:

1. Form or phase distribution (e.g., liquid or solid);
2. Organic or inorganic;
3. Chemical class (e.g., solvents or heavy metals); and
4. Hazardous constituent as it affects treatability (e.g., hexavalent chrome).

This system is simple, yet effective for engineering purposes (e.g., grouping those wastes having similar physical and chemical characteristics and general treatment requirements). Table 1-2 shows the basic classes of an expandable system used in several statewide studies to define the need for new hazardous waste treatment and disposal facilities. The system is expandable.

Example 1-2   Waste categories. How could the major category "Inorganic Aqueous Waste" listed in Table 1-2 be expanded?

Solution. It could be split up among the following subcategories:

- Acid—inorganic acids; aqueous liquid with pH values of 2 or lower;
- Acid with metals—acidic wastes with metal contamination (e.g., ferric chloride solution and spent pickle liquor);
- Alkali—inorganic alkalis; aqueous liquids with pH of 12.5 or higher;
- Alkali with metals—alkali wastes with metal contaminants;
- Cyanide—spent cyanide solutions from plating, stripping, and cleaning; and
- Hexavalent-Chrome—aqueous liquids containing hexavalent chromium in excess of a specified limit.

1-6   GENERATION

Hazardous wastes can originate from a wide range of industrial, agricultural, commercial, and household activities. They are generated by manufacturers of many everyday products, by manufacturers of specialty articles, by both service and wholesale trade

TABLE 1-2
Engineering classification system for hazardous waste

Major category	Characteristics	Examples
Inorganic aqueous waste	Liquid waste composed primarily of water but containing acids/alkalis and/or concentrated solutions of inorganic hazardous substances (e.g., heavy metals, cyanide).	— Spent sulfuric acid from galvanizing — Spent caustic baths from metal finishing — Spent ammoniacal etchants from manufacturing electronic components — Rinse water from electroplating — Spent concentrates from hydrometallurgy
Organic aqueous waste	Liquid waste composed primarily of water but containing admixtures or dilute concentrations of organic hazardous substances (e.g., pesticides).	— Rinse water from pesticide containers — Washings of chemical reactors and formulation tanks
Organic liquids	Liquid waste containing admixtures or concentrated solutions of organic hazardous substances.	— Spent halogenated solvents from metal degreasing and dry cleaning — Distillation residues from production of chemical intermediates
Oils	Liquid wastes comprised primarily of petroleum-derived oils.	— Used lubricating oils from internal combustion engines — Used hydraulic and turbine oils from heavy equipment operations — Used cutting oils from machinery manufacture — Contaminated fuel oils
Inorganic sludges/solids	Sludges, dusts, solids and other non-liquid waste containing inorganic hazardous substances.	— Wastewater treatment sludge from mercury cell process of chlorine production — Emission control dust from steel manufacture and smelters — Waste sand from coking operations — Lime sludge from coking operations — Dust from deburring of chromium parts in fabricated metal industry
Organic sludges/solids	Tars, sludges, solids and other non-liquid waste containing organic hazardous substances.	— Sludges from painting operations — Tar residues from production of dyestuff intermediates — Spent filter cake from production of pharmaceuticals — Distillation bottom tars from production of phenols — Soil contaminated with spilled solvents — Slop oil emulsion solids

companies, as well as universities, hospitals, government facilities, and households. After a waste is generated, the generator can either manage the waste on site or transport it off site for treatment, disposal, or recycling, typically to a commercial hazardous waste facility. Hazardous waste managed on the site where it is generated is termed **on-site waste**. Waste managed at a site other than where it is generated is termed **off-site waste** and requires, in the United States, the use of a document termed a **manifest** for tracking its transport on a "cradle-to-grave" basis (see Fig. 2-5).

National Picture of Waste Generation

In the United States, generators are required to submit biennial reports of the generation, management, and final disposition of hazardous waste. The most recently analyzed data are those for 1987.[18] The data provide an interesting overview of the hazardous waste generation and management dynamic as follows:

- 17,677 large quantity generators (those generating at least 1000 kilograms of waste per month) filed a report for 1987.
- These establishments generated about 238 million tons of hazardous waste.
- More than 90% of all hazardous waste was categorized as hazardous wastewater (it was aqueous liquid waste). Most of the hazardous wastewater is regulated as a hazardous waste merely because it is corrosive (having a pH less than or equal to 2.0 or greater than or equal to 12.5). Typically, these wastes pose a low degree of hazard, and almost all of it is treated on-site, frequently by neutralization, and discharged to either a surface stream or a municipal sewer.
- Only 2% of the generators accounted for 98% of the waste. More specifically:

  5 largest generators: >57% of the waste
  50 largest generators: >90% of the waste

- About 70% of the generators produced less than 100 tons each in 1987, "cumulatively accounting for only 0.1% of the nationwide generation." This excludes the even greater number of generators, referred to as **small quantity generators**, who are exempt from the biennial reporting requirements because they generate less than 1000 kilograms per month (about 13 tons per year).
- The nation's hazardous waste was managed at 3,308 treatment, storage and disposal facilities. About half of these facilities had capability only for storage. A vast majority of the facilities were on-site facilities.
- Of the total quantity of hazardous waste, 13% was disposed of in deep underground injection wells (mostly hazardous wastewater), 1% in landfills, and 0.4% by incineration, with the remainder mostly discharged to sewers or water bodies after treatment.
- Only 3% of the generated hazardous waste was managed off-site, virtually all at commercial facilities. Assuming that all of the hazardous wastewater was managed on-site, the split for the non-wastewater wastes was 70% on-site management and 30% off-site (see Fig. 1-4).

70%
On-site
management

30%
Off-site
management

10%
Hazardous waste other
than hazardous wastewater

90%
Hazardous wastewater

FIGURE 1-4
Generation of hazardous waste in the United
States in 1987 (by weight).

- About 46% of the off-site quantities were transported to facilities in a state other than where the waste was generated.

The high volume of the hazardous wastewater category distorts the national picture. Most public concern is focused on the non-wastewater types of waste, particularly that portion shipped off-site to commercial facilities. These facilities typically accept waste from out-of-state generators. In response to the concerns of their citizens, some state governments have attempted to restrain the import of waste; however, the interstate commerce clause of the U.S. Constitution has prevented these actions.

Overview of Sources and Types

As part of an effort to determine the need for new off-site facilities, many state governments have analyzed in detail the waste that is generated within their borders and managed off-site. The data developed in these efforts provide a relatively accurate and detailed basis for examining waste generation patterns. Table 1-3 summarizes the specific industrial sources of waste manifested in two states with large amounts of hazardous waste generation. Table 1-4 examines the types of waste manifested in these same states in terms of the engineering classification system presented in Table 1-2, which proves to be a very suitable scheme for depicting the dynamic situation of hazardous waste generation.

The two tables demonstrate the complex aspects of hazardous waste generation even within two states that border one another. Motor oils and other oils are generated in both states. However, they are regulated as hazardous waste in New Jersey but were exempt from regulation in Pennsylvania at least for 1987. Hence, the quantities of oil wastes manifested in Pennsylvania are nil.

Pennsylvania has a larger population and a larger industrial base than New Jersey and hence generates more non-oil hazardous waste. However, the types of waste generated and the sources of the waste vary considerably. New Jersey contains more

TABLE 1-3

Industrial sources of hazardous wastes generated and manifested in 1987 (percent of statewide total)

Standard industrial classification code	Industry	New Jersey[19]	Pennsylvania[20]
22	Textile mill products	0.1	0.1
24	Lumber and wood products	0.3	<0.1
25	Furniture and fixtures	<0.1	0.2
26	Paper and allied products	0.4	0.3
27	Printing and publishing	0.3	0.2
28	Chemicals and allied products	18.3	15.0
29	Petroleum and refining	7.3	1.5
30	Rubber and miscellaneous products	0.3	0.4
31	Leather and leather goods	0.1	< 0.1
32	Stone, clay, glass, and concrete	0.7	1.5
33	Primary metals	5.0	39.5
34	Fabricated metals	6.1	5.8
35	Machinery, except electrical	2.0	1.0
36	Electrical and electronic machinery	1.7	14.7
37	Transportation equipment	3.0	0.8
38	Instrumentation	0.9	2.7
39	Miscellaneous manufacturing	0.6	0.2
40–47	Transportation services	5.6	–
48	Communications	0.2	<0.1
49	Electrical and gas services	1.8	0.2
50–59	Wholesale and retail trade	6.9	0.6
73	Business services	9.0	0.3
76	Miscellaneous repair services	12.8	7.9
–	Other/undetermined	16.5	6.9

chemical plants and refineries. Thus, the chemical products group and the petroleum and refining group are the two manufacturing industries with the largest percentage of the state's manifested waste (total of 25%). Pennsylvania, famous for its steel industry, has 40% of its manifested waste generated by the primary metals industry alone. The standard industrial classification (SIC) code 76 (miscellaneous repair services) includes commercial hazardous waste facilities, and Table 1-3 indicates that this group is the second largest source in New Jersey and fourth in Pennsylvania. The wastes manifested by this group are either treatment sludges being sent to other facilities for final disposal, or they are wastes that have been consolidated and reprocessed at transfer stations prior to treatment or disposal. The industrial categories of retail trade and business services in New Jersey may include tank cleaning and remediation services and are accounted for differently in Pennsylvania.

Different categories of waste generators produce different types of waste. The large organic chemical industry based in New Jersey results in a larger portion of the waste in that state being organic, whether liquid, sludge or solid. The large primary metals industry in Pennsylvania results in a different situation. In Pennsylvania, nearly

TABLE 1-4

Types of hazardous wastes generated and manifested in 1987 (quantities rounded to thousands of tons)

	New Jersey[21]	Pennsylvania[22]
Aqueous liquids		
Corrosive with metals	11	179
Contains cyanide	3	3
Contains hexavalent chrome	2	17
Corrosive, not otherwise specified (NOS)	16	82
Inorganics, NOS	2	< 1
Organics, NOS	46	< 1
Aqueous liquids, NOS	23	1
Total aqueous liquids	103	282
Organic liquids		
Non-halogenated solvents	13	56
Halogenated solvents	6	14
Solvents, NOS	32	7
Combustible, NOS	52	36
Contains PCBs	3	≤ 1
Total organic liquids	106	113
Oils		
Automotive oils	54	0
Industrial oils	35	1
Fuel oils	21	0
Oils, NOS	12	0
Total oils	122	1
Organic sludges and solids		
With organics	50	8
Paint residues	< 1	0
Oily residues	43	10
Combustible, NOS	9	0
Sludges and solids, NOS	9	0
Total organic sludges and solids	111	18
Inorganic sludges and solids		
With heavy metals	66	192
Corrosive, NOS	3	0
Reactive, NOS	2	0
Toxic, NOS	< 1	–
Sludges and solids, NOS	6	3
Total inorganic sludges and solids	77	195
Others	20	19
Grand total	**539**	**628**

half the wastes are aqueous liquids with most of the aqueous liquid being pickle liquor generated by the steel industry. Of the inorganic solid waste with metals, 44% in Pennsylvania comes from electric arc furnace dust.

Manufacturing Sources

What are the sources of hazardous waste? Most people associate the generation of hazardous waste with large chemical plants. While the chemical manufacturing industry represents a large source, Table 1-3 shows that other manufacturers (i.e., the chemical users) cumulatively produce far more. Primary manufacturers, those who prepare products from naturally occurring materials, may be thought of as the dominant source. Yet, waste generation extends beyond the primary and secondary manufacturing operations to include those companies that assemble and finish the final products demanded by our society for its style of living.

Virtually all manufacturing operations result in the generation of residuals because no production process can transform all input materials into products or services. Dependent upon economics and other factors, these "by-products" and "non-product output" can become wastes. Fashioning the steel, aluminum, plastic, and other components into automobiles and household items will generate abrasives and oils. Covering durable goods with attractive and protective finishes generates cyanides, solvents, concentrated acids, and paint sludges. Production of medicines yields organic solvents and other diverse residues, possibly containing toxic metals. Production of textiles generates heavy metal solutions, dyes, and solvents. The list of examples continues with electronic components, medical equipment, machinery, and publications.

Manufacturing waste may result from many different sources:[23]

- Spent material—an input material that has been used and can no longer serve the purpose for which it was produced without reprocessing;
- By-products—material generated in the specific process of making a product and which has no use in its generated form without further processing;
- Treatment—sludges from treating wastewater, controlling air emissions, or even from treating or recovering other hazardous waste; and
- Commercial chemical products—an actual product that becomes a waste because of any of numerous reasons:

    1. Cleanup of process equipment, sometimes with chemical cleaners, such as alkalis, that are hazardous by themselves;
    2. Failure to meet manufacturer's specifications because of production start-ups/shutdowns, upsets, breakdowns, or other factors;
    3. Accidental spills or leaks of raw material or product;
    4. Residue from containers used for raw material or product; and
    5. Outdated shelf life.

The generation of waste usually correlates with production and technology. For example, in a comprehensive study of hazardous waste generation in Illinois, the following observations were made:[24]

- The manufacture of paint generated 4% to 6% of total production by weight as hazardous waste.

- The manufacture of steel generated 15 to 25 pounds of electric furnace dust per ton of steel produced.
- The manufacture of printing ink generated 1% of total production by weight as hazardous waste.

National data presented previously show that waste generation when including on-site waste is skewed toward the largest generators. This holds true even for manifested wastes. In New York State, an analysis showed that nearly 50% of the total quantity of manifested waste came from just 35 out of about 1,300 generators who manifested waste.[25] Similarly, in Missouri, only 28 generators in the entire state collectively accounted for 72% of the manifested waste.[26] In New Jersey, 93% of the manifested waste came from 7% of the generators in 1990.[27]

Small Quantity Generators

A wide range of waste is generated by service industries such as dry cleaners, automobile maintenance shops, and photographic film processors. Analytical laboratories at research and educational institutions and even the common household generate hazardous waste. The individual quantity of waste generated by such an operation is small. In fact, almost all generate less than 1000 kilograms/month, the level that exempts generators in the United States from most hazardous waste regulations. This exemption applies to both manufacturing sources and service industries, and such generators are referred to as **small quantity generators**.

Waste from a small generator, if mismanaged, has the same potential for harm as does fully regulated waste from larger sources. It is often disposed of in municipal landfills, in sanitary sewers, or in other ways not intended for hazardous waste disposal. It has even resulted in the death of two boys who were playing in an industrial trash storage bin and inhaled toluene generated from cleaning rollers for printing presses.[28]

Surveys of small generators in Missouri and New York provide the comparisons shown in Table 1-5 to indicate the magnitude of the small generator issue. Most of the small quantity generator wastes are from automobile service establishments. In fact, 85% of the total waste in Missouri from small generators consisted of used motor oil and discarded lead-acid batteries, most of which were reclaimed. Other waste included spent solvents, acids/alkalis, and dry cleaning filtration residues. The

TABLE 1-5
Number of generators of small quantities of hazardous waste

Quantity of waste generated	Number of generators of hazardous waste in 1985	
	New York State[29]	Missouri[30]
> 1000 kg/mo	1300	500
100–1000 kg/mo	10,000	4000
< 100 kg/mo	27,000	9000

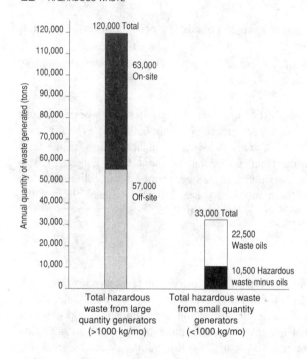

FIGURE 1-5
Comparison of sources of hazardous waste in Missouri.[31]

cumulative quantity of hazardous waste from small generators can amount to more than 10% of the total quality generated in a state, if not greater than 20%, as shown in Fig. 1-5.

Household Hazardous Waste

The sources of hazardous waste extend even to the common household. A wide range of household products, when discarded, have the characteristics of hazardous waste. Pesticides, paint products, household cleaners, hobby chemicals, and automotive products frequently contain hazardous substances. Federal law specifically exempts household hazardous waste from regulation. Nevertheless, some local and state governments have implemented programs to educate the public about household hazardous waste and to operate waste collection programs. Data from various studies suggest that household hazardous waste collectively amounts to a significant quantity, perhaps as much as the cumulative amount from small generators.[32]

1-7 CONTAMINATED SITES

The preceding sections have dealt with that hazardous waste currently being generated. The management of this waste is now subject to extensive regulation; however, such was not always the case. In the past, much waste was dumped indiscriminately or disposed of in inadequate facilities. These problems went ignored, as did spills of product or leaks from tanks. These practices contaminated sites with hazardous substances that can pose a threat to human populations. A typical scenario is that contaminants released from a site migrated into the subsurface where infiltrating rainfall, spilled solvents, or ground water gradually transported the contaminants to points of ground water use or surface discharge.

A great number of such sites exist in the United States. Many took decades to be created, and all have had years of neglect. Defining the problem, especially at sites with multiple contaminants or with a complex environmental setting, can take years and considerable effort before remediation can begin. Remediation can be even more demanding. In a survey of state governments, it was estimated that the average elapsed time per site for the investigation and construction of a remedy is 4.5 calendar years.[33] Implementation of some remedies can take decades. **Collectively, the remediation of contaminated sites dwarfs the management of currently generated hazardous waste on virtually whatever scale one would use.**

The potential extent of the problem is illustrated with a few basic data. The Superfund law requires the EPA to identify contaminated sites and long-closed disposal sites, and to maintain a list (the National Priority List) of the sites deemed to represent the greatest potential threat. As of August, 1990, there were 1187 sites on this list, referred to as "Superfund" sites. The EPA culls its National Priority List on a continuing basis from a much larger list of suspected sites of contamination. This larger list is termed CERCLA Information Service (CERCLIS). The CERCLIS list started with a total of more than 10,000 sites. Because newly discovered sites are added to CERCLIS, it soon grew to 20,000 and was approaching 30,000 as of 1993.

Superfund is just one program for addressing site contamination. Other programs include:

- Sites remediated by private parties to avoid government-regulated programs;
- Sites remediated under state Superfund programs—many states have their own programs for sites not included on the National Priority List;
- Currently operating or closed hazardous waste treatment, storage, or disposal facilities regulated as a hazardous waste facility under RCRA—about 4600 such facilities exist (3000 active and 1600 closed) containing as many as 36,500 separate waste management units;[34]
- Sites remediated under federal programs other than EPA's—both the Department of Defense and the Department of Energy have massive programs for remediating their properties;
- Property transfer initiated remediation—the buyer of property may demand remediation of contamination; some states require this before the legal transfer can occur; and
- Leaking underground storage tanks—there are perhaps as many as one to two million underground storage tanks in the United States and a significant percentage have leaked some of their product. Because these sites represent loss of product and not mismanagement of waste, most are remediated under separate federal and state programs and as such typically are not included on CERCLIS.

TABLE 1-6
Potential number of contaminated sites in New Jersey[35]

Program	Estimated total number of sites/facilities	Potential number of sites/facilities requiring remediation
Superfund's National Priority List	100	100
Current privately funded sites	107	107
Current state-funded sites	23	23
Planned additions to the above programs	71	71
CERCLIS	1239	205
Hazardous waste treatment, storage, and disposal facilities	685	607
Property transfer initiated remediation	–	3000
Underground storage tanks	30,000	9900

The potential size of the total remediation effort is illustrated by the data in Table 1-6 developed by the New Jersey Department of Environmental Protection.

In 1991 a university study estimated the cost of remediating contaminated sites in the United States. The estimate covered all government programs as well as privately funded efforts. The remediation of nuclear weapons complexes was included. The study placed the total cost over the next 30 years at $750 billion. This figure was based on current policies and technologies.[36]

The enormous cost of remediating sites raises an important point about the future extent of remediation programs. The question is not whether contaminated sites exist; literally tens of thousands do exist. What is questionable is the level of risk posed by the sites and the degree of remediation necessary to reduce risks to acceptable levels. There is not a consensus that these sites pose, relative to other environmental problems, a level of risk that warrants the amount of societal investment necessary for their remediation. A policy decision to accept a higher level of risk would probably have two effects: at first, it would accelerate the pace at which sites are remediated, and second, it would eventually result in slackening the program. Nevertheless, site remediation would probably still represent a major environmental program even if it were slackened considerably.

1-8 FUTURE ENDEAVORS

John Quarles, the EPA's first general counsel, recently reviewed the nation's hazardous waste management strategy.[37] He observed that in 1970, when the nation's first Earth Day signaled the arrival of the modern environmental movement, hazardous waste management was an almost totally neglected subject. This has changed remarkably. He

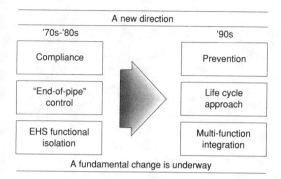

A new direction	
'70s-'80s	'90s
Compliance	Prevention
"End-of-pipe" control	Life cycle approach
EHS functional isolation	Multi-function integration
A fundamental change is underway	

FIGURE 1-6
A new direction for corporations.

noted that the federal hazardous waste laws have ". . . come to dominate environmental legal practice." He goes on to say that "At the current pace Superfund cleanups will require thirty to fifty years at a minimum," and that "Superfund sites are only one part of a much broader national waste contamination problem."

Environmental management, of which hazardous waste management is just one segment, is progressing through a fundamental change as shown in Fig. 1-6. For example, the Environmental, Health, and Safety (EHS) function in most corporations was an isolated entity that primarily dealt with compliance with laws. Today this function typically reports directly to the CEO and is involved in all aspects of corporate activity.

Clearly, the management of hazardous waste represents a challenge and opportunity for the environmental engineering and science professions. While remediation of contaminated sites currently commands the most attention, the increasing costs of managing today's hazardous waste, in combination with the legal foundation that makes generators liable for contamination in the long term, even for past practices, has prompted a new way of managing hazardous waste. Great effort will be expended to reduce, at its source, the quantity of waste generated and its degree of hazard. Looking further ahead, attention should focus on **product life cycle**, a systematic approach for examining the environmental and health consequences of a product at each stage of its life cycle and addressing such consequences in an integrated, cross-functional decision-making process. This results in products that are environmentally sound. Example considerations at each stage are shown in Fig. 1-7. Indeed, hazardous waste management is creating great demand for specialized fields of study within and combining the areas of engineering, geology, chemistry, and biology.

1-9   TOWARDS AN ENVIRONMENTAL ETHIC

Why a section on ethics in a book on hazardous waste? The management of hazardous wastes is not simply a technical problem but is confounded by legal, social, political, and, yes, ethical considerations. Engineers and scientists formulate conclusions and make recommendations that can easily result in *de facto* decisions that affect society as a whole. Embracing ethical standards will ensure that the public trust is not betrayed.

Life cycle stage

	Product design	Raw materials	Manufacture, sales, distribution	Packaging	Use	Final disposal
Example considerations	• Chemical usage • Ease of recycling • Ease of reuse • Ease of disassembly (e.g., modular design) • Size	• Use of renewable resources • Reducing impact from transport	• Pollution prevention, waste minimization • Process hazards management	• Reduced packaging • Vegetable-based inks • Concentrating product	• Aerosol-free cleaning agents • Low-dosage pesticides	• Recycling • Degradability • Combustion products

FIGURE 1-7
Product life cycle considerations.

It is not possible, in this brief section, to provide the tools necessary for resolving the types of ethical problems that tomorrow's hazardous waste manager will confront. Rather our objectives here are more modest: to impress upon the reader that such considerations are of paramount importance and to show that help is available.

Ethical Theories

But where is this help to come from? Philosophers have been studying ethical questions for centuries. While they may not provide simple answers to the type of complex questions likely to be encountered, they do provide help in how to think about ethical questions.

The question of what is and is not ethical is not easily answered. In everyday life, one follows a broad set of personal rules that may include "do not cause pain" and "do not lie." Conflict among the rules may occur which raises an ethical test (e.g., what should an individual do when the only way to avoid causing pain to another is to lie?). How can an individual then do both—be honest and avoid causing pain?

In engineering, rules of conduct are presented as various codes of ethics covering professional practice. For example, the engineer is admonished to be truthful in all dealings with the public. A code will usually express that the engineer is to remain loyal to the client or employer. But what about the situation when the engineer is privy to confidential information about a client's past hazardous waste disposal practices and is asked by a regulatory agency to reveal what he or she knows. How does an individual remain loyal and retain integrity?

The answers to dilemmas such as these have been debated for more than two millennia, without definitive resolution. The search for the solutions to these questions has led to the development of a number of theories of ethics, most of them based on one of two general perspectives:

- Those focusing on the *act,* and
- Those focusing on the *consequences* of the act.

In the former, one is expected to follow the strictures of moral action and never to deviate from them. The consequences are not of any importance. If lying is not accepted, then one should never lie, regardless of the circumstance. Under this type of theory, an individual is acting ethically if he or she does not perform an act that is considered unethical.

Much of the ethical theory based on the importance of the act is based on the work of a German philosopher, Immanuel Kant (1724–1804). Perhaps his most important contribution was his *Categorical Imperative*:

> There is . . . only a single categorical imperative and it is this: Act only on that maxim through which you can at the same time will that it should become a universal law.[38]

Under this theory, engineers would include in their set of ethical values only those rules that they are willing to observe and accept as inviolable (i.e., universal precepts that apply in all circumstances). If one can identify circumstances where a specific rule is inappropriate, the categorical imperative is not met and the rule is defective and lacking in ethical force.

The consequentialist ethical theory is credited in great part to the works of Jeremy Bentham (1748–1832) and John Stuart Mill (1806–1873). In this theory, sometimes termed **utilitarianism**, the consequences of the action are calculated in terms of how much pleasure and pain they produce for all people involved, and the ethical action is the one that has the greatest benefit for the greatest number of people. This is a very comfortable ethical system for engineers who are often placed in situations where it is necessary to calculate the benefits and costs of various projects, making it a "simple" matter to recommend the option having the highest benefit/cost ratio.

Unfortunately, the public does not always appreciate the beauty of utilitarian calculus. For an engineer, it is perfectly reasonable to calculate the public health effect of a hazardous waste incinerator, yielding an estimate that it has the potential to cause one excess case of cancer in a population of one million over a seventy-year period. But is this "cost" acceptable? The public does not care about these numbers. It understands that there will be an additional case of cancer and thinks that this is not right. Who will be the one to bear the unfair burden of this facility? By the public's reasoning, the very *act* of causing one person to get cancer is not ethical, regardless of what the engineers think.

Such a divergence of ethical thinking is very difficult for many engineers and scientists to understand, and they will often attribute the reaction of the public to "technical illiteracy." It is not technical illiteracy that is the problem, but a divergence of value systems. The question is how to make decisions where values conflict. Perhaps some resolution can be found in Hans Jonas's *The Imperative of Responsibility*. He suggests that our first duty as citizens in a technological society is to visualize the long-range effects of technology, and this is followed by a second duty: "Summoning up a feeling appropriate to what has been visualized."[39] These are critical elements in what Jonas calls an "Ethic of the Future."

Environmental Ethics

The previous discussion of ethical considerations addresses our behavior toward other persons, both individually and collectively as a society. But what is environmental ethics? In 1949, Aldo Leopold provided an exposition of environmental ethics in an essay termed "The Land Ethic." In commenting on the conservation attempts in Wisconsin, he bemoaned the fact that conservation decisions were being based solely on the economic value of the species in question:

> One basic weakness is a conservation system based wholly on economic value. Wild-flowers and songbirds are examples. Of the 22,000 higher plants and animals native to Wisconsin, it is doubtful whether more than 5 percent can be sold, fed, eaten or otherwise put to economic use. Yet these creatures are members of the biotic community, and if (as I believe) its stability depends on its integrity, they are entitled to continuance.[40]

Is it sufficient to treat the impact on the environment as essentially value free, unless that impact noticeably affects people? Gunn and Vesilind[41] have termed this approach **speciesism**, a term akin to racism or sexism. This anthropocentric view of the world is probably the most prevalent in today's society, but attitudes towards the intrinsic value of nature are changing. At the time of the first Earth Day in 1970, it would have been impossible to imagine that the construction of a multi-million dollar dam would be held up over concern for an endangered species, a minnow named the snail darter.

Although there is a greater acceptance of the intrinsic value of nature today than in Leopold's time, we probably are still making decisions that Leopold would consider foolish and shortsighted. One reason for this can be found in an essay entitled "The Tragedy of the Commons" by Garrett Hardin.[42] In it Hardin tells the story of a common pasture owned and used by farmers for grazing cattle. Because the commons is the property of all, there is nothing to prevent any farmer from adding an additional animal to the herd. All of the benefits (i.e., the value of an additional fatted calf) accrue to the farmer. The costs (the potential for overgrazing the commons) are not borne by the farmer alone but by all those using the commons. Each farmer being rational decides to maximize personal gain by maximizing the number of animals on the common. The eventual result of these individual decisions is a total destruction of the pasture through overgrazing. While Hardin directed his essay particularly at overpopulation and birth control, there is a clear message for the protection of all environmental resources. Protection is provided by laws; however, as shown earlier, law does not always suffice.

The Law

As will be seen in Chap. 2, much of the decision-making in the hazardous waste field is driven by ever more restrictive laws and regulations. If in fact these laws do represent a set of standards that our society has agreed upon, then why should we be concerned with ethical considerations? Isn't simply obeying the law enough?

The short answer is that the regulatory system presents a moving target. In addition, enforcement of all these laws is highly imperfect. For most corporations simply obeying the law is *not* enough. Informed corporations know that the laws will change and society will hold them to a higher standard. As discussed with site remediation, past practices done in compliance with laws in existence at the time can result in major future liabilities. Major environmental incidents, such as the Exxon Valdez oil spill in Alaska or the release of methyl isocyanate at the Union Carbide Plant in Bhopal, India, threatened the existence of these industrial giants. Smaller corporations know that such incidents could easily bankrupt them.

The current tendency is for corporations to move toward the use of corporate codes of conduct that require more than meeting the law. This is not done out of altruism, but rather due to a heightened environmental awareness on the part of corporate officials and heightened environmental expectation of the public. In short it is simply good business. An example of this trend is illustrated in Sec. 7-3 on Life Cycle Analysis. Using this method, corporations re-evaluate all of their activities from product research and raw materials purchase through the manufacturing and packaging of the final product to determine the impact on the environment.

Codes of Ethics

Engineers and scientists examine the health and environmental impacts of hazardous waste problems and generally have a good understanding of the degree to which mere compliance with the law will address the impacts. Sophisticated clients will appreciate learning whether the law will suffice. What if the client decides to just comply when more seems proper? Help to answer this and earlier ethical questions can be found expressly in formal codes adopted by a number of engineering societies for governing the conduct of their members.

The Code of Ethics of Engineers adopted by the Accreditation Board for Engineering and Technology (ABET) is illustrated in Fig. 1-8. One characteristic of this and most other codes is the first canon: "To hold paramount the safety, health and welfare of the public."

The National Society of Professional Engineers (NSPE) expands on this fundamental canon by noting in Part II of its code (Rules of Practice):

a. Engineers shall at all times recognize that their primary obligation is to protect the safety, health, property and welfare of the public. If their professional judgment is overruled under circumstances where the safety, health, property or welfare of the public are endangered, they shall notify their employer or client and such other authority as may be appropriate.[43]

The language in the ABET, NSPE, and other codes should make the engineers' obligations to public health unequivocal. Ethics are not something that can be forced on a person. However, these codes represent the feelings of individuals who have long ago experienced ethical tests and decided that their profession and society would be better served by such actions.

CODE OF ETHICS OF ENGINEERS
THE FUNDAMENTAL PRINCIPLES

Engineers uphold and advance the integrity, honor and dignity of the engineering profession by:

   I. using their knowledge and skills for the enhancement of human welfare;

  II. being honest and impartial, and serving with fidelity the public, their employers and clients;

 III. striving to increase the competence and prestige of the engineering profession; and

 IV. supporting the professional and technical societies of their disciplines.

THE FUNDAMENTAL CANONS

1. Engineers shall hold paramount the safety, health and welfare of the public in the performance of their professional duties.
2. Engineers shall perform services only in the areas of their competence.
3. Engineers shall issue public statements only in an objective and truthful manner.
4. Engineers shall act in professional matters for each employer or client as faithful agents or trustees, and shall avoid conflicts of interest.
5. Engineers shall build their professional reputation on the merit of their services and shall not compete unfairly with others.
6. Engineers shall act in such a manner as to uphold and enhance the honor, integrity and dignity of the profession.
7. Engineers shall continue their professional development throughout their careers and shall provide opportunities for the professional development of those engineers under their supervision.

FIGURE 1-8
Engineering code of ethics.[44]

DISCUSSION TOPICS AND PROBLEMS

1-1. What are some of the problems encountered in finding a working definition for *hazardous waste*?

1-2. Is the form (e.g., solid, sludge, liquid, or gas) of a waste important in deciding if it is hazardous?

1-3. What is the purpose of grouping the different wastes according to characteristics (e.g., toxic) in classification lists? What are some of the problems with coding systems such as these?

1-4. Why is the generation of hazardous waste not only a result of large-scale industry? What are some other sources?

1-5. Why is it necessary to include a section on ethics in this book? What are some of the ethical considerations regarding decisions engineers have to make when working in the hazardous waste management field? Give examples.

1-6. In manufacturing a generic product, name five potential sources of hazardous waste generation.

1-7. Contrast Hardings's *Tragedy of the Commons* with Leopold's *Land Ethic*. Give an example of how they apply to an environmental problem in the 1990s.

1-8. Why does the definition of *hazardous waste* specifically exclude *radioactive waste*?

1-9. Contrast the incident at Minamata Bay with that Love Canal, noting similarities and differences.

1-10. Describe the impact advances in analytical chemistry have had on the hazardous waste management field.

1-11. In the United States, generators are responsible for hazardous waste they create from "cradle to grave." Explain what is meant by this phrase.

1-12. What factors might explain the dramatic difference between the public's perception of the relative risk of hazardous waste sites and the opinions of an EPA expert panel (Table 1-1)?

1-13. What are the four characteristics that make a waste hazardous in the United States? Can you suggest other characteristics that might be appropriate to such a listing?

1-14. Explain why "degree of hazard" is not a factor in determining whether or not a waste is hazardous in the United States.

1-15. Review the hazardous waste facilities plan for your state and prepare a one-page summary of the most serious problems.

1-16. What are *small quantity generators*?

1-17. Review the summaries of generation in New Jersey and Pennsylvania (Tables 1-3 and 1-4). Explain the differences between these two contiguous states that cause the differences in waste generated.

1-18. Explain briefly the factors that might influence the amount of hazardous waste generated in a given year.

1-19. What is *household hazardous waste*?

1-20. What is the *National Priorities List (NPL)*, and how does it differ from the CERCLIS List?

1-21. Explain what is meant by the *Product Life Cycle* approach to environmental management. Why have many corporations adopted this proactive approach?

1-22. Describe briefly the two general perspectives for dealing with ethical problems.

1-23. When it became known that a new process planned for a chemical plant was expected to produce a highly toxic waste, a plant environmental engineer wrote to the city newspaper expressing opposition to the action. Under what circumstances would the engineer's action be legal and/or ethical?

REFERENCES

1. Batstone, R., J. E. Smith, Jr., and D. Wilson: "The Safe Disposal of Hazardous Waste," World Bank Technical Paper Number 983, Washington, D.C., 1989.
2. Ibid.
3. "Overview of the Environmental Consulting Industry and ERM, Inc.," presentation to Environmental Resources Management, Inc. shareholders, Alan L. Farkas, Farkas, Berkowitz and Company, Washington, D.C., May 15, 1992.
4. Carson, Rachel: *Silent Spring*, Boston, Massachusetts, Houghton Mifflin Co., 1962.
5. Tsu baki, T., and K. Iru Kayama (eds.): *Disease Minimata*, Kodansha Ltd., Tokyo, Elsevier Scientific Publ. Co., Amsterdam, 1977.

6. Dunckel, A. E.: "An Updating on the Polybrominated Biphenyl Disaster in Michigan," *J. Am. Vet. Med. Assoc.*, vol. 167, pp. 838–43, 1975.

7. Brown, Michael: *Laying Waste*, Pantheon Books, New York, New York, 1979.

8. U.S. Environmental Protection Agency: *Report to Congress on Hazardous Waste Disposal*, June 30, 1973.

9. Dowling, Michael: "Defining and Classifying Hazardous Waste," *Environment*, vol. 27, no, 3. April 1985.

10. U.S. Environmental Protection Agency: *Unfinished Business: A Comparative Assessment of Environmental Problems*, Office of Policy Analysis, Washington, D.C., February 1987.

11. Ibid.

12. Ibid.

13. U.S. Environmental Protection Agency: *Explaining Environmental Risk*, Office of Toxic Substances, Washington, D.C., November 1986.

14. Ibid.

15. Dowling, *Defining and Classifying Hazardous Waste*, op. cit., p. 36.

16. U.S. Code of Federal Regulations, 40 CFR261.1, Promulgated May 1980, and amended January 1985.

17. Dowling, *Defining and Classifying Hazardous Waste*, p. 36.

18. U.S. Environmental Protection Agency: "1987 National Biennial RCRA Hazardous Waste Report," Office of Solid Waste, Washington, D.C., July 8, 1991.

19. Environmental Resources Management, Inc.: "New Jersey Hazardous Waste Facilities Plan Update," report prepared for New Jersey Hazardous Waste Facilities Siting Commission, Trenton, New Jersey, October 1989.

20. ERM, Inc. Pennsylvania Hazardous Waste Facilities Plan, Pennsylvania Department of Environmental Resources, 1992.

21. ERM, Inc.: "New Jersey Hazardous Waste Facilities Plan Update," op. cit.

22. ERM, Inc. Pennsylvania Hazardous Waste Facilities Plan, op. cit.

23. U.S. Code of Federal Regulations, 40 CFR261.1, Promulgated May 1980, and amended January 1985.

24. Environmental Resources Management, Inc.: "Statewide Hazardous Waste Generation Study," prepared for Illinois Hazardous Waste Research and Information Center, Champaign, Illinois, HWRIC RR 002, August 1985.

25. ERM-Northeast, Inc.: "New York State Hazardous Waste Facilities Needs Assessment," prepared for New York State Department of Environmental Conservation, Albany, New York, March 1985.

26. Environmental Resources Management, Inc.: "Missouri Hazardous Waste Treatment and Resource Recovery Facility Feasibility Study," prepared for Missouri Environmental Improvement and Energy Resources Authority, Jefferson City, Missouri, January 1985.

27. Personal communication with Susan Boyle, Executive Director, New Jersey Hazardous Waste Facilities Siting Commission, May 27, 1992.

28. Rosen, Marty: "Intoxicating Solvent Cited in Bin Deaths," *St. Petersburg Times*, St. Petersburg, Florida, June 20, 1992.

29. ERM-Northeast, Inc.: "New York State Hazardous Waste Facilities Needs Assessment," op. cit.

30. Environmental Resources Management, Inc.: "Analysis of Hazardous Waste Generation and Management by Small Generators and Households in the State of Missouri," prepared for Missouri Environmental Improvement and Energy Resources Authority, Jefferson City, Missouri, August 1985.

31. Ibid.

32. Ibid.

33. Association of State and Territorial Solid Waste Management Officials: *State Programs for Hazardous Waste Site Assessments and Remedial Actions*, Washington, D.C., June 1987.

34. U.S. General Accounting Office: *Hazardous Waste: Limited Progress in Closing and Cleaning Up Contaminated Facilities*, Report to the Chairman, Environment, Energy, and Natural Resources Subcommittee, Committee on Government Operations, House of Representatives, GAO-RCED-91-79.

35. ERM, Inc., "New Jersey Hazardous Waste Facilities Plan Update," op. cit.

36. "Researchers: Cleanups May Top $1 Trillion," *Superfund*, vol. 5, no. 25, Pasha Publications, Inc., December 13, 1991.
37. Quarles, John: "In Search of a Waste Management Strategy," *Natural Resources and Environment*, vol. 5, Summer 1990.
38. Kant, Immanuel: *The Metaphysic of Morals [1797]*, Bobbs-Merrill Publishers, Indianapolis, 1964.
39. Jonas, Hans: *The Imperative of Responsibility—In Search of an Ethics for the Technological Age*, University of Chicago Press, Chicago, p. 28, 1984.
40. Leopold, Aldo: *A Sand County Almanac, and Sketches Here and There*, Oxford University Press, Cary, North Carolina, 1949.
41. Gunn, A. S., and P. A. Vesilind: *Environmental Ethics for Engineers*, Lewis Publishers, Boca Raton, Florida, p. 22, 1986.
42. Hardin, Garrett: "The Tragedy of the Commons," *Science*, vol. 162, December 13, 1968.
43. Reprinted by Permission of the National Society of Professional Engineers, Code of Ethics for Engineers, NSPE Publication No. 1102, 1993.
44. Accreditation Board for Engineering and Technology, Code of Ethics for Engineers.

ADDITIONAL READING

Attfield, R.: *The Ethics of Environmental Concern*, Columbia University Press, New York, New York, 1983.
Boraiko, A. A.: "Storing Up Trouble . . . Hazardous Waste," *National Geographic*, vol. 167, no. 3, 1985.
Carper, K. L.: "Engineering Code of Ethics: Beneficial Restraint on Consequential Morality," *J. Professional Issues in Engineering*, A.S.C.E., vol. 117, no. 3, pp. 250–57, July 1991.
Dawson, G. W., and B. W. Mercer: *Hazardous Waste Management*, John Wiley & Sons, Inc., New York, New York, 1986.
Epstein, S. S., L. O. Brown, and C. Pope: *Hazardous Waste in America*, Sierra Club Books, San Francisco, California, 1982.
Forester, W. S., and J. H. Skinner: *International Perspectives on Hazardous Waste Management*, Academic Press Inc., London, England, 1987.
Freeman, H. M., ed.: *Standard Handbook of Hazardous Waste Treatment and Disposal*, McGraw-Hill, Inc., New York, New York, 1989.
Hargrove, E.: *Foundations of Environmental Ethics*, Prentice-Hall Inc., Englewood Cliffs, New Jersey, 1989.
Hoffman, W. M., R. Frederick, and E. S. Petry, Jr., eds.: *The Corporation, Ethics and The Environment*, Quorum Books, Westport, Connecticut, 1990.
Katz, E.: "Ethics and Philosophy of the Environment: A Brief Review of the Major Literature," *Environmental History Review*, vol. 15, pp. 79–86, Summer 1991.
Kiang, Y., and A. A. Metry: *Hazardous Waste Processing Technology*, Ann Arbor Science Publishers, Inc., Ann Arbor, Michigan, 1982.
Kolaczkowski, S. T., and B. D. Crittenden: *Management of Hazardous and Toxic Wastes in the Process Industries*, Elsevier Science Publishing Co., Inc., New York, New York, 1987.
Lankford, P. W., and W. W. Eckenfelder, Jr.: *Toxicity Reduction in Industrial Effluents*, Van Nostrand Reinhold Co., Inc., New York, New York, 1990.
Manahan, S. E.: *Hazardous Waste Chemistry, Toxicology and Treatment*, Lewis Publishers, Chelsea, Michigan, 1990.
Martin, E. J., and J. H. Johnson, Jr.: *Hazardous Waste Management Engineering*, Van Nostrand Reinhold Co., Inc., New York, New York, 1987.
Martin, M. W.: *Ethics in Engineering*, McGraw-Hill, Inc., New York, New York, 1983.
Mazmanian, D. and D. Morell, *Beyond Superfailure: America's Toxic Policy for the 1990's*, Westview Press Inc., Boulder, Colorado, 1992.
Nash, R. F.: *The Rights of Nature: A History of Environmental Ethics*, University of Washington Press, 1989.
O'Brien & Gere Engineers, Inc.: *Hazardous Waste Site Remediation—The Engineer's Perspective*, Van Nostrand Reinhold Co., Inc., New York, New York, 1988.
Wentz, C. A.: *Hazardous Waste Management*, McGraw-Hill, Inc., New York, New York, 1989.

White, L.: "The Historical Roots of Our Ecological Crisis," *Science*, vol. 155, pp. 1203–7, March 10, 1967.

Windsor, D.: "Government Business Relationships in Environmental Protection," in *Business, Ethics and The Environment*, W. M. Hoffman, R. Frederick, and E. S. Petry, Jr., eds., Quorum Books, Westport, Connecticut, 1990.

ACKNOWLEDGEMENTS

The authors wish to acknowledge the following individuals who assisted with the preparation of this chapter: Professor A. A. Friedman, *Syracuse University*; David Morell, Ph.D., *ERM-West*; Professor Rita Meyninger, *Brooklyn Polytechnic University*; Bruce Molholt, Ph.D., *ERM, Inc.*; Susan Boyle, New Jersey Hazardous Waste Facilities Siting Commission; Dave Levy, *U.S. EPA's Waste Management Division*; Professor Aarne Vesilind, *Duke University*; Professor Richard Fleming, *Bucknell University*; and Kurt Becker, *Bucknell University*. The principal reviewer for the **ERM Group** was Neil L. Drobny, Ph.D., P.E., *ERM-Midwest*.

CHAPTER

2

THE
REGULATORY
PROCESS

Ignorance of the law excuses no man.
John Selden

'If the law suppose that,' said Mr. Bumble, 'the law is a ass—a idiot.'
Charles Dickens

A successful hazardous waste management program must as a minimum satisfy the environmental laws and regulations. This chapter provides an introduction to the legal setting in which technical problem solving occurs. In the United States, this legal setting includes the U.S. Constitution; federal, state, and local environmental statutes and regulations; and court decisions interpreting these laws.

The chapter begins with a discussion of environmental law in the United States and then introduces the two major federal statutes dealing with hazardous waste: the Resource Conservation and Recovery Act, which deals with the overall hazardous waste program and emphasizes wastes presently being generated, and the Superfund Law, which is primarily concerned with past practices including cleaning up abandoned hazardous waste sites. The regulations derived from these laws are also discussed, followed by a brief description of other environmental statutes of the United States. A final section provides an overview of the regulatory system in other countries.

2-1 ENVIRONMENTAL LAW

To understand the laws governing hazardous waste management in the United States, engineers and scientists must be familiar with the concepts and language of the law. This section summarizes the fundamentals of environmental law and introduces some sources which can help determine how the law applies to a particular problem. For more information, the reader is referred to books on environmental law.[1,2,3,4,5]

The growth in public interest in environmental matters has spurred a concomitant expansion in environmental laws in the United States as illustrated in Fig. 2-1. The rapid proliferation of environmental laws and regulations has resulted in a "regulatory-driven industry."

Case Law vs. Statute Law

When the courts apply a law in rendering a legal decision, they consider two things:

- The actual statute or regulation, or how the law appears in written form as passed by a legislative body or administrative agency, and
- Case law, or how the statute has been interpreted by previous courts (precedent).

Statutes include acts of Congress or of state legislatures. Statutes are referred to by legal designations assigned by the body that passed them. A federal law has several designations that can appear in legal documents or reports. First, when a federal bill becomes a law, it is assigned a Public Law number, which begins with the prefix PL attached to the number of the session of Congress in which the law was passed. A specific number then identifies the law. Thus, the Resource Conservation and Recovery Act (RCRA), passed in 1976 by the 94th Congress, was designated PL94-580. All federal statutory law is codified (i.e., put together in a systematic listing) in the United States Code (USC). The USC is regularly updated with the latest amendments to statutes, and annotations are added describing case law interpretations of the statute. RCRA was amended in 1978, 1980, and 1984. It is referenced as 42 USC 6901 *et. seq.*

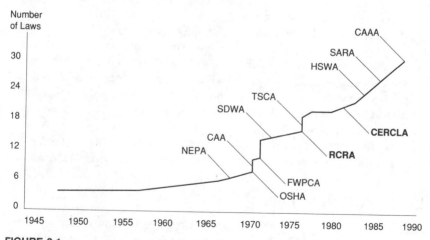

FIGURE 2-1
Growth of environmental laws in the United States.[6]

(Title 42 of the United States Code, section 6901 and the following sections). Finally there is a compendium of U.S. Statutes at Large designated *Stat*, where the Clean Water Act is designated 86 Stat. 880. The statutory law of most states is similarly codified.

When a dispute arises as to the application or meaning of a particular law in a specific instance, the courts interpret which laws apply and how they pertain to the case in question. These court decisions are referred to collectively as **case law**. The basis of a legal decision would include, among other factors, a careful look at cases interpreting the same statutory provision (precedent), legal principles (i.e., contextual and logical analysis), and the intent of the legislature when the statute was enacted. Case law is not codified in any one place, but full text opinions or decisions are reprinted in case law reporters, and summaries of court decisions are published in law reviews and case digests. The full text of recent environmental decisions of note may be found in the Environmental Reporter Cases.[7] Additionally, several computerized services provide easy access.

Common Law and Liability

The term **common law** refers to the body of case law principles that emanate from various rules of law and an intuitive sense of fairness and justice. Common law usually applies to hazardous waste management in cases of tort liability. Civil liability for hazardous waste activity is an extremely complex area of the law, and legal counsel must be sought whenever questions of liability arise.

Common laws "derive their authority solely from usages and customs of immemorial antiquity, or from the judgments and decrees of the courts recognizing, affirming and enforcing such usages and customs; and in this sense particularly the ancient unwritten law of England."[8] Moreover, because of their nature, court decisions provide the primary written statement of common law. Because the factual setting of every case is different, courts look to similar circumstances in previous decisions to provide precedents.

A **tort** may be defined as "A private or civil wrong or injury other than breach of contract, for which the court will provide a remedy in the form of an action for damages. . . . "[9]. The term **civil law** refers to that body of law governing private rights and disputes among individuals as contrasted to **criminal law**, where the state is always a party. Tort actions of importance in the hazardous waste field include nuisance, trespass and negligence. Actions have also succeeded on the basis of strict liability. These concepts are described below.

Nuisance is the "Unreasonable, unwarranted or unlawful use by a person of his own property, working obstruction or injury to the right of another. . . . Nuisance comprehends interference with an owner's reasonable use and enjoyment of his property by means of smoke, odors, noise. . . . "[10] Nuisance suits have traditionally included actions for smoke or noise, but could be brought for virtually any hazardous waste activity perceived as unreasonable by neighbors.

Trespass is defined as "The unlawful interference with one's person, property or rights."[11] The distinction between trespass and nuisance has not always been clear, but generally trespass is considered to be an intrusion on the property of another. Allowing toxic substances to pass to a neighbor's property could result in a civil suit for trespass.

Acting in a careless or unreasonable manner constitutes **negligence**. Courts usually apply the "reasonably prudent person" concept as the standard of care on which to judge the action alleged as negligence. Failing to do something that a reasonably prudent person would do to prevent harm or injury to others (or doing something such an individual would not do) is negligent conduct. Negligence requires four separate elements:

- A **Duty** towards someone (reasonably prudent person standard)
- A **Breach** of that duty
- **Causation** resulting in
- **Damages**

Sullivan has summarized the application of negligence to hazardous waste as follows:

> Persons harmed as the result of careless and improper disposal or handling of hazardous waste can recover for their losses under a negligence cause of action. Indeed, state and federal courts have long recognized this common law theory of recovery against defendants who engage in the negligent disposal of pollutants such as hazardous waste. Where negligence can be established, it is no defense that the negligent action was in full compliance with all government regulations and permit conditions. On the other hand, non-compliance with regulations or a permit in some states may be *prima facie* evidence (that is, proof without any more evidence) of liability.[12]

For a suit for negligence to be successful, the plaintiffs must prove that they were harmed and that the defendant's negligent action (or failure to act) caused the harm. In cases involving hazardous waste, it is usually difficult to associate a specific action (e.g., the release of a carcinogen) to a specific injury (e.g., the onset of cancer). Courts have allowed plaintiffs to recover damages when the plaintiffs present a sufficient body of evidence to allow the court to infer that the defendants were probably responsible, even though there was not sufficient evidence to prove a definitive connection in a scientific sense.

Prosser[13] has summarized six defenses to an action for negligence as follows:

Contributory negligence. The action of the plaintiff was in part responsible for his/her injury and a person should not benefit from his/her own wrong. For example, in cases where the plaintiff had the "last clear chance" to avoid an accident.

Comparative negligence. Where both parties bear some responsibility, there is an apportionment of responsibility between plaintiff and defendant.

Assumption of risk. A person who knowingly assumes a risk may not be permitted to recover costs if injured.

Failure to take precaution against injury. For example, an employee who fails to make use of appropriate safety equipment may reduce his/her ability to sue for negligence if injured.

Statutes of limitations. A law specifying that a claim for negligence must be filed in a specified time period.

Immunity. Under the concept of **sovereign immunity**, tort suits against the U.S. government have been barred for most of its history. The **Federal Tort Claims Act**[14] now permits limited suits against the government for damages caused by federal actions.

While it is ordinarily necessary to show negligence to prevail in court, this was not believed to be strong enough to provide the required protection. Therefore, changes in recent environmental statutes have specified that activities involving hazardous waste should fall under the concept of **strict liability**. Under this principle, proof of negligence is not required. Even if the defendant can demonstrate that the reasonable prudent person standard was exceeded and everything possible was done to properly contain the hazardous substance, if an injury occurred, the individual or company handling the hazardous material can be held liable for the injury. Courts have held that if one engages in a very hazardous activity, that person is strictly liable for any injury to others that may result. This may apply to anyone involved in any aspect of hazardous waste management.

When more than one party is liable, the situation becomes more complex. Where the damage caused by each individual defendant can be reasonably ascertained, each can be held responsible for his/her proportional share. In many environmental cases, it is not possible to apportion damages, and the principle of **joint and several liability** applies. In the case of an abandoned hazardous waste site where a number of generators or transporters have deposited hazardous waste, any or all of the parties may be held liable for clean up of the entire site because the harm is not the waste per se, but the resulting pollution (e.g., contamination of ground water). This means that, in theory, if a company was responsible for depositing only one drum of toxic material at a site where thousands of drums were left, that company could be held liable for the entire cost of removing all of the drums.

Under the tort law concept of **contribution**, a responsible party who is successfully sued for total damages may subsequently recover proportional shares from other parties whose negligence contributed to the harm. Let us examine a Superfund site where parties (companies "A" and "B") are jointly and severally liable. Company A is sued and pays for the clean-up cost. A claim for contribution would enable "A" to recover from "B" *a portion* of the costs paid ("A" recovers less than 100% of what it paid).

Indemnification is a different concept altogether. To indemnify means to secure against loss. A claim for indemnification would enable "A" to recover from "B" *all* of the costs paid. A claim for indemnification can arise in two ways:

- A contract (e.g., an agreement for engineering services) provides that "B" will indemnify "A" for the claim.

- "A" is liable for the claim but was a "passive" responsible party while "B" was the active responsible party (e.g., Engineer A is retained by industry B to study a hazardous waste site, and a neighbor (third party) sues both for perceived damages due to contamination at the site. "A" can seek indemnification from "B.").

Regulations

Because of the technical complexity of environmental issues, environmental laws often specify goals in general terms and leave it up to administrative agencies, such as the U.S. Environmental Protection Agency (EPA), to provide technical details to stipulate the exact requirements. These stipulations are then published as **regulations**. The United States government disseminates its regulations in the *Federal Register*, published daily. In a similar fashion, states publish regulations in a register or bulletin. Normally, regulations are first published as "Proposed," to allow interested parties to comment on the regulations. After a specified comment period, the issuing agency may revise the proposal in light of comments received and then publish a final regulation that is legally binding (see Federal Hazardous Waste Regulations in Sec. 2-3).

The power of a federal agency such as the EPA to write regulations that have the force of law must come from the Congress in an enabling statute. Such regulations must be in accordance with the statute and not be arbitrary and capricious. Practices and proceedings before agencies of the U.S. Government are governed by the **Federal Administrative Procedures Act.**[15]

Federal Rules and Regulations published in the *Federal Register* are codified in the **Code of Federal Regulations** (CFR). Title 40 of this code covers Protection of Environment, and 40 CFR 261 refers to Part 261: Identification and Listing of Hazardous Wastes. The past two decades have seen an explosion of environmental regulations in the United States as illustrated in Fig. 2-2. The distribution of the 291 regulations under development in 1989 is illustrated in Fig. 2-3.

State regulations are usually codified in a similar fashion. For example, the New York regulations covering identification and listing of hazardous waste are designated in the New York Code of Rules and Regulations as 6 NYCRR Part 366. To avoid duplication of administrative and enforcement efforts at both the state and federal level, EPA will often confer **primacy** on a state agency allowing that agency to administer a federal program. Under the concept of primacy, the state has the power to issue federal permits, and acts as an agent of the federal government.

An administrative agency will often provide technical guidelines without conferring the legally binding status of a final regulation. This is accomplished through the publication of **guidance documents**. For example, the EPA published "Preliminary Assessment Guidance"[17] to provide direction to state and federal agencies so that assessment of hazardous waste sites are performed in a consistent manner. To issue guidance documents as regulations would greatly reduce governmental discretion in applying them, because they would have the force of law.

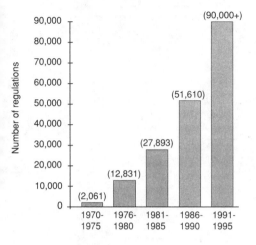

FIGURE 2-2
Changes to U.S. environmental regulations
(federal and state).[16]

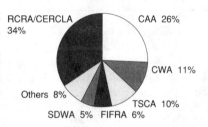

FIGURE 2-3
EPA's short-term agenda—291 regulations under
development.[18]

The extreme complexity and constantly changing nature of environmental regulations have led to the development of several computer databases (e.g., ENFLEX®/INFO, WESTLAW®, or LEXIS®) that allow searching the entire body of federal and state environmental regulations for a particular word or phrase, application, or section. These databases greatly simplify the task of finding the appropriate regulation. The process is illustrated in the following example.

Example 2-1   Searching regulations by computer.   Determine which EPA regulations govern design of landfills using the ENFLEX®/INFO database. The solution is illustrated in the computer screens on page 42.

State and Local Laws

While this chapter deals primarily with the laws of the U.S. federal government, it is important to recognize that very large and complex regulatory programs also exist at the state level, and many local governments have passed laws that deal with hazardous waste. Part of the reason for the separate systems is historical—many states had environmental programs in place before the federal government did.

As you are interested in federal regulations, select the Federal Dataset.

```
HELP   Search  File  Edit  Display  HyperSet  Cursor              14:24:11
ENFLEX INFO                   | Press F10 for menus |        0  | AND|AUTO|

        Select Datasets              F10 - When done

        Federal                   *  ESC - Cancel
        CAA AMENDMENTS
        Arizona                      INS - Select Dataset
        California
        Connecticut                  DEL - Un-Select
        Delaware
        Florida                      + - Select all Data
        Georgia
        Illinois                     - - Un-Select
        Indiana
        Kenty
        Loui
        Mass
        Mich
```

Select the "KEYWORDS" Index.

```
HELP   Search  File  Edit  Display  HyperSet  Cursor              14:25:15
ENFLEX INFO                   | Press F10 for menus |        0  | AND|AUTO|

     Matches          Recor      Choose an Index             Index

                              Section
                              Reference
                              Keywords                    1
                              Full Text
                              Dates
```

Select the two keywords "LANDFILL" and "DESIGN".

```
HELP   Search  File  Edit  Display  HyperSet  Cursor              14:26:05
ENFLEX INFO                   | Press F10 for menus |        0  | AND|AUTO|

     Character               Keywords

     Count    Records    LAND

         1         1     LAER
         7         7     LAMP ENVELOPE MANUFACTURING
        58        58     LAND DISPOSAL
        40        40     LAND TREATMENT
        88        88     LANDFILL                              1
        94        94     LEAD
         5         5     LEAD MONOXIDE
        10        10     LEAD SMELTER
         6         6     LEAD-ACID BATTERY PLANT
       151       151     LEAK
        56        56     LEGAL
```

There are 88 federal regulations with the keyword "landfill" and 400 federal regulations with the keyword "design" — only 13 regulations contain both keywords.

```
HELP   Search  File  Edit  Display  HyperSet  Cursor              14:26:28
ENFLEX INFO                   | Press f10 for menus |       13  | AND|AUTO|

     Matches          Records    Item or Word            Index

        88               88     LANDFILL                Keywords
        13  AND          400     DESIGN                  Keywords
```

This is the third of 13 U.S. EPA regulations that govern the design of landfills.

```
HELP   Search  File  Edit  Display  HyperSet  Cursor              14:27:17
ENFLEX INFO                   | Press F10 for menus |       13  | AND|MANL|
                              (0:0)
SearchSet     Record 3 of 13      Record# 4,859   Dataset-Federal
Search - LANDFILL and DESIGN

Section  :  241.203
Reference:  Title 40 | Part 241 | Subpart B
Keywords :  Waste | Solid Waste | Landfill | Construction |
            Design
Full Text:

    241.203  Design.

    241.203-1  Requirement.

    Plans for the design, construction, and operation of new
  sites or modifications to existing sites shall be prepared or
  approved by a professional engineer. The plans shall be
  submitted to the responsible agency for review and, if
  warranted, approval.

    241.203-2  Recommende

    (a) The types a
```

The conflict among the various levels of government is further complicated by the federal structure of U.S. law. The federal government has a set of limited powers such as entering into treaties, and controlling interstate commerce. The Tenth Amendment to the U.S. Constitution states:

> The powers not delegated to the United States by the constitution, nor prohibited by it to the states, are reserved to the states respectively, or to the people.

Much of the federal power in controlling environmental pollution comes from the interstate commerce clause of the U.S. Constitution which gives Congress the sole power to "regulate commerce ... among the several states." New Jersey's prohibition of out-of-state waste may be cited as an example of where this applies. In defending this statute, New Jersey noted that the state was running out of landfill space, and the continued importation of out-of-state waste would do both environmental and economic harm to the state. In declaring this statute unconstitutional, the Supreme Court noted:

> ... whatever New Jersey's ultimate purpose, it may not be accomplished by discriminating against articles of commerce coming from outside the state. ...[19]

Even with these limitations, individual states play a major role in regulating hazardous waste. Many states have received primacy to administer a federal program such as the Resource Conservation and Recovery Act. A number of recent U.S. environmental statutes specifically allow the states the power to adopt more stringent requirements than those of the federal government. However, under Article VI of the U.S. Constitution, states may not impose regulations that are contrary to federal regulations.

In most instances the role of local government is subservient to the state. A typical example in the hazardous waste field is a local law that prohibits the siting of facilities to serve the needs of the entire state. In this instance the needs of the state would be likely to prevail. In general, local authorities can only pass regulations when there is enabling legislation empowering them to do so. However, local governments typically have a dominant role in protecting the public's health and safety and in making land use decisions. The siting issue is discussed in Secs. 8-4 and 8-5.

While it is absolutely essential that engineers and scientists become familiar with environmental law, it is important to remember that the law will change with time and jurisdiction. Meeting current regulations provides a baseline or minimum standard for the design of solutions to hazardous waste problems. The overall objective should always be protection of human health and the environment through selection of the most cost-effective method (i.e., technology or management practice) best suited to the problem at hand.

2-2 RESOURCE CONSERVATION AND RECOVERY ACT

The Resource Conservation and Recovery Act (RCRA) of 1976 provided for a significant federal role in the management of hazardous waste for the first time.[20] Adopted as a series of amendments to the Solid Waste Act of 1965, RCRA set up a separate Office of Solid Waste within the EPA. This office was charged with establishing a comprehensive regulatory program ranging from identifying which wastes are hazardous to establishing a manifest system for tracking wastes. A major consequence of RCRA was to hold generators responsible for the wastes they produce from "cradle to grave."

Under the **cradle-to-grave** concept, a generator of hazardous waste can no longer avoid liability by contracting with a third party to dispose of the waste. Even if it can be shown that the waste was mishandled through the actions of a third party, the original generator will remain liable for improper disposal. This forces generators to be very careful in the selection of the disposal companies they utilize.

To further assure compliance, Congress provided for **citizen suits**, wherein anyone could bring suit "against any person . . . alleged to be in violation of any permit, standard, regulation, condition, requirement, or order. . . ." Congress specifically permitted suits against the U.S. government and any other government agencies and permitted suits against the EPA for failure to perform any duty under the Act.

RCRA's title clearly announces Congress' intention that future management of hazardous waste should stress conservation and recovery of reusable materials (e.g., recycling) as opposed to disposal. RCRA is directed at the waste presently being generated, whereas the Superfund law is directed at correcting the bad practices of the past (see Sec. 2-4). This section discusses the provisions Congress included in the 1976 RCRA statute or in its major amendment, the Hazardous and Solid Waste Amendments of 1984. Regulations promulgated by the EPA under authority of RCRA will be discussed in Sec. 2-3.

Definitions

The first concern of Congress was to define the term **hazardous waste**. The following definitions are from Sec. 1004 of RCRA:

> The term "**hazardous waste**" means a solid waste or combination of solid wastes, which because of its quantity, concentration, or physical, chemical, or infectious characteristics may—
>
> **(A)** cause or significantly contribute to an increase in mortality or an increase in serious irreversible, or incapacitating reversible, illness; or
> **(B)** pose a substantial present or potential hazard to human health or the environment when improperly treated, stored, transported, or disposed of, or otherwise managed. . . .
>
> The term "**solid waste**" means any garbage, refuse, sludge from a waste treatment plant, water supply treatment plant, or air pollution control facility and other discarded material, including solid, liquid, semi-solid, or contained gaseous material resulting from industrial, commercial, mining, and agricultural operations, and from community activities, but does not include solid or dissolved materials in irrigation return flows or industrial discharges which are point sources subject to permits under section 402 of the Federal

Water Pollution Control Act, as amended (86 Stat. 880), or source, special nuclear, or byproduct material as defined by the Atomic Energy Act of 1954, as amended (68 Stat. 923).

RCRA defined hazardous waste but did not set precise standards for its disposal. Standards for disposal were left to regulatory development by the EPA. For example, RCRA required that the EPA establish criteria for sanitary landfills and gave the following general guidelines: "At a minimum, such criteria shall provide that a facility may be classified as a sanitary landfill and not an open dump only if there is no reasonable probability of adverse effects on health or the environment from the disposal of solid waste at such a facility."

The Hazardous and Solid Waste Amendments

The 1976 Act was comprehensive and led to citizen expectations that hazardous waste problems would be solved in short order. Unfortunately, hazardous waste management problems are too complex to be amenable to simple answers, and the hopes of many were not met. The EPA was blamed by many for failing to carry out the clear mandate of Congress.

The Congressional response to a perceived lack of EPA action was to pass the **Hazardous and Solid Waste Amendments of 1984 (HSWA)**.[21] This statute that amended RCRA was far more detailed than most other pieces of environmental legislation. For example, minimal technical requirements for land disposal facilities were included in HSWA, in contrast to the usual practice of leaving these technical details to the EPA. HSWA also specified standards for the issuance of permits, penalties for violating the law, and controls on underground storage tanks.

Congress, apparently not satisfied that EPA would faithfully carry out its intent as reflected in RCRA, included many "**hammer provisions**" in HSWA. These provisions state that if EPA did not meet specific deadlines for promulgating regulations, then extremely rigid regulations, contained within HSWA, would automatically become enforceable requirements.

A comprehensive summary of HSWA is beyond the scope of this text and the reader is referred to the articles in the references at the end of this chapter.[22,23] The following subsections introduce several of the provisions of HSWA.

Land Disposal Prohibitions

HSWA prohibited the disposal of bulk (non-containerized) liquid hazardous waste in landfills or surface impoundments. Additionally, disposal of containerized liquids is permitted only if the EPA determines on a case-by-case basis that:

- No reasonable alternative is available, and
- Such disposal is environmentally acceptable.

Otherwise, the disposal of containerized liquids is banned. HSWA also prohibits land disposal of specific chemicals (e.g., halogenated organic compounds at concentrations greater than 1000 milligrams of chemical per kilogram of waste). Further, for other wastes not specifically banned, the EPA must issue regulations specifying treatment levels that:

• Diminish the toxicity of the waste, or
• Reduce the likelihood of migration.

Finally, HSWA required that the EPA must review each identified hazardous waste and determine if land disposal should be allowed. Congress mandated that this review must be completed in five-and-one-half years. Failure to meet this deadline results in a complete prohibition. Hammer provisions such as these allow the EPA no discretion to consider other factors, such as EPA staffing available to perform such evaluations or the economic and social implications of such a comprehensive ban. The clear intent was to make land disposal a last resort and to move hazardous waste management toward more "acceptable" technologies.

Minimum Technology Requirements

While Congress stopped short of banning land disposal outright, it determined that if landfills were to exist, they would be state-of-the-art facilities. As an example, all new land disposal facilities must have:

• Double-liners
• Leachate collection systems
• Ground water monitoring
• Leak detection systems

Current practice is to design land disposal facilities with greater redundancy than the law requires (see Sec. 13-3).

New standards for incinerators in HSWA required destruction and removal efficiencies of:

• 99.99% for principal organic hazardous constituents
• 99.9999% for acutely hazardous wastes (e.g., dioxin)

Permitting

The 1984 amendments included many provisions relating to the permitting of treatment, storage, or disposal facilities (TSDFs). Since 1976, such facilities required a RCRA permit to remain in operation. For many manufacturing plants, closing the TSDF means closing the factory, because there is no method to make the products without generating hazardous waste. Therefore, these provisions provide a strong incentive for industry to remain in compliance. Permits issued after HSWA must:

- Incorporate all provisions of HSWA
- Provide for corrective action for releases
- Include a ban on the placement of liquids in landfills
- Include an exposure assessment (i.e., determine the risk involved if a release were to occur)
- Provide for a five-year review of the permit requirements

RCRA Corrective Action Program

HSWA gave the EPA the authority to require corrective action (i.e., clean-up of contaminants in soil and ground water) beyond the boundaries of a manufacturing facility. Similarly, the EPA can require corrective action for continuing releases of pollution, even if the releases do not leave the plant boundaries.

In September, 1988, EPA proposed extensive requirements for corrective action at **solid waste management units** (SWMUs) for facilities permitted under RCRA.[24] These regulations are intended to provide a comprehensive program to correct deficiencies at existing landfills, lagoons, and surface impoundments at manufacturing plants throughout the United States. They require an analysis similar to that conducted at a Superfund site (see Sec. 2-4) whenever contamination was suspected. The basic elements of an RCRA Corrective Action Program are:

- RCRA Facility Assessment (RFA)
- RCRA Facility Investigation (RFI)
- Corrective Measures Study (CMS)
- Implementation

Inspection and Enforcement

While promulgation of laws, rules, and regulations form the foundation of a regulatory system, consistent compliance depends upon adequate enforcement. Federal enforcement is the responsibility of the EPA, although when authorized by the EPA, a state can be responsible for RCRA programs. Additionally, state agencies will enforce state laws and regulations that may be more stringent than federal. In some instances EPA's task is primarily one of education—explaining complex regulations to the regulated community. In other instances, the agency must enforce the law. A Congressional perception that EPA was not aggressive enough in enforcement of the 1976 law led to more stringent enforcement provisions in HSWA.

The 1984 Amendments make it a crime to knowingly:

- Transport hazardous waste without a manifest (see page 53)

- Omit relevant information on a manifest form
- Violate a regulation
- Fail to file a required report or application

The law provides for fines up to $50,000 and prison terms up to two years. HSWA defines **knowing endangerment** as violations that consciously place another person in imminent danger of death or serious bodily injury. Under circumstances constituting knowing endangerment, penalties include fines up to $250,000 and *imprisonment of up to 15 years.*

Enforcement actions include cleanup orders and compliance orders. EPA can order a responsible party to undertake corrective action if a release of hazardous waste has occurred. Compliance orders may be issued for past violations of any requirement. Civil penalties for violations of these orders can range up to $25,000 per day.

Underground Storage Tanks

HSWA requires the EPA to establish a comprehensive regulatory program for underground storage tanks. Underground storage tanks present a unique environmental problem in that their very presence is often unknown to people living in the vicinity until a problem occurs, and then it is too late for preventive measures. An example of the problems created by underground storage tanks is illustrated in Fig. 2-4—the tank being removed has been corroded to the point that large holes were discovered when the tank was removed.

One EPA report indicated that state agencies were aware of more than 12,000 releases from underground tanks covering the period 1970–1984.[25] The majority of these incidents occurred at gasoline stations. Assuming that unreported incidents greatly exceeded those reported, Congress instituted a comprehensive program to assess and control leaking underground storage tanks.

To prevent future leaks, HSWA requires the owner of an existing underground storage tank to provide either a leak detection system or an inventory control with

FIGURE 2-4
Photograph of an underground
storage tank removal.[26]

regular testing of tanks. Owners are required to maintain detailed records of monitoring and tank testing, report releases, and take appropriate corrective action when leaks do occur. Owners are also responsible for maintaining liability insurance to cover the cost of any damage to third parties that might occur if a tank leaks.

2-3 FEDERAL HAZARDOUS WASTE REGULATIONS

In RCRA, Congress required the EPA to institute one of the most far-reaching regulatory programs ever conceived. More than five hundred pages of hazardous waste regulations were published in the *Federal Register* on May 19, 1980.[27] The U.S. Office of Management and Budget estimated that the paperwork burden on industry alone would consume 1.5 million hours annually.[28] These regulations defined which wastes were hazardous; established an EPA notification process for organizations producing waste; and set up detailed regulations covering the generation, transportation, treatment, storage, and disposal of hazardous waste. (Similar provisions are included in the regulations of many developed countries—see Sec. 2-6.) These regulations were modified by HSWA's requirements and continue to be updated regularly. This section discusses some of the more important provisions.

Definitions

What is a hazardous waste? There are three ways that a solid waste (i.e., solid, liquid, semi-solid, or contained gas) can be considered "hazardous" under EPA regulations:

1. The waste is specifically listed in EPA regulations. For example, spent halogenated solvents have been listed and given the EPA hazardous waste number F001.[29]
2. The waste is tested and meets one of four characteristics established by the EPA: **ignitable, corrosive, reactive, or toxic**.
3. The waste is declared hazardous by the generator (the entity producing the waste), based on its knowledge of the waste.

The EPA is responsible for regularly updating the lists of hazardous wastes. Among the criteria utilized in making this determination are the four characteristics mentioned in (2) above and described in detail in the following paragraphs. Additional criteria include materials that have been shown to be fatal to humans in low doses. In the absence of human epidemiological evidence, animal tests may be used to infer human toxicity. This is discussed in greater detail in Chap. 5. The use of lists for many wastes precludes the necessity of expensive laboratory analyses.

The EPA selected the four characteristics of ignitability, corrosivity, reactivity, and toxicity on the basis of tests that met the intent of RCRA law, were reproducible, and were relatively easy to perform.

Ignitable wastes. Ignitable wastes are liquids with a flashpoint below 60°C, or solids capable of causing fire under standard temperature and pressure. Ignitable wastes were assigned EPA Hazardous Waste No. D001.

Corrosive wastes. Corrosive wastes are aqueous wastes with a pH below 2 or above 12.5, or which corrode steel at a rate in excess of 0.25 inches per year. Corrosive wastes are classified D002.

Reactive wastes. Reactive wastes are normally unstable, react violently with air or water, or form potentially explosive mixtures with water. This category also includes wastes that emit toxic fumes when mixed with water and materials capable of detonation. Reactive wastes are classified D003.

Toxicity. The characteristic of toxicity is more difficult to define. The objective of this parameter is to determine whether toxic constituents in a solid waste sample will leach into ground water if the waste is placed in a municipal solid waste landfill. If this is the case, then the waste will be declared hazardous.

The EPA has developed two tests for determining toxicity. The initial test, referred to as the EP Toxicity Test[30] (EP—Extraction Procedure), required grinding of solid samples, followed by rotary extraction in acetic acid for 24 hours. The material would then be filtered and the filtrate analyzed and compared to published standards (generally these were one hundred times the concentration levels acceptable in drinking water for specified metals and organic compounds). If the concentration in the extract exceeded any one of the specified levels, the original solid waste was considered hazardous.

A number of shortcomings with the EP Toxicity Test led the EPA to develop a modified test referred to as the Toxicity Characteristic Leaching Procedure (TCLP).[31,32] The latter test was more specific in many operational parameters, required a zero head-space extractor so that volatile organics would not be lost, and had more stringent quality control measures. The required chemical analyses and regulatory levels are provided in Table 2-1. If even one compound exceeds the level specified in this table, then the original waste is a hazardous waste. For some compounds, e.g., 2,4-Dinitrotoluene and Hexachlorobenzene, the quantitation limit (i.e., the minimum concentration that current laboratory methods can reproducibly measure) exceeds the calculated level based on toxicity. In these instances, the regulatory level is set at the quantitation limit. While TCLP replaced the EP Toxicity Test for most situations, the EP test was still used for several regulatory decisions, leading to a great deal of confusion.[33] A discussion of leaching tests is included in Sec. 11-4.

Generators and Transporters

Under the 1980 regulations, the EPA placed controls on all entities producing or transporting hazardous waste. Any generator producing waste must evaluate its waste and obtain an EPA identification number as a generator if the waste is hazardous. If the waste is stored for more than 90 days, the generator must also obtain a permit as a storage facility. This storage limitation is restricted to 28 days in the United Kingdom. Additional requirements include the use of appropriate containers and labels, maintenance of records of waste generated, training, preparation of contingency plans, and the submission of annual reports.

Trucking companies which transport hazardous waste must obtain EPA identification numbers as transporters. Because trucking companies and railroads were

TABLE 2-1

Toxicity characteristic leaching procedure

Compound	ID No.	Regulatory level in TCLP extract (mg/L)
Arsenic	(D004)	5.0
Barium	(D005)	100.0
Benzene	(D018)	0.5
Cadmium	(D006)	1.0
Carbon tetrachloride	(D019)	0.5
Chlordane	(D020)	0.03
Chlorobenzene	(D021)	100.0
Chloroform	(D022)	6.0
Chromium	(D007)	5.0
o-Cresol*	(D023)	200.0
m-Cresol*	(D024)	200.0
p-Cresol*	(D025)	200.0
2,4-D	(D016)	10.0
1,4-Dichlorobenzene	(D027)	7.5
1,2-Dichloroethane	(D028)	0.5
1,1-Dichloroethylene	(D029)	0.7
2,4-Dinitrotolulene	(D030)	0.13[†]
Endrin	(D012)	0.02
Heptachlor (and its hydroxide)	(D031)	0.008
Hexachlorobenzene	(D032)	0.13[†]
Hexachloro-1,3-butadiene	(D033)	0.5
Hexachloroethane	(D034)	3.0
Lead	(D008)	5.0
Lindane	(D013)	0.4
Mercury	(D009)	0.2
Methoxychlor	(D014)	10.0
Methyl ethyl ketone	(D035)	200.0
Nitrobenzene	(D036)	2.0
Pentachlorophenol	(D037)	100.0
Pyridine	(D038)	5.0[†]
Selenium	(D010)	1.0
Silver	(D011)	5.0
Tetrachloroethylene	(D039)	0.7
Toxaphene	(D015)	0.5
Trichloroethylene	(D040)	0.5
2,4,5-Trichlorophenol	(D041)	400.0
2,4,6-Trichlorophenol	(D042)	2.0
2,4,5-TP (Silvex)	(D017)	1.0
Vinyl Chloride	(D043)	0.2

*If o-, m-, and p-cresol concentrations cannot be differentiated, the total cresol concentration is used. The regulatory level for total cresol (D026) is 200 mg/L.
[†]Quantitation limit.

already regulated by the U.S. Department of Transportation (DOT), the EPA chose to incorporate DOT regulations (49 CFR) into EPA's regulatory scheme, and add additional requirements such as record keeping and cleanup of spills.

To assure that "cradle-to-grave" control of hazardous waste is achieved, Congress mandated that the EPA establish a **manifest system** that uses a form to track all waste sent off-site from the generator through to ultimate disposal. A portion of the manifest form used by New York State is shown in Fig. 2-5. The manifest has two sections, Part A—filled out by generator—and Part B—filled out by transporter or TSD facility. Six copies are distributed:

1. Sent to state agency where disposal occurs
2. Sent to state agency where generation occurs
3. Sent to generator by TSD facility, then retained by generator
4. Retained by TSD facility
5. Retained by transporter No. 1
6. Retained by transporter No. 2 (if required)

The generator is responsible for checking that it receives a copy of the manifest from the disposal facility for each shipment sent. By requiring quarterly and annual reports from both generators and TSD facilities, the system allows the states to track all hazardous waste shipments.

Because of the tremendous cost of paperwork involved in the manifest system, the EPA excluded Small Quantity Generators (i.e., those generating less than one metric ton [1000 kg or 2200 pounds] per month) from many of the 1980 regulations. In 1984 HSWA required the EPA to establish standards for this group. Small quantity generators now must comply with most RCRA regulations, including using the manifest system and shipping only to permitted hazardous waste facilities. Generators producing less than 100 kilograms (approximately one-half of a 55-gallon drum) per month are exempt from most RCRA manifest regulations but are still required to meet minimum standards. This provision is unique to the United States because most developed countries do not provide a small quantity exemption.

Treatment, Storage, and Disposal (TSD) Facilities

Federal regulations covering TSD facilities (40 CFR 264–265) represent the largest part of the 1980 regulatory package. Virtually any facility handling hazardous waste comes under the purview of these standards, including landfills, incinerators, surface impoundments, and storage areas used for more than ninety days.

Facility standards include general requirements for spill prevention procedures, contingency plans, and training programs. There are numerous reporting and paperwork requirements designed to assure compliance. For example, facilities must retain all manifests for three years and keep complete operating records until closure.

Some of the most important provisions of the regulations are those dealing with land disposal (40 CFR 265.300-316.) Among other provisions, operators of landfills

In case of emergency or spill immediately call the National Response Center (800) 424-8802 and the N.Y. Dept. of Environmental Conservation (518) 457-7362

48-14-1 (3/89)—7f

STATE OF NEW YORK
DEPARTMENT OF ENVIRONMENTAL CONSERVATION
DIVISION OF HAZARDOUS SUBSTANCES REGULATION

HAZARDOUS WASTE MANIFEST
P.O. Box 12820, Albany, New York 12212

Please print or type. Do not Staple

Form Approved. OMB No. 2050-0039. Expires 9-30-91

UNIFORM HAZARDOUS WASTE MANIFEST	1. Generator's US EPA No. N｜Y｜D｜1｜2｜3｜4｜5｜6｜7｜8｜9｜3｜8｜7｜5｜2	Manifest Document No.	2. Page 1 of 1	Information in the shaded areas is not required by Federal Law.

3. Generator's Name and Mailing Address
XYZ Corp
1234 Fifth Ave. NY, NY 13201

A. State Manifest Document No,
NY B 203875 2

B. Generator's ID

4. Generator's Phone (212) 555-1234

5. Transporter 1 (Company Name)
ABC Trucking

6. US EPA ID Number
N｜J｜D｜0｜0｜1｜2｜3｜4｜5｜6｜7

C. State Transporter's ID

D. Transporter's Phone (908) 555-1234

7. Transporter 2 (Company Name)

8. US EPA ID Number

E. State Transporter's ID

F. Transporter's Phone ( )

9. Designated Facility Name and Site Address
Chemical Disposal Services, Inc.
Syracuse, NY 13201

10. US EPA ID Number
N｜Y｜D｜0｜0｜0｜1｜2｜3｜4｜5

G. State Facility's ID

H. Facility's Phone
(315) 555-1234

11. US DOT Description (Including Proper Shipping Name, Hazard Class and ID Number)	12. Containers		13. Total Quantity	14. Unit Wt/Vol	I. Waste No.
	No.	Type			
a. Hazardous Solid Waste CEM-E NA 9/88	0｜2｜0	D｜M	｜2｜1｜2｜0	K	EPA ———— STATE B007
b.	｜｜	｜	｜｜｜｜		EPA ———— STATE ————
c.	｜｜	｜	｜｜｜｜		EPA ———— STATE ————
d.	｜｜	｜	｜｜｜｜		EPA ———— STATE ————

J. Additional Descriptions for Materials listed Above
a Soil/PCBs

K. Handling Codes for Wastes Listed Above

15. Special Handling Instructions and Additional Information

16. **GENERATOR'S CLASSIFICATION:** I hereby declare that the contents of this consignment are fully and accurately described above by proper shipping name and are classified, packed, marked and labeled, and are in all respects in proper condition for transport by highway according to applicable international and national government regulations and state laws and regulations.
If I am a large quantity operator, I certify that I have program in place to reduce the volume and toxicity of waste generated to the degree I have determined to be economically practicable and that I have selected the practicable method treatment, storage, or disposal currently avaliable to me which minimizes the present and future threat to human health and the environment; OR if I am a small generator, I have made a good faith effort to minimize my waste and select the best waste management method that is available to me and that I can afford.

Printed/Typed Name
John Smith, Agent for XYZ Corp.

Signature
John Smith

Mo. Day Year
10｜6｜3｜09｜2

17. Transporter 1 (Acknowledgement of Receipt of Materials)
Printed/Typed Name

Signature

Mo. Day Year

18. Transporter 2 (Acknowledgement of Receipt of Materials)
Printed/Typed Name

Signature

Mo. Day Year

19. Discrepancy Indication Space

20. Facility Owner or Operator: Certification of receipt of hazardous materials covered by this manifest except as noted in Item 19.
Printed/Typed Name

Signature

Mo. Day Year

EPA Form 8700-22 (Rev. 9-88) Previous editions are obsolete.

COPY 1—Disposer State—Mailed by TSD Facility

NY B 203875 2

FIGURE 2-5
Hazardous Waste Manifest, N.Y.S. D.E.C.

must construct covers or landfill caps at closure (the end of the useful life of the landfill) that minimize long-term migration of liquids through the closed fill. Additionally,
a ground water monitoring system must be established. Liquids in landfills were conditionally permitted in the 1980 regulations, but Congress banned the disposal of bulk
(i.e., those not in containers) liquids in landfills in the 1984 HSWA.

0-4   SUPERFUND

RCRA established an extensive regulatory program for newly generated hazardous
wastes but did nothing to help correct the results of poor disposal practices and
inadequate technology of the past. In some locations, abandoned plants left drums
and tanks of hazardous waste for others to deal with. (See Fig. 2-6.) In 1978, a
major episode occurred in the section of Niagara Falls, New York, known as the
Love Canal, where toxic chemicals, buried years earlier, were found to be leaking
into the basements of local homes (see Sec. 1-3). This and numerous similar events
led Congress to pass the **Comprehensive Environmental Response, Compensation
and Liability Act**[34] (CERCLA) of 1980. CERCLA was amended in 1986 by the
Superfund Amendments and Reauthorization Act (SARA). The term "**Superfund**"
is used to refer to both these laws and the cleanup program they mandated. A few
of the major provisions of Superfund are described in the following sections. For a
more complete understanding of this law, refer to the references at the end of this
chapter.[35,36,37,38]

The primary purpose of CERCLA was to clean up hazardous waste sites. As the
name indicates, Congress also provided the EPA with the ability to respond to spills
and other environmental incidents, and set up a liability system to recover costs from
those responsible for causing such incidents.

FIGURE 2-6
An abandoned plant becomes a Superfund site (photo courtesy EPA).

CERCLA, 1980

CERCLA established a $1.6 billion fund, derived primarily from feedstock taxes on industry, to implement a massive environmental cleanup program over a five-year period. Generators were required to report to the EPA any facility at which hazardous wastes are, or have been, treated, stored, or disposed. The intent was to identify and clean up hazardous waste sites first, and then to litigate to recover costs. This law gave EPA strong powers to encourage private parties (PRP's) to clean up sites. As in the case of RCRA, Superfund set unrealistically optimistic goals and deadlines and created public anticipation that sites would be rapidly cleaned up.

While the Congressional intent was clear, the incoming Reagan administration had campaigned on a platform that included significantly reducing governmental regulation of industry. The combination of unrealistic expectations and a deregulation attitude had the impact of slowing down an already difficult program. At the end of the five-year period, environmental groups and legislators decried the fact that "only six sites had been cleaned up," and Congress debated whether to reauthorize Superfund at $5 billion or $10 billion.

SARA, 1986

Following nearly two years of debate, SARA[39] was passed in the final days of the 99th Congress in 1986. This statute was a complete rewrite of CERCLA and is four times as long. SARA created an $8.5 billion fund for cleaning up abandoned waste disposal sites and an additional $500 million for cleaning up leaking underground petroleum tanks.

In response in part to the tragedy in Bhopal, India (see box on page 56), the **Community Right-To-Know** provisions of SARA require industries to plan for emergencies and inform the public of hazardous substances being used. It should be noted that the Bhopal tragedy was caused by hazardous materials, *not* hazardous waste. The American right-to-know policy should be contrasted to the European control of "major hazard industries."

SARA requires that the Hazard Ranking System (HRS) (see below) that determines which sites are eligible for cleanup funding be totally revised. Pursuant to SARA, the EPA is mandated to initiate Remedial Investigation/Feasibility Studies (see below) at 650 sites within five years.

The Superfund Process

The procedures followed at a Superfund site are illustrated in Fig. 2-7.

The list of sites eligible for cleanup funding under Superfund is referred to as the **National Priorities List** (NPL). The EPA uses a method called the **Hazard Ranking System** (HRS) to estimate the degree of risk each suspect site poses to human health and the environment. Factors such as proximity to population, nature of the contaminants, and potential pathways are evaluated and then combined to develop a single number—the HRS score—that purports to represent a relative risk. If the

The Bhopal Tragedy[40]

At 1:00 A.M. on Sunday, 3 December 1984, approximately 45 tons of methyl isocyanate (CH_3NCO) (MIC) leaked from a storage tank at the Union Carbide India, Ltd., pesticide plant in Bhopal, India. Before what has been termed "the worst chemical accident in human history" was over, nearly 3,000 people were dead and "about 200,000 injured."[41]

Methyl isocyanate is a precursor chemical used in the manufacture of pesticides. It is highly unstable and needs to be kept at low temperatures. It is extremely toxic, capable of causing severe bronchospasm and asthmatic breathing when inhaled. It is a severe eye, skin, and mucous membrane irritant and can be absorbed through the skin. Exposure to high concentrations can cause blindness, damage to lungs, emphysema, and ultimately death.[42]

The causes of the accident have been ascribed by the media to design flaws, operating deficiencies, maintenance failures, and inadequate training.[43] Union Carbide's internal investigation suggests that the accident was the direct result of employee sabotage.

The plant had a number of safety devices including a vent gas scrubber intended to neutralize any MIC that might escape and a flare tower capable of burning MIC to CO_2, H_2O, and NH_3. A detailed investigation by A. D. Little, Inc. suggests that direct water entry into the MIC tank was the likely cause of the accident.[44] The mixture of water and MIC produced high temperatures and pressures causing MIC gas to be generated at an extremely high rate.

In addition to technological failures, a number of social and political factors contributed to the severity of the incident. Bhopal had grown rapidly, had insufficient low-income housing, and had large numbers of migrants living in shantytowns despite the company's efforts to establish a greenbelt around the plant. Two of these areas were immediately across the street from the chemical plant, even though the area was not zoned for residential use. The state government was reluctant to impose strict environmental and safety standards on industry, for fear of losing badly needed jobs.

The U.S. Environmental Protection Agency was severely criticized because it had not listed MIC as a dangerous pollutant. Congressman Henry A. Waxman, Chairman of the House Health and Environment Subcommittee noted: "It's a terrible thing to think we have to have these tragedies to get industry to come forward and say, 'Yes, maybe we should have thought these things through.'"[45] The fact that so tragic an accident could occur sobered many in Congress and led to the Right-to-Know provisions in SARA.

HRS score exceeds a specified threshold value, the site is placed on the NPL and is eligible for funds for site remediation.

After a site has been listed, the EPA initiates a detailed study of the site—a **Remedial Investigation** (RI). The RI includes a description of the current situation and analysis of air, surface water, ground water, soil, and any waste. Technologies employed in an RI are described in Chap. 15. Concurrent with the RI, EPA officials

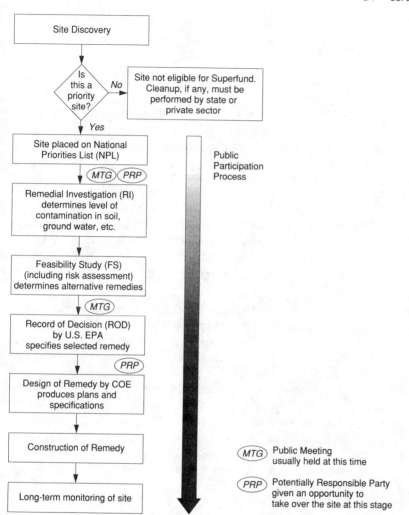

FIGURE 2-7
Activity schematic at a Superfund site.

develop, screen, and evaluate alternative solutions in a Feasibility Study (FS) (see Chap. 17). Nine criteria are used in the Feasibility Study as a basis for selection of the appropriate remedy:

1. Short-term effectiveness
2. Long-term effectiveness
3. Implementability
4. Reduction in toxicity, mobility, and volume
5. Cost-effectiveness

6. Compliance with ARARs (standards)
7. Human health protection
8. State concurrence
9. Local acceptance

Because protection of human health and the environment was specifically spelled out as a primary objective in the original CERCLA statute, this criteria is afforded special significance. Human health protection is typically determined by a quantitative risk assessment (see Chap. 14).

A community participation program provides opportunity for the public in the vicinity of the site to become involved in the process. The study phase of the Superfund process culminates when the EPA issues a **Record of Decision** (ROD), specifying the selected remedy. The remedy is then implemented, usually by private consultants and contractors working under the U.S. Army Corps of Engineers. The overall process is described in more detail in Part IV of this book.

Cleanup Standards

One of the most difficult questions that arises in evaluating remedial measures at NPL sites is "How clean is clean?" For example, what standards should apply to contaminated soil? It is simply not possible to remove contaminants down to the last molecule. The 1986 amendments require remedial actions to meet **Applicable or Relevant and Appropriate Requirements**[46] (ARARs). ARARs are environmental regulations from programs other than Superfund that may be desirable to apply to activities at Superfund sites. For example, RCRA regulations require that when a landfill is closed, it must have a cap less permeable than the liner. Therefore, if waste is left in place at a Superfund site, the design of the cap must meet this RCRA requirement because the regulation is **applicable** to land disposal. Even if a regulation is not legally applicable, the EPA has the authority to decide that it is both **relevant and appropriate** to the situation and to use it in making remediation decisions. There are three types of ARARs:

- Chemical specific (e.g., PCB level in soil < 50 mg/kg)
- Action specific (e.g., If an on-site landfill is proposed, it should meet contemporary landfill standards.)
- Location specific (e.g., prohibition of land disposal in a floodplain)

Where ARARs do not exist for important chemicals at the site, or where the EPA determines an environmental regulation is not appropriate to determine cleanup levels, such levels may be set through the use of quantitative risk assessment (see Chap. 14).

National Contingency Plan

SARA required a total revision of the **National Contingency Plan** (NCP), EPA's blueprint for implementing Superfund. The original NCP provided a framework for the RI/FS process and specified criteria for which sites were eligible for Superfund dollars. The new law addresses the How-Clean-is-Clean? issue, adding more stringent criteria on the extent of cleanup warranted and requiring that preference be given to solutions that permanently diminish the toxicity, mobility, or volume of the hazardous waste.

The EPA published an exhaustive revision of the NCP in March 1990.[47] In addition to incorporating changes mandated by the SARA, the new NCP clarifies language on responsibilities and activities of affected parties and incorporates changes suggested by experience with implementing Superfund.

Liability

The Superfund process includes liability provisions that are intended to replenish the fund and place the burden of cleaning up the site on those entities responsible for the contamination. These provisions have, however, greatly complicated and delayed the cleanup process. Even though the fund comes from a tax on industry, Congress mandated that the EPA should recover costs for site cleanup from **Potentially Responsible Parties** (PRPs). This is difficult because many companies do not agree that they are in fact responsible and are often unwilling to pay costs. Those companies that do acknowledge responsibility frequently question the extent of their responsibility. Both the EPA and the PRPs pursue lengthy litigation to advance their positions.

PRPs include:

- Present and past **owners** of the property in question,
- **Operators** of the facility at the time of disposal,
- **Generators** (i.e., persons whose wastes were disposed of at this site), and
- **Transporters** who conveyed any hazardous substances to the site.

Because of the "cradle-to-grave" nature of hazardous waste liability, generators are responsible for any waste they have produced, even if they contracted with a third party for proper disposal. CERCLA specifically applies this liability to waste produced prior to the enactment of CERCLA. Additionally, because it is rarely possible to ascertain which waste caused which part of the damage to the site, PRPs are jointly and severally liable for cleanup costs (see Sec. 2-1).

EPA may provide PRPs with an opportunity to conduct the Remedial Investigation and Feasibility Study and to undertake the cleanup. This practice would release fund dollars and EPA staff for work on other sites. Generally EPA will not deal individually with entities listed as PRPs, but only with a PRP group. However, PRPs as a group are brought together only because EPA intends to sue them for the cost of cleanup. It sometimes takes years for such a group of unwilling parties to organize themselves to speak with one voice. SARA, however, demands rapid action by the EPA and does not allow the agency to wait for a PRP response. The complex question of corporate liability in the context of Superfund is discussed in the references at the end of the chapter.[48,49]

2-5 OTHER FEDERAL STATUTES

While RCRA and Superfund are the major U.S. laws dealing with hazardous waste, other environmental laws have applications to hazardous waste management. Five of these laws are discussed in the following sections.

Toxic Substances Control Act

Congress passed the **Toxic Substances Control Act** (TSCA) of 1976 to control the manufacture of toxic materials that are not waste products but useful and necessary materials for our industrialized society. Prior to the passage of TSCA, the Federal government could regulate hazardous substances only if they were released to the environment. Whether or not a specific chemical should be produced was entirely the decision of the manufacturer. Congress contended that the public interest would be served if the EPA was involved in these decisions.

TSCA allows the EPA to identify potentially toxic materials before they are manufactured in large quantities. This is accomplished through **Premanufacture Notification** (PMN). Companies are required to give the EPA ninety days' notice before producing a new chemical. EPA can then control the manufacture and distribution of chemicals if the agency believes them to be harmful to human health or the environment.

Chemicals restricted or banned because of TSCA frequently appear in hazardous waste sites. For example, Congress mandated control of polychlorinated biphenyls (PCBs) in TSCA by specifying a regulatory timetable for eliminating production of PCBs:

> Following a number of episodes in Michigan and elsewhere (actually due largely to the unrelated polybrominated biphenyls) Congressman John Dingell (D-Mich.) and other representatives inserted section 6(e) into the final version of TSCA. It provided a schedule that would first stop PCB manufacture, and then gradually curtail its use.[50]

PCBs were in wide use in electrical components because of their excellent electrical properties but were found to be extremely toxic. Most production and use of PCBs were discontinued in 1979. PCBs are still found at many industrial plants and Superfund sites, due to their presence in transformers and capacitors.

The banning of a specific chemical by name, rather than setting broader policy goals and allowing the EPA to make technical decisions, was precedent setting. This precedent was followed in the 1990 Clean Air Act Amendments (see below).

Occupational Safety and Health Act

The **Occupational Safety and Health Act** (OSHA) of 1970 addresses health and safety in the workplace. It created the **Occupational Safety and Health Administration** (also referred to as OSHA) at the same time that the EPA was created. The OSHA law requires the new agency to promulgate standards such that "no employee will suffer material impairment of health or functional capacity even if . . . [exposed] . . .

for the period of his working life."[51] The OSHA law covers all industrial operations, including hazardous waste TSD facilities.

OSHA instituted very stringent requirements for health and safety programs at hazardous waste sites (29 CFR 1910.120). Every hazardous waste disposal facility must now include a medical surveillance program, a site-specific health and safety plan, continuous air monitoring, a minimum of 40 hours of health and safety training for all workers on site, and a decontamination program for workers as they leave the site (see Sec. 15-2). This regulation covers not only treatment, storage, and disposal facilities but also cleanup activities at Superfund sites.

Clean Water Act

The Clean Water Act (CWA), which underwent major amendments in 1972, 1977, and 1987, regulates discharges to surface water in the United States. Among its many sections, this law provided grants to municipalities to build municipal wastewater treatment plants and established a **National Pollutant Discharge Elimination System** (NPDES), which provides for the issuance of permits for discharges to surface water. Discharges covered by the NPDES permit program must meet **effluent limitations** based on available technology.

CWA applies to all but a few categories of point source discharges to surface water. For toxic and non-conventional pollutants (i.e., contaminants different from those found in ordinary domestic sewage), the Act specifies that the **Best Available Technology** (BAT) must be used. To prevent industries from discharging toxic wastes into community sewers, a set of pretreatment standards has been established. These specify the level of treatment for industrial waste prior to discharge to **Publicly Owned Treatment Works** (POTWs). Section 307(a) of CWA requires the EPA to prepare a list of **priority pollutants** (see App. A) subject to control under CWA.

Safe Drinking Water Act

The Safe Drinking Water Act (SDWA) was originally passed in 1974 to ensure that public water supplies are maintained at high quality. Amendments passed in 1986 require the EPA to set **national primary drinking water standards**.

As with the CWA, many provisions of this law have implications for hazardous waste management. One example is the development of drinking water regulations (40 CFR 141) that set **Maximum Contaminant Levels** (MCLs) and **Maximum Contaminant Level Goals** (MCLGs). While the intent was to establish levels that are safe for human consumption, the EPA has determined that they may be used as Applicable or Relevant and Appropriate Requirements (ARARs) (see Sec. 2-4) to specify the cleanup levels at Superfund sites. Cleaning ground water at hazardous waste sites to these levels has resulted in spending limited resources to provide extraordinary treatment at a hazardous waste site. In some instances this has resulted in the product water being discharged to a stream that does not meet such stringent standards. Although remediating ground water to meet drinking water standards is a noble goal, limited resources may be better applied to remediating a greater number of sites to lesser standards.

Clean Air Act[†]

The Clean Air Act (CAA) was originally passed in 1970 and has been amended several times, including extensive amendments in 1977 and 1990. Generally, as amended through 1990, the Clean Air Act consists of major initiatives to attain and maintain National Ambient Air Quality Standards to ensure that all new sources of potential atmospheric emissions are equipped with the "Best Available Pollution Control Technology" and to ensure that emissions of hazardous air pollutants are controlled to the maximum extent possible on both existing and new emission sources. Additionally, the 1990 CAA Amendments mandated an extensive operating permit program applicable to all major emission sources in the country.

National Standards. National Ambient Air Quality Standards represent the cornerstone of clean air programs in the United States. The 1970 CAA required the EPA to investigate and describe the environmental effects of any air pollutant emitted from stationary or mobile sources that may adversely affect human health or the environment. These investigations determine which pollutants should be regulated and are used to establish the **National Ambient Air Quality Standards (NAAQS)**.

An NAAQS value is the concentration that should not be exceeded anywhere in the United States. NAAQS should not be confused with emission standards: NAAQS set the goal for attainment of clean air; **emission standards** regulate source emissions to achieve or maintain the ambient air standards set by the NAAQS.

Criteria Pollutants. Pollutants for which NAAQS are set are classified as **"criteria" pollutants** because they must be set on the basis of health-based or environmental impact-based criteria. The NAAQS specify the allowable concentration of these criteria pollutants in the ambient air. These standards are not directly enforced but are achieved through source emission limitations set within each state and based upon the degree of control predicted to be necessary to meet NAAQS. Pollutants regulated as "criteria" pollutants are shown in Table 2-2.

Air Quality Control Regions. States are divided into regions known as **Air Quality Control Regions (AQCRs)** to determine if the standards for criteria pollutants are being met. If the standards are being met, the area is considered to be in **"attainment"** for the specific NAAQS. If not, the AQCR is in **"nonattainment"** for affected pollutants. If an area is in nonattainment, the regulating authority (either state or local air pollution control agency) must adopt regulations to limit facility emissions sufficient to ensure compliance with the NAAQS by a date specified by the CAA.

Because of the 1970 CAA, most areas of the United States are substantially in attainment with the NAAQS for all the criteria pollutants except ozone and carbon monoxide. The 1990 Amendments include major programs designed to bring

---

[†]This section prepared by Kenneth N. Weiss, P.E., Air Program Director, ERM, Inc.

TABLE 2-2
National ambient air quality standards[52]

Pollutant	Primary standards (protective of health)*
Ozone	0.120 ppm (1-hour average)
Carbon monoxide	9 ppm (8-hour average) 35 ppm (1-hour average)
Particulate matter (PM-10)†	150 μg/m^3 (24-hour average) 50 μg/m^3 (annual arithmetic mean)
Sulfur dioxide	0.140 ppm (24-hour average) 0.03 ppm (annual arithmetic mean)
Nitrogen dioxide	0.053 ppm (annual arithmetic mean)
Lead	1.5 μg/m^3 (arithmetic mean averaged quarterly)

*CAA also requires that EPA establish secondary standards for adverse environmental effects. In most cases the levels are lower than the primary standards.
†Particulate matter 10 microns or smaller in diameter

about ozone attainment through a reduction in the emissions of precursors to ozone: volatile organic compounds (VOCs) and oxides of nitrogen (NO$_x$). Carbon monoxide is addressed through tighter emission standards for motor vehicle tailpipe emissions.

Hazardous Air Pollutants. Section 112 of the Clean Air Act identifies a **hazardous air pollutant** (HAP) as " ... an air pollutant to which no ambient air quality standard is applicable and which ... causes or contributes to ... an increase in mortality or an increase in serious irreversible, or incapacitating reversible illness." Under the 1970 Act the regulation of HAPs was essentially a risk-based approach. After a HAP was listed for regulation by the EPA, the agency was mandated to adopt standards sufficient to protect the public health with an "ample margin of safety" defined by risk-based models. Defining an ample margin of safety for potential carcinogenic emissions proved to be very difficult for EPA. As a result, EPA was only able to develop emission standards for seven HAPs between 1970 and 1990: asbestos, beryllium, vinyl chloride, mercury, benzene, radionuclides, and arsenic.

The new HAP emissions control program is technology-based and requires the installation of state-of-the-art pollution control equipment. The 1990 Amendments list 189 chemicals to be regulated (see App. A). Stringent technology-based standards are to be developed for major sources of these HAP emissions over the next ten years. A major source is defined as any facility which potentially emits 10 tons per year of any single HAP or more than 25 tons per year of a combination of HAPs.

EPA will issue regulations for 174 categories of industrial sources which emit substantial quantities of HAPs. Emissions allowances reflecting **Maximum Achievable Control Technology** (MACT) will be established in each source category regulated. By definition, MACT can include process changes, material substitutions, or air pollution control equipment.

New Source Review. New Source Review (NSR) is the permitting process that governs the construction, modification, or reconstruction of major industrial air pollution emission sources throughout the United States. In areas where the ambient air quality standards are met, the program requires the **Prevention of Significant Deterioration** (PSD) of air quality intended to ensure that air quality is not allowed to degrade in those areas of country where the air quality is already better than required. Where standards are not met, **nonattainment** new source review provisions apply. The PSD provisions and nonattainment provisions apply separately to each regulated pollutant. Thus, facilities may be required to undergo both a PSD review and a nonattainment review if the proposed location is classified as attainment for one pollutant and nonattainment for a different pollutant.

Generally, the 1990 Amendments make EPA's requirements to obtain a permit to construct a new major source or modify an existing major source more stringent in a nonattainment area. New sources or modifications in a nonattainment area are required to install "Lowest Achievable Emission Rate" (LAER) air pollution control technology and to offset remaining permitted emissions after LAER is installed. The concept of "offsetting" pollution is used to assure that the total emissions in a nonattainment area *after* construction of a new facility be *less* than before. This is accomplished by shutting down or adding more efficient air pollution control equipment to existing sources.

The 1970 Clean Air Act also required that EPA promulgate New Source Performance Standards (NSPS) for major emission sources. Although BACT (Best Available Control Technology) and LAER are negotiated on a case-by-case basis, NSPS represent nationwide emission standards for certain significant emission source categories (e.g., Petroleum Refining).

Operating Permits. The 1990 Amendments established a comprehensive new operating permit program to be administered by state agencies in accordance with federal guidelines. The Amendments will require nearly all sources of even minor amounts of air emissions to apply for and obtain operating permits. The new operating permits are intended to lay out in one enforceable document all a facility needs to do to ensure compliance with all air quality regulations.

2-6 INTERNATIONAL PERSPECTIVES[†]

Philosophies of National Regulatory Systems

The alternative philosophies for a national system to control hazardous waste are as diverse as the legal, political, social, and cultural differences of the countries. The control or regulatory system must have three essential components:[54]

- A legal framework (i.e., enabling legislation),
- Implementation and enforcement mechanisms, and
- Adequate facilities with measures to encourage their use.

[†]Section 2-6 is based on the work of Dr. David C. Wilson, ERM-London.[53]

A comparison of the systems is presented in Table 2-3.

A consensus exists among developed countries as to the key elements of a legal framework, although there are differences in emphasis and detail. However, there are as many different implementation and enforcement mechanisms as there are countries. Facility development, particularly which steps will be utilized to advance the use of approved facilities, also vary significantly from country to country. There are several questions useful in classifying regulatory systems:

- At what level of government is control exercised?
- Who selects treatment and disposal methods?
- Who retains long-term liability?
- Who provides facilities?

The issue of devolution of regulatory control from the national government to regional or local authority is very much a function of the cultural considerations and the type of legal system. In the United States, the federal government develops the hazardous waste regulatory system but must delegate implementation and enforcement to those states capable of operating the federal program. States may enact rules and regulations that are more stringent than federal rules, but not less stringent. In the United Kingdom, control is exercised through site licensing and permitting by nearly 200 local authorities with the central government having a largely advisory role. In Japan and Germany, primary responsibility is at the regional level, but the regions have more discretionary power than in the United States.

In countries such as Denmark and Sweden, certain wastes must be shipped to a facility operated by the government. Japan specifies the treatment and disposal routes in regulations, while Austria and Germany use non-mandatory guidance.

The critical question of the selection of treatment and disposal methods is summarized in Table 2-4.

As noted in the earlier sections of this chapter, liability is a major driving force of the hazardous waste system in the United States, mainly due to the somewhat unique nature of the U.S. tort system. This is not the case in most other developed countries. In general, the generator of a waste is free of future liability if the composition of the waste is accurately disclosed and it has been delivered to a disposal facility licensed to accept that type of waste. In Denmark, this indemnification is available only if the waste is delivered to the national treatment facility.

In June, 1990, the European Parliament approved a proposal for a European Community (EC) Directive for civil liability for damage caused by waste.[57] This directive extends liability to the generator, transporter, and disposal facility for damages caused by waste including "impairment of the environment." The EC directive was reviewed by the British House of Lords,[58] who concluded:

- Strict liability should be applied to waste management
- Strict civil liability should cover general environmental harm

TABLE 2-3
Hazardous waste management systems†

	Austria	Canada	Denmark	France	Germany	India	Italy	Japan	Netherlands	Southern Africa†	Spain	Sweden	Taiwan	UK	USA
Status															
Date of main legislation	1983	1987	1972/92	1975	1972/ 86/90	1989 84/86	1982/84	1970	1979	1989 proposed (2)	1986/88	1975	1989	1972/74	1976/
Registration/licensing (3)															
Collector/transporters	L	L	(4)	R	L	L	L	L	No	No	L	L	L	No	R
Treatment/ disposal contractors	L	L	L	R	L	L	L	L	L	Yes	L	L	L	No	R
Transportation															
Manifest system	Yes	Yes	Yes	New	Yes	No	Yes	No	Yes	No	Yes	Soon	Yes	Yes	Yes
Control over import	Yes	Yes	Yes	Yes	Yes	Yes	Yes	Yes	Yes	By sea	Yes	Yes	Yes	Soon	Yes
Control over export	No	Yes	Yes	Soon	Yes	Yes	Yes	Yes	Yes	By sea	Yes	Yes	No	Soon	Yes
Permitting															
Storage	Yes (4)	Yes	Yes	Yes	Yes	Yes	Yes	Yes	Yes	Yes	Yes	Yes	Yes	Soon	Yes
Treatment	Yes	Yes	Yes	Yes	Yes	Yes	Yes	Yes	Yes	Yes	Yes	Yes	Yes	Yes	Yes
Disposal	Yes	Yes	Yes	Yes	Yes	Yes	Yes	Yes	Yes	Yes	Yes	Yes	Yes	Yes	Yes
Have all facilities been permitted?	No	No	Yes	Yes	Yes	No	Yes	Yes	Yes	No	No	No	No	Yes	No

Planning and siting

Is there a national strategy/plan?	Yes	Yes	Yes	No	No	Yes	Yes	No	Yes	Yes	No	Yes	Yes	No	No
Are authorities required to produce a plan?	Yes	Yes (6)	No	Yes	Yes	Yes	No	No	Yes	Yes	No	No	Yes	Yes	(6)
Has this been done?	Yes	No	Yes	No	Yes	No	Yes	Partial	Yes	No	No	No	Partial	Partial	Partial

Abandoned sites

Is there a national inventory	New	Yes	Yes	Yes	Yes (1)	No	No	No	Yes	No	No	Yes	No	No	Yes
Is there a clean-up program?	No	Yes	Yes (5)	(5)	Yes (1)	No	Yes	No	Yes	Yes	No (7)	Yes	No	No	Yes

(1) Partial.
(2) In Southern Africa, the situation varies between countries. In most cases, control is presently informal, in the absence of formal legislation.
(3) L = licensing scheme, implying investigation by authorities; R = registration, implying simply being listed in a register.
(4) Mainly under the Trade Act, not under the Hazardous Waste Act.
(5) Although there is no formal, nationwide cleanup program, the cleanup of individual sites is proceeding.
(6) Provincial or state responsibility.
(7) Being considered by R.O.C. E.P.A.

† In Tables 2-3 through 2-5, Southern Africa refers to the subcontinent south of the Kunene and Zambesi Rivers and comprises at least 17 sovereign states and several geopolitical areas.
‡ Adapted from Wilson and Forester.[55]

- Common interest groups in the U.K. should have the right to recover costs incurred in preventing or cleaning up pollution by waste
- Liability should rest with the person who has control of the waste at the time the damage occurred[59]

In a number of countries such as Denmark, the national government, in association with local government and industry, has taken the initiative in providing treatment and disposal facilities. In Denmark, Sweden, and Finland wastes *must* be delivered to a state-owned company which operates as a monopoly. Only in the United Kingdom and the United States is this left entirely to the private sector, although in the United States, many states have taken on the responsibility for siting facilities (see Sec. 8-4).

Definitions

The term hazardous waste is used in this textbook to describe wastes that meet the American definitions (see Secs. 2-2 and 2-3). Other countries use terms such as special, chemical, difficult, poisonous, or toxic to describe similar wastes. The objectives in having a special category vary. In the United States and Germany, it is to distinguish wastes that cannot be disposed of with municipal solid waste, specifically to keep hazardous waste out of municipal landfills. In Denmark, the objective is to assure the most appropriate treatment.

In the United Kingdom, present policy encourages the disposal of selected hazardous waste in municipal landfills so the American definition is simply inappropriate. Concern in the United Kingdom is to identify wastes that present special hazards during transport. Japan's definition separates those wastes that are difficult to incinerate or treat chemically, and the emphasis is on wastes that require total isolation from the environment.

This lack of standardization of even the purpose of the definition makes international regulation of hazardous waste problematic. There is a move towards an international definition or at least a cross-listing by the Organization for Economic Cooperation and Development (OECD) to ensure compatibility of definitions used in different countries.

A classification of the various definitions used is presented in Table 2-5. While this table provides a simple comparison of some aspects of the definitions, it provides no insight into the breadth of the underlying definitions, the number of substances considered, or the concentrations of various chemicals necessary to make a waste hazardous. All of these factors vary greatly from country to country.

The special rules require some explanation. Under the mixture rule in the United States, a hazardous waste cannot be rendered non-hazardous by mixing it with another material and "diluting" the hazardous constituents. Mixing is similarly prohibited in the Netherlands, while France requires special permission. If the treatment of a hazardous waste results in a residue, that residue is a hazardous waste, by definition, in the United States, unless specifically delisted. France and Denmark have similar provisions for residues.

TABLE 2-4
Control over treatment or disposal options‡

	Austria	Canada	Denmark	France	Germany	India	Italy	Japan	Netherlands	Southern Africa†	Spain	Sweden	Taiwan	UK	USA
Direction of waste															
To particular site(s)	No	No	Yes	No	No	No	No	No	No	No	No	Yes	No	No	No
To particular option(s)	(1)	No	Yes	No	Yes	No	No	Yes	No	No	No	Yes	Yes	No	Yes
Powers exist, in reserve	No	Yes	—	Yes	No	No	No	Yes	No	No	Yes	—	Yes	Yes	No
Prohibition of certain options for particular wastes															
National regulations	(1)	No	Yes	Yes	Yes	No	Yes	Yes	Yes	No	Yes	Yes	Yes	No	Yes
Control via site permits															
Strong national standards mean effective prohibition for certain waste	Yes/No (2)	No	Yes	Yes	Yes/No (2)	Yes/No (2)	Yes	Yes/No (2)	Yes	Yes	Yes/No (3)	Yes	(4)	No	Yes

(1) Recommendations are made, but these are not mandatory.
(2) Strong controls exist in principle, but in practice there are wide local variations in what is/is not permitted at individual sites.
(3) System not yet in place.
(4) Standards exist but are not enforced.

† In Tables 2-3 through 2-5, Southern Africa refers to the subcontinent south of the Kunene and Zambesi Rivers and comprises at least 17 sovereign states and several geopolitical areas.
‡ Adapted from Wilson and Forester.56

Although no acceptable international definition of hazardous waste exists, similarities exist among the definitions of many countries. Contrast the following three definitions with those presented in Secs. 2-2 and 2-3.

Canada 'Hazardous wastes are those wastes which due to their nature and quantity are potentially hazardous to human health and/or the environment and which require special disposal techniques to eliminate or reduce the hazard'
Philippines 'Materials which are inherently dangerous to the human body or to animals, including, but not limited to, materials that are toxic or poisonous; corrosive; irritants; strong sensitizers; flammable; explosive'
United Nations Environment Program 'Wastes other than radioactive wastes which, by reason of their chemical activity or toxic, explosive, corrosive or other characteristics causing danger or likely to cause danger to health or the environment. . . .'[61]

Collection and Transport

It has become common practice in most countries to register transport and disposal firms, but now many countries require a more stringent licensing procedure (see Table 2-3). Similarly a manifest system is used by all of the countries listed in Table 2-3, except Japan, India, and the countries in Southern Africa. The nature of the manifest system does vary considerably. In the United States, primary responsibility for identifying lost shipments is placed on the generator, who must report exceptions to the authorities. In other countries it is the government (local or regional) that does the checking of manifest forms to assure that shipments of waste have not been lost.

In Europe and the United States, collection and transportation services are provided by the private sector, whereas central collection systems are used in Scandinavian countries. In Sweden this is accomplished by local utilities while Denmark has established some 25 central collection points.

In most countries, wastes move freely across internal boundaries. In the United States, states are prohibited from interfering with the transport of waste across their borders under the interstate commerce clause in the U.S. Constitution. Germany is an exception to the free movement of waste inside a country. A special permit, involving a time-consuming process to obtain, is required to cross state boundaries in Germany. The political and social climate in Germany is making it difficult to obtain such permits.

Treatment, Storage, and Disposal

Essentially all developed countries require permitting or licensing of treatment, storage or disposal (TSD) facilities, however, the requirements vary significantly. In the United States, industries may store hazardous waste for up to ninety days without obtaining a permit as a storage facility. In the United Kingdom, this is limited to 28 days and even this exemption is being eliminated in certain instances.

The U.S. TSD permit program is heavily driven by tight restrictions and large penalties (see Sec. 2-2). In many European countries, a more "cooperative" approach is

TABLE 2-5
Definitions of hazardous waste‡

	Austria	Canada	Denmark	France	Germany	India	Italy	Japan	Netherlands	Southern Africa†	Spain	Sweden	Taiwan	UK	USA
Is there a legal definition?	Yes	Yes	Yes	Yes	Yes	Yes	Yes	Yes	Yes	(1)	Yes	Yes	Yes	Yes	Yes
Purpose of definition															
Control over transport	Yes	Yes	Yes	Yes	Yes	Yes	Yes	No	Yes	No	No	Yes	No	Yes	Yes
Control over treatment/disposal	Yes	Yes	Yes	Yes	Yes	Yes	Yes	Yes	Yes	Yes	Yes	Yes	Yes	No	Yes
Type of definition															
List of waste	Yes	No	Yes	No	Yes	Yes	No	Yes	Yes	No	Yes	Yes	No	No	Yes
List of substances	No	No	Yes	Yes	No	No	Yes	Yes	Yes	No	Yes	Yes	Yes	Yes	Yes
List of processes	No	No	No	Yes	No	No	Yes	No	Yes	No	No	No	Yes	No	Yes
Concentrations	No	Yes	No	Yes	Yes	No	Yes	Yes	No	Yes	Yes	Yes	Yes	No	Yes
Criteria															
Toxicity of waste	Yes	Yes	Yes	Yes	Yes	No	Yes	No	No	No	Yes	No	No	Yes	Yes
Toxicity of extract	No	Yes	Yes	Yes	No	No	Yes	Yes		Yes	Yes	No	No	Yes	Yes
Ignitability/flammability	No	Yes	Yes	Yes	No	No	No	No		Yes	Yes	No	Yes	Yes	Yes
Corrosiveness	No	(5)	Yes	Yes	Yes	No	Yes	No	Yes	Yes	Yes	No	No	No	Yes
Reactivity	No	(5)	Yes	Yes	No	No	No	No	No	Yes	Yes	No	No	No	Yes
Special rules															
Mixing rule	Yes	(5)	Yes	Yes	Yes	No	Yes	No	Yes	No	No	No	No	No	Yes
Residue rule	No	(5)	Yes	Yes	No	No	No	No	No	No	No	No	No	No	Yes
Exclusions															
Small generators (2)	No		No	100 kg	No	No	No	Yes	No	No	No	No	No	No	100 kg
Wastewater	Yes	No	No	Yes	Yes	Yes	Yes	Yes	Yes	Yes	Yes	Yes	No	Yes	(3)
Sewage sludge	Yes	No	No	No	No	No	Yes	Yes	No	Yes	Yes	No	No	Yes	No
Mining waste	Yes	No	No	Yes	Yes	No	No	No	No	No	Yes	No	No	Yes	Yes (4)
Agricultural waste	Yes	No	No	Yes	Yes	No	No	Yes	No	No	Yes	No	No	Yes	No

(1) New definition in South Africa only.
(2) Quantity is that per month below which a producer is exempt from the regulations.
(3) Partial exclusion for wastewater treated exclusively in permitted treatment tanks. Wastewater treated in surface impoundments or lagoons is controlled as a hazardous waste.
(4) Mining waste excluded pending further study.
(5) Under consideration in 1992.

† In Tables 2-3 through 2-5, Southern Africa refers to the subcontinent south of the Kunene and Zambesi Rivers and comprises at least 17 sovereign states and several geopolitical areas.
‡ Adapted from Wilson and Forester.[60]

utilized to keep facilities up-to-date and in compliance. For example, France provides subsidies to industries to use the "best practicable" means of treatment and disposal. Belgium provides a zero tax on recycling and a maximum tax on landfills, taxing treatment facilities somewhere in between.

Recent changes in Europe include a gradual move towards the restriction of landfills in favor of more acceptable technologies. Other landfill provisions call for the introduction of stringent provisions for post-closure care. In the United Kingdom, this may mean that the operator of a landfill will not discharge responsibility for the site for a period as long as fifty years after the facility has ceased operation.

Site Cleanup

The failure of past policies to properly manage industrial wastes has left a legacy of abandoned hazardous waste sites that require some form of remedial action. As noted in Table 2-3, most developed countries have some program in place to address these sites. The size of the problem as reported by four countries is shown in Table 2-6.

Extensive remedial action programs, similar to those of the United States, are under way in Denmark and the Netherlands. France and Germany have cleaned up a number of sites on a case-by-case basis, without a national program aimed at all such sites. The costs of cleanup are paid though general revenues (Denmark, Japan, the Netherlands, and Southern Africa) or through a special tax or funding from the chemical industry (France, Germany, United States). Austria, Italy, and the United Kingdom have no special provisions to pay for the cleanup of these sites.

Developing Countries

The regulatory process for controlling hazardous waste in developed countries has evolved over a period of decades. As developing countries increase their economic base through the construction of manufacturing facilities, they are faced with the same basic problems. A challenge facing these countries is to learn from the successes and failures of the developed nations. However, environmental concerns in general and hazardous waste management in particular are not high on the political agendas of most developing countries.

TABLE 2-6
Uncontrolled hazardous waste sites[62]

Country	Number of waste sites	Number containing hazardous waste	Sites requiring immediate action
Denmark	3100	900	110
Netherlands	—	4350	1000
Sweden	3900	500	21
USA	—	20,000	2000

Developing countries have a number of special characteristics that may make improper hazardous waste management a more serious problem than comparable practices in more developed countries:

• Lack of infrastructure that might ameliorate some of the risks of poor practices (e.g., proper treatment of drinking water);
• Lack of education in environmental risks;
• Unique sub-populations at special risk (e.g., scavengers who make their living sorting through refuse); and
• Inadequate health care systems.

All of these attributes argue for a more vigorous regulatory system in these countries. However, economic realities often dictate that environmental concerns not be permitted to stand in the way of development, and many problems are inadequately addressed, or not dealt with at all. Wilson and Balkau[63] identify several common constraints in developed countries, from which the developing countries might learn:

• Inadequate early identification and quantification of waste,
• Inadequate enforcement of existing pollution control laws so that the hazardous waste problem remains hidden,
• The inevitable long delay between recognizing the need for facilities and actually providing them,
• Loss of public confidence leading to opposition to the siting of essential new hazardous waste disposal facilities, and
• The complexity of the administrative response needed to control all parts of the waste cycle.

Another problem facing the less-developed nations is the import of wastes from developed countries. The lack of adequate facilities in most developed countries and the inability to site and build new ones (see Sec. 8-4), has led some generators to export wastes to developing countries. Often the importing countries have no capability to adequately dispose of these wastes but are unable or unwilling to stop the traffic. The response from some importers has been vehement:

> On May 25, 1987, for example, the Organization of African Unity (OAU) unanimously passed a resolution stating that illegal dumping of waste in Africa by other countries "is a crime against Africa and the African people."[64]

In large part due to such strong statements from developing countries, the United Nations Environment Program (UNEP) convened a conference in Basel, Switzerland in 1989. Delegates from 116 nations approved a convention to control import and export of hazardous waste. Among its many provisions, the **Basel Convention**[65] permits export of waste to occur *only* if:

- The exporting country lacks adequate facilities for the proper disposal, or
- The wastes are to be used as raw material in recycling or recovery operations.

Other provisions dictate that permission of the importing country is required and that export is not permitted to countries that have banned imports. The convention makes clear that the exporting country is responsible to make certain that proper disposal occurs and that this responsibility cannot be transferred to the importing country.

DISCUSSION TOPICS AND PROBLEMS

2-1. Contrast the definition of hazardous waste included in RCRA law (Sec. 2-2) with that included in EPA regulations (Sec. 2-3). What are the differences, and why is there any difference at all?

2-2. Contact your congressional representative or U.S. senator and obtain a copy of one federal hazardous waste bill currently being considered. Prepare a brief synopsis of the bill, indicating your opinion as to whether it should be passed.

2-3. Contact your local state legislator and obtain a copy of one hazardous waste bill currently being considered in your state. Prepare a brief synopsis of the bill, indicating your opinion as to whether it should be passed.

2-4. Review recent copies of the *Federal Register* in the library and select a recently promulgated hazardous waste regulation for review. Prepare a brief synopsis of the regulation. Pay particular attention to the preamble and summarize EPA's estimate of the impact of the regulation.

2-5. Repeat problem 2-4 for a recent regulation promulgated in your state.

2-6. Provide a synopsis of a recent court decision involving hazardous waste, based on review of a case in **Environmental Reporter Cases**.

2-7. What is the National Contingency Plan? Where would you obtain a copy of the Plan?

2-8. Explain what is meant by the term Applicable or Relevant and Appropriate Requirements (ARARs).

2-9. Explain the different types of Applicable or Relevant and Appropriate Requirements (ARARs). Provide an example for each.

2-10. Provide an example of how each of the six defenses to an action for negligence cited in Sec. 2-1 might arise.

2-11. One of the three ways for a solid waste to be classified as a hazardous waste is for a company to declare the waste hazardous. Because this will dramatically increase the cost of transportation and disposal, what is the incentive for industries to make such a declaration?

2-12. Using the *Federal Register*—Cumulative List of CFR Sections Affected in the library, determine the most recent changes in regulations for:
   (*a*) Generators (40 CFR 262)
   (*b*) Determining which wastes are hazardous (40 CFR 261)
   (*c*) Landfills (40 CFR 265.300-316)

2-13. Determine what federal regulations would apply to:
   (*a*) Insurance for hazardous waste landfills
   (*b*) Disposal of batteries in landfills

(c) Design and operation of land disposal facilities

(d) Export of wastes containing PCBs

(e) Disposal of PCBs in landfills

(f) A list of extremely hazardous substances

2-14. Repeat problem 2-13 for regulations in your state.

2-15. Define the following legal terms (use a law dictionary) and indicate their relevance to hazardous waste management:

(a) Res ipsa loquitur

(b) Strict liability

(c) Joint and several liability

(d) Sovereign immunity

(e) Negligence

(f) Contributory negligence

(g) Comparative negligence

(h) Ex post facto law

(i) Statute of limitations

2-16. Review the article "Summary and Analysis of the Hazardous and Solid Waste Amendments of 1984," by Beveridge and Diamond (Ref. 22) and prepare a one-page summary of the changes in the 1984 HSWA law that affect land disposal.

2-17. Suggest reasons why (or why not) each of the following might be considered hazardous under EPA regulations:

(a) A railroad tank car of fuming sulfuric acid

(b) Solvents, after use as cleaning agents

(c) Radioactive tracers after use in a university lab

(d) Mercury

(e) Soil that has had gasoline spilled on it

2-18. Review Section 3004(d) of RCRA (42 USC 6924), and prepare a brief synopsis summarizing your understanding of which wastes Congress intended to preclude from land disposal.

2-19. Review OSHA regulations [29 CFR 1910.120(b)-(p)], and determine what health and safety considerations are required at Superfund sites.

2-20. Write a one-page summary of Section 112 of the Clean Air Act.

2-21. In the paragraphs discussing negligence (Sec. 2-1, Common Law and Liability), the concept of a "reasonable prudent person" is mentioned. Are there any instances in which a court might determine a person to be exempt from this standard and therefore not negligent, even if a supposedly negligent event occurred?

2-22. In the discussion on negligence the following statement is made: "Where negligence can be established it is no defense that the negligent action was in full compliance with all government regulations and permit conditions." Explain why this defense may not be valid.

2-23. Suppose that you are an OSHA inspector sent to inspect an industry located in an extremely impoverished area. The industry also employs a large number of local people. In your inspection of the plant, you find a violation of OSHA regulations that could be potentially dangerous to the employees. You are told by the plant manager that correcting the flaw would require the plant to be shut down for several months. This would result in almost all of the company's employees being out of work. What factors should be considered in your response to this information?

2-24. Under the section entitled "Generators and Transporters," the following statement is made: "Companies producing less than 100 kilograms per month are exempt from most RCRA regulations but still required to meet minimum standards." Do you agree with this exemption? If so, why? If not, why not?

2-25. Explain why the statement "CERCLA specifically applies this liability to wastes produced prior to the enactment of CERCLA" does not violate the Constitutional prohibition against ex post facto laws?

2-26. Obtain a manifest form for hazardous waste from your own state regulatory agency. Fill it out for a waste from a local industry.

2-27. Explain how the principle of joint and several liability would apply to potentially responsible parties (PRPs) at a Superfund site.

2-28. What is the Federal Administrative Procedures Act?

2-29. Explain what is meant by "citizen suits."

2-30. Obtain an EPA Guidance Document related to the hazardous waste program from the government documents section of your library and provide a brief abstract of its provisions. Particularly note which laws and regulations provide authority to the guidance document. Finally explain why a guidance document was used rather than a new regulation.

2-31. Review the article by K. E. Aaron and A. J. Slap.[66] Explain how CERCLA may be in conflict with federal bankruptcy laws.

2-32. Review the article by J. Bartels, et al.[67] How does the European system differ from the manifest system in the United States?

2-33. Review the article by M. E. Doyle and C. Sweet.[68] Identify the various methods used in different states to establish cleanup goals.

2-34. Review the article by L. Ford, et al.[69] Explain what happens when the government attempts to regulate pollution at levels that cannot be consistently measured.

2-35. Review the article by J. Josephson.[70] Explain the various phases in the federal Superfund program.

2-36. Explain what is meant by "cradle-to-grave" control of hazardous waste.

2-37. Review the article by J. Nikmanesh, et al.[71] Explain what is meant by a de minimis PRP.

2-38. Review the article by D. C. Wilson and W. S. Forester,[72] noting the differences between European and American approaches to hazardous waste regulations.

2-39. Explain why the American definition of hazardous waste (and the chemical tests derived from that definition) would not be appropriate in Great Britain.

2-40. Contact your state environmental agency and find out what steps are required to obtain a permit for a hazardous waste land disposal facility.

2-41. Determine which departments generate hazardous waste at your university, and where the wastes are sent. Obtain a manifest form for a shipment of hazardous waste from the university.

2-42. Explain the Hazard Ranking System. What is the lowest score a site can have and be included on the NPL?

2-43. Find out from your state's regulatory agency where the nearest Superfund site is located and ascertain the status of cleanup.

2-44. The following is a list of potential hazardous waste generators. Explain the possible hazardous wastes generated from each and how each waste is classified as hazardous.
(*a*) Dry cleaners
(*b*) Automobile service stations
(*c*) "Do-it-yourself" garage mechanics

(*d*) Users of household pesticides (RAID, etc.)
(*e*) Metal finishers (e.g., ball bearing manufacturer)
(*f*) Hospitals
(*g*) Doctors' offices

REFERENCES

1. Bockrath, Joseph: *Environmental Law for Engineers, Scientists & Managers*, McGraw-Hill Inc., New York, New York, 1977.
2. Arbuckle, J. G., et al.: *Environmental Law Handbook*, 10th edition, Government Institutes, Inc., Rockville, Maryland, 1989.
3. Findley, R. W., and D. A. Farber: *Environmental Law—Cases and Materials*, 2nd edition, West Publishing Co., St. Paul, Minnesota, 1989.
4. Rogers, W. H., Jr.: *Environmental Law*, West Publishing Co., St. Paul, Minnesota, 1977.
5. Schoenbaum, T. J.: *Environmental Policy Law*, Foundation Press, Mineola, New York, 1985.
6. *Dupont Directions*, E. I. Dupont deNemours, Inc., vol. III, no. 2, n.d.
7. *Environmental Reporter Cases*, Bureau of National Affairs, Washington, D.C., 1990.
8. Reprinted from *Black's Law Dictionary*, H. C. Black, 5th edition, 1979. With permission of the West Publishing Company.
9. Reproduced with permission. Ibid.
10. Ibid.
11. Ibid.
12. Sullivan, T. F. P.: "Environmental Law Fundamentals" in Arbuckle, J. G., et al., *Environmental Law Handbook*, Government Institutes, Inc., Rockville, Maryland, 1987.
13. Prosser, W. G., J. W. Wade, and V. E. Schwartz: *Cases and Materials on Torts*, 6th edition, The Foundation Press, p. 590ff, 1976.
14. The Federal Tort Claims Act of 1946, 28 U.S.C. §1346 et seq.
15. Federal Administrative Procedures Act of 1946, 60 Stat. 237, 5 U.S.C.A.
16. U.S. Environmental Protection Agency: *EPA Regulatory Agenda*, Washington, D.C., 1989.
17. U.S. Environmental Protection Agency: *Preliminary Assessment Guidance—Fiscal Year 1988*, OSWER Directive 9345.0-01, Office of Emergency and Remedial Response, Jan. 1988.
18. U.S. Environmental Protection Agency.
19. U.S. Supreme Court, *City of Philadelphia v. New Jersey*, 437 U.S. 617, 98 S. Ct., 2531, 1978.
20. The Resource Conservation and Recovery Act of 1976, 42 USC §6901 et seq., 1976.
21. The Hazardous and Solid Waste Amendments of 1984, P.L. 98-616, November 9, 1984.
22. Beveridge & Diamond, P.C.: "Summary and Analysis of the Hazardous and Solid Waste Amendments of 1984," *Hazardous Waste & Hazardous Materials*, vol. 2, no. 1, pp. 113–130, 1985.
23. Environmental Resources Management, Inc.: *The New RCRA—An Introduction to the Hazardous & Solid Waste Amendments Act of 1984*, Exton, Pennsylvania, 1985.
24. U.S. Environmental Protection Agency: *Corrective Action for Solid Waste Management Units (SWMUs) at Hazardous Waste Management Facilities—Proposed Rule*, September 12, 1988.
25. U.S. Environmental Protection Agency: *Summary of State Reports on Releases from Underground Storage Tanks*, EPA/600/M-86/020, August 1986.
26. Courtesy ERM/EnviroClean.
27. Federal Register, *Hazardous Waste and Consolidated Permit Regulations*, 45, no. 98, pp. 3066–3588, May 19, 1980.
28. Deland, M. R.: "EPA Regulates Hazardous Waste," *Environ. Sci. & Tech.*, vol. 14, no. 7, p. 777, July 1980.
29. 40 CFR 261.31.
30. U.S. Environmental Protection Agency: *Test Methods for Evaluating Solid Waste*, SW-846, 3rd edition, Sect. 7.4, November 1986.
31. *Federal Register*, March 29, 1990.

32. Kimmell, T. A.: "Development, Evaluation and Use of the Toxicity Characteristic Leaching Procedure (TCLP)," Chemical and Biological Characterization of Sludges, Sediments, Dredge Spoil, and Drilling Muds, ASTM STP 976, J. J. Lichtenberg, J. A. Winter, C. I. Weber, and L. Fradkin, eds., Amer. Soc. Test. Matls., Philadelphia, Pennsylvania, pp. 129–140, 1988.

33. Anon.: "TCLP vs EP?—Confusion Reigns," *The Hazardous Waste Consultant,* p. 2.2, March/April 1991.

34. The Comprehensive Environmental Response, Compensation and Liability Act (Superfund), 42 USC §9601 et seq., 1980.

35. Hedeman, W. M., Jr., P. E. Shorb, III, and C. A. McLean: "The Superfund Amendments and Reauthorization Act of 1986: Statutory Provisions and EPA Implementation," *Hazardous Waste & Hazardous Materials,* vol. 4, no. 2, pp. 193–210, 1987.

36. Environmental Resources Management, Inc.: *The New Superfund—CERCLA Amendments of 1986,* Exton, Pennsylvania, 1986.

37. U.S. House of Representatives, *Superfund Amendments and Reauthorization Act—Conference Report,* Report 99-962, October 3, 1986.

38. Stoll, R. G.: "The Comprehensive Environmental Response, Compensation and Liability Act (CERCLA or Superfund)" in Arbuckle, J. G., et al., *Environmental Law Handbook,* Government Institutes, Inc., Rockville, Maryland, 1987.

39. Superfund Amendments and Reauthorization Act, P.L. 99-499, October 17, 1986.

40. Shrivastava, Paul: *Bhopal—Anatomy of A Crisis,* Ballinger Publishing Co., Cambridge, Massachusetts, 1987.

41. Diamond, S.: "The Bhopal Disaster: How it Happened," *The New York Times,* January 28, 1985.

42. Sax, N. I., and R. J. Lewis: *Dangerous Properties of Industrial Materials,* 7th edition, Van Nostrand Reinhold, New York, New York, 1989.

43. Diamond, "The Bhopal Disaster."

44. Kalelkar, A. S.: "Investigation of Large Magnitude Incidents: Bhopal as a Case Study," presented at the Institute of Chemical Engineers Conference on Preventing Major Chemical Accidents, London, England, May 1988.

45. "Bhopal Disaster Spotlights Chemical Hazards Issues," *Chemical & Engineering News,* December 4, 1984.

46. Stone, M. E., and A. B. Weissman: "Superfund Compliance with Applicable of Relevant and Appropriate Requirements," Superfund '88, Proceedings of the 9th National Conference, Haz. Matls. Control Res. Inst., Washington D.C., pp. 8–11, 1988.

47. National Oil and Hazardous Substances Pollution Contingency Plan (NCP), Final Rule, 55 FR 8666, March 8, 1990.

48. LaGrega, M. D., A. Ertel, P. A. Marcus, and J. B. McIlvain: "Legal Responsibility of Industry for Superfund Sites," *Journal of Resource Management and Technology,* vol. 16, no. 1, pp. 14–20, April 1988.

49. Mott, Randy M.: *Hazardous Waste Litigation—1986,* Practicing Law Institute, New York, New York, 1986.

50. Miller, M. L.: "Toxic Substances," Chapter 4 in J. G. Arbuckle, et al., *Environmental Law Handbook,* 10th ed., Government Institutes, Inc., Rockville, Maryland, p. 195, 1989.

51. Occupational Safety and Health Act of 1970, PL 91-596, 84 Stat. 1590, section 6(b)(5).

52. Quarles, J., and W. H. Lewis, Jr.: *The New Clean Air Act—A Guide to the Clean Air Program as Amended in 1990,* Morgan, Lewis & Bockius, Washington, D.C., 1990.

53. Wilson, D. C., and W. S. Forester: "Summary and Analysis of Hazardous Waste Management in ISWA Countries" in *International Perspectives on Hazardous Waste Management,* W. S. Forester and J. H. Skinner, eds., Academic Press, San Diego, California, pp. 11–96, 1988.

54. Ibid.

55. Ibid, pp. 17–18.

56. Ibid, p. 24.

57. EC Directive on Civil Liability for Damage Caused by Waste, COM(89)282, June 1990.

58. "Paying for Pollution: Civil Liability for Damage Caused by Waste," U.K. House of Lords—European

Communities Committee, 1990.
59. "UK Report Calls for Strict Civil Liability in Waste Management," *HAZNEWS*, vol. 34, p. 8, January 1991.
60. Wilson and Forester, "Summary and Analysis of Hazardous Waste Management," op. cit., pp. 31–32.
61. Biswas, A. K.: "Environmental Aspects of Hazardous Waste Management for Developing Countries: Problems and Prospects," in *Hazardous Waste Management*, S. P. Maltezou, et al., eds., United Nations Industrial Development Organization, Vienna, Austria, pp. 261–271, 1989.
62. Wilson and Forester, "Summary and Analysis of Hazardous Waste Management," op. cit., p. 71.
63. Wilson, D. C., and F. Balkau: "Adapting Hazardous Waste Management to the Needs of Developing Countries—An Overview and Guide to Action," *Waste Management & Research*, vol. 8, pp. 87–98, March 1990.
64. Huntoon, B. D.: "Emerging Controls on Transfers of Hazardous Waste to Developing Countries," *Law & Policy in International Business*, vol. 21, pp. 247–271, 1989.
65. "Basel Convention on the Control of Transboundary Movements of Hazardous Wastes and Their Disposal," UNEP Document IG.80/3, March 22, 1989.
66. Aaron, K. E., and A. J. Slap: "The Chateaugay Decision: Collision of Bankruptcy and Environmental Law," *Toxics Law Reporter*, Bur. Natural Affairs, June 1990.
67. Bartels, J., et al.: "An Analysis of a European Hazardous Waste Transport Registration System (EARS)," *Waste Mange. Res.*, vol. 7, no. 229, 1989.
68. Doyle, M. E., and C. Sweet: "Soil Cleanup Goals for Total Petroleum Hydrocarbons" in *Superfund '90*, Haz. Materials Control Research Instit., Washington, D.C., pp. 21–24, 1990.
69. Ford, L., et al.: "Environmental Russian Roulette—Compliance At or Near the Detection Level," *Water Environment and Technology*, pp. 58–63, August 1990.
70. Josephson, J.: "Implementing Superfund," *Environ. Sci. Technol.*, vol. 20, no. 1, pp. 23–28, 1986.
71. Nikmanesh, J., et al.: "De Minimis Settlement—A Success Story" in *Superfund '89*, Haz. Materials Control Research Instit., Washington, D.C., pp. 190–193, 1989.
72. Wilson and Forester, "Summary and Analysis of Hazardous Waste Management," op. cit., pp. 11–96.

ADDITIONAL READING

Anderson, D. R.: "Financing Liabilities Under Superfund," *Risk Anal.*, vol. 9, no. 329, p. 615, 1989.
Anon.: "Developments in the Law: Toxic Waste Litigation," *Harvard Law Review*, pp. 1459–1661, May 1986.
Anon.: "The Clean Air Act Amendment of 1990—A Detailed Analysis," *The Hazardous Waste Consultant*, pp. 4.1–4.30, Jan/Feb 1991.
Anon.: *Restatement of Torts*, 2nd ed., West Publishing Co., St. Paul, Minnesota, 1934.
Barr, L. M.: "CERCLA Made Simple: An Analysis of the Cases Under the Comprehensive Environmental Response, Compensation and Liability Act of 1980," *Business Lawyer*, vol. 45, pp. 923–1001, May 1990.
Berman, Eugene, and Jeffrey Howard, "How Should Federal Hazardous Waste Policy Address Potentially Responsible Local Governments," *Environment Reporter*, BNA, pp. 373–376, July 15, 1988.
Cooke, S. M., ed.: *The Law of Hazardous Waste: Management, Cleanup, Liabilities and Litigation*, Matthew Bender, Albany, New York, 1990.
Donovan, P. E.: "Serving Multiple Masters: Confronting the Conflicting Interests that Arise in Superfund Disputes," *Boston College Environmental Affairs Law Review*, vol. 17, pp. 371–403, 1990.
Ferrey, S.: "The Toxic Time Bomb: Municipal Liabilities for the Clean-Up of Hazardous Waste," *The George Washington Law Review*, vol. 57, pp. 197–278, December 1988.
Firestone, D. B., and F. C. Reed: *Environmental Law for Non-lawyers*, Ann Arbor Sci. Publ., 1983.
Forester, W. S., and J. H. Skinner, eds.: *International Perspectives on Hazardous Waste Management*, Academic Press, San Diego, California, 1988.
Frank, William Harris, and Timothy B. Atkeson: "Superfund Litigation and Cleanup," *Environment Reporter*, BNA, June 28, 1985.

Gelwan, L. A.: "PRP Access to Superfund Sites: A Primer," *Virginia Environmental Law Journal*, vol. 10, pp. 77–86, Fall 1990.

Gentry, Bradford S., "Regulation of Hazardous Waste in the United States and Great Britain," *Environment Reporter*, BNA, pp. 2321–2328, February 24, 1989.

Hammitt, J. K., and P. Renten: "Illegal Hazardous Waste Disposal and Enforcement in the United States: A Preliminary Assessment," *J. Hazard. Mater.*, vol. 22, no. 281, 1989.

Howard, J. H., and L. E. Benfield: "CERCLA Liability for Hazardous Waste Generators: How Far Does Liability Extend?" *Temple Environmental Law and Technology Journal*, vol. 9, pp. 33–49, Spring 1990.

Huntoon, B. D.: "Emerging Controls on Transfers of Hazardous Waste to Developing Countries," *Law and Policy in International Business*, vol. 21, pp. 247–271, 1989.

LaGrega, M. D., et al.: "Potentially Responsible Parties and Superfund—An Analysis of Issues," in Proceedings of the 41st Annual Industrial Waste Conference, Purdue University, pp. 370–376, 1986.

Prosser, William: *Torts*, 3rd ed., West Publishing Co., St. Paul, Minnesota, 1964.

Quarles, J.: "In Search of a Waste Management Strategy," *Natural Resources and Environment*, pp. 3–5, Summer 1990.

Ruckelshaus, W. D.: "The Role of the Affected Community in Superfund Cleanup Activities," *Hazardous Waste*, vol. 1, no. 3, pp. 283–288, 1984.

Ryan, M. K.: "The Superfund Dilemma: Can You Ever Contract Your Liability Away?" *Massachusetts Law Review*, vol. 75, pp. 131–137, Summer 1990.

Travis, C. C., and C. B. Doty: "Superfund. A Program without Priorities," *Environ. Sci. Technol*, vol. 23, p. 1333, 1989.

U.S. Government Printing Office: *Code of Federal Regulations — Chapter 40 — Protection of Environment*, 1988.

Wigmore, J.: *Evidence*, 3rd ed., Little, Brown & Co., 1961.

ACKNOWLEDGMENTS

The authors wish to acknowledge the following individuals who assisted with the preparation of this chapter: Thomas F. McCaffery, III, Esq., *SmithKline Beecham*; J. Bradford McIlvain, Esq. and Joseph Dominguez, Esq., *Dilworth, Paxson, Kalish & Kauffman*; Holly Borcherdt, *Bucknell University*; Arnon E. Garonzik, Esq., Ann Sirois, Esq., Kenneth Weiss, and Paul Woodruff, *ERM, Inc.*; David Morell, *ERM-West*; and David C. Wilson, Ph.D., *ERM-London*. The principal reviewer for the **ERM Group** was Peter Schneider, *ERM-New England*.

CHAPTER
3

PROCESS FUNDAMENTALS

Water, water, every where,
And all the boards did shrink;
Water, water, every where,
Nor any drop to drink.
　　　　　　Coleridge

This chapter is intended as an overview of fundamentals that many students using this book will have covered in previous science and engineering courses. For more background on the theory underlying the information presented here, refer to basic texts listed at the end of the chapter.

3-1　A LITTLE CHEMISTRY

Organic chemistry is the study of compounds containing carbon, which make up much of hazardous waste. A working knowledge of the properties and characteristics of organic compounds is essential to the management of hazardous waste. Additionally, a general understanding of chemical nomenclature is critical in the understanding of regulatory documents.

Nomenclature and Structure

When drawing chemical structures for organic molecules, a few rules must be kept in mind. Because the vast majority of chemical bonds in organic molecules are covalent (electrons are shared between atoms), bonds are typically represented by dashes connecting atoms in the molecule. This representation is known as a "dash structure." Because the well-known "octet rule" must be obeyed, atoms in a dash structure must be bonded together by at least one pair (two) of electrons. The rule of thumb is that one dash is used for each pair of electrons in a bond. Therefore, a single bond is represented by one dash, a double bond is represented by two parallel dashes, and a triple bond is represented by three parallel dashes. Another good rule of thumb is that a carbon atom in an organic molecule can only have a total of four bonds with other atoms (i.e., one single–one triple, two single–one double, two double, or four single).

There are several other conventions for drawing organic molecules as seen in Example 3-1. Some are used to illustrate electron pairs; others are used as shorthand representation; still others serve to detail the three-dimensional aspects (stereochemistry) of the molecules.

Example 3-1   Various ways of representing the structure of methane.

Hydrocarbons are a class of organic compounds that are composed solely of carbon and hydrogen. Hydrocarbons are generally divided into two categories, aromatic and aliphatic. **Aromatics** are ring or multiple ring structures containing alternating single and double bonds. **Aliphatic compounds** are those that contain open carbon chains or rings (known as alicyclics) with or without alternating single and double bonds. Aliphatics include alkanes, alkenes, alkynes, and their derivatives.

Alkanes are hydrocarbon compounds that contain only single bonds between carbon atoms. Alkanes have the general formula C_nH_{2n+2} where n is any integer. Several examples of alkanes are listed in Table 3-1.

The names of many organic compounds are derived from the alkane series. Alkyl radicals are alkanes with a terminal hydrogen atom missing. For example, $CH_3–CH_2 \cdot$ is called ethyl, $CH_3–CH_2–CH_2 \cdot$ propyl, etc.

In general formulas, R is used to represent an unspecified alkyl radical (e.g., methyl). An alkane may then be represented by R–H.

Normal alkanes are those that have a straight chain of carbon atoms and the name is proceeded by "n-." A branched alkane is composed of a normal alkane with one or more alkyl groups attached at carbons other than the terminal carbons. In naming a branched alkane, the procedure below must be followed:

1. Look for the longest straight carbon chain. That chain gives the base name.
2. Number the carbon atoms in the base chain. The carbons in this base chain must be numbered in a direction that has the lowest possible numbers assigned to carbons containing any branched alkyl groups.

TABLE 3-1
Alkanes[1]

Name	Chemical formula	Structure
methane	CH_4	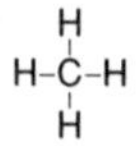 H‑C‑H (with H above and H below)
ethane	C_2H_6	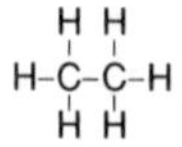 H‑C‑C‑H (with H H above and H H below)
propane	C_3H_8	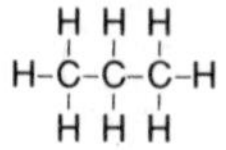 H‑C‑C‑C‑H (with H H H above and H H H below)
butane	C_4H_{10}	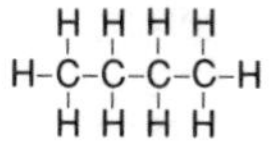 H‑C‑C‑C‑C‑H (with H H H H above and H H H H below)
pentane	C_5H_{12}	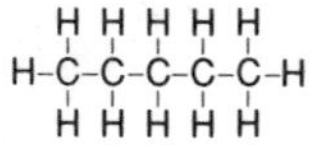 H‑C‑C‑C‑C‑C‑H (with H H H H H above and H H H H H below)

3. Place the appropriate number before the alkyl group's name. Each group must have a number even if two groups are attached to the same carbon atom. When writing the name, numbers are always separated from letters by a hyphen. The name of the last group is joined to the base name as a continuous word and listed in alphabetical order regardless of numerical carbon assignment.

butane:

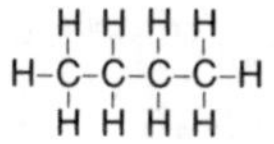

$$
\begin{array}{ccccccc}
 & H & H & H & H & \\
 & | & | & | & | & \\
H- & C & - C & - C & - C & -H \\
 & | & | & | & | & \\
 & H & H & H & H &
\end{array}
$$

3-methyl-hexane:

$$
\begin{array}{c}
H \\
| \\
H-C-H \\
\end{array}
$$

$$
\begin{array}{ccccccc}
H & H & & H & H & H \\
| & | & | & | & | & | \\
H-C & -C & -C & -C & -C & -C-H \\
| & | & | & | & | & | \\
H & H & H & H & H & H
\end{array}
$$

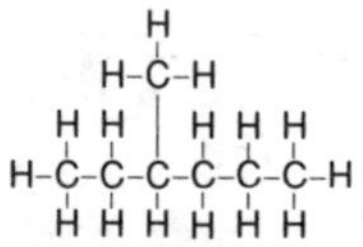

Alkenes contain carbon-carbon double bonds. Alkenes have the general formula C_nH_{2n}. In naming alkenes, the following steps are taken:

1. Look for the longest chain that contains a double bond. Change the suffix of the corresponding alkane of the identical length from "-*ane*" to "-*ene*." This is the base name.

2. Number the chain beginning at the end nearer the double bond. Use the number of the first carbon of the double bond to designate the location of the double bond, and place this number before the base name using a hyphen.

3. Complete the compound name as outlined in Steps 2 and 3 for alkanes above.
4. If a double bond has two identical groups, the prefixes *"cis-"* and *"trans-"* are used. (Cis- denotes two identical groups on the same side of the molecule. Trans- is used if they are on opposite sides.)

cis-2,3-difluoro-2-butene:

trans-2,3-difluoro-2-butene:

5. If more than one double bond is present, indicate the position of each and use diene, triene, etc.

1, 3-butadiene: $CH_2 = CH - CH = CH_2$

Alkynes contain carbon-carbon triple bonds. Alkynes have the general formula C_nH_{2n-2}. Alkynes are named in a manner similar to that used for alkenes except the longest chain must contain the triple bond and the -ane suffix is changed to -yne. There can be no cis- or trans- assignment to a carbon-carbon triple bond.

ethyne (common name – acetylene): $HC \equiv CH$

A somewhat different class of organic compounds are ketones. **Ketones** have the general formula

where R and R' are different alkyl groups. The process of naming ketones involves naming each alkyl group and adding ketone to the end. The alkyl groups are given in alphabetical order.

The second category of hydrocarbons is **Aromatics**. The simplest aromatic ring is benzene:

Nomenclature of benzene is divided into two categories. For certain compounds, benzene is the base name and the group is designated as a prefix.

chlorobenzene:

When two groups are present, the prefixes "ortho-," "meta-," and "para-," or a numerical designation is used to designate position on the ring.

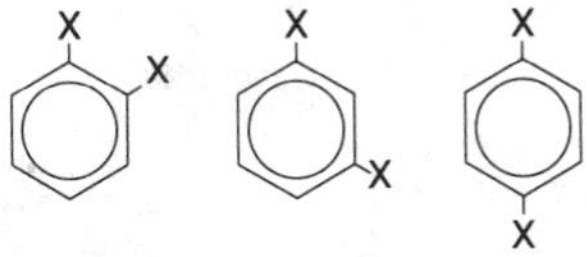

ortho- meta- para-
or 1,2- or 1,3- or 1,4-

When more than two groups are present, numbers must be used to indicate the positions. The ring is numbered so the groups have the lowest possible numbers.

1,2-dinitro-4-chlorobenzene:

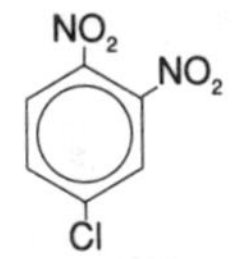

For some compounds, common names can be used.

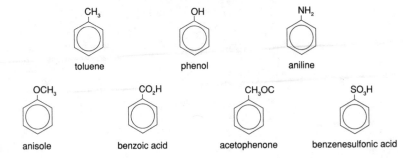

toluene   phenol   aniline

anisole   benzoic acid   acetophenone   benzenesulfonic acid

Dimethylbenzenes are given the common name xylenes.

meta-xylene:

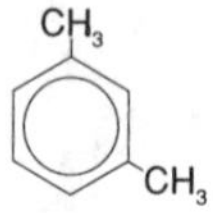

When the benzene ring is a substituent of a larger group, it is called phenyl.

3,4'-dichlorobiphenyl:

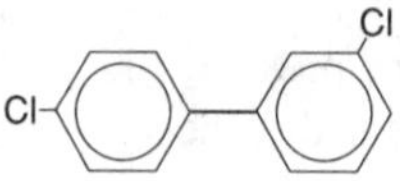

Polyaromatic hydrocarbons (PAHs) are groups of aromatic rings containing only carbon and hydrogen. They may be considered as two or more benzene rings fused together with at least two common carbons.

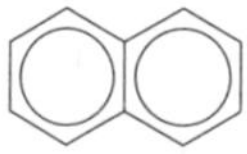

napthalene 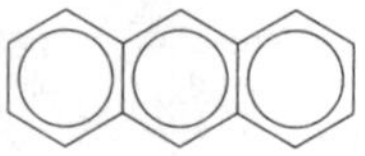   anthracene 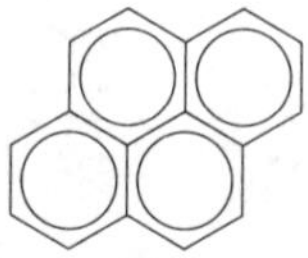   pyrene

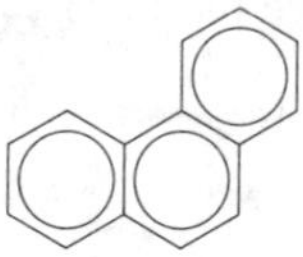

phenanthrene 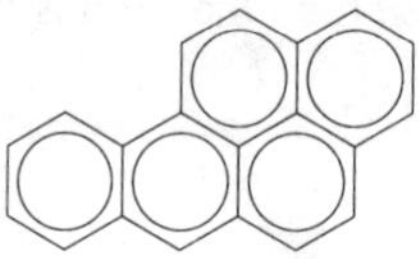   benzo (a) pyrene

Amines are considered organic derivatives of ammonia. They are classified as primary, secondary, or tertiary amines, by the number of attached alkyl groups.

$$R-NH_2 \qquad R-\overset{R'}{\underset{|}{N}}-H \qquad R-\overset{R'}{\underset{|}{N}}-R''$$

Primary amine   Secondary amine   Tertiary amine

A nitrosamine has the nitroso group ($-N=O$), as one of the alkyl groups.

diphenylnitrosamine:

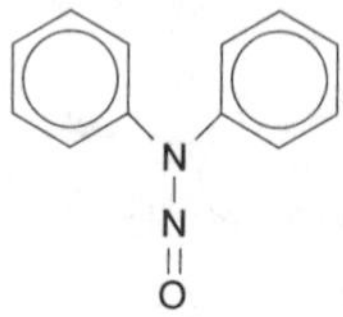

Ethers have the general formula R–O–R or R–O–R′ where R′ indicates a different alkyl group than R. In naming ethers, the two alkyl groups attached to the oxygen are named and the word ether is added.

bis-(2-chloroethyl) ether: Cl–CH$_2$–CH$_2$–O–CH$_2$–CH$_2$–Cl

---

Example 3-2   Structural formulae.   (a) Draw the structure for ethyl propyl ether. (b) What name would you give to the structural formula (CH$_3$)$_2$CHOCH$_2$CH$_3$?

Solution

(a):

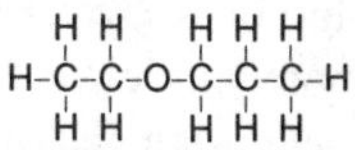

$$\begin{array}{c}\text{H H}\quad\text{H H H}\\ \text{H–C–C–O–C–C–C–H}\\ \text{H H}\quad\text{H H H}\end{array}$$

(b): ethyl isopropyl ether. The prefix "iso" indicates a branched chain.

Esters have the general formula RCO$_2$R′. An ester is formed by reacting a carboxylic acid with an alcohol. To name an ester, the name of the alkyl radical (R') from the alcohol comes first, followed by the name of the carboxylate ion or salt. The carboxylate ion has the general formula R–CO$_2^-$

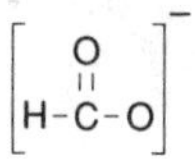

formate ion 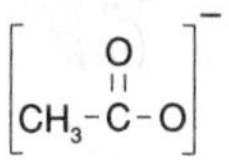   acetate ion 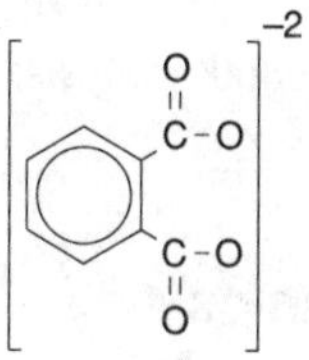   phthalate ion

Phthalate esters contain the phthalate ion and any alkyl groups.

dimethyl phthalate:

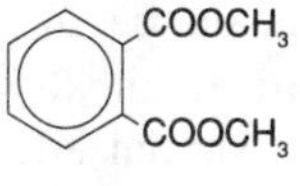

Chemical Characteristics

From the perspective of environmental chemistry, organic chemicals exist in basically three different categories related to volatility: Volatile Organics, Semivolatile Organics, and Nonvolatile Organics. Subcategories can be developed within each of the three main categories as discussed briefly below. It is important to note, however, that no distinct lines can be drawn between what is considered to be volatile or semivolatile, or between semivolatile and nonvolatile. In environmental chemistry, compounds are typically assigned to these broad categories on the basis of which analytical method proves to be the best for the detection of that compound.

Volatile Organics

Volatile organic compounds (VOCs) have a high vapor pressure that allows them to evaporate quickly. In general, they are relatively small compounds. Their physical-chemical properties include high solubility in water, high vapor pressure, high Henry's constant, low organic carbon partition coefficient (K_{oc}), high octanol water partition coefficient (K_{ow}), and low bioconcentration factor (BCF). These properties are described in Sec. 3-2. Examples of volatile organics include:

$$
\begin{array}{cc}
\overset{\displaystyle H}{\underset{\displaystyle Cl}{Cl-\overset{|}{\underset{|}{C}}-H}} & \overset{\displaystyle O}{CH_3-\overset{||}{C}-CH_3} \\[2em]
\text{dichlorethane} & \text{acetone}
\end{array}
$$

Analytically, detection of these compounds is performed by taking advantage of their high vapor pressure. By bubbling a pure gas (e.g. helium) through a water sample (or a soil sample mixed with water), volatile compounds are removed from the sample, are swept away with the evolving gas, and are finally trapped from the gas stream for instrumental analysis. This method of analysis is known as the **purge and trap technique**.

Semivolatile Organics

Semivolatile organics include several broad classes of compounds with different physical and chemical properties. Because of these differences in properties, semivolatile organics are typically subcategorized as **Base/Neutral Extractables** and **Acid Extractables**.

Unlike volatile organics, most of these compounds are not of sufficient volatility for use of the purge and trap technique. Therefore, the chemical analysis of these compounds must rely on the acid/base properties of these compounds to facilitate their isolation from a water sample into a solvent. When the sample is acidic, acid extractable compounds are driven into the solvent. When the sample is basic, base/neutral extractable compounds are driven into the solvent. The insoluble solvent phases are separated from the water sample, and are then analyzed for semivolatile organics. This method of analysis is known as the **solvent extraction** technique.

Base Neutrals

Base neutrals include polyaromatic hydrocarbons (PAHs), nitrosamines, ethers, phthalate esters, and selected aromatics not containing hydroxyl (OH) or carboxyl (COOH) groups. Base/neutrals tend to have the following physical-chemical properties: low solubility, low Henry's constant, low vapor pressure, high K_{oc}, low K_{ow}, and high BCF. These properties are described in Sec. 3-2.

Acid Extractables

In general, acid extractables are aromatic alcohols or phenols. Acid extractables have the following physical-chemical properties that are dependent upon substituent and position: high to low water solubility, moderate vapor pressure, low Henry's constant and variable K_{oc} and BCF. Examples of acid extractables include:

2,4-dinitrophenol   4-chloro-3-methyl phenol
or
4-chloro-3-cresol

Nonvolatile Organics

Nonvolatile organics are typically characterized as organic compounds with little to no volatility even at elevated temperatures. Analytically, these compounds can be isolated from environmental samples using the same solvent extraction technique discussed for semivolatile organics analysis.

Pesticides and Herbicides

Pesticides and herbicides are chemical compounds used to control insects, rodents, plants, etc. These two classes of organics include chlorinated pesticides and organo-phosphorus pesticides (see Tables 3-2 and 3-3).

   Chlorinated pesticides have been widely used as insecticides, fungicides, and herbicides. These pesticides have been discovered as having harmful side effects as they do not readily degrade in nature and tend to accumulate in the fatty tissues of most mammals. Perhaps the most infamous of all chlorinated pesticides is **DDT** (see Table 3-3).

TABLE 3-2
General household and field insecticides[2]

Common name	Remarks
lindane	

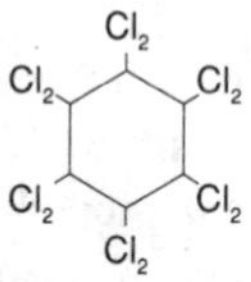

particularly good against the boll weevil, the cotton aphid and the locust

m-delphene

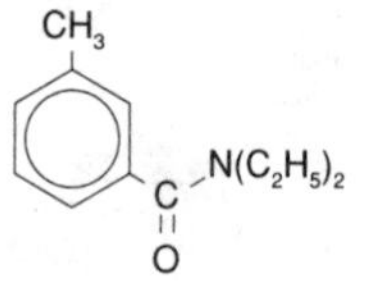

good mosquito repellent

p-dichlorobenzene

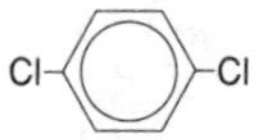

common moth repellent

methoxychlor

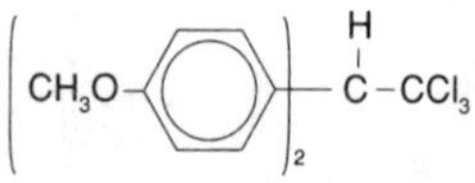

methoxy analog of DDT; can be used in dairies without imparting toxicity or off-taste to milk

Other chlorinated pesticides include lindane, dieldrin, aldrin, chlordane, toxaphene, heptachlor, DDD, and DDE. Virtually all chlorinated pesticides have been scrutinized because of known or suspected hazards to human life.

Organophosphorus pesticides are generally toxic to both humans and animals. One important organophosphorus pesticide is parathion (see Table 3-3).

Other organophosphorus pesticides include malathion, systox, chlorthion, disyston, dicapthon, and metasystox.

Polychlorinated Biphenyls

Polychlorinated biphenyls (PCBs) contain a mixture of biphenyls with chlorine atoms attached at any of the carbons. There are 210 different PCB compounds. Commercial mixtures are designated by their chlorine content; these mixtures typically have 40–60% chlorine by weight.[4]

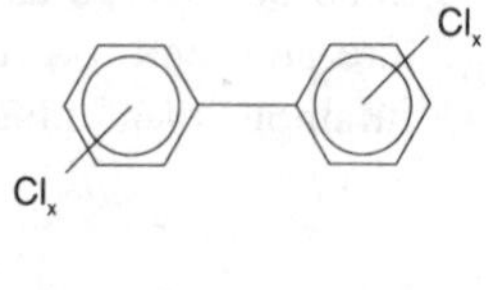

TABLE 3-3
Wide-spectrum insecticides[3]

Common name	Toxicity to mammals
DDT	fairly low
chlordane	moderate
dieldrin:	fairly high
parathion	high
diazinon:	moderate

PCBs are a common pollutant due to their former widespread use in electrical transformers and capacitors. Since 1979, the EPA has banned their manufacture, processing, and distribution.

3-2   PHYSICAL-CHEMICAL PROPERTIES

Physical-chemical properties determine how a specific chemical will interact in the environment. These properties control how chemicals move in the environment, and how effectively they can be removed by various methods.

Solubility

Solubility is the degree to which one substance, the solute, will dissolve into another, the solvent. The solubility of chemicals in water is a function of temperature, and the particular chemical species. Solubility in water is usually expressed in units of milligrams per liter (mg/L), or parts per million (ppm). Because one liter of water weighs one million milligrams, in the dilute solutions typically encountered in environmental work, 1 mg/L = 1 ppm, and 1 μg/L = 1 part per billion (ppb).

All materials are soluble to some degree. Consider the precipitation of calcium hydroxide:

$$Ca^{+2} + 2OH^- \rightarrow Ca(OH)_2 \downarrow \tag{3-3}$$

The **solubility product**[5] for this reaction is:

$$K_{sp} = \left[Ca^{+2}\right]\left[OH^-\right]^2 = 7.88 \times 10^{-6}$$

The square brackets indicate molar concentration (moles/liter).

Example 3-3 Solubility product. Nickel is removed from a plating waste stream by hydroxide precipitation at pH = 10. Determine the solubility of nickel in pure water at 25°C $K_{sp} = 5.54 \times 10^{-16}$.

Solution.

$$Ni^{+2} + 2OH^- \rightarrow Ni\,(OH)_2 \downarrow$$

$$-Log\left[OH^-\right] = pOH = 14 - pH = 4.0$$

$$\left[OH^-\right] = 1 \times 10^{-4} \text{mol/L}$$

$$\left[Ni^{+2}\right]\left[OH^-\right]^2 = 5.54 \times 10^{-16}$$

$$\left[Ni^{+2}\right] = \frac{5.54 \times 10^{-16}}{\left(10^{-4}\right)^2}$$

$$\left[Ni^{+2}\right] = 5.54 \times 10^{-8} \text{mol/L}$$

$$1 \text{ mol Ni} = 58.7 \text{ g} = 58,700 \text{ mg}$$

$$\text{Concentration} = 5.54 \times 10^{-8} \text{ mol/L} \times 58,700 \text{ mg/mol} =$$

$$\underline{\underline{3.25 \times 10^{-3} \text{ mg/L}}} = \underline{\underline{3.25\mu\text{g/L}}}$$

While this is the theoretical solubility at pH = 10, it is not possible to achieve this low value when treating wastes by hydroxide precipitation.

Vapor Pressure

When a liquid is in contact with air, molecules will leave the liquid as a vapor, via evaporation. Equilibrium conditions exist when the rate of molecules leaving the liquid equals the rate of molecules being re-dissolved. The **vapor pressure** of a liquid is the pressure exerted by the vapor on the liquid at equilibrium. Vapor pressure is a function of the temperature of the liquid and is usually measured in atmospheres (atm) (1 atm = 760 mm Hg = 760 torr = 1.0133×10^5 Pa). Vapor pressures vary greatly with temperature, and care must be exercised in using published values. Vapor pressures of various chemicals are tabulated in App. A.

For an ideal binary liquid mixture, equilibrium follows **Raoult's law**:

$$p_a = p_{vp} x_a \qquad (3\text{-}4)$$

where

P_a = partial pressure of contaminant a (atm)

P_{vp} = vapor pressure of the pure compound (atm)

x_a = mole fraction of contaminant a

$$x_a = \frac{\text{mol } a}{\text{mol } a + \text{mol water}}$$

Note that for the dilute solutions typically encountered in environmental work,

$$x_a \approx \frac{\text{mol } a}{\text{mol water}}$$

The partial pressure of this vapor is the pressure it would exert if all other gases were not present. This is equivalent to the mole fraction of the gas multiplied by the entire pressure of the gas. Raoult's law holds only for ideal solutions. In dilute solutions typical in environmental applications, Henry's law (Eq. 3-5) tends to control.

Henry's Constant

Henry's law is used to describe the solubility of a gas in a liquid. When a volatile chemical is dissolved in water, a small amount of the chemical in gaseous form exists in the air immediately above the surface of the water. At equilibrium, as molecules of this gas pass into the water, an equivalent number of molecules of the chemical leave the water phase and become part of the vapor.

Henry's law states: Under equilibrium conditions, the partial pressure of a gas (volatile chemical) above a liquid is proportional to the concentration of the chemical in the liquid:

$$P_g = H C_L \qquad (3\text{-}5)$$

where

P_g = partial pressure of the gas

H = Henry's constant

C_L = concentration of chemical in the liquid

If P_g is expressed in atmospheres (atm), and C_L as a mole fraction, H has units of atm. If C_L is expressed as mol/m^3, H has units of $atm \cdot m^3/mol$. This is the inverse of gas solubility at one atmosphere. From the definition of partial pressure, the Henry's constant also represents the ratio of the concentration in the gas to the concentration in the liquid.

$$H = C_g/C_L \tag{3-6}$$

where C_g = concentration of a specified chemical in the gas phase (air)

C_L = concentration of a specified chemical in the liquid phase (water)

The **Henry's constant** (also referred to as the Henry's law constant) may be expressed in a number of units. When Henry's law is developed for transferring air into water, the constant typically has units of mg/L·atm, or mol/L·atm (the inverse of the Henry's constant used in this book!). It is extremely important to check dimensions carefully before using tabulated values. A dimensionless form of the constant is obtained by:

$$H' = H/RT \tag{3-7}$$

where H' = Henry's constant (dimensionless)

H = Henry's constant (atm·m^3/mol)

R = universal gas constant = $(8.25 \times 10^{-5}$ atm $\cdot$ m^3/mol K$)$

T = temperature (K)

The Henry's constant is significantly affected by temperature and the chemical composition of the water. For example, the Henry's constant for volatile hydrocarbons increases approximately threefold for a 10°C increase in temperature.[6] Regression equations for selected chemicals may be found in App. A. The tabulated values define the Henry's constant as a function of temperature:

$$H = e^{(A-B/T)} \tag{3-8}$$

where H = Henry's constant (atm $\cdot$ m^3/mol)

A, B = regression coefficients

T = temperature (K)

In the absence of tabulated information, the Henry's constant may be estimated as the vapor pressure divided by solubility, provided that both are measured at the same temperature.[7]

$$H = p_{vp}/S \tag{3-9}$$

It should be noted that tabulated H values are normally based on pure water and the values for contaminated water can vary unpredictably.[8]

Example 3-4 Dimensionless Henry's constant. Estimate the Henry's constant for toluene in water at 20°C from vapor pressure and solubility data. Convert this to a dimensionless constant. Compare to the value in App. A.

Solution. From App. A: $p_{vp} = 22.0$ (mm Hg at 20°C)

Solubility at 20°C $= 5.15 \times 10^2$ mg/L

Molecular weight = 92.13 g/mol = 92.13×10^3 mg/mol

$p_{vp} = 22.0$ (mmHg)/760 (mmHg/atm) $= 0.029$ atm

$$S = \frac{5.15 \times 10^2 \text{ mg/L}}{92.13 \times 10^3 \text{ mg/mol}} = 5.59 \times 10^{-3} \text{mol/L} = 5.59 \text{ mol/m}^3$$

From Eq. 3-9:

$H = 0.029/5.59 = \underline{5.19 \times 10^{-3} \text{ atm} \cdot \text{m}^3/\text{mol}}$

From Eq. 3-7:

$$H' = \frac{5.19 \times 10^{-3} \text{ atm} \cdot \text{m}^3/\text{mol}}{\left(8.205 \times 10^{-5} \text{ atm} \cdot \text{m}^3/\text{mol} \cdot \text{K}\right) (20 + 273.2) \text{ K}}$$

$$= \underline{0.216} \text{ (dimensionless)}$$

$H = e^{(A - B/T)}$

$A = 5.13$ (App. A)

$B = 3.02 \times 10^3$ (App. A)

$T = 20 + 273.2 = 293.2$ K

$$H = e^{[5.13 - (3.02 \times 10^3/293.2)]} = 5.68 \times 10^{-3} \text{ atm} \cdot \text{m}^3/\text{mol}$$

Assuming this value to be "correct," the error in estimating H from vapor pressure and solubility data is

$$\frac{5.19 - 5.68}{5.68} = -0.086 \text{ or } 8.6\%$$

Diffusion Coefficient

Contaminants in air or water tend to move from areas of high concentration to areas of lower concentration. **Diffusion** may be defined as the movement of a contaminant under the influence of a concentration gradient. The amount of a contaminant passing through a unit area in a unit time is controlled by **Fick's law**:

$$J = -D(dC/dx) \tag{3-10}$$

where $J =$ flux (mol/cm$^2 \cdot$ sec)

$D =$ diffusion coefficient (cm^2/sec)

$C =$ concentration (mol/cm^3)

$x =$ length in the direction of movement (cm)

Where tabulated values of the diffusion coefficient do not exist, they may be estimated using the methods illustrated below. A more extensive listing of estimation methods may be found in Tucker and Nelken.[9] Additionally, D may be estimated from data on similar chemical species from their molecular weights, M:

$$D_1 = D_2 (M_2/M_1)^{1/2} \tag{3-11}$$

Example 3-5 Estimate of diffusion coefficient. Estimate the diffusion coefficient of ethyl alcohol in water from tabulated values for methyl alcohol and n-butyl alcohol.

Compound	D (cm^2/ sec)^{10}	Mol. weight	Temperature
Methyl alcohol	1.75×10^{-5}	32.05	25°C
n-Butyl alcohol	0.56×10^{-5}	74.12	25°C
Ethyl alcohol	1.24×10^{-5}	46.07	25°C

Solution. Estimates for D for ethyl alcohol:
From methyl alcohol: $D = 1.75 \times 10^{-5}(32.05/46.07)^{1/2} = 1.46 \times 10^{-5}$
From n-butyl alcohol: $D = 0.56(74.12/46.07)^{1/2} = 0.71 \times 10^{-5}$
Average value $= \underline{\underline{1.09 \times 10^{-5}}}$ (12% lower than the tabulated value of 1.24×10^{-5})

The theoretical diffusion coefficient of a contaminant gas in air may be derived from the Boltzmann equation:[11]

$$D = 1.858 \times 10^{-3} \left(\frac{T^{1.5}M'}{P\sigma^2\Omega} \right) \tag{3-12}$$

where $D =$ diffusion coefficient (cm^2/sec)

$$M' = \left(\frac{M_a + M_b}{M_a M_b} \right)^{1/2}$$

$M_a =$ molecular weight of air (28.97 g/mol)

$M_b =$ molecular weight of contaminant (g/mol)

$P =$ pressure (atm)

$T =$ temperature (K)

$\sigma =$ collision diameter (Å)

$\Omega =$ collision integral (dimensionless function of temperature)

This coefficient is termed 'theoretical' to distinguish it from the 'apparent' coefficient measured in the environment. The calculation is dependent on obtaining estimates of σ and Ω, which are available in standard references.[12]

Values of diffusion coefficients in air for a number of compounds may be found in App. A. Diffusion coefficients in air are functions of temperature and pressure and tabulated values must be adjusted to ambient conditions:[13]

$$D_1 = D_2 (P_2/P_1) (T_1/T_2)^m \tag{3-13}$$

where $m =$ theoretically is 1.5 and is measured as 1.75 to 2.0.

The diffusion coefficient in water may be estimated using the Wilke-Chang[14] method:

$$D = \frac{5.06 \times 10^{-7}\ T}{\mu V^{0.6}} \qquad (3\text{-}14)$$

where

D = diffusion coefficient (cm^2/sec)

T = temperature (K)

μ = viscosity of water (centipoises, cP)

V = molal volume of contaminant (cm^3/mol)

Values of the molal volume may be estimated from the methods of LeBas,[15] using Table 3-4.

TABLE 3-4
Additive-volume increments* for the calculation of molal volumes V by method of LeBas [†]

	Increment (cm^3/g-mol)
Carbon	14.8
Hydrogen	3.7
Oxygen (except as noted below)	7.4
In methyl esters and ethers	9.1
In ethyl esters and ethers	9.9
In higher esters and ethers	11.0
In acids	12.0
Joined to S, P, N	8.3
Nitrogen	
Doubly bonded	15.6
In primary amines	10.5
In secondary amines	12.0
Bromine	27.0
Chlorine	24.6
Fluorine	8.7
Iodine	37.0
Sulfur	25.6
Ring, three-membered	−6.0
Four-membered	−8.5
Five-membered	−11.5
Six-membered	−15.0
Naphthalene	−30.0
Anthracene	−47.5

*The additive-volume procedure should not be used for simple molecules. The following approximate values are employed in estimating diffusion coefficients: H_2, 14.3; O_2, 25.6; N_2, 31.2; air, 29.9; CO, 30.7; CO_2, 34.0; SO_2, 44.8; NO, 23.6; N_2O, 36.4; NH_3, 25.8; H_2O, 18.9; H_2S, 32.9; Cl_2, 48.4; Br_2, 53.2; I_2, 71.5.
[†] After Reid[16]

Example 3-6   Estimation of diffusion coefficient. Determine the diffusion coefficient for methanol in water at 25°C.

Solution. Methanol: CH_3OH
$T = 25 + 273.2 = 298.2$ K
Viscosity of Water $= \mu = 0.89$ cP
Molal Volume (from method of LeBas [Table 3-4]) is determined by adding individual increments to each atom:

$$C = 1 \times 14.8 = 14.8$$

$$H = 4 \times 3.7 = 14.8$$

$$O = 1 \times 7.4 = \underline{7.4}$$

$$V = 37.0 \text{ cm}^3/\text{mol}$$

$$D = \frac{5.06 \times 10^{-7} \times T}{\mu V^{0.6}}$$

$$D = \frac{5.06 \times 10^{-7} (298.2 \text{ K})}{(0.89)(37.0)^{0.6}}$$

$$D = 1.942 \times 10^{-5} \text{cm}^2/\text{sec}$$

Partition Coefficients

Partition coefficients are empirical constants that describe how a chemical distributes itself between two media. Three of the coefficients important in waste management are the octanol-water partition coefficient, the soil water partition coefficient, and the vapor liquid partition coefficient.

The **octanol-water partition coefficient** is a dimensionless constant defined by:

$$K_{ow} = C_O/C \tag{3-15}$$

where C_O = concentration in octanol (mg/L or μg/L)

C = concentration in water (mg/L or μg/L)

This coefficient provides a measure of how an organic compound will partition between an organic phase and water. For example, it provides an indication of how much of the chemical will be taken up by aquatic organisms. Values of K_{ow} range over ten orders of magnitude, from 10^{-3} to 10^7. The coefficient is useful in estimating fate and transport of chemicals and is related to soil adsorption coefficients, the bioconcentration factor (BCF), as well as water solubility. Chemicals with low values of $K_{ow}(< 10)$ tend to be hydrophilic and have low soil adsorption and low BCF.

The **soil water partition coefficient**, K_p, measures the tendency of a chemical to be adsorbed by soil or sediment and may be defined as:

$$K_p = X/C \tag{3-16}$$

where

X = concentration in soil (ppb or μg/kg)
C = concentration in water (ppb or μg/L)

Because almost all of the adsorption of organic chemicals by a soil is due to the organic carbon content of the soil, the **organic carbon partition coefficient**, K_{oc}, may be defined:

$$K_{oc} = C_C / C \qquad (3\text{-}17)$$

where C_C = Concentration adsorbed (μg adsorbed/kg organic C, or ppb)

K_{oc} may be estimated from K_p :

$$K_{oc} = K_p / f_{oc} \qquad (3\text{-}18)$$

where f_{oc} = fraction of organic carbon in the soil (dimensionless)

K_p is essentially the same as the Freundlich adsorption coefficient (mg/g) (see Sec. 9-3). K_p and K_{oc} are functions of the organic carbon content but relatively independent of the particular soil or sediment in question. Estimates of K_p and K_{oc} vary greatly, and care must be used in applying these parameters. K_{oc} may be estimated from several other physical-chemical properties as shown in Table 3-5.

The **vapor-liquid partition coefficient** is the ratio of the concentration of a compound in the vapor to the concentration in the liquid at equilibrium. This coefficient is a function of temperature, vapor pressure, atmospheric pressure, the composition of the liquid and vapor, and the specific compound. Raoult's law for ideal solutions and Henry's law for low concentrations are special cases of this partition coefficient.

Bioconcentration Factor

The **bioconcentration factor**, BCF, indicates the amount of a chemical that is likely to accumulate in aquatic organisms. It may be expressed mathematically as:

$$\text{BCF} = C_{org} / C \qquad (3\text{-}19)$$

where C_{org} = equilibrium concentration in organism (mg/kg or ppm)

TABLE 3-5
Correlations between K_{oc} and chemical properties[17, 18]

Class of chemicals	Number of chemicals	Equation	Remarks
Pesticides	45	$\log K_{oc} = 0.544 \log K_{ow} + 1.377$	
Aromatics	10	$\log K_{oc} = 1.00 \log K_{ow} - 0.21$	
Chlorinated hydrocarbons	15	$\log K_{oc} = -0.557 \log S + 4.277$	S in μmol/L
Aromatics	10	$\log K_{oc} = -0.54 \log S + 0.44$	S in mole fraction
Pesticides	106	$\log K_{oc} = -0.55 \log S + 3.64$	S in mg/L
Not stated	—	$\log K_{oc} = 0.681 \log BCF + 1.963$	BCF = bioconcentration factor

The bioconcentration factor is an essential component in risk assessments. The lipid content of animal tissue is an important factor in determining the tendency of a chemical to bioconcentrate. The BCF varies from species to species. For example, fathead minnows tend to have higher values of BCF than trout.[19] Bioconcentration is also affected by metabolism and elimination systems in the test organism.

Example 3-7 Application of the bioconcentration factor. The insecticide chlordane is found in lake water at a concentration approaching its solubility of 560 μg/L. Estimate the concentration in fish tissue.

Solution. From App. A: BCF (Fish) = 14, 000 L/kg.
From Eq. 3-19:
Concentration in Fish = 560 $(\mu$g/L$)\times 14,000$ (L/kg) = 7.84×10^6 μg/kg
= 7,840 mg/kg

BCF is related to solubility, K_{ow} and K_{oc}, with K_{ow} being the most commonly utilized measure from which to estimate BCF. BCF should be distinguished from **biomagnification** which indicates an increase in the concentration of chemical through the food chain. BCF represents the tendency of a chemical to be absorbed by aquatic organisms, and therefore indicates the increase in concentration in the tissues of a single organism.

The BCF in fish is frequently correlated to K_{ow}:

$$\log \text{BCF} = c_1 + c_2 \log K_{ow} \tag{3-20}$$

where c_1, c_2 = empirical constants

Barron[19] indicates that the relationship between BCF and K_{ow} holds over a wide range of K_{ow}; it is not a very strong correlation over a narrow range. This is illustrated in the scatter of Fig. 3-1. Banerjee and Baughman[20] indicate that this relationship does not hold for high K_{ow} compounds. Paustenbach[21] provides several examples of how BCF may be estimated from physical property data:

$$\log \text{BCF} = c_1 - c_2 \log S \tag{3-21}$$

where S = solubility

$$\log \text{BCF} = c_1 \log K_{oc} - c_2 \tag{3-22}$$

It must be emphasized that relationships such as Eq's. 3-20 through 3-22 and Table 3-5 are valid only within the range of experimental conditions for which they were developed and cannot be used generally. Specifically, these relationships do not take into account many factors affecting BCF such as metabolism and excretion.

Sorption

Sorption is the process by which a component (the sorbate or contaminant) moves from one phase to another across some boundary.

Sorption is caused by interactions among three distinct molecules:

- Sorbent (e.g., activated carbon)
- Sorbate (e.g., the contaminant being removed)
- Solvent (e.g., water)

Driving forces behind sorption include the affinity of sorbate for sorbent:

- Electrical attraction
- Van der Waal's Forces
- Covalent bonds
- Hydrogen bonds

An additional driving force is the lyophobic behavior of sorbate (i.e., sorbates of lower solubility tend to sorb more easily).

In **absorption** the sorbate is taken into the sorbent (the sorbing phase). The uptake of contaminants by microorganisms provides an example of absorption (see Chap. 10). Weber, et al.[22] suggest that the partitioning of a contaminant between the solvent (e.g., water) phase and the sorbent phase can be described in terms of the activity coefficient:

$$K = \frac{V_1 f_1}{V_2 f_2} \tag{3-23}$$

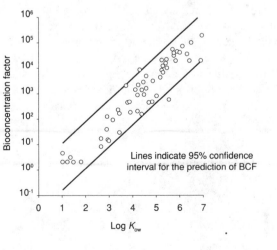

FIGURE 3-1

Relationship between the bioconcentration factor and K_{ow}.[19]

Lines indicate 95% confidence interval for the prediction of BCF

where K = phase partioning coefficient for the specific solute

V_1, V_2 = the molar volume for the solvent and sorbent phases respectively

f_1, f_2 = activity coefficient for the solvent and sorbent phases respectively

In **adsorption** the process takes place at a surface. A typical example of the adsorption process is activated carbon adsorption, discussed in Sec. 9-3.

3-3 ENERGY AND MASS BALANCES[†]

This section briefly presents thermodynamic properties and mass and energy balances that can be conducted to design treatment systems described elsewhere in this book. Additional information can be found in thermodynamics textbooks by Smith and Van Ness,[23] Sonntag and Van Wylen,[24] and Sandler.[25]

The notation for the thermodynamic properties presented in this section is as follows:

Capital letter alone, units of energy or volume (e.g., E_P = potential energy in J);

Lower case letter, units of energy, mass, or volume per time (e.g., q = heat flowrate in J/sec);

Capital letter with a prime symbol, units of energy per mass (e.g., E'_P = potential energy per unit mass in J/kg); and

Capital letter with an asterisk, units of energy or volume per mol of substance (e.g., V^* = molar volume in m^3/gmol).

General Balance Equation

The **law of conservation of energy** states that, with the exception of nuclear reactions, energy can neither be created nor destroyed. Similarly, the **law of conservation of mass** states that, with the exception of nuclear reactions, mass can neither be created nor destroyed. These two principles are used in creating **energy and mass balances** to evaluate the energy requirements of processes and the conversion of reactants to products. In both laws, a **system** is defined as the region where the process takes place, while the **surroundings** are everything outside the system. An imaginary barrier around the system is referred to as the **system boundary**. An **open system** is one where mass is transferred from the system to its surroundings or vice versa. A **closed system** is one where mass does not enter or leave the system boundary, although energy can be transferred from or to the system. Both energy and mass balances are based on the following statement:

$$\text{Input} + \text{Generation} - (\text{Output} + \text{Consumption}) = \text{Accumulation} \qquad (3\text{-}24)$$

[†]Section 3-3 was prepared by Dr. Elsie Millano, *ERM-North Central.*

That is, the difference between the amount of a property that enters and is generated within the system and the amount that leaves and is consumed within the system is accumulated in the system.

Generalized Energy Balance

Energy can only flow through a system boundary as heat or work. **Heat** is energy that flows as a result of a temperature difference or a phase change. **Work** is energy that flows as a result of any other driving force, including a force, a torque, or a voltage. Heat or work are not contained within a system. The **first law of thermodynamics** states that the energy lost by a system during a process must be transferred to the surroundings, as energy is neither destroyed nor created:

$$\text{change in energy of system} - \text{change in energy of surroundings} = 0 \qquad (3\text{-}25)$$

Therefore, when energy disappears in one form, it appears in another. The total energy stored in a system is composed of internal, kinetic, and potential energies. The **internal energy** depends on the random motion of the molecules in a material in the system or on the chemical nature of the material. The **kinetic energy** is a result of the motion of the system. The **potential energy** is a result of the position of the system in a gravity field. The generalized expression for an energy balance, assuming that there is only one phase (i.e., gas, liquid, or solid) in the system, is:[26]

$$\frac{d}{dt}\left\{ U + M\left(E'_{\text{kin}} + E'_{\text{pot}} \right) \right\} = \sum_{k=1}^{K} m_k \left(U' + E'_{\text{kin}} + E'_{\text{pot}} \right)_k + q$$

$$- \left[w_s + P\,\frac{dV}{dt} + \sum_{k=1}^{K} m_k \left(P\,V' \right)_k \right] \qquad (3\text{-}26)$$

where
t = time (sec)
U = internal energy (J)
M = system mass (kg)
E'_{kin} = kinetic energy per unit mass (J/kg)
E'_{pot} = potential energy per unit mass (J/kg)
K = number of entry ports or flow streams
m_k = mass flow rate of the kth flow stream (kg/sec)
U' = internal energy per unit mass (J/kg)
q = rate of flow of heat into the system (J/sec)
w_s = rate of flow of mechanical or electrical
  energy out of the system (J/sec)
P = pressure exerted by the system at its boundaries,
  assumed to be uniform (Pa)
V = system volume (m^3)
V' = system volume per unit mass (m^3/kg)

The left term in Eq. 3-26 is the change in the energy of the system. If there is more than one phase, the total energy is the sum of the energies of each phase.

The first term on the right is the energy flow carried by each fluid element entering (positive) or leaving (negative) the system. The second term on the right, q, is the total rate of flow of heat into the system. The last three terms on the right represent the work done by the system: (1) the shaft work, w_s, which is the rate of flow of mechanical or electrical energy out of the system that occurs without a deformation of the system boundaries; (2) the work resulting from the movement of the system boundaries ($P(dv/dt)$) (e.g., by expansion of the fluid); and (3) the work done by the system due to the movement of the fluid, represented by a change in pressure at constant volume. By convention, the flow of heat into the system is positive, while the work done on the system is negative.

This generalized equation can be simplified depending on the system being evaluated and on whether the instantaneous rate of change or the total change are of importance. For example, for a closed system, $m_k = 0$ and the first and last terms on the right side of Eq. 3-26 drop out. If the process is at steady-state, the left side term is equal to zero, and if no work is done by or on the system, the last three terms on the right side are zero.

Example 3-8 Simplified energy balance. Calculate the change in internal energy resulting when a liquid contained in a tank is stirred by a paddle wheel if the work input to the paddle wheel is W and the heat transferred from the tank is Q.

Solution. Consider the system to be the tank and the liquid. This system is closed because there is no flow of mass into or out of it. Also, there is no change in kinetic or potential energy, as the system is not moving and its elevation with respect to a reference point does not change.

For this system, $m_k = 0, dM/dt = 0, d/dt\left(E'_{kin} + E'_{pot}\right) = 0$, and the total change in internal energy, not the instantaneous rate change, is required. Because the work is done on the system (negative work) and the heat leaves the system (negative heat), Eq. 3-26 reduces to:

$$\underline{\Delta U = +W - Q}$$

Phase Equilibrium and Phase Rule

Thermodynamic properties are either intensive or extensive. **Intensive properties** are independent of the mass of material in the system. **Extensive properties** depend on the mass of material in the system. Extensive properties can be transformed into intensive properties by dividing by the total system mass. Mass and volume are extensive properties, while pressure, temperature, density, volume per unit mass (specific volume), and mass and mole fractions are intensive properties.

Intensive properties can be used to define systems at equilibrium. In thermodynamics, a system is in **equilibrium** if there is no change and no tendency of the properties of the system to change in a macroscopic scale. If a system in an **equilibrium state** returns to the equilibrium conditions after a small disturbance is removed, the system is defined as a **stable system.** State properties are defined as the intensive properties of a system in equilibrium. If a system deviates minimally from thermo-

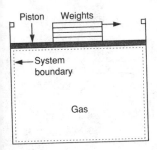

FIGURE 3-2
A system that may undergo a quasi-equilibrium process.[27]

dynamic equilibrium during a process, such that all states through which the system passes can be considered equilibrium states, the process is called a quasi-equilibrium process, and the system is considered to have been in equilibrium throughout the process. For example, for the system shown in Fig. 3-2, if the weights on the piston are small and are taken off one by one, a quasi-equilibrium process will take place.

Intensive properties have been determined and summarized in tables for saturated and superheated pure substances.[28,29,30,31] The **saturation temperature** at a given pressure is the vaporization temperature. A **superheated** vapor has been heated beyond the saturation temperature. Example pages of a table for superheated steam in SI units are included in App. C.

To be able to duplicate an equilibrium condition, a certain number of intensive properties (i.e., temperature, pressure, or the compositions of each phase) must be defined for a specific system, according to the **phase rule** for nonreacting systems:[32]

$$F = C + 2 - \phi \qquad (3-27)$$

where
F = degrees of freedom

C = number of components in the system

ϕ = number of phases in the system

Example 3-9 Use of the phase rule. Determine the degrees of freedom for each of the following systems in equilibrium: (1) pure liquid water; (2) liquid water and saturated vapor; and (3) a liquid solution of water and acetone and saturated vapor.

Solution

(1) Pure liquid water: There is one component ($C = 1$) and one phase (liquid) ($\phi = 1$). Thus, $F = 1 + 2 - 1 = 2$. Therefore, for this system, two intensive properties must be given to define the equilibrium conditions.

(2) Liquid water and saturated vapor: There is one component ($C = 1$) and two phases ($\phi = 2$). The value of F is 1. The tables with liquid-vapor equilibrium data require only one intensive property to define an equilibrium condition.

(3) A liquid solution of water and acetone and saturated vapor: In this case, there are two components ($C = 2$) and two phases ($\phi = 2$). The number of degrees of freedom is two.

To calculate the third intensive property, a relation between concentration and pressure must be used.

Ideal Gas Equation of State

There are equations relating each state variable to two others. One such equation is the **equation of state** for ideal gases. An **ideal gas** is defined as a gas of such low density that the interactions among its molecules are unimportant.[33] The equation of state for ideal gases relates the pressure, temperature, and the molar specific volume of a system, as follows:

$$PV^* = RT \tag{3-28}$$

where
$$P = \text{pressure (Pa)}$$
$$V^* = \text{molar specific volume (m}^3\text{/gmol)}$$
$$R = \text{molar ideal gas law constant} = 8.3144$$
$$\text{Pa} \cdot \text{m}^3/\text{gmol} \cdot \text{K(J/gmol} \cdot \text{K)}$$
$$T = \text{absolute temperature (K)}$$

P, V^*, and T are state properties. The ideal gas equation of state can also be expressed as $PV = NRT$ or $PV = MR'T$, where N and M are, respectively, the number of moles and the mass of the substance in the system, and R' is the mass ideal gas law constant.

The **molar ideal gas law constant**, R, is also referred to as the **ideal gas law constant**, the **gas-law constant**, the **gas constant**, and the **universal gas constant**. Because it is given in units of moles, R is independent of the type of substance and can be used for any substance. Values of R with different units can be found in App. C. When the gas law constant is expressed in terms of mass units (designated R'), it is different for each compound. For example, the values of R' for ammonia and steam are 488 J/kg K and 461.5 J/kg K, respectively.

Example 3-10 Use of the ideal gas equation of state. Calculate the temperature of 10 kg of air (molecular weight = 29) in a 10 m³ container, at a pressure of 100 kPa. Assume that air behaves as an ideal gas under these conditions.

Solution. The number of moles of air in the container is 10,000 g/29 = 345 gmol. Therefore, the molar specific volume of air in the container is 10 m³/345 gmol = 0.029 m³/gmol. Using the ideal gas equation of state:

$$T = PV^*/R = 100,000 \text{ Pa} \times (0.029 \text{m}^3/\text{gmol})/(8.3144 \text{ Pa} \cdot \text{m}^3/\text{gmol} \cdot \text{K})$$

$$T = \underline{\underline{348.8 \text{ K}}}$$

Enthalpy

Certain combinations of thermodynamic properties are frequently found when analyzing processes. One such combination is **enthalpy**, which is defined as:

$$H = U + PV \tag{3-29}$$

where
$$H = \text{enthalpy (J)}$$
$$U = \text{internal energy (J)}$$

Specific enthalpy (i.e., enthalpy per unit mass) is tabulated in thermodynamics tables, as shown in App. C for superheated steam.

Example 3-11 Use of enthalpy data. Water at a flow rate of 1000 kg/hr is evaporated in a boiler by addition of heat at a constant pressure of 100 kPa. The water temperature increases from the saturation temperature (100°C) to 300°C. The system is undergoing a quasi-equilibrium process at steady state. There are no heat losses to the surroundings, and the kinetic and potential energy changes between the entry and exit port are negligible. Calculate the heat transferred to the water.

Solution. The problem will be solved by using Eq. 3-26. The system is the fluid inside the boiler. The term on the left side of Eq. 3-26 is zero because the system is at steady state. Since the changes in kinetic and potential energies are negligible, the first term on the right reduces to $m(U'_{\text{in}} - U'_{\text{out}})$. Heat flows into the system and, therefore, q is positive. The term w_s is zero because there is no work being done that occurs without a deformation of the system boundaries. The work resulting by expansion of the fluid at constant pressure is $P\frac{dV}{dt} = mP(V'_{\text{out}} - V'_{\text{in}})$, and is positive because it is done by the system (i.e., the fluid expands). Finally, the last term on the right hand side of Eq. 3-26, $\left(\sum_{k=1}^{K} m_k(PV')_k \right)$, is zero because there is no work being done by the system due to a change in pressure, as the pressure is constant. The terms remaining in Eq. 3-26 are:

$$0 = m(U'_{\text{in}} - U'_{\text{out}}) + q - mP(V'_{\text{out}} - V'_{\text{in}})$$

Rearranging these terms,

$$q = m\left[(U'_{\text{out}} - U'_{\text{in}}) + P(V'_{\text{out}} - V'_{\text{in}}) \right]$$
$$= m\left[(U'_{\text{out}} + PV'_{out}) - (U'_{\text{in}} + PV'_{\text{in}}) \right]$$
$$= m(H'_{\text{out}} - H'_{\text{in}})$$

From the superheated steam tables (Table C-2), the specific enthalpy (enthalpy per unit mass) for saturated water at 100 kPa and 100°C is 417.5 kJ/kg and for superheated steam at 100 kPa and 300°C, 3074.5 kJ/kg.

Therefore, the heat transferred to the water (positive heat) is:

$$q = 1000 \text{ kg/hr}(3074.5 - 417.5) \text{ kJ/kg} = \underline{\underline{2.657 \times 10^6 \text{ kJ/hr}}} = 2.52 \times 10^6 \text{ Btu/hr}$$

Entropy

A process is **reversible** if it can be reversed such that no change is left in either the system or its surroundings. The process illustrated in Fig. 3-3 is reversible if (1) the piston slides without friction; (2) heat is neither absorbed nor transmitted by the piston or the cylinder; and (3) the work done in taking off the weights is infinitesimal, for example, by having a platform for each weight level such that the transfer of the weight to the platform requires no work.

A **cycle** is a process in which a system undergoes changes of state or individual processes and at the end the system returns to the same initial state. For example, steam that is used to heat a substance, condensed, and then vaporized again to heat more substance undergoes a cycle.

Entropy is a state property defined as always having a positive rate of generation within a system, except when the system is at equilibrium, in which case the internal rate of generation is zero.[35] The entropy balance equation shown below (which can be used to evaluate processes) is a second, independent energy balance based on the **second law of thermodynamics**, which states that no device can operate in a cycle and produce only transfer of heat from a cool body to a hot body. The second law of thermodynamics is based on the fact that spontaneous processes occur only in the direction that dissipates the gradients in a system, toward a state of equilibrium. The entropy balance equation is:

$$\frac{dS}{dt} = \sum_{k=1}^{K} m_k \, S_k' + \frac{q}{T} + S_{\text{gen}} \tag{3-30}$$

where S = Entropy of the system (J/K)

m_k = Mass flow rate of the kth flow stream (kg/sec)

S_k' = Specific entropy of the kth flow stream (J/kg·K)

S_{gen} = Rate of generation of entropy within the system (J/sec·K)

q = Rate of flow of heat into the system (J/sec)

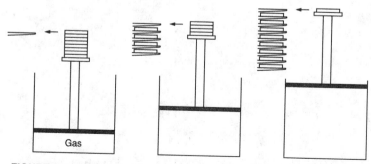

FIGURE 3-3
A system that may be considered to undergo a reversible process.[34]

Example 3-12 Calculation of work using entropy values. Steam enters a turbine at 3 MPa and 500°C and at a flow rate of 3 kg/sec, expands in a reversible, adiabatic (i.e., without gain or loss of heat) process, and exhausts at 300 kPa. Changes in kinetic and potential energy between the inlet and outlet of the turbine are negligible. Calculate the quantity of work produced in the process.

Solution. Assume the process is at steady state ($dS/dt = 0$) and that the system includes the turbine. In addition, from the problem definition, $q = 0$ (adiabatic process), and $S_{gen} = 0$ (reversible and adiabatic process).

From the entropy balance, Eq. 3-30, $S'_{in} = S'_{out}$. The energy balance, Eq. 3-26, becomes $w_s = m(H'_{in} - H'_{out})$, because $dM/dt = 0$ and $dU/dt = 0$ (steady-state process); $m\left[\Delta(E'_{kin} - E'_{pot})\right] = 0$; $q = 0$; $dV/dt = 0$ (constant turbine volume); and $H' = U' + PV'$ (there is a change in pressure).

According to the phase rule (Eq. 3-27), for a system with one component and one phase, two state properties are needed to define the system. In the inlet, T and P are known, and the entropy and enthalpy can be found in the superheated steam tables. In the outlet, P and the entropy are known, since $S'_{in} = S'_{out}$. Thus, the enthalpy at the outlet conditions can also be found by using the superheated steam tables.

From App. C, for the steam entering the turbine, $H'_{in} = 3456.2$ kJ/kg and $S'_{in} = 7.2345$ kJ/kg · K. With the value of $S'_{out} = 7.2345$ kJ/kg · K, the value of enthalpy can be found in the superheated steam tables by interpolation.

For a pressure of 300 kPa, S'_{out} corresponds to a temperature between 175°C ($S' = 7.1990$ kJ/kg · K and $H' = 2813.5$ kJ/kg) and 200°C ($S' = 7.3119$ kJ/kg · K and $H' = 2865.5$ kJ/kg). The enthalpy at the outlet is, therefore:

$$H'_{out} = 2865.5 - (2865.5 - 2813.5) \times [(7.3119 - 7.2345)/(7.3119 - 7.1990)]$$

$$= 2829.9 \text{ kJ/kg}$$

Finally, the work is calculated as:
$$w_s = 3 \text{ kg/sec}(3456.2 - 2829.9) \text{ kJ/kg} = \underline{\underline{1878.9 \text{ kJ/sec}}}$$
Because the work is positive, it is work done by the turbine.

Specific Heats

As mentioned before, several combinations of thermodynamic properties occur frequently in evaluating processes. Enthalpy has already been defined. Other state properties include the specific heat and the Gibbs free energy. The Gibbs free energy will be defined later in this section. **Heat capacity** is defined as the quantity of heat a substance can absorb before a certain rise in temperature occurs:

$$C = dQ/dT \tag{3-31}$$

where

C = heat capacity (J/K)

Q = flow of heat into the system (J)

T = absolute temperature (K)

Two forms of the heat capacity per mol, or **molar specific heats**, are state properties:

$$C_v^* = \left(\partial H^*/\partial T\right)_v \tag{3-32}$$

$$C_p^* = (\partial H^*/\partial T)_p \tag{3-33}$$

where $C_v^* =$ specific heat at constant volume (J/gmol $\cdot$ K)

$C_p^* =$ specific heat at constant pressure (J/gmol $\cdot$ K)

These definitions are based on an energy balance for a closed, steady-state system with no shaft work and occurring at constant volume ($q = N \Delta U^*$) or constant pressure ($q = N \Delta H^*$), respectively, where $N =$ number of moles of the substance. However, because Eq's. 3-32 and 3-33 contain only state properties, the specific heats themselves are also state properties and, thus, these equations can be used for any process. Partial derivatives are shown because the specific internal energy and specific enthalpy are a function of two variables, temperature and volume for U^* and temperature and pressure for H^*.

For the case of an ideal gas, where U^* and H^* are only functions of temperature, the specific heats are also functions of temperature only. For an ideal gas:

$$C_{v0}^*(T) = dU^*(T)/dT \tag{3-34}$$

$$C_{p0}^*(T) = dH^*(T)/dT \tag{3-35}$$

where $C_{v0}^*(T) =$ zero-pressure, constant-volume specific heat (J/gmol $\cdot$ K)

$C_{p0}^*(T) =$ zero-pressure, constant-pressure specific heat (J/gmol $\cdot$ K)

Equations 3-34 and 3-35 are based on the fact that gases approach ideal gas behavior as the pressure approaches zero. Formulas to calculate specific heats for several pure substances based on the temperature of the process are given in Table C-3.

Example 3-13 Use of specific heat value. Air (mol. wt.: 0.029 kg/gmol) enters a compressor at 300 kPa and 300 K at a flow rate of 500 kg/hr and is discharged at an outlet pressure of 3000 kPa. The compression takes place adiabatically and at steady state. The work done on the system is 50 kJ/sec. Assuming that air behaves as an ideal gas under these conditions, and that its constant-pressure specific heat, $C_{p0}^* = 29.3$ J/gmol K, calculate the temperature of the gas at the outlet.

Solution. Assume that there is no change in the kinetic and potential energy of the system, which is the gas contained in the compressor. According to the system description, dU^*/dt and $dN/dt = 0$ (steady state process), $q = 0$ (adiabatic system), and $w = w_s + P \Delta V$ (w_s is negative work, because it is done on the system). In addition, $\Delta U^* + P \Delta V^* = \Delta H^* = C_{p0}^*(T_2 - T_1)$.

Therefore, the energy balance, Eq. 3-26, reduces to:

$$w_s = nC_{p0}^*(T_2 - T_1), \text{ or}$$

$$T_2 = T_1 + w_s/(nC_{p0}^*)$$

where $n =$ molar flow rate, (mol/sec).

$$T_2 = 300 \ K \ + \ \frac{(50 \text{ kJ/sec}) (0.029 \text{ kg/gmol}) (3600 \text{ s/hr})}{\left(0.0293 \frac{\text{kJ}}{\text{gmol} \cdot \text{K}}\right)(500 \text{ kg/hr})} = \underline{\underline{656 \ K}}$$

Latent Heats

Properties used to evaluate heat produced or absorbed during phase changes and reactions have been defined based on a specific process (either a phase change or a reaction) carried out at constant pressure and temperature, in open, steady-state systems where no work is done, and where the kinetic and potential energies are negligible. From the energy balance, Eq. 3-26, the heat produced or absorbed during such processes is equal to the change in enthalpy for each process. The **latent heat of vaporization** and **latent heat of fusion** are the phase change properties most commonly used:

$$\Delta H_{vap}^* = H_{vap}^* - H_{Liq}^* \tag{3-36}$$

$$\Delta H_{fus}^* = H_{sol}^* - H_{Liq}^* \tag{3-37}$$

where $\Delta H_{vap}^* =$ latent heat of vaporization (J/gmol)

$H_{vap}^* =$ specific enthalpy of liquid at the temperature and pressure being evaluated (J/gmol)

$H_{Liq}^* =$ specific enthalpy of liquid at the temperature and pressure being evaluated (J/gmol)

$\Delta H_{fus}^* =$ latent heat of fusion (J/gmol)

$H_{sol}^* =$ specific enthalpy of solid at the temperature and pressure being evaluated (J/gmol)

The latent heats for the reverse processes are defined as the negative of the latent heats listed above (i.e., $\Delta H_{con}^* = -\Delta H_{vap}^*$ where ΔH_{con}^* is the **latent heat of condensation**). The latent heat of vaporization and the latent heat of fusion for several substances at the normal boiling and melting point (i.e., at a pressure of 1 atm, and the boiling or melting temperature of each substance) can be found in Table C-4. The latent heats at other temperatures can then be calculated by adding the changes in enthalpy required to (1) bring the substance from the process temperature to its normal temperature (boiling or melting point), (2) effect the phase change (i.e., the normal latent heat), and (3) bring the substance from its normal temperature back to the original process temperature.

Example 3-14 Calculation of heat of vaporization. Calculate the heat to be added to an open, steady-state system where acetone enters as a liquid at a rate of 100 kg/hr at 40°C and 101.315 kPa (1 atm) and exits as a vapor at 90°C. The pressure is held constant during the process and the kinetic and potential energy changes are negligible. The following data are available from App. C: Molecular weight = 58.08, boiling point at 1 atm = 56°C, latent heat of vaporization at 56°C and 1 atm = 30.2 kJ/gmol, heat capacity of the liquid between −30°C and 60°C = 123.0 + 18.6 ×10^{-2}T J/gmol · °C (T in °C), and heat capacity of the vapor between 0°C and 1200°C = 71.96 + (20.10 × 10^{-2}T) − (12.78 × 10$^{-5}T^2$) + (34.76 × 10$^{-9}T^3$) J/gmol · °C (T in °C).

Solution. The energy balance, Eq. 3-26, reduces to $q = \Delta H$, or

$$q = N(\Delta H^*_{40°-56°C} + \Delta H^*_{vap} + \Delta H^*_{56°-90°C})$$

where $\Delta H^*_{40°-56°C}$ indicate the change in enthalpy when temperature is raised from 40° to 56°

N = number of moles of acetone entering the system

$\quad$ = (100,000 g/hr)/(58.08 g/gmol) = 1721.8 gmol/hr

ΔH^*_{vap} = 30.2 kJ/gmol

From Eq. 3-35, $H = \int C_p dT$

$$\Delta H^*_{40°-56°C} = \int_{40}^{56} (123.0 + 18.6 \times 10^{-2}T) \, dT = 123.0 \, (56 - 40) +$$

$$(18.6 \times 10^{-2})(56^2 - 40^2) \, /2 = 2100 \text{ J/gmol} = 2.1 \text{ kJ/gmol}$$

$$\Delta H^*_{56°-90°C} = \int_{56}^{90} (71.96 + 20.1 \times 10^{-2} \, T - 12.78 \times 10^{-5} \, T^2 + 34.76 \times 10^{-9} \, T^3) \, dT$$

$$= 71.96(90 - 56) + \frac{(20.1 \times 10^{-2})(90^2 - 56^2)}{2} - \frac{(12.78 \times 10^{-5})(90^3 - 56^3)}{3} +$$

$$\frac{(34.76 \times 10^{-9})(90^4 - 56^4)}{4}$$

$$= 3000 \text{ J/gmol} = 3.0 \text{ kJ/gmol}$$

$$q = 1721.8 \text{ gmol/hr} \times (30.2 + 2.1 + 3.0) \text{ kJ/gmol} = \underline{\underline{60,800 \text{ kJ/hr}}}$$

Heats of Reaction, Formation, and Combustion

Additional state properties defined by using enthalpy values include the heats of reaction, formation, and combustion. These properties are used to evaluate the heat produced (negative value, exothermic reaction) or absorbed (positive value, endothermic reaction) in processes where chemical reactions take place under the same system conditions specified above for defining the latent heats.

The **heat of reaction** is, therefore, defined as the change of enthalpy resulting from a reaction occurring at constant pressure in an open, steady-state system, where no work is done, the kinetic and potential energies are negligible, and the reactants and products are at the same temperature. For a similar system, the **heat of formation** of a substance is the change of enthalpy resulting from the reaction that forms that substance from its constituent elements, and the **heat of combustion** of a substance is the change of enthalpy resulting from the combustion of that substance. **Combustion** is defined as the reaction of a substance with oxygen to produce the following set of specified substances: gaseous carbon dioxide, liquid or gaseous water, and gaseous sulfur dioxide and nitrogen (N_2) if the original substance had any sulfur or nitrogen in it.

The heat of formation and the heat of combustion have been tabulated at the **standard states** (see App. C). Standard states have been defined for convenience, to allow definition of only a set of properties (specifically, enthalpy and Gibbs free energy of formation) at specific temperature and pressure that can then be used for calculation of the heat produced or absorbed during reactions at different temperatures and pressures.

To tabulate the heats of reaction, formation, and combustion, the standard state was defined as a pressure of 101.325 kPa (1 atm) and a temperature of 278.15 K (25°C). For gases, the standard state is the ideal-gas state. In the case of the heat of formation, a reference state, at which the value of enthalpy is zero, was defined to allow calculation of the change of enthalpy resulting from a formation reaction. For any atomic species, the reference state is the thermodynamically stable form at 1 atm and 25°C. For example, the thermodynamically stable form of oxygen is O_2, and not O (atomic oxygen) or O_3 (ozone). The following equation can be used to calculate the heat of reaction from the standard heats of formation:[36]

$$\Delta H_r^*(T, P_0) = \sum_p \nu_p \, \Delta H_{f,p}^{*\circ}(T_0, P_0) - \sum_r \nu_r \, \Delta H_{f,r}^{*\circ}(T_0, P_0)$$

$$+ \sum_i \nu_i \int_{T_0}^{T} C_{P_0,i}^{*\circ} \, dT \qquad (3\text{-}38)$$

where $\Delta H_r^*(T, P_0)$ = heat of reaction at the temperature
$\quad\quad\quad\quad\quad\quad$ T and 1 atm (kJ/gmol)

$\quad\quad\quad\quad T$ = reaction temperature (°C)

$\quad\quad\quad\quad P_0$ = standard pressure, 101.315 kPa (1 atm)

$\nu_i, \nu_p,$ and ν_r = stoichiometric coefficient of substance i, product p,
$\quad\quad\quad\quad\quad\quad$ and reactant r in the reaction, respectively (dimensionless)

$\Delta H_{f,p}^{*\circ}(T_0, P_0)$ = standard heat of formation of a product of the reaction
$\quad\quad\quad\quad\quad\quad$ at 25°C and 1 atm (kJ/gmol)

$\quad\quad\quad\quad T_0$ = standard temperature, 25°C

$\Delta H_{f,r}^{*\circ}(T_0, P_0)$ = standard heat of formation of a reactant of the reaction at
$\quad\quad\quad\quad\quad\quad$ 25°C and 1 atm (kJ/gmol)

$\quad\quad\quad C_{P_0,i}^{*\circ}$ = constant-pressure specific heat of
$\quad\quad\quad\quad\quad\quad$ substance i at 1 atm (kJ/gmol · K)

If only the standard heats of combustion are available, the heat of reaction can be calculated from the standard heats of combustion by using an equation similar to Eq. 3-38, except that the heats of combustion of the reactants are positive and the heats of combustion of the products are negative.

Example 3-15 Calculation of standard heat of reaction from standard heats of formation. 50 kg/hr of gaseous ammonia are oxidized in a process following the reaction: $4NH_3(g) + 5O_2(g) \rightarrow 4NO(g) + 6H_2O(g)$. Calculate the standard heat of reaction for this process, given the following data: $\Delta H_f^{*\circ}[NH_3(g)] = -46.19$ kJ/gmol; $\Delta H_f^{*\circ}[O_2(g)] = 0$; $\Delta H_f^{*\circ}[NO(g)] = +90.37$ kJ/gmol; and $\Delta H_f^{*\circ}[H_2O(g)] = -241.83$ kJ/gmol.

Solution. The molar standard heat of reaction can be calculated from Eq. 3-38, by setting the last term on the right hand side to zero (i.e., $T = T_0$):

$$\Delta H_r^{*\circ} = 4\{\Delta H_f^{*\circ}[NO(g)]\} + 6\{\Delta H_f^{*\circ}[H_2O(g)]\}$$
$$- 4\{\Delta H_f^{*\circ}[NH_3(g)]\} - 5\{\Delta H_f^{*\circ}[O_2(g)]\}$$

$$= 4(90.37) + 6(-241.83) - 4(-46.19) - 5(0)$$

$$= -904.74 \text{ kJ/gmol}$$

This is the molar standard heat of reaction for every 4 moles of ammonia that react. The number of moles of ammonia is $n = (50,000 \text{ g/hr})/(17 \text{ g/gmol}) = 2941$ gmol/hr, and the total standard heat of reaction for the process is:

$$\Delta H_r^{\circ} = n(\Delta H_r^{*\circ})/4 = 2,941 \times (-907.74)/4 = \underline{-6.7 \times 10^5 \text{ kJ/hr}}$$

Because the change of enthalpy and, thus, the heat of reaction, are negative, the reaction is exothermic and heat is transferred from the system to its surroundings.

Gibbs Free Energy and Chemical Equilibrium

A frequently used state property is the **Gibbs free energy**, defined as:

$$G = H - TS \qquad (3\text{-}39)$$

where $\qquad G = $ Gibbs free energy (J)

The Gibbs free energy is a property of importance in the evaluation of the thermodynamic equilibrium of systems where chemical reactions take place. For a closed, constant-pressure, constant-temperature system:

$$\Delta G^* = \Delta G_f^{*\circ} + RT \ln K \qquad (3\text{-}40)$$

where $\qquad \Delta G_f^{*\circ} = $ Gibbs free energy of formation,
at the standard state (kJ/gmol)

$K = $ Equilibrium constant (dimensionless)

A negative value of the Gibbs free energy indicates that there is a potential for the reaction to occur as written, while a positive value indicates that the opposite reaction is most likely to occur. The same reference state defined for the enthalpy of formation is used to define the zero Gibbs free energy of formation to allow calculation of the change in Gibbs free energy resulting from a formation reaction. Values of the Gibbs free energy of formation at the standard states can be found in App. C.

The Gibbs free energy for a reaction at any temperature can be calculated from the standard state value as:

$$\Delta G_T^{*\circ} = \sum_p v_p G_{T,p}^{*\circ} - \sum_r v_r G_{T,r}^{*\circ} \tag{3-41}$$

and, for each product p or reactant r

$$G_T^{*\circ} = G_f^{*\circ} + \left(H_T^{*\circ} - H_{T_0}^{*\circ}\right) - \left(T S_T^{*\circ} - T_0 S_{T_0}^{*\circ}\right) \tag{3-42}$$

A combination of Eq. 3-26 and 3-30 for this system at equilibrium gives: $dG^*/dt = 0$, or

$$\Delta G^* = 0 \tag{3-43}$$

Therefore, at equilibrium, Eq. 3-40 becomes:

$$\ln K = -\frac{\Delta G_f^{*\circ}}{RT} \tag{3-44}$$

For a gaseous reaction such as $v_a A + v_b B \rightleftharpoons v_c C + v_d D$, the equilibrium constant relates the molar ratios of the components by the equation:

$$K = \frac{(Y_c)^{v_c} (Y_d)^{v_d}}{(Y_a)^{v_a} (Y_b)^{v_b}} \left(P/P^o\right)^{v_c + v_d - v_a - v_b} \tag{3-45}$$

where $(Y_i)^{v_i}$ = Molar ratio of component i in the gas, (mol_i/total moles)

The equilibrium constants for several gaseous reactions have been tabulated. However, for chemical reactions in non-ideal liquid solutions, the actual equilibrium concentrations are difficult to predict by using the thermodynamic equilibrium constant K. Therefore, in the case of non-ideal liquid solutions, the approach is to determine experimentally the molar concentration equilibrium constant, K_c (which uses the molar concentration of each constituent and, thus, is dependent on the molar concentrations of reactants and solvents), for a certain reaction at a specific temperature and then use it to determine the equilibrium concentrations under a different set of initial concentrations. An example of the calculation of equilibrium concentrations in a gaseous reaction by using the Gibbs free energy is presented in a following section **Mass Balances on Reacting Systems**.

Generalized Mass Balance Equation

Mass balances are based on the law of conservation of mass, defined under **General Balance Equation**, which is represented by Eq. 3-24:

Input + Generation − (Output + Consumption) = Accumulation (3-24)

This equation can be used for any material that enters or leaves the system being evaluated, including compounds, molecules, and atoms. There are three different types of processes that can be evaluated: batch, semibatch, and continuous processes. A **batch process** is one where the constituents are added to the system at the beginning of the process, left in the system for a specified period of time, and then removed from the system. A **continuous process** is one where there is a continuous flow of constituents into and out of the system. A **semibatch process** is one where one or more constituents are continuously added or removed and one or more constituents are added at the beginning of the process and removed at the end of the process (i.e., a hybrid between a batch and a continuous process).

In addition, processes may be operated at steady-state or under transient conditions. The **steady-state** process variables (such as temperature, pressure, flow rates, concentrations, and extent of reaction) do not change appreciably with time, except for minor fluctuations about constant mean values. In a **transient or unsteady-state process**, one or more of the variables change with time. Finally, mass balances can be written as **integral balances** (i.e., determination of conditions over a finite period of time) or as **differential balances** (i.e., determination of instantaneous conditions). In integral balances, each term of the equation is given as a mass, whereas in differential balances, units are mass per unit time. In general, batch processes are evaluated with integral balances and continuous processes are evaluated with differential balances.

Procedure for Conducting a Mass Balance

In conducting a mass balance, the system and its components must be well defined (i.e., in terms of what constitutes the system, how many variables there are, and how many equations can be written to calculate the unknown variables). To avoid going through a lengthy calculation only to find that the system is undefined (i.e., the data provided are not enough to define the system), the problem must be organized to permit an evaluation of its degrees of freedom. The **degrees of freedom** of a system are the minimum number of variables that must be given to be able to define the system, and are calculated as:[37]

$$DF = NV - NE \qquad (3\text{-}46)$$

where DF = degrees of freedom of a system

           NV = number of variables

           NE = number of independent equations relating the variables

In general, to calculate all of the variables in a process, mass and energy balances must be combined. If $DF > 0$, then DF variables must be defined before the rest of the variables can be calculated. If $DF < 0$, the problem is undefined, and either all of the equations counted are not independent, all of the variables were not counted, or additional information is necessary. The procedure to perform a mass balance is as follows:[38]

1. Draw a flowchart of the process, with all of the variables shown. Include the values known from the definition of the problem, and assign names to each unknown variable.

2. Write equations relating the unknown variables, including:
   - Mass balances, that for a system with NV variables can be NV equations if there are no reactions, and NV minus the number of independent chemical reactions if reactions occur;
   - Energy balances;
   - Problem specifications, such as relations between the variables (e.g., a definition that the outlet mass of a component is a fraction of its inlet mass);
   - Physical properties and laws, such as mass/volume relations (specific gravity), concentration/pressure relations (gas laws), or saturation or equilibrium conditions; and
   - Physical constraints, such as the relation between mole fractions (e.g., the sum of the mole fractions in a stream equals 1).

3. Check the degrees of freedom. If DF variables are specified (where $DF > 0$), proceed to solve the problem. If not, evaluate the flowchart and the problem definition again to determine what is wrong and then make the necessary corrections.

4. Convert all given information into the same unit basis, either mass or moles, and the same flow rate basis, if necessary.

5. Solve the equations.

Mass Balances on Nonreacting Systems

In the absence of reactions, the generation and consumption terms are zero and the general mass balance equation for nonreacting systems becomes:

$$\sum_{k_{ii}} \text{Input} - \sum_{k_{oi}} \text{Output} = \text{Accumulation} \qquad (3\text{-}47)$$

where k_{ii}, k_{oi} = streams with component i entering or leaving the system

For batch processes and steady-state continuous processes, accumulation = 0, and the general mass balance equation for nonreacting systems can be simplified as:

$$\sum_{k_{ii}} \text{Input} = \sum_{k_{io}} \text{Output} \qquad (3\text{-}48)$$

Example 3-16 Mass balance on a nonreacting system. 50,000 L/day of ground water contaminated with 5000 mg/L of benzene and 10,000 mg/L of trichloroethylene (TCE) are fed to a system where air is being injected at a rate of 100,000 L/day. The influent air has no organic contaminants and does not react inside the system (i.e., no air dissolves in the water and no water is picked up by the air). The process takes place at 25°C, the same temperature of the influent ground water and air. The benzene concentration in the effluent ground water is 1000 mg/L and the removal efficiency of TCE is 60% of the removal efficiency of benzene.

Assume that the flow of ground water and air does not change as a result of the process, and calculate the mass of benzene and TCE in the effluent air and ground water streams.

A flow chart and mass balance is developed by noting that $M = QC$

$$M = \text{mass mg/day } (\times 10^{-6} =\text{ kg/day})$$

$$Q = \text{flow (L/day)}$$

$$C = \text{concentration (mg/L)}$$

The following subscripts are used:

$$w = \text{water}$$

$$b = \text{benzene}$$

$$t = \text{TCE}$$

$$a = \text{air}$$

$$1 = \text{influent}$$

$$2 = \text{effluent to water}$$

$$3 = \text{effluent to air}$$

The resulting flow chart is shown below.

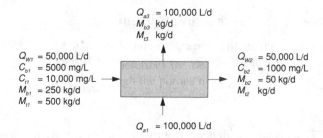

Solution. The problem has 13 variables: influent and effluent ground water and air flow rates, influent and effluent ground water benzene concentrations, influent TCE ground water concentration (the problem does not require the definition of the effluent TCE concentration, but only the mass), influent and effluent benzene and TCE masses in the ground water, and effluent benzene and TCE masses in the air (the influent air has no organics and, therefore, there is no need to define influent benzene and TCE mass variables in the influent air).

There are 8 relations among the variables: the effluent ground water and air flow rates are equal to the influent ground water and air flow rates, respectively; the ground water influent and effluent benzene masses and the influent TCE mass are equal to the respective concentration multiplied by the ground water flow rate; the masses of benzene and TCE in the effluent air are equal to the ground water influent minus the ground water effluent masses; and the efficiency of removal of TCE from the ground water is 60% of the efficiency of removal of benzene from the ground water.

Therefore, the problem has 5 degrees of freedom, and, in fact, 5 variables are given: the influent ground water and air flow rates, the influent and effluent benzene concentrations, and the influent TCE concentration.

The previous flowchart already used some of the relations among variables given. The remaining relations are:

$$(M_{t1} - M_{t2})/M_{t1} = 0.6(M_{b1} - M_{b2})/M_{b1}$$

$$M_{b3} = M_{b1} - M_{b2}$$

$$M_{t3} = M_{t1} - M_{t2}$$

The subscripts b and t refer, respectively, to benzene and TCE. By solving these three equations, the results are:

$M_{t2} = 260$ kg/d, $M_{b3} = 200$ kg/d, and $M_{t3} = 240$ kg/d.

Mass Balance on Reacting Systems

In reacting systems, the mass balance equation takes the form of Eq. 3-24. The evaluation of transient or unsteady-state processes is presented below under **Reactions and Reactors**. For steady-state continuous flow processes (accumulation = 0), the mass balance reduces to:

$$\sum_{k_{ii}} \text{Input} + \sum_{k_{ig}} \text{Generation} = \sum_{k_{io}} \text{Output} + \sum_{k_{ic}} \text{Consumption} \qquad (3\text{-}49)$$

where $\qquad k_{ig}, k_{ic}$ = number of reactions where component i is generated or consumed.

Reactants are present in **stoichiometric proportions** if they are present in the same proportions shown in the balanced reaction equation. A **limiting reactant** is present at concentrations lower than the stoichiometric proportion with respect to the rest of the reactants, which are then **excess reactants**. For example, for the reaction $2A + B \rightarrow 3C + D$, if the influent has 2 moles of A per mole of B, the reactants are present in stoichiometric proportions. However, if the influent has 1 mole of A per mole of B, A becomes a limiting reactant and B an excess reactant. For any constituent:

$$n_{io} = n_{ii} + \beta ER \qquad (3\text{-}50)$$

where $\qquad n_{io}, n_{ii}$ = number of moles of constituent in the effluent and influent, respectively

$\beta = v$, if i is a product; $- v$, if i is a reactant; and 0, if i is inert

v = stoichiometric coefficient of substance

ER = number of moles of the limiting reactant that react

$\quad = f_{lr} \times n_{lri}$ (where the subscript lr = limiting reactant)

f_{lr} = fractional conversion of the limiting reactant (moles reacted/moles fed)

Example 3-17   Mass balance on a reacting process. Ethylene is oxidized to produce ethylene oxide in accordance with the following reaction: $2C_2H_4 + O_2 \rightarrow 2C_2H_4O$. The feed to the reactor contains 100 kmol/hr of ethylene and 200 kmol/hr of molecular oxygen. The reaction proceeds to a fractional conversion of the limiting reactant of 40%. Determine the molar flow rates of all of the product gas constituents if the reaction proceeds under steady-state conditions.

n_{1o} kmol/hr C_2H_4
n_{2o} kmol/hr O_2
n_{3o} kmol/hr C_2H_4O
ER
f_{lr} = 0.4

100 kmol/hr C_2H_4
200 kmol/hr O_2

Solution. The problem has 7 variables and 4 relations (i.e., Eq. 3-48 for each reactant and product, and the relation between the fractional conversion and ER [ER = moles fed $\times$ conversion]). The degrees of freedom are 3, and 3 variables are given. Therefore, the problem is defined.

The equations are:

$$n_{1o} = 100 - 2ER$$

$$n_{2o} = 200 - ER$$

$$n_{3o} = 0 + 2ER$$

$$ER = f_{lr} \times n_{lri}$$

For this problem, ethylene is the limiting reactant, because its ratio with respect to molecular oxygen in the feed is 0.5 and in the reaction is 2.0. Therefore, $n_{lri} = 100$ kmol/hr. Solving the equations results in:

$$ER = 40 \text{ kmol/hr}, n_{1o} = 20 \text{ kmol/hr}, n_{2o} = 160 \text{ kmol/hr}, \text{ and } n_{3o} = 80 \text{ kmol/hr}.$$

The next example illustrates the use of the Gibbs free energy (Eq. 3-39) to calculate: (1) the equilibrium constant for a reaction occurring in the gaseous state, and (2) the equilibrium concentrations based on the equilibrium constant.

Example 3-18   Use of the Gibbs free energy to calculate equilibrium constant and concentrations. Calculate the equilibrium constant, K, for the reaction $2CO_2 \rightleftharpoons 2CO + O_2$ at $T = 2000$ K. All constituents are gases. If 100 kmol/hr of CO_2 enter a system where this reaction takes place essentially to equilibrium at $P = 1$ atm and under steady-state conditions, determine the equilibrium molar flow rate of all of the constituents in the effluent gas. The following data are given for the problem:

Component	$H^{*\circ}_{2000} - H^{*\circ}_{298}$	$S^{*\circ}_{2000}$	S°_{298}	$G^{*\circ}_f$
CO (g)	56,743	258.716	197.651	−137,163
O_2 (g)	59,176	268.748	205.148	0
CO_2 (g)	91,439	309.294	213.794	−394,389

The units of the values given in the table are: kJ/kmol for enthalpy, kJ/kmol·K for entropy, and kJ/kmol for Gibbs free energy of formation.

Solution. Using Eq. 3-41, the Gibbs free energy for the reaction can be calculated as:

$$\Delta G^{\circ}_{2000} = 2G^{*\circ}_{2000}(CO) + G^{*\circ}_{2000}(O_2) - 2G^{*\circ}_{2000}(CO_2)$$

$$G^{*\circ}_{2000} = G^{*\circ}_f + (H^{*\circ}_{2000} - H^{*\circ}_{298}) - (2000S^{*\circ}_{2000} - 298.15S^{*\circ}_{298})$$

Using the data given for the Gibbs free energy of formation, the difference in standard enthalpy, and the absolute entropy for each compound, the Gibbs free energy of each substance is calculated and the Gibbs free energy for the reaction is:

$$\Delta G^{\circ}_{2000} = 2(-538,922) + (-417,155) - 2(-857,795) = 220,591 \text{ kJ/kmol}$$

Equation 3-44 is then used to calculate K:

$$(\ln K)_{2000} = \frac{-\Delta G^{*\circ}_{2000}}{RT} = \frac{-220,591}{(8.3144)(2000)} = -13.2656$$

or $K = 1.73 \times 10^{-6}$

To determine the equilibrium molar flow rate of each constituent, a flowchart of the process is drawn as follows:

$$100 \text{ kmol/hr } CO_2 \longrightarrow \boxed{} \longrightarrow \begin{array}{l} n_{1o} \text{ kmol/hr } CO_2 \\ n_{2o} \text{ kmol/hr } CO \\ n_{3o} \text{ kmol/hr } O_2 \end{array}$$

The molar ratios of the reactants and products are defined as:

$$Y_{1o} = n_{1o}/n_{to}, Y_{2o} = n_{2o}/n_{to}, Y_{3o} = n_{3o}/n_{to},$$

where $n_{to} = n_{1o} + n_{2o} + n_{3o}$. From Eq. 3-45,

$$K = (Y_{2o})^2(Y_{3o})/(Y_{1o})^2, \text{ or}$$

$$\ln K = 2\ln Y_{2o} \ln Y_{3o} - 2\ln Y_{1o}$$

The values n_{io} can be defined in terms of the extent of reaction as:

$$n_{1o} = 100 - ER, n_{2o} = 0 + ER, \text{ and } n_{3o} = 0 + ER/2.$$

Thus, $n_{to} = 100 + ER/2$ and the equilibrium constant equation becomes:

$$\ln K = 2\ln[ER/(100+ER/2)] + \ln[(ER/2)/(100+ER/2)] - 2\ln[(100-ER)/(100+ER/2)]$$

By trial and error, $ER = 1.502$ kmol/hr (i.e., at the indicated temperature, the reaction shifts to the left, and most of the carbon dioxide remains as such in the gas). Therefore, the molar concentrations of the components in the effluent gas are:

$$n_{1o} = \underline{\underline{98.498 \text{ kmol/hr}}}; n_{2o} = \underline{\underline{1.502 \text{ kmol/hr}}}; \text{ and } n_{3o} = \underline{\underline{0.751 \text{ kmol/hr}}}$$

3-4   REACTIONS AND REACTORS

Rate of Reaction

The equilibrium constant (Eq. 3-45) provides no information on how fast a reaction will take place, or how much of the reactants will become products (on a finite amount of time). For a simple irreversible chemical reaction in a constant volume system:

$$A + B \rightarrow AB \tag{3-51}$$

The **reaction rate constant** k may be defined for a constant volume system as:

$$\text{RATE} = -dC_A/dt$$
$$= -dC_B/dt = dC_{AB}/dt = kC_A C_B \tag{3-52}$$

where C_A, C_B = concentration of reactants

C_{AB} = concentration of product

For many reactions of interest in the environment, the reaction rate approximately doubles for every 10°C increase in temperature. Another relationship of importance is the **Arrhenius equation**:

$$k = Ae^{-E/RT} \tag{3-53}$$

where A = frequency factor

E = activation energy (J/mol)

R = universal gas constant (8.314 J/mol · K)

T = temperature (K)

The activation energy represents an energy barrier that must be overcome, if the reaction is to occur. The higher the activation energy, the lower the reaction rate.

Order of Reaction

The reaction rate also depends on the concentration of the chemical species involved. The **order of a reaction** refers to the powers to which concentrations are raised. Suppose A and B are possible reactants and the rate is defined:

$$\text{RATE} = k \times C_A^n \times C_B^m \tag{3-54}$$

The reaction is termed nth order with respect to A, and mth order with respect to B. Typical reaction orders are illustrated as follows:

$$\text{Rate} = dC/dt = k \quad \text{Zero order}$$
$$dC_A/dt = kC_A \quad \text{First order}$$
$$dC_A/dt = kC_A C_B \quad \text{Second order}$$
$$dC_A/dt = kC_A^2 \quad \text{Second order}$$

The rate constant for a first order reaction may be determined by plotting $\ln C$ against time. For a second order reaction, plot $1/C$ against time.

Example 3-19 Determining order of reaction and rate constant. Analyze the following reaction data:

Time (min.)	0	12	24	36	48	60
C (mg/L)	7.5	5.25	3.68	2.48	1.73	1.13

Solution. Calculate the logarithm and inverse of the concentration values:

$\ln C$	2.01	1.66	1.30	0.91	0.55	0.12
$1/C$	0.13	0.19	0.27	0.40	0.58	0.88

Plots of $\ln C$ and $1/C$ versus time are illustrated in Fig. 3-4. The first plot is almost a straight line whereas the second is curved upwards, indicating that the data appear to fit a first order model. The slope of this line provides an estimate of k:

$$k = -[0.12 - 2.01]/[60 - 0] = \underline{\underline{0.032 \text{ min}^{-1}}}$$

Regression analysis is recommended when significant scatter exists in the data.

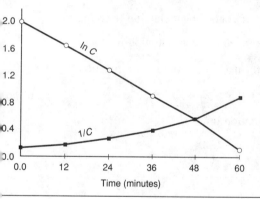

FIGURE 3-4
First and second order plots for Example 3-19.

Reactors

Chemical reactors are designated **continuous flow** or **batch** reactors based on whether a fluid stream continuously flows through the reactor. For hazardous waste applications, the selection is generally based on the nature of the problem being solved. A batch reactor would be appropriate for a one-time cleanup or for intermittent operations, whereas a continuous flow reactor would be more cost-effective to treat a waste that is generated continuously.

The most commonly used reactor is the **continuous flow stirred tank reactor** (CFSTR). This is a **completely mixed** reactor such that the concentration of materials in the reactor equals the concentration in the effluent. However, some of the particles entering the tank will exit immediately, while others will remain in the reactor for a very long time. This variation in particle detention (or residence) may result in effluent concentrations that do not meet environmental standards.

Alternatively, a **plug flow reactor** (PFR) is one in which mixing of the influent with the entire contents of the tank is kept to a minimum. This is often accomplished by long narrow tanks. The actual detention time of all particles is equal to the theoretical detention time (volume divided by flowrate).

The impact the flow regime has on various inputs can be seen through dye tracer studies. The theoretical output (dye concentration) from a PFR and a CFSTR to a step and impulse dye feed is shown in Fig. 3-5. The step input models the start up of a continuous operation, whereas the impulse can be conceived of as a batch dump of material into the reactor influent.

The effluent from a CFSTR may be analyzed by a mass balance as follows. Assuming no reaction takes place and a constant volume reactor:

$$V dC/dt = QC_{in} - QC_{out} \tag{3-55}$$

where

$V =$ volume of mixture in reactor

$dC/dt =$ rate of change of concentration in the reactor

$Q =$ flowrate into reactor

$C_{in} =$ concentration of contaminant entering (step input)

$C = C_{out} =$ concentration leaving $=$ concentration in reactor

Equation 3-55 may be solved for the concentration as a function of time;

where

$$C = C_{in}[1 - e^{-t/td}] \tag{3-56}$$

$t_d =$ detention time $= V/Q$

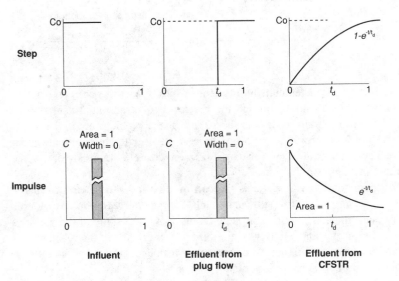

FIGURE 3-5
Response to step and impulse input.

If we now assume an irreversible first order chemical reaction takes place in the reactor:

$$V\,dC/dt = QC_{in} - QC - VkC_{out} \qquad (3\text{-}57)$$

where $\qquad k =$ first order rate coefficient and

$\qquad VkC_{out} =$ the loss due to the chemical reaction

Equation 3-57 may be solved for concentration as a function of time. Equation 3-57 may be written:

$$\frac{dC}{dt} + \beta C = \frac{Q}{V} \qquad (3\text{-}58)$$

where $\beta = k + \dfrac{Q}{V}$

Equation 3-58 is a first order linear differential equation. It may be solved by multiplying both sides by the integrating factor $e^{\beta t}$, and integrating. Noting that at $= 0, C = C_{in}$ yields:

$$C = \frac{QC_{in}}{V\beta}(1 - e^{-\beta t}) + C_{in}e^{-\beta t} \qquad (3\text{-}59)$$

Equation 3-59 is the non-steady state solution to Eq. 3-57.

For many environmental applications, the steady-state $(dc/dt = 0)$ solution is of greater interest:

$$C = \frac{C_{in}}{1 + kt_d} \qquad (3\text{-}60)$$

Many treatment processes involve a recycle line, wherein part of the effluent stream is brought back to the beginning of the process. A schematic showing recycle is illustrated in Fig. 3-6. One result of this modification is that the actual flow rate through the reactor is increased from Q to $Q + R$. If the concentration in the recycle stream is significantly lower than the feed stream, the concentration in the reactor is decreased by dilution.

This chapter has attempted to present a review of some fundamental concepts that are assumed background for many chapters in this book. This review is necessarily brief, and students are referred to basic texts for more detailed information. Particularly recommended are the books by Smith and Van Ness,[39] Lyman, et. al.,[40] and Felder.[41]

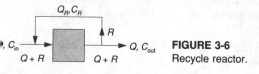

FIGURE 3-6
Recycle reactor.

DISCUSSION TOPICS AND PROBLEMS

3-1. Draw the structure of the following compounds:
(a) benzene
(b) 1,1-dichloroethane
(c) methylene chloride (dichloromethane)
(d) bis (2-chloroethyl) ether
(e) 1,2-dichloro 2-butene

3-2. Name the following structures:

a.
$$H-\underset{\underset{\underset{H}{|}}{\overset{\overset{H}{|}}{C}}}{}-H$$
$$O$$
$$H-\underset{\underset{H}{|}}{\overset{|}{C}}-H$$

b.
COOH

NH$_2$

c.
S−CH$_2$

3-3. Determine the solubility of cadmium hydroxide, $Cd(OH)_2$, as a function of pH. $K_{sp} = 5.33 \times 10^{-15}$.

3-4. The Henry's constant is often tabulated in units of pressure (atm). Derive an expression to convert the value for trichloroethylene, 550 atm at 20°C to a dimensionless Henry's constant.

3-5. Convert the Henry's constant for Benzene, 5.5×10^{-3} atm· m^3 per mole at 20°C, to a dimensionless constant.

3-6. Estimate the Henry's constant for Benzene @25°C from solubility and vapor pressure data in App. A.

3-7. Betterton and Hoffmann report the Henry's constant for formaldehyde in water @25°C as 2.97×10^3M atm^{-1} (moles/atmosphere). Convert this value to units of atm·m^3/mol.

3-8. Estimate the diffusion coefficient for ethylbenzene, from tabulated values for phenol and toluene.

3-9. Estimate the diffusion coefficient for methyl phenyl sulfide in water at 25°C.

3-10. Explain the difference between bioconcentration and biomagnification.

3-11. Estimate the concentration of ethylbenzene in fish tissue in a lake with a concentration in water of 120 mg/L.

3-12. Estimate the organic carbon partition coefficient K_{oc} of benzene from (1) solubility, and (2) Log K_{ow} data in App. A. Compare your results to the measured value in App. A.

3-13. Derive an expression for the change in internal energy of a liquid recirculated by a pump in a tank. (Assume changes in potential energy are negligible and the system is at steady state.) Heat loss to the surroundings is negligible.

3-14. Determine the degrees of freedom for a system at equilibrium with water and benzene over an iron hydroxide layer. Saturated air is above the liquid.

3-15. A decanter for a steam stripper has two liquid layers, primarily toluene and water. How many degrees of freedom does this system have? (Hint: A saturated vapor is above the liquids.)

3-16. A one cubic meter container holds 50g of hydrogen. The temperature is 20°C. Determine the pressure.

3-17. Determine the molar volume of air at 20°C and atmospheric pressure. Compare this value to the molar volume of air at standard conditions (O°C and 1 atm). Assume ideal gas behavior applies.

3-18. A gas stream containing 30 wt % Oxygen and 70 wt % Nitrogen at 120 K and 1 atm enters a process. Determine the work required to raise the temperature to 235 K. Assume the process is adiabatic (i.e., $q = 0$) and the changes in potential and kinetic energy are negligible.

Component	Temperature (K)	Enthalpy (joules/g mol)
Oxygen	120	3480
Oxygen	235	6795
Nitrogen	120	1625
Nitrogen	235	2460

3-19. 100,000 lbm/hr of condensate (liquid water) at 310°F is fed to a wood-fired boiler. The superheated steam leaving the boiler is at 1200 psia and 900°F. How much heat must be transferred into the water in the boiler?

Temperature	Enthalpy of water (Btu/lb mol)°
310°F	280
1200°F	1440.9

3-20. Steam enters a compressor at 101.325 kPA and 350°C and at a flow rate of 100 kg/minute. The steam is compressed to 500 kPA. Changes in kinetic and potential energy between inlet and outlet of the compressor are negligible. Calculate the quantity of work produced in the process. (Assume the system is at steady state. Also assume the process is adiabatic and reversible.)

-21. Determine the heat transferred to water in a boiler given the following data: Temperature is raised from 20°C to steam at 200; flow rate of water $= 1$ m^3/min. Assume atmospheric conditions.

-22. A mixture of air and methane is to be heated from 20°C to 600°C. The mixture is 50 moles air and 100 moles methane. How much energy is needed to achieve this temperature change? What temperature should be used in the temperature-dependent form for the heat capacity?

-23. 50 kg of water is contained in an open tank at room temperature. The goal is to add the steam at such a rate that all of the steam condenses. If the steam supplied is at 2.8×10^5 Pa and 320°C, how much steam (in kg) should be added so that the temperature of the water reaches 90°C? (Assume no heat loss from the water.)

-24. Calculate the energy of formation of aragonite (CaCO$_3$) at 298.15 K, given H $= -1207.0$ kJ/gmol and $S = 7.83$ kJ/K · gmol.

-25. A continuous and steady-state freezing process is used to purify sea water (3 wt % NaCl and 97 wt % water). The liquid brine concentration exiting the process is 10 wt % NaCl. Determine the mass rate of seawater required to produce 300 kg/hr of pure water in the form of ice.

-26. Water enters a 5L tank at a rate of 4.00 kg/sec and is withdrawn from the tank at a rate of 6.00 kg/sec. The tank is initially 4/5 full.

(a) Write a mass balance for this process by assigning a variable function to the mass of water at time t. Describe which terms are present in the balance; give reasons for deletion of any terms.

(b) Solve the balance to find how long it will take the tank to drain completely.

3-27. A distillation column is set up containing A and B. The feed to the column enters a 5000 kg/hr and is in a 1:1 ratio. The top stream out of the distillation column is sent to a condenser at a rate of 3500 kg/hr. At the condenser, A is taken out and separated into a product stream and a recycle stream. The product stream from the condenser contains 95% A. The bottoms stream contains 90% B. Find the ratio of recycle/product at top.

3-28. A reaction with the stoichiometric equation:

$$A + 3B \rightarrow 2C$$

is carried out with 30% conversion of A. The reactor feed stream contains 25% A and 75% B by weight and has a rate of 1000 kg/hr. If the molecular weight of A is 32 and that of B is 2,

(a) Calculate the molecular weight of C.

(b) Calculate the composition of the exit stream on a weight basis.

3-29. (a) Write a balanced reaction for the reaction of 2 moles $CH_3CH_2CH_2NH_3$ with O_2 to produce CO_2, H_2O, and NO_2.

(b) Calculate the rate of production of CO_2, H_2O, and NO_2 if 5 mol/hr of $CH_3CH_2CH_2NH$ is reacted with a stoichiometric amount of O_2.

(c) Calculate the rate of reaction.

3-30. The equilibrium constant K for the formation of nitrogen dioxide: $N + O_2 \rightarrow NO_2(g$ is 8.27×10^{-10} at 298.15 K. Calculate the standard free energy change for the reaction

3-31. Cyclohexane is commercially produced from catalytic, high-pressure and temperatur hydrogenation of benzene. Given the Gibbs Free Energy values for cyclohexane 31.9 kJ/gmol and benzene, 129.72 kJ/gmol at standard temperature and pressure (STP); deter mine whether or not this process is feasible at STP and to what extent.

3-32. Using the heat of combustion (ΔH_c°) value for DDT $(C_{14}H_9Cl_5)$ of -1600 kcal/gmol calculate the heat of formation (ΔH_f°), given the following heats of formation of the combustion products:

Compound:	CO_2	$H_2O(l)$	$HCl(g)$
ΔH_f° (kcal/gmol):	-94.05	-68.32	-22.06

3-33. An environmental scientist wants to prepare some standards to calibrate a portable ga chromatograph to be used to conduct a screening of volatile organic compounds extracte from a soil vapor extraction system. What volume of neat (pure) liquid would be neede to prepare a gas standard containing benzene and tetrachloroethene to produce standar concentrations of 100 μL/L (ppmv) and 500 μL/L (ppmv) respectively? The standard prepared at STP and in a one-liter glass sampling bulb. Use vapor pressure data provide in App. A.

3-34. An impoundment 20×12 feet is contaminated by tetrachloroethylene. The contaminate liquid comes in contact with the groundwater and a concentration gradient of $1.3 \times 10^-$ (mol/cm^3)/cm is established. Use Fick's law to determine the diffusive flux. Assumir diffusion is the only driving force, how many pounds of tetrachloroethylene will mov toward the ground water in one hour?

3-35. Derive Eq. 3-56.

3-36. Derive Eq. 3-59.

REFERENCES

1. Ferguson, Lloyd N.: *Organic Chemistry: A Science and Art*, Willard Grant Press, Boston, Massachusetts, pp. 38–40, 1972.
2. Ibid.
3. Ibid.
4. Solomons, T. W. Graham: *Organic Chemistry*, 4th edition, John Wiley and Sons, Inc., New York, New York, p. 732, 1988.
5. Weast, R.C. (ed.): *CRC Handbook of Chemistry and Physics*, CRC Press, Inc., Boca Raton, Florida, pp. B-219–220, 1983.
6. Kavanaugh, Michael C., and R. Rhodes Trussell: "Design of Aeration Towers to Strip Volatile Contaminants from Drinking Water," *Journal AWWA*, December 1980.
7. Thomas, R. G.: "Volatilization From Water," in Warren J. Lyman, William F. Reehl, and David H. Rosenblatt, *Handbook of Chemical Property Estimation Methods*, Environmental Behavior of Organic Compounds, McGraw Hill, Inc., New York, New York, pp. 15–11, 1982.
8. Yurteri, C., et al.: "The Effect of Chemical Composition of Water on Henry's Law Constant," *J. WPCF*, vol. 59, no. 11, pp. 950–956, 1987.
9. Tucker, W. A., and L. H. Nelken: "Diffusion Coefficients in Air And Water," Chapter 17 in Warren J. Lyman, William F. Reehl, and David H. Rosenblatt, *Handbook of Chemical Property Estimation Methods*, Environmental Behavior of Organic Compounds, McGraw-Hill, Inc., New York, New York, 1982.
10. Reid, R. C., J. M. Prausnitz, and T. K. Sherwood: *Properties of Gases and Liquids*, 3rd edition, McGraw-Hill, Inc., New York, New York, p. 570, 1977.
11. Reid, R. C., J. M. Prausnitz, and B. F. Poling: *The Properties of Liquids and Gases*, 4th edition, McGraw-Hill, Inc., New York, New York, p. 581, 1987.
12. Perry, J. H.: *Chemical Engineers' Handbook*, 6th edition, McGraw-Hill, Inc., New York, New York, pp. 3–285 to 3–287, 1984.
13. Thomas, R. G.: "Volatilization From Soil," in Warren J. Lyman, William F. Reehl, and David H. Rosenblatt, *Handbook of Chemical Property Estimation Methods*, Environmental Behavior of Organic Compounds, McGraw Hill, Inc., New York, New York, pp. 16–10, 1982.
14. Reid, et al., *The Properties of Gases and Liquids*, p. 598
15. Reid, et al., *The Properties of Gases and Liquids*, pp. 52ff.
16. Reid, et al., *The Properties of Gases and Liquids*, p. 53.
17. U.S. Environmental Protection Agency: "Transport and Fate of Contaminants in the Subsurface," EPA/625/4-89/019, Center of Environmental Research Information, Cincinnati, Ohio, September 1989.
18. Dragun, James: "The Fate of Hazardous Materials in Soil," *Hazardous Materials Control*, Hazardous Materials Control Research Institute, vol. 1, no. 5, September-October 1988.
19. Barron, M.G.: "Bioconcentration," *Environ. Sci. Technol.*, vol. 24, pp. 1612–1618, 1990.
20. Banerjee, S., and G. L. Baughman: "Bioconcentration Factors and Lipid Solubility," *Environ. Sci. Technol.*, vol. 25, pp. 536–539, 1991.
21. Paustenbach, D.J.: "A Survey of Health Risk Assessment," in *The Risk Assessment of Environmental and Human Health Hazards: A Textbook of Case Studies*, D. L. Paustenbach, ed., Wiley-Interscience, p. 56, 1989.
22. Weber, W. J., P. M. McKinley, and L. E. Katz: "Sorption Phenomena in Subsurface Systems: Concepts, Models and Effects on Contaminant Fate and Transport," *Water Research*, vol. 25, no. 5, pp. 499–528, 1991.
23. Smith, J. M., and H. C. Van Ness: *Introduction to Chemical Engineering Thermodynamics*, 4th edition, McGraw-Hill, Inc., New York, New York, 1987.
24. Sonntag, R. E., and G. J. Van Wylen: *Introduction to Thermodynamics—Classic and Statistical*, 3rd edition, John Wiley & Sons, Inc., New York, New York, 1991.
25. Sandler, S.I.: *Chemical and Engineering Thermodynamics*, John Wiley & Sons, Inc., New York, New York, 1989.
26. Ibid., p. 33.
27. Sonntag and Van Wylen, *Introduction to Thermodynamics*, p. 20.

28. Smith and Van Ness, *Introduction to Chemical Engineering Thermodynamics*
29. Sonntag and Van Wylen, *Introduction to Thermodynamics.*
30. Sandler, S.I., *Chemical and Engineering Thermodynamics.*
31. Perry, R. H., and D. W. Green: *Perry's Chemical Engineers' Handbook*, 6th edition, McGraw-Hill, Inc., New York, New York, 1984.
32. Smith and Van Ness, *Introduction to Chemical Engineering Thermodynamics*, p. 38.
33. Sandler, S.I., *Chemical and Engineering Thermodynamics*, p. 43.
34. Sonntag and Van Wylen, *Introduction to Thermodynamics*, p. 167.
35. Sandler, S. I., *Chemical and Engineering Thermodynamics*, pp. 85–86.
36. Ibid., pp. 274–277.
37. Felder, Richard M., and Ronald W. Rousseau: *Elementary Principles of Chemical Processes*, 2nd edition, John Wiley & Sons, Inc., New York, New York, pp. 482–483, 1986.
38. Ibid., pp. 103–106.
39. Smith and Van Ness, *Introduction to Chemical Engineering Thermodynamics*
40. Lyman, Warren J., William F. Reehl, and David H. Rosenblatt: *Handbook of Chemical Property Estimation Methods*, Environmental Behavior of Organic Compounds, McGraw-Hill, Inc., New York, New York, 1982.
41. Felder and Rousseau, *Elementary Principles of Chemical Processes.*

ADDITIONAL READING

Alberty, R. A.: *Physical Chemistry*, 7th edition, John Wiley & Sons, Inc., New York, New York, 1987.
Betterton, E. A., and M. R. Hoffmann: "Henry's Law Constants of Some Environmentally Important Aldehydes," *Environ. Sci. Technol.*, vol. 22, p. 1415, 1988.
Christian, G. D.: *Analytical Chemistry*, John Wiley & Sons, Inc., New York, New York, 1986.
Dean, J. A., ed.: *Lange's Handbook of Chemistry*, 12th edition, McGraw-Hill, Inc., New York, New York, 1979.
Ernst, W.: "Determination of the Bioconcentration Potential of Marine Organisms—A Steady-State Approach," *Chemosphere,* vol. 11, pp. 731–740, 1977.
Gossett, J. M.: Measurement of Henry's Law Constants for C1 and C2 Chlorinated Hydrocarbons, *Environ. Sci. Technol.*, vol. 21, pp. 202–208, 1987.
Howe, G. B., M. E. Mullins, and T. N. Rogers: "Evaluation and Prediction of Henry's Law Constants and Aqueous Solubilities for Solvents and Hydrocarbon Fuel Components—Vol I: Technical Discussion," Final Report, Engineering & Services Laboratory, Tyndall Air Force Base, Report No. ESL-86-66, 1987.
Jordan, D. G.: *Chemical Pilot Plant Practice*, Interscience Publishers, New York, New York, 1955.
King, C. J.: *Separation Processes*, McGraw-Hill, Inc., New York, New York, 1980.
Kishi, H., and Y. Hashimoto: "Evaluation of the Procedures for the Measurement of Water Solubility And N-Octanol/Water Partition Coefficient of Chemicals: Results of a Ring Test in Japan," *Chemosphere (G. B.)*, vol. 18, p. 1749, 1989.
Knoche, Herman W.: *Essentials of Organic Chemistry*, Addison-Wesley Publishing Company, Redding, Massachusetts, 1986.
Luyben, William L. and Leonard A. Wenzel: *Chemical Process Analysis: Mass and Energy Balances*, Prentice-Hall, Inc., Englewood Cliffs, New Jersey, E 1988.
Mackay, D. and W. Y. Shiu: "A Critical Review of Henry's Law Constant for Chemicals of Environmental Interest," *Journal Physical Chemical Reference Data*, vol. 10, p. 1175, 1981.
Mailhot, H., and R. H. Peters: "Empirical Relationships Between the l-Octanol/Water Partition Coefficient And Nine Physiochemical Properties." *Environ. Sci. Technol.*, vol. 22, p. 1479, 1988.
Manahan, S. E.: *Environmental Chemistry*, 4th edition, Lewis Publishers, Boca Raton, Florida, 1990.
Manahan, S. E.: *Hazardous Waste Chemistry, Toxicology and Treatment*, 4th edition, Lewis Publishers, 1990.
McCabe, Warren L., Julian C. Smith, and Peter Harriott: *Unit Operations of Chemical Engineering*, 4th edition, McGraw-Hill, Inc., New York, New York, 1985.

Myers, Alan L., and Warren D. Seider: *Introduction to Chemical Engineering and Computer Calculations*, Prentice-Hall, Inc., Englewood Cliffs, New Jersey, 1976.

Nirmalakhandan, N. N., and R. E. Speece: "QSAR Model for Predicting Henry's Constant." *Environ. Sci. Technol.*, vol. 22, p. 1349, 1988.

Prausnitz, J. M.: *Molecular Thermodynamics of Fluid-Phase Equilibria*, Prentice-Hall, Inc., Englewood Cliffs, New Jersey, 1969.

Reklaitis, G. V.: *Introduction to Material and Energy Balances*, John Wiley & Sons, Inc., New York, New York, 1983.

Sawhney, B. L., and K. Brown, eds.: *Reactions and Movement of Organic Chemicals in Soils*, Soil Science Society of America, 1989.

Sawyer, Clair N., and Perry L. McCarty: *Chemistry for Environmental Engineering*, 3rd edition, Mc-Graw Hill, Inc., New York, New York, p. 159, 1978.

Smith, J. H., et al.: "Volatilization Rates of Intermediate and Low Volatility Chemicals from Water," *Chemosphere*, vol. 10, pp. 281–289, 1981.

Snoeyink, V. L., and D. Jenkins: *Water Chemistry*, John Wiley & Sons, Inc., New York, New York, 1980.

ACKNOWLEDGMENTS

The authors wish to acknowledge the following individuals who assisted with the preparation of this chapter: David Gallis, Ph.D., Cheri Pearson, Karen Winegardner, and Jill Sotsky, *ERM, Inc.*, Elsie Millano, Ph.D., *ERM-North Central*, Professor William E. King, Ph.D., and Stephen Galloway, and Lori Hamilton, *Bucknell University*. The principal reviewer for the ERM Group was Roy Ball, Ph.D., *ERM-North Central*.

CHAPTER
4

FATE AND
TRANSPORT
OF
CONTAMINANTS

*I am the daughter of Earth and Water
and the nursling of the Sky;
I pass through the pores of the oceans and shores;
I change but I cannot die.*

Percy Bysshe Shelley

The release of contaminants into the environment is inevitable. Contaminants are released through our manufacture and use of products and as a result of treatment and disposal of our wastes. Upon release to the environment, contaminants move and respond to a number of interrelated natural and manmade factors. They may move quickly or ever so slowly to living receptors in their original or altered form. The pathway may be direct or tortuous. This movement is indicated in Fig. 4-1 for the atmospheric and subsurface environment and in Fig. 4-2 for the aquatic environment.

Understanding how contaminants are released and their transport and fate in the environment is necessary for successful hazardous waste management. In this context, this chapter defines key release mechanisms, presents the fundamental mechanics of contaminant transport, and describes the fate of contaminants in the environment. The reader should recognize that the understanding of contaminant transport builds upon the knowledge of several diverse disciplines including earth science, geology, fluid mechanics, chemistry, physics, hydrology, and biology. The interrelationship of these sciences and the associated mathematical representations of transport phenomena make the understanding of contaminant behavior interdisciplinary in nature.

133

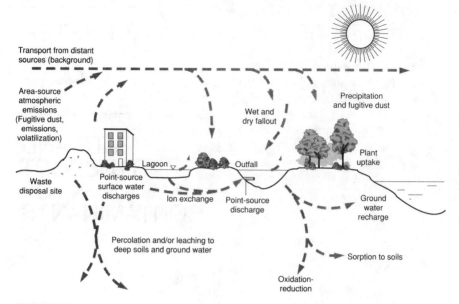

FIGURE 4-1
Fate and transport processes in the subsurface and atmospheric environment.

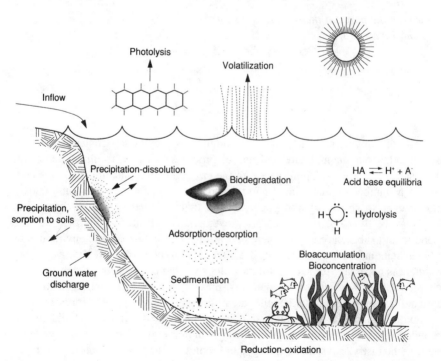

FIGURE 4-2
Fate and transport processes in the aquatic environment.

4-1 CONTAMINANT RELEASE

The release of contaminants from hazardous waste sites occurs in one or all of three phases—liquid, solid, and gas. Liquid releases include such things as contaminated runoff, direct aqueous discharges to surface water, and leachate to ground water. Air emissions can include volatile emissions from lagoons and direct atmospheric discharges from stacks. Stack emissions can include gaseous emissions (e.g., products of combustion such as CO_2 and H_2S and products of incomplete combustion such as trace organic gases) and particulate emissions (e.g., fly ash potentially laden with toxic metals). Releases of contaminants in solid form occur in air (e.g., fugitive dust) and in water (e.g., suspended solids). These releases typically have contaminants adsorbed to solids or dissolved in fluids.

Contaminant releases are either controlled or uncontrolled. Controlled releases are an intrinsic part of our current hazardous waste management practices—part of virtually every manufacturing and waste treatment operation. The notion of controlled releases is that contaminant releases are managed (i.e., controlled) in such a way as to minimize harm to the environment. The predominant mechanisms supporting this approach are dilution and assimilation. The NPDES permits for water quality limited segments of waterways mentioned in Sec. 2-5 are an excellent example of one U.S. regulatory approach to control direct releases to ensure that the receiving waters can assimilate the contaminants released.

Uncontrolled releases are those that are not under the direct management of the unit operators. Some uncontrolled releases may be reasonably foreseeable, such as windblown fugitive dust or "puffs" from incinerators. Others may not, such as a break in a pipeline or a leak from an underground storage tank. Still others derive from closed, inactive sites not having adequate environmental control measures. In many scenarios, uncontrolled releases can result in major pollution incidents as exemplified by the large number of U.S. Superfund sites.

Air Emissions

Air emissions may occur in a variety of forms as point, line, area, volume, or puff (instantaneous) sources. A point source is a well-defined location typically emitting contaminants on a continuous basis. Line, area, and volume sources are considered fugitive in nature. Fugitive emissions are those emissions that could not reasonably pass through a stack, chimney, vent, or other functionally equivalent opening. A puff release is another type of fugitive emission that occurs instantaneously from such sources as spills or other accidental releases. The types of air emissions are illustrated by the following examples:

Point	Incinerator stack; landfill gas vent
Line	Dust from road; vehicle emissions from a roadway
Area	Volatile emissions from a lagoon
Volume	Volatile or particulate emissions from a building with open windows and doors
Puff	Volatile emissions from an accidental spill

Emissions from point sources are expressed in units of mass per time; line sources as mass per time per length of line; an area source as mass per time per two-dimensional area and volume source as mass per time per three-dimensional volume. Instantaneous releases are commonly expressed as the mass of the total release.

Releases can be further classified as gas-phase emissions or particulate matter emissions. Gas-phase emissions consist primarily of organic compounds where the predominant release mechanism is volatilization. Gaseous emissions may also be produced by manufacturing or waste treatment processes. Particulate matter emissions derive principally from combustion, wind erosion, and mechanical processes, releasing fine particles that may have a variety of contaminants. These contaminants can include not only organics but metals and some normally immobile substances (e.g., PCBs and dioxins). Combustion processes can release gases (either as by-products of combustion or as a result of incomplete combustion) and particulate matter in the form of fly ash or uncombusted carbon.

Volatilization. Volatilization is the transfer of a chemical substance from a liquid phase to a gaseous phase. Volatilization is the predominant source of atmospheric emissions at most uncontrolled hazardous waste sites. Aboveground sources include storage tanks, pipe fittings, and the surfaces of lagoons. There are significant subsurface sources as well. It is noted later in this section how the migration of leachate from a landfill or lagoon releases contaminants to the ground water. Organics may volatilize from the leachate and move to the ground surface. Emissions also can come from the volatilization of dissolved chemicals in contaminated ground water, even from a plume that has migrated laterally a considerable distance from a site.

The volatilization rate depends in part upon temperature, vapor pressure of the substance, and the difference in the concentration between the liquid and gas phases. Volatilized contaminants may go directly to the atmosphere, or they may traverse a tortuous path such as contaminants in the subsurface, as depicted in Fig. 4-3. The depicted transfer depends on diffusion (see Sec. 3-2) through a porous medium, and variables such as soil porosity and soil moisture are important.

Volatilization can be measured in the field by various procedures such as through the use of organic vapor analyzers (see Sec. 15-6); however, these instruments may not be capable of differentiating among various organic compounds. Volatile emissions

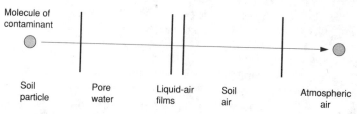

FIGURE 4-3
Contaminant transfer through soil-water-air interfaces.

TABLE 4-1
Volatile air emissions models[1]

Type of source	Number of models presented
Landfills without internal gas generation	5
Landfills with internal gas generation	3
Open landfills	3
Land treatment	2
Non-aerated lagoon	4
Aerated lagoon	2

also can be predicted using mathematical models that have been developed for specific facility types such as landfills, lagoons, or land farms. The basis for most modeling includes Fick's first and second law of diffusion (see Sec. 3-2).

A technical guidance manual, published by the U.S. Environmental Protection Agency for estimating baseline volatile air emissions at Superfund sites describes the number of models available for each of the different types of sources as listed in Table 4-1.

Example 4-1 Volatile emissions. Describe the differences in the rate of volatile emissions from a landfill where co-disposal of municipal solid waste and hazardous waste occurred versus a landfill used for the disposal of hazardous waste only.

Solution. Because of the large organic content of municipal solid waste, co-disposal landfills generate considerable volume of decomposition gases, particularly methane. The convecting sweep of such gas creates a greatly accelerated flux and release of volatile emissions from the co-disposed hazardous waste. Both the soil- and gas-phase diffusion become insignificant when compared with the convecting gas.[2]

At undisturbed, inactive sites, emissions of volatile organic compounds are typically considered insignificant even though the ground water may show the presence of such compounds. Investigations at many Superfund sites have found such compounds in ground water and even surface water but not in the air at detectable concentrations. The results of field measurements of baseline ambient volatile concentrations (prior to disturbances such as those from remediation) from various surface impoundments that have been reported in the literature are shown in Table 4-2.

The remediation of sites typically involves disturbances that may expose waste that was previously covered. Therefore, remediation activities have the potential to substantially increase the release of emissions from an inactive site. The source of the data for TSDF Site N shown in Table 4-2 reported that the flux of organic compounds, exclusive of methane, increased several thousand fold during remediation. Remediation of sites often requires treatment of contaminated ground water and soils. Such treatment may feature pumping, air strippers, oil-water separators, soil venting, and other techniques. Removing volatile contaminants from soil and ground water and allowing them to be released to the atmosphere without controls are simply transferring

TABLE 4-2
Typical volatile organic emissions

Source of data	Identification of surface impoundment	Total emissions of non-methane hydrocarbons prior to disturbance	
		$\mu g/m^2 \cdot$ min	$g/m^2 \cdot$ sec
EPA[3]	NPL Site D	43	7×10^{-7}
	TSDF Site K	120	2×10^{-6}
	TSDF Site L	570	1×10^{-5}
	TSDF Site M	9-31	2×10^{-7} to 5×10^{-7}
	TSDF Site N	630	1×10^{-5}
de Percin[4]	Impoundment 1	690	1×10^{-5}
	Impoundment 2	3400	6×10^{-5}
	Impoundment 3	7700	1×10^{-4}
	Impoundment 4	56–190	9×10^{-7} to 3×10^{-6}

NPL - National Priorities List (Superfund)
TSDF - Treatment Storage and Disposal Facility

the contaminant from one media to another. The ultimate fate of such contaminants needs to be considered. Generally, emissions from sources and the actions to remediate those sources can be estimated with models based on Henry's law constants, standard equilibrium chemistry, and local factors such as soil type, moisture content, wind speed, and lagoon surface area.

Example 4-2 Volatile emission sources. Identify the sources of volatile emissions from a lagoon and a landfill.

Solution. The sources from a lagoon are:

- Lateral migration of volatiles through the sides of the lagoon,
- Lateral migration of volatiles from leachate and contaminated soils in the vadose zone,
- Volatilization from the lagoon surface, and
- Volatilization of dissolved species in ground water.

A landfill would have the same sources. Release from its surface could occur via:

- Release through desiccation cracks in the cover,
- Diffusion through the cover,
- Release through the system used to vent methane generated by the biodegradation of organic waste,
- Release through the system used to collect leachate generated within the landfill, and
- Volatilization of exposed waste prior to covering.

Particulate Matter Emissions. Waste treatment operations generate particulate matter emissions. Incinerators, of course, directly discharge particulate matter. Fugitive particulate matter emissions generally originate from materials handling and surface areas:

- Roadways
- Open waste piles
- Staging areas
- Dry surface impoundments
- Landfills
- Land treatment operations
- Waste stabilization basins

Fugitive dust is usually a key concern at inactive hazardous waste sites. Two key parameters controlling fugitive dust are (1) the wind, which erodes and transports material from the site, and (2) disturbance of the soil by human activities (e.g., vehicular traffic). The dust has the same contaminants as those sorbed to the soil. Remediation of inactive sites generates dust from excavation, grading, and truck traffic. The amount of soil lost due to wind erosion at inactive sites is a function of wind velocity, vegetative cover, soil/waste properties such as texture and moisture, and the area of exposed soil surface. The Soil Conservation Service (SCS) of the U.S. Department of Agriculture has developed a graphical method for predicting annual average wind erosion and resulting soil loss.[5] The method uses annual average climatic data and cannot be used to analyze the impact of climatic extremes. According to the EPA, the model is not reliable for estimating short-term emissions.[6]

Example 4-3   Wind erosion soil loss.   Qualitatively, explain what portion of the soil loss estimated by the SCS wind erosion procedure would represent a release of contaminants that could be transported to off-site receptors.

Solution. The SCS method estimates total soil loss, of which only a portion is **transportable** by the wind and **suspendible**. That transportable and suspendible amount is the portion of the total soil loss represented by particles $\leq 100 \ \mu\text{m}$ in diameter or smaller, dependent upon the location of the receptors (large particles could settle out of the air before moving off site). The quantity transported to off-site receptors would have to be estimated from a grain size distribution of the parent material. If the analysis is concerned with just inhalable sizes, that would limit the analysis to the fraction of particulates $10 \ \mu\text{m}$ in diameter or smaller. Note, however, that contaminants are preferentially adsorbed to the smaller size particles.

The generation of fugitive dust by vehicular traffic is a function of soil properties and vehicular characteristics such as vehicle speed, weight, and number of wheels. Dust generation is particularly sensitive to soil moisture and soil particle size and can be estimated using the empirical formula in Eq. 4-1:

$$E_{VT} = 5.9 \left(\frac{S_p}{12}\right)\left(\frac{V_v}{30}\right)\left(\frac{M_v}{3}\right)^{0.7}\left(\frac{W_v}{4}\right)^{0.5}\left(\frac{365 - D_p}{365}\right) \tag{4-1}$$

where E_{VT} = emission factor (lb/vehicle mile traveled)
S_p = Silt content of road surface material (%)
V_v = mean vehicle velocity (mi/hr)
M_v = mean vehicle mass (tons)
W_v = mean number of wheels
D_p = number of days per year with at least 0.01 inches of precipitation

Again, the percent of the generated dust that is wind transportable and respirable needs to be considered. The release of dust from unpaved roads can be as high as 1 kg/vehicle kilometer (3.6 lb/vehicle mile) if not more.[7]

By far the greatest source of fugitive dust is remediation involving soil handling. Contaminated soils frequently require remediation that may involve either excavation for treatment/disposal or containment using a cap. Excavation entails the following necessary operations, each of which generates large quantities of fugitive dust unless control measures are taken—transportation, dumping, storage in piles, conveyance to treatment unit, and treatment. Even if the soil does not require excavation, usually some form of surface grading is necessary and this generates fugitive dust.

To estimate particulate emissions from materials-handling operations such as soil loading and unloading, the following equation can be used.[8]

$$E = k(0.0032)\frac{\left(\frac{U}{5}\right)^{1.3}}{\left(\frac{M}{2}\right)^{1.4}} \tag{4-2}$$

where E = emission factor (lb of particulates released to air/ton of soil moved)
k = particle-size multiplier (dimensionless)
U = mean wind speed (mi/hr)
M = material moisture content (%)

The emission factor, E, may be multiplied by the tons of material handled to achieve an estimate of total emissions. The particle-size multiplier, k, reflects the aerodynamic particle diameter, as follows:

Particle size	< 30 mm	< 15 mm	< 10 mm	< 5 mm	< 2.5 mm
k	0.74	0.48	0.35	0.20	0.11

The results of applying the preceding methods to estimating generation of fugitive dusts can be considered rough estimates at best. A site should be monitored downwind and upwind of its sources of release if there is any question as to whether the amount of release meets acceptable criteria for health and environmental risk.

Releases to Water

Water is an excellent medium for the transport of contaminants in the environment. How are contaminants released into the water? Controlled releases into surface water (e.g., streams, rivers, and lakes) are quite common. The varieties of aqueous waste streams with direct discharges to surface water are nearly endless. Examples include effluent from treatment works, cooling water scrubber blowdown from incinerators, treated leachate from landfills, and treated wash water from recycling/reuse facilities. Virtually every industrial and commercial facility generates a wastewater. Because our ability to clean up wastewater is and always will be less than 100%, the controlled discharge into surface waters will always be a source, however small, for some contaminant release into the environment.

Controlled releases of hazardous waste to ground water are rare. In fact, most controlled contaminant releases involve materials other than hazardous waste and occur to the land above the ground water. The application of pesticides directly to the land is one such example. Another is the use of on-site septic tanks and drain fields by rural homeowners. Controlled releases of hazardous waste can be rather controversial. One example is the use of deep well injection in which wastes are injected deep beneath the earth into rock formations where permanent containment is anticipated.

Uncontrolled releases occur to both surface and ground water. Leachate and contaminated runoff from landfills are excellent examples of uncontrolled releases to ground and surface water. The releases are a result of natural mechanisms, that is, the impact of precipitation resulting in runoff and/or leachate generation. Uncontrolled releases as a result of human activity, such as spills, also occur.

Table 4-3 lists many principal releases to surface and ground water, contrasting the volume of the release, the concentration of contaminants, and the types of factors affecting the nature of the release. Of these releases, the leakage of leachate through the base of a landfill to ground water is one of the most studied. Characterization of leachate generation is concerned with the volume of liquid generated, its constituents, and the landfill construction.

Landfill Leachate

Generation. The release of leachate first requires the generation of leachate within the landfill. Fig. 4-4 depicts the sources of liquid that may eventually become leachate. Most sources are water such as:

- Precipitation falling directly onto the landfill,
- Surface flow that has run onto the landfill, and
- Ground water inflow through a portion of the landfill lying below the ground water table.

In addition, there is the liquid fraction of the waste disposed in the landfill (past practices sometimes used landfills for disposal of liquid wastes, either in bulk form or in drums). With the exception of the infiltration of direct precipitation, these sources of leachate can be controlled with proper siting and engineering. Infiltration can be minimized by constructing an engineered cover over the landfill as described in Sec. 13-4. Nevertheless, some infiltration inevitably occurs.

The calibration of infiltration through a landfill cover is done with a water budget model such as the widely used model HELP.[9] Such a model quantifies water

TABLE 4-3
Releases of contaminants to surface and ground water

Source	Volume of release	Contaminant concentration	Types of factors affecting release
Transport			
• Spills	Partial to entire contents of transport vessel	High (e.g., often pure product)	Traffic accidents; unloading mishaps
Storage			
• Spills	Partial to entire contents of storage vessels	High (e.g., often pure product)	Structural failure of tank; mishaps in handling
• Leaks	Minimal rate; yet could continue indefinitely, particularly if underground	High (e.g., often pure product)	Frequency of inspections and maintenance; age of storage facilities
Treatment			
• Effluent	Varies, often high	Low (required by regulatory permits)	Influent; design and operation of facility
Landfills			
• Runoff	Possibly large, dependent upon rainfall event	Low; typically contaminated sediment; nil if landfill is capped	Integrity of cap; slope; stormwater retention capacity
• Surface seeps	Minimal rate; yet could continue indefinitely	Medium to high	Characteristics of cap (slope and permeability); disposal of liquids; removal of leachate
• Leachate through base	Minimal to low rate for lined facilities; moderate to high rate for unlined facilities; continues indefinitely	Medium to high	(same as above plus permeability of base)
Lagoons			
• Overflow, washout	Partial to entire contents	High (stored hazardous waste)	Structural failure; flooding
• Leakage	Minimal to low rate for lined facilities, moderate to high for unlined facilities; continues indefinitely	High (stored hazardous waste)	Permeability of base; liquid depth

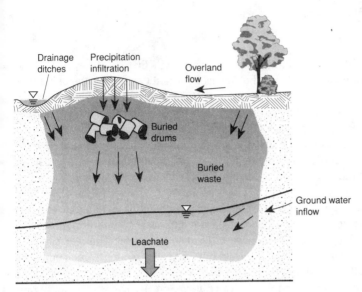

FIGURE 4-4
Sources of fluids for the generation of landfill leachate

movement based on the principle of conservation of mass. It balances the surface water input among the many possible routes and outputs as shown in Fig. 4-5. Given climate, cover characteristics (e.g., slope, thickness, porosity, hydraulic conductivity), and other factors, the model calculates the volumetric flux of water infiltrating into the waste on a time-varying basis. Contaminant transport through landfill barriers is presented in Sec. 13-6.

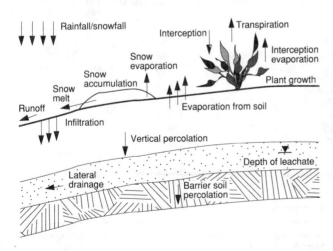

FIGURE 4-5
Water balance variables in the HELP Model[10]

Landfill Leachate

Contaminants. Water infiltrating into waste will first be absorbed by solid waste material if that material is below field capacity. **Field capacity** is the upper limit of a porous material's ability to absorb water. When the waste becomes saturated, water moves through the waste via gravity. This water will become contaminated with waste constituents in a partitioning process. The concentration of a constituent in the water depends upon the amount of leachable material in the waste, the ease of mass transfer, and the ratio of the column depth (i.e., vertical thickness of waste in the landfill) to the infiltration rate. Other factors include contaminant solubility, surface area, contact time, and pH.[11]

For well-engineered landfills in which leachate generation is minimal, the concentration of a waste constituent in the leachate can approach the equilibrium solubility, given the availability of a sufficient mass of leachable material. This is because the residence time in such landfills can be 20 years or more.[12] There are several complicating factors. Cosolvation (mixture of water with a solvent) can result in higher concentrations of contaminants than possible for water alone, greatly exceeding equilibrium solubility. Further, retardation processes and chemical and biological transformations occur within the landfill which alter contaminant types and concentrations.

Because of these many factors, the presence of contaminants and their concentration in leachate varies widely, even within the same landfill. Analytical data collected for leachate in a number of landfills that contained hazardous waste are shown in Tables 4-4 and 4-5 for inorganic and organic substances, respectively. The waste materials, landfill design, and climate varied widely among these landfills;

TABLE 4-4
Selected inorganic constituents detected in leachate from hazardous waste landfills

Constituent	Reported concentration (μg/L)[†]	Number of landfills reported[†]	Reported concentration (μg/L)[‡]	Number of landfills reported[‡]
As	11–< 10,000,000	6	30–5,800	5
Ba	100–2,000,000	5	10–3,800	24
Cd	5–8,200	6	–	–
Cr	1–208,000	7	10–4,200	10
Cu	1–16,000	9	10–2,800	15
Hg	0.5–7	7	0.5–0.8	5
Ni	20–48,000	4	20–670	16
Pb	1–19,000	6	300–19,000	3
Se	3–590	4	10–590	21
CN⁻	–	–	5–14,000	14

[†] "Management of Hazardous Waste Leachate," SW-871 report prepared for U.S. EPA by Shuckrow, Touhill and Associates, Inc., Pittsburgh, Pennsylvania, 1981.
[‡] "Prevalence of Subsurface Migration of Hazardous Chemical Substances at Selected Industrial Waste Land Disposal Sites," EPA/530/SW-634 report prepared for U.S. EPA by Geraghty and Miller, Inc., 1977.

hence, the wide range of concentrations. Although there are analytical and computational procedures for predicting leachate concentrations, the only reliable method is to collect and analyze field samples from the actual site.

Release. A properly designed landfill will provide for leachate collection and removal. Even so, some leachate will leak through the base. For a compacted soil base, the rate of percolation is based on the hydraulic conductivity of the soil material. The case is different for a base that features a flexible membrane liner. Deterioration over time, pinholes, and improperly sealed seams and tears will result in openings in the liner and some leakage. Molecular diffusion will also occur. Even a modern, properly engineered liner system is expected to leak about 22 gal/acre/day through the primary liner (see Sec. 13-3). Much of this is collected in secondary leachate collection systems. A landfill without a liner will pass leachate as rapidly as it is generated.

TABLE 4-5
Selected organic constituents in leachate from hazardous waste landfills*

Substance	Reported concentration (μg/L)	Number of landfills reported
Acetone	0.1–62,000	3
Aldrin	< 2–< 10	2
Benzene	< 1.1–7,370	5
Chlorobenzene	4.6–4,620	5
Chloroform	0.02–4,550	4
Dichlorobenzene	< 10–517	2
1,1-dichloroethane	< 5–14,280	2
1,2-dichloroethane	2.1–4,500	5
Trans-1,2-dichloroethane	25–8,150	2
1,1-dichloroethylene	28–19,850	5
Dichloromethane	3.1–6,570	4
Ethyl benzene	3.0–10,115	4
Hexachlorobutadiene	< 20–109	2
Methylene chloride	< 0.3 mg/L–184 mg/L	3
Methyl isobutyl ketone	2–10 mg/L	2
Perchloroethylene	ND–8,200	5
Phenol	< 3–17,000	4
Tetrachloroethene	< 1–89,155	3
Tetrachloromethane	< 1–25,000	3
TOC	10.9–8,700 mg/L	8
Toluene	< 5–100,000	7
1,1,1-trichloroethane	1.6 μg/L–590 mg/L	5
Trichloroethene	< 3–84,000	4
Trichloroethylene	< 3–260,000	4
Vinyl chloride	140–32,500	1

"Management of Hazardous Waste Leachate," SW-871 report prepared for U.S. EPA by Shuckrow, Touhill and Associates, Inc., Pittsburgh, Pa, 1981.

Example 4-4   Laboratory leaching test. An engineer proposes that metal-plating sludges be disposed of in a municipal waste landfill. To demonstrate the safety of the proposal, the engineer predicted leachate concentrations based on a laboratory test. The test involved drying and pulverizing the sludge, adding distilled water (proportional to the ratio between the mass of sludge to be landfilled and the total pore volume of the landfill), and stirring the mixture for 90 days. Would this test be expected to yield conclusive predictions of worst case leachate concentrations?

Solution. No, for at least two reasons. First, leachate in a municipal landfill will have a lower pH than distilled water due to the presence of acids from biological decomposition of the organic matter in municipal waste. Second, the 90-day test period is far less than the actual leachate residence time for a well-designed landfill. Section 11-4 provides additional detail on leaching tests.

In studies of hazardous waste management it is important to observe civilization's dependence upon the need for potable water. In the United States about 81% of our water supply is from surface water sources, and 19% is from ground water sources.[13] The distribution of water use varies significantly with region and population density. For example, many cities in the West, New Jersey, Florida, and other areas make more use of ground water than surface water. Likewise, rural areas use ground water more extensively than surface water. There is significant concern about the quality of ground water and surface water and their suitability for potable water sources in terms of both quality and quantity.[14]

The fragile nature of our water supplies and the importance of releases to ground water are best illustrated using the results of a calculation from a landfill designed and constructed in Canada.* In this study it was assumed that contaminant transport due to leachate leaking from the landfill could be reduced to zero and the only transport across the landfill liner system would be by diffusion (defined in detail in Sec. 4-? and discussed in the context of landfill design in Sec. 13-6). Even when assuming a very low concentration of benzene in the landfill leachate of one gram per liter (1 g/L), diffusion through the landfill liner would still amount to 19 kg per year. At this rate of 19 kg per year, combined with estimates of ground water flow rates beneath the site, it was calculated that a total of 3.8 billion gallons of ground water would annually be contaminated to the current drinking water standard of 5 parts per billion (5 μg/L) for benzene. It is clear that very small quantities of toxic and hazardous chemicals have a significant impact on the quality of our water supplies.

How do we then protect our ground water and surface water supplies for both our present and future use? The detrimental impact of contaminated ground water as demonstrated above points to the philosophy that, where possible, pollutants should not be generated. If pollutants must be generated, they should be recycled in a safe manner. Where prevention or recycling is not possible, treatment in an environmentally

---

*Johnson, R. L., et al.: "Diffusive Contaminants Transport in Natural Clay: A Field Example and Implications for Clay-lined Waste Disposal Sites," *Environ. Sci. Technol.,* vol. 23, p. 340, 1989.

safe manner should be used. Disposal or other release into the environment should be used only as a last resort, and then only in an environmentally safe manner.

4-2   TRANSPORT OF CONTAMINANTS IN THE SUBSURFACE

Hydrologic Cycle

In the subsurface environment, contaminant transport is strongly dependent upon ground water flow. More precisely, the spatial and temporal distribution of ground water affects contaminant movement. Therefore, it is necessary to begin an understanding of contaminant processes with an understanding of the hydrologic cycle and its effect upon ground water. **Hydrology** is the science that deals with the properties and movement of water. The continuous circulation of water in the atmosphere, in the subsurface, and in surface water bodies is termed the **hydrologic cycle.** Figure 4-6 and the depicted conditions will be used to define a number of important concepts dealing with the water cycle, ground water flow, and contaminant transport.

The hydrologic cycle begins with **precipitation** that lands on the earth's surface and can be in the form of rain, snow, or sleet. From there, the water may spread along the ground surface as **surface water runoff** or **overland flow** or may seep into the ground and become ground water. Overland flow may continue to concentrate as **channel flow,** and progressing as **stream flow,** continuing to concentrate in the form of streams and rivers until ultimately reaching the ocean.

Throughout this cycle the water may enter the subsurface environment through **infiltration** or may be returned to the atmosphere through **evaporation.** Water that does not evaporate or move on the earth's surface as overland flow enters the subsurface environment. Infiltrating water first encounters the **unsaturated zone,** in which the pores (voids) between soil particles contain soil, moisture, and air (a three-phase material). Water in the unsaturated zone may migrate further into the **saturated zone,** or may be held in the form of soil moisture to be returned to the atmosphere by plant and animal life via **evapotranspiration.**

Water reaching the saturated zone in the subsurface will flow from areas of high hydraulic head to areas of lower hydraulic head and may recharge other ground water flow systems. The subsurface stratum, or layer, that readily permits this water movement is termed an **aquifer.** For example, a layer of sand and gravel may provide an excellent stratum to serve as an aquifer. If the upper surface of the saturated zone is free to move up or down (rise or decline) in the aquifer without being physically constrained by a stratum of lower permeability, the aquifer is termed an **unconfined aquifer.** Eventually ground water may discharge to surface water bodies such as streams and lakes. In the migration in the subsurface the ground water may encounter a zone that does not readily transmit water termed an **aquitard** or a **confining layer.** A stratum of clay often serves as a confining layer. An aquifer located between aquitards is termed a **confined aquifer.** It is in the context of this hydrologic cycle that we will study the force systems that cause the migration of water and associated transport of contaminants.

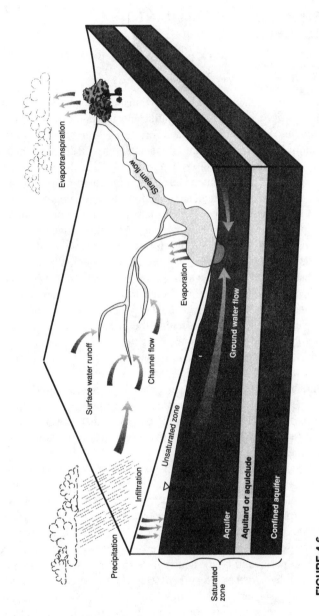

FIGURE 4-6
Hydrologic cycle.

148

Before going further it is necessary to define additional aquifer characteristics. In this context, **soil** will represent all of the unconsolidated sediment or **regolith** above the competent rock. This usage is consistent with that in the soil mechanics field of civil engineering (geotechnical engineering) and encompasses much more material than the term soil denotes in soil science (i.e., that shallow material subjected to biological and physical weathering). A **homogeneous stratum** is one in which the properties of the soil do not vary with location in the layer. Thus, a porous material is homogeneous with respect to permeability if the permeability is the same at all locations within the formation. A stratum whose material properties are different at different locations within the layer is said to be **heterogeneous**. An **isotropic** stratum is one in which the soil properties do not vary with orientation (direction). For example, the materials of a sand and gravel stratum may have a hydraulic conductivity higher in the horizontal direction than in the vertical direction, that is, water may flow through the soils more easily in the horizontal direction. Such a stratum is anisotropic. As an example of the possible combinations, a clay deposit may be homogeneous and isotropic with respect to permeability—the permeability is the same in the vertical and horizontal directions and is the same everywhere in the formation. A clay of lacustrine (lake) origin may be homogeneous and anisotropic with respect to permeability. In this case, the permeability is different (likely lower) in the vertical direction than in the horizontal direction. However, because the example is considered homogeneous and anisotropic, the horizontal hydraulic conductivity is the same everywhere in the formation, and, similarly, the vertical hydraulic conductivity is the same everywhere in the formation (although they are not equal to each other).

In the real world of subsurface conditions, materials are almost always anisotropic and heterogeneous. To strictly account for this heterogeneity and anisotropy theoretically requires an infinite amount of data. Therefore, for the purposes of analysis of ground water flow it is usually necessary to assume that a formation, or a portion of formation, is homogeneous.

The information provided in this section of the chapter allows the student to understand the nature of contaminant fate and transport. It is important to recognize that this chapter on contaminant fate and transport is only an introduction to a complex aspect of hazardous waste management. The readers are referred to several useful texts on ground water flow and contaminant transport including Freeze and Cherry, 1979;[15] Todd, 1984;[16] and Fetter, 1988.[17]

Ground Water Flow

Darcy's Law.   Darcy's law forms the basis for much of our understanding of ground water flow in the subsurface environment. Darcy developed the law in the mid-19th century based on a series of experiments in France to understand water filtration. Figure 4-7 is a schematic of Darcy's experiment to determine the laws of the flow of water through clean sand.

In this experiment Darcy measured the flow of water through a column of sand and found that the flow rate was proportional to the change in hydraulic head and

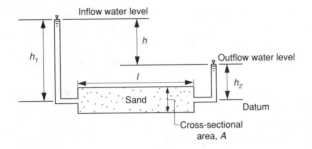

FIGURE 4-7
Darcy's experiment.

the cross-sectional area of flow, and inversely proportional to the length of the sand column. He found that to relate the flow rate to the hydraulic head and area of flow required a constant of proportionality which he termed k. In this text we will refer to k as the hydraulic conductivity. It has units of velocity. Note that the value is a function of both the porous media and the fluid. It is also commonly known as the permeability. His findings describing flow through a porous media are presented in algebraic form as Eq. 4-3:

$$Q = kiA \qquad (4-3)$$

where Q = flow rate (cm^2/sec)
k = hydraulic conductivity (cm/sec)
i = hydraulic gradient (cm/cm)
A = cross-sectional area of flow measured
perpendicular to the flow direction (cm^2)

The **hydraulic gradient**, i, describes the rate of change in which the head (or energy) is lost as water flows through the porous materials and is defined in algebraic form as:

$$i = (h_1 - h_2)/l \qquad (4-4$$

where h_1 = head at location 1 (cm)
h_2 = head at location 2 (cm)
l = length of sand column (cm)

Note the units of $h_1 - h_2$ (change in head) and l (flow distance) are both in units of length resulting in a gradient that is dimensionless.

Darcy's law is empirical—developed as a result of experiments. It has been verified by subsequent experiments and is considered valid over a wide range of subsurface flow conditions. For materials of high hydraulic conductivity and/or with high hydraulic gradients, Darcy's law is valid as long as the flow is laminar. Ground water flow is almost always laminar. For materials with low hydraulic conductivity a threshold gradient may be required before flow can take place. For most of the conditions encountered in the subsurface, Darcy's law has been demonstrated through experience to be valid. Darcy's law is, however, macroscopic; therefore, the parameters and variables are interpreted as averages over the representative elementary volume.

Rearranging terms in Eq. 4-3, Darcy's law can be expressed differently by dividing both sides of the expression by the cross-sectional area of flow to yield Q/A with units of velocity as:

$$Q/A = ki \tag{4-5}$$

where Q/A = specific discharge, Darcy flux, or Darcy velocity, v,

and the equation rewritten as:

$$v = ki \tag{4-6}$$

where v = specific discharge, Darcy flux, or Darcy velocity (cm/sec)

Because gradient is dimensionless, the hydraulic conductivity has units of velocity (cm/sec).

Flow is in the direction of declining hydraulic head and, using sign convention for ground water flow calculations, a negative sign is required. In differential form, Darcy's law can be written as:

$$v = -kdh/dl \tag{4-7}$$

where dh/dl = hydraulic gradient

Example 4-5   Darcy's law.   Using Darcy's law from Eq. 4-6, can we now calculate the actual velocity of a "particle" of water, that is, the seepage velocity? Referring to Darcy's Fig. 4-8 and using the dimensions of $(h_1 - h_2) = 0.1$ m and $l = 0.4$ m, and the hydraulic conductivity is 1×10^{-2} cm/sec or 1×10^{-4} m/sec, (a representative value of hydraulic conductivity for medium to fine sand), what is the Darcy velocity? Is this the seepage velocity?

Solution

$$v = ki$$

$$v = 1 \times 10^{-2}(0.1/0.4) = 2.5 \times 10^{-5} \text{m/sec}$$

Can we assume that the time, t, required for a particle of water to migrate from one end to the other under these hydraulic head conditions is equal to the sample length divided by the velocity?

$$t = l/v = 0.4\text{m}/2.5 \times 10^{-5} \text{ m/sec } = 4.4 \text{ hours? } \textbf{NO!}$$

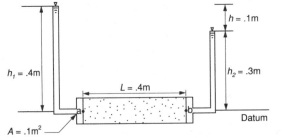

FIGURE 4-8
Schematic of saturated flow
in Darcy's experiment.

The specific discharge or Darcy velocity is *not* the **seepage velocity**. The seepage velocit
is always faster than the Darcy velocity. The travel time cannot be directly computed using th
specific discharge.

Figure 4-8 is a schematic of the saturated water flow in Darcy's experiment
where some measured discharge, Q, flows through some gross cross-sectional area, A
at some measured hydraulic gradient, i. As shown in the section view, cross-section
area of flow, A, includes both the area of solids and voids. The water migrates throug
only the voids and must follow flow paths around the solids. Thus, we can set Q equa
to the velocity of seepage times the area of voids. Invoking the equation of continuit
we have:

$$Q = vA = v_s A_v \qquad (4\text{-}8$$

or rearranging

$$v_s = vA/A_v \qquad (4\text{-}9$$

where Q = volumetric rate of flow(cm^3/sec)
   v = Darcy velocity (cm/sec)
   A = gross cross-sectional area of flow(cm^2)
   v_s = average linear seepage velocity (cm/sec)
   A_v = effective cross-sectional area of flow, area of voids (cm^2)

Now let us look at the relationship between the total area and the area of void
in a soil **phase diagram** as shown in Fig. 4-9 using conventional notation from th
soil mechanics field. Soil in the subsurface is represented by a phase diagram usin
volumetric and gravimetric terms. Soil is a three-phase material consisting of soli
(the soil minerals), liquid (water), and gaseous (air in the voids) phases. If the samp
is saturated, such as the saturated sand in the Darcy experiment, it is a two-phas
material with all the voids filled with water.

Saturated soil is represented on the phase diagram on a volumetric basis as:

$$V_t = V_v + V_s \qquad (4\text{-}10$$

where V_t = total volume (cm^3)
   V_v = volume of voids (cm^3)
   V_s = volume of solids (cm^3)

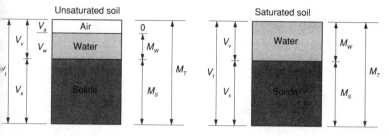

IGURE 4-9
ʰase diagram.

Employing the terms provided on the phase diagram, **porosity**, n, is defined in ercent as the volume of the voids divided by the total volume or:

$$n = (V_v/V_t) \times 100(\%) \tag{4-11}$$

For example, if a soil consists of 3/4 solids and 1/4 voids, the porosity would e 25%. Porosity may also be expressed in decimal form—0.25. The range of "theo-ʰtically" possible values for porosity varies from zero (no void space) to 100% (all oid space, no solids). For the range of soils encountered in the natural environment, ʰorosity is typically in the range of 20% to 70%.

A related parameter is the void ratio, e, defined as:

$$e = V_v/V_s \tag{4-12}$$

ʰhere $V_v =$ volume of voids (cm³)

$V_s =$ volume of solids (cm³)

It is indicated from our phase diagram in Fig. 4-9 that, for a unit dimension of ꞑgth, the porosity is equal to the area of the voids divided by the total area or:

$$n = (V_v/V_t) = (A_v/A_t) \tag{4-13}$$

ʰere $A_v =$ effective area of voids (cm²)

$A_t =$ total area (cm²)

Because the area of the voids is also the seepage area, substitution of this ꞙpression into Eq. 4-9 yields the seepage velocity equal to the Darcy velocity divided ꞙ the porosity as follows:

$$v_s = v/n \tag{4-14}$$

The travel time of a particle of water can be estimated as the linear length of ꞙ flow path, l, divided by the seepage velocity.

ꞙample 4-6 Seepage velocity. Using a porosity of 0.5 and the information from ꞙample 4-5 calculate the travel time, t.

ꞙlution

$$t = l/(v/n) = \frac{0.4 \text{ m}}{(2.5 \times 10^{-5} \text{ m/sec})/(0.5)} \times \frac{\text{min}}{60 \text{ s}} \frac{\text{h}}{60 \text{ min}} = 2.2 \text{ h}$$

Hence, the actual travel time is less than the time predicted using the Darcy velocity. The water particles migrate through the voids and not through the cross-sectional area represented by the Darcy velocity. The estimation of contaminant travel time through porous media uses the seepage velocity as described above.

In summary, Darcy's law is empirical, derived from a series of experiments. Further, it is a porous-media model on a macro-scale. Darcy's law describes flow in the subsurface in a macroscopic way and does not address the tortuous path in which an individual particle of water moves from void to void.

Hydraulic Head. In Darcy's experiment, flow is induced by a change in total hydraulic head. Hydraulic head is an **energy potential**. Examples of other potentials include voltage and temperature. For example, heat moves from a region of high temperature to a region of a lower temperature. The rate at which heat is transmitted is dependent upon the temperature difference, in other words, the temperature potential. A chemical potential, described in more detail later in this section, drives the migration of chemicals from areas of high concentration to areas of low concentration. The hydraulic head is a fluid potential, which drives the flow of fluid from areas of higher hydraulic head to areas of lower hydraulic head. This fluid potential is described in terms of the Bernoulli energy equation as shown in Fig. 4-10:

$$z_1 + p_1/\rho_w g + v_1^2/2g = z_2 + p_2/\rho_w g + v_2^2/2g = \text{constant} \qquad (4\text{-}15)$$

where z_1, z_2 = distance above datum of measuring locations 1 and 2 (m)
  v_1, v_1 = fluid velocity at points 1 and 2 (m/sec)
  p_1, p_2 = fluid pressures exerted by the water column at
    points 1 and 2 (Pa)
  g = acceleration of gravity (m/sec^2)
  ρ_w = mass density of water (kg/m^3)

The Bernoulli equation presented above defines the energy per unit weight of fluid for incompressible fluid and steady-flow (Units: ft·lb/lb or feet; N·m/N or meters). Observe that the terms of the total energy in Eq. 4-15 include contributions from velocity, pressure, and elevation. Further, the Bernoulli equation written in terms of energy per unit weight of fluid as in Eq. 4-15, results in energy expressed in units of length. The three terms, when expressed as energy per unit weight of fluid are

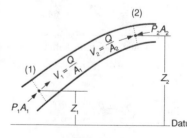

FIGURE 4-10
Bernoulli's equation for flow through a pipe.

ermed the **elevation head, pressure head**, and **velocity head**. Bernoulli's equation Eq. 4-15) can be rewritten for any point in the flow system as:

$$h_e + h_p + h_v = h_t \tag{4-16}$$

where h_e = elevation head (m)
h_p = pressure head (m)
h_v = velocity head (m)
h_t = total hydraulic head (m)

It has been found that for the range of velocity normally encountered in problems dealing with ground water flow, the velocity head is negligible when compared to he pressure and elevation heads. Hence, for ground water flow the problem can be studied in terms of pressure and elevation head components of the total hydraulic head as follows:

$$h_p + h_e = h_t \tag{4-17}$$

The elevation head is the distance above (or below) an arbitrary datum. The pressure head is the pressure exerted by the fluid per unit weight. The **gauge pressure**, p, is he pressure head multiplied by the unit weight of the fluid:

$$p = \rho_w g h_p \tag{4-18}$$

where p = gauge pressure (N/m^2)
ρ_w = fluid density (kg/m^3)
g = acceleration due to gravity (m/sec^2)
h_p = pressure head (L)
$\rho_w g$ = unit weight of the fluid (N/m^3)

Looking again at Darcy's experiment, we can solve for the elevation, pressure, nd total head using the dimensions shown in Fig. 4-11. As shown in the figure, ocations or measuring points placed throughout the flow system in Darcy's experiment ave been identified by the lower case letters a through g. At these locations we can xamine the head conditions in the experiment. First, an arbitrary **datum** (an elevation om which all vertical distances are measured) is assumed through the center line of

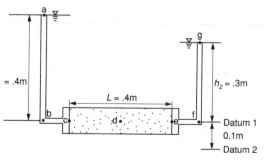

FIGURE 4-11
Hydraulic heads in Darcy's experiment.

the sample in the experiment. This is assigned the notation of Datum 1 to permit later examination of the effects of assuming the datum elsewhere in the system.

A table identifying selected locations in the flow field (measuring points a through g) is created as Table 4-6 as described in the following paragraphs.

In reference to Fig. 4-11, measuring point a is 0.40 m above Datum 1, yielding an elevation head of 0.40 m. Point a is open to the atmosphere so there is no fluid pressure; thus, the pressure head is zero (see Table 4-6). By adding the elevation head and the pressure head, the total hydraulic head at point a is determined to be 0.40 m. At measuring point b, located on datum 1, the elevation head is zero. However, there is a column of water 0.40 m high above measuring point b exerting a fluid pressure; thus the pressure head is 0.40 m. By adding the elevation and pressures head, the total hydraulic head is again found to be 0.40 m. Similarly, at measuring point c, on datum 1, the elevation head is zero. The pressure head (as a result of a column of water above measuring point c of 0.40 m) is 0.40 m. By addition, the total hydraulic head at measuring point c is 0.40 m. It is important to note that the total hydraulic head across the column from measuring points a to c is 0.40 m; there is no change in total hydraulic head. No head is lost in the portion of the column that does not contain soil (the open reservoir portion in the experiment shown in Fig. 4-11). In this experiment, and in ground water flow systems, total hydraulic head is lost as a result of friction on the water flowing through the porous media.

Continuing with the analysis of the downstream head conditions, measuring point g in Fig. 4-11 is examined as representative of the head in the downstream (outlet) portion of the experiment. The elevation of measuring point g is 0.30 m above the datum, that is, the elevation head is 0.30 m. Measuring point g is open to the atmosphere so there is no fluid pressure and the pressure head is zero. Summing, the total hydraulic head is equal to 0.30 m. Similarly, points f and e have a total hydraulic head equal to 0.30 m. On the outlet, or downstream, side of the experiment, the total hydraulic head is found to be constant and equal to 0.30 m.

Now let us determine hydraulic head conditions within the sand of the experiment. To calculate the head within the soil column, such as at measuring point d, it is first necessary to calculate the hydraulic gradient. In this case the change in hydraulic

TABLE 4-6
Values of hydraulic head from Darcy's experiment—Datum 1

Location	h_e	+	h_p	=	h_t
a	0.40		0.00		0.40
b	0.00		0.40		0.40
c	0.00		0.40		0.40
d	0.00		0.35		0.35
e	0.00		0.30		0.30
f	0.00		0.30		0.30
g	0.30		0.00		0.30

ead across the sample column is the head at measuring point c (h_c) minus the head
t measuring point e (h_e) or $h_c - h_e$ (0.40 − 0.30 equals 0.10 m). The linear length of
low, L, through which the head changes is equal to 0.40 m. The gradient, $(h_c - h_e)/L$,
s thus equal to 0.25. The hydraulic head at measuring point d is thus equal to the
ydraulic head at point c minus the head loss as water flows a distance of 0.20 m
at a head loss rate of 0.25 m/m) from measuring points c to d. Therefore the total
ydraulic head at point d is (0.4) − (0.25 × 0.2) = 0.35 m. Having found the total
ydraulic head at measuring point d to equal 0.35 m and knowing the elevation head,
e, to equal 0 m, it is possible to solve for the pressure head to equal 0.35 meters, as
hown in Table 4-6.

These calculations show that the energy lost due to flow is energy measured in
erms of total hydraulic head. The hydraulic gradient must be calculated in terms of
otal hydraulic head.

The hydraulic head values were also determined in Darcy's experiment using
atum 2 shown in Fig. 4-11. The results are presented on Table 4-7.

What can we learn about the nature of hydraulic head by these calculations
sing datum 1 and datum 2? First, note that the values of total hydraulic head for each
easurement in Table 4-7 are different from those in Table 4-6. The total hydraulic
ead is dependent upon the selection of the datum. However, the change in total
ydraulic head from point c to d to e is the same (independent of the datum) as is the
otal hydraulic head difference causing flow (the head difference between measuring
oints). Thus, although the values of total hydraulic head are dependent upon the
atum, the head difference is independent of datum and as a result, the gradient is
independent of datum. The elevation head is dependent upon the datum. Importantly,
e pressure head is independent of the datum.

wo-Dimensional Flow.   To illustrate the principles and terms associated with
vo-dimensional flow, Fig. 4-11 is redrawn with the dimensions shown in Fig. 4-12,
d measuring points r through z have been labeled. Using the Bernoulli equation
d the concepts of one-dimensional flow previously described, the elevation head,

\BLE 4-7

**ydraulic heads from Darcy's experi-
ent—Datum 2**

Location	h_e	+	h_p	=	h_t
a	0.50		0.00		0.50
b	0.10		0.40		0.50
c	0.10		0.40		0.50
d	0.10		0.35		0.45
e	0.10		0.30		0.40
f	0.10		0.30		0.40
g	0.40		0.00		0.40

pressure head, and total hydraulic head have been determined and are shown on Table 4-8.

Measuring points selected along cross sections are labeled r-s-t, u-v-w, x-y-z. On examination of the data generated in Table 4-8, it is found that the measuring points on each of the cross sections, r-s-t, u-v-w, and x-y-z, have a constant total hydraulic head (e.g., the total hydraulic head is equal to 0.30 at x-y-z, 0.375 at u-v-w). Contouring the values of equal total hydraulic head in Fig. 4-12 results in Fig. 4-13 showing the contours of equal total hydraulic head as dotted lines.

These lines of equal total hydraulic head are known as **equipotentials**, that is, the potential energy at any point on one of these lines is the same. **Flow lines**, by definition, are perpendicular to the equipotentials and are shown in Fig. 4-13. The Laplace equation will be derived later in this chapter as the solution for flow in isotropic, homogeneous porous media.

Although we can draw flow lines and equipotentials for the Darcy experiment, it should be noted that all flow is along the axis of the tube of porous media. In this case, all flow is in the x-direction with the flow in the y-direction is equal to zero. The more general case of two-dimensional flow is illustrated by an example of flow through a dike surrounding a liquid impoundment shown in Fig. 4-14.

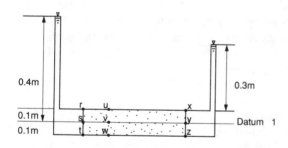

FIGURE 4-12
Darcy's experiment revisited.

TABLE 4-8
Hydraulic heads from Darcy's experiment—Datum 1

Location	h_e	+	h_p	=	h_t
r	0.10		0.30		0.40
s	0.00		0.40		0.40
t	−0.10		0.50		0.40
u	0.10		0.275		0.375
v	0.00		0.375		0.375
w	−0.10		0.475		0.375
x	0.10		0.20		0.30
y	0.00		0.30		0.30
z	−0.10		0.40		0.30

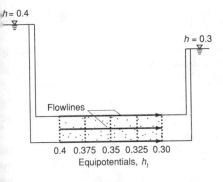

FIGURE 4-13
Flow lines and equipotentials.

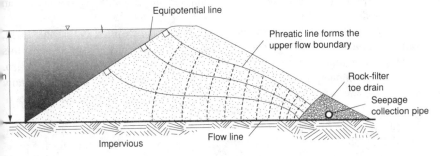

FIGURE 4-14
Flow net for steady-state flow through a homogeneous embankment (after Cernica, 1982).[18]

Hydraulic Conductivity of Geologic Materials

The hydraulic conductivity, k, in Eq. 4-3 is a measure of the capability of a material to transmit water. It is an "order of magnitude" parameter. The most impermeable unconsolidated material (unconsolidated material is synonymous with soil in civil engineering terminology) may have a hydraulic conductivity of 1×10^{-9} cm/sec. At the other extreme, open-worked clean gravel may have a hydraulic conductivity of 1×10^5 cm/sec. Stated another way, the gravel is 100,000,000,000,000 times more permeable than clay. Presented on Table 4-9 are the range of expected values for the hydraulic conductivity of various geologic materials. The values provide a starting point for the assessment of ground water flow. There are two facts to remember about the values in this table.

First, under the best investigatory conditions, hydraulic conductivity can be estimated only within order of magnitude accuracy. The ramifications of being an order of magnitude off in a ground water flow calculation must be considered. For example, if the flow to a ground water extraction well used in site remediation (see Sec. 16-5) is calculated to be 10 gal/min in a clean sand estimated to have a permeability of 1×10^{-2} cm/sec, the flow could be 100 gal/min should the actual permeability of the formation be 1×10^{-1} cm/sec. Could the remedial ground water pumping and treatment system accommodate this variation? If not, extreme care and adequate testing must

TABLE 4-9
Typical ranges of hydraulic conductivity for various soils

Soil type	Hydraulic conductivity (cm/sec)
Clean gravel	1×10^5 to 1.0
Clean sand or sand + gravel mixtures	1.0 to 1×10^{-3}
Fine sands and silts	1×10^{-2} to 1×10^{-6}
Silty clay and clay	1×10^{-5} to 1×10^{-9}

be employed to reduce the risk of overestimating or underestimating the hydraulic conductivity (see Sec. 15-6).

The second aspect of the values in Table 4-9 to bear in mind is that the value of hydraulic conductivity for a given material type may be quite different than that for the entire formation. For example, consider that a sample of glacial clay may have a hydraulic conductivity of 1×10^{-7} cm/sec. However should the clay contain interconnected seams of sand, the formation hydraulic conductivity may be several orders of magnitude higher, even if the sand seams constitute only a minute percentage of the overall deposit. Similarly, should the clay be highly fractured, as glacial tills often are, the secondary structure (i.e., the fractures) would allow for considerably greater flow than the portion of the clay that is not fractured. Given the heterogeneity and anisotropy that exist in the natural subsurface environment, it is little wonder that the accurate prediction of ground water flow and contaminant transport in the subsurface environment requires considerable investigation, analysis, skill, experience, and luck.

Darcy's law (Eq. 4-2) shows that the flow is proportional to both the hydraulic conductivity of the medium as well as the cross-sectional area of flow. It is convenient (for both mathematical modeling and analysis of field data) to define the **transmissivity** of the formation as the hydraulic conductivity times the thickness of the formation. In equation form, we have

$$T = kt \tag{4-19}$$

where T = transmissivity (cm^2/sec)
k = hydraulic conductivity (cm/sec)
t = aquifer thickness (cm)

Thus, transmissivity is defined as the rate of water flow through a vertical section of an aquifer, (over its full saturated thickness, one unit wide) under a unit hydraulic gradient.

To better understand the interaction between hydraulic head and flow from an aquifer it is necessary to describe the storage capacity of the aquifer. As water is pumped from an unconfined aquifer and the water table drops, the saturated portion of the aquifer becomes unsaturated. Unsaturated does not imply "dry." An unsaturated material is one in which the void space is not completely filled with water. The volume of water that drains from the saturated soil pores as the water table drops is termed the **specific yield.** For unconfined conditions as just described the specific yield is synonymous with the **storativity.**

The storativity for a confined aquifer situation is more complex than for the unconfined case described above. As water is pumped from a confined aquifer, the aquifer remains saturated. However, as the hydraulic head decreases, the **effective stress** increases causing a small but real compression in the aquifer. In addition, at the lower water pressure level there is a small, but real, expansion in the water (although we often assume water is incompressible, a given mass of water will expand or contract with changes in pressure). The combination of these two phenomena is responsible for the storativity of confined aquifers.

Coupled Flow Phenomena.   In preceding sections, flow induced by hydraulic gradients has been described. In some cases, ground water may flow in response to gradients other than, or in addition to, hydraulic gradients. These other gradients include voltage, chemical concentration, and temperature gradients. Flow induced by more than one gradient is termed coupled flow. A general flow law would include a specific discharge which includes flow in response to all applicable gradients.[19] The importance of coupled flow processes with respect to hazardous waste management is illustrated by the proposed use of electrokinetics for the cleanup of contaminated ground water. [20] The derivation and a complete discussion of coupled flow phenomena are beyond the scope of this text, although the description and use of electrokinetics for ground water remediation are subsequently detailed.

Flow in the Unsaturated Zone

The foregoing discussions describe flow in the zone below the water table. How do the processes differ in the unsaturated zone? First, a few terms need to be defined:

Water table A surface on which the fluid pressure (pressure head) is exactly atmospheric.

Phreatic zone The zone located below the water table where the fluid pressure is positive (in compression).

Vadose zone The zone located above the water table where the fluid pressure is negative (in tension). The zone is divided into a saturated portion, due to capillary rise, and an unsaturated or partially saturated zone.

In the vadose zone, the pressure head is termed the **suction head**, ψ, and is negative. The total hydraulic head remains the sum of the elevation head and the pressure head. As noted above, there is some water in the unsaturated zone, and the pressure head is dependent upon this soil moisture content, θ. Further, the relationship between θ and ψ is hysteretic, that is, it is different depending upon whether the

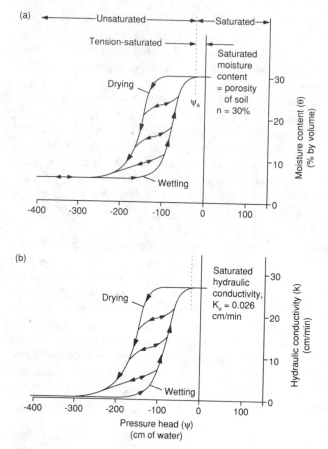

FIGURE 4-15
Characteristic curves relating hydraulic conductivity and moisture content to pressure head for a naturally occurring sand soil (after Liakopoulos).[21]

soil is being wetted or dried. The hydraulic conductivity in the unsaturated zone is similarly affected. A typical relationship is shown in Fig. 4-15. As a result of the changing hydraulic conductivity and pressure head with changing water content, the flow models are considerably more complex for the unsaturated zone.

Contaminant Transport Mechanisms

Contaminants are found in the subsurface environment in various forms (or phases). Contaminants that are dissolved in water are the **solutes**, and water is the **solvent**. Water having dissolved contaminants (solutes) flowing in a porous media is schematically shown in Fig. 4-16. As the water flows, the dissolved contaminants are transported with the water. This method of contaminant transport is termed **advection**. Advection is the transport of solutes along stream lines at the average linear seepage flow

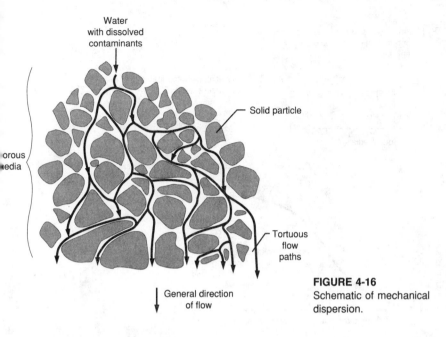

Water
with dissolved
contaminants

Solid particle

orous
media

Tortuous
flow
paths

General direction
of flow

FIGURE 4-16
Schematic of mechanical
dispersion.

velocity. On a macroscopic scale, the porous media model governs the flow rate and direction. However, on a microscopic scale, the porous media is composed of discrete solid particles and pore spaces or voids. Water flows, not through the solid particles but around them through the interconnected pore space. These concepts are shown schematically in Fig. 4-16.

Flowing water, upon encountering the solid particles of the porous media, must alter its course by flowing around the solids, either to the right or left of the solid or by spreading and migrating to both sides. This process is repeated millions of times by millions of water "particles." The result is a mixing of the flowing water by mechanical means termed **mechanical dispersion** (also known as **hydrodynamic dispersion** or just **dispersion**). The most important effect of dispersion is to spread the contaminant mass beyond the region it would occupy without dispersion.

The water in Fig. 4-16 can be assumed to enter the porous medium at a discrete location with known concentration of contaminants, C_o. This mass of water and dissolved contaminants is termed a **slug**, and the discrete location is termed a **point source**. The contaminant concentration will decrease in the slug as it migrates through the formation as a result of the mechanical dispersion spreading the mass of contaminants over a larger and larger volume and mixing with water without the contaminants. The result is a **dilution** or reduction in contaminant concentration. Thus, contaminants are transported by primarily advection, and their concentration changes as a result of dispersion.

To investigate further the transport of contaminants, a laboratory experiment similar to the Darcy experiment could be setup as shown in Fig. 4-17. In this con-

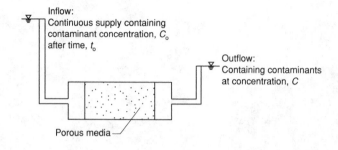

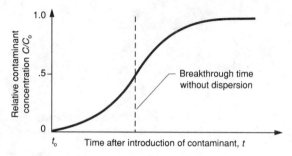

FIGURE 4-17
Effect of dispersion on contaminant transport.

figuration the contaminant is introduced, not as a one-time slug described above, but continuously and at the same concentration. This is termed a **continuous source** and is representative of releases such as from an underground storage tank with a small leak which goes undetected. The concentration of the contaminant can be measured as a function of time at the downgradient end of the column experiment. Without mechanical mixing (dispersion), the water would be free of contaminants until the contaminant **front** reaches the outflow end of the sample. The time required would be determined by the average linear seepage velocity previously defined. However, mechanical dispersion does occur and as a result, contaminant arrival at the outflow (termed **breakthrough**) occurs more rapidly than the seepage velocity would indicate. However, at the time the contaminant is first detected, the concentration is much lower than the input concentration. The concentration at the outflow, C, is presented in Fig. 4-17 as a ratio to the input concentration as C/C_o. The contaminants arrive at the outflow faster than indicated by the average linear seepage velocity. The time required for the outflow concentration to equal the inflow concentration is much longer than predicted directly from the average linear seepage velocity.

The concept of dispersion can be further explained by examining the nature of the contaminant distribution with time in a homogeneous and isotropic formation. The distribution and extent of contaminants migrating in the subsurface is termed the **plume**. Shown in Fig. 4-18 is the plan view of a plume at various times. This figure then depicts the effects of dispersion upon contaminant transport for two separate

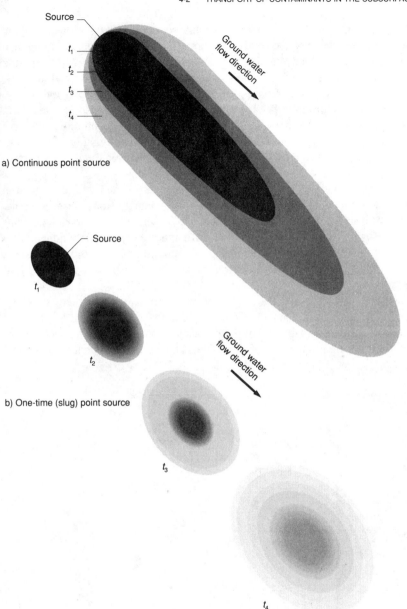

FIGURE 4-18
Plume migration affected by dispersion and source type.

cases. In case (a) the source is a continuous point source loading such as a leaking underground storage tank described earlier. At time t_1, the contaminants have migrated downgradient and have spread laterally due to mechanical dispersion. With increasing times, the plume has migrated farther downgradient and continues to spread laterally. Alternatively, a plume resulting from a one-time event (a slug) such as a spill is also shown in Fig. 4-18. In this case the slug migrates downgradient spreading laterally to encompass a larger and larger area, albeit at lower and lower concentrations. Observe that dispersion spreads the plume both laterally and longitudinally, although the longitudinal dispersion is masked by the advective transport. The longitudinal dispersion is also illustrated by Fig. 4-17.

Contaminants (particularly ionic and molecular constituents) will also move in response to their chemical kinetic activity. The movement is from areas of higher concentration to areas of lower concentration and is termed diffusion. The terms **molecular diffusion, ionic diffusion,** or **self-diffusion** are often used synonymously with diffusion. The difference in contaminant concentration taken with respect to the distance over which the concentration difference occurs is referred to as the **concentration gradient** (see Eq. 3-10).

Shown in Fig. 4-19 is the Darcy laboratory experiment where there is no hydraulic head difference from one end of the sample to the other. At time t_o, there is a concentration C_o at the left hand end of the porous media column and a concentration equal to zero at the right hand end of the column, thus creating a concentration dif-

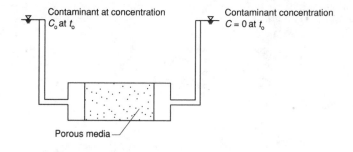

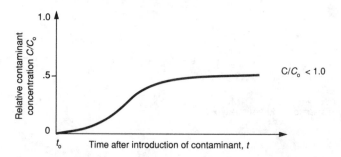

FIGURE 4-19
Effect of diffusion on contaminant transport with no advective transport.

ference from one end of the sample to the other. Because there is no hydraulic head difference, the hydraulic gradient, Darcy flux, and advective transport are all equal to zero. Yet, as shown in Fig. 4-19, contaminants migrate through the sample in response to the chemical concentration gradient. The final concentration ratio at the right-hand side of the column is less than 1.0. The concentration ratio depends upon the initial concentration and the total volume of the sand and water in the laboratory experiment. If contaminants were added to the left-hand side of the sample during the diffusion to maintain the initial concentration, C_o, then the equilibrium concentration ratio would be 1.0 on the right.

It can be generally stated that advective transport and the associated mechanical dispersion processes dominate the contaminant transport system in formations of medium-to-high hydraulic conductivity. In formations of low hydraulic conductivity, including clay liners in waste management facilities, diffusive transport is frequently the controlling mechanism.

Non-Idealized (Real-World) Contaminant Transport

The basic concepts of contaminant transport in the subsurface environment have been described. However, there are many mechanical, chemical, and biological processes by which the contaminant transport differs from that predicted employing the concepts and assumptions described above. Some processes will retard the rate of contaminant transport whereas others can speed up the rate. Some natural processes affecting contaminant transport are summarized in Table 4-10. The chemical and biological processes are discussed in more detail in Sec. 4-3, Fate of Contaminants in the Subsurface.

TABLE 4-10
Some natural processes affecting contaminant transport

Process type	Process
Mechanical (Physical) Processes	Advection
	Dispersion
	Diffusion
	Density stratification
	Non-aqueous phase fluid flow
	Fractured media flow
Chemical Processes	Oxidation-reduction reactions
	Ion exchange
	Complexation
	Precipitation
	Immiscible phase partitioning
	Sorption
Biological Processes	Aerobic degradation
	Anaerobic degradation
	Cometabolism
	Biological uptake

There are also some mechanical processes and special conditions that depart from idealized flow including fractured media, heterogeneity, and non-aqueous phase liquid flow.

Fractured Media Flow. Up to this point, the porous media model has been used as the basis for our understanding of ground water flow and contaminant transport. Indeed, many sites can be adequately characterized by a porous media model. However, there are also many sites where the flow is better characterized as fractured flow. For example, Fig. 4-20 illustrates the subsurface stratigraphy at a site showing the bedding planes and vertical fractures in the underlying limestone bedrock.

Flow and associated contaminant transport will occur predominantly along the joints and fractures and not in the bulk rock. In essence the flow is more like pipe flow than porous media flow. Efforts are currently underway to develop readily usable techniques to enable predictions of ground water flow and contaminant transport under fractured flow conditions although they have not as yet been widely employed in the hazardous waste management field. Present approaches to describe flow and transport in fractured rock systems are:

1. Fractured rock system as an equivalent porous media (EPM) or
2. Discrete fracture approach where channels are represented as channels with parallel sides and individual fractures are combined into fracture networks

Heterogeneity. Heterogeneity in porous media also alters the idealized transport processes described above. As defined previously, heterogeneity is a spatial variation in hydraulic and transport properties originating from the nature of the geologic properties. Figure 4-21 illustrates the effect of a lens of higher hydraulic conductivity within an otherwise homogeneous and isotropic formation. The Darcy flux is greater in the zone of higher hydraulic conductivity than in the adjacent zones under the same gradient. Correspondingly, the linear seepage velocity and the rate of advective transport are greater in the more permeable lens.

Nonaqueous Phase Liquid (NAPL). The problems of contaminant transport predictions are further complicated by the presence of **nonaqueous phase liquid** (NAPL). Many organic liquid compounds are characterized as **immiscible**, that is, they have a very low **solubility** in water. As a result these nearly insoluble organic compounds frequently exist as a separate liquid phase in the subsurface. This separate liquid phase typically has a different density and viscosity than water. A NAPL with a density greater than water is termed a sinker (dense nonaqueous phase liquid [DNAPL]) (e.g., chlorinated solvents) and a NAPL with a density lighter than water is termed a floater (light nonaqueous phase liquid [LNAPL]) (e.g., petroleum products). The transport of NAPL is influenced by gravitational forces as well as differences in hydraulic head. As with fractured flow we are only beginning to understand transport of NAPL,[22] and analytical techniques have not been widely employed. The migration of LNAPL is largely dependent upon the slope of the water table. The migration of DNAPL is strongly dependent upon the subsurface stratigraphy, particularly the distribution of

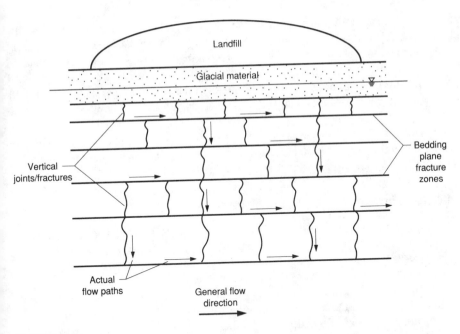

FIGURE 4-20
Schematic of fractured flow.

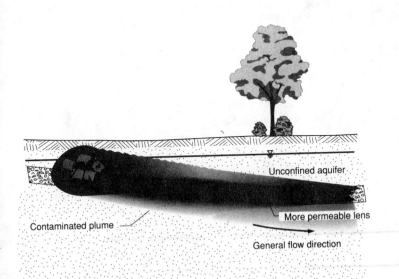

FIGURE 4-21
Effect of high-permeability zone on contaminant transport.

zones of low permeability that act as barriers. Typically, the volume of a ground water plume contaminated with dissolved NAPL constituents is many times greater than a NAPL plume although the greater mass of contaminants may be in the NAPL plume. Fig. 4-22(a) illustrates a schematic of the distribution of DNAPL from a leaking underground storage tank near the surface. Although the hydraulic gradient is from left to right, the DNAPL has predominantly migrated downward and down slope (even if hydraulically upgradient).

Figure 4-22(b) illustrates a schematic of the distribution of LNAPL from a leaking underground storage tank near the surface. As shown, the LNAPL floats on the ground water and spreads laterally as well as being transported downgradient.

Selected Flow and Transport Equations

In this section of the chapter, we will look at the equations which govern ground water flow and contaminant transport. Derivations, where they enhance our understanding of the processes, will also be presented using simplifying assumptions, and the solutions are extended to more complex boundary conditions.

Figure 4-23 illustrates a two-dimensional control volume. The Darcy flux into the control volume in the x and y directions are v_x and v_y, respectively. The dimensions of the element are dx and dy. Assuming no change in the control volume, the **equation of continuity** (what goes in, must come out) tells us:

$$q_{in} = q_{out} \tag{4-20}$$

where q_{in} = flow rate into control volume
q_{out} = flow rate out of control volume

The volumetric flow rate is equal to the Darcy flux times the area of flow as:

$$q_{in} = V_{in} A \tag{4-21}$$

where q_{in} = volumetric flow rate into control volume (cm^3/sec)
V_{in} = flux into control volume (cm/sec)
A = area of flow (cm^2)

Thus, the flow into the control volume is:

$$q_{in} = v_x dy + v_y dx \tag{4-22}$$

and the flow out of the control volume is:

$$q_{out} = \left(v_x + \frac{\partial v_x}{\partial x} dx \right) dy + \left(v_y + \frac{\partial v_y}{\partial y} dy \right) dx \tag{4-23}$$

where $\partial v_x / \partial x$ = the rate of change in the Darcy flux in the x direction
$\partial v_y / \partial y$ = the rate of change in the Darcy flux in the y direction

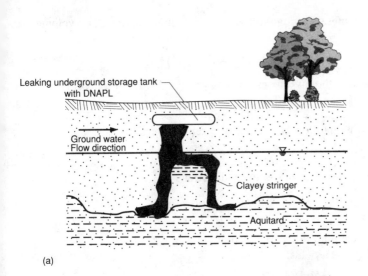

(a)

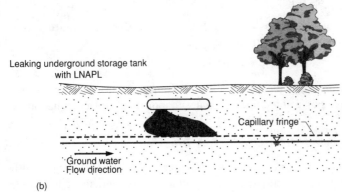

(b)

FIGURE 4-22(a) and (b)
Movement of NAPL.

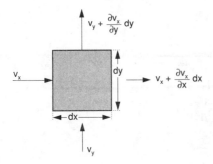

FIGURE 4-23
Two-dimensional control volume.

With no net change in our control volume, the inflow and outflow volumes are equal resulting in:

$$0 = (\partial v_x/\partial x)dxdy + (\partial v_y/\partial y)dydx \tag{4-24}$$

For a square element with unit dimensions $(dx = dy = 1)$, Eq. 4-23 reduces to:

$$0 = \partial v_x/\partial x + \partial v_y/\partial y \tag{4-25}$$

Darcy's law can be written for the flux in the x and y directions as:

$$v_x = k_x \partial h/\partial x \tag{4-26}$$

where k_x = hydraulic conductivity in the x direction

and

$$v_y = k_y \partial h/\partial y \tag{4-27}$$

k_y = hydraulic conductivity in the y direction
Substituting into Eq. 4-25 yields:

$$0 = k_x \partial^2 h/\partial x^2 + k_y \partial^2 h/\partial y^2 \tag{4-28}$$

Finally, for isotropic flow conditions $k_x = k_y$, Eq. 4-28, the **Laplace equation**, for a two-dimensional flow field can be written as:

$$\partial^2 h/\partial x^2 + \partial^2 h/\partial y^2 = 0 \tag{4-29}$$

The Laplace equation describes the value of hydraulic head at any point in the two-dimensional flow field. For a three-dimensional situation the Laplace equation becomes:

$$\partial^2 h/\partial x^2 + \partial^2 h/\partial y^2 + \partial^2 h/\partial z^2 = 0 \tag{4-30}$$

Note that the value of hydraulic head is a function only of position in the flow field. The value of hydraulic head is not a function of the hydraulic conductivity under isotropic flow conditions. Implicit in the derivation of the Laplace equation is the assumption that the flow is **steady** (i.e., not changing with time). The equations of ground water flow are considerably more complex for **unsteady** or **transient** flow conditions. The Laplace equation can be rewritten in three dimensions for unsteady flow in a saturated confined aquifer as follows:

$$\partial^2 h/\partial x^2 + \partial^2 h/\partial y^2 + \partial^2 h/\partial z^2 = (\rho g(\alpha + n\beta)/k)(\partial h/\partial t) \tag{4-31}$$

where x, y, z = positions in the flow system
h = hydraulic head (cm)
ρ = density of water (g/cm^3)
g = acceleration of gravity (cm/sec^2)
α = aquifer compressibility (cm^2/N)
n = aquifer porosity (dimensionless)
β = compressibility of water (cm^2/N)
k = hydraulic conductivity (cm/sec)
t = time (sec)

An examination of the transient form of the Laplace equation reveals that the head is a function of four variables (x, y, z, and t) as well as six constants (ρ, g, n, k, a, β)

The Laplace equation for horizontal unsteady flow in a saturated unconfined aquifer is:

$$\partial^2 h/\partial x^2 + \partial^2 h/\partial y^2 + \partial^2 h/\partial z^2 = (S/T)(\partial h/\partial t) \qquad (4\text{-}32)$$

where x, y, z = spatial positions in the flow system
h = hydraulic head (cm)
S = aquifer storativity (g/sec)
T = aquifer transmissivity (cm^2/sec)
t = time (sec)

Solutions to the Laplace Equation.   The Laplace equation describes the steady-state hydraulic head conditions throughout the ground water flow regime. The solution of this partial differential equation can be by analytical, graphical, electrical analog, or numerical methods. Analytical methods are possible for only relatively simple boundary conditions. Graphical methods (flow nets) provide good, relatively rapid approximations to the flow conditions and are useful for first approximations or "back-of-the-envelope" checks of the results from more sophisticated numerical methods. Electrical analog methods are time-consuming and expensive.

Most complex site and subsurface conditions must depend upon the use of a numerical method that employs a system of equations solved by an iteration method. The solution method yields, by successive approximation, an approximate solution to the system of mathematical equations. The two commonly employed numerical approximation methods are finite element and finite difference methods. In the finite element approach, the partial differential equations are approximated with integral representations. The finite difference method replaces the continuous flow field with discrete units and describes the governing partial differential equations by methods used in differential calculus.

The graphical method (flow nets) and the finite difference method are described below. For additional information regarding these and other numerical methods as applied to ground water modeling, the reader is referred to the References section.[23,24,25,26]

Flow Nets.   Flow nets are two-dimensional graphical representations of the hydraulic head conditions in the subsurface. A flow net is composed of equipotentials (lines of equal hydraulic head) and flow lines (lines of flow normal to equipotentials representing the average path a particle of water takes as it flows in the subsurface). The flow net is a solution to the two-dimensional steady state flow conditions under the specified boundary conditions. Why draw a flow net in this age of computer solutions? For the inexperienced, construction of a flow net develops a sense or appreciation of the nature of ground water flow and, for the experienced, a flow net provides an independent validation of results obtained from more sophisticated methods.

Perhaps the easiest example to visualize is the flow through a homogeneous, isotropic embankment on an impervious foundation surrounding a liquid impoundment as shown earlier in Fig. 4-14. The figure illustrates several principles regarding flow net construction. First, the flow lines are normal to the equipotentials resulting in "squares" bounded by equipotentials and flow lines. Also, the impervious boundary is a flow line and the free water boundary is an equipotential. A flow channel is that flow path bounded by two adjacent flow lines. The fundamental properties of flow nets can be summarized as follows:[27]

- The head difference between any pair of adjacent equipotentials is the same as any other pair.
- Flow lines intersect equipotentials at right angles.
- Figures enclosed by adjacent pairs of equipotentials and flow lines are essentially squares.
- The spacing of equipotentials is inversely proportional to the hydraulic gradient (and thus the Darcy velocity).
- Every flow channel transmits the same quantity of seepage.

It is possible to determine the hydraulic gradient and the quantity of seepage from a well-constructed flow net. The seepage quantity under homogeneous, isotropic flow conditions can be shown to be:[28]

$$q = khn_f/n_d \tag{4-33}$$

where
q = flow rate per unit width (cm^3/sec $\cdot$ cm)
k = hydraulic conductivity (cm^2sec)
h = hydraulic head (cm)
n_f = number of flow channels
n_d = number of equipotential drops

The hydraulic gradient between two equipotentials can be determined as:

$$i = (h/n_d)/l \tag{4-34}$$

where
i = hydraulic gradient
h = hydraulic head (m)
n_d = number of equipotential drops
l = distance along flow path between the two equipotentials under consideration (m)

In summary, flow nets can provide a relatively rapid means to examine accurately the hydraulic head and flow conditions in the subsurface. Flow nets can be constructed for two-dimensional flow through a cross section or in plan view. The following example illustrates a flow net constructed in plan and demonstrates the calculations for seepage quantity and gradient.

Example 4-7   Flow net for a water table aquifer.   The following figure is a flow net constructed in plan view for a water table aquifer discharging to a stream (looking down on the stream valley from above).[29] Calculate the subsurface flow per meter of stream using the entire flow net shown in the figure and assuming $k = 1 \times 10^{-2}$ cm/sec.

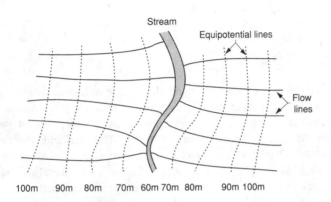

Solution. $q = khn_f/n_d$
By counting from the figure we find $n_f = n_d = 4$ substituting and converting cm to m we have:
$k = 1 \times 10^{-2}$ cm/sec
$h = 100$ m $- 60$ m $= 40$ m
$n_f = 4$
$n_d = 4$ (100 m to 90 m = 1 drop) 100 m to 60 m = 4 drops
$q = 1 \times 10^{-2}$ cm/sec $\times 40$ m $\times 4/4 \times 1$ m/100 cm $= 4 \times 10^{-3}$ m^3/sec $\cdot$ m
The units may appear confusing, but in fact represent the volumetric flow rate per meter of stream length.

Finite Difference Method.   Figure 4-24 illustrates a numerical grid representing a two-dimensional flow regime. The abscissa and ordinate axes of the finite difference grid use i, j notation. Each nodal point is modeled to represent the average head for the area enclosed by the dotted square.

The finite difference method further assumes that the head at a particular node, $h_{i,j}$, is the average of the head at the four surrounding nodes. For an interior node, the

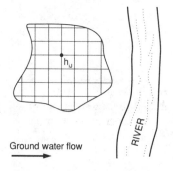

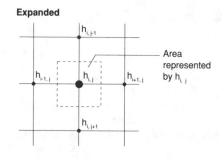

Expanded

Area represented by $h_{i,j}$

Ground water flow

FIGURE 4-24
Finite difference.

head at any point is:

$$h_{i,j} = (h_{i,j-1} + h_{i+1,j} + h_{i,j+1} + h_{i-1,j})/4 \qquad (4\text{-}35)$$

The finite difference numerical technique first requires establishing the boundary conditions, either as **constant head** boundaries or as **no-flow** boundaries. Numerical values for head are then assigned. Next, the heads are recalculated systematically using the finite difference equation for each node. This trial and adjustment method is the **relaxation technique.** The head at each node is then recalculated throughout the system; each pass through the system is known as an **iteration.** The iterations are repeated until the change in head from one iteration to another is less than the **tolerance** that is selected by the ground water modeler and may be on the order of 2 cm for typical hazardous waste sites. The technique is illustrated for the Darcy experiment in Ex. 4-8.

Example 4-8   Finite difference solution for Darcy's experiment. For the boundary conditions in Fig. 4-11 use the finite difference technique to determine the hydraulic heads in the flow regime.

Solution.

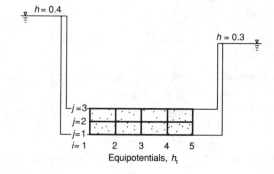

Equipotentials, h_t

Initial head matrix in terms of variable name, $h_{i,j}$

$$
\begin{matrix}
h_{1,3} & h_{2,3} & h_{3,3} & h_{4,3} & h_{5,3} \\
h_{1,2} & h_{2,2} & h_{3,2} & h_{4,2} & h_{5,2} \\
h_{1,1} & h_{2,1} & h_{3,1} & h_{4,1} & h_{5,1}
\end{matrix}
$$

To begin the process, enter the values of head into the initial head matrix under the known boundary conditions and estimate the head at intermediate points. In our example we know that h = 0.4 at the upstream boundary and h = 0.3 at the downstream boundary. Thus, we know that the head is < 0.4 and > 0.3 in the remainder of the flow regime. Therefore, we may make the initial estimate of the head matrix as:

$$
\begin{matrix}
j = 3 & 0.4 & 0.35 & 0.35 & 0.35 & 0.3 \\
j = 2 & 0.4 & 0.35 & 0.35 & 0.35 & 0.3 \\
j = 1 & 0.4 & 0.35 & 0.35 & 0.35 & 0.3 \\
 & i = 1 & i = 2 & i = 3 & i = 4 & i = 5
\end{matrix}
$$

The next step is to iterate the matrix by systematically recalculating the heads using equation 4-35: $h_{i,j} = (h_{i,j-1} + h_{i+1,j} + h_{i,j+1} + h_{i-1,j})/4$

Beginning at $h_{2,1}$ we have

$$h_{2,1} = (h_{1,1} + h_{3,1} + 2h_{2,2})/4$$

$$h_{2,1} = (0.5 + 0.35 + 2[0.35])/4 = 0.3875$$

and

$$h_{2,2} = (h_{1,2} + h_{3,2} + h_{2,3} + h_{2,1})/4$$

$$= (0.4 + 0.35 + 0.35 + 0.3875/4 = 0.3719$$

and

$$h_{2,3} = (h_{1,3} + h_{3,3} + 2(h_{2,2})/4$$

$$= (0.4 + 0.35 + 2[0.3719]/4 = 0.3735$$

Continuing through the matrix in this fashion, the values for head at the end of the first iteration are:

$$
\begin{matrix}
j = 3 & 0.4 & 0.3735 & 0.3598 & 0.3334 & 0.3 \\
j = 2 & 0.4 & 0.3719 & 0.3578 & 0.3369 & 0.3 \\
j = 1 & 0.4 & 0.3875 & 0.3594 & 0.3399 & 0.3 \\
 & i = 1 & i = 2 & i = 3 & i = 4 & i = 5
\end{matrix}
$$

The iterations continue (normally using a computer, spreadsheets work very well) until the change in head between successive iterations is less than a specified value; in this case a value of 0.001m would be reasonable. Thus, the final values of head are:

$$
\begin{matrix}
j = 3 & 0.4 & 0.375 & 0.35 & 0.325 & 0.3 \\
j = 2 & 0.4 & 0.375 & 0.35 & 0.325 & 0.3 \\
j = 1 & 0.4 & 0.375 & 0.35 & 0.325 & 0.3 \\
 & i = 1 & i = 2 & i = 3 & i = 4 & i = 5
\end{matrix}
$$

Ground Water Modeling. The attractiveness of ground water modeling is in its use in evaluating alternatives. As described in Chap. 16, the remediation of contaminated hazardous waste sites frequently means the remediation of contaminated ground water, and there are numerous alternatives to ground water remediation including pump and treat, in situ bioremediation, and containment. In selecting remedial measures, questions arise regarding the performance of potentially suitable alternatives. How is the ground water flow and contaminant transport altered by the selected remedial alternative? Which alternative provides the best control of ground water flow and contaminant transport at a particular site? If a pumping system is proposed, is it more effective to use 10 wells pumping at 50 gpm or 20 wells pumping at 25 gpm? What should the well spacing be? These and many other questions need to be answered in the evaluation of remedial alternatives. Ground water modeling can help the decision maker by providing insight into the relative impacts of various alternatives on ground water.

What is ground water modeling? A model is a replica of some real-world object or system. A ground water model must therefore simulate the actual ground water conditions. This simulation is generally macro scale over a large area. This simulation is often a graphical representation allowing a visualization of the actual conditions. A simulation may also be a complex digital three-dimensional ground water model solvable only through the application of numerical methods. Most applications of ground water modeling use a mathematical model. A mathematical model tries to simulate the actual behavior of a system through the solution of mathematical equations. Specifically, a numerical ground water model solves for the hydraulic head conditions throughout the flow regime (governed by the Laplace equation) using numerical techniques.

How is a numerical ground water model employed for hazardous waste management applications? A complete treatise is beyond the scope of this chapter but general guidance is provided. Upon the selection of a numerical ground water model, the required data regarding subsurface conditions and physical properties are **digitized**. In this step physical parameters, such as hydraulic conductivity, porosity, aquifer thickness, and extent, are converted into numerical inputs. For example, the location and elevation of the ground water are input as x, y, and z coordinates. The next phase of ground water modeling is termed **calibration**. Calibration of the numerical model is the process of adjusting model inputs until the resultant predictions give a reasonably close fit to the observed data. For example, water level elevations may be known at selected locations in the subsurface flow regime. The known observed water levels are compared with the simulated water levels predicted by the numerical model. If the observed and predicted conditions compare reasonably well, the model is said to be calibrated. If not, the modeler must examine the material properties and boundary conditions and make adjustments accordingly. A simplified flow diagram representing the general steps in developing a ground water and contaminant transport model is shown in Fig. 4-25.

Having calibrated the model, it can be used in hazardous waste management as follows:

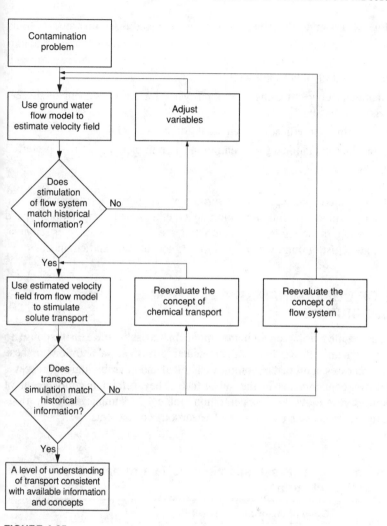

FIGURE 4-25
Simplified flow diagram representing general process of developing a transport model.

- As a tool in understanding ground water flow and contaminant transport processes at a site,
- As a tool for evaluating the adequacy of the available field and laboratory data and for identifying deficiencies in the data base,
- As a tool for estimating flow and transport parameters,
- In the prediction of future patterns of ground water flow and contamination migration, and
- For assessment of ground water flow and contaminant migration patterns in response to proposed remedial actions.

Ground water modeling for remedial investigations and feasibility studies could be used to:

- Guide the placement of monitoring wells,
- Predict contaminant concentrations in ground water for an assessment of present and future risks at the site,
- Assess the feasibility and efficacy of remedial alternatives, and
- Predict the residual contaminant concentration after implementation of preferred remedial action.

Ground water modeling represents a formidable challenge for data collection, parameter estimation, model formulation, numerical methods, and computational speed and power. As a result, modeling is an art combining the skills of a computer scientist, mathematician, geologist, ground water hydrologist, geochemist, and engineer.

4-3 FATE OF CONTAMINANTS IN THE SUBSURFACE

Section 4.2 has detailed how contaminants move in the subsurface in response to advection, hydrodynamic dispersion, and molecular diffusion. In addition to these three transport processes, a myriad of complex chemical and microbiological processes affects the **fate** of contaminants in the subsurface. They **retard** the movement of contaminants if not **attenuate** their concentration. Table 4-11 summarizes the various effects of subsurface processes on the fate of hazardous constituents.

TABLE 4-11
Summary of natural processes affecting the fate of hazardous constituents in the subsurface

Process	Class of chemical	Effect
Sorption	Organic	Retardation
Precipitation	Inorganic	Retardation
Ion exchange	Inorganic	Retardation
Filtration	Organic/inorganic	Retardation
Chemical oxidation-reduction	Organic/inorganic	Transformation/retardation
Biological uptake	Organic/inorganic	Retardation
Biodegradation	Organic	Transformation
Hydrolysis	Organic	Transformation
Volatilization	Organic	Elimination by intermedia transfer
Dissolution	Organic/inorganic	Mobility enhancement
Co-solvation	Organic	Mobility enhancement
Ionization	Organic	Mobility enhancement
Complexation	Inorganic	Mobility enhancement
Immiscible phase	Organic	Various partitioning

Retardation refers to processes that impede the transport of contaminants by removing or immobilizing them from a free state (i.e., aqueous solution or vapor). Examples of chemical reactions that result in retardation include sorption and precipitation. It is important to note that in retardation processes the immobilized contaminants are not transformed and the processes are reversible. Reversal tends to occur when the concentration of a contaminant in the ground water declines and the quantity of retarded contaminants is high. Reversal would return retarded contaminants to solution over an extended period, giving the plume of contamination a long "tail" over time.

Attenuation refers to two types of processes: (1) irreversible removal, and (2) transformation. Removal by an attenuation process differs from retardation in that it reduces the mass of a substance (i.e., the process functions as a sink). A common example is a process that transfers the contaminant to another media (e.g., volatilization). The more common type of attenuation process is one that transforms the molecular structure of the substance, an example being oxidation-reduction reactions.

Some processes actually increase the mobility of chemical substances in the subsurface. Examples include dissolution of organic substances and complexation of metallic ions. Such processes are categorized as **mobility enhancement** phenomena.

Subsurface Environment

Understanding the fate of contaminants in the subsurface requires a more exact depiction of the subsurface environment than used to describe the movement of ground water in response to advection and hydrodynamic dispersion. The model presented in Sec. 4.2 depicts the subsurface as a porous medium through which ground water flows. Describing the subsurface simply as a porous medium is a "macroscale" approach when "microscale" is needed.

The porous medium consists of both consolidated (rock) and unconsolidated geologic formations. The uppermost covering mantle of material is commonly referred to as **soil**. The formation of soil begins with weathering of parent rock or unconsolidated sediment followed by the transport, deposition, and accumulation of this material. Frost action, geochemical processes, water movement, and biological activity cause further change. The physical and chemical characteristics of soil vary widely with location, depth, and time, dependent primarily upon the parent material, climate (temperature, wind, and water), and topography (relief and slope). Note the difference between the material described here as soil and that term used earlier in the context of soil mechanics.

Soil is a mixture of different inorganic and organic materials. The inorganic fraction consists mostly of fine mineral grains, subdivided according to size as shown in Table 4-12. The percentages by weight of gravel, sand, silt, and clay provide a basis for classifying soil by texture. The mineralogy of clays varies, but most are formed from silicates. The most common clays are termed kaolinite, illite, and smectite (montmorillonite). Individual, plate-shaped, colloidal clay particles typically agglomerate to form clay aggregates.

The predominant inorganic elements in soil are silicon, aluminum, and iron, with a great number of micro and trace elements. It is interesting to note that some

TABLE 4-12
Classification of mineral grains in soil

Classification	Description	Size
Clay	Microscopic mineral particles of a colloidal nature, laminated as layers of plates	< 0.002 mm diameter (effective diameter)
Silt	Fine particles composed of minerals from the parent formation	> 0.002 mm diameter < 0.075 mm diameter
Sand	Granular particles composed of minerals from the parent formation	> 0.075 mm diameter < 2 mm diameter
Gravel	Granular particles composed of minerals from the parent formation	> 2 mm diameter < 75 mm diameter

of the naturally occurring and commonly found trace elements in soils are hazardous substances. Average concentrations in soils for such examples are as follows:[30]

Element	Average concentration in soil (μg/kg)
Arsenic	6
Cadmium	10
Nickel	40
Lead	10
Selenium	0.2

These naturally occurring elements differ from the same elements introduced with hazardous waste in that natural elements are essentially in an insoluble form and release very slowly with weathering. Waste constituents frequently are in a soluble form.

Soils typically contain appreciable amounts of organic material, basically consisting of decomposed plant matter or humus. Consequently, organic content normally decreases with depth, varying more in the vertical than horizontal dimension. The organic content of most soils typically would fall in the range of 0.2 to about 3%[31] and is composed of a wide mixture of organic chemicals (e.g., humic acid). The organic fraction can have a colloidal nature and has been described as organic polymer matrices.[32]

The organic matter in soil usually acts as a stabilizer such as to bind inorganic particles together as **aggregates**. Thus, the soil mantle consists of aggregates, each containing sand, silt, clay, and organic matter. It should be noted that the size and structure of aggregates can vary considerably (see Fig 4-26).

With depth, a transition occurs from soil to the underlying geological formations whether unconsolidated deposits or rock. The soil mantle, the underlying permeable formations, and transitional material between the two constitute the porous medium. The three are sometimes referred to collectively as the sub-surface matrix. Movement of ground water through porous media, whether saturated or unsaturated, thus is combination of transport both through the interstitial openings between aggregate and through the pores *within* an aggregate.

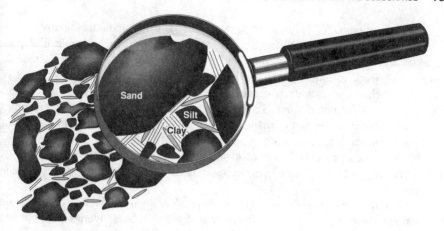

FIGURE 4-26
Soil aggregates in subsurface domain.

Within aggregates are found much of the colloidal inorganic and organic surfaces, the most active portions of the subsurface matrix.[33] It is within this microscale domain where much of the retardation and attenuation processes occur. These processes depend on intra-aggregate transport of ground water (i.e., through pores within the aggregate to reach internal surfaces). The effective diameter of pores within the crystalline mineralogical structure of clay can be 100 to 1000 nm (10^{-6} to 10^{-7} inches), thus severely restricting ground water advection and limiting intra-aggregate contaminant transport mostly to molecular diffusion.[34] Nevertheless, the structural arrangement of clay provides for an extremely large internal surface area, much greater than its external area,[35] and certainly great in relation to its mass.

Retardation Processes

Retardation processes consist of sorption of organic substances and ion exchange and precipitation of inorganic substances. These reactions result in a reduced rate of movement of contaminants compared to the average rate of ground water flow.

Sorption of Organics. Sorption is the partitioning of **sorbate** (chemical contituents) among phases (i.e., movement of **solutes** from a **solvent** to a **sorbent** [or orbing material]), as explained in Sec. 3-2. The sorbent does not have to be a solid; an example of sorption is the dissolution of an immiscible liquid such as oil into an aqueous phase. However, for the specific case of the fate of hazardous waste contaminants in the subsurface, sorption is primarily the accumulation of organic chemicals at soil surfaces.[36] More specifically, it is the partitioning of dissolved organic contaminants (the solutes) from ground or surface water (the solvent) to natural subsurface materials (the sorbent). An important example is the adherence of organic molecules to naturally occurring humic matter in soil.

The partitioning of solutes between phases depends upon the relative affinity of a solute for the solvent and sorbent. Affinity is primarily a molecular phenomenon and is a function of multiple chemical, physical, and electrostatic mechanisms. Most work to attract a solute; others repel a solute. Important examples of attractive mechanisms include hydrogen bonding (chemical), van der Waals forces (physical), and Coulombic attractive forces (electrostatic).[37,38] Two other examples are charge transfer and ligand exchange with metal ions.[39] An important example of a repulsion mechanism is between ground water and hydrophobic nonpolar organics. This example of a repulsion mechanism, referred to as hydrophobic sorption, is true partitioning, while the attractive mechanisms can be considered more as surface adsorption.

Any of these mechanisms may operate dependent upon the nature of the organic chemical and the properties of the subsurface materials. In fact, several could operate simultaneously. The sum of all mechanisms and all active subsurface materials account for the effect of sorption. Cumulatively, these mechanisms can easily yield a non-linear relationship between solute and sorbed concentrations. Figure 4-27 depicts both a linear and non-linear relationship.

Depending upon the mechanism and the organic chemical, the rate of sorption can be rather rapid with much of the sorption occurring within 48 hours. A two-stage model is usually considered for this non-linear relationship, consisting of a rapid, short-lived stage, followed by a slower, longer stage approaching equilibrium (see Fig. 4-28).

The rate of sorption can be slow given the tortuous path of diffusion from the bulk aqueous solution to pore water and then through the pores of the aggregate across the fixed water film to sorption sites in organic or mineral matter. In fact, many contaminants of environmental concern have sorption equilibrium times of days, weeks, or even years as for the case of polychlorinated biphenyls (PCBs).[40]

It is important to note that sorption is reversible (a process termed **desorption**). This can occur when the concentration of solute has decreased and the quantity of chemicals sorbed to the sorbent is high. A hysteresis effect, as shown in Fig. 4-29, can occur. Desorption can take an extremely long time, thus giving the plume shape

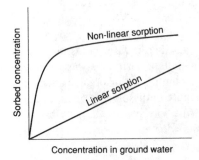

FIGURE 4-27
Partitioning of sorbate between solvent and sorbent.

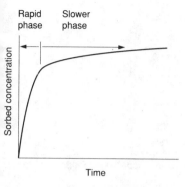

FIGURE 4-28
Two-stage sorption model.

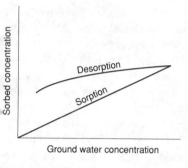

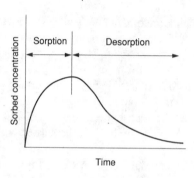

FIGURE 4-29
Desorption of sorbate.

a "long tail of decreasing contamination."[41] This phenomena limits the use of "pump and treat" cleanup technology discussed in Sec. 16-7. Upon cessation of pumping, concentrations in ground water will return to equilibrium levels.

As explained earlier in this section, a microscopic examination of the porous medium composing the subsurface shows that it has both organic and mineral surfaces. Many organic chemicals of environmental concern are hydrophobic. At most hazardous waste sites, nonpolar species such as trichloroethene, benzene, and chlorinated hydrocarbon pesticides account for much of the risk to health. For hydrophobic organic compounds, several researchers have shown that the dominant sorption mechanism is the affinity of such solutes for natural organic material in the aquifer.[42] In fact, researchers have correlated the sorption capability of various soils with their organic matter content.[43] The importance of the organic matter content of the surrounding geologic material is shown in Fig. 4-30. This figure shows the amount of lindane sorbed by a soil under two conditions: natural soil and the same soil stripped beforehand of its organic matter.

It should be noted that hydrophobic organics can adsorb to mineral surfaces, especially to clay-sized particles. Clay particles have a net negative charge due to isomorphous substitutions on the internal surfaces between the plates of silicate clay crystals as well as the external surfaces.[44] Examples of organic chemicals adsorbing to these sites include those with polar organic functional groups (via ion-dipole

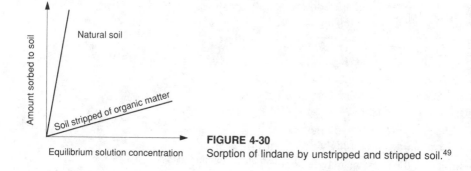

FIGURE 4-30
Sorption of lindane by unstripped and stripped soil.[49]

forces) and large organic molecules (via hydrogen bonding).[45] Cation exchange is also possible.[46] Even hydrophobic organics can accumulate on clay surfaces.[47] This can be significant, particularly for soils with organic matter content below one percent.[48]

It is clear that a variety of molecular mechanisms can act to sorb organic chemicals to natural materials in the subsurface. The magnitude of the mechanisms varies dependent on the nature of the solute, solvent, and sorbent. Their combined net effect determines the distribution between phases. Dependent upon conditions, the degree of sorption can be high, with the sorbed phase exceeding the concentration in ground water by a ratio as high as 5 to 10:1, if not higher. The ratio can also be less than 1.0.

The potential retention capability, which includes sorption, of soil can be estimated by saturating undistributed soil samples with a liquid contaminant and then allowing the sample to drain. The retained contamination is termed residual saturation or retention capacity and is shown in Table 4-13 for fuels. Obviously, retention capacity for fuels represents partitioning from the bulk phase as occluded droplets rather than as solutes to the soil matrix. It is not true molecular sorption as discussed here; however, the values indicate the degree to which bulk contaminants can partition to soil.

Example 4-9 Retention capacity. Based on the data provided in Table 4-13, what are two physical characteristics that control retention capacity?

Solution. For a given soil type, retention of a fluid increases as the viscosity of the fluid increases. For a given contaminant, retention by soil increases as the texture of the soil becomes finer.

TABLE 4-13
Retention capacity of various soil types for common fuels (ppm of fuel in soil)[50]

Fuel	Gravel	Coarse sand	Medium sand	Fine sand
Gasoline	1000	1600	3200	5000
Diesel fuel	2000	3200	6400	10,000
No. 6 fuel oil	4000	6400	12,800	20,000

The properties of the sorbent are also very important. Key factors influencing sorption include the following:

- Molecular size (generally the larger the molecule, the more readily it is adsorbed)
- Hydrophobicity (the adsorption of non-ionic organic chemicals onto soil particles correlates well as an inverse function with water solubility and octanol-water partition coefficients, two surrogates for hydrophobicity)
- Molecular charge (some organics can be an uncharged polar molecule at a certain pH and an anion at a higher pH[51]).
- Structure (one isomer may be more readily adsorbed than another)[52]

Suffice it to say, the mass transfer and reaction rate interrelationships of the various forces are very difficult to formulate. Furthermore, in the field, subsurface materials exhibit a high degree of complexity and heterogeneity. This necessitates that sorption be determined based on tests of the contaminant and subsurface materials and the net effect be expressed using simplified, empirical models.

Linear Sorption Model. For saturated conditions and non-polar organic constituents, sorption from the aqueous phase to the porous media of the subsurface can be "... treated as an equilibrium-partitioning process. ... When solute concentrations are low (e.g., either $\leq 10^{-5}$ molar, or less than half the solubility, whichever is lower) partitioning often can be described using a linear isotherm"[53] as discussed in Secs. 3-2 and 9-3. The linear isotherm relates the concentration of the organic substance in ground water and the sorbent at constant temperature as follows:

$$S = K_d C \tag{4-36}$$

where S = mass sorbed per mass of sorbent (mg/kg)
K_d = partition or distribution coefficient
C = concentration in ground water at equilibrium (mg/L)

Such a linear relationship was shown earlier in Fig. 4-27. The slope in this figure represents K_d. It represents the ratio of the mass concentration of a contaminant sorbed to soil to its concentration in the surrounding ground water. It would vary dependent upon the specific chemical constituent and the subsurface material. The slope can be determined from batch equilibrium tests conducted with samples of actual subsurface material. As discussed earlier, such tests frequently show that the greater the organic content of the aquifer, the greater the sorption potential. Thus, sorption can be characterized in many cases, particularly for hydrophobic organics, based on the organic content of the porous medium of the aquifer by using the following linear expression (Eq. 3-18):

$$K_d = K_{oc} f_{oc} \tag{4-37}$$

where f_{oc} = organic carbon fraction of the porous medium
   K_{oc} = organic carbon partition coefficient of the chemical constituent

Experience has shown that this method produces a reasonable estimate of sorption only for subsurface materials having the following organic content:

Minimum: "Greater than 0.1%"[54]
            "Few tenths of a percent"[55]
            1.0%[56]
Maximum: 20%[57]

As stated earlier, the organic content of most soils falls in the range of 0.2 to about 3%. Thus, this approach may not be applicable for many situations. Where applicable, the variable f_{oc} can be determined from field sampling and analyses, and the variable K_{oc} can be found in App. A for many chemicals.

When unknown, K_{oc} sometimes can be derived by correlation with a property indicating the degree of the chemical's hydrophobic nature (e.g., octanol-water partition coefficient).[58,59] A commonly used correlation is as follows:[60]

$$K_{oc} = 0.63 K_{ow} \qquad (4\text{-}38)$$

where K_{ow} = octanol-water partition coefficient for the chemical

However, the actual value can vary considerably from this approximation as shown by various studies which have developed similar regression equations for different classes of compounds (see Table 3-5). These studies note that some large errors are possible, particularly for a class covering a broad range of chemicals.[61] When using these correlations it is important to consider factors such as chemical structure similarity, molecular weight, organic carbon content of soil, contribution of clay mineral surfaces, solute concentration, and others.[62]

When applicable, an estimate of linear sorption can be developed simply based on Eq. 4-36, the correlation K_{oc}, and the determination of the organic fraction of the porous medium. This linear sorption equilibrium model thus provides a rather simple method for estimating sorption. However, it is not adequate for many conditions, and probably best applies when the concentration of the solute is "very low,"[63] at least not higher than 10% of the maximum solubility of the contaminant in water.[64] A high concentration may saturate if not exceed the sorption capacity of the subsurface. Thus, application of the linear model to any situation requires care.

Example 4-10 Sorption. Ground water near a leaking underground tank contains 0.5 mg/L of benzene. What is the expected concentration of benzene sorbed to a silty soil containing 2% organic matter? Assume sorption follows a linear model.

Solution. From App. A, K_{oc} for benzene is 83 mL/g.

Using Eq. 4-37,

$$K_d = K_{oc} f_{oc}$$
$$= (83)(0.02) \frac{mL}{g}$$
$$= 1.66 \frac{mL}{g}$$

Using Eq. 4-36, the sorbed concentration is estimated as follows:

$$S = K_d C$$
$$= (1.66)(0.5) \text{ mg/kg}$$
$$= \underline{0.83 \text{ mg/kg}}$$

Non-Ideal Sorption Conditions. There are a number of conditions for which sorption does not follow the ideal pattern as expressed by the linear equilibrium approach. Some example conditions are as follows:

- The rate of ground water movement may be rapid such that the relatively slower rate of sorption becomes limiting, thereby negating the assumption of equilibrium.
- Sorption may follow second-order kinetics, necessitating a non-linear model, particularly at high solute concentrations.
- Organic chemicals may adsorb to suspended microparticles of organic matter in the ground water; as such, adsorbed organic chemicals would move with ground water and yield less sorption than predicted.
- In ground water with multiple organic chemicals, competition for limited sorption sites will occur, thereby yielding less than the additive amount of sorption determined for an individual species.

The linear partitioning model assumes equilibrium between sorbed constituents and solute. This is deemed adequate for many hydrophobic organic chemicals if the solvent-sorbent contact time is on the order of "days to months."[65] Nevertheless, for some compounds, such times may not suffice for reaching equilibrium.

Non-linear sorption is frequently observed. For example, in one case involving the sorption of trichloroethene (TCE) onto a glacial till, the K_d decreased from 110 at 1 ppm of TCE to less than 10 at 4 ppm (see Fig. 4-31). There are two models frequently used to describe non-linear adsorption, the Langmuir model (see Eq. 9-34) and the largely empirical Freundlich isotherm (see Eq. 9-33). The latter is used more often. It is very similar to the linear sorption model with the addition of the empirical constant, $\frac{1}{n}$, as follows:

$$S = K_d C^{\frac{1}{n}} \qquad (4\text{-}39)$$

The use of such models requires that they be carefully calibrated using data from batch if not flow through tests of the specific contaminants with samples of the subsurface materials. Even then, it is important to have verified the selected model for the conditions measured in the field at the site under investigation.

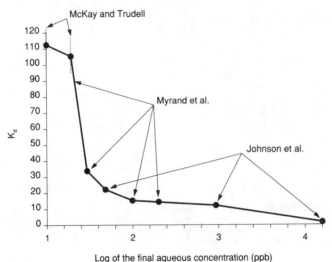

FIGURE 4-31
Variable sorption of trichloroethene on glacial till.[66]

Ion Exchange. Ion exchange involves the sorption of ions in solution onto oppositely charged, discrete sites on the surface of a soil particle. It is driven by the attractive force of maintaining electrostatic neutrality: the electric charges of the surface of the soil are balanced by equivalent free ions of opposite charge.[67] As such, a previously held ion of weaker affinity is exchanged by the soil for the ion in solution. For example, calcium ions in groundwater may be exchanged onto the clay surface, replacing existing sodium ions.

Ion exchange can be considered as a subcategory of sorption involving electrostatic interactions. However, a separate distinction is made in that ion exchange is generally thought to apply to metals while sorption applies to organics. Most metals in aqueous solution occur as charged cations. Thus, metal species adsorb primarily in response to electrostatic attraction, but such is only one type of interaction resulting in the sorption of organics.

In contrast to sorption, which occurs mostly upon organic surfaces, both organic and inorganic surfaces are important with ion exchange. The capability of a soil to retain and exchange cations is quantified as the **cation exchange capacity**. For example, the cation exchange capacity of humus and the two clay minerals, vermiculite and montmorillonite, are in the order of 200, 150, and 100 meq per 100 grams.[68] Although the organic and inorganic fractions together account for the cation exchange capacity of a soil, the clay fraction is usually more important because soils frequently have a far greater clay than organic content.

Clay has a much higher cation exchange capacity than other inorganic particles because of its extremely large surface area, abounding with negative charge sites. The exposed surface area of a unit mass of colloidal clay is at least 1000 times that of coarse sand.[69] It should be noted that the cation exchange capacity of clays varies

TABLE 4-14
Cation exchange capacity of clay minerals

Clay mineral	Cation exchange capacity (meq/100 g)
Kaolinite	3–15
Illite	10–40
Smectite (montmorillonite)	80–150
Attapulgite	20–30

considerably dependent upon the clay mineralogy. Table 4-14 presents the cation exchange capacity of commonly encountered clay minerals.

In the process of ion exchange, ions compete for the exchange sites and displace a previously held cation, termed cation exchange selectivity. Typically displaced cations include sodium and calcium. Exchange affinity depends on electric charge, hydrated radius, and molecular configuration.[70] Generally, divalent cations are more strongly adsorbed than monovalent ions, and smaller cations tend to replace larger cations. Cations are typically replaced in the following order:[71]

$$Na^+ < Li^+ < K^+ < Rb^+ < Cs^+ < Mg^{2+} < Ca^{2+} < Ba^2 < Cu^{2+} < Al^{3+}$$

$$< Fe^{3+} < Th^{4+} \qquad (4\text{-}40)$$

These guidelines are not absolute. For example, at high concentrations a cation with low replacing power such as Na^+ can replace a cation higher in the series.

Ion exchange capacity is very dependent upon pH. As the pH is lowered below neutrality, hydrogen ions readily replace metal ions. As the pH increases, the hydrogen ions are less resistant to replacement and the net result is an increase in the exchange capacity.

A key point is that ion exchange of metallic ions with soil may be partially reversible; saturated exchange sites can release cations as concentrations of contaminants decline in ground water[72] and in response to pH changes. Thus, ion exchange represents a retardation rather than an attenuation process. Further, the exchange capacity of a subsurface material can be saturated such that eventually the transport of contaminants is unaffected by ion exchange and goes unretarded.

Precipitation. Precipitation is the converse of dissolution: the concentration of a solute exceeds the solubility of that particular compound, and any excess solute changes to a solid and thus precipitates out of the solution. That part which separates out of solution is termed **precipitate**. The process is reversible. If the concentration of a solute drops below its solubility, dissolution of previous precipitate could occur. Precipitation is particularly applicable to heavy metals such as nickel, mercury, chromium, and lead.

Precipitation can occur when a chemical reaction transforms a solute to a much less soluble form, typically by mixing a precipitant with the solution (e.g., mixing of a lead solution with high carbonate waters can produce the relatively insoluble form $PbCO_3$). Also, changes in oxidation state can promote precipitation (e.g., oxidation of ferrous solutes will yield the much less soluble ferric forms).

Precipitation depends greatly upon pH. Most metals precipitate at high pH levels as hydroxides. However, continued elevation of pH will increase the solubility of amphoteric metals such as nickel. This is shown in Fig. 4-32.

Filtration.   Filtration is a physical form of retardation, resulting from the clogging of pore spaces. This occurs from solid particles becoming trapped in the pores. It also occurs due to precipitation and accumulation of dissolved matter.

Attenuation Processes

The significant **attenuation** processes that transform substances are the oxidation-reduction reactions, whether initiated via chemical or biological pathways; hydrolysis which can also transform substances; and volatilization which results in transfer of chemical substances to the atmosphere.

Chemical Oxidation-Reduction.   Chemical reduction-oxidation (redox) reactions can occur with organic and inorganic chemicals and classically involves the gain or loss of oxygen. A reaction that adds oxygen to a compound is oxidation (e.g., $FeSO_4$ changes to $Fe_2(SO_4)_3$). Likewise, loss of oxygen (frequently with the addition of hydrogen) is termed reduction (e.g., the ion NO_3^{+1} changes to NH_4^{-1}). However, oxidation-reduction reactions do not necessarily involve oxygen.

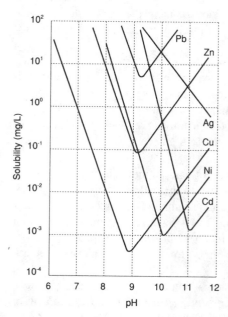

FIGURE 4-32
Solubilities of metal hydroxides as a function of pH.

For organic chemicals, oxidation is defined as a reaction that results in the loss of electrons by a chemical (the chemical donating the electron is oxidized). The converse, gaining an electron, is reduction. For inorganic solutes, oxidation again involves the loss and acceptance of electrons, but is more accurately defined as a reaction that increases the oxidation state of an atom (i.e., the charge on the atom becomes more positive). For the example mentioned in the preceding paragraph, Fe^{+2} is oxidized (i.e., donates electrons) to become Fe^{+3}. The converse is reduction such as the following example in which Cr^{+6} is reduced to Cr^{+3} due to the acceptance of electrons.

$$2H_2CrO_4 + 3SO_2 \rightarrow Cr_2(SO_4)_3 + 2H_2O \qquad (4\text{-}41)$$

Ions having multiple oxidation states (i.e., more than one unit of charge, as with the example of Cr^{+3} and Cr^{+6} above) are termed **redox couples**. Dependent upon the oxidation state, the implications for fate in the subsurface can vary widely as indicated by the examples shown in Table 4-15.

For natural waters, oxidation and reduction of inorganic solutes of hazardous substances may occur over geologic time but is usually negligible for shorter time periods. Oxidation-reduction would occur with the addition of reagents. The presence of chemical oxidizing and reducing reagents can be indicated by the pH and redox potential (measured as E_h or pE, the latter being the negative logarithm of electron activity) of the ground water.

Oxidation and reduction of organic chemicals is principally a result of biological activity (as discussed in Chap. 10) rather than chemical reactions. Nevertheless, some chemical oxidation and reduction reactions occur between organic chemicals and soil materials. The role of soil materials in the oxidation of organic chemicals is complex and not well known. Researchers have reported the free radical oxidation of substituted aromatic chemicals at soil and clay surfaces, catalyzed by adsorbed oxygen and trace metals, particularly in aerated, unsaturated soil.[74] However, the same research suggests that many chlorinated organics likely will not undergo such oxidation, and for organics that do, "not all soils and clays will catalyze these reactions." Clay can also initiate some reduction reactions, particularly in anaerobic conditions. General

TABLE 4-15
Multiple oxidation states[73]

Element	Ion	Comment
Iron	Fe^{+3}	Low solubility, and hydroxide precipitate can adsorb other metals
	Fe^{+2}	"Very soluble"
Chromium	Cr^{+6}	Toxic; "relatively mobile"
	Cr^{+3}	Precipitated as relatively insoluble $Cr(OH)_3$ or other chemical; "strongly adsorbs to surfaces"
Selenium	Se^{+6}	More "mobile, but less toxic"
	Se^{+4}	"More toxic, but less mobile"

shows that redox conditions can vary widely over short distances, with saturated conditions being more homogeneous than unsaturated conditions.

Biological Oxidation-Reduction.   As discussed in Chap. 10, most organic substances in the subsurface will undergo transformation to smaller molecules via oxidation and reduction mechanisms induced by the metabolic activity of native microorganisms. Such transformation is termed **biodegradation.** Due to the small amount of oxygen in the subsurface, most transformation occurs via reducing pathways of anaerobic processes. Anaerobic biodegradation occurs at a very slow rate in the subsurface; however, it favors dehalogenation of chlorinated compounds that typically resist aerobic degradation.

It is important to note that anaerobic degradation may not always transform organic compounds to less toxic or less mobile forms. This is illustrated by the well documented case of the anaerobic degradation of tetrachloroethylene in which vinyl chloride is produced as an intermediate as chlorine atoms are cleaved in the reduction process (see Fig. 4-33). Vinyl chloride is more toxic and has a much lower K_{oc} than the initial compound tetrachloroethylene. The reduced oxidation states of some organics may therefore be more toxic than the oxidation states stable in the presence of oxygen.[75]

Hydrolysis.   Chemical substances may react with water molecules, a chemical process termed hydrolysis. The exothermic reaction of some types of reactive waste is an example. Hydrolysis is frequently depicted as an exchange between OH⁻ and an anionic group X of the chemical compound, resulting in decomposition of the compound as follows:

$$RX + HOH \longrightarrow ROH + HX \tag{4-42}$$

For most chemicals, hydrolysis has a relatively insignificant effect compared to other processes for transforming organic chemicals (e.g., biodegradation). However, for

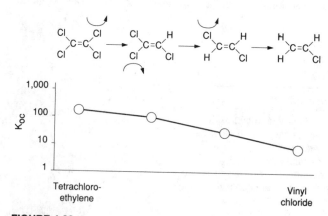

FIGURE 4-33
Degradation of tetrachloroethylene to vinyl chloride.[76]

chlorinated organics, which typically are not readily transformed by biodegradation, hydrolysis may be significant dependent upon other factors.[77] Hydrolysis of chlorinated organics involves exchange of the hydroxyl group with an anionic X on a carbon atom.[78] The reaction typically forms an alcohol or alkene as follows:[79]

$$\overset{\displaystyle H \quad X}{\underset{\displaystyle}{C-C}} \xrightarrow[\text{OH}^-]{\text{H}^+ \text{ or}} C = C + HX \tag{4-43}$$

Actually, the hydroxyl reactant may come from a water molecule or another source of a hydroxyl ion. The reaction can be catalyzed by an acidic or basic intermediate.[80] Therefore, the rate of hydrolysis depends upon the pH and temperature of the water as well as the chemical compound. Figure 4-34 shows the effect of pH on the hydrolysis of two alkyl chlorides. This figure also illustrates the strong dependence of the rate of hydrolysis on the compound structure.

A simple but not always realistic approach is to depict hydrolysis as a first order reaction under constant pH and temperature using a form of Eq. 3-54 as follows:

$$\frac{dC}{dt} = -KC \tag{4-44}$$

where C = concentration
  t = time
  K = first order reaction rate at constant temperature and pH

The value of K can be determined in the laboratory. Nevertheless, it is difficult to develop reliable data because competing reactions mask the individual effect of hydrolysis.

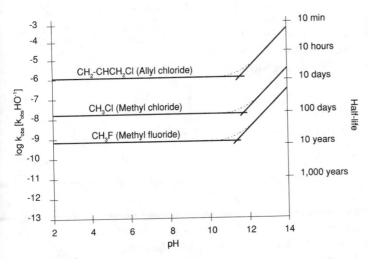

FIGURE 4-34
Hydrolysis of chlorinated alkyl compounds.[81]

Volatilization. Volatilization is the conversion of volatile chemical constituents in ground water to vapor, which ultimately results in the transfer to the atmosphere. This process was covered earlier in Sec. 4-1.

Volatilization can occur from three sources in the subsurface—free product, vadose water, and ground water. The key variables are vapor pressures and the area of contact between soil air and the three sources. The porosity of the subsurface materials less moisture (i.e., the volume of soil air) provides a reasonable indication of the degree of contact. Thus, vadose water and residual free product trapped in the unsaturated zone both have a much greater area of contact with soil air than ground water, limited to the boundary at the ground water table. Low-density immiscible fluids that move to the ground water table and float there in pools because of lower specific gravity would also result in a large relative area of contact with soil air. However, the area is perhaps not as significant as the residual fluid trapped in and on soil particles in the unsaturated zone.

Mobility Enhancement

Retardation and attenuation processes alter the transport of contaminants to yield a lower concentration than what would be predicted based on advection and dispersion alone. Some processes actually work counter to these processes to increase the mobility of a substance. These processes principally consist of cosolvation, ionization, dissolution, and complexation.

Cosolvation. Cosolvation is the result of the introduction of bulk quantities of an organic solvent in the subsurface. The properties of an organic solvent differ from water such that a mixture of the two behaves differently than water alone. The presence of bulk solvents promotes increased interaction between a solute and the solvent than would occur with water. The resultant mixture can dramatically increase the mobility of constituents in comparison with the case in which water is the sole solvent. In particular, the solubility of an organic constituent can increase and the sorption capacity of the subsurface materials can decrease.

Organic solvents affect the sorption capacity of subsurface materials in various ways:[82]

- Competition for sorption sites
- Removal of organic matter from the aggregate, which is the primary site for sorption to occur.
- Physical alteration of soil, thereby changing intra-aggregate flow patterns to bypass sorption sites.

Figure 4-35 shows how adding the cosolvent methanol to three different soils reduces sorption capacity. Besides reducing sorption, solvents may alter the properties of soil such as to increase the hydraulic conductivity of the subsurface.[83]

The effect of cosolvation on solubility can be just as dramatic. The introduction of solvents in water "in the range of 20 percent by volume or greater" can increase

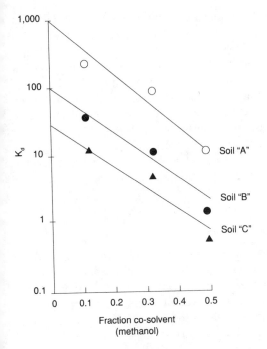

FIGURE 4-35
Reduction of sorption by cosolvation.[86]

solubility of hydrophobic compounds by more than one order of magnitude.[84] The literature is replete with examples where the presence of solvents have prompted the migration of compounds that otherwise are deemed relatively insoluble.

Ionization. Organic acids (e.g., phenols and aliphatic acids) have the ability to donate protons when in aqueous solutions. In a process termed ionization, acids losing a proton would become anions and their solubility in water would increase significantly. An example is 2,4,5-trichlorophenol for which the K_{oc} would "decrease from 2330 for the phenol, to near zero for the phenolate."[85]

Dissolution. Dissolution is the dissolving of chemical substances from free product or solid forms of hazardous waste to solutes in the ground water. The solutes can be inorganic cations and anions, polar organics, and non-polar organics. Leachate, as discussed in Sec. 4-1, is an example of a liquid which dissolves constituents as it percolates through a landfill.

Complexation. Complexation, or chelation, is the formation of a coordinate bond between a metal ion and an anion known as a ligand. The ion-ligand arrangement is termed a complex in which the ligands surround the ion. Complexation increases the potential mobility of a metal because a) the complexed metal is more soluble, and b) the **complex** ties up what would otherwise be free metallic ions, thus decreasing the opportunity for adsorption or precipitation of such ions. An example is mercury and chloride salts; increasing chloride concentrations may drive mercury from Hg^{2+} to a neutral complex.[87]

A ligand is often composed of multiple elements functioning together. Ligands can be inorganic (e.g., many of the common anions such as CO_3^{-2} and OH^{-1}) or organic.[88] Some synthetic organic ligands may be present in ground water from waste sources. Natural organic ligands may also be present, having derived from humic materials. It is noted that in shallow aquifers, organic ligands from humic materials can be present in significant concentrations and dominate metal chemistry.[89] Complexed metals formed with organic ligands can be fairly stable, requiring a low pH (or for some metals a high pH) to break the bond. Because of the variety of ligands that can be found in ground water, a metal can be complexed in different ways; it is difficult to predict the outcomes.

Immiscible Phase Partitioning

Immiscible fluids, such as gasoline, that have migrated into the subsurface may exist in three phases: (1) free product, (2) sorbed to soil particles and interstitial openings in the subsurface matrix, and (3) dissolved in ground and soil water. The amount in each phase depends upon the properties of the subsurface and the immiscible substance. The partitioning to the sorbed phase follows the same principles as explained previously for sorption of organics. Likewise, the partitioning to the solute phase is similar to dissolution of organics. In most cases, the mass in the solute phase is small in comparison with the mass in either the sorbed or free product phases. As such, the latter two represent an extended and continuous source of release of contaminants to the ground water. The free product phase is termed **non-aqueous phase liquid** (NAPL).

Inclusion of Fate in Transport Equations

Some retardation processes, particularly sorption and ion exchange, can have a significant effect on transport rates. The approach to these processes is to represent them in ground water transport equations by the addition of a retardation factor on an empirical basis.

Recall in Sec. 4-2 that the transport of contaminants in the subsurface was in response to hydraulic gradients (advection) and chemical gradients (diffusion). The one-dimensional advection-dispersion equation for non-reactive dissolved constituents in a homogeneous, isotropic aquifer under steady-state, saturated flow conditions is:

$$D_\ell \frac{\partial^2 c}{\partial \ell^2} - \bar{v}_\ell \frac{\partial c}{\partial \ell} = \frac{\partial c}{\partial t} \qquad (4\text{-}45)$$

where D_ℓ = longitudinal coefficient of hydrodynamic dispersion
  c = solute concentration
  $\bar{v}_\ell$ = average linear ground water velocity
  ℓ = distance along the flowline
  t = time

The coefficient of hydrodynamic dispersion, D_ℓ, is a result of two mechanisms, mechanical dispersion and molecular diffusion, and can be expressed as

$$D_\ell = \alpha_\ell \bar{v} + D^* \qquad (4\text{-}46)$$

where $\alpha_\ell =$ longitudinal dispersivity
$D^* =$ coefficient of molecular dispersion

At this juncture, the transport equation has not considered the attenuation, retardation, and mobility enhancement processes discussed above in Sec. 4.3. It is clear that these phenomena collectively can reduce the concentration of inorganic and organic constituents in ground water to levels less than that estimated based on merely mechanical forces. The converse is also true. Each phenomenon is individually complex, much more so if considered in a combined and interdependent manner. Because of this mosaic of great complexity, the processes are typically accounted for in a varied manner. The general, simple approach is to quantify the influence of these processes in a collective, gross manner as a retardation factor, R, expressed as:

$$R = 1 + \frac{\rho_b}{n} K_d \qquad (4\text{-}47)$$

where $\rho_b =$ bulk mass density of the porous medium
$n =$ porosity
$K_d =$ distribution coefficient

Eq. 4-45 can be rewritten to accommodate retardation as follows:

$$\frac{1}{R}\left(D_\ell \frac{\partial^2 c}{\partial \ell^2} - \bar{v}_\ell \frac{\partial c}{\partial \ell}\right) = \frac{\partial c}{\partial t} \qquad (4\text{-}48)$$

How then does retardation alter the contaminant plume? Shown in Fig. 4-36 is a section through a shallow ground water system with a continuous source of contaminants and steady-state flow. In Fig. 4-36(a) and (b) there is no retardation whereas the retardation increases (as K_d goes from 0 to 1 to 10) in Figs. 4-36(b), (c) and (d). The effect is dramatic. After a transport time of 60 years there is little contaminant migration beyond the source for the retardation case ($K_d = 10$) compared with the case having no retardation ($K_d = 0$).

By using only the variable K_d, this approach is limited to sorption and ignores other fate processes. Nevertheless, this approach has been shown to estimate the velocity of the transport of organic contaminants within a factor of two to three of the actual velocity observed in the field.[90] Thus, this approach can provide reasonable values. Part of the success of this approach derives from the fact that sorption can be the dominant process in many cases. There are other mathematical approaches, including monographs, to predicting the mobility of organic chemicals.[91]

With knowledge of site conditions and the individual chemical species present, some attenuation processes can be examined fairly reliably on an individual basis. Mobility enhancement processes need to be considered carefully and individually.

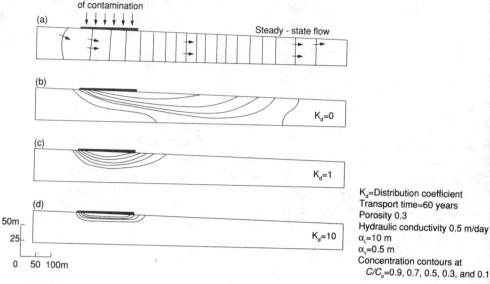

FIGURE 4-36
Effect of the distribution coefficient on contaminant retardation during transport in a shallow ground water flow system. (Adapted from Freeze and Cherry[92])

4-4 ATMOSPHERIC TRANSPORT OF CONTAMINANTS

Releases of gaseous and particulate emissions to the atmosphere disperse at varying rates in response to numerous natural factors controlling their transport. These factors can be grouped as **plume properties**, **wind**, and **atmospheric turbulence**. Their influence on dispersion has been well theorized by engineers and scientists with no significant changes to the fundamental concepts developed in the 1950s and 1960s. These concepts often involve empirical relations and statistical representations coupled with a scientific basis. Their development has focused on the classic case of dispersion of an elevated release (i.e., a plume from a stack) which is only one of the possible types of releases from hazardous waste sites. Nevertheless, the fundamental concepts can be applied to many types of releases.

Application of these concepts can yield reasonable approximations of dispersion within certain limitations and when care is taken in the selection of parameters. However, for receptor locations in very close proximity to sources or at great distances from sources, the predictions are less reliable. Thus, use of these concepts requires special care at hazardous waste sites where maximum ground level concentrations typically occur either on-site or in its immediate vicinity. As such, the results should incorporate substantial safety factors and, in many cases, require verification by air monitoring.

The traditional approach of examining transport of contaminants in the atmosphere is divided into two steps. The first step is to estimate **plume rise** and the second

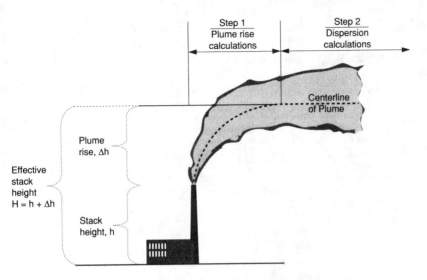

FIGURE 4-37
Two steps of estimating dispersion.

is to estimate **dispersion** of the plume beginning after it has reached its maximum height, as depicted in Fig. 4-37.

Plume Rise

The key concern with airborne hazardous substances is their resultant concentration at ground level, where potential receptors could be exposed. Therefore, the height to which a release (i.e., a plume) rises before bending horizontally and dispersing is important because, obviously, the higher the rise, the lower the contaminant concentration at ground level downwind from the source (i.e., the greater the dilution).

The effective stack height, H, is the height of the stack, h, plus plume rise (see Fig. 4-37), or:

$$H = h + \Delta h \tag{4-49}$$

A plume rises in response to its momentum and buoyancy. The exit velocity from a stack imparts **momentum** to the plume. Heated stack gases, especially those from combustion processes, are considerably warmer than ambient air, giving the plume a lower density and thereby making it **buoyant** in ambient air. Momentum and buoyancy propel the plume upwards. The force of any wind counteracts this upward motion by bending the plume to the horizontal. Ambient air also mixes with the plume in a process called **entrainment** due to phenomena such as turbulence and velocity shear.[93] Entrainment speeds the cooling of the plume, thus dissipating its buoyancy and impeding its rise. As a plume becomes horizontal, the momentum and buoyancy flux dissipate and atmospheric turbulence becomes the dominating factor in the plume dispersion.

Plume rise is measured typically using the centerline of the plume as a reference. The rise of a plume from a stack, except on calm days, follows an elliptical path, and may climb as much as 1000 meters or more, before fully reaching horizontal. In early models, momentum was considered the dominating factor in determining the ultimate height to which a plume could rise. However, with work done in the 1960s by Briggs, it became clear that for stacks from combustion processes, buoyancy has more of an effect than momentum.[94]

Briggs found that the bulk of the plume rise data for windy conditions fits the following formula (simplified from a more theoretical basis):[95]

$$\Delta h = 1.6 \frac{(Fx)^{\frac{2}{3}}}{u}$$ (4-50)

where Δh = rise of plume centerline above top of center stack (m)
   F = buoyancy flux, or $3.7 \times 10^{-5} Q_H$ where Q_H
      equals heat emission rate (cal/m/sec^2)
   x = horizontal distance downwind of stack (m)
   u = average wind speed at top of stack (m/sec)

This equation disregards the contribution of momentum to plume rise. This omission is reasonable because, for at least these common conditions, momentum is negligible. It is seen in Eq. 4-50 that plume rise is an inverse function of wind speed.

Many emissions from hazardous waste sites do not exit from a stack. Rather, they occur at ground level, with minimal momentum and a nominal difference in temperature. Although there are significant differences in molecular weight between that of air and some released gases, the concentration of the gases is too low for density differences to give them a buoyancy flux. Hence, plume rise for most ground level releases at hazardous waste sites is negligible.

Dispersion

Wind has an obvious major effect in advecting airborne pollutants. In addition to controlling the direction of pollutant movement, wind dilutes continuous releases: the greater the wind velocity, the more volume of incoming air per unit mass of pollutants released and the greater the dilution. In fact, gaseous pollutant concentration is inversely proportional to wind velocity, all else being equal. Wind also keeps particulate matter suspended that would otherwise settle sooner to the surface.

Analysis of the dispersion of airborne pollutants is far more complicated than an analysis of just wind velocity and direction. The mechanical agitation imparted by turbulence has a major effect because it promotes vertical and lateral motion components in addition to the advective component of wind. In periods of high wind, ground level concentrations may actually be greater than in calm periods because the plume can reach the ground sooner due to downward motions imparted by turbulence.

Turbulence does not follow a uniform pattern. It has immense temporal and spatial variations. It can consist at one time and place as tight eddy currents and can change to large circling motions, only later to become calm. Tangential wind shear also can occur with strong motions. Turbulence derives from two sources: the wind (as it traverses over the irregular surface of the earth) and atmospheric stability.

Atmospheric stability is a surrogate meteorological term for turbulence, and is dependent primarily upon the atmospheric thermal gradient, as well as surface wind velocity and surface roughness. Stable air is frequently thought of as a uniform decrease in temperature versus elevation and unstable air as having variant temperature gradients. However, atmospheric stability is described by examining what would happen to a parcel of air if it were elevated and allowed to expand due to the decreased atmospheric pressure. If this occurred adiabatically (with no exchange of heat between the parcel and surrounding air), the temperature of the parcel would cool. If the resultant temperature were less than that of the surrounding air at the higher altitude, the parcel would sink. This is defined as stable air (i.e., the atmospheric thermal gradient is such as to retard upward motion). Unstable conditions exist when the converse occurs (i.e., the adiabatic temperature of the parcel is higher than the atmospheric temperature, thus tending to induce vertical motion). Neutral conditions occur when the adiabatic temperature of lifted air equals the temperature of the surrounding air.[96] The adiabiatic lapse rate is the thermal gradient for the condition of neutral stability and equals about 1°C/100 m (5.4°F/1000 ft) for dry air. Vertical stability is related to the actual lapse rate as follows:

When actual lapse rate...   Conditions are...
  = adiabatic lapse rate   Neutral
  > adiabatic lapse rate   Unstable (or superadiabatic)
  < adiabatic lapse rate   Stable (subadiabatic)

A classic study done by Briggs depicted these relationships by showing the effects of various conditions of atmospheric stability on a plume (see Fig. 4-38). The effects can progress through a diurnal sequence in clear weather as discussed in Table 4-16 and shown in Fig. 4-38.

Solar energy is the source of these changes, and the amount reaching the earth's surface depends on the angle of the sun and the cloud cover. Wind can also break up inversions and result in a well-mixed atmosphere. Pasquill developed a chart for classifying atmospheric stability based on incoming solar radiation, wind speed, and other basic observations typically reported by meteorological stations (see Table 4-17). It can be seen from Table 4-17 that a clear day yields less stable atmospheric conditions. Table 4-17, according to Turner, "will give representative indications of stability over open country or rural areas, but are less reliable for urban areas."[97] In urban areas the greater heat flux and the greater surface roughness results in a "dome" of air that is less stable than in surrounding open areas. The dome may extend to 200 to 300 meters in height.[98]

Mathematical Expressions

Although some dispersion of the plume occurs as it rises, practical approaches to calculating plume dispersion assume that dispersion begins upon the plume reaching

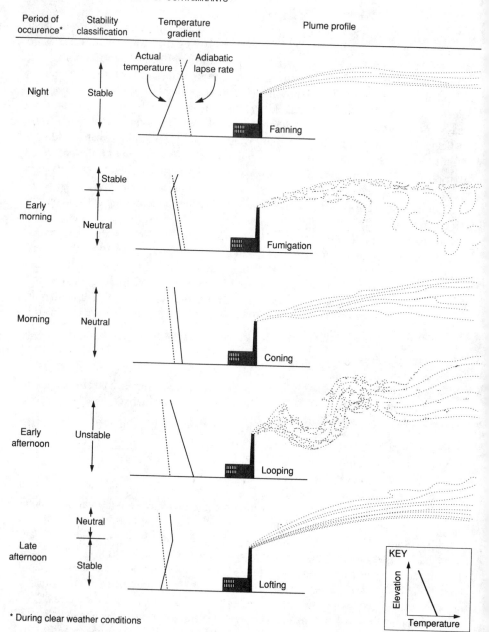

* During clear weather conditions

FIGURE 4-38
Plume profile for various atmospheric stability classifications (after Briggs).[99]

TABLE 4-16
Diurnal sequence in clear weather

Time of day	Phenomenon	Plume diffusion pattern
Night	Air near the ground is cooled as ground loses heat, forming an inversion.	Fanning
Early morning	Solar energy heats the ground, and warmed air rises as cool air replaces it in a convective eddy motion.	Fumigation (downward eddies bring part of the plume to the ground)
Morning	Convective eddies eventually penetrate inversion.	Coning
Early afternoon	Intensified heating may create large, convective eddies.	Looping
Late afternoon	Convection diminishes and inversion builds from the ground up.	Lofting (ideal as stable air below prevents downward diffusion)

the effective stack height. That is, plume rise and dispersion are calculated separately in two steps as depicted in Fig. 4-37. After the plume rise is calculated, any further effects of buoyancy and momentum flux are neglected with the plume dispersing only in response to wind velocity and atmospheric turbulence.

Work done by Pasquill and Gifford around 1960 advanced the theory that turbulence under any condition of atmospheric stability is random such that the concentration of pollutants around the centerline of the plume has a Gaussian (i.e., normal) distribution as depicted in Fig. 4-39.

On the basis of the Gaussian distribution, Pasquill derived the following equation, as modified by Gifford, for the concentration of pollutants in a plume traveling along the direction of wind (i.e., the X-axis).[100]

$$C(x, y, z) = \frac{Q}{2\pi \sigma_y \sigma_z u} \exp\left[-\frac{1}{2}\left(\frac{y}{\sigma_y}\right)^2\right] \left\{\exp\left[-\frac{1}{2}\left(\frac{z-H}{\sigma_z}\right)^2\right]\right.$$

$$\left. + \exp\left[-\frac{1}{2}\left(\frac{z+H}{\sigma_z}\right)^2\right]\right\} \quad (4\text{-}51)$$

where C = concentration of pollutants at coordinates $x, y, z\,(\text{mg/m}^3)$
Q = emission rate of pollutants (mg/sec)
σ_y, σ_z = horizontal (crosswind) and vertical standard deviations
of pollutant concentration along the centerline of the plume
(i.e., dispersion coefficient) (m) (See Figs. 4-40 and 4-41)
u = mean wind velocity (m/sec)
x = downwind distance along the centerline of the plume (m)
y = horizontal distance from the centerline of the plume (m)
z = vertical distance from ground level (m)
H = plume height (m)

TABLE 4-17

Pasquill chart for classifying atmospheric stability (after Turner)[101]

Surface wind speed at 10 m (m/sec)	Day			Night		
	Incoming solar radiation			Mostly overcast or ≥ 4/8 cloud cover	Mostly clear ≤ 3/8 cloud cover	Overcast (day or night)
	Strong	Moderate	Slight			
< 2	A	A-B	B			D
2-3	A-B	B	C	E	F	D
3-5	B	B-C	C	D	E	D
5-6	C	C-D	D	D	E	D
> 6	C	D	D	D	D	D

Key:
A Extremely unstable
B Unstable
C Slightly unstable
D Neutral
E Slightly stable
F Stable to extremely stable

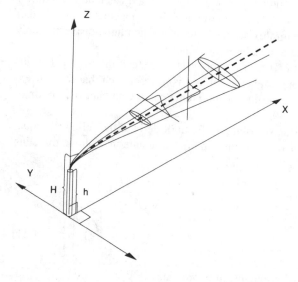

FIGURE 4-39
Assumed Gaussian distribution of pollutant concentration (after Turner).[102]

This equation relies on several assumptions:

- Wind velocity and direction are constant over height and over the averaging period.
- The emission rate is constant.
- The plume reflects completely at the ground (i.e., no deposition).
- No diffusion occurs in the direction of the plume travel.

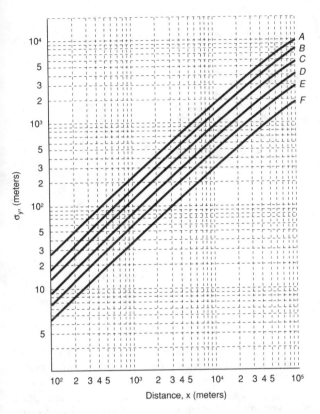

FIGURE 4-40
Standard deviation, σ_y, in the crosswind direction, as a function of distance downwind.[103]

Example 4-11 Gaussian dispersion equation. Simplify Eq. 4-51 to estimate concentration along the plume centerline at ground level.

Solution. In this case, y and z both equal 0. Eq. 4-51 thus reduces to the following form:

$$C_{x,0,0} = \frac{Q}{\pi \, \sigma_y \, \sigma_z \, u} \exp\left[-\frac{1}{2} \left(\frac{H}{\sigma_z} \right)^2 \right] \qquad (4\text{-}52)$$

Example 4-12 Gaussian dispersion equation. Further simplify Eq. 4-52 to describe a ground level source with no thermal or momentum flux, the typical release that occurs at a hazardous waste site. In this situation the effective plume rise, H, is essentially 0.

Solution. Equation 4-52 thus reduces to the following form:

$$C_{x,0,0} = \frac{Q}{\pi \, \sigma_y \, \sigma_z \, u} \qquad (4\text{-}53)$$

The values of σ_y and σ_z vary as a function of atmospheric turbulence and downwind distance and may be estimated from Figs. 4-40 and 4-41. Again, atmospheric

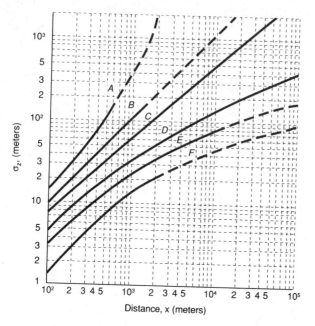

FIGURE 4-41
Standard deviation, σ_z, in the vertical direction as a function of distance downwind.[104]

stability is used as an indicator of turbulence. Figures 4-40 and 4-41 provide values for σ_y and σ_z as a function of downwind distance from the source and the atmospheric stability classification.

Example 4-13   Estimation of downwind concentration. Two acres of soil are contaminated with toluene and the estimated release (flux) from the soil to the atmosphere is 5.7×10^{-6} g/cm$^2 \cdot$ s. The maximum exposed individual is 2 km downwind of the center of the exposed area. Prevailing wind velocity is 5.0 m/sec, and the area is characterized by slight incoming solar radiation during the day. Determine if the contamination at the exposure point will exceed a state standard for toluene of 3770 μg/m^3.

Solution

Flux	$= 5.7 \times 10^{-6}$ g/cm$^2 \cdot$ sec
Area	$= 2$ acres $= 2 \times 4.04687 \times 10^8 = 8.09 \times 10^8$ cm^2
Q	$= (5.7 \times 10^{-6})(8.09 \times 10^8) = 4.61 \times 10^3$ g/ sec
	$= 4.61 \times 10^9 \mu$g/ sec
x	$= 2000$ m
Stability class D	
σ_y	$= 150$ m (from Fig. 4-40)
σ_z	$= 50$ m (from Fig. 4-41)
μ	$= 5.0$ m/sec
C(2000,0,0)	$= \dfrac{4.61 \times 10^9}{\pi(150)(50)(5)} = 3.91 \times 10^4$
C	$= \underline{\underline{39,100 \ \mu\text{g/m}^3}}$

This value exceeds the state standard of 3770 μg/m^3 by more than an order of magnitude.

Estimation of Contaminant Concentrations

The accuracy of the concentrations estimated by the dispersion method presented in Eq. 4-51 is best for medium-range distances, beyond the boundary of most sites. Aerodynamic effects due to site features such as buildings and trees can create localized downwash effects that can result in much higher concentrations than those estimated by Eq. 4-51. Further, plume rise occurs over some distance downwind and is not immediate as assumed in many mathematical approaches. For long distances, the accuracy of Eq. 4-51 lessens as well. Turner states that "several fold errors in estimate of σ_z can occur for longer travel distances;" however, it is "expected to be correct within a factor of 2 . . ." within the following maximum downwind distances:[105]

- "Few hundred meters" regardless of stability,
- "Few kilometers" for neutral to moderately unstable conditions, and
- "10 kilometers or more" for unstable conditions with an **inversion** at 1000 meters or less.

Under these conditions and considering other errors, Turner states that the ground level centerline concentration ". . . should be correct within a factor of 3."[106] Although this may be considered a large error, it is as accurate as results obtained from ground water modeling.

Equation 4-51 can be used to show relative ground level concentration with respect to distance as in Fig. 4-42 (actual concentration is estimated by multiplying ordinate values by Q/u). The effect of different stability classes is shown in Fig. 4-43 where apparently the less stable the atmosphere, the higher the ground level concentration within close proximity to a source. The influence of various effective stack heights is shown in Fig. 4-44. It is apparent from the figure that effective stack height has a great influence on maximum ground level concentration.

Maximum Ground Level Concentrations.   As stated previously, most concern focuses on ground level concentrations, particularly the maximum concentration and

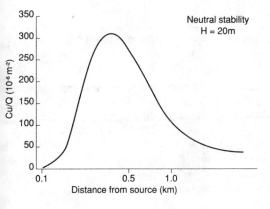

FIGURE 4-42
Relative ground level
concentration vs. distance.

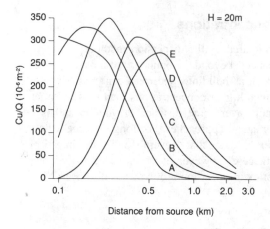

FIGURE 4-43
Relative ground level concentration vs. distance for various stability classes.

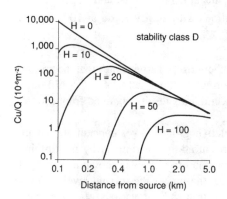

FIGURE 4-44
Relative ground level concentration vs. distance for various effective stack heights.[107]

its location and duration. The maximum ground level concentration is referred to as MGLC. For ground level releases, stable atmospheric conditions obviously create the greatest concentrations, all else being equal. The situation is more complicated for elevated releases. For an elevated release of given rate and duration, the absolute value of the maximum ground level concentration depends on effective stack height, atmospheric stability, and wind velocity.

Example 4-14 Variables affecting maximum concentration. Using Figs. 4-43 and 4-44, determine the relative influences on MGLC of the variables—wind velocity, stability class, and stack height?

Solution. Doubling the effective stack height from 10 m to 20 m and from 50 m to 100 m results in a 75% and 84% reduction in the MGLC, respectively. In comparison, doubling wind velocity for a constant effective stack height would reduce MGLC by 50%. Thus, *effective* stack height has a greater effect than wind velocity. However, effective stack height equals stack height plus plume rise, and the latter is inversely proportional to wind velocity. A doubling of wind velocity for a constant stack height would thus result in <u>more</u> than a 50% reduction in MGLC. An increase in wind velocity also tends to stabilize the atmosphere. Thus, it is unclear from these figures which has a greater effect—wind velocity or stack height. It is clear that both have a much greater effect than stability class.

Given values for these three variables (effective stack height, wind speed, and stability class) and an emission rate, the MGLC and its distance from a source can be calculated using the previously presented equations. Even though wind dilutes continuous releases, maximum ground level concentrations for releases with thermal or momentum flux tend not to occur at low wind velocities because such conditions promote a higher plume rise. Turner states that maximum ground level concentrations "occur at intermediate wind speed" which represents a balance between the countervailing effects of wind on dilution and plume rise.[108] It should be noted that these equations cannot be used to estimate *extreme* values that occur for a relatively short duration and which exceed the calculated MGLCs. Extreme concentrations for elevated releases tend to occur in unstable air when atmospheric turbulence may carry a plume to the ground almost immediately, well before it is more completely dispersed by a longer travel distance. In contrast, models based on the preceding equations calculate MGLCs for durations of 10 minutes to 1 hour, and results are extrapolated to longer averaging periods. Table 4-18 presents the situations that create the maximum ground level concentrations for elevated sources.

TABLE 4-18
Maximum ground level concentrations for elevated sources[109]

Type of turbulence	Downwind distance	Duration
Aerodynamic downwash	1–3 stack heights	Very short
Unstable	1–5 stack heights	Few minutes
Fumigation	30–40 kilometers for tall effective stack heights	30 minutes
Neutral or unstable with inversion	—	Varies

Mathematical Modeling.   The mathematical expressions used to estimate air dispersions are complex. Considering the degree to which wind speed, wind direction, and atmospheric stability can vary, and adding complicating factors such as topography and wake effects, the calculation of contaminant concentrations clearly requires the use of computer models. The EPA states, " . . . Gaussian dispersion models can and have been successfully applied to the types of sources encountered at Superfund sites; but the number of applicable models is limited."[110]

In order to use most models to estimate the impact of a stationary point source on air quality, certain characteristics of the source must be known. At a minimum, the following information is required to execute a model:

- Pollution emission rate
- Stack height
- Stack gas temperature and volume flow rate (for plume rise calculations)

- Location of the emissions point with respect to surrounding topography, and the character of that topography
- Detailed descriptions of all structures in the vicinity of, or attached to, each stack

Also, similar information from other significant sources in the vicinity should be known. This information is used in conjunction with the modeling of the source in question to account for all contributions to ambient concentrations.

Modeling typically proceeds through two levels of sophistication. A screening model is usually run first. These models employ conservative assumptions and default values that tend to overstate ground level concentrations. Therefore, if the results show that the impacts do not pose significant risk even under these conservative conditions, there is usually no need to execute a refined model. A refined model should use extensive meteorological data (e.g., at least several years of hourly meteorological data). They typically can analyze multiple sources and variable source configurations.

Air dispersion models can account for deposition, downwash effects, and other complicating factors. These models can be configured to calculate concentration levels for all areas surrounding the source based on hourly average values or longer-term average values for wind speed, direction, atmospheric stability, and ambient temperature. Results can be determined for durations as short as one hour or as long as the length of the meteorological records, providing estimates of both acute and chronic exposure periods. Calculation of downwind concentrations for intervals less than hourly intervals requires collection of on-site meteorological data. Estimates of ground level concentrations are provided by most models for downwind receptors, allowing the determination of exposure levels and critical locations for air monitoring stations. The models can also analyze worst case scenarios and be used to facilitate decisions on safety precautions and exposure assessment.

Special Topics

There are a number of situations in which the mathematical expressions presented in this subsection may not apply, even with modification. These include dominant surface features, inversions, chemical transformations, and deposition.

Dominant Surface Features. Topographical features on the earth's surface influence turbulence and in turn affect dispersion. Ridges, hills, and mountains divert the wind, thereby altering wind direction and increasing wind turbulence. Releases from elevated locations such as from a ridge result in higher effective stack heights in relation to downwind receptors at lower elevations. Releases in large valleys can be confined by the surrounding ridges, causing the plume to disperse horizontally more in one direction and less in another than if the ridges had not existed. Receptors located on valley walls may experience higher concentrations if releases occur in the valley. Large water bodies tend to foster stable air because water cools and heats slower than land. However, this creates the familiar alternating, diurnal wind directions.[111]

The features at a site can have a major influence on dispersion and create higher concentrations at ground level than otherwise expected. A classic example is the aerodynamic effect of a building or grove of trees creating a downwash. This easily can bring the plume immediately to the ground and fill the air cavity behind the obstacle with concentrated pollutants. In such cases, the height of a stack and the effluent velocity are important considerations. Figure 4-45 illustrates building wake effects.

Inversion.   An inversion is the reverse of the normal progression to lower atmosphere temperatures at higher altitudes. Two types of inversions can occur: ground (low-level) and elevated. A ground inversion lasts less than a day and occurs frequently in response to the sun warming the ground surface (see Fig. 4-38). This type of inversion is short-term and usually breaks up as the day progresses. In contrast, an elevated inversion occurs for a longer term and on a regional scale when a stable, warm air mass overlies a colder layer. The interface between the two air masses may be 1500 to 4000 feet above the ground in the eastern part of the United States and only 1000 feet along the Pacific Coast.[112] Inversions along the Pacific Coast result from specific topographic, climatic, and oceanographic features, while those in the eastern United States are caused by stable high-pressure systems and are referred to as frontal inversions.

An elevated inversion can create severe conditions by trapping airborne pollutants and allowing buildup of contamination, especially in confined topographic situations (e.g., a deep valley). Even plumes rising from tall stacks may not break through as the stable and warm overlying air dissipates buoyancy. The impacts associated with an elevated inversion are regional in nature. The ground inversion poses concern in and around hazardous waste facilities having elevated releases because convective eddy currents, formed as the ground surface warms to a temperature higher than the

Emissions from roof-top exhaust

Emissions from short stack

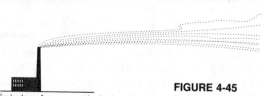

Emissions from properly designed stack

FIGURE 4-45
Building wake effects.[113]

air, can pull an elevated plume downwards. This is termed **fumigation** as depicted in Fig. 4-38 and differs from the stagnation caused by elevated inversions.

Turner presents a mathematical approach for calculating ground level concentrations during fumigation. This approach assumes that the plume is emitted entirely into a stable layer and then calculates the portion of the plume pulled downwards with some adjustment for increased horizontal mixing. Smith suggests that the various formulas proposed for the described fumigation " . . . may easily be in error by as much of a factor as 5."[114]

Chemical Transformation.   The significance of chemical and biological processes in transforming hazardous substances in the subsurface were discussed earlier in this chapter (Sec. 4-3). Chemical transformations likewise occur in the atmosphere. The depletion of stratospheric ozone and the formation of acid rain are excellent examples. Even though releases from hazardous waste sites can consist of reactive constituents, chemical transformation is often of little significance because of the relatively quick travel time to receptors, usually not exceeding one or two hours[115] and considerably shorter in many cases.

Deposition.   Although precipitation may absorb gases and function as an aid to deposition, the effect of this mechanism at hazardous waste sites is insignificant. However, deposition of particulate matter due to the force of gravity can be significant. The Gaussian-based equations ignore gravity and are assumed to apply only to gases and particles less than 20 μm in diameter, a reasonable assumption for treated stack gases. In contrast, hazardous waste sites can release larger particles as dust that settle due to gravitational force, and may be resuspended later by higher wind speed. Equation 4-51 does not apply to such cases because it assumes full reflection at the ground surface. It would tend to underestimate concentrations and accumulations of contaminants at ground locations near the source and overestimate the impact at more distant locations.

The EPA has presented guidelines for the deposition of fugitive dust according to particle size. The source of the dust is considered to be a release at ground level without buoyancy or momentum flux. The expected travel distances are shown in Table 4-19.

Several mathematical concepts have been proposed to account for the effect of deposition of fine particles with most defining a deposition velocity and multiplying it times the concentration to yield a deposition flux. Deposition velocity is a function of several variables with a default value of 0.1 cm/sec acceptable as a maximum upper bound.[116] Such a low velocity, especially for an upper bound, shows that particles tend more to move in a suspended manner rather than fall freely.

Example 4-15   Dual stacks.   The gases from two proposed side-by-side furnaces are each to be treated by a specific technology that is considered the "best available." The treated gases are then to be released from dual 80-foot stacks, one for each furnace. The resultant maximum ground level concentrations (MGLCs) are estimated to exceed allowable standards. The height of the stacks cannot be increased due to land use restrictions. What can be done other than additional treatment (which perhaps cannot be accomplished except at prohibitive cost) to reduce MGLCs?

Solution. Combining the two stacks as one will increase the effective stack height. Even if the cross-sectional area of the new stack equals the sum of the two proposed stacks, the circumference of the plume upon release will be less. A smaller circumference will result in less horizontal dispersion and less entrainment of ambient air, thereby slowing the cooling of the plume. The plume will retain its buoyancy longer and thereby rise higher. Other alternatives to achieve a similar effect would include increasing the exit velocities from the stack(s) or heating the stack gas.

TABLE 4-19

Guidelines for deposition of fugitive dust according to particle size[117,118]

Size of particle	Travel distance
> 100 μm	5-10 meters from source
30–100 μm	within 100 meters from source except for cases of "high atmospheric turbulence"
15–30 μm	transportable "considerable distances downwind"
< 15 μm	"likely to stay adrift"

DISCUSSION TOPICS AND PROBLEMS

4-1. Rewrite Eqs. 4-20, 4-21, and 4-26 for three dimensions.

4-2. Determine the elevation, pressure, and total hydraulic heads at all points (a–g) labeled in Fig. P-1 using datum #1.

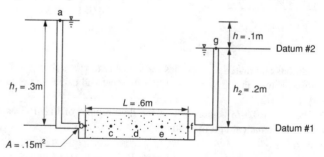

FIGURE P-1

4-3. Determine the hydraulic conductivity for the experiment in Fig. P-1 using datum #1 if the measured flow rate is 100 ml in 15 minutes.

4-4. Redo problems 2 and 3 using datum #2.

4-5. Draw a flow net for the sample and head conditions in problem 2.

4-6. Calculate the rate of flow through the sample shown in Fig. 4-8 assuming the hydraulic conductivity is 5×10^{-3} cm/sec.

4-7. Using the phase diagram presented as Fig. 4-9 of the chapter, develop a relationship for the void ratio, e, in terms of the porosity, n.

4-8. Perform a triaxial permeability test on a cylindrical compacted clay sample that is 2.8 in. in diameter by 6 in. high. It is planned to use a backpressure of 50 psi to aid in saturation and average effective stress of 10 psi. The inflow and outflow burettes have an inside diameter of 0.5 cm.

(a) What is the cell pressure you will use in this test?

(b) If the hydraulic head causing flow is to be applied by elevating the water level in the inflow burette a distance of 20 cm above the water level in the outflow burette, what is the gradient?

(c) If the hydraulic head causing flow applied by elevating the water level in the inflow burette a distance of 10 cm above the water level in the outflow burette after 8 hours, what is the hydraulic conductivity of the sample?

4-9. A laboratory experiment similar to the one conducted by Darcy is shown in Fig. P-2.

(a) What is the pressure head at point A in sand of the laboratory experiment shown below?

(b) Assuming a porosity of 0.5, how long will it take a molecule of water to pass through from one end of the sand to the other?

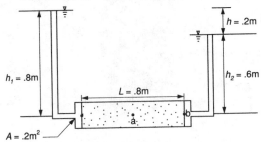

$h_1 = .8m$ $L = .8m$ $h = .2m$ $h_2 = .6m$ $A = .2m^2$

FIGURE P-2

4-10. The boundary conditions and material properties for a flow regime are shown in Fig. P-3.

(a) For the conditions shown in Fig. P-3, sketch the flow net and calculate the flow rate through the system.

(b) Solve for the heads on 0.5 m centers within the flow regime of Fig. P-3 using the finite difference mathematical approximation.

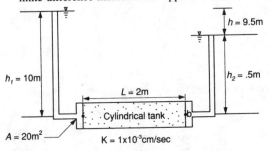

$h_1 = 10m$ $L = 2m$ $h = 9.5m$ $h_2 = .5m$ $A = 20m^2$ Cylindrical tank $K = 1\times10^{-3}cm/sec$

FIGURE P-3

4-11. Differentiate between hydrogeology and ground water hydrology.

4-12. Discuss your opinion about the use of the terms aquitard and aquiclude in the context of hazardous waste management from the perspective of a(n):

(a) landfill operator

(b) state environmental regulator

(c) environmental organization spokesperson

(d) John Q. Public

4-13. Differentiate between water table and piezometric surface.

4-14. The derivation of the Laplace equation assumed steady flow and homogeneous, isotropic conditions and showed that hydraulic head depended only upon position within the aquifer. What would the dependent variables be for:

(a) anisotropic conditions

(b) heterogeneous conditions

(c) transient conditions

4-15. From a contaminant transport viewpoint, describe the advantages and disadvantages of selecting in situ bioremediation to clean a ground water supply aquifer contaminated with petroleum hydrocarbons from a leaking underground storage tank. Benzene is the compound of concern while xylene and ethylbenzene are present at lower concentrations.

4-16. Critically comment on the negative aspects of the use of computer models.

4-17. Sample A has a porosity of 40%. Sample B has a porosity of 8%. Show the relationship between seepage velocities for samples A and B under the same gradient.

4-18. Using Darcy's law and porosity, show that the seepage velocity is not accurate at extreme values.

4-19. Determine the pressure, elevation, and total heads at points a, b, and c in Fig. P-4.

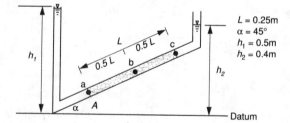

$L = 0.25\text{m}$
$\alpha = 45°$
$h_1 = 0.5\text{m}$
$h_2 = 0.4\text{m}$

FIGURE P-4

4-20. Determine the pressure, elevation, and total heads at points A, B, and C in Fig. P-5.

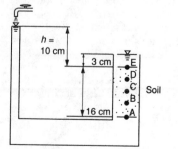

FIGURE P-5

4-21. Given a region found to have a glacial till overlying a thin lacustrine clay layer overlying a competent sandstone, how would you model ground water flow through the system?

4-22. What provides soil the capacity to buffer acidic water?

4-23. For what conditions would ideal sorption not apply?

4-24. What are two properties that can be measured in a laboratory that indicate the "hydrophobicity" of an organic chemical?

4-25. What are five important characteristics of soil affecting its molecular sorption capacity?

4-26. Which of the five characteristics named above apply more to inorganics than organics?

4-27. What is the difference between retention capacity of a soil and its molecular sorption capacity?

4-28. What indirect effect would depth to ground water expect to have on the sorption of organic contaminants?

4-29. How do the releases from an explosion differ from a continuous release? How would this be handled mathematically?

4-30. Frequent shifts of wind direction would cause a plume of air contaminants to do what?

4-31. A stable layer of air above an unstable layer would have what effect on the dispersion of a ground release of air emissions? On the dispersion of emissions from a stack sufficiently tall to penetrate into the stable layer?

4-32. Eq. 4-51 ignores diffusion in the direction of transport (x-axis). What assumption regarding emission duration must this be based upon?

4-33. What differences would occur in effective stack height for two releases, both at the same temperature but one containing water droplets from having passed through a cooling tower?

4-34. Determine the concentration of TCE 5.0 km downwind of a point source at ground level that releases 3,000 g/sec TCE on a continuous basis. The following atmospheric conditions are given: wind velocity $= 4.0$ m/sec; stability class B.

REFERENCES

1. U.S. Environmental Protection Agency: *Procedures for Conducting Air Pathway Analyses for Superfund Applications*, Volume II, Office of Air Quality Planning and Standards, Research Triangle Park, NC, January 1989.

2. U.S. Environmental Protection Agency: *Superfund Exposure Assessment Manual*, EPA/540/1-88/001, Office of Emergency and Remedial Response, Washington, D.C., April 1988.

3. U.S. EPA, *Procedures for Conducting Air Pathway Analyses for Superfund Applications*, Volume II.

4. de Percin, P. R.: "Air Pollution from Land Disposal Facilities," *Standard Handbook of Hazardous Waste Treatment and Disposal*, H. M. Freeman, ed., McGraw-Hill Book Company, New York, NY, p. 10.53–10.102, 1988.

5. Skidmore, E., and N. Woodruff: *Wind Erosion Forces in the United States and Their Use in Predicting Soil Loss*, Agriculture Handbook No. 346, U.S. Department of Agriculture, Washington, D.C., 1968.

6. U.S. EPA, *Procedures for Conducting Air Pathway Analyses for Superfund Applications*, Volume II.

7. U.S. Environmental Protection Agency: *Procedure for Conducting Air Pathway Analyses for Superfund Applications*, Volume III, Office of Air Quality Planning and Standards, Research Triangle Park, NC, January 1989.

8. U.S. Environmental Protection Agency: "Compilation of Air Pollutant Emission Factors," Manual AP-42.

9. Schroeder, P. R., A. C. Gibson and M. D. Smolen: *The Hydrologic Evaluation of Landfill Perfomance (HELP) Model*, EPA/830-SW-84-010, 1984.

10. Eastern Research Group, Inc.: "Design and Construction of RCRA/CERCLA Final Covers," prepared for CERI, U.S. EPA, EPA/625/4-91/025, Arlington, MA, May 1991.
11. "Management of Hazardous Waste Leachate," report prepared for U.S. EPA by Touhill, Shuckrow and Associates, Pittsburgh, PA, September 1980.
12. U.S. Environmental Protection Agency: *Superfund Exposure Assessment Manual*, U.S. EPA, EPA/540/1-88/001, Washington, D.C., April 1988.
13. Murray, C.L.: "Water Use, Consumption, and Outlook in the U.S. in 1970," *J. American Water Works Association,* vol. 65, pp. 302–308, 1973.
14. Pye, V. I., R. Patrick, and J. Quarles: *Groundwater Contamination in the United States,* University of Pennsylvania Press, PA, 1983.
15. Freeze, R.A., and Cherry, J.4.: *Groundwater,* Prentice-Hall, Inc., Englewood Cliffs, NJ, 1979.
16. Todd, D. K.: *Groundwater Hydrology,* 2nd ed., John Wiley and Sons Inc., New York, NY, 1980.
17. Fetter, C. W.: *Applied Hydrogeology,* 2nd ed., Merrill Publishing Co., Columbus, OH, 1988.
18. Cernica, J. N.: *Geotechnical Engineering,* CBS College Publishing, 1982.
19. Domenico and Schawartz: *Physical and Chemical Hydrogeology,* John Wiley and Sons, Inc., New York, NY, 1990.
20. Acar, Y.B.: "Electrokinetic Soil Processing (a Review of the State of the Art)." *Geotech. Special Publication No. 30,* ASCE, New York, NY, pp. 1420–1432, 1992.
21. Liakopoulos, A. C.: "Variation of the Permeability Tensor Ellipsoid in Homogenous Anisotropic Flow," *Water Resources Res.,* vol. 1, pp. 135-141, 1965.
22. Palmer, C.D., and R. L. Johnson: "Physical Processes Controlling the Transport of Non-Aqueous Phase Liquids in the Subsurface," *Transport and Fate of Contaminants in the Subsurface,* EPA/625/4-89/019.
23. Bear, Jacob, and Arnold Verruijt: *Modeling Groundwater Flow and Pollution,* D. Reidel Publishing Company, Dordrecht, Holland, p. 414, 1987.
24. Huyakorn, Peter S., and George F. Pinder: *Computational Methods in Subsurface Flow,* Academic Press, Orlando, FL, p. 472, 1983.
25. Walton, William C.: *Analytical Groundwater Modeling,* Lewis Publishers, Chelsea, MI, p. 173, 1989.
26. Walton, William C.: *Numerical Groundwater Methodology,* Lewis Publishers, Chelsea, MI, p. 272. 1989.
27. Spangler, M. G., and R. L. Hardy: *Soil Engineering,* 4th ed., Harper & Row, New York, NY, 1982.
28. Cedergren, H. R.: *Seepage, Drainage and Flow Nets,* 3rd ed., John Wiley & Sons Inc., NY, NY, 1989.
29. Driscoll, F. G.: *Groundwater and Wells,* 2nd ed., Johnson Division, St. Paul, MN, 1986.
30. *McGraw-Hill Encyclopedia of Science and Technology,* McGraw-Hill Book Co., New York, NY, 1982.
31. Brubaker, G. R.: "Lecture No. 1—Introduction to In Situ Bioremediation and Contaminant Transport," *HMCRI Superfund Conference,* November 28, 1990.
32. Weber, W. W: "Sorption Phenomena in Subsurface Systems: Concepts, Models, and Effects on Contaminant Fate and Transport," *Water Research,* Pergammon Press, vol. 25, no. 5, 1991.
33. Brady, N.C.: *The Nature and Properties of Soils,* 8th ed., McMillan Publishing Co., Inc., New York, NY, 1974.
34. Pignatello: "Sorption Dynamics of Organic Compounds in Soils and Sediments," *Reactions and Movement of Organic Chemicals in Soils,* edited by B. L. Sawhney and K. Brown, Soil Science Society of America, Inc., Madison, WI, 1989.
35. Brady, N.C.: *The Nature and Properties of Soils.*
36. Dragun, James: "The Soil Chemistry of Hazardous Materials," Hazardous Materials Control Research Institute, Silver Springs, MD, 1988.
37. Voice, T. C.: "Activated-Carbon Adsorption," *Standard Handbook of Hazardous Waste Treatment and Disposal,* Harry M. Freeman, ed., McGraw-Hill Book Company, New York, NY, 1989.
38. Dragun, "The Soil Chemistry of Hazardous Materials."
39. Pignatello, "Sorption Dynamics of Organic Compounds in Soils and Sediments."
40. Ibid.
41. U.S. Environmental Protection Agency: "Superfund Exposure Assessment Manual."
42. U.S. Environmental Protection Agency: "Transport and Fate of Contaminants in the Subsurface," EPA/625/4-89/019, Center for Environmental Research Information, Cincinnati, OH, September 1989.

43. Dragun, "The Soil Chemistry of Hazardous Materials."
44. Brady, N.C.: *The Nature and Properties of Soils.*
45. *McGraw-Hill Encyclopedia of Science and Technology.*
46. Pignatello, "Sorption Dynamics of Organic Compounds in Soils and Sediments."
47. Dragun, "The Soil Chemistry of Hazardous Materials."
48. Piwoni, M. D., and J. W. Keeley: "Basic Concepts of Contaminant Sorption at Hazardous Waste Sites," EPA/540/4-30/053, U.S. EPA, Robert S. Kerr Environmental Research Laboratory, Ada, OK, October 1990.
49. U.S. EPA: "Transport and Fate of Contaminants in the Subsurface,"
50. Brubaker, G. R.: "Lecture No. 1—Introduction to In Situ Bioremediation and Contaminant Transport," *HMCRS Superfund Conference*, November 28, 1990.
51. Piwoni, M. D., and J. W. Keeley: "Basic Concepts of Contaminant Sorption at Hazardous Waste Sites," EPA/540/4-30/053, U.S. EPA, Robert S. Kerr Environmental Research Laboratory, Ada, OK, October 1990.
52. Dragun, "The Soil Chemistry of Hazardous Materials."
53. U.S. EPA, "Transport and Fate of Contaminants in the Subsurface."
54. Ibid.
55. Pignatello, "Sorption Dynamics of Organic Compounds in Soils and Sediments."
56. Dragun, "The Fate of Hazardous Material in Soil
57. U.S. EPA, "Superfund Exposure Assessment Manual."
58. U.S. EPA, "Transport and Fate of Contaminants in the Subsurface."
59. Dragun, "The Fate of Hazardous Materials in Soil."
60. Brubaker, "Lecture No. 1—Introduction to In Situ Bioremediation and Contaminant Transport.
61. U.S. EPA, "Transport and Fate of Contaminants in the Subsurface." op. cit.
62. Dragun, "The Soil Chemistry of Hazardous Materials."
63. Weber, W. W. "Sorption Phenomena in Subsurface Systems: Concepts, Models, and Effects on Contaminant Fate and Transport." op. cit.
64. Brubaker, "Lecture No. 1—Introduction to In Situ Bioremediation and Contaminant Transport."
65. Wu, S. C. and P. M. Gschwend: "Sorption Kinetics of Hydrophobic Compounds to Natural Sediments," *Environmental Science and Technology*, American Chemical Society, vol. 20, no. 7, 1986.
66. U.S. EPA "Transport and Fate of Contaminants in the Subsurface," op. cit.
67. Dragun, "The Fate of Hazardous Materials in Soil," p. 154.
68. Brady, *The Nature and Properties of Soils.*
69. Ibid.
70. Weber W.W., "Sorption Phenomena in Subsurface Systems." op. cit.
71. Grim, R. E., Clay Mineralogy, McGraw-Hill, New York, NY, 1960.
72. U.S. EPA, "Transport and Fate of Contaminants in the Subsurface."
73. Ibid.
74. Dragun, "The Fate of Hazardous Materials in Soil," op. cit. p. 307.
75. Bohn, H. L., B. L. McNeal, and G.A. O'Conner: *Soil Chemistry,* John Wiley & Sons, Inc., New York, NY, 1979.
76. U.S. EPA, "Transport and Fate of Contaminants in the Subsurface," p. 46.
77. U.S. EPA, "Superfund Exposure Assessment Manual."
78. U.S. EPA, "Transport and Fate of Contaminants in the Subsurface."
79. Dragun, "The Fate of Hazardous Materials in Soil."
80. Ibid.
81. U.S. EPA, "Transport and Fate of Contaminants in the Subsurface."
82. Dragun, "The Fate of Hazardous Materials in Soil."
83. Brown, K.W., and D.C. Anderson: "Effects of Organic Solvents on the Permeability of Clay Soils," EPA 600/2-83-016, U.S. EPA, Cincinnati, OH, p. 45, 1983.
84. U.S. EPA, "Transport and Fate of Contaminants in the Subsurface," p. 45.
85. Ibid.
86. Ibid.

87. Piwoni, M.D., and J.W. Keeley, "Basic Concepts of Contaminant Sorption at Hazardous Waste Sites."
88. Ibid.
89. Freeze, R. A., and J. A. Cherry: *Groundwater*, Prentice-Hall, NJ, 1979.
90. Dragun, "The Soil Chemistry of Hazardous Materials."
91. Ibid.
92. Freeze and Cherry, *Groundwater*, op cit. p. 406.
93. Briggs, G. A.: *Plume Rise*, U.S. Atomic Energy Commission, Oak Ridge, TN, November 1969.
94. Ibid.
95. Ibid.
96. Shrock, J. E.: "Overview of Dispersion Models and Applications," *Air Pollution: Environmental Issues and Health Effects*, Pennsylvania Academy of Science, Easton, PA, 1991.
97. Turner, D. Bruce: *Workshop of Atmospheric Dispersion Estimates,* U.S. Environmental Protection Agency, Office of Air Programs, Publication No. AP-26, Research Triangle Park, NC, 1970.
98. Ibid.
99. Briggs, *Plume Rise.*
100. Pasquill, F.: "The Estimation of the Dispersion of Windborne Materials," *Meteorology Magazine,* 1961.
101. Turner, *Workshop of Atmospheric Dispersion Estimates.*
102. Ibid.
103. Ibid.
104. Ibid.
105. Ibid.
106. Ibid.
107. Ibid.
108. Ibid.
109. Ibid.
110. Smith, Maynard: *Recommended Guide for the Prediction of the Dispersion of Airborne Pollutants.* American Society of Mechanical Engineers, New York.
111. U.S. EPA, *Procedures for Conducting Air Pathway Analysis for Superfund Applications*, Vol. IV.
112. Briggs, *Plume Rise.*
113. U.S. EPA, Fluid Modeling Facility, Research Triangle Park, NC
114. Smith, Maynard: *Recommended Guide for the Prediction of the Dispersion of Airborne Pollutants.*
115. U.S. Environmental Protection Agency: *Procedures for Conducting Air Pathway Analyses for Superfund Applications*, Volume IV, Office of Air Quality Planning and Standards, Research Triangle Park, NC, July 1989.
116. Shrock, "Overview of Dispersion Models and Applications."
117. U.S. EPA, *Procedures for Conducting Air Pathway Analyses for Superfund Applications*, Vol. III.
118. U.S. Environmental Protection Agency: *Superfund Exposure Assessment Manual,* Office of Emergency and Remedial Response, Washington, D.C., April 1988.

ADDITIONAL READING

Bahr, L. M.: "Analysis of Nonequilibrium Desorption of Volatile Organics during Field Test of Aquifer Decontamination," *J. Contaminant Hydrology*, vol. 4, no. 3, p. 205, 1989.

Boucher, P.: "Ecological Assessment and Modeling of a Contaminated Wetland," *Superfund '90*, Haz. Materials Control Research Instit., Washington, D.C., pp. 148–152, 1990.

Charbeneau, R. J., et al.: "Kinematic Modeling of Multiphase Solute Transport in the Vadose Zone," NTIS PB89-207948, Robert S. Kerr Environ. Res. Lab., Ada, OK, 1989.

Cooper, C. D., and Alley, F. C.: *Air Pollution Control*, PWS Publishers, Boston, MA, 1986.

Devinny, J., et al.: "Subsurface Migration of Hazardous Wastes," Van Nostrand Reinhold, New York, NY, 1989.

Dupont, R. R., et al.: "Evaluation and Modeling of Volatile Organic Vapor Transport in the Unsaturated Zone for Groundwater Quality Protection," *Water Resources Division*, NTIS PB90119736, U.S. Geologic Survey, Reston, VA, 1989.

Hauptmann, M. G., et al.: "Use of Groundwater Modeling During Superfund Cleanup," *Superfund '90*, Haz. Materials Control Research Instit., Washington, D.C., pp. 110–116, 1990.

Hosseinipour, Z., and K. Brown: "Application of Hydrodynamic and Water Quality Models at a Superfund Site in Montana," *Proc. 1989 Natl. Conf. Hydraul. Eng.*, ASCE, New York, NY, p. 466, 1989.

Johnson, R. L., et al.: "Diffusive Contaminants Transport in Natural Clay: A Field Example and Implications for Clay-lined Waste Disposal Sites," *Environ. Sci. Technol.*, vol. 23, p. 340, 1989.

Kapila, S., et al.: "Field and Laboratory Studies on the Movement and Fate of Tetrachlorodibenzo-p-dioxin in Soils," *Chemosphere*, vol. 18, p. 1297, 1989.

Kincare, K. A., and S. M. Aulenbach: "Geostatistical Decision Making Process for Plume Modeling in Cadillac, MI," *Superfund '89*, Haz. Materials Control Research Instit., Washington, D.C., pp. 146–151, 1989.

Koch, D., et al.: "Using a Three-Dimensional Solute Transport Model to Evaluate Remedial Actions for Groundwater Contamination at the Picatinny Arsenal, NJ," *Superfund '89*, Haz. Materials Control Research Instit., Washington, D.C., pp. 152–156, 1989.

Li, W. W., et al.: "Modeling of On-Site Air Concentrations at Superfund Sites," *Superfund '90*, Haz. Materials Control Research Instit., Washington, D.C., pp. 117–122, 1990.

McCarty, J. F., and J. M. Zachara: "Subsurface Transport of Contaminants," *Environ. Sci. Technol.*, vol. 23, p. 496, 1989.

Militana, L. M., and S. C. Mauch: "Statistical Modeling of Ambient Air Toxics Impacts During Remedial Investigations at a Landfill Site," *Superfund '89*, Haz. Materials Control Research Instit., Washington, D.C., pp. 157–162, 1989.

Olsen, R. L., and A. Davis: "Predicting the Fate and Transport of Organic Compounds in Groundwater," *Superfund '89*, Haz. Materials Control Research Instit., Washington, D.C., p. 145, 1989.

Ozbilgin, M. M., et al.: "Evaluation of Groundwater Extraction and Treatment Scenarios Using Quasi-Three-Dimensional Transport Model," *Superfund '90*, Haz. Materials Control Research Instit., Washington, D.C., pp. 386–397, 1990.

Reinhart, D. R.: "Fate of Selected Organic Pollutants During Landfill Codisposal with Municipal Refuse," *Diss. Abstr. Int. B*, vol. 50, p. 1830, 1989.

Ridgway, H. F., et al.: "Investigation of the Transport and Fate of Gasoline Hydrocarbon Pollutants in Groundwater," NTIS PB89151450, U.S. Geologic Survey, Reston, VA, 1989.

Roffman, A., and R. Stoner: "Air/Superfund National Technical Guidance Study Series. Volume 4. Procedures for Dispersion Modeling and Air Monitoring for Superfund Air Pathway Analysis." Office of Air Qual. Plann. and Stand., NTIS PB90113382, U.S. EPA, Research Triangle Park, NC, 1989.

Rudy, R., et al.: "RANDOM-WALK Modeling or Organic Contaminant Migration from the Theresienfeld Landfill Located in the Vienna Basin Aquifer of Austria," *Superfund '89*, Haz. Materials Control Research Instit., Washington, D.C., pp. 163–180, 1989.

Stein, C. L., and D. F. McTigue: "Chromium Distribution and Transport Beneath a Contaminated Site," NTIS DE89009696, U.S. Department of Energy, Washington, D.C., 1989.

Stevens, D. K., et al.: "Simulation of Oxygen Dynamics in the Vadose Zone," *Environ. Eng. Proc. 1989 Specialty Conf.*, ASCE, New York, p. 694, 1989.

Stevens, D. K., et al.: "Sensitive Parameter Evaluation for a Vadose Zone Fate and Transport Model," NTIS PB89-213987, Robert S. Kerr Environmental Research Laboratory, Ada, OK, 1989.

Stoner, R.: "Air/Superfund National Technical Guidance Study Series. Volume 1. Application of Air Pathway Analyses for Superfund Activities." Office of Air Qual. Plann. and Stand., NTIS PB90-1 13374, U.S. EPA, Research Triangle Park, NC, 1989.

Study of Dredging and Dredged Material Disposal Alternatives, Report 5, Evaluation of Leachate Quality, *Army Eng. Waterways Exp. Sta. Rep. No. WES/TR/EL-88-15-5*, 1989.

Thomann, R. V.: "Bioaccumulation Model of Organic Chemical Distribution in Aquatic Food Chains," *Environ. Sci. Technol.*, vol. 23, p. 699, 1989.

Thorne, D. J., and A. S. Rood: "Surface Water for Site Closure of a Remediated Dioxin (TCDD) Site," *Superfund '89*, Haz. Materials Control Research Instit., Washington, D.C., pp. 117–121, 1989.

Ulirsch, G. V., et al.: "Role of Environmental Fate and Transport Data in Health Assessments: Four Case Studies," *Superfund '90*, Haz. Materials Control Research Instit., Washington, D.C., pp. 128–132, 1990.

Walton, J. C., et al.: "Model for Estimation of Chlorinated Solvent Release from Waste Disposal Sites," J. *Hazardous Mater.*, vol. 21, no. 1, p. 15, 1989.

Wark, K., and Warner, C. F.: *Air Pollution: Its Origin and Control*, Harper & Row, New York, 2nd ed., 1981

Whetzel, N.: "SLAPMAN Model for Groundwater Pathway," Office of Water Regulations and Standards, NTIS PB89-138747, U.S. EPA, Washington, D.C., 1989.

ACKNOWLEDGMENTS

The authors wish to acknowledge the following individuals who assisted with the preparation of this chapter: Paul Gruber, *ERM-South*; Nick DeSalvo, Kathryn Z. Klaber, Ken Weiss, John Harsh, *ERM, Inc.*; Mike De Busschere, *ERM-Midwest;* Tim Mead, William Reichardt, Charles Johnson, Michael Costa, and Kevin Grega, *Bucknell University;* and David Garg, *Pennsylvania Department of Environmental Resources.* The principal reviewer for the **ERM Group** was Roy O. Ball, Ph.D., P.E., *ERM-North Central.*

CHAPTER

5

TOXICOLOGY

Was ist das nit gifft ist?
Alle ding sind gifft und nichts ohn gifft
Allein die dosis macht das ein ding fein gifft ist.

What is there that is not poison?
All things are poison and nothing without poison.
Only the dose determines that an agent is not a poison.
                                        Paracelsus, c. 1538

A waste is considered hazardous if it exhibits any of four characteristics—reactivity, corrosivity, ignitability, or toxicity. Of these, the potential for toxicity, particularly to humans, has caused the greatest public concern and has prompted the massive regulatory initiatives in hazardous waste management. It follows that the management of hazardous waste has as its fundamental objective the protection of human health by reducing the risk, if not the toxicity, of the substances in hazardous wastes to acceptable levels. An understanding of toxicology is therefore needed to determine whether the management of hazardous waste or the remediation of contaminated sites meets this objective.

Chapter 4 covered the fate and transport of substances from their release into the environment up to their contact with receptors. This chapter addresses the toxic effects caused by hazardous substances upon gaining access to the human body and other living organisms. Toxic effects can range from mild allergic reactions to death. Indeed, the constituents of some hazardous waste can kill humans outright as evidenced by graphic accounts. Some constituents can cause death in insidious ways such as cancer.

What is toxicology? Toxicology deals with the adverse effects caused by exposure of living organisms to chemical substances. Although poisonous plants and animal venoms have been known since early human history, the emergence and recognition of toxicology as an independent scientific discipline did not occur until well into the 20th Century. It has evolved as a discipline that incorporates areas of several other basic sciences such as physiology, pharmacology, biochemistry, molecular biology,

225

and epidemiology. Being basically a conglomerate of biological disciplines, toxicology is, thus, not an exact science in the same sense as physics or chemistry. Much remains unknown, particularly the basic mechanisms by which chemicals produce their deleterious effects.

What is the single, most important thing for an engineer to know about toxicology? The engineer cannot and should not practice toxicology. An engineer's education and experience falls far short of what is needed to make toxicological evaluations. However, a general understanding of how a toxicologist makes decisions will open meaningful communication between the two disciplines which share the responsibility to decide whether an engineering solution offers adequate protection.

What is the key limitation of applying toxicology? Because the fundamental mechanisms that cause toxic responses are not fully understood, toxicological findings are largely based on observations, only partially derived directly from humans. Toxicology is to a large extent supported by information obtained from experiments on animals. When extrapolated to humans, such test data carry a considerable degree of uncertainty. This lack of precision is not only evident when using information obtained in one animal species (e.g., the mouse) to predict toxic action in another (e.g., humans), but also when extrapolating from the high exposures used in an animal experiment to the low-dose situation usually encountered with respect to exposure of the general population. Depending upon the mathematical method used for such extrapolations, the risk estimates obtained may differ by orders of magnitude. Thus, the quantification of the toxicity of a chemical has associated with it a large degree of uncertainty, typically dwarfing the uncertainty inherent in a scientific analysis of the fate and transport in the environment of the same chemical.

Having said this about the uncertainty associated with toxicological data, it should also be stressed that, for many chemicals, the toxicological database is sufficient to predict the risks associated with their presence in the environment. For chemicals without a sufficient database, generally accepted approaches exist to make a reasonable approximation to provide the basis for selecting appropriate strategies protective of human health.

5-1 PHARMACOKINETICS

Chapter 4 covered the release of hazardous waste constituents and their fate in the environment. The series of steps starting with a release of waste constituents and concluding with the interface with a human body is referred to as an **exposure pathway**. Gaining entry into the body is referred to by toxicologists as the **exposure route**. There are three exposure routes by which environmental contaminants may enter the body: **inhalation** (respiratory tract), **ingestion** (gastrointestinal tract), and **dermal contact** (skin) (see Fig. 5-1).

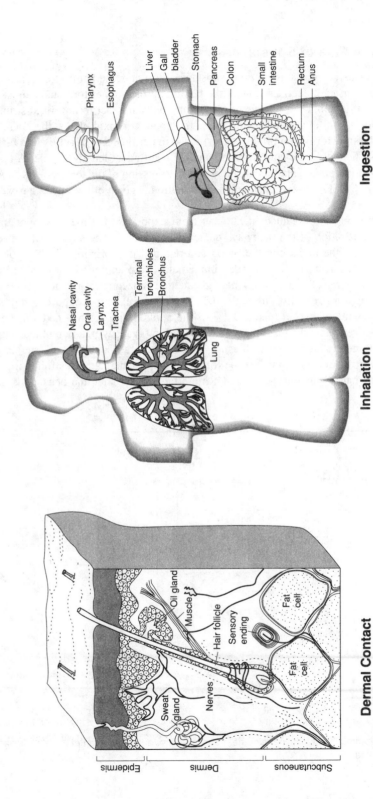

Ingestion

- Pharynx
- Esophagus
- Liver
- Gall bladder
- Stomach
- Pancreas
- Colon
- Small intestine
- Rectum
- Anus

Inhalation

- Nasal cavity
- Oral cavity
- Larynx
- Trachea
- Terminal bronchioles
- Bronchus
- Lung

Dermal Contact

- Oil gland
- Muscle
- Hair follicle
- Sensory ending
- Fat cell
- Fat cell
- Nerves
- Sweat gland
- Epidermis
- Dermis
- Subcutaneous

FIGURE 5-1

Exposure routes for chemical agents in hazardous waste.[1]

227

Overview of Fate of Substances

Except for corrosive agents, most toxic substances do not cause harmful effects at the point of entry. Instead, exposure marks the beginning of a rather tortuous path as the physiological (metabolic) processes of the human body interact to absorb, distribute, store, transform, and eliminate a substance. To produce a toxic effect, the chemical agent or its biotransformation products must reach the critical site of action in a **target organ** at a sufficiently high concentration and for a sufficient length of time. A target organ is the "preferential anatomical site" for the expression of the toxic effects caused by a substance.[2] In Fig. 5-2 a schematic representation is given of the fate of a toxicant upon uptake.

Distribution data obtained in humans from exposure to benzene vapors[3] may serve as an illustration of the pathways depicted in Fig. 5-2. The amount of benzene taken up through the lungs into the blood and retained is approximately 30% of the inhaled dose. The rest is exhaled, 50% immediately and another 20% excreted unchanged via the lungs. Studies in rodents indicate that benzene is also readily (> 90%) taken up by ingestion. Finally, although much less effective, dermal exposure may also give rise to some absorption (about 0.2% of the applied dose). Once taken up, over half of the benzene absorbed is rapidly translocated to organs with a rich blood supply like the liver and the kidneys as well as to tissues rich in fat like adipose tissue, brain, and bone marrow. Benzene may also be found in the placenta and fetus following exposure of pregnant women to benzene. Of importance is the fact that benzene is only

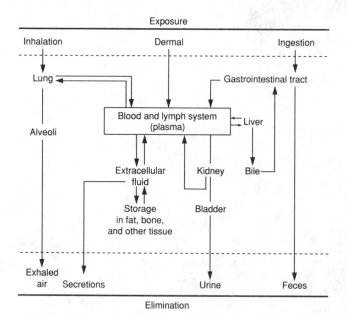

FIGURE 5-2
Overview of absorption, distribution, storage, transformation, and elimination of a toxic substance in the human body.[4]

slowly released from fatty tissues. Coincidentally, complex transformation processes convert the non-polar, fat-soluble benzene into a host of polar metabolites that are excreted in the urine. For several organic compounds there is a substantial database relative to metabolism, but for most substances our knowledge is incomplete.

Absorption

Epithelial layers of cells (membranous tissue) cover the mammalian body at any port of entry regardless of whether the exposure route is inhalation, ingestion, or dermal contact. Having gained entry to the human body, the toxic agent must pass through this interface to enter the blood stream. If ingested, toxic agents must pass through the mucosa (mucous membrane) which lines the gastrointestinal tract. If inhaled, they must pass through the thin lining of cells covering the inside of the alveoli in the lungs, representing the barrier between blood and the air which may carry gaseous toxicants. In the case of dermal exposure, the agents must pass through the stratified layers of the skin. The transport across any body barrier (i.e., skin, lung cells, the lining of the gastrointestinal tract) is referred to as **absorption**.

Whether in the lungs, skin, or elsewhere, these barriers consist basically of cellular membranes having a similar structure of a two-molecule thick, inner layer of various fatty substances (lipids) blanketed between two outer layers of proteins. However, the chemical composition and physical properties of epithelial linings can vary dependent upon the function of the organ they cover. They can be almost fluid-like at some sites, possessing pores of 4 to 70 angstroms in diameter.[5]

The amount of a particular toxic agent to which a body is exposed is termed the **administered dose**. Although the administered dose is important, the amount of the toxic chemical that becomes available to the body (i.e., absorbed by the body) and the amount reaching the target organ are of greater concern. The former is the **intake** or **uptake dose** and the latter can be termed the **target** or **effective dose**. The effective dose depends on how the body's systems interact with the compound. Figure 5-3 illustrates these definitions for contaminated soil.

The same processes that govern the absorption of nutrients, oxygen, and other elements and chemicals vital to life also govern the absorption of toxic agents. The key variables are the properties of the compound, the exposure route, and the susceptibility (or physiological state) of the individual. Absorption through the skin, for example, depends very much upon the medium in which the toxic substance is contained. Soil contaminated with organics would result in less of an intake dose than the same organics in a solvent base. Other factors include the humidity of the skin, the location on the body (penetration by chemicals is especially rapid on the scrotum), and whether the skin is damaged.

In many cases, absorption by more than one exposure route must be considered. For example, about 30-50% of airborne particulate lead inhaled by an adult can be deposited in the respiratory tract, and virtually all deposited lead is absorbed over time. For ingested lead, absorption is about 50% in children and 8–15% in adults with solubility of the particular lead salt in gastric acid being an important variable. Absorption of lead via dermal contact is insignificant except at extremely high concentrations.[6]

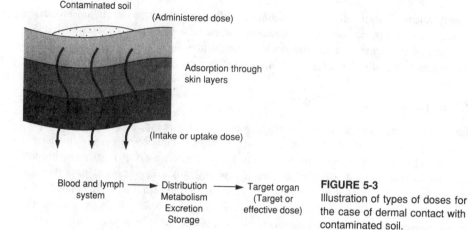

Contaminated soil

(Administered dose)

Adsorption through skin layers

(Intake or uptake dose)

Blood and lymph system ⟶ Distribution Metabolism Excretion Storage ⟶ Target organ (Target or effective dose)

FIGURE 5-3
Illustration of types of doses for the case of dermal contact with contaminated soil.

However, some other substances readily penetrate the skin (e.g., 15–60% of the total applied dose of lindane is taken up by the skin[7]). This route would thus become important.

Absorption mechanisms. Chemical substances may pass in different ways through the cellular linings (epithelium) that make up the protective external and internal barriers of the organism: (a) passage through cellular membranes by passive transport (e.g., simple diffusion) or by active transport, or (b) passage through pores or channels present between the epithelial cells. Most toxic agents pass through cell membranes by simple diffusion.

Diffusion. The rate of diffusion depends upon the chemical and physical properties of a toxic agent, particularly its degree of ionization, lipid solubility, protein binding, and water solubility. For passing through cell membranes, lipid solubility is the most important property. Lipid solubility is a measure of the affinity of a chemical substance for fat-like solvents (e.g., lipids) versus water-like solvents (e.g., blood, urine). The property is related to the polarity of a chemical substance. Polar substances readily dissolve or ionize in water and thus can be thought of as **hydrophilic**. Similarly, non-polar substances can be termed hydrophobic, or **lipophilic**. The property of lipid solubility is indicated by an octanol/water partition coefficient (see Sec. 3-2) above 1.0 (the higher the coefficient, the greater the affinity for lipids). Many organic chemicals (e.g., benzene) are non-polar and highly soluble in lipids. Thus, the lipid-rich composition of a membranous lining facilitates rapid penetration and transport of non-polar compounds; the higher the octanol/water partition coefficient, the more rapid the movement through membranes to enter the blood stream. However, increasing molecular weight reduces the movement through membranes.[8]

It follows that the lipid structure of cellular membranes may inhibit the movement of polar or charged particles. Even for cellular membranes, small molecules of polar compounds may still move as ions in channels with a diameter on the order of 4 angstroms, allowing for passage of only molecules with molecular weights below 200.[9]

The description of the properties of cell membranes given above will to a large extent also hold for the so called "blood-brain" barrier, a complex structure involving several cell types that separates the central nervous system from the circulating blood and lymph. Although most analgesics, anesthetics, and tranquilizers penetrate readily, the blood-brain barrier is much less permeable to toxic substances than other areas of the body. However, it is important to point out that the nature of biological membranes from one part of the body to another may vary as does the composition of the cellular layers that make up the epithelial linings of the organism. The permeability properties of the different parts of the gastro-intestinal tract, for example, present a complex picture, allowing for the uptake also of larger polar molecules (e.g., of alkaloids like codeine sulfate with molecular weights in the range 300–400). In the kidneys the membranous lining is very porous in some places allowing for passage of molecules with molecular weights up to about 60,000. Here, passage occurs through interstitial pores between cells of the membrane.

Specialized absorption. In addition to the primary mechanism of diffusion, some absorption may take place by specialized and complex transport systems. A variety of such mechanisms exists and works selectively, for example, to absorb sugars and other hydrophilic, nutritional compounds. The same systems can also selectively transport some toxicants. Some of the specialized systems are important to the elimination of toxic agents from the body.

Examples include processes such as phagocytosis in which the cell membrane flows around and engulfs particles. This type of absorption is important for the removal of particles from alveoli by specialized white blood cells, so called alveolar macrophages, but may also occur elsewhere to a limited extent (e.g., in the duodenum)[10] accounting for the uptake of high molecular weight toxins like botulinum toxin.

Absorption via the ingestion route. Absorption of toxic substances may occur along the whole length of the gastrointestinal tract, but tends to differ in the stomach compared to the intestine due to the lower pH in the former. Both passive diffusion as well as absorption by specialized transport systems occur. Many factors may alter gastrointestinal absorption. As examples, the presence of chelating agents may increase absorption of heavy metals, the presence of zinc may decrease the uptake of cadmium, and decreased gastric mobility has a general promotive effect on absorption.

Absorption via the inhalation route. In the gaseous state, the inhaled compound is transferred into the liquid layer lining the airway wall by diffusion. Transfer through this layer depends on the gas-diffusion coefficient, layer thickness, and the gas concentrations at the boundaries of this layer. Capillary blood flow removes the dissolved gas on the other side of the liquid and the separating tissue layers. Alveolar uptake of gases is governed by air-to-blood partitioning, ventilation/perfusion, and air and blood concentrations. For readily soluble gases, uptake is linearly related to solubility. In general, the tissue-gas partition coefficient linearly correlates with its fat-gas and blood-gas partition coefficients, providing an approach for estimating these parameters. However, the relation between absorption in the lung and air concentrations *per se* may be non-linear as is the case for a poorly soluble gas like butadiene.[11]

The site of deposition of inhaled particles varies and depends largely on particle size. Particles with a diameter of 2 μm and above are deposited in the upper respiratory tract, removed by the mucous present in the airways, and swallowed. Particles of smaller diameter penetrate to the alveoli of the lung where they may be deposited.[12]

Different animal species when exposed to the same levels of particles or gas in the inhaled air will not receive the same doses in comparable regions of the respiratory tract. Thus, the derivation of human equivalent doses on the basis of animal experiments is particularly difficult, especially for aerosols. The regional deposition pattern is determined by anatomical-physiological features specific to various mammalian species, as well as the physico-chemical characteristics of the inhaled agent. For particles, deposition is controlled by impaction, sedimentation, diffusion, and electrostatic precipitation; for gases the important parameters are convection, diffusion, chemical reaction, and solubility. Retention is governed by the relative rates of deposition and clearance. Whereas deposition data are usually obtainable, the determination of clearance rates is technically difficult to perform and often influenced by the methodology used. Clearly, species' differences in alveolar ventilation and cardiac output are critical determinants of the absorbed dose.[13]

Absorption via the dermal contact route. Skin is an excellent barrier to many toxic compounds. The body has the potential to tolerate 100 to 1000 times the exposure amount by dermal contact versus ingestion or inhalation. Many exceptions exist, however, as exemplified by corrosive agents like phenol and a number of lipophilic compounds like carbon tetrachloride and lindane. Also, certain relatively hydrophilic organophosphorus compounds like the pesticide parathion are highly toxic upon contact with the skin.

Effect of exposure route on toxic response. Clearly the exposure route has a significant effect on uptake. This and subsequent factors manifest themselves such that the magnitude and type of toxic effect will show great variation, dependent on route of exposure. For example, silica can cause a pulmonary disease known as "silicosis" after deposition of dust particles over a protracted period of time, but is virtually harmless upon ingestion. Other examples are asbestos, chromium, and nickel, which act as carcinogens if inhaled but do not seem to induce tumors in humans by the oral route. Many substances are much more toxic by the inhalation route than by oral or dermal administration. This is due to the rapid uptake in the lung that is directly transferred into the blood stream. Calculated per amount absorbed, the toxicity by this route is often comparable to that seen upon intravenous administration. If a substance is found to have a low toxicity by oral administration, but is significantly more toxic by inhalation in the same animal, one might suspect that the low oral toxicity may be due to inactivation or poor uptake in the gastro-intestinal tract.

Distribution and Storage

Upon reaching the blood stream, the toxic agents can move virtually throughout the body. In fact, few toxic agents attack locally at their portal of entry; most are "systemic" and rely on blood flow to reach other organs and tissues. Many factors influence distribution. These factors are absorption, perfusion, exposure route, and tissue affinity.[14] As a result, toxic compounds distribute partially and unequally to multiple "compartments" of the body rather than totally to one or equally to all. Absorption is again important because the same absorption processes which govern the passage of the toxic agent into blood now govern its ability to move from the blood into the tissue of the organs as well as into the cells. Another factor is perfusion—the movement of blood through organ tissue. The liver is well perfused; therefore, its total potential uptake is high. Although the brain is well perfused, it is to a large extent protected by the blood-brain barrier.

The point of absorption, partly determined by the exposure route, has an influence on distribution. One point of absorption may allow the toxicant to bypass the liver, the body's primary detoxification site. For example, toxicants absorbed through the lungs, skin, mouth, and esophagus may temporarily bypass the liver, while those absorbed through the stomach and intestines will follow the blood's direct path to the liver.

Because of their affinity for various tissues, many substances may accumulate at sites other than the target organ to be released over an extended period of time. This is termed **storage**, and typically occurs without any adverse effect on the storage organ. An organochloride pesticide like lindane may accumulate in fat without any adverse effect on the fat cells themselves.

Storage is usually thought of as concentration at a site or sites other than the target organ. The concentrations in these storage sites can be high, as high as in the target organ, if not higher. The fraction that is stored and the particular storage site depend on the characteristics of the chemical substance. Again, polarity and the affinity of the chemical structure for the storage tissue are the dominant factors. Important storage sites include the following examples:

- Fat for non-polar (lipophilic) compounds (e.g., organochloride pesticides, PCBs)
- Blood plasma for compounds bound by blood proteins (e.g., mercuric ions)
- Bone for lead, radium, and fluoride
- Kidneys for cadmium

Many contend that storage is one of the body's defense mechanisms (i.e., it stores compounds which otherwise could not be eliminated quickly enough to prevent damage to a target organ). It should be noted, though, that storage acts in equilibrium with other processes and can be reversed, thus making elimination of stored compounds possible but over an extended period of time even after cessation of environmental exposure. Dieting, stress, and other use of fat reserves can induce rapid release of the stored substance, causing a toxic reaction. In some cases, the site of accumulation defines the point of toxic action. Thus, inorganic mercury accumulates in the kidneys causing severe functional impairment.[15] Above an accumulated total concentration of cadmium in the kidney cortex of 100–200 ppm, kidney damage has been shown to occur.[16]

Biotransformation and Elimination

Upon reaching an organ, three events other than storage can occur, as follows:

1. Biotransformation—enzyme-rich organs metabolize the toxicant to other molecular species, termed metabolites, which may not necessarily be less toxic than the parent.
2. Elimination—toxicants which are not stored, as well as most metabolites, are eliminated from the body.
3. Formation of a chemical-receptor complex—typically, toxic agents attack only one or perhaps a few organs, referred to as target organs.

The body has a remarkable capacity for elimination of a vast array of chemical agents absorbed through dietary and environmental exposures. A toxic agent, particularly polar compounds, may pass from the body in urine, bile, feces, or secretions. Elimination of non-polar, non-volatile compounds is difficult and usually can occur only following their metabolic transformation (biotransformation) by the organism to more polar, and thus more water soluble, metabolites that can be excreted by urine.

Biotransformation of non-polar, non-volatile toxicant molecules can be described as a two-phase biochemical reaction.[17] In the first phase, the body's enzyme system will introduce a polar functional group to the toxicant. Examples of the first-phase reaction include oxidation, reduction, hydrolysis, and dehalogenation. The second phase typically involves the linkage of polar molecules (conjugation) to yield products which are more water-soluble, thus making them more easily eliminated. Conjugating agents are endogenous to the body, and key examples are glucuronic acid and sulfate.

The two most important organs for biotransformation and elimination of toxicants are the liver and the kidneys. Blood from the stomach and intestines flows to the liver, carrying toxicants which have been absorbed through the gastrointestinal tract. The liver is rich in biotransformation enzymes and can quickly transform substances it absorbs. Some substances are rendered water-soluble and transported by blood flow to the kidneys where the kidneys filter or otherwise remove polar toxicants and metabolites from blood and excrete them as concentrated solutes into urine for elimination. Some toxic agents like lead, mercury, and some other heavy metals, as well as many organic substances, are excreted in the bile, often at concentrations much higher than in the blood, to the intestinal tract for elimination in the feces. Still, some hydrophobic toxicants, or their biotransformation products, remain concentrated in the liver.[18]

Considering benzene as an example,[19] it is metabolized primarily in the liver where much of it is oxidized to benzene oxide. Benzene oxide is unstable and converts to phenolic compounds, which are excreted in urine together with other metabolites. Analysis of specific metabolites in the urine are frequently used as quantitative indicators of exposure to substances. This is termed **biological monitoring** and the metabolite a **biomarker**. Phenol is thus a biomarker for benzene exposure. Figure 5-4 shows the direct increase in urinary phenol as the concentration of benzene in inhaled air increases.

Biotransformation does not necessarily reduce the toxicity of a substance. On the contrary, some metabolites are many times more toxic than the parent compound. For example, benzene must undergo metabolic transformation to produce its toxic effects in bone marrow. As another example, benzo(a)pyrene, a substance in tobacco smoke, is non-carcinogenic; however, it is biotransformed to its diol epoxide which is a potent carcinogen. Other examples of such a toxic transformation is the conversion of parathion to paraoxon and of the pesticide DDT to DDE (responsible for eggshell thinning).

Pharmacokinetics

The rates at which toxicants are absorbed, metabolized, and eliminated are very critical. If the body eliminates a particular toxicant rapidly, it may tolerate an otherwise toxic dose when received in fractionated administrations. If the body eliminates a toxicant fairly slowly, a low dose over a long period could result in the accumulation of the toxicant to a critical concentration. For these reasons, the duration, concentration, and frequency of exposure are important.

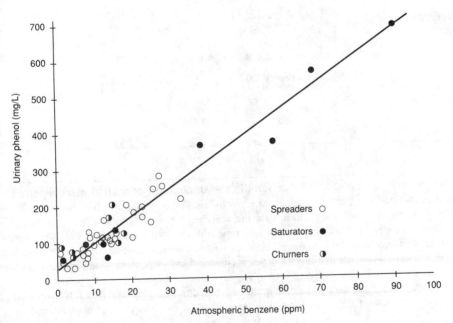

FIGURE 5-4
Correlation of atmospheric benzene concentrations with urinary phenol levels in workers in a rubber coating plant.[20]

Pharmacokinetics is the quantitative description of the time course for uptake, transformation, storage, and transport of compounds (e.g., a hazardous substance).[21] Sometimes referred to as **toxicokinetics,** it is the quantitation in time of all the processes that take place subsequent to uptake in the organism. The various steps and processes involved between exposure and reaching the target organ can often be predicted in a general manner by **pharmacokinetic modeling.** This approach offers a simpler (and less accurate) alternative to the cumbersome analytical and biochemical techniques involved in the elucidation of the pharmacokinetics of a toxicant. Many of the pharmacokinetic models of such actions rely on zero-order, first-order, or second-order reaction kinetics combined with other mathematical approaches familiar to engineers. Such calculations yield values for parameters, such as adsorption rate, half-life of substance, and metabolic saturation concentration. By pharmacokinetic evaluation the intake (uptake) can be related to the target or effective dose. For the example of air pollutants, it would consist of defining the relation between concentrations in inhaled air to the amounts actually taken up and transformed by the organism and, ultimately, its relation to the dose delivered to the sensitive target. Simple models are illustrated in Examples 5-1 and 5-2.

Example 5-1   Elimination of ethanol. Metabolism of ethanol is a zero-order reaction (i.e., elimination is constant regardless of concentration). If an average human is able to metabolize 10 mL of ethanol per hour,[22] how long a time is required to eliminate a six-pack of beer containing 3% alcohol?

Solution

$$t = \text{total dose/elimination rate}$$

$$= \frac{(6 \text{ beers})(400 \text{ mL/beer})(3\% \text{ alcohol})}{10 \text{ mL/hr}}$$

$$= \underline{7.2 \text{ hours}}$$

Example 5-2   Calculation of the half-life for elimination of methyl mercury from the human body. Many metabolic processes follow first-order kinetics (i.e., the formation or elimination of a substance depends on the absorbed concentration) and the level of the substance changes by some fraction per unit time, the first order rate constant, k. The time required for the concentration to decrease by one-half is known as the half-life ($t_{1/2}$). The half-life of a chemical obeying first-order kinetics is constant and independent of administered dose. From determination of the levels of methyl mercury in an individual surviving poisoning by a fungicide in Iraq,[23] the average excretion rate was found to be 1.75% of the total body burden per day. What value for the half-life can be determined on the basis of these observations?

Solution. From Sec. 3-4, a first order reaction may be written as:

$$C = C_0 e^{-kt} \tag{5-1}$$

where C = concentration at time t (mg/L)

t = time (hours)

C_0 = initial concentration (mg/L)

$$t = -\frac{\ln \; C/C_0}{k}$$

$$t_{1/2} = -\frac{\ln(0.5)}{0.0175/\text{day}} = \underline{\underline{39.6 \; \text{days}}}$$

Pharmacokinetic modeling often presents a formidable challenge because most metabolic reaction sequences are exceedingly complex. Many critical reaction variables exist which can only be roughly approximated considering the complexity of the human body and how its components function as a living system in a dynamic environment. Pharmacokinetic modeling is also complicated by the fact that the body cannot be considered as one single compartment with respect to the absorbed compound. This is the case for many lipophilic solvents (e.g., methylene chloride, chloroform, and trichloroethylene) which are characterized by a complex distribution pattern and metabolism. The fate of such compounds can be predicted by using physiologically based pharmacokinetic (PB-PK) models, where attempts are made to identify the various distribution compartments and the different kinetic components in biologically relevant terms.[24]

Exposure Period

The period over which a dose is administered is obviously important. In humans the following criteria have been used for characterizing exposure:

Acute	One day
Subacute	Ten days
Sub-chronic	Two weeks–seven years
Chronic	Seven years–lifetime

When an acute toxic dose is fractionated into smaller portions and administered over a longer period of time, the toxic effect usually decreases. This will be true if there is an appreciable metabolic detoxification and/or excretion of the agent between exposures. If the deactivation/elimination is sufficiently rapid in comparison with the interval between doses, no toxic effects will be noted.

Figure 5-5 illustrates the case of repeated exposure to a compound at two different administered doses, neither dose being sufficiently high to cause a toxic effect after a single exposure. The interval between exposures is the same for both doses. If the elimination rate allows the toxic chemical to decline to pre-exposure levels, as in the case of the lower doses, it prevents the total body burden from reaching the toxicity threshold. However, if the elimination rate cannot keep pace with uptake, as in the case of the higher exposure, this toxicity threshold is eventually reached.

Occupational exposure to TCE constitutes an illustration of the situation depicted in Fig. 5-5. The slow elimination from fat explains why repeated fluctuating

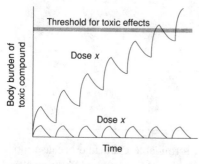

FIGURE 5-5
Effect of dose fractionalization on accumulation of a toxic compound.

concentrations at moderate exposure levels (100 ppm) during a workday, 5 days/week, results in an accumulated concentration as TCE from each day's new exposures adds to the residual concentration from previous exposures until a steady state is reached. This equilibrium level attained towards the end of the week represents a relatively high total body burden of TCE of toxicological significance and has led to a revision of previously adopted standards in occupational hygiene. Assuming first-order elimination kinetics, the "average" body burden at steady state (X_ϕ) after repeated exposure may be approximated by the following equation:[25]

$$X_\phi = 1.44 \times t_{1/2} \times f \times D/\tau \tag{5-2}$$

where $t_{1/2}$ = half-life of the substance (days)
   f = fraction absorbed
   D = dose taken up during each exposure interval (mg/kg of body weight)
   τ = constant time interval between exposures (days)

Regardless of the elimination rate, toxic effects will eventually occur with repeated small doses if either: (a) there remains a "residual injury" after each exposure, or (b) there is a gradual accumulation of the toxicant. As an example of the first, repeated exposure to tri-o-cresyl phosphate at doses below the acutely toxic level will result in accumulation of irreversible nerve damage which will later show up clinically.[26]

5-2 TOXIC EFFECTS

The events leading to a toxic effect can be grouped into three phases: **exposure**, **pharmacokinetic**, and **manifestation**. The previous pages have covered the first two which conclude with a toxic substance or its metabolite reaching a target organ. This marks the start of the third phase, manifestation, which eventually results in the clinically overt symptoms of toxic action. The general sequence of this phase is depicted in Fig. 5-6.

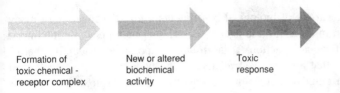

Formation of
toxic chemical -
receptor complex

New or altered
biochemical
activity

Toxic
response

FIGURE 5-6
Sequence of events leading up to a toxic response.

The mechanisms and specific sequence of actions which result in a toxic effect are often not well understood. Generally, it is believed that following the movement of a toxic chemical from the point of absorption to the target organ, the toxic chemical joins, if not binds, with a molecule or group of molecules from a cell of the target organ.[27,28] The affected molecule or group of molecules from the target organ is termed the **receptor**. The typical receptor is an enzyme. However, it can also be the cell membrane, some other molecule from the cell, or DNA, as depicted in Fig. 5-7.

The target molecule(s) may be a specific receptor for endogenous or "natural" substrates located on an enzyme catalyzing a specific biochemical reaction; it (they) may be a receptor on a cell membrane designed to combine with a "signal substance" like a hormone having a central regulating function for various metabolic processes. If the target molecule is DNA, this may result in hereditary alterations (mutation) if this occurs in sex cells (germinal cells), or initiation of cancer in any other cell (somatic cell).

Reactions at the target may also be unspecific in nature. For example, toxic agents like arsenic and some heavy metals bind to reactive groups, like sulthydryl-

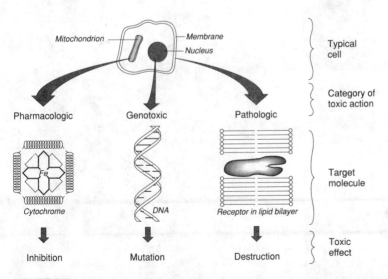

FIGURE 5-7
Representative cellular targets for toxic action.

groups (SH), in a wide variety of biological constituents. Since the function of many enzymes depends on an intact SH, these agents will act as general enzyme inhibitors. Except for a few biochemical functions it is virtually impossible to identify the affected target molecules that are involved relative to the wide range of toxicological effects eventually seen clinically. A common sequel to unspecific toxic reactions is cell death.

Concept of Structural Affinity

A primary factor governing toxic action is the affinity of a substance for a specific receptor in the cells of an organism, or sometimes lack thereof. This affinity derives from the structure of the compound (i.e., its three-dimensional spatial arrangement of atoms or its stereochemical molecular configuration). An affinity exists if the structure of the toxicant is complementary with that of a cellular receptor. Basically, an affinity shared between the structure of a cellular component and that of a xenobiotic molecule causes the two to join, if not bind, together upon contact and recognition of this affinity (see Fig. 5-8). An analogy is the "key within a lock."

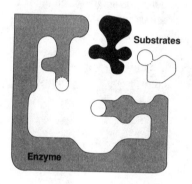

Enzyme and substrates in reaction

FIGURE 5-8
Lock-and-key model for toxic action.[29]

A mechanism of many toxic chemicals for toxic action is to bind to stereospecific clefts in receptor macromolecules such that the receptors no longer can perform their normal cellular function. This binding can result in interference with enzymatic function, or with some other process. Binding does not necessarily result in toxic action; it can, for example, prompt passage through a membrane, storage in particular organ tissue, or induction of enzymes which transform the toxicant. In some cases, such as the binding of carbon monoxide to hemoglobin with 40 times the affinity of oxygen, the "key" fits the same "keyhole" used by the normal substrate, oxygen. Thus, the toxic chemical competes with, if not displaces, beneficial substances. In other cases, insertion of the "key" takes place elsewhere, distorting the receptor molecule sufficiently to make its "keyhole" inoperative. Hypothetically, the more exact the fit, the more potent the toxic response.[30] Xenobiotics not sharing an affinity with cellular molecules are likely non-toxic and mostly excreted; however, exceptions exist for non-specific agents.

The existence of structural affinity is indicated by examples of numerous isomers (isomers are compounds having the same number and kind of atoms but different structure) exhibiting much different toxicity. An example is represented in Fig. 5-9 which shows two isomers of hexachlorinated biphenyl, a species of polychlorinated biphenyls (PCBs) with 6 chlorines. Normally, the two phenyl rings of PCBs are free to rotate about the carbon-carbon bond that joins them. However, chlorination of the internal biphenyl carbons (o-position), as shown in the upper molecule, inhibits rotation such that the two phenyl rings cannot achieve coplanarity. Because coplanarity is required for binding to certain cellular receptors and toxicity, the upper PCB isomer is much less toxic than the lower by a ratio of about 20.[31]

The concept of a toxic chemical-receptor molecule complex based on molecular structure has been the focus of considerable research in the past. When better understood, by means of so-called **quantitative structural-activity relationships (QSAR)**,[32] it should eventually be possible to predict the toxicological properties of new substances before submitting them to full laboratory tests.

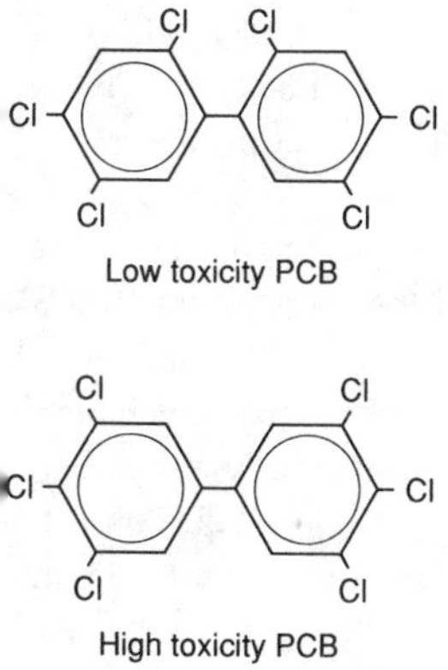

Low toxicity PCB

High toxicity PCB

FIGURE 5-9
Example of compounds with identical molecular formula but different toxicity.[33]

Mechanism of Toxic Action and the Manifestations of Toxic Effects

As discussed above, the binding of toxic substances to enzymes, or to other types of receptors, initiates a chain of events that are not fully understood. Nevertheless, they interfere with the normal beneficial biochemical reactions of the human body or initiate abnormal metabolic reactions, resulting in toxic response. The response may consist of only short-term effects such as headaches or nausea. They could also be fatal such as the case of cyanide which decreases oxygen exchange at the cellular level to the point of disrupting the nervous system function, leading to respiratory failure.

The mechanisms of toxic action and the manifestation of toxic effects can be illustrated by examples. The toxic effects of organophosphorous pesticides can be adequately explained in terms of their primary mechanism of action as acetylcholinesterase inhibitors. At nerve endings, the release of acetylcholine triggers constriction of muscle fibers, but also activates certain parts of the autonomous nervous system.[34] To prevent prolonged and excessive action, the enzyme acetylcholinesterase present at the nerve endings cleaves acetylcholine into inactive acetic acid and choline. Exposure to an organophosphorus compound, like paraoxon (a metabolite of parathion), can inhibit the action of acetylcholinesterase, (AChE), thus allowing the accumulation of acetylcholine at nerve endings. This can result in bronchoconstriction, increased salivation, contraction of smooth muscle (i.e., those performing involuntary functions), and twitching and cramps of skeletal muscles.[35] This is illustrated in Fig. 5-10, showing normal hydrolysis of ACh and the inhibition by the pesticide paraoxon.

Allergic reactions, although only occasionally lethal (in association with anaphylactic shock), nevertheless represent the most widespread manifestation of toxicity in humans that are caused by chemical agents. It is basically a dysfunction of the immune system, and the target organs usually include the respiratory tract—for so called immediate type of reactions—or the skin for contact allergies (delayed). Whereas high molecular weight agents like polysaccharides (e.g., pollen, animal hair) are most prone to induce immediate allergies, a number of low molecular substances, including some metal ions (nickel, chromium, beryllium, mercury, platinum), can cause contact allergy. To produce allergic sensitization, low molecular substances must first bind to an endogenous protein to form an antigen that can activate the memory cells of the immune system. Subsequent exposure to the chemical will then induce an antigen-antibody reaction that elicits the typical symptoms of an allergic reaction.

Even when the primary molecular receptor of a toxic substance is known, the sequence of events that results in the clinical manifestation of toxicity is rarely well understood. Although arsenic acts as an inhibitor of enzymes that depend on free-SH groups for their function, the plethora of divergent toxic effects elicited by this element is difficult to explain on the basis of just this basic inhibitory mechanism. Arsenic's toxic effects include nausea and vomiting, fever, insomnia, progressive general weakness, liver swelling, disturbed heart function, peripheral nervous disturbances, hyperkeratosis of palms and soles as well as melanosis, cancer, anemia, and leukopenia.[36]

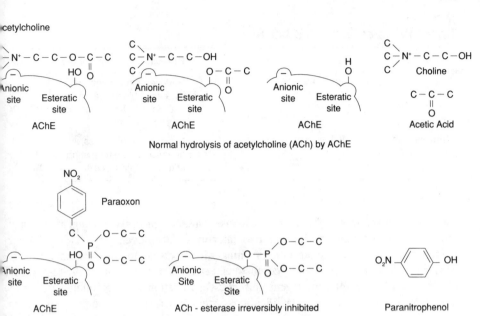

FIGURE 5-10

Inhibition of enzyme acetylcholinesterase (AChE) by organophosphorus.[37]

The biochemical role of riboflavin (vitamin B_2) in intermediate metabolism is well characterized. However, the symptoms of riboflavin deficiency induced by vitamin B_2 antagonists—inflammation of the tongue, lesions of the mucocutaneous juncture of the mouth, corneal vascularization, photophobia, etc.—likewise, cannot be readily related to the primary receptor.[38]

This lack of understanding of basic mechanisms underlying toxic responses seem less surprising when one considers that the mechanism of action of many commonly used beneficial drugs for which a long history exists (e.g., sodium acetylsalicylate, aspirin) are not well understood either. Suffice it to state that toxicologists cannot delineate the whole sequence of events but can identify the organs that are affected by toxic chemicals, the symptoms expressed, and the range of the severity. of the symptoms in response to the different doses.

Classification of Toxic Actions and Effects

Toxic responses are manifested in behavioral and physiological terms and can range from headaches and nausea to convulsions and death. Obviously, depending on the context, there are a number of ways in which a toxic chemical can be classified.

Classification by end point. A simple but apt and commonly used way of classifying the toxic action of a chemical substance is according to toxic end point. Be-

TABLE 5-1
Toxic effects of lead[39] and benzene[40]

Toxic substance	Carcinogenic effects	Non-carcinogenic effects
Lead	Kidney tumors (in test animals)	Reduced birth weight, anemia, increased blood pressure, brain and kidney damage, IQ impaired, decreased learning
Benzene	Leukemia (in humans)	Drowsiness, dizziness, headaches, anemia, suppressed immunity, fetotoxicity

cause of the inherently different dose-response relationships in risk management, toxic effects are often subdivided into the two categories: **carcinogenic** effects with tumor induction as the end point and **non-carcinogenic** effects comprising all other effects. This classification creates some confusion because it means that non-carcinogenic effects would encompass mutagenic effects, yet mutation is more related to tumor induction. Nevertheless, this classification system is widely used, and it provides a way to illustrate toxic effects for lead and benzene (see Table 5-1) and other hazardous substances (see Table 5-2). The term "**genotoxic**" more precisely groups carcinogenic and mutagenic effects. This term would include all chemical substances that can alter DNA or chromosomes, possibly initiating cancer in somatic cells (all cells other than reproductive cells) or leading to heredity changes due to action on germ cells (reproductive cells).

TABLE 5-2
Example of adverse effects caused by selected toxic substances

Carcinogenic chemical	Potential carcinogenic effects	Potential non-carcinogenic effects
Metals (by inhalation) Arsenic	Lung cancer	Liver damage, lung fibrosis, neurological damage
Cadmium	Lung cancer (in test animals)	Kidney damage, osteoporosis, anemia
Chromium	Lung cancer	Bronchitis, liver and kidney damage
Organophosphorous pesticides		Neurotoxicity, systemic poisoning
Chlorinated organic compounds	Liver cancer (in test animals)	Liver damage, neurological effects (in test animals)
Polycyclic aromatic hydrocarbons	Lung cancer, stomach cancer (ingested), skin cancer (dermal contact)	Liver damage, dermatitis

The toxic effects occurring from fertilization until sexual maturation to an adult can be grouped as **developmental effects** with special subcategories. Adverse effects associated with the process of reproduction starting from the production of germ cells via fertilization, implantation, and development of the embryo to fetal growth and ending with birth are usually grouped under **reproductive** effects. **Teratogenicity** (from the Greek work "teras" connoting monster) refers specifically to the subcategory of pathological effects during the sensitive development phase after implantation of the embryo up to the first three months of pregnancy in humans when major tissues and organs differentiate and develop. **Fetotoxicity** represents various disturbances, like growth depression, that may occur to a developing fetus during the later stages of pregnancy. No major skeletal malformations or malformations of organs can occur during this period.

Postnatal development is another period of concern because the rapid growth of a child makes it more susceptible to some toxic agents than the adult body. Lead is an example, with its capacity to retard precognitive development in young children. Exposures of moderate duration, or even a single event, during gestation or prior to sexual maturation could result in adverse effects that would rarely occur in an adult.

Classification by target organ. In some cases, a specific target organ, or tissue, may be distinguished and used as a basis for classification (e.g., the kidney in case of cadmium toxicity, the bone marrow for benzene, the brain for methyl mercury, the liver for carbon tetrachloride, the lung for the pesticide paraquat, the eye for the antimalarial drug chloroquine).

It should be stressed, however, that it may be difficult to single out a characteristic target for toxic action in some cases. Several substances induce generalized symptoms of intoxication, where nausea with abdominal distress, loss of appetite, headache, drowsiness, etc., may be the only obvious symptoms. Oral intake of toxic, but non-lethal, concentrations of inorganic arsenic compounds provides an illustrative example.

Toxicological effects may also be characterized in the following manner:

- Immediate versus delayed effects
- Irreversible versus reversible effects
- Local versus systemic effects

Classification by immediate versus delayed effects. The ingestion of a lethal dose of a soluble cyanide results in death within a few minutes, providing illustration of an immediate toxicological action. In contrast, carcinogenic agents induce tumors in humans only after a latency period possibly extending 10–30 years.

A way of characterizing toxic action according to the latency periods is to use the terms acute and chronic. **Acute or immediate toxicity** results shortly after a single exposure; the typical concern is with a single exposure of a magnitude to overcome existing protective mechanisms. **Delayed toxic reactions** feature long latency periods, perhaps measured in years. It should be noted that an acute exposure may result in a delayed toxicological effect or chronic illness.

Classification by irreversible versus reversible effects. The beneficial pharmacological action of drugs is usually reversible. This may not be the case for non-therapeutic, toxic effects, depending on the nature of the action and capacity for the target tissue to regenerate. Skin and liver have a high-repair capacity; consequently, moderate damage induced at these sites is often reversible. Injuries of the central nervous system caused by chemicals are, on the other hand, mostly irreversible due to the slow or non-existing regenerative power of these tissues.

Classification by local versus systemic effects. In toxicology a distinction is made between local toxic effects and systemic action. Reactive chemicals, such as acid anhydrides, epoxides, strong acids and bases, acyl chlorides, and corrosive agents, cause toxic effects, such as cell necrosis, immediately when coming into contact with tissue (e.g., skin and eyes, mucous membranes of the respiratory tract). However, such agents are usually not taken up appreciably in the organism. In contrast, systemic poisons may exert their toxic action at a point distant from the site of absorption (e.g., tri-o-cresyl phosphate, which induces damage to the peripheral nervous system upon ingestion of small amounts over extended periods of time).

5-3 DOSE-RESPONSE RELATIONSHIPS

A number of factors influence the degree of toxic action. It is readily apparent that the dose represents the major determinant of the extent of toxic action. A relationship exists that the greater the dose, the more severe the response (i.e., the greater the number of chemical-receptor complexes, the more severe the response).

Dose-Response Relationships

Whether a toxic effect occurs, and how severe it is, depends not only upon dose but also upon the susceptibility of the exposed individual. The susceptibility of an individual is primarily a function of the individual's age, sex, diet, health status, genetics, prior exposures to the agent, and exposures to other agents. Even when variations of these key variables are reduced such as in the case of carefully in-bred laboratory animals, other differences will yield variable responses among individuals.

When the incidence (or frequency) of a specified toxic end point among individuals (e.g., proportion of animals with tumors, proportion of dead animals) of a population is recorded as a function of dose, this correlation is called a **dose-response relationship**. The classic example of a dose-response relationship derives from exposing a population of living organisms to increasing doses of a toxic compound. The exposed population will typically experience no deaths at low doses, a few deaths as the dose increases, and more deaths with higher doses until all the organisms die. Most plots of log dose versus cumulative mortality will show the classic non-linear S-shaped curve illustrated in Fig. 5-11a. This concept was first presented in the literature in 1927[41] to introduce the term **median lethal dose** (LD_{50}). This is the dose, usually expressed as mg per kg body weight, at which only 50% of the organisms remained alive. It is sometimes referred to as a quantal dose-response relationship because the elicited response is an all-or-none effect (i.e., dead or alive). The equivalent for inhaled substances is the **median lethal concentration** (LC_{50}), where the dose is expressed as the concentration of the substance present in a volume of inhaled air.

To be able to make comparisons of the acute toxicity of chemicals on a relative scale, the acute oral LD_{50} in rodents is commonly used. In Table 5-3 the oral LD_{50} for the rat is shown for selected chemicals.

Due to the wide variation of susceptibility among individuals, dose-response relationships are frequently plotted with dose on a log-scale and response as mortality frequency on an arithmetic scale as depicted in Fig. 5-11. The bell-shaped form, as shown in Fig. 5-11b, is that of a log-normal distribution. This is typical for the variation in susceptibility of a population with respect to a majority of toxicological effects.

Developing a dose-response relationship first requires selecting a toxicological end point which is caused by exposure to the selected chemical. Preferably, the end point should be clear, definitive and directly measurable such as death or organ dysfunction. Having to use an equivocal end point, such as retardation of mental development, makes the interpretation of the results more susceptible to question.

The ordinate (i.e., response) of a dose-response relationship is usually measured as incidence of the selected end point (e.g., percent mortality as in Fig. 5-11). Similarly, it could be reported as the probability that the end point will occur. This latter approach is used more for carcinogens which are typically reported in terms of probability of cancer over a lifetime of exposure.

Clinical chemistry data, when available and well understood, can serve as indicators of toxic action. The selected parameter must be specific, quantifiable and functionally related to the toxicological action of the chemical. In humans, lead causes anemia by inhibiting the synthesis of hemoglobin from various precursors. This results in the accumulation of the precursor protoporphyrin in the erythrocytes (red blood cells). Thus, erythrocyte protoporphyrin (EP) in blood can be used as a biomarker

TABLE 5-3
Acute oral LD_{50} in the rat for selected compounds

Substance	LD_{50} (mg/kg)
Ethanol[43]	13,000
Table salt[44]	3800
Malathion (insecticide)[45]	2800
Aspirin[46]	1500
Lindane (insecticide)[47]	88–270
Sodium fluoride[48]	180
DDT[49]	113–118
Nicotine[50]	50–60
Sodium arsenite[51]	40
Parathion[52]	2
Dioxin (TCDD)[53,54]	0.02–0.05

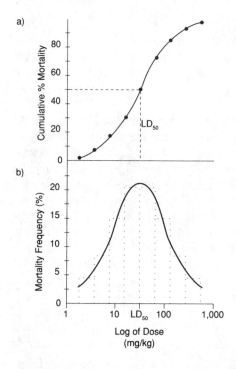

FIGURE 5-11
Dose-response relationship (dose versus mortality).[42]

for measurement of the "toxic" action of lead and has been found to constitute a sensitive and reliable indicator of low-level exposure to lead. In Fig. 5-12 the dose-response relationship for the levels of EP as a function of the concentration of lead in blood derived from inhalation exposure is given for different segments of a human population.[55]

The abscissa (i.e., dose) of a dose-response relationship is straightforward. Doses are most frequently expressed on a body weight basis (e.g., mg or mole of chemical per kg of body weight). Reporting on a body size basis allows the extrapolation from small test animals to humans. The abscissa may not be expressed as a dose in some cases but as an ambient concentration; this is particularly valid when evaluating exposure by inhalation or in testing the response of fish to aquatic contaminants.

Dose-Effect Relationship

A toxic chemical will cause adverse health effects at doses significantly below the lethal level. When instead of incidence among a population of individuals, the intensity (or degree) of the toxic effect in a single individual (e.g., increased blood pressure, retardation of growth, increase in plasma levels of indicators for liver damage) is plotted as a function of dose, the correlation is customarily termed a **dose-effect relationship**. Glycol ethers (an industrial solvent) produce testicular damage in rodents. Figure 5-13 shows the dose-effect relationship between orally administered ethylene glycol monomethyl ether (EGM) and the relative organ weights for the testes and prostate in rats after 11 days of treatment.[57] The dose-effect graph of Fig. 5-13 can be

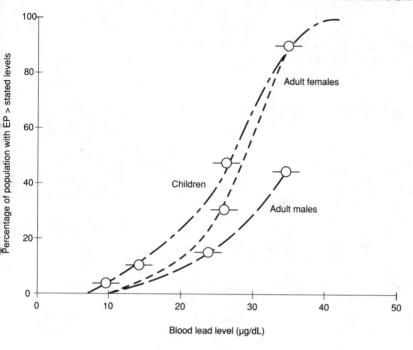

FIGURE 5-12

Dose-response relationship for erythrocyte protoporphyrin (EP) in blood as a function of the blood lead level in subpopulations.[56]

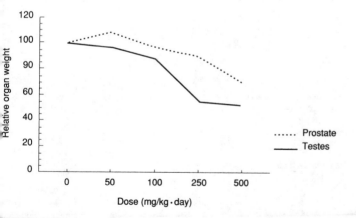

FIGURE 5-13

Effect of EGM on the relative organ weights of testes and prostate.[58]

converted into a dose-response curve by first determining a cut-off value for defining a significant effect, and then plotting the percent of the population which meets this value for each dose tested.

Figures 5-14 and 5-15 provide an alternative manner for demonstrating in a condensed fashion the dose ranges for induction of different toxicological responses by lead and benzene in humans.

Toxicological Data—Human Populations

Toxicological data are based largely on observations of a) human populations which have been exposed to a particular substance, typically as part of their workplace, and b) laboratory animals which have been tested under controlled conditions. Of these, epidemiologic studies of human exposure to a toxic substance represent very convincing evidence for establishing a positive association between a toxic substance

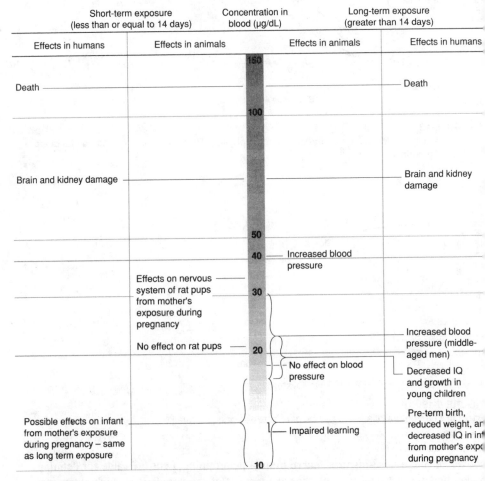

FIGURE 5-14
Health effects from breathing and/or ingesting lead.[59]

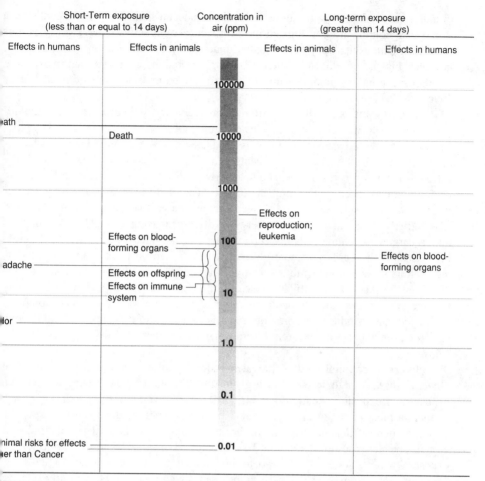

FIGURE 5-15
Health effects from breathing benzene.[60]

and adverse human health effects. For ethical reasons, the cases embraced by these studies have been limited to occupational exposures or to single catastrophic exposure incidents involving the general population.

Types of epidemiological studies.   Out of several types of epidemiological study designs, **cohort** studies, **case-control** studies, and **cross-sectional** studies will be mentioned here. There are two types of cohort studies, prospective and retrospective. In **prospective cohort studies**, a defined population, or cohort, is followed with respect to exposure and a particular toxic end point. This type of study design has the advantage that definite dose-response relationships can be established. On the other hand such studies require a long period of time to complete, and they are characterized by a low power of resolution; unless the exposures are high, a large number of individuals will have to be studied in order to detect a statistically significant effect. When such cohorts are investigated in retrospect (**retrospective cohort study**), dose-response relationships may be estimated under favorable circumstances. Frequently, however, bias is introduced by underreporting, difficulties in reconstructing exposure situations, not having complete information for mixed exposure situations, and other factors.

In **case-control studies**, a different approach is used. In retrospect, individuals with a specific disease are selected from a larger population and the exposure to a suspected causative agent is evaluated for these individuals. Each person with the specific disease represents a case. The incidence of exposure to the selected agent for cases is then compared with the exposure in controls who are healthy with respect to the studied disease. The finding of a more frequent occurrence of exposure to a certain agent among cases than among controls could thus indicate a causal association between disease and exposure. This method requires careful matching with respect to other factors, such as age, sex, and smoking. Because the cases and controls may be drawn from a very large population, the methodology is much more sensitive than cohort studies, but suffers from the disadvantage that bias is difficult to avoid. Case-control studies are in general only feasible when the studied toxicological end point is relatively rare. When properly conducted, such studies may provide important clues as to the causes of human disease (i.e., the etiology), but rarely afford reliable estimates of dose-response relationships.

In **cross-sectional studies** (prevalence studies) information on the prevalence of a certain disease or toxic effect at a given time is collected and related to a certain exposure, or exposure situation. For instance, the following question could be raised: Is anemia more common among workers in copper smelters using sulfide ores with a high content of arsenic than among healthy males in the general population? The current arsenic exposure could here be estimated by measuring arsenic in urine, and exposures over a longer time by determination of arsenic in hair and/or nails. As for other types of epidemiological studies, controls will have to be carefully selected by random sampling and be comparable with respect to age, smoking habits, dietary factors, health status, etc.

Example 5-3 Epidemiological studies at a hazardous waste site. How would an epidemiologist approach the study of a site historically contaminated by illegal dumping of hazardous waste? What types of problems would such a study face?

Solution. The epidemiologist would use the retrospective or case-control method. He/she would first try to obtain data on the extent of environmental contamination to assess potential human exposure to specific chemical contaminants at representative sampling sites, review the literature to determine the types of health effects expected for exposure to such substances at the levels found, and survey the exposed population for such effects. In another approach, the epidemiologist could identify any unusually high incidence of a disease in the surrounding community and then seek to establish a relationship between exposure and the disease by means of a case-control or a cross-sectional study.

It should be realized, however, that an epidemiological investigation of human populations living close to a hazardous waste site represents a formidable task. One of the most difficult problems will be to secure a satisfactory assessment of exposure which, most certainly, will not be confined to one substance. Also, it will not be sufficient to obtain representative analytical data at the time of the investigation; an assessment will also have to be made with respect to previous exposures. This requires examining historical records on morbidity and mortality of residents, as well as tracing residents who have moved from the area, both contributing further to the complexity of the problem. Finally, the population size may be too small to produce statistically valid results.

Limitations of epidemiological studies.   It is important to keep in mind that statistically based epidemiological studies by themselves can never prove the existence of a cause and effect relationship. However, such observations may be used to generate or to test a hypothesis. Many possibilities exist for introducing bias in this type of investigation, and statistical epidemiological correlations are fortuitous. It is, therefore, of utmost importance that any such correlation is supported by a biological plausibility based on existing toxicological data and other information. Here, common sense is a logical first step. Even for the case of a well established cause-and-effect relation, confounding factors often impede the establishment of a reliable dose-response (effect) relationship.[61]

Epidemiological investigations of the general population are especially difficult to conduct. This is particularly true when an investigation of incidence of disease (e.g., cancer) is based on simple comparisons of morbidities for exposed populations of a certain geographical area against "normal" incidences for the general population. The main objection typically raised with such studies is the bias introduced due to inappropriate matching between exposed individuals and controls. Important confounding factors are smoking and other "lifestyle" associated factors which influence exposure to a melange of chemicals as well as the quality of the diagnostic routines employed in various hospitals. For these reasons, only a limited number of toxic substances have sufficient human data to support the development of a quantitative relationship between dose and incidence of adverse health effects. This situation is illustrated in Fig. 5-16 for benzene, a substance which has been the focus of numerous and extensive epidemiological studies.[62]

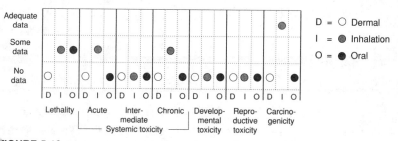

FIGURE 5-16
Adequacy of human exposure data on health effects of benzene.[63]

Toxicological Data—Experimental Animals

Because of the lack of sufficient human data, most determinations of the risk of toxic chemicals must depend on testing of experimental animals, particularly small mammals. In many cases, experimental animals constitute adequate models for the prediction of adverse reactions induced by chemicals in humans, making them indispensable in toxicological and pharmacological research. The interpretation of data from animal studies requires high-level expertise mainly in the fields of biochemistry, toxicology and veterinary pathology, and this can encounter substantial differences of opinion among similarly trained professionals. Hence, peer review of such information by a qualified panel of experts is critical. For much of the published information, this has, unfortunately, not always been the case.

Problems associated with animal test data interpretation arise especially in the following areas:

- Interpretation of the significance of certain types of pathological findings for humans
- Low degree of statistical resolution due to the limited number of animals used per dose group
- Design of carcinogenicity studies
- Extrapolation of dose-response relationships for certain types of effects, in particular carcinogenic effects and induction of allergic reactions

Interpretation of findings. As to the first item, a recent international evaluation concluded that a) the introduction of international guidelines for the conduction of toxicological testing (e.g., the OECD [The Organization for Economic Co-operation and Development] Guidelines[64]) has streamlined the production of experimental data, and b) interpretations of most findings are generally consistent among qualified toxicologists. Disagreement does occur. Notable examples involve pathological changes that are notoriously difficult to classify (e.g., adrenal medullary tumors) and certain minor pathological findings for which it is not clear whether they should be interpreted as signs of toxicity, or only as signs of exposure (e.g., induction of certain liver enzymes).[65]

Limited number of animals. The OECD Guidelines for testing of chemicals, which are legally binding for all of the OECD countries (including the United States), require a minimum of 50 animals per sex and dose group for long-term toxicity testing in rodents. The statistical limitations with respect to the possibility of detecting a significant increase of a certain toxic effect in a test group of 50 animals, using control groups of varying size, are demonstrated in Table 5-4. As is generally the case, the controls also exhibit a varying incidence of the toxicological end-point being studied.

The severe consequences of missing an incidence of a severe toxic reaction (like cancer) induced by chemical exposure of the general population, and which could occur at a possible incidence of 5% in the general population, can readily be visualized. To cope with this problem, the doses used in such experiments as described

TABLE 5-4
Smallest percentage of treated animals exhibiting a toxic effect (F_t) that is considered statistically significant when compared to control incidences (F_c); based on Fisher's Exact Test at $p \leq 0.05$ (one-sided)[66]

Incidence (%) of effect in controls (F_c)	Number of animals in controls	Number of animals in treated group	Smallest statistically significant incidence (%) of effect in treated animals (F_t)
0	50	50	10
	500	50	4
10	50	50	26
	500	50	20
30	50	50	50
	500	50	44
50	50	50	70
	500	50	64

above are, therefore, increased to the so called **Maximum Tolerated Dose (MTD)**, a dose that is high enough to induce minor signs of toxicity in the animals, but not so high as to increase mortality.[67]

Selection of surrogate species. Toxic effects can vary considerably between species even after adjustment for body weight. This derives from differences in metabolism and pharmokinetics. For example, absorption after ingestion may vary between species, thus necessitating caution when extrapolating an oral LD_{50} to humans. In the absence of information from humans or other primates, it is difficult to know which of the common experimental animals will resemble humans the most: the mouse, rat, rabbit, or guinea pig. The toxicity of dioxin (2,3,7,8-tetrachlorodibenzo-p-dioxin, TCDD) is shown in Table 5-5 for different animal species to demonstrate the unusual differences in sensitivity found for this substance. Further, the toxicity of TCDD to three strains of rat shows that similar variations also occur between strains of the same species.

TABLE 5-5
LD_{50} for dioxin (TCDD)

Animal	LD_{50} (mg/kg)
Guinea pig[68]	0.001
Rat (Sprague Dawley or Long-Evans)[69,70]	0.02–0.05
Rat (Han/Wistar)[71]	> 3
Hamster[72]	5

Other tests. Today's legal requirements stipulate for chemicals such as pesticides and pharmaceuticals that, in addition to investigations of toxic reactions in whole animals, studies on metabolism and pharmacokinetics be performed. Such studies are preferably carried out in whole animals, but tissue cultures of perfused isolated organs (e.g., liver) are sometimes utilized. Such data are indispensable for assessing parameters such as uptake, accumulation, and other factors. Although rarely performed, comparative studies of human metabolism and the metabolism of an animal which has shown susceptibility to a toxic agent will further facilitate the interpretation of available animal toxicity data. In some instances, experiments in whole animals may be difficult to perform, and microorganisms and mammalian cell cultures may provide important additional information, especially in the field of mutagenicity testing.

Adequacy of Data

In most cases, the scientific database is not adequate to describe, with a high degree of confidence, the quantitative risk of all health effects from exposure to toxic substances. Figure 5-16, presented earlier, showed that even for benzene, one of the more thoroughly studied toxic substances, human data are inadequate for most exposure situations. Many inadequacies still exist even after including animal test data, as depicted in Fig. 5-17 for benzene. Toxicologists, when assessing toxicity, should bracket their estimates to reflect the inherent uncertainty of such endeavors. On the other hand, it is important to realize that depending on the situation, not all toxicological effects are equally essential to making decisions in the hazardous waste management field. For episodes involving spills, acute hazards are usually the most important. In the remediation of a hazardous waste site, where continuous low-level exposure is generally of highest concern, efforts should be directed towards identifying and defining dose-response relationships for those serious chronic effects that appear at the lowest doses.

When reviewing toxicological information, it is very important to make a clear distinction between toxicological data appropriate for risk assessment and administrative "standards" that are associated with risk management. The former is a scientific procedure. Risk management, on the other hand, comprises a range of actions taken

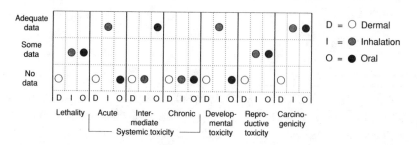

FIGURE 5-17
Adequacy of animal test data on health effects of benzene.[73]

to minimize, reduce, or otherwise control specific risks posed by a certain situation; it contains elements of policy. An LD_{50} derived from an animal experiment is an example of basic information that can be used in risk assessment. In contrast, a threshold limit value (TLV) for occupational exposure, although partly based on scientific data, represents an administrative decision made by regulatory agencies. Their decisions have taken into account, to a greater or lesser extent, a number of non-scientific factors, such as technical feasibility, economic impact, and other factors. Such decisions thereby involve policy judgments.

5-4 NON-CARCINOGENS

Non-carcinogenic effects literally include all toxicological responses other than the induction of tumors. The toxic end points of non-carcinogenic agents vary widely, as do mechanisms by which they cause the toxic effects. Chemical agents may combine with cell membranes, thereby reducing the permeability of the membranes. Some can chemically displace elements important to a cell performing its normal activity. Nevertheless, the most prevalent and typically the most toxic non-carcinogenic responses are those in which an agent affects enzymes. All enzymes are predisposed to perform specific physiologic functions. However, binding with a toxic substance may interfere with, or block, these normal functions, prompting an altered series of biochemical activities that eventually expresses itself as a toxic response.

Concept of Threshold

The assessment of the toxicity of non-carcinogenic chemical substances is based on the concept of threshold. That is, for toxicological end-points other than mutations and cancer, the dose-effect or dose-response relationship is generally characterized by a threshold below which no effects can be observed on the cellular, subcellular, or molecular level. Most structures which have critical functions in the organism exist in a large number of identical copies. The destruction of a few units (e.g., enzymes) will have little effect on overall performance, especially when considering the presence of efficient repair mechanisms. Only when a significant fraction of the targets have been eliminated by toxic action, that is, above a certain threshold for the target dose, will a toxic effect occur.

The toxic effects of lead provide an example. At a blood level in the range of 10–30 μg/100 mL, lead can cause inhibition of precognitive development in children. Until this threshold concentration of lead is reached, the developing brain will function satisfactorily. As another example, carbon tetrachloride will result in pathologic damage to liver tissue; however, the liver will continue to function satisfactorily and its remarkable ability to grow new cells will eventually replace the lost cells. At some threshold, the liver will become dysfunctional, and the damage may not be reversible. Therefore, the body can tolerate a wide range of exposures between negligible intake and the threshold value for a discernible effect.

No Observed Adverse Effect Level (NOAEL)

The threshold value for a toxic substance cannot be identified precisely. Instead, it can only be bracketed based on analysis of epidemiological data and animal tests. In practice, hazard evaluations are often based on controlled animal experiments, where the highest dose that will not cause an effect, the "No Observed Effect Level" (NOEL) is sought. Because we are usually only interested in toxicologically significant effects, a variant of NOEL termed the "No Observed Adverse Effect Level" (NOAEL) is often utilized. Two other terms used in this context are:

- LOEL (Lowest Observed Effect Level)—the lowest dose tested for which effects were expressed; typically used when an effect is expressed at all doses, and
- LOAEL (Lowest Observed Adverse Effect Level)—a stricter version of LOEL to address only adverse effects.

When using data like LOELs (LOAELs) or NOELs (NOAELs), it is important to be aware of the limitations of such values.[74] A NOEL, for instance, is determined simply by selecting the lowest dose in a study that does not seem to cause any adverse effect. Thus, in deriving such a parameter, statistical uncertainty due to the limited number of animals used in the critical study is not accounted for. As mentioned before, the variation in sensitivity in a population for a majority of toxicological effects follows a log-normal distribution.

A NOEL (as well as an LD_{50}) gives no indication of such individual variations in susceptibility. However, without performing any kind of sophisticated analysis, a qualitative judgment can sometimes be made simply by determining the slope of the dose-response curve; a steep curve usually indicates a uniform response curve (e.g., cyanide) with little variation between individuals, and the NOEL may be determined with good precision. A flattened response-curve, as for ethylene glycol, tends to reveal large differences in susceptibility among individuals, thus decreasing the confidence in the NOEL.

Acceptable Daily Intakes and Reference Doses

Instead of establishing a threshold value, toxicologists use the term **acceptable daily intake (ADI)** to represent the level of daily intake of a particular substance which should not produce an adverse health effect. ADIs are based on NOAELs and should not be construed as a strict physiological threshold which when exceeded will result in adverse health effects; an ADI includes safety factors to reflect variable susceptibility in the human population and other uncertainties. An ADI is much smaller than the theoretical threshold.

For regulatory purposes, the accepted manner to establishing safe levels of most chemicals in food and drinking water is to divide the NOEL (or NOAEL) by safety (uncertainty) factors to derive a value for ADI. By international agreement an arbitrarily fixed uncertainty factor of 10 is used to compensate for different sensitivity between individuals, and an additional factor of 10 for extrapolating from an experimental animal to humans. This simplified approach is used in risk assessment by organizations like the World Health Organization (WHO) as well as by government regulatory agencies all over the world.[75] Its main justification for continued use lies largely in the fact that the ADIs have been used historically with no obvious negative consequences for human health.

The reference dose (RfD) is a contemporary surrogate used by the EPA instead of the ADI. The development of an RfD follows a somewhat stricter procedure than used for an ADI, sometimes resulting in lower values for acceptable intake. The standard procedures for establishing an RfD consists of the following steps:[76]

1. Select the most sensitive species for which adequate studies are available.

   This should always be done unless compelling evidence shows that the most sensitive species does not adequately reflect the situation in humans and a better animal model is available. When available, human data are always given priority, taking into account the limitations in epidemiological studies.

2. Select the principal or critical studies using the appropriate route of exposure.

   RfDs are route-specific (i.e., they refer specifically either to inhalation or to oral intake). Thus, a study based on intravenous injection would not be appropriate for the establishment of an RfD for oral intake, or for inhalation. Ideally, the critical study should be sufficiently well conducted to permit the establishment of a NOAEL with acceptable accuracy.

3. Select supporting studies.

   Investigations from a wide variety of sources may provide additional aid in interpreting the results from the critical study. Comparisons of metabolism and pharmacokinetics of the toxic chemical in the animal used for the critical study with such data derived from humans are of particular value.

4. Identify the NOAEL, or if such data are not available, the LOAEL for the most sensitive end point.

   Chemical agents frequently have multiple toxicological effects. For example, arsenic induces a large number of chronic toxic effects upon ingestion: intestinal disturbances, anemia, neurological damage, peripheral vascular disorders, and others. However, based on epidemiological data, the induction of skin disorders and vascular complications is the most sensitive toxicological end point with a LOAEL of 14 μg/kg·day.

5. The NOAEL for the most sensitive end point is adjusted downward by orders of magnitude to reflect uncertainty as follows:

   - Reduce the NOAEL found in humans by an uncertainty factor of 10 to account for variation in the general population, thus protecting the most sensitive subpopulations (e.g., children, elderly);
   - Reduce the NOAEL by an additional uncertainty factor of 10 when extrapolating from animals to humans;

- Reduce the NOAEL by an additional uncertainty factor of 10 if the data are derived from a sub-chronic instead of a chronic study; and
- If the test data do not show a NOAEL (i.e., adverse effects were expressed at all dose levels), the LOAEL is selected and reduced by an additional factor of 10 to account for the uncertainty introduced by this extrapolation.

The uncertainty factors are applied sequentially. The EPA also applies a "modifying factor," which ranges from 1 to 10, to reflect a qualitative professional judgment of uncertainties which are not accounted for by the preceding uncertainty factors. The derivation of an RfD is illustrated in Example 5-4.

Example 5-4   Derivation of an RfD. In a subchronic oral toxicity study in mice, a lowest observed adverse effect level (LOAEL) of 5 mg/kg·day was determined for a specific agent. The quality of the data is given a high rating by the expert evaluating the data. What is the RfD?

Solution

Area of uncertainty	Uncertainty factor
Variation within a population	10
Extrapolation from animals to humans	10
Extrapolation from sub-chronic to chronic	10
Extrapolation from LOAEL to NOAEL	10
Modifying factor	1

$$\text{RfD} = \frac{5 \text{ mg}/\text{kg} \cdot \text{day}}{10 \times 10 \times 10 \times 10 \times 1} = 0.5 \ \mu\text{g}/\text{kg} \cdot \text{day}$$

Similar approaches are taken for the development of an RfD for other exposure durations (e.g., acute, chronic). Another example is that of "developmental exposures RfD." A developmental exposures RfD covers the special case when exposure of short duration, even a single event, prior to conception, during gestation, or prior to sexual maturation, could result in adverse effects that would not occur in an adult. For example, investigations have revealed that precognitive development in young children may still be sensitive to the RfD of 0.0014 mg/kg·day for lead. Hence, the EPA at one time considered lowering the RfD for lead by a factor of 10.

Apart from the arbitrary assignment of safety factors, ADIs, RfDs, and related values are also associated with the same uncertainty as those inherent in the NOAELs or LOAELs from which they are derived. Above all, the misuse of such data as some kind of physical constants is not recommended. The following note in IRIS (the EPA Integrated Risk Information System) for arsenic amply illustrates this point:[77]

Note: There was not a clear consensus among Agency scientists on the oral RfD. Applying the Agency's RfD methodology, strong scientific arguments can be made for various values within a factor of 2 or 3 of the currently recommended RfD value, i.e., 0.1 to 0.8

μg/kg·day. **It should be noted, however, that the RfD methodology, by definition, yields a number with inherent uncertainty spanning perhaps an order of magnitude** [emphasis added]. ... Risk managers should recognize the considerable flexibility afforded them in formulating regulatory decisions when uncertainty and lack of clear consensus are taken into account.

For most non-carcinogenic effects a considerable body of data exist that will permit adequate high-to-low dose extrapolation to be made on the basis of dose-response or dose-effect relationships. However, the establishment of such relationships with respect to immunological effects poses special problems. The characterization of such relationships for pulmonary (and other) sensitizers remains largely an unresolved issue, and constitutes one of the most enigmatic problems in the risk assessment of air pollutants.

Toxicological Databases

A number of databases exist which document RfDs, ADIs, and other toxicological data for both non-carcinogenic and carcinogenic substances. Some of these are accessible as computerized databases maintained by government agencies.

An example is the Integrated Risk Information System (IRIS), the EPA's preferred source for toxicity information for hazardous wastes typically encountered in site remediation projects. IRIS is updated monthly. A compilation of IRIS data is included in App. B, and Fig. 5-18 shows a sample output from IRIS. The complete basis for these data is not shown but is also available from IRIS.

Other databases and references include the following:

1. Health Effects Assessment Summary Tables (HEAST)
   - Updated quarterly
   - Includes interim RfD and CPF values
   - Prepared by EPA's Environmental Criteria and Assessment Office (ECAO)
2. Toxicological profiles prepared by U.S. Agency for Toxic Substances and Disease Registry (ATSDR)
3. The IPCS Environmental Health Criteria documents published by the World Health Organization (WHO), Geneva, Switzerland
4. The Joint FAO/WHO Meeting on Pesticide Residues (JMPR) published by Food and Agriculture Organization (FAO), Rome, Italy
5. Numerous database vendors that compile various other sources of data (e.g., National Library of Medicine, Toxnet, Syracuse Research Center)

One of the more comprehensive sources of information is the set of ATSDR toxicologic profiles listed as Item 2 above. U.S. federal law (SARA, 1986) mandated that the ATSDR prepare profiles for the toxic substances which are most commonly found at Superfund sites. Each profile can be as much as 100 pages in length and presents the following information:

1. Assessment of all relevant, peer-reviewed toxicological testing and information.

Chemical Name	CASRN	Oral RfD mg/kg/day	Date	Inhalation RfD mg/kg/day	Date	Oral CPF 1/mg/kg/day	Date	Inhalation CPF 1/mg/kg/day	Date	Carc. Class.
Dimethylamine	124-40-3	pending		withdrawn	1/1/91	no data		no data		
n-n-Dimethylaniline	121-69-7	2.00E-03	3/1/88	no data		no data		no data		
3,3-Dimethylbenzidine	119-93-7	no data		message	7/1/91	no data		no data		
n,n-Dimethylformamide	68-12-2	no data		8.58E-03	10/1/90	no data		no data		
2,4-Dimethylphenol	105-67-9	2.00E-02	11/1/90	no data		no data		no data		
2,6-Dimethylphenol	576-26-1	6.00E-04	9/7/88	no data		no data		no data		
3,4-Dimethylphenol	95-65-8	1.00E-03	1/1/89	no data		no data		no data		
Dimethylphthalate	131-11-3	pending		message	1/1/92	no data (NA)		no data (NA)		D
Dimethylsulfate	77-78-1	no data		pending		no data (NA)		no data (NA)		B2
Dimethylterephthalate (DMT)	120-60-6	1.00E-01	3/1/88	no data		no data		no data		
m-Dinitrobenzene	99-65-0	1.00E-04	8/22/88	no data		no data (none)		no data (none)		D
o-Dinitrobenzene	528-29-0	no data		no data		pending	5/1/92	pending	5/1/92	
4,6-Dinitro-o-cyclohexylphenol	131-89-5	2.00E-03	1/1/89	no data		no data		no data		
2,4-Dinitrophenol	51-28-5	2.00E-03	7/1/91	message	10/1/91	no data		no data		
2,4-Dinitrotoluene	121-14-2	2.00E-03	6/1/92	message	3/1/91	no data		no data		

CASRN   Chemical Abstract Service Registration Number
RfD   Reference Dose
CPF   Carcinogenic Potency Factor (also called a slope factor, see Sec. 5-5)
Pending   A numerical value is currently under review
Message   A message regarding this value is available
Withdrawn   A previous data value is no longer considered valid

FIGURE 5-18
Sample IRIS output (July 1993).

2. Determination of whether adequate information exists to define levels of exposure that present a significant risk to human health, and if it does not exist, the additional testing needed.

3. Presentation of the known acute, sub-acute, and chronic health effects at various levels of exposure.

4. Description of the substance's relevant pathologic properties, such as pharmacokinetic data and environmental fate information.

Calculation Procedure for Assessment of Non-Carcinogenic Risk

The preliminary assessment of non-carcinogenic risk of a hazardous waste site as recommended by the EPA is typically calculated in four major steps:[78]

1. Identify discrete exposure conditions:
   - Exposure route
   - Frequency
   - Duration
   - Administered dose

2. Derive appropriate reference doses for each discrete set of conditions.

3. Evaluate hazard for noncarcinogenic effects as a ratio of exposure dose to the recommended RfD.

4. Aggregate hazard for multiple chemical agents and exposure pathways as a hazard index, where appropriate.

Application of this four step process is explained in detail in Chap. 14. The ratio referred to in the third step is utilized to quantify risk from non-carcinogens. As the fourth step indicates, the hazard index for individual chemicals may be summed for chemicals affecting a particular target organ or acting by a common mechanism in order to provide a final measure of non-carcinogenic toxic risk. If the sum of hazard indices is less than one, then the risk of adverse health effects is considered acceptable. This additive method is illustrated in Example 5-5.

Example 5-5   The use of RfDs to evaluate non-carcinogenic hazard. Due to contamination from hazardous waste originating from a metal plating facility, the water from a nearby community water supply well was shown to contain cyanide at a concentration of 30 μg/L, nickel at 120 μg/L, and chromium(III) at 12,400 μg/L. The daily water intake is assumed to be 2L, and the body weight of an adult human 70 kg. Do these exposures indicate an unacceptable hazard?

Solution. The dose for cyanide is:

$$\frac{0.03 \text{ mg/L} \times 2 \text{ L/day}}{70 \text{ kg}} = 8.57 \times 10^{-4} \text{ mg/kg} \cdot \text{day}$$

The remaining calculations are shown in the following table, with RFD values from the IRIS database:

Substance	Concentration C (mg/L)	Dose mg/kg·day	RfD[79] (mg/kg·day)	Hazard ratio dose/RfD
Cyanide	0.03	8.57E-4	0.02	0.04
Nickel	0.12	3.43E-3	0.02	0.17
Chromium(III)	12.4	0.35	1.0	0.35
Total				0.56

The individual ratios are well below 1, and this preliminary evaluation does not indicate an unacceptable hazard.

Strictly speaking, additivity assumes similar type of action. It is justified, for example, when dealing with mixed exposure to cadmium, mercury, uranium, and chromium, all of which affect the kidneys. However, adding the hazard indices for separate compounds in the manner demonstrated may overestimate risk if they induce toxic effects by different mechanisms. Thus, for this example, the compounds present in the mixture do not affect a common target organ or act by a common mechanism and should be segregated by critical effect, and separate hazard indices calculated for each toxicological end point. One approach is to sum hazard indices on an organ-specific basis. This should only be done when the mechanisms of the noncarcinogenic toxicants are well-known and the effects are virtually mutually exclusive.

5-5 CARCINOGENS[†]

Cancer is a disease marked by cellular abnormalities in which a normal cell becomes altered and divides uncontrollably. The definition of cancer favored by standard textbooks in pathology is:

> an abnormal mass of tissue, the growth of which exceeds and is uncoordinated with that of the normal tissue and persists in the same excessive manner after cessation of the stimuli which evoke the change.[80]

Cancer is characterized by tumors or neoplasms (meaning "new growth"); however, not all tumors are cancerous. Tumors can be divided into **benign** and **malignant** types, where the latter are commonly known as cancer. Malignant tumors from muscle are designated sarcomas, and those derived from epithelial cells, carcinomas.

The most important difference between benign and malignant tumors is that a malignant tumor will invade surrounding structures and spread (metastasize) to distant sites; in contrast, a benign tumor will not. Malignant growths also tend to

---

[†]Section 5-5 was prepared by Drs. Bruce Moholt and Robert Nilsson.

Stage	Normal cell	Carcinogenic agent

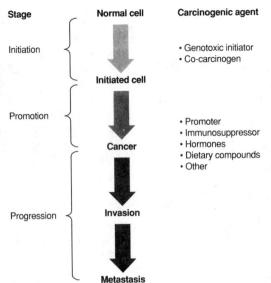

FIGURE 5-19
Three stages in carcinogenesis.

grow more rapidly than benign neoplasms and are composed of highly atypical cells. In a malignant tumor the cells have become altered in such a way that they no longer respond to regulatory signals that coordinate the rate of cell division to prevent invasion of surrounding tissues.

The multi-step process which occurs between initial exposures to carcinogens and the frank development of malignancy is termed **carcinogenesis**. In humans, the process as induced by radiation or chemicals takes about 10–40 years on average dependent upon the nature of the carcinogenic agent and other factors such as heredity.

Within the last decade the multiple steps of carcinogenesis have begun to be understood. Current theory recognizes three stages:[81]

- Initiation—the alteration of a normal cell to a cancer cell through mutation.
- Promotion—multiplication and selection of initiated cells.
- Progression—expansion of the cancer cell line by invasion of local tissue and metastasis to distant organ sites.

These intermediate steps in carcinogenesis are summarized in Fig. 5-19.

Initiation of Cancer

What alterations are required to convert a normal cell into a cancer cell? Because a tumor can arise from one single transformed cell, the expression of cancer is carried over from one cell generation to the next. Thus, the first change which must occur is genetic, that is, alteration of the structure and function of the DNA of a cell. Another clue indicating that DNA is the critical target of carcinogens is provided by the observation that patients with diseases like xeroderma pigmentosa or Bloom's syndrome have a predisposition for developing cancer. In these diseases DNA repair is deficient.[82]

In many cases, transformation to a cancer cell has been associated with changes in specific **oncogenes** into a form that has an altered function in comparison with its natural counterpart (often called **protooncogenes**).[83] One normal function of oncogenes appears to be growth control during development of the human embryo and fetus when proliferation of tissues must proceed at breakneck pace to make a fully formed human within just nine months of gestation. An oncogene can regulate growth by controlling growth factors directly or by suppressing other oncogenes.

Upon birth, oncogenes are repressed (turned off) and normally remain repressed permanently or possibly are reactivated only transiently. However, mutation or other genetic transformation can fully activate an oncogene. There are evidently several ways in which an oncogene may be activated. As in human Burkitt's lymphoma, the change of position of a protooncogene by translocation (swapping of "arms" of chromosomes) to a more active region may result in a greatly enhanced expression of the gene. Successive duplications (amplification) of a protooncogene may, likewise, give an enhanced expression. Another way of oncogene activation involves the change of individual bases in the DNA of a protooncogene (point mutation) that will result in the synthesis of a protein with a changed structure.

If an oncogene is activated, the affected cell of the child or adult individual is completely capable of autonomous growth and replication because it makes its own "hormone" and has a receptor for it. In such cases the mutated cells contain new antigens on their surface which stimulate the production of antibodies that may destroy the mutated cells. However, if these cells are not recognized as foreign by the immune system, and continue to replicate abnormally, they will become the stem cell population for that malignant clone called cancer.

It is established, however, that a single mutated protooncogene cannot cause cancer. At least two changes in the same cell are necessary (i.e., cancer can only result when the second copy undergoes mutation). A mutated oncogene is rarely inherited. If inherited, the person would have a genetic predisposition to cancer because rather then requiring two changes within the same cell, only an additional single change would be necessary to initiate cancer. The resulting tumor would normally develop during childhood because much less time is required for a single oncogene change. However, most cancers develop in adults and the incidence of cancer in children is low. This suggests that mutation of genes after birth causes cancer in most cases.

Only a rare subset of DNA mutations, or chromosonal rearrangements, activates oncogenes. Because there are roughly 300,000 human genes, out of which only a very minor portion involve protooncogenes, the majority of DNA mutations have no effect on carcinogenesis. Furthermore, even within oncogenes, the vast majority of mutations will inactivate rather than activate the oncogene.

Having made these statements about oncogenes, it should be pointed out that in many aspects the role of oncogenes in the process of carcinogenesis remains obscure, largely due to the fact that the mechanisms that regulate normal cell division and tissue differentiation are largely unknown. Further, for many sites, activated oncogenes have been found only in a minor fraction of the tumors; thus, there are likely other routes to cancer, including as yet unidentified oncogenes. Regardless of what role oncogenes may play in carcinogenesis, the initiation of cancer is clearly genetic. Those agents which cause mutation of genes are referred to as **genotoxic**. Such agents consist of many carcinogenic chemicals and ionizing radiations.

Promotion of Cancer

The second stage of carcinogenesis is an increase in the number of initiated cells which may develop into "true" cancer cells.[84] This step is referred to as **promotion**, and it is triggered by promoting carcinogens (promoters). Promoters, in general, are non-genotoxic. Although they act in a wide variety of ways, basically they increase the replication rate of initiated cells, or they cause the cells to undergo differentiation which involves the switching on of new genes. Well-known examples in experimental animals are croton oil with respect to skin cancer, and phenobarbital for promotion of liver tumors. In humans, high-fat diets are considered to promote breast cancer in women and colon cancer in both sexes. A promoter is not usually the same carcinogen which initiated carcinogenesis.

Cells which have undergone two or more genetic changes necessary for initiation and which have been promoted still do not normally progress to frank malignancy. Because these transformed cells display foreign ("non-self") antigens on their outer surfaces, they may be recognized and rejected by the same components of the immune system (T-cells) which reject invading parasites. However, the immune system can be inhibited in a number of ways (e.g., by the AIDS virus). Some toxic chemicals may also act to suppress the body's ability to recognize and reject promoted cells. Such chemicals are termed immunosuppressants, some of which are used clinically in organ transplantations.

Progression of Cancer

Following promotion, the final two steps necessary for malignancy consist of **invasion** of adjacent tissues and **metastasis** (i.e., movement) to distant organ sites. Invasion and metastasis are collectively known as oncogenic **progression**.[85]

For invasion to occur, localized clumps of cancer cells require enzymes not normally available to dissolve the sheath which surrounds the organ of origin. For metastasis, the localized tumor must next invade the circulatory system for movement to another organ. At the distant organ site, these far flung cells must be able to grow in a foreign environment (i.e., without the support of complementary tissues from the organ of origin). These are not normal capabilities of cells and can only come about by further genetic change. Hence, the development of invasive and metastatic cancer requires additional mutations beyond that necessary to initiate carcinogenesis.

Molecular genetic examination of tumors has found progressive cancers to contain many more mutations than the two required for initiation. The promotion and progression of a clone of initiated cells are accompanied by a cascade of genetic alterations, pointing to a genetic instability, the nature and significance of which remain unclear.[86] Nevertheless, further exposure to genotoxic agents seems necessary for the progression of initiated and promoted cells. This requirement constitutes the basis for the rapid decrease of cancer risk in smokers after cessation of smoking.

Role of Carcinogenic Chemicals

Most of our understanding of human carcinogenesis by exposure to carcinogenic chemicals comes from the workplace or from experiments in test animals. Certain types of cancer—like breast cancer in women, prostrate cancer in men, and lung cancer in smokers—are common causes of death. To investigate the role of chemicals in the etiology of such cancers is difficult because epidemiological investigations require very large cohorts, unless the effect is very striking. The chances of detecting an effect improve if the chemical agent in question induces a specific type of cancer that is not common in the general population. Some of these specific cancers with known etiology are angiosarcoma of the liver caused by vinyl chloride exposure, oats cell carcinoma of the lung caused by exposure to bis (2-chloromethyl) ether, and cancer of the scrotum of chimney sweepers induced by exposure to polynuclear aromatic hydrocarbons in soot.

It is clear that exposure to carcinogenic chemicals can result in cancer via different mechanisms. Genotoxic carcinogens may prompt the initiating and progressing mutations necessary for the full-fledged development of human cancer, while non-genotoxic carcinogens speed up carcinogenesis through promotion or immunosuppression.

This distinction serves to divide carcinogenic chemicals into **genotoxic** and **nongenotoxic (epigenetic)** agents.[87] Since the implications with respect to hazard and risk assessment are profound, the distinction between the two groups serves a useful purpose. However, whereas government agencies in Germany, the Netherlands, and the United Kingdom have accepted the concept,[88] the EPA with one exception[89] has been reluctant to make use of this distinction.

Genotoxic carcinogens. With respect to genotoxic carcinogens, toxicologists have found that chemical substances which are known to be active in carcinogenic initiation and progression also cause mutation in bacteria and other short-term test organisms. Because the structure of DNA is the same for all organisms, these test reveal some theoretical information about how specific chemicals prompt alterations of genetic material in humans.

Certain alkylating agents, like ethylene oxide and methyl methane sulphonate, react directly with DNA and can cause initiation. However, most initiating carcinogens are not active as such but require metabolic activation.[90] One example is benzo(a)pyrene, one of the major carcinogens of tobacco smoke and coke oven emissions. Initiation of carcinogenesis by benzo(a)pyrene first requires its metabolic activation to benzo(a)pyrene diol epoxide (Fig. 5-20). The diol epoxide forms a covalently bonded adduct with guanine in DNA,[91] which appears to swivel the base about the glycosidic bond such that the guanine now appears as thymine during DNA

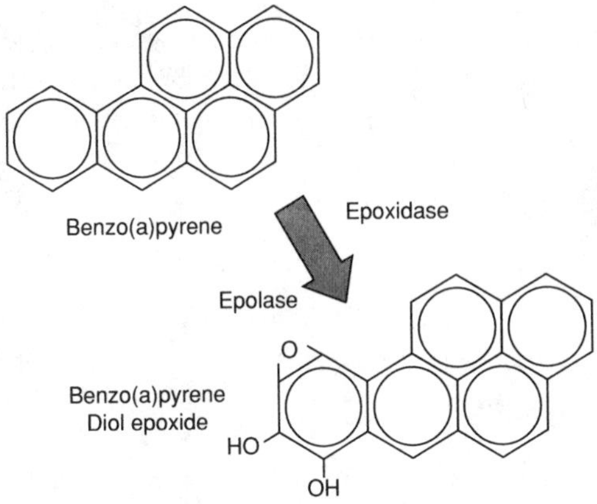

Benzo(a)pyrene

Epoxidase

Epolase

Benzo(a)pyrene
Diol epoxide

FIGURE 5-20
Metabolic activation of benzo(a)pyrene.

replication or repair. In this manner, benzo(a)pyrene causes a so-called point mutation in DNA (Fig. 5-21). It is of interest that the first human oncogene which was isolated and sequenced from the lethal bladder cancer of a heavy smoker contained a guanine-thymine transversion at the 35th base of the H-ras oncogene as its sole mutation.

A point mutation can include base-pair substitution or frameshift mutation. A base-pair substitution involves the replacement of one base pair by another. A frameshift mutation is the case of adding or deleting one or more base pairs of DNA. DNA altered in such a fashion mostly becomes dysfunctional. However, it may continue to function but in a grossly different manner than normal. Some mutagenic agents cause breakage of the backbone of the DNA molecule—events that may show up as chromosome breaks, translocations, or deletions. Also, certain genotoxic agents interfere with the replication of chromosomes and cause an uneven number to be distributed between the daughter cells.

Epigenetic carcinogens. Since the term "epigenetic" refers to a perceived absence of a genotoxic effect, its indiscriminate use has been a legitimate cause of concern. Several compounds thus labeled have simply not been sufficiently well studied with respect to genotoxicity. Further, the mechanism of action for most epigenetic carcinogens is not known. It is important to note that a large portion of regulated carcinogens is believed to be epigenetic.

Complete carcinogens. Some carcinogens can interact at more than one step in the carcinogenic process. For example, benzo(a)pyrene is a polycyclic aromatic hydrocarbon that can act both as an initiator and as a promotor of carcinogenesis. Carcinogens that both initiate and promote carcinogenesis are known as **complete carcinogens**. Complete carcinogens exhibit wide variations in chemical structure. How-

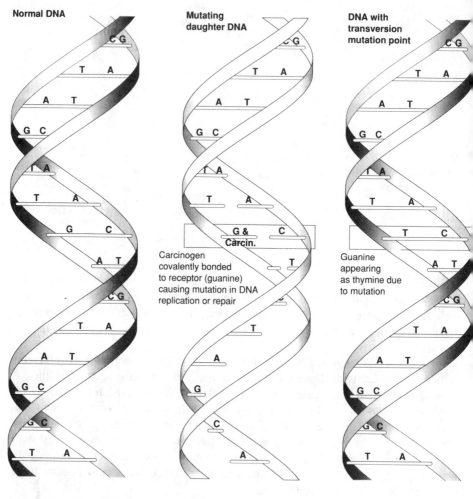

Normal DNA

Mutating daughter DNA

Carcinogen covalently bonded to receptor (guanine) causing mutation in DNA replication or repair

DNA with transversion mutation point

Guanine appearing as thymine due to mutation

G - Guanine   A - Adenine
T - Thymine   C - Cytosine

FIGURE 5-21
Example of a point mutation through base pair substitution.

ever, it has been possible to roughly classify most of these agents with respect to functional group, or other structural characteristics. The most important main categories with some examples are presented in Table 5-6.

Other carcinogens. The mechanism of action of inorganic carcinogens is enigmatic. For arsenic[92] and possibly also for other metals, interference with DNA repair may be involved. Several polychlorinated dibenzo(p)dioxins have been found to induce tumors in rodents at low doses. However, they seem to lack genotoxic action, possibly acting only as promoters and by immunosuppression. Because they are both

TABLE 5-6
Chemical classes of complete carcinogens

Chemical class	Examples
Directly alkylating agents	
Epoxides	Ethylene oxide
	Butadiene epoxide
Ethylenimines	Ethylenimine
	Triethylene melamine
Alkyl halogenides	Methyl bromide
Halogenated ethers	Bis-(chloromethyl)ether
Mustards	Bis-(2-chloroethyl)-sulfide
	Bis-(2-chloroethyl)amine
	Cytoxan (endoxan)
Nistrosamides, nitroso-ureas	N-methylnitrosourethane
	N-methylnitrosourea
Agents requiring metabolic activation	
Aromatic amines	2-Naphthylamine
	o-Toluidine
	3,3'-Dichlorobenzidine
Nitrosamines	Dimethylnitrosamine
	Diethylnitrosamine
	Nitrosopiperidine
Azo dyes	o-Aminoazotoluene
	4-Dimethylaminoazo-benzene
	Direct blue 6
Hydrazines	1,2-Dimethylhydrazine
	Cycasin (natural product)
Polycyclic aromatic hydrocarbons	Benzo(a)pyrene
	Methylcholanthrene
	Dibenz(a,h)acridine
	Dibenz(a,h)anthracene
Alkaloids (natural products)	Aflatoxins
	Pyrrolizidine alkaloids
Inorganic carcinogens	
Metals	Beryllium
	Nickel
	Chromium(VI)
	Cadmium
Metalloids	Arsenic
Solid state carcinogens	Asbestos

powerful promoters and highly immunosuppressive, dioxins appear to be potentially potent carcinogens, especially when chlorinated at the 2, 3, 7, and 8 positions (see Fig. 5-22).

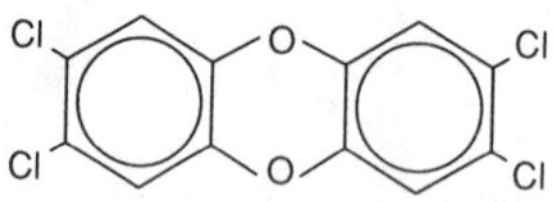

FIGURE 5-22

2,3,7,8-tetrachlorodibenzo(p)dioxin   2,3,7,8-tetrachlorodibenzo(p)dioxin.

Co-carcinogens are agents that enhance carcinogenesis when administered simultaneously with a carcinogen. Although sometimes acting in the same way as promoters, they may also function by affecting the metabolism of a carcinogen in such a way as to increase the level of the active intermediate. Phenol is an example of a co-carcinogen that is not a promoter.

Affinity. An affinity between a carcinogen and a receptor is very important, as discussed in Sec. 5-2. An example of the importance of affinity is the case of polychlorinated biphenyls (PCBs) which were produced commercially by chlorinating benzene-containing precursors. During commercial production the rate and extent of precursor molecule chlorination are virtually impossible to control. Hence, these organochlorines were produced and now exist in the environment as complex mixtures of hundreds of individual **congeners** (isomers). Although each congener of these complex organochlorines has a similar molecular formula, each has its own toxicity properties when tested in experimental animals.

At one time a uniform carcinogenic potency was assigned for all PCBs, but over the past decade it has become clear that risk assessment for entire categories of these complex organochlorines is meaningless. A constructive approach was established first for dioxins (polychlorinated dibenzo(p)dioxins or PCDDs) where toxic equivalency factors (TEFs) were established for classes of polychlorination. Later, in line with international recommendations concerning dioxin toxicity, the EPA embraced a more useful approach to TEFs, which is dependent upon whether chlorination sites include the 2, 3, 7, and 8 positions, inclusively, of PCDD molecules. According to the latest information available, a similar relationship holds among PCB congeners such that pentachlorobiphenyls (PeCBs) are the most toxic class of congeners followed by hexachlorobiphenyls (HxCBs) and tetrachlorobiphenyls (TeCBs). This information suggests the relative carcinogenic potencies shown in Table 5-7.

TABLE 5-7
Relative carcinogenic potencies of PCB classes

PCB class	Number of chlorines	Relative potency
PePCBs	5	1.0
HxPCBs	6	0.5
TePCBs	4	0.1
>HxPCBs	>6	<0.1
<TePCBs	<4	<0.1

Classes of Carcinogens

How do we know which chemical substances are carcinogenic and just how potent they are? The preferred basis for making this decision is the case where sufficient human epidemiologic data exist, mainly from studies of workers, to make a clear causal connection between the exposure to specific substances and subsequent induction of human cancers. Although human epidemiologic studies are severely limited, as discussed in Sec. 5-3, for some carcinogens, the causal relationship is clearly supported by strong statistical correlations. For example, fully half of the known cases of angiosarcoma of the liver have occurred in vinyl chloride workers. Similarly, fully 50 percent of those workers exposed five years or more to naphthylamine dyes or who have worked in uranium mines have succumbed to bladder cancer or lung cancer, respectively.

The number of clear human epidemiologic studies is small. A total of only about 50 compounds (e.g., benzene, vinyl chloride) and complex exposures (e.g., aluminum production, tobacco smoke) have sufficient data available to permit their classification as human carcinogens.[93] The most potent human carcinogens known, the aflatoxins, are of natural origin. Their presence in food products through infestation by toxin-producing fungi constitute a serious problem in several tropical and subtropical countries.

The EPA utilizes the following description for Group A - Human Carcinogens: [Compounds for which] "there is sufficient evidence from epidemiologic studies to support a causal association between exposure to the agent and cancer." This designation is based on the weight of evidence. The designation as a human carcinogen represents a consensus opinion with an expert group rather than a scientific fact. The weight of evidence system is shown in Table 5-8.

Because of limited epidemiological data, most of our understanding about the potential for a toxic substance to cause cancer is based on animal testing. Substances

TABLE 5-8
EPA weight-of-evidence classification system for carcinogenicity

Group	Description
A	Human carcinogen
B1 or B2	Probable human carcinogen
	B1 indicates that limited human data are available.
	B2 indicates sufficient evidence in animals and inadequate or no evidence in humans.
C	Possible human carcinogen
D	Not classifiable as to human carcinogenicity
E	Evidence of noncarcinogenicity for humans

Of the approximately 50 human carcinogens presently known, 13 are contaminants found in the environment (see Table 5-9). The remainder are primarily pharmaceutical products (e.g., azathioprine, an immunosuppressant).

TABLE 5-9
Human environmental carcinogens

Carcinogen	Cancer	Genotoxin
Arsenic	Lung, skin	Yes
Asbestos	Lung, mesothelioma	Rarely
Benzene	Leukemia	Rarely
Benzidine	Bladder	Yes
Bis-chloromethyl-ether	Lung	Yes
Chromium, hexavalent	Lung	Yes
Diethylstilbestrol	Female	No
2-Naphthylamine	Bladder	Yes
Nickel	Lung	Yes
Plutonium-239	Lung	Yes
Radium-226	Lung, osteosarcoma	Yes
Radon-222	Lung	Yes
Vinyl chloride	Liver, angiosarcoma	Yes

not having an apparent epidemiological association with human cancer but which have induced cancer in test animals are referred to as **potential** human carcinogens. Several hundred substances for which there is inadequate human epidemiological data have been assigned as potential human carcinogens based upon testing data in laboratory animals, usually rats and mice. Many of these substances are commonly found in hazardous waste sites. In fact, of the 717 hazardous substances regulated under CERCLA, the EPA has identified 191 substances as "potential carcinogens or as substances having carcinogenic potential."[94]

Testing for Carcinogenicity

Testing of chemical carcinogens primarily involves animal experiments, particularly lifetime studies in rodents. In these tests, doses are generally administered to animals at levels at or below the maximum dose that neither reduces lifespan (other than via tumor induction) nor results in clinical symptoms of toxic effects (frequently referred to as Maximum Tolerated Dose or MTD). Testing at these maximal doses is done for the purpose of increasing the likelihood of inducing cancer.

Such studies are very resource-demanding and can only be carried out for a limited number of economically important chemical products. Although such information is indispensable, many pitfalls exist with respect to its interpretation. The limitations are basically of three kinds: (a) statistical limitations posed by the limited size of the test groups used, (b) difficulties in interpreting the relevance of certain types of tumors for humans, and (c) uncertainties associated with extrapolation from high-to-low dose.

Limited size of test groups. The first issue has already been discussed under the section dealing with animal data in general (Table 5-4). In a human population, a 1% increase of a serious disease is of major concern. To ensure that an excess tumor incidence of 1% over a background spontaneous rate of 10% can be detected, theoretically, about 20,000 animals must be used. Actually, such a "mega mouse experiment" has indeed been performed for a cost of $15 million in 1980 using the potent experimental carcinogen 2-acetylaminofluorene.[95] The purpose was mainly to study the dose-response relationship in the low-dose region. Testing for a 0.0001% excess cancer rate is clearly impractical; however, this level of protection is a common regulatory goal in the United States.

Relevance of tumors. Relative to the significance of certain types of tumors found in animals, there has been considerable controversy with respect to liver tumors found in the mouse, kidney tumors in male rats, bladder tumors induced by non-genotoxic carcinogens in the bladders of rodents, and certain other experimental neoplasms. Primary liver cancer is rare in humans, whereas in the particular mouse strain mostly used in the United States for such tests (the B6C3F1 mouse) the spontaneous incidence is in the range 18 to 47% in males and 3 to 8% in females.[96] This incidence is significantly increased by a number of unspecific factors like hormonal treatment or by dietary changes. Thus, when the test animals are given chemicals near the maximum tolerated dose (MTD) which usually is several orders of magnitude above expected exposures for humans, there is a concern that it may trigger unphysiological processes (e.g., overloading of detoxification mechanisms) that do not occur at lower doses.[97,98]

Many if not most of the substances classified as potential carcinogens received such a classification solely because they induced liver tumors in these select mouse and rat strains. In contrast to the United States, three European regulatory agencies have discounted the finding of liver tumors at high doses in rodents (e.g., induced by the insecticide lindane) on grounds of a non-genotoxic mechanism of action which has no relevance to humans at realistic exposures.[99]

Bladder tumors in rodents illustrate another controversial situation. Although it is true that the background spontaneous incidence of bladder tumors in rats is low, chronic exposure to a number of agents that cause irritation or trauma to the bladder epithelium may stimulate cell division with the subsequent formation of tumors. Examples of such promoting agents are foreign solid bodies, alteration of urinary pH as well as infectious agents. Thus, many compounds inducing such tumors in rodents, like the food additives cyclamate, saccharin, o-phenylphenol as well as butylated hydroxyanisole (BHA), are now recognized as promoters.[100,101]

The difference of opinion about interpretation of cancer studies has resulted in the unsatisfactory situation that a compound may be considered a carcinogen in one country but not in another. At least with respect to judgment of the quality of an animal study as such, the **WHO International Agency for Research on Cancer (IARC)** is able to provide guidance that is universally accepted. These evaluations are published regularly in the series "IARC Monographs on the Evaluation of the Carcinogenic Risks to Humans."

Notwithstanding the issues discussed above, for many compounds rats and mice do seem to represent adequate models for humans. This is demonstrated by comparative dosimetry in animals and humans for those carcinogens which have adequate human data. In these cases, concordant animal data are remarkably similar (within an order of magnitude) in simulating the observed human carcinogenic response.

Other tests. As stated earlier, toxicologists have found that chemical substances which are known to be active in carcinogenic initiation also cause mutation in bacteria and other short-term test organisms. Because the structure of DNA is the same for all organisms, these tests provide a basis for prediction of carcinogenicity. However, the correlation between mutagenicity and carcinogenicity has only been found to be relatively good for initiators or complete carcinogens but not for compounds with a promoter-like action. The following end points have been used for prediction of potential carcinogenic effects:

- Point mutations (base pair substitutions, frame shift mutations, small additions or deletions)
- Disturbances of DNA repair (unscheduled DNA synthesis)
- Clastogenic effects (chromosomal breaks resulting in fragments, losses, rearrangements of chromosomes)

The numerous test systems that now are available range from bacterial cells (the Ames' test), yeasts, cultured mammalian cells, to whole animals (fruit flies, mice). Usually, results from representative test systems for all three categories are preferred before making an evaluation. Short-term *in vitro* tests are easy to perform and have resulted in the generation of a tremendous amount of data of varying quality. Pundits have suggested that the definition of a "non-mutagen" is a chemical that has not yet been tested in a sufficient number of test systems. When using this kind of data, it is important not to rely solely on the results from a single test, but to assess the weight of evidence obtained from several sources. Finally, a positive result from mutagenicity testing only gives an indication of a potential for carcinogenicity and should only be used as supporting information.[102]

Dose-Response Relationships for Carcinogens

The previous pages have explained how the assessment of human cancer risk based on animal tests requires two major considerations:

1. Is the tested chemical a carcinogen?
2. How are data from experimental animals applied to humans?

These interpretations are difficult, but far less difficult than deciding the potency of a carcinogen. This is an important concept because carcinogens vary greatly in regard to their potency. To predict the likely outcome from human exposure to carcinogenic chemicals requires the development of dose-response relationships. To develop these relationships, toxicologists ordinarily depend upon animal data. In Fig. 5-23,

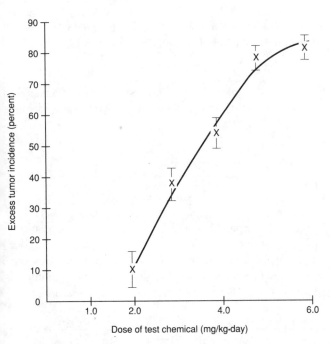

FIGURE 5-23
Hypothetical dose-response curve for a typical complete carcinogen.

a typical dose-response curve in the high dose region is given for a hypothetical complete carcinogen.

The approach to developing dose-response relationships for carcinogens differs somewhat from the approach for noncarcinogens. For ease of comparisons, doses (the abscissa) are almost always plotted as the daily administered dose per unit of body weight (mg/kg·day). Dose is plotted against response (the ordinate) which is the incidence of the total number of tested animals with tumors (dimensionless). The curve in Fig. 5-23 represents the best fit of mean values. It should be noted that the ordinate is reported as **excess incidence of cancer,** or the incidence in the exposed animals minus the normal spontaneous (background) incidence in the control population.

When assessing lifetime cancer risk to humans, it is widely accepted that carcinogenesis works in a manner such that it is possible, however remote, that exposure to a single molecule of a genotoxic carcinogen could result in one of the two mutations necessary to initiate cancer (i.e., genotoxins do not exhibit thresholds). In theory, therefore, the dose-response curve is asymptotic to "zero" incidence. Because there is no threshold, there is no safe level, only "acceptable levels." The public has expressed concern with any level, and regulatory agencies tend to establish a regulatory goal of one-in-one million (1×10^{-6} or 0.0001%) excess lifetime cancer risk.

As stated before, it is not practical to test in the range of such a low incidence. In fact, laboratories conduct animal tests at high doses to increase the likelihood of inducing cancer in a portion of the test population. Thus, to apply the results of animal tests to human exposures, the data from the high doses used in the tests must be extrapolated to the low doses of public concern.

Example 5-6 Comparison with 10^{-6} goal. For the dose-response relationship shown in Fig. 5-23, what is the minimum dose tested? What is the mean incidence induced by this dose? How does this compare with the 10^{-6} goal?

Solution. The minimum dose tested is 2.0 mg/kg·day, yielding a mean excess tumor incidence of about 10%. This incidence is 10,000 times (0.10/0.000001) greater than the 10^{-6} goal.

The calculation of carcinogenic risk involves the use of a **carcinogen potency factor** (CPF). A CPF is basically the slope of the dose-response curve at very low exposures, and is now referred to as **slope factor**. The dimensions of a slope factor are expressed as the inverse of daily dose $(mg/kg \cdot day)^{-1}$. Having derived a slope factor, the calculation of carcinogenic risk is straightforward. Quantification of carcinogenic risk of an exposure simply requires converting the dose to the appropriate terms (mg/kg·day) and multiplying it by the slope factor as demonstrated in Example 5-7:

Example 5-7 Use of slope factors. What are the maximal number of excess lifetime cancer cases expected for a population of 100,000 adults with a daily intake of 0.14 mg of benzene?

Solution

1. The slope factor for benzene is 0.029 $(mg/kg \cdot day)^{-1}$ taken from App. B.
2. Assuming an adult weight of 70 kg, the lifetime cancer risk is calculated as:

$$\text{Individual cancer risk} = (0.14 \text{ mg/day}) \frac{(1)}{(70 \text{ kg})} \frac{(0.029)}{(mg/kg \cdot day)}$$

$$= 0.000052 \text{ or } 0.0052\%$$

$$\text{Maximal cases} = \text{risk} \times \text{exposed population}$$

$$= 0.000052 \times 100,000$$

$$= 5 \text{ excess lifetime cancer cases}$$

Low-Dose Extrapolation Models

To make the necessary extrapolations to low doses, toxicologists use various mathematical models. Current models are usually divided into two types: **tolerance distribution models** and **mechanistic models.**[103]

Tolerance distribution models. Tolerance distribution models assume the existence of a threshold as described in Sec. 5-4. Some of these models (e.g., the

Mantel-Bryan model)[104] can be used for promoters, non-genotoxic carcinogens, and for certain tumor incidence data that exhibit threshold-like appearance for certain genotoxic carcinogens (e.g., the bladder tumor data for 2-acetylaminofluorene in the mega mouse experiment cited above).[105]

It should be kept in mind that most of the modeling carried out for low-dose extrapolation of promotive effects does not consider the following additive theory: the population is already subjected to a high level of promotive stress, to which an additional promotive insult may result in a dose-response relationship that lacks a dose threshold. However, such additivity presupposes a common mechanism of action.

Mechanistic models. In mechanistic models, it is assumed that a certain number of reactions, events, or "hits" (concept derived from radiation biology), or transition stages, related to a critical target in the cell (DNA), are necessary to transform a normal cell to a cancer cell. These models are important because there seems to be a general consensus that tolerance distribution models (e.g., Mantel-Bryan) should not be used. The exception is when available data are sufficiently accurate to exclude models involving a linear component in the low-dose region, or when a genotoxic mechanism might be involved.

The **multistage model**, first proposed by Armitage and Doll in 1954,[106] is an extension of the one-hit model developed for ionizing radiations. One generalized version of this model where at least one of the stages is assumed to be dose-related takes the form:[107]

$$P(D) = 1 - \exp\left[-\left(q_0 + q_1 D + q_2 D^2 + \ldots + q_k D^k\right)\right], \ k \geq 1 \qquad (5\text{-}3)$$

where $P(D)$ = the probability of cancer at dose D
   q_k = coefficients to best fit to the data
   D^k = the applied dose raised to the kth power
   k = the number of stages (usually set arbitrarily
      at the number of dose levels minus one)

The most likely estimate at very low doses becomes increasingly unstable with a small change in the response at experimental doses. Therefore, a further development in modeling was the replacement of the linear term in the polynomial function by its 95% confidence limit to achieve more stable estimates of risk above background than are obtained for the most likely estimates. This so-called **linearized multi-stage model** is currently used routinely by the EPA and has, therefore, become the most widely used model for estimation of cancer risk.[108, 109] At low doses the function becomes essentially linear. The multistage model is very flexible in fitting data sets because it is a polynomial function of dose. The so-called carcinogen potency factors provided by the EPA have been based on this model and are calculated from animal data using commercially available computer programs.[110]

The multistage model is favored by the EPA because it generally gives conservative risk estimates for low exposures, but the model has several weaknesses.[111, 112]

As already pointed out, the model is less well suited for promoters such as PCBs, greatly overestimating risk by not considering a threshold dose. For formaldehyde a much better fit to the experimental data was found using a five-stage model rather than the EPA version with the conventional restrictions.[113] Equation 5-3 given earlier predicts that the probability of tumor induction will eventually approach unity when the dose is sufficiently high. Thus, a good fit will not be obtained for data sets where the dose-response function rises steeply and then reaches a plateau (strongly concave, Michaelis-Menten kinetics), (e.g., for vinyl chloride). To cope with such a situation, a regression procedure is used for computing the maximum likelihood polynomial function from a data set. Whenever the model does not fit the data sufficiently well, data at the highest dose are deleted, and the model is refitted to the rest of the data. This is continued until an acceptable fit is obtained. A more recent development is the **Moolgavkar-Knudson model** that is based on cellular dynamics and transformations and that incorporates time-to-tumor data.[114]

Extrapolation error. It is very possible that the extrapolation of data from the 10–90% carcinogenesis range of test animals to 0.0001% carcinogenesis yields results which are conservative by several orders of magnitude for carcinogenic initiators and are incorrect for carcinogenic promoters and immunosuppressors. The range of error in experimental dose-response extrapolation for low carcinogenic frequencies in potential exposed human populations is indicated in Fig. 5-24 which shows the results from an experiment with 2-acetylaminofluorene.

The EPA is well aware of the problems associated with overly conservative risk estimates, and has repeatedly stressed[115] that the unit cancer risk estimate only provides a plausible *upper* limit for a risk that can very well be much lower. The problem is that, in reality, official EPA unit risk estimates are widely used, more or less, as absolute standards.

The adherence to one single model to cover all situations without taking due consideration to all relevant biological data (epigenetic/genotoxic action, promotive mechanisms, etc.), and sometimes even ignoring available data points in the high dose part of the dose-response curve when trying to obtain a fit to the model used (e.g., the successive exclusion of high dose data points in the application of the linearized multistage model), may have given mathematical modeling an undeserved bad reputation among biologists. In particular, the EPA's use of the linearized multistage model for compounds like chlorinated dioxins and arsenic has been criticized. Although the data sets derived from conventional carcinogenicity studies in animals are limited, the intelligent use of mathematical procedures can—at least in some cases—result in the rejection of a certain type of model for that particular situation, and thereby improve the confidence in low-level risk extrapolation.

Toxicological databases for carcinogens. Most of the toxicological databases discussed in Sec. 5-4 also apply to carcinogens and some also provide slope factors (CPFs). IRIS (see App. B and Fig. 5-18) is a prime example. Section 5-1 discussed the effect that the route of exposure can have on a substance's toxicity. This also holds true for some carcinogens. Thus, the selection of a CPF from a database

Dose response

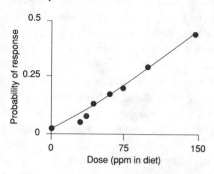

Low dose extrapolation

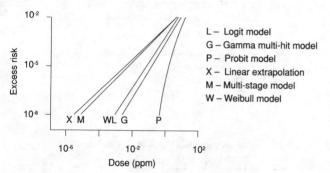

L – Logit model
G – Gamma multi-hit model
P – Probit model
X – Linear extrapolation
M – Multi-stage model
W – Weibull model

FIGURE 5-24

Extrapolation of dose-response relationship for liver tumors induced by 2-acetylaminofluorene in the low-dose range.[116]

must consider the exposure route. For example, vinyl chloride is 8 times more potent by ingestion than inhalation, whereas with arsenic it is the other way around, the ingestion risk being only 3% of that by inhalation (see Table 5-10).

For reasons that are not apparent, most metallic lung carcinogens, such as cadmium, hexavalent chromium, and nickel, do not appear to be carcinogenic by any other route of administration than inhalation. Thus, care must be exercised when defining exposure and in selecting the appropriate slope factor.

TABLE 5-10

Carcinogenic potencies depend upon exposure route

Carcinogen	Inhalation CPF	Ingestion	Ingestion/inhalation ratio
Vinyl chloride	0.295	2.30	7.8
Arsenic	50	1.75	0.035

Mixed exposure to carcinogens. The conservative nature of the EPA's linear-dose response model suggests that the government's approach is overprotective of public health. However, the approach is underprotective from another perspective. To explain, the data used by the model stem from testing of animals exposed to a single compound when, in fact, it is rare that a given human population will receive potential exposure to only one carcinogenic substance in the environment. The normal human is exposed to a complex mixture of substances, including carcinogens which are both initiators and promoters or immunosuppressors. Consequently, the EPA approach to considering multiple exposures has been considered underprotective by some scientists.

At present, the simplified risk assessment methodology for dealing with mixed exposure to carcinogens is that overall carcinogenic risk is equivalent to the sum of individual risks of each substance. This is referred to as an additive model, and is illustrated in the following hypothetical example for an individual exposed to three carcinogens:

Carcinogen	Lifetime risk for specified exposure
Benzene	0.0005
Cadmium	0.0008
Vinyl chloride	0.0025
Total	$0.0038 = 3.8 \times 10^{-3}$, excess cancer probability of approximately four in one hundred

These classic additive models may not be appropriate for multiplicative or antagonistic interactions which may occur either within or as a result from exposure to complex mixtures. The dosimetry of potential human responses to mixtures of different types of carcinogens is more complex, as revealed from epidemiological studies of shipyard workers and uranium miners regarding the impact of cigarette smoking. For example, if the relative risk for lung cancer from tobacco smoking is 10 (ten times the lung cancer risk for nonsmokers), and the relative lung cancer risk from asbestos exposure is also 10, the classic additive model would conclude that the lung cancer risk from exposure to both tobacco smoke and asbestos would be 20 $(10 + 10)$. However, it has been found that the lung cancer risk among asbestos workers who smoked was nearly 100 times that of workers in other industries who did not smoke.[117] This is a prototypic case of carcinogenic synergism in which the ultimate risk of exposure to an initiator and a promoter is the product $(10 \times 10 = 100)$ of relative risks rather than the sum $(10 + 10 = 20)$ of these relative risks.

Background and Lifestyle Carcinogenesis

Twenty-five percent of us will contract cancer in our lifetimes. Toxic substances in the environment do contribute to the carcinogenic burden on humans. It is widely accepted, however, that natural radiation and lifestyle account for far more cancer deaths than toxic chemicals in the environment.

Background carcinogenesis. There is a background rate of carcinogenesis which has very little to do with human behavior, including where we live and what we eat or do. We live within a universe which is rather hostile to DNA and as such are constantly bombarded with cosmic radiation which is rather uniform over the Earth's surface, although slightly more pronounced at higher altitudes and over the poles. In addition there are naturally occurring radionuclides, including radon gas, which are associated with the composition of the Earth's crust and provide further sources of radiation exposure. This cosmic ray and radionuclide flux, at approximately 150 mrem per year at sea level in North America, is one background genotoxic influence on human cells that is responsible for tens of thousands of DNA repair events per day per cell—or an astronomical 10^{20} (100 quintillion) DNA repair events per person per year. Fortunately, the DNA repair process replaces the damaged DNA base with the same base. However, the DNA repair process can go awry, leading to a mutation. If this mutation occurs in or near an oncogene in a somatic cell, it can constitute one of the two mutations required for carcinogenic initiation, or it can be one mutation needed for progression.

Although each of us begins life at conception as a single cell, in the adult body there are approximately 100 trillion cells, each at risk of becoming cancerous. Every cancer begins as a clone of cells, all of which are derived from the same originally transformed single cell which has gained the ability to replicate faster than its neighboring cells. Although a very frequent disease of humans at the level of a cell, cancer initiation is an extremely rare event. Because cancers are clonal and one out of four humans, each with 100 trillion cells, develops a malignancy in a lifetime, the frequency of cancer initiation on a cellular basis may be calculated as 1 per 400 trillion cells, or 2.5×10^{-15}.

It is thought that the balance between faithful DNA repair and more mistake-prone DNA repair was a driving force in evolution, yet with minimal damage to the species. Less-than-perfect DNA repair enzymes allow for a low rate of accumulated mutations both in somatic cells (potentially cancer initiating) and germinal cells (leading to changes in the next generation, a minority of which may be more fit for a changing environment). In retrospect, had cancer been a lethal disease prior to puberty, chances are that this mechanism for DNA repair would not have survived in its present form. If so, we less well-adapted survivors of a less dynamic evolutionary lineage would carry a more perfect DNA repair system today. In that case, it would be doubtful if any of us could read these words, let alone reason—a dubious trade-off for a slightly errant DNA repair system.

Lifestyle and occupational carcinogenesis. In addition to persistent background mutagenesis, we encounter a wide variety of chemicals in our daily lives as part of our diet and lifestyle, having nothing to do with hazardous wastes. Many plants contain natural pesticides which are potentially carcinogenic. Some of us favor foods rampant with naturally formed carcinogens: mushrooms (hydrazines), celery (psoralens), peanut butter (aflatoxin), and charcoal-broiled steaks (benzo(a)pyrene). Such known sources pale, however, in comparison with the widespread habit of wrapping shredded leaves of Mother Nature's produce into small bundles and smoking them. The inhalation of tobacco smoke has been a major pathway for delivery of carcinogens into *Homo sapiens*, and responsible for more lung, bladder, and possibly also pancreatic and breast cancer than all other artificial sources of carcinogenic insult combined.

In addition to background sources of carcinogenic exposure such as cosmic rays, natural radionuclides, and lifestyle-associated sources such as food and cigarettes, many workers are exposed to carcinogens occupationally. Obviously, an awareness of this source of exposure has driven many regulations by the U.S. Occupational Safety and Health Administration (OSHA), such that the workplace is much safer today than ever from the standpoint of potentially carcinogenic insult.

"Safe" levels of carcinogenic chemicals in hazardous wastes. The tools for predicting the potential carcinogenic burden for an exposed human population have become more precise over the past decade; still, there are many uncertainties which severely limit the process. Carcinogenic risk assessment is best at comparing risks and allowing a prioritization of risks rather than quantifying actual risks. As explained in this section, extrapolations from high doses in animals to low doses in humans are fraught with statistical uncertainty and conservative assumptions and are probably incorrect for non-genotoxic carcinogens. Nonetheless, a somewhat inaccurate estimation of risk may be better than no estimation at all, and errors at present are in the direction of human health protection.

In the future, it is expected that improved approaches will be developed such as the use of pharmacokinetic models to account for many of the variables in scaling animal test data to estimate human risk. Nevertheless, the toxicologists will be able to determine only the likelihood of adverse effects; definitive answers will remain elusive.

Given these scientific limitations, what levels of carcinogenic exposure from hazardous wastes (or other sources) should be considered acceptable by society? At present, the EPA and many state agencies have an upper bound "safe" guideline of one-in-ten-thousand (1×10^{-4}) lifetime cancer risk to persons who might be maximally exposed to carcinogens from hazardous wastes. With this upper bound, EPA has defined acceptable risks for carcinogens as the range of one-in-one-million to one-in-ten-thousand (10^{-6} to 10^{-4}) lifetime cancer risk. The figure "one-in-one million" (1×10^{-6}) excess lifetime cancer risk is used as a general regulatory goal (see Sec. 14-1).

5-6 ECOTOXICOLOGY[†]

As discussed in the preceding sections, the mechanisms by which toxic substances affect mammalian species (viz. humans) has historically been the focus of considerable research. In contrast, the assessment of chemical releases on other components of

---

[†]Section 5-6 was prepared by David J. Stout.

the biosphere commenced with the study of "fish kills" in the early 20th century, and extensive research on hazards to non-humans has only recently gained prominence. Rachel Carson's documentation of ecological damage from agricultural pesticide application in *Silent Spring* is widely recognized as a driving force behind heightened public awareness about "ecotoxicity" which emerged during the 1960s.[118] With this came the recognition of the need to protect and manage a multitude of feral (i.e., not domesticated) species, interacting as communities within complex ecosystems, thus providing an impetus for the science of ecotoxicology.

Ecotoxicology is a relatively young scientific discipline which extends the principles of toxicology to natural systems to evaluate the potential impact from releases of toxic materials. While the focus of traditional mammalian toxicology is the effects of chemical exposure on individual species, primarily humans, ecotoxicology addresses effects on groups of populations interacting with the physical environment. In other words, a chemical which kills a percentage of individuals in a population is a catastrophe to humans; however, a similar effect in an ecosystem may be of little ecological significance compared to chemicals which are not directly lethal to an individual, but decrease primary productivity or reduce energy transfer which, in turn, affect system function directly and affect system structure indirectly. Therefore, it is important to understand not only the direct effects of a chemical on an organism, but also how the chemical interacts with the organism's environment (e.g., physical alteration of habitat by a chemical spill; loss of a food source; biological or chemical degradation).[119] As examples of the latter, an accidental oil spill can produce a variety of effects including reduction of photosynthesis from decreased light penetration, lethality to aquatic organisms from exposure to lighter aromatic hydrocarbons, and the transfer of heavier hydrocarbons through the food chain from accumulation in tissue.

It is important to note that the principles of exposure, absorption, metabolism, and manifestation used for the study of humans are also germane to ecotoxicology. Although the specific steps can vary widely (e.g., absorption for non-predaceous fish occurs predominantly in the gills), the pharmacokinetic fundamentals remain the same. The concepts of structural affinity and chemical activity are likewise important to ecotoxicology. Nevertheless, the application of these principles and fundamentals to diverse populations of feral species interacting as communities in natural systems presents difficult challenges for scientists and engineers involved in decisions regarding the management of hazardous waste. This section will highlight the general techniques used to evaluate toxic effects and how the results of these tests are used in the fields of aquatic and terrestrial toxicology.

Evaluation of Toxic Effects

Ecotoxicologists answer numerous questions regarding constituent toxicity using a variety of test systems which differ in complexity, predictive capacity, and cost. A number of factors frustrate the applications of these tests and the use of test results.

The origin of laboratory toxicity testing traces to pre–World War II investigations of the cause of fish kills and the use of primitive test methods often referred to as "one-fish-in-a-jar." The subsequent pursuit of an optimal test organism, synonymous to the "white rat," led to the testing of a wide variety of species ranging from primary producers (e.g., algae) to invertebrates (e.g., water fleas) to ultimate food chain predators (e.g., largemouth bass). Although standard toxicity test methods have been established for numerous species, the lack of scientific agreement on the species most sensitive to a contaminant and/or representative of community structure continues to pace the search for the elusive "white rat."

Some of the ecotoxicology tests, in the sense that they are designed to monitor changes at the subcellular, cellular, and organism level, resemble those developed for estimating toxicity to humans. In many cases, however, evaluation at the organism and sub-organism level does not reveal the extent of a toxic effect on an ecosystem. For example, measurement of contaminant levels in edible fish tissue represents an excellent indicator of individual exposure, but does not convey information about the possible loss of a food source in an aquatic food chain.

Ecotoxicity test methods attempt to bridge the information gaps between measurable subcellular effects and shifts in structure and function of an ecosystem. Such testing encounters increasing complexity at ascending levels of organization: population, community, and ecosystem. This is suggested by Fig. 5-25, which depicts the relationship between responses at levels of biological organization and the time scales for these responses.

Testing and evaluating toxic effects are also frustrated by interactions among chemicals released to the environment. Environmental releases of an industrial nature

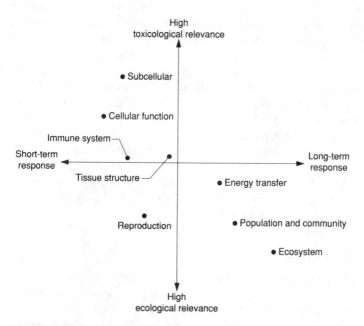

FIGURE 5-25
Toxicological/ecological relevance and time scale of toxic responses for levels of biological organization.[120]

seldom consist of an individual chemical compound; typically, they include a complex mixture of chemicals. In fact, the toxicity of relatively few of the multitude of chemicals released to the environment are understood, and minimal information exists to discern the effects of a complex melange of hazardous effluents and emissions released from active facilities and contaminated sites.

These and the other complicating factors frustrate the practice of ecotoxicology. Notwithstanding these complications, well-founded approaches have been developed for evaluating toxic effects, providing reliable data for hazardous waste management decision-making.

Toxic End Point

The selection and design of an appropriate test method first requires the specification of the toxic effect (**end point**) to be evaluated. This is often accomplished in ecotoxicity evaluations by using two distinct but complementary types of end points: assessment and measurement. **Assessment end points** are ecological characteristics, which, if found to be significantly affected, would indicate a need for mitigative action.[121] An assessment end point for an ecological risk assessment performed at a hazardous waste site could be the discernible adverse effect on populations of commercial or aesthetic value. For example, the potential effects of a discharge to surface water on the survival and abundance of a population of native trout presents an assessment end point of ecological and recreational value.

To be useful, an assessment end point must be measurable. Therefore, a **measurement end point** is selected which provides a quantitative expression (i.e., a measurable characteristic) related to the assessment end point.[122] For the example assessment end point, the measurement end point may quantify the frequency of mortality within this trout population using either toxicity tests or field monitoring of changes in species abundance relative to the discharge location and dilution. Some assessment and measurement end points used in ecological evaluations at hazardous waste sites are provided in Tables 5-11 and 5-12.

TABLE 5-11
Potential assessment end points[123]

Population	Ecosystem
Extinction	Productive capability
Abundance	
Yield/production	
Age/size class structure	
Massive mortality	

Community	Human Health Concerns
Market sport value	Contamination
Recreational quality	Gross mortality
Change to less useful desired type	

TABLE 5-12
Potential measurement end points[124]

Individual	Community
Death	Number of species
Growth	Species evenness/dominance
Fecundity	Species diversity
Overt symptomology	Pollution indices
Biomarkers	Community quality indices
Tissue concentrations	Community type
Behavior	

Population	Ecosystem
Occurrence	Biomass
Abundance	Productivity
Age/size class structure	Nutrient dynamics
Reproductive performance	
Yield/production	
Frequency of gross mortality	
Frequency of mass mortality	

Bioassay

To evaluate the hazards associated with chemical exposure of non-humans, ecotox-icologists have historically relied on the combination of short-term and long-term **bioassays**. A bioassay is a technique by which organisms, biological systems, or biological processes are used to measure effects of chemical exposure.

Results from short- and long-term toxicity tests are often used to establish water quality criteria, to estimate the killing or knock-down power of pesticides, to calculate permit limitations for effluent discharge, and to project the potential impacts of catas-trophic spills and chemical accidents, and the actual and potential effects of hazardous waste disposal and cleanup. The integration of toxicity tests in waste management problems facing engineers can provide significant advantages in the decision-making process to achieve long-term solutions which are protective and cost-effective.

In the context of hazardous waste management, bioassays may be defined as the laboratory exposure of organisms to field-collected environmental samples for the purpose of identifying actual or potential toxic effects on resident species. Typically, such bioassays involve laboratory testing of soil, sediment, or water samples using a standard array of test organisms under controlled conditions.[125] A simple exam-ple is pumping contaminated surface water through a tank containing juvenile trout. Simulating exposure and interpreting effects on communities and ecosystems is far more demanding than measuring exposure and effects on populations of individuals exposed in laboratory systems.

Laboratory simulation of ecosystems has been attempted and frustrated by prob-lems of scale. **Microcosms** have been used to imitate ecosystem structure and function on a reduced scale in the laboratory. For example, terrariums are common microcosms used as surrogates of simplified terrestrial ecosystems. To reduce difficulty presented by scale and lessen uncertainty in data extrapolation, macro model systems known as **mesocosms** have been developed as isolated portions or replications of natural settings. While microcosms and mesocosms are time- and cost-intensive and not commonly applied to hazardous waste management decision-making, these tests are often used to evaluate the fate, transport, and effects of new chemicals seeking registrations.

The following discussions will examine the development and use of these toxicity tests, but an engineer should consult an ecotoxicologist and/or refer to guidance documents, as well as the scientific literature.[126, 127]

Short-Term Toxicity Tests

Short-term toxicity tests, often referred to as **acute** toxicity tests, are used to evaluate severe effects from exposures spanning relatively short periods of time. While any response to a stressor represents a measurement end point, acute toxicity tests are most often used to evaluate percent affected by a chemical compound or perturbation. As in tests to predict potential human toxicity, the LD_{50} is used as an empirical estimate of the dose associated with a 50% lethal response in the test population (see Fig. 5-11). Similar to an LD_{50}, the LC_{50} is used to measure lethality via inhalation or exposure in an aqueous solution. Short-term toxicity tests used to determine doses associated with sub-lethal responses provide an effective concentration, or EC_{50}. Figure 5-26 presents an idealized schematic of the effect of dose on organism response as measured by some physiological change, including percent death.

Acute toxicity tests are commonly designed to elicit a quantal response (i.e., an all-or-none effect: dead or alive). The relation between the concentration of the toxic substance and the percent of exposed organisms affected can be determined as a dose-response curve as discussed in Sec. 5-3. It is sometimes referred to as a dose-mortality curve.[128] The standard test for measuring acute lethality is designed to produce a 50% lethal response in a test organism after a 96-hour exposure duration. While responses for shorter exposure durations are often cited, the 96-hr LC_{50} is recognized as the standard.

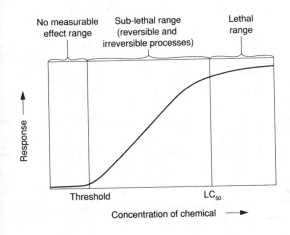

FIGURE 5-26
Idealized plot of dose-response relationship.[129]

TABLE 5-13
Test concentrations and mortality of sheepshead minnows exposed to sodium selenite[130]

Average measured concentration, selenium ion (mg/L)	Percent mortality per exposure period			
	24-hr	48-hr	72-hr	96-hr
Control	0	0	0	0
2.0	0	0	0	0
3.6	0	5	10	20
7.1	0	25	35	45
14	10	30	60	75
27	40	100	100	100

Data for an acute toxicity test for sheepshead minnows exposed to sodium selenite are tabulated in Table 5-13 and the estimated LC_{50} values are presented in Fig. 5-27.

An acute toxicity test can take many forms. For some organisms (e.g., invertebrates) death is not easily discernible; therefore, the 96-hr EC_{50} is cited to reflect end points such as immobilization or loss of equilibrium which are considered surrogates of lethality. In an acute toxicity test, an organism may be exposed by one of four standard techniques: static, recirculation, renewal, or flow-through.[131] Essentially, these methods differ in regard to the introduction and maintenance of the test material in the exposure medium. Other water quality parameters (e.g., dissolved oxygen, hardness) may also be manipulated to evaluate their influence on toxicity.

Acute toxicity tests can be designed to answer numerous questions about waste management such as the toxicity of an individual chemical, mixture of chemicals,

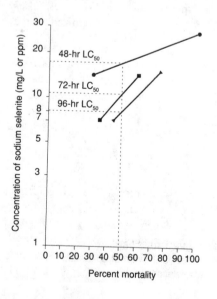

FIGURE 5-27
Graphical interpolation of time-dependent LC_{50} values for sodium selenite.[132]

leachate, or other concentrated waste streams. Also, tests can be used to locate toxic "hot spots" at waste sites, and to evaluate the effectiveness of waste treatment technologies. Selecting the proper test organism, exposure system, and boundary conditions is important to producing meaningful results. Evaluation of the toxicity test results combines statistical procedures with interpretation based on experience.

Long-Term Toxicity Tests

Long-term toxicity tests, commonly referred to as **chronic** toxicity tests, are used to evaluate whether extended laboratory exposure of different life stages (e.g., embryos, juveniles, and adults) to a constituent will adversely affect individuals, species, or populations. These tests may measure effects on development, homeostasis, growth, and reproductive potential to establish benchmark concentrations for the onset of long-term effects. Such benchmark concentrations include the lowest observed effect level (LOEL), which represents the lowest concentration associated with a measurable effect, and the no observed effect level (NOEL) which denotes a concentration which did not produce a measurable effect, as discussed in Sec. 5-4.

Since chemical concentrations associated with chronic effects are typically lower than those that produce acute responses, chronic toxicity usually provides a more sensitive measurement of toxicity. To develop acceptable exposure concentrations (e.g., water quality standards), ecotoxicologists use a sub-category of chronic toxicity tests referred to as **life-cycle** tests. These specific tests evaluate the responses of individuals from a species over a complete life cycle to evaluate effects on survival, growth, and reproduction.[133] For example, a life-cycle test for a freshwater fish may include exposures of embryos and juvenile fish at various stages of maturation, and adults continuing through reproduction and hatching of the next generation. An old tenet in aquatic toxicology is that the early life stages are the most sensitive to toxicants. An evaluation of data from 56 complete or partial "life-cycle" toxicity tests showed that embryo-larval and early juvenile life stages were the most sensitive in most cases.[134]

The problems associated with chronic toxicity tests are the time and costs for conducting such tests. While life-cycle tests provide useful information for decision-making, these tests are limited in that (1) not all species will reproduce in a laboratory setting, (2) they are unnecessary and incredibly expensive to conduct, (3) the toxicant is introduced at a constant rate for a fixed duration whereas in nature the magnitude, frequency, and duration of exposure can vary considerably, and (4) only a small set of all toxic chemicals released to the aquatic environment can feasibly be evaluated.[135]

Biomarkers

The **biomarker** approach to identifying exposure or physiological effects differs from the more traditional laboratory methods of environmental testing (e.g., chemical measurements, toxicity tests) by using naturally occurring organisms as markers of environmental response.[136] The underlying concept for biomarkers is that selected measurement end points, typically comprised of biochemical and physiological responses, can provide sensitive indices of exposure or effects other than death (non-lethal stress).[137] Confronted with the numerous complexities frustrating the measurement of environmental exposure and effects in dynamic, natural systems, there is increasing desire to evaluate exposure and effects of chemicals using biomarkers. These measurements of body fluids, cells, or tissues can indicate in cellular or subcellular terms the relative magnitude of chemical exposure or organism response.[138]

While extensive research continues to refine existing methods and to develop new markers of exposure and effect, experience has identified several biomarkers of value to hazardous waste management decisions. Measures of exposure to chemicals which bioaccumulate in tissue is easily accomplished using an analysis of tissue residue. In the absence of tissue residues, empirical constants often derived and used by ecotoxicologists to predict bioaccumulation include the **octanol-water partition coefficient** (K_{ow}), which approximates a chemical's behavior between the lipid and water phase, and the **bioconcentration factor (BCF)**, which describes the relationship between the chemical concentration in an organism and the concentration in the exposure medium (typically water). These parameters were discussed in Sec. 3-2. The BCF is empirically expressed as[139]

$$C_{org} = (BCF)(C_w) \text{ or } BCF = C_{org}/C_w \qquad (5\text{-}4)$$

where C_{org} = concentration in biota (mg/kg)
   BCF = bioconcentration factor (mg/kg in tissue/mg/L in water, or L/kg), and
   C_w = concentration in water (mg/L)

Table 5-14 provides physico-chemical parameters for some typical compounds which bioaccumulate. Additional data may be found in App. A.

TABLE 5-14
Selected physicochemical parameters and bioconcentration factors for typical compounds which bioaccumulate[140]

Compound	Log K_{ow}	Log BCF (fish)	Log K_{oc}	S (mg L^{-1})
Chlordane	5.15	4.05	4.32	0.056
DDT	5.98	4.78	4.38	0.0017
Dieldrin	5.48	3.76	4.55	0.022
Lindane	4.82	2.51	2.96	0.150
Chlorpyrifos	4.99	2.65	4.13	0.3
Cyhexatin	5.38	2.79	>3.64	<1.0
2,4D	1.57	1.30	1.30	900
2,4,5T	0.60	1.63	1.72	238
TCDD	6.15	4.73	5.67	0.0002

K_{ow} = octanol-water partition coefficient (unitless)
BCF = bioconcentration factor (L/kg)
K_{oc} = organic carbon partition coefficient
S = water solubility

Example 5-8   Predicting bioconcentration. Routine monitoring of sediments and surface water downstream from a hazardous waste site indicate that reportable concentrations of the pesticide DDT are present. On the average, sediments and surface water contain 33 mg/kg and 0.25 μg/L, respectively. Considering that the U.S. Food and Drug Administration has identified levels of 5 ppm or higher of DDT as requiring action, is there a reason to be concerned about potential accumulation of DDT in fish tissue near this site?

Solution. Since the routine monitoring does not provide information regarding DDT residues in fish tissue, we refer to Table 5-14 and bioconcentration factor (BCF). Transforming the Log BCF of 4.78, we estimate a BCF of 60,256 (mg/kg DDT in tissue per mg/L DDT in water, or L/kg).

Substituting our monitoring data and the empirical BCF for DDT in Eq. 5-5,

$$C_{org} = (60,256 \ (mg/kg)/(mg/L)) \times (0.00025 \ mg/L) = 15.06 \ mg/kg$$

we estimate a DDT tissue residue of 15 mg/kg. Based on this estimate, it is probable that DDT residues in tissue are an issue that should be evaluated.

A chemical which bioaccumulates and/or bioconcentrates can increase in tissue concentration as the chemical is transferred up through trophic levels in the food chain in a process known as **biomagnification.** Biomagnification in two simple aquatic food chains is depicted in Fig. 5-28 for two organochlorine pesticides: DDD and toxaphene.

Tissue residues alone do not convey information about biological responses to chemical exposure. Furthermore, measuring tissue residues is not feasible with chemicals that do not readily bioaccumulate (compounds easily eliminated, for example) or with complex mixtures that require time and cost intensive analyses that may not identify all toxic chemicals. In such cases, other indirect measures may be preferred to indicate a biological response to exposure that is of toxicological significance.[142] A biomarker which provides information about both exposure and non-lethal stress is the enzyme δ-aminolevulinc acid dehydratase (δ-ALA-D). This enzyme catalyzes a reaction involved in heme synthesis and is very sensitive to inhibition by inorganic lead.[144, 145] δ-ALA-D provides an inexpensive measurement of lead exposure, which eschews the possible problem of lead contamination of environmental samples. At waste sites where lead is of particular interest, δ-ALA-D measurements may provide useful information about lead exposure and non-lethal organism response,[146] but its utility is reduced at hazardous waste sites contaminated by numerous trace metals. Another common biomarker for trace metal exposure is the metal-binding protein metallothionen. Metallothionen functions as a regulator for normal zinc and copper metabolism as well as providing a mechanism for metal detoxification. Changes in metallothionen activity provide a sensitive marker of trace metal exposure.

Other biomarkers of environmental exposure and effects include induction of metabolic (detoxifying) enzymes such as mixed function oxidases (MFO) or cytochrome P450 by organic contaminants; inhibition of cholinesterase enzymes in the brain by insecticide compounds such as organophosphates; and non-specific effects such as skeletal abnormalities. Additional information about the methods and uses

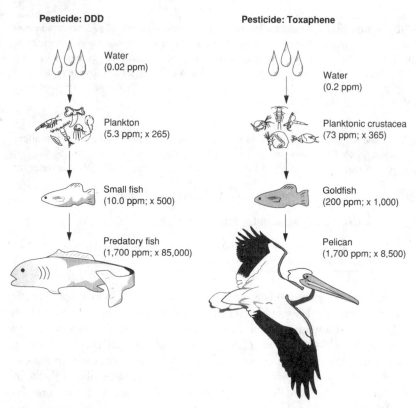

Pesticide: DDD

Water
(0.02 ppm)

Plankton
(5.3 ppm; x 265)

Small fish
(10.0 ppm; x 500)

Predatory fish
(1,700 ppm; x 85,000)

Pesticide: Toxaphene

Water
(0.2 ppm)

Planktonic crustacea
(73 ppm; x 365)

Goldfish
(200 ppm; x 1,000)

Pelican
(1,700 ppm; x 8,500)

FIGURE 5-28
Biomagnification of pesticides in aquatic food chains.[143]

of other specific biomarkers to evaluate exposure to hazardous waste is available in various references.[147,148]

Aquatic Toxicity of Trace Metals

Trace metals naturally cycle through and between the lithosphere, atmosphere, and hydrosphere. Enriched levels of trace metals may occur naturally due to the weathering of a geologic substrate or volcanic emissions, or from anthropogenic sources such as industrial effluent, or non-point sources such as urban storm runoff and atmospheric washout of particulate or gaseous emissions. Mining, fossil fuel combustion, smelting, plating, and other historical anthropogenic activities have altered the biogeochemical cycling of trace metals and frequently enriched the levels of trace metals in the environment. The release from a hazardous waste site may represent a significant local source, but rarely the only source of exposure to local biota.

In aquatic systems, these trace metals may be present as both soluble and insoluble forms within the water column, sediments, and biotic tissues. Trace metal toxicity occurs when a receptor is exposed to an available form of a trace metal for a sufficient duration at a concentration which elicits an adverse response. The potential for trace metal toxicity to aquatic organisms depends on numerous factors, including the chemical and physical characteristics of the water and sediment, composition and health of the biological communities present, and the magnitude and availability of the trace metal. Some of the factors which influence trace metal toxicity in aquatic systems are outlined in Table 5-15.

Many trace metals are essential micronutrients for the maintenance of aquatic life, and toxic only when available to organisms at levels which exceed nutritional requirements. Trace metals such as copper, iron, zinc, manganese, cobalt, and selenium are essential to metabolism; yet, exposure of aquatic organisms to these same elements at elevated levels can adversely affect development and health, or cause death. Other

TABLE 5-15
Factors influencing the toxicity of trace metals in the aquatic environment[141]

Form of metal in water	inorganic / organic	soluble	ion / complex ion / chelated ion / molecule
		particulate	colloidal / precipitated / adsorbed
Presence of other metals or poisons	joint action / no interaction / antagonism		more-than-additive / additive / less-than-additive
Factors influencing physiology of organisms and possibly form of metal in water	temperature / pH / dissolved oxygen / light / salinity		
Condition of organism	stage in life history (egg, larva, etc.) / changes in life cycle (e.g., moulting, reproduction) / age and size / sex / starvation / activity / additional protection (e.g., shell) / adaptation to metals / altered behavior		

trace metals such as lead, cadmium, and mercury do not play an essential role in daily maintenance, but adversely affect an organism if available at toxic levels.

Upon exposure, uptake occurs by three main processes: from the water column through respiratory surfaces (e.g., gill); adsorption from water onto body surfaces; and from ingested food, particles, or water through the digestive system.[150] After intake, as discussed in Sec. 5-1, organisms may absorb, transform, excrete, or store (bioaccumulate) metals. A schematic depicting contaminant exposure and uptake for a fish is presented in Fig. 5-29.

Many free-swimming or mobile aquatic organisms (e.g., fish) actively excrete trace metals, particularly essential nutrients, through the gills, gut, feces, and urine, while less mobile or sessile organisms (e.g., plants and bivalves) are less capable of regulating internal metal concentrations. Once the processes which regulate metals have been saturated, bioaccumulation within tissue or other tolerance mechanisms may preclude an adverse effect. However, these protective mechanisms are often too transient or insufficient to preclude toxicity.

The mechanisms of metal toxicity are divided into three general categories: blocking of essential functional groups of biomolecules (e.g., protein and enzymes); displacing the essential metal ion in biomolecules; and modifying the active conformation of biomolecules.[151] These mechanisms of action may cause death at high level exposures, but also may manifest non-lethal effects on structure (histology/morphology), function (growth, development), biochemistry (blood and enzymes), and reproduction.

Example of Aquatic Toxicity of Copper to Fish

To illustrate the toxic mechanisms and effects of trace metals on fish, the example of copper is selected and discussed thoroughly in the following pages. Copper, a micronutrient, is a necessary component of many enzymes and proteins. Ceruloplasmin,

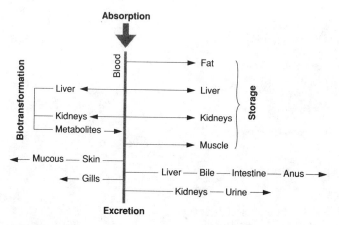

FIGURE 5-29
Possible movement and fate of a contaminant after absorption into the bloodstream.[149]

a protein located in blood serum, accounts for 95% of the circulating copper in mammals. Copper, mercury, and silver vary in the magnitude of their toxicity to aquatic plants, invertebrates, and fish, but as shown in Table 5-16 these metals are commonly the most toxic to aquatic biota. The sensitivity of aquatic organisms to copper is extreme for those species having a high surface to volume ratio, high respiratory flows, plus an extensive, highly permeable gill surface that facilitates rapid uptake of large amounts of copper.[153]

The principal toxic copper species is cupric ion (Cu^{+2}), although two ionized copper hydroxides ($CuOH^+$ and $Cu_2OH_2^{+2}$) have also proven to be toxic.[154] The toxicity of copper in water is controlled primarily by alkalinity (which includes hardness) and pH. Copper is less toxic at higher alkalinities that accompany higher hardness values which reduce bioavailability through formation of copper carbonate complexes.[155] Copper toxicity increases by reductions in alkalinity (hardness), dissolved oxygen, chelating agents, humic acids, pH, and suspended solids. For example, a low level of dissolved oxygen in the water column would require an increase in the ventilation rate of aquatic organisms. The elevated ventilation rate would then increase the flow of water containing copper, facilitating the uptake of copper and increasing the overall body burden.

Upon uptake and distribution, the maximum fish tissue copper concentrations are found in the liver, followed by the gill, kidney, and muscle. In studies which measured tissue concentrations, the liver was the principal accumulator of copper,[156–162] accumulating approximately ten times the amount measured in muscle tissue.[163] The liver retains copper for the longest duration[164] and serves as the best indicator of copper exposure. The copper residing in the liver associates with metalloenzymes at concentrations that could potentially disrupt normal metabolic activity, hence producing deleterious effects in reproduction and development.[165] In Table 5-17, the distribution of copper in freshwater fish tissues is listed as measured by investigators in several studies.

TABLE 5-16

Range of acute toxicity (48- to 96-h LC$_{50}$ or EC$_{50}$ μg/L) data for commonly used freshwater test organisms[152]

Test organism	Copper	Mercury	Lead	Zinc
Arthropoda (crustaceans)	5–3000	0.02–40	—	30–9000
Annelida	6–900	—	—	—
Mollusca	40–9000	90–2000	—	500–20,000
Vertebrata(fish)				
Salmonidea	10–900	3–20,000	1000–500,000	50–7000
Centrachidae	700–10,000	—	20,000–400,000	1000–20,000
Cyprinidae	23–2000	3000–10,000	2000–500,000	400–50,000
Algae				
Chlorophyta	1–8000	<0.8–2000	500–1000	30–8000
Chrysophyta (diatoms)	5–800	—	—	—

TABLE 5-17
Distribution of copper in fish tissues[166]

Species	Cu source	Copper (ppm dry weight)				References
		Gills	Liver	Kidneys	Muscle	
Brook trout	Tank, Cu = .09 ppm	8.2	238	16.5	17	McKim & Benoit (1971)[156]
Brook trout	Tank, Cu = .003 ppm	5.0	239	19.7	15.3	McKim & Benoit (1971)[156]
Brown bullhead catfish	Tank, Cu = .027 ppm 30 days	9.4	33	10.0	—	Brungs et al. (1973)[1]
Brown bullhead catfish	Tank, control	5.7	24	8.0	—	Brungs et al. (1973)[1]
Brown bullhead catfish	Tank, Cu = .027 ppm 20 months	6.9	11	10.0	—	Brungs et al. (1973)[1]
Brown bullhead catfish	Tank, control	3.8	29	7.8	—	Brungs et al. (1973)[1]
Brown bullhead catfish	French R. Cu = 2–9 ppm	1.8	30.3	2.5	1.3	Hutchinson and Fitch (1974)[167]
Perch	D'Alene- Spokane R. (polluted)	—	2.8–44.5	—	.4–1.06	Funk et al. (1975)[168]
Perch	Tank, Cu = .044 ppm 20 days	19.0	191	—	—	Collvin (1984)[160]
Perch	control	3.0	3.0	—	—	Collvin (1984)[160]
Bluegill sunfish	Tank, Cu = .040 ppm 24 months	5.0	61	13.0	—	Benoit (1975)[158]
Bluegill sunfish	control, Cu = .003 ppm	3.0	7.0	22.0	—	Benoit (1975)[158]

Acute and chronic studies have been performed on copper for a variety of freshwater fish species at different life stages. In chronic studies encompassing a complete life cycle, all species tested were found to be most sensitive to copper in their early life stages (larval and juvenile).[169-171] In the critical early stages copper did not produce embryo mortality, but it has been postulated that sublethal effects occurred in the embryo which ultimately influenced the success and survival of larvae and juveniles.[172]

In addition to the obvious effects on survival, copper also produces deleterious results in growth and reproduction. Inhibition of spawning, marked decreases in egg hatchability, and reductions in growth have been observed in a number of species over a wide span of copper concentrations. [173-178] Similar to the effects on survival the greatest reductions in growth occur in the early life stages.

At the lower levels of organization copper produces changes in blood chemistry, including increases in red blood cell count, hematocrit, hemoglobin, plasma glucose, and lactic dehydrogenase.[179] Although these changes in blood chemistry are good indicators of environmental stress their transient nature prevents their measurement as definitive biomarkers.

Similar to combinations of many toxicants, additional effects could result from the association of copper with other contaminants. A combination of copper and other heavy metals, or xenobiotics, could produce synergistic, antagonistic, and additive effects in toxicity. For example, the combination of zinc and copper yielded an additive toxic response,[180, 181] and a mixture of zinc, copper, and cadmium elicited an antagonistic response.[182]

Aquatic Toxicity of Hydrocarbons

The large universe of hydrocarbons naturally present and resulting from anthropogenic releases is addressed as two principal subgroups: aliphatic and aromatic. In the aquatic environment, the fate and persistence of a hydrocarbon are important parameters relative to potential aquatic exposure and associated toxicity. As discussed in Sec. 4-3, aromatic hydrocarbons which tend to partition to organic matter, resist degradation, and persist in the aquatic environment possess a greater potential to affect aquatic life than the aliphatic compounds which are generally less persistent because of greater volatility and solubility. Further, hydrocarbons which are substituted with halogens, nitrogen or other subgroups are generally more persistent and toxic than an unsubstituted, parent hydrocarbon. Table 5-18 presents aquatic toxicity data of some common monocyclic aromatic hydrocarbons for algae, invertebrates, and fish.

Example 5-9   Patterns in aquatic toxicity for organic compounds. A chemical manufacturing facility produces chlorinated benzenes. As part of their effluent monitoring program they conduct acute toxicity tests. Assuming the toxicity data presented in Table 5-18 are representative of the results for their toxicity testing, can you identify a pattern relative to the toxicity of these chlorinated benzenes?

Solution. Based on the data presented in Table 5-18, one can compare the relative sensitivities of algae, invertebrates, and fish; identify potential differences between freshwater and saltwater species; and compare the relative toxicities of the mono-, di-, tri-, tetra-, and penta-chlorinated benzenes. Focusing on any differences in toxicity related to chlorine substitution, the concentration associated with acute toxicity in 20 of the 23 cases decreases for the following comparison: mono to di, di to tri, tri to tetra, and tetra to penta. This indicates that the acute toxicity of chlorinated benzenes increases as the chlorine content increases. This is no coincidence, as a common trend for the toxicity of halogenated hydrocarbons is that the acute toxicity increases as the degree of halogen substitution increases.

An important, pervasive class of hydrocarbons which persist in the aquatic environment is the polycyclic aromatic hydrocarbons (PAHs). Approximately, 200,000 metric tons of PAHs are released annually to the aquatic environment from sources such as petroleum spillage, deposition of particulates from fossil fuel combustion as

TABLE 5-18
Acute toxicity (96-h LC$_{50}$, mg/L^{-1}) of monocyclic aromatic hydrocarbons to algae, invertebrates, and fish[184]

Compound	Algae		Invertebrates		Fish	
	Fresh-water	Salt-water	Fresh-water*	Salt-water	Fresh-water	Salt-water
Benzene	525	—	200	39	22.5	33
Toluene	>433	>433	310	56	13	280–480
Chlorobenzene	228	342	86	16.4	16	10
1,2-Dichlorobenzene	95	44	2.4	2.0	5.6	9.7
1,3-Dichlorobenzene	64	52	28	2.9	5.0	7.8
1,4-Dichlorobenzene	97	57	11	2.0	4.3	7.4
1,2,4-Trichlorobenzene	36	8.9	—	0.5	3.4	21
1,2,3,5-Tetrachlorobenzene	17	0.7–7.1	9.7	0.3	6.4	37
1,2,4,5-Tetrachlorobenzene	50	7.3	—	1.5	1.6	0.8
Pentachlorobenzene	6.7	2.1	5.3	0.2	0.25	0.8
Ethylbenzene	>438	>438	75	88	150	280
Nitrobenzene	44	9.6	27	6.7	43	59
Dinitrotoluene	1.5	0.4	0.7	0.6	0.3	—
2,4,6-Trinitrotoluene	5	50	—	—	—	2.6

*48-h LC$_{50}$.

well as other anthropogenic and natural processes. The hydrophobic character of PAHs leads to rapid association with particulate matter and partitioning to either sediments or tissue. PAHs are acutely toxic to aquatic organisms at concentrations ranging from 0.2 to 10 mg/L,[183] and acute toxicity tends to increase with increasing molecular weight. However, higher molecular weight PAHs (i.e., larger than pyrene, which has a molecular weight of 202) are not considered acutely toxic to aquatic organisms, apparently because solubility is less than the concentration required to elicit a response.[186] Of greater interest are the potential effects resulting from non-lethal, chronic exposure. Although PAHs are hydrophobic, their metabolites are more water soluble and reactive. As discussed earlier for human toxicity, these reactive metabolites can bind to protein, DNA, and other macromolecules, leading to cell damage, mutagenesis, and possible cancer. While PAH-induced cancerlike growths or developmental anomalies have been described in species of aquatic organisms, the role of PAH in carcinogenesis in aquatic organisms is uncertain.[187]

Aquatic Toxicity of Pesticides

The expansion of the synthetic organic chemical industry after World War II led to rapid development of synthetic compounds to control species of insects, plants, rodents, and other "pests." These chemicals were intended to be persistent and to render an adverse affect on a **target** species of "pest" without damaging **non-target** organisms. However, historical data for compounds such as DDT, Parathion, and other persistent pesticides provide clear evidence about deleterious effects on non-target organisms of widespread pesticide use.

The effect of these organic compounds on aquatic organisms is related to several properties fundamental to their use as pesticides. First, pesticides are chemical compounds capable of affecting all types of biota, including non-target species, to varying degrees dependent on physiological and ecological factors. Second, many pesticide compounds are, by design, resistant to degradation to enhance their persistence and effectiveness. Finally, pesticides are usually applied in bulk quantity and dispersed over large areas which results in atmospheric transport to natural systems and run-off contamination of receiving surface water bodies.[188]

For our discussion of aquatic toxicity, the pesticides of interest are primarily insecticides and herbicides. Three important classes of insecticides are the organochlorines (e.g., DDT, methoxychlor, toxaphene, aldrin, dieldrin), organophosphates (e.g., parathion, malathion, diazinon), and the carbamates (e.g., sevin, temik). Most insecticides act by interfering with nervous system function. These neuroactive insecticides are believed to involve effects on either nerve impulse transmission; the receptor for acetylcholine, a neurotransmitter; or acetylcholinesterase, an enzyme[189] (see Fig. 5-10). Phenoxy acid herbicides (2,4-D and 2,4,5-T) effectively eradicate plants by growth disruption.

As shown in Table 5-19, the acute aquatic toxicity of organochlorine compounds is generally greater than organophosphates, and insecticides are relatively more toxic than herbicides.

TABLE 5-19
Relative toxicities (measured as induced mortalities) of pesticides to aquatic animals[185]

Pesticide type	Plankton	Shrimp	Crab	Oyster	Fish
Herbicide	1	1	1	1	1
Organochlorine compounds	×3	×300	×100	×100	×500
Organophosphorus compounds	×0.5	×1000	×800	×1	×2

Non-lethal effects of pesticide exposure are numerous and diverse, and related to a broad spectrum of metabolic and behavioral responses, including alterations in enzyme production, growth, reproduction, activity, production of tumors, and teratagenic effects.[190] Some sublethal effects of pesticide exposure on aquatic organisms are provided in Table 5-20.

Water Quality Criteria

To regulate the discharge of pollutants to surface water and to protect the use of such water by indigenous aquatic biota and humans, the Clean Water Act requires that the EPA and the States develop water quality criteria. These criteria are developed based on the measured toxic effects on aquatic biota and humans of exposure to chemicals in surface water. For aquatic life, these criteria specify a pollutant concentration which, if not exceeded, should protect most, but not all, aquatic life.[191] To develop water quality criteria, the EPA requires data in four areas: acute toxicity to animals (eight data points), chronic toxicity to animals (three data points), toxicity to plants, and tissue residues.[192,193] The specific data requirements for each of the four areas, and the methods used to reduce and analyze the data are provided in the EPA guidance for deriving water quality standards.[194,195,196]

It is important to note that for several toxicants the influence of factors which mitigate toxicity are accounted for in the development of criteria. For example, the acute and chronic water quality criteria for several trace metals depend on the measured water hardness value. The chronic and acute criteria for copper recognize that freshwater aquatic organisms should not be adversely affected if the 4-day average concentration (in μg/L) of copper does not exceed the value given by $e^{(0.8545[\ln(\text{Hardness})]-1.465)}$ more than once every 3 years on the average and if the 1-hour average concentration does not exceed the value given by $e^{(0.9422[\ln(\text{Hardness})]-1.464)}$ more than once every 3 years on the average.[197] Therefore, at a water hardness level of 50 and 200 mg/L $CaCO_3$, the acceptable 4-day average concentrations are 6.5 and 21 μg/L, respectively, and the acceptable 1-hour concentrations are 9.2 and 34 μg/L. Additional equations to estimate freshwater aquatic life criteria for metals are provided in Table 5-21.

TABLE 5-20
Sublethal effects of pesticides on aquatic organisms[198]

Test species	Test material	Observed effect	Effective concentration
American oyster	Chlordane	Reduced shell growth	6.2 μg/L
Four marine algae	Kepone	Effect on growth of cultures	350–600 μg/L
Fathead minnow	Kelthane	Alteration in normal	218 μg/L
	Dursban	schooling behavior after	44 μg/L
		24 h	
	Disulfoton		413 μg/L
	Pydrin		3.6 μg/L
	Permethrin		7.2 μg/L
Yellow perch	Methoxychlor	Effect on respiration	5.0 μg/L
Juvenile blue crabs	Mirex	Paralysis	Mirex bait
Pinfish	Naled (Dibrom)	Brain acetylcholinesterase inhibition	84–99% inhibition at lethal concentrations
Brook trout	DDT	Prevents the establishment of a visual conditioned avoidance response	20–60 μg/L
Bluegill sunfish	Parathion	Effect on locomotor orientation	10–25 μg/L
Large-mouth bass Grass shrimp	Methyl or ethyl Parathion	Impairment of antipredation behavior	0.1 μg/L
Sheepshead minnow	Kepone	Scoliosis (curvature of spine)	0.8 μg/L

Example 5-10 Ecotoxicology of zinc. Downstream of a metal-plating lagoon holding wastewater and sludge, surface water monitoring data indicate 4-day and 1-hour average zinc concentrations of 14.2 μg/L and 27 μg/L. An average water hardness value of 100 mg/L $CaCO_3$ was also measured for the stream. Based on the measured concentrations for zinc and water hardness, and the water quality criteria for zinc, is there reason to be concerned regarding the long-term protection of aquatic life?

Solution. Using Table 5-21 the equation to estimate the zinc criterion for long-term (chronic) protection of aquatic life is as follows:

$$\mu g/L \ Zn = e^{(0.8473[\ln(\text{hardness})]+0.7614)}$$

Substituting the measured hardness value of 100 mg/L $CaCO_3$, we estimate that the criterion for the 4-day average zinc concentration as:

$$Zn = e^{(0.8473[\ln(100)]+0.7614)} \mu g/L, \ \text{or}$$

$$\underline{\underline{Zn = 106 \ \text{mg/L}}}$$

Because the measured 4-day average zinc concentration is an order of magnitude less than the criterion, there is no empirical reason for concern.

While water quality criteria continue to help ensure the maintenance of designated water body uses and the protection of water column aquatic life, these numerical criteria were not intended to protect organisms which inhabit the sediment. The evaluation and protection of sediment quality is an arena of aquatic toxicology which is rapidly expanding, including the development of supplemental information pertaining to sediment toxicity and quality analyses.[200,201]

Early environmental statutes regulated the release of chemicals to water; therefore, the advances in and data available from measuring toxicity to aquatic ecosystems far exceeds the data base for other branches of ecotoxicology. However, of the tens-of-thousands hazardous materials produced and ultimately released to the environment, the EPA has developed water quality criteria to protect freshwater aquatic life for less than thirty pollutants.[202]

TABLE 5-21
Hardness-dependent aquatic toxicity criteria[199]

	Acute criterion = $\exp(m_A[\ln(\text{hardness})]+b_A)$		Chronic criterion = $\exp(m_C[\ln(\text{hardness})]+b_C)$	
	m_A	b_A	m_C	b_C
Cadmium	1.128	−3.828	0.7852	−3.490
Copper	0.9422	−1.464	0.8545	−1.465
Chromium (III)	0.8190	3.688	0.8190	1.561
Lead	1.273	−1.460	1.273	−4.705
Nickel	0.8460	3.3612	0.8460	1.1645
Silver	1.72	−6.52	—	—
Zinc	0.8473	0.8604	0.8473	0.7614

Terrestrial Toxicology

As discussed earlier, ecotoxicology initially focused on aquatic species. A second group of organisms to gain research attention were birds, particularly large birds of aesthetic and recreational value whose abundance visibly declined over a short period of time due to widespread use of several chemical substances, most notably DDT.

Only within the past ten to twenty years has ecotoxicology began to examine the effects of chemical releases on terrestrial species. With the exception of effects on sessile, terrestrial plants, the effects of non-lethal chemical exposure on terrestrial organisms are not as easily measured because organisms are extremely mobile over large areas, exposed to numerous natural stressors such as disease, predation, and habitat loss, and demonstrate a penchant for avoidance, rather than confrontation.

Feral Plants and Mammals

The media attention afforded to the effects of acid deposition on forests in Germany, Canada, and the U.S. has increased public awareness regarding chemical effects on plants. Herbicide use in agriculture to inhibit "weed" growth, in Vietnam to eradicate jungle foliage, and along roadways and utility lines to regulate plant growth has also increased public interest about potential adverse effects on both target and non-target receptors.

Another contemporary debate continues about the application of domestic sewage sludge as a soil enhancement to land used for agriculture, recreation, and/or residential gardening. The land treatment of hazardous waste sludge is a related example. The concentrations of trace metals in sludge raises concern about potential plant phytotoxicity (inhibition of photosynthesis) and/or reductions in plant community productivity. The presence of the mobile lighter organic compounds in sludge could cause potential foliar damage such as chlorosis (loss or damage to chlorophyll), whereas heavier, less mobile organics may accumulate in plant tissue and be transferred through the food chain.

While many mammalian species have been evaluated in laboratory toxicity tests as surrogates for humans, information is limited regarding effects on feral species. An exception is the effects of non-lethal organohalogen exposure on reproduction. It is well established that PCBs can be found in wildlife in many areas, and studies have attempted to correlate the high tissue residues with breeding deficiencies in mammals.[203] While some disagreement remains among scientists regarding the mechanism of action, the diminished capacity of an organism to reproduce apparently is definitive.

Avian Toxicology

The visible effects of chemical use on avian species of recreational and aesthetic value has increased research in avian toxicology. The database for avian effects of chemical exposure has expanded significantly due to regulatory requirements for avian toxicity testing as part of new chemical registration. Similar to the earlier discussion for aquatic organisms, the majority of avian research has focused on the effects of several trace metals and the organohalogens, most notably pesticides and PCBs.

The intake of spent lead shot by waterfowl and the consumption of methylmercury by fish-eating birds are well-researched cases of avian poisoning though trace metal exposure. To maintain efficient grinding of food, most waterfowl ingest sand, small pebbles and other forms of grit. It is during this routine event that many species ingest spent lead shot. Upon ingestion the lead shot is slowly eroded, becoming available for internal transport and distribution. In many cases it results in death from tissue hypoxia. While lead poisoning from spent shot presents an interesting case study, it is important to note that trace metals do not selectively affect waterfowl. Other species of insect-, seed-, and flesh-eating birds are also susceptible to trace metal poisoning from direct exposure to contaminated soil and water and indirect exposure via the food chain.

Avian toxicity research regarding the effects of hydrocarbon exposure has focused primarily on organochlorine and organophosphate pesticides and the PCBs. Although other hydrocarbons have not been as thoroughly scrutinized in toxicity tests, additional information regarding exposure via the food chain is available as tissue residues and physical impairment resulting from oil spills. Because pesticides generally accumulate in the food chain, species positioned at the top of a food chain are often susceptible to significant tissue burdens from long-term, low dose exposure. As energy demands increase during migration, reproduction, and other stressful activities, these materials are often liberated from fatty tissue deposits, and distributed to target organs where a variety of adverse responses occur. From these examples, it is clear that isolated contamination, transported by both physical and biological processes, can affect local receptors as well as migratory species.

DISCUSSION TOPICS AND PROBLEMS

5-1. Which are the main **routes of exposure** by which the organism can be exposed to chemical substances present in the environment?

5-2. The LD_{50} of a certain substance is appreciably higher by inhalation than by ingestion. What conclusions may be drawn from this observation?

5-3. What are the consequences with respect to uptake by the mammalian organism after inhalation of an aerosol with a mean aerodynamic particle diameter of
  (a) 1.5μ and
  (b) 10μ?

5-4. Which properties favor accumulation of a substance in the mammalian body?

5-5. Identify substances that tend to accumulate in
  (a) fatty tissues,
  (b) bone, and
  (c) the kidney.

5-6. What is the difference between dose expressed as administered (or inhaled) dose and target dose? Which is the more appropriate to use for determination of a dose-response relationship? Why?

5-7. How is an accumulating lipophilic compound handled by the mammalian organism to facilitate elimination in urine?

5-8. What are the differences between acute and chronic exposure, and how may these differences relate to acute and chronic health effects?

5-9. Give an example of a chemical compound that specifically induces damages to
(*a*) the kidney,
(*b*) the liver,
(*c*) the bone marrow,
(*d*) the central nervous system, and
(*e*) the testicles.

5-10. By which type of mechanism do neurotoxic organophosphorous compounds act?

5-11. What is the difference between a dose-response and a dose-effect relationship?

5-12. What is the fundamental difference between risk assessment of carcinogenic and mutagenic agents and the compounds that induce effects like malformations or neurotoxicity?

5-13. The NOEL for the herbicide atrazine was found to be 0.5 mg/kg · day based on induction of cardiotoxic effects (EKG alterations) in beagle dogs in a long-term feeding study.
(*a*) Assuming the quality of the study to be high, what is the ADI (RfD)?;
(*b*) Can this ADI (RfD) be used to determine an acceptable occupational exposure level for farm workers when the intake occurs mostly by inhalation?

5-14. What type of disease could be induced by mutation in
(*a*) somatic cells and
(*b*) sex (germinal) cells?

5-15. Why would a laboratory test that may identify a mutation in a single gene in a bacterium be of interest to human toxicology?

5-16. (*a*) List the different stages of carcinogenesis.
(*b*) What characterizes a cancer promoter?
(*c*) What is the difference between an epigenetic and a genotoxic carcinogen?

5-17. Which evidence is available to support the idea that the induction of malignant tumors involves changes in the hereditary material (i.e., DNA)?

5-18. What are the limitations of using laboratory animals in identifying human carcinogens?

5-19. What is the difference between a cohort study and a case control study, and what are the strengths and weaknesses inherent in their design, respectively?

5-20. In 1982 homes were constructed in the close vicinity of an abandoned hazardous waste site. In 1984 sampling indicated that the site was heavily contaminated with the human bladder carcinogen, beta-naphthylamine. According to a cross-sectional epidemiological investigation (completed by the end of 1986) of the area residents, an increased incidence of cancer of the prostate, bladder, lung, and rectum was found in comparison with a control population. The conclusion was that chemical contaminants from the hazardous waste caused the observed increase in cancer. Why was this conclusion unwarranted?

5-21. From epidemiological studies the carcinogenic potency factor (slope factor) for benzene was estimated by the EPA in 1988 to be 0.03 (mg/kg · day)$^{-1}$. Using this estimate, what would be the hypothetical **yearly** increase in the cancer incidence of the U.S. *population* from benzene in gasoline due to exposure during filling of gasoline in motor vehicles? Use the following assumptions:
- Thirty (30) million car owners fill the gasoline tank twice a week at self-service stations.
- Filling averages 3 minutes.

- The concentration of benzene in the breathing zone is 0.5 ppm.
- The volume of inhaled air is 14 L/min.
- The degree of absorption of benzene in the lungs is 50%.
- The annual cancer incidence for the entire U.S. population can be estimated to be about 1 million cases.

5-22. What would be the closest equivalent for aquatic organisms to an oral LD_{50} in a rodent?

5-23. What is a life-cycle test?

5-24. Why is it not enough to investigate the acute toxic effects of chemicals in a specific species of fish to evaluate the impact on this species of the release of the chemical to the aquatic environment? What alternatives exist to test for exotoxicologic effects in a few selected species of organisms?

5-25. Given the fact that MCPA ([4-chloro-2-methylphenoxy]acetic acid) and certain arsenic compounds are used as weed killers, and their toxicity to fish is relatively low, which segment of the aquatic ecosystem would you expect to be the potential main target of their adverse action?

5-26. (*a*) What physical property would you use to predict the BCF in a semiquantitative way? (*b*) How would you define "biomagnification"?

5-27. Which terrestrial animal species would you use to test for the potential environmental impact of DDT?

5-28. Which regulatory standards in the United States have protection of aquatic life as one of their main goals? Are these standards designed to protect all forms of aquatic organisms?

5-29. Name one important environmental factor that will influence aquatic toxicity of several heavy metal ions to such an extent that it influences regulatory standards. Why?

5-30. The maximum permitted levels of pesticide residues in food products are set by tolerance levels. Assuming that the current tolerance level of lindane of 4 ppm for hog fat would also apply to fish, would trout from a lake with a concentration of 0.1 ppm of lindane in water (caused by accidental contamination) be considered fit for human consumption? Base your estimate on a log BCF value for lindane of 2.51 mg/kg tissue per mg lindane/L water.

REFERENCES

1. Keetan, William T.: *Biological Science*, 2nd ed., W. W. Norton & Co., Inc., pp. 93, 167, 1972.
2. Gute, David M., and N. B. Haner: "An Applied Approach to Epidemiology and Toxicology for Engineers," Instructor's Resource Guide under review by U.S Department of Health, Education and Welfare, Cincinnati, Ohio, 1991.
3. U.S. Department of Health and Human Services: "Toxicological Profile for Benzene," Public Health Service, Agency for Toxic Substances and Disease Registry, ATSDR/TP/-88/03, Atlanta, 1989.
4. Klaassen, C. D., and K. Rozman: "Absorption, Distribution, and Excretion of Toxicants," *Casarett and Doull's Toxicology—The Basic Science of Poisons*, M. O. Amdur, J. Doull, and C. D. Klaassen, eds., 4th ed., Pergamon Press, Inc., Elmsford, New York, p. 51, 1991.
5. Klaassen and Rozman, "Absorption, Distribution, and Excretion of Toxicants."
6. U.S. Department of Health and Human Services: "Toxicological Profile for Lead," Public Health Service, Agency for Toxic Substances and Disease Registry, ATSDR/TP-88/17, Atlanta, 1990.
7. International Programme on Chemical Safety (IPCS): "Environmental Health Criteria No. 24, Lindane," World Health Organization, Geneva, 1991.

8. Klaassen C. D. and D. L. Eaton: "Principles of Toxicology," *Casarett and Doull's Toxicology—The Basic Science of Poisons,* 4th ed., M. O. Amdur, J. Doull, and C. D. Klaassen, eds., Pergamon Press, Inc., Elmsford, New York, pp. 12-50, 1991.

9. Klaassen and Rozman, "Absorption, Distribution, and Excretion of Toxicants."

10. Ibid, p. 54.

11. U.S. Environmental Protection Agency: "Interim Methods for Development of Inhalation Reference Doses," Office of Health and Environmental Assessment, EPA/600/8-88/066F, Washington, D.C., August 1989.

12. Klaassen and Rozman, "Absorption, Distribution, and Excretion of Toxicants."

13. U.S. EPA, "Interim Methods for Development of Inhalation Reference Doses."

14. Klaassen and Eaton, "Principles of Toxicology," pp. 12–49.

15. Clarkson, T. W.: "Effects—General Principles Underlying the Toxic Action of Metals," *Handbook on the Toxicology of Metals,* L. Friberg, G. F. Nordberg, and V. B. Vouk, eds., 2nd ed., Elsevier, Amsterdam, p. 135, 1986.

16. World Health Organization: "Recommended Health-Based Limits in Occupational Exposure to Heavy Metals," Report of a WHO Study Group, WHO, Geneva, 1980.

17. Manahan, Stanley E.: *Toxicological Chemistry and Guide to Toxic Substances in Chemistry,* Lewis Publishers, Inc., Chelsea, Michigan, 1989.

18. Sipes, I. G., and A. J. Gandolfi: "Biotransformation of Toxicants," *Casarett and Doull's Toxicology—The Basic Science of Poisons,* M. O. Amdur, J. Doull, and C. D. Klaassen, eds., 4th ed., Pergamon Press, Inc., Elmsford, New York, pp. 88–126, 1991.

19. U.S. Department of Health and Human Services: "Toxicological Profile for Benzene," Public Health Service, Agency for Toxic Substances and Disease Registry, ATSDR/TP/-88/03, Atlanta, 1989.

20. Ibid.

21. Klaassen and Eaton, "Principles of Toxicology."

22. Klaassen and Rozman, "Absorption, Distribution, and Excretion of Toxicants."

23. Clarkson, "Effects—General Principles Underlying the Toxic Action of Metals."

24. Gargas, M. L.: "Chemical-specific Constants for Physiologically-based Pharmacokinetic Models," *CIIT Activities,* vol. 11, no. 3, March 1991.

25. Klaassen and Rozman,"Absorption, Distribution, and Excretion of Toxicants," op. cit., p. 82.

26. Ecobichon, D.: "Toxic Effects of Pesticides," *Casarett and Doull's Toxicology—The Basic Science of Poisons,* M.O. Amdur, J. Doull, and C.D. Klaassen, eds., 4th ed., Pergamon Press, Inc., Elmsford, New York, p. 583, 1991.

27. Hathway, D. E.: *Molecular Aspects of Toxicology,* The Royal Society of Chemistry, Burlington House, London, 1984.

28. Kamrin, Stanley E.: *Toxicology,* Lewis Publishers, Inc., Chelsea, Michigan, 1988.

29. Bailey, James E., and David F. Ollis, : *Biochemical Engineering Fundamentals,* 2nd ed., McGraw-Hill Inc., New York, New York, p. 94, 1986.

30. Kamrin, *Toxicology.*

31. Goldstein, J. A.: "Structure-activity Relationships for the Biochemical Effects and the Relationship to Toxicity," *Halogenated Biphenyls, Terphenyls, Naphthalenes, Dibenzodioxins and Related Products,* R. Kimbrough, ed., Elsevier/North Holland Biomedical Press, Amsterdam, p. 175, 1980.

32. Wold, S., W. J. Dunn, and S. Hellberg: "Toxicity Modeling and Prediction with Pattern Recognition," *Env. Health Perspectives,* vol. 61, pp. 257–268, 1985.

33. Goldstein, "Structure-activity Relationships for the Biochemical Effects."

34. Schmidt, R. F., and G. Thews: *Human Physiology,* Springer Verlag, Berlin, 1983.

35. Ecobichon, "Toxic Effects of Pesticides."

36. Goyer, R. A.: "Toxic Effects of Metals," *Casarett and Doull's Toxicology—The Basic Science of Poisons,* M. O. Amdur, J. Doull, and C. D. Klaassen, eds., 4th ed., Pergamon Press, Inc., Elmsford, New York, pp. 643–645, 1991.

37. Ariëns, E. J., et al.: *Introduction to General Toxicology,* Academic Press, New York, p. 38, 1976.

38. Schmidt and Thews, *Human Physiology.*

39. Goyer, "Toxic Effects of Metals."

40. U.S. Department of Health and Human Services, "Toxicological Profile for Benzene."

41. Trevan, J. W.: "The Error of Determination of Toxicity," *Proc. Roy. Soc. London (Biol.)*, vol. 101, pp. 483–514, 1927.

42. Klaassen and Eaton, "Principles of Toxicology."

43. Merck & Co., Inc.: *The Merck Index*, Merck & Co., Inc., Rahway, New Jersey, 1983.

44. Ibid.

45. The British Crop Protection Council: *The Pesticide Manual—A World Compendium*, 9th ed., C. R. Worthing, ed., Unwin Brothers Ltd., Old Woking, Surrey, U.K.

46. Merck & Co., Inc., *The Merck Index*.

47. The British Crop Protection Council: *The Pesticide Manual—A World Compendium*.

48. Merck & Co., Inc.: *The Merck Index*.

49. The British Crop Protection Council, *The Pesticide Manual—A World Compendium*.

50. Merck & Co., Inc., *The Merck Index*.

51. Ibid.

52. The British Crop Protection Council, *The Pesticide Manual—A World Compendium*.

53. Schwetz, B. A., J. M. Norris, G. L. Sparschu, V. K. Rowe, P. J. Gehring, J. L. Emerson, and C. G. Gerbig: "Toxicity of Chlorinated Dibenzo-p-dioxins," *Env. Health Perspectives*, vol. 5, pp. 87–99, 1973.

54. Pohjanvirta, R., and J. Tuomisto: "Han/Wistar Rats are Exceptionally Resistant to TCDD," II. *Arch. Toxicol. Suppl.*, vol. 11, pp. 344–347, 1987.

55. U.S. Department of Health and Human Services, "Toxicological Profile for Lead."

56. Roels, H., J.-P. Buchet, and R. Lauwerys, et al.: "Impact of Air Pollution by Lead on the Heme Biosynthesis Pathway in School-age Children," *Arch. Environm. Health*, vol. 31, pp. 310–316, 1976.

57. Foster, P. M., D. M. Creasy, J. R. Foster, L. V. Thomas, M. W. Cook, and S. D. Gangolli: "Testicular Toxicity of Ethylene Glycol Monomethyl and Monoethyl Ethers in the Rat," *Toxicol. Appl. Pharmacol.*, vol. 69, pp. 385–399, 1983.

58. Ibid.

59. U.S. Department of Health and Human Services, "Toxicological Profile for Lead."

60. U.S. Department of Health and Human Services: "Toxicological Profile for Benzene."

61. International Programme on Chemical Safety (IPCS): "Environmental Health Criteria No. 27, Guidelines on Studies in Environmental Epidemiology," World Health Organization, Geneva, 1983.

62. U.S. Department of Health and Human Services, "Toxicological Profile for Benzene."

63. Ibid.

64. OECD: "OECD Guidelines for Testing of Chemicals," OECD, Paris, 1987.

65. International Programme on Chemical Safety (IPCS): "Principles for the Toxicological Assessment of Pesticide Residues in Food," *Environmental Health Criteria*, vol. 104, WHO, Geneva, pp. 49–53, 1990.

66. OECD, "OECD Guidelines for Testing of Chemicals."

67. International Programme on Chemical Safety (IPCS), "Principles for the Toxicological Assessment."

68. Schwetz, et al., "Toxicity of Chlorinated Dibenzo-p-dioxins."

69. Ibid.

70. Pohjanvirta and Tuomisto, "Han/Wistar Rats are Exceptionally Resistant to TCDD."

71. Ibid.

72. Henck, J. W., M. A. New, R. J. Kociba, and K. S. Rao: "2,3,7,8-Tetrachlorodibenzo-p-dioxin: Acute Oral Toxicity in Hamsters," *Toxicol. Appl. Pharmacol.*, vol. 59, pp. 405–407, 1981.

73. U.S. Department of Health and Human Services, "Toxicological Profile for Benzene."

74. Barnes, D. G., and M. Dourson: "Reference Dose (RfD): Description and Use in Health Risk Assessments," *Regulatory Toxicol. Pharmacol.*, vol. 8, pp. 471–486, 1988.

75. Hutt, P. B.: "Use of Quantitative Risk Assessment in Regulatory Decision-making under Federal Health and Safety Statutes," *Risk Quantitation and Regulatory Policy*, D. G. Hoel, et al., eds., Banbury Report No. 19, Cold Spring Harbor Laboratory, New York, 1985.

76. Barnes and Dourson, "Reference Dose (RfD): "Description and Use in Health Risk Assessments."

77. IRIS: U.S. EPA Integrated Risk Information System: "Data File for Arsenic," updated January 10, 1991.

78. U.S. Environmental Protection Agency: "Superfund Public Health Evaluation Manual," Office of Emergency and Remedial Response, Washington, D.C., October 1986.
79. IRIS: "Data File for Cyanide, Nickel, and Chromium III."
80. Robbins, S. L., and R. S. Cotran: *Pathological Basis of Disease*, 2nd ed. W.B. Saunders Co., Philadelphia, 1979.
81. Weisburger, J. H., and G. M. Williams: "Chemical Carcinogens," *Casarett and Doull's Toxicology—The Basic Science of Poisons*, 4th ed., M. O. Amdur, J. Doull, and C. D. Klaassen, eds., Pergamon Press, Inc., Elmsford, New York, p. 129, 1991.
82. Ibid.
83. Ramel, C.: "Mutation Spectrum in Carcinogenicity," *Mechanisms of Environmental Mutagenesis-Carcinogenesis*, A. Kappas, ed., Plenum Press, New York, 1990.
84. Weisburger and Williams, "Chemical Carcinogens," p. 131.
85. Ibid., pp. 131–132.
86. Ramel, "Mutation Spectrum in Carcinogenicity."
87. Weisburger and Williams, "Chemical Carcinogens."
88. Nilsson, R., R. B. Jaeger, and M. Tasheva: "Why Different Regulatory Decisions When the Scientific Information Base Is the Same. I—Human Risk Assessment," *The Role of Science in Pesticide Management, Regulatory Toxicol. Pharmacol.*, in press, Stockholm, May 20–22, 1992.
89. Paynter, O. E., G. J. Burin, R. B. Jaeger, and C. A. Gregorio: "Goitrogens and Thyroid Follicular Cell Neoplasia: Evidence for a Threshold Process," *Regulatory Toxicol. Pharmacol.*, vol. 8, pp. 102–119, 1988.
90. Weisburger and Williams, "Chemical Carcinogens," p. 129.
91. Gelboin, H. V., and P. O. P. Ts'o, eds.: *Polycyclic Hydrocarbons and Cancer*, Academic Press Inc., New York, vol. 1, 1978.
92. Rossman, T. G., M. S. Meyn, and W. Troll: "Effects of Sodium Arsenite on the Survival of UV-irradiated *Escherichia coli:* Inhibition of a RecA-dependent Function," *Mut. Res.*, vol. 30, pp. 157–162, 1975.
93. IARC (WHO): *IARC Monographs on the Evaluation of the Carcinogenic Risks to Humans—Overall Evaluations of Carcinogenicity: An Updating of IARC Monographs*, Lyon, vol. 1 to 42, suppl. 7, 1987.
94. Office of Technology Assessment Task Force: *Identifying and Regulating Carcinogens*, Lewis Publishers, Chelsea, Michigan, 1988.
95. Littlefield, N. A., J. H. Farmer, P. W. Gaylor, and W. G. Sheldon: "Effects of Dose and Time in a Long-term, Low-dose Carcinogenesis Study," *J. Environm. Pathol. Toxicol.*, vol. 3, pp. 17–34, 1979.
96. Doull, J., B. A. Bridges, R. Kroes, L. Golberg, I. C. Munro, O. Paynter, H. C. Pitot, R. Squire, G. Williams, and W. Darby: "The Relevance of the Mouse Liver Hepatoma to Human Carcinogenic Risk," a report to the International Expert Advisory Committee to the Nutrition Foundation, Washington, D.C., 34 pp., 1983.
97. Weisburger and Williams, "Chemical Carcinogens."
98. Doull, et al., "The Relevance of the Mouse Liver Hepatoma."
99. Nilsson, et al., "Why Different Regulatory Decisions."
100. Weisburger and Williams, "Chemical Carcinogens."
101. Nilsson, et al., "Why Different Regulatory Decisions."
102. *Federal Register*: Guidelines for Carcinogen Risk Assessment, U.S. Environmental Protection Agency, vol. 51, no. 185, pp. 33993–34003, September 24, 1986.
103. Johannsen, F. R.: "Risk Assessment of Carcinogenic and Non-Carcinogenic Chemicals," *Critical Reviews in Toxicology*, vol. 20, p. 351, 1990.
104. Mantel, N., and W. Bryan: "Safety Testing of Carcinogenic Agents," *J. Natl. Cancer Inst.*, vol. 27, pp. 455–470, 1961.
105. Littlefield, et al., "Effects of Dose and Time."
106. Armitage, P., and R. Doll: "The Age Distribution of Cancer and a Multistage Theory of Carcinogenesis," *Brit. J. Cancer*, vol. 8, pp. 1–12, 1954.

107. Crump, K.: "An Improved Procedure for Low-dose Carcinogenic Risk Assessment from Animal Data," *J. Env. Pathol. Toxicol.*, vol. 5, pp. 339–348, 1982.

108. *Federal Register*, "Guidelines for Carcinogen Risk Assessment."

109. Crump, "An Improved Procedure for Low-dose Carcinogenic Risk Assessment."

110. *Tox Risk*, K. S Crump Division of Clement International Corp., Ruston, LA.

111. Zeise, L., R. Wilson, and E. A. C. Crouch: "Dose-response Relationships for Carcinogens: A Review," *Environm. Health Persp.*, vol. 73, pp. 259–308, 1987.

112. Carlborg, F. W.: "Multistage Dose-response Models and Carcinogenesis," *Food Cosm. Toxicol.*, vol. 19, pp. 361–365, 1981.

113. Ibid.

114. Moolgavkar, S. H., and A. G. Knudson: "A Model for Human Carcinogenesis," *J. Natl. Cancer Inst.*, vol. 66, pp. 1037–1052, 1981.

115. *Federal Register*, "Guidelines for Carcinogen Risk Assessment."

116. Littlefield, et al., "Effects of Dose and Time."

117. Selikoff, I. J., E. C. Hammond, and J. Churg: "Asbestos Exposure, Smoking and Neoplasia," *J. Am. Med. Assoc.*, vol. 204, pp. 106–112, 1968.

118. Carson, Rachel: *Silent Spring*, Houghton Mifflin, New York, 1962.

119. Connell, D. W., and G. J. Miller: *Chemistry and Ecotoxicology of Pollution*, John Wiley & Sons, Inc., New York, 1984.

120. McCarthy, J. F., and L. R. Shugart: "Biomarkers of Environmental Contamination," *Biomarkers*, J. F. McCarthy and L. R. Shugart, eds., Lewis Publishers, CRC Press, Inc., Boca Raton, Florida, p. 10, 1990. (With permission.)

121. Suter, G.: "Ecological Endpoints," Ecological Assessment of Hazardous Waste, U.S. Environmental Protection Agency, Office of Research and Development, Environmental Research Laboratory, Corvallis, OR, EPA 600/3-89/013, 1989.

122. Suter, "Ecologial Endpoints."

123. Ibid.

124. Ibid.

125. Athey, L. A., J. M. Thomas, J. R. Skalski, and W. E. Miller: "Role of Acute Toxicity Bioassays in the Remedial Action Process at Hazardous Waste Sites," U.S. Environmental Protection Agency, Office of Research and Development, Environmental Research Laboratory, Corvallis, Oregon, EPA 600/8-87/044, 1987.

126. Ibid.

127. American Society for Testing and Materials: *Standard Practice for Conducting Toxicity Tests with Fishes, Macroinvertebrates, and Amphibians*. Philadelphia, PA, 1980.

128. Parrish, P. R.: "Acute Toxicity Tests," *Fundamentals of Aquatic Toxicology*, G. M. Rand and S. R. Petrocelli, eds., Hemisphere Publishing Corporation, Washington, 1985.

129. Heath, A. G: *Water Pollution and Fish Physiology*, CRC Press, Inc., Boca Raton, Florida, p. 7, 1987.

130. Ibid., p. 49.

131. Ibid.

132. Ibid.

133. McKim, J. M.: "Early Life Stage Toxicity Tests," *Fundamentals of Aquatic Toxicology*, G. M. Rand and S. R. Petrocelli, eds., Hemisphere Publishing Corporation, Washington, D.C. 1985.

134. Ibid.

135. Heath, A. G., *Water Pollution and Fish Physiology*, CRC Press, Inc., Boca Raton, Florida, p. 7, 1987 (With permission.)

136. Adams, S. M., L. R. Shugart, and G. R. Southworth: "Application of Bioindicators in Assessing the Health of Fish Populations Experiencing Contaminant Stress," *Biological Markers of Environmental Contamination*, J. F. McCarthy and L. R. Shugart, eds., Lewis Publishers, CRC Press, Inc., Boca Raton, Florida, 1990.

137. DiGuilio, R. T.: "Biomarkers," *Ecological Assessment of Hazardous Waste*, U.S. Environmental Protection Agency, Office of Research and Development, Environmental Research Laboratory, Corvallis, Oregon, EPA 600/3-89/013, 1989.

138. McCarthy and Shugart: "Biomarkers of Environmental Contamination."
139. Connell, D. W.: *Bioaccumulation of Xenobiotic Compounds,* CRC Press, Inc., Boca Raton, Florida, 1990.
140. Ibid.
141. Connell, D. W. and G. J. Miller: *Chemistry and Ecotoxicology of Pollution,* John Wiley & Sons, Inc., New York, New York, p. 319, 1984.
142. DiGuilio, "Biomarkers."
143. Connell, *Bioaccumulation of Xenobiotic Compounds.* Ibid., p. 169.
144. Goyer, R. A.: "Toxic Effects of Metals," *Casarett and Doull's Toxicology–The Basic Science of Poisons,* M. O. Amdur, J. Doull, and C. D. Klaassen, eds., 4th ed., Pergamon Press, Inc., Elmsford, New York, NY, pp. 643–645, 1991.
145. DiGuilio, "Biomarkers."
146. Ibid.
147. Ibid.
148. McCarthy and Shugart, "Biomarkers of Environmental Contamination."
149. Heath, A. G.: *Water Pollution and Fish Physiology,* p. 62.
150. Ibid.
151. Connell and Miller, *Chemistry and Ecotoxicology of Pollution.*
152. Leland, H. V., and J. S. Kuwabara: "Trace Metals," *Fundamentals of Aquatic Toxicology,* G. M. Rand and S. R. Petrocelli, eds., Hemisphere Publishing Corporation, Washington, 1985.
153. Hodson, P. V., U. Borgmann, and H. Shear: "Toxicity of Copper to Aquatic Biota," *Copper in the Environment, Part II. Health Effects,* J. O Nriagu, ed., Wiley and Sons, Inc., New York, New York, 1979.
154. Howarth, R.S., and J.B. Sprague: "Copper Lethality to Rainbow Trout in Waters of Various Hardness and pH," *Water Research,* vol. 12, pp. 455–462, 1978.
155. Stiff, M. J.: "Copper/Bicarbonate Equilibria in Solutions of Biocarbonate Ion Similar to Those Found in Natural Water," *Water Research,* vol. 5, pp. 171–176, 1971.
156. McKim, J. M., and D. A. Benoit: "Effects of Long-Term Exposure to Copper on the Survival, Growth and Reproduction of Brook Trout (*Salvelinus fontinalis*)" *Journal of the Fisheries Resource Board of Canada,* vol. 28, pp. 655–662, 1971.
157. Brungs, W. A., E. A. Leonard, and J. M. McKim: "Acute and Long-Term Accumulation of Copper by the Brown Bullhead, *Ictalurus nebulosus,*" *Journal of the Fisheries Resource Board of Canada,* vol. 30, pp. 583–586, 1973.
158. Benoit, D. A: "Chronic Effects of Copper on Survival, Growth, and Reproduction of the Bluegill (*Lepomis macrochirus*)," *Transaction of the American Fisheries Society,* vol. 104, pp. 353–358, 1975.
159. Phillips, G. R., and R. C. Russo: "Metal Bioaccumulation in Fishes and Aquatic Invertebrates: A Literature Review," Ecological Research Series, EPA 600/3–78/103, 1978.
160. Collvin, L.: "Uptake of Copper in the Gills and Liver of the Perch, *Perca fluviatilis,*" *Ecological Bulletins,* vol. 36, pp. 57–61, 1984.
161. Norris, R. H., and P. S. Lake: "Trace Metal Concentrations in Fish from the South Esk River, Northeastern Tasmania, Australia," *Bulletin of Environmental Contamination and Toxicology,* vol. 33, pp. 23–27, 1984.
162. Stout, D. J.: "Copper and Its Distribution in Fish Muscle Tissue in North Carolina," Master's Project. Duke University School of Forestry and Environmental Studies, Durham, North Carolina, 1986.
163. Leland, H. V., S. N. Luoma, and J. M. Fielden: "Bioaccumulation and Toxicity of Heavy Metal and Related Trace Elements," *Journal of Water Pollution Control Federation,* vol. 51, pp. 1592–1616, 1979.
164. Phillips and Russo, "Metal Bioaccumulation in Fishes and Aquatic Invertebrates."
165. Harrison, F. L.: "Effects of Copper Speciation on Aquatic Ecosystems," Workshop Proceedings: The Effects of Trace Elements on Aquatic Ecosystems, 1984.
166. Stout, "Copper and Its Distribution in Fish Muscle Tissue."
167. Hutchinson, T. C., and J. Fitchko: "Heavy Metal Concentrations and Distributions in River Mouth Sediments Around the Great Lakes," *Proceedings of the International Conference on Transport of*

Persistent Chemicals in Aquatic Ecosystems, National Research Council of Canada, Ottawa, Canada, 1974.

168. Funk, W. H., et al.: *An Integrated Study on the Impact of the Metallic Trace Element Pollution in the Couer-D'Alene Spokane River Drainage System,* Washington State Univ. and Univ. Idaho Report to OWRR, B-044 Washington and B-015 IDA., 1975.

169. Mount, D. I.: "Chronic Toxicity of Copper to Fathead Minnows (*Pimephales promelas, rafinesque*)," *Water Research,* vol. 2, pp. 215–223, 1968.

170. McKim and Benoit, "Effects of Long-Term Exposure to Copper."

171. Brungs, et al., "Acute and Long-Term Accumulation of Copper."

172. McKim, J. M., J. G. Eaton, and G. W. Holcombe: "Metal Toxicity to Embryos and Larvae of Eight Species of Freshwater Fish II: Copper," *Bulletin of Environmental Contamination and Toxicology,* vol. 19, pp. 608–616, 1978.

173. Mount, "Chronic Toxicity of Copper to Fathead Minnows."

174. McKim and Benoit: "Effects of Long-Term Exposure to Copper."

175. McKim, J. M., and D. A. Benoit: "Duration of Toxicity Tests for Establishing 'No Effect' Concentrations for Copper with Brook Trout (*Salvelinus fontinalis*)," *Journal of the Fisheries Resource Board of Canada,* vol. 31, pp. 449–452, 1974.

176. Brungs, et al., "Acute and Long-Term Accumulation of Copper."

177. Benoit, D. A, "Chronic Effects of Copper on Survival, Growth, and Reproduction of the Bluegill (*Lepomis macrochirus*)," *Transaction of the American Fisheries Society,* vol. 104, pp. 353–358, 1975.

178. McKim, et al., "Metal Toxicity to Embryos and Larvae."

179. Nem'cok, J., and L. Borross: "Comparative Studies on the Sensitivity of Different Fish Species to Metal Pollution," *Acta Biologica Academie Scientifica Hungarie,* vol. 33, pp. 23–27, 1982.

180. Lloyd, R. "The Toxicity of Mixture of Zinc and Copper Sulphate to Rainbow Trout (*Salmo gairdneri, Richardson*)," *Annals of Applied Biology,* vol. 49, pp. 535–538, 1961.

181. Sprague, J. B.: "Lethal Concentrations of Copper and Zinc for Young Atlantic Salmon," *Journal of the Fisheries Resource Board of Canada,* vol. 21, pp. 17–26, 1964.

182. Roch, M., R. N. Nordin, A. Austin, C. P. J. McKean, J. Densiger, R. D. Kathman, J. A. McCarter, and M. J. R. Clark: "The Effects of Heavy Metal Contamination on the Aquatic Biota of Buttle Lake and the Campbell River Drainage (Canada)," *Archives of Environmental Contamination and Toxicology,* vol. 14, pp. 347–362, 1985.

183. Neff, J. M.: "Polycyclic Aromatic Hydrocarbons," *"Fundamentals of Aquatic Toxicology,"* G. M. Rand and S. R. Petrocelli, eds., Hemisphere Publishing Corporation, Washington, 1985.

184. Moore, J. M., and S. Ramamoorthy: *Organic Chemicals in Natural Waters, Applied Monitoring and Impact Assessment,* Springer-Verlag, New York, Inc., New York, New York, 1984.

185. Connell, D. W., and G. J. Miller: *Chemistry and Ecotoxicology of Pollution,* John Wiley & Sons, Inc., New York, 1984.

186. Neff, "Polycyclic Aromatic Hydrocarbons."

187. Ibid.

188. Connell, *Chemistry and Ecotoxicology of Pollution.*

189. Ibid.

190. Ibid.

191. *Federal Register:* "Water Quality Criteria Documents; Availability," U.S. EPA, vol. 45, no. 231, pp. 79318–79379, 1980.

192. Ibid.

193. U.S. Environmental Protection Agency: "Quality Criteria for Water 1986," U.S. Environmental Protection Agency, Office of Water Regulations and Standards, EPA 440/5-86/001, Washington, D.C., 1986.

194. *Federal Register,* "Water Quality Criteria Documents; Availability."

195. U.S. Environmental Protection Agency: *Water Quality Standards Handbook,* U.S. Environmental Protection Agency, Office of Water Regulations and Standards, Washington, D.C., 1983.

196. U.S. EPA, "Quality Criteria for Water 1986."

197. Ibid.

198. Nimmo, D. R.: "Pesticides," *Fundamentals of Aquatic Toxicology.* G. M. Rand and S. R. Petrocelli, eds., Hemisphere Publishing Corporation, Washington, p. 359, 1985.
199. *Federal Register*, vol. 56, no. 223, p. 58444, 1991.
200. Burton, G. A., ed.: *Sediment Toxicity Assessment*, Lewis Publishers, Inc., Chelsea, Michigan, 1992.
201. Baudo, R., J. Giesy, and H. Mantau: *Sediments: Chemistry and Toxicity of In-Place Pollutants*, Lewis Publishers, Inc., Chelsea, Michigan, 1990.
202. U.S. EPA, Amendments to the Water Quality Standards Regulation to Establish the Numeric Criteria for Priority Toxic Pollutants Necessary to Bring All States Into Compliance with Section 303-c)(2)(B). Proposed Rule 40 CFR Part 131.FR 56 (223): 581/20-58478, 1991.
203. Fuller, G. B., and W. C. Hobson: "Effects of PCBs on Reproduction in Mammals," *PCBs and the Environment*, Volume II, J. S. Waid, ed., CRC Press, Inc., Boca Raton, Florida, 1986.

ADDITIONAL READING

Ames, B. N.: *Dietary Carcinogens and Anticarcinogens, Science*, vol. 221, pp. 1249-1264, 1983.
Annual Review of Pharmacology and Toxicology, Annual Reviews, Inc., Palo Alto, California.
Carson, R.: *Silent Spring*, Houghton Mifflin, New York, New York, 1962.
Casarett and Doull's Toxicology–The Basic Science of Poison, 4th edition, M. O. Amdur, J. Doull, and C. D. Klaassen, eds., Pergamon Press, Inc., Elmsford, New York, 1991.
Clayton, G. D., and F. E. Clayton: *Patty's Industrial Hygiene and Toxicology*, 3rd edition, John Wiley and Sons, New York, New York, 1982.
Congress of the United States, *Identifying and Regulating Carcinogens*, Office of Technology Assessment, U.S. Government Printing Office, Washington, D.C., 1987.
Doll, R., and R. Peto: "The Causes of Cancer: Quantitative Estimates of Avoidable Risks of Cancer in the United States Today," *J. National Cancer Institute*, vol. 66, no. 6, pp. 1192-1308, June 1981.
Efron, E.: *The Apocalyptics–Cancer and the Big Lie*, Simon and Schuster, New York, New York, 1984.
Haley, T. J.: *Toxicology*, Hemisphere Publishing Co., Harper & Row, 1987.
Hamilton and Hardy's Industrial Toxicology, 4th edition, rev. by Asher Finkel, PSG Publishing Co. Inc., Littleton, Massachusetts, 1983.
Hayes, A. W., ed.: *Principles and Methods of Toxicology*, 2nd edition, Raven Press, New York, New York, 1989.
Hayes, W. J., and E. R. Laws, eds.: *Handbook of Pesticide Toxicology*, Academic Press, New York, New York, 1991.
Hynes, H.B.N.: *The Biology of Polluted Waters*, Liverpool University Press, Great Britain, 1971.
Johannesen, F. R.: "Risk Assessment of Carcinogenic and Noncarcinogenic Chemicals," *Critical Reviews in Toxicology*, CRC Press, Pearl River, New York, vol. 20, issue 5, pp. 341-367, 1990.
Johnson, D. W., and M. T. Finley: *Handbook of Acute Toxicity of Chemicals to Fish and Aquatic Invertebrates*, U.S. Department of Interior, Fish and Wildlife Section Resource Publication 137, 1980.
Lu, F. C.: *Basic Toxicology*, Hemisphere Publishing Corp., Washington, D.C., 1985.
Maughan, J. T.: *Ecological Assessment of Hazardous Waste Sites*, Van Nostrand Reinhold, New York, New York, 1993.
Nebert, D., and F. J. Gonzalez: *P450 Genes: Structure, Evolution and Regulation, Annual Review of Biochemistry*, vol. 56, pp. 945-993, 1987.
Paustenbach, D. J., ed.: *The Risk Assessment of Environmental Hazards: A Textbook of Case Studies*, John Wiley and Sons, New York, New York, 1989.
Rodricks, J. V.: *Calculated Risks: Understanding the Toxicity and Human Health Risks of Chemicals in Our Environment*, Cambridge University Press, Cambridge, U.K., 1992.
Sax, N. I., ed.: *Dangerous Properties of Industrial Materials*, 7th edition, Van Nostrand Reinhold, New York, New York, 1989.
Suter, G. W., ed.: *Ecological Risk Assessment*, Lewis Publishers, Boca Raton, Florida, 1993.
Tannock, I. F., and R. P. Hill: *The Basic Science of Oncology*, Pergamon Books, Inc., 1987.
Tardiff, R. G., and J. V. Rodricks, eds.: *Toxic Substances and Human Risk—Principles of Data Interpretation*, Plenum Press, New York, 1987.

Zeise, L., R. Wilson, and E. A. C. Crouch: "Dose-Response Relationships for Carcinogens: A Review," *Environmental Health Perspectives*, vol. 73, pp. 259-308, 1987.

ACKNOWLEDGMENTS

The authors wish to acknowledge the following individuals who assisted with the preparation of this chapter: Sec. 5-5, Carcinogens, was prepared by Dr. Bruce Molholt, Principal Toxicologist for *Environmental Resources Management, Inc.*, and Dr. Robert Nilsson, Professor of Toxicology, *Royal University of Stockholm*. Dr. Nilsson contributed extensively to the entire chapter. Sec. 5-6, Ecotoxicology, was prepared by David Stout, Manager of Environmental Science for Blasland, Bouck and Lee, Inc. The principal reviewer for the **ERM Group** was Dr. Richard Shank, *ERM-Midwest*.

CHAPTER

6

ENVIRONMENTAL AUDITS

The greatest of faults, I should say, is to be conscious of none.
Carlyle

6-1 INTRODUCTION

The dire consequences of mismanaging environmental responsibilities have prompted the use of environmental audits as a preventative tool. An environmental audit derives its name from "financial audit" with which it bears some resemblance. The concept of a financial audit, developed largely in the 1800s as a means for detecting fraud, consists of an independent systematic examination of accounting records to verify that financial reports reflect generally accepted accounting principles. Similarly, an environmental audit is an independent, systematic method of verifying that environmental regulations, internal policies, and good operating practices are being followed.

The concept of an environmental audit had its beginnings in the late 1970s.[1] In comparison with a financial audit, an environmental audit has broader objectives. In fact, environmental auditing can take many forms, each with its own set of objectives. The valuable purposes that an audit can serve prompted one periodical to report in 1983 that the market for environmental audits had "exploded."[2] Since then, demand for audits has continued to increase, environmental requirements have become ever more numerous and complex, and auditors have developed new uses for audits as tools to improve company operations. In summary, environmental auditing has matured and come into its own as a specialized field.

Types of Audits

The EPA defines environmental auditing as a "systematic, documented, periodic, and objective review by a regulated entity of facility operations and practices related to meeting environmental requirements."[3] Such a definition stresses regulatory compliance, and verification of compliance was the driving force behind the initial developments in environmental auditing. Today, it still remains one of the main reasons for conducting audits. Audits with this objective are termed **compliance audits**.

The need for compliance audits is clear. The body of environmental law and regulations has increased in size and complexity over the past two decades, seemingly without bounds (see Fig. 6-1). Violation of these requirements could result in imposition of criminal penalties, civil penalties, and injunctive relief. Additionally, enforcement by regulatory agencies is increasingly under great public scrutiny, with pressure to do even more. In summary, the requirements are great, the price of non-compliance is high, and the chances and consequences of getting caught have increased. Thus, determining what specific regulatory requirements are imposed upon an operation, finding out whether the operation is in compliance, and pinpointing possible violations in time to take proactive measures are the major goals of compliance auditing, but not the only useful purpose which an audit can serve.

A compliance audit is merely a "snapshot" of plant operations and procedures, identifying instances of either compliance with regulations or violation of them. A compliance audit is quantal and superficial. More is needed as industrial companies have gained experience with audits, the companies have realized that the auditing process is very useful in extending beyond the mere identification of regulatory violations to an analysis of both the root causes of any identified violations, and even the potential for future problems. This perspective, termed a **management audit**, determines whether an adequate compliance management "system" is established, implemented, and used correctly to integrate environmental compliance into everyday operating procedures. Such an audit examines cultural, management, and operational elements to

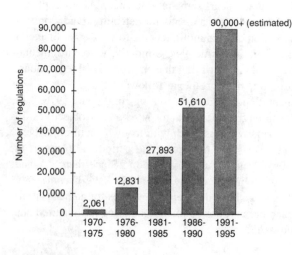

FIGURE 6-1
Environmental compliance regulations for federal and state governments in the United States.[4]

include internal policies, human resources, training programs, budgeting and planning systems, monitoring and reporting systems, and information management systems. A management audit detects potential systematic breakdowns that could manifest themselves as an environmental problem.

Another type of environmental audit is a **liability definition audit**. These are typically done for prospective buyers of real estate and for proposed mergers and acquisitions. Such audits identify environmental problems that could reduce the value of a property or expose the buyer to liability. Not conducting such audits before acquiring property is done at the buyer's peril. The term "due diligence" implies gathering facts and information in a prudent manner and applying technical knowledge to identify where problems exist. The fact-finding of an audit helps to establish due diligence.

Other types of environmental audits are waste contractor, risk definition, and waste minimization. The **waste contractor audit** uses features of both a compliance audit and a liability definition audit to analyze commercial facilities used to store, treat, and dispose of hazardous waste. The **risk definition audit** analyzes the operations of facilities that handle hazardous materials and substances. These audits serve to help in obtaining insurance coverage and are required by some governments as part of catastrophe prevention planning. The results also provide environmental guidance for hazards not specifically addressed in current environmental regulations. **Waste minimization audits** examine waste generated by a facility with the objective of identifying viable actions to reuse, recycle, or otherwise reduce the quantity and toxicity of each waste stream (see Sec. 7-2).

Features of Effective Environmental Auditing

A review of mature auditing programs being conducted by both the public and the private sector has identified many common features deemed to be an important part of an effective auditing program:[5]

- Explicit top management support for environmental auditing and commitment to follow-up on audit findings;
- An environmental audit team separate from and independent of the persons and activities to be audited;
- Adequate team staffing and auditor training;
- Explicit audit program objectives, scope, resources, and frequency;
- A process which collects, analyzes, interprets, and documents information sufficient to achieve audit objectives;
- A process which includes specific procedures to promptly prepare candid, clear, and appropriate written reports on audit findings, corrective actions, and schedules for implementation; and
- A process which includes quality assurance procedures to ensure the accuracy and thoroughness of environmental audits.

These features show that conducting an audit is a process rather than a mechanical application of a rigid, absolute methodology. Although an audit is systematic, as its procedures occur in a methodical manner, the procedures cannot be mechanical because much of the audit involves interpretation. The key findings of an audit typically spring from the ability of an auditor to probe in an open-ended manner, well beyond any checklist, in search of basic facts and specific patterns. In its best form, an audit is part of a continuous improvement program, with the findings of an audit triggering a number of follow-up actions including subsequent audits.

An audit may seem to the casual observer to be an assessment. An audit is different than an assessment; an audit implies statistical verification, direct checking, and an adequate level of detail. An assessment implies an overview with less direct checking and less detail. Although both require professional judgment, an audit demands more. As an example, it is significant whether an audit collects its own samples or simply reviews existing data from previous sampling. Collecting new samples is essentially the only way to directly verify the existing sampling data. This point is so significant as to classify programs which do not sample as an assessment or review rather than true audits. Experience has shown that most programs do not collect samples and, therefore, are reviews and assessments rather than formal audits.

Regardless of the type of audit being performed, it typically would unfold in three stages as follows:

1. Audit program planning (including pre-visit data collection).
2. On-site activities.
3. Evaluation of audit data and reporting of findings.

The next three sections of this chapter describe the approach taken in each stage of conducting a compliance audit of a plant, stressing those aspects which address hazardous wastes. As stated earlier, a compliance audit is just one type of audit, and its selection in this chapter for purposes of illustration does not diminish the importance of the other types of audits. The other types are contrasted with compliance auditing in the closing section of this chapter.

6-2 PROGRAM PLANNING

A compliance audit program consists of many elements, including a visit to the plant. Prior to the actual visit, it is essential to plan the overall program. Planning would include obtaining commitment by management, defining the objectives and requirements, and developing an information management system. It is also important to organize the audit team and make sure that the program has the resources and tools necessary to address the issues that need to be investigated. In addition, one must consider the legal measures which may be appropriate to protect the results of the program. In summary, the auditing program must be tailored to meet the needs of the facility being audited.

Commitment by Management

No compliance audit program can be successful without the clear commitment of senior management. This commitment must come before the start of the process in two ways. First, management must provide the program with the necessary resources and if necessary, direct plant personnel to provide access and to cooperate with the auditors during the process. Second, it is critical that management articulates an explicit written commitment to follow-up an audit finding and correct the problems that the audit will uncover. Undertaking an audit program without a commitment to cure the ills that it may uncover is probably worse than not conducting an audit at all. Not correcting such faults suggests to employees that compliance is not a high-priority item. Further, leaving them uncorrected, perhaps with statements in a file that they were known, is viewed by a regulatory agency as flagrant disregard for the environment and will likely result in severe penalties. Finding a violation, correcting it, and reporting it to the appropriate agencies (i.e., self-correction) is viewed by agencies as one of the strongest benefits of a compliance audit program.

The value of correcting faults is illustrated by the true case of two manufacturing plants with the same type of violation involving the Form R Chemical Release Report (a provision of SARA). One company had retained a consulting firm to conduct an audit, and the auditors discovered that the plant had filed incorrect reports for each of the three previous years; the reports had submitted miscalculated information for three chemicals and had omitted two chemicals which had substantially exceeded the reporting threshold. The plant self-reported the violations to the EPA and filed correct reports within 30 days of discovery. The plant received a fine of $2000. In contrast, a second plant operating in the same state was inspected by the EPA in the same year. The EPA inspector discovered that the information for one chemical had been substantially misreported, and the plant subsequently received a $60,000 fine. The difference derives from the fact that one plant self-reported and the other was caught by a regulatory agency.

Definition of Requirements

The process of performing an audit begins by deciding what type of audit is needed (i.e., is the goal to ensure compliance, to minimize waste generation, or to look for liabilities that might be there from past practices?). After selecting the needed type of audit (in this case, a compliance audit), the auditor must define the scope of the audit. Are all areas of environmental management to be covered or only one specific part (e.g., air quality or hazardous waste)? A compliance audit could cover all or just part of the various environmental management topics identified in Table 6-1. A comprehensive compliance audit would address all of these topics and perhaps related health, safety, and transportation requirements.

After selecting the type of audit and its scope, other decisions need to be made: How will successful performance be measured over time? How frequently will the audits be done? How will the results be reported? Each of these is important to evaluate prior to designing the actual audit process. Defining these objectives in the beginning

TABLE 6-1
Topics of an environmental compliance audit

Hazardous waste regulations	Identification Packaging Storage Waste minimization Emergency preparation Employee training Waste disposal (current) Record keeping and reporting
Past disposal of hazardous waste	Release notification procedures Remedial investigations and actions
Emergency planning requirements	Emergency response planning Emergency release reporting Hazardous chemical inventory Toxic chemical releases
Air pollution regulations	Permitting Record keeping and reporting
Water pollution regulations	Spill prevention and control plans Facility controls Discharge permitting Sampling/analysis
Underground storage tanks	Registrations, notifications Testing Inventory control Incidents/leakage
Solid waste regulations	Permitting Record keeping and reporting

makes a major difference regarding the types of resources and techniques needed to conduct the actual audit.

Confidentiality

It may be appropriate to conduct an audit under legal arrangements to protect confidentiality. It is important to plan this carefully because such legal protections may withstand challenge only if they are built into the auditing process. Why should one consider holding an audit confidential? The reasons go beyond the direct connection of finding violations. For example, the audit report may describe proprietary business information. Also, there may be potential for an adverse community reaction to sensitive information if not communicated clearly in an appropriate forum. There is even the potential for self-incrimination if the disclosed violations carry criminal penalties. Many of these concerns will disappear by developing an effective strategy for correcting problems noted by the audit. Oral reporting will eliminate concern with disclosing highly sensitive issues. For information which clearly needs protection from public disclosure (e.g., proprietary data on processes), it is desirable to use legal protective methods.

One basis for protecting information from the requirement that it be disclosed in connection with litigation is the attorney-client privilege. Attorney-client privilege applies to communications between a client and his or her attorney. If an attorney is involved in the auditing process or if corporation counsel is available to the auditor, information may be communicated directly to the attorney. If it is anticipated that an interview may reveal sensitive information, it may be advisable to have an attorney conduct the interview.

A second basis for protecting information from discovery is the attorney work product privilege, which is a conditional privilege. At least three specific circumstances must be met to establish this privilege. First, legal counsel must have a direct involvement in conducting the audit. Second, the attorney work product is prepared in anticipation of litigation. This does not necessarily mean that litigation is ongoing or even imminent; however, litigation must be more than a remote possibility. The documentation that establishes this privilege should indicate a clear reason for the anticipation of litigation. Third, the distribution of documents prepared under attorney work product privilege must be very limited. The reports of the auditor must first go to the attorney for distribution on a limited basis. The marking of the documents as "Privileged and Confidential–Attorney Work Product" is critical, and all individuals in the process must be educated on the confidentiality and the clear requirement not to reveal the protected information to third parties. In fact, even accidentally revealing the contents of such a document may constitute a waiver of the confidentiality. The disclosure of the existence of an audit document does not constitute a waiver; however, the disclosure of any of its contents may.[6,7]

As a final point, confidentiality applies only to the reports and to the interpretations of the audit results as contained in the reports. Disclosure-limiting methods such as attorney-client privilege cannot protect the underlying facts from disclosure. For example, attorney-client privilege can protect an audit report from being disclosed and subpoenaed for use in litigation. On the other hand, detailed laboratory reports, faulty inspection records, faulty manifests, and similar items cannot be protected under attorney-client privilege. These are, in a sense, facts; they existed prior to the audit and are facts a regulatory agency has a right to see which cannot be protected under a disclosure-limiting method.

Example 6-1   Confidentiality. Company "A" is planning an initial public offering of its stock, and prior to preparing disclosure statements in connection with its offer to sell stock to the public, it conducts a compliance audit. Company "B" conducts a routine compliance audit program, reviewing every site every two years. Company "C" discovers a leaking underground storage tank as the result of an audit. Can any of these results be held confidential?

Solution. Company A may not be able to hold the results confidential. The audit may discover some environmental liabilities that would be required to be revealed by the rules of the United States Securities and Exchange Commission. While this is a complex area of law, in general a prospective buyer has a right to know about potential liabilities, including environmental liabilities.

Company B also may have some difficulty in holding the results confidential. As a matter of course, many companies attempt to protect the results of routine compliance audits from discovery by a regulatory agency or a party suing the company in the absence of specific statutory requirements to reveal the information.

The case of Company C is more straightforward because in certain states the company has a clear obligation to immediately report leaking underground storage tanks to the appropriate governmental authority; failure to do so could result in civil or even criminal sanctions.

Confidentiality and privilege are complex areas of the law, and legal counsel should be consulted when these questions arise.

Organization of Auditing Program

After the objectives are defined for the audit program, the next step is to organize the program. There are many questions to be answered: What is the frequency of reaudits, if any? How to schedule an audit of multiple sites? Will the audit program be conducted by environmental consultants? Or will it be done by the corporate staff as part of their normal jobs? A third alternative is to have the audit managed by a small corporate oversight group with the audit function being conducted by staff from the operating divisions. Each of these three groups (consultants, corporate staff, and operating staff) has advantages and disadvantages in its approach to various issues. Mature, successful auditing programs tend to integrate participation by all three groups.

Example 6-2   Source of audit team.   What are the comparative capabilities and limitations of conducting an audit of a manufacturing plant with consultants, corporate staff, or operating staff?

Solution

Issue	Consultant	Corporate Staff	Operating Staff
Objectivity	High	Partial loss	Questionable
Auditing expertise	High	Varies	Typically nominal
Technical expertise	Broad and specialized	Industry-specific	Plant- or corporate-specific
Regulatory expertise	Broad and specialized	Industry-specific	Varies
Memory	None	Corporate memory	Corporate and plant memory
Familiarity with operations	Low	Moderate	High
Participation in solution	Nominal	Indirect	High and direct

Team of Auditing Specialists

Conducting a successful audit requires a team effort, directed by a team leader and supported by auditor specialists. The selection of the team has a major influence on the success of the audit. The size and makeup of the team depend on the size of the facility and the complexity of the issues that must be addressed.

Looking at the types of issues typically covered in an audit reveals that the area of knowledge required of the auditors is very wide. Auditors need to be "expert" in federal and state environmental regulations and sometimes local or county regulations. They need to know the specific regulations that apply to the facility's particular operations and also possess a knowledge of peer facilities. For a manufacturing facility, the team needs to know production (i.e., how a facility operates, how its manufacturing processes function, where "things" could be hidden, and how "things" are most likely to fail). Auditors need to have a knowledge of waste treatment and pollution control technology. The team needs to have a variety of scientific disciplines. They need a knowledge of information management systems and general management techniques for organizing staff and programs. In summary, a comprehensive audit team would offer specific experience in engineering, regulatory analysis, regulatory enforcement, chemistry, production, and perhaps geology and community relations. An audit could function with a team of just two experienced individuals, a regulatory specialist and an engineer. More complex facilities require additional team members. For example if specialized issues such as contamination of soil and ground water are to be examined, additional auditors may be required.

In addition to knowledge in specific technical areas, the auditing team should have both training and experience in conducting audits. The auditors should be independent (i.e., free and capable of investigating the topic thoroughly and completely, probing appropriately through document reviews and interviews). Finally, the team members should have sensitivity, with the ability to communicate accurately and appropriately with different levels of personnel.

Pre-Visit Data Collection

An audit usually requires facility staff to complete and return a pre-visit questionnaire that familiarizes the audit team with operations of the facility, its regulatory standing, and the environmental setting. Entering the facility without this information will result in wasting valuable on-site time instead of evaluating specific issues. It is essential to have received a completed pre-visit questionnaire before finalizing the audit team. The responses will identify whether special expertise may be needed on the audit team (e.g., air specialists, hydrogeologists) as well as special equipment and materials.

Alerting the auditors to the specific circumstances that they will encounter at the facility, the auditors to focus their preparations on potential problem areas that deserve particular attention. The questionnaire also identifies the type of information needed by the auditors, which the facility can begin to compile and organize as part of its preparations. Based on the responses to the completed pre-visit questionnaire, the auditing team will identify the key people to be interviewed and the key documents to be examined. The audit visit can then be scheduled and an agenda planned.

What type of information is typically solicited by the pre-visit questionnaire? It is the critical information necessary to plan the audit, including the following:

1. A management chart showing key personnel and their line and staff responsibilities.
2. A site layout showing the location of various operations, particularly the key items which should be examined (e.g., waste storage areas, sewer locations and drainage systems, underground storage tanks, and ground water monitoring wells).
3. A brief description and block diagram of operating processes and any pollution control units, showing recent or planned modifications. Block diagrams are very useful because they allow the audit team to anticipate the types of chemicals handled by the facility and the environmental emissions and wastes generated by the process. The team can preview pertinent environmental regulations and compile data on specific chemicals, appropriate maintenance programs and other information which will help them to be more fully prepared for the on-site audit.
4. A list of all major types of waste generated and how each is managed.
5. A list of all environmental regulatory permits of the site, with copies. Permit information tells what type of reporting is required, and what limits are in place. The lack of permits for certain activities would alert the auditor to potential omissions based on knowledge of similar operations.
6. All major environmental plans and policy manuals. For example, spill prevention and control plans provide relevant information about the facility and the issues related to storage of waste and hazardous materials (e.g., inspection programs and structural and non-structural controls that should be in place to prevent spills).
7. Information about non-compliance history of any sort. This is particularly useful because it allows the auditor to concentrate on areas that have attracted regulatory attention and would likely receive close regulatory scrutiny in the future.

The comprehensiveness of the questionnaire should conform to the size of the facility, the sophistication of facility management, and the type of support provided by the management for the auditing efforts. In any event, the questionnaire should be of moderate length (8–12 pages) and directed only at key factors. Examples of key questions under selected headings other than hazardous waste are as follows:

General Issues

- What pollution control equipment is used at the plant?
- Have you been inspected by any regulatory agency in the last year?
- Have you been cited for non-compliance in the last year?
- What is the acreage of the facility?
- How long has the facility been in operation?
- What is the past use of the site?

Hazardous Materials

- What is the latest hazardous materials inventory?
- What quantity of petroleum products are stored on site?
- Do any underground tanks exist on site?

Wastewater

- Where are wastewater discharge locations (point and non-point)?
- What permits are in place?
- Has there been any correspondence with the local sewer agency or the water board concerning compliance within the last two years?

Ground Water

- Does this plant have any required ground water monitoring programs?

Air Quality

- What air permits does this plant have? For what operations?

The plant should submit any available lists of items (e.g., underground storage tanks) with the completed questionnaire. Additionally, the plant staff should assemble backup information for the above items (e.g., ground water monitoring data) and have it available when the auditors arrive.

A review of the pre-visit questionnaire completed by facility personnel will probably raise some questions that necessitate telephone interviews with facility management.

Example 6-3   Hazardous waste questions. What are some examples of key questions and solicitations regarding hazardous waste which need to be included in a pre-visit questionnaire sent to a manufacturing plant?

Solution

- List monthly volumes and types of hazardous and non-hazardous solid waste generated at the facility.
- Please have available all manifests generated within last six months.
- List and briefly describe each waste storage area.
- How long are wastes stored?
- How are wastes stored?
- Are wastes treated or disposed of on-site?

Audit Protocol

One of the most critical audit tools is the audit protocol or the questionnaire used at the facility during the on-site audit. It is a guide to the types of questions that must be asked; it is a "road map." The use of an audit protocol ensures consistency and comprehensiveness in implementing the scope of the audit. It forms a record of the audit procedures, becomes a basis for development of the audit findings, and also serves as a basis of any future critique of the audit. It is much more than a data inventory form.

The many different types of audit protocols range from general instructions and checklists to detailed yes/no guides that may even include scoring mechanisms. Table 6-2 describes the various types of protocols and the advantages and disadvantages of each. Figure 6-2 presents examples from three types of protocols.

All protocols are prepared with separate sections for air, water, waste, and whatever else is deemed important. Each type of protocol offers its own advantages and disadvantages. Regardless of the type, a protocol basically uses either of two approaches: **checklist** or **review and evaluation**. A checklist must be designed so that the question can be answered without leading or prompting the respondent. This forces a rigid structure which may miss a subjective and critical fact. The review and evaluation approach is commonly used by more experienced auditors. This protocol is basically organized as a general topical outline with reminders of important points. It allows open, narrative answers with entries made to identify the basis for the determination (e.g., visual observation, interviewer's statements).

Schedule

Following the preparation of a protocol, the site agenda can be prepared. The agenda lists the main issues that the audit will cover, the key personnel to be interviewed, a schedule for conducting the visit of the facility, and a list of documents and other information that the team wishes to review. Typical schedule items would start with the opening interview and the overview presentation, proceed through the site tour

Type of audit protocol	Example
General instruction	Determine whether laboratory analytical procedures comply with 40 CFR, Part 136
Yes/no guide	Does the facility store oil or oil products above ground? *Yes* → Is total storage capacity greater than 660 gallons? (*No* → No further analysis required; *Yes* ↓) *No* → No further analysis required
Yes/no checklist	Does the site have a written Spill Prevention Control and Countermeasure Plan? Does the plan include...

FIGURE 6-2
Examples of protocol entries.

TABLE 6-2
Types of audit protocol

Type	Description	Advantages	Disadvantages
General instruction list	Divides audits into sequence of audit steps; very general in approach	Very flexible; gives broad freedom to auditors	Requires very experienced auditors; hard to "score"
Review and evaluation list	General topics "reminder" within a given area; details not included	Very flexible; gives broad freedom to auditors	Requires experienced auditors; hard to "score"
Detailed instruction checklist	Summarizes details of specific regulations and standards; aimed as method to walk auditor through steps	Consistency; fills in lack of knowledge	Bulky/long; too much detail; may miss critical points
Yes/no checklist	Organized summary of almost every regulation; common in companies doing first audits	Consistency; easy to use; easy synopsis	Long; tends to lead to "wrong" questions; poor on systems
Yes/no guide	More structured version of above	Consistency; very easy to use	Very long
Expanded analysis checklist	Usually added to "detailed instructions" or "yes/no" checklists to capture more detail	Allows extended response for questions which don't fit into checklist format	Long
Scored checklist	Add a weighted rating system to "yes/no" checklist	Allows comparison of similar facilities	Leads to "passing score" mentality
Verification checklist	Annotates checklists with instructions on verification techniques	Increases reliability of audit	Very long
Procedure-driven checklist	Addresses internal approaches/policies and systems in place to ensure compliance rather than snapshot of current compliance	Gives best analysis of overall potential for compliance	Requires experienced auditors

and each of the environmental media specific audits, and close with the exit interview. Excerpts from a typical audit schedule are as follows:

Day 1—Monday

9:00–10:00 AM — Meet with host Facility Manager and Environmental Manager to make introductions and review purpose of the audit. Refine schedules for interviews, tours, and special requirements. Receive overview presentation of facility operations.

10:00–11:30 AM — General tour of the facility to provide Audit Team with orientation of overall facility. This is a brief "ride through" of key areas of the facility.

11:30–12:00 AM — Conduct briefing for Facility Manager.

1:00–5:00 PM — . . .

Day 2—Tuesday

8:00 AM–5:00 PM — Conduct audit of hazardous waste management compliance. Physically inspect hazardous waste generation points (shops, process areas, storage yards, warehouses, treatment facilities). Inspect waste shipment manifests, waste analysis records, regulatory permits, ground water monitoring, and scheduled inspection plans. Conduct interviews with key facility staff.

Day 3—Wednesday — . . .

Day 4—Thursday

8:00 AM–2:00 PM — . . .

2:00–5:00 PM — Conduct audit of fuel/oil storage and spill response compliance. Physically inspect oil storage areas, tank farms, and drum storage areas. Review oil spill contingency plans, spill reports, and spill prevention and control plans. Conduct interviews with key staff.

Day 5—Friday

8:00 AM–10:00 AM — . . .

10:00–11:00 AM — Audit Team makes an informal briefing to key staff on preliminary findings of audit.

11:00 AM–12:00 PM — Conduct formal exit interview with Facility Manager.

A site agenda serves many purposes. It points out time conflicts among the members of the audit team. Sending the agenda to the facility in advance of the visit is imperative to reduce the disruption of operating activities and to ensure that the facility personnel are prepared for the visit (e.g., the facility can collect and organize the needed records in advance so that the audit team does not experience unnecessary delays while the information is being pulled from files).

6-3   ON-SITE AUDIT

Following the acceptance of the site agenda by the facility, the team members visit the site. The primary purpose of the visit is to review documents maintained at the facility, to interview facility employees, and to inspect thoroughly all relevant operations conducted at the facility. This is an intensive and thorough period of collecting data.

Orientation

The first step upon arrival at the site is an orientation and initial interview, typically with the facility manager. The environmental coordinator and, in some cases, many of the facility staff involved in the audit may also participate. It is important to gain early rapport with these individuals, to start on a positive note, and to ensure that the planned approach to the audit will make the most efficient use of everyone's time. Facility personnel should clearly understand that the audit is an improvement tool, not a policing effort.

The auditor should prepare handouts to describe the scope of the audit and should clearly explain that the efforts of facility staff will be needed to conduct the audit. The audit will likely disrupt the facility's schedule, and the audit team needs to acknowledge that this may represent an inconvenience, if not a problem, to the facility staff. Of course, it is important to point out the reasons for the audit and its benefits to them. This initial interview should encourage questions from the facility staff to assure that the staff members have a thorough understanding of the audit goals and the organized approach.

Several other issues must be discussed prior to proceeding with the audit. The first issue is whether the audit will be kept confidential as part of the attorney-client work product privilege. Similarly, it is important to clarify how the audit and its results will be revealed to employees and how this information will be used. Later, as the audit team proceeds through the audit of the facility, discussing issues and inspecting operations, employees will question the team as to what they are doing. The facility employees may assume, unless told otherwise, that an auditor is a regulator. Alternately, they may think an auditor is a company spy trying to check up on them. Such perceptions may evoke less than ideal responses as the audit team solicits information. Therefore, it is important to have cleared with the facility manager an agreement as to how this audit will be discussed and how information will be used. If the information will be used without attribution, facility personnel need to know this so they can be frank. On the other hand, if attribution will be made, personnel need to know this too. Also, it is important to decide with the facility manager whether personnel will receive any feedback about the results of the audit.

Safety measures and escort issues must be reviewed during orientation. In many circumstances, access to certain parts of a facility may require entry into restricted work areas or may require certain protective equipment (e.g., respirators, protective hats, and boots). This requirement must be understood prior to entering such areas and, in some cases, prior to arriving at the facility. It is important to understand what kind of emergency response signals may be given so that the audit team will respond appropriately and evacuate if necessary.

Having discussed the logistical and overview issues, the audit team should review the results of the pre-visit questionnaire. This is the time to expand the team's understanding of facility operations, particularly the operational setup and chemical inventory. Finally, the facility manager will need to define who will be available and their schedules for interviews, inspections, and other elements of the audit. This may necessitate some adjustment to the previously prepared schedule.

Having completed the orientation, the audit team should take a quick tour of the facility. The goal is simply to obtain an overview of the facility, not to get into details. This tour may suggest further refinement of the previously planned schedule and logistics. After making these adjustments, the auditors should assign themselves specific responsibilities for interviews, document reviews, visual inspections, and other audit activities.

Interviewing Facility Employees

Much of the information gathered during the audit will emerge during discussions with facility personnel—as part of the protocol and as questions prompted by points discovered during visual inspections. These interviews are a particularly valuable method for obtaining subjective information. The success of an interview, based on the quantity and quality of the information obtained, depends largely upon the techniques used by the auditor.

One necessary step of conducting a successful interview is to plan for the interview in advance. Prior to interviewing a person, the interviewer needs to organize thoughts, to identify issues to be discussed with the person, and to define what the desired outcome will be in terms of the types of information needed to draw conclusions.

When conducting the actual interview, the opening moments are critical to establishing both a relaxed atmosphere and a professional relationship with the person:

- Introduce yourself, your background, and your objectives.
- Ensure that the person has the appropriate amount of time available and is prepared to spend that time with you discussing the issues.
- Explain to them how the information will be used during the audit. Particularly, establish whether the information will be used anonymously or will be attributed. This may strongly affect the way that the person will respond.
- After the above explanation is completed, begin the information-gathering stage by requesting an overview of the interviewee's job. The response may be surprising! People will describe their job in many ways that are different from what one might expect from reading a job description or analyzing an organization chart.

As the interview progresses, it is important to continue to use the proper amount of respect. Do not prod; placing someone in a position of having to defend what they have done or just said will not reveal all the needed information. Instead, try to probe constructively with open types of questions. Do not ask questions that are just "yes/no." Instead, try to have open-ended but concrete questions that require the interviewee to express himself/herself so details will be given that otherwise could be overlooked by a yes/no checklist. Make sure that you are not using leading questions which would telegraph evaluations. Some key words in asking questions are "what," "who," and "how" (e.g., "how do you know... ," "how do you ensure..."). Silence is important in letting the person think through the response.

It is vital to provide feedback during the interview. Confirm that you have understood correctly what the person has told you by summarizing for them the information you are learning. Do not jump to conclusions; analysis and conclusions come later. When the interview is over, make sure you end on a positive note. Thank the person for the time and effort, and indicate that you will call if you have additional questions. An interview should ideally last about 20–30 minutes and never longer than 45–60 minutes. Clearly, this requires preparation and organization.

Document Review

Complying with environmental regulations, particularly those pertaining to hazardous waste, requires extensive record keeping and other documentation such as permits, emergency plans, training and monitoring programs, and manifests. These documents contain information on the various characteristics of the facility and its operations, as well as the controls used to ensure compliance. As such, the documents warrant a thorough analysis as part of a compliance audit.

The most important documents to review are permits, because a major component of an audit is to compare operations against applicable permits. Thus, an auditor examines each permit for more information than simply that needed to ensure it has not expired. Nearly every regulatory program requires permitting. A major facility may have many permits for different activities—air quality, discharges of wastewater and stormwater to natural water bodies and public sewers, currently generated and previously generated hazardous waste, other solid waste, underground tanks, ground water, pesticides, and others. A permit covers a number of issues, including:

- A description of relevant facility equipment and operations,
- The specific start and stop dates and the date for renewal,
- The quantitative limitations on discharges or use, and
- Monitoring and reporting requirements (e.g., quarterly reports on actual discharges compared to that permitted).

The review of a permit should include examining the application prepared by the facility prior to obtaining the permit. A permit application typically specifies equipment and procedures to be used by the facility. In many cases, these have changed since filing the original application. Not only may the current equipment and procedures not be correctly described in the application, but the facility may never have applied for a permit for the changes and thus would not have complied with the typical requirements of most regulations. It is important to determine whether the proper person from the facility, in terms of both level and authority, signed the permit application. In some circumstances, the person who did sign may not have had the proper authority.

A permit typically specifies limitations on discharges to the environment, the use of particular equipment, or some other activity. The permit requires periodic reports or at least some documentation of compliance with these limitations (e.g., monitoring data). An audit should ascertain what types of reports and record keeping are necessary to document compliance, whether the facility has met these documentation requirements, and whether the substance of the documentation shows that the facility is in compliance.

The need to use judgment when examining documented data, and not to take them at face value, is illustrated by an example of pH monitoring at an oil terminal. The facility had recorded daily for over a year that the pH of its discharge was always 7.0. The auditor asked the technician to take another measurement. The technician carefully obtained a representative sample in an especially clean bottle and used pH paper to measure the pH. The auditor was puzzled, however, when the technician entered 7.0 in the log book when it was obviously 6.2. It turned out that the technician was color blind and did not know it!

Some records such as manifests are not associated with a particular permit, yet are just as important. These too require auditing. In auditing any record, it is necessary not only to examine it for completeness or timely submittal but also to check the accuracy of the data. This may necessitate having to go back to original data (e.g., log books, laboratory results, strip chart, or whatever originated the record). A random sampling of the arithmetic calculations used to make the record should also be recomputed. The records for some facilities can be too extensive for manual review, necessitating the auditor to use information systems and database tracking.

After analyzing the permits and record keeping, the next step is to review various types of documentation required as part of planning emergency responses. Typical of what one might find at a facility in the United States would be the Spill Prevention Control and Countermeasure (SPCC) Plan, a contingency plan for hazardous material and hazardous waste releases, a preparedness prevention plan, an OSHA 1910.120 emergency action plan, or a written plan for collecting inventory and emissions for Title III of SARA. Similarly, many regulations require written training programs and, in many cases, documented records to verify that the training has taken place at the proper time intervals. In many circumstances training programs have not been updated and thus the auditor may find that the equipment, techniques, and chemicals specified in the written training programs no longer suffice for existing operations at the facility.

The documents will list numerous items that are to be kept and maintained at the facility. For example, the contingency plan and the preparedness prevention plan will list specific items required for emergency response such as evacuation plans and certain types of emergency response equipment such as fire extinguishers, eye washes, and protective gear. The auditor should prepare a list of these items while reviewing the documentation. When conducting the physical inspection, the auditor should verify the presence of each of these items and the equipment maintenance schedules to determine whether they are in good repair.

Example 6-4 Storage facility. In conducting a compliance audit of a manufacturer, the pre-visit questionnaire indicated that spent solvents from a cleaning operation are accumulated and stored in an aboveground tank at the corner of the property. After a long period, the waste is loaded into a rail tank car and shipped to a recycling facility. The waste is definitely hazardous because of its ignitability, and storage is greater than 90 days; thus, a permit is required. The employees show you a permit and a permit application. You examine inventory records, manifests, inspection records, and other documents. All seems in order until employees are unable to answer your questions regarding leak detection and fire precaution measures. What example Resource Conservation and Recovery Act (RCRA) requirements are at issue and what information as an auditor must you obtain and analyze to evaluate the issue?

Solution
RCRA Issue: Have the employees received proper training? Training records must be checked and staff questioned as to their knowledge of pertinent regulatory requirements.
RCRA Issue: Are leaks covered in the various contingency plans under the titles such as Preparedness Prevention and SPCC? These plans must be reviewed and the issue of secondary containment evaluated.
RCRA Issue: Has the past response to any leaks complied with these plans? Inspection records must be thoroughly reviewed and the rail car loading areas carefully examined for stains.
RCRA Issue: Is the storage in compliance with labeling requirements? The storage area must be inspected and markings on the tank (i.e., ignitability labels) examined.
RCRA Issue: Is the tank properly located? Inspect the tank to determine if it is 50 feet from a property line (a rule for ignitable waste) and away from sparks or other ignition sources.

Data Sampling

Proper environmental management results in too many records for an auditor to check all of them. The auditor, therefore, must sample from the whole in a representative manner. The sample size and the sample itself can have a statistical basis or be based on the auditor's judgment. The method depends upon the objective of the protocol.

If the objective is to define the percentage of a population which is not in compliance, the sample selection must be made in a statistically valid manner reflecting the potential for bias to enter the sampling process. For example, if the population is distributed randomly, a random number table can be used. However, if the population is stratified (e.g., a facility has three separate operations each of which files manifests), the sampling should be stratified accordingly.

If the objective is to discover any violation, the sampling may be concentrated away from subgroups expected to be in compliance. In any event, the auditor should collect the sample and not depend upon the facility to do it. The sampling process and the sample must, of course, be thoroughly documented.

Visual Inspections

After the review of the various documents has been completed, the next step in an audit is to conduct the detailed visual inspection of the constructed systems (e.g., treatment plant, containment devices) and the operating procedures (e.g., routine monitoring). The use of a detailed checklist of environmental inspection items is not recommended. Such a list is unwieldy and voluminous. Besides, it promotes a checklist mentality which experience has shown cannot capture every possible excursion from good management practices (see Secs. 7-4 through 7-6).

A more fundamental approach, open to the full capabilities of human analysis and deduction is needed. The approach starts by considering hazardous materials as well as hazardous waste because hazardous materials, if not properly managed, might lead to some hazard to employees or the environment. In fact, many hazardous wastes started out as mismanaged hazardous materials. In inspecting the use of hazardous materials, the auditor is essentially looking for good management practices, whether required by law or not, which prevent these materials from affecting human health or the environment. In many circumstances, complying with just the specific requirements of regulations will not prevent hazards. For example, a chlorinated solvent storage area is not governed by hazardous waste regulations; however, if the base of the storage area is constructed of asphalt, leaking solvents still could pass through to the soil. Over a period of time it is inevitable that leakage would occur, which would lead to soil contamination.

The safe facility is one that goes beyond regulations and institutes management practices appropriate to the hazardous material being handled, the operation which uses it, and the physical and environmental setting. Therefore, when visually inspecting a facility, the auditor must make two specific sets of considerations: whether the facility complies with regulations and whether the facility uses good management practices. Compliance inspections certainly are important because the potential for fines and other problems in not complying with regulations is obviously great. However, following good management practices in the absence of regulations can be just as important. Many Superfund sites started life as waste disposal operations that had complied with applicable regulatory requirements in effect at the time.

The art of conducting an inspection is a process of recognition and evaluation. The following is a simple summary for an audit that focuses on hazardous waste:

1. Look everywhere for hazardous wastes and hazardous materials.
2. Every time a hazardous waste or hazardous material is recognized, ask:
   a. How is it handled?
   b. Why?
   c. What could go wrong?
   d. Are the necessary precautions in place to prevent a hazard?

Again, compliance audits typically encompass all media and not just hazardous waste. In such cases, inspections go well beyond hazardous waste and hazardous materials. The inspection will cover air pollution control devices, examining them for water discharges (from scrubbers) or solid waste generation (from baghouses). The inspection covers transformers, storm drain inlets, and a host of other features.

Example 6-5   Staining process. When conducting a compliance audit of a manufacturer of sporting equipment, you discover a staining process. Substances to be dyed are dipped into four tanks in the following sequence:

Tank No. 1	Methylene blue dye in alcohol
Tank No. 2	6N HCl in ethanol
Tank No. 3	Distilled water
Tank No. 4	Tap water

The spent dye bath is accumulated in 55-gallon drums, then taken to a non-permitted storage area and shipped off site as a non-hazardous waste. The acid-ethanol mixture is drained to the municipal sewer system, as is the distilled water and tap water.

As an auditor, you must determine if the wastes are hazardous and, if so, is the method for managing the waste in compliance with hazardous waste regulations. What example RCRA requirements are at issue and what information as an auditor must you obtain and analyze to evaluate the issue?

Solution

RCRA Issue: Is the waste properly classified? A review of Material Safety Data Sheets (MSDS) and any test results to identify waste characteristics is required.

RCRA Issue: Assuming that the waste is hazardous, is the satellite storage in compliance with 90-day time limits and other requirements? Inspect the storage area and check inventory records.

RCRA Issue: Is the waste disposed of improperly to the sewer system? Ask the employees to show the written permission from the local sewer authority for this discharge. Obtain a copy from the local authority of their industrial discharge requirements and compare their requirements (e.g., minimum pH) with the waste.

Example 6-6   Catalyst reclaimer. A small facility reclaims waste catalysts, most of which are listed as a hazardous waste under RCRA. In reviewing the accumulation records, some materials are noted that have been on-site for over three years without reclamation. When questioned, the manager says that due to depressed economic conditions, reclamation of these particular catalysts is not profitable. He intends to hold them until the market improves. What RCRA issue does this raise and how would you evaluate it?

Solution. Many so-called recycling facilities have become Superfund sites because they merely stored materials and ultimately went bankrupt. Because of this history, RCRA has specific provisions to preclude "speculative accumulation." Receipts, inventory records, and sales records must be examined to determine if the facility has met the criterion of 70% sales in a year.

Inspection of Storage Facilities

The improper storage of hazardous material and hazardous waste is one of the leading sources of environmental contamination. This section describes how storage operations should be inspected as an example of how all inspections can be approached. A methodical inspection of a storage operation is needed to verify whether storage containers are leaking, containers are properly sealed, drip pans are used with absorbent to immediately soak up spills, and containers are clearly marked so that the contents are quickly identified and characterized in case of an emergency or a spill. Other concerns include checking to see that incompatible wastes are stored separately from each other, any flammables in a confined space have proper sprinkler protection, ignitable material is properly grounded, drains do not represent a ready pathway for release of leaking or spilled contents, and so forth.

Next, the auditor should evaluate the storage area itself: Is it properly secure? Are there signs posted to indicate that it is a hazardous waste or hazardous material storage area? Is the base impervious, with no cracks? Is the material used to construct the base compatible with the types of materials stored in the area? Is the emergency equipment operable and in its proper location? Is secondary containment necessary?

What are the most common sins in storing hazardous materials? The private inspections conducted by ERM during 1988–1990[8] found many examples of not complying with regulations or not employing good management practices. Table 6-3[9] summarizes these examples.

One common problem is the existence of storm drains in liquid storage and transfer areas or drains that connect to wastewater sewers. Almost 20% of the sites inspected had this problem. Another frequent problem is improper handling of empty drums (e.g., regarding drums with 2–3 inches of material on the bottom as empty, and allowing the contents to leak onto the ground). Containers for storage of hazardous

TABLE 6-3
Most common deficiencies in storing hazardous wastes and hazardous materials

Deficiency	Percent of audited facilities having this deficiency
Inadequate containment/drum/tank storage areas	68%
Lack of integrity testing	65%
Lack of or inadequate inspection program for storage	48%
Improper labeling	44%
Inadequate training program	40%
Lack of or inadequate spill prevention, control, and contingency plan	34%
Incomplete wastewater analysis	24%
Lack of or inadequate hazardous waste contingency plan	18%
Inadequate handling of materials containing PCB	18%
Storm drains in liquid storage and transfer areas	18%
Inadequate handling of empty drums	17%

waste were not always in good condition and suffered from bulges and corrosion which can quickly lead to leakage. In many cases, drums containing ignitible waste were not grounded; sparks touched off during container filling and moving or due to static buildup might set the facility ablaze. Many inspected facilities had improper base material or otherwise inadequate containment for the wastes and hazardous materials stored in drums or tanks.

Containers are supposed to be stored completely closed. In many circumstances, employees have left funnels in the top of containers, believing that this would count as closed. However, unless funnels screw into the top and have self-closing mechanisms, this is actually a violation of regulations and is dangerous, because if such a container were tipped, it would immediately leak. Drums are supposed to be stored off the ground on pallets or at least stored stably with adequate aisle space. Because some facilities fail to do this, not all drums can be reached for inspection and leaks cannot be responded to quickly.

The storage of hazardous waste must comply with a fairly complete series of regulations. For example, it is important that waste be properly classified as hazardous or non-hazardous. In many circumstances, facilities have misunderstood the classification system or do not correctly evaluate the waste. States may regulate some waste as hazardous which would otherwise be considered non-hazardous. Consequently, some facilities have put regulated hazardous waste into ordinary trash containers yet have classified some waste as hazardous which could have been properly disposed of as non-hazardous. The auditor should examine the "paper trail" prepared by the facility to record the decisions for classifying a waste and the basis for those decisions whether tests, interpretations, or outside advice. The conscientiousness with which the facility made its determination is an important consideration of the auditor because the answer as to whether a waste is hazardous is not always clear.

Another problem with the storage of regulated hazardous waste is improper labeling. Hazardous waste containers require labels which clearly specify the contents and the start date for accumulation and storage. At least 40% of the facilities privately inspected have incomplete or improperly completed labels on some drums. One of the most common problems is an incorrect or lack of start date for accumulation.

The lack of proper inspection programs for hazardous waste storage is another frequently encountered problem. Hazardous waste regulations require a fairly formalized self-inspection program of storage operations. Typical problems noted in inspections included no programs at all or lack of proper documentation. It is incredible that some self-inspection programs have never recorded errors or any problems at all over a period of years. No storage operation is perfect, and it is highly unlikely that an adequate self-inspection program would go more than two to three months without finding at least some problem that would require an actual correction. In such cases an auditor should review the self-inspection program as documented and walk through the program with the facility employee responsible for the self-inspection, waiting for the employee to identify potential problems. In many cases, the auditor will find that the employee is overlooking important points and simply not doing an appropriate and proper inspection.

Example 6-7   Drum storage facility. One of the conditions listed on a permit for a hazardous waste storage facility states that the facility shall store no more than 2000 drums. What actions should be undertaken at the on-site audit to confirm compliance with this requirement?

Solution. The facility is expected to maintain an inventory of drums with weekly if not daily records. The inventory sheets need to be examined and the data analyzed for randomly selected days to ensure the total number of drums stored on those days is less than 2000. The employee who conducts the inventory needs to be questioned to ascertain if his/her techniques are adequate. The auditor should count or estimate the drums actually in storage on the day of the audit and compare this to the current inventory sheet.

6-4   EVALUATION AND PRESENTATION

The audit data are evaluated during two periods while still at the site. The first occurs in the form of immediate feedback as the auditor collects information. The second occurs as the audit team performs a comprehensive analysis upon completion of all physical inspections, interviews, and document reviews. The evaluation of data is such that the findings for the subsequent written report are developed before the audit team leaves the site.

Immediate Feedback

As an auditor gathers information, it is important to provide immediate and on-going feedback to facility personnel. This involves communicating frequently with the personnel and identifying positive points and exceptions when found. Audits should not be conducted in an adversarial manner. The auditor should use the opportunity to explain the violation/exception to the person in charge of that situation so that he or she is able to learn the proper procedure and correct a problem. Even if facility personnel are able to correct a problem immediately, the auditor still should record the discrepancy as well as the response. An experienced auditor should consider internal procedures when reviewing documents, inspecting facilities, and interviewing personnel, specifically, whether the procedures lead to proper types of operations.

Although feedback should be immediate, it is important not to draw hasty conclusions. Frequently, as an auditor gathers more information, the initial conclusion may prove incorrect or there may be a reasonable explanation. This is particularly true for conclusions drawn only on the basis of a single interview. Facility personnel may contradict each other regarding procedures, thus requiring the auditor to sort out the differences. The auditor should not only acknowledge the rationale provided by a facility employee for an exception, but listen to it carefully.

Working Papers

The evaluation of the data collected as part of an audit is a major and complex effort that needs to be completed in a compressed period of time while still at the site. To facilitate this evaluation, it is critical that the extensive notes from the interviews, document reviews, and inspections be organized carefully. Most auditors refer to the organization of these notes as **working papers**.

Working papers are a method for ensuring that all the steps have been taken and for organizing the information obtained from each source. They explain the rationale used for checking various items, and provide a basis for review by the audit team leader, or for others who may follow, as to the source of information and how supportive it is of the conclusions.

An auditor should include the following information in working papers—background information, all descriptive information gained from interviews and inspections, any analysis and supplementary information used to reach any conclusions. The information is documented in the working papers as collected in the form of field notes. When recording field notes, several factors are critical:

1. Write legibly—transcribe your field notes in such a way that you can clearly identify what your thoughts were and what you gained from each interview or each step of your evaluation process.
2. Initial and date all pages.
3. Document the source of information.
4. Relate the information to the individual parts of the protocol which require such data.
5. Summarize findings and draw conclusions even if they are only on an interim basis.
6. Identify other questions prompted by an interview or inspection that require follow-up.

The purpose of these working papers is to provide an exhibit which supports all conclusions, whether positive or negative. Working papers are a factual written record that is made to support final conclusions. They should not contain hearsay statements (i.e., when the individual being interviewed provides information he or she has heard from others but does not know to be true based on personal observation). When such statements raise serious questions, interviewing individuals who do have knowledge should follow up to obtain the facts to address the compliance issues.

After the working papers are completed, they should be read to make sure they would make sense out of context. At some point in the future, someone else may have to refer to them. Some large and complex facilities may have such extensive data to warrant a specialized database to track all information gathered.

Comprehensive Evaluation

The comprehensive evaluation is an integrated analysis of all information, in which all members of the audit team participate. The team as a whole reviews each step of the protocol, ensuring that each step is completed and that appropriate data-gathering techniques have been used. For example, if data gathering has required taking a sample (e.g., selecting and reviewing 100 of the 1000 manifests generated by the facility during a year), the team makes sure that the individual auditor selected an appropriate sample which correctly represents the whole.

Having reviewed each protocol step for completeness, the individual auditors should draw appropriate conclusions for their areas of the protocol and summarize each exception and observation supporting that conclusion. After pulling together all findings and exceptions, the team should critically review each of the findings: Did the auditor talk to the correct employees? Have all the equipment or other items that might apply been identified? Is the regulatory requirement fully understood by the auditor? Each finding must be carefully examined by the entire team to ensure that it is substantiated. The team should develop a complete list of audit exceptions and observations, then organize and summarize the list to group common exceptions and to look for patterns and trends. Team interaction is vital to stimulate the level of thinking necessary to establish patterns and to identify causes.

Exit Interview

Having completed the comprehensive evaluation, the audit team conducts an exit interview with the key employees of the site. One of the common pitfalls in an exit interview is focusing only on the deficiencies and failing to complete each step of the process. To prevent this, the audit team should prepare a discussion sheet covering all points to be addressed at the exit interview.

Typically, the exit interview will be conducted with the same group, if not an expanded group, that the audit team met upon arrival at the facility. The exit interview must allow time for full discussion; it must address all findings no matter how minor. A report will eventually follow and it should not contain any surprises.

When conducting the exit interview, the audit team should set the stage. It is important to acknowledge the cooperation and the aid that the facility has provided. Rapport can be established by a willingness to discuss all the issues thoroughly and to listen to any explanation provided by facility personnel. The exit interview goes through the findings one at a time, making sure to avoid any personal attack, instead discussing methods that might be used to solve the problems. Nevertheless, a philosophy of zero tolerance for problems, however minor, should prevail. Frequently, conflicting opinions can occur. If this happens, the audit team and the facility personnel should talk the dispute out completely and document how the conflict has occurred and how to resolve it.

The final topic of the exit interview is describing the next step in the audit process—the report. The team defines how it will report the results of the audit and discusses the specific format of the report. The team should also explain how facility personnel will be able to challenge and to explain their responses to the findings of the audit team.

Audit Report

After departing the facility, it is important to develop the audit report quickly. In developing the report and report format, style is important. The following guidelines should apply to any audit report:

- Be clear and concise.
- Be sure to distinguish between isolated incidents and chronic problems.
- State the facts as discovered.
- Do not draw unsubstantiated conclusions: limit findings to the facts.
- State the nature of the problem clearly and exactly without using generalities.
- Give regulatory or management practice citations, very clearly citing where requirements have not been met; however, be very careful not to draw legal conclusions. **In particular, do not state that there is a violation.**

Two examples of an acceptable writing style are as follows:

1. A written waste analysis plan is required (40 CFR 265.13(b)(1–4)) for all hazardous wastes treated on site or stored on site. The facility's plan does not include the following two requirements listed in the regulation:
   a. Rationale for selection of parameters to be analyzed—parameters are listed but not justified.
   b. Frequency of review and repeat analysis—none is specified.
2. In reviewing a random sample of pH monitoring records for 15 days of the previous year for discharge #1, three excursions were identified which exceeded permit limits. We were not able to confirm that these incidents were properly reported.

An audit is an **improvement tool**. Therefore, the audit team should present recommendations in addition to a set of findings. It may be appropriate to present recommendations in a separate report so that they could be protected under the attorney work product privilege. It is helpful to present any recommendations in a "value-added format"—one that will provide the foundation for implementing corrective actions and ultimately designing an efficient, manageable facility compliance system. For each negative finding, the report should provide:

- A description of the specific requirements,
- Potential liabilities associated with the negative finding, and
- A recommended corrective action.

Although the recommendations should clearly outline the type of action needed to resolve any negative findings, they should not represent complete work plans. A typical example would start with a finding, identify the requirements, estimate the potential liability, and present a recommendation as follows:

Finding: Incomplete RCRA employee training records

Requirements: RCRA requires that employees receive annual training for emergency procedures and the hazards associated with handling of hazardous wastes. RCRA requires the facility to maintain a copy of the facility's training plan and to maintain detailed training records for a minimum of 3 years.

Liability: Injury of uninformed employee; EPA fine

Recommendation: Review facility's hazardous wastes and develop a training program that addresses potential hazards; incorporate facility's standard emergency response plan into a training document; designate a trainer; develop a record keeping system; update Material Safety Data Sheets (MSDS) system.

Finally, the audit report and its distribution must adhere to the disclosure requirements established early in the process (e.g., confidentiality or use of attorney work product privilege). Even if confidentiality is required, the auditor must assume that the report may be made public at any time because this does happen. An uninformed public can easily misinterpret results or take something out of context. A subsequent investigation may use the report as evidence. Therefore, it is important to be candid, but circumspect.

Action Plan

It can be seen that an audit program helps to troubleshoot potential violations of regulations and excursions from good management practices. However, without an action plan to address the problem and its cause, the next audit may find the same shortcoming. Therefore, after an audit report is completed, the facility (not the audit team) must develop an action plan—a written response to the audit that specifies a clear set of actions and tasks with deadlines. It also clearly identifies the persons responsible for implementing each action. A system of tracking deadlines to see that they are met should be established. Follow-through on audit findings is very important. If a regulatory agency discovers that a facility has conducted an audit and did not attempt to resolve the findings, the agency's response will probably be much harsher than if the audit had never taken place.

In summary, the goals of environmental compliance audits are:

- To obtain a clear understanding of the environmental status, risk and liability of a facility;
- To allow timely, proactive correction of potential sources of problems within the budget process rather than having to respond to deficiencies uncovered by regulatory agencies or emergencies.

The audit brings an increased attention to environmental issues, prompting facility personnel to pay more attention to these issues, and elevating the issues to the level of consciousness and concern that they deserve. With a good audit program, a facility should experience a decreasing pattern of serious discrepancies. This will provide the facility and corporate management increased confidence in the facility's control of environmental issues, including anticipating new issues as new regulations occur.

6-5 OTHER TYPES OF AUDITS

Other types of audits include management audits, waste management contractor audits, property transfer audits, and waste minimization audits.

Management Audits

The traditional audit evolved as a method to verify that a facility has instituted adequate procedures, dedicated sufficient resources, and installed the necessary systems to comply with environmental regulations. It is a mere "snapshot" of compliance. Yet, the overall environmental objective is compliance on a continuing, long-term basis. This is a function of organization, guidance, controls, and communications. Each of these functions is the subject of a management audit whose purpose is not the identification of specific violations but the detection of activities likely to experience problems and their root causes.

Specifically, a management audit addresses a number of interpretive questions under each of several broad issues as follows:

1. Clearly defined responsibilities—How are they established and communicated? How are assignments reinforced? Are any assignments overlapping, shared, or conflicting?
2. Adequate system of authority—How is it granted? Who grants exceptions? How are authorizations and exceptions recorded? Is there a real understanding of the issues?
3. Division of duties—What potential exists for conflict of interest?
4. Trained and experienced personnel—What methods are used to determine who needs training? What training have personnel had to help them understand and complete their duties?
5. Documentation—What records are developed? How are records maintained? Do type and amount of documentation correspond to the importance of the activity? How are exception reports developed? Are the records reliable? Accurate?
6. Internal verification—How does management review and evaluate environmental success? What is the accountability of managers for adherence to policies and procedures?
7. Protective measures—What engineered controls are in place to prevent noncompliance? What procedures?

A management audit detects areas likely to experience compliance problems and, more importantly, the root causes of these problems. A key component of any management audit is an analysis of how organizational resources could be used more effectively to assure compliance.

Example 6-8   Training programs. How would an auditor, in conducting a management audit of a California facility handling oil, hazardous waste, and hazardous materials, evaluate the adequacy of training programs?

Solution. First, the auditor must review the regulations which apply to the facility and define their training requirements (see Fig. 6-3). The auditor would then review the jobs and training records of selected employees who perform tasks covered by these regulations to ascertain data such as the following:

1. Besides meeting the basic requirements, does the level of training match the degree of hazard posed by the various jobs?
2. How are employee training needs identified on an annual basis? After changing positions?
3. How is employee knowledge tested?
4. Are employee records kept properly, and where are they located?

Waste Management Contractor Audits

Some generators ship a significant amount of hazardous waste off site for treatment and disposal by private contractors. A fundamental principle of hazardous waste law in the United States is that of joint and several liability for management of hazardous waste—the liability for remediation of problems created by a waste treatment and disposal contractor can rest with the generator (see Sec. 2-1).

When remediation is required, it can cost several orders of magnitude greater than the originally contracted cost for the treatment and disposal of the waste. In addition, generators may also be financially responsible for damages to third parties derived from waste transport, treatment, and disposal operations. As a result of these potential liabilities, hazardous waste generators exercise great care in the selection of waste contractors. One tool to use in the selection of contractors, and in the verification that the selected contractor continues to operate in a prudent manner, is to audit the contractor's facilities.

A waste contractor audit follows closely a facility compliance audit. It uses the same three phases (e.g., pre-audit planning). The compliance issues are the same and it places the same emphasis on document review, interviews, and inspections. However there are some key differences. A waste contractor audit extends beyond compliance to liabilities (e.g., potential for damages to third parties and potential for remediation). A waste contractor audit also analyzes the financial strength of the company which owns the facility; a company possessing sufficient assets and insurance probably can absorb the costs of remediation without having to pass them to generators.

Another difference between compliance and waste contractor audits is the manner in which the on-site interviews and inspections are conducted—a waste contractor audit typically provides little ongoing feedback to facility personnel. Unlike compliance audits, improvement is not the primary objective. The objective is to identify risks, typically on a basis comparable with other facilities. Hence, a final difference between compliance and waste contractor audits is that the latter may tend to apply uniformly a detailed worksheet type of protocol in which the answers to a long checklist may possibly be scored.

Property Transfer Audits

Transactions involving real property can convey the liabilities associated with contamination present in the soil or ground water. Lenders, too, place their secured capital

Regulatory training requirements for Example 6-9.

| | **Regulations** | | | |
Requirements	CWA Spill prevention control and countermeasure plan	RCRA Contingency plan and emergency procedures	OSHA Hazard communications	California business plan requirements
Who must be trained?	Persons who operate or maintain equipment that may discharge oil	Persons who handle hazardous waste	Employees exposed to hazardous substances in their work area	Employees responsible for responding to hazardous emergencies
Who can do the training?	A designated person who is accountable for oil spill prevention	A person trained in hazardous waste management practices	A person familiar with workplace hazards and self-protection measures	A person with a higher level of training than the level being taught
What must be included?				
Proper hazardous materials handling procedures	✓	✓	✓	✓
Applicable laws, rules, regulations, and rights	✓	✓	✓	✓
Operation and maintenance of routine equipment	✓	✓		✓
Proper use of emergency equipment, procedures, systems, and plans		✓	✓	✓
Proper waste management procedures			✓	✓
Methods for recognizing the release of hazardous chemicals			✓	✓
Physical and health hazards in the work area			✓	✓
Location and use of chemical and hazard information (labeling, MSDSs, plans, etc.)			✓	
When must training occur?	When person begins job; then periodically	Within 6 months after employment or new job; then annually	At time of initial assignment; then whenever new hazard is introduced	Annually; schedule specified in Business Plan
What records must be kept?	None required; however, documentation is a good management practice	Written job title & description; written description of the type & amount of training to be given; records documenting that training was completed	Written description of the type and amount of training that will be given; records documenting that training was completed	None required; however, documentation is a good management practice

FIGURE 6-3

at risk if collateralized property is contaminated. Certain states require audits prior to the transfer of industrial property (e.g., New Jersey's Environmental Cleanup Responsibility Act of 1984 [ECRA]).

The liability for the costs of remediating a contaminated property has surprised many landowners, tenants, and lending institutions as illustrated in the following real examples:

1. A purchaser of land finds that the property was once used for a landfill; the purchaser must pay $6 million to clean up ground water in the area.
2. A tenant moves from a site which they had used for 6 years for an electronics manufacturing operation. Eight years later, ground water contamination is discovered. The owner successfully sues the tenant to pay for cleanup, claiming that the manufacturer was the "only tenant likely to have used a sufficient quantity of the chemical known to have caused the contamination."
3. A bank lends money to a company. When the business started to have financial difficulties, the bank stepped in and ran part of it. The bank ultimately foreclosed. Subsequently, the property was found to be contaminated. The bank was found to be liable for a major part of the remediation. The cost of remediation was twenty-five times the amount of the original loan.

Because the value of property can be greatly diminished by the potential for having to remediate the site, the need to know the potential liabilities is crucial to both the buyer and seller of a property. In fact, the costs of removal of asbestos insulation or remediation of soil and ground water contamination resulting from historic uses can greatly exceed the "pristine" value of a property. Lending institutions may lose the property's value as collateral and possibly become liable for the cleanup costs upon foreclosure. For these reasons, all prudent buyers of property and virtually all commercial lending institutions require an evaluation of environmental risks before completing the mortgage transactions. In fact, property transfer audits have become a routine part of commercial real estate transactions. The purchaser of a property may establish an "innocent landowner defense" if they conduct appropriate inquiries at the time of purchase that would establish that they had "no reason to know" that site contamination existed prior to the purchase. A protocol for conducting a property transfer audit has been developed and adopted for use by prospective purchasers of commercial real estate, and adherence to this protocol may minimize the purchaser's liability, under a number of laws including CERCLA, should prepurchase site contamination be identified in the future.

Example 6-9 Property transfer audit. What parties may require a property transfer audit and why?

Solution. The buyer and seller need to know the liabilities in negotiating a price, and the lender must assure that the value of the property as collateral is free of major environmental liabilities. A tenant or landlord should evaluate liabilities, both upon coming onto a property and upon leaving it, to settle any future claims. Finally, a regulatory agency as part of its objective to protect the environment may bar the transfer of a property to its new owner until an audit is completed and the site remediated.

Elements of property transfer audit.   A property transfer audit differs in two key ways from a compliance audit. It is usually conducted in a phased approach, and it focuses on historic practices. The largest liabilities typically derive from historic rather than current practices, and defining liabilities encounters much uncertainty because of difficulties in detecting such problems. For example, unrecorded waste disposal operations, use of contaminated fill material, and underground storage tank leaks can easily have occurred decades ago at industrial properties. Such activities are much more difficult to discern than a determination of whether a facility has all necessary regulatory permits. Further, the emphasis to complete the transaction quickly makes time a critical factor, compressing the schedule to a matter of two to four weeks for larger transactions and often just a few days for smaller ones. The combination of tight schedules and difficult-to-define liabilities clearly reduces the certainty of the findings of a property transfer audit.

The decided advantage of having full knowledge of such liabilities during negotiations over the selling price will compel some parties to seek a thorough audit. The level of detail to which such an audit is conducted depends upon the characteristics and history of the site and time constraints set by the transaction schedule. Underdeveloped or residential property obviously requires a lower level of scrutiny than industrial facilities. For the latter, a thorough audit will include the following issues:

1. Soil/ground water contamination.

2. Presence of hazardous waste and hazardous substances.

3. Waste discharges to receiving waters.

4. Proper permits and compliance.

5. Presence of underground tanks.

6. Special issues (e.g., asbestos, radon).

Emphasis on ground water and soil contamination.   Of these previously discussed issues, identifying whether the soil and ground water are contaminated and defining the extent of such contamination will require the greatest effort. While a visual inspection of a property and its buildings, manufacturing processes, and hazardous material and waste handling procedures may give some indication of contamination, this is not considered the primary source of information. Instead, it is necessary to conduct an in-depth review of historical usage of the property to include specifically:

Past chemical usage, particularly hazardous substances and materials and how they were stored and handled;

Past waste management and disposal practices;

- Past spills, leaks, or pollution incidents;
- Past wastewater sewer lines, lagoons, leach fields, underground tanks; and
- Past property development involving fill material.

The buyers and sellers of property frequently have not disclosed the proposed transaction to employees at the time of an audit, severely restricting the use of employee interviews to obtain this information. Therefore, the auditors must compile and analyze the following types of data which typically are not addressed as part of a compliance audit:

1. Past aerial photographs.
2. Engineering archives (e.g., construction drawings).
3. Public records regarding enforcement and inspection activities on the site and adjacent properties by federal, state, or local environmental officials.
4. Title searches for utility rights-of-way and previous owners.
5. Review of federal and state lists and databases of contaminated sites reported for the property or nearby areas.
6. Interviews with regulatory agency personnel with inspection authority.
7. Any existing site studies.

An audit should also address the surrounding properties if the possibility exists that their historical activities may have resulted in ground water contamination.

Another difference between property transfer audits and compliance audits is that the former are usually conducted in a phased approach. The preceding steps are usually considered the first phase. Each successive phase, if needed, is done with increasing rigor to increase confidence that potential liabilities have been identified and evaluated as shown in Fig. 6-4.

For example, the first phase may include soil gas screening. If the results of the first phase show some indications of possible contamination, the audit would progress to a site investigation to include appropriate soil and ground water sampling as discussed in Chap. 15. The results of the site investigation provide a basis for estimating probable remedial actions and their costs in subsequent phases.

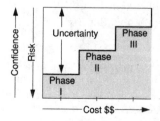

FIGURE 6-4
Phased property transfer audit.

Related Audits

The type of audits performed for property transfers is done for two other cases:

• Mergers, acquisitions, and divestitures
• Financial reporting

The full costs of environmental management are often not represented on the financial statements of target acquisitions. Acquired companies can saddle the parent with unexpected liabilities. The costs of site remediation, required capital improvements, and fines can dramatically undermine the value of the acquisition if not factored into the transaction in the early stage. A due diligence audit addresses this concern by investigating the following:

• **Environmental Compliance.** What are the liabilities of violations and the engineering, legal, and capital costs of bringing a facility into compliance?
• **Ability to Expand Production.** Are there environmental limits or constraints, such as local air pollution restrictions and permit requirements, that will prevent a company from expanding production levels?
• **Liabilities for Remediation.** To what extent could remediation be required in the future?

Financial liabilities arising from environmental non-compliance and site remediation frequently represent material exposures requiring disclosure under public securities laws in the United States. The obligation for public companies to report such financial exposures may arise from specific environmental disclosure requirements adopted by the U.S. Securities and Exchange Commission, financial statement disclosures within the context of accounting for financial contingencies, and general accounting standards. The net effect is an obligation to disclose environmental-related exposures which may have a material effect upon the capital expenditures, earnings, and competitive position of the corporation or its subsidiaries. An audit will identify current, contingent, and future environmental exposures; anticipate the magnitude and timing of the costs; and provide a substantiated basis for adequate disclosure.

Waste Minimization Audits

Waste minimization is the reduction of any solid, liquid, or gaseous waste at its source of generation. It is a noble goal and certainly the preferred method for managing all waste. Accordingly, this book devotes the entire next chapter (Chap. 7) to the subject of pollution prevention. The first steps to successfully achieving the goal of waste minimization is an audit, consisting of planning the entire program and fully characterizing all wastes. From there, waste minimization options are identified, evaluated, selected, implemented, monitored, and communicated.

DISCUSSION TOPICS AND PROBLEMS

6-1. Prepare a protocol and conduct a compliance audit of one of the laboratories at your university.

6-2. You are auditing a refinery site to determine the likelihood of a release from its existing tanks and pipes. The site has more than 150 tanks, 75 miles of pipes, and more than 2000 valves. You have one day to inspect this part of the site. Devise and justify a sampling plan to determine the condition of the tanks, pipes, and valves.

6-3. Please correct or state what is wrong with the following examples of bad audit report writing:

  (a) The audit team inspected 20 transformers and 10 capacitors of which 3 showed signs of leaks.

  (b) Several of the drums at the hazardous waste area had no labels, as required by 40 CFR 262.

  (c) The fire training area has been contaminated by waste solvents. The audit team recommends that the facility excavate the contaminated soils.

  (d) In a review of the manifests generated by the site, 20 were found to be incorrectly completed.

6-4. You have been asked to audit PCB compliance at a very large site containing 65 PCB transformers or capacitors and 110 non-PCB transformers. They are spread out over the site. You wish to inspect to determine if the transformers are properly labeled as PCB.

  (a) How would you select your sample?

  (b) If three of the first five transformers you checked were not labeled (and should be), would you change your plan?

6-5. Review the conditions listed below for a hazardous waste storage permit and indicate the specific items you would expect to review and analyze to determine compliance with the permit. Please provide your methods by individual permit condition.

  (a) The facility shall store no more than 5000 drums.

  (b) The drums shall be segregated by waste type in a manner so that any spillage of incompatible materials will not mix.

  (c) For the flammable materials stored in drums, proper fire protection methods will be used.

  (d) No drum shall be stored longer than 45 days.

  (e) Inspections for spillage, leakage, and poor drum quality will take place daily. Repairs will be recorded.

  (f) No explosives or reactives may be stored for more than 10 days.

6-6. What is wrong with the following statements:

  (a) "OK, we'll do an audit program, but we can't give a copy of the audit report to the plants!"

  (b) "We don't need a spill plan. We don't have any spills!"

  (c) "Well, how good did we do? What was my score?"

  (d) "We send the solvent off to a recycler so they can reuse the hazardous waste. The sludge in the degreaser? Oh, that just goes in the dumpster."

6-7. Electrical transformers are often oil-filled, sometimes with PCB-containing oils. A common occurrence is the location of transformers near storm drains where a transformer leak or fire could result in spillage to the storm drain. What might the plant do to correct this deficiency?

6-8. Hazardous waste container storage areas must be designed to provide secondary containment. How should a facility address and handle rainwater which collects in the diked areas? One facility opened the valve and let it drain out. Is this permissible?

6-9. When asked if a facility had a copy of its air permits, a facility representative answered, "Yes," claiming that the facility was in full compliance. When the facility was asked if it had a copy of the air permit application referenced in the issued permit, a facility representative said no one had any knowledge of such a document. Why is it necessary for the plant to have the air permit application even when the facility has been issued a permit?

6-10. A facility which operates in New Jersey services vehicles and equipment, changes the oil, and repaints them. During the audit, the facility manager stated he checked the EPA regulations, and, based on this information, none of the waste which was generated at the facility was hazardous. What is the deficiency in this approach, and what else should he do to arrive at a correct waste identification determination?

6-11. One facility had undergone some improvements in the process which it had used for 30 years to manufacture the same product. When asked about the air permits for the improvements, the facility manager stated, "The raw materials were the same, the products were the same; we just moved some equipment around. And besides, we're grandfathered." What is incorrect about this assumption?

6-12. A municipal sewage treatment plant produces a sludge containing toxic metals at concentrations in excess of TCLP limits. After dilution with stormwater runoff, the concentration is less than the TCLP limits. Is the sludge a hazardous waste?

6-13. A light manufacturing facility produces 125,000 gallons per week of sanitary sewage and uncontaminated wash water which are discharged to a municipal sewer system. A small parts rinsing operation flushes about 12 ounces per week of trichloroethylene down the sink. Does the facility have to manage their wastewater as a hazardous waste (see 40 CFR 261.3)?

6-14. Prepare a yes/no checklist for auditing a facility's contingency plan/preparedness prevention plan (see 40 CFR 265.50-265.56).

6-15. What would a waste management contractor audit cover in its scope that a compliance audit would not?

6-16. Describe what is meant in Ex. 6-2 when the consultant is described as having no memory.

REFERENCES

1. Cahill, L. B., and R. W. Kane: *Environmental Audits*, Government Industries, Inc., Rockville, Maryland, October 1989.
2. Anon.: "The Exploding Market for Environmental Audits," *Business Week*, pp. 92–96, June 1983.
3. *Federal Register:* "Environmental Protection Agency's Environmental Auditing Policy Statement," vol. 51, no. 131, July 9, 1986.
4. ERM Computer Services, Inc., Exton, Pennsylvania, 1991.
5. *Federal Register*: "Environmental Protection Agency's Environmental Auditing Policy Statement."
6. Frost, E. B. and S. Seigel: "Environmental Audits: How to Protect Them from Disclosure," *Toxic Law Reporter*, p. 1211, February 17, 1991.
7. Reed, P. D.: "Environmental Audits and Confidentiality: Can What You Know Hurt You as Much as What You Don't Know?" Environmental Law Institute, p. 303, October 1983.
8. Roig, Randy, ERM-West, Inc: Personal communication, August 1990.
9. Source: Audits of private facilities conducted by the ERM Group.

ADDITIONAL READING

Blakeslee, W. H., and T. M. Grabowski: *A Practical Guide to Plant Environmental Audits*, Van Nostrand Reinhold Co., New York, 1989.

Friedman, F.: "Why Not to be a Manager? Is This Job Really Worth It?" *Environmental Forum*, p. 20, May/June 1991.

Friedman, F. B.: *Practical Guide to Environmental Management*, Environmental Law Institute Monograph, 1990.

Frost, E. B.: "Voluntary Environmental Compliance Audits: A DOJ Policy Failure," *Toxic Law Reporter*, BNA, p. 499, September 18, 1991.

Greeno, J. L., et al.: *The Environmental Health and Safety Auditor's Handbook*, Arthur D. Little, 1988.

Moore, J., et al.: "Why Risk Criminal Charges by Performing Environmental Audits?" *Toxic Law Reporter*, BNA, p. 499, September 18, 1991.

U.S. Department of Justice: "Factors in Decisions on Criminal Prosecutions for Environmental Violations in the Context of Significant Voluntary Compliance or Disclosure Efforts by the Violator," Washington, D.C., July 1, 1991.

ACKNOWLEDGMENTS

The authors wish to acknowledge the following individuals who assisted with the preparation of this chapter: Dr. Randy Roig, *ERM-West, Inc.*; Brian Bennett and Ann Sirois, Esq., *ERM, Inc.* The principal reviewer for the **ERM Group** was Howie Wiseman, *ERM-Northeast, Inc.*

CHAPTER
7

POLLUTION
PREVENTION

Conspicuous consumption of valuable goods is a means
of reputability to the gentleman of leisure.
The Theory of the Leisure Class, 1899, Thorstein Veblen

In the previous chapter, methods were presented to identify environmental problems at the origin of waste production. This chapter examines practices that reduce or eliminate hazardous waste at this source.

7-1 GENERAL CONSIDERATIONS

Definitions

Pollution prevention consists of all those activities that reduce the generation of hazardous waste. Many terms are used to describe these activities: waste minimization, waste reduction, source reduction, waste diversion, pollution prevention, recycling, and reuse.

An overview of techniques used to reduce pollution is illustrated in Fig. 7-1. This chart provides a clear establishment of priorities in hazardous waste management, ranging from changing products to ultimate disposal of hazardous wastes that cannot be avoided.

In a report to Congress, the U.S. Environmental Protection Agency (EPA) defined waste minimization as the reduction of volume or toxicity of waste:

... the reduction, to the extent feasible, of hazardous waste that is generated or subsequently treated, stored, or disposed of. It includes any source reduction or recycling

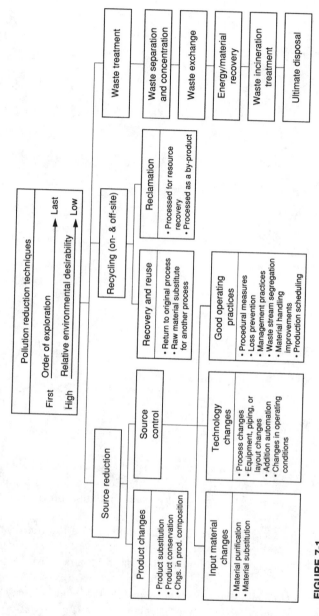

FIGURE 7-1
Pollutant reduction techniques.

356

activity undertaken by a generator that results in either (1) the reduction in total volume or quantity of hazardous waste, or (2) the reduction of toxicity of hazardous waste, or both, so long as the reduction is consistent with the goal of minimizing present and future threats to human health and the environment.[1]

In a recent policy statement,[2] EPA suggested the following hierarchy for management of wastes:

- Source reduction
- Recycling
- Treatment
- Disposal

There is considerable controversy as to what activities should be considered as prevention, in part due to recent U.S. requirements that generators demonstrate that they are doing everything possible to reduce waste. The congressional Office of Technology Assessment defines waste reduction as:

In plant practices that reduce, avoid or eliminate the generation of hazardous waste so as to reduce risks to health and the environment. Actions taken away from the waste generating activity, including waste recycling or treatment of wastes after they are generated, are not considered waste reduction. Also, an action that merely concentrates the hazardous content of a waste to reduce waste volume or dilutes it to reduce degree of hazard, is not considered waste reduction.[3]

In this chapter the term **waste minimization** applies to any management technique or process modification that ultimately reduces the mass or toxicity of waste sent on to treatment, storage, and disposal facilities. The somewhat broader term **pollution prevention** has an evolving definition that includes:

- Managing chemicals to reduce risk
- Identifying and estimating all releases
- Waste minimization

Historical Overview

A common perception is that pollution prevention efforts began in the 1980s. Actually there are numerous examples from the nineteenth century of industrial practices that reduced waste generation. At the time it was characterized as simply good engineering economics to create a saleable product rather than a waste or to reduce the cost of raw materials. The early twentieth century contains numerous examples of waste reduction efforts.[4] A book on waste utilization was published in London in 1915.[5] The technical literature of the mid-twentieth century includes many articles on waste reduction, such as those by Trembler[6] and Besselievre[7] to cite two examples. In a

1964 paper on industry's approach to pollution abatement, Rocheleau and Taylor of E. I. du Pont de Nemours & Co. wrote:

> Emphasis has been placed on in-plant control measures, because operation of a wastewater treatment works is only part, often a small part, of industrial control problems. . . . It is now standard policy with nearly all chemical companies to apply effective wastewater controls as soon as possible in the development of a new or improved process. If a potential waste problem can be recognized early enough, it can often be prevented from becoming a problem, or at least it can be minimized.[8]

It is somewhat ironic that industries that were early pioneers in waste reduction are now being penalized for being too forward thinking. As an example, "save-alls" were originally installed in paper mills in the 1940s as a way of recovering fibers, and preventing them from becoming a source of pollution. They became so effective and widespread that they are now considered part of the manufacturing process, and no credit would be given today to this device as a waste minimization technique, because it comes as a standard part of paper manufacturing equipment.

What has changed today are the dramatic increases in the cost of disposal, legal and regulatory incentives, and public demands that industry simply stop producing hazardous wastes.

Economic Incentives

It is common practice today to think of wastes simply as resources out of place. It is often the case that a second look at the manufacturing process can identify additional ways that wastes can be reduced, sometimes at little or no cost. A frequently cited example is the 3M Company's 3P (Pollution Prevention Pays) program. A recent article reports the following amounts of waste eliminated due to their source reduction programs:

Waste type	Eliminated	(U.S. only)
Air pollutants (tons)	121,000	(110,000)
Water pollutants (tons)	14,600	(13,500)
Sludge/solid waste (tons)	314,000	(303,000)
Wastewater (10^6 gallons)	1600	(1000)

Since its inception in 1975, "The processes and product changes made to achieve these reductions have yielded cumulative first year savings of $420 million (U.S. savings: $350 million)." These savings were achieved in such diverse ways as savings on capital costs of unneeded pollution control facilities and sales from new products from former wastes.[9]

Several changes in the economic environment justify a second look at the way a manufacturer handles waste. The most dramatic is the sharp increase in the cost of hazardous waste disposal. A recent report from DuPont suggests that the cost of managing a pound of hazardous waste is escalating at a rate of 20–30% annually.[10]

Product design	Raw materials	Manufacture, sales, and distribution	Packaging	Use	Final disposal
• Chemical usage • Ease of recycling • Ease of reuse • Ease of disassembly (e.g., modular design) • Size	• Use of renewable resources • Reducing impact from transport	• Pollution prevention, waste minimization • Process hazards management	• Reduced packaging • Vegetable-based inks • Concentrating product	• Aerosol-free cleaning agents • Low-dosage pesticides	• Recycling • PET versus HDPE

Management and organization	Cross-functional management strategies, integrating environmental expertise into decision-making processes
Technologies	Creating new technologies that are environmentally sound

FIGURE 7-2
Adding value from an environmental standpoint: competitive advantage.

Recycling proposals that were economically infeasible only a few years ago now look very cost-effective.

An important aspect in evaluating the economics of environmentally desirable proposals is to examine the "big picture" and not simply focus on micro-economics at the plant level. This is illustrated in Fig. 7-2, which examines the competitive advantages that can accrue to corporations from waste minimization activities. This big picture approach is discussed in detail in Sec. 7-3, Life Cycle Analysis.

Legal and Regulatory Incentives

A second motivation for conducting waste minimization evaluations in the United States is that they are required by law. In the 1984 amendments to RCRA, congress defined as national policy that "... the generation of hazardous waste be reduced or eliminated as expeditiously as possible." In order to ship hazardous waste to a disposal facility, generators are now required to certify that they have a waste minimization program in place. While this requirement lacks a strong method of enforcement, new legislation including Pollution Prevention and Air Toxics legislation and RCRA re-authorization are expected to include regulatory "hammers" that will force recalcitrant industries into waste minimization programs.

Perhaps the most effective legislation does not deal with waste minimization at all. Under the Right-To-Know provisions of SARA (see Sec. 2-4), industries are required to furnish information on toxic waste releases to the local community, providing data that was not generally known before. Not surprisingly, many communities have reacted negatively to the knowledge that a local plant is releasing toxic compounds into the air that they breathe.

The 1990 Pollution Prevention Act in the United States was designed to reduce the generation of waste by a range of institutional measures including:

- Establishing a national pollution prevention strategy
- Providing grants to states to establish programs to promote source reduction
- Establishing a source reduction clearinghouse

This Act requires industries to provide detailed annual reports on the effectiveness of source reduction for all chemicals that they must already report under the community right-to-know provisions of SARA. Because these reports will become public information, this Act provides another strong incentive to develop an effective source reduction program.

In February 1991, EPA instituted a voluntary 33/50 program which had an overall national goal of 50% reduction in the generation of 17 high-priority toxic chemicals by 1995. The program has an interim goal of 33% reduction by the end of 1992. The 17 chemicals listed in Table 7-1 accounted for 1.4 billion pounds of releases and transfers within the United States in 1988.

Another legal inducement is the potential future liability for any hazardous wastes produced. As pointed out in Chap. 2, an industry that generates a hazardous waste is responsible for that waste forever. Because these future liabilities are of unknown magnitude, many industries do not want to ship any wastes to what may become future Superfund sites. Finally, the cost of insurance, when it is available at all, has risen dramatically.

TABLE 7-1
Target chemicals covered by EPA's 33/50 program[11]

Chemical	Quantity released in 1988 (million pounds)
Benzene	33.1
Cadmium	2.0
Carbon tetrachloride	5.0
Chloroform	26.9
Chromium	56.9
Cyanide	13.8
Dichloromethane	153.4
Lead	58.7
Mercury	0.3
Methyl ethyl ketone	159.1
Methyl isobutyl ketone	43.7
Nickel	19.4
Tetrachloroethylene	37.5
Toluene	344.6
1,1,1-Trichloroethane	190.5
Trichloroethylene	55.4
Xylene	201.6

Public Demands

Another important incentive is that the public will no longer tolerate the continued production of hazardous waste. This intolerance manifests itself in many forms, including the new legislation cited previously and consumer boycotts of products from companies perceived not to be environmentally responsible. As will be seen in Sec. 8-5, it has become virtually impossible to site new hazardous waste disposal facilities. Existing facilities are closing and no new facilities are opening. Much of the opposition to proposed new facilities is generated by genuine, even if mistaken, fears of excessive health risks. Another major source of opposition is the perception that industry has no need to produce these wastes in the first place and the proposed facility is simply not necessary. One of the more effective groups at preventing the siting of new facilities notes in its newsletter:

> By slowing and stopping the siting process, communities are forcing industry and government to look hard at why people are opposing existing treatment and disposal methods. ... After all, what makes more sense than to avoid generating waste in the first place.[12]

It is a sine qua non with such groups that all hazardous waste generation can and should be eliminated. While most environmental engineers and industrial managers would consider this a clearly unachievable objective, the minimization of waste is unquestionably an appropriate goal.

7-2   MANAGEMENT STRATEGIES

Overview

Given the goal of minimizing the generation of hazardous waste at a manufacturing facility, how does one proceed? The following outline is suggested:

1. Planning and organization
2. Characterization of waste and losses
3. Development of waste minimization options
4. Technical, regulatory, and economic feasibility
5. Implementation (including training)
6. Monitoring and optimization

The first step includes obtaining a commitment from management and establishing a clear set of goals. Typically this step would also include identifying individual responsibilities and organizing the group of individuals who will establish the Waste Minimization Plan.

The second step is to understand the types and sources of wastes—an outcome of the audits discussed below. Steps three and four analyze this information in a framework that includes a range of perspectives: environmental, economic, legal, political, and technical. The end result of this analysis should be a comprehensive Waste

Minimization Plan. Implementation of this plan must concern virtually all aspects of the company including research and development, marketing, production, training, maintenance, raw materials procurement. Finally, these efforts should be examined to ensure that they are accomplishing the objectives in an optimal way. EPA has divided these activities into four phases as illustrated in Fig. 7-3.

Audits

The steps for successfully achieving the goal of waste minimization begin with an audit following the methods outlined in Chap. 6. The waste minimization audit consists of planning the entire program and fully characterizing all wastes. From there, waste minimization options are identified, evaluated, selected, implemented, monitored, and communicated.

The planning of a waste minimization program is critical to its success. Planning starts with obtaining commitment of top management, setting goals, establishing incentives, and budgeting financial and human resources. Planning also includes com-

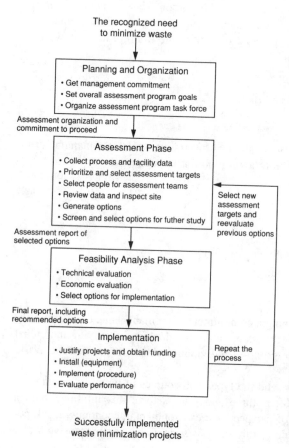

FIGURE 7-3
EPA's waste minimization assessment procedure.[13]

municating the policy and goals to facility employees, designating a person to lead the program, and organizing a team to support it. It also includes preparing schedules, worksheets, and other necessary tools.

The primary functions of a waste minimization audit are to characterize waste generation and to identify obstacles to source reduction. The formality of a waste minimization audit depends not only on the complexity of the facility and its wastes but also on the stage of waste reduction that the facility has already implemented. Hirshorn suggests four stages:[14]

1. Common sense waste reduction—immediate opportunities without having to change production processes. Examples are reducing the amount of water used in cleaning equipment and ceasing the commingling of hazardous and non-hazardous waste.

2. Information-driven waste reduction—relatively easy opportunities resulting from simple changes in production prompted by information on successful cases in similar industries. Examples are changing raw materials (e.g., replacing an organic solvent with a water-based product, or installing a closed-loop recycling unit).

3. Audit-dependent waste reduction—opportunities requiring capital investment and production changes which seem to have a questionable economic payback without the detailed information of a formal audit.

4. Research-and-development-based waste reduction—opportunities facing severe technical obstacles which require extensive study if not research to overcome.

The first and second stages can be implemented with a fairly simple audit using a checklist. The third stage requires a formal audit. A formal waste minimization audit surveys each operation that generates waste to determine how the waste is generated, what are its characteristics, how it is managed, and what are the costs. This provides baseline conditions for evaluating progress toward meeting waste minimization goals. It also provides the necessary data to prioritize waste streams and to identify alternative options for minimizing the high-priority wastes.

The waste minimization audit includes a thorough survey to provide the following data for each hazardous waste stream:

- Specific source(s)
- Generation rate
- Physical characteristics
- Chemical characteristics
- Production rate of process/operation generating the waste
- Current management methods from generation to final disposition
- Costs of each management method

The first step in the waste survey is to assemble and organize existing information on all wastes, not just hazardous waste. This can come from a variety of sources to include mandatory reports required on a one-time-only or periodic basis by regulatory agencies (e.g., annual summaries, waste shipment manifests, contracts for disposal and tolling services, permits and permit applications, laboratory data, and other similar sources).

The assembled sources of information provide a basis for identification and initial characterization of each waste stream. The second task is to relate each generated waste directly to the specific operation or process responsible for its generation. Necessary sources of information to complete this include:

- Process flow diagrams
- Material and heat balances (both design and actual)
- Operating manuals and process descriptions
- Equipment lists
- Equipment specification and data sheets
- Piping and instrument diagrams
- Plot and elevation plans
- Equipment layouts and work flow diagrams

The characteristics of a waste depend not only on the process design but how the process is operated and what raw materials are used as inputs to the process. Therefore, other supporting sources of information include:[15]

- Product composition and batch sheets
- Material application diagrams
- Material safety data sheets
- Material purchase records
- Product and raw material inventory records
- Operator data logs
- Operating procedures
- Production schedules
- Production records
- Maintenance procedures
- Maintenance records

Methods

Perhaps the simplest waste minimization efforts involve housekeeping changes. Such items as segregating hazardous waste, sweeping floors prior to washing, and training employees to be sensitive to the implications of their actions can dramatically reduce waste generation. Only after these relatively inexpensive measures are undertaken should production changes be contemplated.

Most industries change production methods only when forced to by circumstances. There is an inertia that surrounds the large amount of capital invested in present methods and a reluctance to change profitable systems or make changes that may impair product quality. Many of the processes in use today were developed prior to the present concern with hazardous waste. The range of production changes that may be appropriate include:

- Changes in process operating variables (e.g., pressure)
- Changes in feedstocks
- New process chemistry
- Equipment changes
- Product elimination

In the case of operating variables, many processes were established to optimize production or minimize operating cost. Unless the optimization calculations were performed quite recently, it is unlikely that the costs of waste disposal, if they were considered at all, are appropriate today. Feedstocks and process chemistry should be evaluated in terms of the waste produced. It is not unusual for a large volume of waste to be hazardous due to trace amounts of a chemical *not used* in the plant. This might point to a trace contaminant in one of the raw materials, which does not adversely affect the manufacturing process. Converting to a higher quality or simply a different source may reduce or eliminate the problem. It is important to keep in mind that manufacturing calculations involve percent concentrations, not the ppm or ppb concentrations typically used in waste calculations.

Often the cost of waste treatment has risen so dramatically that it is the single factor that causes an industry to upgrade the process through the modification, addition, or total replacement of equipment. (e.g., land disposal costs in California rose 445% in the three-year period 1983–1986).[16]

Any proposed alternative must be evaluated in terms of its technical, regulatory, and economic feasibility. In some instances, this results in no feasible alternative other than eliminating the product entirely, because there is no alternative process that can reduce unacceptable by-products. Unfortunately, in the case of facilities that are already marginal economically, this may mean a reduction in work force or the closing of an entire plant.

Microeconomic Analysis

Where capital expenditures are contemplated, it must be recognized that pollution prevention projects must compete with other proposed capital projects for limited company funds. Capital costs for new waste minimization equipment are often considered a drain on a company's profitability. In order to fully appreciate the actual cost of such equipment, it is necessary to look at the savings resulting from reduced cost of waste disposal. Economic methods applied to such decisions include estimates of annualized cost savings, internal rate of return, and present worth analysis.

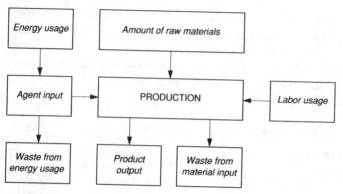

FIGURE 7-4
An economic production model.

A theoretical econometric model developed by Anderson[17] incorporates the analysis of energy and material flows with waste generation to provide a method of evaluating approaches to waste reduction (see Fig. 7-4).

The total cost of production is estimated by:

$$C = C_1 R + C_2 E + C_3 W_m + C_4 W_e + C_5 L \tag{7-1}$$

where C_i = appropriate unit costs
R = amount of raw material
E = energy usage
W_m = waste from material input
W_e = waste from energy usage
L = labor usage

Anderson provides generalized production curves, based on mass and energy balances, showing the relationship between mass of raw material (R, abscissa), agent input (e.g., capital, labor, etc.) (K, ordinate) and the mass of product output (Q, curve parameter). The optimization is illustrated in Fig. 7-5. The equation for the curves is:

$$Q = \frac{vR}{m} \tag{7-2}$$

where Q = maximum amount of output that could be produced given R
units of input and a mass per unit of output of m
v = fraction of input mass R that becomes product

The concept of agent input is an economic term to account for the non-material input to production. It includes capital, labor, and the associated energy. The cost of K would include the cost of the agent's services plus the energy costs.

Using this terminology, waste production is estimated as $(1 - v)R$, implying that waste production is proportional to the material input to the process. If a given production level is selected (e.g., Q^*), moving to the left along the curve reduces W_m,

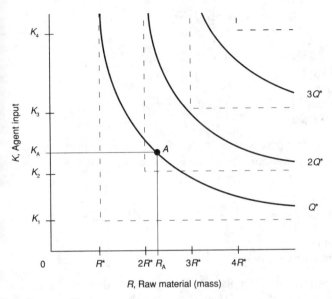

FIGURE 7-5[18]
Optimization of economic production model.

but increases W_e. At point A in the figure, Q^* units of product are produced, using R_A units of material and K_A units of agent input. The curve implies that it is not possible to produce Q^* units of product with less than R^* mass of raw material, no matter how much K is increased. The waste production at point A is $W_m = R_A - vR^*$, which can be divided into by-product waste $[(1 - v)R_A]$ and economic waste $[v(R_A - R*)]$.

Anderson's conclusions regarding policy are:

> ... W_e is an inevitable flow from any production process so that policy directed at the elimination of this flow is doomed to failure. Incentives are instead required to induce firms to use agent-energy combinations which have high thermal efficiencies ... and/or result in flows of a less harmful nature ... **for by-product wastes, waste reduction can only be achieved by a reduction of the output produced (assuming the nature of the output and material input remain unchanged).** [emphasis added] Finally, while policies aimed at reducing economic waste may be effective in doing so, they are likely to result in increased flows of W_e if output levels are to be maintained. ...[19]

While the above theoretical analysis is somewhat pessimistic regarding the ability to reduce wastes without curtailing production, this chapter will examine techniques to change the nature of the output and material input, while still maintaining productive manufacturing facilities. To start the process, microeconomic calculations used to evaluate minimization alternatives will be presented.

Present worth analysis is used to bring all future costs to an equivalent amount of money which, if invested today, would produce the needed future sums. For example, if 10% interest is assumed, one dollar today is effectively worth $1.10 in one year and $1.00 (1.1) (1.1) = $1.21 in two years. In general:

$$PW = \frac{F}{(1+i)^n}$$ (7-3)

where PW = present worth
  F = future sum
  i = interest rate (10% : $i = 0.1$)
  n = number of years

Present worth analysis is typically applied considering all future operating costs as a uniform series. The present worth for a uniform series of annual payments is given by:

$$PW = A \times PWF = A \times \frac{[(1+i)^n - 1]}{i(1+i)^n}$$ (7-4)

where A = annual cost
  PWF = present worth factor

The process is illustrated by Example 7-1. This example does not take into account changes in annual costs due to inflation and other factors. Because it is typical for waste treatment costs to rise much faster than other manufacturing costs, the above analysis is conservative. The selection of an interest rate is normally based on the present cost of money for the industry in question. Design life would be based on the expected life of the equipment in question, perhaps modified by considerations of potential changes in regulations or other factors that may influence how long the new process will remain in place. Example 7-2 illustrates how the effect of inflation may be incorporated into the calculation.

Example 7-1 Present worth analysis. A manufacturing process has present operating costs of $10.9 million annually and waste disposal costs of $3.3 million. The plant manager wishes to upgrade the process by installing new equipment at a capital cost of $23.3 million. This will reduce the annual operating and waste disposal costs to $7.1 and 1.4 million respectively. Compare the two processes on a present worth basis, assuming no salvage value. Use 12% interest for 10 years as a basis for calculation.

Solution. The present worth factor is:

$$PWF = \frac{[(1+i)^n - 1]}{i(1+i)^n} = \frac{[(1.12)^{10} - 1]}{0.12(1.12)^{10}} = 5.650$$

For the present process:
$PW = (10.9 + 3.3)(5.650) = \$80.233$ million
For the new process:
$PW = 23.3 + (7.1 + 1.4)(5.650) = \$71.327$ million
On a present worth basis, the new process saves $8.9 million.

It should be noted that this answer does not consider other important factors such as income taxes, investment tax credits, and inflation. The latter is addressed in Example 7-2.

Example 7-2 Inflation analysis. Revise Example 7-1 to account for a 5% estimated inflation rate for manufacturing costs and a 12% rate for waste disposal costs. Use 12% interest for 10 years as a basis for calculation.

Solution. Calculations are shown in Table 7-2.
These calculations can also be performed by applying equation 7-3 to each future cost. For example the cost of the present process in year 2 is $16.16 million. The present worth of this figure at 12% is:

$$\frac{16.16}{(1.12)^2} = \frac{16.16}{1.25} = 12.93$$

For the present process:

$$PW = (15.14/1.12) + (16.16/1.25) + (17.25/1.40) + (18.44/1.57) +$$

$$(19.73/1.76) + (21.12/1.97) + (22.63/2.21) + (24.47/2.48) +$$

$$(26.06/2.77) + (28.00/3.11)$$

$$= 110.75$$

For the new process:

$$PW = \text{Capital cost} + \sum (\text{years } 1{-}10)$$

$$PW = (23.30) + (8.50/1.12) + (9.02/1.25) + (9.58/1.40) + (10.19/1.57) + (10.83/1.76) +$$

$$(11.53/1.97) + (12.28/2.21) + (13.09/2.48) + (13.96/2.77) + (14.90/3.11)$$

$$= 84.03$$

New process savings $= 110.75 - 84.03 = \underline{\underline{\$26.72 \text{ million}}}$

Equations 7-3 and 7-4 may be inverted to determine a future value or annual payment corresponding to a specified present worth. A final compound interest equation determines the annual payment necessary to produce a specified future amount:

$$A = \frac{F \times i}{[(1 + i)n - 1]} \tag{7-5}$$

While present worth analysis provides a simple mechanism for comparing costs, it does not take into account many factors that should influence waste minimization decisions. One modification of the analysis is to use a **benefit/cost analysis**. While this is often thought of as another form of economic analysis, the values used are usually arrived at by present worth analysis. In addition to the costs identified above, benefits might include the value of goods produced and reduction in paperwork and other transactional costs associated with waste disposal. Additionally one could estimate the value of benefits such as reduction in future liabilities and other less tangible items such as corporate public relations and potential public benefits. Sometimes negative

TABLE 7-2
Inflation calculations for example 7-2 (in millions)

Year n	Present process			New process			Difference	PW
	Manufac-turing*	Waste disposal**	Total T_p	Manufac-turing*	Waste disposal**	Total T_n	$\Delta = T_p - T_n$	$\Delta/(1+i)^n$
0	10.90	3.30		23.30		23.30	−23.30	−23.30
1	11.45	3.70	15.14	7.10	1.40	8.50	6.64	5.93
2	12.02	4.14	16.16	7.46	1.57	9.02	7.13	5.69
3	12.62	4.64	17.25	7.83	1.76	9.58	7.67	5.46
4	13.25	5.19	18.44	8.22	1.97	10.19	8.26	5.25
5	13.91	5.82	19.73	8.63	2.20	10.83	8.89	5.05
6	14.61	6.51	21.12	9.06	2.47	11.53	9.59	4.86
7	15.34	7.30	22.63	9.51	2.76	12.28	10.35	4.68
8	16.10	8.17	24.27	9.99	3.09	13.09	11.19	4.52
9	16.91	9.15	26.06	10.49	3.47	13.96	12.10	4.36
10	17.75	10.25	28.00	11.01	3.88	14.90	13.11	4.22

Total Savings $= \sum PW = \$26.72$ M

*increase at 5%
**increase at 12%
PW Present Worth at $i = 0.12$ (12%)

benefits are termed **disbenefits** (i.e., hardships created by the alternative). This leads to two common equations used to calculate the **benefit cost ratio** (B/C):

$$\frac{(B - D)}{C} \tag{7-6}$$

$$\frac{B}{(C + D)} \tag{7-7}$$

where B = benefits
        C = costs
        D = disbenefits

As illustrated in Example 7-3, the selection of the B/C equation may provide slightly different values of the ratio.

Example 7-3   Benefit/cost analysis. Calculate the B/C ratio given the following present worth data about a waste minimization alternative:
Benefits (Million $):
   Reduction in Future Liabilities $6.0
   Elimination of Potential Fines estimated at $2.0
   Savings in Waste Treatment Costs estimated at $4.5
Total Benefits: $12.5
Disbenefits: Increased production costs: $7.5
   Cost of New Equipment $4.97
Eq. 7-6: (B-D)/C = (12.5 − 7.5)/4.97 = 1.01
Eq. 7-7: B/(C+D) = 12.5/(4.97 + 7.5) = 1.00

A modification of benefit/cost analysis is termed **cost-effectiveness analysis**. In this case, no attempt is made to put a price on intangibles. Instead all monetary costs are evaluated for various alternatives using present worth analysis. Other factors are evaluated on an effectiveness scale. Effectiveness might include safety considerations, potential effects on production (delays, product quality, etc., that cannot easily be quantified), hidden costs (paperwork, monitoring), public relations considerations, ease of implementation and future liabilities. Cost-effectiveness is a particularly useful technique when evaluating factors for which it is simply inappropriate to assign dollar values. For example society deems it indelicate to assign a dollar value to human life or an additional cancer case, even when the objective is to save lives or reduce the incidence of cancer. In the latter cases, risk is used as the measure of effectiveness (see Fig. 17-7).

A plot of effectiveness versus cost provides decision-makers with a graphical illustration of the decision to be made. The process is illustrated in Example 7-4. It must be emphasized that this is simply a way of presenting information to decision-makers. It is only of value to the extent that the selection of factors, weights, and relative effectiveness scores are credible to management. The most appropriate use of cost-effectiveness analysis is to screen a large number of potential pollution prevention alternatives to determine which ones are appropriate candidates for detailed feasibility studies.

Example 7-4   Cost-effectiveness analysis. A manufacturing plant is faced with the loss of a hazardous waste disposal site. They are evaluating three alternative strategies:

A. Ship waste to a site in a distant state.

B. Construct major in-plant changes to reduce the production of hazardous waste.

C. Re-build the manufacturing facility using an entirely new process that does not produce any hazardous waste.

Solution. To begin an analysis, management identifies six factors which they believe cover the intangibles (see Table 7-3). The factors are assigned weights which represent management's value judgment as to the relative importance of each factor. A further subjective evaluation assigns a relative effectiveness score to each alternative for each factor. The raw score is simply the weighted sum of the individual values, and the effectiveness is obtained by subtracting the raw score of the present alternative (the do-nothing alternative provides a baseline on which to judge the others).

From the results in the table, it is seen that Alternative B is twice as effective at achieving objectives, but only adds about 30% to the cost. Alternative C is actually less effective (see Fig. 7-6).

It is possible to use this type of analysis simply to deal with intangibles, in advance of any attempts at estimating cost. For example, an industry with dozens of process lines may wish to prioritize them to determine which processes should be evaluated first.

TABLE 7-3
Cost-effectiveness data for Example 7-4

Factor	Weight	Effectiveness Present	A	B	C
Short-term liability	2	−3	−2	−1	−1
Long-term liability	3	−5	−3	−2	−1
Compliance	5	−2	+1	+2	+2
Production impact	5	0	−2	−1	−3
Public relations	1	0	+2	+2	+4
Regulatory impact	1	−3	−2	−1	0
Raw Score		−34	−18	−2	−6
EFFECTIVENESS		0	16	32	28
COST (millions $)*		0	8.6	10.7	52.4

*Present worth of all identifiable capital and operating costs less savings when compared to present operation

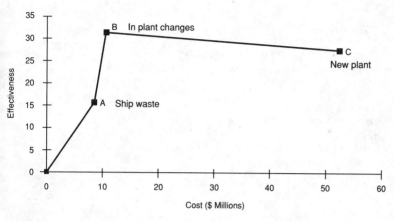

FIGURE 7-6
Cost-effectiveness plot.

Table 7-4 was used by Intel Corp. to prioritize waste streams. For toxicity information the lowest available median lethal dose (LD_{50}) for human oral toxicity was used. Where no human data existed, animal data was used and in the absence of oral data from animal studies, other routes (e.g., dermal) were utilized.

For waste streams with multiple components, the higher rating is selected. Total ratings are determined for all waste streams and used as a basis for selecting which waste streams to study first.

Financing

Major pollution prevention programs may require substantial capital investment. Potential sources of capital include:

- Corporate resources
- Commercial bank loans
- Business development corporations
- Trade associations
- State and Federal governments

Where corporate and commercial bank funding is not available, Business Development Corporations (BDCs) may be able to assist. These institutions were established to assist the private sector in retaining and expanding employment. BDCs exist in most states and were specifically set up to assist businesses to secure funding that would not normally be available through established lending institutions.

Trade associations may be able to secure funding for a group of industries with a common problem and centralized solution. In New Jersey, metal finishers established

TABLE 7-4
Waste stream prioritization[20]

Prioritize waste streams using the following parameters:

A. Monthly Amount Produced

Amount > 5000 lb	= 4
5000 lb $\geq$ Amount > 1000 lb	= 3
1000 lb $\geq$ Amount > 500 lb	= 2
500 lb $\geq$ Amount > 50 lb	= 1
Amount < 50 lb	= 0

B. Toxicity Rating (LD_{50})

LD_{50} < 10 mg/kg	= 3
10 mg/kg $\leq LD_{50}$ < 100 mg/kg	= 2
100 mg/kg $\leq LD_{50} \leq$ 1000 mg/kg	= 1
LD_{50} > 1000 mg/kg	= 0

C. Land Disposal Ban Applicability (LD)

Yes	= 4
No	= 0

D. Disposal Cost (DC)

Cost > $1500/ton ($6.25/gallon)	= 3
$1500/ton $\geq$ Cost > $1000/ton ($4.17/gallon)	= 2
$1000/ton $\geq$ Cost $\geq$ $500/ton ($2.08/gallon)	= 1
Cost < $500/ton	= 0

E. Flash Point (FP)

Flash point < 20°F	= 4
20°F $\leq$ flash point < 80°F	= 3
80°F $\leq$ flash point < 140°F	= 2
140°F $\leq$ flash point $\leq$ 200°F	= 1
Flash point > 200°F	= 0

F. Reactivity (R)

Yes	= 1
No	= 0

G. Carcinogenicity, Mutagenicity, and/or Teratogenicity (CMT)

Yes	= 4
No	= 0

Total rating = sum of seven factors

a non-profit research foundation that examined the feasibility of a centralized waste treatment facility:

> Ninety members of the Master's Association of Metal Finishers each paid $2,500, and additional assistance was obtained through grants ... a conventional treatment system would cost each company approximately $200,000 to purchase and install at their own plants ... the centralized system would save companies at least $14,000 a year.[21]

In order to encourage industries to implement waste reduction programs, state and federal governments have instituted programs to provide assistance. This assistance ranges from technical assistance on how to accomplish waste minimization to financial assistance in paying for the changes. Illinois' program included the development of an interactive computerized database that provides industry with access to the technical literature on waste minimization.[22] Technical assistance programs are operated by many states as part of their economic development efforts. These programs provide information on regulatory compliance and technologies for achieving waste reduction. Small business development centers and state funded university research programs are other ways government is providing direct technical assistance to industry. In general, these programs are aimed at small businesses, which are unlikely to have the required specialized knowledge in-house.

Funding for pollution prevention efforts is also available from many states. While the requirements and level of funding vary from state to state, it is important to obtain this information before analyzing a proposal. Even if a state program merely provides low interest loans for waste minimization projects, the difference in the interest rate may be sufficient to change the way corporate management will view the proposal.

Obstacles

While the previous introductory sections would lead to the conclusion that every industry should adopt wide-ranging waste minimization programs, there are a number of impediments to such action. Hirschorn and Oldenburg characterize this as a serious national problem:

> ... nearly every part of American society is mentally locked into the established, institutionalized pollution control culture, a paradigm that defines environmental protection in terms of what is done to wastes and pollution *once they are produced*. ... The switch from pollution control to pollution prevention is a classic example of a paradigm change—a truly profound change in the way people think about something. Changing to a new paradigm takes time.[23]

In addition to a change in mind-set, there are a number of institutional impediments to waste reduction. These include some governmental policies, "short-term" management strategies, information and educational needs, and economic disincentives.

Recent changes in federal regulations have created some disincentives for waste reduction programs. For example, RCRA specifically prohibits issuance of tax free industrial bonds for funding pollution control equipment to achieve compliance. Additionally, some waste reduction efforts may require a permit under state or federal regulations (e.g., certain types of recycling activities will require a "Part B" permit under RCRA). The added cost of permit preparation and the possibility that the permit might not be granted, may be sufficient to stifle some waste minimization initiatives.

As noted above, the two most obvious sources for these funds are corporate resources or banks. Unfortunately, lending institutions are reluctant to provide commercial loans for such projects, unless it can be shown that they will also improve profitability.[24]

A similar impediment in some corporations is the inability of management to take a "long-term" view of corporate decisions. Often the payback period for waste reduction efforts is many years. If the plant manager is evaluated solely on the next quarter's bottom line, purchase of equipment that does not add to production is undertaken at personal cost.

An effective pollution prevention effort must include research and development, process design, production management, and other corporate functions in order to incorporate meaningful changes into the manufacturing process. Information needs range from a real understanding of the costs (long-term as well as short-term) of waste management practices to newly developed technologies that will enable a plant to dramatically reduce waste generation. Although the information is available, corporate management needs strong incentives to devote the necessary resources to acquiring it. Government applied incentives are often reduced to more forms to fill out to demonstrate pollution prevention efforts, which may result in less prevention.

Because many waste minimization procedures involve changes in operation rather than equipment, it is important that a training component be included in any waste minimization plan. Workers need to know how their actions affect both productivity and the environment.

7-3 LIFE CYCLE ANALYSIS[†]

The environmental impact of all phases of industrial activity, from raw materials acquisition and R&D to the final disposition of a product and its packaging, is having a far-reaching effect on environmental quality and on public health. As a result, industry, environmental groups, and governments are attempting to identify a systematic means of evaluating and minimizing the environmental impact of products and processes. One of the more promising systematic approaches for identifying and evaluating opportunities to improve the environmental performance of industrial activity is termed Life Cycle Analysis (LCA) (see Fig. 7-7). LCA provides an analytical framework for investigating the entire range of environmental impacts (e.g., air emissions, wastewater, solid and hazardous waste, renewable resources and energy utilization).

Employing LCA methods, corporations are identifying opportunities to reduce environmental impacts, while maintaining or gaining competitive advantage. The use of life cycle methods at the design stage of a product or process is particularly noteworthy. This is true whether the concern is manufacturing operations, product use, or disposal. For example, the typical car contains about 180 pounds of plastic. However, when a car is disposed of, the different plastics in the car cannot be easily sorted. To counter this life cycle problem, BMW introduced a new series of small cars in the United States in June 1992. Plastic parts are stamped by type during fabrication so they can be sorted quickly and accurately. In addition, the cars are designed for disassembly. Design for disassembly is one of the many life cycle oriented "Design

---

†Section 7-3 prepared by Marc Wendell, ERM, Inc.

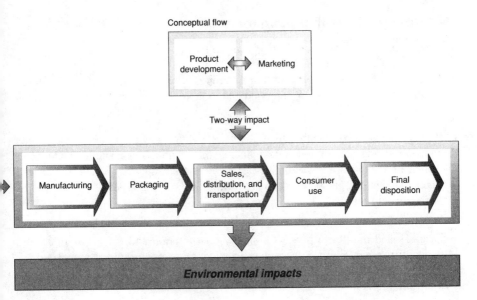

FIGURE 7-7
Product life cycle.

for" techniques being embraced by the auto and other industries. Design for ease of assembly, design for field repair, and design for transportability are other examples.[25]

The remainder of this section provides a brief overview of Life Cycle Analysis, focusing on its use in addressing hazardous waste issues.

Background

Although LCA has been used in various forms over the past several decades in Europe and in the United States, it remains a relatively new concept, the methodology of which is still in the developmental stages. In the United States the origin of LCA can be traced back, prior to its environmental usage, to the defense industry. Defense systems analysts used life cycle methodology to account for the previously neglected operating and maintenance costs associated with owning equipment and systems. The differences in life-cycle costs among systems were factored into the analysts' procurement decision-making processes. This allowed low-maintenance systems with high-capital cost to compete with lower capital cost systems that required extensive maintenance.[26]

Paralleling the concern of defense systems analysts, the environmentalists, policy makers, and corporations of today are employing LCA in an attempt to take into account the previously neglected environmental costs (and benefits) of products, processes, and other business activities.

Performing LCA within Industry

Private corporations are the most likely group to be performing LCAs in the market-place today. These companies are using LCA methods to compare design variations for a single product or to compare new or proposed products to their existing line(s). From this vantage point, LCAs are valuable tools to:

- Provide a means of identifying trade-offs associated with product modifications and design alternatives (such as reducing the weight of packaging material and increasing the use of recycled material), raw material sourcing (e.g., agricultural vs. fossil fuel derived), technology changes (in process or materials) and various waste management options;

- Serve as a baseline for comparison with future modifications; and

- Quantify and monitor a product's energy consumption and emissions in an effort to identify potential resource and waste reduction opportunities associated with specific manufacturing operations (e.g., raw material extraction, fabrication, and transportation).

Some common characteristics exist in all industrial LCA programs regardless of the products manufactured. In particular, LCA methodologies generally involve some variation of the following three-phase process:

Inventory $\Rightarrow$ Impact (or interpretation) $\Rightarrow$ Improvement

One common approach being used in industry to perform LCAs is outlined in Table 7-5.

TABLE 7-5
The life cycle analysis model

Phase 1 Inventory	Phase 2 Impact analysis	Phase 3 Improvement analysis
• Organize LCA teams	• Assess compliance with company requirements	• Assess alternatives for improvement
• Determine priorities		
• Target products for LCA	• Assess risk, where appropriate	
• Diagram the LCA		
• Determine usage and generation rates (of raw materials, wastes, and energy)		

Inventory Phase

Organize Teams. A project coordinator is usually designated to manage an LCA. The coordinator then assembles one or more teams, each of which assesses a limited number of targeted products or activities.

The teams include those personnel that have a significant stake in the outcome of process or product changes. They are cross-functional groups with broad practical experience, including personnel from areas such as operations and maintenance; manufacturing; product/design engineering; finance and accounting and corporate environmental, health and safety (EHS). By bringing together a diverse team that includes all parties with a stake in the outcome, the team can overcome resistance to change, facilitate the sharing of ideas, and address conflicts among departments.

Target Products/Processes for Analysis. Even though LCA was conceived for the purpose of investigating the full range of impacts across the entire life cycle of a product, in actual practice it is financially impractical for an organization to perform detailed life cycle analyses on all their activities. Instead, it is usually necessary for the corporation to target those issues requiring detailed life cycle analysis. For example, a specific product may be targeted because of concern over its usage of toxic materials, air emissions, etc.

The targeting process usually requires a preliminary assessment of products, processes, chemical usage, waste generation, and emissions by experienced company personnel, particularly those with a broad base of experience with product and environmental issues. It is not necessary to compile exacting quantitative information during targeting. Instead, it is better to develop an initial, qualitative understanding of the environmental concerns of the company (e.g., its products, processes, and activities). Such targeting can be performed using readily available sources, such as operating logs, and from the practical knowledge of experienced production and process design personnel. Outside consultants with broad regulatory experience and an understanding of regulatory trends can also be a useful source of information, providing an outsider's "objective" insight.

Diagram the Life Cycle. After an activity has been targeted, the life cycle can be diagrammed using simple box flowcharts (see Fig. 7-8). The scope of the diagram should be thought through carefully before it is developed since the diagram will set the limits of the investigation. The analysis might begin with the acquisition of raw materials and end with disposal to the municipal waste stream, or it might extend to the production of raw materials and the decomposition products of the waste (see Fig. 7-8).

The actual scope, or boundary conditions, of a particular analysis depends on the needs of the investigation, which in turn depends on the issue(s) of concern. For example, if waste generation from the manufacture of a product is of concern, it might be possible to limit attention to key manufacturing operation, and the treatment, storage, disposal, and recycling (TSDR) facility that ultimately receives the waste.

In contrast, if the entire product life cycle is of concern, the "boundary conditions" of the analysis might need to be expanded to include all of the following life

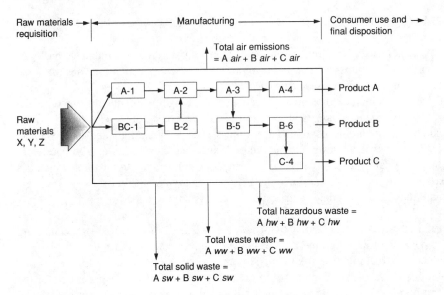

FIGURE 7-8
A manufacturing plant with products A, B, and C (note several stages of the product life cycle are not shown in this particular diagram).

cycle stages: marketing, research and development, raw material acquisition, material manufacture, final product fabrication, packaging, distribution, consumer use, and disposal.

Identify Input and Output. Following the completion of the diagram, the location and amounts of materials and wastes entering and leaving each stage of the life cycle is identified. Raw materials and intermediates and the emissions, wastes, and products should be identified for each stage of the life cycle. The practical knowledge of experienced personnel can be used to accelerate this evaluation.

If the main concern of the LCA is the reduction of hazardous waste (e.g., hazardous waste and hazardous air pollutants [HAPS]), emphasis should be placed on inventorying these wastes at each stage of the life cycle. For example, not only are hazardous wastes generated as a direct result of the production stage, they also result indirectly, from the maintenance activities associated with raw materials processing, manufacturing operations, and transportation equipment. In addition, returned goods can be a significant source of waste from outdated or damaged goods. All possible generation points need to be evaluated for possible inclusion in the final life cycle diagram.

The final diagram(s) and accompanying documents should contain the following information:

- **Targets** (e.g., product A; process B; and/or activity C)
- The **location** of waste emission or energy or material usage (e.g., a process vent, wastewater stream, and/or equipment cleanout)

- The type of **media** transfer (e.g., air emissions, wastewater, and solid or hazardous waste)
- The **amount** of material or waste transferred (and/or energy utilized).

Readily available sources of information for estimating the rates of material usage and waste generation include sales figures; material safety data sheets (MSDSs); operating and maintenance records and manuals; permit information; vendor information; and facility (e.g., purchasing) records. An example of a process assessment worksheet is shown in Fig. 7-9.

The diagramming activity is designed to promote improvement by providing a visual framework for systematic analysis of the life cycle, including the identification and tracking of root causes of waste. For example, the worksheets aid in "normalizing" waste generation rates to the production rates of finished goods (e.g., pounds of hazard waste per thousand "widgets" produced). The normalized waste factors help to distinguish between real reductions in waste and the effects that accompany fluctuations in production rates. In doing so they provide a solid basis for setting environmental goals and tracking performance.

Impact (Interpretation) Phase

After the inventory has been developed, it is necessary to interpret the impacts of waste emissions and energy usage. There are a number of alternative strategies for interpreting life cycle impacts, most of which are still in the developmental stages (e.g., regulatory, hazard based, energy utilization, risk).

Regulatory Strategy. After an inventory has been developed, a review of regulated chemicals, wastes, emission, etc., should be performed to determine whether (1) any of the life cycle stages are regulated or (2) further characterization of a material, waste, air emission, or process is necessary to assess regulatory applicability. The resultant "hit list" of regulated materials, wastes, and/or processes forms one important basis for investigating improvement opportunities.

Non-regulated Targets. Corporations can promote the improvement of EHS performance by setting priorities beyond compliance, either expanding efforts beyond regulated materials, wastes, and processes or increasing the force behind specific regulatory items. For example, many corporations are accelerating their rate of chlorofluorocarbon (CFC) replacement beyond that mandated by law.

The Clean Production Program of EPA's Office of Toxic Substances (OTS) uses three primary criteria to set (non-regulatory) priorities for addressing particular processes:

- Potential for reducing risk
- The significance of the process or use to multiple program initiatives across EPA, both regulatory and non-regulatory

Process Assessment

Process: _Chemical Manufacturing_   Work sheet #: _1 of 1_

General mass balance equation: **Input + Generation = Output + Destruction**

	Operation 1	Operation 2	Operation 3	Operation 4	Operation 5	Operation 6	Outputs
	Sampling & testing: RM & packaging components	Raw Materials — movement to formulation	Compounding	Qality Control batch components	Filling — filler lines	Filler Lines — pump insert	
Raw materials _See formula sheets, function spec sheets_							Finished product and associated packaging
Energy		Energy loss through pumping RM into vessels			Mechanical movement: compressed air	Mechanical energy used - compressed air	
Air (ID#)	VOC fugitive whilst open hauler (will be quantified when instr. installed	NO / closed system	VOC emiss. (fugitive) (can be quant. — see air monitoring done)	NO / small sample testing	VOC emissions extractions to atmosphere	NO VOC	
Wastewater (ID#)	Some washout in lab; 20–40 kg RM liquid from sump each time of sampling tankers	NO	Yes, wash out tanks and pipe work (will be quantifiable)	Lab samples + reagents to drain Neutralize + H2O (policy) (see reagent tests)	Spills, etc.	NO	
Solids (ID#)	As sampling procedure	NO	NO solids	NO	Bottles (can be quantified) Packaging SW, recyclable cardboard, shrink film	Some pumps bottles; Packaging SW, recyclable cardboard, shrink film	
Hazardous (ID#)	NO generation NaOH HCl ⎤ Neutral before drain	NO unless spilled	Yes — if bad batch — incineration	NO	shrink film NO	NO	

• The importance of the process or chemical to agency voluntary pollution prevention efforts, including the ongoing 33/50 Program (See Table 7-1)

While the above example draws from the EPA's prioritization strategy, the same basic strategy can be used by industry. Corporations can set priorities as to which products, processes, and/or activities receive in-depth life cycle analysis.

As the above example suggests, life cycle inventory information provides a basis for assessing compliance with regulatory, company, and customer requirements and developing a strategy to meet these requirements. If, for example, a facility is found to be nearing the large quantity generator status for hazardous waste, an array of regulatory requirements may have to be met (e.g., recordkeeping, reporting and training requirements), entailing significant costs. The company may, however, decide instead to implement a pollution prevention program to reduce the potential regulatory burden, opting to remain a small quantity generator, or become a conditionally exempt small quantity generator. Again, it is the life cycle analysis information that can provide support for strategic decision making.

Finally, LCA inventory information may trigger the need for further in-depth investigation (e.g., risk assessment) of materials, processes, and management practices.

Improvement Phase

The improvement analysis generally consists of two parts—the identification of improvement options (to address each area of concern identified during the impact analysis), followed by a comparative evaluation of set of options. For example, the reduction of hazardous waste generation from maintenance activities might be accomplished by the redesign of manufacturing equipment, changes in maintenance procedures, or a combination of the two. The process of identifying the best options usually begins with the applications of quality management tools (e.g., pareto and cause-effect diagrams).

The comparative evaluation requires a comparison of the technical, economic, and regulatory feasibility of each option and is beyond the scope of this brief section. Instead, the remainder of this section will present some examples of improvement strategies that are currently being employed within industry.

Use Clusters

For example, the new Clean Production Program developed by the EPA's Office of Toxic Substances (OTS) employs the idea of "use clusters"—groups of chemicals used for the same or similar purposes, such as paint strippers—for purposes of targeting. Examining use clusters, rather than single chemicals, OTS has evaluated the potential risks and identified safer substitutes and cleaner technologies for the chemical substances and technologies associated with a cluster.

Compliance Alternatives. Compliance with even the most stringent regulatory or corporate requirements can often be achieved using different strategies and technologies. Similarly, environmental risks can be managed using alternative approaches.

From a company's viewpoint, the aim should be to minimize risk to meet the company's requirements, ensure compliance, and do so at a minimum cost. LCA can provide the information needed for such decision making.

As an example, the 1990 Clean Air Act Amendments will cost industry billions of dollars and fundamentally impact how manufacturing facilities can operate in the future. However, there are opportunities, through emissions reduction, chemical substitution, and other process changes, to avoid or delay regulation under the Clean Air Act. LCA can be employed to gain competitive advantage from such regulations by providing a framework for the strategic evaluation of compliance alternatives.

Design Alternatives

Case Study 1: Evaluating Packaging Designs.   One of the most important uses of LCA is to provide companies with a method of ensuring that new product formulations or packaging options result in meaningful environmental improvements. In the following simplified example (Table 7-6), an LCA was conducted on various packaging alternatives for a liquid fabric softener. Only four of the several possible alternatives were studied. The data are based upon 1,000 liters of a single-strength product. The table indicates *percent decreases* for each energy and emission category as compared to a non-recycled, 64 oz. bottle made of virgin, high-density polyethylene (HDPE).

The results suggest that the *source reductions* associated with *product concentration* offer greater reductions in life cycle energy requirements and emissions than

TABLE 7-6
Strategies for packaging improvement[27]

Strategy	Percent decrease in energy use			Percent decrease in emissions		
	Process	Trans-port	Feed stock	Solid	Water-borne	Air-borne
1. Incorporate 25% recycled HDPI into the virgin HDPE bottle	3	0	9	9	(+4)	4
2. Assume 25% consumer recycling of the virgin HDPE bottle	3	2	11	11	(+4)	5
3. Use a concentrated (3×) product in smaller, virgin HDPE bottles	55	53	56	55	54	55
4. Use a concentrated (3×) product in virgin paper carton to refill HDPE bottle	53	58	94	91	40	62

either incorporation of 25 percent recycled HDPE (from other sources) or a 25 percent recycling rate for the bottle itself. Product concentration *plus the reuse of existing containers* provides a margin of additional benefits.

Relative to the non-recycled virgin bottles, each of the four packaging options studied offered a reduction in energy requirements and solid waste emissions. Note that the emissions associated with each packaging option differ in composition, and this may require a more detailed examination for meaningful interpretation. This case study illustrates the merits of LCA in (1) providing a baseline as a reference for improvement, (2) guiding product development efforts, and (3) lending better understanding of the unique environmental attributes of different strategies.

Case Study 2: Evaluating Chemical Designs.[28]   The traditional approach to synthesizing chemicals is to produce the greatest yield at the least cost. However, many of the synthetic steps, or synthetic pathways, that produce high yields also generate toxic byproducts or use high-risk solvents and catalysts.

These toxins could be reduced or "designed out" of the manufacturing process if alternative synthetic sequences were found. This is a relatively new life cycle design concept that EPA's Office of Toxic Substances is encouraging among researchers in the field of synthetic organic chemistry.

The following sections explore additional improvement strategies in more detail, illustrating specific methods to reduce the volume and toxicity of waste and to eliminate waste generation altogether. Secs. 7-4 through 7-6 are not mutually exclusive but rather provide a convenient method of organizing the information. There is a significant amount of overlap because many methods effect a reduction in both volume and toxicity of waste.

7-4   VOLUME REDUCTION

An appropriate place to initiate waste minimization investigations is to examine ways to reduce the volume of hazardous waste. This can be accomplished by a number of methods including modifying production processes, segregation, and re-use. It should be noted that under some regulatory schemes, simply reducing the volume of the waste without an accompanying reduction in toxicity would not be considered "waste minimization." For example, under California's Hazardous Waste and Management Review Act of 1989,[29] "actions that merely concentrate the constituents of a hazardous waste to reduce its volume" are explicitly excluded from the definition of "source reduction."[30]

Process Modifications

Process modifications include changes in:

- Raw materials
- Equipment
- Operating procedures

- Materials storage
- End products

Raw material substitution includes such simple items as cleaning materials. In the printing industry, the common practice of using organic solvents for cleaning presses has been replaced with water based cleaners. Merck, Sharp & Dohme report replacing organic solvents with inorganic acids and bases, resulting in the reduction in the emission of 300,000 gallons of hexane.[31] Sometimes conversion to a higher quality raw material can eliminate the generation of hazardous waste where the compound causing the waste to be considered hazardous is due to contamination of the raw material.

In a typical metal plating process, a metal piece is submerged in the bath, then removed and allowed to drain prior to rinsing. The piece must be moved away from the bath during the drain step to allow the next piece to be submerged. This results in dragout of plating bath. Simple equipment modifications such as the drain boards and drip tanks as shown in Fig. 7-10 can significantly reduce waste.

A major source of dragout occurs in the rinse tanks used in most plating operations to remove plating solution from the part being plated. A significant reduction in dragout to the rinse tank can be achieved through the use of a double dragout tank not connected to the final flowing rinse tank (see Fig. 7-11). In addition to significantly reducing the volume of wastewater, electroplaters have experienced up to 50% reduction in use of process chemicals.

Hunt[32] suggests the following additional measures to reduce dragout:

- Increase drain time
- Air knife (low pressure air to blow solution off parts)
- Spray rinsing over process tank
- Minimize concentration of metal in plating bath
- Rack parts to maximize drainage
- Use drip bars to hold racked parts over plating tanks

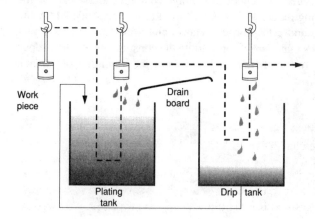

FIGURE 7-10
Dragout recovery devices.[33]
(*Adapted from Higgins with permission.*)

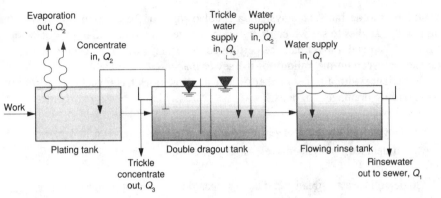

FIGURE 7-11
Providence method of double dragout tank in plating operations.[34]

- Maximum hole size on barrels
- Rotate barrels above plating tank

A final important aspect of any process change is the need to coordinate with production management. They must be an integral part of the planning, design and implementation of any waste minimization efforts. Production supervisors must be committed not only to the changes in equipment, but to training staff as to the reasons for the changes. Creatively designed waste minimization efforts can fail due to a lack of understanding on the part of production staff as to the need for the changes.

Segregation

A primary tenet of source reduction is to avoid mixing wastes. As noted in Chap. 2, a mixture of a small amount of hazardous waste with a larger amount of non-hazardous waste creates a large amount of material that must be treated as a hazardous waste! Another basic rule—don't make it a liquid if it is dry! Housekeeping operations as simple as sweeping prior to washing floors can substantially reduce waste volumes.

Keeping wastes segregated greatly facilitates any required treatment. Proper labeling of all lines and containers will greatly increase the likelihood that plant personnel will follow any changes in practices intended to enhance segregation of wastes. Where hazardous waste treatment is required, segregation permits treating smaller quantities of waste.

Another area where substantial volume reduction can be achieved is keeping non-contact cooling water separate from waste streams. Many manufacturing facilities are in buildings that were constructed long before waste management was a major concern and there is often a single sewer system for such buildings. In such buildings, all aqueous wastes drain to a common sewer, and separating process wastes through the installation of new piping systems is often the only way to properly handle wastes.

In order to design such a system, a detailed study of the existing sewer system, including dye studies to see which drains are connected, is required. Samples of wastes should be taken and a mass balance performed either on water or a specific chemical parameter to ascertain the validity of the sewer map.

Experience with a large number of industrial facilities has revealed two general rules regarding drainage diagrams:

1. No plant has an accurate sewer map—there is always unknown information.

2. After a detailed (and expensive!) survey of sewers, Rule No. 1 still applies.

Some solid waste streams can be segregated effectively through minor changes in equipment. A major source of hazardous waste at a number of industries is baghouse dust emanating from air pollution control equipment. As with sewer systems, common dust collectors were utilized for different production areas, resulting in a mixing of different types of dust and precluding recycling. Lewis[35] reports that FMC Corporation was able to modify dust collectors as illustrated in Fig. 7-12, so that each compartment served a single source.

Major impediments to waste segregation programs are materials that go into plant trash but don't belong there. Examples include laboratory samples, which must be disposed of as hazardous wastes if they have any of the four hazardous waste characteristics (corrosive, ignitable, reactive, or fail the TCLP test). Other materials include solvents and pigments, for which special receptacles must be provided.

Re-use

Many materials that are disposed of as hazardous waste have a potential to be re-claimed for another application. In some instances, contaminated materials may be of adequate quality to serve as a solvent or cleaning material for a less sensitive application. In the printing industry, toluene is often used as both a cleaning agent for presses and a thinning agent for the inks, prior to their being placed on the presses. If the toluene used for cleaning is limited to cleaning one specific color ink, it may be possible to re-use that toluene to thin the same color ink.

The acid recovery process used by the steel industry provides an example of re-use of a liquid waste stream. Iron scale is removed by an acid cleaning process referred to as "pickling." The spent pickle liquor is a hazardous waste that must be neutralized

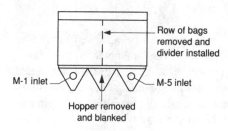

FIGURE 7-12
Modification of baghouse to segregate dust.
(*Reprinted with permission from Air and Waste Management Association.*)

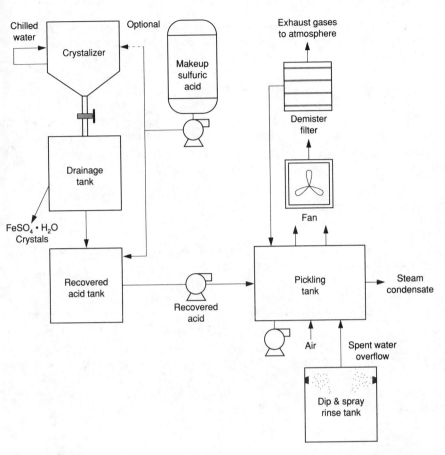

FIGURE 7-13
Acid recovery process for steel pickle liquor.[36]

and the resulting sludges are still considered hazardous and require further treatment or secure land disposal. The recovery process shown in Fig. 7-13 is reported[37] to reduce virgin sulfuric acid use by as much as 40%. The capital cost of recovery facilities exceeds the cost of neutralizing the waste. However, when the savings associated with reduction in purchase of acid and elimination of waste treatment are included, a 100,000 ton/year facility realizes a net annual savings of $350,000.

Another source of hazardous wastes is raw materials stored past their expiration date. Such dates are often established conservatively and it may be possible to re-assay and requalify the material for use in production. A recent report by the U.S. Navy suggests that a total evaluation of specified shelf lives may be appropriate:

Shelf life term is specified by Item Managers who are not usually chemists and, therefore, technically unqualified to make such an assessment. As a result, the suppliers are relied upon to specify a "safe" shelf term. The manufacturers have found that shelf term is inversely proportional to Navy sales and, accordingly, often recommend shelf lives that are unrealistically short. ... It is possible that some goods may not even require specification of a shelf life.[38]

7-5 TOXICITY REDUCTION

A number of waste minimization techniques reduce the concentration of contaminant in a liquid or solid waste stream, without necessarily diminishing the volume of wastes produced. It is often possible to lessen the toxic characteristics sufficiently so that the remaining waste is no longer considered a hazardous waste.

Process Modification

Process modifications were discussed in Sec. 7-4. Some specific examples of modifications that result in a less toxic waste stream include dry powder painting and solvent recovery methods.

Powder coating technology has reduced the need for solvent-based paints in many industrial painting applications. The process consists of spraying a heat fusible powder on a metal surface. An electrode in the spray gun charges the powder particles. The metal to be coated carries an opposing charge and the particles are attracted to the metal surface. Applying heat to bind the paint to the metal completes the process. Overspray is collected and recycled. In addition to eliminating a hazardous waste stream, this process reduces material, labor, and energy costs when compared to conventional solvent based painting.[39] It also reduces air emissions. Where solvents must be used they should be recycled (see Sec. 7-6).

Equipment Modification

As in the discussion of volume reduction, many pieces of equipment can be modified to reduce waste characteristics. Fairly innocuous waste streams can become hazardous wastes through spills or intentional batch dumps of concentrated liquids. Simple devices such as alarms on plating baths can avoid batch dumps and ensure that spent plating baths are transferred to the proper containers for reclamation of useful metals.

Where small quantities of powdered chemicals are used, it is common practice to purchase these materials in 50-pound bags. Initially an operator would simply cut the bags open and dump the powder into a hopper. Health and safety concerns subsequently required that the operating area where the bags are cut and dumped be properly ventilated to protect workers. Then air pollution control equipment in the form of a wet venturi scrubber was added to the ventilating system to avoid violating air pollution laws. This resulted in a liquid hazardous waste (the spent scrubber solution) being produced. The existing system was not designed as a whole, but rather, parts were added to meet changing regulatory requirements. The installation of a small packaged bag dumping station, where bags are placed in a sealed contained box and then cut open, eliminates a major waste source and a lot of unneeded equipment. What is required is a new look at the equipment.

Housekeeping Practices

Perhaps the simplest and most inexpensive methods of reducing hazardous wastes at the source is through revision of housekeeping and maintenance procedures. Table 7-7 provides a list of items to look for in revising such procedures.

Materials handling procedures provide a good place to initiate the search for opportunities to reduce waste. The design of storage areas should allow for easy inspection of potential sources of leaks. For example, in a drum storage area, aisles should be wide enough to permit access for routine inspection. Specifying the inspections and not permitting the inspector to see potential problems is counter-productive. Adequate lighting must be provided in the design of storage areas. Other practices include keeping aisles clear of debris and keeping different chemicals separate to preclude contamination or reaction. Drums that remain on concrete pads often corrode due to "concrete sweating." Use of pallets avoids this.

Training must be a major element of housekeeping practices. It is essential to have an on-going education program to account for turnover of production staff. The best designed waste minimization plan will not work without adequate instruction of the personnel involved. Employees should be taught the implications of their actions, and their thoughts on how to reduce problems should be solicited. For example, while the disposal of spent solvent in a cooling water sump may be clearly inappropriate to an environmental engineer, operating personnel will not know this unless someone takes the time to tell them. They will not take seriously rules about where to dump waste, unless the reasons are explained. Soliciting waste minimization suggestions from operating staff often identifies worthwhile ideas. Some industries provide cash awards for waste minimization suggestions.

Labeling is a simple housekeeping task that can markedly reduce the unnecessary generation of hazardous waste. A specific example is the use of solvents. Plants where both chlorinated and non-chlorinated solvents are used have found that clearly marking disposal storage tanks and sensitizing staff to the need to keep these materials separate has greatly reduced the generation of hazardous waste.

Finally, once a new housekeeping program is underway, it should be monitored to determine that the objectives are being met. A check on housekeeping practices should be part of routine inspections of the operating units. Employees should be encouraged to suggest additional practices that will further reduce waste generation.

Material Substitution

Most manufacturing processes were designed with product quality and profitability as the primary design criteria. The changing economics of waste management are forcing a new look at production formulations. Among the many material changes that have proven to be cost-effective and reduce wastes are the use of non-cyanide baths in metal plating operations and the use of less toxic solvents in cleaning operations. Industries producing machine parts use large quantities of lubricating oils that become contaminated with metals and must be disposed of as a hazardous waste. Synthetic lube oils cost more but last longer than petroleum-based materials, and can be returned

TABLE 7-7

Good operating practices†

Practice	Program elements
Waste minimization assessments	Form a team of qualified individuals Establish practical short-term and long-term goals Allocate resources and budget Establish assessment targets Identify and select options to minimize waste Periodically monitor the program's effectiveness
Environmental audits/reviews (see Chap. 6)	Assemble pertinent documents Conduct environmental process reviews Carry out a site inspection Report on and follow up on the findings
Loss prevention programs	Establish Spill Prevention, Control, and Countermeasures (SPCC) plans Conduct hazard assessment in the design and operation phases
Waste segregation	Prevent mixing of hazardous wastes with non-hazardous wastes Isolate hazardous wastes by contaminant Isolate liquid wastes from solid wastes
Preventative maintenance	Use equipment data cards on equipment location, characteristics, and maintenance Master preventive maintenance (PM) schedule Deferred PM reports on equipment Maintain equipment breakdown reports Keep vendor maintenance manuals handy Maintain computerized repair history file

Training/awareness-building	Provide training for:
	  Safe operation of the equipment
	  Proper materials handling
	  Economic and environmental ramifications of hazardous waste generation and disposal
	  Detecting releases of hazardous materials
	  Emergency procedures
	  Use of safety gear

Employee participation	"Quality circles" (free forums between employees and supervisors) to identify ways to reduce waste Solicit employee suggestions for waste reduction ideas
Production scheduling/planning	Maximize batch size Dedicate equipment to a single product Batch sequencing to minimize cleaning frequency (e.g., light-to-dark batch sequence) Schedule production to minimize cleaning frequency
Cost accounting/allocation	Cost accounting done for all waste streams leaving the facilities Allocate waste treatment and disposal costs to the operations that generate the waste

†Adapted from EPA.[40]

to the manufacturer for recycling. The analysis to evaluate the efficacy of such a proposal is to compare a low cost–high volume material against a high cost–low volume substitute.

Polaroid Corporation has a Toxic Use and Waste Reduction Program that groups chemicals into 5 categories:

I and II	Carcinogens, highly toxic materials, etc. (Category I chemicals are those selected for special priority in reduction)
III	Suspected carcinogens, moderately toxic materials, eye irritants, corrosives, etc.
IV	All other chemicals
V	All other wastes

Example 7-5 indicates how they applied their waste reduction program to the analysis of a new dye.

Example 7-5 Polaroid Corporation. A new dye. "In 1987 Polaroid's research division began evaluating new dyes to replace an existing material used for many years. After completing the research stage, several candidate dyes had been identified that delivered the desired product improvement (a spectrum that better matched the requirements of the photographic use). Each candidate also had a higher molar absorptivity than the existing material, so less dye per picture was required.

"The candidate dyes were evaluated by the process research and development group for 'manufacturability.' Due to considerations of ease of production, the choices were reduced to two. Serious process evaluation on the finalists began. One of the key process steps of each new candidate dye, as well as of the existing product, was an oxidation step using Cr (VI) in aqueous pyridine solvent. Cr (VI) is a Category II material in Polaroid's system, and the workup of the oxidized product was complicated by the mixed solvent, which was also difficult to recycle.

"A review of the literature led us to investigate possibilities of catalyzed air oxidations. Conditions were ultimately found to successfully implement this technology, completely eliminating the Category II Cr (VI) compound, and to link it to readily recoverable solvents. Figure 7-14 plots the impact of this new dye on toxic use and by-product generation per film pack. The combination of increased molar absorptivity and greatly reduced process waste results in an 80 percent reduction in the waste being generated using the new dye, compared with the old. In addition, the wastes that are generated are more readily treated, so disposal costs are lower per pound of waste generated. Overall, this substitution is saving Polaroid about $1 million a year. For most chemical processes, our experience is that the minimum waste process is usually the minimum net cost process."[41]

7-6 RECYCLING

When it is not possible to reduce the volume or toxicity of a waste, it may be possible to recycle it to another process or another plant. Other factors being equal, on-site recycling is preferable because shipping hazardous waste off-site, even for recycling, carries the liability that the waste might be mishandled.

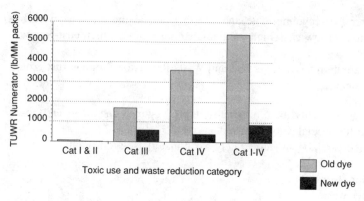

FIGURE 7-14
Toxic use and waste reduction—old dye compared to new dye. (*Reprinted with permission from POLLUTION PREVENTION REVIEW, All rights reserved.*)

Water

Water is in most instances the easiest material to recycle, and this is the first place one should look when evaluating prospects for recycling. In terms of hazardous waste, aqueous waste streams that meet the characteristics of hazardous waste (e.g., toxic, corrosive, reactive) must be treated as a hazardous waste. If these waste streams can be recycled to some useful purpose, a hazardous waste is effectively eliminated.

In the instance of an extremely acidic line, re-use for waste treatment in neutralization of caustic wastes is an obvious solution. Rarely does waste acid exactly balance with alkali, but the total cost of chemicals can be cut by maximizing the recycle of high and low pH lines.

It is often possible to recycle process water to a less sensitive use. Often a high quality water will become degraded in the manufacturing process and no longer suitable for use in that process, but may be adequate for another purpose. Exxon Corp., at its Baytown, Texas, refinery, recycles part of its wastewater to wet gas scrubbers.

Another potential source of recycle water is non-contact cooling water. Often such water is simply discharged as a heated but uncontaminated waste stream. If this source could be recycled, it would reduce the need for raw water make-up in some other part of the plant and perhaps some of the energy in the heated water could be conserved as well.

Solvents

Solvent recycling is common practice in many industries and a wide range of solvents are currently recycled. Some of the more common solvents currently being recycled are shown in Table 7-8. Azeotropic distillation is commonly used to enhance solvent recovery. An azeotrope or azeotropic mixture is a mixture of liquids that behave as a single substance when boiled (i.e., both the vapor and the liquid have the same composition). Azeotropic distillation consists of adding a substance (typically steam) to form an azeotrope with the solvent to be recycled. The azeotrope will have a lower boiling point than the original mixture and recovery of the solvent will be facilitated. Azeotropic boiling points of compounds not listed in Table 7-8 may be found in standard references.[42,43]

Packaged solvent distillation units are available from a variety of manufacturers and usually consist of simple distillation units applicable to low boiling point solvents. More sophisticated units are now becoming available that utilize vacuum distillation to handle solvents with higher boiling points (above 400°F). While package units are available at sizes as small as three gallons, for very small operations it may be more appropriate to return spent solvents to the original vendor or to a commercial recycle operation.

TABLE 7-8
Commonly recycled solvents[†]

Solvent	Boiling point, °F	
	Atmospheric	Azeotropic
Aliphatic hydrocarbon		
Hexane	157.0	142.9
Heptane	209.0	174.8
Aromatic hydrocarbon		
Benzene	176.0	157.0
Aromatic naphtha		
Toluene	232.0	185.0
Xylene	261–318	202.1
Turpentine		
Chlorinated hydrocarbon		
Trichloroethylene	189	163.8
Perchloroethylene	249	189.7
Methylene chloride	104	101.2
Alcohols		
Ethyl	173	172.7
Isopropyl	180.1	176.5
Ketones		
MEK	175	164.1
MIBK	241	190.2
Esters		
Ethyl acetate	170.9	158.7
Butyl acetate	259.2	194.4
Mixtures of solvents		
Toluene/xylene		
Ketones		
Alcohols		
Phenols		

[†] Adapted from Freeman[44] and Higgins.[45]

Another effective way to recycle solvents is **cascade re-use**, where a solvent is used once for high quality cleaning and then used again for cleaning operations that do not require a pure solvent.

Higgins[46] presents a case study illustrating the use of distillation to recover de-ionized (DI) water and spent solvent from electronics manufacturing. N-methyl-2-pyrrolidone (NMP) is used to clean semiconductors. NMP becomes contaminated with tin, lead, oil, and grease. The cleaning process is followed by a DI water rinse, and the water is degraded with NMP and its contaminants. A wide range of purification processes were evaluated and laboratory studies demonstrated that vacuum distillation provided the required recovery of both NMP and DI water. The system is illustrated in Fig. 7-15.

Oil

Oil that has become contaminated with hazardous materials may require disposal as a hazardous waste. Often this material can be recycled rather than shipped for off-site disposal.

Adolph Coors Co. in Golden, Colorado, reports a 40% increase in oil reclamation using ultrafiltration (see Sec. 9.7) when compared to dissolved air flotation.[48]

The energy content of spent oil can be reclaimed in many instances, through burning in industrial boilers or furnaces. This is a common practice at power plants in California where state law permits the mixing of oil recovered from wastewater treatment with fuel oil for burning in boilers. Similarly, most refineries recycle slop oil from wastewater treatment plants.

Solids

A wide range of solid materials, including paper, metals, and plastics, are amenable to recycling. The recycling of paper and paperboard has become routine at most industries. Scrap metal such as swarf from machining operations is also a common candidate for recycling. The recycling of plastics is not as widespread. However, as the following example illustrates, it is a feasible alternative to incineration or land disposal of these materials.

Polystyrene (PS), $(C_6H_5CHCH_2)_n$, represents a major component in plastic waste. Kampouris, et al.[49] present laboratory studies of a solvent method of recovering PS scrap. The method consists of five steps:

1. Dissolution with solvents.
2. Filtering.
3. Re-precipitation of PS.
4. Washing of polymer grains.
5. Drying.

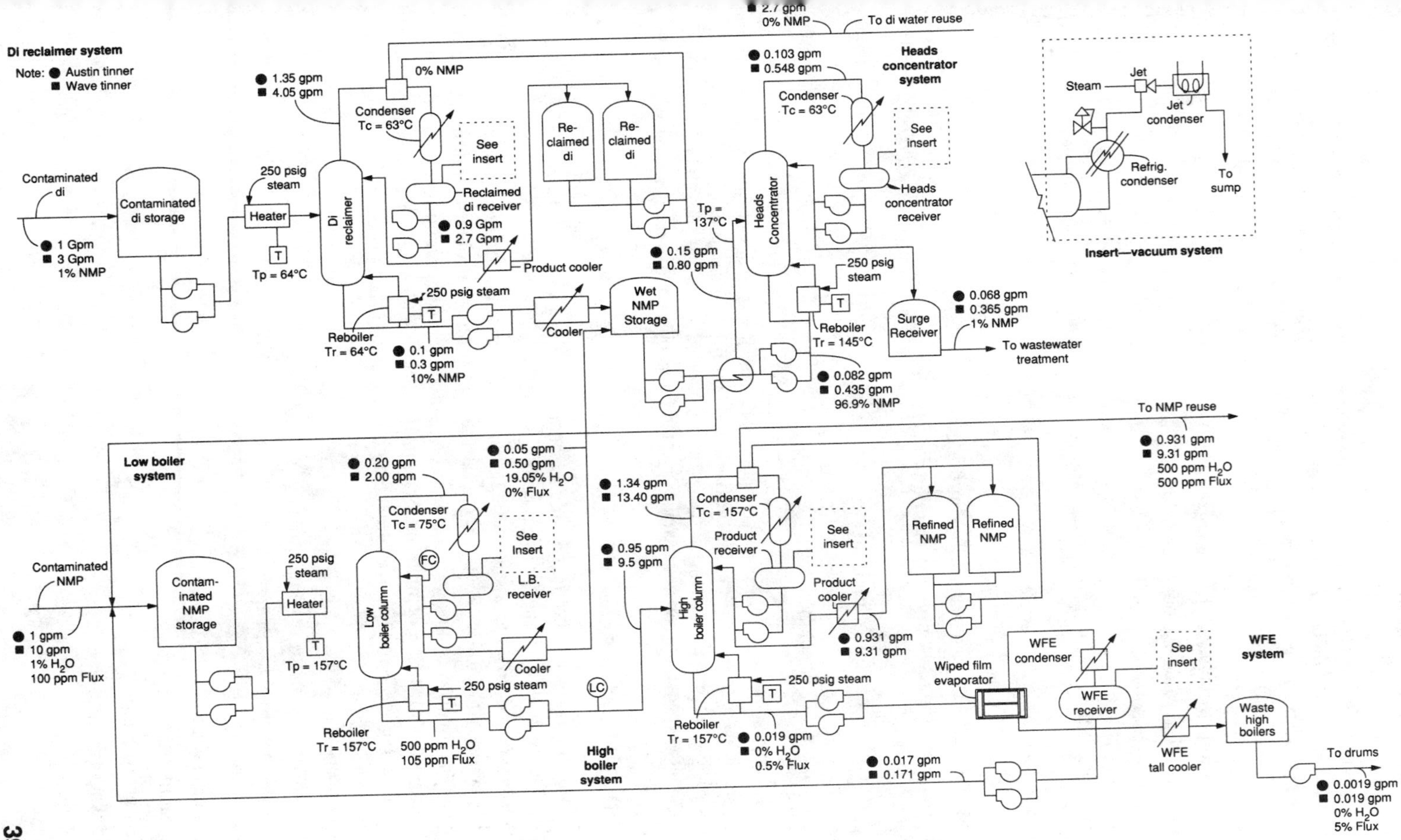

FIGURE 7-15
Vacuum distillation system.[47]

Di reclaimer system
Note: ● Austin tinner
■ Wave tinner

Contaminated di
● 1 Gpm
■ 3 Gpm
1% NMP

Contaminated di storage

250 psig steam
Heater
T
Tp = 64°C

Di reclaimer

● 1.35 gpm
■ 4.05 gpm
Condenser Tc = 63°C

To di water reuse

0% NMP

See insert

Reclaimed di receiver
● 0.9 Gpm
■ 2.7 Gpm

Re-claimed di
Re-claimed di

250 psig steam
Reboiler Tr = 64°C
T
● 0.1 gpm
■ 0.3 gpm
10% NMP

Product cooler

Cooler

Wet NMP Storage

● 0.15 gpm
■ 0.80 gpm

Tp = 137°C

Heads concentrator system

● 0.103 gpm
■ 0.548 gpm
Condenser Tc = 63°C

See insert

Heads concentrator receiver

2.7 gpm
0% NMP

Heads Concentrator

250 psig steam
Reboiler Tr = 145°C
T
● 0.082 gpm
■ 0.435 gpm
96.9% NMP

● 0.068 gpm
■ 0.365 gpm
1% NMP

Surge Receiver

To wastewater treatment

To NMP reuse

● 0.931 gpm
■ 9.31 gpm
500 ppm H_2O
500 ppm Flux

Insert—vacuum system

Jet
Steam
Jet condenser
Refrig. condenser
To sump

Low boiler system

Contaminated NMP
● 1 gpm
■ 10 gpm
1% H_2O
100 ppm Flux

Contaminated NMP storage

250 psig steam
Heater
T
Tp = 157°C

Low boiler column

● 0.20 gpm
■ 2.00 gpm
Condenser Tc = 75°C

FC

See Insert

L.B. receiver

● 0.05 gpm
■ 0.50 gpm
19.05% H_2O
0% Flux

Cooler

250 psig steam
Reboiler Tr = 157°C
T
500 ppm H_2O
105 ppm Flux

LC

High boiler system

● 1.34 gpm
■ 13.40 gpm

● 0.95 gpm
■ 9.5 gpm

High boiler column

Condenser Tc = 157°C

See insert

Product receiver

Product cooler
● 0.931 gpm
■ 9.31 gpm

250 psig steam
Reboiler Tr = 157°C
T
● 0.019 gpm
0% H_2O
0.5% Flux

● 0.017 gpm
■ 0.171 gpm

Refined NMP
Refined NMP

Wiped film evaporator

WFE condenser

WFE receiver

WFE tall cooler

WFE system

See insert

● 0.931 gpm
■ 9.31 gpm
500 ppm H_2O
500 ppm Flux

Waste high boilers

To drums
● 0.0019 gpm
■ 0.019 gpm
0% H_2O
5% Flux

397

The dissolution step employs miscible solvent/non-solvent mixtures. Solvents utilized included MEK and p-xylene, and non-solvents included methanol and n-heptane. These investigations also evaluated the separation of solvents for re-use.

The effectiveness of recycling was determined based on three criteria:

- Rheological behavior of the solvent-PS solution,
- Yield of useable PS polymer, and
- Adequate solvent separation.

The authors report several advantages of solvent systems to recover plastics:

- Volume of plastic refuse is reduced.
- A range of plastics are amenable to this type of recycling.
- The process removes impurities in the non-solvent phase.
- The re-precipitated polymer can be produced in many forms (e.g., granules, fibrils, or powders).
- The end product is of high quality.

This study concludes that

> ... the recovery of PS from foam waste by means of a solvent technique is feasible and even in the case of feeding with dirty stock, removal of contaminants can be ensured. Obviously, several parameters, such as waste transportation, solvent losses and solvents separation energy, have to be taken into account for an optimized operation of the process.[50]

Waste Exchanges

While reusing a waste product inside a manufacturing facility is the most desirable form of recycling, it is not always possible to find another department or process that can effectively utilize the waste. An alternative is to locate another company that can make use of your waste. A **waste exchange**[51] is a regional clearinghouse for such transactions. Waste exchanges maintain computer databases and/or publish periodic lists of wastes available or materials sought by various industries. A typical listing is illustrated in Fig. 7-16. The waste exchange information base would include:

- Company ID Code
- Category (e.g., acid, solvent, etc.)
- Primary usable constituents
- Contaminants
- Physical state
- Quantity
- Geographic area
- Packaging

Phenol Formaldehyde Resin
NE:A05-0242

Phenol formaldehyde resin from vat coating of fiberglass. Solution contains 54% phenolic resin in 23% ethanol with 1% formaldehyde by net weight. Currently 5 drums available. Thereafter 2500 lbs/qtr. No restrictions on amounts. Sample Available. MD

FIGURE 7-16
Waste exchange listing.[52]

DISCUSSION TOPICS AND PROBLEMS

7-1. Discuss how you would develop a waste minimization program for a small metal plating operation.

7-2. Discuss how you would identify potential changes in production methods that might be appropriate for the management of a large chemical plant to consider.

7-3. Discuss some of the impediments to implementing an effective waste minimization program.

7-4. Write a report identifying housekeeping and other activities that might be appropriate in minimizing the hazardous waste from a shop or laboratory identified by your instructor. Identify institutional impediments to implementing your solution.

7-5. Contact a waste exchange in your region and identify the types of wastes that have been successfully exchanged.

7-6. Identify institutional impediments to implementing a waste minimization program at a small manufacturing plant.

7-7. Identify institutional impediments to implementing a waste minimization program in your state.

7-8. Identify institutional impediments to implementing a waste minimization program for a manufacturing facility in a third world country.

7-9. Derive Equation 7-5 from 7-4 and 7-3.

7-10. A firm needs to expand its manufacturing capacity. The present facility could be expanded to meet the needs at a cost of $1.2 million. Waste treatment costs for the expanded facility would be about $184,000 per year. An entirely new facility that would reduce waste production could be built for approximately $3.35 million. The new facility would have an estimated annual waste treatment cost of $56,000. Assume a 20-year useful life for the facilities and a zero salvage value for both. If the prevailing interest rate is 10%, which alternative is preferable?

7-11. A company is faced with the decision of selecting either hazardous waste minimization process A or B. Process A involves purchasing machinery that costs $1.2 million with annual maintenance expenses of $150,000 for the first 10 years and $180,000 for the following 10 years. Process B calls for buying machinery that costs $1.5 million and has maintenance charges of $100,000 for the 20-year period. Both sets of machinery have zero salvage value. Which process is more economical for the company at a 15% rate of interest?

7-12. An industry is considering the construction of a hazardous waste treatment facility of 21 MGD (million gal/day) with the following annual costs and disbenefits:

Capital cost	$20.2 million
Annual O&M cost	$19,000/MGD

Transportation cost $375,000 per year

Loss of agricultural land $0.8 million per year

Benefits for the facility are:

Health benefits $0.64 million per year

Environment and ecosystem $2.25 million per year

Reuse of chemicals $0.6 million per year

Compute the B/C ratio. Calculate the present worth using interest rates of 10% and 15% and compare the results. The expected life for the facility is 10 years.

7-13. The following information is provided for two alternative liquid hazardous waste treatment facilities:

Plant	Flow rate (10^6 gal/day)	Capital costs (10^6)	Operating costs (10^6/yr.)
X	4.5	10.7	0.85
Y	9	16.8	1.63

Assuming a constant inflation rate of 9% over the 20-year life of each plant, determine which of the two alternatives is more economically feasible. Use an internal rate of return of 7%.

7-14. The benefits and disbenefits accruing from a proposed waste minimization project that would involve an initial cost of $7 million are the following:

Reduced air pollution $50,000 per year

Reduced future fees $0.3 million

Increased public image $0.4 million

Disbenefits:

Increased maintenance $1.5 million

Compute the B/C ratio of this project based on a 10% rate of interest for a 20-year project life.

7-15. Redo problem 7-11 assuming an inflation rate of 5% for Process A and 8% for Process B. How do your results differ from the previous computation?

7-16. A management review of the information in Table 7-3 indicates that another major factor, safety, should be included in the analysis. Management rates this as important as compliance or production impact. This assessment reveals that Alternative B should be rated −3 for safety. In the process of this assessment, the rating for production impact for alternative B is lowered from −1 to −3. Revise Example 7-4 to account for these changes. Indicate how this might affect the waste minimization decision.

For problems 7-17 through 7-26, use the manufacturing flowsheet from *Shreve's Chemical Process Industries* (5th edition, McGraw-Hill, 1984), and develop an outline of a waste minimization plan for the manufacturing process specified.

7-17. By-product Coking (Shreve's: Fig. 5.3, p. 76).

7-18. Hydrogen Production (Fig. 7.3, p. 109).

7-19. China Clay (Fig. 9.1, p. 152).

7-20. Portland Cement (Fig. 10.1, p. 175).

7-21. Superphosphate (Fig. 16.3, p.275).

7-22. Sulfuric Acid (Fig. 19.3, p. 330).

7-23. Nitrocellulose (Fig. 22.2, p. 397).

7-24. Photographic Film (Fig. 23.3, p. 419).

7-25. Water-Based Paint (Fig. 24.4, p. 432).

7-26. Styrene (Fig. 36.3, p. 696).

7-27. Prepare a one-page summary of the article by G. J. Hollod, et al.[53]

7-28. Prepare a one-page summary of the article by Y. Isooka, et al.[54]

7-29. Prepare a one-page summary of the article by S. T. Johnnie.[55]

7-30. Prepare a one-page summary of the article by L. R. Martin.[56]

7-31. Prepare a one-page summary of the article by L. L. Zelnio.[57]

7-32. Visit the plant manager of a local manufacturing facility suggested by your instructor. Prepare a brief summary of its current waste minimization program.

7-33. Contact your state environmental agency and determine what programs your state has to encourage hazardous waste minimization.

REFERENCES

1. U.S. Environmental Protection Agency: *Report to Congress: Minimization of Hazardous Waste*, Volumes I and II, EPA/530-SW-86-033A, Office of Solid Waste, p. ii, October 1986.
2. *Federal Register:* "Pollution Prevention Policy Statement," U.S. EPA, vol. 54, Jan. 26, 1989.
3. U.S. Congress: *Serious Reduction of Hazardous Waste: For Pollution Prevention and Industrial Efficiency*, OTA-ITE-317, Office of Technology Assessment, Washington, D.C., September 1986.
4. Colten, C. E.: "Historical Development of Waste Minimization," *The Environmental Professional*, vol. 11, pp. 94–99, 1989.
5. Koller, T.: *The Utilization of Waste Products*, Scott Greenwood & Son, London, 1915.
6. Trembler, H. A.: "Waste Savings by Improvements in Milk Plant Equipment," in *Proceedings of the Industrial Waste Conference*, Purdue University, pp. 6–21, November, 1944.
7. Besselievre, E. B.: "Industries Recover Valuable Water and By-Products from their Wastes," *Wastes Engr.*, vol. 30, p. 760, 1959.
8. Rocheleau, R. F., and E. F. Taylor: "An Industry Approach to Pollution Abatement," *Journal, Water Pollution Control Federation*, vol. 36, pp. 1185–1194, October 1964.
9. Bringer, R. P., and D. M. Benforado: "Pollution Prevention as Corporate Policy: A Look at the 3M Experience," *The Environmental Professional*, vol. 11, pp. 117–126, 1989.
10. Hollod, G. J., and R. F. McCartney: "Hazardous Waste Minimization: Part I. Waste Reduction in the Chemical Industry—DuPont's Approach," *J. Air Pollut. Control Assoc.*, vol. 38, no. 2, pp. 174–179, 1988.
11. U.S. EPA *Pollution Prevention News*, p. 8, Jan./Feb. 1991.
12. Lester, S. U.: "EPA says: 'Build a New Superfund Site or Lose Money to Clean up the Ones You've Got,'" *Everyone's Backyard*, vol. 6, no. 1, Citizens Clearinghouse for Hazardous Waste, Inc., Washington, D. C., pp. 6–8, Spring 1988.
13. U.S. Environmental Protection Agency: *Waste Minimization Opportunity Assessment Manual*, EPA/625/7-88/003, p. 4, 1988.
14. Hirschorn, J. S., and K. U. Oldenburg: "Waste Reduction Audit: Matching Types to Stages" in *Hazardous and Industrial Solid Waste Minimization Practices*, ASTM, STP 1043 (R.A. Conway, et al., eds.), Amer. Soc. for Testing and Matls, Philadelphia, PA, pp. 41–47, 1989.
15. U.S. Environmental Protection Agency, *Waste Minimization Opportunity Assessment Manual.*
16. Ghassemi, M.: "Waste Reduction: An Overview," *The Environmental Professional*, vol. 11, pp. 100–116, 1989.
17. Anderson, C. L.: "The Production Process: Inputs and Wastes," *J. Environ. Econ. Mgmt.*, vol. 14, no. 1, p. 11, 1987.
18. Ibid.
19. Ibid.
20. Smith, Dennis: "Intel Corp. Waste Minimization Program," Univ. Calif. at Davis, 1989.
21. *Hazardous Waste Minimization Manual for Small Quantity Generators*, Center for Hazardous Materials Research, Univ. of Pittsburgh, p. 10-2, October 1989.

22. Thomas, D. L., et al.: "A Technical Assistance and Waste Reduction Program for Illinois," *Waste Minimization–Proceedings of A&WMA Specialty Conference*, SP67, Air & Waste Management Assoc., pp. 128–141, Pittsburgh, 1988.

23. Hirschorn, J. S., and K. U. Oldenburg: "The Obstacles to Waste Reduction," *Chem. Engr. Progress*, pp. 31–35, June 1989.

24. *Hazardous Waste Minimization Manual for Small Quantity Generators.*

25. Greenwood, Mark A.: "Design for the Environment: DfE Concepts Are Being Integrated into OTS Activities, Chemicals in Progress Bulletin, Office of Toxic Substances, U.S. EPA, vol. 12, no. 4, December 1991.

26. Bailey, Paul E.: "Life-Cycle Costing and Pollution Prevention," *Pollution Prevention Review*, vol., 1, no. 1, Winter 1990/91.

27. Data compiled by Franklin Associates, Ltd., and provided courtesy of the Proctor & Gamble Company.

28. Ibid.

29. Calif. Health and Safety Code, Div. 20, Chapter 6.5, Article 11.9.

30. Calif. Code of Regulations (CCR) Title 22, Chapter 30, Article 6.1.

31. U.S. Congress: *Serious Reduction of Hazardous Waste: For Pollution Prevention and Industrial Efficiency*, OTA-ITE-317, Office of Technology Assessment, Washington, D.C., September 1986.

32. Hunt, G. E.: "Waste Reduction in the Metal Finishing Industry," *Journal, Air Poll. Control Assn.*, vol. 38, pp. 672–680, 1988.

33. Higgins, T. E.: *Hazardous Waste Minimization Handbook*, Lewis Publishers, Chelsea, Michigan, p. 81, 1989.

34. Thibault, J. E.: The Costs and Benefits of Source Reduction in Metal Finishing," *Massachusetts Hazardous Waste Source Reduction Conference Proceedings*, MA Dept. of Environmental Management, October 1983.

35. Lewis, D. A.: "Waste Minimization in the Pesticide Formulation Industry," *Journal, Air Poll. Control Assn.*, vol. 38, pp. 1293–1296, 1988.

36. U.S. Environmental Protection Agency: "Recovery of Spent Sulfuric Acid from Steel Pickling Operations," U.S. EPA, 625/2-78-017, 1978.

37. Ibid.

38. Roberts, R. M., J. L. Koff, and L. A. Karr: "Hazardous Waste Minimization-Initiation Decision Report," Tech. Memorandum TM 71-86-03, Naval Civil Engineering Laboratory, Port Hueneme, California, pp. 2–13, January 1988.

39. Higgins, *Hazardous Waste Minimization Handbook*, pp. 119ff.

40. U.S. Environmental Protection Agency, *Waste Minimization Opportunity Assessment Manual*, p. E-4.

41. Ahearn, J., H. Fatkin, and W. Schwalm: "Case Study: Polaroid Corporation's Systematic Approach to Waste Minimization," *Pollution Prevention Review*, pp. 257ff., Summer 1991.

42. R. C. Weast, ed.: *CRC Handbook of Chemistry & Physics*, 64th edition, CRC Press, Boca Raton, Florida, pp. D–9 to D–23, 1984.

43. Perry, R. H.: *Chemical Engineers' Handbook*, 6th edition, McGraw-Hill, New York, New York, pp. 13–57ff, 1984.

44. Freeman, H. M.: *Hazardous Waste Minimization*, McGraw-Hill, New York, New York, p. 328, 1990.

45. Higgins, *Hazardous Waste Minimization Handbook*, p. 56.

46. Ibid., pp. 65ff.

47. Ibid., p. 69.

48. McTeer, P., W. Word, and K. Lewandowski: "A New Application for Ultrafiltration Wastewater Treatment in the Industrial Environment," *Proceedings from 39th Industrial Waste Conference*, Purdue University, 1985.

49. Kampouris, E. M., C. D. Papaspyrides, and C. N. Lekakou: "A Model Recovery Process for Scrap Polystyrene Foam by Means of Solvent Systems," *Conservation & Recycling*, vol. 10, no. 4, pp. 315–319, 1987.

50. Ibid.

51. Slough, W.: "Waste Exchanges: Catalysts for Innovation in Waste Minimization," *J. Air Poll. Control Assn.*, vol. 38, p. 744, 1988.

52. Listings Catalog, Northeast Industrial Waste Exchange, Inc., Issue No. 36, p. 20, May 1990.
53. Hollod, G. J., and R. F. McCartney: "Hazardous Waste Minimization: Part I. Waste Reduction in the Chemical Industry—DuPont's Approach," *J. Air Pollut. Control Assoc.*, vol. 38, no. 2, pp. 174–179, 1988.
54. Isooka, Y., Y. Imamura, and Y. Sakamoto: "Recovery and Reuse of Organic Solvent Solutions," *Metal Finishing*, vol. 82, no. 6, pp. 113–118, June 1984.
55. Johnnie, Susan T.: "Waste Reduction In The Hewlett-Packard, Colorado Springs Division, Printed Circuit Board Manufacturing Shop," *Hazardous Waste Hazardous Mater.*, vol. 4, no. 1, pp. 9–22, 1987.
56. Martin, L. R.: "Demanding Waste Reduction: The Roles of Public Interest Organizations in Promoting the Institutionalization of Waste and Toxics Reduction," *The Environmental Professional*, vol. 11, pp. 132–141, 1989.
57. Zelnio, L. L.: "Reclamation of Metalworking Fluids from Individual Machine Tools," *Lubrication Engineering*, January 1987.

ADDITIONAL READING

Bridges, J. S., et al : "Results from a Cooperative Federal, State, and Trade Association Waste Minimization Research Program," *Hazard. Waste Hazard. Mater.*, vol. 6, no. 17, 1989.

Brinkman, D. W.: "Technologies for Re-Refining Used Lubrication Oil," *Lubrication Engineering*, vol. 43, no. 5, pp. 324–328, May 1987.

Caldart, C. C., and C. W. Ryan: "Waste Generation Reduction: A First Step Toward Developing a Regulatory Policy To Encourage Hazardous Substance Management Through Production Process Change," *Hazardous Waste Hazardous Mater.*, vol. 2, no. 3, pp. 309–331, 1985.

Cranford, B.: "Federally Sponsored Waste Minimization Research and Development for Hazardous and Non-Hazardous Wastes," *J. Air Pollut. Control Assoc.*, vol. 39, no. 34, 1989.

Davies, R., R. Curtis, and R. Laughton: "Recycling Metal Stamping Plant Wastes," *Water and Pollution Control*, vol. 123, no. 5, pp. 26–27, September/October 1985.

deYoung, R.: "Exploring the Difference Between Recyclers and Non-Recyclers: The Role of Information," *J. Environ. Sys.*, vol. 6, p. 341, 1989.

Freeman, H. M.: *Hazardous Waste Minimization-Industrial Overviews*, JAPCA Reprint Series-RS14, Air & Waste Management Assoc., Pittsburgh, Pennsylvania, 1989.

Johnson, J. C., et al.: "Metal Cleaning by Vapor Degreasing," *Metal Finishing*, September 1983.

Kalnes, T. N., et al.: "Recycling Waste Tube Oils for Profit (UOP Direct Hydrogeneration Process)," *Hazard. Waste Hazard. Mater.*, vol. 6, no. 51, 1989.

Kenson, R. E.: "Recovery and Reuse of Solvents from VOC Air Emissions," *Environmental Progress*, vol. 4, no. 3, pp. 161–165, August 1985.

Kikendall, T. R.: "Converting from Conventional to Compliance Coating Systems," *Industrial Finishing*, vol. 58, no. 6, pp. 40–42, June 1982.

Kolaczkowski, S. T., and B. D. Crittenden: *Management of Hazardous and Toxic Wastes in the Process Industries*, Elsevier Applied Science, 1987.

Loehr, R.: *Reducing Hazardous Waste Generation*, National Academy of Sciences, National Academy Press, 1985.

Junno, T. J., and H. M. Freeman: "Minimizing Plating Bath Wastes in the Electronics Products Industry," *J. Air Poll. Control Assn.*, vol. 37, pp. 723–729, 1987.

Overby, C.: "Design for the Entire Life Cycle: A New Paradigm?" in *Proceedings of the International Conference on Pollution Prevention* (EPA/600/9-90/039), June 1990.

Overcash, M. R.: *Techniques for Industrial Pollution Prevention*, Lewis Publishers, Chelsea, Michigan, 1986.

Sarokin, D., W. R. Muir, C. G. Miller, and S. Sperber: *Cutting Chemical Wastes*, Inform, Inc., New York, New York, 1986.

Theodore, L. and McGuinn, Y. C., *Pollution Prevention*, Van Nostrand-Reinhold, New York, New York, 1992.

Tsai, E. C., and R. Nixon: "Simple Techniques for Source Reduction of Wastes from Metal Plating Operations," *Hazard. Waste Hazard. Mater.*, vol. 6, no. 67, 1989.

U.S. Environmental Protection Agency, *Waste Minimization Issues and Options*, Volumes I, II, and III, EPA/530-SW-86-041, Office of Solid Waste, Washington, D. C. (NTIS:PB87-114351, PB87-114369 and PB87-114377), October 1986.

U.S. Environmental Protection Agency: *Proceedings of the International Conference on Pollution Prevention*, EPA/600/9-90/039, June 1990.

U.S. Environmental Protection Agency: *Report to Congress: Minimization of Hazardous Waste*, Volumes I and II, EPA/530-SW-86-033A, Office of Solid Waste, Washington, DC (Available from NTIS:PB87-114336 & PB87-114344), October 1986.

U.S. Environmental Protection Agency: *Seminar Publication—Solvent Waste Reduction Alternatives*, EPA/625/4-89/021, September 1989.

Waste Minimization—Proceeding of A&WMA Specialty Conference, SP67, Air & Waste Management Assoc., Pittsburgh, Pennsylvania, 1988.

ACKNOWLEDGMENTS

The authors wish to acknowledge the following individuals who assisted with the preparation of this chapter: Mark Wendell, Ph.D., and Brian Bennett, P.E., *ERM, Inc.*; Donald Grogan, *ERM-Taiwan*; Donald Landeck, *ERM-West*; and Sharmistha Law, *Bucknell University*. The principal reviewer for the **ERM Group** was Amir Metry, P.E., Ph.D.

CHAPTER
8

FACILITY DEVELOPMENT AND OPERATIONS

When society requires to be rebuilt, there is no use attempting to rebuild it on the old plan.

John Stuart Mill

A **facility** is defined as the contiguous land, structures, and other improvements and appurtenances used for storing, recovering, recycling, treating, or disposing of hazardous waste. This chapter covers the steps in the development and operation of hazardous waste management facilities.

This chapter begins with an explanation of facility types and their operations. Next, the chapter briefly highlights the step of assessing the need for facilities, thus providing a lead to the controversial subject of siting. This is followed by a review of how the public can participate in siting, including a case study of a successful project to site a major facility. Although this review of public participation addresses just the subject of siting, the general concepts of public communication are important to virtually all aspects of hazardous waste management. The chapter concludes with a discussion of permitting.

8-1 FACILITY TYPES

The concept of a facility dedicated to the management of hazardous waste is not new. Long before the enactment of U.S. hazardous waste laws, many generators recognized the need for specialized treatment and disposal of these wastes. Many generators constructed and operated their own captive facilities referred to as **on-site**

405

facilities. For example, Dow Chemical constructed a rotary kiln incinerator at its Midland, Michigan plant in 1954 for the primary purpose of thermal destruction of organic waste from the production of a great number of organic chemicals. Hundreds of plants developed simple, inexpensive surface impoundments.

Other generators, not having a suitable site or not generating a sufficiently large volume of waste to justify the investment in an on-site facility, transported their waste off site to specialized facilities for treatment and disposal. Such facilities are referred to as commercial, **off-site facilities**. The commercial hazardous waste management industry began developing off-site facilities in the late 1960s.They collected and transported waste to specialized off-site facilities for treatment and disposal. Some of these early facilities, such as the Rollins Environmental Services facility built in Bridgeport, New Jersey, in 1971, are still operating today, at the same sites but with vastly modernized equipment and operations.

Just as there are many types of hazardous waste, there are many ways in which hazardous wastes can be managed. In fact, there are at least 50 commercially proven technologies for the recovery and treatment of hazardous waste. A hazardous waste facility may function with just one technology, or it may combine multiple technologies, particularly if it is a commercial facility serving a number of generators.

The predominant types of facilities, other than storage facilities, are depicted in Fig. 8-1 under the following major categories:

- **Recovery/recycling facilities:** recover material as a salable product (typically solvents, oils, acids, or metals); some recover energy values in waste.
- **Treatment facilities:** change the physical or chemical characteristics of a waste, or degrade or destroy waste constituents, using any of a wide variety of physical, chemical, thermal, or biological methods.

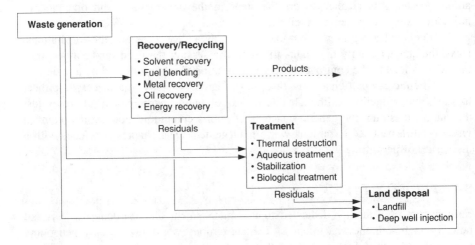

FIGURE 8-1
Hazardous waste recovery, treatment, and disposal facilities.

- **Land disposal facilities:** permanent emplacement of waste on or below land surface.

There are some differences between a commercial, off-site facility and a captive, on-site facility. Public reaction to their siting contrasts remarkably. This derives in part from the fact that an off-site facility accepts waste from outside the community, while an on-site facility handles only that waste generated by what could be a long-standing and important economic activity in the community. From a technical perspective, the off-site facility generally handles a wider range of waste types and is typically larger and more complex. Ignoring this, however, the risks posed by off-site and on-site facilities are comparable, and depend far more upon types of waste received, design, operation and other site-specific factors. The two must meet the same hazardous waste regulations. Thus, no further technical distinction is made in this chapter.

Numerous references are available that describe in detail the various types of facilities.[1] In fact, some books may be dedicated to a single type.[2] The major types of facilities are discussed in Chaps. 9 through 13 of this textbook. The following summaries are developed from a report intended to introduce the reader to basic categories of facilities.[3] Within each category, many subcategories exist dependent upon a particular technology employed.

Recovery Facilities

Solvent recovery. These facilities separate contaminants from waste solvents and thus restore the solvent to its original quality or to a lower grade product (e.g., lacquer thinner). This practice is fairly widespread for halogenated solvents. Distillation (batch, continuous, or steam) is used by most commercial solvent processors, and typically recovers about 75 percent of the waste solvent (see Sec. 7-6). The residue, known as still bottoms, can be a liquid or a sludge, depending upon a number of conditions, and typically requires management as a hazardous waste. Other separation technologies used by solvent processors include filtration, simple evaporation, centrifugation, and stripping.

Oil recovery. Used lubricating oils can be recovered to a quality essentially equal to virgin lubricating oils. This process is typically referred to as oil re-refining. The term oil recovery excludes the larger practice of fuel blending, which does not recover lubricating oils. Two methods may be employed in waste oil re-refining: the acid/clay method or the capital-intensive distillation method. Because the former produces a large amount of spent acidic oily clay residue, distillation is preferred.

Acid regeneration. Acid recovery usually involves the separation of unreacted acid from an acid waste such as spent pickle liquor generated by the steel industry. One method, used in the steel industry, is to cool sulfuric acid to precipitate ferrous compounds. In another method, acid can be regenerated by injecting it into a spray roaster.

Metals recovery.   The technologies for recovering metals can be grouped as pyrometallurgical or hydrometallurgical. Pyrometallurgy uses the differences in melting and boiling properties to separate metals at high temperatures. The heat is typically provided by roasting or smelting. The hydrometallurgy technology extracts and concentrates metals from liquid waste using any of a variety of processes such as ion exchange, electrodialysis, reverse osmosis, membrane filtration, adsorption, sludge leaching, electrowinning, solvent stripping, and precipitation. Non-liquid wastes first require dissolution.

Fuels blending.   Waste-derived fuels can be made from waste oils, solvents, and distillation bottoms by either blending different wastes that have high BTU values or by mixing wastes with coal or fuel oils. Wastes, particularly oils, normally need pretreatment to remove bottoms, sediments, and water. This is achieved through separation and dehydration. Fuels blending is a simple process that is well developed.

Co-incineration in industrial kilns/furnaces.   The manufacturers of cement, light aggregate, and some other products use a special type of rotary kiln. The destruction of hazardous waste by co-incineration with primary fuels and raw materials in such industrial kilns has been practiced for a long time. Cement kilns operate at high temperatures (2600°F to 3000°F) and provide very long residence times, which make possible very high destruction of Principal Organic Hazardous Constituents (POHCs, see Sec. 12-2). The waste thus represents a supplementary fuel. The waste may be blended at the site where the waste is burned or blended at a separate facility and shipped to the site of the industrial kiln.

An example of a recovery facility is shown in Fig. 8-2. This particular facility has the capability to recover waste solvents and oils.

Treatment Facilities

Thermal destruction.   Exposing waste to high temperatures in the presence of oxygen results in either partial or complete destruction of organic waste materials. The most common application is an incinerator, which is an "enclosed device using controlled flame combustion, the primary purpose of which is to thermally break down hazardous waste."[4] Existing federal regulations specify that incineration must achieve a destruction and removal efficiency (DRE) of at least 99.99%. Chap. 12 describes several technologies for achieving thermal destruction.

Aqueous treatment.   An aqueous waste treatment system removes and/or detoxifies hazardous constituents that are dissolved or suspended in water. The selection and sequence of unit processes are determined by the characteristics of incoming wastes and the required effluent quality. Figure 8-3 shows the unit process diagram for destruction of aqueous waste containing cyanide. (Cyanide destruction is discussed in Sec. 9-5.)

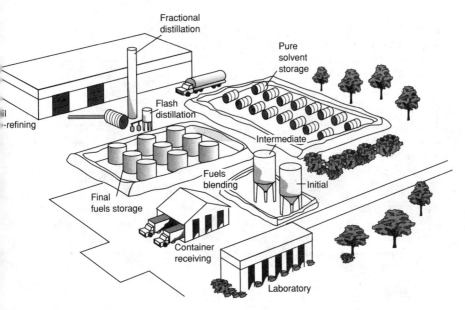

FIGURE 8-2
Liquid organics recovery facility.[5]

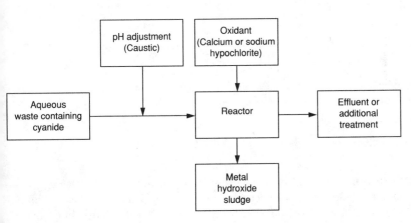

FIGURE 8-3
Cyanide destruction process diagram.

An example of a full-service aqueous treatment facility is shown in Fig. 8-4. Typically, it would feature most, if not all, of the following unit processes and could treat essentially any aqueous waste:

- Cyanide destruction
- Chrome reduction
- Two-stage metal precipitation

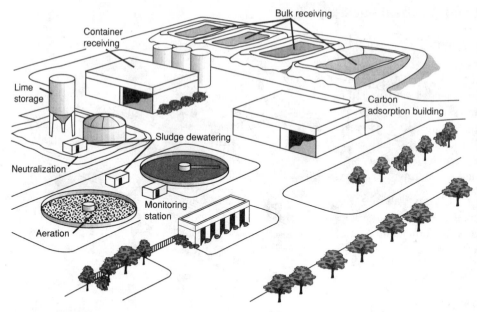

FIGURE 8-4
Full-service aqueous treatment facility.[6]

- pH adjustment (neutralization)
- Solids filtration
- Biological treatment
- Carbon adsorption
- Sludge dewatering

Stabilization. Stabilization involves admixing materials with wastes to improve the handling and physical characteristics of the wastes and to reduce mobility of contaminants in a landfill environment. Stabilization may either induce a chemical change to produce an essentially insoluble form or chemically bind the waste in a matrix of crystal lattice having a very low solubility. Several types of stabilization processes have been developed, and these are described in Chap. 11.

Biological treatment. Biological treatment systems utilize microorganisms to degrade organic wastes. Organic substances are either mineralized or transformed into compounds having a lower molecular weight. The primary process variables are the method of contacting the waste with biomass (microbes), the moisture content of the waste (e.g., liquid, slurry or solid), and the method and degree of aeration. Biological methods are explained in Chap. 10.

Land Disposal Facilities

Landfill. A landfill involves permanent emplacement of hazardous waste. In many instances the waste will undergo some form of pretreatment (e.g., stabilization) prior to land disposal. Emplacement would occur either in relatively shallow trenches or in mounded cells or vaults that extend mostly, if not totally, above grade. A landfill facility has several protective elements (e.g., liners), both design and operational, to contain emplaced wastes, to minimize the generation of leachate, and to remove leachate as it is generated. It is important to note that landfills should be operated as containment facilities with necessary provisions for virtually perpetual maintenance and monitoring after closure. Landfills are described in Chap. 13.

Deep well injection. This disposal method features injection of liquid waste into a deep, porous subsurface geological formation containing salt water. The receiving formation is well below potable aquifers and should have an impermeable confining layer above it. The injection well consists of the injection pipe surrounded by a series of concentric casings with the annuli containing cement and noncorrosive fluids to detect and prevent leakage into shallow formations.

Other. Surface impoundments are used for the retention of waste liquids and sludges. They require liners and many of the other protective elements associated with a landfill, particularly if they are closed with some waste left in place. A land farm is a biological treatment method in which soil is the treatment medium; however, some waste constituents are not degraded but incorporated into the soil, thus representing a form of disposal.

Fully Integrated Facility

Some large commercial facilities employ aqueous treatment, incineration, land disposal, and possibly other components to form a fully integrated facility as depicted in Figs. 8-5, 8-6, and 8-7.

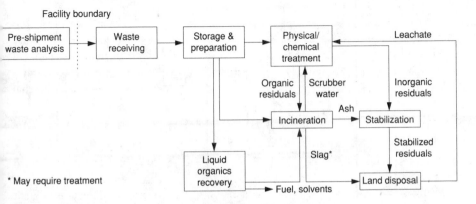

FIGURE 8-5
Waste management flow diagram for fully integrated hazardous waste management facility.

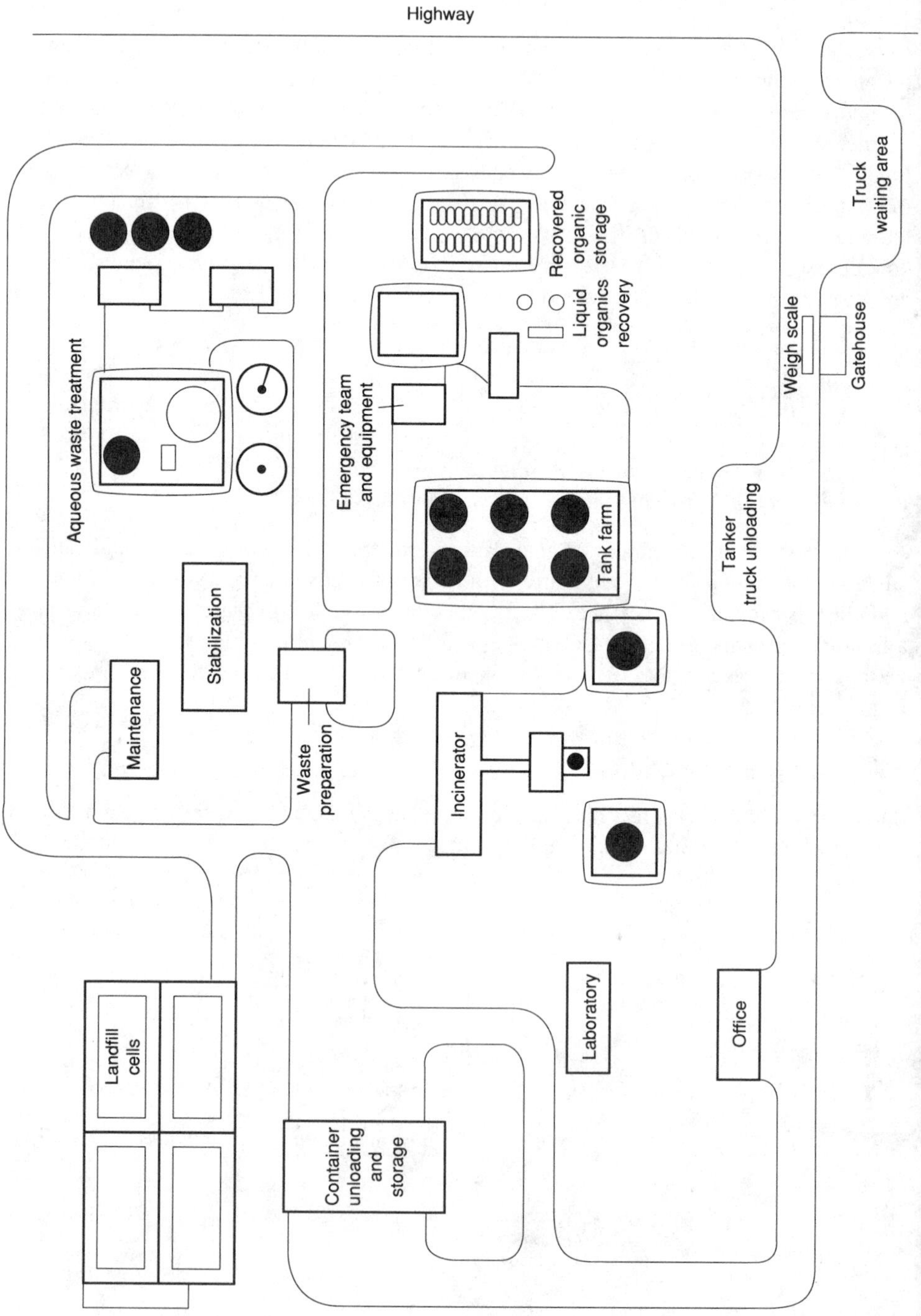

Highway

Truck waiting area

Aqueous waste treatment

Recovered organic storage

Liquid organics recovery

Emergency team and equipment

Maintenance

Stabilization

Weigh scale

Gatehouse

Tank farm

Waste preparation

Tanker truck unloading

Incinerator

Landfill cells

Laboratory

Office

Container unloading and storage

412

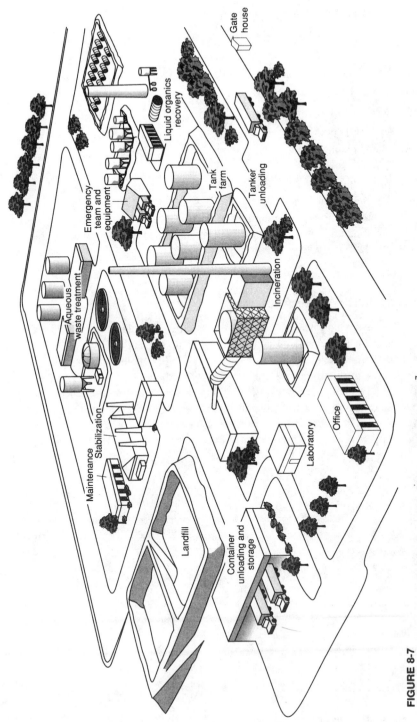

Gate house

Liquid organics recovery

Emergency team and equipment

Tank farm

Tanker unloading

Aqueous waste treatment

Incineration

Stabilization

Maintenance

Office

Laboratory

Landfill

Container unloading and storage

FIGURE 8-7
A hypothetical fully integrated hazardous waste management facility.[7]

413

8-2   FACILITY OPERATIONS

The operations at a hazardous waste facility consist of five subsystems:

1. Pre-shipment waste analysis
2. Waste receiving
3. Waste storage and preparation
4. Waste treatment
5. Residuals management

Figure 8-8 matches these subsystems with the components of the fully integrated facility depicted in the preceding figures. It is important to note that all components operate under an umbrella of a number of special measures. These special precautionary measures include security, inspections, maintenance, training, incident prevention, emergency planning, safety, monitoring, and auditing.

Pre-Shipment Waste Analysis

A waste analysis plan is a critical part of a facility. The plan specifies the parameters for which each waste will be analyzed, the sampling and analytical methods to be used, and the frequency of analysis. Before a facility treats, stores, or disposes of a waste, it must profile the waste, including a detailed chemical and physical analysis of a representative sample of the waste (see Fig. 8-9). Commercial facilities require this full characterization prior to shipment by the generator. Representative sampling of a waste shipment is conducted upon arrival at the facility to verify that the composition of the shipped waste matches the fully characterized waste. The purpose of the full

Facility components	Operations subsystems				
	Pre-shipment waste analysis	Waste receiving	Waste storage & preparation	Waste treatment	Residuals management
Analytical laboratory	✓	✓			
Truck waiting area		✓			
Gate house		✓			
Weigh scale		✓			
Drum unloading & storage		✓	✓		
Tank farm		✓	✓		
Bulk waste strg. & waste prep.		✓	✓		
Physical / chemical treatment				✓	✓
Incinerator				✓	✓
Stabilization plant				✓	✓
Landfill cells					✓

FIGURE 8-8
Operations subsystems and their components.

Waste Profile Form

Generator Name _____

Address _____

_____

City, State, Zip _____

Technical Contact _____

       (Name)   (Title)

Area Code ( ) Telephone # _____

Facility EPA ID # _____

Common Name of Waste _____

Billing Address _____

City, State, Zip _____

Technical Contact _____

       (Name)   (Title)

Area Code ( ) Telephone # _____

Business SIC Code _____

Duns # _____

Generating Process _____

Chemical Composition (Totals must add up to 100%)

_____ _____ %

_____ _____ %

_____ _____ %

_____ _____ %

_____ _____ %

_____ _____ %

_____ _____ %

_____ _____ %

_____ _____ %

_____ _____ %

_____ _____ %

EPA Haz

Waste # _____

      (Quan)   (Units)   P (Time Interval)

                        E

Rate of Generation _____

      (Quan)   (Units)   R

Volume in Storage _____   I (Container)

                        N

Is Waste DOT Hazardous ☐ (Yes) ☐ (No)

Proper DOT Shipping Name _____

Hazard Class _____ ID # _____

Transportation Equipment _____

Placarding _____

Metals (mg/l or ppm)

	Total	Leachate		Total	Leachate
Arsenic (As)			Selenium (Se)		
Barium (Ba)			Silver (Ag)		
Cadmium (Cd)			Copper (Cu)		
Chromium (Cr)			Nickel (Ni)		
Chromium, Hex			Zinc (Zn)		
Lead (Pb)					
Mercury (Hg)					

Inorganics (mg/l or ppm)

Total CN			Bromide		
Free CN			Iodide		
Sulfide			Asbestos		
Bisulfite					
Sulfite					
Sulfate					
Phosphate					
Fluoride					
Chloride					

Organics (mg/l or ppm)

Endrin			BOD		
Lindane			COD		
Methoxychlor			Organohalide		
Toxaphene			Organo-sulfur		
2,4-D			Mercaptans		
2,4,5-T					
Phenol (ics)					
PCB					
TOC					

Physical Description

Physical State ☐ Liquid ☐ Semi Solid ☐ Solid

Phases/Layering ☐ None ☐ Bilayered ☐ Multilayered

Total Solids (wt %) _____ Suspended Solids (wt %) _____

Type of Solids ☐ Organic ☐ Inorganic ☐ Mixed

Specific Gravity _____

Viscosity ☐ High ☐ Medium ☐ Low

Flash Point (°F) _____ Type _____

Boiling Point (°C) _____ Freezing Point (°C) _____

Vapor Pressure (mm Hg @ 25°C) _____

BTU/lb _____ % Ash Content _____

pH (Avg) _____ (Range) _____ To _____

Total Alkalinity/Acidity (%) _____

Odor _____

Color _____

Hazardous Properties ☐ None

☐ Ignitable   ☐ Corrodes Steel ☐ Toxic Vapor

☐ Reactive   ☐ Pyrophoric   ☐ Shock Sensitive

☐ Explosive   ☐ Water Reactive ☐ Radioactive

☐ Biological   ☐ Pathogen   ☐ Etiological

☐ Pesticide Residuals   ☐ Other _____

NFPA   **Toxicity**

Hazard   Oral _____

Identification   Dermal _____

                   Inhalation _____

Please attach all material safety data sheets, handling precautions, additional hazard information, support data & comments.

_____

_____

_____

_____

I believe that the above information is true, accurate and complete to the best of my knowledge and, that all known and suspected hazards have been disclosed.

_____ By _____   _____ _____

Date   Name   Title   Signature

FIGURE 8-9
Waste profile form.

characterization before shipment is to satisfy the following requirements:

1. Determine if the waste is acceptable for receipt at the facility in terms of (a) the facility's permit and (b) the capability of the facility to treat or dispose of the waste.
2. Identify the inherent hazards of the waste so that appropriate precautions can be taken during its handling and storage at the facility to prevent incidents.
3. Determine the physical characteristics and chemical constituents of the waste to allow selection of effective waste processing and disposal methods.
4. Select the verification parameters to be tested upon arrival at the facility. These parameters would ensure that each shipment of waste is the same type as the fully characterized waste.
5. Select any treatability parameters to be tested that could vary so as to influence how waste processing would be programmed.
6. Develop an estimate of the cost of treatment or disposal.

Waste Receiving

Waste shipments typically arrive by truck at a facility's gatehouse. Upon accepting the waste, the facility signs the manifest and sends a copy to the generator. At that point, the facility shares liability with the generator and the transporter. Thus, it is critical that the pre-shipment waste analysis has already been completed and the shipment scheduled. Without prior scheduling of the incoming shipment or if the shipment is improperly documented, the gatehouse will refuse entry of the truck.

Scheduled and properly documented shipments are directed to the receiving station where any packaging is checked, the loaded truck is weighed, and representative samples are collected for testing the verification parameters. The waste may arrive as bulk liquids in a tank truck, containerized liquids or sludges in drums, bulk shipments of contaminated soil in dump trucks, or by a number of other methods. Collecting a representative sample can pose a difficult task considering that a waste may be in multiple phases and states or have pockets of high contamination. The receiving station must use previously established procedures for each situation to ensure the collection of a representative sample.

Upon collection of the sample, the laboratory analyzes a portion for the verification parameters and retains the remainder for subsequent testing of treatability parameters. Upon verification of the waste shipment, the truck is directed to an unloading area where it is emptied and then reweighed before it leaves the facility.

The mere "emptying" of a truck can pose a difficult challenge if the waste has stratified, a container has leaked, or a solidification reaction has occurred. It is important that facilities have planned procedures and are prepared with special equipment to resolve such problems. The truck may need to be decontaminated to remove any trace residues.

Waste Storage and Preparation

Upon unloading, the wastes are moved into storage that can consist of tanks or impoundments for bulk liquids, hoppers for solids and sludges, or pads and warehouses for containers. The objectives of storage and preparation are fourfold: (a) store the waste safely before introduction as feed into the system of unit treatment and disposal processes, (b) provide adequate accumulation time during periods when treatment and disposal process systems are out of service, (c) facilitate mixing, blending, and repackaging of waste as deemed necessary, and (d) allow staged input of various wastes with reagents to the subsequent unit treatment processes.

An obviously important safety consideration is fire prevention and protection. The storage of certain types of hazardous waste requires automatic alarms and possibly sprinklers. The facility must provide adequate water supply for extinguishing fires plus the capability to collect and store fire water runoff. The storage or treatment of any water-reactive waste necessitates an alternative type of fire protection system.

A key issue in providing safe storage is compatibility. This has two independent considerations:

1. The compatibility of the waste with the material used to construct the container, tank, or liner in contact with the waste (e.g., certain solvents should not be stored in plastic containers).

2. The compatibility of the waste with other wastes stored together (e.g., containers of cyanide waste should not be located near acid waste).

The key is to segregate incompatible wastes by placing them in separate areas constructed of suitable materials. If stored together, incidents such as leaks could result in mixing of incompatible wastes. Different chemical reactions could occur. Some reactions could produce excessive pressure, thus posing fire or explosion hazards. Others could produce toxic fumes or gases. Figure 8-10 presents a compatibility chart and indicates that careful planning must be given to chemical storage.

Example 8-1   Waste compatibility. Given the following list of wastes stored in containers in a common area, what could happen in the event of a fire in which high-pressure water was used to extinguish it? To provide safe segregation, what is the minimum number of separate storage areas required? Does this list provide sufficient information to make this determination?

Waste A—Spent pickle liquor from a steel manufacturer

Waste B—Metal-plating rinsewater from cyanide bath

Waste C—Cadmium-bearing sludges from a metal plater

Waste D—Spent degreaser from an electroplater

Waste E—Solvent-based paint sludges from a metal-finishing operation

Waste F—Distillation bottoms from a phenolics manufacturer

Waste G—Laboratory packs from a university

Waste H—Pesticides with an expired shelf life

Waste I—Soil contaminated with a spill of diesel fuel

Waste J—Alkali cleaners

Waste K—Off-spec sodium product

Waste L—Arsenical wastewater treatment sludges from veterinarian product manufacturer

Waste M—Drum-reconditioning waste

No.	Reactivity group Name	1	2	3	4	5	6	7	8	9	10	11	12	13	14	15	16	17	18	19	20	21	22	23	24	25	26	27	28	29	30	31	32	33	34	101	102	103	104	105	106
1	Acids, minerals, non-oxidizing	1																																							
2	Acids, minerals, oxidizing		2																																						
3	Acids, organic		G H	3																																					
4	Alcohols & glycols	H	H F	H P	4																																				
5	Aldehydes	H P	H F	H P		5																																			
6	Amides	H	H GT				6																																		
7	Amines, aliphatic & aromatic	H	H GT	H		H		7																																	
8	Azo compounds, diazo comp. & hydrazines	H G	H GT	H G	H G	H			8																																
9	Carbamates	H G	H GT						H G	9																															
10	Caustics	H	H	H		H				H G	10																														
11	Cyanides	GT GF	GT GF	GT GF					G			11																													
12	Dithiocarbamates	H GF F	H GF F	H GF GT		GF GT		U	H G				12																												
13	Esters	H	H F						H G		H			13																											
14	Ethers	H	H F												14																										
15	Fluorides, inorganic	GT	GT	GT												15																									
16	Hydrocarbons, aromatic		H F														16																								
17	Halogenated organics	H GT	H F GT					H GT	H G		H GF	H						17																							
18	Isocyanates	H G	H F GT	H G	H P			H P	H G		H P G	H G	U						18																						
19	Ketones	H	H F						H G		H	H								19																					
20	Mercaptans & other organic sulfides	GT GF	H F GT						H G									H	H	H	20																				
21	Metals, alkali & alkaline earth, elemental	GF H F	GF H F	GF H F	GF H F	GF H F	GF H	GF H	GF H	GF H	GF H	GF H	GF GT H	GF H				H E	GF H	GF H	GF H	21																			
22	Metals, other elemental & alloys as powders, vapors or sponges	GF H F	GF H F	GF					E F GT	U	GF H							H E	GF H		H GF F		22																		
23	Metals, other elemental & alloys as sheets, rods, drops, moldings, etc.	GF H F	GF H F						H F G									H F						23																	
24	Metals & metal compounds, toxic	S	S	S			S	S			S														24																
25	Nitrides	GF H F	H F E	H GF	GF H E	GF H			U	H G	U	GF H	GF H	GF H				GF H	U	GF H	GF H	E				25															
26	Nitrites	H GT GF	H F GT	H							U											H P			S	GF H	26														
27	Nitro compounds, organic		H F GT			H					H E											H GF E				H GF E		27													
28	Hydrocarbons, aliphatic, unsaturated	H	H F			H																	H E						28												
29	Hydrocarbons, aliphatic, saturated		H F																											29											
30	Peroxides & hydroperoxides, organic	H G	H E		H F	H G		H GT	H F E	H F GT		H E GT	H F GT					H E	H	E	H F GT	H E	P G		H G	H GF E	H P GT		H P		30										
31	Phenols & cresols	H	H F						H G										H P			GF H				GF H					H	31									
32	Organophosphates, phosphothioates, phosphodithioates	H GT	H GT						U		H E											H									U		32								
33	Sulfides, inorganic	GT GF	HF GT	GT		H			E										H												H GT			33							
34	Epoxides	H P	H P	H P	H P	U		H P	H P		H P	H P	U								H P	H P	H P		H P	H P					H P	H P	U	H P	34						
101	Combustible & flammable materials, misc.	H G	H F GT																			H G F				H GF F					H F GT					101					
102	Explosives	H E	H E	H E					H E		H E			H E								H E	H E	H E	E	E					H E	H E		H E	H E	H E	102				
103	Polymerizable compounds	P H	P H	P H					P H		P H	P H	U									P H	P H	P H	P H	P H					P H	P H		P H			H E	103			
104	Oxidizing agents, strong	H GT		H GT	H F	H F	H F GT	H F GT	H E	H F GT		H E GT	H F GT	H F	H F		H F	H GT	H F GT	H F	H F GT	H F E	H F E	H F		H F E	H F GT	H E	H F	H F	H G	H F	H F GT	H F GT	H F G	H F G	H E	H F GT	104		
105	Reducing agents, strong	H GF	H F GT	H GF	H GF F	GF H F	GF H	H GF	H G				H GT	H F				H E	GF H	GF H	GF H						H GF	H E			H E	GF H	GT GF H		H	GF H	H E	H P GF	H F E	105	
106	Water & mixtures containing water	H	H						G										H G			GF H	GF H		S	GF H								GT GF						GF GT	106
107	Water reactive substances	Extremely Reactive! Do Not Mix with Any Chemical or Waste Material!																																							

KEY

Reactivity code	Consequences
H	Heat generation
F	Fire
G	Innocuous and non-flammable gas generatio
GT	Toxic gas generation
GF	Flammable gas generation
E	Explosion
P	Violent polymerization
S	Solubilization of toxic substances
U	May be hazardous but unknown

Example: H F GT — Heat generation fire, and toxic ga generation

FIGURE 8-10
Compatibility chart for storage of hazardous waste.[8]

Solution. These limited descriptions do not characterize the wastes sufficiently for the purposes of storage. For example, drum reconditioning wastes could include acid and alkaline rinses, spent solvents, and paint sludges. Laboratory packs could include any of innumerable chemicals used in research laboratories. Additional information may be obtained from the waste profiles needed for facilities to accept these wastes. In this example, it is assumed that the waste profiles would identify that the compatibility groups for the purposes of storage are as follows for each waste:

Waste	Reactivity group	Name
A	2, 24	Oxidizing mineral acids, toxic metal compounds
B	11	Cyanides
C	24	Toxic metal compounds
D	17	Halogenated organics
E	16	Aromatic hydrocarbons
F	31	Phenols and cresols
G	16	Aromatic hydrocarbons
H	32	Organophosphates
I	29	Saturated aliphatic hydrocarbons
J	10	Caustics
K	107	Water reactive substances
L	24	Toxic metal compounds
M	10	Caustics

A number of events could occur if a high-pressure water hose were used in an attempt to extinguish a fire among the above wastes stored together in individual containers. For example, the force of the water stream could tip over the containers of mineral acids and cyanides, thereby creating highly toxic hydrogen cyanide gas. If the containers leaked, the elemental sodium could come into contact with water, liberating hydrogen gas, which spontaneously ignites. Similarly, acids could react violently with alkalis, releasing heat. Violent reactions in surrounding containers could release the non-halogenated solvents, that would form an explosive mixture with air.

Using the information in Fig. 8-10, a total of four separate areas are needed for the compatible storage of these wastes as follows:

Area	Wastes
1	A, C, L
2	B, J, M
3	D, E, F, G, H, I
4	K

Certain types of wastes will require prior mixing. For example, the concentration of waste constituents can vary considerably because of large differences in incoming waste strength. This is particularly true at most commercial treatment facilities. Mixing can control such variations to a range that will not upset performance of the subsequent unit treatment processes. Some waste must undergo other forms of preparation before feeding to the waste treatment processes. For example, contaminated soil can contain

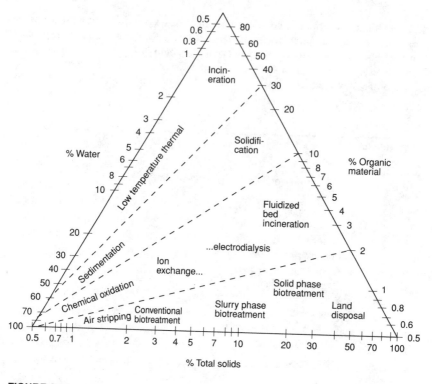

FIGURE 8-11
General suitability of various treatment methods versus physical and chemical characteristics of waste.

rocks, bricks, and reinforcing rods that could damage materials handling equipment if not removed.

Waste Treatment

While the waste is maintained in storage, a treatment schedule is developed that will identify the waste to be treated, its storage location, any necessary preparations, the method of treatment, and the rate at which the waste is fed. Upon commencement of waste treatment operations, the waste is typically fed by bulk materials handling systems such as pipelines or conveyors to the equipment used to perform the prescribed treatment steps.

Treatment operations may be carried out on a batch or continuous basis. A facility must monitor operations carefully to assure that its performance attains the desired results. Operational monitoring is done with instrumentation, direct human observation, and chemical analysis. This typically involves extensive record keeping using a combination of computers, chart recorders, and manually entered paper logs.

Hazardous waste can be treated using any of a large number of commercially proven unit processes. Many of these methods are discussed in Chaps. 9 through 12 of this textbook. The treatment methods fall into four categories:

- Phase separation (e.g., sedimentation, steam stripping)
- Component separation (e.g., ion exchange, electrodialysis)
- Chemical transformation (e.g., chemical oxidation, incineration)
- Biological transformation (e.g., fixed film aerobic treatment)

The selected method of treatment not only depends upon the type of waste but on the waste's individual physical and chemical characteristics and the specifications for the treated waste. Fig. 8-11 presents a diagram showing how basic characteristics (e.g., percent organics and percent moisture) can influence the selection of particular treatment methods. The unit treatment processes can be interconnected to attain more efficient and more effective treatment as shown in Fig. 8-12.

Residuals Management

Each waste treatment process produces gaseous emissions, wastewater effluents, or residuals requiring subsequent management if not additional treatment. An incinerator, for example, produces combustion gases that require scrubbing that in turn produces an acidic washwater requiring wastewater treatment. Incineration also produces fly ash and bottom ash requiring disposal, if not treatment. The unit treatment processes are not the only operations that generate residuals. Spillage and runoff from storage areas require treatment. The opening of containers may need to be done under negative atmospheric pressure with the fumes collected and treated. A full-service facility can usually provide all necessary treatment of residuals. Smaller facilities may have to collect the residuals and transport them as hazardous waste to another facility capable of treating them.

Special Measures

A hazardous waste facility needs to take a number of special precautionary measures for all of its day-to-day operations. These measures can be quite complex for full-service, commercial facilities and much less involved for single purpose, on-site facilities. Nevertheless, both types of facilities need all of these measures to prevent incidents. In fact, the permitting process requires that the permit application provide detailed plans and procedures on how the facility will implement precautionary measures. These measures can be listed as follows:

- Security
- Inspection and maintenance
- Incident prevention
- Emergency planning

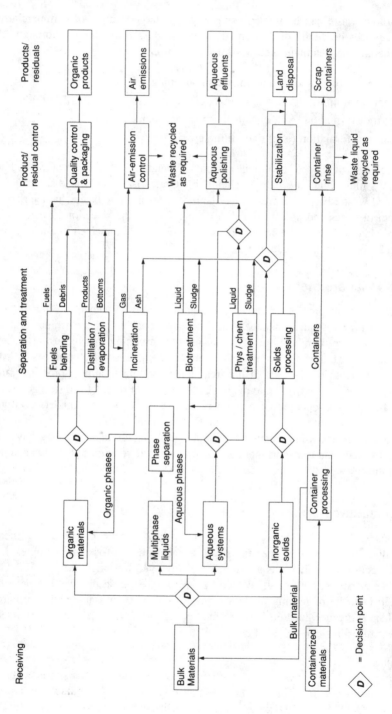

FIGURE 8-12
Integrated hazardous waste treatment processes.[9]

422

- Employee training
- Safety
- Monitoring
- Audits

A facility needs to provide strong **security** to prevent vandals and inadvertent intruders who could become exposed from contact with waste, damage the equipment, or cause illicit dumping. Most facilities use a combination of security guards, total enclosure (usually with fences), controlled entry points, adequate lighting, proper warning signs, and 24-hour surveillance. The guards operate the gatehouse where they prevent entry of unscheduled trucks and monitor entry of visitors.

A facility needs regular and thorough **inspection** of its equipment, structures, and other appurtenances with preventive and corrective **maintenance** taken. This requires preparing a list of items to be inspected, a schedule, and the typical problems that may be encountered. The inspection should examine process equipment, storage areas, emergency equipment, monitoring equipment, and security devices. Basically, the inspection should check for equipment malfunctions, structural deterioration, operator errors, and discharges that could lead to the release of hazardous waste constituents. Some critical components of a facility require daily inspection. For example, the operator should inspect storage areas, looking not only for leaks but for signs of corrosion and mismanagement (e.g., loose lid on a drum). The unloading area is critical to inspect because of the higher potential for spills.

Example 8-2   Inspection of unloading area. What potential problems could an inspector find in an unloading area?

Solution. The spill containment structure may have signs of deterioration. There may be spillage outside the spill containment structure. The loading dock equipment may be inoperable or show signs of corrosion. The eyewash may not have sufficient water pressure. Absorbent material used to control spills may be low if not out of stock. Access for fire protection equipment may be obstructed. Incoming trucks may be leaking waste.

It is important to implement **incident prevention** measures to minimize the possibility of a fire, explosion, spill, or any other unplanned, sudden release of hazardous waste constituents. Such a program begins by conducting a risk analysis to identify hazards and potential incidents. The risk analysis would indicate the routine activities and waste processing operations that have a higher probability of initiating an incident. In this way, preventive measures can be instituted and appropriate responses planned in the event one of the more probable incidents were to occur. Fire, for example, is a major concern. Having alarms, fire control equipment, water of adequate pressure, and personnel trained to fight fires is important, but so is maintaining adequate aisle space to allow unobstructed movement of fire protection equipment. Of course, the inspection and compatibility considerations discussed earlier in this section are important to prevent incidents.

If an incident occurs, the facility should follow a set of procedures prepared as part of an **emergency plan** that describes the actions that facility personnel should take in response to fires, explosions, or other incidents. The plan should provide for interaction with local police departments, fire departments, hospitals, and other appropriate emergency and community services. The plan should also describe evacuation procedures for facility personnel. Practice drills are important. One facility employee should be appointed as emergency coordinator and take leadership responsibility for implementing the plan.

It is important that the facility train its employees to perform their duties effectively and safely and to know how to respond to an emergency. The **training** usually consists of both classroom instruction and supervised on-the-job training. Each employment position should have a description of its responsibilities and duties, the content of the necessary introductory and continuing training, and a frequency for review sessions (usually annually).

One of the most important measures is **environmental monitoring**—collecting samples of the environmental media and testing for the presence of hazardous substances that may have been released by the facility. The objective is to detect potential problems before they impact human health and the environment. Early detection should allow sufficient time for adequate warning of potentially impacted individuals and effective implementation of remedial measures. Important monitoring points are ground water wells for storage and land disposal facilities (see Sec. 15-6) and air monitoring stations at critical locations around the facility. Monitoring could also include surface water, employees (e.g., blood samples), and flora and fauna.

Regulatory compliance could be thought of as a separate function, yet this is essentially the driving force for the entire facility. The market for facilities is heavily influenced by dynamic regulatory programs that continue to undergo significant change. The operation of a facility is thus geared around managing waste in a manner that meets, if not exceeds, environmental regulations. This is no easy matter. For example, the record keeping requirements for documents and reports alone represent a major undertaking because there are nearly 200 U.S. Environmental Protection Agency (EPA) regulations specifying retention periods.[10]

Audits are an excellent tool for ensuring that the facility operating staff is complying with all standard operating procedures, best management practices, and regulatory requirements. Actually, diverse groups from outside the facility conduct periodic inspections and audits from varied perspectives. Federal and state agencies inspect the facility to determine compliance with regulatory requirements. Many clients (i.e., generators of waste) audit facilities to ensure that the facility's practices for managing wastes do not pose future liabilities for the generators. Some facilities may even allow local citizen groups to tour the facility to review particular operations important to the community (e.g., truck traffic and monitoring data). The corporate parent of the facility may also conduct periodic audits. For these reasons, some facilities conduct their own internal audits. Chap. 6 is devoted to a discussion of audits.

Closure

Some of the nation's Superfund sites were originally hazardous waste management facilities that were not closed properly when operations ceased. The importance of a proper closure is such that a closure plan is a necessary part of a permit application, thus providing a plan for final closure prior to startup of operations. A closure plan provides a clear and orderly set of steps and methods to be followed upon cessation of all operations at a facility. The steps are designed to ensure that the closed facility (a) poses a minimal risk to the environment and human health and (b) requires minimal post-closure maintenance.

A closure plan requires assurance that funds are available to close the facility even if the facility owner enters bankruptcy proceedings. This assurance can be in the form of a bond, corporate guarantee, or some other financial instrument. The monetary amount is determined based on a cost estimate prepared as part of the closure plan. To satisfy the EPA regulations, the cost estimate must equal the maximum costs of closing all waste management units ever activated at the facility.

Closure of a storage or treatment facility requires removal of all remaining waste to another facility. All equipment and structures that had been in contact with waste must be decontaminated. This may entail removal of concrete pads used to hold waste containers as well as contaminated soil where leaks had occurred.

Example 8-3   Estimate of closure costs. What key activities should be included in the closure cost estimate for a treatment facility?

Solution. EPA regulations specify the following key elements: on-site treatment of the maximum possible inventory of waste, removal and transportation to an off-site facility of all wastes that cannot be treated on site, decontamination of all equipment and other facility components, monitoring and security during closure, and certification of closure. This listing ignores the cost to remediate ground water if such were to occur.

The closure of land disposal facilities, where waste has been emplaced, differs from closure of a treatment facility. In the former case, waste is not to be removed but contained, necessitating installation of cover systems plus implementation of a long-term plan referred to as a post-closure plan. This plan provides for leachate management, monitoring, maintenance, security, and other measures typically for a period of 30 years, if not longer.

8-3   NEEDS ASSESSMENT

The first step in developing a facility, even prior to the step of selecting a site, is to decide what type of facility is needed and its capacity. It is readily apparent that the ideal characteristics for a site depend upon the type and scale of facility to be located there. The process for making such a decision is termed a **needs assessment**.

In its logical sense, a needs assessment involves an analysis of the following:

1. The current and future types and quantities of hazardous waste generated.
2. The current methods for managing the generated waste.

3. The methods expected or desired in the future for managing the generated waste.
4. The capacity of existing facilities providing such methods.

The need for new facilities is thus determined by comparing future waste generation (No. 1) with capacity (Nos. 3 and 4). A shortfall in capacity indicates a need for new facilities. Economists would equate this with a comparison of "supply" and "demand."

Need is a concept that has different dimensions dependent upon one's perspectives. A private sector developer would examine need on a market basis (i.e., "supply" and "demand"). From this perspective, waste might be diverted from facilities now in operation to new types of facilities. The capacity shortfall and the potential economic return must be sufficiently large to justify the financial risk of the investment. A state government may perceive need as sufficient facility capacity to manage all waste generated within the state, thus avoiding having to depend on facilities in other states. With this view, possible diversion from existing facilities is excluded from the calculation. An environmental group may perceive need as that capacity needed only after implementation of extensive waste minimization and source reduction programs by waste generators.

Example 8-4   Different assessments of need. Which needs assessment made by the following entities would likely indicate the greatest need for new facility capacity: a private developer considering market conditions, a state government seeking independence from other states, or an environmental group emphasizing waste minimization? Which is correct?

Solution. The state government would probably show a greater need for new facility capacity. A state, by seeking independence, would ignore the capacity of facilities in other states. Yet, the state must recognize that the capacity of commercial facilities within its own borders is not totally available because interstate commerce provisions allow the facilities to serve generators in other states. In contrast, a private developer would consider a new facility as having to operate in an interstate market; facilities in other states could compete with the new facility. A developer, thus, may show less need for additional capacity unless the developer projects that waste now going to competing out-of-state facilities might be sent instead to a new facility. In any event, the perspective of an environmental group emphasizing waste minimization could likely be that no additional facilities are needed. All three might be "correct."

8-4   SITE SELECTION

A primary objective of a site selection process is to assure that new facilities are located at intrinsically superior sites that, by virtue of their natural features and land use setting, provide a high degree of protection to public health and the environment. In this concept, the land use setting and natural features function as an additional line of defense if the facility operations do not perform as planned.

A site selection process usually proceeds through a phased approach, as depicted in Fig. 8-13. It begins with the use of regional screening techniques to reduce a large study area, such as an entire state, to a manageable number of discrete search areas.

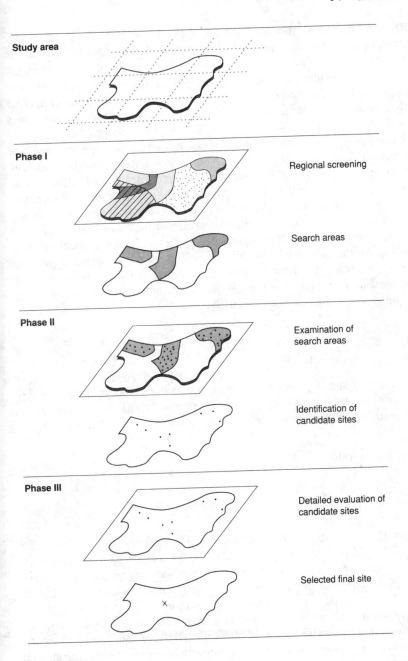

Study area

Phase I — Regional screening

Search areas

Phase II — Examination of search areas

Identification of candidate sites

Phase III — Detailed evaluation of candidate sites

Selected final site

FIGURE 8-13
Phased site selection process.

Computerized geographic information systems (GIS) are available to assist in this task (see Sec. 15-5). Because of this screening, the search areas should have higher probabilities of containing suitable sites. The next phase evaluates the discrete search areas in more detail and identifies candidate sites within them. These candidate sites are then evaluated in even more detail at a site-specific level of analysis to provide the basis for selecting a site for the facility. "The overall site selection process is thus one of increasingly intensive analysis of progressively smaller areas."[11]

A phased approach is usually employed by a governmental agency when the selection process is under public scrutiny. Another approach to site selection, and one frequently implemented by the private sector, is to preempt a phased approach and simply select from an inventory of a few specific sites. The inventory, for example, may consist of those sites advertised as "for sale" or sites already owned by the facility developer. Defenders of such an approach contend that if the characteristics of a site selected in such a manner meet all regulatory requirements, the site is suitable. Thus, it should still be considered suitable regardless of whether a phased approach, starting with regional screening, may have located an environmentally superior site. Even so, sites selected in a preemptive approach may necessitate more expensive design features to mitigate less than ideal characteristics.

Site Screening Methods

There are alternative methods used to integrate and evaluate data in the phased approach of progressing from a study area to the selection of the final site. The following list contains four examples of screening methods (i.e., methods for data synthesis).

- Intuitive
- Stepped-down exclusion
- Scaling
- Criteria combinations

The **intuitive** method is one in which the decision maker examines all data as a whole and judges whether the site is acceptable. The argument for using this method is that the site characteristics are too functionally interdependent with each other and with the design and operation of the proposed facility to support an assessment derived from an examination of individual factors. This is akin to Gestalt psychology.[12]

The **stepped-down exclusion** method examines each siting factor sequentially, determines a criterion (definitive threshold of acceptability) for the factor, and applies the criterion to eliminate areas from further consideration. A criterion may be thought of as an environmental, technical, social, economic, or political constraint to siting a facility. The process continues by selecting another factor and applying a criterion for it. The method progresses through the entire sequence of siting factors to identify "not unacceptable areas." The process may stop short of applying all criteria if application of the previous factors had accomplished the siting objective.

The **scaling** method is one of weighting the site data. That is, the data are modified from their original form by replacing them with numerical surrogates having a common scale. For example, permeability of the formation below a site is a typical consideration in selecting a site for a land disposal facility. Clay is an indicator of low permeability, and a site underlain by 10 feet of clay may have a value of 10, while one with sandy silt may be assigned a value of 3. Similar values are assigned for other data describing site suitability. The next step in scaling is to assign weights to indicate the relative importance of the factors (e.g., permeability and depth to ground water may be judged to have a weight of 8 and 4, respectively). After assigning weights, all factors are combined in a multiplication-summation step to yield a numerical score for each site. The selected site would be the one with the best score.

The **criteria combination** method can use either the stepped-down exclusion or scaling technique. Instead of working with a single set of criteria or scaling values, alternatives are developed. This involves selecting a particular combination of criteria and identifying sites satisfying them. The process is then repeated for another combination. For example, one combination may stress that the facility be located over thick deposits of clay, and it may give less consideration to factors such as proximity to population. Another combination of criteria may relax the requirement for clay but require that the facility be located on remote, public-owned land. The criteria combination method results in the selection of a set of starkly different sites representing different perspectives about what constitutes a better site.

Example 8-5   Comparison of site-screening methods.

What are the capabilities and limitations for the four site-screening methods listed previously?

Solution

Method	Capabilities	Limitations
Intuitive	Possibly the only way to consider public input and subjective information when selecting from otherwise technically superior sites.	Difficult to justify decisions, especially at regional level.
Stepped-down exclusion	Works well in regional screening and in application of mandatory criteria.	Not appropriate for preferential criteria because it may unnecessarily exclude a site that could rate "high" in subsequent evaluations.
Scaling	Works well if the number of sites is small and the differences among sites are broad.	May introduce biases. Incorrectly treats each factor as wholly independent. Significance of the quantitative output is not readily apparent.
Criteria combinations	Only method that takes into account (a) the interdependence among factors and (b) differences in professional opinion. Excellent for encouraging public participation.	Does not rank sites. Data management difficult for a large number of sites.

Siting Criteria

Most screening methods require siting criteria. Siting criteria stem from siting **factors**. A siting factor is associated with a specific consideration important in judging the suitability of a site. For example, the potential for leachate to migrate to ground water is a consideration with land disposal facilities. Accordingly, it is desirable that a land disposal site have one or more of the following characteristics: the permeability of the uppermost formation is low, the depth to ground water is high, and the thickness of any clay deposits is high. Thus, the pertinent siting factors are permeability, depth to ground water, and thickness of clay. The factors applicable to a treatment facility may be very different.

A **criterion** is a definitive threshold of acceptability for a siting factor. For example, a criterion for permeability (hydraulic conductivity) could be "maximum of 1×10^{-7} cm/sec." Sites having a value higher than this would be deemed unsuitable. Similarly, a criterion for thickness of clay could be "minimum of 10 feet."

A great number of siting factors can be applied when selecting the site for a hazardous waste facility. Table 8-1 lists some example factors frequently used in such endeavors.

Siting criteria can be divided into two types: mandatory and discretionary. **Mandatory** criteria represent a legal requirement, a regulatory standard, or some other aspect deemed so important that it cannot be violated under any circumstance. Most siting criteria are **discretionary**—they represent desirable but not mandatory features. Discretionary criteria reflect preferences and value judgments.

The development of siting criteria acceptable to the public requires much input from the public. An example of a process to develop siting criteria is one used by the state of New Jersey. A state-appointed advisory council, representing a broad range of interests, used a six-step process:

1. Make basic assumptions.
2. Identify types of facilities.
3. Analyze facility failure possibilities—types and consequences.

TABLE 8-1

Example hazardous waste facility siting factors[13]

Category	Example siting factors
Surface water	Flood hazard areas, drinking water supplies, reservoirs
Ground water	Hydraulic conductivity, depth to ground water, thickness of clay deposits, aquicludes/aquitards, recharge areas, proximity to wells, karst area, ground water flow direction
Environmentally sensitive lands	Wetlands, habitat for endangered species, parks
Population	Proximity to dwelling units, proximity to schools, population density

4. Establish objectives of siting criteria.

5. Select siting factors.

6. Determine siting criteria.

The first two steps resemble a needs assessment. Based on prior studies, the advisory council assumed that: "(1) New Jersey needs new hazardous waste facilities, (2) one or more of these new facilities will be sited within the State's borders, and (3) all facilities engender some degree of risk and, therefore, siting criteria must be developed to help minimize the risk to public health and the environment."[14] Any approach to the development of siting criteria must consider what types of facilities are needed because the possible failure events vary dependent upon type of facility. An analysis of failure events indicates potential consequences from which a set of siting objectives can be established. For example, the site selected for a hazardous waste land disposal facility should have intrinsic features that satisfy the following objectives:

- Provide structural stability
- Protect surface water quality
- Protect ground water quality
- Protect environmentally sensitive areas

Meeting each objective requires identifying pertinent siting factors and then determining a criterion for each factor. An example follows:

- Objective: provide structural stability
- Siting factor: riverine flood hazard areas
- Siting criterion: all facilities prohibited in flood hazard areas

---

Example 8-6 Narrative and numerical siting criteria. What are examples of narrative and numerical siting criteria for protecting ground water from the possible subsurface migration of leachate from a hazardous waste landfill?

Solution. The protection of ground water is a hydrogeologic issue. The relevant siting factors, as discussed in Chap. 4, could include:

- Permeability
- Ground water flow velocity
- Depth to bedrock
- Karst topography
- Time of travel

The criteria must be carefully selected to reflect the geology of the study area. A narrative criterion for permeability and time of travel could read as follows:

Allowed only in areas having a sufficient thickness of naturally occurring geologic materials of low permeability (e.g., fine-grained geologic deposits, dense bedrock) to assure that effective remedial action can be taken.[15]

If a numerical format is desired, the criterion could specify a minimal thickness (e.g., 20 feet) of regionally continuous clay having a permeability no greater than 1×10^{-7} cm/sec. The EPA has determined that an appropriate criterion is that the time for ground water to travel 100 feet along the flow line originating at the base of a hazardous waste unit is to be greater than 100 years.[16]

Example 8-7 Siting criteria sensitive to public's concern. The types of criteria produced in Example 8-6 were presented and discussed in great detail in a series of well-attended public meetings in New Jersey in 1982. The public objected to the narrative criteria because qualified descriptors such as "sufficient" give too much discretion and flexibility, opening the door to finesse and compromise. The public also eschewed arbitrary numbers such as 100 years, 20 feet, and 10^{-7} cm/sec. Of most concern, the public noted that criteria stressing permeability would drive the site search towards regional clay deposits that happen to overlie New Jersey's major aquifers used by a large percentage of its population for drinking water. The failure of a facility in such a location could have major consequences. What are examples of siting criteria sensitive to the public's concerns listed above?

Solution. This task begins by examining risk. Risk involves two independent considerations— **likelihood** and **consequential**. That is, risk is composed of the potential for an event to occur and the severity of the consequences if it does occur. Theoretically, risk is the product of multiplying the two. In this instance, the risk to ground water can be reduced by applying criteria that address either the likelihood of contaminant release or the consequences if such a release were to occur.

Factors such as permeability provide for containment. Containment is a measure of the potential for release; it ignores the consequences of such a release. The public did not accept the position that clay prevents release of toxic pollutants; therefore, a criterion requiring clay does not meet the public's concerns. The public in New Jersey asked for criteria that would protect the ground water even if the facility operation and clay did not function as planned. The criteria, therefore, should address consequences of failure, however remote.

One option is to prohibit the siting of facilities except above ground water not presently used for water supplies and so severely contaminated from past practices or natural salinity as to make it highly unlikely that the ground water would ever be remediated. The consequences of failure would be nominal as additional contamination would not affect current or future users. The public again expressed disapproval, claiming that (a) this would represent a license to pollute and (b) this "writes off" major aquifers when the policy eventually should be to remediate the past contamination and allow use of the ground water.

Developing criteria sensitive to the public's concerns requires consideration of the entire ground water flow regime: the criteria should eliminate recharge areas of water supply aquifers, particularly areas where the uppermost aquifer is used for drinking water supplies downgradient from a facility. Also, the site conditions should foster the early detection of leakage, if it occurs, and its remediation. The following ground water flow regime would meet these requirements: the ground water under the site should flow either horizontally (parallel to the water table) or upwards toward the water table (discharge area) until discharging to surface water. In such an area the recharge to deep, regional aquifers would be nominal, possibly nonexistent. In areas with downward flow, in contrast, any migrating pollutants could bypass monitoring wells, and remediation would be more difficult.

Thus, the above requirements can be captured in the following criterion: "the flow of ground water in the uppermost saturated unit is predominantly parallel to or upwards toward the water table and the predominant ground water flow direction is toward the nearby surface water bodies without any intermediate withdrawals from the uppermost saturated zone for public or private water supply."[17] This criterion is only effective in conjunction with other criteria concerned with the complexity of the geology, ground water travel time between the waste and property boundaries, and uses made of the surface water into which the ground water discharges. Such requires a site-specific assessment, with some flexibility to assure the spirit of the criteria are met without eliminating the entire study area.

Value Judgments and Trade-offs

A site selection process will involve value judgments because:

1. Criteria are based on an "acceptable risk" concept (there are *no* circumstances for which risks, however remote, are zero), and
2. The line between "acceptable" and "unacceptable" is typically unclear, and often varies from one point of view to the next.

More significant than these two reasons, value judgments are required because the ideal site with optimum conditions for all siting considerations (e.g., hydrogeology, land use, and transportation) does not exist. The siting process must, therefore, examine trade-offs in its determination of what constitutes "better" conditions, and this will inevitably encounter differences in judgments, even among leading authorities in the same field.

Individuals responsible for siting facilities often make value judgments too early—when establishing criteria rather than when the criteria are applied. Establishing inflexible or narrow criteria could preempt important decisions that should be deferred until after the application of the criteria and the collection of site-specific data. Example 8-7 illustrates the limitation of requiring that a facility be located in areas having regional deposits of clay. As another example, a decision to require a facility be located only in remote areas would immediately eliminate areas near cities where suitable sites may exist. In fact, upon detailed examination, sites near cities may prove to be technically superior in key ways. For example, risks from accidents involving the transportation of hazardous waste might be reduced at urban sites because transportation distances are shorter. Such sites may also be more politically viable because most generators are located near cities. As a result, "equity-based" siting criteria can be applied. Yet, decision to consider only sites in remote areas preempts the application of the stringent criteria to urban sites.

Site selection processes that start with inflexible and narrow criteria typically conclude with a trade-off analysis of nothing other than trivial differences among sites. It is important to provide a wide choice and keep many options open as part of the final site selection. This facilitates a trade-off analysis of stark differences that should encourage participation by affected parties. Such an objective can be met by establishing different combinations of criteria and selecting a set of alternative sites for each combination. This progressively screens the study area without reducing value judgment flexibility until the end when different sites epitomizing different opinions can be contrasted and more data can be obtained.

8-5 PUBLIC PARTICIPATION

The process of selecting sites for new facilities and obtaining permits for them is commonly referred to as "siting." It can cost tens of millions of dollars and take a decade or more of effort before a definitive decision is received from the permitting agency. That decision is frequently a denial. The Province of Ontario started a process in the early 1980s to site a full-service facility. In 1990, after the selection of a site, public hearings commenced that were expected to continue for another three-and-one-half years.[18] The cost in Ontario of siting a facility may exceed the cost of constructing the facility, that is, if one is ever built. In a recent private sector proposal, Clean Harbors, Inc., spent $12 million in a now terminated attempt to site an incinerator in Massachusetts.[19]

Even if a permit is granted, subsequent litigation can delay placing the facility into operation and add to the mounting costs. Almost all efforts to develop new, modern facilities have failed because of an obstacle much greater than market forces: public opposition. While every aspect of developing and operating a facility is conducted under close scrutiny, nothing seems to raise as much opposition as siting a new facility. The situation has given rise to a new acronym, NIMBY (Not in My Back Yard).

Root Causes of Public Opposition

Why do hazardous waste facilities arouse so much public opposition? Why can the opposition become so vehement? The answers stem from many causes including risk perception, public mistrust, inequities in risk sharing, and other factors.

As discussed in Chap. 1, the typical layperson perceives a facility in his or her community as (1) imposed upon the community (involuntarily encountered), (2) having no real benefit, and (3) representing an unknown but substantial risk. To a public already cautioned by the label "hazardous," all three perceptions can prompt severe public response. While public concern with hazardous waste spurred regulatory initiatives that created and drove the expanding market for hazardous waste management services, the same concern prompts opposition to new facilities for treating and disposing of waste.

Compounding these problems is the general observation in society today that there is not enough public trust upon which to promote understanding and acceptance. The engineering and science professions believe they have developed the capabilities required to construct, operate, and monitor safe waste management facilities within technically acceptable risk limitations. However, the public does not necessarily share the same belief. The link between risk and technology is perceived in a climate of public distrust of both industry and government. This is part of the legacy of Love Canal and similar episodes that were deemed at their time to provide adequate disposal. Underdesigned, undercapitalized, and underregulated, many of the early hazardous waste facilities have, in fact, ceased to operate and eventually became Superfund sites. Although government is the watchdog to ensure that these episodes will not be repeated, the public lacks confidence in the adequacy of the laws and regulations and the capabilities and willingness of regulatory agencies to enforce them.

Once formed, public opposition is not known to lessen measurably. A number of different countervailing programs have been attempted, without success, to "overcome" opposition. As if public mistrust and the heightened risk perceptions were not enough, new factors can emerge to harden opposition. One of these factors is inequity in risk sharing. There is no real link between waste management at a facility in one community and waste generation in another. The host community bears the risk of waste treatment and disposal while other communities benefit from the jobs, local purchases, and other economic activities associated with the industries that generate the waste.

Another factor behind public opposition is the perceived loss of a community's sacred right to shape its own future.[20] Opposition stems in part from how the public first becomes aware of a proposal to site a facility. Most of the public, even in a process that has been conducted openly, will learn of the project after a site has been identified in its community and publicized. The *post-facto* revelation of a siting decision as a *fait accompli*, particularly without proper explanation, can prompt an intense feeling that citizens have no way of influencing other than to oppose it, a decision that will affect their well-being. Getting the public involved early is a difficult undertaking. When the public has been given an opportunity to participate in "front-end" planning, the public usually fails to perceive its stake and typically will not participate. Only after sites are identified do citizens in the potential host communities get involved, marked by a rapid "yawn to yell" transition. Upon reaching this stage, discussions between the facility developer and the public make little progress toward finding common ground. How can there be when the developer, having to site a facility somewhere, will not consider any option that involves siting the facility elsewhere, and the public will not consider any option that involves siting the facility in its community? Satirists refer to these discussions as a "dialogue of the deaf."

Another compounding factor is that some well organized political activist groups exist that oppose all new hazardous waste facilities in general. These groups will support local communities in their opposition to a proposed facility. Support can consist of an information clearinghouse and expert witnesses. The positions of these groups usually disagree sharply with those of facility proponents, thus adding a degree of confusion for laypersons struggling to understand a technical issue.

Public Involvement

With all these obstacles, facilities have been successfully sited where success can be measured as (1) selecting a technically suitable site, (2) identifying public concerns, and (3) addressing them. Some facility developers have even gained public acceptance (see Case Study of Siting a Hazardous Waste Facility). If one analyzes the procedure that was followed for the successful permitting of these new facilities, it is clear that in each case the community participated extensively in the process and had an influential and constructive effect. Simply stated, public involvement was the key to success, fostering trust upon which understanding and acceptance could build. It is not sufficient to merely profess to have adequate technology and expect public trust. For the public to have confidence, it must participate in the process of making decisions on issues perceived as having great effect on its future.

Public involvement is a political art, possibly even foreign to the professional scientist and engineer who typically respond only with technical information. Some such response is not the answer. The public is skeptical of technical experts. In fact, technical information may even be counterproductive; observers of opposition to nuclear power facilities have noted an inverse relationship between knowledge and public support.

The triad of the facility developer, the local community, and the regulatory agency is closely interrelated in the issue of siting new facilities (see Fig. 8-14). State governments have assumed active, non-traditional roles. One role may be to ensure that local authority does not override what is otherwise a suitable, needed site. Some state governments, however, are actually leading the search for new sites, a role that governments do not necessarily relish. Other state governments are taking a less active but supportive role, endeavoring to use institutional measures to promote negotiation between the community and facility developer in the common interest of all its citizens.

In the supportive role, government initiatives include a credible waste management plan justifying the need for a new facility, assistance grants to proposed host communities, extensive public information, public participation in selection of siting criteria if not the site itself or at least the type of facility developed at the site, and a negotiated siting agreement between the facility developer and the local community. To the facility developer, this level of public involvement may seem excessive, if not impossible, to deal with. However, without it, experience shows that an expensive and long siting process and the high probability of regulatory denial are inevitable.

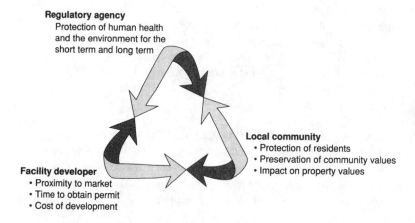

Regulatory agency
Protection of human health
and the environment for the
short term and long term

Local community
• Protection of residents
• Preservation of community values
• Impact on property values

Facility developer
• Proximity to market
• Time to obtain permit
• Cost of development

FIGURE 8-14
Key issues among triad of participants in siting a facility.

Case Study of Siting a Hazardous Waste Facility

It is reasonable to estimate that siting a new facility has a success rate of less than one percent. Almost all failures derive from public opposition, with many proposals ending in ugly incidents: facility proponents wearing bullet proof vests before an alarmed public, or police using stun guns on protesters.[21] Contrary to this trend, a recent headline captures the spirit underlying one of the very few success stories of siting a new hazardous waste facility.

Small Town in Alberta Embraces What Most Reject: Toxic Waste[22]

The Province of Alberta in Canada sponsored a siting process that eventually prompted two separate communities to vote overwhelmingly to accept a large, full-service waste management facility. Because only one facility would be constructed, the two communities competed actively for the facility and even lobbied Parliament to be selected.

Why did the Alberta process achieve success when almost all others have failed? The case is documented in several references[23, 24, 25] and presented here as an example of the cornerstone effect of an open public participation program that educates the public and involves citizens fully in the decision making process.

As background, a private proposal to site a facility in Alberta had failed in 1979 amid strong public opposition.[26] This prompted the Alberta government to develop a special (i.e., hazardous) waste management plan. One element was to form a committee of representatives from government and the public to develop and implement a site selection process consisting of two parts:

1. The development and scientific application of technical siting criteria to ensure the intrinsic suitability of the selected site.
2. The resolution of sociological issues such as to foster *acceptance* by the host community.

The application of siting criteria followed a sequential progression, as discussed earlier in this chapter, to yield large regions of land free of technical constraints (e.g., steep slopes, significant ground water recharge). The application of such criteria is a fundamental land use planning technique. It is a necessary part of siting facilities; however, it is not the key to success. In fact, it is reasonable to state that many failed siting proposals had undertaken a similar screening of constraints and identified technically suitable if not superior sites. The Alberta success hinged upon the approach taken after the screening—soliciting volunteer communities rather than naming sites and threatening the use of eminent domain. That is, Alberta did not plan to preempt or override the traditional preeminence of local government in land use decisions, even though the facility was needed for the benefit of the whole province. In fact, Alberta's approach was termed the "invitational process"[27] because the Alberta government had to be invited by the local government.

In 1981, after preparing the regional screening maps but before naming specific sites, the Alberta task force met with counties, municipalities, and other local governments having constraint-free land within their borders. The meetings provided an open forum to explain the need for a facility, to discuss the planned siting process, and to receive local input. After these introductory meetings, the local governments had to decide whether to request further consideration to site the facility in their jurisdiction or to withdraw from the site selection process. Opting for further consideration did not forfeit their eventual right to reject the facility. Nearly 75% of the local governments took the next step by inviting the Alberta task force to conduct an environmental analysis of potential sites in their area.[28]

The next step in the process included more local meetings and workshops to educate the community. Local review committees were formed. This public communication program had two objectives:

1. Engender a sense of individual responsibility for overall industrial, municipal, and household waste problem of which hazardous waste is one small part.
2. Demonstrate that the risks of a facility were no greater than those posed by other industries.

The Alberta task force did not identify the potential sites. Instead, the local governments reviewed the constraint maps and delineated specific sites within the constraint-free areas. Following that, the task force conducted environmental analysis of the sites, including geologic drilling to verify the constraint mapping.

Some communities were eliminated due to environmental constraints revealed by the environmental analysis. For communities having sites that satisfied the environmental constraints, local government again could elect to be removed from further consideration. Many did drop out because of public opposition. Many communities, however, perceived that a facility would benefit their local economy and that the risks posed by a facility were manageable. About 25% of the communities studied decided to continue with the process.[29]

Several factors obviously had much to do with gaining such a high degree of public acceptance. For example, the institutional arrangement that the Province would own the site was important to ensure perpetual care. Clearly, the local government retaining the sole right to accept or reject a facility without intervention by a higher government fostered a high level of trust and respect. The communication program built upon this trust and was able to educate the public on the highly emotive issue without forming and hardening opposition. This has proven very difficult to achieve in most other siting cases.

The communication program worked with the media but also had a number of other communication vehicles:

- Materials to assist at public meetings
- Displays and participation at local conferences and conventions
- A print campaign of brochures and handout materials
- A literature review on progress made in other jurisdictions
- A publication and newspaper-clipping inventory

- Preparation of key officials for on-the-spot radio, television, and newspaper interviews
- An educational program for schools
- Development of the mapping program for presentation in public meetings
- A newsletter outlining the issues, activities, and progress[30]

In 1983 the Province had narrowed the list of candidate sites for the facility to five communities. This marked the beginning of detailed site investigations and a series of intensive educational seminars. Community opinion leaders, including facility opponents, were sent on informational tours of active facilities in Europe and the United States where they visited and talked with the host communities about their experiences.

A necessary requirement for siting a facility was that the community supported it. To gauge support, each community held a referendum in 1984. Seventy-nine percent of the voters in a community named Swan Hills voted in favor of the facility. Another community had a 77% favorable vote.[31]

Soon after the plebiscites, the Provincial cabinet selected a site near the community of Swan Hills as the location for the facility. The town had a population of about 2500, with oil and gas production representing the dominant industry. Many residents worked in the oil and gas fields and were already familiar with hazardous operations. The community leadership was concerned with its dependence on the oil industry, and it envisioned a facility as a progressive and needed step toward diversifying its economic base.

It should be noted that no "host fees" were paid to Swan Hills.[32] Host fees are typically proposed in the United States, not so much as an inducement but as a means to compensate a community for accepting a facility that will benefit many other communities. Although frequently offered by facility developers, host fees can backfire and harden opposition if not carefully presented; they can be easily misconstrued as a bribe instead of compensation for accepting an unwanted facility.

The facility opened in 1987, employing 94 workers,[33] and featuring an incinerator, a landfill, a deep well, and physical-chemical treatment units[34] (see Fig. 8-15). The facility manager has operated the facility in the manner in which it was sited—including an "open door" policy with full disclosure of information. A key feature of the operation is that the public has ready access to facility personnel to seek additional information. Community support continues, with "virtually all residents" backing the facility's plans for a $60 million expansion that would quadruple its capacity.[35]

Public Participation in Other Activities

This section has focused on public participation in the site selection process. However, the public is keenly interested not just in siting but in all aspects of facility development. In particular, the public has expressed concern with the permitting of facilities, the operation and closure of facilities, and the transportation of waste. Activities other than the development of facilities also encounter public concern. For example, the remediation of contaminated sites can alarm the surrounding community who perhaps

FIGURE 8-15
Alberta Special Waste Treatment Centre at Swan Hills (Alberta Special Waste Management Corporation)[36]

believes that the planned remediation will stop short of what is necessary or that the proposed remediation techniques will place short-term human health at risk. The same fundamental concepts discussed in this section generally apply to the other areas (e.g., getting the public involved as early as possible).

8-6 PERMITTING

Permitting is the action of regulatory agencies to authorize construction and operation of a facility. It is based upon finding that the facility site, design, and operations do not violate health and environmental standards.

A hazardous waste treatment and disposal facility is probably the most closely regulated non-nuclear or non-medical activity in the United States today. A new term, "Part B permit," has entered the language, and the RCRA permit application alone can require hundreds of pages of documentation. This permitting requirement may be just one of a host of federal, state, and frequently, local requirements that must be met. In one instance, for example, a commercial hazardous waste treatment company was required to obtain 24 separate permits to open an integrated treatment and landfill facility. Only one permit addressed hazardous waste regulations; the balance of permits dealt with transportation, stormwater, zoning, and a host of other concerns.

By any measure, the EPA permit application process for a hazardous waste facility is lengthy, expensive, and burdensome. In many states, a facility developer must also submit a separate permit application to satisfy the state's particular requirements. Preparing the permit application takes considerable effort and time. Federal regulations (primarily 40 CFR Parts 264 and 270) specify the technical requirements of a permit application. The major elements of a typical application are listed in Table 8-2.

TABLE 8-2

Elements of a permit application for a hazardous waste facility

Major element	Example subjects
Facility description	Location and configuration of structures, improvements and appurtenances; traffic patterns
Waste characteristics	Chemical and physical analyses of all waste types, waste analysis plan (e.g., sampling and testing methods)
Process information	Container management practices, control of run-on, tank system integrity, piping and instrumentation diagrams, spill prevention and control, liner system compatibility, process engineering analyses, waste compatibility
Ground water monitoring	Demonstration of migration potential, description of wells, sampling and analysis plan, corrective action program
Procedures to prevent hazards	Security, inspection schedule, communication and emergency equipment, preventive procedures, management of reactive and ignitable wastes
Contingency plan	Responses to fires, explosions, or other unplanned release of hazardous waste
Personnel training	Programs to prepare personnel to operate and maintain facility
Closure and post-closure requirements	Waste inventory, equipment decontamination, final cover, long-term monitoring, inspection and maintenance schedule, financial assurance, liability requirements
Other federal laws	Demonstration of compliance with other applicable federal laws (e.g., Endangered Species Act)

Extensive detail is required for any element. For example, in "Process information," a facility having tanks must provide the following information for each tank:

- Dimensions and capacity
- Assessment of tank system integrity
- External control protection
- Tank system installation and testing plans and procedures
- Description of feed systems, safety cutoff, bypass systems, and pressure controls
- Diagram of piping, instrumentation, and process flow
- Design, construction, and operation of secondary containment system
- Leak detection requirements
- External liner, vault, double-walled tank, or equivalent device
- Containment and leak detection requirements for ancillary equipment

There are extensive requirements for technical information for unit waste treatment or disposal operations. Some types of treatment or disposal have to meet **performance standards**. For example, an incineration facility must conduct a trial burn to demonstrate 99.99% destruction and reduction efficiency (DRE) for principal organic hazardous constituents (POHCs) (see Chap. 12). DREs are higher for a select few compounds, such as PCBs. Concern with products of incomplete combustion (PICs) from incineration has led some states to establish a 99.9% combustion efficiency as an additional requirement. Some states further require three-dimensional dispersion modeling of air emissions to calculate concentration of contaminants at the location of receptors (see Chap. 4). The calculated concentrations are compared with toxicology data to assess health risks.

Preparing a permit application has many pitfalls that may have adverse consequences for years to come if not addressed effectively. For example, it is important that the applicant provide the details that allow the permit writer (i.e., the regulatory agency) to extract specific terms and conditions. Permits written in general terms are subject to a variety of interpretations and may result in unexpected and unwarranted enforcement actions that may threaten the facility's viability. On the other hand, a commercial facility designed to accept waste from industry-at-large will need the flexibility to operate in an efficient and cost-effective manner and to accept particular waste types that were not addressed when writing the original permit.

The technical requirements are fairly clear but still leave open many decisions that, in aggregate, constitute a judgment as to what is an "acceptable" level of risk. However minutely small the calculated risk is, it is not zero. This contributes to public concern and becomes a controversial point of conflict between the public and the facility developer. Lengthy delays on the part of regulatory agencies have occurred as they endeavor to judge the "bottom line" acceptability of a facility—without definitive standards to guide their decision but with the pressure of a constituency bitterly opposing the facility.

The public's concern with hazardous waste facilities has prompted some state governments to involve citizens more fully in the review of a permit application. Table 8-3 shows what one state, Pennsylvania, has proposed as a schedule of review steps in its permitting procedure. This schedule assumes no Notice of Deficiency (a finding that the information in the application is either incomplete or identifies an operational or design feature not in compliance with regulatory requirements) on the review for completeness, no Notice of Deficiency on the technical review, no new issues as the result of the public hearing on the draft permit, only five days for the public hearing, and a technical review in 180 days. Experience shows that each assumption is highly unlikely. A Notice of Deficiency would extend the permitting period for several months or a year, if not longer. Additionally, changes in facility design, configuration, or even ownership will delay the process even longer. The permitting process itself includes several points where additional public hearings can be convened and decisions appealed. As a result, it is not unusual to take five years or more to issue a permit.

Outside of the permitting process, facilities can face challenges in the form of local zoning restrictions, litigation by affected parties, or simply political intervention. These actions string out the construction and operation of the facility and add to the mounting costs. The developer must "carry" the costs of facility development without the prospect of revenue generation on the horizon, making the project less

TABLE 8-3
Proposed schedule for review of permit application in Pennsylvania

Step	Duration
Pre-application conference	Optional
Site investigation by state	Optional
Notify host community application has been submitted (within 10 days)	
Publish notice in state bulletin and twice in local paper	15 days
State reviews application for completeness	30 days
State publishes notice of start of technical review	15 days
Public review and comment period	60 days
Technical review by state	180 days
If requested, state holds fact-finding meeting with the public	45 days
State prepares draft permit and fact sheet	45 days
EPA review of draft permit and fact sheet	60 days
State publishes notice that draft permit has been prepared	15 days
Public review and comment period	45 days
If requested, public hearing is held	60 days
State can extend the public review and comment period if new issues are raised	?
State prepares a responsiveness summary on the permit and publishes comments	45 days
State prepares and issues the permit	30 days
	645 days (21.5 months)

attractive to potential investors. These delays and cost increases can be enough to make undercapitalized ventures withdraw applications for permits and cease when investors realize that the risk of denial makes competitive investments more attractive.

The facts speak for themselves. The typical schedule *intended* by state government for reviewing and approving a permit application may take from six months to two years. Actual experience shows that only from one-third to one-half the number of permit applications received during a seven-year period (1980–1986) were completely processed. Of the applications submitted, only 53 facilities were permitted, of which 46 were existing facilities undergoing some form of expansion or modernization. About half were commercial facilities, the remainder being on-site operations.[37]

One critic of the particular procedure for reviewing a permit application claims that it "reverses George Orwell's *1984*" and it is "a prescription for governmental paralysis."[38] While the schedule may be lengthy, it does respond to a need and a growing outcry to involve the public more in siting. The question remains of how to make this involvement meaningful and productive.

DISCUSSION TOPICS AND PROBLEMS

8-1. Which types of facilities do not generate residual waste?

8-2. What types of facilities could recover or treat the residual wastes from a solvent recovery facility?

8-3. Briefly describe ten measures for ensuring the safe operation of a hazardous waste facility.

8-4. Define five criteria for siting a hazardous waste incinerator? Should any of these be mandatory?

8-5. How could a computer prove useful in siting hazardous waste facilities?

8-6. Name a positive and negative way in which public concern with hazardous waste impacts the development of facilities to treat or dispose of the waste.

8-7. Name five root causes of the general reaction of the public to oppose the siting of hazardous waste facilities in its community.

8-8. What did the Province of Alberta do to address these root causes? Which was/were the most important?

8-9. Would one expect public opposition to differ dependent upon the type of facility proposed?

8-10. Identify the representation one would expect to find on a public advisory committee to a statewide site selection process.

8-11. What are Part A and Part B permit applications, and what are their main elements?

8-12. What are the implications of the term "acceptable risk" as it applies to the siting and permitting of hazardous waste facilities?

8-13. Why is a "needs assessment" important to the development of a facility?

8-14. Determine the minimum number of separate storage areas necessary to provide safe segregation of the following wastes: plating sludges containing Cd; cleaning agents containing caustics; chemical wastes (aromatic hydrocarbons); waste cresol from a wood-treating operation; sludge from plating operation containing cyanides.

8-15. Under what conditions can a surface impoundment be considered a land disposal facility?

8-16. What are two fundamental measures for preventing a deep injection well from contaminating ground water?

8-17. What is the difference between a waste profile and a waste analysis plan?

8-18. Name five media/receptors that could be monitored for release of hazardous waste constituents from an incineration facility.

8-19. Which is more important in selecting a site for a new hazardous waste facility: applying siting criteria in a scientific fashion to select a technically superior site or involving the public in site selection?

8-20. What is the process difference between an aqueous waste treatment facility and a municipal sewage treatment plant?

REFERENCES

1. Freeman, H. M., ed.: *Standard Handbook of Hazardous Waste Treatment and Disposal*, McGraw-Hill, Inc., New York, New York, 1988.
2. *Incinerating Hazardous Wastes*, edited by H. M. Freeman, Technomic Publishing Co., Lancaster, Pennsylvania, 1988.
3. ERM, Inc.: "Hazardous Waste Management Technologies—An Assessment," report to Pennsylvania Department of Environmental Resources, 22 February 1983.
4. 25 PA Code §75.260.
5. Clark-McGlennon Associates, Inc.; "An Introduction to Facilities for Hazardous Waste Management,' report to the New England Regional Commission, Boston, Massachusetts, November 1980
6. Ibid.
7. Ibid.
8. Hatayama, H. K., et al.: "A Method for Determining the Compatibility of Hazardous Waste," EPA 600/2-80-076, U.S. Environmental Protection Agency, Cincinnati, Ohio, April 1980.

9. Freeman, H. M., ed.: *Standard Handbook of Hazardous Waste Treatment of Disposal*, p. 11.10, 1988. McGraw-Hill, Inc., New York, New York.
10. Ackerman, S. A., "Do I Need to Keep All These Records?" *Environment Today*, newsletter by Montgomery McCracken, Walker, and Rhoads, Philadelphia, Pennsylvania, Summer 1991.
11. Buckingham, P. L.: "Regional Screening Approaches to Selecting Sites for New Secure Landfills," *Proceedings of Fourth Annual Madison Conference of Applied Research & Practice on Municipal & Industrial Waste*, University of Wisconsin, Madison, Wisconsin, September 28–30, 1981.
12. Rogers, Golden, and Halpern, Inc.: "Regional Screening and Conflict Resolution," report prepared for Maryland Department of Natural Resources, 1977.
13. ERM-New England, Inc.: "Development and Application of Criteria for Siting Hazardous Waste Facilities in Vermont," report prepared for Vermont Agency of Natural Resources, February 1991.
14. Buckingham, P. L.: "Protecting Ground Water with Siting Criteria," *Proceedings of Hazardous Materials Management Conference*, Philadelphia, Pennsylvania, July 14, 1983.
15. New Jersey Department of Environmental Protection: "Basis and Background Document for N.J. Major Commercial Hazardous Facility Siting Criteria," Trenton, New Jersey, February 1983.
16. McCoy and Associates: "Locational Criteria for Siting TSD Facilities May Have Broad Impact," *The Hazardous Waste Consultant*, November/December 1986.
17. New Jersey Department of Environmental Protection, "Basis and Background Document."
18. ERM-New England, Inc., "Development and Application of Criteria."
19. McGlennon, J. A. S., "Siting New Facilities: Problems and Possibilities," presented to the American Institute of Chemical Engineers, August 25, 1986.
20. Ibid.
21. Tomsho, Robert: "Small Town in Alberta Embraces What Most Reject: Toxic Waste," *The Wall Street Journal*, vol. CXXV, no. 126, p. A1, December 27, 1991.
22. Ibid.
23. Ibid.
24. McQuaid-Cook, J., and K. S. Simpson: "Siting a Fully Integrated Waste Management Facility," *Journal of the Air Pollution Control Association*, vol. 36, no. 9, p. 1031, September 1986.
25. Kupchanko, E. E.: "Case Study of a Successful Facility Siting," *Standard Handbook of Hazardous Waste Treatment and Disposal*, H. M. Freeman, ed., McGraw-Hill, Inc., New York, New York, p. 3.39, 1988.
26. McQuaid-Cook and Simpson, "Siting a Fully Integrated Waste Management Facility."
27. Ibid.
28. Ibid.
29. Ibid.
30. Kupchanko, E. E. "Case Study of a Successful Facility Siting."
31. McQuaid-Cook and Simpson: "Siting a Fully Integrated Waste Management Facility."
32. Tomsho, "Small Town in Alberta Embraces."
33. McQuaid-Cook and Simpson, "Siting a Fully Integrated Waste Management Facility."
34. Ibid.
35. Tomsho, "Small Town in Alberta Embraces."
36. Alberta Special Waste Management Corporation, Edmonton, Alberta, Canada.
37. National Governors Association: "Siting New Treatment and Disposal Facilities," Washington, D.C.
38. Kury, F. L.: comments on draft report of "The Proposed Public Participation Program for Hazardous Waste Management in Pennsylvania," submitted to Pennsylvania Department of Environmental Resources by Rogers, Golden, and Halpern, February 8.

ADDITIONAL READING

Andrews, R. N. L., and F. M. Lynn: "Siting of Hazardous Waste Facilities," *Standard Handbook of Hazardous Waste Treatment and Disposal*, H. M. Freeman, ed., McGraw-Hill, Inc., New York, New York, pp. 3.3–3.16, 1989.

Andrews, R. N. L., and T. K. Pierson: "Hazardous Waste Facility Siting Processes: Experience from Seven States," *Hazardous Waste*, vol. 1, no. 3, pp. 377–386, 1984.

Bacow, L. S., and J. F. Milkey: "Overcoming Local Opposition to Hazardous Waste Facilities: The Massachusetts Approach," *Harvard Environmental Law Review*, vol. 6, pp. 265–305, 1982.

Baker, R. D., et al.: "Management of Hazardous Waste in the United States," *Hazardous Waste and Hazardous Materials*, vol. 9, pp. 37–59, 1992.

Bingham, G., and T. Mealey: *Negotiating Hazardous Waste Facility Siting and Permitting Agreements*, The Conservation Foundation, Washington, D.C., 1988.

Clark-McGlennon Associates: *A Decision Guide for Siting Acceptable Hazardous Waste Management Facilities in New England*, New England Regional Commission, Boston, Massachusetts, 1980.

Daley, Peter: "Comprehensive Hazardous Waste Treatment Facilities," *Standard Handbook of Hazardous Waste Treatment and Disposal*, H. M. Freeman, ed., McGraw-Hill, Inc., New York, New York, pp. 11.1–11.18, 1989.

Dawson, G. W., and B. W. Mercer: *Hazardous Waste Management*, John Wiley and Sons, New York, New York, 1986.

Freeman, H. M. ed.: *Standard Handbook of Hazardous Waste Treatment and Disposal*, McGraw-Hill, Inc., New York, New York, 1989.

Greenberg, M. R., R. F. Anderson, and K. Rosenberger: "Social and Economic Effects of Hazardous Waste Management Sites," *Hazardous Waste*, vol. 1, no. 3, pp. 387–396, 1984.

Hadden, S. G., J. Veillette, and T. Brandt: "State Roles in Siting Hazardous Waste Disposal Facilities: From State Preemption to Local Veto," Chapter 10 in J. P. Lester and A. O'M. Bowman, eds., *The Politics of Hazardous Waste Management*, Duke University Press, Durham, North Carolina, pp. 196–211, 1983.

Hirschhorn, J. S.: "Siting Hazardous Waste Facilities," *Hazardous Waste*, vol. 1, no. 3, pp. 423–429, 1984.

Kupchanko, E. E.: "Case Study of a Successful Facility Siting," *Standard Handbook of Hazardous Waste Treatment and Disposal*, H. M. Freeman, ed., McGraw-Hill, Inc., New York, New York, pp. 3.39–3.49, 1989.

Morell, D. L., and C. Magorian: *Siting Hazardous Waste Facilities: Local Opposition and the Myth of Preemption*, Ballinger, Cambridge, Massachusetts, 1982.

National Audubon Society: *Siting Hazardous Waste Facilities: A Handbook*, National Audubon Society Tavernier, Florida, 1983.

O'Hare, M., L. S. Bacow, and D. R. Sanderson: *Facility Siting and Public Opposition*, Van Nostrand Reinhold, New York, New York, 1983.

Seley, J. E.: *The Politics of Public Facility Siting*, Lexington Books, Lexington, Massachusetts, 1983.

Smith, M. A., F. M. Lynn, and R. N. L. Andrews: "Economic Impacts of Hazardous Waste Facilities," *Hazardous Waste*, vol. 3, no. 2, 1986.

Smith, M. A., F. M. Lynn, R. N. L. Andrews, R. Olin, and C. Maurer: *Costs and Benefits to Local Government Due to Presence of a Hazardous Waste Management Facility and Related Compensation Issues*, Governor's Waste Management Board, Raleigh, North Carolina, pp. 195–204, 1985.

Susskind, L. E.: "The Siting Puzzle: Balancing Economic and Environmental Gains and Losses," *Environmental Impact Assessment Review*, vol. 5, pp. 157–163, 1985.

U.S. EPA: "Hazardous Waste Treatment, Storage and Disposal Facility Report for 1987," U.S. Environmental Protection Agency, Washington, D.C., 1987.

ACKNOWLEDGEMENTS

The authors wish to acknowledge the following individuals who assisted with the preparation of this chapter: Alan S. MacGregor, *ERM Program Management Company*; Ruth Baker, *ERM, Inc.*; and David Morell, *ERM-West*. The principal reviewer for the **ERM Group** was John A. S. McGlennon, President, *ERM-New England*.

CHAPTER
9

PHYSICO-
CHEMICAL
PROCESSES

If seven maids with seven mops, Swept it for half a year,
Do you suppose', the Walrus said, 'That they could get it clear?'
I doubt it' said the Carpenter, and shed a bitter tear.

Lewis Carroll

The physico-chemical processes described in this chapter include technologies that can be used for hazardous waste treatment and recycling, as well as for ground water or soil remediation. Each section presents a description of a technology, the theory behind its application, and the procedure for designing the physical facility to apply the technology. This chapter presents a selection of available technologies as space limitations preclude in-depth treatment of all technologies currently available.

9-1 AIR STRIPPING

Air stripping is a mass transfer process that enhances the volatilization of compounds from water by passing air through water to improve the transfer between the air and water phases. Air stripping is one of the most commonly used processes for remediating ground water contaminated with volatile organic compounds (VOCs) such as solvents. The process is ideally suited for low concentrations (<200 mg/L). Air stripping can be performed by using packed towers, tray towers, spray systems, diffused aeration, or mechanical aeration. Packed towers are generally used for the specific application of remediating ground water. This section describes packed air stripping towers, although the theory can be applied to other types of air stripping systems.

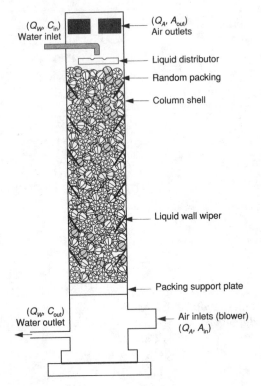

FIGURE 9-1
Packed tower air stripper.

Process Description

The process consists of counter-current flow of water and air through a packing mate rial. The packing provides a high surface area for VOC transfer from the liquid phas to the air.

The process is illustrated in Fig. 9-1. A contaminated water stream is brough in at the top of the stripper and distributed evenly over the packing, while the a stream is brought in at the bottom of the column. Typical packing materials consi of plastic shapes that have a high surface-to-volume ratio and provide the necessar transfer surface to allow volatile components to move from the liquid stream to th air stream.

The air stream leaves the column at the top; the water stream at the bottom A mass balance equation for the column may be written by setting the mass contaminant removed from the water equal to the mass added to the air:

$$Q_W \left[C_{in} - C_{out} \right] = Q_A \left[A_{out} - A_{in} \right] \tag{9-}$$

where Q_W = water flow rate (m³/sec)
   C = concentration in water (kmol/m³)
   Q_A = air flow rate (m³/sec)
   A = concentration in air (kmol/m³)

Assume that the influent air contains none of the contaminant ($A_{in} = 0$) and the effluent water also is free of contamination ($C_{out} = 0$). This latter assumption implies the stripper will work at 100% efficiency. Finally, the concentration in the effluent air, A_{out}, is determined from Henry's law (see Sec. 3-2) by using the dimensionless Henry's law constant (H'), and assuming equilibrium:

$$A_{out} = H'C_{in}$$

$$Q_W[C_{in}] = Q_A\left[H'C_{in}\right]$$

$$Q_W = Q_A\left[H'\right] \tag{9-2}$$

or

$$H'(Q_A/Q_W) = 1 \tag{9-3}$$

This theoretical relationship is true for ideal equilibrium and perfect stripping. The expression $R = H'(Q_A/Q_W)$ is termed the **stripping factor**. Values of this factor must be greater than one for stripping to occur. Q_A/Q_W is termed the air to water ratio.

Air stripping systems are limited to VOC concentrations of less than a few hundred mg/L. Non-volatile compounds (i.e., those with $H' \leq 0.01$) will not be removed effectively. However, this process is applicable to a wide range of VOCs and is capable or removing contaminants to extremely low levels.

Stripping Theory

The transfer of a volatile organic compound from water to air follows the two-film theory (Fig. 9-2) covering transfer from:

- Bulk liquid to liquid film
- Liquid film to air film
- Air film to bulk air

Bulk air

Air film
Liquid film

Bulk liquid   **FIGURE 9-2**
Two-film theory.

The rate at which a contaminant is transferred from water to air, for low solubility contaminants, is represented by an overall transfer rate constant, $K_L a(\sec^{-1})$ which is the product of two variables: K_L, the liquid mass transfer coefficient (m/sec), and a, the area-to-volume ratio of the packing (m^2/m^3). For design purposes, $K_L a$ should be determined experimentally. However, for dilute solutions, $K_L a$ can be estimated from the Sherwood and Holloway[1] equation:

$$K_L a = \alpha \times D_L \times \left(305 \; \frac{L}{\mu}\right)^{1-n} \left(\frac{\mu}{\rho D_L}\right)^{0.5} \qquad (9\text{-}4)$$

where D_L = liquid diffusion coefficient $(m^2/\sec)$
   L = liquid mass loading rate $(kg/m^2 \sec)$
   μ = liquid viscosity $(1.002 \times 10^{-3} \text{ Pa·sec at } 20°C)$
   ρ = density of water $(998.2 \text{ kg/m}^3 \text{ at } 20°C)$
   α, n = constants from Table 9-1

Diffusion coefficients may be estimated by the Wilke-Chang method (see Sec. 3.2). Mathematical correlations developed by Onda[3,4] also provide values of $K_L a$ that are valid over a range of contaminants and are preferred for preliminary design. The Onda correlations are:[5]

$$k_L \left(\frac{\rho_L}{\mu_L g}\right)^{1/3} = 0.0051 \left(\frac{L}{a_w \mu_L}\right)^{2/3} \left(\frac{\mu_L}{\rho_L D_L}\right)^{-0.5} \left(a_t d_p\right)^{0.4} \qquad (9\text{-}5)$$

and

$$\frac{a_w}{a_t} = 1 - \exp\left[-1.45 \left(\frac{\sigma_c}{\sigma}\right)^{0.75} \left(\frac{L}{a_t \mu_L}\right)^{0.1} \left(\frac{L^2 a_t}{\rho_L^2 g}\right)^{-0.05} \left(\frac{L^2}{\rho_L \sigma a_t}\right)^{0.2}\right] \qquad (9\text{-}6)$$

$$\frac{k_G}{a_t D_G} = 5.23 \left(\frac{G}{a_t \mu_G}\right)^{0.7} \left(\frac{\mu_G}{\rho_G D_G}\right)^{1/3} \left(a_t d_p\right)^{-2} \qquad (9\text{-}7)$$

TABLE 9-1
Constants for equation 9-4[2]

Packing	Size (mm)	α	n
Raschig Rings	12	920	0.35
	25	330	0.22
	38	295	0.22
	50	260	0.22
Berl Saddles	12	490	0.28
	25	560	0.28
	38	525	0.28
Tile	75	360	0.28

Note: millimeters × 0.0394 = inches

The overall mass transfer coefficient may be found by:

$$\frac{1}{K_L a} = \frac{1}{H' k_G a} + \frac{1}{k_L a}$$ (9-8)

Example units for the Onda correlations are:

k_L liquid-phase mass transfer (m/sec)

ρ_L liquid density (kg/m^3) = 998.2 at 20°C

μ_L viscosity of water (kg/m·sec) = 1.002×10^{-3} at 20°C

g gravity (m/sec^2) = 9.81

L liquid mass loading rate (kg/m^2·sec)

a area/volume ratio; use a_w

a_w wetted packing area (m^2/m^3) (used as equal to a in $K_L a$)

D_L liquid diffusivity (m^2/sec) (see Sec. 3.2)

a_t total packing area (m^2/m^3) (see Table 9-2)

d_p nominal packing diameter (m)

σ_c critical surface tension of packing material (N/m) = 0.033 for polyethylene; = 0.040 for PVC

σ surface tension of water (N/m) = 0.073 at 20°C

k_G gas-phase mass transfer (m/sec)

D_G gas diffusivity (m^2/sec) (see App. A and Sec. 3.2)

G gas mass loading rate (kg/m^2· sec)

ρ_G density of air (kg/m^3) = 1.205 at sea level at 20°C

$K_L a$ overall mass transfer (sec^{-1})

H' Henry's constant (dimensionless) (see App. A and Sec. 3.2)

TABLE 9-2
Packing characteristics

Size in mm	Intalox saddle		Raschig ring		Pall ring		Berl saddle		Tri-pack	
	F	a	F	a	F	a	F	a	F	a
13	200	623	580	364	—	—	240	466	—	—
25	98	256	155	190	52	206	110	249	28	279
50	40	118	65	92	25	102	45	105	16	157
75	22	92	37	62	—	—	—	—		
88	—	—	—	—	16	85	—	—	12	125

F = Packing factor (dimensionless); a = area/volume ratio (m^2/m^3). Data for ceramic shapes except Pall Ring and Tri-pak® data for plastic shape.

Example 9-1 Onda correlations. Given the following characteristics of an air stripping column to remove chloroform ($CHCl_3$) from ground water: Flow = 170 gpm; Column diameter = 3 ft; Packing = 2 in polyethylene Tri-paks®, $D_L = 6 \times 10^{-6}$ cm²/s; Air-to-water ratio = 100:1. Determine the overall mass transfer coefficient, $K_L a$, using the Onda correlations. Assume 20°C (293.2 K).

Solution

$$\frac{\sigma_c}{\sigma} = \frac{0.033}{0.073} = 0.45$$

diameter = 3 ft = 0.91 m; area = $\pi (0.91)^2 / 4 = 0.650$ m²

unit weight of water = 8.34 lb/gal

$$L = \frac{170 \frac{gal}{min} \times \frac{1\ min}{60\ sec} \times 8.34 \frac{lb}{gal} \times \frac{1\ kg}{2.204\ lb}}{0.650\ m^2} = 16.49\ kg/m^2 \cdot sec$$

$a_t = 157$ m⁻¹ (Table 9-2)

$D_G = 0.09404$ cm²/sec (see App. A) $\times 10^{-4}$ m²/cm² $= 9.404 \times 10^{-6}$ m²/sec

$$Q_W = 170 \frac{gal}{min} \times \frac{ft^3}{7.48\ gal} \times \frac{min}{60\ sec} = 0.379\ ft^3/sec$$

$$Q_A = 100 \times Q_W = 37.9\ ft^3/sec$$

$$G = \frac{37.9 \frac{ft^3}{sec} \times 0.02832 \frac{m^3}{ft^3} \times 1.205 \frac{kg}{m^3}}{0.650\ m^2} = 1.99\ kg/m^2 \cdot sec$$

$$D_L = 6 \times 10^{-6} \frac{cm^2}{sec} \times \frac{m^2}{10^4\ cm^2} = 6 \times 10^{-10} m^2/sec$$

$$d_p = 2\ in \times \frac{m}{39.37\ in} = 0.051\ m$$

From App. A for chloroform: A = 9.84; B = 4.61 $\times 10^3$

$$H = \exp\left[9.84 - \left(4.61 \times 10^3 / T\right)\right]$$
$$= \exp\left[9.84 - \left(4.61 \times 10^3 / 293.2\right)\right]$$
$$= 2.786 \times 10^{-3}\ atm \cdot m^3/mol$$

From Eq. 3-7:

$$H' = \frac{H}{RT} = \frac{2.786 \times 10^{-3}\ atm \cdot m^3/mol}{\left(8.205 \times 10^{-5}\ atm \cdot m^3/mol \cdot K\right)(293.2\ K)} = 0.116\ \text{(dimensionless)}$$

Reynolds No.:

$$\frac{L}{a_t \mu_L} = \frac{16.48\ kg/m^2 \cdot sec}{157\ m^{-1} \times 1.002 \times 10^{-3}\ kg/m \cdot sec} = 104.76\ \text{(dimensionless)}$$

Froude No.:

$$\frac{L^2 a_t}{\rho_L^2 g} = \frac{\left(16.48^2\right)(157)}{\left(998.2^2\right)(9.81)} = 4.36 \times 10^{-3} \text{ (dimensionless)}$$

Weber No.:

$$\frac{L^2}{\rho_L \sigma a_t} = \frac{16.48^2}{(998.2)(0.073)(157)} = 0.0237 \text{ (dimensionless)}$$

From Eq. 9-6:

$$a_w = 157 \left\{1 - e\left[-1.45\,(0.45)^{0.75}\,(104.76)^{0.1}\left(4.36 \times 10^{-3}\right)^{-0.05}(0.0237)^{0.2}\right]\right\}$$

$$= 85.6 \text{ m}^2/\text{m}^3$$

From Eq. 9-5:

$$k_L\left(\frac{998.2}{\left(1.002 \times 10^{-3}\right)(9.81)}\right)^{1/3} = 0.0051\left(\frac{16.48}{(85.6)\left(1.002 \times 10^{-3}\right)}\right)^{2/3} \times$$

$$\left(\frac{1.002 \times 10^{-3}}{(998.2)\left(6 \times 10^{-10}\right)}\right)^{-0.5} ((157)(0.051))^{0.4}$$

$$k_L = 2.00 \times 10^{-4} \text{ m/sec}$$

From Eq. 9-7:

$$\frac{k_G}{(157)\left(9.104 \times 10^{-6}\right)} =$$

$$5.23\left(\frac{1.99}{(157)\left(1.81 \times 10^{-5}\right)}\right)^{0.7}\left(\frac{1.81 \times 10^{-5}}{(1.225)\left(9.4 \times 10^{-6}\right)}\right)^{1/3} ((157)(0.051))^{-2}$$

$$k_G = 1.374 \times 10^{-2} \text{ m/s}$$

$$\frac{1}{K_L a} = \frac{1}{H' k_G a} + \frac{1}{k_L a} =$$

$$\frac{1}{(0.116)\left(1.374 \times 10^{-2}\right)(85.6)} + \frac{1}{\left(2.05 \times 10^{-4}\right)(85.6)} = 64.32$$

$$\underline{\underline{K_L a = 0.0155 \left(\text{sec}^{-1}\right)}}$$

In order to develop a design equation, a section of the stripping tower is selected (see Fig. 9-3). The section has a superficial surface area B, and a differential thickness dz.

The mass transferred per unit volume of the tower is:

$$M = (Q_W dC)/(B dz) = (Q_W/B)(dC/dz) \tag{9-9}$$

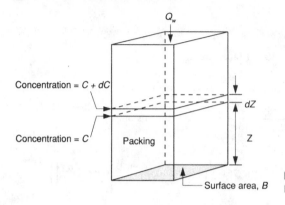

FIGURE 9-3
Differential section.

where
M = mass transfer rate, (kmol/sec·m^3)
Q_w = liquid flow rate (m^3/sec)
C = contaminant concentration (kmol/m^3)
B = surface area (m^2)
z = depth in column (m)
dC/dz = concentration gradient (kmol/m^4)

The mass transfer can also be defined by examination of the concentration gradient across the air/water interface:

$$M = K_L a[C - C_{eq}] \qquad (9\text{-}10)$$

where
C_{eq} = concentration in water in equilibrium with the air at a specific point
$[C - C_{eq}]$ = the degree to which the system is out of equilibrium

From the definition of Henry's law (Eq. 3-6):

$$C_{eq} = A/H' \qquad (9\text{-}11)$$

where
A = concentration in air (kmol/m^3)
H' = dimensionless Henry's constant (see Sec. 3-2).

The concentration imbalance $[C - C_{eq}]$ is the driving force for transfer. Setting Eq. 9-10 equal to Eq. 9-9, and converting to molar units by substituting L/M_w for Q_w/B:

$$(L/M_w)(dC/dz) = K_L a[C - C_{eq}] \qquad (9\text{-}12)$$

where
L = liquid molar loading rate (kmol/sec · m^2)
M_W = molar density of water (55.6 kmol/m^3) = 3.47 lb mol/ft^3

Equation 9-12 is a first order linear differential equation. Separating the variables:

$$dz = \frac{L}{M_W K_L a} \times \frac{dC}{[C - C_{eq}]} \tag{9-13}$$

The first term on the right is independent of C, and is designated HTU, the height of a transfer unit:

$$HTU = \frac{L}{M_W K_L a} \tag{9-14}$$

Integrating from the bottom of the column to the top, we obtain the required depth of packing in the column:

$$Z = HTU \times NTU \tag{9-15}$$

where NTU (the number of transfer units) is:

$$NTU = \int_{C=C_{out}}^{C=C_{in}} \frac{dC}{[C - C_{eq}]} \tag{9-16}$$

The equilibrium concentration, C_{eq}, may be determined from a mass balance from the bottom of the column up to the differential section, dz. Let A and C be the concentrations in the air and water at the section, respectively. The mass gained by the air must equal the mass lost by the water:

$$Q_A(A - A_{in}) = Q_w(C - C_{out}) \tag{9-17}$$

From Eq. 9-11: $A = C_{eq}H'$ and $A_{in} = 0$, therefore

$$Q_A C_{eq} H' = Q_w(C - C_{out}) \tag{9-18}$$

Solving for the equilibrium concentration:

$$C_{eq} = Q_w(C - C_{out})/(Q_A H') = (C - C_{out})/R \tag{9-19}$$

Substituting Eq. 9-19 into Eq. 9-16 and integrating produces:

$$NTU = \left(\frac{R}{R-1}\right) \ln\left(\frac{(C_{in}/C_{out})(R-1)+1}{R}\right) \tag{9-20}$$

where R = stripping factor, $H'(Q_A/Q_w)$ (dimensionless)
   C_{in} = concentration in the influent water (kmol/m^3)
   C_{out} = concentration in the effluent water (kmol/m^3)

Since the units for concentration cancel, C_{in} and C_{out} may be expressed in mass concentration units (mg/L or μg/L).

Practical values of R range from 2 to 10 or greater. The stripping factor is one of the primary design variables in the system. Realistic values of the air-to-water ratio (Q_A/Q_w) range from 5 to several hundred.

It should be noted that in the chemical engineering literature the stripping factor is typically defined as[6]

$$R = H'(G/L) \qquad (9\text{-}21)$$

where H' = Henry's law constant (dimensionless)
   G = molar flow rate of air (kmol/hr)
   L = molar flow rate of water (kmol/hr)

At the low concentrations encountered in environmental work, Eqs. 9-21 and 9-20 produce approximately the same value for R.

Example 9-2   Preliminary design of air stripping column. A ground water supply has been contaminated with ethylbenzene. The maximum level of ethylbenzene in the ground water is 1 mg/L and this must be reduced to 35 μg/L using an air stripping column. The following data are provided:

$K_L a = 0.016 \text{ sec}^{-1}$; $Q_W = 7.13$ L/sec; Temperature, $T = 20°C = 293.3$ K

$H = 6.44 \times 10^{-3} \text{atm} \cdot \text{m}^3/\text{gmol}$

Select: Column diameter = 2.0 ft (0.61 m); air-to-water ratio $(Q_A/Q_W) = 20$

Determine: Liquid loading rate, L

             Stripping factor, R

             HTU, NTU, and height of packing in column

Solution. From Eq. 3-7: $H' = H/RT = 6.44 \times 10^{-3}/(8.205 \times 10^{-5} \times 293.3) = 0.27$

1. Liquid loading rate:
   cross-sectional area of column = 0.292 m^2
   mass rate = 1.0 kg/L $\times$ 7.13 L/sec = 7.13 kg/sec
   mass loading = 7.13/0.292 = 24.4 kg/sec $\cdot$ m^2
   $L = (24.4 \text{ kg/sec} \cdot \text{m}^2)(10^3 \text{ g/kg})(1/18 \text{ mol/g}) = \underline{1360 \text{ mol/sec} \cdot \text{m}^2}$

2. Stripping factor: $R = H'(Q_A/Q_W) = 0.27(20) = \underline{5.4 \text{ (dimensionless)}}$

3. Height of transfer unit: $\text{HTU} = \dfrac{L}{M_W K_L a} = \dfrac{1360}{(55,600)\,(.016)} = \underline{1.53 \text{ m}}$

4. Number of transfer units: $\text{NTU} = \left(\dfrac{R}{R-1}\right) \ln\left(\dfrac{(C_{in}/C_{out})\,(R-1)+1}{R}\right)$

   $$\text{NTU} = (1.23) \ln\left(\dfrac{(1,000/35)\,(5.4-1)+1}{5.4}\right)$$

   $$= \underline{3.88 \text{ transfer units}}$$

5. Height of packing in column $Z = \text{NTU} \times \text{HTU} = (3.88)(1.53) = \underline{5.93 \text{ m} = 19.45 \text{ ft}}$

In an actual design, a safety factor of 20% would be added and the height of packing raised to the next whole number:

$$Z = 19.45(1.2) = 23.34\text{ft}$$

Use $\underline{\underline{Z = 24 \text{ ft}}}$

Design Considerations

In practice, stripping towers have diameters of 0.5 to 3 m and heights of 1 to 15 m. The air-to-water ratio ranges from as low as 5 to several hundred and is controlled primarily by flooding and pressure drop considerations.

As the air flow in a tower is increased, it will ultimately hold back the free downward flow of water and cause flooding of the tower. The smaller size packings are more susceptible to flooding. Channeling occurs when water flows down the tower wall rather than through the packing. Distribution plates should be placed approximately every 5 diameters to avoid this. Using a smaller size packing will also reduce the tendency of flow to channel.

It is sometimes necessary to collect air emissions and remove organics from the air stream prior to release to the atmosphere. This can be accomplished by passing the air stream through activated carbon, but requires piping the air stripper to collect off-gases (see Fig. 9-4).

The pressure drop in a tower should be between 200 and 400 N/m^2 per meter of tower height (0.25 to 0.5 in H_2O/ft) to avoid flooding.[7] The pressure drop appears to be the parameter that most directly affects the operating cost of the tower. Capital cost is a function of the tower volume.

FIGURE 9-4
An air stripping unit.

Flooding and pressure drop calculations are simplified through the use of Fig. 9-5. The notation for Fig. 9-5 is:

G = air loading rate $(\text{lb}/\text{ft}^2\text{sec})$

L = liquid loading rate $(\text{lb}/\text{ft}^2\text{sec})$ $(= \text{kg}/\text{sec} \cdot \text{m}^2 \times 0.2048)$

F = packing factor (dimensionless, see Table 9-2)

ρ_{air} = air density $(\text{lb}_f/\text{ft}^3)$ $(= \text{kg}/\text{m}^3 \times 0.06243)$

ρ_W = water density $(\text{lb}_f/\text{ft}^3)$

g = gravitational constant $(32.17 \text{ ft}/\text{sec}^2)$

Because the ordinate axis of Fig. 9-5 is not dimensionless, the English units shown must be used. Figure 9-5 may be used to estimate the pressure drop for a specified design. Additionally, it may be used to modify a design by moving to a different acceptable pressure drop and determining a new value for G from the ordinate axis.

Characteristics of several typical packing shapes are shown in Table 9-2. Additional information is available in several references.[9, 10, 11]

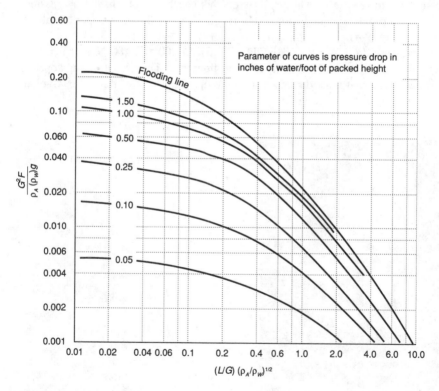

FIGURE 9-5
Generalized pressure drop correlations.[8]

Example 9-3   Design of air stripping tower. Using the data from Example 9-2, determine the pressure drop through the tower. Examine the impact on effluent quality of varying the air-to-water ratio (A/W) and the packing height.

Solution

1. Select 3″ (75 mm) ceramic Raschig Rings as packing.

2. Calculate pressure drop. The parameters for Fig. 9-5:

$$L = 24.4 \frac{kg}{sec \cdot m^2} \times 0.2048 = 4.997 \frac{lb}{sec \cdot ft^2}$$

$$Q_a = Q_W(A/W) = 7.13(20) = 142.6 \text{ L/sec} = 0.1426 \text{ m}^3/\text{sec}$$

$$\rho_a = \text{air density} = 1.205 \text{ kg}/m^3 = 0.075 \text{ lb}/ft^3 \ (@20°C)$$

$$\rho_W = \text{water density} = 62.3 \text{ lb}/ft^3 \ (@20°C)$$

$$G = \frac{(0.1426 \text{ m}^3/\text{sec})(1.205 \text{ kg}/m^3)}{(0.292 \text{ m}^2)} = 0.5885 \frac{kg}{sec \cdot m^2}$$

$$= 0.5885 \frac{kg}{sec \cdot m^2} \times (0.2048) = 0.1205 \frac{lb}{sec \cdot ft^2}$$

$$F = 37 \text{ (Table 9-2)}$$

$$\text{Ordinate} = G^2 F/\rho_a(\rho_w)g = (.1205)^2(37)/(.075)(62.3)(32.17) = \underline{0.0036}$$

$$\text{Abscissa} = (L/G)(\rho_a/\rho_w)^{.5} = (4.997/.1205)(.075/62.3)^{.5} = \underline{1.44}$$

This point intersects the curves of Fig. 9-5 at dP = 0.1 inches of water per foot of packing depth (82 Pa/m). This pressure drop is below the recommended range of 0.25 to 0.5 inches of water per foot of packing depth, indicating that the air flow rate could be increased significantly without causing flooding.

$$\text{Pressure Drop} = (19.45)(0.1) = 1.95 \ \underline{2 \text{ inches of water}} = \underline{500 \text{ pascals.}}$$

3. Evaluate impact on effluent quality. It should be noted that column diameter, A/W, and packing shape are design parameters to be selected to provide a cost-effective design. The following output from the AIRSTRIP Program[12, 13] provides a range of designs. Note that at $A/W = 20$ and $Z = 6$ m, the effluent concentration approximates that of Example 9-2. For $A/W = 20$, the pressure drop is 106 pascals/meter. For six meters this yields a pressure drop of 636 pascals, about 25% greater than the calculation shown above, but within the accuracy of Fig. 9-5.

Contaminant: Ethylbenzene	Concentration In: 1000.0 μg/L
C-Raschig Rings 76.2 mm	Atmospheric Pressure: 101.3 kPa
Temperature: 20.0°C	Liquid Loading Rate: 24.4 kg/m^{2}sec
Min Packing Depth: 2.0 meter	Minimum A/W Ratio: 10.0
Max Packing Depth: 6.0 meter	Maximum A/W Ratio: 50.0

Effluent Concentration (μg/L)					
Packing Depth (meter)	$A/W = 10$	$A/W = 20$	$A/W = 30$	$A/W = 40$	$A/W = 50$
2.0	359.1	310.6	292.5	82.9	276.8
3.0	231.1	180.2	162.7	153.7	148.2
4.0	152.1	105.8	91.2	84.0	79.7
5.0	101.5	62.5	51.3	46.1	30.0
6.0	68.4	37.1	29.0	25.3	23.2
R	2.7	5.4	8.1	10.8	13.5
dP (Pa/m)	<50	106	208	371	612

9-2 SOIL VAPOR EXTRACTION[†]

Soil Vapor Extraction (SVE) is a relatively new remedial option utilized to remove volatile organic compounds (VOCs) from soils in the vadose zone (the unsaturated zone above the ground water table) or from stockpiled, excavated soils. The SVE process consists of passing an air stream through the soil, thereby transferring the contaminants from the soil (or soil/water) matrix to the air stream.

Modifications of the process are distinguished by the location of the treatment system (i.e., in situ or above ground) and the method for developing air flow. SVE systems are implemented by installing vapor extraction wells or perforated piping in the zone of contamination and applying a vacuum to induce the movement of soil gases (refer to Fig. 9-6). SVE systems typically include knockout drums to remove moisture from the soil gases followed by vapor phase treatment prior to discharge to the atmosphere. The effective radius for extraction wells is reported to vary from 6 m to over 45 m (20 to 150 feet) depending on soil conditions.[14] SVE systems have also been demonstrated to depths of 7 m (23 feet) in soils with an average permeability (or hydraulic conductivity) of 10^{-4} cm/s.[15] SVE systems may be enhanced through the addition of alternative options/enhancements.

Examples of these system options/enhancements include:

- Installation of ground water extraction pumps into the vapor extraction wells to either lower the ground water table (and therefore enlarge the vadose zone) and/or concurrently remove contaminated ground water.

[†]Sec. 9-2 was prepared by James D. Fitzgerald, P.E. and Kevin E. McHugh, ERM-New England, Inc.

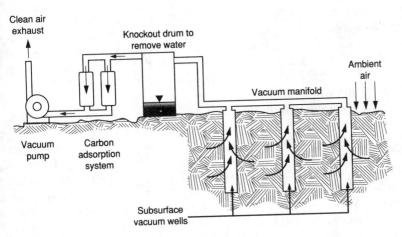

FIGURE 9-6
Schematic of SVE system.

- Placing an impermeable barrier over the surface to minimize short-circuiting of air flow from the surface. (Under these conditions the effective radius for the extraction wells has been reported to extend to 90 m [300 feet].[16])

- Installing air recharge wells around the zone of contamination to enhance movement of soil gases through contaminated soils.

- Providing a compressor to force clean air into the recharge wells enhancing soil gas movement. (Applied pressures are small and typically range from 3500 to 85,000 Pa [0.5 to 12.5 psi].)

- Installing wells into the zone of contaminated ground water and blowing air through the ground water (air sparging). The induced air flow enhances the volatilization of the contaminants (especially compounds with low solubility in water), causing upward migration to the vadose zone where it is captured by the SVE system.[17] (Care should be taken to also install hydraulic controls with the air sparging system to ensure that ground water mounding does not result in migration of contaminants into uncontaminated zones.)

The above modifications represent a partial listing of the various alternatives that can be incorporated into an SVE system. Depending on site-specific conditions, the appropriate system might include various combinations of the above items.

Only SVE treatment of vadose zone contaminants will be discussed in detail in this section. However, the reader should be aware that the basic theory describing movement of soil gases and contaminants through the vadose zone applies to the various permutations of the SVE system presented above.

Theory of Soil Aeration

A VOC spill or release to the ground can be represented using a three-phase model of the vadose zone including: the soil matrix (solid phase), the soil moisture that is located in the voids between and on the surface of the soil matrix (liquid phase), and the soil gas that constitutes the remaining portion of the voids between the soil particles (gas phase). After a release has occurred, VOC contamination in the vadose zone will be distributed among all three phases. The mass distribution of contamination between the phases will depend on the soil conditions (e.g., water content, organic content) as well as the specific contaminants of concern.

Contaminant movement among the various phases in the vadose zone is ongoing with the subsurface system moving toward equilibrium conditions. The removal of VOCs from the vadose zone may be modeled as the five-step process illustrated in Fig. 9-7. Volatile compounds will desorb from the soil particle surface, transfer to the soil water (moisture in the vadose zone) and volatilize to the soil gas. This soil gas then migrates through the void spaces until reaching the ground surface and is finally released to the atmosphere. Assuming localized equilibrium (that is, the concentration of a volatile contaminant in the soil gas at any point in the system is in equilibrium with the chemical concentration in the stationary phase) may lead to an improper mass distribution.[18] However, equilibrium conditions are assumed on an area-wide scale in order to facilitate remedial design.

The movement of contaminants in the soil gas through the soil media can be described by two processes: advection and diffusion. Advection is the movement of a vapor phase contaminant with the bulk airflow through the soil media and best describes the flow through permeable soils within the unsaturated zone. Diffusion is the movement of a contaminant through the soil medium via concentration gradients.[19] Diffusion tends to control in soils with lower permeability.

Spills or leaks of VOCs to the environment may result in surface or subsurface contamination. Provided the leak is of sufficient quantity, the VOC contaminants will tend to migrate downward through the unsaturated zone, leaving globules, films, and small droplets of the released material. Low-density contaminants will tend to collect in the capillary fringe or float on the ground water surface. Dense contaminants (following transport through the vadose zone) will tend to pass through the ground water until encountering an impermeable layer (see Fig. 4-22).

A release of contaminants will result in residual contamination in the soil pores. This residual material in the unsaturated zone is the target contamination for cleanup through SVE (with the addition of further processes to the SVE system, contaminants

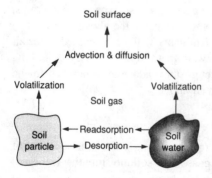

FIGURE 9-7
Schematic of soil/soil vapor model.

trapped in the capillary fringe and on the ground water surface may also be remediated). The type (low- or high-density) of chemical released and therefore the location where the contaminants accumulate will affect the volatilization and movement of the contaminants and the design of a remediation system.

A SVE system removes volatile organic contaminaton by evacuating the soil gases, thereby enhancing the movements of contaminants from the solid and liquid phase to the gas phase as shown in Fig. 9-7. The success of SVE remedial systems is therefore based on the effectiveness (both cost and performance) of moving the soil gases through the contaminated zone. Before the volatile compounds can be carried via advective flow to the SVE system for treatment, the contaminants must flow from lower permeability to higher permeability soils by diffusion. Therefore, diffusion may be the rate-limiting step for mass transfer.[20] It has been estimated that 40 to 60% of contaminant removal is controlled by advective flow, while the remaining amount is controlled by diffusion.[21]

Contaminant transport between different phases is affected by certain physical, environmental, and chemical properties as listed in Table 9-3. Physical properties that control SVE system performance include: moisture content of the soil in the vadose zone, organic content of the soil, grain size distribution of the soil, and porosity of the soil in the vadose zone. Environmental properties that affect SVE system performance include: area rainfall, air temperature, and ground cover. Chemical properties affecting SVE system performance include: the volatility of the contaminant (the contaminant's tendency to transfer from the liquid to the vapor phase), the relative polarity between the contaminant and the soil particles, and the density and viscosity of the liquid phase contaminant. The latter properties dictate where the liquid phase contaminants will accumulate in the subsurface. For example, dense, non-aqueous phase liquids (DNAPLs) will migrate downward through the ground water. The physical-chemical properties that control contaminant movement in the environment were presented in Sec. 3-2, and contaminant transport is discussed in Sec. 4-2.

While theoretical models are available to attempt to predict the performance for each of the individual parameters identified in Table 9-3, actual soil conditions are too complex and non-homogeneous for these models to be used in the design of SVE systems. Current practice is to utlilize empirical models to select the most appropriate mechanical system and then to use field data to refine system design.

TABLE 9-3
Properties affecting soil aeration systems

Soil properties	Contaminant properties	Environmental properties
Permeability	Henry's law constant	Temperature
Porosity	Solubility	Humidity
Grain size distribution	Adsorption coefficient	Wind speed
Moisture content	VOC concentration in soil	Solar radiation
pH	Polarity	Rainfall
Organic content	Vapor pressure	Terrain
Bulk density	Diffusion coefficient	Vegetation

As previously discussed, movement of VOCs through the soil is controlled in part by diffusion (see Sec. 3-2). Fick's law best describes this type of vapor transport (vapor flux) through the soil media:

$$J = -D_v \frac{dC}{dz}$$

(9-22)

where J = vapor flux (mol/cm$^2 \cdot$ sec)
   D_v = diffusion coefficient in the vadose zone (cm^2/sec)
   C = concentration of vapor in soil air (mol/cm^3)
   z = distance in the direction of diffusion (cm)

Diffusion coefficients for pure compounds in air are tabulated in App. A. These values should be modified to account for the fact that air movement in a soil matrix is affected by soil properties. Grasso[22] suggests the Millington and Quirk[23] relationship:

$$D_v = \frac{a^{10/3} D}{\phi^2}$$

(9-23)

where a = portion of voids filled with air (dimensionless)
   ϕ = soil porosity (volume voids/total volume)
   = a + (portion filled with water) + (portion filled with immiscible fluid)
   D = diffusion coefficient of chemical in free air (cm^2/sec)

Dragun[24] suggests the following modification of the diffusion coefficient to account for soil properties:

$$D_v = T(\phi - \phi_B)D$$

(9-24)

where D_v = diffusion coefficient to be used in soil air
   T = correction factor due to tortuous flow path
   ϕ_B = volume of blocked pores (fraction)

Experimental values of T and ϕ_B are highly variable. Dragun quotes the following experimental values:

T Range from 0.6 to 0.7; median 0.66

ϕ_B Range from 0.0 to 0.25; median 0.1

The organic carbon partition coefficient (K_{oc}) (see Sec. 3-2) expresses a chemical's tendency to be adsorbed to soil particles in the soil (or soil/water) matrix. K_{oc} may be thought of as the ratio of the amount of chemical adsorbed per unit weight of organic carbon (oc) in the soil to the concentration of the chemical in the water phase at equilibrium.[25]

$$K_{oc} = \frac{\mu g \text{ adsorbed/g organic carbon}}{\mu g/\text{mL of solution}}$$

(9-25)

Soils with relatively high-organic content (e.g., peat) will tend to more readily adsorb organics than those soils with low-organic content (e.g., silts). K_{oc} values of some volatile contaminants may be found in App. A.

The partition coefficient (K_p) for a particular compound refers to the preference of a contaminant for the soil or water phase. K_p was defined in Sec. 3-2 as:

$$K_p = \frac{X}{C} \tag{9-26}$$

where
- $X =$ contaminant concentration on soil surface
  (μg contaminant/kg soil) at equilibrium
- $C =$ contaminant concentration remaining in water phase
  (μg contaminant/liter) at equilibrium

A higher K_p value indicates that a contaminant is more likely to remain on the soil and not be transmitted through soil moisture movement.

Gas and liquid phase equilibrium can be described for dilute aqueous solutions by Henry's law and for concentrated non-aqueous phase liquid (NAPL) solutions by Raoult's law (see Sec. 3-2).

To determine the effect of Henry's law on the effectiveness of SVE, an expression relating the total concentration and the concentrations of the individual soil, water, and air phases is used along with the concept of partitioning:[26]

$$C_T = d_B C_a + V_w C_l + V_a C_v \tag{9-27}$$

where
- $C_T =$ total concentration of contaminant (mass/volume)
- $d_B =$ soil dry bulk density (mass/volume)
- $C_a =$ mass of contaminant adsorbed per mass of soil
- $V_w =$ volumetric fraction of water in the soil (volume/volume)
- $V_a =$ volumetric fraction of air in the soil (volume/volume)
- $C_l, C_v =$ concentration of contaminant associated with liquid
  and vapor phase respectively (mass/volume)

By substituting equilibrium equations into Eq. 9-27, the vapor partition coefficients can be derived.

Two equilibrium equations describe the phase relationships between chemicals that exist in the soil matrix-water system and the relationship between total concentrations of a compound and the concentration of that compound in each phase. These relationships are as follows:

$$C_a = K_p C_l \quad \text{(a linear adsorption model)} \tag{9-28}$$

where
- $C_a =$ mass adsorbed per mass of soil
- $K_p =$ soil water partition coefficient
- $C_l =$ concentration in water (mass per volume)

$$C_v = H'C_l \quad \text{(Henry's law)} \tag{9-29}$$

where C_v = concentration in air (mass per volume)
H' = Henry's constant (dimensionless, see Sec. 3-2)

By rearranging Eqs. 9-28 and 9-29 and substituting, Eq. 9-27 becomes:

$$C_T = d_B \left(K_p C_l \right) + \frac{V_w \left(C_v \right)}{H'} + V_a C_v = d_B \frac{\left(K_p C_v \right)}{H'} + \frac{\left(V_w C_v \right)}{H'} + V_a C_v$$

$$C_T = \left[d_B \frac{K_p}{H'} + \frac{V_w}{H'} + V_a \right] C_v$$

$$C_T = R_v C_v \tag{9-30}$$

where R_v = vapor partition coefficient

The liquid phase equilibrium equations are the same as the vapor phase equations above. By substitution, Eq. 9-30 becomes:

$$C_T = d_B K_p C_l + V_w C_l + V_a H' C_l$$

$$C_T = \left[d_B K_p + V_w + V_a H' \right] C_l$$

$$C_T = R_l C_l \tag{9-31}$$

where R_l = the liquid partition coefficient

Example 9-4   Soil vapor extraction. Three areas of a large site have been contaminated by similar concentrations of different compounds (Area 1: ethylbenzene; Area 2: dichlorobenzene; Area 3: 1,1,1, trichloroethane). Assuming the physical and environmental conditions are the same at each area, which area would be best suited to vadose zone treatment by SVE? Explain.

Solution. From App. A:

Compound	Solubility (mg/L)	Vapor Pressure (mm Hg)	Henry's Constant @20°C (atm · m³/mol)
Ethylbenzene	152	7.0	7.03×10^{-3}
Dichlorobenzene	79	2.28	2.86×10^{-3}
1,1,1, Trichloroethane	4400	100	1.34×10^{-2}

From Eq. 9-29 through 9-31, compounds with higher values of H' will have a greater affinity for the vapor phase. Based on this information, Area 3 would appear to be the best candidate for SVE.

Thomas[27] summarizes a number of methods for estimating the volatilization of chemicals from soil with the caveat that no single method currently addresses all of the complexities of the process. The Hartley method provides an estimate based on the heat balance between the contaminant and air:

$$J = \frac{A_{sat}\frac{(1-h)}{\delta}}{\frac{1}{D} + \frac{\lambda^2(A_{sat})M}{kRT^2}}$$ (9-32)

where J = flux (g/cm^2 · sec)
  A_{sat} = saturation concentration in air (at temperature of outer air) (g/cm^3)
  h = relative humidity ($0 < h < 1$)
  δ = thickness of stagnant boundary layer (cm)
  D = diffusion coefficient in air (cm^2/sec)
  λ = latent heat of vaporization (cal/g)
  M = molecular weight (g/mol)
  k = thermal conductivity of air (61×10^{-6} cal/sec·cm·K)
  R = universal gas constant (1.987 cal/mol·K)
  T = temperature (K)

The second term in the denominator indicates the resistance to volatilization due to thermal considerations and may be neglected for compounds significantly less volatile than water.

Example 9-5 Volatilization of TCE from soil. A tank containing 30,000 gallons of trichloroethylene (TCE) ruptures contaminating 2,000 ft^2 of soil. Use Hartley's equation to estimate the volatilization of TCE to the air and the time required for the entire 30,000 gallons to evaporate. Comment on the limitations of the method.

The following data are provided:

TCE saturation in air, $A_{sat} = 415$ g/m^3 (Verschueren)[28]
Diffusion coefficient, $D = 0.070$ cm^2/sec   (Lyman, et al.)[29]
Humidity, $h = 0.3$;   Temperature, $T = 20°C = 293°K$
Latent heat of vaporization, $\lambda = 7.52$ kcal/mol   (Dean)[30]
Thickness of stagnant boundary layer, $\delta = 0.3$ cm
Specific gravity of TCE = 1.46   (Verschueren)[31]

Solution

1. Flux:

$$A_{sat} = 415 \text{ g/m}^3 \times 10^{-6} \text{ m}^3/\text{cm}^3 = 4.15 \times 10^{-4} \text{ g/cm}^3$$

$$M = 131.5 \text{ g/mol}$$

$$\lambda = \frac{7520 \text{ cal/mol}}{131.5 \text{ g/mol}} = 57.2 \text{ cal/g}$$

$$k = 61 \times 10^{-6} \text{ cal/sec} \cdot \text{cm} \cdot \text{K}$$

$$J = \cfrac{\cfrac{4.15 \times 10^{-4} \, (\text{g/cm}^3)(0.7)}{0.3 \text{ cm}}}{\cfrac{1}{0.07 \text{ cm}^2/\text{sec}} + \cfrac{(57.2 \text{ cal/g})^2 (4.15 \times 10^{-4} \text{ g}/\text{cm}^3)(131.5 \text{ g/mol})}{(61 \times 10^{-6} \text{ cal/sec·cm·K})(1.987 \text{ cal/mol·K})(293 \text{ K})^2}}$$

$$= 3.08 \times 10^{-5} \text{g/cm}^2 \cdot \text{sec}$$

2. Area and mass of chemical:

$$\text{area} = 2{,}000 \text{ ft}^2 \times 929.034 \text{ (cm}^2/\text{ft}^2) = \underline{1.858 \times 10^6 \text{ cm}^2}$$

$$\text{volume} = 3 \times 10^4 \text{ gal} \times .003785 \text{ (m}^3/\text{gal)} = 113.55 \text{ m}^3$$

$$\text{mass} = 113.55(1.46)(1000 \text{ kg/m}^3)$$

$$= 165{,}783 \text{ kg} = \underline{1.658 \times 10^8 \text{ g}}$$

3. Time required:

$$\text{time} = \text{mass}/(\text{area} \times \text{flux})$$

$$= \frac{1.658 \times 10^8 \text{g}}{\left(1.858 \times 10^6 \text{ cm}^2\right)\left(3.08 \times 10^{-5} \text{ g}/\text{cm}^2 \cdot \text{sec}\right)}$$

$$= 2.897 \times 10^6 \text{ sec} = 33.5 \text{ days}; \underline{\text{approximately 5 weeks}}$$

4. Limitations of method: Hartley's method accounts for only a few of the parameters listed in Table 9-3. Specifically, this problem does not consider soil properties, depth of contamination and many of the environmental properties. These limitations indicate that the results are probably accurate only to an order of magnitude.

Design Considerations

SVE's primary benefit is as an in situ remediation tool for the removal (and treatment) of VOCs from soils in the vadose zone. By utilizing in situ remediation, soil excavation and disposal costs, as well as site disruption that may be associated with large scale excavation activities, are minimized. Also, the long-term liability which may result due to off-site disposal of contaminated soils will be eliminated. SVE may be used to remediate soils beneath structures; does not require reagents; and employs conventional equipment, labor, and materials.

SVE is not an applicable remedial alternative at all locations. Its usefulness will be limited at sites with low-permeability soil, where contaminants with low vapor pressures (and low values of Henry's constant) are present, and at sites with high ground water tables. Furthermore, due to the number of uncertainties that are inherent with in situ remediation, the time required to achieve the clean-up levels is unpredictable. However, through the collection of site-specific information, the anticipated effectiveness of SVE can be evaluated.

Process Description

A typical SVE system consists of the following components:

Infrastructure:

- One or more vapor extraction well(s).
- Manifold piping connecting extraction wells with the vacuum source.
- Vacuum monitoring wells.
- Vacuum gauges and flow control valves at each monitoring and extraction well (optional).
- Impermeable cover over the system site to minimize the downward infiltration of air and water (optional).
- One or more vent wells to enhance air movement through the vadose zone (optional).

Extraction Equipment:

- Vacuum/blower unit.
- Moisture knockout drum (optional).
- Off-gas treatment.

A schematic diagram of a typical SVE system was previously provided as Fig. 9-6. A photo of an SVE system under construction is shown in Fig. 9-8.

Three variables control the performance of an in situ SVE system and must be considered during system design: (1) the extraction well spacing, (2) the air flow rate induced in the subsurface remedial zone, and (3) the subsurface pressure. Addi-

FIGURE 9-8
An SVE system under construction.

tional variables relating to the specific contaminant which must be evaluated for the design include: (1) pressure gradients, (2) identity and concentration of VOCs in soil, (3) extracted air VOC concentration, (4) extracted air temperature, (5) extracted air moisture, and (6) power usage.[32]

Consideration must also be given to the size and type of blower/vacuum unit which produces the vacuum pressure. Typical blower/vacuum units generate a negative pressure of between 0.5 inches to 30 inches of mercury (Hg). The blower/vacuum units may be regenerative centrifugal (the most common type which provides a wide operating range), multi-stage centrifugal, rotary positive displacement, and liquid ring. In situ pilot studies have been conducted to evaluate the effect of site-specific variables on system performance.

Example Pilot Studies

Results of in situ pilot studies to evaluate the design variables on a site-specific basis have been summarized in the literature.[33] These studies were intended to determine the effects of soil type and varying source vacuum pressures on system performance (i.e., what is the effect of various source pressures on the vadose zone?). Two example pilot studies are presented below.

The results of one study are provided in Fig. 9-9. The figures reveal that the change in negative pressure relative to distance from the extraction well is a decreasing exponential relationship. The pilot test consisted of the installation of one extraction well and five monitoring wells at the test site. The extraction well (VES-1) was installed in the area of highest-known soil VOC concentration, while the monitoring wells (VES-2 through VES-6) were installed at radial distances from VES-1. To evaluate the effects of different soil types (high- and low-permeability) on the SVE system, the monitoring wells were installed in two different hydrogeologic zones. Wells VES-1 through VES-4 were installed in sandy soils, while wells VES-5 and VES-6 were installed in an area of tight clay soils. All wells ranged in depth from 6 feet to 11 feet.

The results of this and similar studies indicate that SVE is highly dependent on specific site conditions. The results further reveal the importance of conducting an in situ pilot test prior to the design and implementation phases of a SVE project to ensure optimal well layout and blower selection.

A second study was conducted at the Twin Cities Army Ammunition Plant in New Brighton, Minnesota.[34] The study consisted of a 14-week field program to evaluate the feasibility of in situ air stripping as a remedial alternative. The target contaminant of the study was trichloroethylene (TCE). However, the information collected is applicable to VOCs with similar vapor pressures. It should be noted that the information gathered from the study is site-specific and should not be applied directly to other sites. Once again, pilot studies are typically performed at each site to evaluate the effectiveness of soil venting and to develop specific design information.

The Twin Cities pilot study consisted of the installation of three rows of vapor extraction wells around four air injection wells at two adjacent locations at the site. The extraction wells were manifolded to the vacuum intake of a vacuum/blower unit.

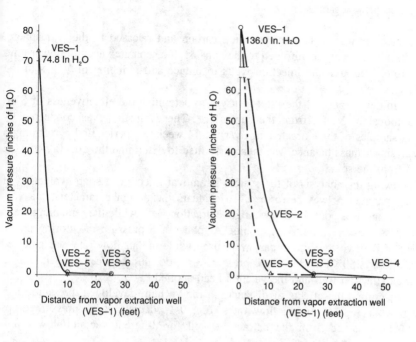

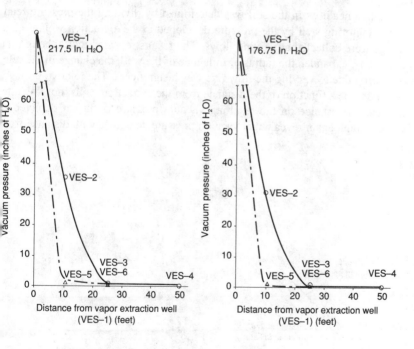

Note

VES 1 through VES 4: Sandy soils
VES 5 and 6: Tight clay soils

FIGURE 9-9
Effect of distance from vapor extraction well. [35]

The extracted air was treated by activated carbon and released to the atmosphere. The injection wells were manifolded to the exhaust of a separate vacuum/blower unit. Air was forced through the injection wells to induce soil stripping of VOCs in the subsurface.

The initial purpose of the pilot test was to determine the effectiveness of SVE on the removal of VOCs from the subsurface. The evaluation was conducted by collecting samples of the extracted air over the 14-week test period and performing a TCE analysis. A mass balance was then performed to determine the quantity of TCE removed from the soil.

Following the analysis of the system's removal effectiveness, an evaluation of the design parameters was performed. To evaluate the design parameters, the vacuum/blower unit was connected to a single extraction well, while the remaining wells were used as pressure monitoring points. A total of 26 pressure monitoring points (inclusive of the two remedial areas) were used while the air flow rate was adjusted between 50 and 250 CFM. Fig. 9-10 presents a relationship between the withdrawal rate at the extraction well and the vacuum head applied.

To determine the effects of distance on the vacuum developed, pressure readings were collected at varying flow rates. Fig. 9-11 depicts vapor pressure drops at increasing distance from the extraction well for different vacuum/blower unit flowrates.

The relative headloss in the soil was determined by dividing the pressure readings at each monitoring well by the pressure developed in the extraction well. Again, these readings were collected at varying flows. The results revealed that although the increased flow rate expanded the radius of influence of the well, the change in the relative pressure drop due to varying the flow rate was insignificant. The relative pressure drop was found to be a function of the distance from the extraction wells, not flowrate.

The effect of distance on the resulting vacuum pressure is further depicted in Fig. 9-12. This graph again reveals that vacuum pressure decreases with distance from

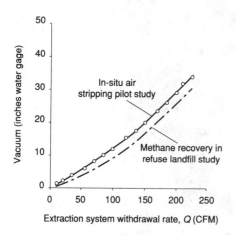

FIGURE 9-10
Twin Cities army ammunition plant ratio of withdrawal rate vs. vacuum head.[36]

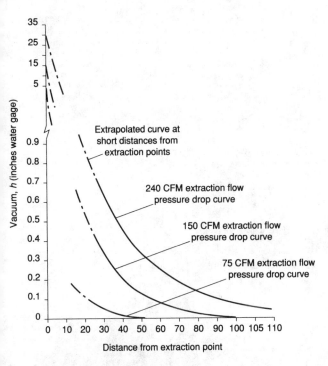

Note: Soil conditions consist of sandy
materials with traces of silt or gravel
and occasional gravelly layers.

FIGURE 9-11
Twin Cities army ammunition plant vapor pressure drop vs. distance.[37]

the extraction well. This graph would be prepared using site-specific pilot test data
and would then be used to design the full-scale remedial system.

Buscheck and Peargin[38] recommend that SVE well spacing be determined as
the radial distance corresponding to a percentage of the projected operating vacuum as
interpolated from the fitted vacuum distribution line. The percentage used to determine
well spacing will be dependent on how representative the pilot test data are of site
conditions. That is, if subsurface site conditions are well known and consistent, then a
higher confidence in the accuracy of the pilot test data will allow a higher percentage
to be used in determining well spacing. Conversely, if site conditions are variable or
unknown, then a lower percentage will be used in spacing extraction wells.

SVE studies have become less expensive through the use of trailer-mounted
units containing vacuum pumps, controls, recorders and air pollution control equipment. A single vapor extraction well is drilled near the center of the contaminated
area. Monitoring wells are placed at several distances along perpendicular axes, and
varying vacuum levels are applied. The resulting data are used to determine well
spacing for site remediation.

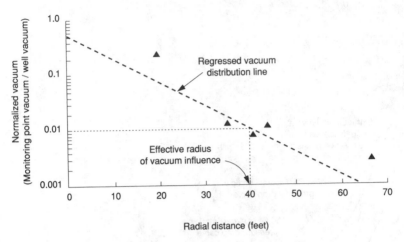

FIGURE 9-12
Effective radius of vacuum-influenced normalized pilot-test vacuum data plot.[39]

FIGURE 9-13
Trailer-mounted soil vapor extraction unit. (*Photo courtesy of ERM-Northeast*).

Example 9-6   Preliminary design of SVE system.   An underground storage tank (UST) at a gasoline station is leaking and the plume is about to migrate off the property. The plume currently extends south to a distance of 230 feet and is 100 feet wide at this distance. The depth of the plume is 12 feet and the seasonal high ground water table is at a depth of 21 feet. Provide a preliminary layout of an SVE system to remediate the site.

Specifically determine:

1. The required number of extraction wells and their layout
2. The volumetric capacity in cubic feet per minute (CFM) and operating system pressure of the blower
3. Equipment layout and any required ancillary facilities

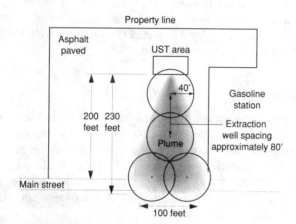

Assume the curves in Figs. 9-10 and 9-11 are applicable to this site. Further assume that the minimum measurable vacuum pressure at the radius of influence of the extraction wells will be 0.5 inches water gage (125 Pa).

Solution

1. Using Fig. 9-11, a vacuum of 0.5 inches water gage can be obtained from either a 150 CFM blower, with 25-foot radius of influence per extraction well, or a 240 CFM blower with 40-foot radius of influence. Selecting a 240 CFM blower yields an extraction well spacing of 80 feet. Four extraction wells should be adequate to cover the plume.
2. Using Fig. 9-10, a 240 CFM blower per extraction well must be capable of generating a vacuum of 34 inches of water gage. This does not include any losses in the piping system, which would have to be considered after the piping system is designed. Because the blower(s) will be manifolded to all four extraction wells, the system must be capable of producing $4 \times 240 = 960$ CFM at 34 inches of water gage (8450 Pa).
3. Ancillary equipment required includes manifold piping, a knockout drum to remove any water removed by the vacuum system, and a carbon adsorption system or other means of removing organics form the air before release to the atmosphere. A schematic of the equipment layout is shown in Fig. 9-6. Many of the system options/enhancements discussed above may be considered in this preliminary design. Because the site is shown as paved, it would be inappropriate to suggest placing an impermeable barrier over the site, but air recharge wells may enhance the system.

9-3 CARBON ADSORPTION

Adsorption is a process in which a soluble contaminant (the adsorbate) is removed from water by contact with a solid surface (the adsorbent). The adsorbent most widely used in environmental applications is carbon that has been processed to significantly increase the internal surface area (activated carbon). Use of different raw materials and processing techniques results in a range of carbon types with different adsorption characteristics. Activated carbon is available in both powdered and granular form. Granular activated carbon (GAC) is most commonly used for removal of a wide range of toxic organic compounds from ground water and industrial waste streams and is discussed in this section. Powdered activated carbon is often used in biological treatment systems and is briefly discussed in Sec. 10-5.

Process Description

A typical activated carbon contactor is illustrated in Figs. 9-14 and 9-15. Carbon is held in place by a plenum plate. Contaminated water enters the top of the column, is contacted with the carbon, and exits through an underdrain system at the bottom. Typical applications require facilities for air scouring and back washing the carbon to avoid the build up of head loss due to the accumulation of solid particles present in the influent. Additionally, the system must allow for the removal of spent carbon for regeneration and the addition of new carbon.

Carbon adsorption systems are typically continuous flow columns, set up in series so that the final column in the system is in effect a polishing unit (see Fig. 9-16). Because the downflow beds also act as filtration units, they must be periodically backwashed.

Other modifications of the process include downflow columns in parallel, upflow-expanded bed systems in series and moving bed systems where new carbon is continually added to the system. The moving bed is operated in a counter-flow mode wherein regenerated carbon is continuously added at the top and spent carbon is continuously removed at the bottom for regeneration, while the contaminated water enters at the bottom and exits at the top.

Upflow expanded beds are used when the influent contains significant amounts of suspended matter. Use of an upflow mode eliminates the filtering of solids within the carbon system, which can be more efficiently removed with sand filters rather than carbon.

Activated carbon adsorption is a well developed technology that is capable of effectively removing a wide range of soluble organic compounds. It is able to produce a very high-quality effluent, and is utilized in drinking water systems as well as for pollution control.

Adsorption Theory

Sorption is the process by which a component moves from one phase to another across some boundary. In adsorption the process takes place at a surface. Examination of a

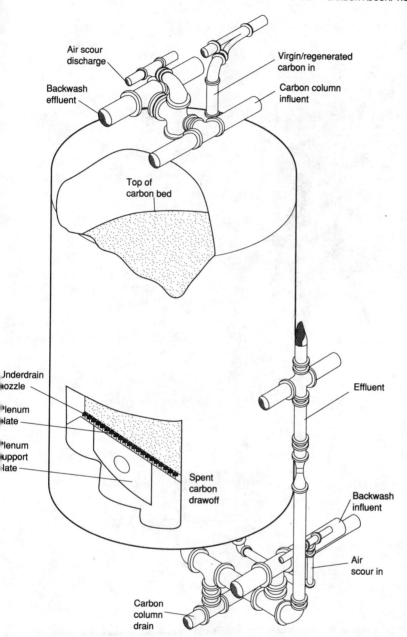

Air scour
discharge

Virgin/regenerated
carbon in

Backwash
effluent

Carbon column
influent

Top of
carbon bed

Underdrain
nozzle

Effluent

Plenum
plate

Plenum
support
plate

Spent
carbon
drawoff

Backwash
influent

Air
scour in

Carbon
column
drain

FIGURE 9-14
Schematic of carbon contactor.[40]

FIGURE 9-15
Carbon contactors. (*Courtesy of Calgon Carbon Corp.*)

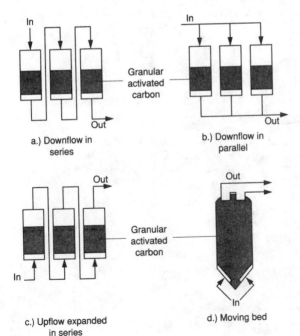

a.) Downflow in series

b.) Downflow in parallel

Granular activated carbon

c.) Upflow expanded in series

d.) Moving bed

Granular activated carbon

FIGURE 9-16
Schematic of activated carbon adsorption systems.[41]

microscopic cross-section of activated carbon reveals a porous structure with a large internal surface area. As illustrated in Fig. 9-17, movement of an organic molecule to a surface site requires four separate transport phenomena: bulk fluid transport, film transport, intraparticle (or pore) diffusion, and the actual physical attachment.

Film and intraparticle transport are normally the slowest and therefore the rate limiting steps which control design. The rate of diffusion tends to increase with increasing solute concentration and decrease with increasing temperature and pH. Where intraparticle diffusion controls, the rate decreases with increasing molecular weight of the solute and decreasing pore size of the carbon.

The driving forces that control adsorption include electrical attraction, a chemical affinity of the particular organic molecule for the adsorbent, van der Waal's forces (weak attractive forces acting between molecules) and the hydrophobic nature of the organic. A summary of some of the factors that affect the adsorption capacity of carbon for organic compounds is listed in Table 9-4.

A plot of the amount of contaminant adsorbed per unit mass of carbon X/M against the concentration of contaminant in the bulk fluid C is referred to as an adsorption isotherm. There are several different mathematical forms of isotherms. The Freundlich isotherm is an empirical model expressed mathematically:

$$X/M = KC_f^{1/n} \qquad (9-33)$$

where X = mass of contaminant adsorbed (mg) = $(C_i - C_f) \times V$
 V = volume of solution (L)
 C_i = initial concentration of contaminant in solution (mg/L)
 C_f = final concentration of contaminant in solution (mg/L)
 M = weight of carbon (mg)
 K, n are empirical constants (see App. A)

K and n can be determined from laboratory studies by a logarithmic plot, as illustrated in Example 9-7.

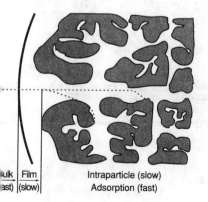

| Bulk (fast) | Film (slow) | Intraparticle (slow) Adsorption (fast) |

FIGURE 9-17
Transport processes.[42]

TABLE 9-4
Factors affecting carbon adsorption

Factor	Effect
Solubility	Less soluble compounds are adsorbed more easily than more soluble compounds.
Molecular structure	Branch-chain organics are more easily adsorbed than straight-chain organics.
Molecular weight	Larger molecules are generally more easily adsorbed than smaller molecules. However, when pore diffusion governs the adsorption process, the adsorption rate decreases with molecular weights above values specific to each type of carbon and within a given class of organics.
Polarity	Less polar (or weakly ionized) organics are more easily adsorbed than polar (or strongly ionized) organics.
Hydrocarbon saturation	Double- or triple-carbon bond (unsaturated) organics are adsorbed more easily than single-carbon bond (saturated) organics.

Example 9-7 Determining isotherm constants. 100 mL of a 600 mg/L solution of xylenes are placed in each of five containers with different amounts of activated carbon and shaken for 48 hours. The samples are filtered and the concentration of xylene measured, yielding the following analyses:

Container:	1	2	3	4	5
Carbon (mg):	600	400	300	200	50
Xylenes (mg/L):	25	99	212	310	510

Determine the Freundlich constants, K and n, and plot the isotherm. Initial contaminant $C_i = (600 \text{ mg/L})$; Volume, $V = 0.1$ liter

Solution

M (grams)	C_f (mg/L)	$X = (C_i - C_f)V$ (mg)	X/M (mg/g)	$\text{Log}_{10}(X/M)$	$\text{Log}_{10}(C_f)$
0.6	25	57.5	95.8	1.98	1.40
0.4	99	50.1	125.3	2.10	2.00
0.3	212	38.8	129.3	2.11	2.33
0.2	310	29.0	145.0	2.16	2.49
0.05	510	9.0	180.0	2.26	2.71

Freundlich constants are determined by linear regression on the last two columns:
Constant = 1.71; Slope = 0.187

$$K = 10^{1.71} = \underline{51.3 (\text{mg/g})}$$

$$1/n = \underline{0.187}$$

$$X/M = 51.3\, C_f^{0.187}$$

A plot of the adsorption isotherm is shown here:

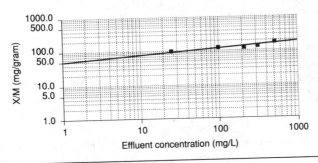

Where a single contaminant is involved, isotherm data may be more closely approximated by the Langmuir equation:

$$X/M = abC/(1 + bC) \tag{9-34}$$

where $a, b = $ constants

The Langmuir equation is developed by assuming that a fixed number of adsorption sites are available, and that the adsorption is reversible. The Langmuir equation can be linearized by plotting the inverse of X/M against the inverse of C.

Example 9-8   Use of adsorption isotherm data.   Use the isotherm data from Example 9-7 to determine the amount of carbon required to treat 10,000 gal/day of water contaminated with 600 mg/L xylenes. Assume a required effluent of 10 mg/L and that the facility will operate at the same temperature and pH for which the isotherm was developed. Determine daily carbon usage for a batch reactor.

Solution.   After the required contact time, the carbon and solution should reach equilibrium with a concentration of xylenes of 10 mg/L. From the Freundlich equation developed in Example 9-7: $X/M = 51.3 \, C_f^{0.187}$
If $C_f = 10$ mg/L
$X/M = 78.9$ mg xylenes/g carbon
Carbon required = 0.0789 g/g = 0.0789 lb/lb
Xylenes removed = 10,000 gal/day $\times$ (600-10) mg/L $\times 10^{-6}$ L/mg $\times$ 8.34 lb/gal
$= 49.21$ lb xylenes per day
Carbon required:
(49.21 lb xylenes/day)/(0.0789 lb xylenes/lb carbon) = <u>624 lb/day</u>

Design Considerations

While isotherm information is useful in evaluating the effectiveness of adsorption, isotherms have a number of limitations. By their very nature, isotherm tests reach equilibrium, a condition that is usually not met in full scale units. With carbon beds in series and a contained adsorption zone (AZ, see below), the isotherm dosage can be approached. Additionally, the carbon is not exhausted (i.e., not all the possible sites are utilized), and the effects of biological activity and carbon regeneration are not seen in these tests.

The design of adsorption units requires column tests that simulate the actual operation of full scale units. In the laboratory, small diameter (e.g., two-inch diameter) columns are filled with the carbon to be tested and the contaminated ground water or wastewater run through the columns. The column effluent is monitored for the contaminant(s) of interest.

The volume within the carbon bed where adsorption takes place is referred to as the **adsorption zone** (also known as a mass transfer zone). As the carbon becomes exhausted (i.e., all of the adsorption sites are filled), the adsorption zone moves down the column. This is illustrated in Fig. 9-18 where the effluent concentration from a column is plotted against time (or volume of water treated, V). At $V = V_1$, the effluent concentration is very small. A small portion of the carbon at the top is totally exhausted (dark shading) and the adsorption zone (light shading) is near the top of the column. At breakthrough, V_b, the adsorption zone reaches the bottom of the column. Breakthrough is typically defined as the point where a specified amount of the influent is detected in the effluent (usually 5 or 10%). After breakthrough the effluent concentration begins to rise rapidly, but the carbon in the adsorption zone at breakthrough is not yet exhausted. Operating columns in series allows complete exhaustion of the first carbon column without releasing significant amounts of contaminants in the effluent.

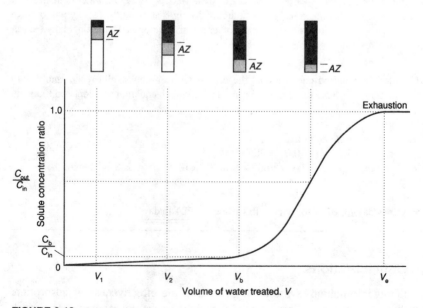

FIGURE 9-18
Carbon adsorption breakthrough curve showing movement of an adsorption zone.[43]

The results of laboratory column studies with three 5 cm (2 in) diameter columns in series, each containing a carbon depth of 2.3 m (7.5 ft), are plotted in Fig. 9-19. The effluent concentration, C_{out}, is divided by the influent concentration, C_{in}, to provide the fractional breakthrough of contaminant versus time for each column. Breakthrough does not occur in the second column until the carbon in column one is almost fully exhausted.

Hutchins[44] presented a simplified way of examining such data by evaluating the bed-depth service time (BDST). A horizontal line is drawn through Fig. 9-19 at $C_{out}/C_{in} = 0.9$, representing the time in each column where breakthrough = 90% of the influent concentration. The line intersects the curve for column one at time = 22 days. Selected data for the three columns are presented in Table 9-5. Bed-depth service time curves are plotted as Fig. 9-20. The lower line represents 90% removal of organics. This line can be represented by the Bohart-Adams[45] equation.

$$t = aX + b \qquad (9\text{-}35)$$

where $X =$ depth in column (m)

$$a = \frac{F_1 N}{C_{in} V} \qquad (9\text{-}36)$$

where $a =$ slope (hr/m)
$F_1 =$ conversion factor for units $= 10^3$ for metric units shown
$\quad\ = 1998$ if N has units of lb/ft^3 and V has units of (gal/min)/ft^2
$N =$ adsorptive capacity of carbon (mass of contaminant removed per volume of carbon in the column, kg/m^3)
$C_{in} =$ influent contaminant concentration (mg/L)
$V =$ superficial velocity through column (m/hr) (m^3/hr per m^2 of column)

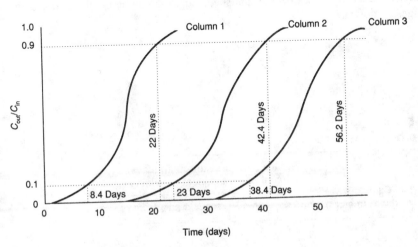

FIGURE 9-19
Laboratory breakthrough curves.

$$b = (F_2/KC_{in}) \times \ln\{(C_{in}/C_{out}) - 1\} \qquad (9\text{-}37)$$

where b = intercept (hr)

      F_2 = conversion factor = 10^3 for the units shown

      = 16,025 if K has units of ft^3/lb-hr

      K = adsorption rate constant required to move an adsorption zone through the critical depth (m^3/kg-hr) (m^3 of liquid treated per kg impurity fed per hr)

      C_{out} = contaminant concentration at breakthrough (mg/L)

This technique provides carbon dosage based on a single fixed bed. The actual carbon usage rate will be lower for beds in series.

At $t = 0$:

$$X(0) = (F_2/F_1) \times (V/NK) \times \ln\{(C_{in}/C_{out}) - 1\} \qquad (9\text{-}38)$$

The abscissa intercept, $X(0)$, is the critical bed depth (i.e., the minimum to obtain satisfactory effluent at time zero). The ordinate intercept, b, measures the time required for an adsorption zone to pass through the critical depth. The slope term, a, provides a measure of the velocity of the adsorption zone (i.e., the speed at which carbon is exhausted). The velocity of the adsorption zone is $1/a$.

The rate of carbon utilization may be determined by using this velocity:

Carbon utilization = area $\times$ (1/a) $\times$ unit weight

After a BDST curve has been established, Eq. 9-35 can be modified to account for changes in flowrate:

$$t = a'X + b \qquad (9\text{-}39)$$

$$a' = a(V/V') \qquad (9\text{-}40)$$

TABLE 9-5
Bed-depth service time

Column number	Cumulative depth (meters)	Service time (breakthrough) (days) percent of contaminant remaining				
		10%	60%	70%	80%	90%
1	2.3	8.4	12.3	13.8	14.5	22.0
2	4.6	23.0	27.7	29.3	30.8	42.4
3	6.9	38.4	41.2	43.3	44.8	56.2

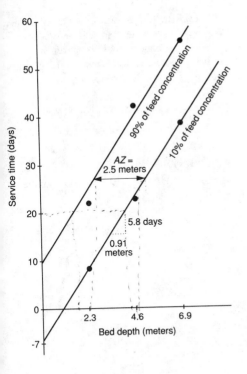

FIGURE 9-20
Bed-Depth Service Time curves
corresponding to Table 9-5.

where V' = the new flowrate (m³/hr per m² of column)

Similarly, the BDST equation is easily modified for changes in the feed concentration:

$$a' = a \times C_{in}/C'_{in} \tag{9-41}$$

$$b' = b \times \frac{C_{in}}{C'_{in}} \times \frac{\ln\left\{\left(C'_{in}/C'_{out}\right) - 1\right\}}{\ln\left\{(C_{in}/C_{out}) - 1\right\}} \tag{9-42}$$

where C'_{in} and C'_{out} represent the influent and effluent concentrations under the new operating conditions.

The horizontal distance between the curve representing exhaustion (90% in Fig. 9-19) and breakthrough (10% in Fig. 9-19) represents the height of an adsorption zone. Where a number of columns in series are used, the total number of columns is related to the height of an adsorption zone by:

$$n = (AZ/d) + 1 \tag{9-43}$$

where n = number of columns in series (rounded up to a whole number)
   AZ = height of an adsorption zone
   d = height of a single column

Example 9-9 Preliminary design of carbon adsorption system. A ground water flowing at a rate of 0.145 m^3/min (55,000 gal/day) requires treatment to reduce organic concentration from 89 mg/L to 8.9 mg/L (90% removal). Laboratory studies are run in 2.3 m (7.5 ft) long by 0.051 m (2 in) diameter columns at a flow rate of 0.5 L/min (190.2 gal/day). The results are shown in Figs. 9-19 and 9-20. Assume a unit weight of carbon of 481 kg/m^3 (30 lb/ft^3).

Determine: The height of the adsorption zone, number and size of columns required, and the carbon usage rate.

Solution.

1. Height of adsorption zone, $AZ = \underline{\underline{2.5 \text{ m}}}$ (8.25 ft) (from Fig. 9-20)
2. Number and size of units: $n = (AZ/d) + 1 = (2.5/2.3) + 1 = 2.09$; therefore, number of units = 3 columns.

   Area of lab columns = 2.043×10^{-3} m^2 (3.14 in^2)
   Loading rate in laboratory columns:

$$V = (5.0 \times 10^{-4} \text{ m}^3/\text{min})/2.043 \times 10^{-3} \text{ m}^2$$
$$= 0.245 \text{ m}^3/\text{m}^2 \cdot \text{ min (6.14 gal/ft}^2 \cdot \text{min)}.$$

   Using the same loading rate for the full scale unit yields:

$$\text{Area} = Q/V = \frac{0.145 \text{ m}^3/\text{min}}{0.245 \text{ m}^3/\text{m}^2 \cdot \text{min}} = 0.59 \text{ m}^2 \text{ (6.32 ft}^2)$$

   Diameter = $\underline{0.867 \text{ m}}$ (2.84 ft)

3. BDST equation for 90% removal (use curve for 10% of feed concentration in Fig. 9-19)
   Slope $a = 5.8$ days/0.91m = 6.34 days/m (1.93 days/ft)
   Intercept b: -7 days
   Equation of line: $t = 6.34X - 7$
   Velocity of adsorption zone = $1/a = 0.158$ m/day (0.52 ft/day)
   Carbon utilization = Area $\times (1/a) \times$ Unit weight = 0.59 m$^2 \times 0.158$ m/day $\times 481$ kg/m^3 = $\underline{\underline{44.8 \text{ kg/day}}}$ (131 lb/day)

9-4 STEAM STRIPPING[†]

Process Description

Steam stripping is utilized for the removal of volatile and sometimes semivolatile compounds from ground water or wastewater. This process is capable of reducing volatile organic compounds (VOCs) in water to very low concentrations. Table 9-6 shows the low treatment limits that the U.S. EPA established for the Organic Chemical,

---

[†] Sec. 9-4 was prepared by David Potter, P.E., Science Applications International Corp.

Plastics, and Synthetic Fibers (OCPSF) category based on the use of steam stripping as the model technology.[46,47] In some cases, discharge VOC concentrations from steam strippers can approach or be below analytical detection limits.

Both steam and air strippers are based on the transfer of organics from the liquid phase to the gas. However, as discussed in this section, the higher concentration of organics in a steam stripper requires more complex process design techniques than for an air stripper. The functional differences between steam stripping and air stripping are:

- Steam, rather than air, is used as the stripping gas,
- The stripping gas, steam, is infinitely soluble in the liquid phase, water,
- Steam strippers operate at much higher temperatures than air strippers,
- The organics in the water are recovered as a separate liquid phase.

TABLE 9-6
Best Available Technology Economically Achievable (BATEA) effluent limitations set for OCPSF products and production based on steam stripping technology (40 CFR Part 414.101)

Contaminant	Maximum concentration for any one day (μg/l)	Maximum concentration for monthly average (μg/l)
Benzene	134	57
Carbon tetrachloride	380	142
Chlorobenzene	380	142
1,2,4-Trichlorobenzene	794	196
Hexachlorobenzene	794	196
1,2-Dichloroethane	574	180
1,1,1-Trichloroethane	59	22
Hexachloroethane	794	196
1,1-Dichloroethane	59	22
1,1,2-Trichloroethane	127	32
Chloroethane	295	110
Chloroform	325	111
1,2-Dichlorobenzene	794	196
1,3-Dichlorobenzene	380	142
1,4-Dichlorobenzene	380	142
1,1-Dichloroethylene	60	22
1,2-trans-dichloroethlyene	66	25
1,2-Dichloropropane	794	196
1,3-Dichloropropylene	794	196
Ethylbenzene	380	142
Methylene chloride	170	36
Methyl chloride	295	110
Hexachlorobutadiene	380	142
Tetrachloroethylene	164	52
Toluene	74	28
Trichloroethylene	69	26
Vinyl chloride	172	97

A schematic diagram of a steam stripper operating at atmospheric pressure is presented in Fig. 9-21, and a schematic diagram of a steam stripper operating under vacuum is presented in Fig. 9-22.

In both Fig. 9-21 and Fig. 9-22, the bottom portion of the steam stripper is stripping wastewater in a manner similar to an air stripper. This is known as the stripping section of the column. The top section of the column, above the feed point, is known as the rectifying section. This section of the column enriches the organic content of the steam to a point where a separate organic phase can be achieved in the overhead decanter. The steam is enriched because it is (theoretically) in equilibrium with the saturated liquid that is fed to the top of the column. The combination of the stripping and rectifying sections is a mass transfer process known as distillation in the chemical process industry. Distillation has many other production applications in the chemical industry in addition to its specialized application to wastewater treatment as steam stripping.

In Fig. 9-21, the contaminated water feed to the steam stripper is preheated to near-boiling temperature by exchanging heat with the stripped water exiting from the bottom of the stripping column. The contaminated water enters the column at the feed point, and the water flows downward through the stripping section of the column. Steam passes counter-currently up through the column. The column operates at a temperature that is slightly higher than the normal (one atmosphere) boiling point of water, usually in the range of 215° to 220°F. The temperature difference between the

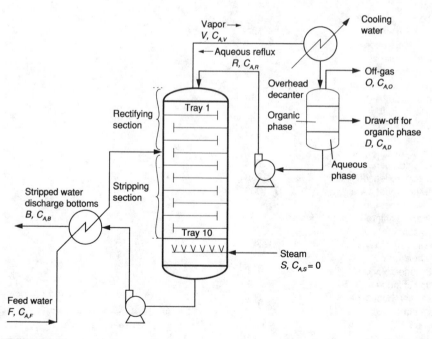

FIGURE 9-21
Atmospheric pressure steam stripping column.

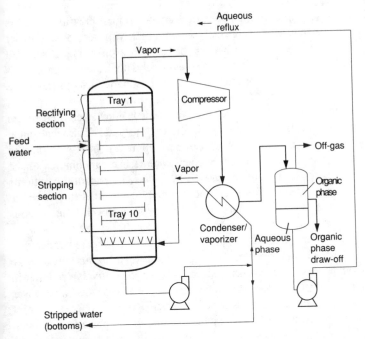

FIGURE 9-22
Vacuum steam stripper.

top and the bottom of the column is modest, generally on the order of a few degrees Fahrenheit.

At the elevated temperature inside the column, the volatile organics in the water exert a higher vapor pressure than at ambient conditions. As the organics vaporize in the column, they are transferred from the liquid phase into the gas phase. As the steam travels up the column, the concentration of organics increases in the stripping steam.

The steam exits the top of the column where it undergoes a phase change to a liquid in the overhead condenser. (Condenser is a term describing a heat exchanger where a gas is cooled to effect a change to the liquid phase.) This liquid is supersaturated with the organics so a separate organic layer forms in the decanter. This organic phase can have a density greater than or less than the aqueous phase in the decanter, so this organic phase will float or sink in the decanter. In Fig. 9-21, this is shown as a floating "organic phase."

The aqueous phase from the decanter is returned to the top of the column where it flows down the column. This aqueous phase is saturated with the organics because the organic concentration in the aqueous phase is in equilibrium with the separate organic phase in the decanter. An analysis of phase-equilibria thermodynamics and a mass balance will show that, because this feed to the top of the column is saturated with the organics, the steam leaving the top of the column will contain a high enough concentration of organics to form a separate "organic" phase when the steam is condensed to a liquid phase (see Example 9-10).

Figure 9-22 shows a steam stripper operating under vacuum conditions in the column. A vacuum stripper has a stripping section and a rectifying section and many similarities to an atmospheric stripper. However, because it is operating at reduced pressure, the boiling point of water is reduced. Consequently, the entire vacuum stripping column operates at a lower temperature than an atmospheric stripping column. Typical operating values are 140° to 180°F and 0.2 to 0.5 atmosphere.

In a vacuum steam stripper, the overhead steam is mechanically compressed to an elevated pressure that will cause the gas to condense to a liquid. The mechanical compression of the gas also creates a vacuum on the compressor inlet side that, in turn, creates vacuum conditions in the steam stripper. The mechanical compression of a gas gives off a great deal of heat that is recovered by vaporizing part of the bottoms flow to create the stripping steam. The overhead mixture is decanted into separate organic and aqueous phases as in an atmospheric pressure stripper, and the aqueous phase is returned to the top of the stripping column as in an atmospheric stripper.

Steam strippers can also operate at elevated pressures. The operating temperature of the stripper will be higher than the normal boiling point of water because the column is under pressure. Operation of a pressurized steam stripper is relatively common in refineries where hydrogen sulfide and ammonia are stripped from the wastewater before it is discharged to a biological wastewater treatment plant. However, use of pressurized steam stripping is not a normal practice outside of the petroleum refining industry because the higher pressure complicates the operation of a steam stripper.

A steam stripper could be applied to water containing volatile organic compounds that are highly soluble in water (e.g., acetone, ethanol). In such an application, a separate organic phase would not form in the overhead decanter. The purpose of such an application would be to increase the concentration of VOCs in the water and thereby reduce the volume of water for a subsequent treatment unit. Although this is theoretically sound, as a practical matter the extra capital and O&M costs of a steam stripper outweigh the benefits gained by reducing the volumetric rate to downstream unit operations.

Design Theory

Development of column design theory occupies several chapters in mass transfer textbooks.[48,49] Consequently, the detailed process design of steam strippers is a complicated mass transfer problem that requires engineers who are experienced in column design. However, an understanding of process design fundamentals can be applied to engineering feasibility studies without the benefit of detailed process design experience. In this section, the basic design theory underlying design of distillation columns is presented on a conceptual level, including:

- Thermodynamics
- Material balance
- Mass transfer
- Process design selection

These subsections are discussed on a conceptual basis to provide familiarity with the process design principles and practices involved in column design. A list of fundamental chemical engineering references is provided at the end of this chapter for more detailed discussion of these topics.[50,51,52,53,54]

Thermodynamics. The conceptual principles of the complex liquid-vapor-equilibrium and phase separation processes involved in the design of a steam stripper are presented in this section. The fundamental differences and similarities between liquid-vapor equilibrium design for steam strippers and air strippers should then be readily apparent.

Up to this point, the equilibrium between an organic in the vapor phase and an organic in the liquid phase has been represented using a simple Henry's law model (Eq. 3-5):

$$p_g = HC_l \qquad (9\text{-}44)$$

where p_g = partial pressure of gas
H = Henry's constant
C_l = concentration in liquid

However, the Henry's constant is applicable to a limited temperature range (usually 0° to 35°C), to simple binary systems (usually water and a single organic), and to very dilute systems. These simplifications are not valid within a steam stripper. In an air stripper, the feed water concentration is generally less than 100 mg/L, and the concentration of organics decreases in the liquid phase as the water is processed through the air stripper. In contrast, the feed water to a steam stripper generally is 100 mg/L or more, and the concentration of organics in the liquid phase increases above the feed point. Consequently, the assumption of a very dilute system is not valid in a steam stripping column. This means the interaction between various organics in the system must be considered when estimating liquid-vapor equilibrium.

Activity coefficient models were developed to consider such interactions among components in a mixture. In fact, Eq. 3-5 is a generally accepted simplification of the more complex activity coefficient model of liquid-vapor equilibrium:

$$p_{g,A} = \frac{a_{A,B}}{\delta_{A,B}} X_A P_{vp,A} \qquad (9\text{-}45)$$

where $p_{g,A}$ = partial pressure of component A above the specified system of A and B
$a_{A,B}$ = activity coefficient of component A in component B (unitless)
$\delta_{A,B}$ = fugacity coefficient of gas A at the system total pressure (unitless)
X_A = mole fraction of component A in the mixture of A and B
$P_{vp,A}$ = saturation vapor pressure of component A at system temperature

The fugacity coefficient and the activity coefficient in Eq. 9-45 represent non-ideal gas and liquid behavior, respectively. In a steam stripper, the modest operating pressures usually result in a fugacity coefficient near one.

Several models exist for estimating the activity coefficient for a given system. Two examples are the Non-Random Two Liquid (NRTL) model and the Universal Quasi Chemical (UNIQUAC) model,[55] both of which are generally available on computerized thermodynamics packages. To develop activity coefficient models, the mutual solubility of two or more components is experimentally measured. These data are then reduced to models that utilize other thermodynamic properties, mole fractions, and absolute temperature.

When more than two components are present in a mixture, the binary activity coefficients between all components are calculated using published data and models such at NRTL or UNIQUAC. An additional set of correlations is utilized to account for the interactions between the various components in the mixtures that can increase or decrease the binary activity coefficients. Finally, a special case activity coefficient is utilized to estimate the activity coefficient at infinite dilution. This infinite dilution activity coefficient is not applicable in the overhead decanter where a separate organic phase is formed. In the overhead decanter, a separate set of liquid-liquid activity coefficients must be estimated to calculate the expected phase separation.

There is much fundamental data available to determine activity coefficients for binary organic/water mixtures as well as data for ternary and quaternary water/organic mixtures. The generally accepted published source for this data are the Dechema series.[56,57,58] In addition, many companies and vendors in the chemical process industries have their own internal databases developed for materials they process. These databases are developed from empirically measured parameters, but they are not published because they represent proprietary process information. Models to account for multi-component interactions and to estimate activity coefficients are available on the open market, and there are a number of in-house modifications of these models that are also considered proprietary information in the chemical process industry.

An activity coefficient model can also be used to predict the separation of the "organic phase" and the "aqueous phase" in the overhead decanter. This would, however, be a liquid-liquid equilibrium model. Ideally, the composition of the organic and aqueous phases would be a simple phase-equilibria calculation based upon the solubility of organics in water. However, in actual practice there is some water in the organic phase. Co-solvent effects can increase the solubility of a given organic in water, and a strong hydrophobic organic can extract highly water-soluble organics from the aqueous phase.[59] The proper way to model this separation is very similar to the above methods discussed for liquid-vapor equilibria. This includes a combination of measured water-organic and organic-organic activity coefficients[60] and computer modeling of the component interactions in a multi-component two-phase liquid mixture. The NRTL and UNIQUAC models are often used for steam stripper design because these methods support estimation of both liquid-vapor and liquid-liquid equilibria.

Material Balance. A schematic diagram of an atmospheric steam stripper is presented in Fig. 9-21, with notation for a material balance. For purposes of illustration only, this discussion will assume that the feed wastewater is a simple binary water-organic mixture. As previously noted, this is not the case for most applications of steam strippers. The student can develop a more complex material balance, as needed, using these same principles.

The material balance consists of two balances—one for the organic and one for water. Because there are no chemical reactions taking place in the steam stripper, the accumulation and generation terms are zero for both balances. Using the generalized material balance approach for non-reacting steady-state processes presented in Eq. 3-48, the material balance for the organic component A is expressed as:

$$\text{Input} = \text{Output} \tag{9-46}$$
$$FC_{A,F} = BC_{A,B} + DC_{A,D} + OC_{A,O}$$

where
F = feed water mass rate (lb/hr)
$C_{A,\_}$ = concentration of component A in the various streams (weight percent)
B = bottoms discharge mass rate (lb/hr)
D = overhead organic phase withdrawal mass rate (lb/hr)
O = off-gas (from overhead decanter) mass rate (lb/hr)

Eq. 9-46 can be simplified by the recognition that the concentration of component A out through the bottoms discharge and the mass rate of stream O are almost always negligible terms in this mass balance:

$$FC_{A,F} = B\cancel{C}_{A,B} + DC_{A,D} + \cancel{O}C_{A,O}$$
$$\qquad\quad \text{neg.} \qquad\qquad \text{neg.} \tag{9-47}$$
$$= DC_{A,D}$$

Example 9-10 shows the basis for considering the $C_{A,B}$ and O terms to be negligible. Although this is almost always true, the assumptions presented in Example 9-10 should be confirmed before using the simplified form of Eq. 9-46. It is also worth noting that the overhead phase withdrawn from the decanter is not assumed to be a pure organic phase ($C_{A,D} \neq 1.0$). In actual practice, this organic phase can contain a very high dissolved water fraction, on the order of up to fifty percent.

Example 9-10. Demonstrate the basis for the negligible terms in Eq. 9-47. Discuss limitations of this simplified water mass balance.

Given: 1. Non-reactive system described in Fig. 9-21
2. Feed organics = 230 mg/l
3. Organics removal efficiency = 99.4%
4. Decanter temperature = 35°C

Solution. The generalized water mass balance on the non-reactive system described in Fig. 9-21:

$$\text{Inlet} = \text{Outlet}$$

$$F(1 - C_{A,F}) + S(1 - C_{A,S}) = D(1 - C_{A,D}) + O(1 - C_{A,O}) + B(1 - C_{A,B})$$

terms which may be considered negligible are:

$C_{A,F}$ is 230 mg/l or 2.30×10^{-4} weight fraction. So $1 - C_{A,F} = 0.999770$. Within the accuracy of these types of engineering calculations, at most three figures are significant. Therefore, 0.999770 is rounded to 1.00 and $C_{A,F}$ is neglible in terms of the water mass balance calculations.

$C_{A,S}$ is theoretically zero. In actual practice, plant steam will contain less than 10 mg/l of water treatment chemicals. By the same logic applied to $C_{A,F}$, $C_{A,S}$ may then be considered megligible.

$C_{A,B}$ may be calculated as (1 − removal efficiency/100) $C_{A,F} = (1 - 0.994)(230 \text{ mg/l}) = 1.38 \text{ mg/l} = C_{A,B}$. $C_{A,B}$ may then be considered negligible since $C_{A,B} << 1$. (The same logic as used for $C_{A,F}$).

O is the overhead gas rate. This gas is almost entirely composed of non-condensible gases (which are oxygen and nitrogen). The non-condensibles come from air dissolved in the feed water. In example 9-11 it is shown that the solubility of air in water is 0.0022 wt.% at 25°C and atmospheric conditions. Therefore, the first estimate of O is based on non-condensible flow only:

$$O_{NC} = (0.0022/100)F = 2.2 \times 10^{-5}F$$

The non-condensible gases flow over the liquid surface in the decanter and then to the atmosphere. This liquid surface can be the organic layer or the aqueous layer, depending upon the relative densities of the aqueous and organic layers.

The highest water content in the non-condensible gas flow occurs if the top layer is a pure water layer. This case is assumed for the rest of this discussion. At 35°C, pure water exerts a vapor pressure of 91 mm Hg or .12 atmospheres. The total flow of water and non-condensible gas flow is:

$$O_{tot} = O_{water} + O_{NC}$$

and in terms of pressure:

$$P_{tot} = 1 \text{ atmos} = P_{water} + P_{NC}$$

$$= 0.12 \text{ atmos} + P_{NC}$$

$$P_{NC} = 0.88 \text{ atmos}$$

Therefore:

$$O_{water} = O_{NC}(0.12/0.88)$$

$$= 0.136 \, O_{NC}$$

$$O_{tot} = O_{water} + O_{NC} = 0.136 \, O_{NC} + O_{NC}$$

$$= 1.14 \, O_{NC}$$

As discussed above, $O_{NC} = 2.2 \times 10^{-5} F$.

So $O_{tot} = 1.14 \, (2.2 \times 10^{-5} F)$

$\qquad = 2.51 \times 10^{-5} F$

Since $O_{tot} << F$, O_{tot} may be treated as negligible in the water mass balance calculations.

D is the flow of the organic layer, which often contains a significant water fraction. For purposes of this calculation, assume a very high water fraction in the organic layer, eg., 50% water, by weight.

The organic layer total mass flow is composed of an organic flow and a water flow:

$$D_{tot} = D_{water} + D_{organic}$$

The pure organic portion is the outlet term of an organic mass balance. For simplification, assume 100% of the organic in the feed water goes into the organic decant layer:

$$D_{organic} = (C_{A,F})F$$

$$= (2.30 \times 10^{-4})F$$

$$D_{water} = (.50/.50)D_{organic}$$

$$= D_{organic}$$

$$D_{total} = D_{water} + D_{organic}$$

$$= \left(D_{organic}\right) + D_{organic}$$

$$= 2D_{organic}$$

$$= 2 \, (2.30 \times 10^{-4})F$$

$$= 4.60 \times 10^{-4} F$$

Conclusion $D_{total} << F$ and therefore may be treated as negligible in the overall water mass balance.

Using the generalized material balance approach presented in Eq. 3-24, the material balance for water is expressed as:

$$F \left(1 - C_{A,F}\right) + S \left(1 - C_{A,S}\right) = D \left(1 - C_{A,D}\right) + B \left(1 - C_{A,B}\right) + O \left(1 - C_{A,O}\right) \tag{9-48}$$

where
- F = feed water mass rate (lb/hr)
- $1 - C_{A,\_}$ = concentration of water in the various streams (weight percent)
- S = steam mass rate (lb/hr)
- D = overhead organic phase withdrawal mass rate (lb/hr)
- B = bottoms discharge mass rate (lb/hr)
- O = off-gas (from overhead decanter) mass rate (lb/hr)

This generalized material balance can be simplified by recognizing that the concentration of water is near unity for the steam, feed, and bottoms streams, and that the water mass rate of the organic phase withdrawal and the off gas is a negligible term in the water mass balance (demonstrated in Example 9-10):

$$F(1 - \underset{neg.}{\cancel{\phi}_{A,F}}) + S(1 - \underset{neg.}{\cancel{\phi}_{A,S}}) = \underset{neg.}{\cancel{\phi}}(1 - C_{A,D}) + B(1 - \underset{neg.}{\cancel{\phi}_{A,B}}) + \underset{neg.}{\cancel{\phi}}(1 - C_{A,O})$$

$$F + S = B$$

$$(9\text{-}49)$$

As can be confirmed from this balance, essentially all of the condensed water vapor from the steam used in the stripping column ends up in the bottoms discharge flow. This can easily result in a net increase of ten percent in the water flowrate after water is treated in a stripping column. This must be considered when sizing downstream equipment or subsequent treatment processes.

The above simplified material balances should be used for purposes of estimating material balances around a steam stripper. The exact form of these equations are used for purposes of detailed process design. The concentration of organic A at various points in the process is often negligible in terms of the overall material balance, but these concentration terms are essential for calculation of mass transfer rates and in the detailed process design of a steam stripper.

Example 9-11   *Air Solubility.*

Calculate the weight percent of air that is soluble in water at 25°C, 1 atmos., and equilibrium conditions.

Assume   1. Ideal gas behavior
         2. Ideal liquid behavior

Solution. Ideal gas and liquid behavior is assumed, so the Henry's law expression is applicable:

$$P = H_X$$

where $P =$ partial pressure, atmospheres
        $H =$ Henry's Constant, atmospheres
        $X =$ mole fraction solute in solvent

Per Perry's *Chemical Engineers Handbook*, 5th edition, p3–96:
for air/water, at 25°C, $H = 7.20 \times 10^4$ atmos.

$$X = P/H$$

$$= 1 \text{ atmosphere}/7.20 \times 10^4 \text{ atmosphere}$$

$$= 1.39 \times 10^{-5} \text{ moles air/mole water}$$

$$\text{wt \% air} = (\text{mole fraction air/water}) \left(\frac{\text{MW air}}{\text{MW water}} \right) \times 100\%$$

$$= (1.39 \times 10^{-5}) \left(\frac{\text{moles air}}{\text{moles water}} \right) \left(\frac{28.9\text{g air/mole air}}{18\text{g water/mole water}} \right)$$

$$\times 100\%$$

$$= 0.0022\%$$

Conclusion, air is nearly insoluble in water.

Mass Transfer. Mass transfer design of a steam stripper is fundamentally different than the mass transfer design of an air stripper. These different design approaches are a result of the different organic concentrations in an air stripper and a steam stripper as well as the different solubilities of the stripping gases in water.

In the case of an air stripper, air is so sparingly soluble in water that air is treated as a water-insoluble gas for mass transfer purposes. In addition, the ratio of the air flow to liquid flow is so high that there is no significant approach to equilibrium conditions between the organic concentrations in the gas phase (air) and the liquid phase (water). The design of an air stripper is consequently based on an estimation of the resistances to mass transfer (e.g., the two-film theory).

In a steam stripper, the stripping gas is entirely soluble in water (steam is water vapor!), and the concentration of organics at the top (rectifying) section of a steam stripper can easily be several percent. This is four to five orders of magnitude higher than the liquid concentrations in an air stripper. Consequently, operating gas-liquid conditions do significantly approach liquid-vapor equilibrium conditions. This applies to both the water-steam and the organic components. As a result, a steam stripper is designed as an equilibrium stage process; mass transfer resistances are not directly considered in this design approach.

The equilibrium stage approach is illustrated in Fig. 9-23. As the gas travels up the column, it passes through each tray and through the liquid that is on top of the tray (say tray $n + 1$). The gas then exits the liquid, picking up organics from the liquid on the tray $(n + 1)$, and proceeds upwards to the next tray $(n + 2)$. Under ideal conditions, the gas leaving the top of the tray $(n + 1)$ would be in equilibrium with the liquid on the tray $(n + 1)$. In actual practice, however, the mass transfer only approaches equilibrium conditions.

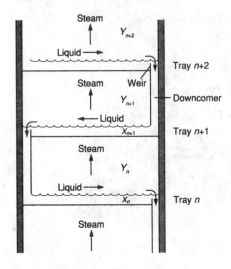

FIGURE 9-23
Mass transfer across trays.

The mass transfer at tray $n + 1$ may be described by the following mass balance and equilibrium equations:

$$\text{Input} = \text{Output}$$

$$Gy_{A,n} + LX_{A,n+2} = Gy_{A,n+1} + LX_{A,n+1} \tag{9-50}$$

where G = steam molar rate (lb-mole/hr)
L = water molar rate (lb-mole/hr)
y_A = mole fraction of organic component A in the gas above the designated tray number (lb-moles A/lb-moles gas)
X_A = mole fraction of organic component A in the liquid on the designated tray number (lb-moles A/lb-moles water)

The equilibrium conditions between the liquid at and the gas above each tray are determined from the thermodynamic liquid-vapor equilibria considerations previously discussed. A tray efficiency is used to estimate the approach to these equilibrium conditions.

$$E = \frac{\text{actual vapor pressure of component A}}{\text{equilibrium vapor pressure of component A}}$$

$$E = \frac{(Y_{A,n})(P_{T,n})}{(a_{A,n}/\delta_{A,n})(X_{A,n})(P_{vp,A})} \tag{9-51}$$

or

$$(Y_{A,n}) = E(a_{A,n}/\delta_{A,n})(X_{A,n})(P_{vp,A})/(P_{T,n})$$

where $\quad E$ = tray efficiency (unitless fraction)

$Y_{A,n}$ = mole fraction of organic component A in the gas above tray n (lb-moles A/lb-moles gas)

$a_{A,n}$ = activity coefficient of organic component A in liquid on tray n (unitless)

$\delta_{A,n}$ = fugacity coefficient of organic component A in gas above tray n (unitless)

$X_{A,n}$ = mole fraction of organic component A in gas above tray n (lb-moles A/lb-moles liquid)

$P_{vp,A}$ = vapor pressure of organic component A at conditions on tray n

$P_{T,n}$ = total pressure of system above tray n

Obviously, a tray with a higher efficiency is more desirable for mass transfer purposes. Tray efficiency is a function of both the mechanical design of the tray and the process conditions for operating the tray. Tray efficiencies of 50 to 65 percent are common for steam strippers.

Packed column steam strippers are actually designed as equilibrium stage (or tray) columns. Mechanical and process considerations are then utilized to estimate the Height Equivalent to a Theoretical Plate (HETP). HETP is expressed in feet of packing depth, which is equivalent to one theoretical equilibrium stage. The calculation of HETP is also treated as a proprietary trade secret in the chemical process industry, although there are correlations available in published literature.

The calculation of the number of theoretical trays required to meet a given set of process requirements is now routinely performed on computers. However, a review of an older technique, the McCabe-Thiele diagram, is useful for conceptually understanding the design of a steam stripper. In Figs. 9-24 and 9-25, a McCabe-Thiele diagram is presented for a simple binary water/organic mixture. The diagram in Fig. 9-24 consists of:

- An equilibrium curve
- The rectifying section operating line
- The stripping section operating line
- The bottom (X_B), feed (X_F), and the overhead vapor (X_O) organic mole fractions

In Fig. 9-25, the number of theoretical equilibrium stages (also called theoretical trays or theoretical plates) is stepped off from X_O to X_B. The underlying principle in this approach is that the liquid leaving a theoretical tray is in equilibrium with the vapor leaving that same tray. Also, the slopes of the rectifying section and stripping section operating lines are functions of the ratio of the liquid-to-gas (steam) mass flowrates (L/G).

For example, in Fig. 9-25, set tray n equal to tray 5, tray $n+1$ equal to tray 6, and tray $n+2$ equal to tray 7. The equilibrium conditions on tray 6 are the point, on Fig. 9-25, on the equilibrium curve where step 6 is indicated. The liquid then

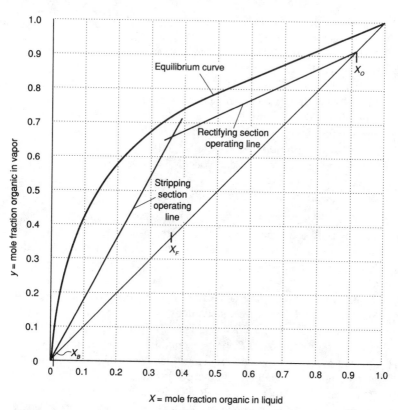

FIGURE 9-24
Construction of McCabe-Thiele diagram.

flows down from tray 6 to tray 7 on the straight vertical line. Similarly, the vapor from tray 6 steps off against the operating line to tray 5. At step 5, the operating lines for the rectifying section and stripping section intersect. Consequently, in this example, tray 5 is the feed tray. The rectifying section is located above tray 5 and the stripping section is located below tray 5. The preceding discussion is presented only as an example, (for purposes of explaining the equilibrium stage design concept); McCabe-Thiele diagrams are generally not used for the design of steam strippers. The limitations of the McCabe-Thiele diagram in a steam stripper application are:

- The method is applicable to pure binary mixtures (not multicomponent mixtures), and
- The method does not graphically recognize the proper mass balances for the top and bottom trays in a steam stripper.

   Current process design methods generally utilize a computerized process simulator package, as discussed later in this section. For more information on the McCabe-

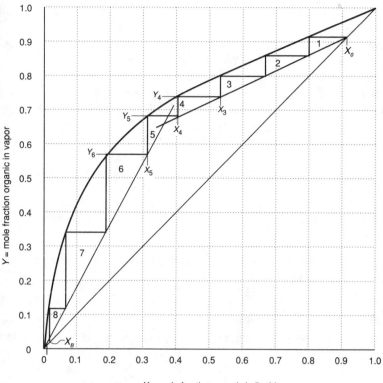

FIGURE 9-25
Completed McCabe-Thiele diagram.

Thiele design method, the reader is referred to Treybal,[48] McCabe and Smith and Harriott,[49] and Wankat.[61]

In actual practice, the design of a column is much more complicated than the simple graphical method presented in Figs. 9-24 and 9-25. However, the concept of steps and equilibrium stages is an important one. In actual process design, process simulation software is used to calculate the complex equilibrium relationships between vapor and liquid at a given tray along with mass and energy balances. The vapor then travels up to a higher tray, the liquid flow down to the next lower tray, and another set of complex equilibrium calculations must be performed for each tray. Although these calculations are too complex to perform by hand, the equilibrium stage concept is still fundamental to the design of distillation columns such as a steam stripper.

Curve of Constant Bottoms Concentration. For a given water flowrate and a given set of feed and bottoms concentrations, there is a large number of potential column designs that will satisfy the process requirements. By increasing the steam rate, a smaller number of theoretical trays would be required to meet the specified design performance requirements. This would be reflected on the McCabe-Thiele diagram by a lower slope on the operating lines in the stripping and rectifying sections of the column. A number of these solutions can be prepared to develop a curve of constant bottoms concentration shown in Fig. 9-26. The curve presented in Fig. 9-26, based on actual pilot plant data, shows a number of solutions to a set of design constraints in terms of number of theoretical trays versus steam flowrate.

The prudent design of a steam stripper will be to the left of the inflection point on this curve. In this manner, if process conditions shift the design curve, a small change in the steam rate will bring the operation back onto the design curve. Contrast this to the portion of the design curve to the right of the tangent point. In this area, a shift in the design curve will require a large change in steam flowrate in order to return the design curve. Such a large increase in steam rate may be beyond the steam supply capabilities of a given installation or exceed the capacity of a given tray hydraulic design, as well as causing a large increase in steam use and operating costs.

Design Considerations. The initial analysis of the feasibility of a steam stripper must consider two key factors:

- The strippability of the organics, and
- Whether the organics will form a separate organic phase in the overhead decanter.

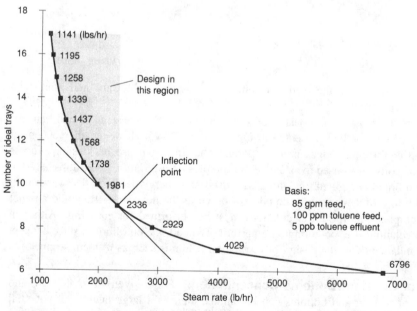

FIGURE 9-26
Design curve—constant bottoms concentration.

For a single organic in wastewater, this is a relatively easy analysis. However, for water containing a mixture of organics, this often requires the use of a computerized process simulator to assess the thermodynamics of the interactions between the various organics in the wastewater. However, there are some general considerations that can be examined before utilizing a process simulator. These types of considerations can be used to eliminate obviously poor candidates for steam stripping.

As a general rule of thumb, any priority pollutant that is analyzed by direct injection on a gas chromatograph can be considered a good candidate for high-efficiency steam stripping. This can easily be determined by a detailed reading of the analytical reports for a water sample. Another rule of thumb is that any compound with a boiling point below 150°C is a potential candidate for steam stripping,[62] and the lower the boiling point the more easily a compound would be expected to be steam stripped. A key factor not included in this rule of thumb is what effects the presence of other organics will have on the activity coefficient of a given organic. A large increase in the activity coefficient will effect an increase in the strippability of a given organic compound.

The second feasibility factor is whether a separate organic phase can be formed in the overhead condenser. This analysis involves looking at the solubility limits of the various organics in water. If even one organic compound has a low solubility limit (e.g., less than 1%), then there is a basis to expect a separate phase to form in the decanter. Some organic compounds are infinitely soluble in water, and will not form a separate organic phase, by themselves, in the overhead decanter. Additional treatment of the aqueous phase would be necessary if a separate organic phase is not formed. Treatment options other than steam stripping are then, usually, more economical. Some examples include simple alcohols such as methanol and ethanol. Generally speaking, the infinitely soluble organic compounds are simple low molecular weight organics.

Two example water/organic mixtures are shown in Table 9-7. Mixture A contains organic compounds that are all infinitely soluble in water. Consequently, this ground water is not a candidate for steam stripping. Mixture B contains the same organic compounds along with toluene and xylene, both of which are sparingly soluble in water. The Mixture B ground water is actually being treated using a steam stripper. When the toluene forms an insoluble phase in the overhead decanter, the other organics partition to the organic phase to a high degree (e.g., in excess of ninety percent). The point of this illustration is to recognize that only one sparingly soluble organic need be present to make steam stripping feasible. As a general rule, the organics will preferentially partition to the organic phase created by the single sparingly soluble organic.

After an initial desk top feasibility analysis has been performed, further analysis of promising candidates should be considered. Such an analysis would be performed using a process simulator. A positive result from a process simulator would then normally be followed by pilot-scale tests using samples of the candidate wastewater or ground water. These pilot-scale tests may be a simple two- to three-day lab program using a single water sample. Or the pilot-scale tests may involve a field pilot plant that is installed for several weeks.

TABLE 9-7
Example water mixtures for steam stripping feasibility analysis

Mixture A	37 mg/L methanol
	194 mg/L ethanol
	114 mg/L n-butanol
Mixture B	37 mg/L methanol
	194 mg/L ethanol
	114 mg/L n-butanol
	110 mg/L toluene
	14 mg/L xylene

The key factor in determining the appropriate pilot program is the variability of the water composition, including the presence of inorganic salts that will also have an effect on activity coefficients. A properly designed steam stripper can handle water composition variations of less than one order of magnitude. Consequently, simple lab scale programs are appropriate for these types of candidate waters. This typically includes ground waters and wastewaters from continuously operated manufacturing processes that produce a single product.

Where large day-to-day variations in composition are expected or a number of batch or campaign (switching products every few weeks or months) production units are utilized, a field pilot plant may be advisable. The purpose of such a field pilot plant would be to ascertain the variability in the activity coefficients over a sustained period of time. Based upon the range of activity coefficients, a proper stripping column could be designed along with an agitated equalization tank that would dampen composition variations to a level that is acceptable for operation of a steam stripper. This type of variability in composition is usually only present in industrial wastewater from large complexes or from batch production plants.

Field pilot plant trials may also be appropriate where fouling or plugging of the trays is a concern. An example would be a wastewater containing monomers that polymerize at elevated temperature.

Mechanical Design. The design of a steam stripper can utilize any of several types of internals to provide the necessary gas-liquid contact time for effective mass transfer of the organics. For steam strippers, the practical choice is usually one of three types of internals:

- Random packing
- Valve trays
- Sieve trays

An important concept in the design of tray towers is the **turndown ratio**. This is the ratio of the maximum rate of flow of either the gas or liquid to the corresponding minimum rate of flow. This is particularly important in environmental applications because of the possibility of significant variability in the feed flow rate. As an example, if a steam stripper with a turndown ratio of 3 were designed to handle contaminated ground water at a maximum flow of 100 gpm, it could not function properly if the flow dropped below 33 gpm.

Random packing (see Sec. 9-1), has the advantage of providing the highest mass transfer capability for a given column height. Random packing also provides for a very high turndown ratio of steam rates—on the order of a factor of ten. However, random packings are limited in their ability to pass solids in the ground water. Valve trays (Fig. 9-27) have a high turndown ratio that approaches random packing performance and are capable of passing modest suspended solids in the water. Sieve trays (Fig. 9-27) have the best solids-passing capabilities, but the steam turndown ratio is on the order of a factor of four.

In addition to turndown ratios and solids-passing capabilities, a number of site-specific items need to be considered such as inspection, maintenance, and the availability of packings in the more exotic materials of construction (see the following section, Other Design Considerations). Concerns over unacceptable downtime can drive internals selection to a less efficient but more easily maintained or repaired design. Unacceptable downtime can shut down a production process or lead to high fines from regulating authorities because the authorities expect virtually 100% compliance with discharge limitations.

In addition to the selection of internals, a basic decision must be made regarding atmospheric operation or vacuum operation of a steam stripping column. The choice of a vacuum stripper may be determined by a limited or nonexistent steam supply at a proposed site. Other factors include the availability of operators who are trained in mass transfer processes and the need to maintain a mechanical vapor compressor.

All of the above factors are site specific and should not be considered to be an exhaustive list of factors that can affect the mechanical design of a steam stripping column.

Other Design Considerations. Ground waters containing high concentrations of iron are commonly found in many parts of the United States. Iron will be oxidized in the high-temperature environment of a steam stripping column, and a ferric hydroxide precipitate will form. This can easily overwhelm the solids-passing capabilities of

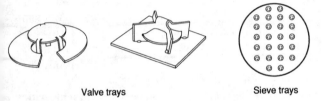

Valve trays   Sieve trays

FIGURE 9-27
Plate types.

even the best mechanical design of column internals. Consequently, it is sometimes necessary to add a pretreatment step to precipitate iron and other ions that can cause excessive scaling or precipitation in a steam stripping operation.

The elevated temperature of a steam stripping operation often requires exotic materials of construction. The presence of free chlorides and elevated temperatures will attack the common austenitic stainless steels (e.g., 304, 316), so the use of Monel™, titanium, or other specialty alloys is often indicated for wetted parts where high temperature is present. Other exotic materials of construction could include a carbon steel column lined or coated with corrosion-resistant materials. The exact materials of construction are usually determined on a case-by-case basis.

9-5   CHEMICAL OXIDATION

In general, the objective of chemical oxidation is to detoxify waste by adding an oxidizing agent to chemically transform waste components. For example, an organic molecule can be converted to carbon dioxide and water or to an intermediate product that may be less toxic than the original. This intermediate might be amenable to further treatment by the biological methods presented in the next chapter.

Chemical oxidation of wastes is a well-established technology that is capable of destroying a wide range of organic molecules, including chlorinated VOCs, mercaptans, and phenols, and inorganics such as cyanide. This section focuses on the oxidizing agents most commonly used for hazardous waste treatment: ozone, hydrogen peroxide, and chlorine. Ultraviolet (UV) light is usually added along with ozone and/or hydrogen peroxide to accelerate the oxidation of chlorinated VOCs.

Oxidation and reduction reactions occur in pairs to comprise an overall redox reaction. In chemical oxidation for hazardous waste treatment, an oxidizing agent is added to oxidize the waste components of concern, which serve as the reducing agents. Oxidizing agents are nonspecific and will react with any reducing agents present in the waste stream. Therefore, these processes are most economical when organics other than the ones of concern are in low concentration. For example, in the oxidation of cyanides, the presence of large quantities of other organic molecules will require excessive amounts of oxidizing agent to be used. Additionally, reactions between some oxidizing agents (e.g., chlorine) and some organics (e.g., hydrocarbons) may result in chlorine substitution rather than destruction of the organic. The resulting chlorinated hydrocarbon may be more toxic than the original waste.

Although chemical oxidation is typically applied to liquid hazardous wastes and contaminated ground water, soils may also be amenable to these processes. Contaminated soils can be excavated and treated in a slurry form in reaction vessels. However, because excavation is an expensive process, the tendency in soil cleanup technology is to use in situ processes. The feasibility of in situ chemical oxidation depends upon the ability to get reagents in contact with contaminants, which is a function more of the soil properties such as permeability, than of the process chemistry. In remediation technologies employing addition of ozone or hydrogen peroxide (two of the most common oxidizing agents in hazardous waste treatment) to soils in situ, enhanced volatilization of contaminants and biological activity of native microorganisms, rather than chemical oxidation, are the primary mechanisms of remediation. Therefore, in situ soil aeration and bioremediation have been more widely developed and applied for site remediation. These two technologies are discussed in Sec. 9-2 and Sec. 10-6, respectively.

Process Description. Chemical oxidation is typically conducted in completely mixed tanks or plug flow reactors. The contaminated water is introduced at one side of the tank and the treated water exits at the other side. The oxidizing agent is either injected into the contaminated water just before it enters the tank or is dosed directly into the tank. Complete mixing of the water with the oxidizing agent must be provided, either by mechanical agitation, pressure drop, or bubbling into the tank. Complete mixing, which prevents short-circuiting in the tank, is necessary to ensure contact of the contaminants with the oxidizing agent for a minimum period of time, and, thus, reduce the chemical dosage required to obtain a specific effluent concentration.

Ozone is produced from atmospheric oxygen using electrical energy to split the oxygen into two oxygen radicals ($O\cdot$), which readily combine with other oxygen molecules (O_2) to form ozone (O_3). Ozone is unstable under normal environmental conditions and readily decomposes back to oxygen. Ozone is added to the liquid waste as a gas, either through porous diffusers at the bottom of a tank, or through an injector where the pressure drop produced draws ozone into the injector and mixes it with the liquid.

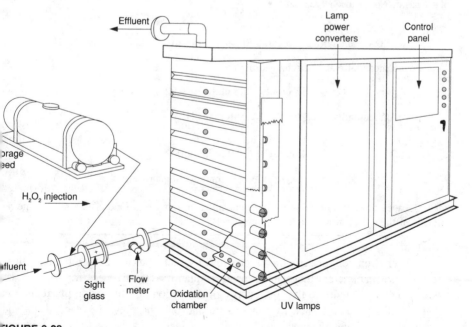

FIGURE 9-28
Equipment arrangement for a typical UV/H$_2$O$_2$ system.[64]

Hydrogen peroxide (H_2O_2) and UV light are generally used together. A typical arrangement is shown in Fig. 9-28. Hydrogen peroxide is stored on site as a 35 to 50% solution. The high solubility of H_2O_2 in water obviates the need for mixing equipment, reducing the number of moving parts in the system. The size of the combined H_2O_2/UV system can be significantly reduced through the use of high-energy UV lamps, capable of imparting 500 Watts/Liter to the water.[63]

The last oxidizing agent discussed in this section, chlorine (Cl_2), is a dense gas that readily liquefies under pressure. Liquid chlorine is commercially available in 150-lb and 1-ton cylinders. Chlorine is also available in solid form such as calcium hypochlorite [$Ca(OCl)_2$]. For aqueous waste treatment applications, liquid chlorine is evaporated to a gas and then mixed with water in a solution feeder. This concentrated solution is then added to the liquid waste. The dry calcium hypochlorite is also dissolved in water prior to application to the liquid waste.

Some common oxidation and reduction reactions are listed in Table 9-8.

Example 9-12 Chemical quantity determination. A waste stream of 20,000 gal/day contains 270 mg/L of cyanide as NaCN. Determine the theoretical (stoichiometric) amount of chlorine required daily to destroy this waste.

Solution. Atomic weights: Na = 23, C = 12, N = 14, Cl = 35.45

From reaction 3-1 (see also Table 9-8): $2NaCN + 5Cl_2 + 12NaOH \rightarrow N_2 + Na_2CO_3 + 10NaCl + 6H_2O$ This implies that 2.5 moles of Cl_2 are required for each mole of NaCN. The total mass of NaCN discharged daily is:

$$(20,000 \text{ gal/day})(8.34 \text{ lb/gal})(270 \text{ parts/} 10^6 \text{ parts}) = 45.04 \text{ lb/day NaCN}$$

Pound moles of NaCN discharged per day:

$$(45.04 \text{ lb/day})/(49 \text{ lb/lb mol}) = 0.92 \text{ lb mol/day}$$

Cl_2 required daily = 2.5 (0.92) = 2.3 lb mol Cl_2

$$= 2.3 \text{ lb mol } Cl_2 \text{ (70.9 lb/lb mol)}$$

$$= \underline{163 \text{ lb/day chlorine required}} \text{ (stoichiometric)}$$

Because of short-circuiting and/or incomplete reaction time, more than stoichiometric amounts of the chemical are required in practice.

The environment under which redox reactions take place greatly influences their effectiveness. Parameters that can be used to quantify effectiveness for use in hazardous waste treatment are the free energy of a reaction and the oxidation reduction potential (ORP). Factors such as temperature and pH, the presence of catalysts and concentration of other reactants, will influence selection of the oxidizing agent and will determine whether cost-effective treatment is feasible.

Reagent	Contaminant	Reaction	Comments	Reference
Ozone	Cyanide	$NaCN + O_3 \rightarrow NaCNO + O_2$		65
Ozone	Phenol	(See Fig. 9-31)		66, 67
Hydrogen peroxide	Cyanide	$NaCN + H_2O_2 \rightarrow NaCNO + H_2O$	pH 9.5–10.5	68
	Sulfide	$H_2O_2 + H_2S \rightarrow 2H_2O + S$ $H_2O_2 + S^= \rightarrow SO_4^= + 4H_2O$		69
Chlorine	Ferrous to ferric	$2Fe^{+++} + HOCl + 5H_2O \rightarrow$ $2Fe(OH)_3 + Cl^- + 5H^+$		70
Chlorine	Cyanide to cyanate	$NaCN + Cl_2 \rightarrow CNCl + NaCl$ $CNCl + 2NaOH \rightarrow$ $NaCNO + H_2O + NaCl$ $2NaCNO + 3Cl_2 + 4NaOH \rightarrow$ $N_2 + 2CO_2 + 6NaCl + 2H_2O$	CNCl is a toxic gas	71
Chlorine dioxide	Cyanate to CO_2 & N_2 Manganous ion to manganese dioxide	$Cl_2 + 2NaClO_2 \rightarrow 2ClO_2 + 2NaCl$ $Mn(NO_3)_2 + 2e^- \rightarrow MnO_2 + 2NO_2^-$	MnO_2 insoluble	72, 73
Potassium permanganate	Sulfide	$4\ KMnO_4 + 3H_2S \rightarrow$ $2K_2SO_4 + S + 3MnO + MnO_2 + 3H_2O$		
Ozone/UV	Organics	$CH_3CHO + O_3 \rightarrow CH_3COOH + O_2$		
UV/peroxide	Organics (CH_2Cl_2)	$CH_2Cl_2 + 2H_2O_2 \rightarrow CO_2 + 2H_2O + 2HCl$	Forms H_2O_2	75
Oxygen	Organic matter (CH_2O)	$CH_2O + 1/2\ O_2 \rightarrow CO_2 + H_2O$		
Oxygen	Cyanide	$2\ CN^- + O_2 \rightarrow 2CNO^-$ $CNO^- + 2H_2O + 2H^+ \rightarrow CO_2 + NH_4^+ + H_2O$	Requires activated C and Cu catalyst	76
Sulfur dioxide	Chromium (VI)	$3SO_2 + 3H_2O \rightarrow 3H_2SO_3$ $2CrO_3 + 3H_2SO_3 \rightarrow Cr_2(SO_4)_3 + 3H_2O$	Cr(III) is then precipitated as	77
Ferrous sulfate	Chromium (VI)	$2CrO_3 + 6FeSO_4 + 6H_2SO_4 \rightarrow$ $3Fe_2(SO_4)_3 + Cr_2(SO_4)_3 + 6H_2O$	$Cr(OH)_3 \downarrow$ by adding lime	78
Sodium borohydride	Copper	$NaBH_4 + 8Cu^+ + 2H_2O \rightarrow$ $8Cu + NaBO_2 + 8H^+$		79, 80

Oxidation

Reduction

The power of an oxidizing or reducing agent is measured by its electrode potential. An indication of how a reaction will proceed can be determined from the free energy considerations (see Sec. 3-3). The free energy of a reaction, $\Delta G°$, is related to the standard electrode potential $E°$, and the equilibrium constant K by:

$$\Delta G° = -nFE° = -RT\ln K \qquad (9\text{-}52)$$

where $\Delta G°$ = free energy of a reaction (kcal/gmol)
  n = number of electrons exchanged during the reaction per gmol
  F = Faraday constant 23.062 kcal/(volt-equiv.)
  $E°$ = standard electrode potential (volts)
  R = gas constant = 1.98×10^{-3} kcal/gmol·K
  T = temperature (K)

$E°$ may be determined from tabulated half reactions (see Table 9-9). To assemble a complete reaction from two half reactions, an oxidation half reaction is added to a reduction half reaction. Water, H^+, OH^-, etc., occurring on both sides of the reaction are removed to simplify. The process is illustrated by Example 9-15.

It is possible to measure the ORP directly by means of a galvanic cell made up of a gold or platinum anode and a reference electrode (the cathode). This galvanic cell instrument is capable of measuring only those potentials from species that react with the electrode and, thus, does not provide a direct measure of $E°$. However, for some reactions of interest in waste treatment, the ORP value does provide a useful measurement, as shown in Fig. 9-32 for hydrogen peroxide.

The **Nernst equation** may be derived by substituting the definition of K from Chap. 3. This equation allows E to be determined at conditions other than standard:

$$E = E° - (RT/nF)\ln Q \qquad (9\text{-}53)$$

$$Q = \frac{\Pi\,(\text{products})^x}{\Pi\,(\text{reactants})^y} \qquad (9\text{-}54)$$

where x,y represent the coefficients of the products and reactants in the original equation and Q is termed the reaction quotient. Note that at equilibrium, $Q = K$.

For the reaction $aA + bB \rightarrow cC + dD$:

$$Q = \frac{[C]^c\,[D]^d}{[A]^a\,[B]^b} \qquad (9\text{-}55)$$

where $[C]$ = molar concentration of component C
  c = the coefficient of component C in the reaction

In simplified form, at 25°C, Eq. 9-53 reduces to

$$E = E° - (0.059/n)\log_{10}Q \qquad (9\text{-}56)$$

TABLE 9-9
Standard electrode potentials at 25°C[81,82,83,84]

Reaction	E° (volts)
$Cl_{2(g)} + 2e^- \rightarrow 2Cl^-$	1.36
$HOCl + H^+ + 2e^- \rightarrow Cl^- + H_2O$	1.49
$ClO^- + H_2O + 2e^- \rightarrow Cl^- + 2OH^-$	0.90
$NH_2Cl + H_2O + 2e^- \rightarrow Cl^- + NH_3 + OH^-$	0.75
$NHCl_2 + 2H_2O + 4e^- \rightarrow 2Cl^- + NH_3 + 2OH^-$	0.79
$NH_3Cl^+ + H^+ + 2e^- \rightarrow Cl^- + NH_4$	1.40
$NHCl_2 + 3H^+ + 4e^- \rightarrow 2Cl^- + NH_4$	1.34
$O_3 + 2H^+ + 2e^- \rightarrow O_2 + H_2O$	2.07
$O_3 + H_2O + 2e^- \rightarrow O_2 + 2OH^-$	1.24
$H_2O_2 + 2H^+ + 2e^- \rightarrow 2H_2O$	1.78
$HO_2 + 2e + H_2O \rightarrow 3HO^-$	0.85
$ClO_2 + 2H_2O + 5e^- \rightarrow Cl^- + 4OH^-$	1.71
$MnO_4 + 4H^+ + 3e^- \rightarrow MnO_2 + 2H_2O$	1.68
$MnO_4 + 8H^+ + 5e^- \rightarrow Mn^{2+} + 4H_2O$	1.49
$MnO_4 + 2H_2O + 3e^- \rightarrow MnO_2 + 4HO^-$	0.58
$O_2 + 4H^+ + 4e^- \rightarrow 2H_2O$	1.23
$O_2 + 2H_2O + 4e^- \rightarrow 4OH^-$	0.40
$HBrO + H^+ + 2e^- \rightarrow Br_- + H_2O$	1.33
$MnO_2 + 4H^+ + 2e^- \rightarrow Mn^{2+} + 2H_2O$	1.21
$ClO_2 + e^- \rightarrow ClO_2$	1.15
$Fe(OH)_3 + e^- + 3H^+ \rightarrow Fe^{2+} + 3H_2O$	1.01
$Fe^{3+} + e^- \rightarrow Fe^{2+}$	0.77
$ClO_2 + 2H_2O + 4e^- \rightarrow Cl^- + 4OH^-$	0.76
$ClO_3 + H_2O + 2e^- \rightarrow ClO_2 + 2OH^-$	0.35
$S_{(s)} + 2H^+ + 2e^- \rightarrow H_2S$	0.14
$NO_3 + H_2O + e^- \rightarrow NO_2 + 2HO^-$	0.01
$NO_2 + 2OH^- \rightarrow NO_3 + H_2O + e^-$	-0.01
$1/4\ CO_{2(g)} + H^+ + e^- 1/24(glucose) + 1/4H_2O$	-0.20
$H^+ + e^- \rightarrow 1/2H_{2(g)}$	0
$Na^+ + e^- \rightarrow Na_{(s)}$	-2.72
$Mg^{2+} + 2e^- \rightarrow Mg_{(s)}$	-2.37
$Cr_2O_7^{2-} + 14H^+ + 6e^- \rightarrow 2Cr^{3+} + 7H_2O$	+1.33
$Cr^{3+} + e^- \rightarrow Cr^{2+}$	-0.41
$MnO_{2(s)} + 4H^+ + 2e^- \rightarrow Mn^{2+} + 2H_2O$	+1.23
$Fe^3 + e^- \rightarrow Fe^{2+}$	+0.77
$Fe^{2+} + 2e^- \rightarrow Fe_{(s)}$	-0.44
$Cu^{2+} + e^- \rightarrow Cu^+$	+0.16
$Cu^{2+} + 2e^- \rightarrow Cu_{(s)}$	+0.34
$Ag^{2+} + e^- \rightarrow Ag^+$	+2.0
$Ag^+ + e^- \rightarrow Ag_{(s)}$	+0.8
$AgCl_{(s)} + e^- \rightarrow Ag_{(s)} + Cl^-$	+0.22
$Au^{3+} + 3e^- \rightarrow Au_{(s)}$	+1.5
$Zn^{2+} + 2e^- \rightarrow Zn_{(2)}$	-0.76
$Cd^{2+} + 2e^- \rightarrow Cd_{(2)}$	-0.40

TABLE 9-9
Standard electrode potentials at 25°C (*continued*)

Reaction	E° (volts)
$Hg_2Cl_{2(s)} + 2e^- \rightarrow 2Hg_{(l)} + 2Cl^-$	+0.27
$Mn(OH)_3 + e \rightarrow Mn(OH)_2 + OH^-$	0.15
$Mn(OH)_3 + 2e \rightarrow Mn + 2OH^-$	−1.56
$2Hg^{2+} + 2e^- \rightarrow Hg_2^{2+}$	0.91
$Al^{3+} + 3e^- \rightarrow Al_{(s)}$	−1.68
$Sn^{2+} + 2e^- \rightarrow Sn_{(2)}$	−0.14
$PbO_{2(s)} + 4H^+ + SO_4^{2-} + 2e^- \rightarrow PbSO_{4(s)} + 2H_2O$	+1.68
$Pb^2 + 2e^- \rightarrow Pb_{(s)}$	−0.13
$NO_3^- + 2H^+ + 2e^- \rightarrow NO_2^- + H_2O$	+0.84
$NO_3^+ + 10H^+ + 8e^- \rightarrow NH_4 + 3H_2O$	+0.88
$N_{2(g)} + 8H^+ + 6e^- \rightarrow 2NH_4^+$	+0.28
$NO_2^- + 8H^+ + 6e^- \rightarrow NH_4^+ + 2H_2O$	+0.89
$2NO_3^- + 12H^+ + 10e^- \rightarrow N_{2(g)} + 6H_2O$	+1.24
$O_{3(g)} + 2H^+ + 2e^- \rightarrow O_{2(g)} + H_2O$	+2.07
$O_{2(g)} + 4H^+ + 4e^- \rightarrow 2H_2O$	+1.23
$O_{2(aq)} + 4H^+ + 4e^- \rightarrow 2H_2O$	+1.27
$SO_4^{2-} + 2H^+ + 2e^- \rightarrow SO_3^{2-} + H_2O$	−0.04
$S_4O_6^{2-} + 2e^- \rightarrow 2S_2O_3^{2-}$	+0.18
$S^{(s)} + 2H^+ + 2e^- \rightarrow H_2S_{(g)}$	+0.17
$SO_4^{2-} + 8H^+ + 6e^- \rightarrow S_{(s)} + 4H_2O$	+0.35
$SO_4^{2-} + 10H^+ + 8e^- \rightarrow H_2S_{(s)} + 4H_2O$	+0.34
$2HOCl + 2H^+ + 2e^- \rightarrow Cl_{2(aq)} + 2H_2O$	+1.60
$Cl_{2(g)} + 2e^- \rightarrow 2Cl^-$	+1.36
$Cl_{2(g)} + 2e^- \rightarrow 2Cl^-$	+1.39
$2HOBr + 2H^+ + 2e^- \rightarrow Br_{2(l)} + 2H_2O$	+1.59
$Br^2 + 2e^- \rightarrow 2Br^-$	+1.09
$2HOI + 2H^+ + 2e^- \rightarrow I_{2(s)} + 2H_2O$	+1.45
$I_{2(aq)} + 2e^- \rightarrow 2I^-$	+0.62
$I_3^- + 2e^- \rightarrow 3I^-$	+0.54
$ClO_2 + e^- \rightarrow ClO_2^-$	+1.15
$CO_{2(g)} + 8H^+ + 8e^- \rightarrow CH_{4(g)} + 2H_2O$	+0.17
$6CO_{2(g)} + 24H^+ + 24e^- \rightarrow C_6H_{12}O_6(glucose) + 6H_2O$	−0.01
$CO_{2(g)} + H^+ + 2e^- \rightarrow HCOO(formate)$	−0.31
$H_2BO_1 + 5H_2O + 8c \rightarrow BH_4^- + 8OH^-$	−1.24
$H_2BO_1 + H_2O + 3c \rightarrow B + 4OH^-$	−1.79
$H_2BO_1 + 3H^+ + 3c \rightarrow B + 3H_2O$	−0.8698
$Ba(OH)_2 + 2e^- \rightarrow Ba + 2OH^-$	−2.99
$Cd^{2+} + 2e^- \rightarrow Cd(Hg)$	−0.3521
$Cd(OH)_2 + 2e^- \rightarrow Cd(Hg) + 2OH^-$	−0.809
$Fe^{2+} + 2c \rightarrow Fe$	−0.447
$Fe^{1+} + 3c \rightarrow Fe$	−0.037
$Fe^3 + c \rightarrow Fe^-$	0.771
$H_2O_2 + 2H^- + 2c \rightarrow 2H_2O$	1.776
$Ni(OH)_2 + 2e \rightarrow Ni + 2OH^-$	−0.72

Example 9-13 Calculation of free energy. Determine the free energy of the oxidation of soluble manganous nitrate to insoluble manganese dioxide:

$$Mn(NO_3)_2 + 2e^- \rightarrow MnO_2 + 2NO_2^-$$

Solution. The half reactions from Table 9-9:

(Note that $2H_2O + 4H^+$ appear on both sides when the two half reactions are combined.)

$$E° \text{ (volts)}$$

Oxidation:

$$Mn^{++} + 2H_2O \rightarrow MnO_2 + 4H^+ + 2e^- = -1.23 \text{ V}$$

Reduction:

$$2\left[NO_3^- + 2H^+ + 2e^- \rightarrow NO_2^- + H_2O\right] = 2\ [+0.84V] = +1.68 \text{ V}$$
$$E° = \text{Total electrode potential} = \quad 0.45 \text{ V}$$

From Eq. 9-52:

$$\Delta G° = -(2 \text{ equiv/gmol})(23.062 \text{ kcal/V-equiv})(0.45 \text{ V}) = -20.76 \text{ kcal/gmol}$$

The equilibrium constant may also be determined from Eq. 9-52:

$$\ln K = -\Delta G°/RT = \frac{(20.76 \text{ kcal}/\text{gmol})}{(1.98 \times 10^{-3} \text{ kcal}/\text{gmol } K)(298 \text{ K})} = 35.01$$

$$K = 1.60 \times 10^{15}$$

The high value of K indicates that the reaction should proceed considerably to the right. However, it should be noted that for many reactions, kinetics rather than thermodynamic considerations control.

Design Considerations

Ozone. Ozone has long been used as a disinfectant in water and wastewater applications. Ozone is a blue gas at normal temperatures and pressures and has a recognizable pungent odor. It is irritating to mucous membranes and is a significant surface air pollutant in many areas. Ozone is a powerful oxidant and readily reacts with most toxic organics. Most of the reaction products are less toxic and many are biodegradable.

Reactions between ozone and organic molecules include the insertion of oxygen into a benzene ring, the breaking of double bonds, and the oxidation of alcohols. Ozonolysis of the C=C bond produces aldehydes and ketones:[85]

Cleavage products
(Aldehydes and ketones)

Alkene   Molozonide   Ozonide

(9-57)

Alcohols react with ozone to form organic acids:

$$3RCH_2OH + 2O_3 \rightarrow 3RCOOH + 3H_2O \qquad (9\text{-}58)$$

Aromatic compounds may be converted to toxic phenolics at pH greater than 9 and in the presence of redox salts such as iron or manganese.[86] Ozone reacts with a wide range of organic compounds as shown in Fig. 9-29. Note the dramatic reduction in the time required by using ultraviolet light with ozone.

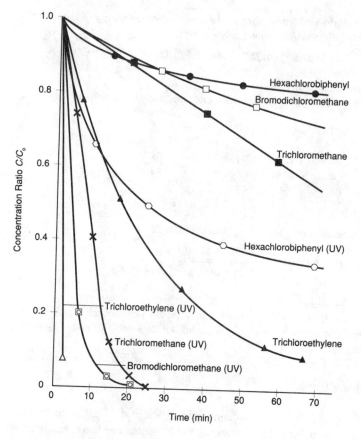

* At pH 6–7, ozone dose rate = 1.0–1.4 mg/L min: UV frequency is 254 nm from a low pressure mercury lamp, with a flux of 0.42 W/L.

FIGURE 9-29

Destruction of chlorinated organics by ozone with and without UV radiation.[87]

The primary advantage of ozone as an oxidizing agent is that it has a very high free energy, which indicates that the oxidation reaction may proceed to completion. Glaze[88] reports a value of $\Delta G° = -400 \text{KJ/mol}$ for the reaction of ozone to form water and oxygen:

$$O_3 + 2H^+ + 2e^- \rightarrow H_2O + O_2 \tag{9-59}$$

However, kinetics indicate that the reaction rates for ozone oxidation are not always high. A disadvantage of ozone is its rapid dissociation to oxygen in the atmosphere. This dissociation can compete with any chemical oxidation reactions during waste treatment, and therefore, ozone must be generated and immediately applied at the treatment location.

The transformation of compounds in aqueous solutions by ozone via its decomposition in water is due to a chain reaction involving the ·OH (hydroxyl) radical.[89]

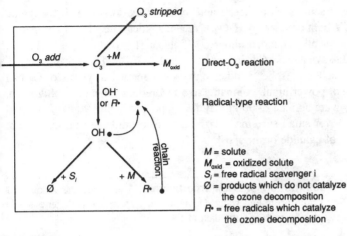

FIGURE 9-30
Reaction of ozone in aqueous solution.[91]

The reaction rate constant for destruction of organics by ·OH is typically 10^7 to 10^9, many orders of magnitude greater than for O_3 alone.[90] A schematic indicating the various types of O_3 reactions is presented in Fig. 9-30.

Phenol is readily degraded by ozone. Early work by Eisenhauer[92] indicated that the destruction of phenol by ozone follows the pathways indicated in Fig. 9-31.

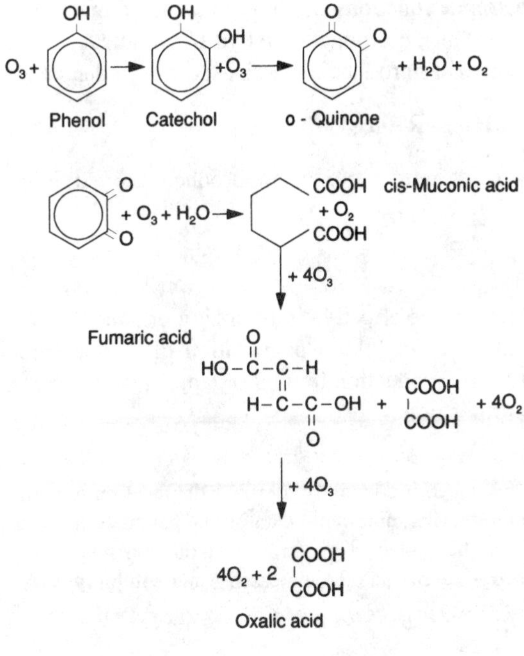

FIGURE 9-31
Degradation of phenol by ozone.[94]

Because simple organic acids such as acetic and oxalic acids do not readily degrade unless catalyzed by UV light or aided by H_2O_2,[93] recent applications of O_3 to oxidation of toxic wastes have typically included either UV light or H_2O_2.

The principal disadvantages of using ozone are the cost and inability to inject the gas efficiently. Posselt and Weber[95] indicate that only about 150 grams of O_3 can be produced per kWh of power input, compared to a theoretical value of 1058 g/kWh. Ozone generation costs are about \$2,400/kg/day for plants producing 900 kg/day, and are dramatically higher for small systems.[96] Ozonation of cyanide is reported to cost \$0.14–0.64 per pound of cyanide destroyed.[97]

Example 9-14   Ozone destruction of trichloroethylene.   Determine the amount of ozone required to treat 760 L/min of ground water contaminated with trichloroethylene at 150 μg/L. Assume 90% removal is required, and the data of Fig. 9-29 apply.

Solution.   The figure indicates an ozone dose rate of 1.0 to 1.4 mg/L·min. Select the highest value (1.4 mg/L·min). From the figure, 90% removal requires approximately 1 hour of contact at 1.4 mg/L-min.

Required dose = 1.4 mg/L-min×60 min = 84 mg/L
Mass of ozone required daily:
(760 L/min)(1440 min/day)(84 mg/L) = 9.193 × 10^7 mg/day = 91.9 kg/day

Note that using ozone combined with ultraviolet light at 0.42 watts/L the contact time is reduced to approximately one minute.

Hydrogen peroxide.   Hydrogen peroxide functions similarly to ozone in that in the presence of an iron catalyst (Fenton's reagent), it produces the hydroxyl radical (·OH). This radical reacts with organic compounds to produce a reactive organic radical (·R)

$$\cdot OH + RH \rightarrow \cdot R + H_2O \tag{9-60}$$

The organic radical can again react with peroxide to produce an additional hydroxyl radical:

$$\cdot R + H_2O_2 \rightarrow \cdot OH + ROH \tag{9-61}$$

Hydrogen peroxide has been shown to be effective in oxidizing organics in soil through in situ treatment. Watts, et al.[98] report 99.9% degradation of pentachlorophenol (PCP) contaminated soils, noting that reduction in total organic carbon closely followed the removal of PCP. They further note:

The phenomenon that organic carbon was removed rapidly after PCP degradation suggests that hydroxyl radical attack on the products is more rapid than on the parent compound. High rates of product degradation may be explained by the lower oxidation state of the ring as it is hydroxylated [i.e., hydroxyl added] and the increased water solubility of the products.[99]

Schmitt[100] reported on the use of hydrogen peroxide to treat a paint stripping liquid waste containing the following contaminants:

Methylene Chloride: 50–1,000 mg/L
Phenol: 500–6,000 mg/L
Chromium (VI):   15–400 mg/L

Initial pilot studies showed that 99% of the phenol could be destroyed following Cr^{6+} reduction and precipitation, at a dosage of 2.5g H_2O_2 per g phenol, with an iron catalyst level of 0.05 mg Fe^{2+}/mg H_2O_2. However, a considerable amount of color remained. Color could be removed at a peroxide/phenol dosage of 7:1, but this released O_2 gas and raised the temperature of the waste more than 60°F.

Further studies demonstrated that H_2O_2 treatment should precede Cr^{6+} reduction. In order to ensure phenol destruction and avoid overdosing, an ORP control system was recommended following the data in Fig. 9-32. Not only was the phenol effectively oxidized, but the reduction and subsequent precipitation of chromium was enhanced.

As illustrated in Fig. 9-33, the effectiveness of H_2O_2 may be greatly enhanced by UV light. The application of H_2O_2 + UV light is a well-established technology with operating costs ranging from $1 to $10 per thousand gallons treated.[102] When UV is used to catalyze peroxide reactions, the ability of the water to transmit light becomes of consequence, and information must be obtained on light transmissibility. Pre-treatment to remove color, iron, or suspended solids may be required.

The effectiveness of treatment by H_2O_2 + UV is dependent on which organic compounds are to be destroyed. This is illustrated for halogenated aliphatics in Fig. 9-34.

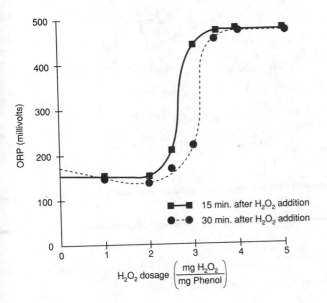

FIGURE 9-32
Control of H_2O_2 dosage by ORP.[101]

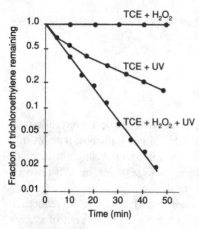

Initial TCE = 58 ppm
Initial H_2O_2/TCE = 4.5 mol/mol
UV lamp output = 2.5 watts @ 254 nm

FIGURE 9-33
Decomposition of trichlorethylene with H_2O_2, UV and H_2O_2+UV.[103]

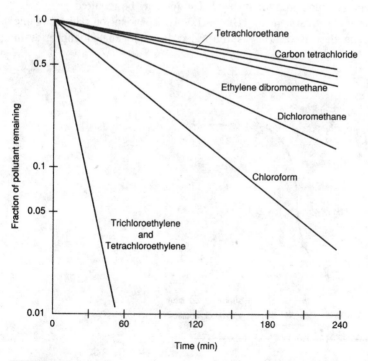

FIGURE 9-34
Destruction of halogenated aliphatics by H_2O_2 + UV at 20°C.[104]

Example 9-15 Removal of phenol with H_2O_2 and iron catalyst. Calculate the amount of H_2O_2 and iron catalyst (as Fe^{2+}) required to remove 99% of the phenol in a 10,000 L/day (2,640 gal/day) waste stream containing 6,000 mg/L of phenol. Use the dosages found by Schmitt[105] for oxidation of phenol after hexavalent chromium reduction.

Solution. The required dosages for 99% removal are 2.5 g H_2O_2/g phenol and 0.05 mg Fe^{2+}/mg H_2O_2.

Mass of phenol to be treated per day: 10,000 L/d × 6,000 mg/L = 6×10^7 mg/day = 60 kg/day

Mass of H_2O_2 required daily = 60 kg/day × 2.5 g H_2O_2/g phenol = $\underline{150 \text{ kg/day}}$ (330 lb/day)

Mass of Fe^{2+} required daily = 150 kg/day × 0.05 mg Fe^{2+}/mg H_2O_2 = $\underline{7.5 \text{ kg/day}}$ (16.5 lb/day)

Chlorine. Chlorine and chlorine compounds such as chlorine dioxide are frequently utilized as chemical oxidizing agents in waste treatment applications. For aqueous waste treatment applications, chlorine is evaporated to a gas and mixed with water to provide a hypochlorous acid (HOCl) solution:

$$Cl_2(aq) + H_2O \rightarrow HOCl + H^+ + Cl^- \tag{9-62}$$

Hypochlorous acid ionizes as follows:

$$HOCl \rightarrow H^+ + OCl^- \tag{9-63}$$

Eq. 9-63 is very pH dependent, with the dissociation constant of hypochlorous acid being approximately 7.6 at ambient temperatures.[106] The oxidizing power of chlorine tends to increase with pH. At pH values higher than 7.5, OCl^- is the predominant species in Eq. 9-63.

A classical application of oxidation with chlorine is the destruction of cyanide.[107] Under alkaline conditions cyanide is converted to the less toxic cyanate:

$$CN^- + OCl^- \rightarrow CNO^- + Cl^- \tag{9-64}$$

Caution must be exercised to maintain pH above 10 for reaction 9-64 to prevent cyanogen chloride, a toxic gas, from being produced:

$$NaCN + Cl_2 \rightarrow CNCl + NaCl \tag{9-65}$$

Under caustic conditions, cyanogen chloride is converted to sodium cyanate:

$$CNCl + 2NaOH \rightarrow NaCNO + H_2O + NaCl \tag{9-66}$$

Cyanate is destroyed by further alkaline reaction with chlorine:

$$2NaCNO + 3Cl_2 + 4NaOH \rightarrow N_2 + 2CO_2 + 6NaCl + 2H_2O \tag{9-67}$$

In practice, excess chlorine must be added to ensure that cyanogen chloride is not produced.

The above reactions provide information to determine the stoichiometric amount of chlorine needed. However, the actual amounts required for real treatment systems would depend on other constituents in the waste such as metals and reducing agents that create a chemical demand for chlorine. Additionally, chlorine is not stable in aqueous solutions and some will be lost to decomposition. When the cyanide is in the form of iron or nickel complexes, destruction of cyanide by chlorine is more difficult. Ferrocyanide ($[Fe(CN)_6]^{4-}$) is converted to ferricyanide ($[Fe(CN)_6]^{3-}$) and additional destruction by chlorine is ineffective. For Ni, an excess dosage of approximately 20% is recommended.[108]

Example 9-16 Chemical oxidation of cyanide. A waste stream contains 85 kg of CN^- daily. Determine the stoichiometric amounts of chlorine (Cl_2) and NaOH required to: (1) oxidize the cyanide to cyanate; and (2) oxidize the cyanide to N_2. Disregard the NaOH required to maintain the pH = 10.0 to prevent reaction 9-65.

Molecular Weights: CN^-: 26; Cl_2: 70.9; NaOH: 40

Solution. The conversion of CN^- to N_2 occurs through Eqs. 9-64 and 9-67.

1. Oxidation of cyanide to cyanate.
   From Eq. 9-64, the required molar ratio $OCl^-/CN^- = 1$. Based on Eq. 9-62, $Cl_2/OCl^- = 1$. Therefore, the molar ratio $Cl_2/CN^- = 1$ and the mass ratio $Cl_2/CN^- = 1 \times 70.9/26 = 2.73$ kg/kg. The mass of Cl_2 required daily is 85 kg/day × 2.73 kg/kg = 232.1 kg/day (510.6 lb/day).
   Mass of NaOH required daily = None according to Eq. 9-64.

2. Oxidation of cyanide to N_2.
   Step 1—As calculated above, 232.1 kg/day of Cl_2 are required.
   Step 2—From Eq. 9-67, the molar ratio $Cl_2/CNO^- = 3/2 = 1.5$, and the mass ratio is $1.5 \times 70.9/26 = 4.09$ kg/kg. The mass of Cl_2 required daily is 85 kg/day × 4.09 kg/kg = 347.7 kg/day (764.9 lb/day).
   Total mass of Cl_2 required = 232.1 + 347.7 = 579.8 kg/day (1,275.6 lb/day).
   Mass of NaOH required daily =
   Step 1—None
   Step 2—From Eq. 9-67, the molar ratio $NaOH/CNO^- = 4/2 = 2$ and, from Eq. 9-64, the molar ratio $CNO^-/CN^- = 1$. Therefore, the molar ratio $NaOH/CN^- = 2$ and the mass ratio $NaOH/CN^- = 2 \times 40/26 = 3.08$ kg/kg. The mass of NaOH required daily is 85 kg/day × 3.08 kg/kg = 261.8 kg/day (576 lb/day).
   Total mass of NaOH required = 261.8 kg/day (576 lb/day).

In a continuous treatment system, destruction of cyanide by chlorine can be controlled by oxidation reduction potential as illustrated in Fig. 9-35.

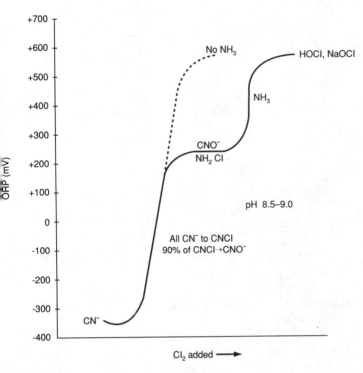

FIGURE 9-35
Control of cyanide destruction by ORP.[109]

9-6 SUPERCRITICAL FLUIDS

Supercritical fluids (SCF) extraction and supercritical water oxidation (SCWO) represent emerging technologies in hazardous waste management. Supercritical fluids are materials at elevated temperature and pressure that have properties between those of a gas and a liquid. In SCF extraction, organics in soils, sediments, or water are dissolved in the fluid at elevated temperature and pressure conditions and released from the SCF at lower temperatures and pressures. In supercritical water oxidation, air and contaminated water are brought together above the critical point of water where complete oxidation of organic components occurs rapidly.

Process Description

In SCF extraction, the contaminated stream (liquid or solid) is introduced into an extraction vessel into which the extraction fluid, pressurized and heated to the critical point in a compressor, is continuously loaded (see Fig. 9-36).[110] The organic contaminant in the contaminated stream dissolves in the SCF. The SCF is then expanded by passing it through a pressure reduction valve. The expansion of the SCF lowers the solubility of the organic contaminant in the SCF and, thus, results in the separation

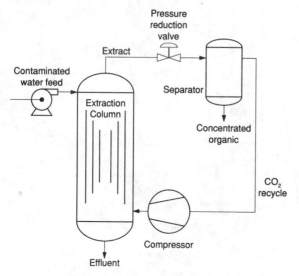

FIGURE 9-36
Schematic diagram of an SCF process for extracting organic compounds from water.[111]

of the organic contaminant from the extracting fluid. The SCF is then recompressed and recycled to the extraction vessel.

Alternatively, instead of reducing the pressure of the exiting SCF, the temperature can be reduced at a constant pressure by passing the SCF through a heat exchanger. The end result is also a reduction in the solubility of the organic contaminant in the SCF and the separation of both components. In this case, a blower is used instead of a compressor, and the separated extracting fluid entering the process vessel is heated in the same heat exchanger used to cool the exiting SCF.

In supercritical water oxidation, pressurized and heated water (at supercritical conditions) containing the organic contaminants is mixed with compressed air in a reactor (oxidizer). Figure 9-37 shows a schematic of the process.[112] The oxidation of the organics in the reactor takes place in a short time (residence times of less than one minute at 600–650°C [1,112–1,200°F] have been reported to produce removal efficiencies of 99.9999%).[113] From the oxidizer, the mixture passes through a solids separator (the solubility of inorganic compounds in water decreases by about three to four orders of magnitude when the water is at supercritical conditions) and through high- and low-pressure vapor separators to separate the gases (mainly CO_2 and N_2) from the effluent water. A portion of the energy recovered from the high-pressure separator is used to compress the influent air.

Theory of Supercritical Fluids Applications

Fluids are normally divided into two phases: liquids and gases. However, at elevated temperatures and pressures, a point is reached where this distinction is no longer discernible. This is termed the **critical point**. An SCF is a single-phase fluid at temperature and pressure above the critical point. An SCF has properties between those of the liquid and gas phases (i.e., densities approaching those of the liquid phase and

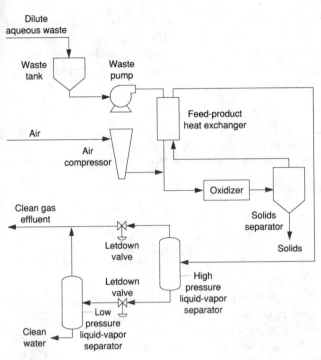

FIGURE 9-37
Process schematic diagram for SCF oxidation of an aqueous waste.[114]

diffusivities and viscosities approaching those of the gas phase). As a result of these properties, organic compounds are highly soluble in SCFs and can easily transfer to the SCF from their original medium. Critical conditions for a number of SCF solvents are given in Table 9-10.

Phase diagrams like the pressure-temperature diagram for water shown in Fig. 9-38 aid in illustrating the relationships between a compound's properties. Similar diagrams can be constructed for other pure compounds. Measurements represent the

TABLE 9-10
Critical constants[115]

Solvent	Temperature (°C)	Pressure (atm)	Density (g/cm³)
Carbon dioxide	31.1	73.0	0.460
Water	374.15	218.4	0.323
Ammonia	132.4	111.5	0.235
Benzene	288.5	47.7	0.304
Toluene	320.6	41.6	0.292
Cyclohexane	281.0	40.4	0.270

pressure at which the compound is present in the solid phase (from point A up to the triple point) and the pressure at which the compound is present in the liquid phase (from the triple point up to the critical point). The solid liquid equilibrium relationship is illustrated by the line from the triple point through point B. At the **triple point**, all three phases co-exist in equilibrium.

Example 9-17   Degrees of freedom.   Use the phase rule to determine the degrees of freedom (i.e., number of variables that can be selected independently) at the triple point of water.

The phase rule (Eq. 3-27) is:

$F = C + 2 - P$

$C = 1$ component (i.e., water)

$P = 3$ phases (solid/liquid/vapor)

Therefore, $F = 0$ independent variables, and the triple point is considered invariant.

As shown in Fig. 9-38, the **critical point** (T_C, P_C) represents the highest values of T and P where vapor/liquid equilibrium can exist. The material in the area labeled supercritical fluid is neither a liquid nor a gas. For example, a liquid can be vaporized (i.e., instantaneously converted to a vapor) if the pressure is reduced at constant temperature. An SCF undergoes a gradual change in properties from those of a liquid to properties of a vapor with no instantaneous phase change.

Properties are commonly expressed in terms of reduced temperature and pressure (i.e., the actual value divided by the critical value). The basis for this is the **theorem of corresponding states**:

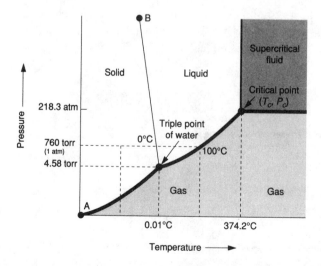

FIGURE 9-38
Pressure temperature phase diagram for water.[117]

All gases when compared at the same reduced temperature and reduced pressure, have approximately the same compressibility factor and all deviate from ideal gas behavior to about the same degree.[116]

The compressibility factor, Z, is simply the inverse of the density, ρ.

Figure 9-39 shows isotherms of reduced temperature (T/T_c) on a plot of reduced density (ρ/ρ_c) (ordinate) versus reduced pressure (P/P_c) (abscissa). All three variables are termed "reduced" because values of T, ρ, and P have been divided by their critical values.

Figure 9-39 illustrates one of the properties of SCFs. Near the critical point (CP), a small change in pressure (from $P_R = 1$ to $P_R = 1.1$) can create a large change in density (from $\rho_R = 1$ to $\rho_R = 1.6$) at constant temperature ($T_R = 1$). At the higher density the SCF acts as a more effective solvent.

In most instances, the viscosity changes rapidly near the critical point. Curves of generalized reduced viscosity as a function of temperature and pressure are illustrated in Fig. 9-40.

For CO_2, the viscosity at 250 atm and 37°C is about 0.09 cP, an order of magnitude below typical organic solvents. Diffusivities of organic constituents in supercritical fluids are typically orders of magnitude higher than is usual for organic liquids at

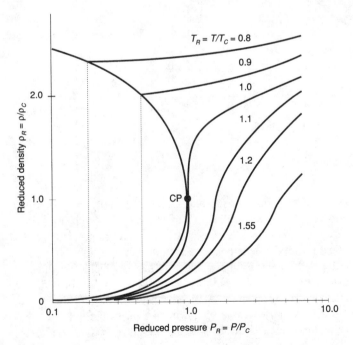

FIGURE 9-39
Variation in reduced density (ρ_R) near the critical point for a pure compound.[118]

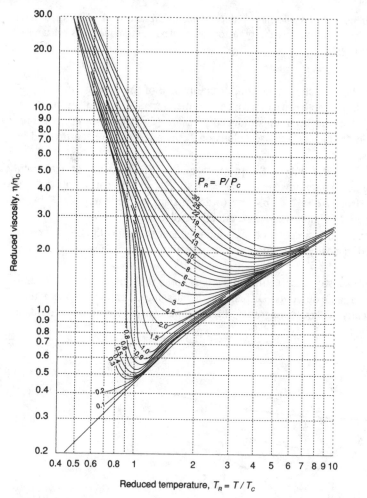

FIGURE 9-40
Generalized reduced viscosities.[119]

normal conditions. This lower solvent viscosity and higher diffusivity permits much higher mass transfer rates of organic contaminants into SCFs.

Properties of supercritical water are illustrated in Fig. 9-41. As the temperature is raised to near the critical point, the density changes from that of a liquid to that of a vapor, without going through an abrupt phase change. The high dielectric constant of water at ambient conditions is due to strong hydrogen bonds which provide the very high solubility for inorganic compounds. At supercritical conditions, the dielectric constant can drop to 2.5. As an example of the impact on solubility, NaCl has a solubility in water at ambient conditions of 37% by weight (370,000 mg/L). At supercritical conditions, this drops by almost four orders of magnitude to about 100 mg/L. Many organic compounds, on the other hand, become essentially miscible in

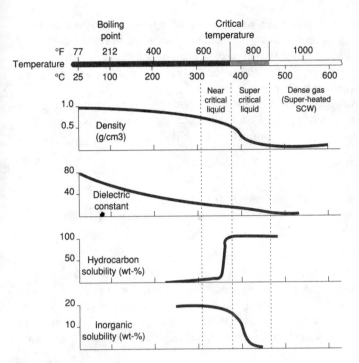

FIGURE 9-41
Properties of water at 250 atm.[120]

supercritical water. For example, benzene is miscible in supercritical water, but has a solubility of only 1700 mg/L at ambient temperature and pressure.

Design Considerations

Among the many decisions to be made in designing an SCF extraction system is the selection of solvent. Treybal[121] suggests the factors shown in Table 9-11 as appropriate for selecting solvents for solvent extraction.

TABLE 9-11
Factors for selection of solvents

Distribution coefficient	Recoverability
Density	Critical temperature and pressure
Toxicity	Chemical reactivity
Interfacial tension	Cost
Hazard	

It is essential to be able to re-use the solvent in order for these systems to be cost-effective. Separation of the SCF from the feed stream depends on the difference in density of the two fluids. For SCF extraction, reactions between the solvent and solute would be undesirable. However, for oxidation of organics in supercritical water, reactivity is essential. Concern with solvent toxicity has led to increased use of CO_2 as an SCF. Potential hazards with SCF extraction technology would include corrosive characteristics and flammability. The lower the critical temperature and pressure of the SCF, the less expensive the process is.

The distribution coefficient (partition coefficient) of a compound between two phases can be calculated by using Eq. 9-68. In this equation, the extract is the extraction solvent leaving the reactor, and the raffinate is the effluent.

$$K = C_e/C_r \qquad (9\text{-}68)$$

where K = distribution coefficient (dimensionless)

C_e = concentration of organic in the extract at equilibrium (weight %)

C_r = concentration of organic in the raffinate (effluent) at equilibrium (weight %)

The distribution coefficient is used to determine the *minimum* quantity of extraction solvent needed to extract a compound from a feed. For example, a value of $K = 4$ indicates that *at least* 1 kg of extraction solvent would be required to extract all of the compound from 4 kg of feed.[122] In actual application, the quantity of extraction solvent would be a function of the removal required and substantially above the minimum.

In supercritical water oxidation (SCWO), organic compounds are destroyed rather than removed. Reactions occur in a single homogeneous phase, characteristic of SCFs, and produce the end products shown in Table 9-12

TABLE 9-12
SCWO end products

Initial chemical	Resulting end product
Organic compounds	Carbon dioxide
Chlorine	Chlorides
Nitrogen compounds	Nitrates
Sulfur	Sulfates
Phosphorous	Phosphates

The process is reported to be cost-effective for aqueous wastes with organic concentrations in the range of one to twenty percent by weight. In this range the heat of combustion raises the SCF temperature from the critical temperature of water (374°C) to 600—650°C.[123]

Additional design considerations include: (1) the ability of the waste to be pressurized, (2) the potential to form char, and (3) the removal of the solids produced.[124] If the waste is a solid, slurry, or suspension, the solid particles must be reduced to less than 100 μm in diameter, which is the particle size limit for high-pressure pumps. Charring may occur when organics are heated slowly to the supercritical water conditions. The formation of char is undesirable, as the char is oxidized more

TABLE 9-13
Results of bench-scale evaluation tests of SCF oxidation[†]

Compound	Temp. (°C)	Time (min)	Destruction efficiency (%)	Reference
Aliphatic hydrocarbons				
Cyclohexane	445	7.	99.97	126
Aromatic hydrocarbons				
Biphenyl	450	7.	99.97	127
o-Xylene	495	3.6	99.93	128
Halogenated aliphatics				
1,1,1-Trichloroethane	495	3.6	99.99	129
1,2-Ethylene dichloride	495	3.6	99.99	130
1,1,2,2-Tetrachloroethylene	495	3.6	99.99	131
Halogenated aromatics				
O-Chlorotoluene	495	3.6	99.99	132
Hexachlorocyclopentadiene	488	3.5	99.99	133
1,2,4-Trichlorobenzene	495	3.6	99.99	134
4,4-Dichlorobiphenyl	500	4.4	99.993	135
DDT	505	3.7	99.997	136
PCB 1234	510	3.7	99.99	137
PCB 1254	510	3.7	99.99	138
Oxygenated compounds				
Methyl ethyl ketone	460	3.2	99.96	139
Methyl ethyl ketone	505	3.7	99.993	140
Dextrose	440	7.	99.6	141
Organic nitrogen compounds				
2,4-Dinitrotoluene	457	0.5	99.7	142
2,4-Dinitrotoluene	513	0.5	99.992	143
2,4-Dinitrotoluene	574	0.5	99.9998	144

[†] Source: Modell.[145]

slowly than solubilized organics, resulting in the need for longer detention times in the oxidizer.

The separation of inorganic salts precipitated during the process may be a problem. These solids may adhere to the reactor walls decreasing the effective volume available for the reaction and, thus, the residence time in the reactor. Also, the solids may adhere to the pipelines downstream of the reactor, eventually plugging the lines. Solids problems can be avoided by periodically washing the reactor and pipelines. The precipitated solids present a removal and disposal problem due to the potential for either handling a concentrated, corrosive brine or a dilute brine that has to be further treated for removal of toxic metals prior to disposal.

Another major design consideration for both SCF extraction and supercritical water oxidation is that the materials of construction must be able to withstand the extreme pressures and temperatures required by SCF systems and the corrosive charcteristics of the supercritical streams. SCF extraction vessels are typically constructed of stainless steel or glass, while supercritical water oxidation has been carried out in Hastelloy C-276 and Inconel 625 (high-nickel alloys) vessels.

As noted in the introduction, SCF processes are emerging technologies and few full-scale applications exist. Generally accepted design procedures do not as yet exist and current practice is to perform bench scale studies in conjunction with SCF equipment manufacturers to ensure that the process is both feasible and cost effective for a specific application. Table 9-13, from Modell,[125] provides the results of such tests for a range of organic compounds, indicating the potentially wide applicability of SCF processes.

9-7 MEMBRANE PROCESSES

The application of membranes to separate water from contaminants is a well-established technology in industry. For example, a membrane process (reverse osmosis) is used to produce industrial water of extremely high quality in the manufacture of semiconductors, for which the mineral content of ordinary drinking water is too high. The term **membrane** usually refers to a barrier to flow which will allow the passage of water, ions, or small molecules (**semi-permeable membrane**). However, membrane processes do not operate as a conventional filtration process. In most applications, the solution flows in parallel to the membrane and the transfer of solute or solvent occurs through application of direct electrical current (electrodialysis) or high pressures (reverse osmosis and ultrafiltration) to the solution. The membrane may be made of a solid matrix or swollen gel.[146]

The membranes used in industrial processes are subject to fouling and degradation, which limit their applicability in many waste treatment situations. The application of these processes to hazardous waste management is generally limited to those instances where extremely toxic materials cannot be removed by more cost-effective technologies. Research is being conducted to produce membranes resistant to fouling or new cleaning procedures used to prevent fouling.[147]

This section discusses three membrane processes of interest in managing hazardous wastes: **electrodialysis, reverse osmosis,** and **ultrafiltration.**

Process Description.   Electrodialysis, reverse osmosis, and ultrafiltration have different modes of operation and, thus, use different membrane types. Each of these processes is described in the following paragraphs.

Electrodialysis.   Electrodialysis consists of the separation of ionic species from water by applying a direct-current electrical field. The ion-selective membranes used in electrodialysis (cation and anion exchange membranes) permit the passage of cations through the cation exchange membrane and of anions through the anion exchange membrane. By alternating cation and anion exchange membrane between two electrodes, alternate dilute and concentrated cells are created (see Fig. 9-42). This results in the separation of the ionic species from water. Newer cell design incorporates the periodic reversal of current to reduce fouling.

Electrodialysis is used in the electroplating industry for the recycling of metal and purified water. Because a single pass through an electrodialysis cell usually removes only 30 to 60 percent of the metal, recycling of the produced brine to the feed inlet is necessary until the desired metal concentration is achieved.[149] This brine is returned to the plating bath, while the treated water is returned to the rinse tanks.

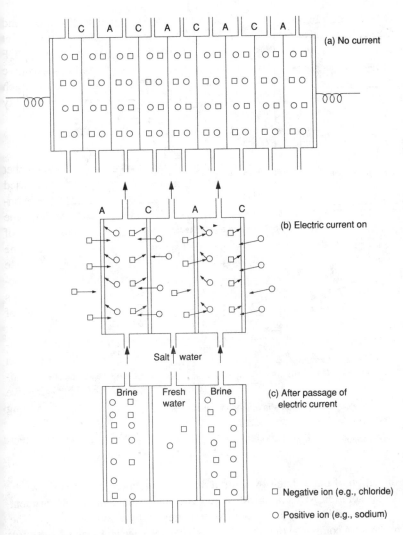

FIGURE 9-42
Operation of electrodialysis systems.[148]

As shown in Fig. 9-43, hundreds of individual cells can be combined between two electrodes.[150] The membranes are about 0.5 mm thick and the spacers that distribute the liquid over the surface of the membrane are 1 mm thick. The electrodes must be continually rinsed to prevent buildup of hydrogen gas at the cathode and oxygen or chlorine at the anode.

Typical electrodialysis units operate at pressures in the order of 276 to 414 kPa (40 to 60 psi). Generally, 90% of the feed is transformed to product water, while approximately 10% remains as the concentrate.[151]

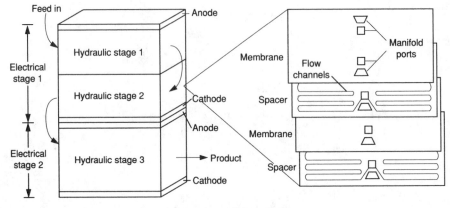

Typical electrodialysis stack

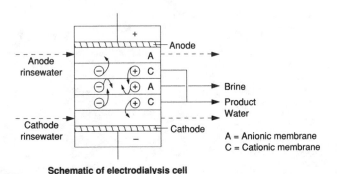

Schematic of electrodialysis cell

FIGURE 9-43
Schematic of electrodialysis system.[152,153,154]

A modification of the technology using a three-compartment cell[155] is shown in Fig. 9-44. A salt solution MX (cation M, anion X), enters the central compartment and is separated into acid (HX) and basic (MOH) streams by electrical energy. The bipolar membrane is a composite of an anion selective membrane and a cation selective membrane with an interface. Water in the interface is dissociated electrically to form H^+ and OH^-.

This three-compartment cell has been applied to the regeneration of spent pickle liquor. The process starts with pretreatment consisting of neutralization with KOH to form a KF/KNO_3 solution which is then filtered using diatomaceous earth. An electrodialysis cell then regenerates hydrofluoric and nitric acids, which are returned to the pickling process. A conventional electrodialysis unit is used to recover water from the filter cake wash. The process is illustrated in Fig. 9-45.

Reverse Osmosis. In reverse osmosis, a solvent is separated from a solution by applying a pressure greater than the osmotic pressure, thus forcing the solvent through a semipermeable membrane.[157]

Standard mode

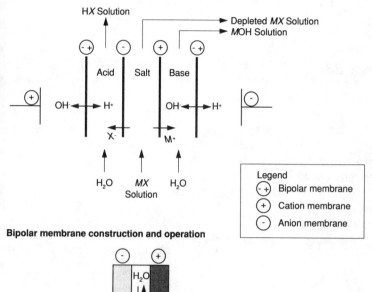

Bipolar membrane construction and operation

FIGURE 9-44
Three-compartment electrodialysis cell.

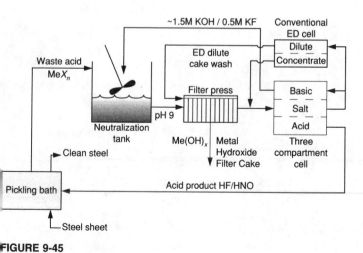

FIGURE 9-45
Pickle liquor recovery process.[156]

Osmosis is the transport of a solvent such as water through a semi-permeable membrane via diffusion. The driving force is the solute concentration gradient in the reverse direction of solvent flow. The process is illustrated schematically in Fig. 9-46. In the left diagram, water flows from the pure water side to the salt side because the concentration of salt is higher on the right and the system tends to move towards an osmotic equilibrium state. The semi-permeable membrane will allow water molecules to pass, but only a small fraction of salt molecules. This results in an increase in pressure in the salt solution until the osmotic pressure is reached (center diagram). At this point, equilibrium exists and the net transport of solvent is zero. **Reverse osmosis** (RO) occurs if mechanical pressure greatly in excess of the osmotic pressure is applied to the salt solution. This results in the transfer of water from the salt solution through the membrane into the pure water side (right diagram).

A reverse osmosis unit consists of several modules to which the feed is added through a high-pressure pump. The feed must be pumped at a pressure high enough as to maintain the design pressure difference in the last module.[159] The difference between the system and osmotic pressure is usually around 1,000 kPa (145 psi) with a maximum system pressure of 5,500 kPa (800 psi).

Reverse osmosis is used in the electroplating industry to recover both metals and water. Figure 9-47 presents a flow diagram of the process. The rinsewater from the first tank is passed through a reverse osmosis unit to recover a concentrated metal solution that is recycled to the plating bath and a purified water stream that is recycled to the third rinse tank. Figure 9-48 shows a photograph of a typical unit.

Reverse osmosis has been applied to both organic and inorganic hazardous waste problems. Martin, et al.[161] report on EPA research where a reverse osmosis unit was operated at 0.69 to 2.07 MPa (100–300 psi]) to remove organics. Reported rejection rates were 97% for dimethylphthalate, and 98% for naphthalene. Higgins[162] reports that reverse osmosis can concentrate divalent metals in plating rinse water to a 10–20% solution, greatly facilitating the recycling of these materials.

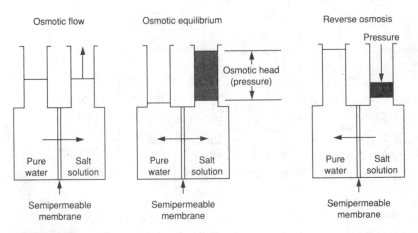

FIGURE 9-46
Pictorial explanation of reverse osmosis.[158]

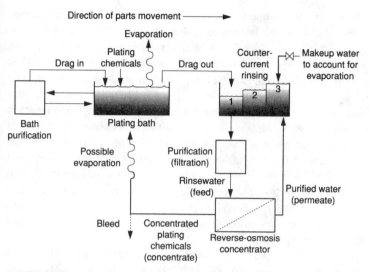

FIGURE 9-47
Flow diagram for a reverse osmosis system treating plating rinsewater.[160]

FIGURE 9-48
Reverse osmosis unit (courtesy of Process Science, Inc.).

An important RO design consideration is the need to dispose, recycle, or reuse both the permeate and concentrate streams, since complete separation of solute from solvent is not possible.

Ultrafiltration. The third membrane process to be discussed in this chapter, ultrafiltration, separates solutes from a solvent on the basis of molecular size and shape by passing the solution through a membrane module where a pressure difference is maintained across the membrane. Water and small molecules move through the membrane to the lower pressure side, while the larger molecules are retained by the membrane. To prevent fouling, the solution is passed at a high velocity over the membrane, resulting in small separation efficiencies. To improve the separation efficiency, the feed is recycled through the ultrafiltration unit several times or passed through several modules in series.[163]

In ultrafiltration, solutes of molecular weight greater than about 500 and less than 500,000 can be separated from a solution. The lower limit is related to the pore size of commercially available membranes. Solutes above the upper limit of molecular sizes are no longer separated from solution by ultrafiltration, but by conventional filtration.[164] Ultrafiltration membranes generally allow the passage of water and most ionic molecules while retaining nonionic molecules. Because of the lower osmotic pressures of the high molecular weight solutes separated by ultrafiltration membranes, the operating pressure differences are lower than those used in reverse osmosis, in the order of 35 to 690 kPa (5 to 100 psi).[165]

Ultrafiltration can be operated as a batch, single-stage continuous or feed-and-bleed, and multistage continuous operation. Figure 9-49 depicts the batch and continuous operation modes. In all operation modes, the feed is passed through ultrafiltration

(a) Batch ultrafiltration

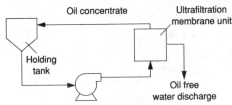

(b) Continuous ultrafiltration

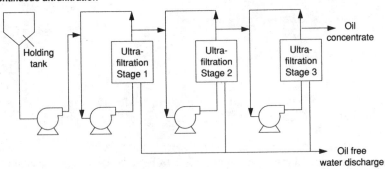

FIGURE 9-49
Batch and continuous ultrafiltration operations. [168]

membrane modules several times to improve the solute separation efficiency. Ultrafiltration units have been used in the recycle of electrophoretic paints, the removal of starches and enzymes from wastewaters, the recycling of polyvinyl alcohol in the textile industry, and the separation of oil-water emulsions.[166] In most applications, both the water and the concentrated solute are reused.

Theory

Electrodialysis. The current required to operate an electrodialysis unit can be calculated by using Faraday's law, which indicates that to transfer one gram equivalent of a substance by electrical force, a single cell will require one Faraday of electrical current (96,487 coulombs). In n cells, the required current is:[167]

$$I = (FQN/n) \times (E_1/E_2) \tag{9-69}$$

where I = current (amperes)
F = Faraday's constant = $96,487$ coulombs/g-equivalent (or amp·sec/g-equivalent)
Q = flow rate (L/sec) [gal/min $\times 0.06308$ = L/sec]
N = normality of solution (g-equivalents/L)
n = number of cells between electrodes
E_1 = removal efficiency (fraction)
E_2 = current efficiency (fraction)

The current efficiency, E_2, represents the number of g-equivalents of product actually produced per Faraday applied. Current efficiency is adversely affected by inefficient selectivity of the electrodialysis membranes and diffusional losses due to concentration gradients.

The voltage may be determined from Ohm's law:

$$E = IR \tag{9-70}$$

where E = voltage requirement (volts)
R = resistance through the unit (ohms)

The required power is:

$$P = I^2 R \tag{9-71}$$

where P = power (watts)

The **current density**, CD, is defined as the current passing through a unit area of membrane (amp/m^2). The ability of the solution to carry current is a function of the normality, N, of the solution. The ratio CD/N provides an important characterization of the system. A large value can result in too much current for the ions available in solution. This can lead to concentration polarization (i.e., regions of low ion concentration) which will result in an increase in resistance and a decrease in system efficiency, as described in the section Design Considerations.

Example 9-18 Electrodialysis of nickel-bearing waste. An electroplating plant has 1,700 m^3/day of a nickel-bearing waste. The nickel concentration is 15,400 mg/L as $NiSO_4$. Assume the following characteristics of the system:

Resistance through unit: 10.5 Ohms; Current Efficiency = 90%; Maximum CD/N ratio = 6,000 (A/m^2)/(g-equivalent/L); Membrane area = 1.0 m^2. Provide a preliminary design of a system to produce 90% removal of Ni. Determine the number of membranes; power required and the annual electrical cost @ $0.05/kWh. (Molecular weight of $NiSO_4$ = 154.7 g/gmol)

Solution

1. Normality:

$$\text{Equivalent Weight of } NiSO_4 = \frac{\text{Molecular Weight}}{\text{g-equivalent/gmol}}$$

$$= \frac{154.7 \text{ g/gmol}}{2 \text{ g-equivalent/gmol}}$$

$$= 77.4 \text{ g}\big/\text{g-equivalent}$$

Concentration = 15,400 mg/L = 15.4 g/L
Normality = 15.4/77.4 = <u>0.2N</u>

2. Maximum current density = 6000(0.2) = 1200 Amps/m^2
   Current = 1200 Amps/m^2 × 1.0m^2 = <u>1200 amperes</u>

3. Number of cells: $I = (FQN/n) \times (E_1/E_2)$; but $E_1 = E_2 = 0.9$
   $n = (FQN/I) =$

$$\frac{96,487 \left(\frac{\text{amp} \times \text{sec}}{\text{g-equivalent}}\right) \times 1.7 \times 10^3 \left(\frac{m^3}{\text{day}}\right) \times 10^3 \left(\frac{L}{m^3}\right) \left(\frac{0.2 \text{ g-equivalent}}{L}\right)}{1,200 \text{ amps} \times 86,400 \text{ seconds/day}} = 316.4 \text{ cells}$$

No. of membranes = 2 × 316.4 = 633 ∼ <u>640 membranes in 320 cells</u>

4. Power = I^2R = (1,200 amps)^2 × 10.5 ohms = 1.51 × 10^7 watts
   = <u>1.51 × 10^4 kW</u>

5. Annual Electrical Cost = 1.51 × 10^4 kW ×(365 × 24) hr × $0.05/kWh =
   <u>$6.62 million</u>

Reverse Osmosis. In reverse osmosis systems, the osmotic pressure, solute rejection, and water and solute fluxes are of interest. The flow across the membrane exerts an osmotic pressure, π determined by the **van't Hoff equation:**[169]

$$\pi = \phi_c N C_s RT \tag{9-72}$$

where π = osmotic pressure (atm)
  ϕ_c = osmotic pressure coefficient
  N = number of ions in each molecule
  C_s = concentration (gmol/L)
  R = universal gas constant = 0.082 atm·L/gmol·K
  T = absolute temperature (K)

Example 9-19   Osmotic pressure. Determine the osmotic pressure for a sodium chloride brine solution with concentration = $23,400$ mg/L at 25°C, given $\phi_c = 0.92$.

Solution

$$\pi = \phi_c N C_s RT$$

$$N = 2; (NaCl = Na^+ + Cl^-); \text{ molecular weight} = 58.45 \text{ g/gmol}$$

$$C_s = 23{,}400 \text{ mg/L} = (23.4 \text{ g/L})/(58.45 \text{ g/gmol}) = 0.4 \text{ gmol/L}$$

$$\pi = (0.92) \times 2 \times (0.4 \text{ gmol/L}) \times (0.082 \text{ atm} \cdot \text{L/gmol} \cdot \text{K}) \times (298.2 \text{ K})$$

$$= \underline{18.01 \text{ atm}} \text{ (264.7psi)}$$

Membrane rejection of contaminants is defined in terms of influent and effluent concentrations:

$$R = 100 \times (C_{in} - C_{out})/C_{in} \tag{9-73}$$

where R = rejection (%)
  C_{in} = concentration in feed (mg/L or gmol/cm^3)
  C_{out} = concentration in permeate (mg/L or gmol/cm^3)

The flux of water through a semi-permeable membrane is proportional to the difference in the operating pressure and the osmotic pressure.[170]

$$J_w = (D_w C_w V_m / RT \Delta Z)(\Delta P - \Delta \pi) \tag{9-74}$$

where J_w = water flux through the membrane (gmol/cm^2·sec)
  D_w = diffusivity of water in membrane (cm^2/sec)
  C_w = concentration of water (gmol/cm^3)
  V_m = molar volume of water = 0.018 L/gmol = 18 cm^3/gmol
  R = universal gas constant = 82.057 (atm· cm^3)/(gmol·K)
  T = absolute temperature (K)
  ΔZ = membrane thickness (cm)
  ΔP = pressure differential across membrane = $P_{in} - P_{out}$ (atm)
  $\Delta \pi$ = osmotic pressure differential across membrane
   = $\pi_{in} - \pi_{out}$(atm)

Equation 9-74 applies only if D_w, C_w, and V_m are independent of pressure. This is a valid assumption at pressures of up to 14,200 kPa (2,050 psi). The term $(D_w C_w V_m / RT \Delta Z)$ is referred to as the coefficient of water permeation, W_p, and is a characteristic of the particular membrane utilized with units of gmol/cm^2·sec·atm:

$$J_w = W_p \times (\Delta P - \Delta \pi) \qquad (9\text{-}75)$$

Similarly, the salt flux through the membrane is driven by the concentration gradient:

$$J_s = (D_s K_s / \Delta Z)(C_{in} - C_{out}) \qquad (9\text{-}76)$$

where J_s = salt flux through the membrane (gmol/cm^2·sec)
  D_s = diffusivity of the solute in the membrane (cm^2/sec)
  K_s = solute distribution coefficient (dimensionless)
  C = concentration (gmol/cm^3)

The term $(D_s K_s / \Delta Z)$ is referred to as the coefficient of permeability, K_p, for the particular contaminant. The coefficient of permeability is also a characteristic of the particular membrane and has units of cm/sec:

$$J_s = K_p \times (C_{in} - C_{out}) \qquad (9\text{-}77)$$

Example 9-20   Selection of reverse osmosis system.   A wastewater stream at 25°C is to be treated using a reverse osmosis system. The wastewater has a flowrate of 50 gpm and a concentration of 5000 mg/L of sodium chloride. A 75% recovery rate is required.
  A reverse osmosis vendor has recommended a membrane with the following characteristics:

$W_p = 1.9 \times 10^{-6} \dfrac{\text{gmol}}{\text{cm}^2 \cdot \text{sec} \cdot \text{atm}}$
Area for a bundle = 300 ft^2
50% recovery rate
Optimal pressure differential across the membrane = 500 psi

95% rejection rate of the salt

Determine: 1.   The osmotic pressure of the solution.
2.   The water flow through the membrane.
3.   Number of units required for 75% recovery.

Solution

1. The osmotic pressure of the solution using the van't Hoff equation

$$\pi = \phi_c N C_s R T$$

$$\phi_c = 0.92, \text{ from Example 9-21}$$

$$N = 2$$

$$C_s = 5,000 \text{ mg}/L = \frac{5 \text{ g}/L}{58.45 \text{ g}/\text{mol}} = 0.09 \text{ gmol}/L$$

$$\pi = (0.42)(2)(0.09)(0.082)(298.3) = 4.05 \text{ atm} = \underline{\underline{59 \text{ psi}}}$$

2. The water flow through the membrane

$$J_w = W_P(\Delta P - \Delta \pi)$$

$$J_w = 1.9 \times 10^6 \frac{\text{gmol}}{\text{cm}^2 \cdot \text{sec} \cdot \text{atm}} (500 \text{ psi} - 59 \text{ psi}) \left(\frac{1 \text{ atm}}{14.7 \text{ psi}} \right)$$

$$J_w = \underline{\underline{5.7 \times 10^{-5}}} \text{ gmol/cm}^2 \cdot \text{sec}$$

3. To treat 50 gpm of water

$$50 \frac{\text{gal}}{\text{min}} \times 3.79 \frac{\text{liters}}{\text{gallon}} \times \frac{1 \text{ min}}{60 \text{ sec}} \times \frac{1000 \text{ grams}}{\text{liter}} \times \frac{\text{gmol}}{18 \text{ grams}} \times \frac{\text{cm}^2 \cdot \text{sec}}{5.7 \times 10^{-5} \text{ gmol}}$$

$$= 3.1 \times 10^6 \text{ cm}^2 \text{ of membrane is required}$$

Each bundle of reverse osmosis membrane has 300 ft^2 of area. The number of bundles for the first phase is

$$3.1 \times 10^6 \text{ cm}^3 \times \frac{\text{bundle}}{300 \text{ ft}^2} \times \frac{(3.28 \text{ ft})^2}{\text{m}^2} \times \frac{1 \text{ m}^2}{(100 \text{ cm})^2} = \underline{\underline{11.1}}$$

Therefore, 12 bundles are necessary for the first phase.

Each pass has a 50% recovery rate. In order to obtain a 75% recovery rate, the reject stream must be passed through a second reverse osmosis unit. A 95% rejection rate of the salt is expected. Therefore, the concentration in the reject can be calculated

$$\text{Concentration reject} = 0.95 \times 5000 \text{ mg/L} \times 2 = 9,500 \text{ mg/L}$$

Determine the osmotic pressure of the reject stream of the first reverse osmosis unit using the van't Hoff equation.

$$\pi = \phi_c N C_s RT$$

$$C_s = 9,500\text{mg}/\text{L} = \frac{9.5\text{g}/\text{L}}{58.45 \text{ g}/\text{mol}} = 0.16 \text{ gmol}/\text{L}$$

$$\pi = 0.92 \times 2 \times 0.16\frac{\text{gmol}}{\text{L}} \times 0.082\frac{\text{atm} \cdot \text{L}}{\text{gmolK}} \times 298.2 \text{ K} = 7.2 \text{ atm}$$

$$7.2 \text{ atm} \left(\frac{14.7 \text{ psi}}{\text{atm}} \right) = \underline{\underline{106 \text{ psi}}}$$

The water flow through the membrane is calculated by

$$J_w = w_p(\Delta P - \Delta \pi)$$

$$= 1.9 \times 10^6 \frac{\text{gmol}}{\text{cm}^2 \cdot \text{sec} \cdot \text{atm}} (500 \text{ psi} - 106 \text{ psi}) \left(\frac{1 \text{ atm}}{14.7 \text{ psi}} \right)$$

$$= \underline{\underline{5.1 \times 10^{-5}}}\text{gmol/cm}^2 \cdot \text{sec}$$

To treat the 25 gpm reject stream the number of bundles required is found by the following calculations:

$$25\frac{\text{gal}}{\text{min}} \times 3.79\frac{\text{liters}}{\text{gallon}} \times \frac{1 \text{ min}}{60 \text{ sec}} \times \frac{1000 \text{ grams}}{\text{liter}} \times \frac{\text{gmol}}{18 \text{ grams}} \times \frac{\text{cm}^2 \cdot \text{sec}}{5.7 \times 10^{-5}\text{gmol}}$$

$$= 1.54 \times 10^6 \text{cm}^2 \text{of membrane is required}$$

$$1.54 \times 10^6 \text{ cm}^3 \times \frac{\text{bundle}}{300 \text{ ft}^2} \times \frac{(3.28 \text{ ft})^2}{\text{m}^2} \times \frac{1 \text{ m}^2}{(100 \text{ cm})^2} = \underline{\underline{5.52}}$$

Therefore, 6 bundles are necessary for the second pass.

Ultrafiltration.   In ultrafiltration, because solute is rejected (and accumulated) at the membrane, a gel layer may form a secondary membrane. The solvent flux may be expressed as:[171]

$$J_w = \frac{(\Delta P - \Delta \pi)}{(R_g + R_m)} \tag{9-78}$$

where J_w = water flux through the membrane (gmol/cm^2 · sec)
   ΔP, $\Delta \pi$ have units dynes/cm^2
   R_g = resistance due to the gel layer (g · cm)/(gmol · sec)
   R_m = membrane resistance (g · cm)/(gmol · sec)

Osmotic pressure can usually be neglected and the equation reduced to:

$$J_w = \Delta P /(R_g + R_m) \tag{9-79}$$

For dilute solutions of large solutes, R_g tends to be much less than R_m and may be neglected. When this is not the case, however, concentration polarization can ccur and R_g will control (see Design Considerations for a discussion of concentration olarization).

Design Considerations

'oncentration polarization is a phenomenon that occurs in all three membrane proesses described in this section. In general, concentration polarization results in an icreased power requirement without an appreciable performance enhancement.

In electrodialysis, there is a value of current that produces a depletion of ions n the feed side of the membrane because the ions are being transferred through the iembrane faster than they can be transferred from the membrane to the solution. An icrease in resistance occurs not only because there are no ions to transfer the curnt at the membrane boundary, but also because the excess current is used to ionize ater ("water splitting") into H^+ and OH^- ions and the increased OH^- concentraon (increased pH) on the concentrating side of the membrane results in scaling by recipitation of pH-sensitive substances, such as calcium carbonate.[172]

In reverse osmosis, concentration polarization occurs because the solute builds ɔ on the feed side of the membrane, thus exceeding the bulk concentration. As a .sult, the local osmotic pressure increases, necessitating higher system pressure to 'oduce the same water recovery, and the transfer of solute through the membrane ineases, producing a less pure water. In addition, the membrane may deteriorate faster certain substances accumulate on the membrane surface and the water recovery may crease if precipitation of sparingly soluble salts occurs on the membrane surface.[173]

Finally, concentration polarization occurs in ultrafiltration because of the buildup ' macrosolutes on the membrane surface. These macrosolutes have very low diffusion efficients and, thus, their diffusion back into the bulk liquid is slow. Eventually, a gel yer develops on the feed side of the membrane and the process becomes controlled ' mass transfer and independent of the applied pressure.[174]

The methods to control concentration polarization include the use of a highed velocity or turbulent flow (all membrane processes), lower current density to feed ncentration ratio (electrodialysis), lower recovery (reverse osmosis), and higher feed nperature (ultrafiltration).[175]

Fouling of membranes can be reduced by periodic cleaning with alkalis or acids d, in the case of electrodialysis and reverse osmosis, reversal of flow. Pretreatment the feed is also useful to reduce membrane fouling and may include filtration; ftening; adsorption; pH control; and dispersants or surfactants addition to prevent aling or wetting of membranes by oil emulsions, respectively.

Conlon[176] notes several design restrictions for electrodialysis systems:

The ion concentration ratio (concentrate versus product water) should be below 150.

The voltage applied to the membrane stack should be below 80% of the limiting voltage (i.e., where excess current will travel through adjacent membranes and generate enough heat to damage membranes).

The maximum current density should be 70% of the limiting current density.

- Reversal of the direct current on a timed cycle (30 seconds every 20–30 minutes) will reduce fouling of the membranes.

Other requirements for electrodialysis systems include the use of membranes and electrodes that function adequately in both current flow directions. In addition, because the size of the unit and the energy consumption are dependent on the total dissolved solids (TDS) concentration, electrodialysis systems are generally used for low TDS concentration feeds (less than 5,000 mg/L), unless the recovered product is valuable enough to offset the costs.[177]

Additional design considerations for reverse osmosis systems include the feed temperature and pH. High feed temperatures and pH values may result in membrane hydrolysis and, thus, reduced performance. However, within the operating range of temperature and pH values, increased temperature results in an increased water flux through the membrane with a constant solute rejection.[178]

For effective separation in ultrafiltration, the solute molecule should be at least one order of magnitude larger than the solvent molecule. Ultrafiltration membranes are specified in terms of a molecular weight cutoff (MWC), a value representing the molecular weight that results in 90% of the solute of that size in a very dilute solution being retained by the membrane.[179] However, the MWC is simply a theoretical value useful to rank membranes. Other solute characteristics (i.e., shape and affinity for the membrane) and the operating conditions (i.e., whether a layer of solute has formed at the membrane surface or fouling has occurred) also play a role in the retention capability of the membrane.[180]

DISCUSSION TOPICS AND PROBLEMS

9-1. Derive Eq. 9-20 starting with Eq. 9-16.

9-2. Provide a preliminary design of an air stripping column to remove toluene from ground water. Levels of toluene range from 0.1 to 2.1 mg/L and this must be reduced to 50 μg/L. A hydrogeologic study of the area indicates that a flow rate of 110 gal/min is required to ensure that contamination not spread. Laboratory investigations have determined the overall transfer constant, $K_L a = 0.020$ sec^{-1}. Use a column diameter of 2.0 feet and an air to water ratio of 15. Specifically determine: Liquid loading rate, stripping factor, height of the tower and provide a sketch of the unit indicating all required appurtenances.

9-3. Using the data in problem 9-2, determine the pressure drop through the tower.

9-4. Using the data in problem 9-2, determine the impact on effluent quality by varying the air to water ratio and the packing height.

9-5. Determine the overall mass transfer coefficient, $K_L a$, given the following data from an air stripping column: contaminant = tetrachloroethene (C_2Cl_4); liquid mass loading rate = 10.2 kg/m^{2}s; packing = 1$''$ polyethelene Tri-paks®; air mass loading rate = 1.5 kg/m^{2}s; temperature = 6°C; diffusion coefficient in water = 1.1 $\times 10^{-6}$ cm^2/s

9-6. Recalculate $K_L a$ from Example 9-1 incorporating the following changes: H' is given as 0.0704 (dimensionless) at a temperature of 6°C. $a_t = 138$ m^2/m^3.

9-7. Design an air stripping column to remove TCE from water. the initial concentration of TCE is 1.3 mg/L and this must be reduced to 75 μg/L. Use the following criteria: water flow rate: 350 gal/min; water temperature: 16°C.; air temperature: 23°C; packing: use 2" Intalox saddles

9-8. Demonstrate that the stripping factor used in this book is numerically similar to that in common use by chemical engineers:

$R = H'G/L$

$H' = $ Henry's constant (dimensionless)

$G = $ gas loading rate (kmol/hr)

$L = $ liquid loading rate (kmol/hr)

9-9. Soil on a five-acre site has become contaminated with a mixture of volatile organic compounds. Prepare a brief outline indicating what type of data you would need to collect to begin analyzing remedial measures.

9-10. Use Hartley's method to estimate the rate of volatilization of trichloroethane. Use an estimated diffusion coefficient of 0.80 cm^2/s

9-11. Repeat problem 9-10 for benzene. Use an estimated diffusion coefficient of 0.087 cm^2/s

9-12. Repeat problem 9-10 for toluene. Use an estimated diffusion coefficient of 0.078 cm^2/s

9-13. Briefly describe the alternative processes that could be considered for removing volatile organics from soils.

9-14. An underground storage tank (UST) containing trichloroethane at a chemical plant has ruptured. The plume currently extends west to a distance of 345 feet and is 120 feet wide at this distance. The depth of the plume is 12 feet and the seasonal high ground water table is at a depth of 7 feet. Provide a preliminary design and layout of a soil vapor extraction system. Specifically determine:

(*a*) The required number of extraction wells and their layout.

(*b*) The volumetric capacity in cubic feet per minute (CFM) and operating system pressure of the blower.

(*c*) Equipment layout and any required ancillary facilities.

State any assumptions made.

9-15. Identify additional parameters other than those presented in Table 9-3 that might affect SVE system performance. Further, describe how these parameters influence system performance. (Refer to Lyman, et al., Reference 25, for additional information.)

9-16. 100 mL of a solution with a TOC (total organic carbon) concentration of 0.5% is placed in each of five containers with activated carbon and shaken for 48 hours. The samples are filtered and the concentration of TOC measured, yielding the following analyses:

Container:	1	2	3	4	5
Carbon (grams):	10	8	6	4	2
TOC (mg/L):	42	53	85	129	267

Determine the Freundlich constants, K and n, and plot the isotherm.

9-17. 200 mL of a solution with a para-xylene concentration of 500 mg/L is placed in each of six containers with activated carbon and shaken for 24 hours. The samples are filtered and the concentration of p-xylene measured, yielding the following analyses:

Container:	1	2	3	4	5	6
Carbon (grams):	24	20	16	12	8	4
p-Xylene(mg/L):	10.7	14.6	23	29	48	107

Determine the Freundlich constants, K and n, and plot the isotherm.

9-18. Estimate the daily carbon utilization to remove chlorobenzene from 1.0 MGD of ground water saturated with chlorobenzene. Assume a chlorobenzene concentration of 5 mg/L is acceptable for discharge to a POTW.

9-19. Provide a preliminary design of a carbon adsorption system for removal of 2,4,6 trichlorophenol from 250,000 gal/day of water. The following data is provided: Bohart-Adams model: $a = 2.3$ days/ft; $b = -10$ days in laboratory tests where trichlorophenol concentration was reduced from 395 mg/L to 10 mg/L at a loading of 4.0 gal/ft^2·min. The adsorption zone was 19.0 feet.

9-20. A ground water aquifer is contaminated with benzene. Contrast the relative effectiveness of air stripping to activated carbon adsorption by providing a preliminary design of each process.

9-21. Using the information in Example 9-9, determine the Bohart-Adams coefficients N and K.

9-22. Revise Example 9-9 for a flow rate of 250,000 gallons per day.

9-23. Revise Example 9-9 for a contaminant concentration in ground water of 239 mg/L.

9-24. Derive the conversion factors F_1 and F_2 in the Bohart-Adams model.

9-25. Derive a simplified Nernst equation (Eq. 9-56) from Eq. 9-53.

9-26. A plating waste with a flow of 50 L/min contains 274 mg/L of NaCN. Determine the amounts of chlorine as $Ca(OCl)_2$ and NaOH required to:

(*a*) Oxidize the cyanide to cyanate; and

(*b*) Oxidize the cyanide to CO_2 and H_2O.

9-27. A plating waste with a flow of 40,000 gal/day contains 274 mg/L of NaCN. Determine the amounts of chlorine as Cl_2 and NaOH required to oxidize the cyanide to CO_2 and N_2.

9-28. Demonstrate that reactions 9-65 and 9-66 are equivalent to reaction 9-64.

9-29. An industrial waste stream contains 650 mg/L of hexavalent chromium at a flow rate of 35 L/min. Determine the amount of sulfur dioxide required to reduce the chromium to the less toxic trivalent form.

9-30. A waste contains 150 mg/L of copper at a flow rate of 75 L/min. Determine the amount of sodium borohydride required to treat this waste.

9-31. A waste contains 250 mg/L of cadmium and 450 mg/L lead at a flow rate of 45,000 gal/day. Determine the amount of sodium borohydride required to treat this waste.

9-32. Given the following octanol water partition coefficients, K_{ow}, indicate for which compounds vacuum extraction is a candidate process for removing the compound from the vadose zone of contaminated soil.

Compound	Log K_{ow}
Trichloroethene	2.38
Pyrene	4.88
Polychlorinated biphenyl congeners	6.01
Vinyl chloride	1.38
Ethyl benzene	3.15
Phenol	1.46
Lindane	3.90
DDT	6.19
2-Butanone (methyl ethyl ketone)	0.26

9-33. Provide a preliminary design of an electrodialysis unit to treat a plating waste with the following characteristics: Ni concentrations = 2.1% as $NiSO_4$; flow rate = 20 gpm. Determine the number of membranes, power required, and annual electrical cost @ $0.07/kWh. The effluent Ni concentration must be less than 500 mg/L.

9-34. Determine if the following mixtures of organics in wastewater are good condidates for steam stripping. State reason. Apply the OCPSF discharge limitations as criteria for the steam stripper bottoms concentrations.

Mixture A	Mixture B	Mixture C
93 mg/L benzene	133 mg/L phenol	56 mg/L ethanol
127 mg/L acetone	27 mg/L nitrotoluene	182 mg/L acetone
	42 mg/L methanol	158 mg/L methyl ethyl ketone

9-35. Show, through material balances, that the ratio of liquid to gas (L/G) in the stripping section of a steam stripper is higher than in the rectifying section. State any necessary assumptions or givens.

9-36. Using the equilibrium curve in Fig. 9-24, and given the following equations for operating lines:

$$\text{Stripping section: } y = 0.73x$$
$$\text{Rectifying section: } y = 1.65x + 0.39$$

Use the McCabe-Thiele method to estimate the number of theoretical trays. Assume: $x_B = 0.05$; $x_F = 0.35$; $x_O = 0.9$

REFERENCES

1. Perry, R. H., and D. W. Green, eds.: *Perry's Chemical Engineers' Handbook*, 6th ed., McGraw-Hill, Inc., New York, NY, pp. 18–32, 1984.
2. Ibid., pp. 18–33.
3. Ibid., pp. 18–34.
4. Onda, K., H. Takeuchi, and Y. Okumoto: "Mass Transfer Coefficients Between Gas and Liquid Phases in Packed Columns," *Journal Chemical Engineering of Japan*, Vol. 1, p. 58, 1968
5. Cornwell, D. A.: "Air Stripping and Aeration," in *Water Quality & Treatment* (F. W. Pontious, ed.), McGraw-Hill, 1990.
6. McCabe, W. L., J. C. Smith, and P. Harriot: *Unit Operations of Chemical Engineering*, 4th ed., McGraw-Hill, Inc., New York, NY, 1985, p. 466.
7. Ibid., p. 622.
8. Eckert, J. S.: "Selecting the Proper Distillation Column Packing," *Chem. Engr. Prog.*, vol. 66, pp. 39–44, 1970.
9. Haarhoff, J., and D. Schoeller: "Theory and Design of Countercurrent Packed Aeration Towers," Airstrip, Ames, IA, 1988.
10. Perry, R. H., and D. W. Green, eds.: *Perry's Chemical Engineers' Handbook*, pp. 18–21 to 18–23.
11. Treybal, R. E.: *Mass Transfer Operations*, McGraw-Hill, Inc. New York, NY, pp. 196ff, 1980.
12. Haarhoff, J., and D. Schoeller, "Theory and Design of Countercurrent Packed Aeration Towers"
13. Haarhoff, J., and J. L. Cleasby: "Evaluation of Air Stripping for the Removal of Organic Drinking Water Contaminants," *Water South Africa*, vol. 16, pp. 13–22, 1990.
14. Hiller, D., and H. Gudeman: "Analysis of Vapor Extraction Data from Applications in Europe," Proceedings of the Third International Conference on New Frontiers in Hazardous Waste Management, U.S. EPA, pp. 434–441, September 1989.
15. Applegate, J., John K. Gentry, and James J. Malot: "Vacuum Extraction of Hydrocarbons from Subsurface Soils at a Gasoline Contamination Site," in Superfund '87, proceedings of the 8th National Conference, The Hazardous Materials Control Research Institute, Greenbelt, MD, pp. 273–279, 1987.
16. Hiller, D., and H. Gudeman: "Analysis of Vapor Extraction Data from Applications in Europe."
17. Martin, L. M., Richard J. Sarnelli, and Matthew T. Walsh: "Pilot Scale Evaluation of Groundwater Air Sparging: Site-Specific Advantages and Limitations," Proceedings of the National R&D Conference

on the Control of Hazardous Materials, Hazardous Materials Control Research Institute, pp. 318–327, February 4–6, 1992.

18. Anon.: "In Situ Vapor Stripping: Avoiding Problems Due to Simplifying Assumptions," *The Hazardous Waste Consultant,* pp. 1–5, September/October 1990.

19. Grasso, Dominic: "Soil Vapor Extraction (Module)," *Hazardous Waste Site Remediation,* The University of Connecticut under a grant from the U.S. Environmental Protection Agency, January, 1993.

20. Multch, Robert D, Jr., Ann N. Clarke, James H. Clarke, and David J. Wilson: "In Situ Vapor Stripping: Preliminary Results of a Field-Scale U.S.EPA/Industry Funded Research Project," Superfund '89, proceedings of the 10th National Conference, The Hazardous Materials Control Research Institute, Greenbelt, MD, November 1989.

21. Grasso, Dominic, "Soil Vapor Extraction (Module)."

22. Ibid., p. 9.

23. Millington, R. J., and J. M. Quirk: "Permeability of Porous Solids," *Trans. Faraday Soc.,* vol. 57, p. 1200, 1961.

24. Dragun, James: *The Soil Chemistry of Hazardous Materials,* Hazardous Materials Control Research Institute, Washington, D.C., p. 271, 1988.

25. Lyman, Warren J., Warren F. Reehl, and David H. Rosenblatt: *Handbook of Chemical Property Estimation Methods,* McGraw-Hill Book Company, New York, NY, 1982.

26. SITE Program Demonstration Test, EPA 540 5–89-003a, U.S. Environmental Protection Agency, April, 1989.

27. Thomas, R. G.: "Volatilization from Soil," in W. J. Lyman, W. F. Reehl, and D. H. Rosenblatt, *Handbook of Chemical Property Estimation Methods,* McGraw-Hill Book Company, New York, NY, 1982.

28. Verschueren, Karel: *Handbook of Environmental Data on Organic Chemicals,* 2nd ed., Van Nostrand Reinhold Co., Inc., New York, NY, p. 1132, 1983.

29. Lyman, Warren J., William F. Reehl, and David H. Rosenblatt: *Handbook of Chemical Property Estimation Methods,* Environmental Behavior of Organic Compounds, McGraw Hill, Inc., New York, NY, pp. 16–35, 1982.

30. Dean, J. A. (ed.): *Lange's Handbook of Chemistry, 12th Ed.* McGraw-Hill, 1979.

31. Verschueren, Karel: *Handbook of Environmental Data on Organic Chemicals,* 2nd ed., Van Nostrand Reinhold Co., Inc., New York, NY, 1983.

32. Fiss, Edward C., and Michael O. Smith: "The Application of Vapor Extraction for the Removal of Volatile Organic Compounds in Soils," presented at Aquifer Reclamation and Source Control Conference, New Jersey Institute of Technology, Newark, NJ, June 1991.

33. Ibid.

34. U.S. Army Toxic and Hazardous Materials Agency: "Installation Restoration General Environmental Technology Development Contract DAAK11-82-C-0017, Final Report," Task 11, In Situ Air Stripping of Soils Pilot Study, October 1985.

35. Fiss, E. C. and M. O. Smith, "The Application of Vapor Extraction for the Removal of Volatile Organic Compounds in Soil."

36. Anon.: "In Situ Air Stripping of Soils Pilot Study, Installation Restoration General Environmental Technology Development Contract DAAK11-82-C-0017, Final Report," prepared for U.S. Army Toxic and Hazardous Materials Agency, October 1985.

37. Ibid.

38. Buscheck, T. E., and R. G. Peargin: "Summary of a Nationwide Vapor Extraction System Performance Study," Proceedings National Water Well Association., Houston, pp. 205–219, 1991.

39. Chevron Corp: "General Procedures and Data Interpretation for Vapor Extraction System Pilot Tests," Chevron Corp., Houston, 1992.

40. Metcalf and Eddy, Inc.: *Wastewater Engineering,* McGraw-Hill, Inc., New York, NY, p. 316, 1990.

41. Anon.: "Activated Carbon Application Bulletin 20-4b," Calgon Corp., 1973.

42. Weber, Jr., W. J.: "Evolution of a Technology," *Journal of Environmental Engineering,* ASCE, vol. 110, no. 5, pp. 899–917, 1984.

43. Metcalf & Eddy, Inc., *Wastewater Engineering,* p. 321.

44. Hutchins, R. A.: "New Method Simplifies Design of Activated Carbon Systems," *Chem. Engr.*, vol. 80, pp. 133–138, Aug. 23, 1974.
45. Bohart, G. S., and E. Q. Adams: "Some Aspects of the Behavior of Charcoal with Respect to Chlorine," *J. Amer. Chem. Soc.*, vol. 42, pp. 523–544, 1920.
46. U.S. EPA: "Development Document for Effluent Guidelines and Standards for the OCPSF Source Category, Volume III (Best Available Technology)," EPA 440/1-83/009b.
47. U.S. EPA: "Organic Chemicals and Plastics and Synthetic Fibers Category Effluent Limitations Guidelines, Pretreatment Standards, and New Source Performance Standards," 40 CFR Parts 414 and 416, *Federal Register*, vol. 52, no. 214 pp. 42522–42584, November 5, 1987.
48. Treybal, R. E.: *Mass Transfer Operations*, Second Edition, McGraw Hill Book Company, New York, NY, 1968.
49. McCabe, W. L., J. C. Smith, and P. Harriott: *Unit Operations of Chemical Engineering*, 4th ed., McGraw Hill, Inc., New York, NY, 1985.
50. King, C. J.: *Separation Processes*, McGraw Hill, Inc., New York, NY, 1980
51. Perry & Green: *Perry's Chemical Engineers' Handbook*.
52. Smith, J. M., and H. C. Van Ness: *Introduction to Chemical Engineering Thermodynamics*, 4th ed., McGraw-Hill, New York, NY, 1987.
53. Balzhiser, Samuels, and Eliassen: *Chemical Engineering Thermodynamics*, Prentice-Hall, Inc., Engelwood Cliffs, NJ, 1972.
54. Reid, R. C., Prausnitz, and Poling: *The Properties of Gases and Liquids*, 4th ed., McGraw Hill Book Company, New York, NY, 1987.
55. Ibid, p. 254 ff.
56. DECHEMA Chemistry Data Series: *Vapor-Liquid Equilibrium Data Collection, Volume 1, Parts 1 and 1a*, DECHEMA, Frankfurt, 1981.
57. DECHEMA Chemistry Data Series: *Vapor-Liquid Equilibrium Data Collection, Volume I, Part 1b*, DECHEMA, Frankfurt, 1986.
58. DECHEMA Chemistry Data Series: *Activity Coefficients at Infinite Dilution, Volume IX, Parts 1-2*, DECHEMA, Frankfurt, 1986.
59. Pinal, R., Suresh, Rao, Lee, Cline, and Yalkowsky, "Cosolvency of Partially Miscible Organic Solvents on the Solubility of Hydrophobic Organic Chemicals," *Environmental Science and Technology*, vol. 24, No. 5 (1990), pp. 639–647.
60. DECHEMA Chemistry Data Series: *Liquid-Liquid Equilibrium Data Collection, Volume V, Part 1*, DECHEMA, Frankfurt, 1979
61. Wankat, Phillip C.: *Equilibrium Staged Separations*, Elsevier, New York, NY, pp. 157–198.
62. U.S. Environmental Protection Agency: "Technical Resource Document—Treatment Technologies for Halogenated Organic Containing Waste," U.S. EPA PB88-131271, Dec. 1987.
63. Bernardin, F. E., and E. Froelich: "Chemical Oxidation of Dissolved Organics Using Ultraviolet-Catalyzed Hydrogen Peroxide", Superfund '90, Haz. Materials Control Research Instit., Washington, D.C., pp. 768–771, 1990.
64. Ibid., p. 770.
65. Fochtman, E. G.: "Chemical Oxidation And Reduction," *Standard Handbook of Hazardous Waste Treatment and Disposal*, Harry M. Freeman, ed., McGraw-Hill, Inc., New York, NY, pp. 7.41ff, 1989.
66. Glaze, W. H.: "Chemical Oxidation in American Water Works Association," *Water Quality and Treatment*, 4th ed., F. W. Pontius, ed., McGraw-Hill, Inc., New York, NY, pp. 747ff, 1990.
67. Eisenhauer, H. R.: "The Ozonation of Phenolic Wastes," *J. WPCF*, vol. 40, pp. 1887–1889, 1968.
68. Fochtman, E. G., "Chemical Oxidation And Reduction."
69. Eckenfelder, W. W.: *Industrial Water Pollution Control*, McGraw-Hill, Inc., New York, NY, p. 301, 1989.
70. Glaze, W. H.: "Chemical Oxidation in American Water Works Association."
71. Eckenfelder, W. W.: *Industrial Water Pollution Control*, op. cit., p. 306.
72. Posselt, H. S., and W. J. Weber: "Chemical Oxidation," in W. J. Weber, *Physicochemical Processes for Water Quality Control*, Wiley-Interscience, New York, NY, pp. 398, 391, 1972.
73. Patterson, J. W.: *Industrial Wastewater Treatment Technology*.

74. Posselt, H. S., and W. J. Weber: "Chemical Oxidation."
75. Sundstrom, D. W., et al.: "Destruction of Halogenated Aliphatics by Ultraviolet Catalyzed Oxidation with Hydrogen Peroxide," *Hazardous Waste & Hazardous Materials*, vol. 3, pp. 101–110, 1986.
76. Weber, W. J., and O. Corapcioglu: "Catalytic Oxidation of Cyanides," Proceedings of the 36th Industrial Waste Conference, Purdue University, Ann Arbor Sci. Publ., Ann Arbor, MI, 1982.
77. Fochtman, E. G.: "Chemical Oxidation and Reduction."
78. Ibid, p. 7.43.
79. Lindsay, M. J., M. E. Hackman, and R. J. Tremblay: "Sodium Borohydride Reduces Hazardous Waste," Proceedings of the 40th Industrial Waste Conference, Purdue University, Ann Arbor Sci. Publ., Ann Arbor, MI, p. 479, 1985.
80. Fochtman, E. G.: "Chemical Oxidation and Reduction," p. 7.49.
81. Weast, R. C., ed.: *Handbook of Chemistry and Physics*, CRC Press, p. D-156ff, 1984.
82. Snoeyink, V., and D. Jenkins: *Water Chemistry*, Wiley, pp. 328–329, 1980.
83. Posselt, H. S., and W. J. Weber: "Chemical Oxidation," in Weber, W. J., *Physiocochemical Processes for Water Quality Control*, Wiley-Interscience, New York, NY, pp. 398, 391, 1972.
84. Weber, W. J., and O. Corapcioglu: "Catalytic Oxidation of Cyanides," in Proceedings of the 36th Industrial Waste Conference, Purdue University, Ann Arbor Sci. Publ., Ann Arbor, MI, 1982.
85. Morrison, R. T., and R. N. Boyd: *Organic Chemistry*, 3rd ed., Allyn & Bacon, p. 218, 1973.
86. Eckenfelder, W. W., *Industrial Water Pollution Control*, op. cit., p. 301.
87. Glaze, W. H.: "Drinking Water Treatment with Ozone," *Environ. Sci. Tech*, vol. 21, p. 227, 1987.
88. Ibid.
89. Staehelin, J., and J. Hoigne: "Decomposition of Ozone in Water in the Presence of Organic Solutes Acting as Promoters and Inhibitors of Radical Chain Reaction," *Environ. Sci. Tech.*, vol. 19, p. 1206, 1985.
90. Glaze, W. H.: "Drinking Water Treatment with Ozone."
91. Hoigne, J., and H. Bader: "Rate Constants of Reactions of Ozone with Organic and Inorganic Compounds in Water," *Water Research*, vol. 17, pp. 173–183, 1983.
92. Eisenhauer, H. R.: "The Ozonation of Phenolic Wastes," *J. WPCF*, vol. 40, pp. 1887–1889, 1968.
93. Miller, G. W., et al.: *An Assessment of Ozone and Chlorine Dioxide Technologies for Treatment of Municipal Water Supplies*, EPA 600/2-78-147, [NTIS No. PB-285972], 1978.
94. Eisenhauer, H. R.: "The Ozonation of Phenolic Wastes."
95. Posselt, H. S., and W. J. Weber: "Chemical Oxidation," op. cit., p. 368.
96. Glaze, "Drinking Water Treatment with Ozone."
97. Patterson, J. W., *Industrial Wastewater Treatment Technology*, p. 14, Butterworths Publishers, Stoneham, MA, 1985.
98. Watts, R. J., M. D. Udell, and P. A. Rauch: "Treatment of Pentachlorophenol-Contaminated Soils Using Fenton's Reagent," *Hazardous Waste & Hazardous Materials*, vol. 7, pp. 335–345, 1990.
99. Ibid.
100. Schmitt, R. J.: "Combining Chromium Treatment with Hydrogen Peroxide Oxidation of Phenolics," Proceedings of the 36th Industrial Waste Conference, Purdue University, Ann Arbor Sci. Publ., Ann Arbor, MI, pp. 375–382, 1982.
101. Ibid., p. 379.
102. Barich, J. T., and J. D. Zeff: "A Review of Ultrox, Ultraviolet Oxidation Technology," Superfund '89, Haz. Materials Control Research Instit., Washington, D.C., pp. 264–266, 1989.
103. Sundstrom, D. W., et al., "Destruction of Halogenated Aliphatics by Ultraviolet Catalyzed Oxidation with Hydrogen Peroxide," *Hazardous Waste & Hazardous Materials*, vol. 3, pp. 101–110, 1986.
104. Eckenfelder, W. W., *Industrial Water Pollution Control*, op. cit., p. 308.
105. Schmitt, R. J.: "Combining Chromium Treatment with Hydrogen Peroxide Oxidation of Phenolics."
106. Haas, C. N.: "Disinfection," in American Water Works Association," *Water Quality and Treatment*, 4th ed., F.W. Pontius, ed., McGraw-Hill, Inc., New York, NY, p. 883, 1990.
107. Milne, D.: "Chemistry of Waste Cyanide Treatment," *Sewage & Industrial Wastes*, vol. 23, p. 174, 1951.
108. Posselt, H. S., and W. J. Weber: "Chemical Oxidation," op. cit., p. 396.

109. Eckenfelder, W. W., *Industrial Water Pollution Control*, op. cit., p. 310.

110. Thomason, T. B., and M. Modell: "Supercritical Water Destruction of Aqueous Wastes," *Hazardous Wastes*, vol. 1, p. 459, 1984.

111. McHugh, M. A., and V. J. Krukonis: *Supercritical Fluid Extraction*, Butterworths Publishers, Stoneham, MA, p. 131, 1986.

112. Modell, M.: "Supercritical-Water Oxidation," *Standard Handbook of Hazardous Waste Treatment and Disposal*, Freeman, H.M., editor in chief, McGraw-Hill, Inc., New York, NY, pp. 8.153 to 8.168, 1989.

113. Thomason, T. B., and M. Modell: "Supercritical Water Destruction of Aqueous Wastes."

114. Modell, M.: "Supercritical-Water Oxidation," p. 8.159.

115. Perry, R. H., and D. W. Green, eds.: *Perry's Chemical Engineers' Handbook*, op. cit., p. 3–111.

116. Smith, J. M., and H. C. Van Ness: *Introduction to Chemical Engineering Thermodynamics*, 4th ed., McGraw-Hill, Inc., New York, NY, p. 86, 1987.

117. Josephson, J.: "Supercritical Fluids," *Environ. Sci. & Tech.*, vol. 16, p. 549A, 1982.

118. McHugh, M. A., and V. J. Krukonis: *Supercritical Fluid Extraction*, Butterworths Publishers, Stoneham, MA, p. 5, 1986.

119. Reid, R. C., J. M. Prausnitz, and T.K. Sherwood: *The Properties of Gases and Liquids*, 3rd ed., McGraw-Hill, Inc., New York, NY, p. 422, 1977.

120. Thomason, T. B., and Modell, M., "Supercritical Water Destruction of Aqueous Wastes," p. 459.

121. Treybal, R. E.: *Mass Transfer Operations*, 3rd ed., p. 488, McGraw-Hill, New York, NY.

122. Perry, R. H., and D. W. Green, eds.: *Perry's Chemical Engineers' Handbook*, op. cit., p. 15-5.

123. Thomason, T. B., and M. Modell, "Supercritical Water Destruction of Aqueous Wastes," p. 462.

124. Modell, M., op cit, p. 154.

125. Modell, M.: "Supercritical-Water Oxidation," *Standard Handbook of Hazardous Waste Treatment and Disposal*, Freeman, H. M., editor in chief, McGraw-Hill, Inc., New York, NY, pp. 8.153 to 8.168, 1989.

126. "Detoxification and Disposal of Hazardous Organics by Processing in Supercritical Water," Final Report, U.S. Army Contract No. DAMD 17-800-C-0078, prepared by MODAR, Inc., Houston, Texas, 1987.

127. Ibid.

8–138. Modell, M., *Processing Methods for the Oxidation of Organics in Supercritical Water*, U.S. Patent 4,543,190, September 1985.

139. "Detoxification and Disposal of Hazardous Organics by Processing in Supercritical Water," Final Report, U.S. Army Contract No. DAMD 17-800-C-0078, prepared by MODAR, Inc., Houston, Texas, 1987.

140. Modell, 1985.

141. "Detoxification and Disposal of Hazardous Organics by Processing in Supercritical Water," Final Report, U.S. Army Contract No. DAMD 17-800-C-0078, prepared by MODAR, Inc., Houston, Texas, 1987.

142. Thomason, T. B., and M. Modell, "Supercritical Water Destruction of Aqueous Wastes," *Hazardous Waste*, vol. 1, no. 4, p. 453, 1984.

143. Ibid.

144. Ibid.

145. Modell, M.: "Supercritical Water Oxidation," op cit., p. 8.163.

146. Gregor, H. P., and C. D. Gregor: "Synthetic Membrane Technology," *Scientific American*, pp. 112–128, July 1978.

147. Palmer, S. A. K., M. A. Breton, T. J. Nunno, D. M. Sullivan, and N. F. Surprenant: "Technical Resource Document: Treatment Technologies for Metal/Cyanide Containing Wastes, Volume III," U.S. Environmental Protection Agency, NTIS PB88-143896, pp. 6–9, 1987.

148. Perry, R. H., and D. W. Green, eds.: *Perry's Chemical Engineers' Handbook*, pp. 17–37.

149. MacNeil, J., and D. E. McCoy: "Membrane Separation Technologies," in *Standard Handbook of Hazardous Waste Treatment and Disposal*, Freeman, H.M. (ed.), McGraw-Hill, Inc., New York, NY, p. 6–103, 1989.

150. "Membrane Separation Technology: Applications to Waste Reduction and Recycling," *The Hazardous Waste Consultant*, vol. 3, no. 3, McCoy & Assoc., Lakewood, CO, p. 4–21, 1985.

151. MacNeil, J., and D. E. McCoy, "Membrane Separation Technologies."

152. "Membrane Separation Technology: Applications to Waste Reduction and Recycling," *The Hazardous Waste Consultant*.

153. Applegate, Lynn E., "Membrane Separation Processes," *Chemical Engineering*, June 11, 1984, pp. 64–89.

154. Katz, William E., "State-of-the-Art of the Electrodialysis Reversal (EDR) Process," *Industrial Water Engineering*, vol. 21, no. 1, pp. 12–20.

155. Anon.: "Aquatech Acid Recovery System," *Chemical Engineering*, December, 1989.

156. Mani, K. N.: "Electrodialysis Water Splitting Technology," Aquatech Systems, 1985.

157. Eykamp, W., and Steen, J.: "Ultrafiltration and Reverse Osmosis," in *Handbook of Separation Process Technology*, Rousseau, R.W. (ed.), Wiley-Interscience, New York, NY, p. 836, 1987.

158. Conlon, W. J.: "Membrane Processes," *Water Quality and Treatment*, 4th ed., American Water Works Assoc., p. 714, 1990.

159. Eykamp, W., and J. Steen, "Ultrafiltration and Reverse Osmosis," p. 838.

160. MacNeil, J., and D. E. McCoy, "Membrane Separation Technologies," p. 6–97.

161. Martin, J. F., M. J. Stutsman, and E. Grossman: "Hazardous Waste Volume Reduction Research in the U.S. E.P.A.," *Hazardous and Industrial Solid Waste Minimization Practices, ASTM STP1043*, Amer. Soc. for Testing & Materials, pp. 163–171, 1989.

162. Higgins,T.: *Hazardous Waste Minimization Handbook*, Lewis Publ., Chelsea, MI, p. 99, 1989.

163. Perry, R. H., and D. W. Green, eds.: *Perry's Chemical Engineers' Handbook*, pp. 17–32.

164. Weber, W. J.: *Physicochemical Processes for Water Quality Control*, John Wiley & Sons, Inc., New York, NY, p. 330, 1972.

165. Perry, R. H., and D. W. Green, eds.: *Perry's Chemical Engineers' Handbook*, pp. 17–28.

166. Ibid. pp. 17–32.

167. Rich, G. L.: *Unit Processes of Sanitary Engineering*, John Wiley & Sons, Inc., New York, NY, p. 130, 1963.

168. Perry, R. H., and D. W. Green, eds. *Perry's Chemical Engineers' Handbook*.

169. Perry, R. H., and D. W. Green, eds.: *Perry's Chemical Engineers' Handbook*, op. cit., pp. 17–23.

170. Weber, W. J., and E. H. Smith: "Removing Dissolved Organic Contaminants from Water," *Env. Sci. & Tech*, vol. 20, pp. 970–979, 1986.

171. Perry, R. H., and D. W. Green, eds.: *Perry's Chemical Engineers' Handbook*, op. cit., pp. 17–29.

172. Klein, E., R. A. Ward, and R. E. Lacey: "Membrane Processes—Dialysis and Electrodialysis," in Rousseau, R. W., ed., *Handbook of Separation Process Technology*, Wiley-Interscience, New York, NY, pp. 972–975, 1987.

173. Weber, W. J.: *Physicochemical Processes for Water Quality Control*, pp. 322–323.

174. Eykamp, W., and J. Steen, "Membrane Separation Technologies," p. 827.

175. Weber, W. J.: *Physicochemical Processes for Water Quality Control*, pp. 342, 323, 334.

176. Conlon, W. J.: "Membrane Processes," *Water Quality and Treatment*, 4th ed., American Water Works Assoc., pp. 735ff, 1990.

177. MacNeil, J., and D. E. McCoy: "Membrane Separation Technologies," pp. 6.102 and 6.104.

178. Weber, W. J.: *Physicochemical Processes for Water Quality Control*, p. 320, 319.

179. Eykamp, W., and J. Steen, "Membrane Separation Technologies," p. 829.

180. Ibid.

ADDITIONAL READING

Anon., "Advances in Air Sparging Design," *The Hazardous Waste Consultant*, Jan/Feb, 1993.

Angell, K. G.: "In Situ Remedial Methods: Air Sparging," *The National Environmental Journal*, pp. 20–23, January/February 1992

Applegate, Lynn E.: "Membrane Separation Processes," *Chemical Engineering*, June 11, 1984, pp. 64–89.

Baker, E. G., et al.: "Catalytic Destruction of Hazardous Organics in Aqueous Wastes: Continuous Reactor System Experiments," *Hazard. Waste Hazard. Mater.*, vol. 6, no. 1, p. 87, 1989.

Branscome, M. C., et al.: "Field Assessment of Steam Stripping Volatile Organics from Aqueous Waste Streams," Proceedings of the Thirteenth Annual Research Symposium; Land Disposal, Remedial Action, Incineration and Treatment of Hazardous Wastes, EPA 600/9-87/015, Cincinnati, OH, 1987.

Breton, M., et al.: "Technical Resource Document—Treatment Technologies for Solvent Containing Wastes," EPA/600/2-86/095, October 1986.

Chemical Treatment," vol. 2 in *Innovative Site Remediation Technology*, W. C. Anderson, ed., American Academy of Environmental Engineers, 1994.

DeFillippi, R. P., and M. E. Chung: "Laboratory Evaluation of Critical Fluid Extractions for Environmental Applications," EPA/600/2-85/045 (NTIS No. PB 85-189843), April 1985.

Dobbs, R. A., and J. M. Cohen: "Carbon Adsorption Isotherms for Toxic Organics," EPA-600/8-80-023, 1980.

Fair, G. E., and F. E. Dryden: "Comparison of Air Stripping Versus Stream Stripping for Treatment of Volatile Organic Compounds in Contaminated Ground Water," Superfund '89, Haz. Materials Control Research Instit., Washington, D.C., pp. 558–561, 1989.

Freeman, Harry M., ed.: *Innovative Hazardous Waste Treatment Technology Series—Vol. 2, Physical/Chemical Processes,* Technomic Publishing Co., Lancaster, PA, 1990.

Freeman, Harry M., ed.: *Standard Handbook of Hazardous Waste Treatment and Disposal,* Mc-Graw Hill, Inc., New York, NY, 1989.

Helling, R. K., and J. W. Tester: "Oxidation of Simple Compounds and Mixtures in Supercritical Water: Carbon Monoxide, Ammonia, and Ethanol," *Environmental Science & Technology,* vol. 22, no. 11, pp. 1319–1324, 1988.

Lyandres, S., et al.: "Evaluation of Membrane Processes for the Reduction of Trace Organic Contaminants," *Environ. Prog.,* vol. 8, p. 239, 1989.

Lyman, Warren J., William F. Reehl, and David H. Rosenblatt: *Handbook of Chemical Property Estimation Methods*, Environmental Behavior of Organic Compounds, McGraw Hill, Inc., New York, NY, 1982.

McCabe, W. L., J. C. Smith, and P. Harriot: *Unit Operations of Chemical Engineering*, 4th ed., McGraw-Hill, Inc., New York, NY, 1985.

Niramalakhandan, N., Y. H. Lee, and R. E. Speece: "Designing a Cost-Efficient Air-Stripping Process," *Journal AWWA,* vol. 79, no. 1, p. 56, 1987.

Raghaven, R., et al.: "Cleaning Excavated Soil Using Extraction Agents: A State-of-the-Art Review," Rep. No. EPA/600/2-89/034, 1989.

Shanableh, A., and E. F. Gloyna: "Supercritical Water Oxidation—Wastewaters and Sludges," *Water Science & Technology,* vol. 23, pp. 389–398, 1991.

"Soil Vapor Extraction," vol. 8 in *Innovative Site Remediation Technology*, W. C. Anderson, ed., American Academy of Environmental Engineers, 1994.

"Soil Washing/Soil Flushing," vol. 3 in *Innovative Site Remediation Technology*, W. C. Anderson, ed., American Academy of Environmental Engineers, 1994.

"Solvent/Chemical Extraction," vol. 4 in *Innovative Site Remediation Technology*, W. C. Anderson, ed., American Academy of Environmental Engineers, 1994.

Streebin, L. E., H. M. Schornick, and A. M. Wachinski: "Ozone Oxidation of Concentrated Cyanide Wastewater from Electroplating Operations," Proceedings of the 35th Industrial Waste Conference, Purdue Univ., Ann Arbor Sci. Publ., Ann Arbor, MI, 1983.

Verschueren, Karel: *Handbook of Environmental Data on Organic Chemicals*, 2nd ed., Van Nostrand Reinhold Co., Inc., New York, NY, 1983.

Yu, X., et al.: "Supercritical Fluid Extraction of Coal Tar Contaminated Soil," *Environmental Science & Technology,* vol. 24, no. 11, pp. 1732–1738, November 1990.

ACKNOWLEDGMENTS

The authors wish to acknowledge the following individuals who assisted with the preparation of this chapter: Sec. 9-2, Soil Vapor Extraction, was prepared by James D. Fitzgerald and Kevin E. McHugh of *ERM-New England*. Sec. 9-4, Steam Stripping, was prepared by David Potter, Science Applications International Corp. Review assistance was provided by Karen Winegardner, Deborah Watkins, William Murphy, Richard Dulcey, Jeffrey Case, John Haasbeek, and Ruth Baker, *ERM*; Amy Wong, ERM-New England; Profs. W. C. King and M. Prince and Steven Galloway, *Bucknell University*, and Wayne Schuliger, *Calgon Carbon Corporation*. The principal reviewers for the **ERM Group** were James Fitzgerald, *ERM-New England, Inc.* and Dr. Elsie Millano, *ERM-North Central, Inc.*

Bacterium ... : any of a class of microscopic plants ...
often aggregated into colonies ... living in soil, water, organic ...
and important to man because of their chemical effects ...
　　　　　　　　Webster's *New Collegiate Dictionary*

Biological treatment is the degradation of organic waste by the action of microorganisms. Degradation alters the molecular structure of organic compounds, and the degree of alteration determines whether **biotransformation** or **mineralization** has occurred. **Biotransformation** refers to the simplification of an organic compound to a daughter compound. **Mineralization** is the complete breakdown of organic molecules into cellular mass, carbon dioxide, water, and inert inorganic residuals. That is, biotransformation is partial degradation and mineralization is complete.

　　Biological treatment of almost any organic hazardous waste can be accomplished because virtually all organic chemicals can be degraded if the proper microbial communities are established, maintained, and controlled. With a sound understanding of microbiology and the same engineering principles which have governed the treatment of industrial wastewater, an engineer can design systems which create the environment necessary for effective treatment of hazardous waste.

10-1　BASIC MICROBIOLOGY

A number of microbiological factors affect biological treatment. These factors can be grouped under the following categories:

- Energy and substrate sources

- Enzymatic processes
- Substrate biodegradability
- Inhibition and toxicity
- Microbial community

Energy and Substrate Sources

Any form of living matter requires energy and carbon for growth and maintenance. The particular sources from which microorganisms derive their energy and cellular carbon provide a basis for their classification as shown in Table 10-1.

Biological treatment is the direct result of heterotrophic metabolism in which organic waste is the substrate (i.e., the waste serves as a source of carbon and energy to living organisms). Of the many different organisms involved in biological treatment of hazardous waste, the most commonly relied upon are heterotrophic bacteria. Nevertheless, fungi and higher living forms can be effective in specific applications, and the latter as predators can affect biotreatment results in the mixed biological populations that occur naturally outside the laboratory.

Enzymatic Processes

Waste constituents can occur in a size as small as a single molecule or as large as a droplet, if not a clump. For degradation to occur, the waste, as substrate, must first come into contact with a bacterial cell's outermost coating. This action initiates a series of metabolic steps in the degradation of organic waste as depicted in Fig. 10-1.

The first step consists of transport of the substrate into the cell. This can occur via three methods:

- Extracellular enzyme complexation
- Liquefaction
- Direct transport

Typically, upon contact of bacteria with substrate, extracellular enzymes produced by the bacteria will form complexes with substrate molecules. These complexes

TABLE 10-1
Types of microorganisms

Classification of microorganism	Energy source	Carbon source (substrate)
Autotrophic		
Photoautotrophic	Light	Carbon dioxide
Chemolithotrophic	Oxidation-reduction reactions of inorganic compounds	Carbon dioxide
Heterotrophic	Oxidation-reduction reactions of organic compounds	Organic carbon

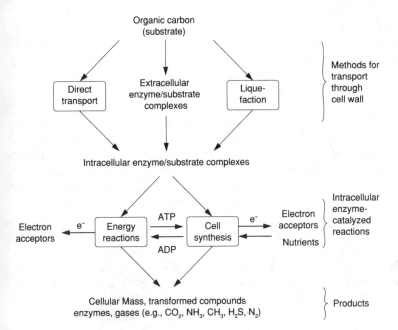

FIGURE 10-1
Metabolic steps in degradation of waste.

allow the substrate to pass through the cell wall. Once inside the cell, intracellular enzymes will further complex with the substrate to catalyze other reactions necessary to obtain energy and to build new cellular material. Some organic chemicals are too large to be transported into the cell, and liquefaction (i.e., enzyme-mediated transformation) takes place on the surface of the cell. Also some organic chemicals can pass into bacterial cells directly without complexing with enzymes.

Enzymes are large protein molecules composed primarily of amino acids twisted into complex shapes by peptide links and hydrogen bonding. Their role in metabolism is extremely complex. Essentially, they lower the energy required to activate a reaction and thereby speed up biological activity. This capability is due to evolution-induced stereochemistry (i.e., a structural affinity between the enzyme and the substrate as discussed in Sec. 5-2). Because enzymes function as catalysts, they are not used up in the process but can be reutilized.

A myriad of different enzymes, not necessarily from the same species of bacteria, typically act in a sequential manner to degrade organic waste to successively simpler compounds (e.g., hydrocarbons to alcohols to organic acids to carbon dioxide). The specificity of enzymes for a particular substrate varies widely. There are some organic compounds with which only one known enzyme, produced by not more than two or three different species of bacteria, will complex. Similarly, there are some bacteria which can use only a single specific organic chemical as substrate.

A cell operates two critical types of metabolic processes, referred to as **anabolic** (cell building) and **catabolic** (energy releasing) processes. The oxidation-reduction reactions which release energy involve a transfer of electrons from organic carbon as it is oxidized to a higher state. To complete the reaction, electron acceptors are needed. These can be oxygen, nitrate, sulfate, or others. The energy released by catabolic processes is captured and transformed by the adenosine triphosphate and adenosine diphosphate (ATP-ADP) energy coupling cycle to support cell maintenance and cell building activities.

The anabolic processes produce protoplasm as an end product which is composed of proteins, carbohydrates, DNA and other cellular components. Bacterial protoplasm is 75–80% water. The solid material in the protoplasm is 90% organic and 10% inorganic. An empirical formula for the organic portion can be approximated as $C_{60}H_{87}O_{23}N_{12}P$, or more simply as $C_5H_7O_2N$.[1] The inorganic cellular material contains numerous elements including phosphorus, sodium, sulfur, potassium, calcium, magnesium, iron, and various trace metals.

In summary, microorganisms can metabolize organic waste via a series of low-temperature (i.e., ambient) enzyme-catalyzed reactions. Metabolism releases energy from the organic waste and uses the carbon to build new cellular material. As a side result, the organic waste is degraded. Degradation either transforms the waste to simpler organic compounds or, if complete, mineralizes it to carbon dioxide, water, and cell mass.

Substrate Biodegradability

Experience and research show that most synthetic organics are biodegradable, making biological treatment a technically feasible alternative. Nevertheless, the literature is replete with cases where specific compounds have resisted degradation (such compounds are termed **recalcitrant** or **refractory**), or their degradation occurs so slowly as to make biological treatment inefficient (**persistent** compounds). Even with recalcitrant compounds, continuing research has often been successful in identifying some microorganisms, or consortium of microorganisms, which degrade them. Nevertheless, some noteworthy compounds that have been subjected to considerable research still remain labeled as recalcitrant or persistent. Examples include TCDD (dioxin) and PCBs.

What makes one organic chemical persistent and another not? The inherent biodegradability of a chemical depends to a large extent upon its molecular structure. Unsubstituted hydrocarbons, except for multiple ring polycyclic aromatics (PAHs), generally degrade rapidly. Yet certain substituents will resist biodegradation. Small changes in the configuration of a biodegradable molecule may result in persistence (see Fig. 10-2).

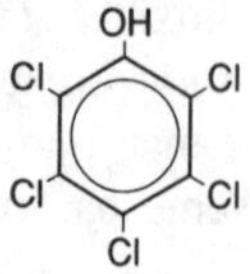

Pentachlorophenol
(readily degradable)

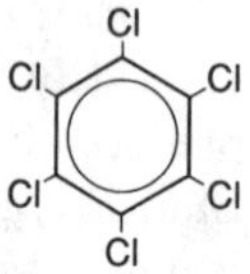

Hexachlorobenzene
(persistent)

FIGURE 10-2
Similarly structured compounds having different biodegradability.

Attempts to correlate degradability with various structural parameters indicate that the following conditions, either individually or in combination, are frequently associated with recalcitrant or persistent compounds:

1. Halogenation
2. Large number of halogens
3. Highly branched
4. Low solubility in water
5. Atomic charge difference[2]

The conditions listed above should be considered as general guidelines rather than definitive relationships. A more specific assessment is provided in Table 10-2, which lists types of structural features associated with persistence. This particular work tested 287 chemicals and found 81 to be persistent as defined by half-lives greater than 15 days using the standard biochemical oxygen demand (BOD) procedure.

Some general conclusions regarding the ability of microorganisms to degrade specific classes of organic chemicals have been reported by researchers.[3,4,5,6] These can be summarized as follows:

- Typically, straight-chain aliphatic compounds are easily degraded. However, introduction of extensive branching of alkane substituents results in sterically hindered configurations that often are resistant to degradation. Also, unsaturated aliphatics are less readily transformed than the saturated analogues.

- Simple aromatic compounds are usually degradable by several mechanisms of ring cleavage. The introduction of halogens will result in decreased biodegradability because halogens stabilize the ring. The degree of decreased biodegradability usually is a function of the number of halogens incorporated because biodegradation will require dehalogenation at some stage. However, microorganisms capable of utilizing haloaromatic as well as haloaliphatic compounds as the sole source of carbon and energy have often been isolated from natural environments.

- Even highly chlorinated aromatic compounds like PCBs can be degraded, albeit slowly. Highly substituted PCB isomers are degraded much more slowly than less halogenated derivatives, and catabolic enzymes preferentially attack and cleave the ring carrying the lowest number of chlorines. For halogenated substances, degradation of isomeric substrates may require distinctly different pathways.

- Biodegradation of nitrogen and sulfur containing compounds is often linked to their utilization as nutrient sources. Branched chain alkyl or aryl sulphonates are often slowly degraded.

- Polymeric materials are among the most resistant to microbial attack. Nevertheless, microorganisms have been isolated that will utilize nylon polymer or, to a more limited extent, polystyrene, as the sole source of carbon and energy.

Many exceptions exist to these guidelines. An engineer considering biological treatment of a hazardous substance should depend on literature studies or testing of that

TABLE 10-2

Structural features associated with chemicals that were persistent in a standard biochemical oxygen demand test[7]

Structural feature	Half-life (days)	Number of chemicals tested with this feature	Number that were not persistent
Presence of at least one *tertiary*-butyl terminal branch	> 15	11	0
Epoxides	> 20	6	0
Aliphatic chemicals with fused rings and no branches	> 35	3	0
Two terminal isopropyl subgroups for a non-cyclic chemical	> 35	3	0
Aliphatic cyclic chemicals without branches	> 40	5	0
One or more halogen substitutions on a branched, noncyclic, or cyclic chemical	> 15	12	1
At least one isopropyl or dimethyl amine substitution on a cyclic chemical without other "degradable" substitutions	> 25	4	2
Two halogen substitutions on an unbranched, noncyclic chemical	> 15	3	0
More than two hydroxy substitutions on an aromatic ring	> 15	2	0
Two or more rings (except two aromatic rings)	> 20	8	0
Two terminal diamino groups on a noncyclic chemical	> 35	1	0
More than 1 amino branch on a ring with nitrogen being a member of the ring	> 100	2	0
Two terminal double-bonded carbons on an unbranched chemical	> 100	1	0
Benzene ring with > 2 substitutions (non-hydroxy) and K_{ow} > 2.18	> 100	1	0
Cyano group on a chain consisting of > 8 atoms	> 100	1	0
Highly branched chemicals	> 100	1	0

specific substance. The literature serves as an excellent starting point. For substances whose biodegradability has not been reported in the literature, tests include 5-day biochemical oxygen demand (BOD_s), ultimate oxygen demand, respirometry studies, and batch or continuous flow laboratory or pilot scale treatability studies. These tests are usually conducted in a tiered sequence, starting with simple laboratory measures and advancing to more involved studies only if the screening tests prove positive.

Inhibition and Toxicity

Many organic and inorganic waste materials move through a progression of effects as the concentration increases (similar to that of a dose-effect relationship discussed in Sec. 5-3). An organic substance that is biodegradable at one concentration can become persistent at higher concentrations by inhibiting the growth of the microbial culture. At even higher concentrations, the substance can become toxic to the culture. These effects derive presumably from the degree to which the substance overwhelms the enzymatic systems which normally degrade it. Many of the inorganic nutrients necessary for the synthesis of cellular mass are toxic to the same organisms at higher concentrations.

Microbial Community

Biological treatment involves complex interaction of mixed biological populations. Growth rates and utilization of substrate is frequently higher in mixed enrichment cultures than in pure cultures isolated from the mixture. In mixed microbial communities, not only are organisms that can initiate catabolism important, but so are secondary utilizers (i.e., organisms that utilize intermediates derived from transformation of the original substrate). Also, secondary utilizers may support the growth of primary utilizers (e.g., by providing a specific factor for growth, by removing toxic products, by concerted metabolism [cometabolism], as well as by transfer of genetic material between strains). In some cases primary utilizers may be completely dependent on secondary utilizers for growth.

A microbial consortium, in which each species serves a vital link in a chain if not web of transformation and mineralization steps, may be necessary for many organic compounds. In one instance,[8] a mixed microbial community of three species of bacteria was found to be able to degrade 3-chlorobenzoate. However, none of the species could, by themselves, utilize the chemical. The initial step as shown in Fig. 10-3 is the reductive dechlorination by a specific bacterium to yield benzoate. This dechlorinating bacterium receives no carbon source from this step. Another species degrades the benzoate to yield acetate and H_2. This second species, termed the benzoate oxidizer, is not known to use any other chemical for substrate. Because of the production of H_2 and the need to keep its levels below a threshold to enable benzoate oxidation, a third bacterial species is needed. This is a methanogen which uses the H_2 with CO_2

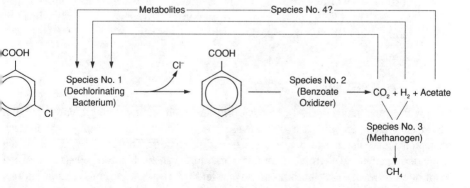

FIGURE 10-3
Consortium of three bacteria species for degrading chlorobenzoate.[9]

to produce methane. A portion of the H_2 is returned to the dechlorinating bacterium for the reduction step. Presumably, the dechlorinating bacterium receives some form of carbon from one or more of the metabolites produced in the subsequent oxidation stages for acetate, meaning that still another species of bacteria is possibly required.

When naturally occurring organisms are exposed to organic contaminants, they tend to develop, by **adaptation** an increased ability to degrade them as substrate. The rate at which the microorganisms can adapt will, to a great extent, determine how quickly the concentration of contaminants diminish in water and soil. Adaptation results in the natural selection of primary utilizers. Such adapted strains may arise by selection among strains already present, but also by mutation and gene transfer (in the latter case, e.g., by plasmids). In this way novel enzymatic activities may appear in a population. It is also possible that the gene encoding for such enzymes was initially present in a non-expressed form. Since adaptation requires a long period of time, one approach for enhancing the potential of microbial breakdown of chemical wastes has been to introduce previously adapted species by inoculation.

Co-metabolism is an important example of a microbial community at work. It features the transformation of one compound (the secondary substrate) by enzymes from microorganisms routinely degrading another compound (the primary substrate or co-metabolite). The microorganisms derive neither a source of carbon nor energy from the secondary substrate; its degradation is serendipitous and fortuitous. The co-metabolite induces the enzymes needed for transformation of the secondary substrate. Such enzymes usually have a low specificity. Although the secondary substrate does not enter the catabolic and anabolic pathways of the microorganism degrading the cometabolite, other microorganisms may be able to use the transformation products for substrate.

Co-metabolism is a potential alternative for compounds which otherwise would degrade very slowly, such as trichloroethylene (TCE). It has been well documented[10,11] that functioning methanotrophs (a group of heterotrophic bacteria that use methane as their source of carbon and energy) will indirectly oxidize TCE and other chlorinated aliphatics as secondary substrate in the presence of oxygen. Methanotrophs produce the enzyme methane monooxygenase (MMO) when metabolizing methane, the primary substrate. MMO has broad substrate specificity[12] and can transform TCE to a TCE epoxide. The TCE epoxide readily decomposes in water to alcohols which are more readily degraded by other species in the environment.

10-2 ENGINEERING FACTORS

Biological treatment consists of promoting and maintaining a microbial population (biomass) that metabolizes a target waste. A number of factors influence the rate at which metabolism, and hence biodegradation, occurs. Through investigations and engineering of these factors, favorable conditions can be identified, established and controlled to assure that the biomass grows and flourishes. If so, biodegradation may occur at a sufficiently robust rate, such that the cost of biological treatment can be competitive with physical and chemical treatment alternatives. The engineering factors affecting biological treatment include:

- Electron acceptor
- Moisture
- Temperature
- pH
- Total dissolved solids
- Nutrient availability
- Reactor design
- Alternative carbon source

Electron Acceptor

The catabolic reactions involve a transfer of electrons from the waste to an electron acceptor, and the biological process by which this occurs is termed **respiration**. In **aerobic** respiration, bacteria utilize oxygen as the terminal acceptor of electrons removed from oxidized organic compounds. Typically, in the case of biotransformation, oxygen is added to the organic molecule and hydrogen removed (e.g., oxidation of an alcohol to an acid). Upon complete mineralization, oxygen is reduced to water and organic carbon is oxidized to carbon dioxide. The mass of oxygen required by aerobic systems can be calculated based on stoichiometry or laboratory determinations. For example, the following stoichiometric equation for the theoretical mineralization of branched alkene in catabolic reactions skips the intermediate steps of oxidation to an alcohol and then to an acid:

$$C_7H_{12} + 10\ O_2 \rightarrow 7CO_2 + 6H_2O$$

The oxygen required to completely mineralize one milligram of alkene is calculated as follows:

$$\frac{10\,(2 \times 16)}{(7 \times 12) + (12 \times 1)} = 3.33 \text{ mg O}_2/\text{mg alkene}$$

Example 10-1 Oxygen requirements. What are the oxygen requirements of the anabolic metabolism of alkene?

Solution. The stoichiometric equation for the assimilation of alkene to build new cellular material is as follows:

$$C_7H_{12} + 5O_2 + NH_3 \rightarrow C_5H_7O_2N + 2CO_2 + 4H_2O$$

The new cellular material, if mineralized in a process termed endogenous respiration, also exerts oxygen demand as follows:

$$C_5H_7O_2N + 5O_2 \rightarrow 5CO_2 + 2H_2O + NH_3$$

Summing these two equations yields:

$$C_7H_{12} + 10O_2 \rightarrow 7CO_2 + 6H_2O$$

which is the same as for catabolic reactions. It can be noted in this case that not having to mineralize the cellular mass would reduce the oxygen demand for anabolic metabolism by half.

Because the solubility of oxygen in surface water is in the range of 8 mg/L, this example illustrates that even small concentrations of organic waste necessitate that oxygen be supplied to aerobic treatment operations. To provide a safety factor in aerobic treatment, the generally accepted practice is to maintain continuously a minimum dissolved oxygen concentration of 2 mg/L.

Anaerobic processes may be defined as treatment that occurs in the absence of oxygen. This definition greatly simplifies a complicated system of multiple reactions resulting in the conversion of hydrolyzed organics to higher organic acids. These are subsequently degraded to acetic acid, hydrogen and carbon dioxide which are metabolized by methanogenic bacteria (methanogens), thereby producing methane. Unlike aerobic processes, oxygen is not the terminal electron acceptor. Instead, the terminal electron acceptor in anaerobic respiration can include any of several inorganic, oxygen-bearing compounds in the following order of preference:

Nitrates	Reduced to nitrogen; sometimes referred to as anoxic rather than anaerobic conditions
Sulfates	Reduced to hydrogen sulfide
Carbon dioxide	Reduced to methane

An example of anaerobic degradation is shown as Fig. 10-4. The rate of anaerobic degradation is usually lower than aerobic degradation. However, anaerobic degradation offers key advantages when degrading high-strength wastes. By not requiring the supply of free oxygen, the expense of aeration is eliminated. By converting more of the organic substrate eventually to methane, the process produces less biomass per unit of organic waste removed.

Conversely, several wastes have been found that undergo anaerobic transformation more efficiently than under aerobic conditions. Of particular interest with hazardous waste is the highly specialized ability of anaerobic processes to transform halogenated compounds. A chemical such as tetrachloroethene (a dry cleaning fluid also known as perchloroethylene or PCE) is too highly oxidized for efficient degradation by aerobic processes. Yet, reductive dechlorination occurs in a subsurface, anaerobic environment. The concept of reductive dehalogenation involves the cleaving of a chlorine atom or other halogen from an organic compound by oxidation-reduction. This is illustrated in Fig. 10-5 for PCE.

In this concept, a reduced organic compound loses an electron and thus becomes oxidized. A mediator transfers the electron to the halogenated compound, thus

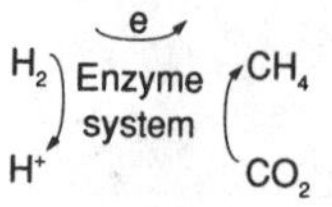

FIGURE 10-4
Example of electron transfer in anaerobic degradation.

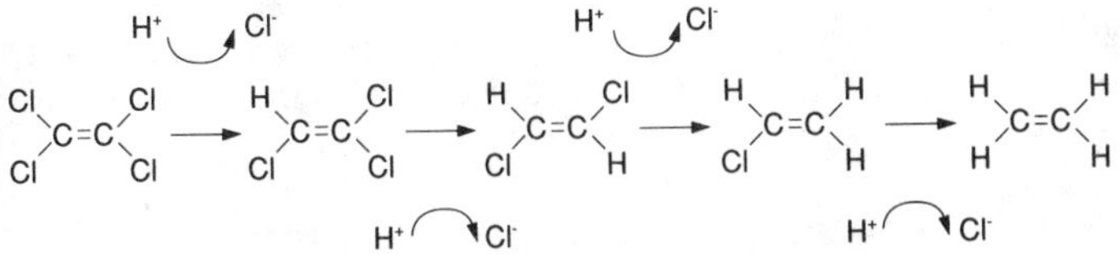

FIGURE 10-5
Anaerobic reductive dehalogenation.

cleaving the halogen from the compound. This concept of reductive dehalogenation was discussed in Sec. 4-3 in explaining the natural degradation of trichloroethylene to dichloroethylene and eventually vinyl chloride.

Moisture

Biodegradation requires moisture for two reasons:

1. For cellular growth because cellular tissue is 75–80% moisture.
2. As a medium for movement of the microorganisms to the substrate, or vice versa, for non-motile species.

Biodegradation in soil systems can occur at moisture levels well below saturation. One reference obscurely states that the optimum moisture level is only one bar.[13] Singleton and Sainsbury[14] indicate that most bacteria fail to grow if the water content of the medium falls below 92% relative humidity. However, it is generally accepted that the minimum moisture content necessary for treatment of wastes such as contaminated soil is 40% of saturation.

Temperature

Temperature has a major influence on growth rate. Cellular activity, particularly enzyme systems, responds to heat so that the rate of cell growth increases sharply with increasing temperature until the optimum is reached. Increases in temperature just a few degrees above the optimum can slow growth dramatically by inactivating enzyme systems and reducing reproductive capability.[15] Continued exposure to high temperature may melt membrane lipids, resulting in cell death.[16] Unlike high temperatures, low temperatures are usually not lethal. Instead, the cells eventually become dormant. Activity decreases when dropping below the optimum because of reduced enzyme activity and a loss of the fluidity of the cell membranes, thereby restricting transport of substrate molecules.[17] A sudden decrease to a low temperature produces a much greater reduction in cell activity than gradual decreases to the same temperature; gradual decreases enable the microorganisms to acclimate.

Figure 10-6 shows how cellular activity observed below the optimum temperature can more than double with a rise of about 10°C.

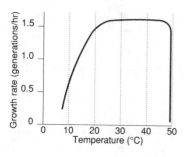

FIGURE 10-6
Effect of temperature upon growth rate of a particular species of bacteria.

The effect of temperature on cellular activity, and thus the rate of biological degradation, can be expressed simply as follows:

$$\frac{r_T}{r_{20}} = \theta^{(T-20)}$$

(10-1)

where
r_T = activity rate at $T°C$
r_{20} = activity rate at 20°C
θ = temperature-activity coefficient
T = temperature (°C)

In conventional wastewater treatment systems, values for θ can range from 1.0 to 1.14.[18]

Figure 10-6 shows bacterial growth rate dropping drastically above 45°C. Actually, viable species of bacteria have been found in thermal springs and glaciers. Temperature provides a basis for classification of the typical ranges in which bacteria function. Within this range there exists an optimum for which the maximum cell growth rate occurs, all else being equal. This classification system is shown in Table 10-3.

pH

Enzyme activity depends on pH. The amount of a particular enzyme that is in a catalytically active state changes as pH changes, with the maximum amount occurring at the optimum pH. Accordingly, the growth of bacteria depends on pH. Most bacteria

TABLE 10-3
Classification of bacteria based on temperature

Classification	Functional range	Optimum
Psychrophilic	−2 to 30°C	12 to 18°C
Mesophilic	20 to 45°C	25 to 40°C
Thermophilic	45 to 75°C	55 to 65°C

grow best in a relatively narrow range around neutrality (i.e., in a range of pH of 6–8). Die-off typically occurs below a pH of 4–5 and above a pH of 9–9.5; nevertheless, there are strains which thrive outside these limits.

Microbial activity can alter the external pH. Examples include anaerobic fermentation which converts organic waste to organic acids, depressing pH. Nitrification (conversion of NH_4^+ to NO_3^-) also lowers pH, as does the carbon dioxide produced by aerobic degradation. The breakdown of organo-nitrogen compounds can raise pH by releasing NH_4^+. If any of these effects are not buffered, the altered pH can inhibit, if not kill, the microbial population.

Total Dissolved Solids

The concentration of dissolved solids can affect liquid-based biological treatment. If the concentration (as measured by total dissolved solids or TDS) is too high, microbes die due to osmotic rupture of the cell membrane. If the TDS varies widely, the activity of the microbial population can decline. Environmental engineers have developed practical rules of thumb based on experience:

- TDS should not exceed 40,000 mg/L.
- TDS should not vary by more than a factor of 2.0 over a period of a few days.

Nutrients

Cellular mass contains carbon and numerous other elements. Metabolism requires these elements as nutrients in addition to organic carbon as substrate. Phosphorus and nitrogen are referred to as **macronutrients** because the synthesis of cellular tissue requires much more of these than other nutrients. Frequently, nitrogen and phosphorus are not available in sufficient amounts in hazardous waste and must be added, usually as ammonia and orthophosphate.

The empirical formula for the organic portion of bacterial protoplasm, given earlier as $C_{60}H_{87}O_{23}N_{12}P$, provides a basis for calculating the theoretical amount of nitrogen and phosphorus required for synthesis of cellular mass from waste as shown in Table 10-4.

The TOC:N:P ratio for anabolic reactions is thus 100:23:4.3. Environmental engineers commonly use, based on general experience, a rule-of-thumb TOC:N:P ratio

TABLE 10-4
Nitrogen and phosphorus requirements for anabolic reactions

	Number of atoms	Atomic weight	Component of empirical formula weight	Ratio
Total organic carbon	60	12	720	100
Nitrogen	12	14	168	23
Phosphorus	1	31	31	4.3

of 20:5:1. It should be noted that the ratio does not apply to the total organic carbon measured in a waste because not all carbon metabolized by bacteria is synthesized to new cellular material. As much as half, if not more, of the metabolized carbon may be oxidized by catabolic reactions to CO_2 without the need for nutrients.

Example 10-2 Nutrient requirements. What is the ratio of the nutrient requirements per unit of biochemical oxygen demand (BOD) for the aerobic treatment of alkene in which all cellular mass is physically removed without endogenous respiration (see Ex. 10-1)?

Solution. As calculated earlier, the mineralization of alkene requires 3.33 mg of oxygen per mg of alkene for catabolic reactions. The same is also required for anabolic reactions, assuming endogenous respiration of the cellular mass. However, if such were to occur, it would free the nutrients for synthesis of new cells and, theoretically, no additional nutrients would be required. Because in this example all cellular mass is physically removed per mg of alkene.

In the absence of experimental data, environmental engineers customarily assume that half of an organic substrate in a hazardous waste is converted to cellular mass and half oxidized for energy. If so, the total biochemical oxygen demand would be $3.33/2 + 1.67/2 = 2.5$ mg O_2 per mg of alkene as follows:

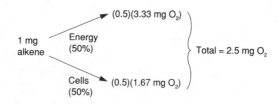

The synthesis of 0.5 mg alkene would involve the following amount of carbon:
Carbon into cellular mass $= (0.5$ mg$)(84/(84 + 12)) = 0.4375$ mg
The previously calculated theoretical ratio of TOC:N:P shows that 0.5 mg of alkene converted into cells would represent a need for the following amounts of nutrients:
N needed $= (0.4375$ mg$)(23/100) = 0.201$ mg
P needed $= (0.4375$ mg$)(4.3/100) = 0.019$ mg
Therefore, the theoretical BOD:N:P ratio for treatment of alkene is calculated as follows:
BOD:N:P $= 2.5:0.201:0.019 = 100:8:0.8$
Environmental engineers have commonly used a rule-of-thumb BOD:N:P ratio of 100:5:1 based on general experience. The rule-of-thumb approximates the theoretical ratio for alkene, and the true ratio would depend on the split between catabolic and anabolic reactions, the degree of endogenous respiration and other factors.

Rather than calculating theoretical requirements or using a general rule of thumb, the need for nutrients should be determined as part of laboratory treatability tests. The specific requirements are very much dependent upon the waste, the availability of nutrients in the waste, the yield of biomass, and the design of the treatment system. The rules of thumb provide a basis for calculating ranges to be tested or for conducting economic evaluations. The objective is to ensure that carbon is the limiting nutrient.

Micronutrients needed to support metabolism include sulfur, potassium, calcium, magnesium, iron, and others. To provide physical movement of ions across the cell-wall membrane into the cell cytoplasm, the micronutrients should have a minimum concentration in water of 1 to 100 mg/L.[19] Metabolism also requires a number of trace nutrients such as nickel, copper, zinc, various vitamins, and others. The concentration of these **trace nutrients** can be less than 1 mg/L. In most cases, all micronutrients and trace nutrients can be obtained by bacteria from the environment and do not have to be added, particularly if the waste is a contaminated soil or one that has been in contact with soil (e.g., ground water). However, an industrial process wastewater may require their addition.

Reactor Design and Operation

The particular physical design of the reactor within which biodegradation occurs has a major effect on treatment efficiency and economics. Important design factors for a reactor include equalization, mixing regime, biomass retention time, and hydraulic detention time. By controlling these factors, biodegradation can be enhanced to reduce contaminant concentrations to required levels. Optimization of these variables will reduce economic costs and increase system reliability.

Flow Equalization.   Theoretical descriptions of biological treatment systems typically assume steady-state conditions. In reality, this is not true as evidenced by numerous examples in which the nature and concentration of contaminants, if not the rate of flow, can vary with time: waste streams from industrial processes, a ground water plume undergoing recovery, and sludge removed from an inactive lagoon. The microbial population must adjust to these variations in a process known as **acclimation**. Acclimation takes time, and shock loads can upset the biomass. As discussed earlier, high concentrations are even capable of toxic and inhibitory effects. While biological systems have a significant "buffering capacity," equalization is necessary for wastes with temporal variations that, if not dampened, would upset if not destroy the biomass. For the case of a liquid waste with a highly variable rate of flow, equalization consists of passing the influent into a tank where the depth (and hence the volume) is permitted to vary. The waste is pumped from this tank at a constant rate of flow.

Mixing Regime.   Complete mixing provides internal equalization to compensate for variations in concentration of waste constituents. The opposite regime is termed plug flow, and it provides for more efficient removal of waste constituents if the influent flow and concentration have nominal variation. Another factor is that the reactor must provide sufficient mixing for the waste to come into contact with the biomass. This holds obvious importance when treating waste having a great amount of inert solid material (e.g., contaminated soil).

Solids Retention Time. The population in a reactor of waste-degrading microorganisms, measured as biomass, is important because the rate of degradation shows a linear dependence upon the concentration of biomass. The concentration of biomass can be increased by recycling, a process in which the microorganisms (or biological solids) generated in the bioreactor are withdrawn with the effluent and allowed to settle in a clarifier, with the sludge returned to the bioreactor (see Fig. 10-7).

Thus, the biomass retention time (referred to as solids retention time, or SRT) can exceed the hydraulic retention time. An obvious relationship exists that the greater the SRT, the greater is the concentration of biomass. It follows that removal efficiency increases as SRT increases. A key advantage of recycling other than to increase the concentration of biomass is that it serves to retain the specific "degraders" that have acclimated to the particular organic chemicals being treated. If the SRT is too short (i.e., less than the time to generate new cells), the biomass will diminish and wash out.

Hydraulic Retention Time. The hydraulic retention time of a reactor is equal to its volume divided by the rate of flow. Obviously, costs decrease as hydraulic retention time decreases. Minimizing the hydraulic retention time is constrained by the fact that as it reduces, the ability to handle shock loads decreases. Also, as the hydraulic retention time is decreased the biomass must increase to achieve effective biodegradation. Experience shows that extremely high biomass levels are unacceptable in liquid treatment systems because of potential problems with overloading the subsequent clarification process, and the inability to transfer adequate amounts of oxygen in a limited volume.

Other Factors

Other factors can influence waste treatment efficiency but not to the degree represented by the key factors previously discussed. Examples of these other factors include alternative carbon sources, a concept in which a waste contains multiple organic compounds capable of serving as sources of carbon for the biomass. Research shows that because of preferential differences, the biomass may degrade the compounds sequentially, possibly delaying the degradation of a target compound. However, further research shows that in systems with a long SRT, the compounds are removed simultaneously, each at a different rate.[20]

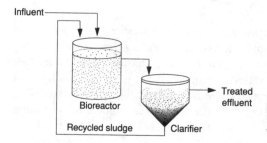

FIGURE 10-7
Recycling of biomass in liquid-phase treatment system.

10-3   GROWTH KINETICS

The performance of biological waste treatment can be measured by the rate at which microorganisms metabolize the waste which, in turn, is directly related to their rate of growth. Bacteria reproduce by binary fission, and the time required to divide is referred to as generation or doubling time, which can be as little as 15 to 20 minutes. Although individual cells reproduce at an approximately constant linear rate in optimum conditions, the total population of biomass grows at exponential rates.

Batch Culture

A classic laboratory study is the batch liquid culture which starts with a small inoculation of bacterial cells into a solution containing a high concentration of a single substrate. The pure culture is then incubated under optimum conditions. The pattern for population growth of such a culture follows five distinct phases as follows (see Fig. 10-8).

Lag phase. Cells require a period to adapt to the new environment before beginning their growth; pre-adapted cells experience a much shorter lag phase.

Exponential-growth phase. The biomass grows exponentially as cells divide at a constant rate which is the maximum growth rate for the established laboratory conditions.

Stationary phase. Upon approaching exhaustion of the substrate, the number of dying cells offsets the number of new cells.

Death phase. Without the addition of substrate, dying cells outnumber the new cells. The rate can be as great as the "inverse" of the exponential growth phase.

Endogenous phase. In some cases, the rate of population decline slows as dead cells supply the substrate needed for new cells.

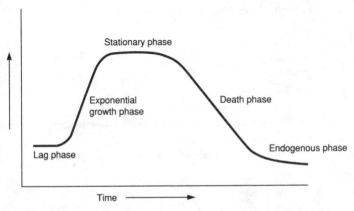

FIGURE 10-8
Growth of bacteria in batch culture.

The buildup in the microbial population, or the biomass, corresponds to a removal of substrate (degradation of organic contaminants) as shown in Fig. 10-9.

Kinetics of Batch Cultures

The rate at which cells divide is the **specific growth rate** or μ. It is measured as generations per unit time, or the inverse of the doubling time. The term provides a starting point for deriving a mathematical formulation of growth kinetics. Without die-off, the bacterial cells accumulate in the batch culture at an exponential rate which can be defined as follows:

$$\frac{dX}{dt} = \mu X \tag{10-2}$$

where X = concentration of biomass, mass/unit volume
μ = specific growth rate (time^{-1})
t = time

In the exponential-growth phase, cell division is unrestrained and occurs at a constant specific growth rate which is the maximum observed for batch conditions. As shown in Fig. 10-8, the growth in biomass slows upon approaching the stationary phase because the specific growth rate begins to decrease in response to the diminishing concentration of substrate. A classic relationship exists between bacterial growth rate and substrate concentration when all other conditions are kept constant. This is depicted in Fig. 10-10 for a hypothetical example. The specific growth rate at lower concentrations of substrate is almost linearly dependent upon substrate concentration. As the concentration of the substrate increases, the specific growth rate eventually reaches a maximum that is maintained independently of higher concentrations. The actual maximum specific growth rate depends upon the nature of the substrate and other factors but not upon substrate concentration.

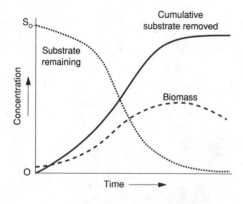

FIGURE 10-9
Degradation of organics and corresponding buildup in biomass.

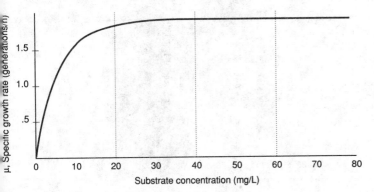

FIGURE 10-10
Effect of substrate concentration on bacterial growth rate (hypothetical example).

Environmental engineers and scientists have traditionally formulated the relationship between specific growth rate and concentration according to the Monod equation first proposed in 1942.[21,22]

$$\mu = \mu_{max} \frac{S}{K_s + S} \tag{10-3}$$

where μ = specific growth rate (time^{-1})
μ_{max} = maximum specific growth rate (time^{-1})
S = concentration of substrate in solution (mass/unit volume)
K_s = half-velocity constant (i.e., substrate concentration at which the specific growth rate is one-half μ_{max}) (mass/unit volume)

Although an oversimplification, the Monod equation reasonably depicts the general observation that the specific growth rate is almost linearly dependent on substrate concentration at the lower range of concentrations and approximately independent at higher concentrations. The **half-velocity constant**, K_s, represents an approximate division between the two ranges (see Fig. 10-11).

The Monod equation implies that some growth occurs even at extremely low concentrations of substrate, a condition not only unproven but probably untrue. The equation also may not describe reasonably well the conditions which yield very rapid growth.[23] Nevertheless, the equation represents an acceptable starting point for describing kinetics of growth.

Substituting Eq. 10-3 in 10-2 yields the following expression:

$$\frac{dX}{dt} = \frac{\mu_{max} S X}{K_S + S} \tag{10-4}$$

It is generally accepted that the mass of new cells synthesized per unit mass of substrate removed is constant for a given substrate and bacterial species. The ratio of the increase in cellular mass to the decrease in mass of substrate is termed the **growth**

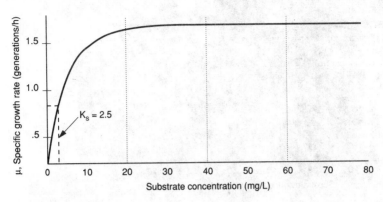

FIGURE 10-11
Monod half-velocity constant (hypothetical example).

yield coefficient, Y, and can be expressed as follows:

$$Y = \frac{dX/dt}{dS/dt} \tag{10-5}$$

Substituting Eq. 10-5 in Eq. 10-4 results in the following relationship:

$$\frac{dS}{dt} = \frac{\mu_{max} S X}{Y(K_S + S)} \tag{10-6}$$

The term $\frac{\mu_{max}}{Y}$ can be replaced by the term k which is referred to as the **degradation rate constant** or maximum rate of substrate removal per unit weight of biomass. This results in the following expression for the rate of substrate degradation:

$$\frac{dS}{dt} = \frac{k\,SX}{K_S + S} \tag{10-7}$$

Similarly an expression for the rate of biomass formation is as follows:

$$\frac{dX}{dt} = \frac{Yk\,SX}{K_S + S} \tag{10-8}$$

Actually, in any biological treatment system, a portion of the cells are not proliferating but are in a maintenance if not death phase. The net yield in biomass is thus the hypothetical amount of new cell formation represented by Eq. 10-8 less any reduction due to death and maintenance. The reduction can be grouped and defined as endogenous decay. Inclusion of decay in Eq. 10-8 yields the following:

$$\frac{dX}{dt} = \frac{Yk\,SX}{K_S + S} - bX \tag{10-9}$$

where b = endogenous decay constant (time^{-1})

These basic equations form the foundation for any detailed analysis of the kinetics of biological treatment systems. There are four important kinetic parameters which can be determined only through treatability tests (see Table 10-5).

ABLE 10-5

mportant kinetic parameters

Term	Definition	Significance
k	Degradation rate constant (time^{-1})	Indication of the degradability of an organic compound; for biological treatment to be effective, high values are preferred.
K_s	Half-velocity constant (mass/unit volume)	Indication of the efficiency with which degradation will occur; to degrade organic compounds to low concentrations, low values of K_s are necessary. Determining K_s is very difficult at sites with complex contamination (e.g., variable concentrations of multiple compounds).
Y	Yield of biomass generated per unit mass of substrate removed (mass/unit mass)	Indication of utilization of substrate for the production of biomass.
b	Endogeneous decay constant time^{-1})	Indication of the rate of loss of biomass due to natural death and decay.

Starting with these basic equations for a batch culture system, equations can
e derived to describe the kinetics of biological treatment reactors at steady state. A
umber of books on wastewater engineering present these derivations.[24,25,26] Table 10-
lists example steady-state derivations which are commonly used in the engineering
esign of liquid- and slurry-phase treatment systems. These equations introduce two
w variables:

$$\theta_h = \text{hydraulic retention time} = \text{volume/flow rate}$$

$$\theta_c = \text{solids retention time (SRT)}$$

BLE 10-6

ommonly used steady-state equations for liquid- and slurry-phase treatment
stems

	System		
etermination	Complete mix system without solids recycle	Complete mix system with solids recycle	Plug flow system with solids recycle
ffluent substrate ncentration	$S = \dfrac{K_S(1 + b\theta_c)}{\theta_c(Yk - b) - 1}$	$S = \dfrac{K_S(1 + b\theta_c)}{\theta_c(Yk - b) - 1}$	
oncentration of omass	$X = \dfrac{Y(S_0 - S)}{1 + b\theta_c}$	$X = \dfrac{Y(S_0 - S)}{1 + b\theta_c}\left(\dfrac{\theta_c}{\theta_h}\right)$	$\bar{X} = \dfrac{Y(S_0 - S)}{1 + b\theta_c}\left(\dfrac{\theta_c}{\theta_h}\right)$
ritical solids tention time	$\dfrac{1}{(\theta_c)^M} = \dfrac{YkS_0}{K_S + S_0} - b$	$\dfrac{1}{(\theta_c)^M} = \dfrac{YkS_0}{K_S + S_0} - b$	
oncentration of ert material (I) reactor	$I = I_0$	$I = I_0\dfrac{\theta_c}{\theta_h}$	$I = I_0\dfrac{\theta_c}{\theta_h}$

^{m} denotes minimum SRT.

In designing and operating a conventional treatment system, there are many objectives other than to reduce the concentration of an organic chemical to an acceptable level. A key economic objective is to maximize the rate of degradation at minimum hydraulic retention times. Other objectives include minimizing the quantity of sludge requiring disposal and minimizing the cost of chemical additives and oxygen supply. Maximizing reliability represents still another objective. The design of conventional systems to achieve treatment objectives is covered in detail in the same standard textbooks that elaborate upon derivation of the equations depicting treatment kinetics. There are limitations to these equations as discussed by Grady.[27] Furthermore, determination of kinetic coefficients requires bench-scale and pilot-scale studies, in which test units are operated over a range of solids retention times, influent concentrations, food:microorganism ratios, and other conditions. The application of the kinetic equations can represent one of the more mathematically rigorous designs to be undertaken by an environmental engineer.

Solids Retention Time

In designing biological reactors, a key factor is solids retention time (SRT), also referred to as mean cell residence time or θ_c. The implication is simple: within limits, a longer value of θ_c results in more efficient degradation, smaller reactor size, and lower cost. If the SRT drops below the cell regeneration time, biomass will wash out faster than it forms new cells. Without complete data, conceptual screening studies for activated sludge use a rule of thumb of an SRT of about 20–30 days with X not exceeding 5000 mg/L. A higher value of X will likely result in the clarifier failing.

The SRT can be calculated as follows:

$$\theta_c = \frac{XV}{\text{Rate of biomass wastage}} \tag{10-10}$$

where θ_c = solids retention time (time)
   V = volume of reactor (volume)

Example 10-3 Solids retention time. A 5000 gallon bioreactor operates at a biomass concentration of 2000 mg/L, measured as mixed liquor volatile suspended solids (MLVSS), and treats 10,000 gallons per day of liquid waste containing 1000 mg/L of total organic carbon (TOC). The suspended solids are separated in a clarifier following the bioreactor with recycle of separated sludge. The recycle flow rate is 5000 gallons/day. Each day, 300 gallons of recycle are wasted. The effluent from the clarifier contains 40 mg/L MLVSS. What is the solids retention time? How does that compare with the rule of thumb? If too short, how can the SRT be increased?

Solution. The conceptual flow and solids balance for a system with recycle is shown as Fig. 10-12. SRT is calculated for a recycle system as depicted in Fig. 10-12 as follows:

$$\theta_c = \frac{XV}{(Q - Q_W) X_E + Q_W X_U} \tag{10-10}$$

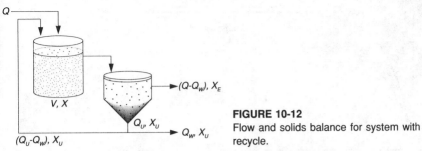

FIGURE 10-12
Flow and solids balance for system with recycle.

For this example, all variables are known except for X_U. To determine X_U, a solids balance must be done around the clarifier. For conservation of matter,

$$(Q + Q_U - Q_W) X = (Q - Q_W) X_E + Q_U X_U$$

Solving for the underflow sludge concentration, X_U,

$$X_U = \frac{(Q + Q_U - Q_W) X - (Q - Q_W) X_E}{Q_U}$$

$$= \frac{(10{,}000 + 5000 - 300)(2000) - (10{,}000 - 300)(40)}{5000}$$

$$= 5800 \text{ mg/L}$$

Now solving for θ_c,

$$\theta_c = \frac{(2000)(5000)}{(10{,}000 - 300)(40) + (300)(5800)}$$

$$= 4.7 \text{ days}$$

This is a very short SRT, probably too short for efficient treatment. It needs to be increased. Not wasting any recycle (i.e., $Q_W = 0$) would yield an SRT as follows:

$$\theta_c = \frac{(2000 \text{ mg/L})(5000 \text{ gal})}{(40 \text{ mg/L})(10{,}000 \text{ gal/day})}$$

$$= 25 \text{ days}$$

To increase the SRT even more would require improving clarification such as to reduce the solids in the effluent from the clarifier to below 40 mg/L.

Toxic Inhibition

As discussed previously, many hazardous wastes can inhibit their own degradation at increased concentrations (see Fig. 10-13). Two cases are shown, one for an organic chemical with no toxic inhibition (the same as shown earlier in Fig. 10-10) and the other for an organic chemical whose toxic nature inhibits growth at higher concentrations. In both cases, the specific growth rate at lower concentrations increases in a nearly linear relationship. However, as the concentration increases further, the specific

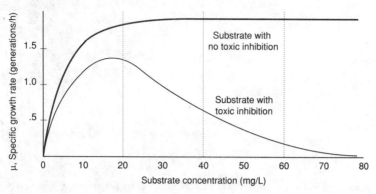

FIGURE 10-13
Effect of toxic inhibition on bacterial growth rate (hypothetical example).

growth rate of the toxic substance slows, eventually reaching a threshold concentration above which growth ceases. An even greater concentration than that shown could result in die-off.

If the concentration of a waste with a toxic nature is sufficiently low such that the effect of inhibition is negligible, the equations which have been derived from Monod kinetics are still applicable. However, for higher concentrations, the equations must be modified. The effect of toxic inhibition, as demonstrated in Fig. 10-13, can be described by Andrews kinetics (also referred to as Haldane equation)[28] as follows:

$$\frac{dS}{dt} = k \frac{SX}{K_S + S + \dfrac{S^2}{K_i}} \tag{10-11}$$

where K_i = inhibition coefficient (mass/unit volume)

From this expression, the biokinetics for various treatment reactors can be derived. The expressions become particularly complicated when dealing with mixtures of wastes in which one waste may inhibit the degradation of another.

For any organic chemical exhibiting toxic inhibition, such as shown in Fig. 10-13, concentration is critical. The maximum μ, or μ^*, for such a substance occurs at concentration S^* (see Fig. 10-14). If the concentration in the reactor exceeds S^*, inhibition or die-off can commence on a continuing, irreversible basis, resulting in loss of the reactor's treatment capability. Because of the great number of potentially toxic substances and the synergism among them, determination of the concentration that results in toxic inhibition requires specific testing. Even when the waste proves toxic to the biomass, certain operational measures such as acclimation of the microbes, introduction of hardier species, and dilution of the waste can possibly mitigate toxic inhibition.

Minimum Substrate Concentration

Many cases of soil and ground water contamination involve dilute concentrations of hazardous waste constituents (e.g., parts per billion). While such a low waste

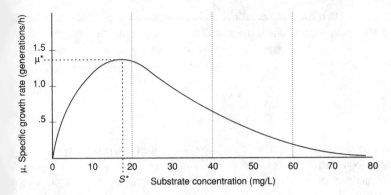

FIGURE 10-14
Critical concentration for inhibitory substance.

concentration may still pose a harmful risk to receptors, it may not support a viable biomass needed for treatment. The rate of growth decreases as substrate concentrations diminish, eventually reaching a point where growth ceases. The biomass may still survive, but not grow, at low concentrations if the diffusion of substrate molecules is sufficient to meet the energy needs for maintenance. At even lower concentrations, a threshold will be reached where not even energy needs are met, resulting in loss of the biomass.

Explanations for this threshold include the following: the flux of energy from degradation of insufficient substrate cannot sustain the biomass; the concentration may not be sufficient to induce enzyme production; and spores may form that cause microorganisms to become dormant.[29,30] This theoretical concentration behavior is noted universally in laboratory cultures, in full-scale treatment systems, and in natural water. It may explain the persistence of low levels of otherwise biodegradable substances in natural water.[31]

The kinetic expressions provide a basis for examining the minimum substrate concentration required to sustain a viable biomass. Equation 10-9 shows that the minimum sufficient substrate would occur when formation of new biomass equals loss by endogenous decay, or as expressed in the following equation:[32]

$$\frac{Yk\, S_{min}\, X}{K_S + S_{min}} = bX \qquad (10\text{-}12)$$

Solving for S_{min}:

$$S_{min} = \frac{bK_S}{Yk - b} \qquad (10\text{-}13)$$

Researchers have reported typical S_{min} concentrations for aerobic treatment systems in the range of 0.1 to 1.0 mg/L. In contrast, for the specific conditions of many contaminated sites, some toxic organics are considered to pose an unsafe carcinogenic risk at lower concentrations, as low as 0.01 mg/L, if not lower. When concentrations of toxic organics do not exceed their S_{min}, it is necessary to induce cometabolism or add a primary substrate. Otherwise, effective biodegradation may not occur.

It should be noted that this mathematical derivation pushes the Monod equation perhaps beyond its limit. Approached from another perspective, the concept of a minimum substrate concentration is refuted mathematically by examining Eq. 10-7 for the case when S approaches zero. Eq. 10-7 reduces to the following as $S \ll K_s$:

$$\frac{dS}{dt} = \frac{k}{K_S} X S \qquad (10\text{-}14)$$

Researchers have noted bacteria with very low K_s values in the open ocean and very low microbial concentrations (as low as 0.001 mg/L) in the subsurface.[33] Both cases represent situations in which low concentrations of organics are present. This observation and Eq. 10-14 suggest that organic substrates may indeed degrade at trace concentrations, albeit at extremely low rates perhaps measured in months if not years.

The theoretical S_{min} is also the minimum effluent concentration that can eventually be achieved when treating any substrates. The same equation as Eq. 10-13 can be derived by starting with the following equation from Table 10-6:

$$S = \frac{K_S (1 + b\theta_c)}{\theta_c (Yk - b) - 1} \qquad (10\text{-}15)$$

Rearranging,

$$S = \frac{K_S + K_S b\theta_c}{(Yk - b)\theta_c - 1}$$

As $\theta_c \to \infty$ for the case of extended treatment,

$$S_{max} = \frac{K_S b\theta_c}{(Yk - b)\theta_c} \qquad (10\text{-}16)$$

$$= \frac{K_S b}{Yk - b}$$

Mathematical Approximation for Hazardous Waste Cases

The application of the classic wastewater treatment equations in the field of hazardous waste is sometimes limited because of multiple complexities: the need to use non-conventional treatment methods, the toxic inhibition effects of some wastes, the competitive nature of some mixed wastes, the dilute concentrations of substrate, the difficulty in accounting for abiotic losses which can be significant, and concepts new to the environmental engineering field such as cometabolism and primary and secondary substrate. Much research is being conducted to formulate degradation kinetics for hazardous waste treatment; however, these efforts have not yet derived practical mathematical expressions. In the field of hazardous waste, the classical equations are applicable primarily to conventional biological treatment of aqueous waste.

For other types of biological treatment or other types of hazardous waste, a simple first order approximation has been used, in many cases with reasonably accurate results, to describe degradation of organic waste. This approximation stems from Eq. 10-7 and is based on two key assumptions concerning the concentrations of substrate and biomass. For the typical hazardous waste application, in which the target organic chemical is at a very low concentration, the numerical value of S is much smaller than K_s. Therefore, assuming S is insignificant by comparison, Eq. 10-7 reduces to Eq. 10-14 as explained previously:

$$\frac{dS}{dt} = \frac{k}{K_S} X S \qquad (10\text{-}14)$$

As a second assumption, if the biomass concentration X changes little with time, it can be considered a constant. Integration of Eq. 10-14 yields:

$$\ln\left(\frac{S}{S_o}\right) = -\frac{k}{K_S} X t \qquad (10\text{-}17)$$

where t = duration of treatment (time)
S_0 = initial substrate concentration (mass/unit volume)

The constant X can be grouped with the two degradation coefficients as follows:

$$k' = \frac{k}{K_S} X \qquad (10\text{-}18)$$

where k' = first order degradation rate constant (time^{-1})

Substituting Eq. 10-18 in Eq. 10-17 and solving for S yields the following expression for determination of contaminant concentration after treatment for the time interval t:

$$S = S_0 \, e^{-k't} \qquad (10\text{-}19)$$

In most cases where investigators have reported hazardous waste degradation rates in the literature, they have reported it as this first order degradation rate coefficient or in terms of half-life. Half-life can be determined using Eq. 10-19, yielding:

$$k' = \frac{0.633}{t_{1/2}}$$

where $t_{1/2}$ = half-life (days)

Example 10-4   Half-life calculation.   What is the half-life of a toxic compound for which a laboratory treatability test yielded a first-order degradation rate constant of 0.02 day^{-1}?

Solution. Substituting values of 0.5 for $\frac{S}{S_0}$ and 0.02 day^{-1} for k' in Eq. 10-19 and solving for t, yields:

$$t = -\frac{\ln 0.5}{0.02 \text{ day}^{-1}}$$

$$t = 35 \text{ days}$$

10-4 TREATMENT SYSTEMS

Biological treatment systems for hazardous wastes can be classified in many ways. Based on flow configuration, a system can be batch or continuous flow with or without recycling of biomass. A continuous flow system has subcategories to reflect complete mixing or plug flow. Whether the biomass is suspended (e.g., activated sludge) or attached (e.g., trickling filter), and whether free oxygen is maintained (i.e., aerobic) or not (i.e., anaerobic), also represent other classifications.

The preceding classifications traditionally apply to treatment of municipal and industrial wastewater, and likewise would apply to the treatment of both diluted and concentrated liquid hazardous waste. However, biological processes can also treat organic sludges and solids, as well as soil contaminated with organics. These non-liquid wastes necessitate a different classification system. For purposes of presentation in this book, the classification of treatment systems is made as shown in Table 10-7.

A key technical factor for selecting the appropriate system is the capability to provide proper contact between the organic constituents of the hazardous waste and

TABLE 10-7
Classification of biological treatment systems

System	Form of waste	General aspects
Conventional liquid phase	Liquid	Attached or suspended growth, with or without recycle of cellular mass, and continuous or batch flow; principally aerobic, but anaerobic can play key roles.
In situ	Liquid and contaminated soil	Treatment of ground water in its natural state below the surface and treatment of subsurface sources of ground water contamination.
Slurry-phase	Sludge or solid (e.g., contaminated soil)	Similar to conventional systems for liquid treatment except that non-volatile solids content in the reactor may range from 5% to over 50%.
Solid-phase	Sludge or solid (e.g., contaminated soil)	Unsaturated conditions if not minimal free moisture; land treatment, composting, and heaping are prime examples.

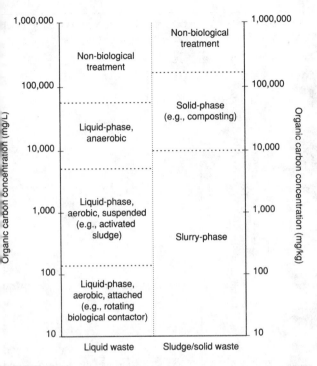

FIGURE 10-15
Appropriate treatment system as a function of concentration and form of waste.

the microbial population. This capability depends primarily on the state of the waste and its concentration. An approximate delineation of the system offering the most capability as a function of these two variables is shown as Fig. 10-15.

10-5   CONVENTIONAL TREATMENT

Process Description

Conventional liquid-phase treatment consists of technologies, or variants, originally developed for treatment of industrial wastewater. The typical hazardous wastes treated by this method include contaminated ground water and industrial process wastewater containing toxic organic substances. The method consists of passing aqueous hazardous waste through a reactor containing either suspended or attached biomass of highly active and acclimated microorganisms. Flow can be continuous or batch, and the reactor can be operated under aerobic or anaerobic conditions. Oxygen is added in aerobic systems by diffused or surface aeration. The liquid waste receives treatment both before and after biological treatment. The sequence of a general system is depicted simply in Fig. 10-16.

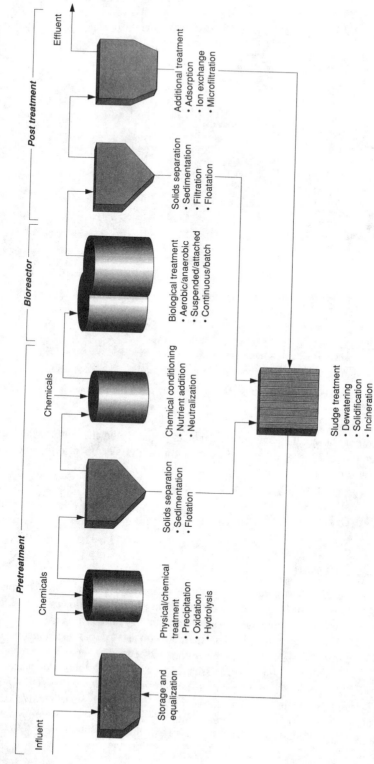

FIGURE 10-16
Conventional liquid-phase treatment.

Pretreatment could consist of several steps dependent upon the types of waste:

1. **Equalization**—to dampen/modulate hydraulic surges and variable organic loadings in continuous flow systems.
2. **Chemical Treatment**—typically to precipitate toxic metals, if present, but could involve other steps such as breaking of emulsions.
3. **Physical Separation**—sedimentation of metallic precipitates, removal of floating material, etc.
4. **Conditioning**—typically to supply nutrients and to adjust pH to optimum range.

After pretreatment, the liquid waste flows into the bioreactor where the dissolved organics are metabolized by the biomass with a resulting growth of cellular mass. This can be achieved only by biodegradation. The actual yield of biomass depends on numerous factors. As stated earlier, not all utilization of substrate results in synthesis of cells; some substrate is oxidized to produce energy.

The removal of a portion of the organic waste occurs by methods other than biological (i.e., abiotic losses). Volatilization can result in significant removal of some organics, especially in aerated systems. Some organics may not be metabolized but adsorbed with colloidal contaminants onto the biomass. In sum, abiotic losses can account for a significant amount of the organic waste removed with no reduction of the toxic nature of the waste. The chemical constituents have merely been transferred to other media which may require their own particular treatment.

Suspended Growth Systems

Figure 10-17 depicts a typical suspended growth system in which the bioreactor is mixed continuously, thereby facilitating contact between the substrate (i.e., organics in the liquid waste) and the suspended biomass. The effluent washes the suspended biomass out of the bioreactor to a separation step where the biomass is separated from the treated effluent, typically by sedimentation. Subsequent steps (e.g., filtration) may also be utilized dependent upon the regulatory limitations on the quality of the effluent.

A major portion of the settled biomass (i.e., sludge) is returned to the bioreactor to maintain the proper solids retention time (i.e., sludge age) and the proper ratio between substrate and acclimated microorganisms (i.e., food:biomass ratio). Recycling increases the solids retention time beyond the simple hydraulic retention time. This speeds degradation and reduces the size of the needed reactor. Excess sludge must be removed for treatment and disposal. The typical sludge age of an effective system is about 20 to 30 days, and no system should have a sludge age outside the range of 10 to 50 days. A lower sludge age would tend to result in system washout, instability, or filamentous organism growth. A higher sludge age could produce non-settling, pinpoint floc. An additional consideration is that the biomass MLSS should exceed 1000 mg/L to promote proper flocculation.

The suspended growth process described above is a conventional system; there are many variations. The incorporation of hydraulic design measures can create plug-flow conditions, unlike the completely mixed conditions of the conventional system. Plug-flow systems theoretically can be more efficient than continuous-flow systems;[34]

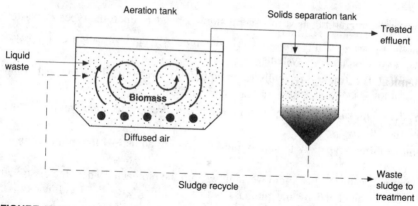

FIGURE 10-17
Completely mixed, suspended growth treatment system.

however, they are more easily upset by rapid changes in loadings, particularly of toxic compounds. Completely mixed suspended growth systems withstand shock loadings of non-degradable toxic compounds better than most other systems, because influent concentrations are diluted almost immediately throughout the reactor.

In another variation, powdered activated carbon is added to the bioreactor to remove organic compounds not metabolized by the biomass. It also adsorbs organics which otherwise could be toxic to the biomass. The largest application of this technology is the 40-MGD activated sludge system operated by Du Pont's Chambers Works in Deepwater, NJ. A discussion of this treatment plant can be found in a current book on hazardous waste technologies.[35] This technology has been patented as PACT and is depicted in Fig. 10-18.

Suspended growth systems are best suited to wastes containing moderate to high concentrations of organics, as high as 5000 mg/L of TOC. Low concentrations of organics do not yield enough growth to provide a properly flocculated biomass. This produces poor final settling of biomass such that new cells cannot replace those lost through washout and endogenous decay.

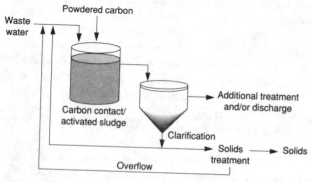

FIGURE 10-18
Zimpro PACT process (powdered carbon-activated sludge system).

Example 10-5   Biomass calculations. An industrial plant generates a process wastewater stream which has averaged 440 gal/min and 4,800 mg/L of COD over the past three years. The regulatory agency has stated that the organic chemical constituents are hazardous and must be reduced to 100 mg/L measured as COD. The stormwater runoff from the site also contains the organic chemicals of concern, and all runoff from an annual average rainfall of 60 inches must also be treated.

The industrial plant conducted some studies and found the following:

- Rainfall/runoff measurements over three months reported a total of 12 inches of rainfall over that period which yielded 3,300,000 gallons of runoff containing 300 mg/L of COD.
- Treatability tests of the process wastewater reported the following data:
  - BOD:COD::1:2
  - Biomass production rate $(Y) = 0.7$ mg/mg BOD removed
  - Endogenous decay rate $(b) = 0.05$ day^{-1}

The plant is evaluating the feasibility of treating the wastewater and runoff in a full-scale completely mixed, suspended growth system with solids recycle. Based on the above information, could a sufficient biomass concentration be maintained to prevent settling problems?

Solution. Calculate average runoff:

$$\text{Runoff} = (3.3 \times 10^6 \text{ gal}/ 12 \text{ in}) \times (60 \text{ in/yr}) = 16.5 \times 10^6 \text{ gal/yr}$$

$$\underline{= 45,200 \text{ gal/day}}$$

Calculate average daily flow:

$$\text{Flow} = (440 \text{ gal/min} \times 1440 \text{ min/day}) + 45,200 \text{ gal/day}$$

$$= 633,600 \text{ gal/day} + 45,200 \text{ gal/day}$$

$$= 678,000 \text{ gal/day}$$

Calculate average BOD of influent:

$$\text{Avg. COD} = \frac{(633,600 \times 4800 \text{ mg}/\text{L}) + (45,200 \times 300 \text{ mg}/\text{L})}{678,800}$$

$$\underline{= 4500 \text{ mg/L}}$$

$$\text{Avg. BOD} = 0.5 \times 4,500 \text{ mg/L}$$

$$\underline{= 2250 \text{ mg/L}}$$

Estimate suspended biomass (MLSS):

The steady-state equations include a number of interdependent variables. Use of these equations may necessitate assuming values for select variables. These assumed values represent starting points from which refinements may be necessary. For this example, it is first necessary to assume a sludge age (θ_c) and hydraulic detention time (θ_h) as follows:

$$\theta_c = 50 \text{ days}$$

$$\theta = 7 \text{ days}$$

With this starting point, the suspended biomass can be calculated from an equation provided in Table 10-6 for this type of system:

$$X = \frac{Y(S_0 - S)}{(1 + b\theta_c)} \frac{\theta_c}{\theta}$$

$$= \frac{0.7[2250 - (0.5 \times 100)]}{(1 + 0.05(50))} \frac{50}{7}$$

$$= 3100 \text{ mg/L}$$

This is a reasonable level for the suspended solids in a reactor. It is within the range defined by a minimum of 1000 mg/L and a maximum of 5000 mg/L as discussed earlier.

Attached-Growth Systems

Attached-growth (or fixed film) systems rely on the ability of microorganisms to attach to surfaces of inert media. Contaminated water is passed through a bioreactor which houses the media. The resulting microbial growth attaches to the media and forms a thick film. The biomass remains in the reactor except that which sloughs off the supporting media. Part of the effluent and biomass may be recycled.

Attached-growth systems can develop high concentrations of biomass in relatively small reactors because relatively little biomass is lost with the effluent. This results in a low food:microorganism ratio and the long solids retention time necessary to foster slow-growing microorganisms. Both enhance degradation of hazardous contaminants, particularly wastes with low concentrations of organics. Attached-growth systems require that the microbes have the ability to attach to the appropriate media; such ability varies among strains of microbes and must be examined during the design phase.

The plug-flow nature of attached-growth systems would seemingly make them more vulnerable to shock loadings. It has been shown, however, that significant biodegradation can occur in attached-growth systems at substrate concentrations which would inhibit suspended growth reactors.[36] It is believed that the outer layers of the biofilm protect the inner layers from toxic effects.

The field of municipal sewage treatment historically has relied extensively on trickling filters, a form of attached-growth treatment using rock as the inert media. For a period, numerous rotating biological contactor systems were constructed for biomass attainment. These rotating contactors consisted of closely spaced vertical discs mounted to a rotating shaft suspended horizontally just above the surface of the liquid undergoing treatment. The circular discs provide a large surface area for supporting the biofilm. By rotating the discs, the biomass is aerated and treatment occurs. A pilot-scale unit of such a system was reported to have reduced an already low concentration of chlorinated phenolic compounds by 36–68%.[37]

Other applications feature columnar beds (or towers) packed with synthetic porous media having a high area to volume ratio. The influent is normally sparged with air before moving as plug flow through the columns. These units are usually constructed in modular form to provide the necessary treatment volume (i.e., treatment time). Such a system was used to treat ground water contaminated with ethylene dichloride (EDC).[38] The system reduced concentrations of EDC from 560 mg/L in the influent to about 1 mg/L in the effluent. A check of volatilization indicated that 99% of the removal of EDC was due to degradation. Batch tests indicated an initial degradation rate of 60 mg/kg/hr, dependent of course on the substrate.

Another innovation is the fluidized bed reactor. A fluidized bed is achieved by growing the biofilm not on a fixed media but on particles of sand or other inert media. The influent is introduced from below the bed in an upflow reactor; the upflow velocity must be sufficiently high to keep the particles suspended. Withdrawal of treated effluent from the top of the reactor maintains the upflow velocity. A portion of the inert media is withdrawn periodically as the bed expands. The virtually complete mixing reduces the susceptibility to shock loadings. Further, these systems can remove more organic loading per unit reactor volume than other fixed-film processes.[39]

Two examples of modern attached-growth systems are shown in Fig. 10-19.

Current research has studied the use of rotating biological contactors to treat trichloroethylene (TCE) and related volatile compounds in a cometabolic process.[40] An enclosed pilot-scale unit was fabricated to provide a gas-tight reactor which: (1) prevented volatilization of contaminants and (2) maintained methane (the co-metabolite or primary substrate) at fixed concentrations. After developing a steady-state biomass,

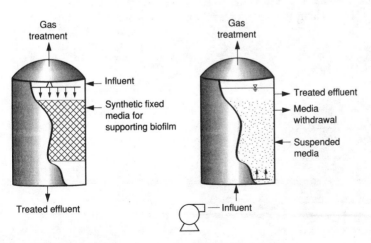

Downflow, fully-packed reactor   **Fluidized-bed reactor**

FIGURE 10-19
Examples of attached-growth systems.

batch tests were conducted of water contaminated with 700 mg/L each of the chlorinated organics: tetrachloroethylene (TECE), TCE and dichloroethene (DCE). TCE and DCE degraded at first-order kinetic rates, with faster rates observed at the lower concentrations of methane tested.

Aerobic Batch Reactor Systems

A batch reactor combines equalization, biotreatment, and sedimentation in a single tank. The three steps are conducted sequentially (see Fig. 10-20), hence the term sequencing batch reactor (SBR) is used to describe these systems. The biomass can be maintained by discharging only the clarified effluent after the sedimentation step. The time allocated to each step can be adjusted within the constraint of the incoming flow. This provides much greater flexibility than continuous flow systems, allowing SBRs to achieve high performance when treating variable wastewaters. The simplicity and flexibility of the SBR approach makes it especially appropriate for small-scale processes.

If the flow to be treated is continuous, at least two SBRs are needed: one to fill while the other sequences through the treatment steps. Placement of SBRs in series helps to ensure that the biomass in each tank is acclimated to the transformed compounds being discharged from the previous tank. This can speed the sequential transformation of waste. A commercial hazardous waste treatment facility has used SBRs to remove 75% of the organic contamination in leachate and other aqueous wastes.[41]

A demonstration project treated ground water contaminated with PAHs from wood-preserving wastes in a two-stage batch reactor system. The first tank operated for 4 days and the second 8 days. The first tank received concentrations of specialized bacteria for treating low and medium molecular weight PAHs, and the second tank was prepared with bacteria for treating high molecular weight PAHs. The percent removal for low, medium, and high molecular weight PAHs was > 99, > 98, and 98%, respectively. The system reduced pentachlorophenol concentrations by 88%.[42]

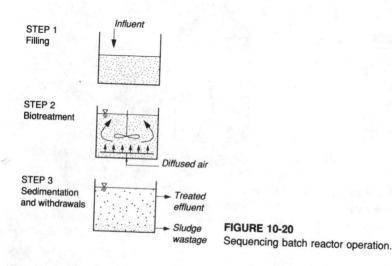

FIGURE 10-20
Sequencing batch reactor operation.

A European research project has shown the potential capability of SBR to foster genetic exchange in the microbial community of a bioreactor. This is importan. because bacteria adapted in a laboratory to a specific waste frequently do not survive in the bioreactor because of predatory activity or because of washout due to failure to floc or attach. The experiment compared rapid (20 minutes) and slow fill rates for an SBR operation, under the following conditions:

Waste influent:	660 mg/L of the pesticide 2,4-D and 270 mg/L of fructose
Fill time:	Fast fill—20 min Slow fill—6 to 24 hr
Reaction time:	Constant
Drain time:	30 min
Volume exchange per cycle:	50%
Inoculation:	Donor bacteria with laboratory-developed capability of degrading 2,4-D

Under "fast fill" conditions, transconjugants (i.e., microorganisms that have acquired genetic material for another microbe) acquired the capability to degrade 2,4-D from the inoculated donors and increased significantly in population. The "feast or famine" conditions created by the fast fill cycle "... resulted in selective pressure on the transconjugants to preserve the newly acquired capability for ..." degrading 2,4-D. In the slow fill cycle, they "abandoned" this capability. Longer reaction times also resulted in the loss of this capability.[43]

Anaerobic Systems

The treatment of municipal sewage has long depended on anaerobic systems to digest excess sludge to methane and carbon dioxide in an oxygen-free environment. Industries with high-strength, readily degradable wastewater (e.g., food products, pharmaceutical, pulp and paper, and chemical), have increasingly used anaerobic systems because of the savings realized with energy conservation and greatly reduced sludge production. The specific systems used in these applications typically feature attached growth processes similar to those shown in Fig. 10-19 with modifications to maintain anaerobic conditions.

There have been few applications of anaerobic treatment of hazardous waste. One application with demonstrated effectiveness is the treatment of phenols, cresols, and other compounds associated with coal gasification.[44] Anaerobic treatment probably has its greatest potential in treating those wastes where reductive dehalogenation can occur more efficiently than by oxidative degradation. Examples of such wastes include volatile, chlorinated hydrocarbons. Due to their volatility and relatively slow degradability under aerobic conditions, wastes such as TCE are not well suited for the aerated processes. Both problems are overcome in a cometabolic approach using an enclosed, attached growth process (e.g., methanogens maintained on a filter of

tment of tetrachloroethene (PCE)

PCE degraded (μmol/bottle)	Transformation product (μmol/bottle)	
	Vinyl chloride	Ethene
55	10	47
3	55	2	58
4	55	0.4	56

inert media through which the TCE-contaminated water is pumped in a methane-rich environment).

One researcher reported the nearly complete reductive dechlorination of tetrachloroethene (PCE) in a methanol-enriched culture. The tests in 160-mL serum bottles resulted in nearly complete transformation of a 94 mg/L solution of PCE to ethene with "relatively little residual vinyl chloride" as shown in Table 10-8 (see also Fig. 10-5). Of interest is that this transformation occurred in the absence of methanogenesis (methane formation). It is surmised that high levels of PCE or its degradation products, including ethene, could have inhibited methanogenesis. Instead, acetogenesis (acetic acid formation) dominated with "about two thirds of the reducing equivalents from methanol" used to produce acetate with the remaining one-third involved in reductive dechlorination. Although this test does not identify the specific dechlorinating microorganism, work such as this indicates the promise of anaerobic treatment for specialized cases.[45]

10-6 IN SITU BIOREMEDIATION

Background

In situ is a Latin phrase meaning "in its original place." In situ bioremediation is the method of treating contaminated ground water and subsurface contaminants where they are found without excavating the overlying soil. That is, essentially all of the treatment actually takes place in the subsurface. As discussed in Sec. 4-1, contaminants that have migrated into the subsurface from sources such as product spills and leaking disposal facilities exist in three phases: (1) free product, (2) adsorbed or otherwise bound to soil particles and interstitial openings in the geologic matrix (sorbed phase), and (3) dissolved in the ground water (solute phase). The amount in any phase varies, dependent upon the characteristics of the contaminants and the subsurface, and can be described by such factors as the soil-water partition and adsorption coefficients (see Secs. 3-2 and 4-3). In most cases, the mass in the solute phase is very small compared with the mass in either the sorbed or free phases; however, the solute phase impacts a much greater volume of the subsurface because of the mobility of ground water. In situ bioremediation can degrade directly the contaminants in all three phases.

Example 10-6   Phase distribution of fuel in subsurface.   A slow leak from a diesel fuel tank has contaminated a site, both the soil below the tank and the underlying aquifer. After removal of free product, testing shows the following information:
Extent of heavy contamination of soil at source (sorbed phase)

$$10 \text{ m} \times 25 \text{ m} \times 3 \text{ m}$$

Areal extent of ground water contamination (solute phase)

$$100 \text{ m (maximum width)} \times 300 \text{ m (length) at 5 m depth}$$

Average concentration of diesel fuel:
      Dissolved in ground water   6 mg/L
      Absorbed in soil at source   4000 mg/kg
Soil characteristics:
      Type   silty sand
      Density   1800 kg/m^3
Aquifer characteristics:
      Average depth contaminated with fuel   5 m
      Effective porosity   30%
      Hydraulic gradient   0.0042 ft/ft
      Hydraulic conductivity   0.02 cm/sec
Based on this information and assuming that ground water contamination plume follows a truncated oblate ellipsoid, what is the phase distribution of fuel?

Solution.   The volume of contaminated ground water can be approximated as a pyramid on its side with base 5 m × 100 m and a height of 300 m. Thirty percent of this volume is void space that can contain ground water.

Volume of contaminated ground water $= 1/3(100 \text{ m} \times 300 \text{ m} \times 5 \text{ m})(0.30)$
                                            $= 15{,}000 \text{ m}^3$

Mass of fuel in solute phase $= 15{,}000 \text{ m}^3 \times 6 \text{ mg/L} \times 10^3 \text{ L/m}^3 \times 10^{-6} \text{ kg/mg}$
                                            $= \underline{\underline{90 \text{ kg}}}$

Mass of contaminated soil at source $= (10 \text{ m} \times 25 \text{ m} \times 3 \text{ m})(1800 \text{ kg/m}^3)$
                                            $= 1.35 \times 10^6 \text{ kg}$

Mass of fuel in sorbed phase $= 1.35 \times 10^6 \text{ kg} \times 4000 \text{ mg/kg}$
                                            $= \underline{\underline{5400 \text{ kg}}}$

Thus, the mass of contaminants in the sorbed phase is 60 times greater than that dissolved in the ground water.

Remediation Alternatives

In some cases (e.g., petroleum products), the free product may float on the ground water table and can be removed. Such removal is dependent upon a number of hydrogeologic and other factors, and a significant amount will still remain in the sorbed phase. This represents a continuing and long-term source of ground water contamination due to desorption by infiltrating precipitation and the seasonal horizontal and vertical movement of ground water.

The contaminated ground water can be intercepted near the point of desorption, pumped to the surface, and treated. This is referred to as "pump and treat" remediation. The time scale for natural desorption and "pump and treat" operations to reduce contamination to low levels can be enormously long because of the slow release of contaminants from the free product and sorbed phases (see Sec. 4-1). Even when contaminants are soluble, other factors act to retard their release. For example, in the frequent contamination case of underground fuel tanks, the hazardous compounds of most concern (i.e., benzene, toluene, and ethyl benzene) are relatively soluble but generally contained within the immiscible bulk fuel phase. Another situation is that of non-aqueous phase liquids with a higher specific gravity than water (dense non-aqueous phase liquids or DNAPLs). DNAPLs are particularly difficult, if not impossible, to remediate because of their propensity to accumulate in isolated but concentrated "pools" in the low spots in an aquifer (see Fig. 4-22a).

There are numerous accounts that document the long time scale involved with desorption. For example, a study of gasoline-contaminated soils showed that infiltration of 46 pore volumes of water (which would take years in some aquifers) removed only 1.6% of the absorbed fraction.[46] An analysis by the EPA showed that "pump and treat" operations rapidly lowered ground water contamination to 10 to 50% of the initial concentrations, but contamination quickly reached a residual concentration which stayed fairly constant with continuous pumping for a long period of time.[47] Leading ground water scientists predict that to lower ground water concentrations by a factor of 100 could require 100-200 years of pumping for ideal conditions,[48] and thousands of years for contaminants with very low solubility.[49] It should be noted that "pump and treat" frequently represents a viable alternative for plume control, and it may even offer an adequate level of treatment if low concentration levels are not required. However, wide experience has prompted scientists to conclude that "pump and treat" operations generally are incapable of achieving drinking water standards.[50]

Recovery of free product and treatment of ground water address two phases of contamination but do not address continuing releases from the sorbed phase. Several physical and chemical methods are presented in Chaps. 9 and 11 for treating the sorbed contaminants; however, these methods, especially if they involve excavation, can be expensive. As an alternative, the sorbed phase can be treated directly by **in situ bioremediation** which reduces contamination in some cases to acceptable levels in periods as short as one or two years. The concept is a simple one; it relies on the natural biodegradation that occurs spontaneously in the subsurface. The difference is that in situ bioremediation enhances the process through stimulation and management of the subsurface microbiological communities. However, its successful implementation requires careful consideration of a complicated set of factors involving not only microbiology, but also hydrogeology, geochemistry, and engineering.

Status of Development

The concept of in situ bioremediation was developed in the 1970s by a subsidiary of Sun Oil Company for the treatment of gasoline, diesel fuels, and other mineral oil hydrocarbons. Few applications had been made until the mid-1980s when technical advancements and heightened interest prompted a much greater level of activity. As of 1987, more than 30 applications of in situ bioremediation have been described in the literature.[51] Project experience still has been confined largely to remediation of simple petroleum hydrocarbons in which first order biodegradation rates as high as 10 mg/kg·day have been recorded.[52] Interest is shifting toward other chemical contaminants with good field results reported for phenols and simple aromatic compounds. Research on in situ bioremediation of chlorinated compounds has intensified in the United States and Germany.

Process Description

In situ bioremediation is practiced in the aerobic mode, with anaerobic processes holding future promise for special applications. Ground water naturally contains low concentrations of oxygen because of the minimal reaeration resulting from its laminar flow below the surface. Thus, even modest biological activity readily depletes the oxygen in the ground water. For example, the oxidation of one mg of gasoline theoretically requires 2.5 mg of oxygen.[53] The subsurface environment is also typically nutrient-starved. The concept of aerobic in situ bioremediation thus depends upon the delivery of oxygen and nutrients to the contaminated subsurface, typically by withdrawing ground water, adding oxygen and nutrients, and reinjecting the enriched water. The injected water with nutrients and oxygen moves through the aquifer and stimulates the growth of native microorganisms, resulting in the degradation of the contaminants.

In situ biodegradation follows the same principles as other biological treatment methods except that the contaminated subsurface, rather than an engineered vessel or lagoon, functions as the bioreactor. The subsurface cannot easily be controlled; the injection of water changes ground water flow patterns which could spread contamination unless injection is linked with ground water withdrawals and containment barriers. The combination of injecting water upgradient from withdrawing ground water increases the hydraulic head between the two points, thus speeding the rate of flow of enriched water through the aquifer. In the design of in situ systems, the hydrogeologic factors discussed in Sec. 4-2 are perhaps even more important than the biodegradation factors.

Typical System

Figure 10-21 depicts a typical contamination problem and the basic elements of in situ bioremediation. The change in ground water flow patterns is also depicted.

Many variants exist to this system including the following:

1. Recovered ground water can be treated by conventional liquid-phase systems before reinjection.

2. Part of the recovered ground water can be discharged elsewhere, if permitted by regulatory agencies.

3. The injected water, usually ground water extracted downgradient from the site, is augmented if necessary by water supplied from outside sources.

4. For shallow ground water conditions, trenches can be used instead of wells (see Fig. 10-22).

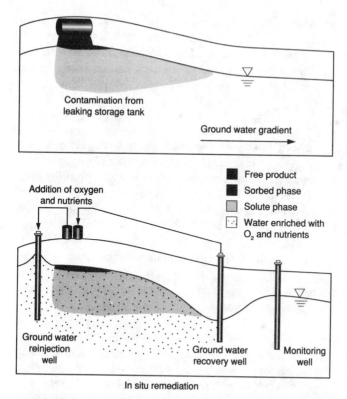

In situ remediation

FIGURE 10-21
In situ bioremediation system.

5. Oxygen can be supplied by sparging the ground water with air or pure oxygen or by adding hydrogen peroxide.

6. To treat the sorbed phase in the vadose zone, the recovered ground water can be applied to an infiltration basin overlying the contaminated vadose zone (see Fig. 10-23). Surfactants might be added as well to help in the desorption of contaminants from the soil.

Theory

The supply of oxygen is almost always the rate limiting factor in operating an in situ bioremediation project. For example, a test performed on gasoline-contaminated soils has shown the dependence of bacterial growth on available oxygen.[54] The experiment involved adding a known amount of gasoline to 50 ml columns of wet sand, adding nutrients, and maintaining different design levels of oxygen for two weeks. At the conclusion, counts of bacterial growth correlated closely with the available oxygen (i.e., dissolved oxygen plus that available from the addition of hydrogen peroxide) as shown in Table 10-9.

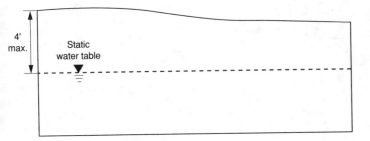

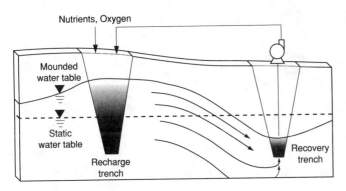

FIGURE 10-22
Use of trenches to recover and reinject ground water.

The test also measured removal of gasoline by biodegradation. The amount of biodegraded gasoline also correlated with the maintained level of oxygen. The amount of gasoline which had been degraded greatly exceeded the amount which, in a control test, flushed out due to dissolution.

The presence of even nominal amounts of organic contaminants will deplete the subsurface of oxygen, thereby creating anaerobic conditions which do not favor the degradation of most compounds as rapidly as sustained aerobic conditions. Thus, in situ bioremediation requires that the subsurface be artificially oxygenated. Sparging the ground water below the surface water with air, or aerating the injected water, rarely can meet the demand for oxygen in biologically active conditions. Sparging with pure oxygen or ozone can increase concentrations of dissolved oxygen to 40 mg/L. Although this is higher than the 8–12 mg/L achievable with aeration, it still may not keep pace with a biologically active system. For these reasons, emphasis has been placed on the use of hydrogen peroxide.

In water, the following hydrogen peroxide solution dissociation occurs:

$$2H_2O_2 \rightarrow 2H_2O + O_2$$

Aerobic bacteria produce an enzyme termed catalase which catalyzes this reaction, thus releasing the oxygen from its chemical bond to support aerobic biodegra-

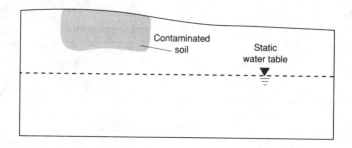

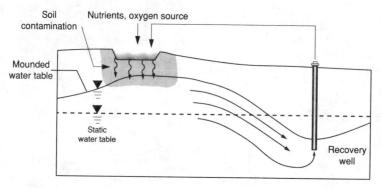

FIGURE 10-23
Use of infiltration basin to treat vadose zone.

TABLE 10-9
Correlation of bacterial growth with oxygen level

Maintained level of oxygen (mg/L)	Heterotrophic bacteria (10^6 colony forming units)	Gasoline-utilizing bacteria (10^6 colony forming units)
8	0.05	0.0001
40	5.5	0.7
112	75	27
200	207	31
Correlation with oxygen level	0.979	0.933

dation. The presence of iron in the subsurface can chemically decompose the hydrogen peroxide molecule, prematurely releasing the oxygen which can be lost to the atmosphere. Further, the released oxygen gas can physically inhibit ground water movement.[55] Addition of phosphate, however, can stabilize the chemical decomposition of peroxide until it is catalyzed by enzymes from the degrading microorganisms.

Besides providing oxygen for biological degradation, the addition of hydrogen peroxide may also chemically oxidize some of the contamination. In the case of some biologically resistant compounds (e.g., trichloroethylene), the oxidation product may be more easily biodegraded than the parent compound.[56]

Because it is possible to obtain concentrations of greater than 1% of hydrogen peroxide in water, hydrogen peroxide serves as a large storage reservoir for continuous release of chemically bound oxygen. Saturated solutions, however, are inappropriate because hydrogen peroxide can have a toxic effect on microorganisms at concentrations as low as 1000 mg/L. Further, high concentrations can be wasteful as excessive degassing of oxygen may occur. For these reasons, the range of 100–500 mg/L in the ground water phase is typically used for in situ applications. To acclimate microorganisms, practitioners suggest starting with 50 mg/L and increasing step-wise to higher concentrations.[57]

Example 10-7   Oxygen demand for in situ treatment. For the contaminated ground water case described in Ex. 10-6, what is the mass of oxygen required to mineralize the gasoline? Which is the best alternative for oxygen supply: natural, sparging, or hydrogen peroxide?

Solution. From the stoichiometry of the degradation of simple hydrocarbons discussed earlier in this chapter, it can be seen that their mineralization requires about 2.5 grams of oxygen per gram of hydrocarbon.

$$\text{Oxygen demand} = (2.5 \text{ gram/gram})(5,400 \text{ kg} + 90 \text{ kg})$$

$$= 13,725 \text{ kg}$$

Natural supply of oxygen relies totally upon movement of upgradient ground water which typically contains 6 mg/L of dissolved oxygen. Using Darcy's law (Eq. 4-6) and Eq. 4-14:

$$v_s = \text{Average linear seepage velocity} = \frac{ki}{n}$$

where $k =$ hydraulic conductivity $= 0.02$ cm/s
$i =$ hydraulic gradient $= 0.0042$
$n =$ effective porosity $= 0.30$

$$v_s = \frac{\left(2 \times 10^{-2} \text{ cm/s}\right)(0.0042)}{0.30}$$

$$v_s = 24.2 \text{ cm/day}$$

Flow through highly contaminated area $= (24.2 \text{ cm/day})(100 \text{ m})(5 \text{ m})(0.30)$

$$= 36,300 \text{ L/day}$$

Natural supply of oxygen $= (36,300 \text{ L/day})(6 \text{ mg/L}) \times 10^{-6} \text{ kg/mg}$

$$= 0.22 \text{ kg/day}$$

At this rate, natural supply of oxygen is not acceptable because it would take more than 170 years to biodegrade the contamination aerobically. Sparging the ground water with pure oxygen would augment the supply of oxygen. Three sparging wells, each with an assumed radial influence of 1 m (3.3 ft) for the full saturated thickness of the contamination, will add oxygen to the following flow:

$$\text{Flow through influence of sparging wells} = 3(24.2 \text{ cm/day})(2 \text{ m})(5 \text{ m})(0.30)$$

$$= 2180 \text{ L/day}$$

Sparging with pure oxygen will yield dissolved oxygen concentrations of 40 mg/L in the sparged water. Thus, the additional supply of oxygen is calculated as

$$\text{Sparged supply of oxygen} = (2180 \text{ L/day})(40 \text{ mg/L})$$

$$= \underline{\underline{0.087 \text{ kg/day}}}$$

Thus, sparging with pure oxygen in three wells does augment the natural supply of oxygen but by only 40%, too little to reduce the treatment time significantly.

A hydrogeologic investigation shows that interception of 300 L/min (79 gal/min) of ground water downgradient of the plume and reinjection through two wells upgradient at the source will hydraulically contain further migration of the plume. Sparging the reinjected ground water with air or pure oxygen will add the following oxygen,

$$\text{Sparging with pure oxygen} = (300 \text{ L/min})(40 \text{ mg/L})$$

$$= 17.3 \text{ kg/day}$$

$$\text{Sparging with air} = (300 \text{ L/min})(8 \text{ mg/L})$$

$$= 3.5 \text{ kg/day}$$

Theoretically, if oxygen supply were the rate limiting factor, surface sparging of reinjected ground water would meet the oxygen demand in 2.2 years using pure oxygen and in 11 years using air.

Because two molecules of hydrogen peroxide will yield one molecule of oxygen, the addition of hydrogen peroxide would supply more oxygen than the 40 mg/L yielded by sparging when H_2O_2 concentrations in the injected water are higher than:

$$\frac{2H_2O_2}{O_2} = \frac{2 \times 34 \text{ g/mol}}{32 \text{ g/mol}} \times 40 \text{ mg/L}$$

$$= 85 \text{ mg/L}$$

Whether to add hydrogen peroxide or sparge with pure oxygen depends largely upon degradation kinetics, the toxicity of H_2O_2 to the microbiota, and economics. Degradation testing suggests a first order degradation rate constant of 0.007 day^{-1}, which yields a half-life for the contamination of 100 days. This suggests that sparging with pure oxygen could not supply the total oxygen requirements. To eliminate oxygen supply as the rate-limiting factor, the use of hydrogen peroxide should be evaluated.

Hydraulic Controls

Supplying oxygen is only one aspect of developing an in situ bioremediation project. For example, the lowering of pH due to the production of CO_2 and the retention of nutrients on soil particles can create problems.[58] Nevertheless, of all such considerations, controlling the hydraulic regime of the subsurface poses the most difficult challenge of in situ bioremediation. It requires that several factors be considered.

One important factor is the hydrogeology of the site. Vertical flow conditions and bedrock formations with complex fracturing or dissolution channels are to be avoided due to the potential for spreading contamination. The formation must be capable of transporting the injected enriched water fairly rapidly in controllable directions. Fine-grained formations restrict transport, and actually they need greater transport of oxygen, because they have a higher capacity for retention of contaminants. Sandy formations are ideal, and practitioners generally state that formations with permeabilities of less than 1×10^{-4} cm/sec are not favorable and possibly not feasible.

Given a site with favorable hydrogeologic conditions, an integrated network of withdrawal and injection wells is still needed to prevent the spreading of contamination. The number of wells depends upon many factors including the transmissivity of the aquifer (see Sec. 4-2) and the dimensions of the contamination. Hydraulic barriers such as slurry trench walls may also be needed (see Sec. 16-5). Trenches may prove more economical and effective for shallow ground water conditions.

Simply injecting water enriched with oxygen and nutrients may not mix the solution adequately with the contaminated ground water. Short circuiting can occur. The design of the injection system must receive careful attention, particularly the three dimensional spatial relationships among points of withdrawal and reinjection.

Even with a carefully designed system of wells, the hydraulic characteristics of an aquifer may change upon implementation of the in situ method. Plugging and fouling of an aquifer can occur due to enriched growth of microorganisms, precipitation of nutrients, and other factors. The results of laboratory column studies show that permeability and porosity can decrease by factors of up to 10^{-3} and 10^{-1}, respectively, and dispersivity can increase by 10^3.[59] The operation and monitoring of in situ bioremediation systems require as much care as the design.

Applications

There is extensive experience with in situ bioremediation of simple hydrocarbons, and the method is widely recognized as an effective option dependent upon site conditions, particularly the hydrogeology. Remediation of chlorinated compounds is another matter. Laboratory tests and some field tests are showing that the method has much future promise. For example, a laboratory test used a reactor filled with glass beads to simulate a porous medium and the in situ bioremediation of 20–25 mg/L of ethylene dichloride (EDC) in ground water. Using hydrogen peroxide as a source of oxygen, a retention time of only 2–3 hours reduced the influent concentration of EDC by 80%.[60]

Transfer of this and other laboratory tests to the field could encounter several problems which would make in situ bioremediation infeasible. These problems include other organic chemicals acting as toxic inhibitors or naturally occurring predators eliminating the dechlorinating microorganisms Nevertheless, some field applications have succeeded in degrading chlorinated compounds.

A pilot-scale test of in situ treatment in Montana demonstrated the effectiveness of bioremediation and the equally important role played by ground water movement.[61] Releases of wood preservatives had resulted in a large (94 hectares, 2 km long) plume of ground water contamination. Total hydrocarbon contaminant concentrations exceeded 500 mg/L. Injection of hydrogen peroxide at concentrations of 50 mg/L created an oxygen-rich front. This front reached a downgradient well located 61 meters from the injection site in 148 days, elevating the dissolved oxygen in this well from 0.5 mg/L to 15–20 mg/L. Concurrently, the concentration of pentachlorophenol and PAHs declined from 3–7 mg/L to less than 20 μg/L, and then steadily declined further for the next 12 months to as low as 1 μg/L.

For any case of subsurface contamination, the challenge consists first of determining whether in situ bioremediation is a feasible option and then testing it in the field. Last, the method requires verification that it, rather than abiotic mechanisms, will reduce contamination to acceptable levels. This becomes an especially complex undertaking for contamination involving a mixture of organic substances. It involves an analysis of hydrogeologic, chemical, engineering, and microbiological factors. It can be approached in five steps as follows:[62]

1. Investigation of degree of contamination and the ground water flow regime.
2. Assessment of the feasibility of in situ bioremediation.
3. Detailed investigation of the site characteristics which affect removal of contamination by biotic and abiotic methods.
4. Differentiation of biotic from abiotic losses based on an analysis of physical-chemical parameters.
5. Verification of the effectiveness of in situ bioremediation based on microbiologic assays.

Emerging Applications

In situ remediation can be achieved by different methods other than the standard method featured on the preceding pages. Research has shown that anaerobic conditions successfully degrade pollutants; one example is alkylbenzene.[63] In an anaerobic system, injection of nitrate or nitrous oxide can serve as the needed alternate electron acceptor. This was demonstrated in a laboratory test which added 25 and 75 mg/L of nitrate as N to two sets of samples of core material contaminated with jet fuel, and incubated the samples anaerobically at 12°C. The samples contained about 8 mg/L of benzene and 5 mg/L each of toluene and ethylbenzene. After a lag period, degradation under these denitrifying conditions occurred at the first order rates of 0.03 day^{-1} for toluene and 0.002 day^{-1} for ethylbenzene. Rates were generally higher with the larger addition of nitrate. Benzene did not degrade within the 60-day test period.[64]

Another emerging application is the co-metabolism of chlorinated aliphatic solvents in the presence of methane. One field demonstration showed that 20%–30% of TCE was biodegraded in such conditions.[65] In another test, ground water was amended with methane, nutrients, and oxygen and reinjected into the aquifer. Downgradient wells showed 95% removal of vinyl chloride and about 20% of the TCE.[66]

In some cases, the microorganisms which can degrade the contaminants may not be naturally present in the subsurface. Successful in situ bioremediation in these cases would require identification and inoculation of exogenous microorganisms which are capable of metabolizing the specific contaminant. This has not yet been demonstrated for in situ bioremediation, partly because of hydraulic obstacles to the injection of large populations of exogenous microorganisms.

Rather than injecting the oxygen-enriched water directly into the ground water, it can be infiltrated through the vadose zone to degrade pollutants above the ground water table. This is not nearly as effective as remediation below the ground water table. Because of preferential pathways for flow in the vadose zone, not all voids are saturated with enriched ground water, which would then fail to contact a significant portion of the mass of pollutants.

In situ bioremediation of contamination in the vadose zone can also be achieved by a "dry" method. Rather than reinjecting ground water, air can be injected in the capillary fringe above the ground water table and exhausted at the ground surface through an air stripping unit. With proper controls, this upflow system can stimulate biodegradation, thereby reducing the treatment requirements of the exhaust air.[67]

10-7   SLURRY-PHASE TREATMENT

In this method, wastes are suspended with water or wastewater in a mixed reactor to form a slurry. Wastes can be sludges, solids, or contaminated soil. The agitation not only homogenizes the slurry but also promotes the following actions:

- Breakdown of solid particles
- Desorption of waste from solid particles
- Contact between organic waste and microorganisms
- Oxygenation of the slurry by aeration
- Volatilization of contaminants

Slurry-phase treatment is similar to conventional, suspended growth biological treatment except that the biomass is not recycled and the suspended solids can frequently be inert. The fundamental treatment steps are mixing/aeration, desorption, and biodegradation. These can occur in an in-place reactor or in a multiple-stage system, and flow can be continuous or batch. A multiple-stage system includes pretreatment steps, such as passing the soil through a sieve or adding surfactants, to enhance desorption and fractionate the waste. The treated soil and other inert solids typically can be left on the site.

Figure 10-24 shows a batch flow, in-place slurry-phase system to treat sludge in its holding lagoon. The mixing action resuspends the settled sludge. Figure 10-25 shows the components of a continuous flow, multiple-stage slurry-phase treatment system for treating contaminated soil.

Slurry-phase treatment degrades waste at a faster rate and requires less land area than solid-phase treatment.[68] This advantage is demonstrated by measuring the popu-

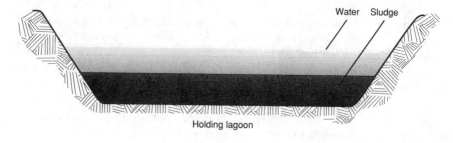

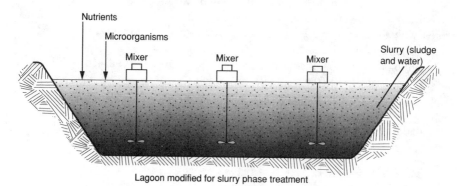

FIGURE 10-24
In-place slurry-phase treatment system.

lation density of microbial colonies. Table 10-10 summarizes the population densities of bacterial cells recovered from assorted samples of treated and untreated soils and sludges. This table demonstrates the capability of slurry-phase treatment for increasing biological activity in comparison with untreated samples and samples from land treatment systems.

The effectiveness of slurry-phase treatment depends upon a number of theoretical factors including pretreatment, desorption, solids concentration, mixer design and retention time.

Pretreatment

A waste can be pretreated in a number of ways to increase the effectiveness and efficiency of the overall system. Pretreatment methods include those which enhance

TABLE 10-10
Bacterial cell density (cells/gram of solids)[69]

Sample source	Contaminated soils	Waste sludge
Untreated	10^6	10^6
Land treatment	10^7–10^8	10^7–10^8
Slurry-phase treatment	10^8–10^9	10^9–10^{10}

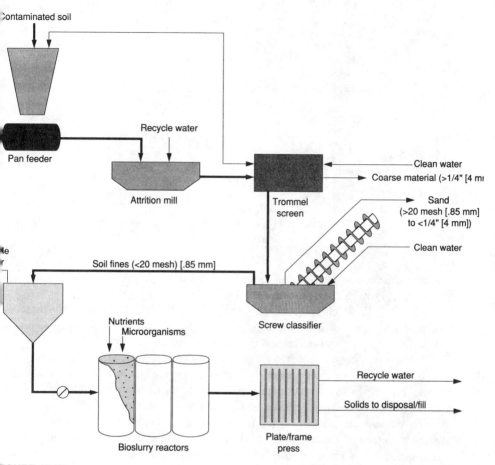

FIGURE 10-25
Multiple-stage slurry-phase treatment system.[70]

desorption (e.g., adding surfactants and reducing soil particle size) and those which concentrate the waste to be treated (e.g., fractionation of soil by particle size). Fractionation holds much merit for treating soil because contaminants preferentially adsorb to finer soil particles. Thus, fractionation can (1) eliminate the heavier particles which require higher energy to keep suspended and (2) concentrate the contaminants prior to degradation. A research project fractionated a soil contaminated with polyaromatic hydrocarbons (PAHs) and demonstrated that the fines, which could easily be kept suspended in a slurry, represented 37% of the dry mass of the soil but contained 94% of the contamination[71] (see Fig. 10-26).

Desorption

Desorption of the contaminant from soil particles must take place before biodegradation can occur. The exception is the case where waste exists as droplets to which

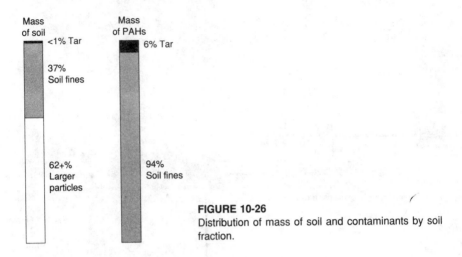

FIGURE 10-26
Distribution of mass of soil and contaminants by soil fraction.

microorganisms have attached. Desorption is particularly important with the soil-water suspensions characteristic of slurry-phase treatment.

The rate at which the pollutant desorbs from the soil particles is influenced by the physical and chemical properties of both the soil and the contaminant. Essentially, the same factors control desorption as control sorption (see Sec. 4-3). A portion of each pollutant is bound in a stable position along micropores within soil aggregates and particles. The relatively large size of microorganisms excludes them from the micropores, preventing their direct access to the sorbed contaminant. However, some microorganisms do attach to the outer surfaces of the soil particles. To be accessible, the contaminant must desorb to the micropore water and then move by diffusion to the outer surface of the soil particle where the solute is either degraded by attached microorganisms or passes to the bulk aqueous phase. Local equilibria may develop within the soil particle and act to slow further desorption. The high affinity of many compounds for soil (e.g., PAHs have a high affinity for clay minerals) will also inhibit desorption.

Investigations in the Netherlands and United States claim that desorption-diffusion from the sorbed phase to the solute phase is the rate limiting factor for slurry-phase treatment.[72,73,74] This conclusion derives from observations in laboratory tests that (1) the rate of degradation is a function of the concentration of the contaminant in solution rather than the total sorbed mass, and (2) microbial populations grow at a linear rather than logarithmic rate. Another investigator observes that this is true for lower molecular weight PAH compounds, even for those with fairly low solubility; however, for higher molecular weight PAH compounds, the degradation rate is the limiting factor.[75] With regard to the higher, unsubstituted PAHs, their solubility may be too low to support significant biological activity, again identifying desorption as the rate limiting factor.

It was noted in the Dutch research projects that when experimenting with different mechanical mixing techniques and the same pollutant, the mixing technique which reduced the soil particle size to 30 microns resulted in degradation rates several times greater than the technique which yielded a particle size of 60 microns.[76] Thus, an important consideration in pretreatment and mixing is the reduction of particle size. Similarly, washing the soil with a surfactant or other suitable chemical could facilitate transfer of contaminants into the aqueous phase.

Concentration of Solids in Reactor

The reactor volume for a given throughput can be reduced by maximizing the solids content. Various applications have reported satisfactory degradation for solids content as low as 5% and as high as 50% (dry weight basis). The maximum solids content depends upon the type of soil and the reactor design. The physical factors involved in keeping a slurry in suspension may establish an upper constraint of 30 to 40% solids.[77] However, the practical limit may be less if a high solids concentration proves toxic to the biomass.

Mixer Design

Mixing and particle suspension are critical to desorption and biological degradation. Keeping large concentrations of sludge and heavy soil particles suspended in a reactor requires high-energy mixers such as turbine mixers or surface aerators. The design of the mixers and reactor is perhaps the most critical factor in slurry-phase treatment. For example, baffle systems are required to improve mixing and prevent settling. Mixing does supply oxygen but may also promote volatilization, possibly necessitating the use of enclosed tanks.

Reactors can be as basic as installing a mixer in a lagoon as shown earlier in Fig. 10-24. However, the importance of the mechanical aspects of the reactor has prompted advanced designs. One concept applies technology developed in the mineral processing industry to improve energy efficiency[78] (see Fig. 10-27).

The dual drive design of the reactor suspends independently the finer and coarser particles found in contaminated soil. The axial flow impeller rotates at 20–30 rpm to agitate the slurry. Other rake arms turn at 2 rpm and push the coarser particles to a central airlift which pumps them to a removal system. Diffusers mounted on the rake arms release fine bubbles of air which keep the finer particles suspended. The manufacturer claims the system reduces operating costs to only 20% of conventional reactors primarily due to energy savings.

Retention Times

As stated before, the biomass grows on the surface of the soil particles. Clearly, it represents a very small portion of the overall solids content of the slurry, and separation and recycling of the biomass has not proven practical. Therefore, compared to liquid treatment methods employing much larger concentrations of biomass, longer hydraulic retention times in the slurry bioreactor are necessary. For a continuous flow process, the rate at which the soil is fed and removed from the bioreactor is critical. If the removal rate is higher than the bacterial generation rate, the biomass will diminish.

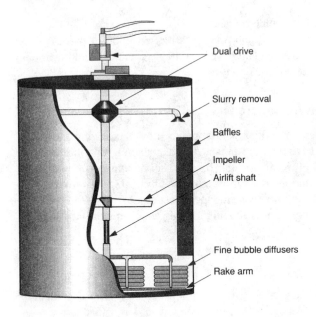

FIGURE 10-27
Eimco Biolift™ slurry-phase
bioreactor.[79]

Applications—Wood Preserving Wastes

Wood preserving wastes contain high concentrations of creosote oil, pentachlorophenol
(PCP), and PAHs (including potentially carcinogenic PAHs). Past practices typically
have placed these wastes in lagoons. The contents of the lagoon are remixed using
pumps, dredges, and aerators, with nutrients added as necessary. Upon completion of
biological treatment, the solids are settled, possibly to be closed in place, and the liquid
decanted for further treatment as necessary. Although treatment in lagoons has some
economic merits, such applications may result in excessive losses to the atmosphere
by volatilization.

As of 1988, there may be as many as 20 to 30 applications of slurry-phase
systems within the wood treating industry alone, primarily to close inactive lagoons.[8]
Practitioners have reported obtaining significant reductions in total mass of wood
preserving wastes as well as in concentrations of contaminants. For example, one
full-scale treatment project treated 100 cubic yards of wood preserving impoundment
sludge in a slurry-phase unit operated on a sequenced batch basis. It reduced total
PCP concentrations from 2600 to 32 mg/L and PAH concentrations from 1200 to 8
mg/L.[81]

The few full-scale projects reported in the literature have not yet presented
scientific biokinetic data. Further, most projects have not distinguished between biotic
and abiotic losses. The few good data in the literature result almost totally from bench-
scale tests. Although these test data do not provide a basis for engineering design,
they indicate the viability of slurry-phase treatment.

As one example of this research, a bench-scale test found that slurry-phase
treatment performed more effectively than solid-phase treatment in degrading more
than 40 targeted constituents in soil and sediments contaminated by wood-preserving

waste.[82,83] The solid-phase treatment was unable to effectively degrade the higher molecular weight PAHs within the 90-day period of the test. In contrast, slurry-phase treatment achieved greater than 50 percent removal of targeted compounds in 3 to 5 days. Pentachlorophenol and high molecular weight PAHs in sediment were more readily removed than the same constituents in surface soil.

Example 10-8 Two-step slurry-phase treatment degradation rates. A laboratory test evaluated a two-step batch process to treat creosote waste: (1) suspending creosote waste with surfactants for seven days at 20% solids in a roughing vessel and (2) transferring the supernatant to a polishing reactor for 14 days of additional biological treatment.[84] The reported reductions in concentrations of selected PAHs are shown in Table 10-11.

Assuming the above bench-scale data are representative of a larger operation and starting at the same initial concentrations, how many days of treatment in a pilot-scale operation would be necessary to degrade each of the constituents to 100 ppm or less?

Solution. The analysis begins by identifying the constituent most resistant to treatment, which is pyrene. Assuming first-order kinetics and accepting the limitations of Eq. 10-19, the degradation rate can be determined:

$$S = S_0 e^{-k't}$$

$$e^{-k'} (21 \text{ days}) = \frac{500}{6000}$$

$$k' = 0.12/\text{day}$$

Again using Eq. 10-19, calculate t for $S = 100$ ppm,

$$S = S_0 e^{-k't}$$

$$100 = 6000 e^{-(0.12/\text{day}^{-1})t}$$

$$t = 34 \text{ days}$$

The pilot-scale test should be designed with a longer test period than 34 days because it involves two extrapolations (i.e., to a larger scale operation and to a final concentration not previously achieved). Practical experience with biological treatment has shown that such extrapolations can fall short of expectations.

TABLE 10-11
Reported reduction in polycyclic aromatics

Constituent	Initial concentration (ppm)	Final concentration (ppm)
Phenanthrene	13,000	< 100
Fluorene	7000	< 100
Fluoranthrene	8000	< 500
Pyrene	6000	< 500
Benzo(a)pyrene	9000	< 100
Benzo(b)fluoranthrene	13,000	< 500
Benzo(a)anthrarene	11,000	< 100

Applications—Petroleum Refining Sludges

Wastes from oil refineries contain large concentrations of readily degradable organics. For example, tests have indicated that oil refinery sludges degrade at rates nearly twice those of wood preserving wastes.[85] Oil refinery sludges have been treated by solid-phase methods (i.e., land treatment) for many years with excellent reductions in oil content and other organic contaminants. Due to the cost of minimizing releases of contaminants from land treatment systems, slurry-phase methods are attracting interest for treatment of oil refinery wastes. Pilot-scale studies comparing land treatment with slurry-phase treatment have reported the following half-lives of non-PAH oily constituents in refinery wastes:[86]

Land treatment	6 to 15 months
Slurry-phase treatment	
Lagoon sludges	2 to 4 weeks
Stockpiled sludges	6 to 14 weeks

Although the results are variable, the tests show that slurry-phase treatment degrades oil sludges at roughly four to 15 times the rate of land treatment. Degradation of PAHs also occurred with slurry-phase treatment, eliminating 30 to 80% of the carcinogenic PAHs and 70 to 95% of the total PAHs over two months.

Example 10-9   Slurry-phase treatment in tanks of petroleum refinery sludge. In a full-scale study, the sludge from an inactive pond and an inactive pit at a petroleum refinery were mixed with dilution water and aerated in two tanks operated in a slurry-phase, batch mode. One tank was operated at 5% solids, the other at 12%. The study reported the data shown in Tables 10-12 and 10-13.[87]

Which of the two tests resulted in faster degradation rates? What factors may explain the differences?

TABLE 10-12
Slurry-phase treatment of petroleum refinery sludge at different solid concentrations

Factor	Pit sludge test	Pond sludge test
Tank diameter	26 ft	40 ft
Tank depth	10 ft	10 ft @ perimeter
		12 ft @ center
Actual operating volume	17,300 gal	70,000 gal
Percent solids (dry)	12%	5%
Surfactant addition	Yes	Yes
Nutrient addition	Yes	Yes
Aeration power	10 HP	40 HP*
Batch operating period	57 days	92 days

*40 HP "was disrupting mixing" and was replaced on day 65 by a 7.5-HP mixer and 10-HP aerator.

TABLE 10-13
Reduction in sludge constituents

Constituent	Pit sludge test (12% solids)	Pond sludge test (5% solids)
Oil & grease		
Initial concentration	27.4%	51.4%
Reduction	50%	63%
Total PAHs		
Initial concentration	1445 mg/kg	1904 mg/kg
Reduction	81%	76%
Carcinogenic PAHs		
Initial concentration	335 mg/kg	532 mg/kg
Reduction	60%	25%

Solution. Assuming first-order kinetics and accepting the limitations inherent with Eq. 10-19, the degradation rate constant can be calculated. For example, k' for the total PAHs in the 12% solids test is calculated as follows:

$$S = S_0 e^{-k't}$$

$$k' = -(\ln S/S_0)(1/t)$$

From Tables 8-12 and 8-13:

$$S_0 = 1445 \text{ mg/Kg}$$

$$\frac{S}{S_0} = 1 - .81$$

$$t = 57 \text{ days}$$

Therefore:

$$k' = -(\ln 0.19)(1/57 \text{ days})$$

$$k' = \underline{\underline{0.029 \text{ day}^{-1}}}$$

Similarly, degradation rates are calculated for other constituents as shown in Table 10-14.

The test with 12% solids degraded PAHs at a faster rate and oil and grease at essentially the same rate as that with 5% solids. It may be that the inherent biodegradability of the samples

TABLE 10-14
Degradation rates for other constituents in petroleum refinery sludge

	Degradation rate constant k' (day^{-1})	
	12% solids test	**5% solids test**
Oil & grease	0.012	0.011
Total PAHs	0.029	0.016
Carcinogenic PAHs	0.016	0.003

differs because they were collected from separate sources. The difference in rates, however, probably derives from reduced mixing effectiveness in the 5% solids tank which resulted from the mixer configuration in the tank (a "disruption in mixing" was noted), its sloping bottom (an arrangement which promotes clarification, not suspension), and the lower horsepower:operating volume ratio.

Example 10-10 Slurry-phase treatment in impoundments of petroleum refinery sludge. Another full-scale study at a refinery/petrochemical plant treated sludge contained in two impoundments. Six mixers (five 25-horsepower surface aerators and one 15-horsepower mixer) were installed in each impoundment to suspend the sludge with the supernatant, yielding a mixed solids concentration of 15% as a minimum. Surfactants, nutrients, pH control chemicals, and an adapted microbial culture were added to enhance degradation. The project reported the results shown in Table 10-15.[88]

What are the degradation rates for oil and grease? What factors may explain the differences? What is the ratio of the mass of oil and grease removed per unit consumption of energy?

Solution. Again, the degradation rate constant k' is calculated using the procedure shown in Example 10-9:

Impoundment No. 1: $k' = \underline{0.046 \text{ day}^{-1}}$

Impoundment No. 2: $k' = \underline{0.033 \text{ day}^{-1}}$

The major factor could be the inherent differences in the degradability of the sludge samples as they were collected from two separate sources. The most notable difference is the operating temperature. Using Eq. 10-1 and assuming all of the difference is attributable to temperature, the temperature activity coefficient, θ, is calculated as follows:

$$\theta^{18-14} = \frac{r_{18}}{r_{14}}$$

$$\theta^4 = \frac{0.046}{0.033}$$

$$\theta = \underline{1.086} \text{ (dimensionless)}$$

TABLE 10-15
Slurry-phase treatment of petroleum refinery sludge in impoundments

Factor	Impoundment No. 1	Impoundment No. 2
Sludge volume	4000 cy	2600 cy
Sludge moisture	66%	62%
Sludge oil & grease concentration (dry wt)	32%	41%
Operating period	21 days	61 days
Operating temperature	18°C	14°C
Reduction in sludge volume	68%	61%
Reduction in oil & grease mass	62%	87%

This value is within the range reported for conventional wastewater treatment systems, thus indicating that temperature accounts for the different rates.

The ratio of mass of oil and grease removed per unit consumption of energy is calculated as follows:

Impoundment No. 1

$$Energy = power \times time$$

$$= 140 \text{ HP} \times 21 \text{ days} \times 24 \text{ hr/day}$$

$$= 71,000 \text{ HP-hr}$$

$$Initial\ mass\ of\ oil\ \&\ grease = sludge\ volume \times density \times (100\% - percent\ moisture)$$

$$\times\ percent\ oil\ \&\ grease$$

$$= (4000 \text{ yd}^3)(1800 \text{ lb/yd}^3)(100 - 66\%)(32\%)$$

$$= 780,000 \text{ pounds}$$

$$Ratio = \frac{(780,000 \text{ lbs})(62\%)}{71,000 \text{ HP-hr}}$$

$$= \underline{\underline{6.8 \text{ lbs oil \& grease removal per HP-hr}}}$$

Impoundment No. 2

$$Energy = 140 \text{ HP} \times 61 \text{ days} \times 24 \text{ hr/day}$$

$$= 200,000 \text{ HP-hr}$$

$$Initial\ oil\ \&\ grease = (2600 \text{ yd}^3)(1800 \text{ lb/yd}^3)$$

$$= (100 - 62\%)(41\%)$$

$$= 730,000 \text{ pounds}$$

$$Ratio = \frac{(730,000 \text{ lbs})(87\%)}{200,000 \text{ HP-hr}}$$

$$= \underline{\underline{3.2 \text{ lb/HP-hr}}}$$

Thus, the higher temperature impoundment is more than twice as energy efficient as impoundment No. 2.

Applications—Contaminated Soil

The use of slurry-phase treatment for contaminated soil has attracted the interest of researchers in Europe and the United States. A laboratory-scale study in the Netherlands indicated that this method biomineralizes soil contaminated with hexachlorocyclohexane (HCH) (400 ppm of α-HCH and 230 ppm of β-HCH) at rates much greater than achieved for solid-phase treatment methods.[89] The mixing of soil slurries in a batch reactor containing 5%–25% solids readily degraded α-HCH at an initial rate of 32 mg/kg of soil·day, much higher than the 2.5 mg/kg·day measured in a land treatment experiment of the same waste. The tests showed that temperature affected the initial degradation rates; the rate at 30°C was about 33% higher than the rate at 20°C, which was about six times higher than that at 10°C. However, given sufficient time, the same degree of degradation was achieved at all three temperatures. The difference observed between degradation at the low and high temperatures is believed to be due less to changes in biological activity and more to the sensitivity of solubility, and hence desorption, to temperature.

The laboratory experiments described above were not able to degrade the β isometer of HCH. This is due to two factors. First, this isomer has a lower solubility which would reduce the rate of desorption. Second, this isomer is believed to be recalcitrant because of its spatial chloride arrangement.

Example 10-11 Slurry-phase treatment of contaminated soil. For the laboratory-scale study described above, a degradation rate of 32 mg/kg of soil·day was reported for α-HCH. Convert this rate to its conventional form.

Solution

$$k' = (32 \text{ mg/kg of soil·day})(1 \text{ kg soil}/400 \text{ mg})$$

$$k' = 0.08 \text{ day}^{-1}$$

Another Dutch experiment[90] treated PAH-contaminated soil in a laboratory reactor under the following conditions:

- Treatment as a slurry (dry solids content of 40%–50%)
- Continuous feed
- Continuous mixing
- Room temperature

For a soil sample containing a low concentration of total PAHs (6 mg/kg) and no fuel oil, no degradation occurred. However, for another soil sample containing a significant amount of fuel oil and a much higher concentration of total PAHs (3900 mg/kg), degradation occurred, but at a linear rate without exhibiting logarithmic growth kinetics typically associated with biological treatment. This linear rate indicates that desorption is the growth-limiting factor.

The lower molecular weight PAH compounds (those having two and three rings) showed rapid degradation with essentially complete mineralization after 43 days or less. Although the higher molecular weight PAHs had slower rates, the experiment was able to degrade the higher PAHs present to less than half of their initial concentration.

10-8 SOLID-PHASE TREATMENT

The term solid-phase treatment represents a broad class of biological methods which feature treatment of sludges, solids, and contaminated soils at conditions of no or minimal free moisture. There are three categories of solid-phase treatment:

1. **Land treatment**—controlled incorporation of waste into the upper soil zone.
2. **Composting**—co-degradation of hazardous waste with larger amounts of supplementary organics mixed together in windrows, piles, or vessels.
3. **Heaping**—static aeration of large mounds of waste, particularly contaminated soil.

Removal of organics can occur by various mechanisms other than biodegradation. Primary examples are volatilization and leaching. Nevertheless, for any method to qualify as "treatment," reduction in organics due to biological degradation should dominate abiotic losses and immobilized amounts.

Land Treatment

Land treatment of wastes involves applying wastes to the land at controlled rates, mixing it with the surface soil, and utilizing the natural physical, chemical, and biological systems existing in the soil for degradation and immobilization of the waste constituents. The practice closely follows agriculture principles, and is even referred to sometimes as "land farming." A number of similar operational steps are taken to stimulate bacterial growth (e.g., aeration, pH adjustment, nutrient addition, moisture control, and mixing). However, a hazardous waste land treatment site is not farmed in an agricultural sense; no agricultural crops are grown because of concern with crop uptake of toxic constituents. In comparison with slurry-phase treatment, land treatment is usually less expensive.

The widest use of land treatment of hazardous waste occurs in the petroleum industry where it has been used at more than 100 locations in the United States primarily to treat API separator sludges and other hazardous wastes generated by petroleum refineries.[91] Land treatment is also used in other industries, such as the wood products industry for treatment of waste containing wood preservatives, primarily creosote and pentachlorophenol.

Even though land treatment has probably been used more than any other hazardous waste bioremediation method, little operating and research data had been reported until recently. Recent investigations of the performance of land treatment facilities have provided data which demonstrate that many types of hazardous waste can be degraded or immobilized at a properly developed and managed land treatment site. However, abiotic losses can represent a problem.

Process description. Wastes undergoing land treatment are applied uniformly to the surface of an engineered site by spreading over the surface or injecting just below the surface. Some type of cultivation equipment, such as a disk harrow or rototiller is then used to incorporate the wastes into the soils (see Fig. 10-28).

The cultivation step tends to equalize the concentration of waste constituents; however, it serves two more important processes. It mixes the soil with waste, thereby maximizing contact among the waste, indigenous microbes, and chemical additives such as nutrients. Cultivation also provides the aeration necessary for biological degra-

FIGURE 10-28
Land treatment system and equipment.

dation. The depth to which the waste is cultivated, typically about 10–30 cm, represents the zone of incorporation (ZOI) (or the "plow zone" in operator's terms).

The soil constitutes the treatment medium. The area where treatment occurs is referred to as the zone of treatment, and it may extend as much as 1.5 m below the soil surface, well below the zone of incorporation. The physical, chemical, and biological properties of both the soil and the waste interact with external variables such as climate to determine the fate of contaminants. As a result, contaminants can be degraded or immobilized. Figure 10-29 shows the possible biological, chemical, and physical mechanisms (see Sec. 4-3) which can affect the fate of various waste constituents after land treatment.

Soil naturally contains a large and diverse microbial community. This community can degrade applied organics, basically as an aerobic treatment process in which virtually all biological degradation occurs within the zone of incorporation. Besides bacteria, other biological forms such as protozoa, mites, and earthworms act to degrade organics.

	Precipitation	Ion exchange	Neutralization	Volatilization	Migration	Fixation	Adsorption	Biological or chemical degradation	Plant uptake
Organics	●	●	●	●	●		●	●	●
Oils				●	●		●		
Metals	●	●			●	●	●	●	●
Inorganic acid, base and salts	●	●	●		●		●		

FIGURE 10-29
Fate of waste constituents in land treatment.

Land treatment also removes waste constituents other than organics. Suspended solids are removed by physical settling and filtration on the soil. Heavy metals are removed from solution by adsorption onto soil particles and by precipitation and ion exchange in the soil. Certain soils have a high sorptive capacity for metals and exhibit high levels of removal. Where a vegetative cover exists, plant uptake can also be a removal mechanism.

No particular type of soil is ideal. The more permeable soils tend to aerate and drain better, allowing larger applications of organics than the fine, low permeability soils. The latter, however, generally have a higher adsorptive capacity for metals. The specific site conditions must be evaluated with regard to the waste characteristics and the operational practices. Land treatment systems are routinely operated with an organic loading as high as 5 to 10% (total mass of waste as a percent of mass soil in the ZOI). Land treatment systems can also be operated for wastes having low concentrations of organics (e.g., contaminated soil and sediment). This latter case may not require cultivation into fresh soil (i.e., the contaminated soil itself can be the treatment medium).

Waste migration pathways.   There are several waste migration pathways (see Fig. 10-30).

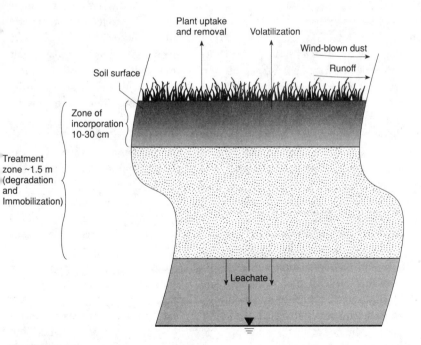

FIGURE 10-30
Pathway for migration of contaminants from land treatment.

Volatilization of organics and wind-blown dust represent air pathways for migration of waste constituents. Runoff can carry contaminated stormwater to surface streams. Infiltration of rainfall can leach waste constituents into ground water; this typically is the major environmental concern with land treatment. A study of several land treatment sites after their closure showed that the concentrations of copper, lead, and zinc in the soils below the zone of incorporation were about the same as observed in background soils. However, concentrations of chromium and nickel appeared greater than background concentrations, indicating that they had migrated.[92] With proper site selection, development, and operation, the potential for adverse environmental impacts can be controlled.

Adequate construction and maintenance of control structures (e.g., berms, runoff channels) can virtually eliminate the potential for surface runoff from the treatment site. Collected runoff can be treated and discharged or reapplied to the soil during dry weather.

Volatilization of organics can be controlled by keeping waste application rates in safe ranges considering waste characteristics and site conditions. In some cases, it may be necessary to enclose the land treatment site to facilitate capture and treatment of the vapor phase. As an added benefit, complete enclosure would enable the elevation of temperature in colder regions. Even under dry conditions, the oil in refinery waste will control wind-blown dust. For non-oily wastes, wind-blown dust can be prevented by adding water to maintain proper soil moisture.

Proper site selection is critical in helping to prevent ground water contamination. Of particular importance is the selection of sites with the hydrogeologic characteristics which offer protection against migration into ground water. The concern with contamination of ground water has prompted some operations to construct a synthetic liner below the zone of treatment, somewhat similar to a landfill liner. Drainage pipes above the liner allow the collected leachate to be removed. A further enhancement involves overhead enclosures which prevent excessive infiltration. Complete enclosure, as in a greenhouse, provides for control of volatile emissions.

Metals pose a special problem. Proper control of soil pH will facilitate their immobilization in the soil. The soil pH can be adjusted to the optimum range (typically no less than pH 6–7) and maintained within such range by application of lime and/or other soil amendments. The presence of amphoteric metals (e.g., Se and Sb) would necessitate a narrow acceptable pH range because their solubilities increase with increasing pH. The immobilization of metals within the upper portion of the treatment zone will result in ever higher concentrations of metals with each application of waste. Eventually the concentrations rise to levels that are toxic to the microbial community. At that point, biodegradation will slow or stop, requiring closure of the site or replacement of the soil with uncontaminated soil. If the soil is replaced, the removed soil must be disposed of, typically by landfilling. Even if the concentrations of metals are not toxic to the microorganisms in the soil, they may still reach phytotoxic levels and inhibit establishment of the planned post-closure vegetative cover. Whether phytotoxic or biotoxic, the build-up of metals typically represents the limiting constraint on the life of the land treatment facility.

Assimilative capacity.   The key to successful land treatment is to apply wastes at a rate and frequency which does not exceed the assimilative capacity of the site. The assimilative capacity must be viewed from three perspectives:[93]

- **Capacity limiting**—addresses conservative, immobile waste constituents (e.g., metals) which accumulate in the soil with successive applications, their concentration eventually increasing to a threshold level requiring the site to be closed. The cumulative amount applied during the life of the facility is critical.

- **Rate limiting**—addresses non-conservative waste constituents which are degraded in the soil over a period of weeks, months or even years (e.g., oil). The cumulative amount applied in comparison with the cumulative amount degraded is critical.

- **Application limiting**—addresses mobile waste constituents which may readily migrate from the site (e.g., volatile organics). It also addresses the maximum hydraulic loading which will infiltrate through the site surface without runoff. The amount applied at any one time is critical.

Each perspective can be evaluated using mathematical models which simulate the physical, chemical, and biological processes of land treatment and which also account for characteristics of the site, characteristics of the waste, environmental parameters, and the design and operating practices. Each constituent is evaluated to determine the amount of land area required for a given application rate and frequency.

Mathematical Models of Land Treatment

Within the zone of treatment, a contaminant moves among four phases—soil grains, pore water, oil or any other immiscible liquid in the pores, and unsaturated pore space—in response to a variety of physical and chemical influences including advection, dispersion, adsorption/desorption, and volatilization. A contaminant may also degrade biologically. Each of these actions can be modeled to mathematically describe the fate and behavior of the contaminant. Such a model can predict the total amount of a pollutant leached below the treatment zone or lost as vapor under varying design and operating practices.

Different models exist for predicting the fate and behavior of waste constituents in a land treatment facility. These models can be highly involved due to the complexity of vadose zone processes. Most of the framework of any model addresses physical and chemical processes of mobilization, and many of these are covered in Chap. 4. Biological degradation is typically approximated by the first order kinetics of Eq. 10-19.

Two of the more widely used mathematical models are VIP[94] and RITZ.[95] RITZ is a plug-flow model which simulates the fate and behavior of organic constituents during the land treatment of oily waste. The model predicts organic constituent losses due to biological degradation, volatilization to the atmosphere, and migration through the treatment zone. As an example of the degree of simplification necessary in virtually any environmental modeling approach, it is interesting to note that RITZ is based on the following assumptions:

• The individual organic constituents are either degraded or migrate from the site because of volatilization (to the atmosphere) or leaching (to the water).

• The soil properties are uniform from the surface to the bottom of the treatment zone and are assumed to be free of residuals from any previous waste application.

• The flux of water is uniform throughout the treatment zone at all times.

• Hydrodynamic dispersion is insignificant and can be neglected.

• Partitioning of a constituent among liquid, soil, vapor, and oil phases is linear at constant temperature.

• Localized equilibrium exists between the phases.

• Degradation of waste constituents and the oil follow first order reaction kinetics, and the degradation constants are assumed to be uniform and constant throughout the zone of incorporation.

Land Treatment—Applicable Wastes

Oily sludges and waste oils.   There are a variety of different types of oily sludges and waste oils generated by the petroleum refining industry. The principal types of oily sludges generated are API separator sludge, slop oil emulsion solids, induced and dissolved air flotation float, and leaded and unleaded tank bottoms. These sludges are biodegradable to varying extents and can be treated by land application techniques. The half-life can range from 200 to 500 days. The limiting parameter is usually the oil loading rate, although the cumulative loading of certain metals contained in the waste (e.g., lead) may be limiting.

Organic sludges and liquids.   The extensive biological treatment capability of natural microbes in the soil makes land application an excellent method for treatment of most organic wastes. Therefore, any waste that is amenable to biological degradation is a candidate for treatment in a land treatment system. However, land application of halogenated solvent wastes must be designed to account for their mobility and their extremely slow biodegradability, with application rates chosen such that the leachate will not contain excessive levels of these solvents.

Land Treatment Case Study

The BP Oil Alliance Refinery near Belle Chasse, Louisiana, conducted a pilot land treatment project for one year on two 50×120 ft. plots to demonstrate that the process does degrade or immobilize its refinery waste.[96] The project also provided data for designing a full-scale facility.

Construction details.   The soil at the test plots is fine grained with a high clay content (Unified Soil Classification System designation of CH and CL). In constructing the plots, a subsurface drainage system, consisting of 4 in. drainage laterals spaced at 20 ft. intervals, was placed on a low permeability clay layer which naturally occurs at an average depth of $5 - 6$ ft. below the existing grade. Soil from outside the plots was added to elevate the plots 3 ft. above the existing grade. This additional soil helped to provide a treatment zone of about 4–4.5 ft. in depth plus a buffer of 3 ft. between the treatment zone and the seasonal high water table. To prevent run-on and contain runoff, a dike was constructed 2 ft. above the land treatment surface around each plot.

Operations. Thickened and filtered sludges from the refinery's API separator units, API skim treatment tank, dissolved air flotation units, and miscellaneous tank bottoms were applied in quarterly batches at an oil loading rate of 827 bbl/acre · year on one plot and 513 bbl/acre · year on the other. The zone of incorporation was tilled weekly, with soil moisture, pH, and nutrient balance monitored periodically. The date of application and the amount of waste applied to the plot receiving waste at the rate of 827 bbl/acre · year are shown in Table 10-16.

Sampling and monitoring. The demonstration project included quarterly sampling of lysimeters, soil cores at various depths, and the subsurface drainage system water. Analysis of samples for all hazardous constituents in the waste materials applied to the plots revealed the following:

1. The applied waste contained 6 volatile organics, 9 semi-volatile organics, and 12 metals.
2. Second quarter sampling of soils showed applied organic waste constituents in the 0–1 ft. interval but none at deeper intervals. Metal concentrations showed a build-up in the 0–1 ft. interval.
3. Waste was not applied between the second and third quarter; therefore, the third-quarter soil coring showed a decrease in organic constituents, with metal concentrations exhibiting substantial variations.
4. Fourth-quarter soil core sampling gave results consistent with the previous results.
5. No organic hazardous constituent present in the applied waste was detected in the soil core liquid.
6. Five metals were detected in the lysimeter samples, and all except for barium were at very low levels.
7. None of the hazardous constituents in the waste were detected in the subsurface drainage system water.

TABLE 10-16
Land treatment of petroleum refinery sludge

Time of application (days after initial application)	Amount of applied waste (tons)	Oil & grease concentration of applied waste
0	20.1	16.5%
110	30.6	13.1%
229	30.6	16.8%
298	24.7	18.5%

Example 10-12 General assessment of treatment effectiveness. Based on the above sampling results, what overall conclusion can be made about the effectiveness of the project? What important information is not provided by the above results?

Solution. The project demonstrated that the operation could treat the waste at the applied rates without migration via ground water during the one-year period of the test. However, the results do not address oil and grease degradation and how that factor limits the rate at which waste can be applied over the long term.

Degradation kinetics. To support making a determination of the rate at which waste can be applied in a full-scale facility over a longer period, it is necessary to calculate oil degradation kinetics. The oil and grease concentrations exhibited significant variation across each plot. This variation was deemed natural and not due to improper operating procedures. To provide a better definition, 32 grab samples were taken of the upper four to six inches of the surface soil, which is the zone of incorporation (ZOI), and analyzed for oil and grease (mass basis). The grab sampling occurred just over one year (368 days) after the initial application; results are shown in Fig. 10-31.

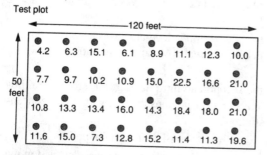

FIGURE 10-31
Final oil and grease concentrations at test plot.

Example 10-13 Degradation rate of land treatment of petroleum refinery sludge.
Based on the results shown in Table 10-16 and Fig. 10-31 in the case study, what are the half-life and degradation rate constants for the oil phase? Assume a soil density of 96 lbs/ft^3.

Solution

1. Calculate oil loadings.

The initial application increased the oil concentration in the ZOI by the following incremental amount:

$$\text{Incremental increase} = \frac{\text{loading of oil}}{(\text{area of plot})(\text{depth of sampling})(\text{density of ZOI})}$$

$$= \frac{(16.5\%)(20.1 \text{ tons})(2000 \text{ lbs/ton})}{(50 \text{ ft})(120 \text{ ft})(4 \text{ in})(1 \text{ ft/12 in})(96 \text{ lbs/ft}^3)}$$

$$= \frac{6600 \text{ lbs}}{192,000 \text{ lbs}}$$

$$= 3.4\%$$

Similarly, the incremental loadings represented by all applications are shown in Table 10-17.

2. Calculate residual oil.

From Fig. 10-31, the following statistics are calculated:

Mean 13.0% (percent oil and grease)
Standard deviation 4.6
Coefficient of variation 35%

3. Calculate degradation rate.

From the above, residual oil at day 368 is 13.0%. Therefore, using Eq. 10-19, the first order kinetics degration rate is as follows:

$$13.0 = \sum S_0 e^{-k'(368-t_0)} \text{ for } t_0 = 0, 110, 229, 298 \text{ days}$$

$$= 3.4e^{-368k'} + 4.2e^{-258k'} + 5.2e^{-139k'} + 4.8e^{-70k'}$$

$$k' = \underline{\underline{0.0016 \text{ day}^{-1}}}$$

4. Calculate half-life.

By definition, half-life is $t_{1/2}$ where $S_t/S_0 = 0.5$. Rearranging Eq. 10-19:

$$0.5 = e^{-k't_{1/2}}$$

$$t_{1/2} = -\frac{\ln 0.5}{0.0016/\text{day}}$$

$$t_{1/2} = \underline{\underline{433 \text{ days}}}$$

To simulate the performance of land treatment in treating the waste, the RITZ model was used in which the degradation rate was assumed to equal 0.0016 day^{-1}. The model results showed little deviation from actual field data, indicating RITZ is a good predictor of actual performance for the one-year land treatment demonstration project. Long-term modeling predicts that if waste were applied every 50 days, the land treatment system will reach a steady state condition, relative to the oil content, within the first six years of operation. Xylene and napthalene will stabilize within the first two years, with no migration. In summary, the modeling predicts that the tested application rates will not limit the life of the facility.

TABLE 10-17
Incremental loadings to land treatment plot

Days after initial application	Applied oil (lb)	Incremental increase to ZOI oil concentration (%)
0	6600	3.4
110	8000	4.2
229	10,000	5.2
298	9200	4.8

Composting

Composting has been used for centuries to degrade organic solids to humic material for use as a soil conditioner. The degradation occurs aerobically at elevated temperatures, usually in the thermophilic range. Composting has a fairly long history in treating municipal solid waste and municipal sewage sludge. When a hazardous waste contains a high concentration of organics, it may be possible to compost it using basically the same technique for composting municipal wastes.

Most hazardous wastes do not contain a sufficiently high concentration of organics to sustain composting; however, they can be composted by mixing a small percentage of hazardous waste with highly biodegradable solids. The biodegradable solids serve as a bulking agent and thermal source. As a bulking agent, they create void space for passage of air needed to supply the necessary oxygen. As a thermal source, their biological decomposition releases a large amount of energy in the form of heat which enhances biological activity, if not achieving the elevated temperatures in which aerobic thermophilic microorganisms thrive.

Two different materials usually are required. Good bulking agents (e.g., wood chips or vermiculite) do not necessarily serve as an adequate thermal source. Likewise, good thermal sources (e.g., dry molasses) may make poor bulking agents. Frequently it is cost-effective to use inert material as a bulking agent, enabling it to be recycled. The thermal source must not represent a preferential carbon source which preempts degradation of the hazardous waste. In some cases, it must serve as a cometabolite.

Temperature is important to the composting of municipal solid waste and sewage sludge because the higher operating temperature of thermophilic bacteria can sterilize the waste of pathogenic bacteria. Such sterilization is not important with hazardous waste and, therefore, mesophilic bacteria may suffice, dependent upon their ability to degrade the waste. If mesophilic conditions are desired, the heat produced by composting requires control.

Composting—Aeration Techniques

Composting of waste will occur readily given the presence of appropriate microorganisms, adequate aeration, optimum temperature, necessary nutrients, pH, and moisture. Aeration is critical, and several techniques have been developed in municipal waste applications to enhance aeration and thereby promote decomposition. Basically there are three types and each can be applied to hazardous waste-organic mixtures as follows (see Figs. 10-32 and 10-33):

1. **Windrows**—In this technique the mixture of waste and organics is piled in long rows and turned once or twice per week, if not daily. The turning aerates the mixture and releases excess heat. Turning also increases the release of volatiles.

2. **Static piles**—This is similar to windrowing except that the piles are not turned. Rather, by building the piles on top of a grid of perforated pipes, air is mechanically drawn or forced through the piles using a vacuum or forced air system. A negative aeration system created by a vacuum at the base of pile enables process

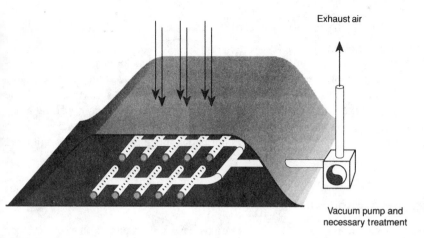

Exhaust air

Vacuum pump and
necessary treatment

FIGURE 10-32
Composting methods—static piles.

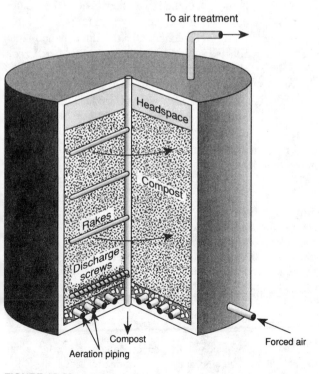

To air treatment

Headspace

Compost

Rakes

Discharge
screws

Compost

Aeration piping

Forced air

FIGURE 10-33
Composting methods—enclosed reactor.[97]

air to be treated before it exhausts to the atmosphere. The process air would be recycled through the pile.

3. **Enclosed reactor system**—By composting within an enclosed reactor, the operation can be optimized to complete the process in as little as three days compared with as much as 30 days for windrows.[98] It also facilitates control of volatile emissions by collecting process air in the headspace above the compost, by a diffuser system below the compost, or by an exhaust system within the compost. A reactor can be operated in a batch or continuous mode, with or without agitation.

Composting Applications

Some bench-scale tests and field demonstrations have been reported on explosive and propellant manufacturing wastes, soils contaminated with coal tar, and other wastes. The test variables and results are shown in Table 10-18.

The test of soil contaminated with coal tar yielded some interesting observations. Although the 5-ring PAHs did not degrade, the data suggest that they became immobilized. The recycling of off-gases proved effective; only 1% of the volatiles were trapped by carbon filters, with the remainder assumed to have been degraded or entrapped in the compost matrix. Although not tested, the researchers believe the residue is suitable for "daily cover in landfills, industrial fill, or underlayment for parking lots."

In addition to the tests conducted with TNT-contaminated sediments and soils, two nearly identical tests were also conducted with nitrocellulose-contaminated sediments. In the first test, the degradation of extractable total explosive wastes proceeded from an initial concentration of 17,900 mg/kg to about 74 mg/kg in 153 days. Little degradation occurred after 80 days. In the second test, extractable nitrocellulose concentrations decreased at a fairly constant rate from an initial value of 3040 mg/kg to 59 mg/kg in 70 days. Minimal reduction occurred after that date through the remainder of the 151-day test. The split between mineralization and transformation to non-extractable compounds was not determined.

Example 10-14 Composting degradation rate. The owner of a property contaminated with petroleum must remediate the site completely in one year. Approximately 40,000 cubic yards of soil contaminated with a mean total petroleum (TPH) concentration of 10,000 mg/kg, ranging from "not detected" to over 98,000 mg/kg, must be remediated to a mean TPH of not greater than 100 mg/kg, nor any sample greater than 500 mg/kg, if it is to be left on site. As an alternative to the costly option of incineration and off-site disposal, the owner has conducted laboratory tests of composting which show the reductions in TPH listed in Table 10-19. Is it possible that composting can meet the clean-up standard? In 90 days?

Solution. The short-term tests show rapid reduction of TPH on a continuing basis, suggesting that the contamination is readily degradable via composting. The question remains if the objectives of 500 mg/kg as a maximum and 100 mg/kg as a mean value can be achieved, because neither has been achieved in the short-term tests. This question can be addressed by analyzing the

Item	Sediments and soil containing explosive and propellant manufacturing waste[99]	Soil contaminated with coal tar[100]	Waste containing ethylene glycol[101]
Waste constituents	76,000 mg/kg total explosives (primarily TNT)	20,000 mg/kg coal tar	1900 mg/kg ethylene glycol 120,000 mg/kg TOC
Aeration methods	Static pile	Static pile	Static pile
Compost pile volume	34 cubic yards	4 cubic yards	32 gallons
Compost pile height		4 feet	
Bulking agent	None (thermal source served as bulking agent)	Not disclosed	Wood chips
Additives	Alfalfa, straw, manure, horse feed, and fertilizer	Not disclosed	Dry molasses, sewage sludge, soil, ammonium nitrate, 10-10-10 fertilizer, lime
Mix ratio of waste to additives	1% (vol.) and 24% (mass)	Not disclosed	2.5%, 5%, 10%, 20% (vol.)
Air supply	Vacuum	Forced air using blower; off-gases recycled through the compost	Forced air using blower; adjusted to maintain minimum of 10% O_2 in exhaust gas
Moisture	Not reported, other than 400 gallons of water added initially	Not reported	40–60%
Operating temperature	Approximately 55°C	65–85°F (18–29°C)	Exceeded 120°F (49°C)
Test period	22 weeks	80 days	30 days
Reduction of waste constituents	99% of total explosives	94% of total hydrocarbon 95% of 2-ring PAHs 94% of 3-ring PAHs 89% of 4-ring PAHs	> 95% of ethylene glycol
Comment	Split between mineralization and transformation to non-extractable compound was not determined.	87% of total hydrocarbon removed in 35 days. 5-ring PAH did not degrade.	Test did not account for abiotic losses.

TABLE 10-19

Composting of contaminated soil

	Concentration of TPH (mg/kg)					
Day	Sample no. 1	Sample no. 2	Sample no. 3	Sample no. 4	Sample no. 5	Sample no. 6
0	18,600	9100	4500	98,300	8100	6900
3	16,400	5500	4200	87,700	5800	2600
6	13,200	2100	3000	55,800	5100	2500
10	11,200	1900	1600	43,600	3200	2100

degradation kinetics. Accepting the limitations of Eq. 10-19, it can be used to determine the degradation rate constants. For Sample No. 1, the constant is calculated as follows:

$$\frac{S}{S_0} = e^{-k't}$$

$$k' = \frac{-\ln(11,200/18,600)}{10 \text{ days}}$$

$$k' = 0.05 \text{ day}^{-1}$$

For all samples, the degradation rate constants are:

Sample no.	Degradation rate constant (day⁻¹)
1	0.05
2	0.16
3	0.10
4	0.08
5	0.09
6	0.12

The mean rate is 0.10 day^{-1}. Again using Eq. 10-19, the number of days to reduce the mean TPH concentration from 10,000 mg/kg to 100 mg/kg is calculated as follows:

$$\frac{S}{S_0} = e^{-k't}$$

$$k' = \frac{-\ln(100/10,000)}{0.10 \text{ day}^{-1}}$$

$$t = 46 \text{ days}$$

Similarly, reducing 98,000 mg/kg to 500 mg/kg would require the following time:

$$\frac{S}{S_0} = e^{-k't}$$

$$k' = \frac{-\ln(500/98,000)}{0.10 \text{ day}^{-1}}$$

$$t = 53 \text{ days}$$

Both time periods are well under the 90-day objective. However, scaling up from bench-scale tests and extrapolating to a longer time period will most likely encounter much slower rates of degradation. Pilot testing over a time period embracing the 90 days is necessary.

Soil Heaping

Heaping is a combination of land treatment and windrow composting. It is an effective method for treating large volumes of contaminated soil and other wastes containing low concentrations of organics, particularly if available land area is limited. The process involves stockpiling contaminated soil in large mounds, typically over an air piping system and an impermeable liner. The mounds can be several meters in height, and they may be covered if necessary.

Rather than aerating the soil by cultivating it as in land treatment, the piping system pulls a vacuum through the heap or blows air through it. The former is preferred as it enables treatment of volatile emissions. By using a bottom liner, leachate can be collected and treated. The operation is monitored for pH, nutrient balance, and moisture with adjustments made as necessary. The method is similar to static pile composting except for the difference in height of the mounds.

The soil-heaping process degrades waste at a somewhat slower rate than land treatment, due to the low concentrations of organics and to the lack of frequent cultivation (e.g., one project reported remediation times from 4–18 months).[102] Reported cost data suggest this process is less expensive than land treatment for large volumes of contaminated soil due, in part, to the reduced need for land area. This treatment method, albeit slower than land treatment, offers the advantage that migration by volatilization and leaching, two primary concerns with land treatment, can be virtually eliminated. The emission of volatile organics is controlled by covering the heap and maintaining a vacuum, with the exhaust from the vacuum passed through an activated carbon filter. Leaching is controlled by a cover and liner. The former eliminates excessive infiltration of precipitation, with any leachate intercepted by the underlying liner.

Soil Heaping—Application to Treat PAHs

A project located near Blackburn, England, used a variant of heaping to treat 30,500 m^3 of soil contaminated by waste from a gas works.[103] The contaminated soil contained coal tar, phenols, and various polyaromatic hydrocarbons (PAHs). After screening, the soil was heaped into large mounds. Rather than pulling a vacuum through the soil, the operators cultivated the soil frequently.

The project featured the identification of "vanguard" microorganisms capable of degrading PAH compounds. These vanguard microorganisms were isolated from the manufacturing site and cultivated in the laboratory to provide large populations for inoculation into the heaps. The presence of the vanguard microorganisms catalyzed the initial stages of degradation of PAHs. From these stages, the natural soil microorganisms could complete the transformation and degradation.

The project began with a series of laboratory tests which showed that this procedure reduced total quantifiable PAHs from 454 to 234 ppm after five weeks of

TABLE 10-20
Heaping of contaminated soil

Additives	Initial concentration	Concentration after 8 weeks
Microbe A plus nutrients	9800 mg/kg	7300 mg/kg
Microbe B plus nutrients	12,500 mg/kg	7600 mg/kg
Microbe B plus nutrients	8000 mg/kg	5100 mg/kg
Nutrients only	3500 mg/kg	5300 mg/kg*
No additives	2900 mg/kg	4300 mg/kg*

*Increase in concentration is attributable to "continual mechanical breakdown of lumps of tarry material."

incubation at 15°C. A controlled experiment, in which microbial activity was inhibited by adding mercuric chloride, showed no reduction in PAHs, thus demonstrating that microbial activity rather than volatilization was responsible for the observed reduction in PAH. Based on laboratory tests, field trials were commenced using two different isolated microbes. The field trials showed reductions in PAH concentrations over an 8-week period listed in Table 10-20.

The field trials showed that effective biological degradation occurred only with inoculation of vanguard microorganisms. A full-scale test was implemented subsequent to the field trials. Although the 11-week period of the full-scale project encountered the low temperatures of winter, the use of vanguard microbes plus nutrients was able to reduce total phenolics to acceptable levels as shown in Table 10-21.

TABLE 10-21
Full-scale results from treatment by soil heaping

Duration from start of test	Concentration (mg/kg)
Start	36.0
6 weeks	24.9
11 weeks	7.0

10-9 EMERGING TECHNOLOGIES

The use of biological processes to degrade hazardous waste is a relatively recent practice. Although it has proven to be an effective and economical treatment alternative for many types of hazardous waste (e.g., light aromatic constituents of distilled petroleum products) there are many others for which biological treatment is still in an early state of development. These can be grouped into two categories:

- Compounds which have been subjected to extensive research but which remain classified as persistent, if not recalcitrant (e.g., dioxin, highly chlorinated PCBs, and dibenzofurans).

- Compounds known to be degradable, but only under conditions not easily replicated in the field (e.g., trichloroethylene and other volatile chlorinated aliphatic hydrocarbons).

Experience with some methods, even for degradable compounds, is not great. Almost every case represents a new undertaking with few standard procedures for evaluating, designing, and operating systems. Further research and development is clearly needed. Many organizations and institutions are identifying and addressing these needs,[104] with much success expected to be reported over the next several years in the following areas:

1. Isolation of specific microbes, or communities of microbes, which degrade persistent compounds.
2. Development of large-scale systems that replicate in the field, if not enhance, the specific laboratory conditions that promote biodegradation.
3. Development of systematic, standardized approaches to the design of the large-scale systems based on treatability data and mathematical expressions of biodegradation kinetics.
4. Integration of biological treatment processes with physical-chemical processes that enhance the overall treatment results.

Isolation of Specific Microbes

Continuing research has often been able to identify microorganisms, or consortia, which degrade compounds previously thought to be recalcitrant. This research is not simple. A key complexity is that the isolation of microbes must be specific. Yet, this is frequently masked by the fact that a myriad of different enzymes, acting in a sequential manner and not necessarily coming from a single species, are needed to degrade a persistent compound. An example is trichloroethylene (TCE). It is normally persistent in the environment; however, it has been demonstrated that methanotrophs will cometabolize TCE by the action of the enzyme methane monooxygenase (MMO) which they produce when metabolizing methane. There are examples of success in this field of research which have selected microbes other than bacteria. The examples include white-rot fungus which has the capability to degrade a wide range of compounds, and algae which remove heavy metal ions from aqueous solutions.

The potential of genetic engineering to go beyond naturally occurring microorganisms creates all sorts of possibilities for degrading recalcitrant compounds. However, the obstacles of public perception, general liability, and regulatory licensing requirements keep the field application of genetically altered microbes, even in an enclosed reactor, a decade or so away at best. Prior to that, one envisions the use of enzyme harvesting in which enzymes are separated from genetically altered microbes capable of degrading specific compounds. The harvested enzymes can be immobilized on resin beads or some other suitable media for subsequent use in the field where the demobilization of the enzymes can be controlled in presence of the waste.

Replication of Laboratory Results in Full-Scale Systems

Many important pollutants may be found biodegradable under laboratory conditions; however, degradation under field conditions is a different matter. It is often observed that metabolic rates are much slower in the field for many reasons. Environmental conditions such as temperature, pH, and oxygen may not be optimal. Microbial cultures isolated in the laboratory may not thrive or even survive in competition with the vast population of microbes that naturally occur in the field. Also, field conditions may favor the production of intermediates not encountered in the laboratory.

Another possible reason for slower rates is that wastes can differ remarkably between field and laboratory conditions. The spatial and temporal concentrations of a target compound at a site to be remediated typically exhibit great variability which usually is not addressed in the laboratory. The waste may contain multiple contaminants other than the target contaminant. Either condition may inhibit degradation of the target compound.

The operation of the reactor presents another set of reasons for slower degradation rates. A physical step such as mixing is not as effective because of differences in scale. The degradation of some persistent compounds may require a sequence of events which may not be easily duplicated in a reactor at a reasonable cost. Control of abiotic losses of the compound (e.g., via volatilization) is another consideration.

Because of these many obstacles, the design of a full-scale system is not simple. It involves a series of distinct tasks, each of which is crucial to the development of a successful operation. The tasks start with literature reviews and desktop evaluations, and progress through a series of laboratory and field treatability tests to an engineering design.

Treatability Testing

Treatability testing typically progresses through three steps as follows:

1. Laboratory screening
2. Bench-scale testing
3. Pilot-scale testing

Laboratory screening is a quick test to establish whether a particular technology has the potential to treat a waste. A biological treatment technology is usually considered a reasonable option if the test results in greater than 20% reduction in the mass or concentration of the target compound, excluding abiotic losses.

Bench-scale treatability tests are moderately quick tests for the purpose of selecting a process for further evaluation and identifying the key parameters which have the greatest influence upon degradation. They usually are conducted in batch reactors with adjustments made in loading rates, chemical additives (e.g., nutrients), physical parameters (e.g., aeration and mixing rates), and other variables to study their relative impact upon the rate of degradation. The data provided by bench testing should lead to the selection of one or two processes for further testing at the pilot-scale level.

The pilot-scale test is done not only to verify the process but to provide kinetic degradation data for final design. A pilot-scale test may require weeks to months, or possibly a year, to complete.

There are few widely accepted standardized treatability tests which support the evaluation of alternative biological treatment processes on a comparable basis. A complete evaluation of biological treatment variables can be a labor intensive undertaking and can involve much engineering judgment. Further innovation is needed in the approach and tools used to evaluate and design biological treatment processes.

One tool for reducing the effort required is that of evaluating biodegradation rates based on automated respirometry. Respirometry, or the measurement of the rate of oxygen uptake, has been largely automated and integrated with ancillary computer and modeling analysis to provide instantaneous data and interpretations.[105] It has long been employed for many applications in the pollution control field, based on the simple fact that oxygen uptake rates provide a rapid indicator of microbial activity. One application is to model the rate of biodegradation. To do this, one must be able to utilize the oxygen uptake data to compute cell growth or substrate utilization rates.[106] A major effort showed that respirometric data can be employed to determine cell growth rates which can be transformed into equivalent growth or substrate utilization data.[107] This is accomplished by using equations which are derived based on the principles of COD and energy balances for aerobic systems.

Several research and field application efforts validated the use of respirometry as a means to obtain values of the biokinetic constants for model calibration.[108] Experience shows that the model predictions which were calibrated with respirometric data accurately described the field performance of aerobic biological treatment systems. This approach saves times and provides kinetically meaningful data for designing and operating biological treatment systems. For example, proactive adjustments can be made as a treatability test progresses, enabling rapid turnaround times (as low as 30 hours from receipt of a sample for a bench-scale test).

Integration of Biological Treatment with Physical-Chemical Treatment Systems

The biological treatment of hazardous wastes can be greatly enhanced with pretreatment steps such as increasing the concentration of the target compound and decreasing the concentration of inhibitors to degradation. In fact, without pretreatment, it may not be possible to initiate biological treatment of wastes. The toxic nature of some compounds may inhibit degradation, or the concentration of substrate may not meet the minimum below which biodegradation cannot occur.

There is much room for innovation in the integration of biological treatment with the many physical and chemical processes (see Chap. 9). An example of what is needed is that depicted in Fig. 10-34. This system features membrane filtration to extract and concentrate contaminants followed by mixing with selectively adapted microorganisms in a reactor suitable for both aqueous and slurry-phase treatment. The system involves two innovative processes: slurry-phase reactor (see Fig. 10-27) and membrane microfiltration (see Sec. 9-7).

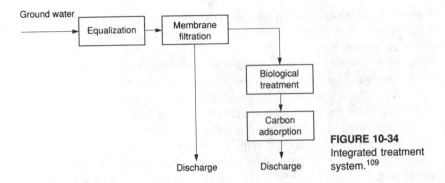

FIGURE 10-34
Integrated treatment system.[109]

DISCUSSION TOPICS AND PROBLEMS

10-1. Name three advantages and three disadvantages of biological treatment in comparison with physical-chemical treatment.

10-2. Land treatment and in situ remediation of the vadose zone both occur in unsaturated soil conditions. Why is the former more effective than the latter?

10-3. Adding nitrate rather than oxygen before reinjecting water as part of in situ remediation would offer what key advantage? Disadvantage? Would nitrous oxide have the same disadvantage?

10-4. What are the advantages of treatment in an enclosed reactor?

10-5. Contrast the three categories of solid-phase treatment.

10-6. Describe a situation in which soil heaping would be preferable to land farming, including consideration of contaminant migration.

10-7. What are the limitations of in situ remediation for a contaminant requiring anaerobic decomposition? Cometabolism? Degradation by non-native bacteria? How might these limitations be addressed?

10-8. Describe one or more shortcomings inherent with the assumptions which the RITZ model makes in predicting land treatment results. How could this affect the modeling results?

10-9. The biomass produced in a reactor is separated in a downstream clarifier and recycled to the reactor. What portion of the biomass must be wasted to yield a solids retention time within the clarifier of 10 days?

10-10. What is the likely reason explaining the observation of a bench-scale test of slurry-phase treatment of contaminated soil that microbial populations grew at a linear rather than logarithmic rate?

10-11. What can be done to increase the rate of degradation by slurry-phase treatment of organics sorbed to soil?

10-12. What kinetic equations demonstrate why the hydraulic retention time for slurry-phase treatment must be greater than for conventional liquid-phase treatment?

10-13. What are the abiotic losses expected in an open land treatment system for organic wastes?

10-14. What type of contaminants limit the long-term life of a land treatment site?

10-15. What effect does the molecular weight and the number of functional groups of an organic chemical typically exhibit on its rate of biodegradation?

10-16. What are the capabilities and limitations of suspended growth versus attached growth in conventional liquid-phase treatment?

10-17. What is the most difficult aspect of in situ bioremediation?

10-18. Which biological treatment method occurs in the thermophilic range?

10-19. Derive a mathematical expression for the rate of biomass formation as a function of substrate concentration, degradation rate, and other appropriate variables.

10-20. An investigation shows that ground water downgradient of a municipal landfill contains low concentrations of halogenated aliphatics, released in the leachate from the landfill. What are alternative methods for treating the ground water using biological processes?

REFERENCES

1. Metcalf & Eddy, Inc.: *Wastewater Engineering: Treatment, Disposal, Reuse*, McGraw-Hill, Inc., New York, New York, p. 379, 1979.
2. Grady, C. P. Leslie, Jr.: "Biodegradation of Hazardous Wastes by Conventional Biological Treatment," *Hazardous Waste & Hazardous Materials*, vol. 3, no. 4, 1986.
3. Niemi, G. J., et. al.: "Structural Features Associated with Degradable and Persistent Compounds," *Environmental Toxicology and Chemistry*, vol. 6, Pergammon Journals, Ltd., Elmsford, New York, 1987.
4. Rochkind, M. L., J. W. Blackburn, and G. S. Saylor: "Microbial Decomposition of Chlorinated Aromatic Compounds," EPA/600/2–86/090, Hazardous Waste Engineering Laboratory, U.S. Environmental Protection Agency, Cincinnati, Ohio, September 1986.
5. Nilsson, Robert: Professor, University of Stockholm, personal communication, 1990.
6. Vogel, T. M. et al.: "Transformations of Halogenated Aliphatic Compounds," *Environmental Science and Technology*, vol. 21, no. 8, pp 722–735, 1987.
7. Niemi, et. al.: "Structural Features Associated with Degradable and Persistent Compounds."
8. Tiedje, J. M., and T. O. Stevens: "The Ecology of an Anaerobic Dechlorinating Consortium," *Environmental Biotechnology Reducing Risks from Environmental Chemicals through Biotechnology*, edited by G. S. Omenn, Plenum Press, New York, p. 3–14, 1990.
9. Ibid.
10. Leahy, M. C., et al.: "Biodegradation of Clorinated Aliphatics by a Methanotrophic Consortium in a Biological Reactor," *Proceedings of the Second National Conference of Biotreatment: The Use of Microorganisms in the Treatment of Hazardous Materials and Hazardous Waste*, HMCRI, Washington, DC, November 27–29, p. 3–9, 1989.
11. Roberts, P. V. et al.: "In-Site Aquifer Restoration of Chlorinated Aliphatics by Methanotrophic Bacteria," Technical Report no. 310, Department of Civil Engineering, Stanford University, June 1989.
12. Ibid.
13. Torpy, M. F., et al.: "Biological Treatment of Hazardous Waste," *Pollution Engineering*, May 1989.
14. Singleton, Paul, and Diana Sainsbury: *Dictionary of Microbiology*, John Wiley & Sons, New York, New York, 1978
15. Rochkind, Blackburn, and Saylor, "Microbial Decomposition of Chlorinated Aromatic Compounds."
16. Gaudy, A. F., Jr., and E. T. Gaudy: *Elements of Bioenvironmental Engineering*, Engineering Press, Inc., San Jose, California, p. 180, 1988.
17. Irvine, R. L., and P. A. Wilderer: "Aerobic Processes," *Standard Handbook of Hazardous Waste Treatment and Disposal*, McGraw-Hill, Inc., New York, New York, p. 98, 1988.
18. Metcalf & Eddy, Inc.: *Wastewater Engineering*, p. 376.
19. U.S. Environmental Protection Agency: "Corrective Action: Technologies and Applications," EPA/625/4-89/020, Cincinnati, Ohio, September 1989.
20. Grady, C. P. L.: "Biodegradation of Toxic Organics: Status and Potential," *Journal of Environmental Engineering*, vol. 116, no. 5, September/October 1990.
21. Bailey, J. E., and D. F. Ollis: *Biochemical Engineering Fundamentals*, 2nd ed., McGraw-Hill, Inc., New York, New York, p. 383, 1986.

22. Monod, J.: Recherches sur la croissance des cultures bacteriennes, Herman et Cie, Paris, 1942.

23. Bailey and Ollis, *Biochemical Engineering Fundamentals*.

24. Metcalf & Eddy, Inc., *Wastewater Engineering*.

25. Gaudy and Gaudy, *Elements of Bioenvironmental Engineering*.

26. Bailey and Ollis: *Biochemical Engineering Fundamentals*.

27. Grady, "Biodegradation of Toxic Organics: Status and Potential."

28. Ibid.

29. Kobayashi, Hester, and Bruce E. Rittman: "Microbial Removal of Hazardous Organic Compounds," *Environmental Science and Technology*, vol. 16, no. 3, 1982.

30. Srinivasan, P. N., and J. W. Mercer: "Simulation of Biodegradation and Sorption Processes in Ground Water," *Ground Water*, vol. 26, no. 4, July 1988.

31. Alexander, Martin: "Biodegradation of Organic Chemicals," *Environmental Science and Technology*, vol. 18, no. 2, 1985.

32. Bhattacharya, S. K.: "Innovative Biological Processes for Treatment of Hazardous Wastes," *Proceedings of the 11th National Conference—Superfund '90*, Hazardous Materials Control Research Institute, Washington, D.C., November 26–28, 1990.

33. Bouwer, E. J., and P. L. McCarty, "Utilization Rates of Trace Halogenated Organic Compounds in Acetate-Grown Biofilms," *Biotechnology and Bioengineering*, Vol. 27, p. 1564–1571, 1985.

34. Metcalf & Eddy, Inc., *Wastewater Engineering*, p. 395.

35. Irvine and Wilderer, "Aerobic Processes."

36. Grady "Biodegradation of Toxic Organics: Status and Potential."

37. Tokuz, R. Y.: "Biotreatment of Hazardous Organic Wastes Using Rotating Biological Contactors," *Environmental Progress*, vol. 10, no. 3, pp. 198–204, August 1991.

38. Friday, D. D., and R. J. Portier: "Modular Reactor Approaches for Remediation of Ground Water: A Case Study with Volatile Chlorinated Aliphatics," *Proceedings of the 10th National Conference—Superfund "89*, Hazardous Materials Control Research Institute, Washington D.C., November 27–29, 1989.

39. Bhattacharya, S. K.: "Innovative Biological Processes for Treatment of Hazardous Wastes," *Proceedings of the 11th National Conference—Superfund '90*, Hazardous Material Control Research Institute, Washington, D.C., November 26–28, 1990.

40. Leahy, et al., "Biodegradation of Clorinated Aliphatics."

41. Irvine and Wilderer: "Aerobic Processes."

42. "Bioremediation of Polycyclic Aromatic Hydrocarbons & Pentachlorophenol," SBP Technologies, Inc., Stone Mountain, Georgia.

43. Wilder, P. A., and M. A. Rubio, "Sequencing Batch Reactors: A Versatile Technology for the Biological Treatment of Hazardous Leachates," *Contaminated Soil '88*, Kluwer Academic Publishers, 1988.

44. Grady, "Biodegradation of Toxic Organics: Status and Potential."

45. DiStefano, T. D., et. al.: "Reductive Dechlorination of High Concentrations of Tetrachloroethene to Ethene by an Anaerobic Enrichment Culture in the Absence of Methanogenesis," *Applied and Environmental Microbiology*, vol. 57, no. 8, pp. 2287–2292, August 1991.

46. Wilson, S. B.: "In Situ Remediation of Ground Water and Soils," *Proceedings of the 10th National Conference—Superfund '89*, Hazardous Materials Control Research Institute, Washington, D.C., November 27–29, 1989.

47. Cannon, J. Z.: "Considerations in Ground Water Remediation at Superfund Sites," memorandum from acting Assistant Administrator of U.S. Environmental Protection Agency, October 18, 1989.

48. Mackay, D. M., and J. A. Cherry, *Environmental Science and Technology*, vol. 23, no. 6, June 1989.

49. Hall, C. W.: "Practical Limits to Pump and Treat Technology for Aquifer Remediation," preconference workshop, Annual Conference of Water Pollution Control Federation, Dallas, Texas, 1989.

50. Travis, C. C., and C. B. Doty: "Can Contaminated Aquifers at Superfund Sites be Remediated," *Environmental Science and Technology*, vol. 24, no. 10, October 1990.

51. Schwefer, H. J.: "Latest Development of Biological In-Situ Remedial Action Techniques Portrayed by Examples from Europe and U.S.A.," *Contaminated Soil '88*, Kluwer Academic Publishers, 1988.

52. Nunno, T. J., J. A. Hyman, and T. Pheiffer: "Assessment of Site Remediation Technology in European

Countries," *Superfund '88*, Hazardous Materials Control Research Ins., pp. 193–198, 1988.

53. Verheul, J. H. A. M., R. Van den Berg, and D. H. Eikelboom: "In-Situ Biorestoration of a Subsurface Contaminated with Gasoline," *Contaminated Soil '88*, Kluwer Academic Publishers, 1988.

54. Brown, R. A., and J. R. Crosbie: "Oxygen Sources for In-Situ Bioremediation," *Proceedings of the 10th National Conference—Superfund '89*, Washington, D.C., November 27, 1989.

55. Brubaker, G. R.: "Lecture No. 1—Introduction to in Situ Bioremediation and Contaminant Transport," In Situ Bioremediation Short Course, Hazardous Materials Control Research Institute, November 28, 1990.

56. Carberry, J. B.: "Enhancement of PCP and TCE Biodegradation by Hydrogen Peroxide," *Proceedings of the 11th National Conference—Superfund '90*, Hazardous Materials Control Research Institute, Washington, D.C., November 26–28, 1990.

57. Thomas, J. M., and C. H. Ward: "In-Situ Biorestoration of Organic Contaminants in the Subsurface," *Environmental Science and Technology*, vol. 23, no. 7, pp. 760–766, July, 1989.

58. Brubaker, "Lecture No. 1."

59. Taylor, S. W., et al.: "Enhanced In-Situ Biodegradation and Aquifer Permeability Reduction," *Journal of Environmental Engineering*, vol. 117, no. 1, January/February 1991.

60. Stucki, G., et. al.: "Biological Degradation of 1, 2-Dichloroethane under Groundwater Conditions," *Water Research*, vol. 26, no. 3, pp. 273–278, 1992.

61. Piotrowski, M. A.: "Bioremediation: Testing the Waters," *Civil Engineering*, August 1989.

62. Madsen, E. L.: "Determining In Situ Biodegradation," *Environmental Science and Technology*, vol. 25, no. 10, pp. 1663–1673, October 1991.

63. Thomas and Ward, "In-Situ Biorestoration of Organic Contaminants."

64. Hutchins, S. R., et al.: "Biodegradation of Aromatic Hydrocarbons by Aquifer Microorganisms under Denitrifying Conditions," *Environmental Science and Technology*, vol. 25, no. 1, 1991.

65. Thomas and Ward, "In-Situ Biorestoration of Organic Contaminants."

66. Scalf, Marion: "Biological Cleanup of TCE, DCE and VC in Ground Water," *Tech Trends*, EPA/540/M-90/008, U.S. Environmental Protection Agency, July 1990.

67. Ostendorf, D. W., and D. H. Kampbell: "Bioremediation Soil Venting of Light Hydrocarbons," *Hazardous Waste & Hazardous Materials*, vol. 7, no. 4, 1990.

68. Ross, D.: "Slurry-Phase Bioremediation: Case Studies and Cost Comparisons," *Remediation*, vol. 1, no. 1, Winter 1990/91.

69. Stroo, H. F.: "Bioremediation of Hydrocarbon-Contaminated Soil Solids Using Liquid/Solids Contact Reactors," *Proceedings of the 10th National Conference—Superfund '89*, Washington, D.C., November 27, 1989.

70. "Eimco Biolift Reactor," Eimco Process Equipment Company, Salt Lake City, Utah.

71. Ahlert, R. C., and D. S. Kosson, "Aerobic Mineralization of Organic Contaminants Bound on Soil Fines," *Third International Conference on New Frontiers for Hazardous Waste Management*, Pittsburgh, Pennsylvania, September 1989.

72. Bachmann, A., and Zehnder, A. J. B.: "Engineering Significance of Fundamental Concepts in Xenobiotics Biodegradation in Soil," *Contaminated Soil '88*, Kluwer Academic Publishers, 1988.

73. deKreuk, J. F., and Annokkee, G. J.: "Applied Biotechnology for Decontamination of Polluted Soils: Possibilities and Problems," *Contaminated Soil '88*, Kluwer Academic Publishers, 1988.

74. Alvarez, L. M., et al.: "The Effects of Sorption on the Biotransformation rate of TCE by Methanotrophs, paper presented at *62nd Annual Water Pollution Control Conference*, October 16, 1989, San Francisco, California.

75. Smith, J. R., J. Lynch, and G. R. Brubaker: "Overview of Selected Bioremediation Technologies," *Hazardous Waste and Hazardous Materials Conference*, April 12–14, 1989, New Orleans, LA.

76. Bachmann, A., and Zehnder, A. J. B.: "Engineering Significance of Fundamental Concepts in Xenobiotics Biodegradation in Soil," *Contaminated Soil '88*, Kluwer Academic Publishers, 1988.

77. Stroo, "Bioremediation of Hydrocarbon."

78. Brox, G. H.: "A New Solid/Liquid Contact Bioslurry Reactor Making Bio-remediation More Cost-Competitive," *Proceedings of the 10th National Conference—Superfund '89*, Washington, D.C., November 27, 1989.

79. "Eimco Biolift Reactor," Eimco Process Equipment Company, Salt Lake City, Utah.

80. Smith, J. R., J. Lynch, and G. R. Brubaker: "Overview of Selected Bioremediation Technologies," *Hazardous Waste and Hazardous Materials Conference*, April 12–14, 1989, New Orleans, LA.

81. Stroo, H. F.: "Bioremediation of Hydrocarbon-Contaminated Soil Solids Using Liquid/Solids Contact Reactors," *Proceedings of the 10th National Conference—Superfund '89*, Washington, D.C., November 27, 1989.

82. Mueller, J. G., et. al.: "Bench-Scale Evaluation of Alternative Biological Treatment Processes for the Remediation of Pentachlorophenol- and Creosote-Contaminated Materials: Slurry-Phase Bioremediation," *Environmental Science and Technology*, vol. 25, no. 6, p. 1055–1061, June 1991.

83. Mueller, J.G., et. al.: "Bench-Scale Evaluation of Alternative Biological Treatment Processes for the Remediation of Pentachlorophenol- and Creosote-Contaminated Materials: Solid-Phase Bioremediation," *Environmental Science and Technology*, vol. 25, no. 6, p. 1045–1055, June 1991.

84. Portier, R. J.: "Examination of Site Data and Discussion of Microbial Physiology with Regard to Site Remediation," *Proceedings of the 10th National Conference—Superfund '89*, Washington, D.C., November 27, 1989.

85. Stroo, "Bioremediation of Hydrocarbon-Contaminated Soil Solids Using Liquid/Solids Contact Reactors," op. cit.

86. Ibid.

87. Kabrick, R. M., et al.: "Biological Treatment of Petroleum Refinery Sludges," *Third International Conference on New Frontiers for Hazardous Waste Management*, Pittsburgh, Pennsylvania, September 10-13, 1989.

88. Christiansen, J., et al.: "Liquid/Solids Contact Care Study," *Proceedings of the 10th National Conference—Superfund '89*, Hazardous Materials Control Research Institute, Washington, D.C., November 27-29, 1989.

89. Bachmann, A., and Zehnder, A. J. B.: "Engineering Significance of Fundamental Concepts in Xenobiotics Biodegradation in Soil," *Contaminated Soil '88*, Kluwer Academic Publishers, 1988.

90. deKreuk, J. F., and Annokkee, G. J.: "Applied Biotechnology for Decontamination of Polluted Soils: Possibilities and Problems," *Contaminated Soil '88*, Kluwer Academic Publishers, 1988.

91. Smith, J. R., J. Lynch, and G. R. Brubaker: "Overview of Selected Bioremediation Technologies," *Hazardous Waste and Hazardous Materials Conference*, April 12–14, 1989, New Orleans, LA.

92. Loehr, R. C., et al.: "Mobility and Degradation of Residues at Hazardous Waste Land Treatment Sites at Closure," U.S. Environmental Protection Agency, EPA/600/S2-90/018, Ada, Oklahoma, July 1990.

93. Brown, K. W., Evans, Jr., G.B., and Frentrup, B.D.: *Hazardous Waste Land Treatment*, Butterworth Publishers, Woburn, Massachusetts, 1983.

94. Grenney, W. J., Caupp, C. L., and Sims, R. C.: "A Mathematical Model for the Fate of Hazardous Substances in Soils: Model Description and Experimental Results," *Hazardous Waste and Hazardous Materials*, vol. 4, no. 3, 1987.

95. Nolziger, D. L., J. R. Williams, and T. E. Short: "Interactive Simulation of the Fate of Hazardous Chemicals During Land Treatment of Oily Waste: Ritz User Guide: EPA/600/8-88/001, U.S. Environmental Protection Agency, Ada, Oklahoma, January 1988.

96. "Land Treatment Demonstration," BP Oil Alliance Refinery, Report Prepared by ERM-Southwest, Inc., Metarie, Louisiana, May 24, 1988.

97. "In-Vessel Composting of Municipal Wastewater Sludge," EPA/625/8–89/016, Center for Environmental Research Information, U.S. Environmental Protection Agency, Cincinnati, Ohio.

98. Smith, J. R., J. Lynch, and G. R. Brubaker: "Overview of Selected Bioremediation Technologies," *Hazardous Waste and Hazardous Materials Conference*, April 12–14, 1989, New Orleans, LA.

99. Williams, R. T., et al.: "Composting of Explosives and Propellant Contaminated Sediments," *Proceedings of the 21st Mid-Atlantic Industrial Waste Conference*, Pennsylvania State University, June 25–27, 1989.

100. Taddeo, A.: "Field Demonstration of a Forced Aeration Composting Treatment for Coal Tar," *Proceedings of the Second National Conference of Biotreatment: The Use of Microorganisms in the Treatment of Hazardous Materials and Hazardous Waste*, Washington, D.C., November 27–29, 1989.

101. Mays, M. K., et al.: "Composting as a Method for Hazardous Waste Treatment," *Proceedings of the*

10th National Conference—Superfund '89, Washington, D.C., November 29, 1989.

102. Hater, G. R.: "Bioremediation of In-Situ and Ex Situ Contaminated Soils," *Proceedings of the Fourth Annual Hazardous Materials Management Conference*, Tower Conference Management Company, 1988.

103. Bewley, R. J. F. and P. Theile: "Decontamination of a Goal Gasification Site Through Application of Vanguard Microorganisms," *Contaminated Soil '88*, Kluwer Academic Publishers, 1988.

104. Grady, "Biodegradation of Toxic Organics: Status and Potential."

105. Rozich, A. F., and A. F. Gaudy, Jr.: *Design and Operation of Activated Sludge Processes Using Respirometry*, Lewis Publishers, Chelsea, Michigan, 1992.

106. Gaudy, A. F., Jr., A. Ekambaram, A. F. Rozich, and R. J. Colvin: "Comparison of Respirometric Methods for Determination of Biokinetic Constants for Toxic and Nontoxic Wastes," *Proceedings, 44th Purdue Industrial Waste Conference*, Lewis Publishers, Chelsea, MI, pp. 393–403, 1990.

107. Gaudy, A. F., Jr., A. F. Rozich, N. R. Moran, S. T. Garniewski, and A. Ekambaram: "Methodology for Utilizing Respirometric Data to Assess Biodegradation Kinetics," *Proceedings, 42nd Purdue Industrial Waste Conference*, Lewis Publishers, Chelsea, Michigan, pp. 573–584, 1988.

08. Rozich, A. F., and R. J. Colvin: "Formulating Strategies for Activated Sludge Systems," *Waste Engineering & Management*, vol. 137, no. 10, pp. 39–41, 1990.

09. U.S. Environmental Protection Agency: *Superfund Innovative Technology Evaluation (SITE) Program*, EPA/540/8-91/005, Spring 1991.

ADDITIONAL READING

Alexander, M.: "Biodegradation of Chemicals at Trace Concentrations," *NTIS AD-A212 409/7*, Army Res. Office, Research Triangle Park, North Carolina, 1989.

Alexander, Martin: "Biodegradation of Organic Chemicals," *Environmental Science and Technology*, vol. 18, no. 2, pp. 106–111, 1985.

Bouwer, Edward J.: "Biotransformation of Aromatics in Strip-Pit Pond," *Journal of Environmental Engineering*, vol. 115, no. 4, August 1989.

Brown, Lewis R.: "Oil-Degrading Microorganisms," *Chemical Engineering Progress*, October 1987.

Brown, Susan Cooper, C. P. Leslie Grady, Jr., and Henry H. Tabak: "Biodegradation Kinetics of Substituted Phenolics: Demonstration of a Protocol Based on Electrolitic Respirometry," *Water Resources*, vol. 24, no. 7, pp. 853–861, 1990.

Casarini, D. C. P., R. M. de Macedo, R. C. de A. Cunha, and J. C. O. Mauger: "The Development of Assessment Techniques to Evaluate the Biodegradation of Oily Sludge in a Land Farming System," *Water Science and Technology*, vol. 20, no. 10, pp. 231–236, 1988.

Coombs, J.: *Dictionary of Biotechnology*, Macmillan Press Ltd., London, 1986.

DiGrazia, Phillip M., et al.: "Development of a System Analysis Approach for Resolving the Structure of Biodegrading Soil Systems," *Applied Biochemistry and Biotechnology*.

George, E. J., and R. D. Neufeld: "Degradation on Fluorene in Soil by Fungus Phanerochaete Chrysosporium," *Biotechnol. Bioeng.*, vol. 33, p. 1306, 1989.

Horvath, Raymond S.: "Microbial Co-Metabolism and the Degradation of Organic Compounds in Nature," *Bacteriological Reviews*, vol. 36, pp. 146–155, June 1972.

Kim, Chung J., and Walter J. Maier: "Acclimation and Biodegradation of Chlorinated Organic Compounds in the Presence of Alternate Substrates." *Journal WPCF*, vol. 58, no. 2, February 1986.

Rhee, G. Y., et al.: "Anaerobic Biodegradation of Polychlorinated Biphenyls in Hudson River Sediments and Dredged Sediments in Clay Encapsulation," *Water Res.*, (G.B.), vol 23, p. 957, 1989.

Robinson, Kevin G., et al.: "Availability of Sorbed Toluene in Soils for Biodegradation by Acclimated Bacteria," *Wat. Res.*, vol. 24, no. 3, pp. 345–350, 1990.

Rogers, L. A., and R. C. Loehr: "Weathering of Constituents at Land Treatment Sites," *Environ. Eng. Proc. 1989 Specialty Conf.*, ASCE, New York, p. 742, 1989.

Sims, Judith L.: "Approach to Bioremediation of Contaminated Soil," *Hazardous Waste and Hazardous Materials*, vol. 7, no. 2, 1990.

Singleton, Paul: *Dictionary of Microbiology*, John Wiley & Sons Ltd., New York, New York, 1978.

Tabak, Henry H., Stephen A. Quave, Charles I. Mashni, and Edwin F. Barth: "Biodegradability Studies with Organic Priority Pollutant Compounds," *Journal WPCF*, vol. 53, no. 10, pp. 1503–1518, October 1981.

Wang, Xiaoping, et al.: "Effect of Bioremediation on Polycyclic Aromatic Hydrocarbon Residues in Soil." *Environ. Sci. Technol.*, vol. 24, pp. 1086–1089, 1990.

ACKNOWLEDGMENTS

The authors wish to acknowledge the following individuals who assisted with the preparation of this chapter: Derek Ross, Ph.D., Al Rozich, Ph.D., P.E., and Jack Harrison, P.E., *ERM, Inc.*; Nicholas Hollingshad of *ERM-Southwest*; and Michael Goldman, *Buchart-Horn Engineers*. The principal reviewer for the **ERM Group** was Brian Flynn, *ERM-Southwest, Inc.*

CHAPTER
11

STABILIZATION
AND
SOLIDIFICATION

There is nothing stable in the world.
John Keats

11-1 INTRODUCTION

Stabilization and solidification have been widely applied in the management of hazardous wastes. The technologies are being applied to 1) the remediation of hazardous waste sites; 2) the treatment of residue from other treatment processes (e.g., the ash from thermal treatment); and 3) the treatment of contaminated land where large quantities of soil containing contaminants are encountered.

In general terms, stabilization is a process where additives are mixed with waste to minimize the rate of contaminant migration from the waste and to reduce the toxicity of the waste. Thus, stabilization may be described as a process by which contaminants are fully or partially bound by the addition of supporting media, binders, or other modifiers.[1] Likewise, solidification is a process employing additives by which the physical nature of the waste (as measured by the engineering properties of strength, compressibility, and/or permeability) is altered during the process.[2] Thus, objectives of stabilization and solidification would encompass both the reduction in waste toxicity and mobility as well as an improvement in the engineering properties of the stabilized material.[3]

Although stabilization and solidification are physiochemical processes, they are discussed separately from those in Chap. 9. This allows for more detailed examination of these widely employed hazardous waste management techniques including presen-

641

tation of stabilization mechanisms, techniques, and design considerations along with the tests used to evaluate the treatment effectiveness.

Definitions

Stabilization is a process employing additives (reagents) to reduce the hazardous nature of a waste by converting the waste and its hazardous constituents into a form:

- To minimize the rate of contaminant migration into the environment, or
- To reduce the level of toxicity.

Fixation is often used synonymously with stabilization. Stabilization is accomplished through the addition of reagents that:[4]

- Improve the handling and physical characteristics of the waste,
- Decrease the surface area across which transfer or loss of contaminants can occur,
- Limit the solubility of any pollutants contained in the waste, and
- Reduce the toxicity of the contaminants.

In contrast, **solidification** is described as a process by which sufficient quantities of solidifying material, including solids, are added to the hazardous materials to result in a solidified mass of material.[5] Solidifying the mass is accomplished through the addition of reagents that:

- Increase the strength,
- Decrease the compressibility, and
- Decrease the permeability of the waste.

Throughout this chapter the term stabilization will be utilized to mean both stabilization and solidification except where specifically noted otherwise.

The potential for contaminant loss from a stabilized mass is usually determined by leaching tests. **Leaching** is the process by which contaminants are transferred from a stabilized matrix to a liquid medium such as water.

During stabilization, certain contaminants may be destroyed such as by the dechlorination of chlorinated hydrocarbons. Other organics may "disappear" as a result of volatilization. However, the stabilization of inorganic contaminants that are already in their atomic form such as cadmium, lead, and other metals should mimic nature. By studying the form in which they occur in nature, we can learn much about the optimum stabilization method. For example, zinc phosphate ore bodies provide the source of zinc for metallurgical processes, and a stabilization technique that precipitates the zinc as zinc phosphate is likely to be the optimum technique in terms of environmental effectiveness. Stabilization must be considered as a waste treatment process that reduces, to an acceptable or *geologically slow* rate, the movement of contaminants into the environment.

Regulations

The cleanup of hazardous waste sites that endanger the public health and welfare and the environment requires the application of remedial technologies that are both effective and permanent. In this regard, stabilization offers technological advantages over alternative remedial options such as landfilling that manages waste by transfering it from one location to another. Stabilization may also be utilized in situations where other treatment methods, such as bioremediation or incineration, are not technically applicable or cost-effective. For long-term solutions it is important to consider more permanent technologies such as stabilization. As a result, stabilization is specifically identified in the U.S. National Contingency Plan.[6]

It has been recognized that the disposal of liquid wastes in landfills is an environmentally unsound practice. Liquids migrate downward through the landfill, assist in the generation of leachate, and potentially migrate through defects in the liner system. As a consequence of the high mobility of liquid hazardous wastes, their disposal in landfills has been prohibited in the United States.[7] This ban has resulted in the increased use of stabilization in order to preclude disposal of liquid wastes. In this application, the stabilization agents must chemically bond the hazardous chemicals in liquid wastes. The agents cannot act simply as absorbents (physically containing the liquids much the same way in which a sponge holds water).

The U.S. regulations have thus brought into focus two distinctly different technological objectives for stabilization: 1) stabilization of materials prior to land disposal and 2) the stabilization of contaminants to be left in place as part of a site cleanup. Stabilization may require a less stringent performance measure (eliminating free liquids). Alternatively, for those situations where the treated material is left in place, the stabilization process may require considerably greater performance (reducing leachability).

Applications

The three major areas of application for stabilization technologies are:

- Land disposal—the stabilization of wastes prior to secure landfill disposal;
- Site remediation—the remediation of contaminated sites; and
- Solidification of industrial wastes—the solidification of non-hazardous, unstable wastes, such as sludges.

Land Disposal. The land disposal of liquid waste increases the likelihood of contaminant migration (and is currently banned under U.S. regulations). Liquid waste, along with wet sludges, must be stabilized prior to landfilling. To effectively stabilize liquids, the stabilization agents cannot be absorbents (such as sawdust). The liquids absorbed by agents could be easily released (desorbed) in the landfill when compressed under additional loads. That is, as more material is landfilled, the weight of the overlying material would squeeze the liquids out of the underlying material. Therefore, the liquids must be chemically and physically bound by the stabilization reagents so that they are not expelled by the consolidation stresses or leached out by the downward percolation of precipitation. Note that under U.S. regulations, incineration is the preferred treatment method for certain organic wastes. However, there are provisions for demonstrating alternative technologies in the land disposal restrictions.

Site Remediation.   The remediation of contaminated sites having organic wastes, inorganic wastes, and/or contaminated soils may be accomplished employing stabilization technology. For site remediation, stabilization is used to 1) improve the handling and physical characteristics of the wastes, 2) decrease the rate of contaminant migration by decreasing the surface area across which the transfer of pollutants can occur and by limiting the solubility of pollutants, and 3) reduce the toxicity of certain contaminants. Stabilization may be considered to be a more permanent remedial solution than other alternatives such as containment. Stabilization is particularly well suited for sites where the hazard involves large quantities of soils contaminated at low levels. In many instances it may not be environmentally sound nor cost-effective to excavate, transport, and landfill or incinerate soils contaminated with low levels of pollutants because of 1) the additional air pollution caused by excavation equipment, trucks, and the exposure of buried contaminated soils to the air, which enhances the volatilization of organics; and 2) the increased risks as a result of traffic accidents.

Example 11-1   Evaluation of landfilling versus stabilization.   The total quantity of contaminated soils at the Pepper's Steel and Alloy site was 120,000 tons.[8] Evaluate the cost of landfilling versus stabilization for the management of the hazardous waste at this site. Each truck can carry 31,500 pounds to the nearest suitable landfill site at Emile, Alabama, at a distance of 850 miles. The trucking cost per mile is $2.00 and the total stabilization cost is $67 per ton. Identify advantages and disadvantages of landfilling for this site.

Solution. Stabilization cost estimate:
    120,000 tons × $67/ton = $8,040,000
Trucking cost estimate:
    (120,000 tons × 2,000 pounds/ton)/31,500 pounds/truck = 7,619 trucks
    7,619 trucks × 850 miles/truck × $2.00/mile = $12,952,381
Because trucking costs alone are almost $13 million, stabilization is certainly more cost-effective.
    Advantages of landfilling: Landfilling permits the actual reclamation of the contaminated land at the Pepper's Steel and Alloy site and eliminates one site from the national inventory of contaminated sites.
    Disadvantages of landfilling: Valuable landfill space is being used by soils amenable to other treatment, precluding the use of that space by wastes better suited for landfill.

Is site remediation employing stabilization widely used? The answer for the United States is increasingly so. Table 11-1 contains a partial listing of waste sites where stabilization was selected and implemented as the hazardous waste management technology of choice. The table also presents information regarding the type and quantity of treated contaminants. Further, studies using four different types of artificially contaminated soils representative of those found at U.S. Superfund sites have demonstrated significant reductions in contaminant leachability.[9]

Case studies for stabilization technologies (modified from U.S. EPA.[10])

Site/contractor	Contaminant (concentration)	Treatment volume	Physical form	Binder	Percentage binder(s) added	Treatment (batch/continuous in situ)	Disposal (onsite/offsite)	Volume increase, %	Scale of operation
Midwest, U.S. Plating Company, Envirite	Cu, Cr, Ni	16,000 yd³	Sludge	Portland cement	20%	In situ	Onsite	> 0	Full scale
Marathon Steel Phoenix, AZ Silicate, Tech.	Pb, Cd	150,000 yd³	Dry-landfill	Portland cement and silicates (Toxsorb)™	Varied 7–15% (cement)	Concrete batch plant	Landfill	NA	Full scale
N.E. Refinery ENRECO	Oil sludges, Pb, Cr, As	100,000 yd³	Sludges, variable	Kiln dust (high CaO content)	Varied, 15–30%	In situ	Onsite	>Varied, ~ 20% average	Full scale
Amoco Wood River Chemfix	Oil/solids Cd, Cr, Pb	90,000,000 gallons	Sludges	Chemfix proprietary	NA, proprietary	Continuous flow (proprietary process)	Onsite	Average 15%	Full scale (site delisted 1985)
Pepper Steel & Alloy, Miami, FL VFL Technology Corporation	Oil sat: soil Pb (1000 ppm) PCBs (200 ppm) As (1–200 ppm)	62,000 yd³ (plus 5000 tons of surface debris)	Soils	Pozzolanic and proprietary	~ 30%	Continuous feed (mixer proprietary) design	Onsite	~ 1%	Full scale
Vickery, OH Chemical Waste Management	Waste acid PCBs (< 500 ppm), dioxins	~235,000 yd³	Sludges (viscous)	Lime and kiln dust	~ 15% CaO ~ 5% kiln dust	In situ	Onsite (TSCA cells)	>~ 9%+	Full scale
Wood Treating, Savannah, GA Geo-Con, Inc.	Creosote wastes	12,000 yd³	Sludges	Kiln dust	20%	In situ	Onsite lined cells	>~ 14%	Full scale
Wyandotte, MI Treatment Plant Chem Met	Various/combined	20 million gal/yr	Various	Lime		Continuous	Offsite (secure landfill)		In-plant process
Chem Refinery, TX HAZCON	Combined metals, sulfur, oil sludges, etc.	90,000 gal. (445 yd³)	Sludges (synthetic oil sludges)	Portland cement and proprietary	NA	Continuous flow	Onsite (secure landfill)	>Estimated 10%	Full scale
Metalplating, WI Geo-Con, Inc.	Al (9500 ppm) Ni (750 ppm) Cr (220 ppm) Cu (2000 ppm)	3000 yd³	Sludges	Lime	10–25%	In situ	Onsite landfill	> 4–10%	Full scale

645

Solidification of Industrial Wastes. As a legacy from our past waste management practices, a wide variety of organic and inorganic industrial wastes can be found in pits, ponds, and lagoons. Solidification (affecting of physical properties such as strength and compressibility) improves the engineering properties and may reduce the rate at which contaminants migrate into the environment. Although many of the non-hazardous waste materials may not endanger the public health or the environment, the materials are frequently structurally unstable, aesthetically unsuitable, and their condition precludes other uses of the site area. Thus, the primary goal of solidification is the improvement of the structural integrity of the material. For such solidification projects, the effectiveness of the solidification process can often be evaluated by measurement of the material's strength.[11]

Figure 11-1[12] depicts a waste solidification project where cement kiln dust is added to neutralize and stabilize an oily sludge. In a project such as this, improving the appearance and stability of the material, coupled with the neutralization, results in an overall improvement to the environment. The solidified mass could then be covered to reduce precipitation infiltration as described in Chap. 13.

11-2 MECHANISMS

Understanding of the fundamental physical and chemical mechanisms that control the effectiveness of the stabilization reagents is essential to the correct implementation of stabilization as a hazardous waste management technology. Further, as new reagents

FIGURE 11-1
Typical waste solidification project. (photo provided by Enreco, Inc.)

are developed or existing reagents are modified and adapted to new and different wastes, fundamental stabilization mechanisms should be considered to evaluate the potential for success or failure of such developments. Successful stabilization employs one or more of the following mechanisms:

- Macroencapsulation
- Microencapsulation
- Absorption
- Adsorption
- Precipitation
- Detoxification

Macroencapsulation

Macroencapsulation is the mechanism by which hazardous waste constituents are physically entrapped in a larger structural matrix,[13] that is, the hazardous waste constituents are held in discontinuous pores within the stabilizing materials. Upon physical degradation (breakdown) of the stabilized material, even to relatively large particle sizes, the entrapped materials are free to migrate. The stabilized mass may break down over time (as measured on a geologic time scale) due to imposed environmental stresses. These stresses include such things as repeated cycles of wetting and drying or freezing and thawing, introduction of percolating fluids, and physical loading stresses. Thus, contaminants stabilized by only macroencapsulation may find their way into the environment if the integrity of the mass is not maintained.

Figure 11-2 is an example of an oily waste that is stabilized, in part, as a result of macroencapsulation. In the stabilized sample shown in this photograph, note the dark nodules of oily waste encapsulated within the light gray matrix of fly ash, lime, and cement binder. The waste in these pores is bound only by physical entrapment in the stabilizing matrix. Should this matrix degrade due to cycles of freezing and thawing or wetting and drying, the stabilized oily waste would be free to migrate into the environment.

The degree of macroencapsulation is enhanced by the type and energy per unit mass of mixing. Laboratory-prepared samples are generally well blended with a high energy per unit mass as compared with field mixing. This gives rise to a greater degree of macroencapsulation in the laboratory (a greater number of particles to be encapsulated) than in the field (fewer, but larger, particles to be encapsulated). Therefore, laboratory studies must be conducted using field mixing condition to the extent practicable.

Microencapsulation

In **microencapsulation**, hazardous waste constituents are entrapped within the crystalline structure of the solidified matrix at a **microscopic** level. As a result, even if the stabilized materials degrade into relatively small particle sizes, most of the stabilized hazardous wastes remains entrapped. However, as with macroencapsulation, because

FIGURE 11-2
Organic waste encapsulated in
binder matrix (photo courtesy
of Debra Cook, Bucknell
University.)[14]

the waste is not chemically altered or bound, the rates of contaminant release from the stabilized mass may increase as the particle size decreases and more surface area is exposed.

Figure 11-2 also provides an example of an oily waste that is stabilized, in part, as a result of microencapsulation. The oily waste is encapsulated within the small pores of the light gray matrix of fly ash, lime, and cement binder. The waste in these very small pores is bound only by physical entrapment in the stabilized mass in a manner similar to macroencapsulation. If the stabilized materials degrade to very small particles over the long term, the stabilized oily waste would be released from the matrix.

As previously described in the macroencapsulation section, laboratory-prepared samples are generally well blended with a high energy per unit mass as compared with field mixing. This gives rise to a greater degree of microencapsulation in the laboratory than in the field, and this must be considered when deciding whether to use solidification or not.

Absorption

As described in Sec. 3-2, absorption is the process by which contaminants are taken into the sorbent in very much the same way a sponge takes on water. As applied in stabilization, absorption requires the addition of solid material (sorbent) to soak up or **absorb** the free liquids in the waste. The process is primarily employed to remove free liquid to improve the waste-handling characteristics, that is, to solidify the waste. The liquids are free to squeeze out of the material should the mass be subjected to consolidating stresses. Thus, the use of absorption is considered only a temporary measure to improve handling characteristics.

The most common absorbents include:

- Soil
- Fly ash
- Cement kiln dust
- Lime kiln dust
- Clay minerals including bentonite, kaolinite, vermiculite, and zeolite
- Sawdust
- Hay and straw

Some of the absorbents previously listed, such as cement kiln dust, have additional benefits due to their pozzolanic (self-cementing) characteristics. The use of these materials as stabilization reagents is discussed later in this chapter.

Adsorption

In addition to the physical entrapment within the stabilized and solidified mass described above, electrochemical interactions may occur.[15] **Adsorption** is the phenomenon by which contaminants are electrochemically bonded to stabilizing agents within the matrix. These are typically considered surface phenomena and the nature of the bonding may be through van der Waal's or hydrogen bonding. Further discussion of adsorption mechanisms is contained in Sec. 3-2. Contaminants that are chemically adsorbed (fixed) within the stabilized matrix are less likely to be released into the environment than those that are not fixed. Unlike microencapsulation and macroencapsulation, where simple particle breakdown may enhance the rate of contaminant migration, additional physicochemical stress is necessary to desorb the material from their adsorbing surfaces. As a result, the treatment is considered more permanent.

The stabilization of organic wastes using organically modified clays illustrates how adsorption can be used in waste stabilization. Organically modified clays are clays which have been altered by replacing exchangeable inorganic cations adsorbed on the clay surfaces with long-chain organic cations, rendering the clays **organophilic**. Organophilic clays have an affinity for organic molecules. Without modification, naturally occurring clays are generally organophobic. Organic waste molecules are then adsorbed to the clay as shown in Fig. 11-3. The adsorption bond strength must then be overcome if the organic waste molecules were to be released to migrate into the environment (see Organically Modified Clays in Sec. 11-3).

Precipitation

Certain stabilization processes will precipitate contaminants from the waste resulting in a more stable form of the constituents within the waste.[16] Precipitates such as hydroxides, sulfides, silicates, carbonates, and phosphates are then contained within the stabilized mass as part of the material structure. This phenomenon is applicable to the stabilization of inorganic wastes such as metal hydroxide sludges. For example,

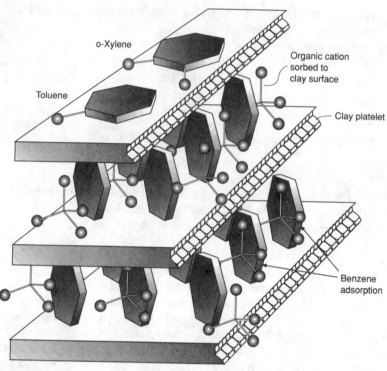

FIGURE 11-3
Organic waste adsorbed to an organophilic clay (modified from reference 17).

metal carbonates are typically less soluble than metal hydroxides. At high pH, the reaction to form a metal carbonate from a metal hydroxide sludge, is as follows:[18]

$$Me(OH)_2 + H_2CO_3 \rightarrow MeCO_3(s) + 2H_2O \tag{11-1}$$

where Me represents a metallic cation.

The permanence of the fixation of a metal as a metallic carbonate depends upon, among other things, the pH. In this example, the metallic carbonate is quite stable except at low pH. Under strongly acidic conditions, the metal may be redissolved and may then be free to migrate as a solute into the environment. Thus, the assessment of the effectiveness of stabilization must properly account for present and future environmental conditions.

Detoxification

Certain chemical reactions taking place during the stabilization process may result in a waste with reduced toxicity. Detoxification is any mechanism that changes a chemical constituent into another constituent (or form of the same constituent) that is either less toxic or non-toxic.

An example of this is the reduction of chromium in the +6 valence state to chromium in the +3 valence state during stabilization with cement-based materials. The trivalent chromium has a lower solubility and toxicity than hexavalent chromium. A number of fixation systems can be used to detoxify chromium by reducing its valence state; ferrous sulfate and a combination of ferrous sulfate and sodium sulfate have been found to be effective.[19] Leaching of the reduced chromium poses a lesser threat to the environment than the leaching of the original hexavalent chromium.

11-3   TECHNOLOGY

A detailed description of stabilization technology can best be compartmentalized by a discussion of the various types of additives (reagents) employed. Although it is recognized that this classification into categories is somewhat arbitrary because more than one reagent may be employed for any given waste stabilization project, this approach allows examination of the mechanisms by which each of the reagents act. Where agents are used in conjunction with each other in a synergistic fashion, it will be noted in the text. The term **binder** is used to denote a reagent that contributes to the strength gain associated with stabilization. The term **sorbent** is used to denote a reagent that primarily contributes to retaining contaminants in the stabilized matrix.

As a first approximation, the effectiveness of each of the stabilization reagents upon selected waste types is shown in Table 11-2. The application of Table 11-2 depends on several factors, such as the concentration of the contaminant, the quantity of the reagent, and the synergistic effects of multiple contaminants and reagents to name a few. Nonetheless, the information of Table 11-2 is useful for preliminary screening and the development of bench scale studies.

There is an extensive range of sorbents and binders available in the worldwide marketplace, including several proprietary reagents. This text describes those that are not proprietary to enable the student to appreciate the complexity of the chemistry involved and to begin to understand the nature of the stabilization process.

Cement

Stabilization of hazardous wastes frequently employs cement as the principal reagent. What is cement? The most common cement is **portland cement** which is made by firing a mixture of limestone and clay (or other silicate) in a kiln at high temperatures. The kiln produces a clinker which is ground to a powder which is a mixture of calcium, silicate, aluminum, and iron oxides. The main constituents are tri- and di-calcium silicates.[20,21] For cement-based stabilization, waste materials are mixed with cement followed by the addition of water for hydration, if necessary because the waste does not have enough water. The hydration of the cement forms a crystalline structure, consisting of calcium alumino-silicate. This results in a rock-like, monolithic, hardened mass. **Concrete,** by definition, is a particulate composite consisting of hydrated cement and aggregate.

TABLE 11-2

Reagent applicability for waste stabilization[22]

Waste component	Cement-based	Pozzolan-based	Thermoplastic	Organic polymer
Nonpolar organics as: oil and grease, aromatic hydrocarbons, halogenated hydrocarbons, PCBs	May impede setting. Decreases durability over a long time period. Volatiles may escape upon mixing. Demonstrated effectiveness under certain conditions.[a]	May impede setting. Decreases durability over a long time period. Volatiles may escape upon mixing. Demonstrated effectiveness under certain conditions.[b]	Organics may vaporize upon heating. Demonstrated effectiveness under certain conditions.[c]	May impede setting. Demonstrated effectiveness under certain conditions.[d]
Polar organics as: alcohols, phenols, organic acids, glycols	Phenol will significantly retard setting and will decrease durability in the short run. Decreases durability over a long time period.[e]	Phenol will significantly retard setting and will decrease durability in the short run. Alcohols may retard setting. Decreases durability over a long time period.	Organics may vaporize upon heating.	No significant effect on setting.
Acids as: hydrochloric acid, hydrofluoric acid	No significant effect on setting. Cement will neutralize acids. Types II and IV portland cement demonstrate better durability characteristics than Type I. Demonstrated effectiveness.[f,g]	No significant effect on setting. Compatible, will neutralize acids. Demonstrated effectiveness.[f,g]	Can be neutralized before incorporation.	Can be neutralized before incorporation. Ureaformaldehyde demonstrated to be effective.[f]
Oxidizers as: sodium hypochlorite, potassium permanganate, nitric acid, potassium dichromate	Compatible	Compatible	May cause matrix breakdown, fire.	May cause matrix breakdown, fire.
Salts as: sulfates, halides, nitrates, cyanides	Increase setting times. Decrease durability. Sulfates may retard setting and cause spalling unless special cement is used. Sulfates accelerate other reactions.	Halides are easily leached and retard setting. Halides may retard setting, most are easily leached. Sulfates can retard or accelerate reactions.	Sulfates and halides may dehydrate and rehydrate, causing splitting.	Compatible[h]
Heavy metals as: lead, chromium, cadmium, arsenic, mercury	Compatible. Can increase set time. Demonstrated effectiveness under certain conditions.[i]	Compatible. Demonstrated effectiveness on certain species (lead, cadmium, chromium).[d,j]	Compatible. Demonstrated effectiveness on certain species (copper, arsenic, chromium).[d]	Compatible. Demonstrated effectiveness with arsenic.[d]
Radioactive materials	Compatible	Compatible	Compatible	Compatible

a Tittlebaum and Seals 1985; Van Keuren et al. 1987; JACA 1985; U.S. EPA 1986; Jones 1986.
b Musser and Smith 1984; U.S. EPA 1984; Kyles, Malinowski, and Stanczyk 1987; Tittlebaum and Seals 1985.
c Tittlebaum and Seals 1985; JACA 1985.
d Tittlebaum and Seals 1985.
e Kylutes and Bishop 1987.
f JACA 1985.
g Van Keuren et al. 1987.
h Musser and Smith 1984.
i Federal Register 1988; Kyles, Malinowski, and Stanczyk 1987.

In its simplest form, the reaction of the tricalcium silicate can be expressed by the reaction equation:

$$2(3CaO \bullet SiO_2) + 6H_2O \rightarrow 3CaO \bullet 2SiO_2 \bullet 3H_2O + 3Ca(OH)_2 \qquad (11\text{-}2)$$

and the reaction of dicalcium silicate by:

$$2(2CaO \bullet SiO_2) + 4H_2O \rightarrow 3CaO \bullet 2SiO_2 \bullet 3H_2O + 3Ca(OH)_2 \qquad (11\text{-}3)$$

Additional reactions occur during the hydration of portland cement as shown in Fig. 11-4. The resulting silicate gel is termed tobomorite. These reactions are quite slow, contributing to the lengthy curing time associated with portland cement concretes.

The most rapid reaction in portland cement is:

$$3CaO \bullet Al_2O_3 + 6H_2O \rightarrow 3CaO \bullet 2Al_2O_3 \bullet 6H_2O + \text{ heat} \qquad (11\text{-}4)$$

This reaction provides portland cement with its initial set. A schematic representation of this reaction is shown in Fig. 11-5.

Cement-based stabilization is best suited for inorganic wastes, especially those containing heavy metals. As a result of the high pH of the cement, the metals are

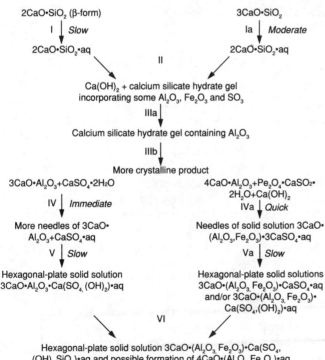

FIGURE 11-4
Portland cement reactions.[23]

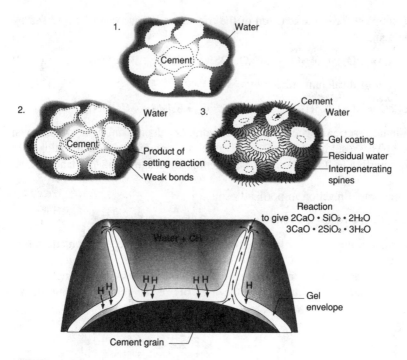

FIGURE 11-5
Schematic representation of cement hydration.[24]

retained in the form of insoluble hydroxide or carbonate salts within the hardened structure.[25,26] Studies have shown that lead, copper, zinc, tin, and cadmium are likely bound in the matrix by chemical fixation, forming insoluble compounds while mercury is predominantly held by physical microencapsulation.[27]

On the other hand, organic contaminants interfere with the hydration process, reduce the final strength, and are not easily stabilized.[28–34] They may also reduce the crystalline structure formation resulting in a more amorphous material.[35] Other additives may be incorporated into the stabilizing mix with the cement to reduce organic contaminant interference with cement hydration and enhance stabilization. These additives may include organically modified and natural clays, vermiculite, and soluble sodium silicates.

It is difficult to determine precisely where a specific contaminant will react within the stabilized cement matrix. In a study of the binding chemistry of hazardous waste in cement, the locations of the heavy metals lead and chromium were studied.[36] The results indicated that the lead precipitated on the outer surface of the hydrated cement particles whereas chromium was widely and more uniformly dispersed through the hydrated cement particles.

Stabilization with cement has shown considerable promise and utilization for the fixation of inorganic wastes such as metal hydroxide sludges from the plating industry. The wide use of stabilization of inorganics derives from (1) the lack of better alternatives (for example, metals do not biodegrade and do not change in atomic structure when incinerated) and (2) available and understood physiochemical mechanisms such as precipitation and adsorption.

There are a number of advantages of cement-based stabilization. The technology of cement is well known including handling, mixing, setting, and hardening. Cement is widely employed in the construction field, and, as a result, the material costs are relatively low and the equipment and personnel readily available. Extensive dewatering of wet sludges and waste is not typically necessary as water is required for cement hydration. In fact, cement-based stabilization may proceed with sufficient water so as to make the material pumpable.[37] The system can handle variations in the chemistry of the waste.[38] Finally, the alkalinity of the cement can neutralize acidic wastes. The principal disadvantage is the sensitivity of cement to the presence of certain contaminants that could retard or prohibit proper hydration and the resulting setting and hardening of the material.

Pozzolans

A **pozzolan** is a material that can react with lime in the presence of water to produce a cementitious material. The reaction of alumino-silicious material, lime, and water results in the formation of a concrete-like product termed **pozzolanic concrete.** Pozzolanic materials include fly ash, ground blast furnace slag, and cement kiln dust. The resulting pozzolanic structures are termed amorphous alumino-silicates.[39] Fly ash, the most commonly employed pozzolan, is typically about 45% SiO_2, 25% Al_2O_3, 15% Fe_2O_3, 10% CaO, 1% MgO, 1% K_2O, 1% Na_2O, and 1% SO_3. In addition, the amount of unburned carbon may vary from source to source.

Like cement-based stabilization, most stabilization applications involving the use of pozzolans are for inorganic materials. The high pH environment is well suited for waste contaminated with heavy metals. An example in the literature shows that the untreated sludge from an Imhoff digester tank was treated with fly ash and lime reagents to stabilize elevated levels of cadmium, chromium, copper, iron, lead, magnesium, manganese, selenium, silver, and zinc.[40] Although the stabilized material had a soil-like consistency, the leaching tests indicated that the stabilization process significantly slowed leaching of several of these components. Unburned carbon in fly ash may adsorb organics from the waste. As a result, a pozzolan such as fly ash may have beneficial effects in the stabilization of both organic and inorganic species.

Example 11-2   Evaluation of stabilization effectiveness. The total mass of contaminants leached in a column leaching test from the unstabilized and pozzolanic reagent stabilized Imhoff sludge example presented above is as follows:[41]

Parameter	Untreated sludge	Treated sludge
Arsenic	0.724	0.049
Cadmium	3.3	0.126
Chromium	0.566	1.16

Copper	3.86	3.2
Iron	48.4	1.57
Lead	0.238	0.019
Manganese	11.5	1.48
Zinc	120	3.02

a.) Calculate the percent reduction in the mass of the leached constituents.

b.) What do you consider the average reduction in leached constituent concentrations?

c.) Which chemical parameter requires further consideration in evaluating the effectiveness of stabilization and why?

Solution.

a.) Reduction (%) = (100) × (Untreated sludge − Treated sludge)/(Untreated sludge) e.g., $R = ((0.724 − .049)/0.724) \times (100) = \underline{93.2\%}$

Continuing in this manner yields:

Parameter	Untreated sludge	Treated sludge	Reduction (%)
Arsenic	0.724	0.049	93.2
Cadmium	3.3	0.126	96.2
Chromium	0.566	1.16	−104.9
Copper	3.86	3.2	17.1
Iron	48.4	1.57	96.8
Lead	0.238	0.019	92.0
Manganese	11.5	1.48	87.0
Zinc	120	3.02	97.5

b.) Because each constituent has a different toxicity and mobility, it is not reasonable to average the value and report the average reduction in contaminant leachability. Each value must be assessed individually; the average is meaningless at best, misleading at worst.

c.) The chromium concentrations actually increased in the leachate of the stabilized sludge over the untreated sludge. The high chromium concentrations suggest a high mobility and therefore a + 6 valence state. The stabilization process should consider first reducing the chromium to a + 3 valence state and then precipitation as chromium hydroxide.

Lime

Stabilization of sludges[42] is frequently accomplished through the addition of calcium hydroxide $Ca(OH)_2$ or **lime** (also called hydrated lime). The resulting reaction with materials included in the waste may result in hydrates of calcium silicate, calcium alumina, or calcium alumino-silicate. These materials are formed from the reaction of calcium in the lime and alumino-silicates in the waste. As with all additives, additional stabilization can be accomplished through the use of other ingredients in smaller quantities. Note that lime may also be added to raise the pH of acidic sludges with other reagents that provide the main stabilization reactions, such as fly ash. Lime-based stabilization is typically best suited for inorganic contaminants and has been widely employed for metal sludges.

Soluble Silicates

The use of siliceous materials in metals stabilization has been long applied in full-scale commercial processes. In one such process, silica reagents are acidified to form a monosilic acid solution to which metal-bearing wastes are added.[43] In another, a combination of liquid soluble silicates and cement form the basis of a process that has been demonstrated to be effective in stabilizing soils contaminated with high concentrations of lead, copper, and zinc.[44]

Organically Modified Clays

In the past, the use of stabilization and solidification for organic wastes has been limited because the ability of traditional stabilization additives to hold the organics in the stabilized matrix was limited. In fact, as pointed out earlier, organics interfere with the cement hydration process. Organically modified clays have been recently employed in conjunction with other stabilization reagents in order to entrap the organic portion of the waste to be stabilized.[45–51]

Organically modified clays are produced when natural clays are organically modified to become organophilic.[52,53] This characteristic is in contrast to their original organophobic nature. The modification process is accomplished through the replacement of inorganic cations within the clay crystalline mineralogical structure with organic cations, typically quaternary ammonium ions. After this replacement process, organic molecules are adsorbed within the crystalline structure of the clay that then swells in the presence of organic contaminants. The clay structure of a montmorillonite, shown on Fig. 11-6, depicts the clay as a tetrahedral layer on each side of an octahedral layer. The components of this two-to-one structure are bound tightly by covalent bonds. However, as a result of isomorphous (same form) substitutions, the clay has a net negative charge that is balanced by exchangeable cations in the intercrystalline region (see Fig. 11-6).

In the production of organically modified clay, the exchangeable inorganic cations shown in the intercrystalline region are replaced with organic cations. The result is a nearly identical clay structure where organic cations occupy the intercrystalline region as shown on Fig. 11-7. In this figure, the organic cations are quaternary ammonium ions. These organic ions attached to the clay readily adsorb other organic species. The effectiveness of organically modified clays in the stabilization of wastes containing organics is due to the adsorption of organic contaminants to the clay which in turn can be encapsulated by cement or other binders.

Organophilic clays typically are added to the waste first and allowed to interact with the organic components.[55,56] Additional agents are added to provide shear strength and solidify the material into a monolithic mass.

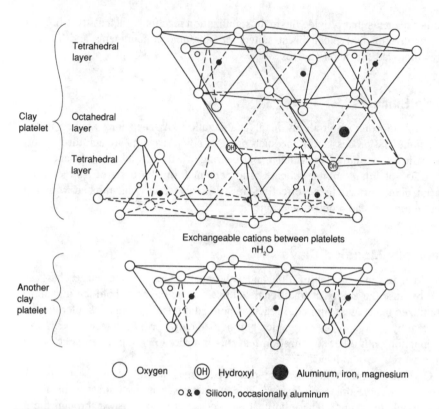

FIGURE 11-6
Montmorillonitic clay structure.[54]

The effectiveness of organically modified clays compared to other sorbents in waste stabilization is illustrated on Fig. 11-8. To stabilize an acidic petroleum sludge, a finely ground cement (MC-500) was used as a binder, and both unmodified and modified clays were used as sorbents. By weight, the sludge-to-sorbent-to-binder ratio was 1.0 to 0.4 to 0.25 for this series.

As shown on Fig. 11-8, the total organic carbon in the leachate from the organically modified clay mixes was one-third less than the bentonite and attapulgite mixes. The unconfined compressive strength for samples mixed with organically modified clays was nearly an order of magnitude stronger than the unmodified clay mixes.

The impact of organically modified clays on the stabilization of organics is further illustrated in Fig. 11-9. In this figure investigating the effect of sorbent-to-waste ratio, the phenol concentration in the leachate decreased as the sorbent quantity increased for two different organophilic clays. Correspondingly, the strength of the stabilized materials increased with increasing organophilic clay content. Detailed studies measuring the partitioning coefficients show that the adsorption of phenols by organophilic clays from water increases with chlorine addition (i.e, phenol < chlorophenol < dichlorophenol < trichlorophenol).[57]

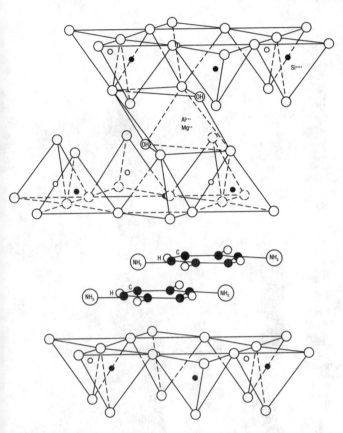

FIGURE 11-7
Organically modified clay structure.[58]

Both the type of quaternary ammonium ion used and the nature of the base clay affect the ability of the organically modified clay to adsorb organics.[59] Figure 11-10 shows the relationship between the trichloroethene concentration in the soil pore water and the adsorption on the organically modified clay. An important fundamental is illustrated in this figure. Utilizing different quaternary ammonium ions with the identical base clay produces two very different organophilic clays as measured by their adsorptive capacities. The example problem demonstrates the efficiency of organically modified clays to adsorb organics.

Example 11-3   Stabilization with organically modified clay. A saturated sandy soil contaminated with trichloroethene is to be stabilized employing one of the organically modified clays illustrated in Fig. 11-10. The saturated soil has a porosity of 50% and a wet density of 2 g/cm^3. The concentration of the trichloroethene is found to be 500 mg/L in the pore water of the saturated soil.

For the two clays on Fig. 11-10, calculate the quantity of organically modified clay to be added to the stabilization mix per ton of contaminated soil to adsorb all of the contaminant.

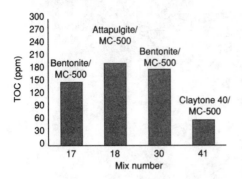

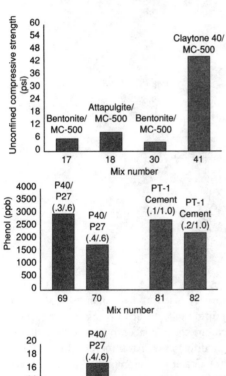

FIGURE 11-8
Stabilization effectiveness of organically modified clay.[60]

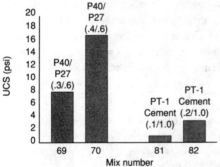

Note: PT-1 and P-40 – organically modified clay from Bentec, Inc.

FIGURE 11-9
Effect of organically modified clay.[61]

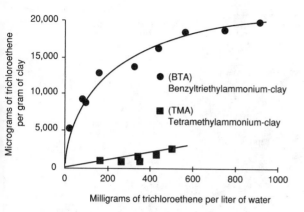

FIGURE 11-10
Effect of quaternary amine on organophilic clay.[62]

Present solution as a percentage of the wet weight of saturated soil as well.

Solution.

a.) From Fig. 11-10, the quantity of each organically modified clay required to adsorb the trichloroethene found in the contaminated soil pore water at 500 mg/L is:

Amount TMA-Clay = 2,750 μg (2.75 mg) of trichloroethene
adsorbed per g of organically modified clay

Amount of BTA-Clay = 17,500 μg (17.5 mg) of trichloroethene
adsorbed per g of organically modified clay

Now calculate the mass of trichloroethene per ton of contaminated soil. The following phase diagram describes the two-phase soil system.

where V_T = total volume
  V_V = volume of void space (filled with water for a saturated soil)
  V_S = volume of soil particles
  M_W = mass of water including aqueous phase contaminants
  M_S = mass of soil particles
  M_T = total mass

Since the mass density, ρ, is given as 2.0 g/cm^3, the total volume per gram of soil can be calculated from:

$$\rho = M_T / V_T$$

Substituting for ρ and M_T, and incorporating units conversion factors, the total V_T is

$$V_T = \frac{1 \text{ ton} \times 2000 \text{ lb/ton} \times 454 \text{ g/lb}}{2 \text{ g/cm}^3}$$

$$V_T = 4.54 \times 10^5 \text{ cm}^3 = 4.54 \times 10^5 \text{ ml}$$

Since the porosity is known to be 50%, the volume of voids can be calculated by:

$$n = \frac{V_V}{V_T} \quad \text{where}$$

$$n = \text{porosity}$$

$$V_V = \text{volume of void space}$$

$$V_T = \text{total volume}$$

Substituting gives:

$$V_V = n \times V_T = 0.5 \times 4.54 \times 10^5 \text{ ml} = 2.27 \times 10^5 \text{ ml}$$

$$= 2.27 \times 10^2 \text{ L}$$

Knowing the contaminated soil is saturated results in:

$$V_V = V_W = 2.27 \times 10^5 \text{ ml} \quad \text{for each ton of soil}$$

The soil pore water is contaminated with trichloroethene at a concentration (C) of 500 mg/L. Thus, the mass of trichloroethene can be found as:

$$M_{\text{trichloroethene}} = C \times V_W = 500 \text{ mg/L} \times 2.27 \times 10^2 \text{ L}$$

$$M_{\text{trichloroethene}} = 1.14 \times 10^5 \text{ mg}$$

The amount of organically modified clay can now be computed:

$$M_{\text{TMA–Clay}} = \frac{1.14 \times 10^5 \text{ mg}}{2.75 \text{ mg/g clay}} = 4.13 \times 10^4 \text{ g} = \underline{41.3} \text{ kg of TMA-clay}$$

$$M_{\text{BTA–Clay}} = \frac{1.14 \times 10^5 \text{ mg}}{17.5 \text{ mg/g clay}} = 6.49 \times 10^3 \text{ g} = \underline{6.49} \text{ kg of BTA-clay}$$

The quantities are per ton of contaminated soil because M_T was set at 1 ton.

b.) To convert the mass of organically modified clay to a % addition:

$$\% \text{ TMA clay} = \frac{\dfrac{4.13 \times 10^4 \text{ g/ton of soil}}{454 \text{ g/lb}}}{2000 \text{ lb/ton of soil}} \times 100$$

$$= \underline{4.5\%}$$

$$\% \text{ BTA clay} = \frac{\dfrac{6.49 \times 10^3 \text{ g/ton of soil}}{454 \text{ g/lb}}}{2000 \text{ lb/ton of soil}} \times 100$$

$$= \underline{0.71\%}$$

Modified Lime

Organically modified lime products have been specifically developed for the stabilization of organic wastes. The lime-based process has been developed to convert toxic organic wastes with significant percentages of hydrocarbon to an inert mass. One system using hydrophobized lime is patented and has been used since 1976 in Germany.[63] The process employs a calcium oxide-based reagent modified with stearic acid. Heat and water-as-steam are given off during the reaction (along with volatile organic contaminants). The waste is converted to a dry water repellent (hydrophobic) powder. The material becomes quite strong and relatively impermeable to water. The resulting material is often considered suitable as a construction material for road and embankment fill.

Thermosetting Organic Polymers

Hazardous wastes may be stabilized through an organic polymer process that involves mixing of a monomer, such as urea-formaldehyde that acts as a catalyst, to form a polymeric material.[64] A sponge-like mass is thereby formed, trapping solid particles of hazardous waste within the matrix (macroencapsulation). This process does, however, leave some wastes untrapped, particularly liquid wastes. Because liquids generally remain after thermosetting organic polymer treatment, the final waste product is often dried and containerized prior to ultimate disposal.

The principal advantage of this process is that it generally results in a low-density material relative to other fixation techniques. Further, only small quantities of additives are required to solidify wastes. Thus, this particular technique is most applicable in the solidification of liquid, non-volatile, organic hazardous wastes. This method has also been employed to solidify low-level radioactive waste prior to ultimate disposal.

These technologies are generally not applicable to site remediation processes due to their cost, fire hazard, production of water from the waste matrix, and volatilization of organics. However, in one study remediation of sediments in the James River near Richmond, Virginia, the sediments were contaminated with an organic pesticide kepone.[65] Bench scale studies determined that an organic epoxy grout was the most effective stabilization agent in treating the kepone-contaminated sediment. Thermosetting organic polymers may ultimately prove useful for remedial action.

Thermoplastic Materials

Hazardous waste may be stabilized by blending molten thermoplastic materials with wastes at high temperatures.[66] Molten thermoplastic materials include asphalt, paraffin, bitumen, polyethylene, polypropylene, or sulfur. When cooled, the solidified material is characterized as a thermoplastically coated waste and typically containerized (put in drums) for ultimate disposal. When bitumen has been employed, the typical waste-to-bitumen ratio has been between 1:1 and 2:1.[67] This technique has been predominantly employed for radioactive waste applications due to its cost. Further, organics are volatilized at the operating temperatures requiring control of air emissions during the process.

The use of thermoplastic stabilization has received attention for mixed waste, that is, waste that is both hazardous and radioactive.[68] The waste is stabilized so that it is no longer considered hazardous, and then it is disposed of in accordance with the requirements for radioactive waste.

A number of research studies have been conducted, indicating the potential for this technique.[69-71] The primary limitations include the presence of materials within the matrix that could deteriorate the thermoplastic material and the presence of organic chemicals that act as solvents to the stabilizing thermoplastic materials. Thus, a potential for long-term degradation exists if these detrimental components are in the waste product. Thermoplastically stabilized waste, however, is quite resistant to leaching and biodegradation.

Vitrification

Vitrification does not require the addition of reagents but is a stabilization technique and, as such, is included under this technology section. Archaeologists estimate that humans have been making glass for more than 7000 years. Glassmaking or *vitrification* involves melting and fusion of materials at temperatures normally in excess of 1600°C followed by rapid cooling into a non-crystalline, amorphous form. The application of vitrification to the management of hazardous wastes has been explored as both an in situ technique and as an in-plant technique. Vitrification is considered a solidification and stabilization procedure because it renders the waste more structurally stable with a reduced potential for contaminant migration into the environment. At the time of this writing (1993) the field scale application of vitrification to the management of hazardous wastes is limited.

In Situ Vitrification. In situ vitrification has been proposed and tested at the pilot scale for the remediation of contaminated soil. In the in situ process, the high temperatures needed to melt the soil into a molten mass are achieved through the application of an electrical current. As the current flows through the soil, heat builds up, eventually causing the soil to melt. Upon melting, the soils become more conductive and the molten mass becomes a heat transfer medium allowing the molten mass to grow. The entire process is begun using a surficial layer of graphite and glass frit to serve as a starter path. Typically, electrodes spaced a maximum of 18 feet apart in a rectangular pattern are employed. Once initiated, the molten mass grows downward and outward to a maximum depth of about 30 feet to encompass a total melt mass of 1000 tons.[72] Schematically, the process stages are shown in Fig. 11-11. The resulting residue bears a strong resemblance to obsidian glass, which is formed by nature's own vitrification process.

The soil is melted at a rate of 4 to 6 tons per hour giving a rate of melt advance of 1 to 2 inches per hour. But what exactly happens to the contaminants during this process? As the temperature increases, organic materials first vaporize and then pyrolyze (break down in the absence of oxygen) into their elemental components. The gases move slowly through the molten mass (which is quite viscous) toward the surface. A portion of the gases dissolves in the molten mass while a portion is emitted to the environment above the molten mass (off-gassed). Those gases which are combustible burn in the presence of oxygen upon off-gassing. All gases are collected (a hood is constructed over the area being processed) and treated to ensure air emission standards are met. Because of the high temperature of the melt (1600° to 2000°C), no residual organic contamination remains in the residue (the glass). Inorganic contaminants

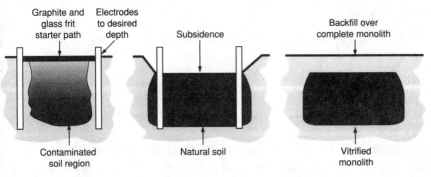

FIGURE 11-11
In situ vitrification.

behave in a similar manner—some decompose while others are dissolved in or react with the melt. For example, nitrates yield gaseous nitrogen (N_2) and oxygen (O_2).

The porosity of soils can vary over a large range, but is commonly in the range of 20% to 40%. That is, soil in its in situ state is a three-phase material with solids, liquids, and gaseous components. The liquids and gases make up the volume of voids. The molten material during the vitrification process is entirely liquid phase made up of melted solids as the original liquids and gases are off-gassed. Thus, upon cooling, there is a net volume reduction in the vitrification process. This is in contrast with the previously discussed techniques where the addition of reagents causes a net volume increase.

Example 11-4   In situ vitrification. A sandy soil with a porosity of 0.3 is contaminated with a wide range of organic and inorganic contaminants to a depth of 3 meters. The site is a candidate for in situ vitrification. Calculate the surface subsidence expected at the completion of vitrification.

Solution. Basis of calculation: 1 square meter of surface.

$$V_T = 3m^3/m^2$$

$$\text{Porosity } n = \frac{V_{\text{Voids}}}{V_{\text{Total}}}$$

$$= 0.3$$

Prior to vitrification:

$$V_V = n \cdot V_T = 0.3 \times 3 = 0.9m^3/m^2$$

$$V_S = V_T - V_0 \Rightarrow V_S = 2.1$$

After vitrification:

$$n = 0 \text{ (porosity is effectively zero after vitrification)}$$

$$V_T = V_S = 2.1$$

$$\Delta V = 3 - 2.1 = 0.9 = \Delta H = \text{surface subsidence}$$

In-Plant Vitrification. The use of contaminated soils as feedstock for glassmaking (vitrification) offers several advantages as a hazardous waste management technique. The technology has the potential to both treat hazardous waste, particularly contaminated soil, and to produce a product that may be usable (e.g., as a road aggregate). The technology employs a glassmaking furnace operating at 1,600°C. A starter mix of recycled glass, fly ash, and limestone is used to initiate the process. The contaminated soil is then introduced into the furnace for the melting and fusion stages that last a minimum of 5 hours.[73] At the time of this writing, the use of glassmaking technologies and facilities for the management of hazardous wastes is not employed on a commercial scale although it has been demonstrated on a pilot scale under the U.S. EPA SITE (Superfund Innovative Technology Evaluation) program.[74]

Additional considerations for this vitrification technology include air emissions and leachability of the residue (the glass). Air emissions can be captured and treated as necessary for environmental protection employing traditional air treatment technologies including ammonia for NO_x, lime for SO_x, and condensation of volatiles in a fractionating pebble bed.[75] As for the in situ vitrification product, leaching tests have shown the glass to be inert. Despite the test results, the heightened environmental awareness may dictate that the product be used in applications that limit its distribution to the general public. For example, significant quantities of clean fill are required for daily cover, intermediate berms, and drainage layers in landfill construction and operation. The use of the vitrified product in these applications would provide an added margin of safety to the process and preserve our mineral resources for other future uses.

Asbestos waste has also been treated through vitrification. A pilot plant with the capacity of 500 kg per day has been operating since 1984 in the United Kingdom. Asbestos, mixed with glass debris and other glassmaking additives, is melted at 1,400°C for approximately 10 hours. During the melt the asbestos becomes amorphous and mixes with the other silicates. Upon cooling, the glass is non-toxic and used as an aggregate filler in concrete.

11-4 TESTING

What does it mean to say that a hazardous waste has been stabilized and solidified? How can the effectiveness of the stabilization and solidification process be measured? How strong does the solidified material need to be? How fast can contaminants leach from the stabilized material? The answers to these questions are complex. The assessment of the effectiveness of stabilization requires the measurement of physical, engineering, and chemical properties of the stabilized material. The measured property depends upon the measurement technique, that is, different tests for the same property may yield different measures of that property. To further complicate the evaluation and design process, it is difficult, if not impossible, to predict the long-term environmental stresses to which these stabilized materials might be subjected—freezing and thawing, wetting and drying, percolation of precipitation, and pressure due to overburden load. Thus, if the range of environmental stresses cannot be predicted, these stresses cannot be simulated in the laboratory. Even if the long-range environmental stresses could be predicted, how can the predicted environmental stresses be simulated in the laboratory?

As a natural outcome of these complexities, a large number of laboratory tests are utilized to evaluate the effectiveness of stabilization. No one test is ideal for all wastes and all applications. Each, when properly interpreted, provides partial insight into the effectiveness of a particular stabilization technique and reagent mix for a specific waste. This section discusses the available tests, how they are conducted, the type of information these tests provide, and the circumstances for which the tests are most applicable. Because it is inevitable that tests and the interpretation of the test results will change with time, the fundamental aspects of the tests are emphasized.

The selection of the appropriate tests and the interpretation of the test results depend upon the objectives of the stabilization program. For example, site remediation was identified earlier in this chapter as a major application area of stabilization technology. The selection of the type and amount of reagents used in stabilization depends upon specific goals for the site remediation program. As described in Sec. 17-6, risk assessment may be used to establish these goals. For example, the calculated risk from ground water might be a function of the amount of arsenic that is estimated to leach daily from a proposed stabilization alternative. In this case, the selected tests must estimate the amount of leachable arsenic.

Extraction and Leaching Tests

In most cases, a reduction in the rate at which contaminants can migrate into the environment is the first and foremost reason for selecting stabilization and solidification as a hazardous waste management technique. As precipitation infiltrates the stabilized waste, contaminants migrate from the stabilized mass into the water (the transport medium) and continue their trek into the environment. Of course, in the short term (geologically speaking) efforts are usually taken to minimize precipitation infiltration and to collect the leachate prior to release into the surrounding environment (see Secs. 13-3 and 13-4).

Leaching tests are listed in Table 11-3. The fluid to which the contaminants are leached is called the **leachant**. After the leachant has become contaminated it is termed the **leachate**. The terms **extraction** and **leaching** are used interchangeably and are, as defined earlier, the process by which contaminants are transferred from a solid or stabilized matrix to the leachant. Finally, the overall ability of a stabilized material to leach contaminants is termed **leachability**.

TABLE 11-3
Leaching test methods

Paint Filter Test
Liquids Release Test
Extraction Procedure Toxicity Characteristics (EPTox)
Toxicity Characteristic Leaching Procedure (TCLP)
Modified Uniform Leach Procedure (ANS 16.1)
Maximum Possible Concentration Test
Equilibrium Leach Test
Dynamic Leach Test
Sequential Leach Test (Sequential Chemical Extraction)
Multiple Extraction Procedure

Do leaching tests have a single purpose? If that were so, it would be much easier to define a single acceptable leaching test protocol. However, there are many reasons to conduct a leaching test. The leaching test may be conducted as a **regulatory test** run to provide the basis for a consistent and uniform decision-making process. In this case a highly reproducible test protocol is needed and the results can be compared to predetermined standards. The stabilized material either passes or fails. Alternatively, a leaching test may be a **predictive test** run to generate data that can be used to model real world contaminant migration. In this case, the data would be employed in subsequent fate and transport analysis (Chap. 4) and risk assessments (Chap. 14). **Investigatory tests** may also be run to study the basic binding mechanisms, interference factors, and the underlying principles of stabilization technology. Because each of these test types has different goals, different test methods have evolved. As the test methods are discussed, consider which purpose is best served by the method described.

How is material leached from a stabilized mass? As the leachant passes through and around, the sample contaminants are transferred from the stabilized mass to the leachant. This may occur as contaminants are dissolved into the leachant, washed from the surfaces of the stabilized material, or as contaminants diffuse from within the stabilized mass to the leachant. Thus, the leachability is dependent upon the physical and chemical properties of both the stabilized material and the leachant. Studies have shown that the main factors affecting leachability are the alkalinity of the stabilized product; the surface-to-volume ratio of the waste; and the tortuosity, a measure of the path length for diffusion.[76] The mechanisms of leaching must be considered in the selection and evaluation of leaching test methods as described below.

The test method affects the leachability of the sample. Specifically, there are a number of test variables that affect the contaminant concentrations in the leachate:

- leachant-to-waste ratio
- surface area of the waste (e.g., grinding the stabilized mass into small particles)
- type of leachant (e.g., distilled water, acetic acid, simulated acid rain)
- pH of the leachant

- contact time
- extent of agitation
- number of replacements of fresh leachant
- extraction vessel
- temperature

The impact of these variables should be self-evident. For example, the greater the ratio of leachant to waste, the greater the amount of contaminant leached from a stabilized mass. As each test method is described, consider the previously listed variables and their impact upon the test result.

Paint Filter Test

The placement of liquid hazardous wastes in landfills in the United States is banned. The Paint Filter Test evaluates the absence or presence of free liquids in bulk and containerized hazardous wastes.[77] This test is rapid, economical, easy to conduct, and easy to evaluate. In this test, wastes are placed in a standard paint filter; if liquid is drained by gravity through the filter within 5 minutes, the hazardous waste is considered to contain free liquids and must be treated prior to landfilling. The test may also be used after stabilization to determine if the stabilization process has been effective in eliminating free liquids from the hazardous waste.

In the strictest of interpretations, the paint filter test is not a leaching test as leachant is not added to the system. The test may, however, yield a leachate. The paint filter test provides results that trigger minimal technology requirements for the stabilization" of the hazardous waste. The test is adequate to evaluate the presence of free liquids in materials prior to disposal in a landfill. The paint filter test, however, is not adequate to evaluate the effectiveness of stabilization for other applications such as site remediation.

The stresses causing liquids to flow are very small and do not simulate the long-term stresses expected in a landfill. An alternative test, the "liquids release test" (see below) has been developed to better account for the overburden stress in a landfill.

Liquids Release Test

The Liquids Release Test[78] determines if a solidified liquid waste is suitable for landfilling. The procedures were developed to preclude the use of sorbents such as sawdust to "solidify" liquid wastes. In this test a "consolidation" stress is applied to test how readily leachate can be squeezed from a solidified mass.

Extraction Procedure Toxicity Test

The EP Toxicity Test, an older regulatory test being phased out in the United States can be used to generate a liquid extract from solid wastes.[79] A particular waste is considered EP toxic if its extract has concentrations of any of the eight regulated metals and six pesticides that are greater than the "standards" previously specified in the federal regulations. In this test, the solidified monolithic block is crushed to pass a 9.5 millimeter sieve. A 0.04M acetic acid (pH = 5) leaching solution is used at a liquid-to-solid ratio of 16:1. The extraction takes place over a period of 24 hours with agitation. The liquid extract is analyzed for specific chemical constituents as shown in Table 11-4. For a stabilized mass, the extraction liquid (leachant) is a mixture of deionized water and acetic acid so the specified pH is achieved. Chemical analyses of the filtered extract are then conducted to determine the concentration of specified organic and inorganic constituents.

As discussed in Chap. 2, the EP Toxicity Test has been extensively used to classify materials as hazardous or non-hazardous. In this sense it is considered a regulatory test. It is not a design test as there is no realistic way to apply the results to any sort of transport, fate, or risk analysis. The measured chemical concentrations are simply artifacts of the test procedure. For example, if the pH is changed the chemical concentrations will change. Similarly, if the liquids-to-solids ratio is changed, the chemical concentrations will change. The test procedure allows the volatile organic contaminants to be released into the atmosphere during the 24-hour extraction period. As a result, a test of this type is not well suited to evaluating wastes with a significant volatile organic component.

Toxicity Characteristics Leaching Procedure (TCLP)

The Toxicity Characteristics Leaching Procedure (TCLP) was adopted on November 7, 1986, by the U.S. EPA under the Hazardous and Solid Waste Amendments of

TABLE 11-4
EP Toxicity Characteristics, Maximum Concentration of Contaminant Stanc

EPA hazardous waste number	Contaminant	Stanc mg
D004	Arsenic	5.
D005	Barium	100
D006	Cadmium	1
D007	Chromium	5
D008	Lead	5.
D009	Mercury	0.
D010	Selenium	1.
D011	Silver	5
D012	Endrin (1,2,3,4,10,10-hexachloro-1,7-epoxy-1,4,4a,5,6,7,8, 8a-octahydro-1, 4-endo, endo-5, 8-dimethano-naphthalene)	0.
D013	Lindane (1,2,3,4,5,6-hexa-chlorocyclohexane, gamma isomer)	0.
D014	Methoxychlor (1,1,1-trichloro-2,2-bis[p-methoxyphenyl-ethane)	10
D015	Toxaphene ($C_{10}H_{10}Cl_8$, technical chlorinated camphene, 67-69% chlorine)	0.
D016	2,4-D (2,4-dichlorophenoxyacetic acid)	10.
D017	2,4,5-TP Silvex (2,4,5-trichlorophenoxypropionic acid)	1.

† Source: U.S. Environmental Protection Agency

1984.[80] This test, a regulatory test, was adopted as a replacement for the EP Toxicity Test to determine if a particular waste meets the applicable technology-based treatment standards to be land-disposed. The TCLP is also widely used to evaluate the effectiveness of stabilization. In this test method, the stabilized material is crushed to a particle size smaller than 9.5 millimeters. The crushed material is mixed with a weak acetic acid extraction liquid, in a liquid-to-solid weight ratio of 20:1, and agitated in a rotary extractor for a period of 18 hours at 30 RPM and 22°C. After 18 hours of agitation, the sample is filtered through a 0.6–0.8 micrometer glass fiber filter and the filtrate is defined as the TCLP extract. This TCLP extract is analyzed for a wide variety of hazardous waste constituents including volatile and semi-volatile organics, metals, and pesticides as shown in Table 2-1.

Special procedures are used to test the sample for the presence of volatile organic compounds within the extract. These special procedures use a zero head space extraction vessel, a cylindrical container with an airtight seal at each end. Extraction in the zero head space extractor maintains the concentration of volatile organics within the fluid to be tested.

The TCLP is subject to criticism with respect to its use for the evaluation of stabilization effectiveness for several reasons. First, a solidified monolithic mass is broken down to pass the 9.5 millimeter sieve, reducing beneficial effects relating to macroencapsulation and microencapsulation. It has been shown that as the particle size reduces, the leachability increases.[81,82] Further, the low pH environment during extraction may not be representative of the real world field conditions although it may be representative in a landfill containing domestic garbage. As a result, proponents argue that those severe steps yield conservative results. However, a stabilized material with a high alkalinity, such as cement-based stabilization, may rapidly increase the pH of the leachant and, in effect, result in leaching under basic rather than acidic conditions. Despite these criticisms, the TCLP is particularly useful in comparing the effectiveness of one treatment technology with another or one stabilization mix or process with another.[83]

It is important to note that the EP Toxicity Test and the TCLP Test yield concentrations for specific constituents. These data are sometimes compared to 100 times the drinking water standard to determine if a sample has "passed" or "failed" the TCLP. The factor of 100 is to account for dilution into the environment. These tests do not yield data on the time rate at which contaminants will be leached into the environment. They simply yield a concentration that can be compared to standards or to leachate concentrations obtained using alternative stabilizing techniques and mixes. In this regard, the TCLP is also a regulatory test.

American Nuclear Society Leach Test (ANS 16.1)

The ANS 16.1 Leach Test,[84] also called the Uniform Leach Test (similar to the International Atomic Energy Agency standard leach test), was developed to determine the leachability of solidified radioactive wastes. This test was designed for use with a solid block of material and, as a result, may not be applicable for soil-like products.

The test requires an accurate calculation of surface area so that diffusivity can be determined and then the rate of loss of contaminants for larger amounts of the waste can be predicted. In the ANS 16.1 test, a solidified test specimen is rinsed in a leaching medium of aerated demineralized water. Following the 30-second rinse, the specimen is left unagitated in a non-reactive container (usually glass) for specific time intervals. The leaching medium is removed and replaced at various time intervals throughout the 14-day test period. The test yields a number termed the **leachability index** that can be related to the diffusion coefficient. The values of diffusivity should be used with caution, and transport analysis must take into account the amount of disposed waste, the annual net infiltration of water or movement of ground water through the landfill, the factors affecting the leaching of the waste, the possible interactions between the waste and the leachate, and the fate of the landfill leachate after it has passed from the waste.

Maximum Possible Concentration Test

The Maximum Possible Concentration Test, originally used in Harwell Laboratories, United Kingdom, is a shake test that attempts to determine the solubility of the contaminants present in the test material by developing a condition of saturation. In this test, distilled water is shaken with a dried, powdered solid sample, and concentrations of selected contaminants are measured. This "quick" test may be used in a screening phase of bench scale studies to provide rapid comparisons of stabilized mixes.

Equilibrium Leach Test

The Equilibrium Leach Test[85] is a batch extraction process that utilizes distilled water as the extraction fluid. In this test, a dried sample, ground to pass through an ASTM No. 100 sieve, is mixed with four times its mass of distilled water (4:1 liquid-to-solid ratio). The mixture is agitated for seven days, and the filtered extract is analyzed for total dissolved solids. Specified chemical analyses are also performed on the filtered extract.

Dynamic Leach Test

The Dynamic Leach Test provides data on the rate of leachability as does the multiple extraction procedure.[86,87] The concentrations obtained from all of the leaching tests previously mentioned except the ANS 16.1 Leach Test, are "equilibrium" or "static" values, and therefore, they do not provide any information regarding the rate of release or its dependence on time. To overcome this deficiency, Dynamic Leach Tests are performed. The Dynamic Leach Test, adapted from the American Nuclear Society Test (ANS) 16.1, is a true leaching test used to determine the mobility of contaminants through a material matrix, by measuring the rate of leachability. In this test, a monolithic solidified specimen is immersed in distilled water at a specific volume-to-surface-area ratio. The specimen is placed in a nylon mesh harness, rinsed, and suspended in containers filled with distilled water. Fresh leaching agents are used after certain liquid-solid contact times have elapsed, and the solid and leachate are separated.[88] Over a period of nine days, the leachant is renewed at intervals calculated according to a diffusion model.[89] The samples may be analyzed for organics and/or inorganics, and the pH of the leachant is measured after each transfer. The amount leached in each interval, relative to the total amount of contaminant initially present, is determined, and an apparent diffusion coefficient can be calculated. With this type of data available, time-related behavior, or diffusive transport, can be evaluated and used for the long-term prediction of leaching rate.

Sequential Leach Test (Sequential Chemical Extraction)[90]

This test attempts to evaluate the leachability of metals from a solid material. Five successive chemical extractions of increasing aggressiveness (pH varying from neutral to very acidic) are performed to separate the contaminants into five fractions: the ion exchangeable fraction, the surface oxide and carbonate bound metal ions, the metal ions bound to iron and manganese oxides, the metal ions bound to organic matter and sulfides, and the residual metal ions.[91] The first three fractions are classified as being "available for short- and medium-term leaching," and the last two fractions are "unavailable for leaching." In this test, a sample is dried in an oven at 60°C and ground to pass through an ASTM No. 325 sieve (45 μm opening size). Then, a 0.5 gram sample is placed in a polysulfone centrifuge tube and subjected to a series of five successive extraction procedures, each suitable for extracting a specific fraction of the metals. In each separate extraction, the specific extraction fluid is added and the mixture is agitated and/or heated for a specific time period before being centrifuged to separate the solid and liquid portions. Chemical analyses are performed on the liquid portion, and the solid portion is rinsed with distilled water, centrifuged, and isolated for use in the next extraction procedure. The procedure is slightly modified when performing a total extraction, in which a new sample is used for each extraction instead of using the material from the previous extraction.

Multiple Extraction Procedure

The U.S. EPA Multiple Extraction Procedure appeared in the *Federal Register* on November 22, 1982, and was required by the EPA for temporary exclusion or delisting petitions. It has subsequently been revised.[92] The test attempts to duplicate leaching conditions that a waste would be subjected to from repetitive contact with acid rain at an improperly designed sanitary landfill. The test is not applicable for treated wastes that will be placed in a hazardous waste landfill, but it does provide severe conditions for the evaluation of the effectiveness of a particular treatment technique.

At the start of this test, the procedures are the same as for the EP Toxicity Test. However, the remaining solid portion is re-extracted nine additional times with a synthetic acid rain extraction fluid, consisting of a 60/40 mixture of sulfuric and nitric acids with a pH of 3.0. In each of the nine additional extractions, the mixture is agitated for 24 hours and separated. The extract analysis is the same as that specified in the EP Toxicity Test. If the contaminant concentrations in the extracts have increased over the seventh and eighth extractions, the extractions are continued, past the ninth repetition, until the concentrations in the extract do not increase.

As in the case of the other tests, these results should not be directly applied to actual field conditions. These results were determined for contaminants released under controlled laboratory conditions that are not directly comparable with conditions at a specific site. In some cases, the laboratory results represent the worst condition and are conservative, while, in other cases, the predicted laboratory leach rate can be less than that which could occur at the site.

Chemical Tests

The chemical analysis of leachate is a complex task. The type of analysis and the analytical procedures are frequently specified in the leaching procedure by citing Standard Methods or U.S. EPA protocols. A discussion of analytical methods is beyond the scope of this chapter. Nonetheless, Table 11-5 contains a list of chemical tests on the leachate or soil frequently employed to evaluate the effectiveness of stabilization.

Much of the focus of chemical testing of stabilized materials relates to the analysis of the extracted fluid or the identification of contaminants in the waste. Because our application of stabilization as a hazardous waste management technology relies upon the fixing of the waste within the matrix, tests which help in our evaluation of that process can be useful. One such test is Fourier transform infrared spectra (FTIR), used to evaluate the presence and type of chemical bonding between the organic contaminant and the binding reagents by interpretation of the shift in the infrared frequencies. Differential scanning calorimetry data are used to determine the excess energy needed to release the contaminant from the stabilized matrix. Finally, gas chromatography and mass spectrometry (GC/MS) techniques can be used to identify products in the stabilized matrix that may result from degradation of the original contaminants. Taken as a group these three tests can give insight into the nature and extent of the chemical bonding of organic contaminants in the stabilized matrix.[93]

TABLE 11-5
Chemical test methods[94]

Total waste analysis
Metals by inductively coupled plasma spectrometry or atomic absorption
Organics by gas chromatography (GC) and mass spectrometry (MS)
Total organic carbon
Loss on ignition
pH
Fourier transform infrared spectra (FTIR)
Differential scanning calorimetry and thermal gravimetric analysis

Physical and Engineering Property Tests

A number of other tests have been adapted from the civil engineering area to evaluate the physical integrity and engineering properties (strength, compressibility, and permeability) of stabilized materials. These tests are described below and are summarized in Table 11-6.

Moisture Content

The Moisture Content Test[95] is used to determine the amount of water (or liquid) in a given quantity of material. For untreated materials, the moisture content aids in the determination of water needs for hydrating reagents. For solidified materials, the moisture content is needed to calculate the dry unit weight (or dry density) of solids given measurements of bulk density. For soil-like materials, the moisture content influences the compactability of the materials. For hardened, concrete-like materials, the moisture content is less significant.

The wet weight of a small (few-gram) sample is first determined. The sample is dried at 105°C for 24 hours and reweighed. The moisture content can then be calculated. Water content on a dry weight basis is presented as a percentage and is calculated as the weight of the water divided by the weight of the dry mass. This

TABLE 11-6
Physical, engineering, and durability property tests

Property	Purpose
Physical Properties	
Moisture content (ASTM D2216)	Phase calculations (saturation, void ratio)
Wet and dry bulk density (unit weight)	Stress and volume calculations
Specific gravity	Phase calculations (saturation, void ratio)
Atterberg limits	Correlates to engineering properties
Particle size distribution	Classification descriptor (e.g., distinguishes sand, silt, clays or)
Laboratory cone index (ASTM D3441)	As-mixed strength
Pocket penetrometer	As-mixed strength
Microstructural examination by:	
X-ray diffraction	Crystallization
microscopy	
SEM (scanning electron microscopy)	
Rate of setting	
Supernatant formation during curing	Excess liquid
Engineering Properties	
Strength	Stability analysis
Compressibility	Settlement analysis
Permeability	Flow and transport calculations
Durability Properties	
Wet/dry	Long-term integrity
Freeze/thaw	Long-term integrity

is numerically different from the moisture content typically used by environmental engineers, which is expressed on a wet weight basis (weight of water/total weight). Some wastes contain more liquid than dry solids, and it is possible to have moisture content on a dry weight basis exceeding 100%. This is a straightforward test, but misleading results may arise when the organic contaminant level is high because organics will volatilize during the drying period.

Wet and Dry Bulk Unit Weight (Wet and Dry Bulk Density)

The bulking (increase in volume) associated with stabilization is evaluated using bulk unit weights that provide information for weight-volume relationships and are used to calculate density and void volume. These data are used to determine space requirements for ultimate disposal, to compute earth pressures in stability calculations.

The bulk unit weight is determined by measuring and weighing the stabilized samples. The bulk unit weight is calculated as the weight of the solids plus pore fluid per unit volume of material. The dry unit weight is determined from the weight of oven-dried solids per unit volume. Typical units are gm/cm^3 or lb/ft^3.

Specific Gravity

Specific gravity and density are measures of the mass-to-volume ratio of the individual solid component of the three-phase material. The density of solids divided by the density of water is the specific gravity. This parameter is needed to calculate certain physical properties such as porosity and unit weight. Specific gravity is numerically equal to the density of solids in S.I. units as water has a density of 1.0 gm/cc.

The procedure to determine specific gravity is difficult to execute with precision and is not generally amenable to a solidified material that contains a number of different materials, each having a different specific gravity. At best the measured specific gravity represents a composite of the individual components.

Particle Size Distribution

The grain-size distribution[96] of the waste or soil breaks the total mass of soil into various fractions by particle size. This particle size distribution is important to the design of stabilization alternatives. Fine-grained wastes are subject to wind dispersion, are difficult to handle during construction, and present problems in producing a high-strength solidified waste. On the other extreme, wastes with a significant coarse fraction may require prescreening to enable mixing equipment to operate and to preclude particle segregation.

To determine the particle size distribution, sieves are employed for the coarse fractions. Gravitational methods employing a hydrometer are used to separate the fine fraction-particles with size ranging from 0.074 mm to micron size, Larger particles settle more quickly through water than smaller particles and a hydrometer allows measurement of the differences in particle size. Because the stabilized mass is often a monolith, particle size distribution is not generally applicable to stabilized materials.

Laboratory Cone Index (ASTM D3441)

The Laboratory Cone Index Test is often conducted on the stabilized material immediately after the reagents are blended and prior to setting. The test provides information on the strength of the as-mixed materials. This information is used in making decisions regarding materials handling and the support of construction traffic. The latter is important in full-scale stabilization in that construction vehicles may have to go over recently stabilized materials.

A standardized cone is dropped into the material from a standard height, and penetration distance is measured. Tests are rapid, simple, and inexpensive and give insight into the strength of the stabilized materials prior to curing.

Pocket Penetrometer

The application of the pocket penetrometer is similar to that for the Laboratory Cone Test Index (i.e., providing an assessment of the as-mixed properties of the stabilized sludge prior to curing). A pocket penetrometer is a standardized cylinder that is pushed into the material to measure the penetration resistance (force), estimating the unconfined compressive strength in tons/ft^2 (see Fig. 11-12). This test is also used in field applications.

FIGURE 11-12
Pocket penetrometer evaluating a stabilized waste (photo courtesy of Debra Cook, Bucknell University).

Microstructural Examination

The stabilized mass can be examined using X-ray diffractometry, optical microscopy, scanning electron microscopy, and energy dispersive microscopy to better understand the nature of the stabilization processes at work. The microstructural analyses are part of basic research into stabilization mechanisms. These techniques allow improved understanding of the mechanisms by which specific contaminants are bound in the matrix. For example, in a laboratory study microstructural examination was made of a heavy metal sludge containing chromium, nickel, cadmium, and mercury stabilized using a combination of ordinary portland cement and fly ash.[97] The microstructural examination revealed that there was a wide variability in the sludge composition and a corresponding wide variability in the chemical composition of the stabilized material. With such variability on a microscopic level, the effect on properties measured on the macroscopic level such as leaching, strength, and durability is unknown. At this point in time, more research is needed to answer this and other questions regarding the details of stabilization for hazardous waste management.

Supernatant Formation during Curing and Rate of Setting

These test methods characterize the performance of treated wastes in the first few hours after mixing. They are particularly suitable for concrete-like (high slump) materials and not suited for soil-like materials. These tests give early information regarding the suitability of the mix.

To determine the supernatant formation, a freshly mixed sample is placed in a one-liter, graduated cylinder, and as the solids settle and supernatant rises, the quantities of each are recorded. Generally speaking, the formation of any supernatant is undesirable.

The set time represents the time required for the fluid-like material to achieve some initial, yet minimal, strength. The set time can be measured with a vicat needle.[98] In this test, a standard "needle" penetration is measured at various times. Set is defined as the time for less than 1/4-inch penetration.

Unconfined Compressive Strength Test

The Unconfined Compressive Strength Test[99, 100] is conducted to determine the strength of cohesive materials. Cohesive materials include materials ranging from soft clays to concrete. In this test, a cylindrical specimen typically 2.8 inches in diameter by 5.6 inches tall is axially loaded to failure as shown in Fig. 11-13. The axial load and corresponding deformation are measured. From these measurements, the applied stress and resulting strain can be calculated and the stress-strain relationship plotted as shown in Example 11-5. The unconfined compressive strength is then the maximum applied stress in the test.

FIGURE 11-13
Unconfined compression strength test (photo
courtesy of Debra Cook, Bucknell University).

Example 11-5   Unconfined compressive strength test.   A sample of sludge was sta-
bilized with an attapulgite clay sorbent and binders of fly ash, lime and cement. The sample
data measured are:
Diameter (d):   2.70 in   7.01 cm
Height (H_o):   5.73 in   14.55 cm
Weight (M):   1.83 lb   829.6 g
The following data were recorded during the test:

Load, P (lb)	Deflection, dial reading (in)
0.0	0.057
231.0	0.068
405.8	0.079
620.5	0.091
861.9	0.102
1131.4	0.113
1408.3	0.124
1651.1	0.136
1814.0	0.147
1905.8	0.158
1969.5	0.170
1634.9	0.181

Compute sample area, volume, wet bulk density, stress, and axial strain. Plot the stress-
strain curve. What is the unconfined compressive strength?

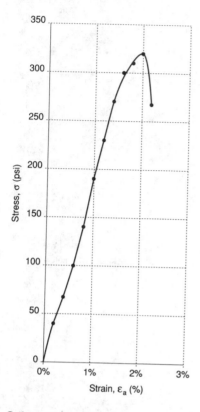

Strain, ε_a (%)

Solution. The initial area, A_o, of the cylindrical test specimen is:

$$A_o = \frac{\pi d^2}{4} = 38.6 \text{ cm}^2 \ (5.98 \text{ in}^2)$$

where $d = $ sample diameter

The volume, V, of the cylindrical test specimen is:

$$V = A_o H_o = 38.6 \times 14.55 = 561.7 \text{ cm}^3 \ (34.27 \text{ in}^3)$$

where $H_o = $ initial height of sample

The density, ρ, is:

$$\rho = M/V = 829.6 \text{ g}/561.7 \text{ cm}^3 = 1.48 \text{ g/cm}^3 \ (92.2 \text{ lb/ft}^3)$$

As shown on the data table, at the start of the test, the applied load, P, is zero. The axial deformation δ must also be zero, although there is an initial dial reading of 0.057 in corresponding to an axial deformation of zero.

Computing the axial strain, ϵ_a, for the first load and deflection dial readings yields:

$$\epsilon_a = \frac{\delta}{H_o} \times 100 = \frac{0.068 - .057}{5.73} = .002$$

or in percent

$$\epsilon_a(\%) = 0.2\%$$

The stress, σ, can now be calculated as force per unit area:

$$\sigma = \frac{P}{A_c} = \frac{P}{\frac{A_o}{1-\epsilon_a}} = \frac{231}{\frac{5.98}{1-.002}} = 38.5 \text{ psi}$$

Note that the area is corrected, A_c, to account for the sample getting "fatter" as it gets "shorter"; remember, there is no net volume change so as the sample is strained, the area increases. Continuing to calculate σ and ϵ_a shows:

P Load (lb)	Deflection (in)	ϵ_a Strain (%)	Corrected area (sq in)	Stress (psi)
0.0	0.057	0.00	5.982	0.0
231.0	0.068	0.20	5.994	38.5
405.8	0.079	0.40	6.006	67.6
620.5	0.091	0.59	6.018	103.1
861.9	0.102	0.79	6.03	142.9
1131.4	0.113	0.99	6.042	187.3
1408.3	0.124	1.19	6.054	232.6
1651.1	0.136	1.38	6.066	272.2
1814.0	0.147	1.58	6.078	298.5
1905.8	0.158	1.78	6.09	312.9
1969.5	0.170	1.97	6.103	322.7
1634.9	0.181	2.17	6.115	267.4

The use of the Unconfined Compressive Strength Test to evaluate the effectiveness of stabilization and solidification has evolved into widespread application as an international standard for the industry. Stabilized and solidified materials must have adequate strength to enable them to support the loads of materials placed over them. Further, the test results would be expected to correlate with the effectiveness of stabilization of inorganic wastes as the inorganics are tied up in the hydrating matrix. This test also offers several other advantages for evaluation of the effectiveness of stabilization. The test is relatively quick (and thus inexpensive), and the required testing equipment is widely available. The test also provides an indication of the ductility/brittleness of the stabilized material. The test can be conducted over time to provide a measure of the improvement of the stabilization process as a function of time. In general, for any given stabilization reagent, the stronger the stabilized hazardous waste, the more effective the stabilization process, particularly for inorganic contaminants. Thus, the unconfined compressive strength is often used as an indicator parameter of the effectiveness of stabilization.

It is generally accepted that an unconfined compression strength of 50 psi represents a suitable strength measure for stabilized materials based upon the calculated typical overburden pressures in a landfill. It is not necessary, however, from a stability standpoint, to achieve a specific unconfined compressive strength to achieve stability in a landfill. Engineering materials can be quite stable and yet may exhibit an unconfined compressive strength of zero (e.g., sand). Alternatively, the strength of engineering materials may be characterized by their frictional properties, termed the angle of internal friction. This is analogous to the sliding block on an inclined plane problem examined in introductory physics. The block does not begin to slide down the inclined plane until the angle of inclination increases to the point that the down plane component of gravity exceeds the frictional force resisting the movement. The frictional force is equal to the normal force times the coefficient of friction between the plane and the block. Similarly, the strength of engineering materials may depend upon the coefficient of friction of the material sliding against itself, thus the term **angle of internal friction**. For example, the strength of sand and gravel is characterized by an angle of internal friction generally greater than 30°, but the unconfined compressive strength of that material is zero. Structures, dams, sand dunes, and other examples of geotechnical stability can be identified for granular materials with unconfined compressive strengths of zero.

In the long term, it appears that the unconfined compressive strength may not be a very reliable indicator for the effectiveness of stabilization,[101] and it is likely that this test will demonstrate its limited usefulness as the science and art of stabilization evolve. Specifically, the effectiveness of stabilization as determined with chemical analysis of leachate, particularly with organic sludges, does not necessarily correlate with unconfined compressive strength as shown in Fig. 11-14. Further, the strength properties may change with time, and the unconfined compressive strength measured at

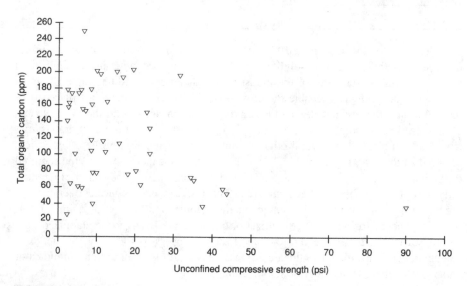

FIGURE 11-14
Unconfined compression and organic waste stabilization (from Evans and Pancoski, 1989).[102]

28 days may not reliably predict long-term strength behavior without due consideration of product durability.

Consolidation

It is widely recognized that some form of consolidation test may be well suited for the evaluation of the settlement of stabilized waste. The consolidation test measures the time rate of one-dimensional compression as well as the relationship between applied vertical stress and total deformation. During a consolidation test, an axial load is applied to the material which is restrained from lateral deformation (see Fig. 11-15). The time rate of deformation (consolidation) is measured for each load applied. The resulting vertical deformation, in terms of axial strain or void ratio, is plotted versus the applied load (see Fig. 11-16). For soft compressible materials, liquids will be squeezed out of the pores as a result of the applied load. The figure shows axial compression with increasing stress as the material is loaded through the recompression range and the virgin compression range. Upon unloading, the material rebounds. The

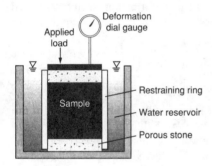

FIGURE 11-15
Consolidation test schematic.

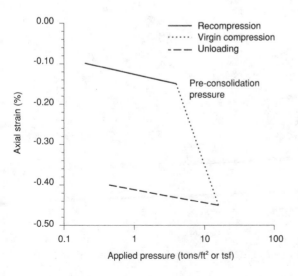

FIGURE 11-16
Idealized consolidation behavior.

liquid that is squeezed out of the pores can be chemically analyzed and used in an assessment of the rate of contaminant migration into the environment.

For well-stabilized, rigid materials the compression will be primarily elastic without the expulsion of fluids from the pores of the stabilized mass. Thus, this test can be used to simulate the field condition of stabilized materials in a large landfill. In the field, stabilized material is loaded by the overlying fill and cover materials. As a result of lateral drainage and anisotropy (i.e., difference in properties such as permeability based on whether flow is horizontal or vertical) due to placement in layers or sections in the field, it is likely that the time rate of consolidation predictions provided by the laboratory consolidation test data will overestimate the time required for consolidation in the field condition.

Shown on Fig. 11-17[103] is the consolidation behavior for three samples of sludge, each stabilized with different proportions of stabilization reagents. The data show that the compressibility of Sample A is about four times greater than the compressibility of Sample B at an applied pressure of 10 tons per square foot. The data also show that Sample B exhibits a substantial preconsolidation pressure as compared with Sample A. Consolidation results such as these are used in evaluations of the suitability of the stabilization process in terms of allowable settlement (deformation) of the stabilized materials.

Hydraulic Conductivity

Hydraulic conductivity, sometimes called permeability, is the ability of a material to conduct or discharge water in response to an applied hydraulic gradient (see Chap. 4). This property is valuable in that it indicates the ease with which water will pass through the material. Hydraulic conductivity is used to estimate the flow rate of fluid through a porous material. As a result, hydraulic conductivity is used to calculate

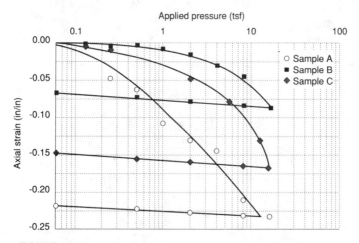

FIGURE 11-17
Consolidation test results.

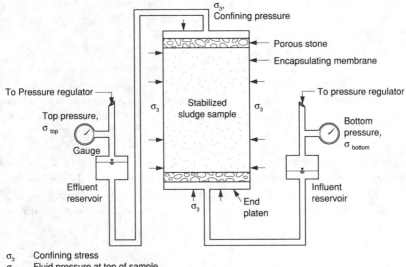

σ₃ Confining stress
σ top Fluid pressure at top of sample
σ bottom Fluid pressure at bottom of sample

FIGURE 11-18
Triaxial permeability test schematic (from Evans and Fang, 1988).[106]

information regarding the rate at which chemicals in the treated waste may migrate into the environment. Some of the factors affecting the hydraulic conductivity of a material include density, degree of saturation, temperature, type of permeating fluid, hydraulic gradient, and particle size distribution. The contaminant concentrations in the effluent can also be measured and values of effective diffusivity determined.

The Triaxial Permeability Test[104, 105] is frequently employed to evaluate the hydraulic conductivity of stabilized materials. This test is also called the Flexible Wall Permeability Test because the sample is encapsulated in a flexible membrane that permits the sample to shrink or swell under the applied confining stresses. A schematic of the Triaxial Permeability Test is shown on Fig. 11-18. Flow is induced by elevating the hydraulic head on one end of the sample with respect to the other. The in situ stress can be simulated by applying pressure to the triaxial cell water surrounding the membrane-encapsulated sample.

Example 11-6 Effective stress for hydraulic conductivity testing. Stabilized sludge will ultimately be disposed in a 5-acre landfill to a maximum depth of 35 feet. The total unit weight of the stabilized sludge is 110 pounds per cubic foot. What confining stress would you select for a triaxial permeability test?

Solution. The confining stress for a triaxial permeability test should simulate the in situ stresses. For the above conditions, the maximum in situ vertical stress would be:

$$\sigma_v(\text{max}) = 35 \times 110 = \underline{3850} \text{ psf}$$

The minimum stress would be at the top of the fill, immediately beneath the cover. Assuming a cover of 5 feet in thickness (see Chap. 13) and with a unit weight of 110 pcf, the minimum vertical stress would be:

$$\sigma_v(\text{min}) = 5 \times 110 = \underline{\underline{550}}\,\text{psf}$$

Because the hydraulic conductivity is stress dependent, it is recommended that hydraulic conductivity tests be conducted over a range of stress levels to a maximum of 3850 psf. For example, for a three-test series, confining stresses of 500, 2175, and 3850 psf could be used.

By monitoring the influent and the effluent quality and quantity, the effectiveness of stabilization can be evaluated along with the effective diffusivity of the contaminant in the stabilized mass. Typical test results for permeation of an acidic oil sludge stabilized with a combination of cement, fly ash, lime, and attapulgite are shown in Fig. 11-19.

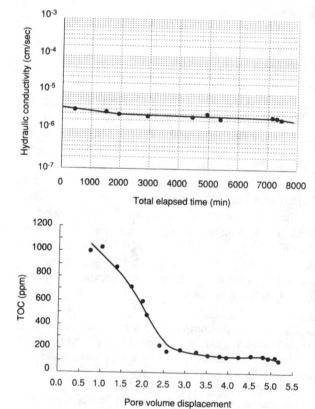

FIGURE 11-19
Triaxial permeability test results.[107]

Example 11-7   Hydraulic conductivity test. The stabilized sludge described in Example 11-6 was tested as per your recommendation. The sample is 14.2 cm long with a diameter of 7.1 cm. The cylindrical sample was tested in a triaxial cell, shown schematically on Fig. 11-18, using a bottom pressure, σ_{bottom}, on the influent reservoir of 55 psi and a top pressure, $\sigma_{top} = 50$ psi. A total flow of 1.76 cm^3 was measured over an elapsed time of 3 hours. Calculate the hydraulic conductivity.

Solution. Darcy's law, $q = K i A$, can be rewritten as $K = \dfrac{q}{i A}$

where K = hydraulic conductivity (cm/sec)
        i = hydraulic gradient (dimensionless)
        A = cross-sectional area in a direction normal to flow (cm^2)
        q = flow rate (cm^3/sec)

The area, A, is:

$$A = \frac{\pi d^2}{4} = \pi \times \frac{7.1^2}{4} = 39.6 \text{ cm}^2$$

The gradient, i, $= \dfrac{\Delta h}{L}$

where Δh = change in hydraulic head
        L = length of flow path

The hydraulic head on the bottom of the sample is:

$$h_{bottom} = \frac{\sigma_{bottom}}{\gamma_w}$$

where γ_w = unit weight of water

$$h_{bottom} = \frac{55 \text{ lb/in}^2}{62.4 \text{ lb/ft}^3} \times \frac{144 \text{ in}^2}{\text{ft}^2} \times \frac{12 \text{ in}}{\text{ft}} \times \frac{(2.54 \text{ cm})}{\text{in}}$$

$$= 3869 \text{ cm}$$

The hydraulic head on the top of the sample is:

where $h_{top} = \dfrac{\sigma_{top}}{\gamma_w} = \dfrac{50 \text{ lb/in}^2}{62.4 \text{ lb/ft}^3} \times \dfrac{144 \text{ in}^2}{\text{ft}^2} \times \dfrac{12 \text{ in}}{\text{ft}} \times \dfrac{2.54 \text{ cm}}{\text{in}}$

$$= 3517 \text{ cm}$$

The gradient is

$$i = \frac{\Delta h}{L} = \frac{3869 \text{ cm} - 3517 \text{ cm}}{14.2 \text{ cm}} = 24.8$$

The flow rate, q, is:

$$q = \frac{1.76 \text{cm}^3}{3\text{hr} \times 60\text{min/hr} \times 60\text{sec/min}} = 1.63 \times 10^{-4} \text{ cm}^3/\text{sec}$$

The hydraulic conductivity may now be computed as:

$$K = \frac{q}{iA} = \frac{1.63 \times 10^{-4} \text{ cm}^3}{24.8 \times 39.6 \text{ cm}^2 \text{sec}}$$

$$= 1.7 \times 10^{-7} \text{ cm/sec}$$

Durability Test Methods

Durability test methods are needed to assess the long-term performance of the stabilized mass. Durability testing is used to evaluate the ability of a material to resist repeated cycles of weathering. A successfully solidified waste material will remain intact throughout these tests, so that the area of the waste directly exposed to leaching is not greatly increased. Experience has shown that most specimens which withstand freeze/thaw cycles also withstand wet/dry cycles, but the reverse is not necessarily true.[108] As a result, the freeze/thaw test, discussed in the following section, was found to be the more severe of the two durability tests.

Wet/Dry Durability (ASTM D4843)

The wet/dry durability test evaluates the resistance of stabilized materials to the natural weathering stresses of repeated wetting and drying cycles. Cured test specimens are subjected to twelve test cycles. Each cycle consists of a period of five hours submerged under water and 42 hours in an oven under low-temperature drying conditions (160°F). The change in volume, moisture content, and weight loss is determined after each cycle. After the twelve cycles, the total sample weight loss is determined or the number of cycles to cause disintegration is determined if the sample is not intact after the twelve cycles are completed.

Freeze/Thaw Durability (ASTM D4842). The resistance of a material to the natural weathering stresses of freezing and thawing are evaluated in the freeze/thaw durability test. Cured test specimens are exposed to twelve cycles, each consisting of a 24-hour freezing period at −20°C and a 24-hour thawing period in water. As in the wet/dry test, performance is evaluated by calculating the weight loss after twelve cycles or the number of cycles that cause disintegration, whichever occurs first. In these durability tests, the weight loss after each cycle can be compared to that of a control specimen.

11-5 FIELD IMPLEMENTATION

The selection of the type of stabilization process depends upon factors such as waste characteristics, materials handling and processing, stabilization objectives, regulatory requirements, and economics.[109] Selection of alternatives for site remediation is discussed in Chap. 17. This section discusses alternative methods for implementing stabilization technologies in the field.

In-drum Mixing Alternatives

To improve the physical characteristics of drummed hazardous wastes prior to land disposal, containers of toxic and hazardous liquids and sludges can be stabilized using in-drum solidification procedures. The drum serves as both the mixing vessel and shipment container for the solidified hazardous waste. However, no credit is taken for the long-term integrity of the drum, assuming the drum will ultimately disintegrate in a landfill.

Reagents are added to the drum containing the waste to be stabilized. Because this may include new drums from a presently operating manufacturing process or old drums from an on-site remediation, it is necessary to evaluate the condition of the drum to determine if it has suitable integrity for the mixing and shipment. Generally, the process involves the addition of the chemical reagents and mixing using a top entering mixing blade. This method of solidification has been used extensively and may be performed remotely through the use of robots.

In situ Mixing Alternatives

The most widely employed method for implementing the stabilization and solidification alternative is in situ mixing. A typical photograph of in situ mixing is shown on Figure 11-1. In situ mixing frequently uses commonly available construction equipment to accomplish the mixing process. In the typical case of the treatment of a large lagoon, this method is often the most economical because it minimizes the handling of wastes. This process was used for the stabilization of a site originally used for oil recovery.[110] A blend of 100 parts sludge to 15 parts lime to 5 parts cement kiln dust was employed. The reagents were added to the lagoons using cranes with 100 foot long booms and mixed with backhoes using a "kneading" action. The completion of fixation was defined according to criteria as follows:

- Visual inspection indicating the material to be solid and earth-like
- No free liquid
- Possible to excavate with material sitting in bucket without overflowing
- Ratio of reagent to fixed material as 1/6

Contractor innovations have improved the thoroughness of mixing compared to that achieved with a backhoe. The backhoe bucket can be replaced with vibratory injectors that simplify the process of adding reagents and increase the mixing energy to obtain a more uniform mix.

For sites requiring relatively deep in situ stabilization or more thorough mixing, modified augers with mixing blades have been employed. These may take two different forms—either large diameter single augers or smaller diameter overlapping augers working in a line. Figure 11-20 (a) and (b) show schematics of these two types.

There are, however, questions as to the thoroughness of an in situ mixing process. In situ mixing is better suited for situations where it is not necessary to blend 100% of the waste material uniformly with the stabilization agents.

FIGURE 11-20a
In situ mixing process with modified augers (photo courtesy of Geo-Con).[111]

Plant Mixing Alternatives

A more thorough way to blend stabilization reagents with the hazardous wastes employs a mechanical mixer using either batch or continuous processes. With this alternative, the hazardous waste is supplied to a continuous or batch mixing process where agents are added and blended. The treated materials are then placed in their ultimate disposal area. A batch process adds fixed amounts of waste and reagents for blending for a fixed contact time (see Fig. 11-21). A continuous process provides for on-going input of waste and reagents and provides the necessary contact time by design of the feed and mixing equipment. These processes can provide significantly improved control over the mix proportions and the thoroughness of blending. Equipment used can include pug mills or extruders. The blended material is then transported for disposal. It is typically placed back on site, compacted, and allowed to cure in place. These processes enable better control over the reagent quantity waste ratios and ensure better and more uniform mixing than in situ techniques. However, even with batch mixing, it may be difficult to ensure the treated materials are adequately blended with the reagents and no unmixed lumps remain in the batch.[112] Plant mixing is typically implemented on site by employing transportable equipment.

Example 11-8. At a secure landfill, wastes with free liquids (failing the Paint Filter Test) are stabilized with fly ash.[113] The alkaline fly ash is derived from power plants burning western coal. The on-site mixing alternative is employed, using a batch process. The wastes are mixed with fly ash by using a front-end loader working on a concrete mixing pad. The front-end loader measures the quantity of fly ash, usually two or three buckets, and places the fly ash on the mixing pad for each batch. Wastes are then added directly from the tank trucks and roll-off containers. Additional fly ash is then added, and each batch mixed using the front-end loader to

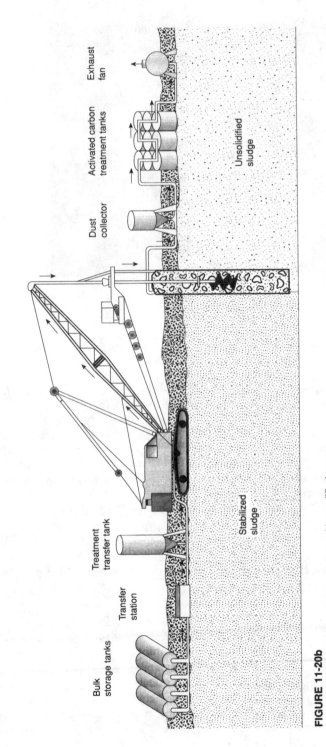

FIGURE 11-20b
Schematic of an in situ mixing process with modified augers

FIGURE 11-21
Remote process batch mixing (photo provided by Enreco, Inc.)

turn and spread the material. The blended material is then transported to the landfill and spread in uniform lifts with a bulldozer.

An oily waste was stabilized using the process described above. The waste, a brown, oil-water emulsion, was analyzed and found to contain 14% oil and grease, 2.2% total organic carbon (TOC), and 72% volatile solids including 180 mg/kg 1,1,1-trichloroethane, 71 mg/kg butylbenzylpthalate, 15 mg/kg cadmium, 150 mg/kg copper, 20 mg/kg lead, and 15 mg/kg zinc. Based upon visual observations, the waste required 1.2 parts fly ash to one part waste by volume for stabilization. After stabilization, testing showed an unconfined compressive strength of 1,758 kg/m² (0.18 tsf) at the time of mixing and 48,828 kg/m² (5.0 tsf) after six days of curing. After stabilization, the material was analyzed and found to contain 89% total solids and 14% volatile solids including 1.5 mg/kg 1,1,1-trichloroethane, 1.1 mg/kg 4-methyl-2-penatone, 12 mg/kg toluene, 14 mg/kg ethyl benzene, and 25 mg/kg total xylenes.

a.) Calculate the reduction in contaminant concentrations due to the dilution only.

b.) Identify other processes that could account for further reductions in contaminant concentrations.

Solution.

a.) $C_0 \times W_w = C_f(W_w + W_{FA})$

where C_0 = original containment concentration
W_w = weight of waste
C_f = final concentration
W_{FA} = weight of fly ash

$$\frac{C_o}{C_f} = \frac{W_w + W_{FA}}{W_w} = \frac{1.0 + 1.2}{1.0} = \frac{2.2}{1} = \underline{\underline{2.2}}$$

$$\frac{C_f}{C_o} = \frac{W_w}{W_w + W_{FA}} = \frac{1.0}{1.0 + 1.2} = \frac{1}{2.2} = \underline{\underline{0.45}}$$

The contaminant concentrations are 45% of original as a result of dilution only; that is a reduction of 55%, assume no change in density.

b.) The metals could react in the pozzolanic reactions of the fly ash.

The volatiles are likely released through volatilization during the extensive handling. The heat of any pozzolanic reaction enhances the rate of volatilization.

Fly ash is a pozzolan, and hydration reactions are time-dependent.

11-6 DESIGN

The selection and design of a stabilization process necessitate characterization of both the waste and the site. A flow chart for these evaluations[114] is presented in Fig. 11-22. The most important lesson regarding design of stabilization projects is that often a poor correlation exists between bench scale studies and full-scale field performance. The variations are frequently related to the thoroughness with which the waste and reagents are mixed; in the laboratory the mix is well blended whereas in the field, such control may be lacking. A full-scale field demonstration (50 to 100 m^3) should be a critical part of the design process.

The selection and design of the appropriate stabilization formulation and technique begin with treatability studies. Treatability studies are conducted to select and develop the appropriate type of binder to eliminate or reduce the hazard of the hazardous waste. Should treatability studies indicate the potential for stabilization, further studies are needed to demonstrate the suitability and to optimize the mix formulation.

The development of the stabilization formulation is an art, not a science. How then does one start the process? The selection of the appropriate formulations is based upon the nature of the waste coupled with judgment incorporating published literature and corporate experience. In addition to consideration of the mechanisms of stabilization discussed in Sec. 11-2, Conner[115] suggest the following additional factors to consider:

- Curing time—field and laboratory
- Volume increase/decrease
- Cost—ingredients, labor, and equipment
- Schedule—project duration, weather restrictions
- Type of mixing—batch or continuous
- Heat generation—heat of hydration and other chemical reactions
- Gas generation—loss of volatiles, material porosity, possible toxicity
- Mixing time—field and laboratory, maximum and minimum
- Other restrictions—flash set, explosion potential, fire potential
- Handling characteristics—waste and reagents

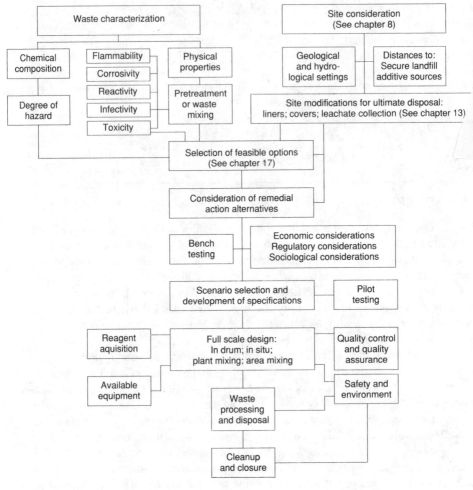

FIGURE 11-22
Flow chart for stabilization process design.[114]

Treatability tests are conducted by mixing different types and quantities of reagent with small (often about 500 g) samples of the waste. The studies are conducted in a phased manner, progressively applying what has been learned to arrive at a suitable solution. The first phase typically includes bench scale mixing and testing of leachability and strength. For example, in one study four variables (sand/cement ratio, water cement ratio, wet sludge quantity and additive dosage) were varied in a statistically designed experiment to determine the variable effect upon the strength and leachability.[116] As the mix proportions and ingredients are refined, additional bench scale testing typically includes more extensive engineering property testing such as hydraulic conductivity and weathering durability. It has been suggested that the mix proportions and test results could be plotted to define an "area" of acceptable test results; in essence, a quality control chart[117] (e.g., a plot of a specific organic of concern in the leachate versus the proportion of cement in the mix). Finally, field pilot scale studies are undertaken to confirm that the laboratory tests can be replicated under field conditions.

Example 11-9   Stabilization cost estimate. Stabilization of sludge from a waste water treatment plant is being considered. The plant treats 0.3×10^6 gal/day (0.3 mgd) with a suspended solids concentration 220 mg/L (ppm) in the wastewater. The treatment plant has a solids removal efficiency of 90%. From bench scale studies, adequate stabilization can be achieved by adding 8% lime by weight. Lime stabilization costs are $80.00 per ton including materials, labor and equipment.

What is the annual cost? How much lime is needed per year?

Solution.

$$\text{Solids in the sludge} = (0.3 \times 10^6 \text{ gal/day}) \times (220 \frac{\text{parts}}{10^6 \text{ parts}}) \times (0.9) \times (8.34 \text{ lb/gallon})$$

$$= 495.4 \text{ lb/day}$$

$$\text{Annual sludge production} = 495.4 \text{ lb/day} \times 365.25 \text{ days per year}$$

$$= 180,944 \text{ lb/year} = 90.5 \text{ ton/year}$$

$$\text{Annual cost} = 90.5 \text{ ton/year} \times \$80/\text{ton} = \underline{\underline{\$7238/\text{year}}}$$

$$\text{Lime usage} = 90.5 \text{ ton/year} \times 0.08 = \underline{7.2 \text{ tons}}$$

11-7   CASE STUDIES

Case Study No. 1

Working under the U.S. EPA's SITE (Superfund Innovative Technology Evaluation) program, an in situ soil stabilization system for PCB contaminated soil was evaluated.[118,119] The system employed a proprietary additive incorporating organically modified clay and deep soil mixing. Initial results were promising as the unconfined compressive strength was high, the permeability low, and the material stable to cycles of wetting and drying. However, many samples subjected to cycles of freezing and thawing showed degradation. The effect of stabilization on the PCB leachate concentration was inconclusive since all analytical results (untreated and treated samples) were near detection limits. The in situ test equipment operated satisfactorily resulting in a dense and homogeneous structure of low porosity. The 1989 cost was estimated at $110 per ton. A follow-up study one-year later showed improvement in the physical properties although the freeze/thaw durability weight loss results were unsatisfactorily high.[120]

Case Study No. 2

It must be emphasized that although theory and bench scale testing can provide tremendous insight into and confidence in stabilization processes, the work in the field must be adequately controlled to yield comparable results. A recent report by Harwell Laboratories[121] described the findings of an investigation of a waste stabilization site that has been operating since 1974. The Leigh facility employs a process which, according to company literature, encapsulates a wide range of organic and inorganic wastes in a "synthetic rock which is safe, impermeable, non-polluting and non-leaching." Testing revealed much of the material was weaker than required and could not support drilling equipment for sampling in one pit. Measured values of hydraulic conductivity in the pit with the lowest strength were generally three orders of magnitude higher than those measured elsewhere. Free liquids were present in one pit with phenol concentrations up to 140 mg/L. The problems were attributed to a decision in the early 1980s to include organically contaminated wastes and use of less cement in the stabilization process. This case points to the need to continually test, evaluate, monitor, and change the stabilization reagents and/or mix proportions as the nature of the waste changes.

Case Study No. 3

A steel manufacturer's chromium wastewater treatment plant generates a filter cake currently classified as Hazardous (EPA code F006). The listed constituents are cadmium, hexavalent chromium, nickel, and cyanide (complexed). A stabilization process employing lime and flyash in a pozzolanic reaction has been shown sucessful in significantly reducing the mobility of the hazardous constituents. The contaminants are fixed in the calcium-aluminosilicate structure formed during the hydration process. For the process to succeed, the hexavalent chromium is first reduced to trivalent chromium. Based upon the results of both TCLP and EP toxicity testing, a petition to delist the stabilized waste has been filed.[122]

DISCUSSION TOPICS AND PROBLEMS

11-1. Define the following terms: (a) solidification, (b) stabilization, (c) fixation, and (d) hydration.

11-2. Define and provide an example of the following mechanisms: (a) macroencapsulation, (b) microencapsulation, (c) adsorption, and (d) absorption.

11-3. Differentiate between an engineering property and a physical property.

11-4. Describe how portland cement can stabilize metallic wastes.

11-5. Identify similarities and differences between the following tests: (a) TCLP and EPtox, (b) paint filter test and liquids release test, and (c) unconfined compression test and one-dimensional consolidation test.

11-6. For each leaching test discussed, identify whether the test is principally a regulatory test, a predictive test, or a investigatory test.

11-7. Construct a table listing leach test variables and each leaching test discussed. Using the EPtox test as a standard, identify whether the contaminant concentrations are expected to be higher or lower as a result of differences in each of the variables.

11-8. Identify three "problems" associated with a stabilization project which might go undetected in the laboratory bench scale studies.

11-9. The data in Table P11-9 were obtained during an unconfined compression test conducted on a stabilized petroleum sludge. Plot the stress-strain relationship and determine the unconfined compressive strength.

11-10. Identify the advantages and disadvantages of the unconfined compression test as a means of evaluating stabilized waste.

11-11. Where did the value of 2175 psf come from in Example 11-6?

11-12. Prove that the specific gravity in English units is equal to the bulk density in SI units.

11-13. Differentiate between batch and continuous processes in waste stabilization.

11-14. The subsurface at the site of toy manufacturing site consists of a well-graded mixture of gravel and sand with 5% silt. Contaminants found in the soil include benzene, toluene, xylene and vinyl chloride to depths of 3 m. The water table is at a depth of 4 m. Select the appropriate stabilization reagents and mixing technique; justify your selection. Suggest other remedial alternatives which might be feasible.

11-15. A lagoon 100 m by 50 m by 3 m deep is found to contain a PCB and metal contaminated sludge at a 10% solids content. Sludge dewatering can be used to increase the solids content to 40% at a cost of $34 per cubic meter of sludge processed. Stabilization of the sludge costs $110 per metric ton of sludge treated. Leachate can be treated at a cost of $20 per cubic meter. By calculating the volume of sludge and leachate generated by dewatering, compare the cost of two alternatives; stabilization of the entire sludge or dewatering and leachate treatment with stabilization of the dewatered sludge. Comment on the overall environmental effectiveness of the two approaches.

11-16. Differentiate landfilling versus stabilization in terms of long-term environmental considerations for the Pepper's Steel and Alloy site in Example 11-1.

TABLE P11-9

Load (kg)	Axial deflection (cm)
0	0
508	.028
893	.056
1364	.086
1896	.114
2488	.142
3098	.170
3632	.200
3991	.229
4191	.257
4332	.287
3597	.315

Original sample length= 15 cm
Original sample diameter= 7.1 cm

REFERENCES

1. Conner, J. R.: *Chemical Fixation and Solidification of Hazardous Wastes*, Van Nostrand Reinhold, New York, NY, 692 pp., 1990.
2. Pojasek, R. B., Ed: *Toxic and Hazardous Waste Disposal*, Ann Arbor Science Publishers, Ann Arbor, MI, 1979.
3. Cullinane, M. J., L. W. Jones, and P. G. Malone: *Handbook for Stabilization/Solidification of Hazardous Waste*, U.S. Environmental Protection Agency Hazardous Waste Engineering Research Laboratory (HWERL), EPA/540/2-86/001, 1986.
4. Malone, P. G., and L. W. Jones: *Survey of Solidification/Stabilization Technology for Hazardous Industrial Wastes*, EPA 600/2-79-056, 41 pp., 1979.
5. Truett, J. B., R. L. Holberger, K. W. Barrett; "Feasibility of *in situ* Solidification/Stabilization of Landfilled Hazardous Wastes." EPA-600/2-83-088, September 1983.
6. *Federal Register*: National Oil and Hazardous Substances Pollution Contingency Plan, Final Rule, vol. 55, p. 8666, March 8, 1990.
7. Environmental Laboratory: "Guide to the Disposal of Chemically Stabilized and Solidified Wastes," EPA-IAG-D4-0569, September 1982.
8. Dole, L.: "In Situ Immobilization of PCBs at the Pepper's Steel and Alloys Site: A Success Story," Company Report, Qualtec, Inc., 1989.
9. U.S. Environmental Protection Agency: *Stabilization/Solidification of CERCLA and RCRA Wastes: Physical Tests, Chemical Testing Procedures, Technology Screening, and Field Activities*, EPA/625/6-89/022, 1989.
10. Weitzman, L., and L. E. Hamel: "Evaluation of Solidification/ Stabilization as a Best Demonstrated Available Technology for Contaminated Soils," EPA/600/2-89/049, U.S. EPA, 1989.
11. Hwang, Daekyoo: "Geotechnical Performance Evaluation of Solidified Wastes," in *Hazardous and Industrial Waste—Proceedings of the 20th Mid-Atlantic Industrial Waste Conference*, Howard University, June 19–21, 1988.
12. Enreco, Inc., Amarillo, Texas
13. Somerville, R. B.: "Encapsulation Solidification of Hazardous Wastes," in *1986 AICHE Symposium—Hazardous Waste Management*, Overland Park, KS, November 14, 1986.
14. Debra Cook, Bucknell University, Lewisburg, Pennsylvania
15. Bricka, R. M., and D. O. Hill: "Metal Immobilization by Solidification of Hydroxide and Xanthate Sludges," in *Proc. 4th International Hazardous Waste Symposium on Environmental Aspects of Stabilization/Solidification of Hazardous and Radioactive Waste*, May 3-6, 1987.
16. Camp, Dresser, and McKee Inc.: "Mobile Treatment Technologies for Superfund Wastes," EPA 540/2-86/003, 1986.
17. Cadena, F.: "Use of Tailored Bentonite for Selective Removal of Organic Pollutants," *J. Environ. Eng. Proc. Am. Soc. Civ. Eng.*, vol. 115, p. 756, 1989.
18. Conner, *Chemical Fixation and Solidification*, op. cit.
19. Carpenter, C. J.: "Ferrous Sulfate/Sodium Sulfide Chromium Reduction Metals Precipitation," *Proceedings of the 5th National Conference on Hazardous Waste and Hazardous Materials*, Las Vegas, NV, pp. 52–56, 1988.
20. Chou, A. C., H. C. Eaton, F. W. Cartledge, and M. E. Tittlebaum: "A TEM Study of the Interactions Between Certain Organic Hazardous Wastes and Type I Portland Cement," Hazardous Waste Research Center, Louisiana State University, Baton Rouge, LA.
21. Walsh, M. B., H. C. Eaton, M. E. Tittlebaum, F. K. Cartledge, and D. Chalasani: "The Effect of Two Organic Compounds on a Portland Cement-Based Stabilization Matrix," *Hazardous Waste and Hazardous Materials*, vol. 3, no. 1, pp. 111–123, 1986
22. U.S. EPA, *Stabilization/Solidification of CERCLA and RCRA Wastes*.
23. Lea, F. M.: *The Chemistry of Cement and Concrete*, Chemical Publishing Company, New York, NY, pp. 727, 1971.
24. Ashby M. F., and D. R. H. Jones: *Engineering Materials 2: An Introduction to Microstructures, Processing and Design*, Pergamon Press, Oxford, UK, 368 pp, 1986.

25. Bishop, P., D. Gress, and J. Olafsson: "Cement Stabilization of Heavy Metals: Leaching Rate Assessment," in *Industrial Wastes—Proceedings of the 14th Mid-Atlantic Industrial Waste Conference*, College Park, MD, 1982.

26. Shively, W., P. Bishop, D. Gress, and T. Brown: "Leaching Tests of Heavy Metals Stabilized with Portland Cement," *J. WPCF*, vol. 38, no. 3, pp. 234–241, 1986.

27. U.S. Army Construction Engineering Research Laboratory: "Critical Review of Cement Based Stabilization/Solidification Techniques for the Disposal of Hazardous Wastes," ADA-184 427, IL, 1986.

28. Walsh, et al: "The Effect of Two Organic Compounds on a Portland Cement-Based Stabilization Matrix," op. cit.

29. Skipper, D. G., Eaton, H. C., Cartledge, F. K., and Tittlebaum, M. E.: "The Microscopic Fracture Morphology of Hardened Type I Portland Cement Paste Containing Parachlorophenol," *Hazardous Waste and Hazardous Materials*, vol. 4, no. 4, pp. 389–402, 1987.

30. Cullinane, M. J., R. M. Bricka, and N. R. Francingues, Jr.: "An Assessment of Materials That Interfere with Stabilization/Solidification Processes," in *Proc. 13th Annual Research Symposium*, Cincinnati, OH, pp. 64–71, 1987.

31. Chalasani, D., F. K. Cartledge, H. C. Eaton, M. E. Tittlebaum, and M. B. Walsh: "The Effects of Ethylene Glycol on a Cement-Based Solidification Process," *Hazardous Wastes and Hazardous Materials*. vol. 3, no. 2, 1986.

32. Cartledge, F. K., H. C. Eaton, and M. E. Tittlebaum: *The Morphology and Microchemistry of Solidified/Stabilized Hazardous Waste*, EPA/600/S2/056, January 1990.

33. Jones, L. W.: *Interference Mechanisms in Waste Stabilization/Solidification Processes*, EPA/600/S2-89/067, April 1990.

34. Montgomery, D. M., C. J. Sollars, R. Perry, S. E. Tarling, P. Barnes, and E. Henderson: "Treatment of Organic- Contaminated Industrial Wastes Using Cement-based Stabilization/Solidification—I. Microstructural Analysis of Cement-Organic Interactions," *Waste Management & Research*, vol. 9, pp. 103–111, 1991.

35. Chou, A. C., H. C. Eaton, F. W. Cartledge, and M. E. Tittlebaum: "A Transmission Electron Microscopic Study of Solidified/Stabilized Organics," *Hazardous Waste and Hazardous Materials*, vol. 5, no. 2, pp. 145–154, 1988.

36. Cocke, D. L.: "The Binding Chemistry and Leaching Mechanisms of Hazardous Substances in Cementitious Solidification/Stabilization Systems," *Proceedings of the 2nd Annual Symposium of the Gulf Coast Hazardous Substance Research Center*, Lamar University, Beaumont, TX, pp. 81–102, February 15–16, 1990.

37. Sams, T. L., and T. M. Gilliam: "Systematic Approach for the Design of Pumpable, Cement-Based Grouts for Immobilization of Hazardous Wastes," *Environmental Aspects of Stabilization and Solidification of Hazardous and Radioactive Wastes, ASTM STP 1033*, P. L. Cote and T. M. Gilliam, Eds., American Society for Testing and Materials, Philadelphia, PA, pp. 15–20, 1989.

38. Daniali, S.: "Solidification/Stabilization Technology for Hazardous Industrial Wastes," *Journal of Hazardous Materials*, vol. 24, no. 2–3, pp. 225–230, 1990.

39. Rademaker, P. D., and R. B. Wiegers: "The Processing of Industrial Waste for Immobilization and/or Recycling Applying Pozzolanic Reactions," presented at *Environmental Technology—2nd European Conference on Environmental Technology*, pp. 411–421, 1987.

40. Malone, P. G., and L. W. Jones: *Solidification/Stabilization of Sludge and Ash from Wastewater Treatment Plants*, EPA/600/S2-85-058, 1985.

41. Ibid.

42. DuPont, A.: "Lime Treatment of Liquid Hazardous Waste Containing Heavy Metals, Radionuclides, and Organics," National Lime Association, August 1, 1986.

43. Ezell, D., and P. Suppa: "The Soliroc Process in North America: A Stabilization/Solidification Technology for the Treatment of Metal-Bearing Wastes with Reference to Extraction Procedure Testing," *Environmental Aspects of Stabilization and Solidification of Hazardous and Radioactive Wastes, ASTM STP 1033*, P. L. Cote and T. M. Gilliam, eds., American Society for Testing and Materials, Philadelphia, PA, pp. 7–14, 1989.

44. U.S. Environmental Protection Agency: "Chemfix Stabilization/Solidification Process, Clackamas,

Oregon," *Superfund Innovative Technology Evaluation*, EPA/540/S5-89/011, December 1990.

45. Beall, Gary W.: "Method of Immobilizing Organic Contaminants to Form Non-Flowable Matrix Therefrom," U.S. Patent Number 4,650,590, March 17, 1987.

46. Evans, J. C., S. E. Pancoski, and G. Alther: "Organic Waste Treatment With Organically Modified Clays." *Third International Conference on New Frontiers for Hazardous Waste Management*, pp. 48–57 Pittsburgh, PA, EPA 600/9-89/072 September 10–13, 1989.

47. Boyd, S. A., J. F. Lee, and M. M. Mortland: "Attenuating Organic Contaminant Mobility by Soil Modification." *Nature*, vol. 3, pp. 345–347, 1988.

48. Boyd, S. A., M. M. Mortland, and T. J. Pinnavaia: "Use of Modified Clays for Adsorption and Catalytic Destruction of Contaminants," in *Proc. 13th Annual Research Symposium*, Cincinnati, OH, 1987.

49. LaGrega, M. D., J. C. Evans, C. O. Acuna, S. J. Zarlinski, and D. S. Hall: "Stabilization of Acidic Refinery Sludges," *Journal of Hazardous Materials*, vol. 24, pp. 169–187, 1990.

50. Montgomery, D. M., C. J. Sollars, R. Perry, S. E. Tarling, P. Barnes, and E. Henderson: "Treatment of Organic-Contaminated Industrial Wastes Using Cement-based Stabilization/Solidification—II. Microstructural Analysis of the Organophilic Clay as a Pre-Solidification Adsorbent," *Waste Management & Research*, vol. 9, pp. 113–125, 1991.

51. Faschan, A., M. Tittlebaum, F. Cartledge, and R. Portier: "The Use of Organoclays to Reduce Organic Waste Volatization during Solidification," *Hazardous Waste & Hazardous Material*, vol. 9, no. 4, 1992.

52. Boyd, S. A., M. M. Mortland, and C. T. Chiou: "Sorption Characteristics of Organic Compounds on Hexadecyltrimethylammonium-smectite," *Soil Sci. Soc. Am. J.*, vol. 52, pp. 652–657, 1988.

53. Boyd, S. A., S. Shaobai, J. Lee, and M. M. Mortland: "Pentachlorophenol Sorption by Organo Clays," *Clays and Clay Minerals*, vol. 36, no. 2, pp. 125–130, 1988.

54. Grim, R. E.: *Clay Mineralogy*, McGraw-Hill, New York, NY, 1968.

55. Alther, G. R., J. C. Evans, and S. E. Pancoski: "Organically Modified Clays for Stabilization of Organic Hazardous Waste," *Superfund '88—Proceedings of the 9th National Conference*, Washington, D.C., November 28–30, 1988.

56. Sheriff, T. S., C. J. Sollars, D. Montgomery, and R. Perry: "The Use of Activated Charcoal and Tetra-Alkylammonium-Substituted Clays in Cement-Based Stabilization/Solidification of Phenols and Chlorinated Phenols," *Environmental Aspects of Stabilization and Solidification of Hazardous and Radioactive Wastes, ASTM STP 1033*, P. L. Cote and T. M. Gilliam, eds., American Society for Testing and Materials, Philadelphia, PA, pp. 273–286, 1989.

57. Mortland, M. M., S. Shaobai, and S. A. Boyd: "Clay-Organic Complexes as Adsorbents for Phenol and Chlorophenols," *Clays and Clay Minerals*, vol. 34, no. 5, pp. 581–585, 1986.

58. Grim, *Clay Mineralogy*.

59. Evans, J. C. and S. E. Pancoski: "Organically Modified Clays," *Transportation Research Record* 1219, Geotechnical Engineering 1989, Transportation Research Board, National Research Council, pp. 160–168, 1989.

60. Evans, J. C., and G. Alther: "Hazardous Waste Stabilization Using Organically Modified Clays," Geotechnical Engineering Congress 1991, ASCE, Geotechnical Engineering Special Technical Publication No. 27, pp. 1149–1162, June 1991.

61. Ibid.

62. Smith, J. A., P. R. Jaffe, and C. T. Chiou: "Effect of Ten Quaternary Ammonium Cations on Tetrachlormethane Sorption to Clay from Water," *Environmental Science and Technology*, vol. 24, no. 8, pp. 1167–1772, 1990.

63. Company Literature: "Monitoring of Potential Leachate Contamination from Contaminated and DCR-Treated Material," Miller Environmental LTD, United Kingdom, undated.

64. Environmental Laboratory: "Guide to the Disposal of Chemically Stabilized and Solidified Wastes." EPA-IAG-D4-0569, September 1982.

65. McNeese, J. A., et al: "Laboratory Studies of Fixation of KEPONE® Contaminated Sediments," in *Toxic and Hazardous Waste Disposal, Robert B. Pojasek, ed. vol. II*, Ann Arbor Science Publishers, Ann Arbor, MI, pp. 219–220, 1979.

66. Tittlebaum, M. E., R. K. Seals, F. K. Cartledge, and S. Engles: "State of the Art on Stabilization

of Hazardous Organic Liquid Wastes and Sludges," *CRC Critical Reviews in Environmental Control*, vol. 15, no. 2, pp. 179–211, 1985.

67. Doyle, R.: "Use of Extruder/Evaporator to Stabilize and Solidify Hazardous Waste," in *Toxic and Hazardous Waste Disposal*, R. Pojasek, ed., Ann Arbor Science Publishers, Ann Arbor, MI, vol. 1, pp. 65–91, 1979.

68. Swindlehurst, D. P., R. D. Doyle, and A. J. Mattus: "The Use of Bitumen in the Stabilization of Mixed Wastes," *Environmental Aspects of Stabilization and Solidification of Hazardous and Radioactive Wastes, ASTM STP 1033*, P. L. Cote and T. M. Gilliam, eds., American Society for Testing and Materials, Philadelphia, PA, pp. 21–27, 1989.

69. Mattus, A. J., M. M. Kaczmarsky, and C. K. Cofer: "Leaching and Comprehensive Regulatory Performance Testing of an Extruded Bitumen Containing a Surrogate Sodium Nitrate-Based, Low-Level Waste," *Environmental Aspects of Stabilization and Solidification of Hazardous and Radioactive Wastes, ASTM STP 1033*, P. L. Cote and T. M. Gilliam, eds., American Society for Testing and Materials, Philadelphia, PA, pp. 28–39, 1989.

70. Unger, S. L., R. W. Telles, and H. R. Lubowitz: "Surface Encapsulation Process for Stabilizing Intractable Contaminants," *Environmental Aspects of Stabilization and Solidification of Hazardous and Radioactive Wastes, ASTM STP 1033*, P. L. Cote and T. M. Gilliam, eds., American Society for Testing and Materials, Philadelphia, PA, pp. 40–52, 1989.

71. Heiser, J. H., E.-M. Franz, and P. Colombo: "A Process for Solidifying Sodium Nitrate Waste in Polyethylene," *Environmental Aspects of Stabilization and Solidification of Hazardous and Radioactive Wastes, ASTM STP 1033*, P. L. Cote and T. M. Gilliam, eds., American Society for Testing and Materials, Philadelphia, PA, 53–62, 1989.

72. Geosafe Corporation: "Application and Evaluation Considerations for In Situ Vitrification Technology: A Treatment Process for Destruction and/or Permanent Immobilization of Hazardous Materials," GSC 1901, April 1989.

73. MacNeill, K. R.: "Recycling Waste Streams Using Glass Making Activities," *Proceedings of Contaminated Land: Policy, Regulation, and Technology*, IBC Technical Services, London, UK, February 1991.

74. U.S. Environmental Protection Agency: "The Babcock and Wilcox Cyclone Furnace Soil Vitrification Technology," EPA/540/F-92/010, September 1992.

75. MacNeill, "Recycling Waste Streams Using Glass Making Activities."

76. van der Sloot, H. A., G. J. de Groot, and J. Wijkstra: "Leaching Characteristics of Construction Materials and Stabilization Products Containing Waste Materials," *Environmental Aspects of Stabilization and Solidification of Hazardous and Radioactive Wastes, ASTM STP 1033*, P. L. Cote and T. M. Gilliam, eds., American Society for Testing and Materials, Philadelphia, PA, pp. 125–149, 1989.

77. *Federal Register*, "U.S. EPA: Free Liquids (Paint Filter)," vol. 50, no. 83, p. 18370, April 30, 1985.

78. *Federal Register*, U.S. EPA: "Liquids Release Test," 40 FR, vol. 51, no. 247, p. 46824, December 1986.

79. U.S. Environmental Protection Agency: "Extraction Procedure (EP) Toxicity Test," *Code of Federal Regulations*, 40 CFR, Part 261, Subpart C, Sec. 261-24, July 1, 1982.

80. *Federal Register*, U.S. EPA, "Toxicity Characteristic Leaching Procedure (TCLP)," 40 CFR, vol. 51, no. 286, Appendix 2, Part 268, Page 40643, November 7, 1986.

81. Brown, T., and P. Bishop: "The Effect of Particle Size on the Leaching of Heavy Metals from Stabilized/Solidified Waste," in *Proc. International Conference on New Frontiers in Hazardous Waste Management*, 1985.

82. Bishop, P. L.: "Leaching of Inorganic Hazardous Constituents from Stabilized/Solidified Hazardous Wastes," *Hazardous Wastes and Hazardous Materials*, vol. 5, no. 2, pp. 129–143, Spring 1988.

83. Thurnau, R. C. and M. P. Esposito: "TCLP as a Measure of Treatment Effectiveness," *Hazardous Wastes and Hazardous Materials*, vol. 6, no. 4, pp. 347–361, November 4, 1989.

84. American Nuclear Society (ANS): "ANSI/ANS 16.1-1986, American National Standard Measurement of Solidified Low-Level Radioactive Wastes by a Short-Term Test Procedure," American Nuclear Society, LaGrange Park, IL, April 14, 1986.

85. Environment Canada and Alberta Environmental Center, "Test Methods for Solidified Waste Charac-

terization," Alberta, Canada, 1986.

86. Cote, P. L. and D. P. Hamilton: "Evaluation of Pollutant Release from Solidified Aqueous Wastes Using a Dynamic Leaching Test," in *Proc. Hazardous Wastes and Environmental Emergencies Conference*, pp. 302–308, 1984.

87. Cote, P. L., and D. Isabel: "Applications of a Static Leaching Test to Solidified Hazardous Waste," presented at the *ASTM International Symposium on Industrial and Hazardous Solid Wastes*, Philadelphia, PA, 1983.

88. Hannak P., A. J. Liem, and P. Cote: "Methods for Evaluation Solidified Waste," Alberta Environmental Centre, Vegreville, Alberta, Canada, and Wastewater Technology Centre, Environmental Protection Service, Environment Canada, Burlington, Ontario, Canada, presented to the *Second Annual Symposium on Solid Waste Testing and Quality Assurance Organized by U.S. Environmental Protection Agency*, Washington, D.C., 14 p., July 15–18, 1986.

89. Stegemann, Julia: "Results of a Cooperative Study of Twelve Test Methods for Solidified Waste Characterization," Environment Canada, Wastewater Technology Centre, Burlington, Ontario, *presented at the 34th Ontario Industrial Waste Conference*, Toronto, Canada, June 1987.

90. Ibid.

91. Shively, W., and S. Dethloff: "Hazardous Waste Treatment Demonstration Guide—Test Methods for Solidification/Stabilization of Hazardous Wastes," EPA seOS8801/44/1, 82 pp, September 14, 1987.

92. U.S. Environmental Protection Agency: *Federal Register*, "Multiple Extraction Procedure," 40 CFR 261, Appendix II, January, 1989.

93. Soundararajan, R., E. F. Barth, and J. J. Gibbons: "Using an Organophilic Clay to Chemically Stabilize Waste Containing Organic Compounds," *Hazardous Materials Control*, vol. 3, no. 1, pp. 42–45, Jan–Feb. 1990.

94. Weitzman, L., and L. E. Hamel: "Evaluation of Solidification/ Stabilization as a Best Demonstrated Available Technology for Contaminated Soils," op. cit.

95. American Society for Testing and Materials (ASTM), "Laboratory Determination of Water (Moisture) Content of Soil and Rock," D2216, Annual Book of Standards, vol. 4.08, pp. 295–298, 1992.

96. American Society for Testing and Materials (ASTM) "Particle Size Analysis of Soils," D422, Annual Book of Standards, Vol. 4.08, pp. 94–100, 1992.

97. Roy, A., H. C. Eaton, F. K. Cartledge, and M. E. Tittlebaum: "Solidification/Stabilization of a Heavy Metal Sludge by a Portland Cement/Fly Ash Binding Mixture," *Hazardous Waste and Hazardous Materials*, vol. 8, no. 1, pp. 33–41, 1991.

98. American Society for Testing and Materials (ASTM), "Vicat Needle" Annual Book of Standards, vol. 4.08, pp. 94–100, 1992.

99. American Society for Testing and Materials (ASTM), "Compressive Strength of Molded Soil-Cement Cylinders" D1633, Annual Book of Standards, vol. 4.08, pp. 295–298, 1992.

100. American Society for Testing and Materials (ASTM), "Unconfined Compressive Strength of Cohesive Soil," D2166, Annual Book of Standards, vol. 4.08, pp. 281–285, 1992.

101. Evans, J. C., and S. E. Pancoski: "Stabilization of Petroleum Sludges" Superfund '89—Proceedings of the 10th National Conference, HMCRI, Washington, D.C., November 27–29, 1989.

102. Ibid.

103. Zarlinski, S. J., and J. C. Evans: "Weathering Resistance of Stabilized Petroleum Sludges," *Proceedings of the 11th National Conference of the Management of Uncontrolled Hazardous Waste Sites*, Washington, D.C., pp. 712–715, November 1990

104. American Society for Testing and Materials (ASTM), "Permeability of Granular Soils (Constant Head)," D2434, Annual Book of Standards, vol. 4.08, pp. 309–313, 1992.

105. U.S. EPA Method 9100.

106. Evans, J. C., and H. Y. Fang: "Triaxial Permeability and Strength Testing of Contaminated Soils," *Advanced Triaxial Testing of Soil and Rock*, ASTM STP 977, American Society for Testing and Materials, Philadelphia, PA, pp. 387–404, 1988.

107. Zarlinski and Evans, "Weathering Resistance of Stabilized Petroleum Sludges," op. cit.

108. Cote, P. L. and D. P. Hamilton: "Evaluation of Pollutant Release from Solidified Aqueous Wastes Using a Dynamic Leaching Test," op. cit.

109. Wiles, C. C., et al.: "Status of Solidification/Stabilization in the United States and Factors Affecting Its Use," EPA/600-D-89/159, U.S. EPA, 1989.
110. Curry, M. F. R.: "Fixation/Solidification of Hazardous Waste at Chemical Waste Management's Vickery, Ohio Facility," *Superfund '89—Proceedings of the 10th National Conference*, November 27-29, 1989.
111. Geo-Con, Monroeville, Pennsylvania.
112. U.S. Environmental Protection Agency: "Technology Evaluation Report: Site Program Demonstration Test, Soliditech, Inc. Solidification/Stabilization Process, Vol. 1," EPA/540/5-89/005a, February 1990.
113. Shively, W. E., and M. A. Crawford: "Extraction Procedure Toxicity and Toxicity Characteristic Leaching Procedure Extractions of Industrial and Solidified Hazardous Waste," *Environmental Aspects of Stabilization and Solidification of Hazardous and Radioactive Wastes, ASTM STP 1033*, P. L. Cote and T. M. Gilliam, eds., American Society for Testing and Materials, Philadelphia, PA, pp. 150–1692, 1989.
114. Cullinane, M. J., and L. Jones: "Solidification and Stabilization of Hazardous Wastes," *HMCRI Journal of Hazardous Materials Control*, vol. 2, no. 2, pp. 22, March–April, 1989.
115. Conner, *Chemical Fixation and Solidification of Hazardous Wastes*.
116. Shin, H-K, N-R Her, and J-K Koo: "Design Optimization for Solidification of Hazardous Wastes," *Hazardous Waste and Hazardous Materials*, vol. 5, no. 3, pp. 239–250, 1988.
117. Rushbrook, P. E., Baldwin, and C. B. Dent: "A Quality-Assurance Procedure for Use at Treatment Plants to Predict the Long-Term Suitability of Cement-Based Solidified Hazardous Wastes Deposited in Landfill Sites," *Environmental Aspects of Stabilization and Solidification of Hazardous and Radioactive Wastes, ASTM STP 1033*, P. L. Cote and T. M. Gilliam, eds., American Society for Testing and Materials, Philadelphia, PA, pp. 93–113, 1989.
118. U.S. Environmental Protection Agency: "Superfund Innovative Technology Evaluation," EPA/540/8-89/010, November 1989.
119. U.S. Environmental Protection Agency: "Superfund Innovative Technology Evaluation, Technology Demonstration Summary, International Waste Technologies In Situ Stabilization/Solidification, Hialeah, FL," EPA/540/S5-89/004, June 1989.
120. U.S. Environmental Protection Agency: "International Waste Technologies/Geo-Con In situ Stabilization/Solidification Update Report," *Superfund Innovative Technology Evaluation*, EPA/540/S5-89/004a, January 1991.
121. Harwell Laboratories: ENDS Report 173, June 1989.
122. Stephanatos, B. N., and A. MacGregor: "Stabilization and Delisting of Hazardous Wastes: An Effective Approach for Reducing High Sludge Disposal Costs," *Proceedings of the 12th National Conference on the Management of Uncontrolled Hazardous Waste Sites*, Washington, D.C., 1991.

ADDITIONAL READING

Conner, J. R.: *Chemical Fixation and Solidification of Hazardous Wastes*, Van Nostrand Reinhold, New York, NY. 692 pp., 1990.
Cullinane, M. J., and L. Jones: "Solidification and Stabilization of Hazardous Wastes, Part 1," *Hazard. Mater. Control*, vol. 2, p. 9, 1989.
Cullinane, M. J., and L. W. Jones: *Stabilization/Solidification of Hazardous Waste*, U.S. Environmental Protection Agency Hazardous Waste Engineering Research Laboratory (HWERL), EPA/600/D-86/028, Cincinnati, Ohio, 1986.
Cullinane, M. J., L. W. Jones, and P. G. Malone: *Handbook for Stabilization/Solidification of Hazardous Waste*, U.S. Environmental Protection Agency (U.S. EPA) Hazardous Waste Engineering Research Laboratory (HWERL), EPA/540/2-86/001, 1986.
Hannak, P.: "Solid Waste Testing: Toxicity Characteristic Leaching Procedures," presented at *Petroleum Waste Management Conference*, January 22–23, 1986.

ACKNOWLEDGMENTS

The authors wish to acknowledge the following individuals who assisted with the preparation of this chapter: Ed VanKeuren, Steve Montagna, James Archer, Daekyoo Hwang, *ERM, Inc*, Nick Hollingshad, *ERM-Southwest, Inc*; Charles Johnson, Eric Smalstig, Marc Smugar, John Floyd, Brian Swanson, Jason Fralick, *Bucknell University*; Dennis Contois, Eric Mortenson, the *Pennsylvania State University–Harrisburg*; Adam Faschan, *Louisiana State University*. The principal reviewer for the *ERM Group* was Dr. Elsie Millano, *ERM-North Central*.

CHAPTER
12

THERMAL METHODS[†]

Lasciate ogni speranza voi ch'entrate!
(Abandon all hope, ye who enter here!)
Dante

12-1 INTRODUCTION

Incineration systems are designed to destroy only organic components of waste; however, most hazardous waste is non-exclusive in its content and therefore will contain both combustible organics and non-combustible inorganics. By destroying the organic fraction and converting it to carbon dioxide and water vapor, incineration reduces the waste volume, and to the extent that the organic components include toxic components, its threat to the environment.

Hazardous wastes come in all physical forms: liquid, solid, and somewhere in between. Commercial hazardous waste incineration systems have to handle the gamut of waste streams (see Sec. 12-6). Nevertheless, many hazardous waste incinerators are concerned only with liquid wastes. Section 12-5 addresses this special type of system. While gaseous waste streams are not regulated as hazardous waste, the combustion of such wastes is certainly a part of the overall incineration process and is addressed in Sec. 12-4. The theory and equipment design for each type of waste is similar but different, and their combination into a single system is often required.

[†]Chapter 12 was prepared by Richard D. Ross and Joseph Santoleri of Four Nines, Inc.

Regulations governing hazardous waste incineration were essentially non-existent until 1980. The Clean Air Act was the only regulatory constraint on incineration of any type before that date. Table 12-1 shows the historical perspective of the regulatory process and some of the process developments along the way.

Incineration is a rather simple chemical process when viewed superficially. Organic compounds are confined to those containing carbon, hydrogen, and sometimes oxygen, with the possible addition of metals and non-metals such as halogens and nitrogen. This should make the chemistry of combustion easy to decipher. Unfortunately, this chemistry applies only when reactions are complete, and in combustion, they often are not.

Good combustion is good oxidation of the organic components—carbon and hydrogen. To achieve this, air, which contains only 21% oxygen by volume, must be thoroughly mixed with the carbon and hydrogen of the fuel (waste) to produce a stoichiometric product of carbon dioxide and water. Unfortunately, air also contains 79% nitrogen, which is inert and gets in the way of the combustion process. Given a completely homogeneous system (a well-stirred reactor) which requires both Time and Turbulence, the complete oxidation of carbon and hydrogen should occur at some prescribed Temperature. Thus, the three "Ts" of combustion affect the reaction.

TABLE 12-1

Historical perspective

Significant legislation	Year	Technology	Waste types
	1955	Open burning	Solids, drummed waste
		Sewer or stream	Liquids
		Grate incinerator	Industrial solids
	1960	Multiple hearth	Sewage sludge
		Fume incineration	Waste gases
		Liquid injection	Non-halogenated solvents
Federal Water Quality Act	1965	Liquid injection	Heavy tars
Clean Air Act	1970	Rotary kiln	Aqueous wastes
		Incinerator APC	Sludges, drummed waste
		Fluid bed	All wastes
		Liquid injection	Solids and sludges
		Liquid injection heat recovery	HCl recovery
			Halogenated liquids
	1975	Rotary hearth	Drummed wastes
Resource Conservation and Recovery Act	1976		
Clean Air Act Amendments	1977		
RCRA Regulation for Hazardous Waste Incinerators	1980	Molten salt process	Halogenated liquids
		Required destruction and removal efficiency	All wastes
	1982	Performance standards	Hazardous wastes
		Huber AER	Contaminated soils
RCRA Amendments	1985		

Decrease one of these factors and the other two must be increased to achieve the same degree of combustion completeness.

For this reason, few combustion reactions are complete at their minimum theoretical temperature or with exactly the theoretical amount of air to provide stoichiometric amounts of oxygen for the carbon and hydrogen in the fuel. As the turbulence (mixing) of the reactor (burner/incinerator) improves, and the time allowed for the reaction to take place increases, the amount of excess air (oxygen) needed in the reaction decreases. If combustion is viewed as an isosceles triangle with time, temperature, and turbulence as the three sides, then a decrease in the length of one side requires a compensatory increase in the other two. This is combustion!

Typical combustion process reactions are

$$C + (O_2 + N_2) \rightarrow CO_2 + N_2 + O_2 + HEAT \tag{12-1}$$

$$H_2 + (O_2 + N_2) \rightarrow H_2O + N_2 + O_2 + HEAT \tag{12-2}$$

$$CH_4 + (O_2 + N_2) \rightarrow CO_2 + H_2O + N_2 + O_2 + HEAT \tag{12-3}$$

Example 12-1   Combustion calculation. A waste mixture of 30% toluene, 65% acetone, and 5% water is to be burned in a liquid injection type incinerator at a rate of 1000 lb/hr using 20% excess air.

Compound	Formula	Heating value (Btu/lb)
Toluene	$C_6H_5CH_3$	18,252
Acetone	CH_3COCH_3	13,120
Water	H_2O	0

The ambient temperature is 60°F and the incinerator is lined with 9″ of refractory having a thermal conductivity of 13 Btu · ft^2·°F/in at 1800°F.

1. What is the total heat release in the incinerator?
2. What is the percent by volume of each component in the flue gas?

Solution.

1. Toluene heat release $=$ $0.30 \times 18,252$ $=$ 5,476 Btu/lb
   Acetone heat release $=$ $0.65 \times 13,120$ $=$ 8,528 Btu/lb
   Water heat release $=$ 0 Btu/lb

   Heat release
   per pound of mixture $=$ 14,004 Btu/lb

   Heat release
   in the incinerator $= 1000$ lb/hr $\times 14,004$ btu/lb $=$ 14,004,000 Btu/hr

2. $C_6H_5CH_3 + 8 O_2 \rightarrow 6 CO_2 + 4 H_2O$
   The molecular weight of toluene is 92
   The weight of toluene is 0.30×1000lb/hr $= 300$lb/hr
   $300/92 = 3.26$ moles/hr of toluene

Constituent	MW	Moles/hr
Toluene	92	3.26
O_2	32	26.08
CO_2	44	19.56
H_2O	18	13.04

The oxygen in the table above is the stoichiometric amount (i.e., that needed to just react with the toluene). An excess amount of 20% has been specified and this excess will be in the flue gas as O_2.

$0.2 \times 26.08 = 5.22$ moles/hr of O_2 in addition to that above.

But air, not oxygen, is used in the incinerator, and air is 79% nitrogen. Therefore, all of the N_2 will be in the flue gas.

$26.08 + 5.22 = 31.3$ moles of O_2 are required or:

$79/21 \times 31.3 = 117.75$ moles of N_2.

Using the same approach for acetone:

$CH_3COCH_3 + 4 O_2 \rightarrow 3 CO_2 + 3 H_2O$

The molecular weight of acetone is 58

The weight of acetone is $0.65 \times 1000lb/hr = 650$ lbs

$650/58 = 11.21$ moles/hr of acetone

Constituent	MW	Moles/hr
Acetone	58	11.21
O_2	32	44.84
CO_2	44	33.63
H_2O	18	33.63

Excess $O_2 = 0.2 \times 44.84 = 8.99$ moles/hr

Total $O_2 = 44.84 + 8.99 = 53.83$ moles/hr

Therefore the nitrogen is $79/21 \times 53.83 = 202.5$ moles/hr

In addition to the above there is 5% water in the waste, or 50 lb/hr, or $50/18 = 2.78$ moles.

By adding the moles of flue gas generated from the three waste components the total moles of each can be determined.

	CO_2	H_2O	O_2	N_2
Toluene	19.56	13.04	5.22	117.75
Acetone	33.63	33.63	8.99	202.50
Water		2.78		
Totals	53.19	49.45	14.21	320.25

Total moles = 437.10

Mole % = Volume %

Flue gas is therefore

12.17%	CO_2
11.30%	H_2O
3.25%	O_2
73.27%	N_2

12-2   REGULATIONS

The design and evaluation of incineration systems, as with most aspects of hazardous waste management, is driven by laws and regulations enacted to assure environmental and human health protection. In the United States the law governing incineration systems is the Resource Conservation and Recovery Act (RCRA)[1] as amended by the Hazardous and Solid Waste Amendments of 1984 (HSWA)[2] This statute was discussed in Sec. 2-2. As noted in Chap. 2, this statute provided far more technical details than other environmental legislation. For example, in contrast to the usual practice of leaving technical details to the EPA, HSWA required specific destruction and removal efficiencies:

> for each incinerator which receives a permit under section 3005(c) after the date of enactment of the Hazardous and Solid Waste Amendments of 1984, the attainment of the minimum destruction and removal efficiencies required by regulations in effect on June 24, 1982[3]

Incineration facilities are regulated by rules promulgated under RCRA. The specific requirements for an operating permit[4] mandate a trial burn and a detailed report including a complete waste analyses quantifying the hazardous constituents in the wastes to be burned, a detailed engineering description of the incinerator, etc.

The Trial Burn Plan, if approved, will identify *Principal Organic Hazardous Constituents* (POHCs) that must be monitored during the *Trial Burn* operation. POHCs are specific organic compounds selected on the basis of their difficulty to incinerate. Facility standards[5] include general requirements for waste analysis, performance standards, operating requirements, etc.

The following Federal performance standards apply to hazardous waste incinerators in the U.S.[6]

- Emissions of particulates—0.08 grains/dry standard cubic foot (dscf) (180 mg/dscm) corrected to 7% O_2 (see Ex. 12-11) in the flue gas. This is a Federal regulation and state regulations may be more stringent.
- Emissions of HCl—4 lb/hr or 99% control. RCRA regulations will probably change to risk based limits for HCl and chlorine.
- Carbon monoxide emissions—100 part per million by volume (ppmv) as a 60 minute rolling average corrected to 7% oxygen measured on a dry basis.
- Metals emissions—Emissions of ten priority metals must meet risk based guidelines.
- Destruction and removal efficiency—The incinerator must demonstrate its capability to achieve a 99.99% DRE on one or more selected Principal Organic Hazardous Constituents (POHCs) during a supervised Trial Burn.

DRE is defined as:

$$DRE = \frac{W_{in} - W_{out}}{W_{in}} \times 100 \tag{12-4}$$

where DRE = Destruction and Removal Efficiency (%)

W_{in} = Mass feed rate of a particular POHC (kg/hr or lb/hr)

W_{out} = Mass emission rate of the POHC (kg/hr or lb/hr)

Example 12-2 Performance standards. The mixture described below is being incinerated at 2,000°F with 50% excess air and a residence time of 2.1 seconds. Principal Organic
Hazardous Constituents (POHCs) for this waste are benzene, chlorobenzene, and toluene. The
flow rate of gas from the incinerator is measured at 12,500 dscfm (dry standard cubic feet per
minute). The O_2 concentration in the flue gas is 7.0%.

Compound	Formula	Mol. wt.	Inlet (lb/hr)	Outlet (lb/hr)
Benzene	C_6H_6	78.11	2015	0.537
Chlorobenzene	C_6H_5Cl	112.5	1150	0.109
Ethylbenzene	C_8H_{10}	106.17	2230	0.757
Toluene	C_7H_8	92.10	637	0.0 22
Xylene	C_8H_{10}	106.17	3040	1.25
Hydrochloric acid	HCl	36.45	0	10.7
Particulates				23.4

1. Calculate DRE for all of the organic compounds
2. Determine if this emission meets requirements for:
   (a.) POHCs
   (b.) Particulates
   (c.) HCl
3. Comment on the results

Solution.

1. For benzene, from Eq. 12-4, DRE= [(2015 − 0.537)/2015] × 100 = 99.9733
   DREs for the other compounds are shown in the table below.

Compound	Formula	Mol. wt.	Inlet (lb/hr)	Outlet (lb/hr)	DRE(%)	Number of chlorines	Molar feed rate (lb mol/hr)	Moles HCl (lb mol/hr)
Benzene	C_6H_6	78.11	2015	0.537	99.9733	0	25.80	0.00
Chlorobenzene	C_6H_5Cl	112.5	1150	0.109	99.9905	1	10.22	10.22
Ethylbenzene	C_8H_{10}	106.17	2230	0.757	99.9661	0	21.00	0.00
Toluene	C_7H_8	92.1	637	0.022	99.9965	0	6.92	0.00
Xylene	C_8H_{10}	106.17	3040	1.250	99.9589	0	28.63	0.00
					Removal efficiency (%)			
HCl		36.45		10.7	97.12829			
Particulates				23.4			Total inlet HCl (lb/hr)	

Particulates out 0.218 (grains/dry scf)

2. **(a.)** The DRE for benzene is less than 99.99, therefore, this emission does not meet the regulatory requirement for DRE.

   (b.) The concentration of particulates leaving the system may be calculated by dividing the outlet mass rate by the stack flow rate:

   $$\frac{(23.4 \text{ lb/hr}) \times (7000 \text{ grains/lb})}{\left(12,500 \text{ dry std ft}^3/\text{min}\right) \times (60 \text{ min/hr})} = 0.218 \text{ grains/dscf}$$

   (c.) Only one compound (chlorobenzene) contains chlorine. The inlet HCl can be determined by multiplying the inlet feed rate (1,150 lb/hr) by the molar ratio of hydrochloric acid to chlorobenzene (36.45/112.5) times the number of chlorine atoms in chlorobenzene (1)

   Inlet HCl = (1,150 lb/hr) × (36.45/112.5) × (1) = 372.6 lb/hr

   The removal efficiency for HCl is:

   [(372.6-10.7)/372.6] ×100 = 97.128%

   The HCl emissions fail the requirements on both bases: the total mass rate exceeds 4 lb/hr and the removal efficiency is less than the required 99%.

3. The emissions from this unit fail to meet any of the three regulations listed, and if these were the results of a test burn, this unit clearly would not pass. It is also interesting to note that the three POHCs listed in the problem do not include xylene, which has the largest mass rate leaving the unit.

12-3   COMBUSTION

Introduction

Combustion of hazardous wastes is not greatly different than the combustion of conventional fuel as discussed in the previous section except that the wastes may contain many different organic compounds. Every organic compound and every waste has a measurable heating value (expressed in Btu/lb) which can be determined experimentally by bomb calorimetry; however, the heating values of most common waste chemicals and mixtures are available from the literature[7] (see Tables 12-2 and App. C).

Excess Air

When organic wastes are burned with a stoichiometric amount of air (oxygen), the products of complete combustion should not include any oxygen. This is known as "perfect combustion," which is not possible in commercial burners or incinerators. Perfect combustion is an infinitely narrow line between excess air combustion on the one side and a deficiency of air (pyrolysis) on the other (see Fig. 12-1). Incinerators must always utilize excess air to achieve combustion; however, they may accomplish this in two stages—the first stage operating pyrolitically and the second with an excess of air. Excess air is also used in incinerators for temperature control because the excess air absorbs heat from the combustion reaction.

TABLE 12-2
Miscellaneous waste data

Waste	Btu value/lb. (as fired)	Weight lb/ft³ (loose)	Weight lb/ft³	Content by weight Ash %	Moisture %
Kerosene	18,900		50	0.5	0
Benzene	18,210		55	0.5	0
Toluene	18,440		52	0.5	0
Hydrogen	61,000		.0053	0	0
Acetic acid	6,280		65.8	0.5	0
Methyl alcohol	10,250		49.6	0	0
Ethyl alcohol	13,325		49.3	0	0
Turpentine	17,000		53.6	0	0
Naphtha	15,000		41.6	0	0
Newspaper	7,975	7		1.5	6
Brown paper	7,250	7		1.0	6
Magazines	5,250	35		22.5	5
Corrugated paper	7,040	7		5.0	5
Plastic coated paper	7,340	7		2.6	5
Coated milk cartons	11,330	5		1.0	3.5
Citrus rinds	1,700	40		0.75	75
Shoe leather	7,240	20		21.0	7.5
Butyl sole composition	10,900	25		30.0	1
Polyethylene	20,00	40–60	60	0	0
Polyurethane (foamed)	13,000	2	2	0	0
Latex	10,000	45	45	0	0
Rubber waste	9,000–11,000	62–125		20–30	
Carbon	14,093		138	0	0
Wax paraffin	18,621		54–57	0	0
1/3 wax–2/3 paper	11,500	7–10		3	1
Tar or asphalt	17,000	60		1	0
1/3 tar–2/3 paper	11,000	10–20		2	1
Wood sawdust (pine)	9,600	10–12		3	10
Wood sawdust	7,800–8,500	10–12		3	10
Wood bark (fir)	9,500	12–20		3	10
Wood bark	8,000–9,000	12–20		3	10
Corn cobs	8,000	10–15		3	5
Rags (silk or wool)	8,400–8,900	10–15		2	5
Rags (linen or cotton)	7,200	10–15		2	5
Animal fats	17,000	50–60			0
Cotton seed hulls	8,600	25–30		2	10
Coffee grounds	10,000	25–30		2	20
Linoleum scrap	11,000	70–100		20–30	1

The above chart shows the various Btu values of materials commonly encountered in incinerator designs. The values given are approximate and may vary based on their exact characteristics or moisture content.
Source: Incinerator Institute of America, New York, N.Y.

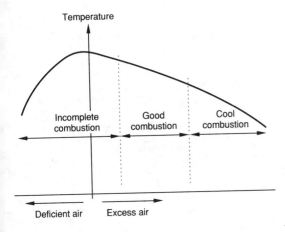

FIGURE 12-1
Excess air temperature curve.

Fuels

Organic wastes often have heating values high enough to support combustion. In such systems auxiliary fuel is needed only for ignition of the waste. In many hazardous waste incineration applications the waste heating value is low so conventional fuels must be used to bring the waste to a temperature where rapid oxidation of the organic fraction of the waste can occur.

The fuels used in an incineration system to provide auxiliary heat may be any commercially available fuel such as natural gas (methane), propane (LPG), light or heavy fuel oil, or possibly a waste fuel such as a spent solvent mixture. Combustion characteristics of some commercial fuels are shown in Table 12-3. Combustion constants for a range of substances are shown in Table 12-4.

There are hundreds of hydrocarbon waste mixtures which have sufficient heating value to be considered a fuel. Primary among these are solvent mixtures which have been used for process washdown or cleanup. Often these mixtures have heating values approaching those of commercial fuels so they may be used in a hazardous waste incinerator as a fuel. For example: consider a waste solvent mixture of two commercial solvents such as toluene and acetone. Toluene has a heating value of 18,440 Btu/lb,

TABLE 12-3
Analysis for common fuels[8]

	Weight fraction (lb element/lb fuel)		
Element	**Residual fuel oil (no. 6)**	**Distillate fuel oil (no. 2)**	**Natural gas**
C	0.866	0.872	0.693
H	0.102	0.123	0.227
N	—	—	0.08
S	0.03	0.005	—
Net heating value	19,500 Btu/lb	18,600 Btu/lb	21,301 Btu/lb

TABLE 12-4

Combustion constants

No. Substance	Formula	Molecular weight*	Lb. per cu ft.[b]	Cu ft per lb.[b]	Sp. gr. air- 1.000[b]	Heat of combustion Btu per cu ft Gross	Net[d]	Btu p[er lb] Gross
1 Carbon	C	12.01	—	—	—	—	—	14.093g
2 Hydrogen	H_2	2.016	0.005327	187.723	0.06959	325.0	275.0	61.100
3 Oxygen	O_2	32.000	0.08461	11.819	1.1053	—	—	—
4 Nitrogen (atm)	N_2	28.016	0.07439e	13.443e	0.9718e	—	—	—
5 Carbon monoxide	CO	28.01	0.07404	13.506	0.9672	321.8	321.8	4.347
6 Carbon dioxide	CO_2	44.01	0.1170	8.548	1.5282	—	—	—
Paraffin series C_nH_{2n+2}								
7 Methane	CH_4	16.041	0.04243	23.565	0.5543	1013.2	913.1	23,879
8 Ethane	C_2H_6	30.067	0.08029e	12.455e	1.04882e	1792	1641	22,320
9 Propane	C_4H_{10}	44.092	0.1196e	3.365e	1.5617e	2590	2385	21,661
10 n-Butane	C_4H_{10}	58.118	0.1582e	6.321e	2.06654e	3370	3113	21,308
11 Isobutane	C_5H_{12}	58.118	0.1582e	6.321e	2.06654e	3363	3105	21,257
12 n-Pentane	C_5H_{12}	72.144	0.1904e	5.252e	2.4872e	4016	37C9	21,091
13 Isopentane	C_5H_{12}	72.144	0.1904e	5.252e	2.4872e	4008	3716	21,052
14 Neopentane	C_5H_{12}	72.144	0.1904e	5.252e	2.4872e	3993	3693	20,970
15 n-Hexane	C_6H_{14}	86.169	0.2274e	4.398e	2.9704e	4762	4412	20,940
Olefin series C_nH_{2n}								
16 Ethylene	C_2H_4	28.051	0.07456	13.412	0.9740	1613.8	1513.2	21,644
17 Propylene	C_3H_6	42.077	0.1110e	9.007e	1.4504e	2336	2186	21,041
18 n-Bulene (Butylene)	C_4H_8	56.102	0.1480e	6.756e	1.9336e	3084	2885	20,840
19 Isobulene	C_4H_8	56.102	0.1480e	6.756e	1.9336e	3068	2869	20,730
20 n-Pentene	C_5H_{10}	70.128	0.1852e	5.400e	2.4190e	3836	3586	20,712
Aromatic series C_nH_{2n-6}								
21 Benzene	C_6H_6	78.107	0.2060e	4.852e	2.6920e	3751	3601	18,210
22 Toluene	C_7H_8	92.132	0.2431e	4.113e	3.1760e	4484	4284	18,440
23 Xylene	C_8H_{10}	106.158	0.2803e	3.567e	3.6618e	5230	4980	18,650
Miscellaneous gases								
24 Acetylene	C_2H_2	26.036	0.06971	14.344	0.9107	1499	1448	21,500
25 Naphthalene	$C_{10}H_8$	128.162	0.3384e	2.955e	4.4208e	5854f	5654f	17,298f
26 Methyl alcohol	CH_3OH	32.041	0.0846e	11.820e	1.1052e	867.9	768.0	10,259
27 Ethyl alcohol	C_2H_3OH	46.067	0.1216e	8.221e	1.5890e	1600.3	1450.5	13,161
28 Ammonia	NH_3	17.031	0.0456e	21.914e	0.5961e	441.1	365.1	9,668
29 Sulfur	S	32.06	—	—	—	—	—	3,983
30 Hydrogen sulfide	H_2S	34.076	0.09109e	10.979e	1.1898e	647	596	7,100
31 Sulfur dioxide	SO_2	64.06	0.1733	5.770	2.264	—	—	—
32 Water vapor	H_2O	18.016	0.04758e	21.017e	0.6215e	—	—	—
33 Air	—	28.9	0.07655	13.063	1.0000	—	—	—

All gas volumes corrected to 60°F and 30 in. Hg dry. For gases saturated with water at 60°F, 1.73% of the Btu va[lue] be deducted.

[a] Calculated from atomic weights given in "Journal of the American Chemical Society," February 1937.

[b] Densities calculated from values given in grams per liter at OC and 760 mm in the International Critical Tables all[ow] the known deviations from the gas laws. Where the coefficient of expansion was not available, the assumed value w[as] as 0.0037 per °C. Compare this with 0.003662 which is the coefficient for a perfect gas. Where no densities were a[vailable] the volume of the mol was taken as 22.4115 liters.

[c] Converted to mean Btu per lb (1/180 of the heat per lb of water from 32°F to 212°F) from data by Frederick D. National Bureau of Standards, Letter of April 10, 1937, except as noted.

[d] Deduction from gross to net heating value determined by deducting 18.919 Btu per pound mol of water in the pr[oducts of] combustion. Osborne, Stimson, and Ginnings, "Mechanical Engineering," p. 163, March 1935, and Osborne, Stim[son, and] Flock, National Bureau of Standards Research Paper 209.

ABLE 12-4

ombustion constants, cont.[†]

	Cu ft per cu ft of combustible						Lb per lb of combustible						Experimental error in heat of combustion percent
	Required for combustion			Flue products			Required for combustion			Flue products			
	O_2	N_2	Air	CO_2	H_2O	N_2	O_2	N_2	Air	CO_2	H_2O	N_2	1 or −
1	—	—	—	—	—	—	2.664	8.863	11.527	3.664	—	8.863	0. 012
2	0.5	1.882	2.382	—	1.0	1.882	7.937	26.407	34.344	—	8.937	26.407	0.015
3	—	—	—	—	—	—	—	—	—	—	—	—	—
4	—	—	—	—	—	—	—	—	—	—	—	—	—
5	0.5	1.882	2.382	1.0	—	1.882	0.571	1.900	2.471	1.571	—	1.900	0.045
6	—	—	—	—	—	—	—	—	—	—	—	—	—
7	2.0	7.528	9.528	1.0	2.0	7.528	3.990	13.275	17 .265	2.744	2.246	13.275	0.033
8	3.5	13.175	16.675	2.0	3.0	13.175	3.725	12.394	16.119	2.92 7	1.798	12.394	0.030
9	5.0	18.821	23.821	3.0	4.0	18.821	3.629	12.074	15.703	2.99 4	1.634	12.074	0.023
10	6.5	24.467	30.967	4.0	5.0	24.467	3.579	11.908	15.487	3.02 9	1.550	11.908	0.022
11	6.5	24.467	30.967	4.0	5.0	24.467	3.579	11.908	15.487	3.02 9	1.550	11.908	0.019
12	8.0	30.114	38.114	5.0	6.0	30.114	3.548	11.805	15.353	3.05 0	1.498	11.805	0.025
13	8.0	30.114	38.114	5.0	6.0	30.114	3.548	11.805	15.353	3.05 0	1.498	11.805	0.071
14	8.0	30.114	38.114	5.0	6.0	30.114	3.548	11.805	15.353	3.05 0	1.498	11.805	0.11
15	9.5	35.760	45.260	6.0	7.0	35.760	3.528	11.738	15.266	3.06 4	1.464	11.738	0.05
16	3.0	11.293	14.293	2.0	2.0	11.293	3.422	11.385	14.807	3.13 8	1.285	11.385	0.021
17	4.5	16.939	21.439	3.0	3.0	16.939	3.422	11.385	14.807	3.13 8	1.285	11.385	0.031
18	6.0	22.585	28.585	4.0	4.0	22.585	3.422	11.385	14.807	3.13 8	1.285	11.385	0.031
19	6.0	22.585	28.585	4.0	4.0	22.585	3.422	11.385	14.807	3.13 8	1.285	11.385	0.031
20	7.5	28.232	35.732	5.0	5.0	28.232	3.422	11.385	14.807	3.13 8	1.285	11.385	0.037
21	7.5	28.232	35.732	6.0	3.0	28.232	3.073	10.224	13.297	3.38 1	0.692	10.224	0.12
22	9.0	33.878	42.878	7.0	4.0	33.878	3.126	10.401	13.527	3.34 4	0.782	10.401	0.21
23	10.5	39.524	50.024	8.0	5.0	39.524	3.165	10.530	13.695	3.317	0.849	10.530	0.36
24	2.5	9.411	11.911	2.0	1.0	9.411	3.073	10.224	13.29 7	3.381	0.692	10.224	0.16
25	12.0	45.170	57.170	10.0	4.0	45.170	2.996	9.968	12.964	3.434	0.562	9.968	—
26	1.5	5.646	7.146	1.0	2.0	5.646	1.498	4.984	6.482	1.374	1.125	4.984	0.027
27	3.0	11.293	14.293	2.0	3.0	11.293	2.084	6.934	9.018	1.922	1.170	6.934	0.030
28	0.75	2.823	3.573	—	1.5	3.323	1.409	4.688 SO_2	6.097	—	1.587	5.511	0.080
29	—	— SO_2	—	—	—	—	0.998	3.287 SO_2	4.285	1.998	—	3.287	0.071
30	1.5	5.646	7.146	1.0	1.0	5.646	1.409	4.688	6.097	1.880	0.529	4.688	0.30
31	—	—	—	—	—	—	—	—	—	—	—	—	—
32	—	—	—	—	—	—	—	—	—	—	—	—	—
33	—	—	—	—	—	—	—	—	—	—	—	—	—

Denotes that either the density or the coefficient of expansion has been assumed. Some of the materials cannot exist as
ases at 60°F and 30 in. Hg pressure, in which case the values are theoretical ones given for ease of calculation of gas
roblems. Under the actual concentrations in which these materials are present their partial pressure is low enough to keep
hem as gases.
From Third Edition of "Combustion."
National Bureau of Standards, RP 1141.
Reprinted from "Fuel Flue Gases," 1941 Edition, courtesy of American Gas Association.
Source: B & W Steam Book

and acetone has a heating value of 13,119 Btu/lb. In a 50/50 weight mixture the combined heating value would be 15,686 Btu/lb which makes an excellent fuel.

Wastes Containing Sulfur, Halogens, Nitrogen, and Inorganics

Hydrocarbon combustion produces CO_2, and water vapor and possibly some CO. The combustion of wastes containing sulfur produces SO_2 and possibly SO_3. Halogen-containing wastes, (i.e., chlorine, fluorine, and bromine), produce the acid halogen gas of each in a combustion reaction—HCl, HF, and HBr. Each acid halogen gas is formed according to the equilibrium conditions present at the time of the combustion reaction. Typical reactions are:

$$CS_2 + 3O_2 \rightarrow CO_2 + 2SO_2 + HEAT \qquad (12\text{-}5)$$

$$CHCl_3 + O_2 \rightarrow CO_2 + HCl + Cl_2 + \ HEAT \cdot \qquad (12\text{-}6)$$

From the heating value of carbon disulfide (6236 Btu/lb) it can be readily seen that it will burn rapidly without the need for auxiliary fuel; however, the chloroform ($CHCl_3$) has a lower heating (1350 Btu/lb) value and will require auxiliary fuel to complete the reaction. In actual combustion situations, some chlorine (Cl_2) will always be formed. This equilibrium is demonstrated by the equilibrium curve shown in Fig. 12-2.

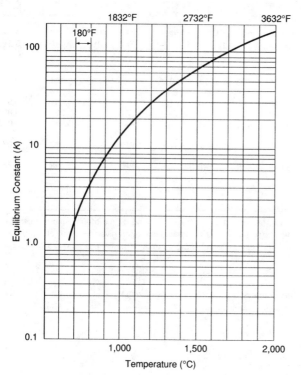

FIGURE 12-2
HCl/Cl$_2$ equilibrium.[9]

For the reaction

$$Cl_2 + H_2O \leftrightarrow 2HCl + 1/2O_2$$

the equilibrium constant is defined by:

$$K = \frac{[P_{HCl}]^2 [P_{O_2}]^{1/2}}{[P_{H_2O}][P_{Cl_2}]} \tag{12-7}$$

P denotes the partial pressures (mole fractions) of each reactant. As shown in Fig. 12-2, the value of K is approximately 1 at 670°C and rises to 100 at 1700°C.

Example 12-3 HCl/Cl$_2$ equilibrium. Chlorinated organics, when burned in air, will produce HCl gas and some chlorine gas in the combustion products. The amount of chlorine in the flue gas is proportional to the incineration temperature and the amount of excess air in the combustion process. It is desirable to maximize the HCl and minimize the Cl$_2$ in the flue gas because HCl is very soluble in water. If much chlorine is formed it will have to be removed with a caustic solution.

The reason that excess air affects the chlorine content is due to the following reaction:

$$2HCl + 1/2O_2 \rightarrow H_2O + Cl_2$$

The equilibrium constant (K_p) for this reaction can be expressed as follows:

$$K_p = \frac{(P_{HCl})^2(P_{O_2})^{1/2}}{(P_{H_2O})(P_{Cl_2})}$$

where P is the partial pressure of the gas in the mixture.

At 1200°C the K_p for the reaction is 30 while at 1000°C it is 20. When burning 200 lb/hr of monochlorobenzene in a liquid injection incinerator with 10 percent excess air, how much more chlorine will be in the flue gas at the lower temperature?

Solution. Monochlorobenzene: Formula: C_6H_5Cl MW = 112.5
Reaction: $C_6H_5Cl + 7O_2 = 6CO_2 + 2H_2O + HCl$

$$\frac{200 \text{ lb/hr}}{112.5/\text{mole}} = 1.78 \text{ moles of monochlorobenzene}$$

The theoretical oxygen required is $7 \times 1.78 = 12.46$ moles and the excess oxygen is $12.46 \times 1.1 = 13.7$ moles so the O$_2$ in the flue gas will be $13.7 - 12.46 = 1.24$ moles.

The molar flue gas composition will then be as follows:

Component			Moles	Mole %
Oxygen			1.24	1.8
Nitrogen	$13.7 \times 79/21$	=	51.50	74.9
CO$_2$	6×1.78	=	10.68	15.5
H$_2$O	2×1.78	=	3.56	5.2
HCl	1×1.78	=	1.78	2.6
Total			68.76	100.00

At 1 atmosphere the mole fraction is numerically the partial pressure, so if $K_p = 20$ and

$$P_{HCl} = 0.026; \qquad (0.026)^2 = 0.000676$$

$$P_{O_2} = 0.018; \qquad (0.018)^{1/2} = 0.134$$

$$P_{H_2O} = 0.52$$

$$P_{Cl_2} = ?$$

$$20 = \frac{(0.000676)(0.134)}{(0.52)(P_{Cl_2})}$$

$$P_{Cl_2} = 0.0000087 \text{ or } 8.7 \text{ ppm } @1000°C$$

If $K_p = 30$, then

$$P_{Cl_2} = 0.0000058 \text{ or } 5.8 \text{ ppm } @1200°C$$

The difference between operation at 1000°C and 1200°C is 2.9 ppm.

Nitrogen compounds in the fuel present a complication because they form various oxides of nitrogen during combustion, especially when excess oxygen is present. The oxides which are of primary concern are NO and NO_2. These two compounds are also formed in the combustion of fuels which do not contain nitrogen by a process which is called "nitrogen fixation." Nitrogen in the air used for combustion is "fixed" as a nitrogen oxide in the combustion process. "Fuel-bound nitrogen" is more easily formed into nitrogen oxides (NO_x) than oxides formed by fixation. NO_2 is not generally present in flue gases above 1200°F. While other oxides of nitrogen are present in the flue gas, NO predominates. Burning ammonia in air is a typical example of the combustion of a nitrogen compound:

$$NH_3 + 1/2O_2 \rightarrow NH_2 + OH^- \qquad (12\text{-}8)$$

The NH_2 then oxidizes or reduces as follows:
Oxidation:

$$NH_2 + O_2 \rightarrow HNO + OH \qquad (12\text{-}9)$$

or Reduction:

$$NH_2 + NO \rightarrow N_2 + H_2O \qquad (12\text{-}10)$$

The oxidation/reduction ratio is 4/1 when oxygen is present in the mixture so the products of equation 12-9 are most likely.

In the oxidation reaction, the HNO reacts with oxygen as follows:

$$HNO + 1/2O_2 \rightarrow NO + OH^- \qquad (12\text{-}11)$$

In actual practice ammonia is burned with an excess of natural gas or steam to provide a reducing atmosphere to reduce the formation of NO (see eq. 12-10).

Metals

Inorganic components of wastes fed to an incinerator cannot be destroyed, only oxidized. Most of the inorganic materials are chemically classified as metals, and enter the combustion process as a component of a waste. Generally, these metals will exit the combustion process as oxides of the metal that enters.

If the metal enters the process as a metal salt which has a boiling point lower than the incinerator temperature, it may vaporize and not oxidize and therefore be present in the flue gas. For example, lead chloride has a vaporization temperature of 950°C (1747°F).[10] Lead which enters an incinerator as a chloride may show up as lead in the flue gas unless it is condensed in the air pollution control equipment. Lead oxide, however, is not volatile and would stay in the bottom ash. Most metal compounds will remain in the incinerator ash, but the volatility of certain metals such as arsenic, antimony, cadmium, and mercury can create problems in the flue gas. High metal content wastes are not good candidates for incineration but appropriate air pollution control equipment can usually remove metals to acceptable flue gas levels for discharge to atmosphere.

Combustion Calculations

To perform basic combustion calculations it is necessary to apply the chemistry of the reaction (e.g., Eqs. 12-5 and 12-6). If the waste chemistry cannot be determined exactly, then a waste fuel analysis similar to that shown on Table 12-5 is required.

Good combustion requires that excess air be present. The percent of excess air may be as low as 5% in a good combustor or burner and as high as 100% in a rotary kiln incinerator to provide good mixing. Excess air may also be used to control the temperature when burning a high heating value waste (see Figs. 12-3, 12-4, and 12-5). Figure 12-3 demonstrates that the adiabatic combustion temperature for a fuel of any given heating value decreases as the excess air increases. Figure 12-4 similarly shows this effect for natural gas and also the effect of preheating combustion air. Figure 12-5 illustrates flame temperatures that can be achieved on various fuels at excess air percentages from 0 to 100%.

Combustion calculations are often based upon the change in the amount of heat content between the reactants and the products of combustion. This heat content is known as enthalpy. (See Sec. 3-3.) Table 12-6 shows the enthalpy of common flue gases and air.

A typical combustion problem is solved in Example 12-4.

TABLE 12-5
Waste fuel analysis

Type of waste	Heating Btu/lb	Volatiles* %	Moisture %	Ash %	Flash point COC °F	Fire point COC °F	Sulphur %	Dry combustible %	Density lbs/ft³
1. Coated fabric–rubber	10,996	81.20	1.04	21.20	265	270	0.79	78.80	23.9
2. Royalite	20,299	81.90	0.37	9.62	270	280	0.04	90.38	24.3
3. Coated felt–vinyl	11,054	80.87	1.50	11.39	165	170	0.80	88.61	10.7
4. Coated fabric–vinyl	8,899	81.06	1.48	6.33	155	175	0.02	93.67	10.1
5. Ensolite–expanded	10,216	88.90	0.35	9.99	265	300	0.32	90.01	5.7
6. Missile–rubber scrap	12,238	71.36	1.69	24.94	340	360	1.17	75.06	28.9
7. Fuel cell spray booth	12,325	79.10	1.74	20.67	125	130	0.25	79.33	9.5
8. Missile–rubber dust	9,761	62.36	0.87	36.42	250	260	1.06	63.58	9.9
9. Banbury–rubber scrap	13,242	60.51	1.74	4.18	145	180	0.53	95.82	34.9
10. Polyethylene film	19,161	99.02	0.15	1.49	180	200	0	98.51	5.7
11. Foam–cloth backed	10,185	55.06	0.37	31.87	215	235	1.44	68.13	11.3
12. Uppers–cloth	7,301	70.73	1.25	24.08	295	300	0.40	75.92	13.5
13. Foam–scrap	12,283	75.73	9.72	25.30	185	240	1.41	74.70	9.1
14. Tape–resin covered glass	7,907	15.08	0.51	56.73	300	330	0.02	43.27	9.5
15. Fabric–nylon	13,202	100.00	1.72	0.13	625	640	0	99.87	6.4
16. Fuel cell bladder and tire cord	15,227	87.35	1.24	3.57	270	290	0.55	96.43	19.5
17. Vinyl scrap	11,428	75.06	0.56	4.56	155	165	0.02	95.44	23.4
18. Liquid waste	13,140	100.00	3.2	1.04	68	68	0.07	95.76	53.0

NOTE: Test for Volatiles was run independently. The difficulty of getting two similar samples accounts for the higher percentage of Volatiles.
Source: Four Nines, Inc.

TABLE 12-6
Enthalpy of combustion gases (Btu/lb)

Temp °F	O₂	N₂	Air	CO	CO₂	SO₂	H₂	CH₄	H₂O*	HCl	HF	HBr	Temp °F
60	0.00	0.00	0.00	0.00	0.00	0.00	0.00	0.00	1060.00	0.00	0.00	0.00	60
100	8.80	9.90	9.60	10.00	8.00	5.90	137.00	21.00	1077.70	7.60	13.20	3.40	100
200	30.90	34.80	33.60	54.90	29.30	21.40	484.00	76.10	1122.40	26.70	45.20	12.00	200
300	53.30	59.90	57.70	59.90	52.00	37.50	832.00	136.40	1168.00	45.80	79.50	20.70	300
400	76.20	85.00	81.80	85.00	75.30	54.40	1182.00	202.10	1213.00	64.90	112.70	29.30	400
500	99.40	110.30	106.00	110.60	99.80	71.80	1532.00	272.60	1260.00	84.10	153.00	38.00	500
600	123.10	136.10	130.20	136.30	125.10	89.80	1882.00	347.80	1307.00	103.00	188.00	46.80	600
800	171.70	187.70	178.90	188.70	177.80	127.00	2584.00	511.20	1404.00	142.00	258.00	64.50	800
1000	221.70	240.70	235.00	242.70	233.60	165.50	3291.00	691.10	1504.00	182.00	329.00	82.90	1000
1200	272.50	294.70	288.50	297.80	290.90	205.10	4007.00	886.20	1608.00	222.00	400.00	101.00	1200
1400	324.30	350.80	343.00	354.30	349.70	245.40	4729.00	1094.00	1715.00	264.00	465.00	119.00	1400
1600	377.30	407.30	398.00	407.50	416.30	286.40	5460.00	1313.00	1827.00	302.00	538.00	138.00	1600
1800	430.70	465.00	455.00	465.30	470.90	327.80	6198.00	1542.00	1947.00	349.00	613.00	157.00	1800
2000	484.00	523.80	513.00	523.80	532.80	369.10	6952.00		2060.00	388.00	688.00	177.00	2000
2200	539.30	583.20	570.70	583.30	596.10	411.10	7717.00		2186.00	433.00	764.00	198.00	2200
2400	594.40	642.30	628.50	643.00	659.20	452.70	8490.00		2303.00	477.00	841.00	218.00	2400
2600	649.00	702.80	687.30	703.20	723.20	495.20	9272.00		2430.00	523.00	920.00	239.00	2600
2800	702.80	763.10	746.60	771.30	787.40	537.50	10060.00		2570.00	568.00	998.00	260.00	2800
3000	758.60	824.10	806.30	832.60	852.00	580.00	10870.00		2700.00	615.00	1085.00	283.00	3000
3200	816.40	885.80	866.00	894.00	916.70	622.50	11680.00		2830.00	665.00	1167.00	304.00	3200
3400	873.40	947.60	925.90	956.00	981.60	665.00	12510.00		2960.00	712.00	1249.00	326.00	3400
3600	931.00	1010.00	986.10	1018.00	1047.00	707.50	13330.00		3100.00	759.00	1331.00	347.00	3600
3800	988.00	1070.00	1046.00	1081.00	1112.00	760.00			3240.00	806.00	1409.00	369.00	3800
4000	1045.00	1132.00	1102.00	1138.00	1177.00	804.00			3380.00	854.00	1499.00	391.00	4000

* Water vapor enthalpy includes latent heat of vaporization.

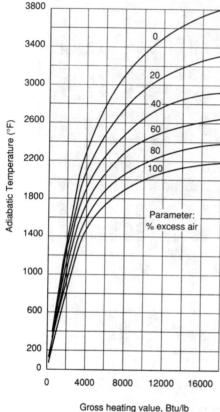

FIGURE 12-3
Adiabatic temperature vs. heating value at various % excess air.

Gross heating value, Btu/lb

Example 12-4   Enthalpy—quench. Many hazardous waste incinerators are followed by an "adiabatic quench" where the flue gases are cooled directly by water sprays to the adiabatic end point (i.e., the lowest temperature that can be achieved in single-stage cooling). Cooling is due to the evaporation of water, usually in a spray tower.

For simplicity, assume that the flue gases are essentially air at 2000°F and contain 20% water by weight. The total gas flow is 100,000 lb/hr.

1. What is the adiabatic saturation temperature?
2. What is the cooling water requirement in GPM?

Solution.

The enthalphy of dry air at 2000°F from Table 12-6 is 513 Btu/lb. The enhalpy of water vapor at the same temperature is 2060 Btu/lb.

$$\text{Heat in the dry air} = 80,000 \text{ lb/hr} \times 513 \text{ Btu/lb} = 41.04 \times 10^6 \text{ Btu/hr}$$
$$\text{Heat in the water vapor} = 20,000 \text{ lb/hr} \times 2060 \text{ Btu/lb} = \underline{41.20 \times 10^6 \text{ Btu/hr}}$$
$$\text{Total heat in the flue gas} = 82.24 \times 10^6 \text{ Btu/hr}$$

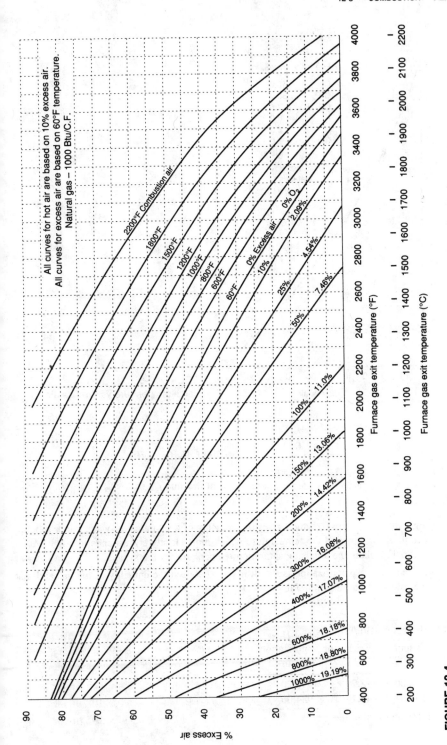

FIGURE 12-4
Excess air vs. temperature.[11]

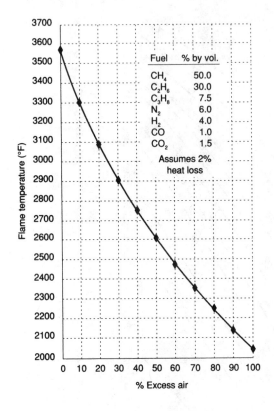

FIGURE 12-5
Flame temperature vs. excess air.

$$\frac{82,240,000}{80,000} = 1028 \text{ Btu/lb of dry flue gas.}$$

This is the enthalpy of the saturated mixture per lb of dry flue gas.

1. Using Table C-7 (App. C) which gives the enthalpy of a saturated air-water vapor mixture in Btu/lb of dry air, find the temperature which most closely corresponds to 1028 Btu/lb. It is 186°F. This is the adiabatic saturation temperature for the system.

2. The latent heat of vaporization is the difference between the enthalpy of the gas and that of the saturated liquid. From Table C-8 (App. C):

Temp	Enthalpy (Btu/lb)		Latent Heat of Vaporization (Btu/lb)
	Vapor	Liquid	
180°	148.00	1138.2	990.20
190°	158.04	1142.1	984.06

Thus, the latent heat of water at 186°F is 986.6 Btu/lb. The heat available to evaporate water in the quench is that contained in the dry flue gas, or 41 million Btu/hr

$$\frac{41,040,000 \text{ Btu/hr}}{986.6 \text{ Btu/lb water}} = 41,597 \text{ lb/hr} = \underline{\underline{83 \text{ gpm of water}}}$$

12-4 GASES AND VAPORS

Even though waste gases and vapors are not covered by federal RCRA regulations because RCRA essentially regulates liquid and solid hazardous wastes, there are a number of thermal processes which are of general interest and should be mentioned.

Waste gases usually consist of a hydrocarbon or a mixture of hydrocarbons in air. They may be in such high concentrations that ignition is not possible until more air is added. Such mixtures are said to be above the higher explosive limit (HEL). Mixtures of hydrocarbon vapors in air that can be ignited are said to be within the explosive range. Mixtures which have such low hydrocarbon concentrations that they cannot be ignited are said to be below the lower explosive limit (LEL). Most hydrocarbon air mixtures are of this type (see Table 12-7).

Example 12-5 Excess air calculations. A liquid injection incinerator has a stack gas which contains 7% oxygen by volume on a wet basis at standard conditions. The incinerator is burning toluene at a rate of 184 lb/hr with air.

a.) What percent excess air is required?

b.) If the stack gas had been on a dry basis, what would the excess air be?

c.) What is the combustion efficiency of the system if the CO content of the flue gas is 500 ppmv?

Solution.

a.) There is no other source of oxygen in the system except the combustion air. Toluene, $C_6H_5CH_3$ has a molecular weight of 92 and a heating value of 18,252 Btu/lb.

$$C_6H_5CH_3 + 9O_2 \rightarrow 7CO_2 + 4H_2O$$

$$\frac{184 \text{ lb/hr}}{92 \text{ lb/lb} \cdot \text{mole}} = 2 \text{ lb moles toluene hr}$$

This requires 18 moles of O_2 or $100/21 \times 18 = 85.71$ moles air on an hourly basis. This is the stoichiometric amount of air required to burn the toluene. Two moles of toluene will produce 14 moles of CO_2 and 8 moles H_2O plus $79/21 \times 18 = 67.7$ moles of N_2. The total moles of flue gas at theoretical air requirements = 14 moles + 8 moles + 67.7 moles = 89.7 moles.

When burning with an excess of theoretical air more nitrogen and some oxygen will be present in the flue gas. Oxygen is present at 7% by volume = 7 mole %

TABLE 12-7

Combustibility characteristics of pure gases and vapors in air

Gas or vapor	Lower limit % by volume	Upper limit % by volume	Closed cup flash point°F
Acetaldehyde	4.0	57	−17
Acetone	2.5	12.8	0
Acetylene	2.5	80	—
Allyl alcohol	2.5	—	70
Ammonia	15.5	26.6	—
Amyl acetate	1.0	7.5	77
Amylene	1.6	7.7	—
Benzene (benzol)	1.3	6.8	12
Benzlyl chloride	1.1	—	140
Butene	1.8	8.4	—
Butyl acetate	1.4	15.0	84
Butyl alcohol	1.7	—	—
Butyl cellosolve	—	—	141
Carbon disulfide	1.2	50	−22
Carbon monoxide	12.5	74.2	—
Chlorobenzene	1.3	7.1	90
Cottonseed oil	—	—	486
Cresol m- or p-	1.1	—	202
Crotonaldehyde	2.1	15.5	55
Cyclohexane	1.3	8.4	1
Cyclohexanone	1.1	—	111
Cyclopropane	2.4	10.5	—
Cymene	0.7	—	117
Dichlorobenzene	2.2	9.2	151
Dichloroethylene (1,2)	9.7	12.8	57
Diethyl selenide	2.5	—	57
Dimethyl formamide	2.2	—	136
Dioxane	2.0	22.2	54
Ethane	3.1	15.5	—
Ether (diethyl)	1.8	36.5	−49
Ethyl acetate	2.2	11.5	28
Ethyl alcohol	3.3	19.0	54
Ethyl bromide	6.7	11.3	—
Ethyl cellosolve	2.6	15.7	104
Ethyl chloride	4.0	14.8	−58
Ethyl ether	1.9	48	−49
Ethyl lactate	1.5	—	115
Ethylene	2.7	28.6	—
Ethylene dichloride	6.2	15.9	56
Ethyl formate	2.7	16.5	−4
Ethyl nitrite	3.0	50	−31
Ethylene oxide	3.0	80	—
Furfural	2.1	—	140
Gasoline (variable)	1.4–1.5	7.4–7.6	−50
Heptane	1.0	6.0	25
Hexane	1.2	6.9	−15

TABLE 12-7
continued

Gas or vapor	Lower limit % by volume	Upper limit % by volume	Closed cup flash point°F
Hydrogen cyanide	5.6	40.0	—
Hydrogen	4.0	74.2	—
Hydrogen sulfide	4.3	45.5	—
Illuminating gas (coal gas)	5.3	33.0	—
Isobutyl alcohol	1.7	—	82
Isopentane	1.3	—	—
Isopropyl acetate	1.8	7.8	43
Isopropyl alcohol	2.0	—	53
Kerosene	0.7	5	100
Linseed oil	—	—	432
Methane	5.0	15.0	—
Methyl acetate	3.1	15.5	14
Methyl alcohol	6.7	36.5	52
Methyl bromide	13.5	14.5	—
Methyl butyl ketone	1.2	8.0	—
Methyl chloride	8.2	18.7	—
Methyl cyclohexane	1.1	—	25
Methyl ether	3.4	18	—
Methyl ethyl ether	2.0	10.1	−35
Methyl ether ketone	1.8	9.5	30
Methyl formate	5.0	22.7	−2
Methyl propyl ketone	1.5	8.2	—
Mineral spirits No. 10	0.8	—	104
Naphthalene	0.9	—	176
Nitrobenzene	1.8	—	190
Nitroethane	4.0	—	87
Nitromethane	7.3	—	95
Nonane	0.83	2.9	88
Octane	0.95	3.2	56
Paraldehyde	1.3	—	—
Paraffin oil	—	—	444
Pentane	1.4	7.8	—
Propane	2.1	10.1	—
Propyl acetate	1.8	8.0	58
Propyl alcohol	2.1	13.5	59
Propylene	2.0	11.1	—
Propylene dichloride	3.4	14.5	60
Propylene oxide	2.0	22.0	—
Pyridine	1.8	12.4	74
Rosin Oil	—	—	266
Toluene (toluol)	1.3	7.0	40
Turpentine	0.8	—	95
Vinyl ether	1.7	27.0	—
Vinyl chloride	4.0	21.7	—
Water gas (variable)	6.0	70	—
Xylene (xylol)	1.0	6.0	63

The total flue gas = 89.7 + moles N_2 + moles O_2
Let X be the moles of O_2 in the flue gas:
$X = 0.07(89.7 + \text{moles } N_2 + X)$
Moles of $N_2 = 79/21 \times X$

Therefore: $\qquad\qquad X = 0.07(89.7 + 79/21X + X)$

$$X = 9.42 \text{ moles } O_2$$

Additional moles of N_2 added with the excess air = $79/21 \times 9.42 = 35.5$
Additional moles of air = $100/21 \times 9.42 = 44.86$
Total air in system = $85.71 + 44.86 = 130.57$
Excess air = excess/theoretical = $\dfrac{44.86}{85.71} \times 100 = \underline{52.34\%}$

b.) From part a.):
On a dry basis the theoretical moles of flue gas must be $89.7 - 8$ moles of water = 81.7 moles.
If X = moles of O_2 in the flue gas, then:

$$X = 0.07 \times (81.7 + 79/21X + X)$$

$X = 8.59$ moles of O_2
Moles $N^2 = 79/21 \times 8.59 = 32.3$ moles
Moles air = $100/21 \times 8.59 = 40.9$
% excess air = $40.9/85.71 \times 100 = \underline{47.7\%}$
The total flue gas will now be 81.7.

c.) 500 parts/million is equivalent to 0.05% by volume CO.
Combustion efficiency % = $\dfrac{CO_2 - CO}{CO_2} \times 100$

$$\text{Combustion efficiency} = \dfrac{14 - 0.05}{14} \times 100 = \underline{99.64\%}$$

Example 12-6 Explosive limits. A film drying system uses hot air to evaporate acetone. The acetone-air vapor mixture is transported in a duct to a thermal type VOC incinerator. Acetone has the following physical and chemical properties:

Molecular weight	54
Formula	CH_3COCH_3
Boiling point	56°C
Heating value	13,120Btu/lb
Auto ignition temperature	537°C
Lower explosive limit	2.5% by volume

Lower explosive limit   2.5% by volume

Higher explosive limit   12.8% by volume

The flow from the dryer is 1000 actual ft^3/min (acfm) at 150°F and 216 lb/hr of acetone are evaporated. The molal volume = 379 ft^3/mole at 60°F.

a.) Determine the concentration of the acetone in the duct.

b.) Is this a dangerous situation? Why?

c.) Insurance regulations require that flammable vapor concentration in ducts be maintained at 25% of the LEL or less. How can the problem be solved?

Solution.

a.) $\dfrac{216\,lb/hr}{54\ lb/mole} = 4$ moles/hr acetone

The molal volume at 150° is:

$$379 \times \frac{460 + 150}{460 + 60} = 445 \ ft^3/\text{mole at } 150°F$$

$$4 \text{ moles/hr} \times 445 \ ft^3/\text{mole} = 1780 \ ft^3/\text{hr of acetone}$$

The acetone is in 1000 acfm or 60,000 acfh total flow; therefore:

$$\frac{1,780}{60,000} = 0.0297 \text{ or } \underline{\underline{2.97 \text{ \% by volume}}}$$

b.) It is dangerous because the acetone concentration in air is above the lower explosive limit (LEL), and it is below the higher explosive limit (HEL). This means that the concentration is in the explosive range and any source of ignition could cause a fire or explosion in the system.

c.) The insurance limit is $0.25 \times 2.5\%$ or 0.625%

2.97/0.625 = 4.75 times higher than required; therefore:

4.75 times the flow is needed to reduce the acetone concentration to this level, or 4,750 acfm. This can be accomplished by adding 3,750 acfm of air after the dryer.

When volatile organic compounds (VOCs) are emitted to the atmosphere, the VOCs may present a problem regarding toxicity, odor, or in the formation of photochemical smog. Most states have regulations that control the emission levels of VOCs, and a states' definitions of what constitutes a VOC vary. There are no explicit federal regulations at present.

This kind of emission is normally extremely dilute. The VOCs are mixed with air or occasionally a carrier gas such as nitrogen. The compounds come from coating and painting operations which employ volatile solvents as a vehicle to carry the coating, and may also result from the venting of storage tanks or from leaks in pumps valves and flanges. High VOC concentrations (over 10,000 parts per million in air, or 1% by volume) are rare. Normal emission levels from most sources are lower. VOC emissions, or gaseous waste streams with high concentrations which will support combustion, are usually burned in burners or flares or boilers. Waste streams with low VOC concentrations may be recovered by condensation or carbon adsorption or burned in a VOC incinerator.

Flares

Flares are used for waste gases that are above the HEL and may be mixed with air, ignited, and burned cleanly. Unfortunately, few waste gases will burn cleanly under these conditions so a steam injection method was conceived to provide for clean burning by providing mixing energy at the flare tip (see Fig. 12-6). Flares can be elevated, i.e., burning from the tip of a flare stack, commonly seen in a refinery, or can be enclosed at ground level and burned within the enclosure. Flares are used primarily to dispose of combustible gases during a process upset condition. Regular emissions of such combustible waste gases could be utilized to recover heat.

Catalytic VOC Incinerators

Low concentrations of VOC in air can be effectively removed by catalytic incineration. The VOC-air stream is heated by direct contact with products of combustion from a fuel burner to a temperature at which the VOC will begin to burn on the surface of the catalyst. This preheat temperature is determined by the VOC, the catalyst employed, and the VOC concentration in the air. The catalyst enhances the combustion reaction and causes oxidation on the catalyst surface with very little heat loss. Therefore, catalytic incineration of VOC-air mixtures usually occurs at lower temperatures than thermal incineration and less auxiliary fuel is used.

The catalyst used for combustion of VOCs is usually a noble metal, e.g., platinum or palladium, although other catalysts are being employed today. Alumina is the most common catalyst support material because it will withstand high temperatures and is easily formed into a variety of configurations. The most popular shape for catalyst support in incinerators has become the honeycomb. The noble metal in a soluble form, often the nitrate, is dried onto the support surface to provide the maximum catalyst surface area and a low air flow pressure drop.

Catalytic incinerators provide excellent destruction of most VOCs; however, care must be exercised in their application. A sudden increase in the VOC content of the VOC-air mixture to the incinerator may cause a temperature rise which will destroy the catalyst and support; therefore, it is best applied to waste gas streams where the VOC content is constant.

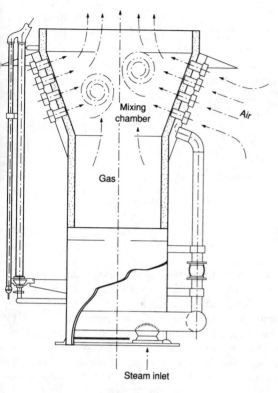

FIGURE 12-6
Flare tip.[11]

Catalysts are also subject to poisoning (which lowers their effectiveness) by a variety of elements and compounds such as chlorine, sulfur, and vanadium.

Heat recovery is a normal part of most catalytic systems. The incoming VOC-air mixture is preheated by the gases exiting the catalyst.

Thermal VOC Incinerators

The thermal incinerator for VOC-air mixtures is a catalytic incinerator without the catalyst. This usually means that the preheater must heat the incoming mixture to the temperature where the VOC will be oxidized. Most VOCs will be removed to desirable levels at exit gas temperatures of 815°C (1500°F).

The thermal incinerator consists of a fuel burner firing into a chamber where the VOC-air mixture can be adequately mixed with the burner combustion products and achieve the desired mixture temperature for at least 1/2 second. Heat recovery from these systems is usually effected by preheating the cold incoming VOC-air stream with the hot effluent gas (see Fig. 12-7).

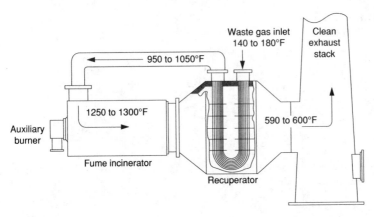

Waste gas inlet
140 to 180°F

Clean
exhaust
stack

950 to 1050°F

1250 to 1300°F

590 to 600°F

Auxiliary
burner

Fume incinerator

Recuperator

FIGURE 12-7
Thermal VOC incinerator.[13]

Example 12-7 VOC incinerator with heat recovery. A steel strip coater/drier emits 1500 ppmv of toluene vapor in a total air volume of 10,000 acfm at 230°F. Design a VOC thermal oxidizer with a heat recuperator schematically similar to the system shown in Fig. 12-7.

The fumes enter the tubular recuperator at 230°F where they are preheated to 900°F by the products of combustion from the incinerator. Natural gas is used as the auxiliary fuel and the preheated fumes are used as combustion air. The thermal oxidizer must operate at an outlet temperature of 1600°F to achieve the required destruction efficiency of toluene. Assume that the natural gas is methane (CH_4) with a heating value of 1000 Btu/ft^3 or 23,875 Btu/lb. Toluene ($C_6H_5CH_3$) has a heating value of 18,252 Btu/lb and a molecular weight of 92 lb/mole.

1. How much natural gas is required assuming a 5% heat loss?
2. What is the stack temperature?
3. To provide a 1/2-second residence time in the incineration chamber, what should its dimensions be? Assume a gas velocity of 20 ft/sec and a residence time of 0.5 sec.

Solution.

1. Determine the flow rate and mass rate:
   1500 ppmv $= 0.15\%$ by volume
   $0.0015 \times 10,000 = 15$ cubic ft of toluene in the mixture but this is at 230°F. At 60°F:

$$15 \times \frac{460 + 60}{460 + 230} = 11.3 \text{ cubic ft/minute}$$

The molal volume at 60°F is 379 cubic ft/mole.

$$\text{Molar flow rate of toluene} = \frac{11.3 \times 60}{379} = 1.79 \text{ moles/hr}$$

$$\text{Mass rate of toluene} = 1.79 \times 92 = 164.7 \text{ lb/hr}$$

$$\text{Heat release from toluene} = 164.7 \text{ lb/hr} \times 18,252 \text{ Btu/lb}$$

$$= 3,006,000 \text{ Btu/hr}$$

The total air flow to the system is

$$10,000 \text{ acfm} \times \frac{460 + 60}{460 + 230} = 7536 \text{ scfm}$$

From Table 12-3, the density of air is 0.076 lb/ft^3

$$7536 \text{ scfm} \times 0.076 \text{ lb/ft}^3 \times 60 \text{ min/hr} = 34,360 \text{ lb/hr}$$

$$\frac{34,360 \text{ lb/hr}}{29 \text{ lb/mole}} = 1185 \text{ moles/hr of air}$$

In order to heat air from 900°F to 1600°F, the heat input must be the difference in enthalpy for air at these temperatures:

At 1600°F the enthalpy $= 398$ Btu/lb (table 12-6)

At 900°F the enthalpy $= 207$ Btu/lb

The enthalpy difference $= 191$ Btu/lb

The total heat required is 191 Btu/lb $\times$ 34, 360 lb/hr $= 6,564,000$ Btu/hr.

Since there is a heat loss of 5% and the toluene contributes 3,006,000 Btu/hr, the natural gas requirement is:

$$1.05(6,564,000) - 3,006,000 = 3,885,000 \text{ Btu/hr}$$

$$\frac{3,885,000 \text{ Btu/hr}}{1000 \text{ Btu/ft}^3} = \underline{\underline{3885 \text{ ft}^3/\text{hr of natural gas}}} \text{ or}$$

$$\frac{3,885,000 \text{ Btu/hr}}{23,875 \text{ Btu/lb}} = \underline{\underline{162.72 \text{ lb/hr of natural gas}}}$$

2. The incinerator will burn both natural gas and toluene.

Natural gas as CH_4: $CH_4 + 2O_2 \rightarrow CO_2 + 2H_2O$

$$\frac{162.72 \text{ lb/hr}}{16 \text{ lb/mole}} = 10.17 \text{ moles of natural gas}$$

The combustion products will be

$$CO_2 = 10.17 \text{ moles}$$

$$H_2O = 20.34 \text{ moles}$$

$$N_2 = 76.52 \text{ moles}$$

Toluene: $C_6H_5CH_3 + 9O_2 \rightarrow 7CO_2 + 4H_2O$

$$\frac{164.7 \text{ lb/hr}}{92 \text{ lb/mole}} = 1.79 \text{ moles toluene}$$

The combustion products will be:

$$CO_2 = 12.53 \text{ moles}$$

$$H_2O = 7.16 \text{ moles}$$

$$N_2 = 60.60 \text{ moles}$$

The total moles of O_2 used for both natural gas and toluene are

$$(10.17 \times 2) + (1.79 \times 9) = 36.45 \text{ moles}$$

This is equivalent to $36.45 \times 100/21 = 173.6$ moles of air.

Since the inlet to the incinerator was 1185 moles of air and 173.6 moles were used in the combustion process this leaves 1011.4 moles of air at the outlet plus the sum of the products above from the combustion of the natural gas and toluene. The outlet gas from the incinerator is made up as follows:

Air	1011.4 moles	$\times$ 29 lb/mole =	29,331 lb
N_2	137.12 moles	$\times$ 28 lb/mole =	3,839 lb
CO_2	22.70 moles	$\times$ 44 lb/mole =	999 lb
H_2O	27.50 moles	$\times$ 18 lb/mole =	495 lb
	1160.48 moles		34,664 lb

The heat needed in the recuperator is the difference in enthalpy between 900°F and 230°F or $(207 - 41) = 166$ Btu/lb. This must equal the heat given up by the flue gas entering the recuperator at 1600°F. The weight fraction of each component in the flue gas times its enthalpy at 1600°F, from Table 12-6, will give the flue gas enthalpy:

Component	Wt. Fraction	Enthalpy, Btu/lb	Weighted Enthalpy
Air	0.846	398	337
N_2	0.111	407.3	45
CO_2	0.029	416.3	12
H_2O	0.014	1827.0	26
Enthalpy of flue gas =			420 Btu/lb

Let the heat in the stack gas be X.

If heat in = heat out in the recuperator then

$$(420 \times 34,664) - X = (166.12 \times 34,364)$$

$$X = 8,850,000 \text{ Btu into the stack}$$

$$\frac{8,850,000 \text{ Btu}}{34,664 \text{ lb/hr}} = 255 \text{ Btu/lb of flue gas in stack}$$

By trial and error the temperature of the stack can be estimated. Assume a stack temperature of 1000°F and use the enthalpies from Table 12-6.

Component	Wt. Fraction	Enthalpy, Btu/lb	Weighted Enthalpy
Air	0.846	235	199
N_2	0.111	241	27
CO_2	0.029	243	7
H_2O	0.014	1504	21
Enthalpy of flue gas =			254 Btu/lb

Since this almost matches the 255 Btu/lb calculated above, the stack temperature is approximately <u>1000°F.</u>

3. The products of combustion from the incinerator total 1160.5 moles/hr, which is approximately:

$$1160.5 \text{ moles/hr} \times 379 \text{ ft}^3/\text{mole} \times 1 \text{ hr}/60 \text{ min.} = 7330 \text{ scfm}$$

The average temperature in the incinerator is

$$\frac{1600 + 900}{2} = 1250°F$$

The average gas flow at 1250°F is:

$$\frac{460 + 1250}{460 + 60} \times 7330 = 24,104 \text{ acfm}$$

The velocity in the incinerator is:

$$20 \text{ ft/sec} = 1200 \text{ ft/min}$$

The area of a chamber is:

$$24,104 \text{ ft}^3/\text{min} \div 1200 \text{ ft/min} = 20.75 \text{ ft}^2$$

A cross sectional area of 20.75 has a diameter of 5.14 ft, say 5 ft diameter. The length of the chamber is:

$$20 \text{ ft/sec} \times 0.5 \text{ sec} = 10 \text{ ft}$$

Overall dimensions: <u>5.0 ft diameter × 10 ft length.</u>

Regenerative Thermal VOC Systems

Regenerative thermal units offer higher fuel efficiencies than either the catalytic or thermal systems but they are larger and usually more expensive in first cost. These units employ a "heat sink"—a chamber filled with heat-absorbing and heat-retaining ceramic materials. The VOC-air mixture passes into a thermal incineration chamber where mixing with the products of combustion from a fuel burner raises the mixture to 1500°F. The hot effluent gases from the incineration chamber pass through the ceramic filled chamber on their way to the stack, preheating this chamber to a high temperature. Valves in the VOC-air inlet line then re-direct the mixture through the heated ceramic chamber before it reaches the combustion chamber effecting significant heat recovery. Regenerative units can achieve 90% heat recovery. These systems are sensitive to high VOC content in the incoming mixture but can solve this problem in bypassing some of the inlet gases directly to the combustion chamber (see Fig. 12-8).

12-5 LIQUID INJECTION INCINERATORS

Introduction

There is more experience with liquid injection (L.I.) incinerators than all other types combined for the incineration of hazardous waste. The greatest proportion of hazardous waste incinerators in operation today are the L.I. type. When the systems covered by the Boiler and Industrial Furnace regulations (BIF) are included, the predominance is even greater. Various types of incinerators burn liquid wastes as well as solids, sludges, and slurries. This section discusses the liquid injection type, which as the name implies is applicable, almost exclusively, to pumpable liquid waste. The waste is burned directly in a burner (combustor) or injected into the flame zone or combustion zone of the incinerator chamber (furnace) through atomizing nozzles. The heating value of the waste is the primary determining factor for the location of the injection point.

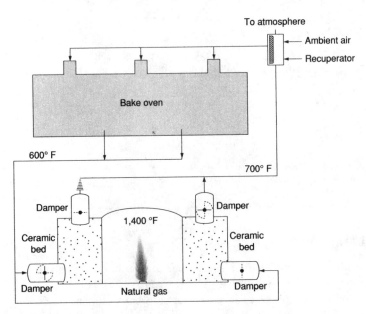

FIGURE 12-8
Regenerative VOC incinerator.[14]

Liquid injection incinerators are usually refractory-lined chambers (horizontal flow or vertical flow either up or down), generally cylindrical in cross section, and equipped with a primary burner (waste and/or auxiliary fuel fired). Often secondary combustors or injection nozzles are required where low heating value materials such as dilute aqueous-organic wastes are to be incinerated.

Liquid incinerators operate at temperature levels from 1,000°C (1,832°F) to 1,700°C (3,092°F). Residence time for the combustion products in the incinerator may vary from milliseconds to as much as 2.5 seconds. The atomizing nozzle in the burner or the incinerator is a critical part of the system because it converts the liquid waste into fine droplets.

The viscosity of the waste determines whether good atomization of a liquid is possible. Two-fluid atomizers, using compressed air or steam as an atomizing fluid, are capable of atomizing liquids with viscosities up to 70 centistokes (2.7 ft²/hr).

When considering the use of a liquid injection incinerator to burn a waste, the ability of the entire system and all of its auxiliary components to meet the federal and state regulations is paramount.

Determination of Waste Properties

The physical, chemical, and thermodynamic properties of the waste must be considered in the basic design of any incinerator system.

Any commercial facility receiving hazardous waste requires a complete analytical laboratory on site. Most commercial operators require a sample of the waste before they will provide a treatment cost to the generator. This sample is then analyzed for the parameters shown below, a treatment and handling regimen is developed and a price for treatment and unit disposal is provided to the generator. If the generator accepts this price then a contract is developed which sets forth the responsibilities of both parties and certain physical and chemical limits for all waste sent to the disposal facility under that contract.

When each shipment of the waste material is received at the facility, the contractor will perform a cursory analysis known as a "fingerprint" analysis to determine that the waste meets the criteria of the contract. For example: if an initial sample of a liquid waste to be treated by incineration indicates a high heating value, say 10,000 Btu/lb, the contractor might offer the generator an attractive disposal price because it will reduce the amount of auxiliary fuel needed in the incinerator. If the actual waste received for disposal has a heating value of only 2000 Btu/lb then the generator has misrepresented the waste and the contractor needs to charge a higher disposal cost. This "fingerprint" is a heating value which can be determined by bomb calorimetry.

Today's commercial laboratories include sophisticated analytical instrumentation such as gas chromatographs coupled with mass spectrometers (GC/MS) which can quickly analyze for thousands of compounds, atomic absorption spectrophotometers for the analysis of metals and many other devices for rapid characterization of waste.

The following information on the waste to be burned in the incinerator is required to design and properly engineer the total system and the auxiliary components.

1. Chemical composition
2. Heat of combustion
3. Viscosity
4. Corrosivity
5. Reactivity
6. Potential for polymerization
7. Ash content
8. Ash fusion temperature

The atomizer design in the burner or incinerator usually dictates the pressure requirements of the transport system. Where a hydraulic (pressure-type) atomizing nozzle is used, requiring high pressures, a pump design with close clearances between the impeller and the housing is needed. High viscosity materials and liquids containing solids will erode the pump surface causing excessive wear and rapid deterioration of pump pressure resulting in a reduction of flow as well as poor atomization. Waste transport is critical in the proper design of any incinerator.

Some liquid wastes and sludges must be maintained at above ambient temperatures to permit pumping. When cooled, these wastes will solidify in the pipeline. Conversely, if these wastes are heated too much, they may polymerize. Once polymerized, it may be impossible to re-liquify the material. With heated liquids, pump bearings may become overheated or centrifugal pumps may cavitate, so pump selection is critical.

Atomizers

The method of injection of the liquid into the burner or incinerator furnace is one of the most important features of a well-designed system. The reasons for injecting the liquid as a fine spray are:

1. To break up the liquid into fine droplets;
2. To develop the desired pattern for the liquid droplets in the combustion zone with sufficient penetration and kinetic energy; and
3. To control the rate of flow of the liquid discharged to the combustion system.

Organic and aqueous liquids pass through three phases before oxidation takes place. (Hydrocarbons ignite at temperatures as low as 20°C and as high as 650°C.) The liquid droplets are heated, vaporized, and superheated to ignition temperature. These droplets must receive heat by radiation and convection from the furnace as rapidly as is practical. At the same time, they must also be in intimate contact with oxygen from the combustion air. If the droplet diameter is large, fewer droplets will be produced per unit flow of waste, and the total surface available for heat transfer is small. In a good atomizer the droplet size will be small providing greater surface area and resulting in rapid vaporization. Note the burning time for a 300-

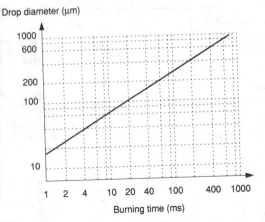

FIGURE 12-9
Burning time for atomized droplets.

micron droplet is 150 milliseconds while only 30 milliseconds for a 125-micron droplet (Fig. 12-9).

There are three basic types of atomizers for liquid wastes. The first is the mechanical or pressure atomizer that is a small orifice through which high-pressure liquid expands as it goes from a high pressure to the low pressure of the combustion chamber. The higher the pressure to the nozzle the better the atomization. Droplet size at full flow is usually on the order of 100 to 150 microns, and the atomization deteriorates as the flow is decreased. The mechanical atomizing nozzle is simple in construction, subject to fouling and plugging by small particles, and has poor turndown characteristics (i.e., atomization deteriorates rapidly below 50% flow capacity).

The second type of atomizer is the two fluid internal mix atomizer where steam or compressed air acts as the second fluid and the waste or fuel as the first. The two fluids are mixed within the atomizer, and the energy of the steam or compressed air achieves the atomization. Droplets down to 50 microns can result from good internal mix nozzle design. This nozzle still has small orifices so it should not be selected for a "dirty" liquid waste. It is an excellent atomizer with good turndown (see Fig. 12-10).

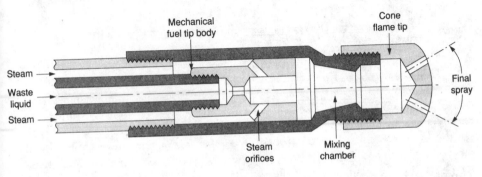

FIGURE 12-10
Internal mix nozzle.[15]

Because most wastes contain solid particles, the two fluid external mix atomizing nozzle is most widely used. Here the atomizing fluid and the waste mix outside the inner nozzle. The atomization may not be as good as in an internal mix nozzle, but because there are no small openings it does not foul or plug easily. In fact, particles as large as 1/4 inch may be tolerated. It is significantly better than the mechanical nozzle in atomization ability (see Fig. 12-11).

With high-viscosity materials, mechanical nozzle orifices have to be made larger to minimize pressure drop, erosion, and blockage. Atomizers must be designed to cause a shearing action of the liquid to break it into many smaller diameter particles. Most atomizers for viscous liquids use a two-fluid nozzle.

Combustion Air Systems

Proper mixing of combustion air with the atomized liquid droplets is very important. As the liquid is vaporized and superheated to ignition temperature, oxygen reacts with the hydrocarbon vapor to produce combustion. As this occurs, there is a sudden rise in temperature increasing the velocity of the gases in the area surrounding the droplet which causes greater mixing and completes the reaction.

As the viscosity of the liquid increases, droplet size tends to get larger. To completely vaporize and superheat the droplet, more time is needed. Increased turbulence created by high-intensity burners (Fig. 12-12) permits this reaction to be achieved rapidly. Greater energy, imparted to the combustion air by a fan or blower, provides higher velocities in the combustion zone improving the mixing of the air and the fuel droplets. In many burners this turbulence causes internal recycling of the hot products of combustion which provides better heat transfer to the atomized droplets achieving the ignition point more rapidly.

If the vaporized liquid contains solids, the incinerator design must allow the particles to be carried into the gas stream without agglomeration. A high swirl or a cyclonic design may cause the solids to be re-agglomerated into larger particles becoming more difficult to burn. Proper design of the air mixing device and the

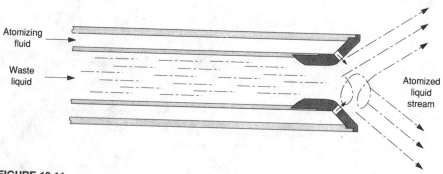

FIGURE 12-11
External mix atomizer.[16]

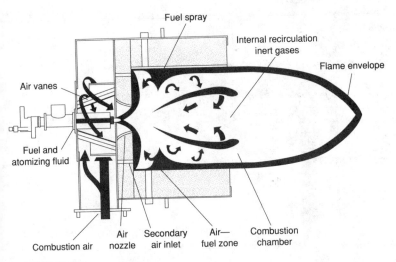

FIGURE 12-12
High intensity burner.[16]

nozzle location are therefore important. Proper oxygen concentration at the surface of these solids is needed to ensure that gradual oxidation will occur. Because they are solid, the particles will burn at the surface, and heat will be transferred inwardly to the core of the particle. Sufficient time must be provided to permit complete burnout of the solid particles in suspension. It is for this reason that coal is finely pulverized before combustion in a fluid system.

Inorganic compounds are carried with the liquid into the gas stream as particulates. Depending on the type of atomizer, the composition of the solid, and the temperature of the primary oxidizer chamber, many particles will be reduced to submicron size and diffused into the gas stream. The heavier particulates may become molten and agglomerate into a molten ash. The combustor must be designed to collect the molten ash without plugging the flow passages of the incinerator and quench system. A slanted horizontal design or a downward vertical orientation is most often used for these types of waste (Fig. 12-13).

Combustors (Burners)

Primary and secondary combustion units are utilized in the Liquid Injection type incinerator systems. Primary units are used to burn wastes which have sufficient heating value to burn without the need for auxiliary fuel. With good burner design, heating values from approximately 2,500 kcal/kg (4500 Btu/lb) and above can be burned without the use of auxiliary fuel. The burner design (e.g., air mixing, turbulence) determines the minimum heating value waste which can be burned without need for auxiliary fuel.

Figure 12-3, Gross heating value vs. adiabatic temperature, provides a guide to the ability of a combustor to burn a particular waste. Low-intensity or laminar

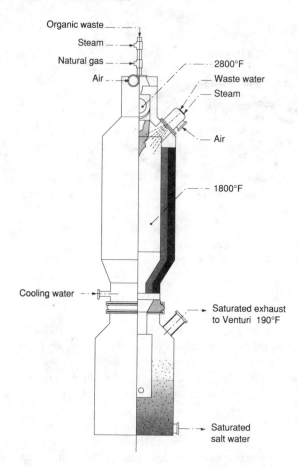

Organic waste

Steam

Natural gas

Air

2800°F

Waste water

Steam

Air

1800°F

Cooling water

Saturated exhaust
to Venturi 190°F

Saturated
salt water

FIGURE 12-13
Down fired combustor.[16]

flame burners where the air pressure drop across the burner varies from 50 mm to 150 mm [2″ water column (w.c.) to 6″ w.c.] utilize high excess air levels (25% to 60%) to provide the mixing of air with the waste fuel. These usually demonstrate low turbulence. High-intensity burners operate with combustion air pressures of 200 to 500 mm w.c. (8″ to 20″ w.c.). As a result, the increased turbulence resulting from the energy imparted by the air allows operation at much lower excess air levels (0% to 20%). The minimum combustion temperature for burning a fuel is approximately $1,200°C$ to $1,315°C$ ($2,200–2,400°F$). Figure 12-3 indicates that a high-intensity burner would be able to oxidize a 2,500 kcal/kg (4,500 Btu/lb) waste material at 2,200°F and 10% excess air. A low-intensity burner operating at 40% excess air and 2,200°F would be limited to a minimum heating value of 3,220 kcal/kg (5,800 Btu/lb). This burner would require auxiliary fuel to properly burn a 2,500 kcal/kg (4,500 Btu/lb) waste.

Other Design Considerations

In the process industries, numerous liquid wastes are generated which are hard to segregate or categorize. These wastes can be defined as organic, aqueous-organic, or aqueous-organic-inorganic.

Wastes which are defined as organic, in most cases, will sustain combustion and may be injected into the primary combustor. Many organic wastes are utilized as a source of fuel in boilers, process furnaces, and cement, lime and aggregate kilns.

Aqueous wastes which contain ash are more difficult to handle than wastes without ash. Non-ashbearing wastes are easily handled in either a horizontally or vertically fired liquid incinerator. The waste is usually injected downstream from the burner and into the flame of the main burner. Care must be taken in the sizing, location, number, and direction of the aqueous waste nozzles.

Because a highly aqueous waste stream is seldom viscous, it may be atomized by hydraulic pressure in a mechanical-atomizing nozzle (or in a dual fluid nozzle by steam or compressed air). Use of compressed air for these nozzles has been very successful. The air serves as combustion air and reduces the heat load on the system that is required with steam. The design must control the proper ratio of liquid to atomizing fluid to ensure the best spray pattern and droplet size through the entire range of flow.

Atomizing nozzles must also be protected from excessive heat in the event of loss of waste flow. This can be achieved by maintaining atomizing fluid flow. In many units, atomizers are designed to be physically removed from the burner or incinerator chamber when not in use to prevent burnout of the nozzle tip.

Sufficient air must be provided at all times in the incinerator to oxidize the organics contained in the waste stream. Incinerators can produce soot when burning aqueous organic wastes unless good air mixing is present at the nozzle.

Aqueous wastes containing organic and inorganic compounds are produced by industries that manufacture herbicides, pesticides, etc. The pharmaceutical industry also produces a number of high water content wastes, some containing phosphorus, chlorine, and sulfur. Aqueous solutions are not ignitable when the water content is greater than 75%. If the waste contains ash (salts), consideration must be given to the maximum operating temperature to minimize refractory spalling problems. Aqueous wastes are normally injected into the incinerator downstream from the primary combustor in a lower temperature zone, rather than directly into the high-temperature zone.

Incinerators designed for the oxidation of high ash-containing wastes are usually arranged for vertical downward flow (see Fig. 12-13). This provides for gravity discharge of molten ash from the waste into the quench zone. Critical to the design and operation of this type of incinerator is the refractory selection. Reactions often occur at elevated incineration temperatures between the metallic oxides formed from oxidation of the waste and the refractory lining (typically alumina and silica). Severe spalling and high maintenance result from improper selection of refractories. Pre-testing of various refractories with the slag that results from the incineration of such wastes at various temperatures can be achieved in a muffle furnace. This often eliminates subsequent refractory problems in a full-scale system.

When burning aqueous wastes and using organic wastes as an auxiliary fuel, the primary purpose of the incineration process is to completely oxidize the organics present in the waste. Therefore, conditions must be established to provide the necessary atomization, mixing, temperature, and time for the various types of wastes to be incinerated. Certain conditions of waste composition may limit the operating temperature, i.e. aqueous wastes with salts. Therefore, it is important to determine how and where each waste may be introduced into the incinerator. Completely organic wastes are usually introduced into the primary combustor to provide the necessary heat input and temperature needed for the low heating value waste streams. If the aqueous waste stream will not cause a problem with refractories at high temperatures, it may also be introduced with organic waste in the primary combustor as long as it does not lower the primary combustor temperature below 2,200°F.

Acid Gas Producing Liquid Wastes

If the waste contains sulfur, phosphorus, chlorine, or another halogen it will produce an acid gas upon combustion. The primary concerns in the hazardous waste industry are with sulfur and chlorine because these are the most frequently encountered acid producers. As outlined in Sec. 12-3, the quantity of acid gases produced will vary depending upon their equilibrium constants at the operating temperature, but they must be reduced to acceptable emission levels before they can be discharged to the atmosphere. Acid gases are generated from many types of hazardous wastes, but acid gas removal methods are essentially the same (see Sec. 12-9).

12-6 SOLID WASTE INCINERATION

Introduction

When solid wastes are to be burned, the combustion must occur in suspension, on a grate, or on a solid hearth. Over the years many different incinerator designs have been employed for burning solid wastes. Suspension systems rely on a relatively uniform feed sizing. For example, lumber and woodworking operations burn sawdust suspended in a high-velocity air stream. The fluid bed incinerator consists of a bed of sand or alumina fluidized with air, into which the waste is injected while the bed is in suspension.

Grate-type systems are suitable for large and irregular wastes which can be supported on a stationary or moving alloy grate which allows combustion air to pass through the grate into the waste. Most municipal solid waste incinerators are of the grate type. Hearth-type incinerators include the two-chamber, fixed-hearth; the rotary kiln; and the multiple-hearth.

Hazardous waste incineration is currently performed in all types of incinerators but mainly in suspension and hearth-type systems. The rotary kiln incinerator accounts for 75% or more of all the hazardous waste incinerators in the United States which handle liquid and solid waste. The two-chamber, fixed-hearth incinerator probably accounts for about 15%, and the multiple-hearth and fluidized bed incinerators combined account for the other 10%. All of these systems also can burn liquid hazardous wastes.

Grate Type Incinerators

Stationary grate type incinerators burn the waste on metal grates providing for air circulation below, above, and through the waste. They are not generally suitable for hazardous waste because temperatures reached in the primary chamber might destroy the grates. Obviously, this type of incinerator depends upon the waste to be of such character that it will be supported on the grate and will not fall through to the ash pit until it is burned. Burning cellulosic wastes (wood and paper) is the most practical use for most grate-type incinerators.

Traveling grate or belt incinerators have also been used for municipal trash but seldom for process waste. One exception is the infrared incinerator which employs an alloy-woven belt or a segmented belt made of alloy castings moving through a radiant electric furnace. This primary furnace is followed by a secondary chamber where additional air and heat are added to complete combustion of the vapors from the primary furnace. Only a few systems have been employed for soil decontamination, and with mixed results.

Hearth-Type Incinerators

Most hazardous waste is burned in hearth-type systems of which there are several basic types:

- The rotary kiln
- A "controlled air" or "two chamber fixed hearth" system
- The multiple hearth incinerator
- The monohearth (seldomly used)

The rotary kiln shown in Fig. 12-14 consists of a refractory-lined cylinder which sits on trunnions and rotates slowly (0.5 to 2 RPM) on its longitudinal axis. The kiln is sloped 1° to 2° from the feed end to the ash discharge end so that the waste moves horizontally as well as radially through the cylinder. Waste burns as it moves toward the ash discharge end. While the ash is discharged at the low end, the flue gases from the kiln pass into a secondary combustion chamber and are heated to an even higher temperature for complete destruction.

Rotary kiln advocates have consistently debated the subject of "slagging" vs. "non-slagging" kiln operation. The slagging technology, of European design and preference, provides for operating the kiln at temperatures where inorganic components of the waste, including steel drums, will be discharged to the ash quench chamber in molten form. Systems in the United States have long employed the non-slagging ash discharge where the ash is kept below its melting point. Both systems are viable.

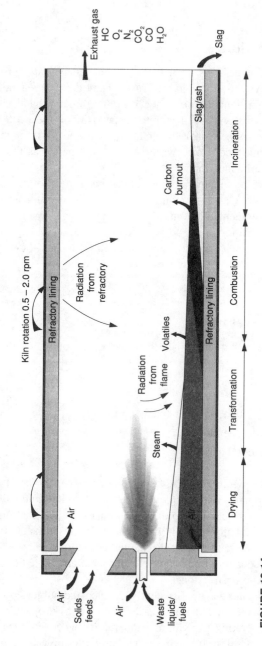

FIGURE 12-14
Rotary kiln incinerator.

An advantage of the slagging system is that it melts everything and discharges a "liquid" ash. It demands more of refractories and a closer temperature control, but it does allow for all types of feed. The non-slagging approach doesn't require the rigorous control nor does it provide the feed flexibility and larger non-combustibles would require removal from the ash discharge.

Real operating difficulty exists between the two temperature conditions because in a system which is neither non-slagging nor slagging there lies a plastic condition where the waste can build a semi-solid ash that cannot be discharged either dry or as a liquid. The solution is temperature control at all times.

The fixed-hearth incinerator shown in Fig. 12-15 consists of a primary chamber which may have either a single-level hearth or a stepped hearth. In smaller units, waste is intermittently charged to the primary chamber, but ash is not removed until the accumulated quantity interferes with normal operation. In larger units a mechanical ram pushes the charge through the incinerator, and the ash is continually removed. Both types have secondary combustion chambers which further heat the flue gases from the first chamber to destroy organic compounds present in the flue gas.

Combinations of liquid and solid wastes are often burned in the same incinerator, and there is nothing to prevent a waste gas from also being burned. Judicious use of the energy from one waste to help burn another is good energy stewardship. Because liquid- or gas-burning equipment is usually employed in a solid waste incinerator to burn auxiliary fuel, it can be employed in much the same way to burn a waste liquid or gas.

The multiple-hearth incinerator shown in Fig. 12-16 was originally utilized in an incineration mode for sewage sludge although multiple-hearth systems had been used in the mining industry for years as ore roasters. It is a complicated, highly mechanical system used to burn sludges. It has seen limited use in hazardous waste incineration because the temperatures required to provide reasonable destruction efficiencies are not compatible with long equipment life. The system consists of two-to-six horizontal hearths in a vertical array. The waste is fed to the top hearth and ignited by fuel burners. An air-cooled central shaft moves rabble arms over the surface of each hearth in a carousel fashion. The ploughs on the arms move the waste across the surface of the hearth until it reaches a pie-shaped opening. The waste drops through the opening to the next lower hearth and continues to burn, is cooled, and discharged as an ash. The flue gases move vertically upwards within the outer shell and into the air pollution control equipment. A variation on this design, often called a "mono-hearth," employs a single hearth which revolves beneath fixed rabble arms which move and level the waste. The mono-hearth also requires a secondary combustion chamber for complete combustion.

Fluidized-Bed Incinerators

The fluidized-bed incinerator shown in Fig. 12-17 utilizes a "fluid hearth" consisting of sand or alumina on which and in which combustion occurs. Imagine a bed of sand sitting on a porous surface. If an air flow from below has sufficient pressure, it will "fluidize" the bed of sand, or hold it in suspension, as long as the velocity of the air is

a.) Horizontal two-chamber controlled air incinerator

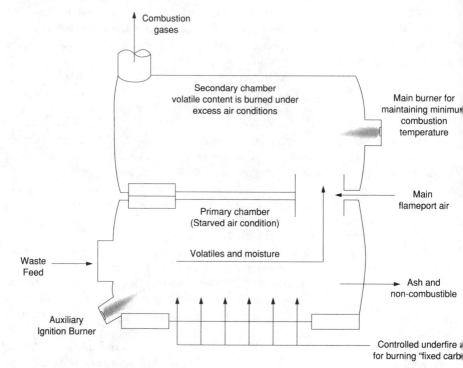

b.) Vertical two-chamber controlled air incinerator

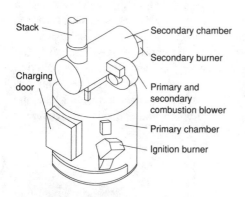

c.) Relationship of temperature to excess air

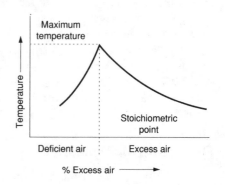

FIGURE 12-15
Fixed hearth incinerator.

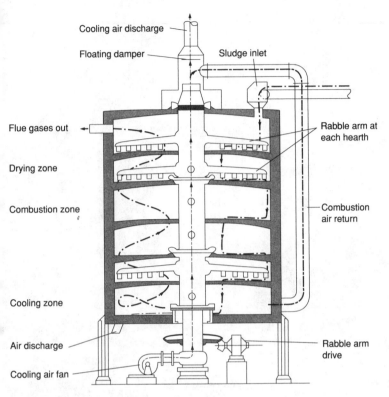

FIGURE 12-16
Multiple-hearth incinerator.

not so great that it transports the sand out of the system. This is a fluid bed in which the particles of the bed are in suspension, but not in flow. Waste is injected into this fluidized bed, either as a liquid, sludge, or uniformly sized solid. The air with which the bed is fluidized is heated to at least the ignition temperature of the waste, and the waste begins to burn (oxidize) within the bed. Most of the ash remains in the bed, and some exits the incinerator into the air pollution control equipment. Heat also exits with the flue gases and can be captured in a boiler or used to preheat combustion air.

There are a number of variations for fluidized-bed systems. One is the circulating fluid bed combustor where the gas velocities are above those required for mere fluidization. The differences between the various technologies are explained in Fig. 12-18.

Oxidation vs. Pyrolysis

There are two distinct operating modes for hazardous waste incinerator systems: oxidizing and pyrolytic. These modes apply to the first combustion stage only because complete oxidation of the flue gases is required to make them acceptable for release to the atmosphere.

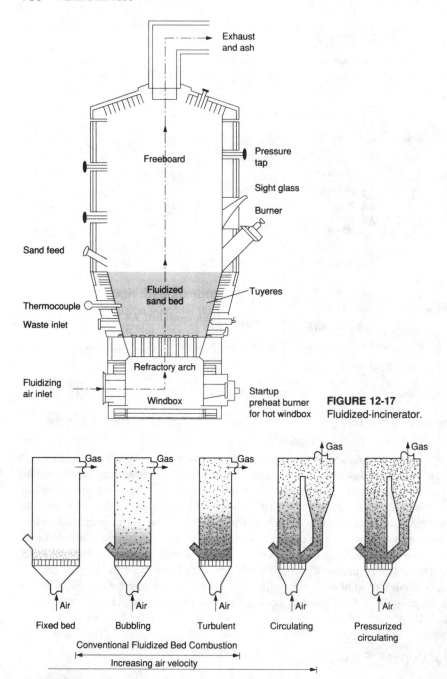

FIGURE 12-17
Fluidized-incinerator.

FIGURE 12-18
Fluid bed incinerator types.

The two-chamber, controlled air type hearth incinerator is the incinerator type most often operated in a pyrolytic mode. While there are rotary kiln systems which claim pyrolytic operation, the actual operation is almost always oxidizing. The two-chamber system is designed to be pyrolytic in the first stage for one reason—to reduce gas velocity—and therefore to reduce particulate carryover. This approach was cogent when it was possible to meet particulate emission regulations without the addition of particulate capture equipment. Today's particulate emission levels make that impossible in most cases, so pyrolytic operation has few advantages. The partially combusted gases from the pyrolytic chamber, consisting of CO and some H_2, are heated and oxidized in the second chamber to CO_2 and water vapor. The ash, however, from many of these units often contained unacceptable levels of hazardous constituents.

Oxidizing systems, which employ an oxidizing atmosphere in both a primary and secondary combustor, are more often used for hazardous waste incineration. The rotary kiln is such a device. Because the kiln is under negative pressure and adequate seals have not been devised to prevent the in-leakage of air, it is almost always in an oxidizing mode.

Secondary Combustion Chamber

The secondary combustion chamber (SCC) is necessary in virtually every hazardous waste incinerator where solid wastes are burned because the primary chamber, whether it is a kiln, fixed-hearth, or multiple-hearth design, does not provide enough time, turbulence, or temperature to destroy the organic components of the waste to the required DRE. Low temperatures in the primary chamber, 705°C to 815°C (1,300°F to 1,500°F), often provide better operation than higher temperatures because they are below the slagging (melting) temperatures of inorganics in the waste. This is especially true when using incinerators for soil decontamination. The primary chamber's function is to volatilize the organic fraction of the waste. The secondary chamber's function is to heat the vaporized organics to a temperature where they will be completely oxidized. This oxidation process occurs at 980°C to 1200°C (1,800°F to 2,200°F). Because purchased fuel is used as the heat source in most SCCs, the lowest the temperature at which the DRE can be achieved will be the most economical. Again, the time, temperature, and turbulence triangle is at work. Most regulatory agencies insist that the flue gas residence time in the SCC be at least two seconds. The destruction of organics in the flue gas is a function of not one, but all three, factors.

The use of a secondary combustion chamber is not usually required where the incinerator burns only liquids because these three "Ts" can be achieved in a single chamber.

The SCC may be either a horizontal or vertical refractory lined vessel with its own liquid or gas burners of sufficient capacity to heat the flue gases from the primary chamber to the desired final temperature where the required destruction of hydrocarbons can be achieved. Vertical vessels are preferred over horizontal when the particulate loading from the primary combustor is expected to be substantial, simply because gravity will cause particulate deposits on the bottom of the chamber which cause a removal problem. Design gas velocities in the SCC are usually 15 to

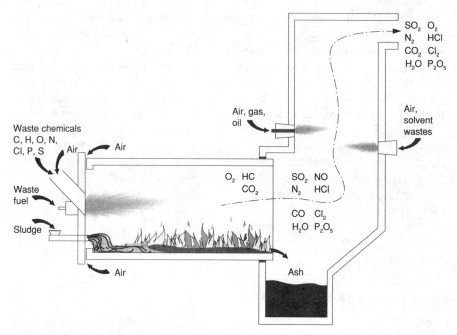

FIGURE 12-19
Rotary kiln with vertical secondary combustion chamber.

25 ft/second at temperatures of 1,800°F or above. Figure 12-19 shows a rotary kiln incinerator with a vertical secondary combustion chamber.

Example 12-8   Rotary kiln—batch feed.   A rotary kiln designed for a nominal heat release of 15,000 Btu/hr·ft^3 has an inside diameter of 8 ft and is 30 ft long.

1. Determine the design heat release in Btu/hr.
2. Normally a waste having a heating value of 750 Btu/lb is burned in the kiln. It is fed continuously through an auger. Occasionally a waste consisting of polyethylene pellets is "batch" fed to the kiln in 30 gallon fiber containers. The pellets have a bulk density of 50 lb/ft^3 and a heating value of 18,350 Btu/lb. A single container is consumed in 6.5 minutes.
   (a) Will the kiln operate within its design parameters?
   (b) What kind of combustion would you expect?
   (c) How would you correct the problem if you were the operator?

Solution.

1. Kiln volume $= 30 \times \dfrac{(8)^2 \pi}{4} = 1508$ ft^3
   The heat release is:
   1507 ft$^3 \times 15{,}000$ Btu/hr · ft$^3 = \underline{\underline{22{,}620{,}000 \text{ Btu/hr}}}$

2. (a) $30 \text{ gal} \times 8.35 \text{ lb}/\text{gal} \times \dfrac{50}{62.4} = 201 \text{ lb}$ in each container.

201 lb $\times 18, 350$ Btu/lb $= 3,688,000$ Btu heat release.

But this will occur in 6.5 minutes, therefore the equivalent hourly heat release is:

$\dfrac{60 \text{min/hr}}{6.5 \text{ min}} \times 3,688,000$ Btu $= 34$ million Btu/hr

This is about 1.5 times the design heat release of 22.6 million Btu/hr, so the kiln will *not* operate within its design parameters.

(b) There will not be sufficient air in the system to burn the polyethylene waste at that rate, so the kiln will produce a dense black smoke.

(c) To correct the situation, package the polyethylene in smaller batches. The batch size will have to be smaller by at least the same factor (1.5). Therefore:

$201/1.5 = 134$ lb/batch

$134 \times 18,350 = 2,458,900$ Btu heat release

$60/6.5 \times 2,458,900 = 22.7$ million Btu/hr, which is almost the same as the design value. Good practice would suggest that a safety margin of at least 10% be used, so about 120 lb/batch is a good size. Also, the containers must be fed at no greater rate than one every 6.5 minutes, say 9 per hour.

Thermal Relief Vent

Most hazardous waste incinerator systems, except liquid incinerators, operate under a negative pressure. If the pressure in the system becomes positive because of failure of the induced draft fan, it is usually because of power failure. If this happens the points in the system which leak air in now may, because of the positive pressure, leak flue gases out. This is obviously undesirable. In a rotary kiln system this occurs at the kiln seals. In addition, many components of the air pollution system may be designed to operate at close to ambient temperatures because their materials of construction are plastics which cannot withstand elevated temperatures. If power fails or water flow decreases, then the plastic components of the system are in jeopardy. Process sensors detect these problems immediately, but it is necessary to prevent either a positive discharge of flue gases through the kiln seals or an over-temperature in the air pollution control system. This is accomplished by installing a thermal relief vent (TRV) at the top of the SCC or between the SCC and the waste heat boiler or quench.

The TRV is a low stack (see Fig. 12-20) which opens automatically when the emergency condition occurs. It provides sufficient draft to maintain a negative pressure condition in the kiln and prevents flow into the air pollution control system. It is downstream from the SCC so the reduced flow from the kiln combined with the high SCC temperature provides for high organic compound destruction efficiencies. The low-flow condition entrains few particulates. Therefore, the threat of environmental contamination is low. While the TRV is not an ideal solution, it is the safest and most practical approach to circumstances which seldom occur but need a solution.

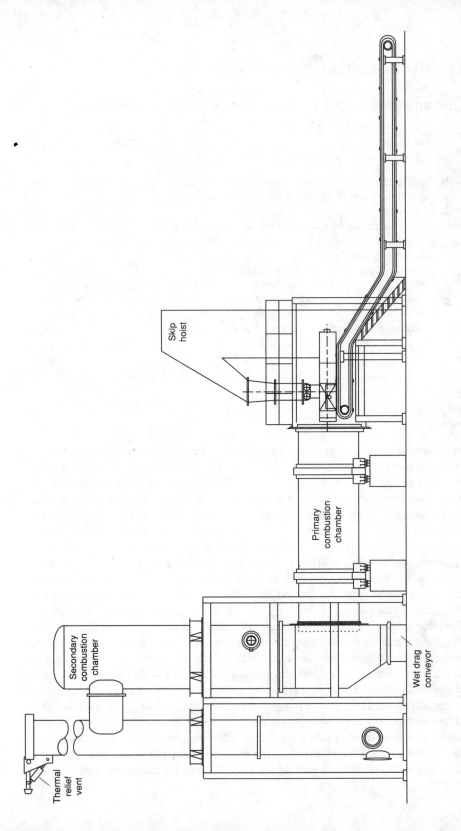

FIGURE 12-20
Partial elevation of an incinerator.

12-7 STORAGE AND FEED SYSTEMS

Liquid and Sludge Storage and Feed Systems

To provide a uniform feed from the storage tanks, good mixing is required. Layering or phase separation often occurs in large tanks, especially where a number of liquid wastes are introduced from different sources. Mixing may be accomplished by internal mixers, external recycle pumps, and air or steam spargers. Proper venting of storage and mixing tanks is also necessary to prevent pressure or vacuum conditions in the vapor space of the tank. Vent gases from storage tanks may be exhausted to the incinerator or controlled by other means such as condensation by refrigeration or carbon adsorption.

The preferred physical design for hazardous waste liquid and sludge storage tanks is a vertical vessel supported on legs and with a maximum capacity of 20,000 gallons. There are three important reasons for selecting these tank parameters:

- Leak detection from a leg-supported (elevated) tank is obviously easier than leak detection from a tank that has a flat bottom sitting on a concrete foundation.
- Twenty-thousand-gallon (20,000) vertical tanks are about the largest size that can be reasonably agitated or mixed by normal methods.
- Hazardous waste tank containment volume must be equal to the largest tank within the containment.

Materials of construction of all components in the storage and feed system—tanks, pumps, mixers, piping, valves, and nozzles—must be compatible with the chemical composition of the waste to prevent corrosion and erosion. If the waste contains suspended solids, which are not combustible, filtration of the waste may be desirable to prevent solids carryover into the incinerator and air pollution control system. If the solids are organic, filtration defeats its purpose so the design must provide for their transport into the incinerator.

High-viscosity wastes (slurries and sludges) and liquids containing solids present feed system handling problems. Various generic pump types are available, and the selection and the use of each is predicated on the properties of the waste stream. Centrifugal pumps often have too great a capacity for incineration systems when waste flows are low. Progressive cavity pumps have been used successfully when the construction materials have been selected properly to prevent corrosion and erosion. Gear pumps, which provide low flows at high pressures with low-to-medium viscosity wastes having some lubricating quality, are also widely utilized. Diaphragm pumps, operated by compressed air, are good for pumping heavy sludges.

In systems where a variety of waste streams must be pumped into the incinerator, it is difficult to select the proper feed system to use with each waste. In some cases, pumps may not be practical due to the high maintenance costs and shutdowns that result. Pressurizing a storage or feed tank with an inert gas (nitrogen) or air where the waste is not highly flammable may be the safest and most economical method.

Solid Waste Storage and Feed Systems

The most difficult design problem in any hazardous waste incinerator is the feed system and primarily the solid waste feed system. Storage of solid wastes is relatively simple. Solids may either be stored in small containers such as drums and boxes, large containers such as 20 cubic yard (or larger) roll-off boxes, or in bulk in some type of solid waste containment. Bulk storage, in other than containers, is not usual in hazardous waste incinerator facilities because it is difficult to contain and difficult to control air emissions. Therefore, small containers, steel drums, fiber and plastic drums, and similar containers are used for most hazardous waste other than heavy sludges or contaminated soil. The latter are usually shipped to the incinerator in roll-off boxes and stored in these containers until processed into the feed system.

Covered, dry storage areas are certainly required for roll-off containers and also for the drum storage and processing facility. What happens to the waste after receipt and storage is a function of the incinerator type. Only the rotary kiln has been widely enough utilized to provide a basis for a discussion of solid feed processing systems because rotary kiln systems constitute the majority of commercial hazardous waste incinerators handling solid wastes.

Solid waste feed systems present the most difficult design problem in any incinerator project, and their proper design is vitally important to the success of the incineration system. Commercial hazardous waste incinerators must provide for feeding:

1. Containerized waste
2. Bulk waste
3. Contaminated soils
4. Oversized items
5. Problem materials

Containers, (i.e., steel, fiber, or plastic drums), or cardboard boxes can be fed directly into a rotary kiln through an airlock door by a hydraulic ram feeder (see Fig. 12-21). Whole steel drums are only fed to slagging-type kilns, but the fiber or plastic drums will burn completely in non-slagging systems except for metal rings or closures. Problems with this type of feed system arise when the container contains a waste which is volatile and has any significant heating value.

For example, if a 50-gallon drum of liquid toluene (HHV = 18,252 Btu/lb and 7.2 lb/gal) is fed into a hot rotary kiln, the potential heat release from that drum is over 6.5 million Btu. It has a vaporization temperature of only 230°F so virtually all of the toluene in the drum will be vapor within a few seconds in a kiln operating at 1,500°F. If vaporization took 15 seconds, the hourly heat release rate would be $4 \times 60 \times 6.5$ million = 1,560 million Btu/hr. It is important to see this as a *rate* of heat release and not an actual release. Most rotary kilns are designed for heat release rates of 20,000 Btu/hr · ft³. This would mean that the kiln would need to have a volume of 78,000 ft³ to be within good design practice. Since a typical rotary kiln is on the order of 12 ft diameter × 30 ft long (3,400 ft³), this is obviously impractical.

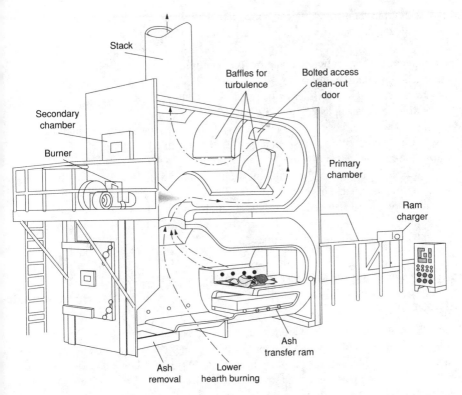

FIGURE 12-21
Hydraulic ram feeder.

It is even more apparent to recognize the combustion consequences of such a single waste charge. There would not be enough combustion air to satisfy the high volume of hydrocarbon vapor in the kiln so smoke or soot would instantaneously appear. The oxygen in the kiln would drop to zero, and the CO concentration would rise to a level which would exceed regulatory limits.

This example is an extreme case for illustrative purposes only, but batch feeding always presents this problem. The higher the heat release in a batch charge of any size, the more difficult is the control of the combustion process.

Containers of all types, containing liquids or sludges, are usually pre-processed by pumping and bulking the waste into storage tanks for continuous feed through lances and atomizers. The containers, either empty or with some residual waste material, can then be batch fed to the incinerator or shredded and fed continuously.

Bulk waste can also be fed by hydraulic ram. It presents the same batch charge problem as containers. It also can be fed by gravity through a chute. Continuous feed is still difficult if not impossible unless there is some "sizing" procedure, such as shredding ahead of the feed system.

The normal heating value for contaminated soils is low, but may occasionally have significant caloric levels as high as 4000 Btu/lb. If they are first processed through a sizing system such as a bar screen or grizzly, they can be fed best by a screw conveyor. This continuous type of feeder provides for uniform but variable feed rates depending upon heating value.

Oversized items, regardless of their heating value, need to be reduced in size by shredding to be accepted by any feed system.

Special wastes, those that are extremely volatile, pyrophoric, or potentially explosive, must be fed through special chutes (often called a bottle drop chute) in quantities small enough to obviate any problem.

The best feed system, from a combustion standpoint, is one that will provide a steady, continuous feed. This is impossible to achieve with types 1, 2, 4, and 5 above unless these types are reduced to a relatively uniform size. This can be accomplished by shredding the waste before it is fed to the incinerator after first removing liquids and pumpable sludges from containers.

12-8 FLUE GAS TEMPERATURE REDUCTION

Introduction

When the flue gases leave the secondary combustion chamber, they are at temperatures usually exceeding $1,800°F$ and at times as high as $2,200°F$. They contain particulate matter and may contain acid gases such as HCl, Cl_2, and other halogens; SO_2 and possibly P_2O_5 which must be removed; and lots of heat, which also must be removed or utilized. This leaves two options—recover the heat or cool the gases without recovering the heat.

Heat Recovery

Heat recovery is a major economic consideration because heat recovery equipment is expensive and there must be a practical use for the recovered heat. In Europe, where fuel cost is high, heat recovery is more widely employed than in the United States. Heat recovery from hazardous waste incinerators in the United States is virtually non-existent in commercial operations because there is little use for steam on-site, the potential of selling steam to off-site customers is minimal (usually because hazardous waste incinerators are not located close to potential users), and the economics of steam-electric power generation are poor. In-plant hazardous waste incinerators may have entirely different economics, and this is where heat recovery often is viable.

The options for heat recovery are the use of boilers to generate steam or hot water, recuperators to preheat combustion air with the incinerator flue gases or to reheat stack gases with flue gases to eliminate the stack water vapor plume after a wet scrubber. Boilers have been widely used in municipal solid waste (MSW) incinerators to produce steam for electric power generation. Here the waste heating value is reasonably constant. This is not always true with hazardous waste incinerators. Also, the size of the MSW incinerator is large compared to the hazardous waste incinerator making power generation economical.

The waste heat boiler (see Fig. 12-22) is a special design for most hazardous waste incinerators. It should be capable of handling a high-temperature, high-particulate, high-acid content flue gas. The particulate in the flue gas from the SCC may be essentially "plastic," i.e., containing inorganics at temperatures above their melting point, or even their vaporization temperature which, upon contacting cold boiler tubes, will deposit a solid residue on the tubes. For this reason, radiant sections of tubes with wide tube spacing are used in the inlet end of the boiler. Some manufacturers have even provided for rapping or shaking of the tubes to release the solid buildup into hoppers below. After the temperature has been sufficiently reduced to eliminate this problem, the remaining boiler can serve as a convective heat transfer device.

Most waste heat boilers on hazardous waste facilities which burn solids or wastes containing solids utilize a specially designed boiler of the "water tube" type.

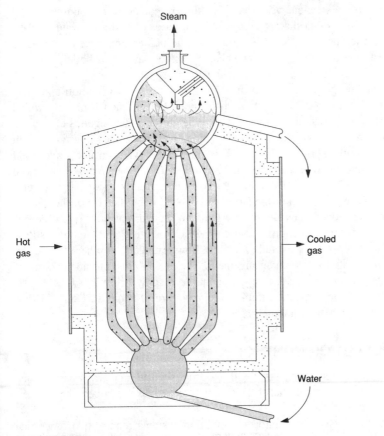

FIGURE 12-22
Waste heat boiler.

Liquid incinerators which burn low ash liquids may use a "fire tube" type boiler. The generated steam pressure is usually nominal, below 100 psig, because there is no application for high-pressure steam in the plant and no customers for the steam external to the plant.

Design considerations when burning wastes which produce acid gases, especially halogen gases, involve operating restrictions more than design restrictions. A cold boiler fired on a waste which will produce HCl upon combustion is a boiler with a short life because hydrochloric acid will condense on the tubes and rapid corrosion will result. Therefore, any boiler handling acid gases must be preheated on a fossil or non-acid fuel to at least its normal operating temperature before acid gas-producing fuel is burned. Similarly, upon shutdown, the acid gas-producing fuel must be stopped and replaced with fossil fuel at least 10 minutes before the boiler fuel is cut off.

Flue Gas Quench

The other option for cooling flue gases is to use the direct injection of air or water. Water injection is more efficient because of its higher specific heat when compared with air. This also means that the net volume of flue gases cooled with water will be lower at any given temperature. Most hazardous waste incineration systems which do not employ a waste heat boiler, and many that do use a waste heat boiler, cool the flue gases either finally or initially by direct injection of water.

Cooling hot flue gases is a tricky problem. It is one of distribution and mixing. There is no such thing as a "dry bottom quench" (i.e., a quench process which evaporates all the water at adiabatic saturation temperatures). The vagaries of cooling a hot gas stream involve both heat and mass transfer between the cooling medium and the flue gases if the coolant is water. This means fine water droplets and good distribution throughout the flue gas stream. The higher the flue gas volume the more difficult this is to achieve. Usually, multiple, two-fluid, atomizing nozzles are employed. The number and location of the atomizers depend upon the size of the quench chamber and the gas flow. Many quench systems are concurrent-flow (i.e., flue gas and water flow in the same direction) because the SCC is vertical and the hot gas flow is downward, but this does not exclude a counterflow arrangement where the equipment design is amenable to such an arrangement. The design must provide cooling of the gases to satisfy inlet temperature requirements to a spray dryer 260°C to 371°C (500°F to 700°F) or to an adiabatic saturation temperature of about 82°C (180°F) for low-temperature downstream equipment.

12-9   AIR POLLUTION CONTROL

Air pollution control systems associated with hazardous waste incinerators are unique in their combination of elements but the elements themselves (scrubbers, baghouses, cyclones, etc.) are not unique.

Most hazardous waste incinerator air pollution systems require two functional elements; one, a system or equipment item which will remove particulates from the flue gas stream; and two, a system or item of equipment for the removal of acid gases. These two functions may be achieved with wet systems or dry systems or with a combination of both.

The particulate removal system must specifically address the removal of the U.S. EPA's list of ten priority metals which have regulated emission levels. These ten metals are: arsenic, beryllium, cadmium, chromium, antimony, barium, lead, mercury, silver, and thallium.

Particulate Removal

Particle density in the flue gas from an incinerator is measured in grains/dscf or mg/dscm. The particle density in the gas varies widely depending upon two factors: gas velocity in the incinerator and actual particle size. The smaller the particle the more easily it is carried by the flue gas at a low gas velocity. The higher the gas velocity, the larger the particle capable of being carried in the flue gas stream and the greater the number of particles of all sizes carried. Current federal regulations specify that particulate emissions from an incinerator shall not exceed 180 mg/dscm (0.08 grains/dscf) corrected to 7% oxygen in the flue gas. These regulations will surely become more stringent, and some states now require levels as low as 0.015 grains/dscf.

Particle removal from gas streams can be effected by gravity separation, interference such as impacting the particles on another object, centrifugal separation, filtration through a filter media, electrostatic separation, or by sufficiently wetting a particle with water so that it is "scrubbed" from the gas stream. Particulate removal can be accomplished by completely dry removal methods, a combination of wet-dry methods, or completely wet methods.

The dry particulate removal methods include:

- Impaction—baffles and screens
- Centrifugal separation—cyclone separators
- Filtration—fabric filters
- Electrostatic—precipitators

Wet methods which employ water as a medium include:

- Impaction—packed and tray columns
- Centrifugal separation—wet cyclones
- Particle wetting—venturis and similar units
- Particle conditioning and wetting—collision scrubber
- Electrostatic—wet ionizer/precipitator

Acid Gas Removal

In an incineration system, especially a hazardous waste incineration system, there is a need to remove acid gases from the flue gas before discharge to the atmosphere. The normal acid gases encountered are HCl, SO_2, and HF. Occasionally HBr will be present. HCl and HF are very soluble in water and HBr slightly less. SO_2 is not as soluble so its removal depends on using a base such as lime, NaOH, or soda ash for neutralization. The same solution can be used to remove HCl and HF. Acid gas removal can be accomplished in either a wet or dry system. The efficiencies of acid gas removal vary between the wet and dry system. The wet system generally has a higher efficiency and uses more nearly stoichiometric amounts of base than the dry system.

The dry removal systems are:

- Dry lime injection—a variety of systems
- Dry scrubber—spray dryer

Dry removal systems are usually followed by a fabric filter to collect the neutralized acid gas as a salt.

The wet systems are:

- Absorption/reaction—packed and tray columns
- Wetting contactors—venturis/Calvert/Hydro-Sonic

In a wet system, the effluent is a salt solution of the acid such as $CaCl_2$, NaCl, $CaSO_3$, or Na_2SO_3. These salt solutions are normally removed from the scrubber continuously and treated in on-site wastewater treatment facilities.

Control System Options

Air pollution control for a hazardous waste facility is a system design problem because of the many possibilities available to remove particulates, including metals and acid gases. For example, many hazardous waste incineration systems today are being designed to eliminate the need for handling wet effluent; therefore, the system is designed to produce only a dry waste product. The flue gases are cooled to about 600°F in a boiler or by direct contact with water or air. They are then exposed to a neutralizing agent such as lime which is air conveyed into the flue gas. The finer the lime particles the better the neutralization efficiency. This is known as a dry-dry injection system (see Fig. 12-23). The lime also might be fed into the hot gas stream as an atomized slurry so that the heat from the flue gas evaporates the water as the acid gases are neutralized by the lime. This is usually accomplished in a spray dryer, sometimes called a dry scrubber. The lime-sulfite mixture leaves the spray dryer as a solid, either at the bottom of the unit or entrained in the flue gases. This is known as a wet-dry system (see Fig. 12-24). Sodium hydroxide or soda ash also might be used, but both are more expensive than lime and because sodium salts are more hygroscopic, lime is preferred.

Both types of dry reaction systems are followed by particle removal devices such as an electrostatic precipitator or fabric filter (baghouse). The operating temperatures for these units are normally about 350–450°F. Baghouses are the preferred

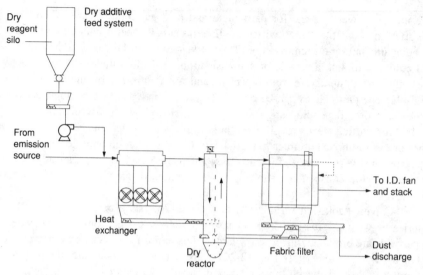

FIGURE 12-23
Dry/dry APC.

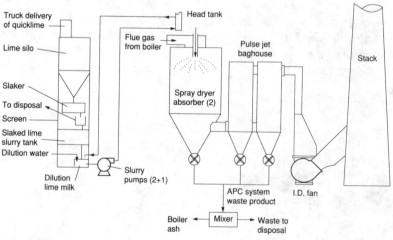

FIGURE 12-24
Wet/dry APC.

removal device in most systems because the acid-base reaction which begins upstream continues on the surface of the bags, increasing acid gas removal efficiency. Both the acid gases and the particulate from the incinerator are removed in the dry scrubber and baghouse. Because the reaction between lime and acid gases is slow and perfect mixing is a problem, the lime-to-acid gas ratio must be 1.5 to 2.0 times the stoichiometric reaction quantity.

Completely wet systems for particulate and acid gas removal are usually employed after an adiabatic quench where the flue gases have been saturated with water at the adiabatic saturation temperature. Then acid gases can be easily removed in a packed column utilizing a caustic or soda ash solution. Unlike the dry neutralization systems, the wet systems are quite efficient and are controlled by maintaining the solution pH. The particulates can be removed from the flue gas before the acid gases or at the same time depending upon the device employed. Venturi scrubbers are one possibility for particulate removal. They are simple and rugged but require a high-pressure drop to remove sub-micron particles to today's required emission levels and are therefore not very energy efficient. They often are equipped with an adjustable "throat" to provide constant pressure drop over a range of gas flow conditions (see Fig. 12-25).

The Calvert "collision scrubber" (see Fig. 12-26) has a better efficiency than a venturi at any given pressure drop because it employs a conditioning tower where small particles are provided time to grow larger, thus requiring less energy for removal. The Hydro-Sonics scrubber (see Fig. 12-27) is also capable of removing fine particles and at pressure drops between those of the venturi and collision scrubbers.

The wet electrostatic precipitator and the ionizing wet scrubber impart an electrostatic charge to the particles in the gas stream, and after the particles are attracted to the collection plates, they are washed off by water. These devices, like the dry electrostatic precipitator, are load-sensitive to the amount of particulate (grains/scf) in the gas stream. If the loading increases for any given design or configuration, the emissions will also increase. By placing a number of units in series, each with the same removal efficiency, the particulate emission level can be controlled for almost any scenario as long as the inlet loading does not exceed the design loading (see Fig. 12-28).

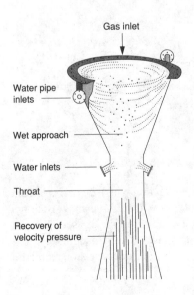

FIGURE 12-25
Venturi scrubber.

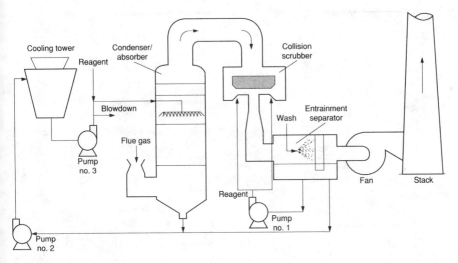

FIGURE 12-26
Calvert collision scrubber.

Packed towers are the most widely used wet contact devices for acid gas removal. In essence, they transfer pollutants from a gas stream to a liquid stream. This liquid is a waste that requires further treatment. Packed towers consist of a simple cylindrical column partially filled by a packing of ceramic or plastic material that has proven to be an efficient medium for gas/liquid contacting (see Fig. 9-1). Ceramic packing is used only when the temperature of the gas phase requires it. Plastic packing is the most widely used for acid gas removal because it is light in weight and inexpensive, and incinerator flue gases are normally cooled sufficiently to be compatible with the plastic. PVC and polypropylene are the most common plastics utilized but others are available. Packed columns are not suitable for particulate removal and are not recommended for use in flue gas streams where the particulate loading is high. Packed towers applied to transferring contaminants from water to air were discussed in Sec. 9.1.

Tray and bubble cap columns are also suitable for acid gas removal but have little advantage over the packed column in efficiency and no advantage in construction materials.

Fans and Blowers

These components are of prime importance to a hazardous waste incinerator because they move the air into the incinerator and the flue gas out of the system. The combustion air for most incinerators is supplied by either a forced draft blower or an induced draft fan or both. Hazardous waste incinerators are ideally operated under a slight negative pressure so that all leakage will be inward, but many liquid injection type incinerators do operate under a positive pressure.

Steam/air ejector drive

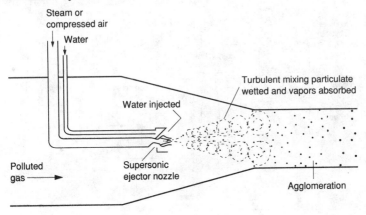

SuperSub fan/ejector drive

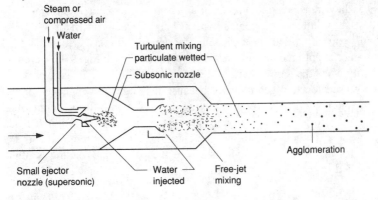

Tandem nozzle fan drive

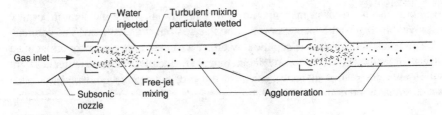

FIGURE 12-27
Hydro-Sonic gas cleaning equipment.

FIGURE 12-28
Ionizing wet scrubber. (courtesy of Ceilcote Co.)

Fans may be driven by electric motors, steam turbines, or hydraulic drives. The speed of the fan may be variable or fixed depending upon the service. Variable speed may be achieved through a variable speed motor or a hydraulic drive. For most incineration systems the fans and blowers are fixed speed motors with a V-belt drive to the fan. Air flow control on these systems is achieved through a mechanical damper on the inlet to the fan.

Coolers and Condensers

A flue gas at adiabatic saturation temperature which has passed through a mist eliminator still remains saturated at a temperature higher than the ambient air. When this flue gas exits the stack and is cooled by the ambient air, a dense steam plume results. Often this plume is conceived to be pollution by the public. To overcome this impression, many incinerator designs employ a condensing step in the process to reduce the dew point of the stack gas and reduce if not eliminate the plume. This can be done with a direct contact system using cooling towers or with a tubular condenser employing either air or water to cool the flue gas. In a geographical area where water is scarce, there is an additional benefit because the condensate may be reused in the process.

Mist Eliminators

Flue gas from an incinerator, especially when it has been adiabatically cooled in a direct water contact system, will contain a significant amount of water. As this gas is further cooled by heat loss in the duct, fine water droplets begin to condense in the air stream. To remove these fine droplets and keep them from entering the stack, a mist eliminator or entrainment separator is used. Mist eliminators are generally impaction type separation devices although sometimes centrifugal separators are used. The impaction separator is either a wire or plastic mesh pad or a series of baffle plates which intercept the droplets and allow them to coalesce so that water may be drained from the system (see Fig. 12-29).

Stacks

At the end of every incinerator there must be a stack to disperse the flue gases to atmosphere. For hazardous waste incinerators the natural draft created by the stack is usually unimportant because the fan forces the flue gas up the stack. The critical design aspect of the stack is not its draft but its ability to disperse the flue gases. The

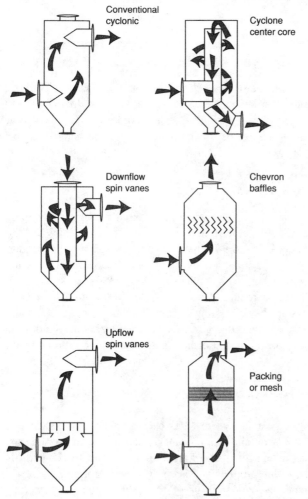

FIGURE 12-29
Entrainment separators.

proper height of the stack depends upon the surrounding terrain, nearby buildings, and prevailing wind direction. Before deciding on the stack location and height, a dispersion analysis of the plume is usually performed. The dispersion analysis estimates the ground concentrations and location of pollutants present in flue gas so that a health risk analysis can be performed (see Sec. 4-4).

The stack must be structurally capable of withstanding the maximum possible wind velocity at its location and its design must incorporate a sampling platform and sampling ports so located that they provide for isokinetic sampling of the flue gas. Ideally this is eight or more stack diameters above the gas inlet and at least five stack diameters from the discharge point. At this sampling level, two or more $4''$ circular sampling ports located at 90 degrees to each other are required.

Example 12-9   Particulate removal. A rotary kiln incinerator and secondary chamber is followed by an adiabatic quench to cool the gases to approximately $185°F$. There are no acid gases present in the flue gas so the only problem is particulate removal. Because the gas has been cooled to saturation, only wet removal systems are practical. This suggests that a venturi scrubber is a reasonable solution.

(1) The flue gas has been examined by an impactor and the results show that 5 weight % of the particles are less than 0.5 micron size while the remaining 95 weight % are larger. The venturi scrubber operates with a $50''$ w.c. gas pressure drop at which it will remove 99% of all particles larger than 0.5 micron.

The stack gas emission limitations are 180 mg/dscm corrected to 7% O_2 in the flue gas. Assuming that the flue gas contains 7% O_2, what is the allowable particle concentration at the inlet to the venturi?

It must be assumed that the venturi scrubber will capture only those particles larger than 0.5 micron and none of those smaller while in actual fact it may capture some small particles.

Solution (1)

Let X = weight of the total particulate into the scrubber.
Capture = $0.99 \times 0.95X = 0.94X$ or 94% of the total in.

$$\text{Therefore: allowable inlet loading} = \frac{180 \text{ mg}/\text{dscm}}{1.0 - 0.94} = 3000 \text{ mg/dscm}$$

(2) For small gas flows the venturi is economical, especially because it is a low capital cost item, but for large flows the horsepower used by the fan would be a high operating cost. An ionizing wet scrubber would only require $4''$ w.c. pressure drop per stage. Each stage is 86% efficient in particulate removal and the ionizing wet scrubber efficiency is independent of particle size. At the same inlet loading, how many ionizing stages are required to meet regulations?

Solution (2)

If each stage is 86% efficient then: $1 - 0.86 = 0.14$ or 14% of the inlet concentration would pass through the scrubber so it would require two stages to reduce 3000 mg/dscm to less than 180 mg/dscm.

(3) Another option for particulate removal is a fabric filter, or baghouse; however, this cannot be used at saturation temperatures because of certain condensation on the filter bags which would make bag cleaning impossible. Therefore, a partial water quench would be utilized to bring the flue gas temperature down to 350°F. Assuming a gas flow of 10,000 scfm at 350°F, and a pulse jet baghouse with a 99% removal efficiency, what filter bag area would you specify? Air to cloth ratios, i.e., ft³/ft² are usually about 4/1 for a pulse jet baghouse.

Solution (3)

$$\text{Gas flow} = 10,000 \times \frac{350 + 460}{60 + 460} = 15,576 \text{ ACFM}$$

If A/C = 4 then: 15,576/4 = 3894 ft² of bag area

(4) If each filter bag is 6″ in diameter and 10 ft long, and the baghouse is divided into four chambers, three of which are "on line" and one is "off line" in the cleaning mode, how many bags are required per section?

Solution (4)

Per bag area = 15.71 ft²
"On line" no. bags = 3894 ft²/15.71 ft² = 248 bags
If three compartments are on line, then 248/3 = 83 bags each plus an off line compartment of 83 bags.
From a practical standpoint, baghouses of this size are usually modular in design so a baghouse of this size or larger would be chosen for this application.

Example 12-10   Acid gas removal. A hazardous waste incinerator, equipped with a waste heat boiler, delivers 30,000 scfm of flue gas at 600°F. The flue gas is contacted with a lime slurry in a dry scrubber where acid gases are partially neutralized and the gases cooled to 320°F. They then pass through a baghouse for particulate removal and finally are cooled to adiabatic saturation and further neutralized by a 10% caustic solution in a conterflow packed tower.

The flue gas contains 700 lb/hr of SO_2 and 500 lb/hr of HCl. The dry scrubber lime feed rate is 1.6× stochiometric, and it is 80% efficient in removing SO_2 and 90% efficient in HCl removal.

1. What is the lime feed rate in lb/hr?

2. How many lb/hr of SO_2 and HCl will be in the flue gas entering the caustic scrubber.

Assume that CO_2 in the flue gas does not react with the lime.
Assuming the lime as CaO
$CaO + SO_2 \rightarrow CaSO_3$
$CaO + 2HCl \rightarrow CaCl_2 + H_2O$
Molecular weights: CaO = 56; SO_2 = 64; HCl = 36.5

Solution.

700 lb/hr/64 lb/mole = 10.9 moles/hr SO_2 in flue gas
500 lb/hr/36.5 lb/mole = 6.85 moles/hr HCl in flue gas

SO_2 requires 10.9 moles/hr CaO × 1.6 = 17.44 moles/hr

HCl requires 6.85 moles/hr CaO × 1.6 = 10.96 moles/hr

Total lime as CaO = 28.4 moles/hr

1. Total lime usage = 28.4 moles/hr $\times$ 56 lb/mole = $\underline{\underline{1,590.4 \text{ lb/hr}}}$

2. 0.2 $\times$ 700 lb/hr = $\underline{\underline{140 \text{ lb/hr } SO_2 \text{ in flue gas}}}$

0.1 $\times$ 500 lb/hr = $\underline{\underline{50 \text{ lb/hr HCl in flue gas}}}$

12-10   INSTRUMENTATION
Introduction

The basic control concepts utilized in a hazardous waste incineration system are essentially the same as those employed in any process. These can be quite simple and basic, electronic or pneumatic, but because they are utilized on a hazardous waste incinerator they must be reliable. They must be able to measure all of the system variables and to shut down the system *if* there is any indication of malfunction which might cause a discharge of the waste to the environment. These specific monitors are designed as automatic hazardous waste feed cutoff devices for the system. The basic measurement parameters are temperature, flow, pressure and pressure differential, pH, and level.

The control strategy for most incinerators is straightforward. The primary combustion chamber exit gas temperature is controlled automatically by the primary combustion air damper, which in turn is linked with the burner fuel control valve to maintain temperature. The secondary combustion chamber exit temperature is automatically maintained by modulation of the combustion air flow rate and the burner fuel control valve. In a wet APC system the variable venturi throat is modulated to maintain constant pressure drop across the venturi. Draft is controlled by a damper on the induced draft (ID) fan inlet. Minor variations occur from plant to plant. For example, some control room operators prefer to control the draft manually while leaving the primary burner fuel and air on automatic control, the secondary burner fuel on automatic control, and the secondary burner combustion air on manual control. These variations occur due to control and incinerator system dynamics.

Temperature

Temperatures are measured by thermocouples installed in standard industrial protection tubes or thermowells. These are to be installed in the combustion gas stream and away from burners to obtain accurate temperature measurement.

Thermocouples are one of the more difficult sensing devices to maintain. Waste streams often create slagging conditions in the incinerator which affect thermocouple readings in both the primary and secondary chambers. Duplicate or dual element thermocouples should be provided in both chambers which will indicate major deviations. Often, a variation in readings is noted before a shutdown occurs and replacement can be made upon shutdown. Manual logging of data on an hourly or once-per-shift basis can alert operators to instrument problems by comparing readings to normal expected levels and/or previous readings.

A poor thermocouple location may also cause problems. In some incinerators, primary chamber exit gas thermocouples are located too close to air seals. An increase in leakage created by kiln seal wear or worn airlocks will result in a low temperature at the thermocouple.

An irregular burner flame in the secondary combustion chamber due to a refractory spalling or a liquid atomizer failure may also affect the thermocouple reading. A daily review of this data will often spot the problem before a serious condition occurs.

Flow

Flow measurements are important for all waste feeds, combustion air, scrubbing solutions in the APC system, and flue gas flow rates.

The feed rate of the waste to the incinerator is monitored by mass flow devices, weigh belt scales for soils, and similar instruments. The readout in the control room gives instantaneous feed rate plus integrated totals.

Gas flow rates are measured by differential pressure across an orifice, pitot tubes, annubars, and other methods. These measurements are also important and require regular calibration of the instrumentation.

Pressure and Differential Pressure

Pressures are indicated on standard industrial pressure and vacuum gauges for low pressures and draft and on industrial Bourdon tube gauges for high pressures. In most waste incinerator systems it is important to maintain a negative pressure throughout. Pressure controllers are used to modulate dampers at the inlet of constant speed ID fans or to vary the speed of variable speed fans to maintain a constant pressure at the inlet to the fan.

Differential pressure transmitters and controllers are employed to modulate throat valves in venturi type scrubbers or to set atomizing steam or air pressures at a constant differential pressure above the waste liquid pressure.

pH

Control of pH in wet scrubbers is important to ensure that the neutralizing agent for acid gas removal is added in sufficient quantities. By maintaining a set pH value, the maximum acid gas content of the flue gas can be virtually guaranteed.

Level

Level control in wet scrubbing systems, liquid feed tanks, and storage tanks is obviously important. It is more difficult with sludges and corrosive liquids than with clean, non-corrosive wastes so the selection of the proper system is important. The options include differential pressure systems such as diaphragms and bubblers, sonic and nuclear sensing units, conductivity sensors, floats, and gauges.

Table 12-8 lists the principal process variables which are monitored in the operation of an incinerator.

Automatic Waste Feed Cutoff

RCRA regulations require that every incinerator provides an automatic cutoff of waste feed streams under conditions which might lead to a discharge of the waste to the environment.

TABLE 12-8
Principal process variables monitored

Process variable	Measurements
Feed rate of waste*	Feed inlet
Temperatures	PCC exit gas*
	SCC exit gas*
	Stack gas
	Venturi scrubber inlet gas or baghouse inlet gas
Pressures	PCC feed end draft
	Venturi scrubber pressure drop or baghouse pressure drop*
Flows	Venturi scrubber water flow rate*
	Stack gas velocity*
Stack gas composition	CO*
	O_2*

*Recorded continuously via strip chart and/or digital data logger.

Instrumentation is provided to monitor process conditions, to provide data for ensuring compliance with regulatory requirements, and to ensure appropriate process response, control, and operational flexibility. The safety interlocks and shutdown features comprise a major portion of the control system. Periodically, the system should be checked to ensure that waste feeds are cut off when one of these shutdown limits has been exceeded. Most operating permits require that the waste feed cutoffs be tested monthly and the information logged for review by the regulatory agency. The conditions under which the automatic waste feed shutoff (AWFSO) system and emergency shutdown (ES) operate are noted in Table 12-9.

12-11   CONTINUOUS EMISSION MONITORS

Parameters most critical to incinerator operation are the combustion chamber temperature, stack carbon monoxide, oxygen, and the combustion gas flow monitors. They are used as indicators to ensure the proper destruction and removal of the toxic components. For a continuously operated system, such as a commercial disposal facility, redundant instrumentation is necessary to minimize emissions of hazardous chemicals and eliminate or minimize waste feed cutoffs and down time. The installation, location, calibration, and maintenance of these instruments are important to the continued performance of the system.

RCRA regulations were designed to ensure that incineration systems that treat hazardous wastes operate in an environmentally responsible manner. If the standards are met, the trial burn should identify the operating conditions necessary to ensure the incinerator's ability to meet or exceed the performance standards throughout the life of the permit.

Currently, monitoring technologies are available to ensure incinerator operation at high combustion efficiency (> 99.9%) by continuous measurement of carbon monoxide (CO) and carbon dioxide (CO_2). Monitors measuring waste stream flow rates, combustion air flow, stack gas flow, and oxygen level in the stack also indi-

TABLE 12-9

Typical events triggering emergency control systems

Item	Control action
High stack temperature	ES
Loss of fan draft	ES
Loss of water flow to wet scrubber	ES
Power failure	ES
High secondary outlet temperature	ES
Low draft in kiln for > 15 seconds	AWFSO
Low absorber pH	AWFSO
Low secondary chamber temperature	AWFSO
Low oxygen (< 3%) in the stack	AWFSO
Low SCC temperature	AWFSO
High CO, > 100 ppm, corr. 7% O_2, 1 hr rolling avg	AWFSO
High gas flow velocity in stack	AWFSO
High waste feed rate	AWFSO

Notes:
1. ES = Emergency shutdown
2. AWFSO = automatic waste feed shutoff
3. When ES occurs, fuel and waste feed are stopped.
4. In addition to automatic operation, both the AWFSO and ES can be activated manually at the operator's discretion.

cate that the total heat release (Btu/hr) from the system is within design and permit conditions.

The following monitors are needed to ensure that DRE is maintained at the levels achieved in the trial burn:

- Temperature in the primary and secondary chambers
- Oxygen in the stack
- Carbon monoxide in stack corrected to 7% O_2 dry volume
- Stack gas volume flow

Instruments which analyze stack gases for CO, CO_2, SO_2, and HCl are referred to as continuous emission monitors (CEMs). The U.S. EPA prefers extractive systems, which condense water vapor before sending the gas to the analyzer. In situ CEMs are less costly but are usually used for process feedback rather than regulatory compliance. Redundant CEM systems are frequently used to allow maintenance, calibration, and/or replacement of the monitor without shutting down the incinerator.

Carbon Monoxide

The EPA has established a level of 100 ppm CO, rolling average corrected to 7% oxygen on a dry basis, as an acceptable level to ensure > 99.99% DRE. This has been determined from many trial burns to be conservative. Accurate CO readings require regularly scheduled calibration of each monitor. The calibration must be done by a qualified instrument technician, trained in this field, and certified calibration gases must be used.

Carbon Dioxide

Carbon dioxide monitors have been added to many incinerators to assess combustion efficiency. The combustion efficiency is calculated as follows:

$$\text{C.E.} = \frac{CO_2}{CO_2 + CO} \times 100 \% \tag{12-12}$$

where CO_2 = volume concentration (dry) of CO_2 (parts per million, volume, ppmv)
   CO = volume concentration (dry) of CO (ppmv)

In some installations, this is continuously recorded. If the combustion efficiency drops below 99.9%, it indicates that CO levels are probably increasing, which alerts the operator to take corrective action before the high CO triggers automatic waste feed cutoff.

Oxygen

Oxygen analysis, when combined with stack gas volume flow, can be used as a means of determining total heat input to the system. An increase in oxygen at the maximum stack gas flow rate indicates a reduction in the maximum Btu/hr heat release. This may be traced to kiln seal wear or leakage to the APC system. Incinerator systems which use ionizing wet scrubbers for particulate removal in the APC system have this air leakage. Also air in-leakage through the ID fan shaft seal which occurs at high-draft conditions as employed with a venturi scrubber will result in an O_2 level in the stack higher than in the SCC outlet. If the O_2 level drops at the maximum stack flow rate, an increase in total heat release is probably occurring, due to a failure in the waste flow monitors, a "hot" charge into the solid waste feed, or failure of an atomizing nozzle. All three of these events place more fuel into the incinerator, which leads to the increase in heat release.

Example 12-11   Oxygen correction. Current hazardous waste regulations require that emissions be corrected to 7% oxygen in the stack on a dry basis.
    The carbon monoxide (CO) concentration in the stack from a hazardous waste incinerator is measured at 20 ppm by volume at a temperature of 175°F. The oxygen concentration in the stack is measured to be 12% by volume on a wet basis. In another test the particulate concentration in the stack is measured at 20 mg/dscf at a stack oxygen concentration of 12% measured on a dry basis at 70°F. The water content of the stack gas is 10 mole %.

1. What is the corrected CO content in the stack gas?
2. What is the corrected particulate concentration?
3. Would this stack meet federal regulations (180 mg/dcsm corrected to 7% O_2)?

Solution.

O_2	12%
H_2O	10%
N_2+CO_2	78% (by difference)

On a dry basis the oxygen content is $12/(1 - 0.10) = 13.33\%$

1. The correction formula provided in the regulations (see 40CFR Part 264.343) is as follows:

$$P_c = P_m \times \frac{14}{21 - Y}$$

where P_c = corrected concentration

        P_m = measured concentration

        Y = O_2 concentration in stack (dry basis)

Therefore:

$$P_c = 20 \times \frac{14}{21 - 13.33} = \underline{36.5 \text{ ppm of CO}}$$

2. Since the O_2 in the second test is measured on a dry basis, no correction for water content is required.

$$P_c = 20 \times \frac{14}{21 - 12} = \underline{\underline{31.1 \text{ mg dscf}}}$$

3. 31.1 mg/ft^3 × 35.3 ft^3/m^3 = $\underline{\underline{1098 \text{ mg/dscm}}}$

No, it would be well in excess of regulations.

Hydrogen Chloride

In the U.S., RCRA regulations require 99% removal (or 4 lb/hr maximum) of HCl. The new guidelines[17] for HCl and Cl_2 emissions are risk-based which may result in lower emission limits. Most incineration systems do not include a continuous HCl monitor. The HCl is sampled and analyzed during the trial burn based on the maximum chlorine content in the feed. APC systems which use wet scrubbers for acid gas control employ pH control of the scrubbing solution to ensure that the HCl is removed.

For the dry APC systems, if the combustion gas flow, neutralizing agent flow and the baghouse/electrostatic precipitator (ESP) gas pressure drop are monitored and within the proper limits and the chlorine level in the feed is not higher than that experienced in the trial burn, proper HCl removal efficiencies are likely.

HCl monitors have been developed in the past few years. They are used routinely in western European countries for monitoring controlled and uncontrolled HCl emissions from municipal and hazardous waste incinerators. The German government has established HCl standards and relies on the CEM data to demonstrate compliance. Experience in the United States, however, is limited.

Total Hydrocarbons

There is currently no U.S. federal limit to total hydrocarbon emissions (THC) from hazardous waste incinerators but some EPA regions and some states are willing to allow operators to use THC as an indicator of proper operation in place of the CO rolling average, especially with batch-fed rotary kilns. When the operator cannot comply with the 100 ppm sixty-minute rolling average, a maximum 20 ppmv THC limit has been acceptable.

Combustion Gas Velocity

RCRA requires that flue gas velocity be monitored and recorded. The velocity is an indirect measure of total heat release, and high gas velocities usually mean higher particulates in the exit gas.

Measuring combustion gas velocity is not a simple task. The usual direct methods are pitot tubes, annubars (which are pitot tubes that average velocity pressure across a duct), and vortex shedding flow meters. These all work well in clean gas at ambient temperature but not as well at high temperatures, corrosive conditions, or in dirty gas streams where they plug and foul. They are also sensitive to disturbances in flow patterns upstream of the probe.

The simplest and most reliable method for measuring combustion gas flow is by the induced draft (ID) fan motor amperage. During stack tests, four or more EPA Method-1[18] velocity traverses are made, each at a different ID fan damper setting, and hence, different gas flow rate. The data is then plotted and correlated with the motor amperage. This method works well for the majority of incinerator installations which use a constant-speed centrifugal fan and an inlet VANE-type damper to control flow. A system using a variable-speed fan would require a more complex algorithm incorporating fan rpm, pressure drop, and amperage to calculate gas volume.

Non-Continuous Parameters

At present, the RCRA standard for particulate emissions from incinerators is 0.08 gr/dscf corrected to 7% O_2. It is well-recognized that particulate emissions may contain trace quantities of toxic metals. Concern about health hazards associated with metal emissions, both to the atmosphere and by ash leaching into the ground water, has led to regulatory "guidelines."[19]

There are no CEMs that adequately provide information related to the quantitative particulate content of a flue gas. Opacity meters are used in the power industry for coal and oil-fired boilers, but are reliable only for dry stack gases. There are no federal requirements for continuous opacity measurement on RCRA incinerators. Some incinerators have installed opacity meters as an additional source of operating information.

12-12   TRIAL BURNS

Trial burns must be run on all hazardous waste incinerators in accordance with 40CFR Part 270.62. The trial burn is run to verify that the incinerator is capable of operating in compliance with performance standards outlined in 40CFR Part 264.343.

An analysis of the flue gas from the operating unit must be made to demonstrate a destruction and removal efficiency of 99.99% or greater for principal organic hazardous constituents contained in the waste.

The flue gas must also be sampled and analyzed to determine the amount of particulate present in terms of weight units per unit of dry volume corrected to 7% oxygen. The federal standard is currently 0.08 grains/dscf. Many states have more stringent emission standards for particulate.

Hazardous waste incinerators must also conform to HCl emission limits of no more than 4 lb/hr or 99% removal of all HCl in the flue gas, but recent guideline changes will place these emissions on a risk basis and include an HCl/Cl_2 total flue gas content.

Carbon monoxide emissions may not exceed 100 ppmv on the basis of a sixty-minute rolling average corrected to 7% oxygen. An exception to this regulation is made when the operator is willing to perform a risk assessment and monitor and maintain total hydrocarbons below 20 ppmv.

Flue gas analysis for oxygen is necessary to provide a correction for the CO and particulate analysis. Flue gas analysis for CO_2 is required for TSCA trial burns (i.e., substances such as PCBs which come under the Toxic Substances Control Act) to determine combustion efficiency.

Ten priority metals[20] have been listed in the EPA guidelines. Permissible feed rates and emission levels have been established for these metals and may not be exceeded unless a dispersion analysis and risk assessment is made for all metals exceeding these levels. Individual state air regulations may require analysis for other metals also.

Sampling and analysis for other pollutants such as sulfur oxides and NO_x may also be required during the trial burn if they are expected to be present in any quantity. In some trial burns, the EPA or the state have required sampling and analysis for dioxins (PCDDs) and furans (PCDFs).

Sampling for all of these parameters is done in the stack of the incinerator. Samples are taken, and analyses are subsequently made in the laboratory except for O_2, CO, and CO_2, which must be made in the field.

12-13 MOBILE SYSTEMS

The idea for a transportable or mobile incinerator grew out of the Superfund program (see Sec. 2-4). It was obvious that incineration was a reasonable method of remediation; however, it was no simple job to dig up an entire site and transport it to one of the few commercial hazardous waste incinerators in the U.S.A. A site not near any commercial disposal facility might spend more in transportation than for any other aspect of the cleanup program.

By placing the incinerator on wheels (mobile system), or making it transportable in modules, the incinerator can be taken to the Superfund site rather than the site being moved to the incinerator. The idea caught on quickly, and a number of companies entered the field with a variety of portable and mobile units.

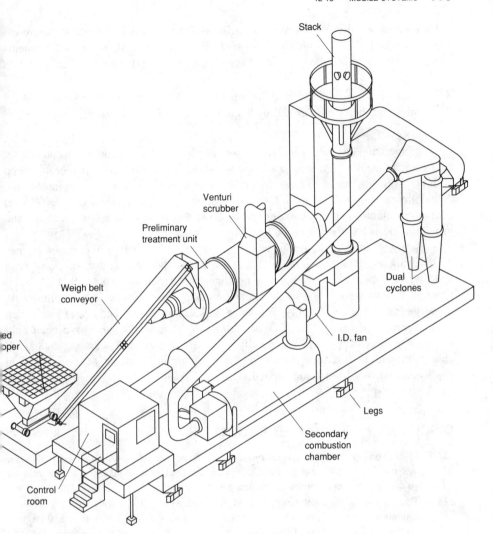

FIGURE 12-30
Mobile (transportable) incinerator.

Today, available transportable systems include rotary kiln systems, fluid bed incinerators, and infra-red units. Most completely mobile units are small rotary kilns. These systems can be set up on-site and ready for operation in a time period as short as one week or as long as several months.

The technology has developed into three basic areas:

• Incineration where the soils are brought to a temperature of more than 1200°F and the subsequent vapors burned in a secondary chamber (see Fig. 12-30).

- Incineration of non-RCRA wastes such as contaminated soil from underground storage tanks (UST), which contain fuel oil or gasoline. Many of these systems merely de-volatilize the soils at lower temperatures and burn the flue gas as a VOC.
- Low temperature de-volatilizers which indirectly heat the waste to vaporize the organics and then condense the hydrocarbon vapors for reclamation.

Incineration has been widely used for site cleanup. For example: a wood chemical treatment facility in Mississippi where telephone and electric poles had been treated with creosote was found to have a significant amount of creosote contaminated soil. A mobile/transportable rotary kiln incinerator was used to process over 9000 tons of the contaminated soil, destroying the creosote and other organic contaminants and providing clean soil that could be landfilled on site.

Many other technologies have been proposed for industrial and commercial use. Many of these are still attempting to find a niche in the marketplace, but most have been rejected because of economics or capacity restrictions. The U.S. EPA SITE program has attempted to give the more promising technologies the opportunity to prove their commercial viability. Few have broken this barrier, and even those which have seen commercial use such as the infra-red incinerator system have been only moderately successful. There are literally reams of literature and operating data available to the interested scholar concerning these "emerging" systems, but their practical application has not been demonstrated.

DISCUSSION TOPICS AND PROBLEMS

12-1. 1500 pound per hour of a waste mixture of 40% benzene, 50% toluene, and 10% water is burned with 25% excess air. Determine the total heat release and the percent by volume of each component in the flue gas.

12-2. The mixture in the table below is being incinerated at 2200°F with 50% excess air, a residence time of 2.3 seconds, and a Stack Flow Rate = 37500 dscfm. Principal Organic Hazardous Constituents (POHCs) for this waste are benzene, tetrachlorophenol, and toluene. Determine if this emission meets requirements and comment on the results. The following table provides incinerator emissions. Oxygen in the flue gas is 7.0%.

Compound	Formula	Mol. wt.	Inlet (lb/hr)	Outlet (lb/hr)
Benzene	C_6H_6	78.11	1025	0.087
Chlorobenzene	C_6H_5Cl	112.5	278	0.034
Ethylbenzene	C_8H_{10}	106.17	780	0.089
Tetrachlorophenol	C_6HOHCl_4	231.9	760	0.056
Toluene	C_7H_8	92.10	756	0.024
Xylene	C_8H_{10}	106.17	168	0.204
HCl		36.45		4.3
Particulates				20.3

12-3. Using the flue gas generated in Example 12-1 calculate the flame temperature in the incinerator if there is a 10% heat loss.

12-4. Good incinerator design provides for a flue gas residence time of 2 seconds in a liquid incinerator and a gas velocity of 20 ft/sec. Using the information provided in Example 12-1, determine the inside diameter and length of the incinerator. Assume an incinerator temperature of 3163°F and a gas velocity of 20 ft/sec.

12-5. Given the following stack data: CO = 17 ppmv at 190°F; O_2 = 10% by volume (wet basis); particulates are 11 mg/dscf. Determine the CO and particulate concentration corrected to 7% O_2.

12-6. Estimate the heat released from a rotary kiln incinerator 12 ft inside diameter × 45 ft long.

12-7. An incinerator which normally burns waste with a heating value of 700—1,000 Btu/lb and is designed to handle 2×10^7 Btu/hr must burn a rubber waste with a heating value of 12,000 Btu/lb. Estimate the maximum size of a waste batch for this system to perform effectively. Assume the residence time in the incinerator is 6 min.

12-8. Determine how much caustic will be required to reduce SO_2 emissions from the incinerator in Example 12-10 to 25 tons/year at 98% efficiency in the packed column scrubber if the operating year is 7200 hours.

12-9. How much additional caustic is required to reduce the HCl emissions from the incinerator in Example 12-10 to less than 4 lb/hr?

12-10. A VOC thermal incinerator operates at 1500°F. How much natural gas at 1000 Btu/ft^3 will be required to incinerate the emissions from the film dryer described in Example 12-6. Assume the diluted mixture of acetone vapor and air is at 80°F.

12-11. Determine the adiabatic saturation temperature and cooling water flow for a spray tower. Assume the flue gas is essentially air at 1800°F and contains 12% water by weight. The total gas flow is 80,000 acfm. The specific weight of air at 60°F is 0.0763 lb/ft^3.

12-12. An incinerator burns 150 lb/hr chlorobenzene and has a stack gas containing 5% oxygen by volume on a dry basis. Determine the percent excess air.

12-13. Discuss briefly the types of particulate removal devices that might be appropriate to a commercial hazardous waste incinerator using a rotary kiln. What factors would control the selection of air pollution control equipment?

12-14. An incinerator release 8,000 scfm of flue gas containing particulates that must be removed by fabric filtration. Assume an "air to cloth" ratio of 4.0 and a baghouse contain six modular units with one always off-line. Determine the number of bags per module.

REFERENCES

1. The Resource Conservation and Recovery Act of 1976, 42 USC 6901 et seq., 1976.
2. The Hazardous and Solid Waste Amendments of 1984 (HSWA) (P.L. 98-616), November 9, 1984.
3. HSWA, op. cit., Sect. 3004(o)(1)(B).
4. Code of Federal Regulations (40 CFR 270.19).
5. Code of Federal Regulations (40 CFR 264, Subpart O).
6. Code of Federal Regulations (40 CFR 264.343).
7. Perry & Green: *Perry's Chemical Engineers' Handbook*, sixth edition, McGraw-Hill, New York, NY, 1984, p. 4–53.
8. Theodore, L., and J. Reynolds: *Introduction to Hazardous Waste Incineration*, John Wiley & Sons, Inc., New York, NY, 1987.
9. Nittetu Chemical Engineering, Ltd., Tokyo, Japan.

10. West, R. C., Ed., *Handbook of Chemistry & Physics*, 64th Ed., CRC Press, 1983.
11. *North American Combustion Handbook*, North American Manufacturing Co., Cleveland, OH, 1979.
12. Flaregas, Inc.
13. Ross, R. D.: "Incineration of Solvent-Air Mixtures," *Chemical Engineering Progress*, August 1992.
14. Reeco, Morris Plains, N.J.
15. National Air Oil Co., Philadelphia, PA.
16. T-Thermal Co., Conshohocken, PA.
17. *Guidance on Metals and Hydrogen Chloride Controls for Hazardous Waste Incinerators*, vol. IV, U.S. EPA 1989.
18. Code of Federal Regulations, Title 40, Parts 60 and 61.
19. *Guidance on Metals and Hydrogen Chloride Controls for Hazardous Waste Incinerators*, vol. IV, U.S. EPA, 1989.
20. Ibid.

ADDITIONAL READING

Bayer, Johan E.: "Incinerator Operations," *Journal of Hazardous Materials*, vol. 22, no. 2, pp. 243–247, Nov. 1989.

Brunner, Calvin: "Industrial Sludge Waste Incineration," *Environmental Progress*, vol. 8, no. 3, pp. 163–6, August 1989.

Carnes, Richard A.: "RCRA Trial Burns: Adventures at Rollins," *Journal of Hazardous Materials*, vol. 22, no. 2, pp. 151–160, Nov. 1989.

Cooper, Eliot D.: "Incineration Permitting: The Long and Winding Road," *Hazardous Materials & Waste Management*, January–February 1986, pp. 10–13.

Cundy, V. A., T. W. Lester, C. Leger, G. Miller, A. N. Montestruc, S. Acharya, A. M. Sterling, D. W. Pershing, J. S. Lighty, G. D. Silcox, and W. D. Owens: "Rotary Kiln Incineration—Combustion Chamber Dynamics," *Journal of Hazardous Materials*, vol. 22, no. 2, pp. 195–219, Nov. 1989.

Cundy, V. A., T. W. Lester, A. M. Sterling, J. S. Morse, A. N. Montestruc, Christopher B. Leger, and S. Acharya: "Rotary Kiln Injection I. An In Depth Study—Liquid Injection," *Journal of the Air Pollution Control Association*, vol. 39, no. 1, pp. 63–75, Jan. 1989.

Freeman, Harry: "Recent Advances in the Thermal Treatment of Hazardous Wastes," *Management of Hazardous and Toxic Wastes in the Process Industries*, vol. 19, no. 4, pp. 34–36, 1987.

Leuser, R. M., L. A. Valazquez, and A. Cohen: "Remediation of PCB Soil Contamination by On-Site Incineration," *Journal of Hazardous Materials*, vol. 25, pp. 375–385, December 1990.

McFee, John, George Ramussen, and Craig Young: "The Design and Demonstration of a Fluidized Bed Incinerator for the Destruction of Hazardous Organic Materials in Soils," *Journal of Hazardous Materials*, vol. 12, no. 2, pp. 129–142, November 1985.

McGowan, T., and R. Ross: "Hazardous Waste Incineration Is Going Mobile," *Chemical Engineering*, vol. 98, pp. 114–123, October 1991.

Munoz, Henry, Frank Cross, and Joseph Tessitore: "Comparisons Between Fluidized Bed and Rotary Kiln Incinerators for Decontamination of PCB Soils/Sediments at CERCLA Sites," *Proceedings of the National Conference on Hazardous Wastes and Hazardous Materials*, pp. 424–245, March 4–6, 1986.

Santoleri, Joseph J.: "Design and Operation of Thermal Treatment Systems for Contaminated Soils," *Proceedings—A&WMA Annual Meeting*, Air & Waste Management Association, pp. 79–93, 1989.

Santoleri, Joseph J.: "Rotary-Kiln Incineration Systems: Operating Techniques for Improved Performance," *Hazardous and Industrial Wastes—Proceedings of the Mid-Atlantic Industrial Waste Conference*, Technomic Publishing Company, Inc., pp. 743–758, 1990.

Stoddart, Terry L., MAJ, USAF, and Jeffrey J. Short: "Dioxin Surrogate Trial Burn: Full Scale Rotary-Kiln Incinerator for Decontamination Soils Containing 2,3,7,8-Tetrachlorodibenzo-P-Dioxin," *Chemosphere*, vol. 18, no. 1–6, pp. 355–361, 1989.

Takeshita, R., and Y. Akimoto: "Control of PCDD and PCDF Formation in Fluidized Bed Incinerators," *Chemosphere*, vol. 19, no. 1–6, pp. 345–352, 1989.

Theodore, L., and J. Reynolds: *Introduction to Hazardous Waste Incineration*, John Wiley & Sons, Inc., New York, NY, 1987.

Tilly, Jean, Gary Dietrich, and Dan Pyne: "Exposure and Risk Assessment for a Proposed Hazardous Waste Incinerator," *Environmental Progress,* vol. 8, no. 3, pp. 207–211, August 1989.

U.S. Environmental Protection Agency, Office of Solid Waste: *Permitting Hazardous Waste Incinerators.* EPA/530-SW-88-024, Washington, D.C., April 1988.

Wiley, S. K.: "Incinerate Your Hazardous Waste," *Hydrocarbon Processing,* vol. 66, no. 6, pp. 51–54, June 1987.

Wood, R. W., R. E. Bastian, et. al.: "Rotary Kiln Incinerators: The Right Regime," *Mechanical Engineering,* vol. III, no. 9, pp. 78–81, September 1989.

ACKNOWLEDGMENTS

Chap. 12 was prepared by Richard D. Ross and Joseph J. Santoleri of Four Nines, Inc. The authors wish to acknowledge the following who assisted with the preparation of this chapter: Jack Riggenbach, *ERM-Southwest*, and Debra J. Brovero, *Bucknell University*. The principal reviewer for **The ERM Group** was Paul H. Woodruff, P.E., *ERM, Inc.*

CHAPTER
13

LAND
DISPOSAL

Now I would give a thousand furlongs of sea for an acre of barren ground.
Shakespeare

Landfills will remain a significant part of our hazardous waste management practice. A landfill is defined as that system designed and constructed to contain discarded waste so as to minimize releases of contaminants to the environment. Landfills are necessary because 1) other hazardous waste management technologies such as source reduction, recycling, and waste minimization cannot totally eliminate the waste generated and 2) hazardous waste treatment technologies such as incineration and biological treatment produce residues. Thus the generation of hazardous waste can not be reduced to zero in the foreseeable future. It is with this inevitability in mind that the topic of land disposal is addressed.

Historically, many land disposal facilities did not provide adequate environmental protection. As a result, the total environmental and societal costs of landfilling were not considered. Society is now paying these costs in the form of cleaning up many old sites, such as those identified by the Superfund program in the United States and by contaminated land inventories in the European Community. This has heightened the public sensitivity to the construction of new storage and land disposal facilities. With this historical perspective, what are the responsibilities of modern managers of hazardous waste? It is important to promote both minimization of the amount of hazardous waste generated at the source and the treatment of the irreducible minimum. For the unavoidable hazardous waste and the residue from waste treatment processes, it is necessary to design and construct safe land disposal facilities. Safe facilities must employ the best of the available technologies, incorporating state-of-the-art design and analysis methods.

When establishing guidelines for the design, construction, and operation of safe disposal facilities, the designers must anticipate imperfection and build physical and institutional redundancies to accommodate inevitable leakage. The goal must be a facility that minimizes releases of contaminants into the environment to an acceptable, perhaps non-detectable, rate. This contaminant release rate could be considered a geologically slow rate. Remember, many contaminants, particularly toxic metals, originated in our natural environment. At low concentrations many of these metals are not only non-toxic but are necessary for the sustenance of life.

Land disposal, as a *principal* hazardous waste management technique, is a vestige of the past; presently, the emphasis on landfilling is reduced. However, land disposal remains a necessary part of any hazardous waste treatment technology for residuals and is an attractive alternative as a component of remediation of uncontrolled hazardous waste sites. Professional waste management requires a decision on the final destination of these materials. Confining our choices to planet Earth, we have only three options: air, water, or land. These three media are interrelated (e.g., what is initially disposed of on the land may end up in the water). Nonetheless, our decisions should be made in full recognition of our options: air, water, or land?

Example 13-1   Justification of land disposal.   Select a hazardous waste treatment technology described in the previous chapters and identify the hazardous waste management technology and the specific need, if any, for land disposal facilities.

Solution. Incineration: The incineration of hazardous waste creates a number of secondary waste products, ash the most obvious. In addition, depending upon the incinerator design and the wastes being burned, residues from air emission controls may require landfilling.

It is useful to differentiate between land disposal and land treatment, often known as land farming. The concept behind **land disposal** is that landfilling represents the ultimate fate of the waste. Land disposal also includes deep well injection, a topic covered in only an introductory manner in this textbook. **Land farming** is a treatment technique wherein biologically degradable wastes are placed onto the land in low concentrations so that microbes can degrade the materials.

It is important to note the difference between land disposal facilities and storage facilities. Land disposal facilities represent a hazardous waste management technique that constitutes a final placement of the waste. In contrast, **storage facilities** represent a temporary management technique where the waste has not yet reached its final destination. Storage facilities hold waste prior to its shipment for treatment such as incineration, physical-chemical treatment, or reuse and recycling.

Examples of storage facilities include pits, ponds, lagoons, tanks, piles, and vaults. Storage facilities are designed such that waste can be retrieved whereas land disposal facilities are designed to function in perpetuity. Storage facilities and disposal facilities employ many of the same precautions in design and construction. This chapter focuses on land disposal; it should be noted that the approach employed for land disposal facilities (to minimize the rate of contaminant migration into the environment) is often the same approach employed for the design and construction of storage facilities.

The design of municipal solid waste landfills will not be covered in the textbook, even though municipal solid waste landfills are used worldwide. However, because municipal solid waste landfills receive significant quantities of household hazardous waste, many of the design concepts and materials employed for hazardous waste land disposal facilities are also employed for municipal solid waste facilities. Note that municipal solid waste landfills can be (but typically are not) designed and operated as bioreactors because much of the waste is decomposable.[1] The overall design of secure land and disposal facilities includes 1) control of the top to minimize air emissions and infiltration of precipitation and 2) control of the bottom to maximize the collection of leachate and minimize contaminant transport through the bottom.

A simplified cross section of a hazardous waste landfill, presented on Fig. 13-1, depicts the major components of the systems engineered to protect the environment from the landfill wastes. The components of this system and their interactions form the basis of design and analysis details presented in this chapter: 1) the components of storage and disposal facilities and their function, 2) the materials and their properties. Land disposal regulations and their history are covered in Secs. 2-2 and 2-3; therefore, the focus of this chapter is only on the technologies employed in the design and construction of land disposal facilities.

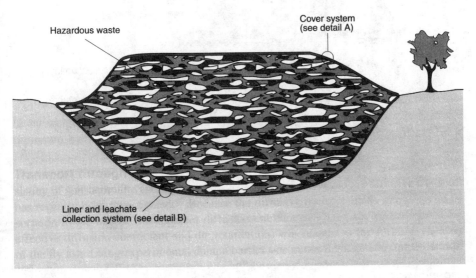

FIGURE 13-1
Cross-section of a hazardous waste landfill.

Detail A – Cover system

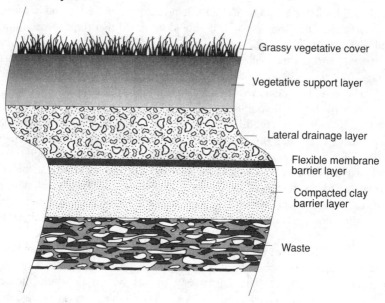

- Grassy vegetative cover
- Vegetative support layer
- Lateral drainage layer
- Flexible membrane barrier layer
- Compacted clay barrier layer
- Waste

Detail B – Linear and leachate collection systems

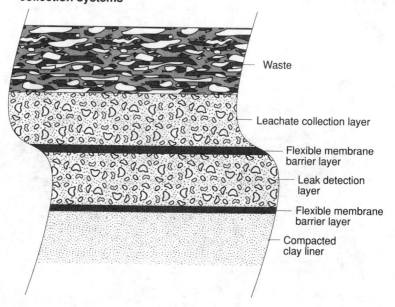

- Waste
- Leachate collection layer
- Flexible membrane barrier layer
- Leak detection layer
- Flexible membrane barrier layer
- Compacted clay liner

FIGURE 13-1 (continued)
Cross-section of a hazardous waste landfill (detail A and B).

13-1   LANDFILL OPERATIONS

The management of hazardous waste in land disposal facilities requires the **tracking** of the waste, that is, the recording of the journey of the waste from the time it is generated to its ultimate disposal site. This tracking extends to the recording of the waste's location within the final disposal site—in smaller mappable units called **cells**. The cells are laid out at a grid. Records are maintained including who provided the waste, what was the nature of the waste, and where and when the waste was landfilled.

The principal reason for the careful tracking and disposal in cells is to ensure **waste compatibility**. Many wastes may react with each other, resulting in the potential for heat, combustion, and/or toxic fumes. By tracking the waste and where it is disposed within the land disposal facility, a check can be made of the waste compatibility, providing safe operating conditions for landfill operators as well a safe long-term disposal with respect to potentially harmful chemical reactions. A detailed compatibility chart is presented as Fig. 8-10. Additional purposes of this detailed tracking include future resource recovery and/or future recovery for alternative treatment, as well as the ability to pinpoint defects in the system should specific contaminants show up in the environment at some time in the future (i.e., identify the source of the leak).

The wastes disposed in a landfill take several forms. Drummed wastes are typically aligned and covered with other wastes, such as contaminated soils or sludges, using care not to damage the drums. Drums may be stacked or may be placed in single lifts. In all cases the wastes are in solid, not liquid, form.

Example 13-2   Land disposal cells.   A shipment of petroleum sludge, generated by a process which employs sulfuric acid to refine crude oils, arrives at Landfills-R-Us for disposal. The operator decides that it would be useful to dispose of this in the area containing sludge from an old storage lagoon from an electroplating wastewater treatment plant. What do you think?

Solution.   Not a good idea because electroplating sludge would be expected to contain, among other things, cyanide. Acidifying the electroplating waste would generate hydrogen cyanide (HCN). HCN, a gas at slightly above room temperature, can enter the body by inhalation and skin absorption. It is highly toxic—lethal concentrations have been reported as low as 10 ppm.

Operational considerations include the application of a **daily cover** placed at the close of each operational day. A daily cover typically consists of soil one foot thick. The purpose is to minimize odor, airborne transport of contaminants, and potential for direct contact, and maximize aesthetics. Although the advantages of a daily cover are clear, it is also clear that daily cover, that is, placement of uncontaminated clean soil on a daily basis in a hazardous waste landfill, is both expensive and uses up valuable landfill space. Further, daily cover creates a highly anisotropic environment in the landfill and, as a result, seepage often exits along the relatively permeable daily cover layers and through the side slopes. Thus, the concept of daily cover for hazardous waste facilities is debatable when balancing the advantages with the disadvantages.

Example 13-3 Land disposal daily cover. Landfills-R-Us asks: "How much material do we need for daily cover for our proposed hazardous waste landfill?" How much of our total volume (air space) will be used by the daily cover during the life of the landfill? and How much longer could the landfill be used if daily cover were not employed?" The landfill is 500 m by 250 m in average plan and 15 m high. Daily filling rates are 15 m by 10 m by 2 m and the daily cover is 0.3 m.

Solution. The total landfill air space is:

$$V_1 = 500 \times 250 \times 15 = 1.88 \times 10^6 \text{ m}^3$$

Each daily layer (lift) has a volume of:

$$V_2 = 15 \times 10 \times 2 = 300 \text{ m}^3$$

The daily cover has a volume of:

$$V_3 = 15 \times 10 \times .3 = 45 \text{ m}^3$$

The volume of air space (in percent) used by daily cover is:

$$V_{p.c.} = (V_3/V_1) \times 100 = \frac{45}{300} \times 100 = 15\%$$

The life of the landfill using daily cover is:

$$t_1 = V_1/(V_2 + V_3) = 5449 \text{ days}$$
$$= 21 \text{ years at 5 days per week}$$

The life of the landfill without daily cover is:

$$t_2 = V_1/V_2 = 6267 \text{ days}$$
$$= 24.1 \text{ years}$$

Therefore the landfill life will be extended by:

$$t_3 = t_2 - t_1 = 818 \text{ days or more than two years}$$

Daily operations of land disposal facilities require the use of heavy equipment, with the proper protective health and safety gear for the operators. The technology for land disposal facility operations is based upon the routine technologies employed for construction with some modifications. Typically, bulldozers, compactors, and trucks developed for heavy construction are used for land disposal operations. Often, modifications are made with respect to equipment weight and blade design to accommodate their use in a hazardous waste landfill. In all cases, modifications are made to accommodate health and safety requirements for the operators.

As a consequence of precipitation while a cell is being filled and of infiltration after a cell is closed, leachate is generated in landfills. This necessitates the collection and treatment of leachate as an integral part of any hazardous waste land disposal facility. The unit processes utilized in leachate treatment are the same as those described in Chaps. 9 and 10 for other liquid waste streams. Typically, the leachate characteristics and flow rates are definable and have characteristics similar to other industrial and hazardous wastewater. The predictability of the leachate characteristics increases for cells designated for specific wastes. Experience has shown that even a mixed waste cell will generate a more predictable leachate than those liquids from uncontrolled hazardous waste sites.

One of the guidelines affecting the design, construction, and operation of land disposal facilities is that long-term monitoring data are needed to provide effective environmental protection. Land disposal facilities include ground water and air quality monitoring along with monitoring of leachate quality and quantity. The monitoring programs are designed and implemented to establish the performance of the landfill with respect to leachate generation and contaminant migration. The monitoring begins prior to landfill construction, and continues during its operation and long after its closure.

13-2   SITE SELECTION

The selection of a hazardous waste disposal site is a highly charged political topic, as discussed in Sec. 8-4. This chapter will briefly focus on some of the technical aspects of storage and disposal facilities. From a technical standpoint, the site selection should incorporate considerations of geology, hydrogeology, hydrology, and transportation. The natural environment of the site should exhibit characteristics that provide **redundancy** to the engineered systems designed to protect the public health and the environment. For example, an area exhibiting an upward gradient from deeper aquifers would, in the case of a spill or unforeseen leakage, carry contaminants away from the aquifer as opposed to into the aquifer. Locating the landfill within a clay formation or the presence of aquitards immediately beneath the site also offers a level of natural protection.

The intrinsic value of the site environment also needs to be considered. For example, there are many areas in the United States where the ground water is unsuitable for use due to background levels of naturally occurring chemical constituents. Sites such as these constitute better sites than those where the underlying ground water aquifers are of high quality.

In addition to the subsurface hydrology and geology, there are surface water hydrology considerations. Undoubtedly the construction of an engineered final cover will increase surface water runoff. Both the capacity and sensitivity of the receiving waters to surface water runoff and to leachate discharge after treatment require consideration. Streams of high value need special consideration, such as those that constitute drinking water supplies and/or those used for sport fishing.

In addition to these issues, the technical issue of transportation (i.e., road capacity and nearness to regional markets) ought to be considered during the site selection. Additional discussion of site selection, including such considerations as adjoining land use, floodplains, wetlands, habitats for endangered species, breeding or stopping areas for migratory birds, prime agricultural land and proximity to human habitats (population densities, schools, hospitals, nursing homes and the like) for hazardous waste facilities, is included in Sec. 8-4.

13-3 LINER AND LEACHATE COLLECTION SYSTEMS

It can be confidently said that more energy, effort, and resources have gone into the development of liner systems for hazardous waste disposal facilities than any other technical aspect of land disposal facilities. The purpose of a **liner** is to provide a barrier to minimize migration of contaminants. A liner with 100% effectiveness will prevent chemical constituents from migrating across the liner system into the environment. Because no liner is 100% effective, the engineered liner systems discussed below offer redundancy to accommodate the imperfections. Because some leachate will inevitably be generated, leachate collection is required. Therefore, the landfill bottom consists of 1) alternating layers of materials to provide barriers to contaminants attempting to migrate from the landfill and 2) layers providing collection of these contaminants through collection systems (see Fig. 13-1).

The complexity of liner systems has evolved as new materials are developed and as preceding generations of liner systems have proven to be inadequate. At the present time, landfill bottom systems have evolved in the United States to the multiple layer system shown in Fig. 13-2.

The concepts behind the performance of the system as shown in Fig. 13-2 are described and the terminology is made a bit more precise in the following paragraphs. The infiltration of precipitation will be inevitable during the operational period and will continue, although to a greatly reduced degree, after landfill closure.

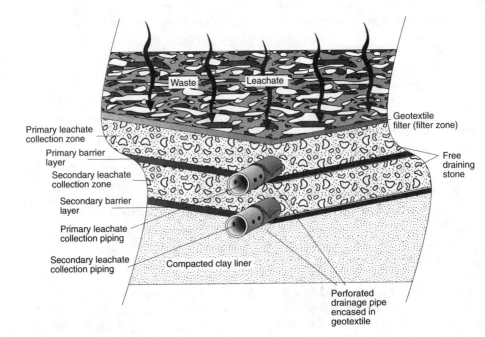

FIGURE 13-2
Schematic of liner and leachate collection systems for a hazardous waste disposal facility.

Leachate is a combination of the direct precipitation infiltration, and any liquids squeezed out as a result of consolidation of landfill waste materials. Leachate seeps, or percolates downward to the base of the landfill due to gravitational forces. The accompanying seepage forces may cause the leachate to carry particulate material in suspension (suspended solids) along with the dissolved constituents. Generally, the landfill environment is considered to be partially saturated with the exception of the very bottom.

The transport of contaminants by leachate through the landfill is by diffusion and advective transport as described in Sec. 4-2. At the base of the land disposal facility, the leachate first passes through a **filter zone** (see Fig. 13-2). The filter zone may be a **geotextile**, and/or a well-graded sand and gravel through which the particulates are filtered out. A **geogrid** may also be included to provide additional structural stability. The filter separates the waste from a relatively free draining zone around the **primary leachate collection piping**. In this primary leachate collection zone the leachate is free to flow to the piping system for removal for treatment. Since flow through the liner is proportional to hydraulic head, it is desirable to minimize this head. The hydraulic head on the liner is minimized through proper design of the leachate collection system including pipe spacing, pipe size, drainage material, and slope.

The **primary barrier layer** underlines the entirety of the primary leachate collection zone. In the United States, the primary barrier layer must be a synthetic material known as a **geomembrane** or **flexible membrane liner** (**FML**). Both geotextiles and

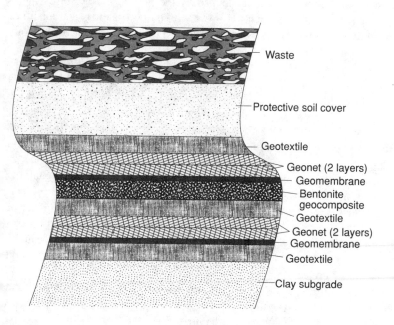

Waste

Protective soil cover

Geotextile

Geonet (2 layers)

Geomembrane

Bentonite
geocomposite

Geotextile

Geonet (2 layers)

Geomembrane

Geotextile

Clay subgrade

FIGURE 13-3
Schematic of geosynthetic liner and leachate collection systems.[2]

geomembranes (as well as other products developed to function in lieu of previously employed naturally occurring materials) are known as **geosynthetics**.

Example 13-4 Leachate generation. As a first approximation, how much leachate, (liters per minute) would be generated during the operation of a Pennsylvania landfill 1000 m by 1000 m in plan? The average annual rainfall is 100 cm/year. Comment on the answer.

Solution. The total rate of infiltration landfill can be conservatively assumed to be equal to the average annual precipitation:

$$L = 1000 \times 1000 \times 1 = 1 \times 10^6 \, m^3 / \, yr$$
$$= 1 \times 10^6 \, m^3/yr/365.25 \, day/yr \; = 2738 \, m^3/day$$
$$= 1 \times 10^6 \, m^3/day/24 \, hr/day \; = 114 \, m^3/hr$$
$$= 1 \times 10^6 \, m^3/hr/60 \, min/hr \; = 1.9 m^3/min$$
$$= (1 \times 10^6 \, m^3/min)/1 \times 10^3 \, l/m^3 = 1900 \, L/min$$
$$= 700,000 \, gal/day$$

Comment. This flow rate is calculated assuming that the waste has no capacity to absorb water (that is, the waste is at field capacity), that no evaporation or surface runoff is taking place, and that the entire landfill area is uncovered. The high rate of leachate generation demonstrates the importance of 1) maintenance of a cover over areas not actively being filled, 2) a low placement moisture content for the waste, and 3) timely placement of intermediate and final cover.

In a perfect system, the leachate is stopped by the primary barrier layer and is collected there for subsequent treatment. However, it has been shown that all liners leak—detailed studies of landfill liner systems built with the best available technology and construction quality control have revealed that a rate of leakage of about 22 gal/acre/day through the primary liner can be expected.[3] At present, the EPA identifies an Action Leakage Rate of about 5 gal/acre/day, below which no corrective action is required.[4] This action rate of leakage of 5 gal/acre/day is much less than the leakage rate of 22 gal/acre/day that can be readily achieved.

The **secondary leachate collection zone** functions much the same as the primary leachate collection system and underlies the primary barrier layer. This second system was originally known as the **leak detection** system. Now it is known as the secondary leachate collection system, recognizing the inevitability of liquids containing contaminants entering this second layer. Given the effectiveness of the primary liner, the secondary leachate collection systems handles a considerably reduced quantity of leachate. Beneath the secondary leachate collection system is yet another geomembrane, the **secondary barrier layer**, that acts in the same manner as the primary barrier layer, serving as a hydraulic barrier preventing downward flow of contaminants and allowing the secondary collection system to collect the leachate.

Underlying the entire system of primary and secondary leachate collection and geomembrane barriers is a third barrier, a layer normally constructed of a "natural material," compacted clay, or clay admixed into the natural underlying soils. The third barrier layer is required to control contaminants that may have passed through the secondary leachate collection system. Conceptually, the redundant barrier layers progressively reduce the rate at which contaminants exit the land disposal facility through hydraulic transport.

Is this the only arrangement for the liner and leachate collection system? No, the combinations and variations are endless. Shown in Fig. 13-3 is a liner system in service at a municipal solid waste and industrial waste landfill that employs exclusively geosynthetics for both drainage and barrier layers. The filters are geotextiles, drainage layers are geonets, and barrier layers are geomembranes and geocomposites. The geocomposite or **geosynthetic clay liner** is a sodium montmorillonitic clay attached to a geotextile to create a thin clay sheet of low permeability.

A European design emphasizes natural materials and minimizes the use of geosynthetics. Shown in Fig. 13-4 is an example of a composite liner employing a geomembrane directly on a clay (mineral sealing) layer. In employing such a design approach, it is recommended that a risk assessment be carried out "to define the degree of security attainable in relation to the various elements of the design and their interaction. Attention should be given to potential contaminant flow paths [to receptors]."[5]

Composite Systems Description.   As illustrated by the examples presented above, the emphasis in landfill liner design has been on the resistance to contaminant transport by hydraulic means (advective contaminant transport). What about the future? Has our technology progressed to its limits? No, the focus has recently been

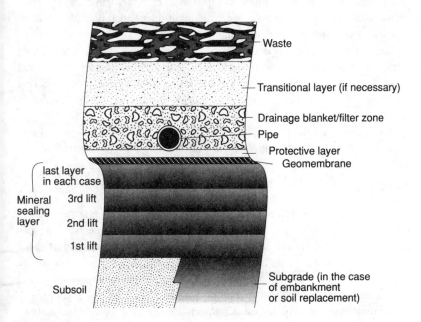

FIGURE 13-4
European liner and leachate collection systems.

drawn to the use of composite liner systems that limit both advective and diffusive contaminant transport.[6]

As described in Sec. 4-2, the rate of contaminant migration is affected by a number of factors including hydrolysis, sorption, cosolvation, ionization, precipitation, and biotransformation.[7] A liner designed to minimize the rate of contaminant migration into the environment must 1) minimize hydraulic conductivity, 2) minimize molecular diffusion rate and 3) maximize retardation. **Composite liners** are made of dissimilar materials, each employed to achieve one or more of these goals. The composite liner materials include bentonite, a zeolite (suitable for the inorganic components of the leachate), and organically modified clays, high-carbon fly ash (suitable for organic constituents of the leachate, and geosynthetics).

The use of materials with high adsorptive capacities to attenuate contaminant migration is well established.[8] Earlier sanitary landfills for municipal waste were designed as attenuating landfills assuming leakage at a low rate. Attenuating landfills function on the premise that the undesirable constituents within the landfill leachate are attenuated by the underlying soils through adsorption, biological uptake, cation and anion exchange, dilution, and precipitation.[9]

The concept of designing liner systems that release contaminants only by molecular diffusion has been examined.[10] In this approach, hydraulic transport is eliminated by hydraulic barrier redundancy and the remaining transport mechanism, diffusion, is the basis for design. This approach can be taken one step further by introducing a composite liner system that seeks to minimize the diffusive transport.

Shown in Fig. 13-5[11] is an example composite liner system that incorporates both the barrier and leachate collection zones in addition to materials that reduce the rate of diffusion through the system. Materials that delay the transport of inorganic contaminants include calcium bentonite, sodium bentonite and zeolite (shown here). Materials that attenuate organic contaminants include high-carbon fly ash and organically modified clay (shown here). Organically modified clay is most effective in retaining higher molecular weight organics whereas the carbon in the fly ash is effective for low molecular weight organics.

The thickness of the attenuating layers depends upon the allowable breakthrough time and steady-state transport rate which in turn are governed by hydraulic conductivity, hydraulic head, chemical potential, effective diffusivity, and retardation factor. The method of calculation can be based upon the analytical solution to the advective-dispersive equation for solute transport (see Sec. 4-2).

The adsorbing barrier materials can be admixed with clay or attached to the geomembranes. At present, admixed soil liners are commonly constructed using native soil with a bentonite admixture to reduce hydraulic conductivity. Admixing attenuating materials can result in material of relatively low hydraulic conductivity ($< 1 \times 10^{-7}$ cm/sec)[12,13] thereby reducing both advective and diffusive transport. Attaching clays to geomembrane barriers is a well-developed manufacturing technique.[14] The use of these materials individually for adsorption is well known. Zeolite, carbon, and organically modified clays (see Fig. 11-7) have been used in wastewater treatment processes.[15]

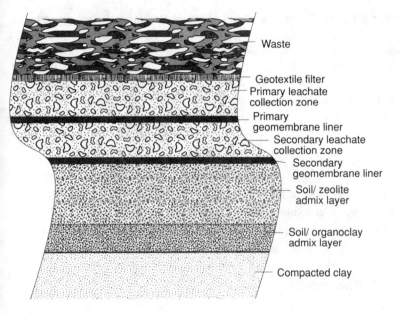

FIGURE 13-5
Schematic of composite liner system for a hazardous waste disposal facility (from Evans, et al., 1990).

Adsorptive layers should be located on the underside of the secondary geomembrane liner because the geomembranes function as the primary barrier against contaminant transport by acting as a hydraulic barrier. In this way the contaminant loading to the adsorptive layers is decreased and their useful lives are extended. Contaminants that migrate through the geomembrane liners (either through diffusion or advection through liner defects) are sorbed by the underlying attenuating materials.

A composite liner that considers both diffusive and advective transport is more effective in minimizing the rate of contaminant transport in comparison to liners that consider only advective transport. The retardation and reduced diffusion rates result from adsorption and cation exchange of inorganics within the zeolite and/or bentonite and natural clays, and adsorption of organics by the organically modified clays and/or carbon-rich fly ash. Further retardation can be achieved employing electrokinetics (described briefly in Sec. 16-8) to reduce the contaminant mobility.[16]

13-4 COVER SYSTEMS

A significant source of leachate generation at hazardous waste disposal facilities is the infiltration of water through the top of the landfill, by either direct precipitation or stormwater runon. The following section describes the controls available to minimize precipitation infiltration and to minimize surface water entering the landfill.

General Description of Cover Systems.   Much attention has been given to the design of the final cover for landfills. The final cover design considers health and safety, aesthetics, and site usage after closure, coupled with engineering requirements for permeability, compressibility, and strength. Because the final cover is expected to remain in service for as long as the waste is present, a thorough design assesses potential failure mechanisms both immediately after construction and throughout the service life (30 years or more), and provides for the following:

- Control water movement into the landfill system to minimize leachate generation.
- Control animals and vectors which can introduce disease into the ecosystem.
- Protect the public from the results of direct contact with waste.
- Control gas movement to avoid a decline in air quality.
- Minimize fire potential to avoid air emissions and damage to landfill components.
- Ensure overall stability of the cover on the landfill slopes where slope instability could result in mass movement of contaminants into the environment.
- Control surface water runoff.
- Resist erosion.
- Control blowing debris.
- Minimize noxious odors.
- Provide a more sightly appearance.

The planned usage of the site after landfill closure may generate additional design considerations. The final cover must provide structural support of the vegetative cover and sustain loadings imposed by site traffic such as maintenance vehicles.

The landfill cover must satisfy the needs of health and safety, aesthetics and site usage after closure while simultaneously providing long-term performance and accommodating potential failure mechanisms. For example, the underlying landfill will subside due to consolidation of the landfill materials: Can the cover accommodate this deformation without cracking and thus compromising its integrity? Will seasonal variations in the weather cause wetting and drying of the cover resulting in desiccation cracking or other degradation? Is the barrier layer of the cover system protected from cycles of freezing and thawing?

A typical cover is shown in Fig. 13-6. The uppermost layer, a **vegetation support layer**, typically consists of an organic silty loam (topsoil) material used to support vegetation. Vegetation provides several important functions in the performance of the landfill cover:

- Reduces erosion.
- Reduces precipitation infiltration.
- Enhances evapotranspiration, returning moisture that has been absorbed into the topsoil layer into the atmosphere to further reduce deeper infiltration.

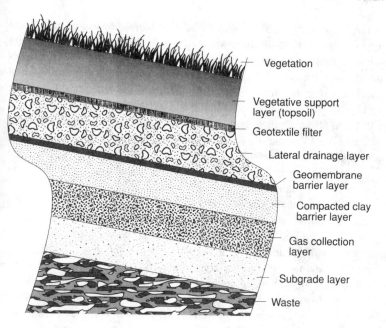

FIGURE 13-6
Schematic of hazardous waste landfill cover.

The lateral drainage layer underlies the vegetative support layer and is a free-draining porous, gravelly material, geonet, or geocomposite. The purpose of this drainage layer is to enhance lateral drainage of any precipitation that infiltrates through the vegetative support zone. With proper grading of the lateral drainage layer, the infiltration can be collected. Collection of water from the lateral drainage layer also results in minimization of hydraulic head upon the underlying barrier layer. As a result, the precipitation infiltration is reduced. The lateral drainage layer and the vegetative support layer function in combination to protect the underlying barrier layers from the environmental stresses of wetting and drying and freezing and thawing.

The lateral drainage layer may also include piping and water collection systems. A **geotextile filter** may be placed beneath the topsoil and above the underlying **lateral drainage layer**. The geotextile serves to maintain separation between the layers and to act as a filter to minimize migration of materials. Should fines from the topsoil migrate downward into the gravel, the topsoil layer will have a reduced capability for supporting vegetation and the gravel will have a reduced capability for lateral drainage. The presence of a geotextile between these layers thus reduces the risk of clogging of the drainage layer with topsoil fines.

Beneath the lateral drainage layer is one or more **barrier layers**. The barrier layers may be composed of any of the barriers previously discussed including geomembranes, natural clays, or admixed materials. The barrier layers represent the final impedance to precipitation infiltration. Maintenance of the integrity of the barrier layer during and after settlement of the underlying landfill is a prime design consideration. The rate of precipitation infiltration and leachate generation must not exceed the ability of the leachate collection system to remove the leachate. If this occurs, leachate levels build up within the landfill much the same was as water fills up a bathtub. Under this condition, the higher hydraulic head induces greater contaminant migration out of the base of the landfill. As a result, landfills lined with geomembrane liner systems are typically covered with geomembrane cover systems. The leachate collection system capacity is usually designed to handle worst case conditions—an open cell with no waste and high rainfall.

Underlying the barrier layer may be a **gas collection layer** used to collect generated gases migrating from the landfill for subsequent venting to the atmosphere. This layer is typically composed of porous sands and gravels and may include gas collection piping. A gas collection layer is virtually always included in the cover system for a landfill containing municipal solid waste or any other biodegradable materials. Gases that may be generated include carbon dioxide and methane from the anaerobic digestion of organic matter as well as volatile organics. In cases where active gas collection systems consisting of wells, laterals, headers, and pumps are employed, the overlying gas collection layer is not needed. Analytical solutions have been developed to determine gas flux in a landfill into which gas collection pipes have been installed.[17] Hazardous waste landfills contain significantly less organic matter than municipal solid waste landfills, thereby making it impractical to collect gases for resource recovery. Nevertheless, some gases can be expected and need to be managed.

The lowermost layer in a landfill cover system may be a **subgrade layer** to accommodate uneven and unstable landfill surfaces. This layer also aids construction of a cover with appropriate contours needed to enhance lateral drainage and minimize hydraulic head.

The cover system may include a geogrid layer to enhance the structural integrity of the cover system. Because some total and differential settlement is inevitable, a geogrid adds tensile capacity to the cover system to redistribute stresses and minimize differential settlements. As a result, the integrity of the cover layers is better protected.

The cover system may also include a biotic barrier layer below the vegetation support zone. This biotic layer could be installed in arid regions to prevent burrowing animals from destroying the integrity of the barrier layer. In humid regions, such a layer serves to prohibit penetration of roots into the barrier layers. The biotic layer is typically constructed of coarse-grained material such as cobbles and coarse gravel.

Cover System Performance

After the components of the cover system and their function are understood, calculations to predict their performance and to establish their design are required. Computer methods have been developed to assist in the evaluation of alternative cover designs such as the widely employed HELP model: Hydrologic Evaluation of Landfill Performance.[18] Using this model, alternative designs can be investigated to study the impact of alternative cover designs upon surface water runoff, lateral drainage above the barrier layer, and infiltration into the landfill.

A sketch for the lateral drainage and barrier layers is shown on Fig. 13-7. As steady-state seepage develops, a hydraulic head builds up on the barrier layer to provide the hydraulic energy for flow to the drain. Note that this schematic also represents the flow of leachate to the underdrains beneath the landfill. The analysis of flow to the drains considers the hydraulic conductivity of the materials, the spacing of the drains, the slope of the underlying barrier layer, and the head buildup between drains. Solutions to this problem, although complex, have been developed.[19]

Surface Water Controls. The construction of a hazardous waste disposal facility alters the natural hydrology of the site. Prior to construction, surface water probably flowed onto the site. Surface water moves first as overland flow and ultimately through streams and larger surface water bodies. Hazardous waste disposal facilities must use surface water controls. The flowing surface water must be controlled to preclude erosion and transport of sediment. A landfill section showing the deployment of surface water controls is shown in Fig. 13-8.

To control surface water, routine technologies can be incorporated into the site development for hazardous waste disposal facilities.[20] The technologies include **surface water diversion and collection systems** to minimize any surface water entering active areas of the site. Surface water diversion technologies include dikes and berms, ditches and drainage ways, terraces and benches, and chutes and downpipes. Surface water can also be controlled by grading and revegetation. Areas need to be revegetated to enhance evapotranspiration and to minimize erosion and sediment transport. The surface water controls and their functions are summarized in Table 13-1. Successful application of surface water controls ensures that surface water originating in areas

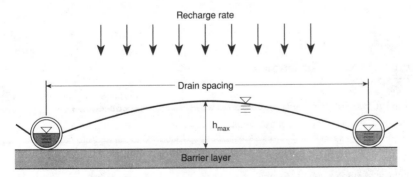

FIGURE 13-7
Schematic of drainage.

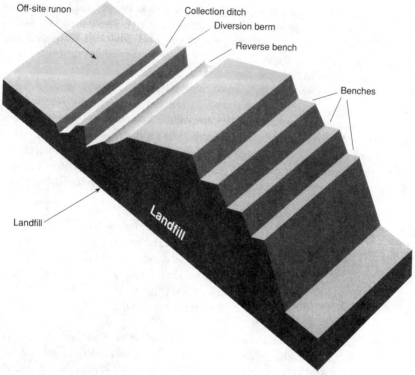

FIGURE 13-8
Landfill section showing surface water controls.

beyond the active portion of the site will not enter the site. If it were to enter the site, it could become contaminated and could contribute to the generation of leachate, and require treatment.

TABLE 13-1
Surface water controls for landfills

Grading	Revegetation	Surface water diversion and collection
Reshapes the surface to enhance runoff	Controls erosion	Dikes and berms
Minimizes infiltration/leachate generation	Enhances evapotranspiration	Ditches and drainage ways
May need periodic maintenance, e.g. annual	Root penetration must be controlled	Terraces and benches
	Maintenance	Chutes and downpipes
		Sedimentation and detention basins

Control of surface water continues when the final cover is constructed. Runoff from the cover system must be managed through the use of selected cover materials, collection channels, and other techniques as shown in Table 13-1 to prevent erosion. Sedimentation and detention basins must be included in the site drainage system to capture eroded sediment from the cover.

Most, if not all, landfills have problems with sedimentation and erosion control. This performance record emphasizes the need for careful attention to what might be considered a mundane component of a landfill system.

13-5   MATERIALS

The cover, liner, and leachate collection system components can be constructed of a wide range of materials. The following sections of this chapter describe the material types and present the test methods available to evaluate the material performance.

Geomembranes.   A geomembrane is an engineered polymeric material (a **geosynthetic**) which is fabricated to be virtually impermeable. A geomembrane can be made of any one of a number of plastic, manufactured materials, such as butyl rubber, chlorinated polyethylene, chlorosulphonated polyethylene (Hypalon®), ethylene-propylene rubber (neoprene), polyethylene (both high-density and low-density polyethylene), and polyvinyl chloride (PVC). A number of different polymers with a wide range of chemical formulations are employed to manufacture geomembranes. Shown in Table 13-2 are the most common of these and their predominant uses. Compatibility studies and studies of liner performance have shown that high-density polyethylene (HDPE) is, at present, the material of choice for the mixed range of wastes typically encountered in land disposal facilities. HDPE can be made relatively thick, 100 mils or more; seams can be readily joined in the field (by extrusion or other welding techniques); and the material offers a high degree of chemical compatibility, resisting degradation to a wide range of chemicals at a wide range of concentrations. The design of geomembrane systems has emerged as a specialty in and of itself.[21]

To evaluate the performance of the geomembrane barrier layers, it is necessary to evaluate a wide range of properties as shown in Table 13-3. Tensile properties include the tensile strength at break, the strain at failure, the stress level that causes 100% and 200% elongations, and modulus of elasticity. The modulus of elasticity is a measurement of the stiffness of the geomembrane and reflects the strains that will be induced by applied stresses. Hardness is a measure of the ability of the material to resist indentations. Tear resistance is a measure of how easily a rip is propagated once it begins. Puncture resistance is a measure of how easily the material is first ripped. It is also necessary to evaluate the strength of the seam as well as the strength of the material itself. The test methods for geosynthetics are presented in Table 13-4.[22]

Measures of the environmental performance include measure of water vapor transmission, which is indicative of how quickly the gaseous phase will migrate through the specimen. Water absorption is indicative of potential property changes when the geosynthetics is in contact with the leachate. The greater the water absorption the more likely the properties of the liner will change. The most common test

TABLE 13-2
Geomembrane types

Type	Use
High-density polyethylene (HDPE)	Excellent resistance to chemical degradation and widely used for municipal and hazardous waste landfills
Polyvinyl chloride (PVC)	OK if buried, subject to ozone and UV degradation, not good for oils
Chlorosulphonated polyethylene	Hypalon[B], fabric reinforced, second most common after PVC, good resistance to corrosive chemicals, not very good for oils and solvents
Chlorinated polyethylene	Chlorinating HDPE, good resistance as above
Low-density polyethylene (LDPE)	Easily punctured, hard to seam, used as a liner for 35 years, deficient for waste containment in chemical resistance
Neoprene	Excellent resistance to oils, weathering, ozone, ultraviolet, puncture, abrasion, and mechanical damage; expensive
Polyester elastomer	Good resistance to oil, fuel, and chemicals
Butyl rubber	Low gas and vapor permeability, thermal stability, ozone and weathering resistance, chemical and moisture resistance, resistance to animal and vegetable oils; swollen by hydrocarbon solvents; more than 35 years in water impoundments
Elasticized polyolefin	Specialized use
Ethylene propylene rubber	Similar in chemistry and behavior to butyl rubber

TABLE 13-3
Geomembrane properties[a]

Physical properties	Thickness
	Density
	Water vapor transmission
	Solvent vapor transmission
Mechanical properties	Tensile behavior
	Seam behavior
	Tear resistance
	Impact resistance
	Puncture resistance
	Friction characteristics
	Stress cracking
Chemical properties	Swelling resistance
	Chemical compatibility
	Ozone resistance
	Ultraviolet light resistance
Biological properties	Resistance to animals
	Resistance to fungi
	Resistance to bacteria
Thermal properties	Warm temperature behavior
	Cold temperature behavior
	Thermal expansion

[a] Properties discussed in detail in ref 21.

TABLE 13-4
Test methods for geosynthetics[†]

Type	Property (units)	Test procedure
Geotextile	Thickness (mils)	ASTM D-1777 (part 32)
	Mass per unit area (oz/sq yd)	ASTM D-3776
	Dimension (cm)	Direct measure machine and cross machine direction
	Grab tensile strength/elongation	ASTM D-4632
	Trapezoidal tear resistance	ASTM D-4533
	Hydraulic burst strength	ASTM D-3786
	Puncture resistance	ASTM D-3787
	Permittivity	ASTM D-4491
	Transmissivity (cm^2/sec)	ASTM D-4716
Geonets	Mass per unit area	Direct measure (sample size $\geq$ 1 sq/ft)
	Volatiles	Appendix G
	Extractables	Appendix E
	Thickness (mils)	ASTM D-1777 (part 32)
	Rib dimensions	Direct
	Aperture size	Direct
	Dimensions of configuration (cm)	Direct measure machine and cross machine direction
	Specific gravity or density	ASTM D-792 Method A or ASTM D-1505 (dry sample before test)
	CBR puncture	GS-1
	Transmissivity (cm^2/sec)	ASTM D-4716
	Compression behavior of geonets	GNI
	Strip tensile strength	Alternative method (see below)
Geogrid	Geogrid rib tensile strength	GG1
	Geogrid node junction strength	GG2
	Creep behavior and long-term design load of geogrids	GG3
	Rib dimension	Direct
	Aperture size	Direct
Pipe	Volatiles @ 105°C	Appendix G
	Extractables	Appendix E

[†] (from Landreth, 1990)

method employed for the measurement of compatibility of the membrane with the wastes is EPA Method 9090.[23] In this method, rectangular geomembrane samples are immersed in the expected leachate for a minimum of 120 days at room temperature of 22°C, and at an elevation temperature of 55°C. Physical properties, including hardness, tensile behavior, puncture resistance, specific gravity, volatile organics content, extractable organics content, hydraulic resistivity, dimensions, and weight are measured after each 30 days, as well as before and after immersion. In addition, semicrystalline geosynthetics should be tested for modulus of elasticity, tear resistance, and environ-

ess cracking whereas fabric-reinforced geomembranes should be tested for
ion. Additional discussion on the limitations of these compatibility tests is
⌐ ⎽y White and Verschoor.[24]

Geotextiles. A **geotextile** is a geosynthetic which is fabricated to be permeable
and has two categories of hydraulic properties:[25]

- Filtration—the removal of suspended solids from the flowing liquid.
- Drainage—the transportation of liquids across the plane of the fabric.

The filtration function of a geotextile has historically been carried out by graded
aggregate filters whereas the drainage function is carried out by sand or gravel layers.
In the liner system of a hazardous waste disposal facility, the liquid of concern is the
landfill leachate.

Geotextiles can be classified by the way they are manufactured as either woven
or non-woven. Woven geotextiles tend to be stronger in tension and have been used for
reinforcement and for silt fences. Non-woven geotextiles have been used for filtration,
for separation, and as a protective layer over geomembranes.

Water flowing perpendicular to the plane of the geotextile is termed cross-plane
flow. The ease by which cross-plane flow occurs is termed permittivity (cross-plane
permeability). The permittivity, Ψ, is defined as:

$$\Psi = K_n/t \tag{13-1}$$

where $\Psi =$ permittivity (s^{-1})
$K_n =$ normal hydraulic conductivity (cm/sec)
$t =$ fabric (geotextile) thickness (cm)

The permittivity can be substituted into Darcy's law (Eq. 4-3) to estimate flow
rates through geotextiles under the imposed hydraulic head conditions. If the geotextile
has inadequate permittivity, its function in providing a filtration media is impaired.

Water can also flow within the plane of the geotextile. The ease in which in-plane
flow occurs is termed the transmissivity, θ, (in-plane permeability). The transmissivity
is defined as:

$$\theta = K_p t \tag{13-2}$$

where $\theta =$ transmissivity (cm^2/sec)
$K_p =$ in-plane hydraulic conductivity (cm/sec)
$t =$ fabric thickness (cm)

This can be substituted into Darcy's law to estimate flow rates within geotextiles
to study their drainage characteristics and to design lateral drainage systems using
geotextiles and geocomposites.[26]

The durability of a geotextile for the long-term demands of a hazardous waste landfill is of critical importance. The properties for consideration of endurance include:[27]

- Ultraviolet light stability
- Abrasion resistance
- Chemical stability
- Thermal stability
- Creep resistance

Generally, the durability of a geotextile is not of concern except for chemical instability. They are protected from ultraviolet light and abrasion because they are buried. They are protected from thermal instability because they are stable over quite a wide range of temperatures. Although we can test the compatibility of geotextiles in accelerated lab tests designed to simulate the in-service environment, there is no substitute for the test of time. Unfortunately, a failure 30 years hence could have serious consequences unless adequate redundancies are built into the system.

A critical issue regarding long-term performance is the issue of clogging. How rapidly will the filter clog with sediment? What sort of biological activity will be present and how will this contribute to clogging? Will chemical precipitation contribute to clogging? These questions have no clear answers.

Compacted Clays.   To supplement geomembrane liners, compacted clay soil is typically used as the tertiary liner system. The native soils typically consist of natural clays, silty clays, sandy clays, and clayey silts. An extensive body of literature has been developed regarding the compaction and resulting permeability in clay properties. The summary of this information constitutes a book in itself.[28] This chapter can only hope to identify the critical issues and present the fundamental geotechnical engineering concepts governing the behavior of clay liners.

The hydraulic conductivity of a clay barrier layer depends upon the type of clay as characterized by the clay mineralogy, grain size distribution, and plasticity limits. However, for a given clay, it has long been established that there is not a unique value of hydraulic conductivity. In one of the early studies of hydraulic conductivity, significant influence of compaction moisture content was established.[29] In fact, as shown on Fig. 13-9, the hydraulic conductivity can vary, dependent upon compaction moisture content, by more than two orders of magnitude for the same clay compacted at the same compaction effort. The tremendous variation in hydraulic conductivity is attributed to differences in the microstructure of the compacted clay with varying compaction moisture content.[30]

Let's assume for discussion purposes that we are able to obtain representative samples of a clay proposed for use in a clay barrier layer and test them in the laboratory. Is it reasonable to assume that the laboratory value of hydraulic conductivity is the same as the hydraulic conductivity of the field constructed barrier layer? Probably

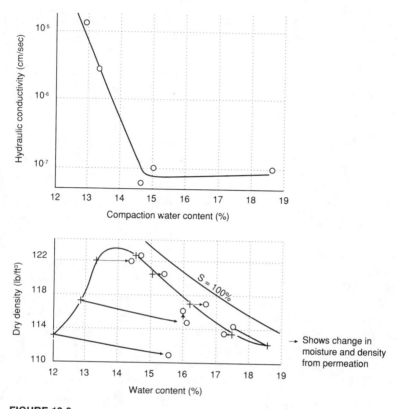

FIGURE 13-9
Influence of compaction moisture content on the hydraulic conductivity of compacted clay (after Lambe, 1958).

not. In one study of liner systems with known high rates of leakage, the field hydraulic conductivity was found to be 10 to 1000 times greater than the laboratory value.[31] Although more difficult and costly to run, field hydraulic conductivity testing may be a more reliable method to determine the hydraulic conductivity of the barrier.[32] It has been shown that the macro-structure of the compacted soil (size of clods and pores between clods) can have a significant influence on the hydraulic conductivity.[33] If the lessons of more than thirty years of research into the hydraulic conductivity of compacted clays can be summarized in a few key points, they would be:

- Compact the soil wet of the optimum moisture content to achieve the desirable microstructure and macrostructure for low hydraulic conductivity.
- Use adequate compaction effort to break down clods into a remolded mass relatively free of large pores between clods and to achieve the desirable microstructure.
- Construct the barrier layer in numerous thin lifts eventually forming a barrier layer that is relatively thick (a meter or more).

How does the hydraulic conductivity of compacted clays change when permeated with landfill leachate? Permeation of compacted clay with concentrated organics, acids, and bases may cause dramatic increases in hydraulic conductivity. For example, shown in Fig. 13-10 are the results of permeability tests where compacted clays were initially permeated with water and then with heptane, $CH_3(CH_2)_5CH_3$. As the curves show, the permeability increased two orders of magnitude or more due to heptane permeation.

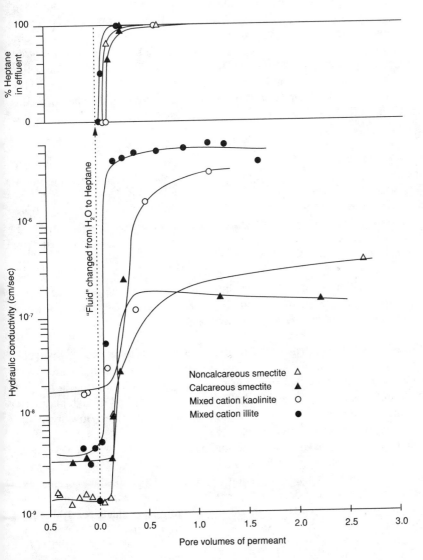

FIGURE 13-10
Influence of organic permeants on the hydraulic conductivity of compacted clay (after Brown and Anderson, 1983).[34]

In contrast, permeation with more dilute solutions generally does not reveal such changes in permeability. An examination of the data from many researchers, coupled with an assessment of expected fundamental physio-chemical changes based upon clay mineralogy and the clay-water-electrolyte system, has led to the conclusions[35] that

- Dilute solutions of organics have essentially no effect upon the hydraulic conductivity of organics; however,
- Dilute solutions of inorganic chemicals may affect the hydraulic conductivity.

A state-of-the-art compacted clay barrier does not stop the transport of contaminants. Containment systems simply delay transport; contaminants will still migrate through the compacted clay liner to the environment beneath the liner. As discussed earlier, even with absolute hydraulic containment, there is diffusive transport. The technical issues are 1) at what rate will the contaminants migrate, 2) when does this migration begin, and 3) will the contaminant concentrations be detrimental to the environment.

In Situ Clay Formations. The principles of clay behavior previously described for compacted clays are, in many regards, similarly applicable to in situ clay formations. However, when assessing the integrity of underlying clay formations as barrier layers we do not have the luxury of inspecting each 6″ (150 mm) lift. Rather, we must evaluate the formation utilizing the tools and methods described in Chap. 15. It is important to state here, however, that hydraulic pathways in clay deposits are not uncommon. Glacial till, a well-graded, dense clayey material deposited on the margins of glaciers, is often considered to be the best natural aquiclude. However, glacial till may be highly fractured. Hydraulic transport through these fractures is essentially unimpeded. Any dependence upon in situ formations of clay requires careful exploration of the bulk properties of the formation.

Bentonite Admixtures. Bentonite is added to natural soils to reduce the hydraulic conductivity and increase the adsorptive capacity of the native materials. Processed sodium montmorillonitic clay (bentonite) is added to the native soil using 5–15% by dry weight. This added clay changes the natural soil to a clayey soil of high plasticity. The soil-bentonite admixture is compacted into a liner of relatively low hydraulic conductivity.

Bentonites are clays having montmorillonite (a smectite) as the major clay constituent (see Sec. 11-3 and Fig. 11-6 for additional detail). Bentonite is further described by the principal exchangeable cation (typically 35–99% of the exchangeable cations) as calcium or sodium bentonite. The cation exchange capacities of calcium and sodium bentonites are approximately 80 and 150 milliequivalents per 100 grams of dry soil, respectively.[36] Calcium bentonite is commonly mined in Mississippi and Alabama, and sodium bentonite is found and mined in Wyoming and South Dakota.

Example 13-5   Bentonite admixed layers. An 8-inch layer of soil-bentonite is planned for placement immediately beneath the secondary geomembrane liner. The layer is to be constructed in two four-inch lifts. The bentonite content is to be 6% by dry weight. The compacted dry unit weight of the soil, including bentonite, has been determined to be 100 pcf. How much dry bentonite needs to be spread for mixing on each lift?

Solution. Each lift is 0.333 ft thick yielding:
$$W_t = 0.3333 \times 100 = 33.3 \text{ pounds per square foot of lift}$$
Of this, 6% is bentonite and the remainder is the base soil yielding:
$$W_t = 0.06 \times 33.3 = 2 \text{ pounds per square foot per lift}$$

Both sodium and calcium bentonite are hydrophylic, although sodium bentonite expands more than calcium bentonite in the presence of water. The monovalent sodium is more readily exchangeable than the divalent calcium. The exchangeable ions (calcium and sodium) in the clay readily exchange with heavy metals such as copper (Cu), nickel (Ni), and lead (Pb). Thus, as leachate migrates through a soil liner containing bentonite, the rate of advective transport is reduced as a result of the lower hydraulic conductivity, and the diffusive transport is reduced due to the exchange and adsorption of metal ions onto the clay surface.

Zeolite. Although not yet used in landfill liners, zeolites are naturally occurring or engineered alumino-silicates with a caged structure as shown on Fig. 13-11. The cation exchange capacity of zeolites is about 250 milliequivalents per 100 grams with sodium and calcium being the major exchangeable cations.[37] The caged structure of the zeolite acts like a molecular sieve. Metal ions that pass into the sieve are trapped and held by ion exchange. The five principal zeolites found in nature are analcite, chabazite, clinoptilolite, erionite, and mordenite. The structure and behavior of zeolites have been studied extensively and are well established.[38] Zeolites already perform a wide range of waste management functions including purification of water contaminated with radioactive ions, removal of ammonium from the effluent of wastewater treatment plants, purification of water in fish tanks, odor control in kitty litter, and adsorption of metals from industrial wastewaters. Among the metal ions, zeolites preferentially adsorb arsenic, lead, and cadmium. The benefits of zeolite in

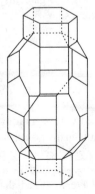

FIGURE 13-11
Zeolite structure.

enhancing landfill liner performance have been demonstrated in a laboratory scale but not in full-scale installations.

Organically Modified Clays. Organically modified clays have been shown to enhance liner performance in the laboratory but have not yet been incorporated into full-scale installations. The modification of naturally occurring clays into organically modified clays was described more than 40 years ago when Jordan published his original work on the subject.[39,40] Since that time, organically modified clays have been used in a number of environmental applications.[41] As described in Sec. 11-3, organically modified clays are naturally occurring clays in which a portion of the inorganic exchangeable cations are exchanged with suitable organic cations such as quaternary ammonium. Quaternary ammonium is an organic nitrogen compound in which a central nitrogen atom is joined to four organic groups along with an acid radical. The reactions that exchange the organic cations with the inorganic cations on the surface of the clay include adsorption, ion exchange, and intercalation. Different organically modified clays result from different base clays (e.g., montmorillonite and attapulgite) and different organic cations used in the exchange. The unmodified clay is hydrophilic whereas the modified clay is organophylic and hydrophobic.

Organically modified clays adsorb organic molecules as a result of adsorbate interactions and adsorbate-solvent interactions[42,43] The adsorption capacity of an organically modified clay is therefore dependant upon both the properties of the organic cation used to make the organically modified clay as well as the properties of the fluids, such as temperature, pH, and organic waste type and concentration. A number of studies have clarified the behavior of organophylic clays.[44,45,46] For example, clays modified with strongly hydrophobic organic molecules tend to be excellent adsorbents of chlorophenols but are less effective for non-chlorinated phenols.

Fly Ash. Fly ash is the by-product of the combustion of coal burned for the production of electricity. The material is a silt-sized particulate, silicious in nature, containing various levels of unburnt carbon. A description of the fly ash properties best suited for landfill liner applications is found in Mott and Weber.[47] The addition of high-carbon fly ash within the barrier results in a substantial delay in breakthrough of organics as discussed later in this chapter. High-carbon fly ash shows an adsorptive preference for low molecular weight organics (similar to activated carbon) in contrast to organically modified clays that preferentially adsorb high molecular weight organics.

13-6 CONTAMINANT TRANSPORT THROUGH LANDFILL BARRIERS

Transport through Liners

Contaminant transport through barriers includes contaminants carried with migrating fluid due to hydraulic gradients (advection) and molecular diffusion in response to chemical concentration gradients. The migration of contaminants by advection and molecular diffusion has been described in Sec. 4-2 and is well documented.[48] In an effort to better understand hydraulic transport, recent studies have included the hydraulic conductivity of clays in both the laboratory and field,[49,50,51] the transport of contaminants through geosynthetic liners,[52] and compatibility between membranes and the contained waste.[53,54] The absolute impermeability of geomembranes is well established as a myth.[55] Imperfections such as pinholes, improperly sealed seams and puncture tears as a result of construction are inevitable. The inevitablility of hydraulic transport of contaminant through barrier layers is presently accounted for by the redundancy of multiple barrier layers. As we are able to build better and better liner systems with lower and lower effective hydraulic conductivities, molecular diffusion becomes the controlling mechanism for contaminant transport.

Amplification of the governing equations, dimensionless charts, and design methodology for contaminant transport through earthen barriers is presented in additional sources.[56] In this chapter, the diffusive transport is further examined, and transport through the liner materials is described.

The importance of contaminant transport through landfill liners by diffusion is illustrated by calculating the release rate for a contaminant diffusing clay liner. Employing data from a landfill study[57] for a 1 meter thick liner, an assumed landfill area of $10^4 m^2$, and a benzene concentration of 1 g/L in the landfill, a steady-state transport rate of 19 kg/yr is computed. This transport rate has the potential to contaminate 3.8 billion liters of water at the EPA drinking water limit of 0.005 mg/L. Diffusion continues even with double liners and leachate collection systems, and the diffusion rate through a double liner system is similar to a single layer barrier of equal thickness.

Molecular diffusion occurs two ways: a) **liquid phase diffusion** through a saturated barrier layer or b) **gas phase diffusion** after **partitioning** from the liquid phase. For liners constructed of thin geomembranes, the contaminant migration rate due to molecular diffusion is surprisingly high. For example, the rate of diffusion of the low-octane gasoline through a 30 mil geosynthetic liner tested according to ASTM D1814-55 Inverted Cup Method was found equivalent to approximately 3000 gallons per acre in 24 hours.[58] Even though gasoline is not stored in geomembrane-lined landfills, this example illustrates the significant rate at which some organic contaminants of low molecular weight (volatile organics) can diffuse through a geosynthetic liner.

To examine liner systems to minimize diffusion, one can assume that the seepage velocity approaches zero and the molecular diffusion rates are characterized by Fick's second law describing one dimensional diffusion of solutes in soil as:

$$\partial c/\partial t = (D^*/R)(\partial^2 c/\partial x^2) \tag{13-3}$$

where c = concentration of a solute (mg/cm^3)
t = time (sec)
D^* = effective diffusion coefficient (cm^2/sec)
R = dimensionless retardation factor
x = direction of diffusion (cm)

According to Fick's law, the driving force (or gradient) for molecular diffusion is the chemical potential. The effective diffusion coefficient attempts to include real-world subsurface conditions including the tortuosity of the flow, temperature variations, and reactions occurring in the subsurface.[59] The apparent diffusion coefficient is defined as D^*/R and includes retardation effects such as those due to adsorption (see Sec. 4-3).

Contaminant partitioning from the water phase to the solid phase is described by the partition coefficient, and retardation is related to the partition coefficient. For fast reversible adsorption with a linear isotherm, the partition coefficient can be replaced by the **distribution coefficient** (see Sec. 4-3). **Retardation**, R, expressed as a dimensionless retardation factor, is a function of the bulk density, ρ, porosity, η, and distribution coefficient, K_d, as:

$$R = 1 + (\rho/\eta)K_d \tag{13-4}$$

An increasing retardation factor results in a decreasing initial rate of diffusive transport. That is, more is adsorbed or otherwise retarded, and the **breakthrough** is delayed. In liners designed with diffusive transport in mind, this retardation is accomplished using materials with higher distribution coefficients.

Transport through Geomembranes. The hydraulic conductivity of the liner along with the diffusion potential through the liner can and needs to be measured. Geomembranes are manufactured as solid homogeneous nonporous materials. As such, they are not porous media materials, and cannot be permeated by liquids in the traditional sense. Geomembranes do, however, contain interstitial spaces between the large polymer molecules. Small molecules can diffuse through the geomembrane within these interstitial spaces or gases. Whereas flow through porous media can be described by Darcy's law (Eq. 4-3), flow through geomembranes is on a molecular level. Thus, even a perfectly constructed geomembrane, free of defects or holes, permits some dissolved constituents of the contained fluids and gases to escape.

Such transport of contaminants through the geomembrane is on a molecular basis by the diffusion process. Liquids per se do not migrate through the geomembrane. The transport involves three steps:[60]

1. The solution or adsorption of the permeant into the upstream surface of the geomembrane (partitioning of molecular from leachate to gases).
2. Diffusion of the contaminant through the geomembrane.
3. Volatilization or desorption of the contaminant at the downstream surface of the geomembrane.

The steady-state transport is described by Fick's law (Eq. 4-15). The permeability of a variety of geomembranes to carbon dioxide (CO_2), methane (CH_4), and nitrogen (N_2) gases is shown in Table 13-5[61].

The migration of "liquids" from one side of a geomembrane to another, by the multi-step, complex process previously described above, can be quantified and expressed as a "permeability" value. This rate of migration can be measured and reported as the permeability (see Table 13-6)[62].

TABLE 13-5
Permeability of polymeric geomembranes to selected gases at 23°C, determined in accordance with ASTM D1434, Procedure V[†]

Geomembrane description				Gas transmission rate mL(STP[b])/m² · d · atm			Gas permeability coefficient ($\bar{P}$) barrer[c]		
Base polymer	Density	Thickness[a] mm	Compound type[d]	CO_2	CH_4	N_2	CO_2	CH_4	N_2
Butyl rubber	...	1.60	XL	512	120	19.7	12.5	2.92	0.480
Chlorosulfonated polyethylene	...	0.82R	TP	122	21.6	26.2	1.52	0.27	0.33
Elasticized polyolefin	...	0.86	TP	418	124	27.1	5.47	1.62	0.36
Ethylene propylene rubber	...	0.58	CX	1450	280	125	12.8	2.47	1.10
	...	0.89R	TP	2720[e]	...	...	36.8[e]	...	...
	...	0.90	XL	5260	1400	314	72.0	19.2	4.30
Neoprene	...	0.71	XL	716	80.9	31.1	9.81	1.11	0.43
Polybutylene	...	0.25	CX	818	248	62.3	8.84	2.68	0.67
Low-density polyethylene[f]	0.921	0.46	CX	6180[e]	1340[e]	...	23.5[e]	5.10[e]	...
Linear low-density polyethylene	0.923	0.61	CX	1370	322	...	9.59	2.25	...
High-density polyethylene	0.945	0.86	CX	729	138	...	6.77	1.28	...
	0.945		CX	467	104	...	6.11	1.28	...
Polyvinyl chloride (plasticized)	...	0.25	TP	7730[e]	1150[e]	...	29.4[e]	4.38[e]	...
	...	0.49	TP	3010	446	108	22.4	3.32	0.81
	...	0.81	TP	2840[e]	285[e]	...	35.0[e]	3.51	...
Polyester elastomer[g]	...	0.022	CX	357	...	...	0.119	...	...

[a] g = fabric-reinforced.

[b] STP = standard temperature and pressure.

[c] One barrer = 10^{-10} mL(STP) · cm/cm² · s · cm Hg.

[d] XL = crosslinked; TP = thermoplastic; CX = semicrystalline.

[e] Measured at 30°C.

[f] Natural resin (no carbon black).

[g] This sample is NBS Standard material 1470. The determination was made at 15.0 psi, under which condition the NBS Certified CO_2 transmission rate was calculated to be 338 mL(STP)/m² · d · atm.

[†] (after Haxo)

TABLE 13-6

Coefficient of permeability (hydraulic conductivity) to deaerated water (hydraulic conductivity) of several geomembranes at various pressures[a†]

Base polymer	Hydraulic pressure, kPa			
	100	400	700	1000
Butyl rubber	22	8.0	4.1	2.8
Chlorosulfonated polyethylene	300	120	77	60
Ethylene propylene rubber	115	57	34	25
Polyvinyl chloride	86	37	20	10

[a] Data is reported in 10^{-15} m/s
[†] (after Haxo)

TABLE 13-7

Permeability of polymeric geomembranes to various solvents, measured in accordance with ASTME96, Procedure B (modified to test solvents)[†]

Polymer[a]	ELPO	HDPE	HDPE	HDPE-A	LDPE	PB
Average thickness, mm	0.57	0.80	2.62	0.87	0.75	0.69
Type of compound	CX	CX	CX	CX	CX	CX
SVT, g m^{-2} d^{-1}						
Methyl alcohol	2.10	0.16	⋯	0.50	0.74	0.35
Acetone	8.62	0.56	⋯	2.19	2.83	1.23
Cyclohexane	7.60	11.7	⋯	151	161	616
Xylene	359	21.6	6.86	212	116	178
Chloroform	3230	54.8	15.8	506	570	2120
Solvent vapor permeability[b], 10^{-2} metric perms · cm						
Methyl alcohol	0.11	0.01	⋯	0.04	0.05	0.02
Acetone	0.23	0.02	⋯	0.09	0.10	0.04
Cyclohexane	0.49	1.05	⋯	14.7	13.6	47.8
Xylene[c]	292	24.6	25.6	262	124	175
Chloroform	103	2.46	2.32	24.6	24.0	82.8

[a] ELPO = elasticized polyolefin; HPDE = high-density polyethylene; HDPE-A = high-density polyethylene alloy; LDPE = low-density polyethylene; PB = polybutylene.
[b] The median thickness value was used to calculate the permeability.
SVT = Solvent vapor transmissibility
[†] (after Haxo)

In an experiment employing various solvents, the transmission rates through selected liners were measured.[63] The results of this experiment are shown in Table 13-7[64]. These data can be used to assess contaminant transport through liner systems.

Transport through Compacted Clay.   The rate of advective flow through clay liners depends upon the hydraulic conductivity, gradient, and liner dimensions. A straightforward analysis of the contaminant loading in the steady state can be made using Darcy's law (Eq. 4-3) and the concentration of the contaminants in the leachate.

Identified in Fig. 13-12 are the critical parameters for the analysis. The hydraulic head is equal to the height of free leachate above the liner plus the thickness of the clay liner, not the height of the leachate alone. The solution then is a direct application of Darcy's law to yield the flow rate. The transport rate can be determined as the product of the contaminant concentration and the flow rate.

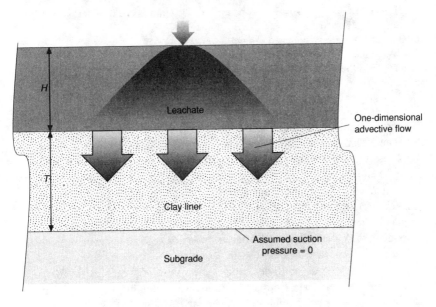

Hydraulic gradient $i = \dfrac{H+T}{T}$

Steady state Darcy flux $V = K_L i$

FIGURE 13-12
Schematic of advective flow and transport.

Example 13-6   Calculation of contaminant migration through a compacted clay liner.   An old landfill, 300 m by 500 m in plan, was constructed using a single clay liner 1 m thick and having a hydraulic conductivity of 5×10^{-7} cm/sec. The average leachate head is 0.5 m. An analysis of the leachate indicates a concentration, C, of carbon tetrachloride at 1.0 mg/L among other contaminants. What is the annual steady-state flow and advective contaminant transport rate from this landfill? How much drinking water per year could be contaminated to the drinking water standard of 0.005 mg/L?

Solution. Calculate the hydraulic gradient:
From Fig. 13-12, $i = (H + T)/T$

$$i = (0.5 + 1)/1 = 1.5 \text{ m/m}$$

Calculate the Darcy flux:

From Fig. 13-12, $v = ki$

$$v = 5 \times 10^{-7} \text{ cm/sec} \times 1.5\text{m/m}/100 \text{ cm/m} = 7.5 \times 10^{-9} \text{ m/sec}$$

Calculate the flow for the entire landfill on an annual basis:

$$Q = vAt = 7.5 \times 10^{-9}\text{m/sec} \times 300 \text{ m} \times 500 \text{ m} \times 3.16 \times 10^7\text{sec/yr}$$

$$Q = 3.6 \times 10^7 \text{ m}^3/\text{yr}$$

Calculate the loading to the subsurface environment:

Loading, $L = Q \times C = 3.6 \times 10^4 \text{ m}^3/\text{yr} \times 1.0 \text{ mg/L} \times 1000 \text{ L/m}^3 = 3.6 \times 10^7 \text{ mg/yr} = 36\text{kg/yr}$

Calculate the quantity of contaminated ground water, V_w:

$$V_w = 3.6 \times 10^7 \text{ mg/yr}/0.005 \text{ mg/L} = 7.2 \times 10^9 \text{ L/yr}$$

More than 7 billion liters per year from about 36 kg.

A comprehensive study employing both field and laboratory data examined the effect of sorption on diffusion rates for natural clay beneath a hazardous waste landfill.[65] The effective diffusion coefficient for the field and lab conditions was from 0.2 to 0.02 less than the theoretical diffusion coefficient for chloride, meaning movement is slower in field tests than in the laboratory. The decreased mobility of chloride through compacted clay results from sorption and molecular weight effects. However, in actual practice, much quicker rates have been noted. This is usually because transport has been through "defects" in the barrier systems, otherwise assumed to be homogeneous in the calculations. Diffusion of organics such as benzene is likely to be further retarded compared with chloride. To calculate the breakthrough time for high toxicity of organics such as benzene, small values of C/C_0 such as 0.001 are suggested. Thus, for the field data from a hazardous waste landfill, the time for benzene to break through a one-meter-thick clay layer (at $C/C_0 = 0.001$), without defects, was calculated to be 70 years.

Transport through Barriers with Bentonite. Bentonite has been found to reduce the rate of contaminant migration by lowering the hydraulic conductivity and increasing the retardation factor. A number of commercial bentonite products are available, and manufacturers provide guidance with respect to application rates. The ability of bentonite to retard contaminant migration is demonstrated by a study that used silica sand/bentonite mixtures and the reactive solute strontium 85. The relationships among the retardation factor R, the distribution coefficient, K_d, and the bulk density/porosity ratio are shown in Fig. 13-13.[66] Increasing the bentonite content in the admixed liner results in a corresponding increase in the sorptive capacity (and thus an increase in K_d). It would be expected that increasing K_d (by increasing the bentonite content) would result in an increasing retardation factor, but countering this expected relationship is the fact that increasing the bentonite content also generally results in a decreasing density/porosity ratio. Thus, as shown in Fig. 13-14[67], the retardation factor reaches a maximum value at a bentonite content between 5 and 10

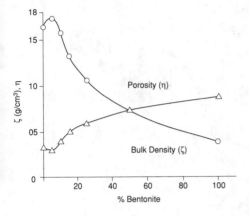

FIGURE 13-13
Strontium-85 coefficients, bulk density to porosity ratios and retardation factors versus bentonite content (from Gillham, et al.).

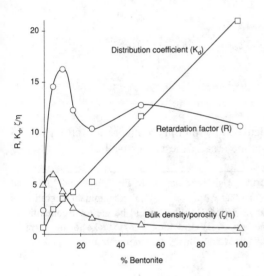

FIGURE 13-14
Strontium-85 diffusion coefficients: experimental and calculated versus bentonite content (from Gillham, et al.).

% by weight. It is noted that a bentonite content of 5–10 % typically results in the minimum hydraulic conductivity in admixed soil liners.

Transport through Barriers with Fly Ash. The effectiveness of a barrier consisting of soil-bentonite and high-carbon fly ash employed to reduce diffusive transport has recently been investigated.[68] Because the diffusive transport rates are hindered by sorption to the media, the apparent diffusion coefficient (which incorporates both the effective diffusion coefficient and the retardation) was related to the sorption capacity of the fly ash. Using experimental data, a barrier one meter thick and having an area of 10^8 cm^2 was modelled by assuming a volatile organic solute at 1 mg/L on the interior and 0.0 mg/L on the outside. These results shown on Figs. 13-15[69] and 13-16[70] are for the two limiting cases with no sorptive capacity and equilibrium sorptive capacity.

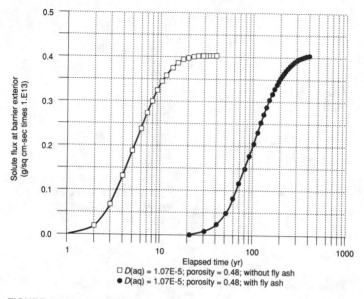

FIGURE 13-15
Model simulations of solute flux (from Mott & Weber).

The total contaminant migration is calculated by multiplying the results on Fig. 13-15 by the actual interior concentration and the area of the barrier wall.

Example 13-7 Contaminant Transport. Using the data from Fig. 13-15 and 13-16 and assuming a liner beneath site of one acre: 1) What would be the total contaminant migration in the first 12 years with a soil-bentonite barrier layer? 2) What would be the total contaminant transport for a composite wall for the same period? 3) How long would it take for the sorption capacity of the composite barrier to be exhausted and for the transport rate to be equal to that of the soil-bentonite barrier?

Solution

1. From Fig. 13-16, the total solute flux through the barrier after 12 years is about $.015$ g/1×10^3 cm^2.

    For one acre we have:

    Total = rate × area

    $$= \frac{.015 \text{ g}}{1 \times 10^3 \text{ cm}^2} \times 1 \text{ acre} \times 43560 \frac{\text{ft}^2}{\text{acre}} \times 1.44 \times 10^6 \frac{\text{cm}^2}{\text{ft}^2} \times \frac{1 \text{ kg}}{1000}$$

    Total = $\underline{941 \text{ Kg}}$

2. From Fig. 13-16, the total flux is essentially zero.

3. From Fig. 13-15, the flux rate for the barrier with fly ash approaches that without fly ash after about 400 years.

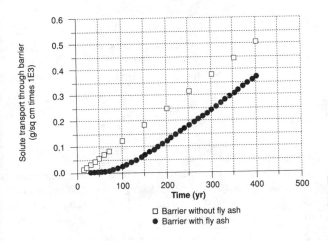

FIGURE 13-16
Model simulations of solute transport (from Mott & Weber).

☐ Barrier without fly ash
● Barrier with fly ash

Transport through Barriers with Organically Modified Clays. The process of sorbing organics onto organically modified clays from water is, in part, a partitioning process. The quaternary ammonium bound to the organically modified clay surface acts as a partitioning media to remove organic molecules from water. When used in a composite liner, these organically modified clays retard the migration of organic contaminants that have diffused or otherwise passed through the geomembrane.

Transport through Barriers with Zeolites. The adsorption characteristics of chabazite (a type of zeolite) as examined in triaxial permeability tests is shown in Fig. 13-17[71]. In this figure, elapsed time is normalized with respect to hydraulic conductivity and presented as pore volume displacement. Pore volume displacement is the ratio between the volume of fluid permeated and the volume of the voids in the soil. Using a solution of lead at 10 mg/L in distilled water, permeability tests on a silty

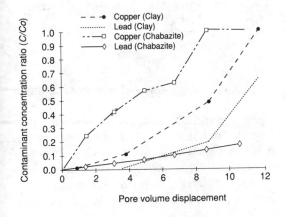

FIGURE 13-17
Permeation of chabazite and silty clay.

clay and a silty clay chabazite composite were conducted. Fig. 13-17 demonstrates the retardation of the lead in the composite liner as compared to the compacted silty clay. Comparison of the results for copper are also shown. Copper reaches breakthrough faster than the lead for both liners. This is due to the difference in retardation factors between copper and lead. Chabazite preferentially adsorbs lead as compared to the silty clay. Chabazite preferentially adsorbs lead over copper. These results emphasize the importance of designing composite liners to sorb specific contaminants of interest.

Example 13-8   Long-term permeability test. A sample of clay barrier soil is permeated in the laboratory to study contaminant transport through the sample. The cylindrical sample is 70 mm in diameter, 70 mm high, and has a porosity of 0.4. What total volume of permeant has passed through the sample to achieve a pore volume displacement of 3? If the test is run at a gradient of 5 and the liner hydraulic conductivity is less than 1×10^{-7} cm/sec, how long will the test take?

Solution

Total sample volume:

$$V_t = \pi \times r^2 \times h = 3.14 \times 3.5^2 \times 7.0 = 269 \text{ cm}^3$$

Void Volume:

$$V_v = V_t \times n = 269 \times 0.4 = 107.6 \text{ cm}^3$$

Permeant volume:

$$V_p = 3 \times V_v = 3 \times 107.6 = 322.8 \text{ cm}^3$$

Test duration:

$$q = ki At$$

$$t \geq q/ki A = 322.8/1 \times 10^{-7} \times 5 \times 3.14 \times 3.5^2$$

$$t \geq 1.68 \times 10^7 \text{ sec}$$

$$t \geq 194 \text{ days}$$

Comment. These long-term permeability tests require considerable time to complete necessitating special test precautions[72] and considerable advance planning.

13-7   LANDFILL STABILITY

A landfill can fail to meet performance requirements for a number of reasons. Many of them pertain to stability.

Failure mechanisms for cover systems include subsidence where total and differential settlements exceed the tolerance of the landfill cover system. In this case, geotextiles or geogrids may be used to span expected zones of subsidence. Changes in the environment in the cover system may also alter its performance. For example, frequent wetting and drying may cause desiccation cracks in any clayey barrier layers. Freezing and thawing may likewise alter the performance of the drainage layer and cover layer. Frequently, therefore, cover systems are quite thick to provide adequate protection to the barrier layers from these environmental stresses. Failure of the cover system may also occur as a result of erosion. A cover is designed to enhance runoff, and the system must be able to withstand the erosional forces associated with this runoff. Of course, the cover materials must remain in place. Cover instability as a result of sliding between the dissimilar materials (such as geotextiles, granular drains, geomembranes, and clay barrier layers) is well documented[73] and the reader is advised to fully consider all potential failure surfaces and the strength along these surfaces.

Example 13-9   Landfill failure. An aboveground ash monofill was constructed over the natural sloping ground with little modification. Perimeter ditches were installed to control surface water. During operation and immediately after a rain, 40,000 m^3 of ash washed out of the landfill.[74] How could this happen?

Solution. The working area was too large, resulting in low runoff and high infiltration. As rainwater entered the ash and migrated downward and outward to drainage systems, the fill became unstable causing the significant loss of ash from the landfill.

Landfills must be stable during construction, during operations, and for many years following closure. Some settlement is inevitable; however, the key measure is whether the waste can be considered **stable** is if it stays within the containment system or, more precisely, if there are no mass movements down slope. Stability, often termed **slope stability**, can be mathematically defined in a number of different ways such as by rotational force equilibrium, rotational moment equilibrium, or sliding wedge force equilibrium. Entire books have been written on the subject of slope stability.[75] A recent conference included five papers on the stability of waste fills.[76] The analysis of slope stability lies in the realm of the geotechnical engineer who is trained and experienced in the selection of both the method of analysis most appropriate and the selection of input parameters to that analysis. The following method of analysis is presented to give the student of hazardous waste management some insight into stability evaluations.

The analysis of stability necessitates the hypothesis of a failure surface, the determination of driving and resisting forces, and the definition of a Factor of Safety employing these force systems. Figure 13-18 depicts a cross section through a landfill. The potential failure surface is often assumed circular, a common assumption which makes the mathematics easier and closely models observed failure surfaces. The driving forces are from the gravitational pull on the landfill materials, and the resisting forces are from the strength of the waste and the underlying soil. The strength of the material is expressed in terms of the sum of cohesion (stress independent) and friction (stress dependent). The strength of waste has received considerably less attention although it can modeled for stability analysis in the same manner as soil. The computed factor of safety will vary with the assumed location of the failure surface. The factor of safety for the landfill is the lowest computed for the various assumed failure surface locations. The location that produces the lowest factor of safety is termed the **critical failure surface.**

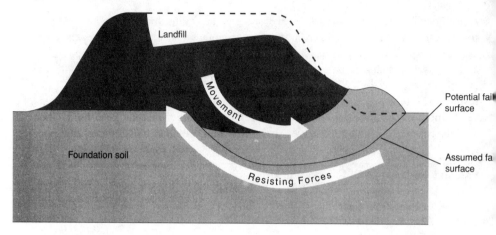

Landfill

Movement

Resisting Forces

Foundation soil

Potential fail
surface

Assumed fa
surface

FIGURE 13-18
Schematic of landfill stability.

In the next several paragraphs, an examination of a limit equilibrium analysis for evaluating slope stability is made. By choosing one relatively simple case, the student can see one approach to slope stability analysis.

The maximum shearing stress that a soil can withstand without failure is defined by the Mohr-Coulomb failure criteria as:

$$T_{ff} = c + \sigma_{ff} \times \tan(\phi) \tag{13-5}$$

where T_{ff} = shear stress at failure on the failure plane (kPa)
c = cohesion (kPa)
σ_{ff} = applied normal stress at failure on the failure plane (kPa)
ϕ = angle of internal friction (degrees)

If the shear stress to failure is applied at a relatively rapid loading rate, it is considered an undrained loading. A landslide at the end of construction is represented by the undrained loading condition. The ability of the mass to remain stable for many years after construction is modeled as a drained loading condition. The analysis of landfill stability must demonstrate stable conditions at the end of construction (undrained shear strength parameters are used) and in the long-term (drained shear strength parameters are used). The terms drained and undrained refer to the absence or presence of excess pore water pressures in the materials undergoing shearing strain. The landfill must be stable for potential failure surfaces that pass only through the waste and pass through both the waste and underlying foundation soil. The stability of the cover must also be considered with respect to the sliding of the vegetative and drainage layers over the barrier layer, particularly when the barrier layer is a geomembrane.

To illustrate just one of the methods of stability analysis, a special case is examined where the cohesion is zero, that is, for cohesionless soils (sands and gravels).

This stability analysis is significantly simplified as compared to that for cohesive soils. For this case, the critical failure surface is an infinitely small distance into the slope and the slope is assumed to be infinitely long. The resulting solution is termed the infinite slope analysis. Force equilibrium equations can be written for the downslope movement of a soil mass infinitely thin. The Factor of Safety (FS) is defined as ratio of the resisting forces, F_r, to the driving forces, F_d, written as:

$$FS = F_r/F_d \tag{13-6}$$

The resulting equation of equilibrium for a slope at angle, β, measured from the horizontal is:

$$FS = \tan(\phi)/\tan(\beta) \tag{13-7}$$

The infinite slope analysis is applicable to fills of cohesionless materials and can be used to examine the potential for sliding between geomembranes and soil zones in cover and liner systems. Example 13-10 illustrates its use.

Example 13-10   Landfill stability. An ash monofill is planned employing fly ash from an incineration process. The angle of internal friction, ϕ, was measured to be 28° and the cohesion, c, was measured to be zero. A factor of safety of 2.0 is desired for stability, given the hazardous nature of the ash and the potential variability in ash properties. What is the maximum slope angle, β, for this ash monofill?

Solution. Rewriting Eq. 13-7 yields:

$$\beta = \tan^{-1}\left(\frac{\tan\phi}{FS}\right)$$

Substituting:

$$\beta = \tan^{-1}\left(\frac{\tan 28°}{2.0}\right) = 14.9° \text{ or}$$

3.76 horizontal to 1 vertical

When developing or reviewing the design of a waste disposal facility, an often overlooked question is "Can it be built?" Instability during construction may render a design inadequate even though, if it could be built, it would have been stable.

Example 13-11   Landfill constructability. A landfill design such as shown on Fig. 13-1 is planned. On behalf of Landfills-R-Us, a consulting firm employed a design using below-grade slopes of 1 horizontal to 1 vertical. The soil at the site is a medium dense, medium-to-fine sand. Is the landfill stable in the final configuration? Can it be built?

Solution. Yes and no. After the waste is placed in the landfill, the side slopes below grade are stable. However, during construction the evacuation slopes are not stable. Assume $\phi \approx 40°$ for a meduim dense sand. The factor of by Eq. 13-7 is:

$$FS = \frac{\tan\phi}{\tan\beta}$$

where $\text{FS} = \tan 40° / \tan 45° = 0.84 < 1$ Unstable!
β = slope angle
ϕ = angle of internal friction

As shown by the preceding sections of this chapter, there is much to consider in the design of waste disposal facilities. Let us imagine a perfect landfill design for a perfect landfill site. What are the chances that the constructed product is exactly as designed? Zero to none. However, with redundancies in design and adequate quality control, the design intent can be fulfilled. It must be first recognized that, although the parameters previously presented were described in a deterministic fashion, they are, in fact, stochastic parameters. To fully understand the results achievable in the field, it is likely that probabilistic methods need to be employed.[77]

13-8 CLOSURE AND POST-CLOSURE CARE

The closure of a hazardous waste facility should "minimize or eliminate, to the extent necessary to protect human health and the environment, post-closure escape of hazardous waste, hazardous constituents, leachate, contaminated runoff, or hazardous waste decomposition products to the ground or surface waters or to the atmosphere..."[78] The environmental protection should be accomplished with minimal need for further maintenance although it must be recognized that landfills require some perpetual care, however minimal. The owners/operators of hazardous waste facilities must be financially responsible for properly closing and maintaining a closed facility as well as pre-closure and post-closure sudden and non-sudden occurrences.

With the successful closing of a land disposal facility and construction of the final cover, monitoring of the system is required. Monitoring of the system includes measurement of the leachate head within the landfill to preclude the **bathtub effect**. Should the precipitation infiltration and leachate generation exceed the ability of the leachate collection system to remove the leachate, leachate levels build up within the landfill much the same as water fills up a bathtub. This higher hydraulic head causes greater contaminant migration out of the landfill.

Another major closure consideration is maintenance of the cover integrity. For example, burrowing animals dig holes through the cover and challenge its integrity. Erosion of slopes due to heavy precipitation events also has been known to occur. Woody vegetation or vegetation with deep tap roots must be controlled so the root systems of vegetation do not challenge the integrity of the system. If the cover is being subject to traffic, the effects need to be monitored.

Finally, monitoring of the ground water and subsurface environment is a necessary component of the final closure and post-closure care. Monitoring wells in the saturated zone and lysimeters in the unsaturated (vadose) zone are required. Discussion of monitoring well installations and their performance is included in Sec. 15-6. Sampling and analysis of ground water from properly sited and installed monitoring wells provide a well-proven method to ensure that ground water beneath a land disposal facility is not contaminated. However, should contaminants be found using this approach, the detection occurs *after* the ground water is contaminated. The approach of using monitoring wells to detect contamination and employing cleanup techniques to remediate the site should contaminants be found is analogous to closing the barn door after the horse is out. To provide an improved level of environmental protection, methods to check the barn door are required.

Air monitoring may also be required. Air emissions can be monitored at the site boundaries to ensure limited risk to the public health and the environment.

The previous discussion of closure relates to landfill facilities closed in place. For facilities such as pits, ponds, and lagoons, closure may also be accomplished by total removal. In this case, liner systems, contaminated soils, and components of the temporary storage facility are excavated and removed for ultimate treatment and/or disposal.

It may be interesting to the student of hazardous waste management that closure and post-closure care has, in the past, not always proceeded along the technological lines previously described. In a 1986 study by the U.S. General Accounting Office (GAO), seventy-four hazardous waste facilities had filed for bankruptcy, hindering the efforts of the government to enforce a proper facility closure.[79] The GAO also found that about 75% of the firms employed the approved financial test required at the time to demonstrate they are financially strong and could pay closure and post-closure costs without set aside (a trust fund) money. However, because corporate financial strengths change rapidly, adequate funds were not available at the time of closure. Since this study, revisions in the financial tests and bonding and insurance requirements have been made to increase the likelihood that monies needed for closure and post-closure care are available.

13-9   OTHER TYPES OF LAND DISPOSAL FACILITIES

Hazardous waste placed in landfill facilities is considered managed. Hazardous waste placed in storage facilities is just passing through on its way to treatment and/or ultimate waste management. Land-based storage facilities include surface impoundments such as pits, ponds, and lagoons.

In a study of the performance of surface impoundments, nine existing facilities were examined in an effort to improve design, construction, and operation of these types of facilities. Presented in Table 13-8[80] is a description of the general features and waste characteristics of the nine surface impoundments selected for detailed examination. This table illustrates to the student of hazardous waste management the range of wastes managed in surface impoundments. The ineffective performance of seven of the nine facilities studied in detail was attributed to inadequacies in design, subsurface investigations, and construction quality control. The reverse was true for the two facilities with successful performance records. It is intuitive that the level of environmental protection provided by surface impoundments depends upon excellence in both design and construction.

TABLE 13-8
Case Study Facilities - General Features and Waste Characteristics†

Case Study No.	Type of Facility	Type of SI (No. and Function)	Year Placed in Service	SI Size (Acres)	Waste Type	Waste Quantity
1	Electrolytic metal refining plant	Small disposal pond	1972	0.4	Acidic process liquor and sludge waste (pH < 2) high in heavy metal content	A total of 843.750 gal in 1982
		Disposal pond	1979	1.1	Acidic process liquor and sludge waste (pH < 2) high in heavy metal content	
2	Pesticide formulation and distribution plant	Pesticide washdown evaporation disposal pond	1979	< 0.1	Pesticide rinsewater	Batch operation (400 gal/day maximum)
		Pesticide rinsewater evaporation disposal pond	1982	< 0.1	Pesticide rinsewater	Batch operation (400 gal/day maximum)
3	Commercial hazardous waste disposal facility	SiteA: 8 impoundments used for settling, storage, and sludge disposal	1951	15	Oily water and brines, alkaline and acid wastes, heavy metals, paint sludge, tank bottom sediments, cyanide, pesticides, and other chemical wastes	A total of 53 million gal in 1982
4	Agricultural fertilizer manufacturing plant	11 settling ponds used to remove gypsum	1965	14	Production water for ammonium phosphate/phosphoric acid plant with pH < 2 and high radionuclides content	20,000 gal/day
		One evaporation pond (treatment)	1976	8	Wastewaters from plant boilers, water treaters, and nitric and sulfuric acid plants (pH < 2, high in radionuclides)	130,000 gal/day
		One cooling pond (treatment)	1976	38	Same as gypsum SI	10,000 gal/day
5	Mineral ore mining/manufacturing plant	5 low-heads solar ponds (treatment and storage)	1972	90	Mineral liquor tailings with high arsenic and boron content	A total of 50 million gal/month to ponds A-E, 4, and 5
		High-head evaporation pond (treatment and storage)	1975	80	Mineral liquor tailings with high arsenic and boron content	
		High-head evaporation pond	1976	100	Mineral liquor tailings with	

6	Commercial hazardous waste disposal facility	Evaporation pond (treatment and storage) high arsenic and boron content	1980	120	Acid plant wastewater with hight arsenic and boron content	12 million gal/month
		Evaporation pond (disposal)	1980	5	Geothermal muds and brines, wastewater treatment sludge, tank bottom sediments, cooling tower blowdown sludge and oil drilling muds	9.85 million gal in 1982
		Disposal pond (currently used for land treatment)	1980	5	Geothermal muds and brines, wastewater treatment sludge, tank bottom sediments, cooling tower blowdown sludge and oil drilling muds	
		Disposal pond (currently)	1981	5	Geothermal muds and brines, wastewater treatment sludge, tank bottom sediments, cooling tower blowdown sludge and oil drilling muds	
7	Agricultural fertilizer manufacturing plant	Cooling pond (treatment)	1974	100	Process water from phosphoric acid plant with pH< 2 and high flouride content	40,000 gal/min (maximum)
		Initial gypsum pond (disposal)	1974	150	Gypsum slurry with pH< 2 and flouride and phosphorus content	No data available
		Expansion gypsum pond (disposal)	1980	200	Gypsum slurry with pH< 2 and flouride and phosphorus content	No data available
8	Chemical production plant	2 equalization/ retention basins (treatment)	1976	35	Wastes high in organic nitrogen content and varying pH, resulting from synthetic fiber production	3,000 gal/min
9	Uranium mining/ milling	Tailings pond (disposal)	1980	64	Acidic tailings slurry containing kerosene and radium 226	No data available

† (after Ghassemi, et al.)[81]

Hazardous waste has been and continues to be disposed of in deep wells. The Environmental Institute for Waste Management Studies concluded in a study of the performance of deep injection wells that most hazardous waste injection wells were properly sited, constructed, and operated.[82] It was found that many of the problems with deep well disposal were operator related and could have been prevented with standardized operating procedures.

The fate of contaminants after disposal in deep wells is not clear, although the combination of favorable chemical reactions and dilution afford protection to ground water and surface water supplies. In fact one facility owner, who operates more than 25% of the Class I injection wells, claims no known water supply contamination over an operational history of more than 35 years.[83]

The injected wastes undergo chemical reactions both between waste constituents and between waste constituents and the formation materials. Reactions include carbonate dissolution in limestone and dolomite, sand dissolution, clay dissolution, hydrolysis, and co-precipitation. The fate of contaminants is predicted by modeling neutralization, hydrolysis, and precipitation reactions using reaction rates and equilibrium constants from the literature for solid-liquid equilibria for aqueous electrolyte solutions. Because these reactions occur deep in the bedrock formations, it is difficult to predict with certainty the extent of unreacted waste and the nature and distribution in the highly anisotropic and heterogeneous subsurface environment.

DISCUSSION TOPICS AND PROBLEMS

13-1. Cite four reasons landfills remain a viable hazardous waste management alternative.

13-2. Redo Example 13-1 for three additional hazardous waste management techniques.

13-3. Explain the relationship between aquifers and the siting of landfills.

13-4. Considering the results of Example 13-3, discuss the advantages and disadvantages of employing 0.2 m of daily cover. Use calculations to support your answer.

13-5. Assume 17% of the infiltration is absorbed by the landfill and daily cover in Example 13-4. What will be the new rate of leachate generation?

13-6. Assume the landfill waste in Example 13-4 consolidates and releases liquids as leachate from the pore spaces equal to 1% of the landfill volume. No infiltration is absorbed by the landfill and daily cover. What will be the new rate of leachate generation?

13-7. Divide the class into two groups: environmental extremists and industrial party liners. Discuss the following questions employing factual information.

(*a*) We are just making our 55-gallon drums bigger.

(*b*) All landfills leak, so they should be banned.

(*c*) When single liners were found inadequate, double liners were used beneath landfills. Now (1992) hazardous waste landfills have three liners, two of geosynthetic materials and one of natural materials. Are landfills now "safe" methods of hazardous waste disposal or will the future prove these systems inadequate?

13-8. Comment on the implications of an action leakage rate of 5 gal/acre/day versus the readily obtainable leakage rate of 22 gal/acre/day.

13-9. What are the factors which affect the rate of contaminant migration.

13-10. Explain the difference between the two types of molecular diffusion.

13-11. You are considering the selection of a geomembrane as secondary containment beneath a buried underground fuel (gasoline) storage tank. PVC, HDPE, and neoprene are being considered. Which would you recommend and why?

13-12. Identify and describe the three steps involved in the transport of a contaminant through a geomembrane.

13-13. Final cover over a hazardous waste landfill is typically installed to minimize the environmental risk by addressing potential contaminant pathways. Identify four of the contaminant pathways mitigated through the use of a cover system.

13-14. Explain the importance of the vegetative support layer in a landfill cover system.

13-15. Identify the three major components of a landfill cover system and describe their functions.

13-16. Sketch your conceptual design of a cover system for four regions of the United States: Minnesota, Arizona, Washington, and South Carolina.

13-17. Water treatment systems often use sand and gravel as filters just as landfills use sand and gravel in leachate collection and filter zones. How are sand filters cleaned in water treatment plants? What are the implications for cleaning sand in leachate collection zones? What are implications for attempting to clean geotextiles?

13-18. A 12-inch layer of soil-bentonite is planned for placement immediately beneath the secondary geomembrane liner. The layer is to be constructed in two six-inch lifts with a bentonite content of 9% by dry weight at a compacted unit weight of 95 pcf. By weight, how much bentonite is needed per square foot of liner per lift?

13-19. Using the data from Table 13-7:

    (*a*) Calculate the total contaminant migration of trichloroethylene through a 10-acre site lined with a single 1.0 mm HDPE geomembrane.

    (*b*) Calculate the number of gallons of ground water that could become contaminated to the drinking water limit for this contaminant in one year.

13-20. Using the data from Figs. 13-15 and 13-16 and assuming a liner beneath site of 10 acres, calculate the number of gallons of ground water that could become contaminated per year to the drinking water limit for the contaminant during the first year, at 10 years, and at 100 years.

13-21. Describe the various ways by which liquids can migrate through a geomembrane liner.

REFERENCES

1. Ham, R. K.: "New Approaches to Technology—Landfills," *Fourth National Conference on Environmental Issues*, Academy of Natural Sciences, Philadelphia, PA, January 1989.

2. Lycoming Landfill, Lycoming Co., PA, 1991.

3. Bonaparte, R., and B. A. Gross: "Field Behavior of Double Liner Systems," *Waste Containment Systems: Construction, Regulation, & Performance*, R. Bonaparte, ed., ASCE Geotechnical Special Publication No. 26, pp. 52–83, November 1990.

4. U.S. Enviromental Protection Agency: *Requirements for Hazardous Waste Landfill Design, Construction, and Closure*, EPA/625/4-89/022, 127 pp., August 1989.

5. German Geotechnical Society, Geotechnics of Landfills and Contaminated Land: *Technical Recommendations, for International Society of Soil Mechanics and Foundation Engineering*, Ernst and Sohn, Berlin, 76 pp., 1990.

6. Evans, J. C., Y. Sambasivam, and S. J. Zarlinski: "Attenuating Materials in Composite Liners," *Waste Containment Systems: Construction, Regulation, & Performance*, R. Bonaparte, ed., ASCE Geotechnical Special Publication No. 26, pp. 246–263, November 1990.

7. Johnson, R. L., C. D. Palmer, and W. Fish,: "Subsurface Chemical Processes," *Transport and Fate of Contaminants in the Subsurface*, EPA/625/4-89/019, September 1989.

8. Bagchi, A.: *Design, Construction & Monitoring of Sanitary Landfill*, John Wiley & Sons, New York, NY, 1990.

9. Bagchi, A.: "Design of Attenuating Landfills," *Journal of Environmental Engineering Div., ASCE*, 110, No. EE-6, 1211–1212, 1984.

10. Daniel, D. E., and C. D. Shackelford: Disposal Barriers that Release Contaminants Only by Molecular Diffusion, *Proceedings of the Oak Ridge Model Conference, Oak Ridge National Laboratory, Oak Ridge, Tenn.*, Vol. I, No. 3, 1987, pp. 315–332.

11. Evans, et al., "Attenuating Materials in Composite Liners."

12. Smalstig, E.: "Comparison Study of Different Clay Liners Used as Barriers to Contaminant Transport," unpublished student independent laboratory report, Bucknell University, May 1990.

13. Mead, T.: "Composite Slurry Walls with Organo-clays" unpublished student independent laboratory report, Bucknell University, May 1990.

14. Paraseal, Paramount Technical Products, Inc., Bloomington, MN, 55431.

15. Evans, J. C., and S. E. Pancoski: "Organic Waste Treatment with Organically Modified Clays," *Proceedings of the 3rd International Conference on New Frontiers for Hazardous Waste Management*, Pittsburgh, PA, EPA/600/9-89/072, pp. 48–57, September 10–13, 1989.

16. Mitchell, J. K., and A. T. Yeung: "Electro-Kinetic Flow Barriers in Compacted Clay," *Transportation Research Record 1288*, Transportation Research Board, National Research Council, Washington, D.C., pp. 1–9, 1990.

17. Young, A.: "Mathematical Modeling of Landfill Gas Extraction," *Journal of Environmental Engineering*, vol. 115, no. 6, pp. 1073–1087, December 1989.

18. Schroeder, P. R., A. C. Gibson, and M. D. Smolen: *The Hydrologic Evaluation of Landfill Performance (HELP) Model*, EPA/830-SW-84-010, 1984.

19. McEnroe, B. M.: "Steady Drainage of Landfill Covers and Bottom Liners," *Journal of Environmental Engineering*, vol. 115, no. 6, pp. 1114–1122, December 1989.

20. U.S. Environmental Protection Agency: *Remedial Action at Waste Disposal Sites*, EPA/625/6-85/006, October 1985.

21. Koerner, R.: *Designing with Geosynthetics*, 2nd ed., Prentice-Hall, Englewood Cliffs, NJ, 652 pp., 1990.

22. Landreth, Robert E.: "Chemical Resistance Evaluation of Geosynthetics Used in Waste Management Applications," *Geosynthetic Testing for Waste Containment Applications*, R. M. Koerner, ed., ASTM STP 1081, pp. 3–11, 1990.

23. EPA Method 9090; "Compatibility Tests for Wastes and Membrane Liners," EPA SW-846, *Test Methods for Evaluating Solid Waste*, U.S. Environmental Protection Agency, Washington, D.C.

24. White, D. F; and K. L. Verschoor: "Practical Aspects of Evaluating the Chemical Compatibility of Geomembranes for Waste Containment Applications," *Geosynthetic Testing for Waste Containment Applications*, R. M. Koerner, ed., ASTM STP 1081, pp. 25–36, 1990.

25. Carroll, R. G., Jr.: "Hydraulic Properties of Geotextiles," *Geotextile Testing and the Design Engineer*, ASTM STP 952, J. E. Fluet, Jr., Ed., American Society of Testing and Materials, Philadelphia, PA, pp. 7–20, 1987.

26. Koerner, R. M., and J. A. Bove: "Lateral Drainage Designs using Geotextiles and Geocomposites," *Geotextile Testing and the Design Engineer*, ASTM STP 952, J. E. Fluet, Jr., Ed., American Society of Testing and Materials, Philadelphia, PA, pp. 33–44, 1987.

27. Hodge, J.: "Durability Testing," *Geotextile Testing and the Design Engineer*, ASTM STP 952, J. E. Fluet, Jr., ed., American Society of Testing and Materials, Philadelphia, PA, pp. 119–121, 1987.

28. U.S. EPA: *Design, Construction, and Evaluation of Clay Liners for Waste Management Facilities*, EPA/530/SW-86/007F, November 1988.

29. Lambe, T. W.: "The Engineering Behavior of Compacted Clay," *J. Soil Mechanics and Foundation Engineering Division, American Society of Civil Engineers*, vol. 84, SM2, May 1958.

30. Lambe, T. W., "The Structure of Compacted Clay," *J. Soil Mechanics and Foundation Engineering Division, American Society of Civil Engineers*. Vol. 84, SM2, May 1958.

31. Daniel, D. E.: "Predicting Hydraulic Conductivity of Clay Liners," *Journal of the Geotechnical Engineering Division, ASCE*, vol. 110, No. 2, February 1984.

32. Daniel, D. E., and S. J. Trautwein: "Field Permeability Test for Earthen Liners," *Use of In Situ Tests in Geotechnical Engineering*, ASCE Geotechnical Special Publication No. 6, pp. 146–160, June 1986.

33. Benson, C. H., and D. E. Daniel: "Influence of Clods on Hydraulic Conductivity of Compacted Clay," *Journal of the Geotechnical Engineering Division, ASCE*, vol. 116, no. 8, August 1990.

34. Brown, K. W., and D. C. Anderson. 1983. "Effects of Organic Solvents on the Permeability of Clay Soils." EPA 600/2-83-016. U.S. Environmental Protection Agency, Cincinnati, OH.

35. Mitchell, J. K., and F. T. Madsen: "Chemical Effects on Clay Hydraulic Conductivity," *Geotechnical Practice for Waste Disposal, ASCE*, Geotechnical Special Publication No. 13, pp. 87–116, June 1987.

36. Grim, R. E.: *Clay Mineralogy*. McGraw-Hill Book Company, Inc., New York, NY, 1968.

37. Ibid.

38. Mumpton, F. A.: *Mineralogy and Geology of Zeolites*, Mineralogical Society of America, Short Course Notes, Southern Printing Co., Blacksburg, Va., vol. 4, 233 pp, 1977.

39. Jordan, J. W.: "Alteration of the Properties of Bentonite by Reaction with Amines," *Mineralogical Magazine and Journal of the Mineralogical Society*, vol. 28, no. 205, pp. 598–605, June 1949.

40. Jordan, J. W.: "Organophilic Bentonites: Swelling in Organic Liquids," *The Journal of Physical & Colloid Chemistry*, Vol. 53, No. 2, pp. 294–306, February 1949.

41. Alther, G. R., J. C. Evans, and S. E. Pancoski: "Organically Modified Clays for Stabilization of Organic Hazardous Wastes," *Superfund '88, Proceedings of the Ninth National Conference, Hazardous Materials Control Research Institute*, pp. 440–445, November 1988.

42. Chiou, C. T., P. E. Porter, and D. W. Schmedding: "Partition Equilibria of Nonionic Organic Compounds between Soil Organic Matter and Water," *Environmental Science and Technology*, 17, pp. 227–231, 1983.

43. Chiou, C. T., L. J. Peters, and V. H. Freed: "A Physical Concept of Soil-Water Equilibria for Nonionic Organic Compounds," *Science*, 213, pp. 684–685.

44. Boyd, S. A., J. F. Lee and M. M. Mortland: "Attenuating Organic Contaminant Mobility by Soil Modification," *Nature*, vol. 33, pp. 345–347, May, 1988.

45. Mortland, M. M.; S. Shaobai, and S. A. Boyd: "Clay-Organic Complexes as Adsorbents for Phenol and Chlorophenols," *Clays and Clay Minerals*, vol. 34, no. 5, pp. 581–585, 1986.

46. Wolf, T. A., T. Demirel, and R. E. Bauman: "Adsorption of Organic Pollutants on Montmorillonite Treate with Amines," *Journal of the Water Pollution Control Federation*, 58, pp. 68–76, 1986.

47. Mott, H. V. and W. J. Weber: "Diffusive Transport and Attenuation of Organic Leachates in Cut-Off Wall Backfill Mixes," Presented at the Twelfth Annual Madison Waste Conference, Department of Engineering Professional Development, University of Wisconsin-Madison, September 20–21, 1989.

48. Freeze, R. A. and J. A. Cherry: *Groundwater*, Prentice-Hall, Englewood Cliffs, NJ, 604 pp, 1979.

49. Daniel, D. E., S. J. Trautwein, S. S. Boynton, and D. E. Foreman: "Permeability Testing with Flexible Wall Permeameters," *Geotechnical Testing Journal*, vol 7, no. 3:113–122, 1984.

50. Dunn, R. J., and J. K Mitchell: "Fluid Conductivity Testing of Fine Grained Soils," *Journal of Geotechnical Engineering*, vol. 110 no. 11:1648–1665, 1984.

51. Evans, J. C. and H. Y. Fang: "Triaxial Permeability and Strength Testing of Contaminated Soils," *Advanced Triaxial Testing of Soil and Rock*, ASTM STP 977, American Society for Testing and Materials, Philadelphia, PA, pp. 387–404, 1984.

52. Haxo, H. E. Jr., J. A. Miedema, and Nelson, N. A.: "Permeability of Polymeric Membrane Liner Systems," *Proceedings of the International Conference on Geomembranes*, Denver, CO, pp. 151–156, 1984.

53. Verschoor, K., L. Britton, and R. Thomas: "Chemical Compatibility Testing of Geosynthetics to be Used as Containment Barriers in Hazardous Waste Management," *Hazardous Waste and Hazardous Materials*, pp. 205–209, 1988.

54. Haxo, H. E., R. S. Haxo, N. A. Nelson, P. D. Haxo, R. M. White, S. Dakessian, and M. A. Fong: *Liner Materials for Hazardous and Toxic Waste and Municipal Solid Waste Leachate*, Noyes Publications, 1985.

55. Giroud, J. P.: "Impermeablilty: The Myth and a Rational Approach," *Proceedings of the International Research Conference on Geomembranes*, Denver, CO, pp. 157–162, 1984.

56. Shackelford, C. D.: "Transit-time Design of Earthen Barriers," *Engineering Geology*, vol. 29, pp. 79–94, 1990.

57. Johnson, R. L., J. A. Cherry, J. F. Pankow: "Diffusive Contaminant Transport In Natural Clay: A Field Example and Implications for Clay-Lined Waste Disposal Sites," *Environ. Sci. Technol.*, vol. 23, no. 3, pp. 340–349, 1989.

58. Seaman Corporation, XR5 Geomembrane literature, Wooster, Ohio.

59. Shackelford, C. D.: "Diffusion of Contaminants Through Waste Containment Barriers," *Transportation Research Record 1219*, Geotechnical Engineering 1989, Transportation Research Board, National Research Council, pp. 169–182.

60. Haxo, H. E.: "Determining the Transport through Geomembranes of Various Permeants in Different Applications," *Geosynthetic Testing for Waste Containment Applications*, R. M. Koerner, ed., ASTM STP 1081, pp. 75–94, 1990.

61. Haxo, "Determining the Transport through Geomembranes."

62. Ibid.

63. August, H., and R. Tatzky: "Permeabilities of Commercially Available Polymeric Liners for Hazardous Waste Landfill Leachate Organic Constituents," *Proceedings of the International Conference on Geomembranes*, Denver, CO, Vol. 1, Industrial Fabrics Association International, St. Paul, MN, pp. 163–168, 1984.

64. Haxo, "Determining the Transport through Geomembranes."

65. Johnson, R. L., J. A. Cherry, J. F. Pankow: "Diffusive Contaminant Transport In Natural Clay: A Field Example and Implications for Clay-Lined Waste Disposal Sites," *Environ. Sci. Technol.*, vol. 23, no. 3, pp. 340–349, 1989.

66. Gillham, R. W., M. J. L. Robin, D. J. Dytynyshyn, and H. M. Johnston: "Diffusion of Nonreactive and Reactive Solutes through Fine-Grained Barrier Materials," *Can. Geotech. J.*, 21, pp. 541–550, 1984.

67. Ibid.

68. Mott, H. V. and W. J. Weber: "Solute Migration in Soil-Bentonite Containment Barriers," *Superfund '89, Proceedings of the Ninth National Conference, Hazardous Materials Control Research Institute*, pp. 526–533, November, 1989.

69. Ibid.

70. Ibid.

71. Evans, J. C., Sambasivam, Y. and Zarlinski, S. J., "Attenuating Materials in Composite Liners," *Waste Containment Systems: Construction, Regulation, & Performance*, ASCE Geotechnical Special Publication No. 26, November, 1990, pp. 246–263.

72. Evans, J. C. and H. Y. Fang: "Triaxial Permeability and Strength Testing of Contaminated Soils," *Advanced Triaxial Testing of Soil and Rock*, American Society for Testing and Materials, Philadelphia, PA, pp. 387–404, 1988.

73. Boschuck, J.: "Landfill Covers, An Engineering Perspective," *Geotechnical Fabric Report*, pp. 23–34, March, 1991.

74. Morris, D. V. and C. E. Woods: "Settlement and Engineering Considerations in Landfill and Final Cover Design," *Geotechnics of Waste Fills—Theory and Practice*, ASTM STP 1070, A. Landva and G. D. Knowles, eds., American Society for Testing and Materials, Philadelphia, PA, 1990.

75. Huang, Y. H.: *Stability Analysis of Earth Slopes*, Van Nostrand Reinhold, New York, NY, pp. 305, 1983.

76. Landva, A., and G. D. Knowles, eds.: *Geotechnics of Waste Fills—Theory and Practice*, ASTM STP 1070, American Society for Testing and Materials, Philadelphia, PA, 1990.

77. Benson, C.: "Reliability Analysis for Time of Travel in Compacted Soil Liners," *Geotechnical Engineering Congress 1991*, ASCE, Geotechnical Engineering Special Technical Publication No. 27, June 1991.
78. U.S. Federal Regulations: "Closure Performance Standard," Sec. 265-111(b), September 23, 1988.
79. General Accounting Office: *Environmental Safeguards Jeopardized when Facilities Cease Operating*, GAO/RCED-86-77-Hazardous Waste, p. 3, February 1986.
80. Ghassemi, M., M. Haro, and L. Fargo: *Assessment of Hazardous Waste Surface Impoundment Technology: Case Studies and Perspectives of Experts*, EPA-600/S2-84-173, January 1985.
81. Ibid.
82. Environmental Institute for Waste Management Studies: "Deep Well Injection Studies," Newsletter, vol. IX, no. 2, June 1989.
83. Scrivner, N. C., et. al.: "Chemical Fate of Injected Wastes," *Ground Water Monitoring Review*, vol. 6, no. 3, pp. 53–58, Summer 1986.

ADDITIONAL READING

Bagchi, A.: *Design, Construction, & Monitoring of Sanitary Landfills*, John Wiley & Sons, New York, NY, 284 pp., 1990.
Daniel, D. E., ed.: *Geotechnical Practice for Waste Disposal*, Chapman & Hall, New York, NY, pp. 632, December 1992.
Linsley, R. K., Kohler, M. A., and Paulhus, J. L.: *Hydrology for Engineers*, 2nd ed., McGraw-Hill, New York, NY, 1975.
Vesilind, P. A., Peirce, T. T., and Weiner, R. F.: *Environmental Engineering*, 2nd ed., Butterworth, 1988, p. 215–234.

ACKNOWLEDGEMENTS

The authors wish to acknowledge the following individuals who assisted with the preparation of this chapter: Craig Benson, *University of Wisconsin*, John Repasky and Robert Owens, *Bucknell University*. The principal reviewer for the **ERM Group** was Alan F. Hassett, P.E., *ERM, Inc.*

CHAPTER 14

QUANTITATIVE RISK ASSESSMENT

Chaos umpire sits,
And by decision more embroils the fray
By which he reigns: next him high arbiter
Chance governs all.

Milton

The purpose of this chapter is to provide an understanding of risk, how to assess it at a specific site, and what the numerical estimates of risk represent. The National Research Council defines risk assessment as:

> ... the characterization of the potential adverse health effects of human exposures to environmental hazards. ... Risk assessment also includes characterization of the uncertainties inherent in the process of inferring risk.[1]

Risk assessment is a tool for understanding the health and environmental hazards associated with hazardous waste and can greatly improve the basis upon which to make hazardous waste management decisions. Risk assessment can also assist the public in understanding the human health implications of the issues and available alternatives.

14-1 RISK

In the classical sense, **risk** is defined as the probability of suffering harm or loss. When the resulting harm is measurable (e.g., person-days lost to accidents), risk may be calculated as the probability of an action occurring multiplied by the severity of the harm if the action does occur:

$$\text{Risk} = (\text{Probability}) \times (\text{Severity of Consequence})$$

Using this calculation suggests that the following two activities have the same risk:

Activity no.	Severity of outcome (person-days lost	Probability of outcome	Risk (person-days lost)
1	20	0.1	2
2	10	0.2	2

Often the undesirable consequence is not a quantifiable matter. An example of such a consequence is death; no one is ever slightly dead. Another example is the onset of cancer. For such consequences, risk is simply defined as the probability of the harm occurring. Because approximately 25% of the U.S. population will experience cancer in their lifetime, the lifetime risk of cancer is approximately 0.25.

An important distinction is whether the risk is background, incremental, or total risk. **Background risk** is what people are exposed to in the absence of the particular source of risk being studied, **incremental risk** is that caused by this source, and **total risk** is simply the sum of the two. The previous reference to 0.25 probability of cancer represents background risk. The U.S. Environmental Protection Agency (EPA) target for Superfund sites of 1×10^{-6} excess lifetime cancer risk represents an incremental risk—the probability of a person developing cancer from exposure to contaminants from a Superfund site is 1×10^{-6} (or "one in a million") in excess of that person's risk of cancer from all other sources combined. It also implies that for every million people potentially exposed to a site, one additional cancer case over their lifetimes would be expected. If the population exposed to a site is 10,000, this number can be interpreted as 0.01 additional occurrences of cancer.

Table 14-1 provides some estimates of some commonplace risks (death in the case of accidents and cancer for the other actions). These represent background risks on an annual basis.

TABLE 14-1
Common risks[2]

Action	Annual risk	Uncertainty
Motor vehicle accident		
Total	2.4×10^{-4}	10%
Pedestrian	4.2×10^{-5}	10%
Home accidents	1.1×10^{-4}	5%
Cigarette smoking (one pack per day)	3.6×10^{-3}	Factor of 3
Peanut butter (4 teaspoons per day)	8×10^{-6}	Factor of 3
Drinking water with EPA limit of TCE	2×10^{-9}	Factor of 10

The terms risk and hazard are often confused. **Hazard** is a descriptive term, referring to the intrinsic capability of the waste to cause harm; it is the source of the risk. The hazard posed by a waste is a function of such variables as its toxicity, mobility, and persistence as well as how it is contained. The release or threatened release of a hazardous waste thus represents a hazard—it is a source of risk; however, the waste does not represent a risk unless exposure has occurred or the possibility exists for future exposure.

Emergence as a Systematic Methodology

Undoubtedly, the concept of risk has a long history of influencing environmental policy making. However, the initial efforts relied more on intuition than on the scientific principles of toxicology, chemistry, and fate and transport modeling inherent in modern risk assessments. Only in the past few years has a *science-based* risk assessment played a major role in environmental decision making.

In this chapter we will use the term *quantitative* risk assessment to describe the process of using scientific principles to calculate quantitative estimates of risk. The widely recognized approach is a four-stage process:

1. Hazard identification (which chemicals are important)
2. Exposure assessment (where do the chemicals go, who might be exposed. and how)
3. Toxicity assessment (determining numerical indices of toxicity for computing risk)
4. Risk characterization (estimating the magnitude of risk, and the uncertainty of the estimate)

Risk assessment, as performed under most hazardous waste regulatory programs, employs this concept of conducting the risk assessment process in these four stages. The U.S. National Academy of Sciences created the four-stage process and eventually the EPA codified it. The EPA first applied the concept in 1984 for carcinogens following a suit brought by the Natural Resources Defense Council regarding ambient water quality criteria. In response to this suit, the EPA issued carcinogenic potency factors (CPFs) (also called slope factors) that could be utilized in a predictive capacity to estimate carcinogenic risk to potentially exposed populations. A similar approach was then applied to non-carcinogens by setting acceptable daily intake (ADI) levels, that are analogous to reference doses (RfDs).

Purposes of Risk Assessment

Using the estimates calculated by a risk assessment as a basis for making decisions is termed risk management. **Risk management** may be defined as:

> ... evaluating alternative ... actions and selecting among them ... entails consideration of political, social, economic, and engineering information with risk-related information to develop, analyze, and compare regulatory options and to select the appropriate reg-

ulatory response to a potential chronic health hazard. The selection process necessarily requires the use of value judgments on such issues as the acceptability of risk and the reasonableness of the costs of control.[3]

In the hazardous waste field, risk assessment provides information to decision makers as to the consequences of possible actions. Important decisions that could use risk estimates include selecting waste treatment/disposal options, remediating contaminated sites, minimizing waste generation, siting new facilities, and developing new products. It should be emphasized that risk estimates are only one type of information used, and hazardous waste decisions are often driven by political, social, economic, or other factors.

Risk assessment plays a major role in the decision making for the remediation of contaminated sites. It provides an important part of the basis for selecting from alternative remedies (see Sec. 17-6). Risks from exposure are calculated for each alternative under the assumption that its remedial actions have been implemented fully. This enables a balanced comparison of the effectiveness of each alternative at reducing risks.

Another application in site remediation is the establishment of cleanup standards. In many cases, definitive numerical standards for specific contaminants in soil or ground water simply do not exist. Risk assessments are used to determine "How clean is clean?"[4] In this instance the calculation procedure is reversed. The process starts with a numerical definition of acceptable risk and works back to the level of contamination that will produce the acceptable risk level. An example would be a determination at a specific site that any soil containing greater than 588 mg/kg of toluene would pose an unacceptable risk and thus has to be remediated; soils containing smaller concentrations can be left in place.[5]

The EPA has proposed guidelines for the preparation of hazardous waste site risk assessments in a number of publications. The initial publications, *Endangerment Assessment Handbook*[6] and the *Superfund Public Health Evaluation Manual*,[7] have been superseded by *Risk Assessment Guidance for Superfund*,[8] the *Superfund Exposure Assessment Manual*,[9] and the *Exposure Factors Handbook*.[10] These risk assessments are conducted to quantify potential human health and ecological risks stemming from toxic contaminants that may be transported to potential receptor populations in both present and future use scenarios. Human health risks for carcinogens are calculated as the probability of developing cancer in a lifetime and for non-carcinogens as hazard indices. In addition, an ecological assessment is attempted to quantify risk to on-site and off-site biota.

The following four sections describe each of the four stages of the risk assessment process as an overview of the state of the art in risk assessment. The methods in general follow those recommended by the EPA[11] because much of the risk assessment currently conducted adheres to these procedures. However, it must be noted that these procedures are constantly being revised, and it is our objective to explain the science of risk assessment and not to provide guidelines for regulatory compliance. To provide material for illustrating the process, the examples of exposure from a hazardous waste site are used.

14-2   HAZARD IDENTIFICATION

It is not rare to detect as many as 100 different chemicals at a contaminated site. The hazard identification stage examines the data for all contaminants detected at a site and consolidates the data to stress the chemicals of concern (i.e., those representing the vast majority of risk posed by the site).

Risk assessment requires a clear understanding of what chemicals are present at a site, their concentration and spatial distribution, and how they could move in the environment from the site to potential receptor points. The types of data typically sought at a hazardous waste site to address such needs are shown in Table 14-2, and the collection of such data is detailed in Chap. 15. A site investigation easily can produce a huge amount of data, necessitating that steps be taken in the hazard identification stage to facilitate subsequent analysis.

One approach is to examine data for all chemical contaminants detected in any media and select a subset of chemicals, consisting of the specific chemicals of concern and representative of all detected chemicals. These selected chemicals thus represent indicators or surrogates for all chemicals detected at the site. Their purpose is to limit the number of chemicals that must be modeled in fate and transport analyses and to focus efforts on the most significant hazards. The advent of modern computer methods that allow a large number of chemicals to be modeled simultaneously may obviate the need for a surrogate chemical selection step in many instances. Nevertheless, the EPA endorses the data consolidation approach for sites having a large number of chemicals detected in different media; however, professional judgment is emphasized. In general, most constituents at hazardous waste sites should be carried through a quantitative risk assessment.

The surrogate chemicals are selected on the basis of which compounds best represent the risk posed by the site:

- The most toxic, persistent, and mobile
- The most prevalent in terms of spatial distribution and concentration
- Those involved in the more significant exposures

TABLE 14-2
Data needs

Site history
Land use
Contaminant levels in media:
    Air, ground water, surface water, soils and sediments
Environmental characteristics affecting chemical fate and transport:
    Geologic
    Hydrogeologic
    Atmospheric
    Topographic
Potentially affected population
Potentially affected biota

The list of surrogate chemicals should encompass those chemicals that are estimated to account for 99% of the risk at the site. It should contain compounds that will support an adequate evaluation of both carcinogenic and non-carcinogenic risk.

Initial Screening

One approach to selecting this subset from all detected chemicals begins with an initial screening of data which may be summarized as follows:

1. Sort the contaminant data by medium (e.g., ground water, soil, etc.) for both carcinogens and non-carcinogens.
2. Tabulate for each detected chemical the mean and range of concentration values observed at the site.
3. Identify the reference doses for non-carcinogens and slope factors for carcinogens (see Chap. 5) for each potential exposure route.
4. Determine the toxicity scores for each chemical in each media. For non-carcinogens:

$$TS = C_{max}/RfD \qquad (14\text{-}1)$$

where TS = toxicity score
C_{max} = maximum concentration
RfD = chronic reference dose (i.e., an estimate of acceptable daily intake) (see Sec. 5-4 and App. B)

For carcinogens:

$$TS = SF \times C_{max} \qquad (14\text{-}2)$$

where SF = slope factor (also called carcinogen potency factor) (see Sec. 5-5 and App. B)

5. For each exposure route, rank the compounds by toxicity scores.
6. For each exposure route, select those chemicals comprising 99% of the total score.

The process is illustrated by the following two examples. Data for surface soils appears in Table 14-3. Reference doses (RfD) and slope factors may be found in App. B.

TABLE 14-3
Chemical concentrations at the ABC landfill

Chemical	Air Mean (mg/m^3)	Air Maximum (mg/m^3)	Ground water Mean (mg/L)	Ground water Maximum (mg/L)	Soil Mean (mg/kg)	Soil Maximum (mg/kg)
Chlorobenzene	4.09E − 08	8.09E − 08	2.50E − 04	1.10E − 02	1.39E + 00	6.40E + 00
Chloroform	1.12E − 12	3.12E − 12	4.30E − 04	7.60E − 03	1.12E + 00	4.10E + 00
1,2-Dichloroethane	1.40E − 08	2.40E − 08	2.10E − 04	2.00E − 03	ND	ND
BEHP	3.29E − 07	8.29E − 07	ND	ND	1.03E + 02	2.30E + 02

ND—Not detected.

Example 14-1   Selection of non-carcinogenic chemicals. Rank the chemicals for soil in Table 14-3.

Chemical	C_{max}(mg/kg)	RfD*	Toxicity score	Rank
Chlorobenzene	6.40E + 00	2.00E − 02	3.20E + 02	3
Chloroform	4.10E + 00	1.00E − 02	4.10E + 02	2
1,2-Dichloroethane	ND	NA	NA	NA
BEHP	2.30E + 02	2.00E − 02	1.15E + 04	1

NA–Not applicable
*Source: IRIS (App. B)

Example 14-2   Selection of carcinogenic chemicals. Rank the chemicals for soil in Table 14-3.

Chemical	C_{max}(mg/kg)	SF*	Toxicity score	Rank
Chlorobenzene	6.40E + 00	NA	NA	NA
Chloroform	4.10E + 00	6.10E − 03	2.50E − 02	2
1,2-Dichloroethane	ND	NA	NA	NA
BEHP	2.30E + 02	1.40E − 02	3.22E + 00	1

NA–Not applicable
*Source: IRIS (App. B)

Note in Examples 14-1 and 14-2 that BEHP (bis-2-ethyl hexyl phthalate) and chloroform have both carcinogenic and non-carcinogenic toxicity so that it appears in both rankings.

Further Screening of Chemicals

The ranking of toxicity scores as illustrated by the previous examples indicate which compounds pose the greatest hazard based solely on their maximum concentration and toxicity. The selection of surrogate chemicals requires further evaluation to consider for each chemical its range of concentrations, its mobility in the environment, and other issues. These additional considerations include:

- Mean concentration
- Frequency of detection
- Mobility
- Persistence in the environment
- Chemicals associated with site operation
- Treatability

In general, chemicals with a low frequency of detection or extremely low mean concentration are not deemed sufficiently significant to receive direct evaluation. The evaluation of mobility in the environment uses information such as half-lives and physical-chemical properties (Henry's law constant, vapor pressure, octanol-water partition coefficient and solubility) (see Sec. 3-2). Chemicals whose persistence and mobility make them significant risks would remain in the list of surrogate chemicals. Also, the risk analyst can selectively add compounds that are degradation products of others on the list (e.g., vinyl chloride).

Although the list of surrogate chemicals should account for about 99% of the risk, it may account for less than 10% of the detected chemicals. It is not infrequent to reduce a total of 100 detected chemicals to 10–15 surrogate chemicals.

14-3 EXPOSURE ASSESSMENT

The second stage of a quantitative risk assessment consists of estimating the exposure to the chemicals by the populations potentially at risk. To provide a comprehensive understanding of the sources of contamination, this stage begins with a delineation of the sources (e.g., sludge lagoons, contaminated soil) and the spatial distribution of contaminants at the site. The exposure assessment continues by analyzing how the contaminants might be released. Given the release of a chemical, it is necessary to estimate how it may migrate (e.g., via ground water) to a potential receptor (e.g., users of ground water). Having identified current and potential receptor points, considerable attention is given to (a) identification of general and sensitive populations of current and potential receptors and (b) estimation of both short- and long-term exposures in terms of doses by exposure route.

Environmental Pathways

Much of this stage consists of a fate and transport analysis as presented in depth in Chap. 4, and only a brief summary will be presented here. A chain of events must occur to result in exposure. This chain in a collective sense is termed an environmental **pathway** (i.e., the environmental routes by which chemicals from the site can reach receptors). A pathway, consisting of the following elements, thus defines the framework of a fate and transport analysis:

- Source (e.g., a lagoon)
- Chemical release mechanism (e.g., leaching)
- Transport mechanisms (e.g., ground water flow)
- Transfer mechanisms (e.g., sorption)
- Transformation mechanisms (e.g., biodegradation)
- Exposure point (e.g., residential well)
- Receptors (e.g., residential consumers of drinking water)
- Exposure route (e.g., ingestion)

An example of two simple but common pathways, one involving ground water and the other atmospheric transport, is shown in Fig. 14-1. Other examples of potential pathways, some of them quite complex, are discussed in Chap. 4 with various useful approaches for estimating fate and transport. It is necessary to establish potential pathways between sources and receptors for each chemical of concern. Figure 14-2 provides a flow chart indicating the types of considerations necessary in analyzing potential soil and ground water pathways.

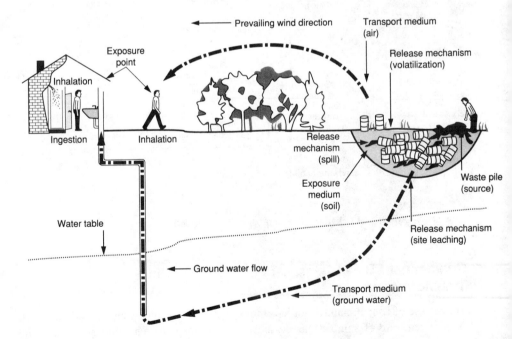

FIGURE 14-1
Examples of exposure pathways.[12]

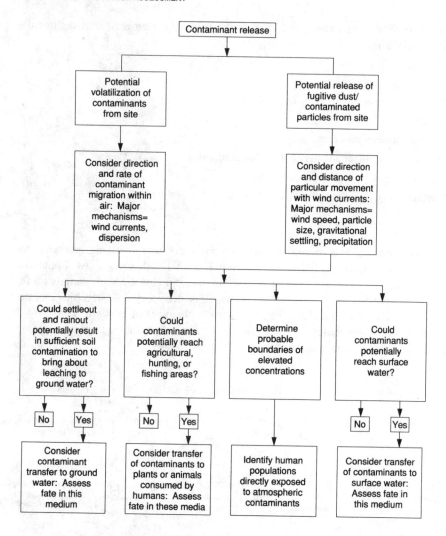

Environmental fate and transport assessment:
Atmosphere

FIGURE 14-2a
Flow chart—Analysis of potential soil and ground water pathways.[13]

Contaminant Release, Transport, Transfer, and Transformation

The release of contaminants from hazardous waste sites results from natural processes, such as leaching of soluble chemicals to ground water; human activity, such as the construction of site drainage channels; and accidents such as chemical spills. Table 14-4 summarizes some potential mechanisms for release of chemicals.

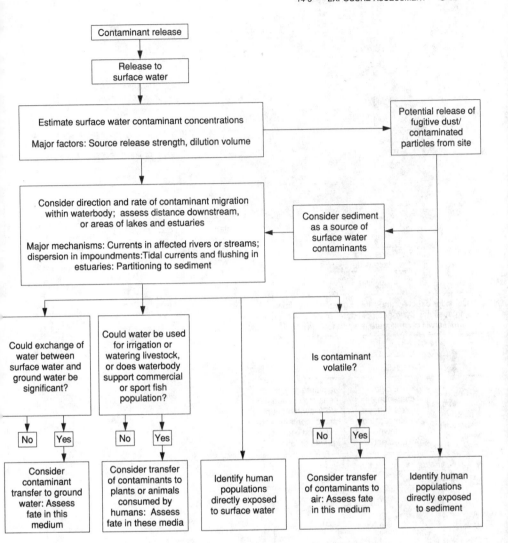

**Environmental fate and transport assessment:
Surface water and sediment**

FIGURE 14-2b
Flow chart—Analysis of potential soil and ground water pathways (continued).

In addition to determining what chemicals are released from a source, it is important to determine what happens to these chemicals: how they are transported, transferred, or transformed. Transport involves movement by advection and diffusion. A great number of mechanisms could act to transfer the contaminants to another medium or to storage sites (e.g., sorption onto soils). In some instances, organic contaminants transform under environmental mechanisms to CO_2 and water, while in

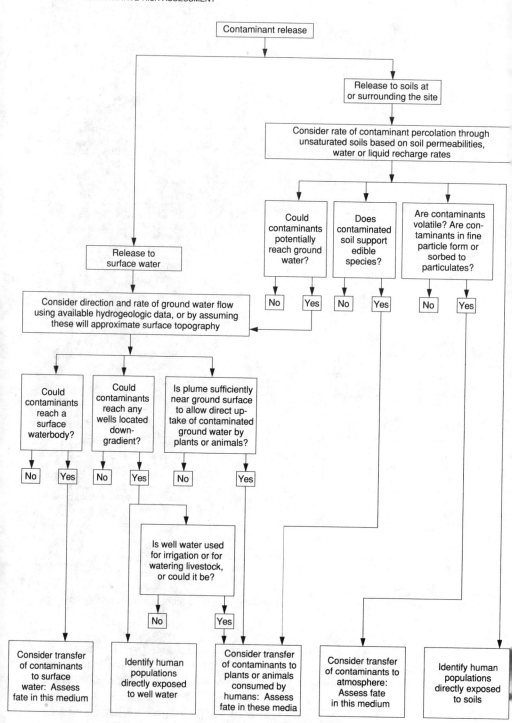

**Environmental fate and transport assessment:
Soils and ground water**

FIGURE 14-2c
Flow chart—Analysis of potential soil and ground water pathways (continued).

TABLE 14-4
Contaminant release mechanisms

Media	Mechanism	Time frame
Air	Volatilization	C
	Fugitive dust generation	C, E
	Combustion	E
Soil	Erosion	C, E
	Leachate generation	C
	Spills	E
Surface and ground water	Leachate generation	C
	Spills	E

C—Chronic   E—Episodic

other cases the degradation products may be more toxic than the original chemical (e.g., vinyl chloride). Example mechanisms for the transfer and transformation of chemicals are presented in Chap. 4, and several examples are shown in Table 14-5.

Potentially Exposed Populations

The next step in the exposure assessment is to determine potentially exposed populations. These would include the following:

- Present population in vicinity of the site
- Future population in vicinity of the site
- Sub-populations of special concern (e.g., young children in the case of lead contamination)
- Potential on-site workers during any remediation

The present population may be initially identified as those living within specified distances from site boundaries (e.g., 1 and 3 miles). However, any distance could be

TABLE 14-5
Mechanisms affecting environmental transfer and transformation of chemicals[†]

	Fate mechanism	
Media	**Transfer**	**Transformation**
Water	Volatilization	Biodegradation
	Adsorption	Photochemical degradation
Soil	Uptake by plants	Biodegradation
	Dissolution in rainwater	
Atmosphere	Washout by rain	Oxidation by ozone
	Gravitational deposition	

[†]Adapted from R. A. Conway.[14]

specified depending on the results of the fate and transport analysis. Determining potentially exposed populations is based on the surrounding land use and documented sources of demographic information such as summarized in Table 14-6.

After a preliminary overview of documents has been made, a thorough site visit to "ground truth" the current situation is critical. Patterns of human activity associated with various land uses should be determined, including answering questions such as:

- What sort of sensitive public facilities are located near the site (e.g., schools, hospitals, day care centers, nursing homes)?
- What outdoor activities occur (e.g., parks, play grounds, recreational facilities nearby)?
- What fraction of time do various sub-populations spend in potentially contaminated area (e.g., children in a grade school)?
- What is likely to change (e.g., seasonally, immediate future, long term future)?
- What secondary exposures are possible (e.g., contamination of crops at nearby farms)?

Certain sub-populations may require special consideration because of their higher sensitivity to toxic substances—children, pregnant women, the elderly, and people with chronic illnesses. An initial step in identifying such sub-populations is to locate sensitive facilities such as schools and nursing homes. It is critical to evaluate future populations to address how changes could affect the estimates of risks. For example, if ground water is not currently utilized as a drinking water source but could be used in the future, this may dramatically affect future risk calculations.

Development of Exposure Scenarios

After potential populations and exposure pathways have been defined, it is then necessary to characterize the conditions under which the populations may be potentially exposed. This involves an evaluation of both current and reasonable future uses of the site to establish a credible set of conditions under which the exposure could occur. After the scenarios have been established, then the specific parameters governing exposure (e.g., exposure frequency, duration, and intake rate) can be selected.

Some examples of commonly used exposure scenarios include a worker scenario, a trespasser scenario, a residential use scenario, a recreational use scenario, and a

TABLE 14-6
Sources of demographic information

Topographic maps	Tax maps
Census reports	County/regional demographic studies
Municipal zoning maps	Land use data
Projected future land use	Site visit
Human activity patterns	Aerial photos

construction scenario. Each of these is discussed in the following listing, together with some pertinent questions for the risk assessor to consider when determining whether site conditions warrant evaluation of the given scenario.

- **Worker scenario:** Is the site currently used for industrial activities? Are workers exposed to site-related constituents under normal conditions? Could workers be exposed in the future, either because of a change in use of the site or because the workers would be involved in remedial activities? It should be noted that workers are strictly protected under OSHA regulations. The estimation of potential risks for workers is usually done as a hypothetical evaluation to determine whether the potential exists for unacceptable risks, rather than as a mandated exercise designed to ensure that workers are not placed at risk during the normal course of their activities.

- **Trespasser scenario:** Is there evidence that trespassing may routinely occur at the site? Is there a fence that would limit access to the site? If so, is the fence in good condition? Have other measures been taken to limit access to the site? Is the site close to a school, shopping, or residential area where area children would have reason or inclination to play at the site? Is the site attractive to children?

- **Residential use scenario:** Is the site currently used for residential purposes? Will it or could it be used for residential purposes in the future? Are there any zoning or deed restrictions that would limit its use for residential purposes? Are the residences single-family dwellings? Is there the potential for residential use of ground water? A residential scenario is frequently evaluated as a hypothetical condition in an effort to estimate worst-case risks. Exposure under a residential scenario is generally the most restrictive condition for any age group, thus resulting in the greatest potential exposures and the highest potential risks.

- **Recreational use scenario:** This scenario is particularly applicable for evaluation of potential risks associated with surface water bodies where people may fish, swim, canoe, wade, etc. Observations made during site investigation activities can be useful in the development of a recreational exposure scenario. It may also be important to confirm the regulated use and classification of the water body with the appropriate state or federal agency.

- **Construction scenario:** Are construction activities planned or likely at the site? Will the construction result in potential exposures for both on-site receptors (e.g., direct contact with soils by construction workers) and off-site populations (e.g., exposure by off-site residents or workers to fugitive dust and volatiles released as a result of earth-moving activities)?

The development of scenarios involves making assumptions and must include subjective decisions. Ideally, this should produce credible scenarios, but in an effort to ensure conservative risk assessments, unrealistic scenarios have been used at some hazardous waste sites.

As an example, pica behavior is a real phenomenon and should be considered, but the assumption that children will eat 200 mg/day of soil for a 70-year lifetime is not realistic. Non-degradation policies, in many states, prohibit the discharge of contaminants into streams, independent of the present water quality in the stream. That policy has been used to suggest that a stream heavily contaminated with acid mine drainage be considered a primary drinking water source for the purpose of developing a credible scenario for a risk assessment. The point is that while non-degradation may be an appropriate public policy, drinking acid mine drainage is not realistic.

Exposure points define the locations of the receptors for the various scenarios. They are identified for each exposure scenario simply by overlaying the demographic information with the exposure pathway. An exposure point may be as close as the sources of waste at the site itself (e.g., the trespasser scenario) or at a considerable distance, particularly for pathways involving the food chain.

Exposure Point Concentrations

The risk analyst must next estimate the concentration of contaminants at the exposure points, including all pathways—air, ground and surface water, soils, sediments, and food (e.g., plants and fish). For present exposures, actual monitoring data at the exposure point should be used wherever possible. For example, contaminant concentrations should be obtained for drinking water wells in the vicinity of a site. For on-site exposure points, representative concentrations in soil or ground water may be calculated as the arithmetic or geometric mean (depending on the statistical distribution of the site analytical concentration data).

Future conditions can differ starkly. A plume may not have yet migrated to a potential exposure point. Remediation will, of course, drastically reduce migration. In a comprehensive risk assessment, the exposure concentrations for each for remedial alternative would be estimated including the present situation (baseline conditions or no remedial action). Determining the concentration of a contaminant at the exposure point for future conditions often requires the use of fate and transport modeling methods of Chap. 4 and standard references.[15,16] The major effort with these models is calibration; once calibrated, successive runs can be made relatively easily to estimate concentrations for a range of conditions and assumptions.

For ground water contaminants, hydrogeologic models can be used to estimate the future concentration at a downstream well. For volatile organic compounds released to the atmosphere, a Gaussian diffusion model can be employed to estimate future downwind concentrations. In general, the level of effort employed in data collection and modeling will depend on the estimated severity of the risk. Nominal risks do not warrant the same level of analysis as the clearly significant risks.

All mathematical models require the making of assumptions. It is essential that the appropriateness of assumptions be reviewed carefully. The impact of assumptions on the exposure point concentrations and, ultimately, the risk values should be examined through a sensitivity analysis. A **sensitivity analysis** identifies which of the myriad of input variables have the most significant impact on the resulting risk value.

Receptor Doses

The final step in the exposure assessment stage is to estimate the doses of the different surrogate chemicals to which receptors are potentially exposed at the exposure points. As with the previous stages, three exposure routes are considered—ingestion, inhalation, and dermal contact. Also, there are three types of doses—the administered dose (the amount ingested, inhaled, or in contact with the skin), the intake dose (the amount absorbed by the body), and the target dose (the amount reaching the target organ).

For purposes of calculating risks, the dose should be in the same form as that of the dose-response relationship reported for the specific chemical and the exposure route under study. This will almost always be either administered dose or absorbed dose. Given the concentration of the contaminant at the exposure point, the calculation of administered dose is straightforward. In contrast, the calculation of absorbed dose based on administered dose requires consideration of some complex factors (see Sec. 5-1). The key factors influencing the uptake of contaminants by the body are simplified as follows:

Ingestion	Contaminant concentration in the ingested media
	Amount of ingested material
	Bioavailability to the gastrointestinal system
Inhalation	Concentration in air and dust
	Particle size distribution
	Bioavailability to the pulmonary system
	Rate of respiration
Dermal contact	Concentration in soil and dust
	Rate of deposition of dust from air
	Direct contact with soil
	Bioavailability
	Amount of skin exposed

Other factors to be considered in determining the intake of contaminants include considerations of life style, frequency and duration of exposure (e.g., chronic, sub-chronic, or acute), and the body weight of the receptor. In the majority of hazardous waste sites, long-term (i.e., chronic) exposures are frequently of greatest concern.

The calculation of an administered dose is summarized in the following generic equation:[17]

$$I = \frac{(C \times CR \times EF \times ED)}{(BW \times AT)} \tag{14-3}$$

where
I = intake (mg/kg of body weight·day)
C = concentration at exposure point (e.g., mg/L in water or mg/m^3 in air)
CR = contact rate (e.g., L/day or m^3/day)
EF = frequency (day/year)
ED = exposure duration (yr)
BW = body weight (kg)
AT = averaging time (days)

Eq. 14-3 is typically modified for specific exposure pathways. For example, the intake dose from the inhalation of fugitive dust may be calculated as:

$$I = \frac{(C \times CR \times EF \times ED \times RR \times ABS)}{(BW \times AT)} \tag{14-4}$$

where
RR = retention rate (decimal fraction)
ABS = absorption into bloodstream (decimal fraction)

For fugitive dust, the concentration in the air is determined by:

$$C = C_s \times P_c \tag{14-5}$$

where C_s = concentration of chemical in fugitive dust (mg/mg)
P_c = concentration of fugitive dust in air (mg/m^3)

Often, appropriate parameters may be found in the literature. Considerable research has been done in recent years to define many basic parameters, such as skin surface areas, soil ingestion rates, and inhalation rates. Other parameters, such as exposure frequency and duration, are often based on site-specific information (if available) or professional judgment. For example, in the evaluation of a trespasser scenario, observations of trespassers during site investigation activities may dictate the values used for exposure frequency (e.g., number of days per year or number of events per year) and exposure duration (e.g., number of years that the activity occurred). Common sense also plays an important role in the selection of exposure parameters. For example, if one were evaluating residential use of two sites, one in North Dakota and one in Florida, it would not be reasonable to assume the exposure frequency (e.g., days per year that an activity occurred) would be the same in the two risk assessments for a child potentially exposed to surface soils as a result of their playing outdoors in the backyard.

Averaging time (AT) is another important parameter which must be defined in the intake equation. The averaging time selected will depend on the type of constituent being evaluated. For example, to assess long term or chronic effects associated with exposure to noncarcinogens, the intake is averaged over the exposure duration (expressed in days). Exposure to carcinogens, however, is averaged over a lifetime (assumed to be 70 years or 25,550 days), to be consistent with the approach used to develop slope factors.

An example of the parameters used for this type of calculation are shown in Table 14-7. It should be noted that these values can vary greatly depending on the assumed exposure conditions (i.e., the selected exposure scenario). As an example, the air breathing rate for adult males is 0.83 m^3/hr in Table 14-7. However, this rate can vary by an order of magnitude from 0.6 m^3/hr at rest to 7.1 m^3/hr for vigorous physical exercise.[18]

TABLE 14-7

Standard parameters for calculation of dosage and intake determined for the ABC landfill

Parameter	Adults	Child age 6–12	Child age 2–6
Average body weight (kg)	70	29	16
Skin surface area (cm²)	18,150	10,470	6,980
Water ingested (L/day)	2	2	1
Air breathed (m³/hour)	0.83	0.46	0.25
Retention rate (inhaled air)	100%	100%	100%
Absorption rate (inhaled air)	100%	100%	100%
Soil ingested (mg/day)	100	100	200
Bathing duration (minutes)	30	30	30
Exposure frequency (days)	365	365	365
Exposure duration (years)	30	6	4

Example 14-3a   Calculation of contaminant intake.   Determine the chronic daily inhalation intake, by adults, of a noncarcinogenic chemical as a function of concentration in fugitive dust at the ABC Landfill (Table 14-3).

Solution.   For an adult exposed to a noncarcinogenic constituent, the intake (I_N) may be calculated from Eq. 14-3:

$$I_N = \frac{(C \times CR \times (EF \times ED) \times RR \times (ABS)}{BW \times (AT)}$$

From Table 14-7, air breathing rate for adults = 0.83 m³/h.
$CR = 0.83 \times 24 = 19.92$ m³ day
$EF = 365$ days
For chronic exposure to non-carcinogens, $ED = 30$ years
In the absence of better information, a conservative approach would assume the retention rate (RR) and the absorbtion into bloodstream would both equal 1.0.
$BW = 70$ kg (Table 14-7)
$AT = 365$ days $\times$ 30 years

$$I_N = \frac{(C \times 19.92 \times (365 \times 30) \times 1.0 \times 1.0}{70 \times (365 \times 30)}$$

$$I_N = \underline{0.285 \text{ m}^3/\text{kg} \cdot \text{day} \times C}$$

C = Exposure point concentration (mg/m)³

Example 14-3b   Calculation of contaminant intake.   Determine the chronic daily inhalation intake in children age 6–12 of a carcinogenic chemical (i.e. calculate I_C).

Solution

$$I_C = \frac{(C \times CR \times EF \times ED \times RR \times ABS)}{(BW \times AT)}$$

From Table 14-7:
CR $= 0.46 \text{ m}^3/\text{h} \times 24 \text{ h}/\text{d} = 11.04 \text{ m}^3/\text{d}$
EF $= 365$ days/yr
ED $= 6$ years
RR $= ABS = 1.0$ as in Ex. 14-3a
BW $= 29$ kg (Table 14-7)
The averaging time (AT) for determining carcinogenic risk is 70 years or 25,550 days. As noted in the text, this is to be consistent with the approach used to develop carcinogenic slope factors.

$$I_C = \frac{C \times 11.04 \times 365 \times 6 \times 1.0 \times 1.0)}{29 \times 25,550}$$

$$I_C = 3.26 \times 10^{-2} \text{ m}^3/\text{kg·d} \times C$$

C = Exposure point concentration (mg/m^3)

Example 14-4a Average daily intake from dermal contact with soil. Determine the average daily intake of chlorobenzene over one year of exposure for on-site workers from dermal contact of the soils in Table 14-3.

Assume the following additional parameters:

A = skin exposed $= 20\% = 0.2 \times 18,150 \text{ cm}^2 = 3630 \text{ cm}^2$

DA = dust adherence $= 0.51 \text{ mg/cm}^2$

ABS = skin absorption rate $= 6\%$

SM = effect of soil matrix $= 15\%$ (i.e., due to soil matrix, only 15% of contamination is actually available for contact)

EF = two exposure events per day; 156 exposure days per year

$ED = 1$ year

$BW = 70$ kg

$AT = 365$ days

Solution

$$I_N = \frac{C\left(\frac{mg}{kg}\right) \times A \text{ cm}^2 \times \frac{DA}{\text{Exp. event}} \frac{mg}{cm^2} \times ABS \times SM \times \frac{2 \text{ Exp. events}}{day} \times \frac{156 \text{ days}}{year} \times ED \times \frac{kg}{10^6 \text{ mg}}}{(BW \times AT)}$$

$$I_N = \frac{C \times 3630 \times 0.51 \times 0.06 \times 0.15 \times 2 \times 156 \times 1 \times 10^{-6}}{70 \times 365}$$

$$I_N = 2.03 \times 10^{-7} \times C \text{ (mg/kg·day)}$$

From Table 14-3, the average concentration of chlorobenzene in soil is 1.39 mg/kg. The daily intake of chlorobenzene is

1.39 mg/kg × 2.03 × 10^{-7}

Thus,

$$I_N = 2.82 \times 10^{-7}\ \mathrm{mg/kg \cdot day}$$

Example 14-4b   Average daily intake from dermal contact with soil.   Determine the average daily intake of chloroform for on-site workers from dermal contact of the soils in Table 14-3. Assume that the intake estimate will be used to assess the carcinogenic effects of chloroform (i.e., calculate I_C).

Solution

$$I_C = \frac{C \times A \times \frac{DA}{\text{Exp. event}} \times ABS \times SM \times \frac{2\ \text{Exp. events}}{\text{day}} \times \frac{156\ \text{days}}{\text{year}} \times ED \times \frac{\text{kg soil}}{10^6\ \text{mg}}}{BW \times AT}$$

$$I_C \frac{= C \times 3630 \times 0.51 \times 0.06 \times 0.15 \times 2 \times 156 \times 10^{-6}}{70 \times 25{,}550}$$

$$I_C = 2.91 \times 10^{-9} \times C$$

Using the data from Table 14-3, the daily intake of chloroform is 1.12 mg/kg × 2.91 × 10^{-9}

Thus,

$$I_C = 3.26 \times 10^{-9}\ \mathrm{mg/kg \cdot day}$$

It should be noted that unlike the oral and inhalation routes, there are no generally accepted reference doses or slope factors for the dermal route. In lieu of these, oral RfDs and SF values may be utilized, but the advice of a qualified toxicologist should be sought.

Equations similar to those developed in Examples 14-3 and 14-4 are used in spreadsheets to calculate administered/absorbed doses for all surrogate chemicals for all pathways. These are repeated for each exposure scenario.

14-4   TOXICITY ASSESSMENT

This stage of the risk assessment process defines the toxicity (i.e., the dose-response relationship) for each surrogate chemical. The output takes the form of mathematical constants for insertion into risk calculation equations. In addition to providing a set of mathematical constants for calculating risk, the toxicological assessment should also analyze the uncertainty inherent in these numbers, and describe how this uncertainty may affect the estimates of risk.

Chapter 5 explains in detail the scientific basis underlying the development and application of dose-response relationships. This section highlights some of the important concepts and quantitative methods involved in such undertakings and how they apply to the calculation of the risk from exposure to hazardous waste constituents.

Carcinogens vs. Non-Carcinogens

For the purposes of quantifying human health risks, chemicals are characterized as carcinogens (i.e., those with demonstrated propensity for cancer induction) and non-carcinogens. Some chemicals behave as both carcinogens and non-carcinogens and, hence, will appear in both types of calculation of potential human health risk.

Carcinogens tend to dominate public concerns about health risk; however, this is not the reason for distinguishing between chemicals which induce cancer and those which do not. The distinction is important because the two elicit toxic responses in different ways, giving rise to two quantitatively different models for induction of toxic response as a function of dose. Thus, the mathematical expressions of risk differ for the two categories of chemicals. Non-carcinogens have thresholds below which they fail to induce any discernible adverse health effect. In contrast, a linear non-threshold model has been adopted for all carcinogens by the EPA (see Fig. 5-24). According to this model, some risk is assumed for carcinogens at any dose, regardless of how small. There is increasing evidence, however, that some carcinogens do have threshholds.

Sources of Toxicity Data

The experimental research effort involved in developing a new dose-response relationship for a toxic substance takes considerable time, much longer than associated with starting and completing a study of most hazardous waste issues. Rather than conduct experimental research, the risk analyst defers to existing data found in standard sources of toxicological data, and selects from them the appropriate mathematical descriptors of toxicity.

Perhaps the most used source is IRIS (Integrated Risk Information Systems). The toxicologic indices in the IRIS database are updated monthly and available on-line. The IRIS database contains both qualitative and quantitative data regarding non-carcinogens as well as carcinogens. A sample printout from the IRIS database is included as App. B.

It should be noted that regulatory agencies will frequently specify the particular mathematical constants to be used. Whether taken from regulatory guidance or other sources, the toxicity constants must apply to the range of doses predicted by the exposure assessment. Also, because toxic response for the same dose can vary dependent upon the exposure route, a separate toxicity constant frequently must be selected for each exposure route predicted by the exposure assessment.

Slope Factors for Carcinogens

Dose-response relationships for carcinogens are conventionally reported as incidence of lifetime cancer (i.e., probability) versus dose (see Fig. 5-23). The slope, known as the **slope factor**, represents the carcinogenic potency for the chemical. It is calculated as the 95 percent upper confidence limit of the cancer dose-response curve and is expressed as the inverse of dose (e.g., $[mg/kg \cdot day]^{-1}$) (see Sec. 5-5). Thus, the probability of cancer for a given exposure is calculated by multiplying the slope factor times the dose.

Substances are assigned carcinogenic status dependent upon the weight of evidence from human epidemiologic studies along with data from laboratory experiments in test animals. Substances are classified by EPA as **known, probable** or **possible** human carcinogens on the basis of their apparent epidemiologic association with human cancer, induction of cancer in multiple species of test animals or induction of cancer in one species, respectively. As the strength of a possible casual association declines from human epidemiology to carcinogenesis in a single species of test animal, so does the certainty of cancer risk assessment in exposed populations.

Of all carcinogens, only 14 have adequate epidemiologic evidence in order to be classified as known human carcinogens (Class A). The vast majority of EPA's Priority Pollutants List (approximately 70%) are Class B2 (probable human carcinogens— insufficient human epidemiologic evidence but adequate evidence of carcinogenesis in at least two species of experimental test animals).

Reference Doses for Non-Carcinogens

Unlike carcinogens, non-carcinogens exhibit a threshold effect. Namely, below a specific dose, they fail to induce any adverse health effect in exposed populations. This threshold is defined as the **reference dose** (RfD) which is the estimated daily intake that is not believed to be associated with adverse health effects. The toxic endpoint varies from chemical to chemical as does the reference dose.

Because a non-carcinogenic chemical can induce several different toxic endpoints, the RfD is set for the most sensitive toxic endpoint. For example, carbon disulfide, a commonly used industrial solvent, may affect the central nervous system, the cardiovascular system, and the auditory and visual systems. Effects on these target organs occur at high levels of exposure. A more sensitive endpoint, however, is fetal development. Long-term inhalation studies at lower levels of exposure to carbon disulfide demonstrated fetal toxicity and malformation. Therefore, this more sensitive endpoint provides the basis for a chronic oral RfD of 0.1 mg/kg·day. Exposure below this level would protect against not only developmental toxicity, but also against the other target organs for carbon disulfide.

The hazard index is the standard unit for quantifying non-carcinogenic risk of a particular chemical. This simply is the ratio of the estimated daily intake of the toxic substance to its recommended RfD. An index greater than 1.0 would indicate the possibility for an adverse health effect from the exposure.

Protective Nature of EPA Approach

Although based upon the best available information at present, many of the toxicologic indices utilized in calculations of carcinogenic and non-carcinogenic risks are highly uncertain. In a prudent approach to public health protection, the EPA has built several safety factors into its methods for establishing reference doses and carcinogenic slope factors. Virtually all errors of uncertainty are made in the direction of public health protection to insure that risks are over-estimated rather than underestimated. Four examples of this protective approach follow:

- For non-carcinogens, extrapolation of animal reference doses to humans utilizes two safety factors of ten; one for animal-to-human extrapolation, a second for variation for toxic sensitivities within the human population.

- For carcinogens, the linearized multistage model assumes the upper-bound 95 percent confidence level of extrapolated data (i.e., it is likely to be conservative 19 out of 20 times).

- For carcinogens, the linearized multistage model extrapolates data from the 10-90% carcinogenesis range observed in experimental animals to the regulatory target of 0.0001% carcinogenesis, a step which could overstate risk by several orders of magnitude.

- Although evidence indicates that, like non-carcinogens, nongenotoxic carcinogens have thresholds below which they fail to influence cellular differentiation or division, they are treated mathematically like genotoxic carcinogens according to the linearized multistage dose-response model. (Genotoxic carcinogens are discussed in Sec. 5-5.)

14-5 RISK CHARACTERIZATION

The final stage of a four-stage human health risk assessment is to estimate risks. This consists in part of calculating quantitative estimates of both the carcinogenic and non-carcinogenic risks to receptors for all exposure scenarios considered. Estimates are typically calculated for all three exposure routes and for the maximally exposed individual (MEI) as well as the most probable exposed population. Such calculations are straightforward, and yield quantitative estimates of risk. The challenge lies not in making the calculations but in interpreting the results such that they are applied properly in decision making. The overall effort is referred to as risk characterization.

Risk for Average and Maximum Exposures

In the exercises required for Superfund sites, both average and maximum exposure point concentrations are used to estimate risk. Performing a specific risk calculation using both values permits the estimation of a range of potential risks, which can frequently be useful in providing perspective regarding the potential hazards associated with a particular set of exposure conditions. The significance of either measurement depends, of course, on the amount of data and the associated confidence in that data. However, in general, calculation of potential risk using an average concentration permits a better estimate of risk associated with chronic exposures, since the average value represents a more likely estimate of the exposure point concentration to which a receptor would be exposed over time. Use of a maximum value is best in the estimation of shorter-term, subchronic risks, although its use can provide a useful upper bound estimate of potential risk.

It should be noted that current risk assessment guidance[19] emphasizes the use of a single, upper bound estimate of exposure point concentration in the calculation of potential risks. This value is often taken to be the 95th percentile upper confidence limit of the arithmetic mean. Use of this number generally provides a worst-case estimate of risk, and can result in a significant overestimate of potential risk, especially when this value is used in combination with other worst-case assumptions to define a reasonable maximum exposure.

Carcinogenic Risk

Carcinogenic risk may be defined as the chronic daily intake dose (developed in the exposure assessment) multiplied by the carcinogenic slope factor (selected by the toxicity assessment). The product is a real term: the probability of excess lifetime cancer from exposure to this chemical. The computation is as follows:

$$\text{Risk} = \text{CDI} \times \text{SF} \qquad (14\text{-}6)$$

where   CDI = chronic daily intake (mg/kg·day)
   SF = carcinogen slope factor (kg·day/mg)

Example 14-5   Calculation of carcinogenic risk.   Calculate the average carcinogenic risk from chloroform due to dermal contact of soil by on-site workers at the ABC Landfill.

Solution
From Example 14-4: $I_C = 2.91 \times 10^{-9} \times C$, where C = exposure point concentration
From Table 14-3: $C = 1.12$ mg/kg
$I_C = 3.26 \times 10^{-9}$ mg/kg·day
SF for chloroform $= 6.10 \times 10^{-3}$ (App. B)

$$\text{Risk} = (3.26 \times 10^{-9})(6.10 \times 10^{-3}) = \underline{1.99 \times 10^{-11}} \qquad (14\text{-}7)$$

This would be termed the most probable risk since the average concentration value was used. If $C_{\max} = 4.10$ mg/kg is used, then the maximum carcinogenic risk $= 7.27 \times 10^{-11}$

Characterization of carcinogenic risk first involves performing calculations similar to this example for each chemical appropriate to the exposure route and pathway under study. The classical methodology for risk assessment assumes additivity of risks from individual toxicants. For carcinogens this means that the total carcinogenic risk equals the sum by exposure route of carcinogenic risks from all individual substances as shown in Table 14-8. This is repeated for each exposure scenario and exposed population. It must be emphasized that slope factors are specific to the exposure route (e.g., oral) and may only be used when the exposure data is for the same route.

TABLE 14-8
Carcinogenic risks—additive model

Carcinogen	Lifetime risk
Benzene	0.0005
Cadmium	0.0008
Vinyl chloride	0.0025
Total	$0.0038 = 3.8 \times 10^{-3}$

Non-Carcinogenic Risk

Non-carcinogenic risk is normally characterized in terms of a **hazard index**. This index is simply the ratio of the estimated intake dose from exposure to the reference dose (RfD). Reference doses are dependent on the route of exposure and may only be used with exposure data for the same route. The hazard index is calculated as follows:

$$HI = CDI/RfD \qquad (14\text{-}8)$$

where   HI = hazard index (dimensionless)
           CDI = chronic daily intake (mg/kg·day)
           RfD = reference dose (mg/kg·day)

If the acceptable level of intake is deemed to equal the reference dose, then by definition, a hazard index less than 1.0 is acceptable. An exposure typically involves multiple chemicals, and an index must be calculated for each surrogate chemical for all pathways and exposure routes. For exposure to multiple non-carcinogens, the hazard index scores for all non-carcinogens normally are summed to provide the final measure of the risk for non-carcinogenic toxic effects, as shown in Table 14-9. It should be noted that the acceptable target for the sum of hazard indices remains as less than 1.0.

The standard additive model should not necessarily be applied to all contaminants as a single group because many toxic substances show a predilection for only one or two organ sites. If the mechanisms of toxic action of different non-carcinogenic chemicals are well known, it is preferable to sum hazard indices on an organ-specific basis (e.g., all compounds having a toxic effect on the liver [hepatotoxic compounds]

TABLE 14-9
Non-carcinogenic risks—additive model

Non-carcinogen	Hazard index
Acenaphthene	0.05
Cyanide	0.15
Mercury	0.03
Total	0.23 which is < 1.0

would be added together but separate from those affecting the brain). Again, this approach requires considerable understanding of the toxic mechanisms of the significant chemicals.

The importance of this approach is illustrated by the example of a hypothetical exposure yielding a hazard index of 0.7 for cyanide and 0.6 for cadmium. If summed in the conventional manner, this would yield an overall hazard index of 1.3, thereby indicating an adverse health effect. However, since cyanide affects the brain and cadmium is mainly toxic to kidneys, it is unlikely that the sum of their two hazard indices is, in fact, a viable measure of overall toxicity.

Example 14-6   Calculation of non-carcinogenic risk. Determine the hazard index for chloroform, based on the intake calculated in Example 14-3.

Solution
From Example 14-5: $CDI = 2.27 \times 10^{-7}$ mg/kg · day
$RfD = 1.00 \times 10^{-2}$ mg/kg/day (App. B)
Most probable hazard index $= (2.27 \times 10^{-7})/(1.00 \times 10^{-2}) = \underline{\underline{2.27 \times 10^{-5}}}$

Background Risk

When characterizing the risk from exposure involving hazardous waste, one consideration is whether to include or exclude background risk. This depends upon the decision to be made. In comparing a set of remedial alternatives, the background risk of cancer is not relevant. It dwarfs the incremental risk and remains constant regardless of the remedy.

On the other hand, background risk is important in deciding what maximum allowable level of soil contamination is necessary to protect aquatic life in a nearby stream. In this latter case, the impact on aquatic life is determined by the total exposure, independent of the source. Total risk rather than the incremental risk drives the decision.

Iterative Nature of Risk Assessment

It is important to keep in mind that risk assessment is an iterative process. The four stage process necessitates a number of assumptions. An analysis of the sensitivity of the magnitude of calculated risk to these assumptions will indicate whether the risk analyst should revisit these assumptions and refine them. This may even require additional data because the health risks are initially calculated from available soil, ground water and surface water monitoring data. Additional sampling may be required to fill data gaps or to present a more credible worst-case assessment.

Uncertainty Inherent in Calculated Risks

Upon calculating risks, it is essential to examine and explain the degree of uncertainty associated with these estimates. In many instances, this variability is not properly presented to decision makers who use the risk estimates. When risk numbers are reported in the public press, the uncertainty is rarely reported, much less explained.

The methods used in each of the four stages of risk assessment have deficiencies that can introduce a high degree of uncertainty and thus impair the validity of the results. The hazard identification stage is based on data for which detection, identification and quantification limits could introduce errors. Selecting carcinogens as surrogate chemicals requires much judgment and is made difficult by the fact that carcinogenic initiators are treated the same as promoters and experimental data is the basis for classifying all but 14 carcinogens.

Exposure assessments of future conditions depend heavily on fate and transport models, estimates of the performance of remedial alternatives, projection of land use as well as assumptions about the frequency and duration of the exposure. Each is a potential source of uncertainty. For example, calibrating models to site-specific situations is a difficult task.

The toxicity assessment stage has a very high degree of uncertainty associated with the slope factors and reference doses. For many contaminants, there is a dearth of information regarding carcinogenicity for certain exposure routes. Even for those chemicals where data exist, extrapolation introduces a large measure of uncertainty (e.g., extrapolation from animal tests to human exposures and particularly extrapolation to the range of a 0.0001% carcinogenesis response).

Finally, the computation of risk is an exercise in applied probability of extremely rare events. It is not possible to enumerate every conceivable outcome, and credible worst-case exposure scenarios are used. This introduces an inherent conservatism which often results in assessing scenarios that will never be experienced.

In summary, the process by design is very protective of human health by ensuring that potential exposures and risks are not understated. Nevertheless, there are some sources of uncertainty which possibly are not fully taken into account from a conservative perspective. An example is that existing toxicity data typically derive from experiments with exposure of a single organism to a single chemical under a specified set of environmental conditions. Such data do not account for confounding factors such as synergism, unique species characteristics and the multitude of factors that define a specific environment. Each factor serves to make the application of toxicity data to the complexity of real hazardous waste sites problematic. Simply stated, the classic additive models may understate the risk of exposure to complex mixtures of toxic substances. Another factor that could prompt underestimating actual risk is the presence of sensitive subpopulations.

Mathematical techniques such as Monte Carlo simulation may be utilized to assess both uncertainty and the sensitivity of final answers to individual input parameters when estimating exposure. This is particularly important for situations with many pathways, such as found at hazardous waste sites. Monte Carlo methods are discussed in Sec. 14-8.

Quantitative assessment of risk is still in the developmental phase. As the field matures, some of these limitations, such as our knowledge of carcinogenicity, will become less severe through research. Other constraints, such as our inability to communicate risk, may prove more challenging.

14-6　RISK COMMUNICATION

Having characterized the risk of exposure to constituents of hazardous waste and the degree of uncertainty associated with the risks, the next step is to use the information to improve the basis for making decisions. This often involves the public, or at least embraces public concerns and attitudes. It requires an examination of the question, "What is acceptable risk?" It enters the area of perception, and while not leaving the world of science, it is a much different world and one not relished by many scientists and engineers.

What Is Acceptable Risk?

At present the EPA has defined acceptable risks for carcinogens as within the range of 10^{-4} to 10^{-6} excess lifetime cancer risk and for non-carcinogens as a hazard index of less than 1.0. As discussed in Sec. 5-5, there have been precedents with other issues which have defined "acceptable risk" outside this range. The U.S. FDA deemed that less than 10^{-7} cancer risk for saccharin was "acceptable," and local citizens were willing to accept a risk much greater than 10^{-4} for a copper smelter. Clearly, "acceptability" is a personal concept and demands that the public, which ultimately must have jurisdiction over what level of risk is acceptable, be informed. At many sites it is ultimately the public which determines by its influence which level of potential health risks are acceptable.

The U.S. EPA uses 10^{-6} excess lifetime cancer risk as a point of departure, meaning that a higher risk may be deemed acceptable only if there was special extenuating circumstances. For purposes of comparison, examples of actions that would increase the risk of death by a probability of 1×10^{-6} are illustrated in Table 14-10. Almost all of the actions shown in Table 14-10 seem highly acceptable if not commonplace; yet, the magnitude of their risk compares with that of the U.S. EPA's target risk for hazardous waste sites. It should be noted that risk of cancer is not the same as risk of death because not all cancer cases result in death. Still, cancer is the second greatest cause of death in the United States.

Another comparison is that of incremental risk and background risk. The 10^{-6} target represents an incremental risk of 0.0001% probability, an especially small level in comparison to the 25% background risk of the very same disease for which this regulatory target is directed. The total risk to an individual exposed at EPA's target would increase from 25% to 25.0001%. This increase is hardly meaningful from a scientific perspective, especially considering that the exposed population is not the whole nation but isolated pockets.

Risk Perception

While an increase in cancer risk by an increment of 10^{-6} may not be significant from a scientific viewpoint, it easily can alarm the community near a hazardous waste

TABLE 14-10
Actions increasing risk of death by one in a million[20]

Action	Nature of risk
Smoking 1.4 cigarettes	Cancer, heart disease
Drinking 0.5 liter of wine	Cirrhosis of the liver
Spending 1 hour in a coal mine	Black lung disease
Spending 3 hours in a coal mine	Accident
Living 2 days in New York or Boston	Air pollution/heart disease
Traveling 6 minutes by canoe	Accident
Traveling 10 miles by bicycle	Accident
Traveling 30 miles by car	Accident
Flying 1000 miles by jet	Accident
One chest x-ray taken in a good hospital	Cancer caused by radiation
Living 2 months with a cigarette smoker	Cancer, heart disease
Eating 40 tablespoons of improperly stored peanut butter	Liver cancer caused by Aflatoxin B
Drinking heavily chlorinated water (e.g., Miami) for 1 year	Cancer caused by chloroform
Drinking 30 12-oz. cans of diet soda	Cancer caused by saccharin
Living 5 years at site boundary of a typical nuclear power plant in the open	Cancer caused by radiation
Living 150 years within 20 miles of a nuclear power plant	Cancer caused by radiation
Eating 100 charcoal broiled steaks	Cancer from benzo(a)pyrene

site. The reason is perception. Some of the factors that affect the perception of risk are indicated in Table 14-11 and are discussed in Sec. 1-4. The factors work in such a manner that, for example, an action *voluntarily* undertaken by an individual is perceived as posing a smaller risk than one *imposed* upon that individual, all else being equal. Almost all of the "more risky" items are associated with hazardous waste. Clearly, the four-stage risk assessment process described on the preceding pages can not incorporate such factors in a precise quantitative calculation.

A concept accompanying public perception is the notion of acceptable risk. Fischoff, et al.[21] identified five generic complexities that affect acceptable risk:

TABLE 14-11
Risk perception

Less risky	More risky
Voluntary	Involuntary
Familiar	Unfamiliar
Controlled by self	Controlled by others
Chronic	Acute
Natural	Artificial
Fair	Unfair
Detectable	Undetectable
Not memorable	Memorable

- Defining the decision problem
- Assessing the facts
- Assessing relative values
- Addressing the human element in decision making processes
- Assessing the quality of the decisions

With each of these, there is a perception gap between the decision makers and the local citizens. Starting with the first item, a decision maker's responsibility in site remediation climaxes with the evaluation of several remedial alternatives and the selection of one. The public likely will view this evaluation as an unnecessary exercise. They will ask, "Why deliberate when it has been intuitively clear from the onset that the safest decision is to excavate the waste and haul it out of their community?"

The difficulties that surround assessing the facts are primarily those of dealing with the large magnitude of uncertainty inherent in any risk assessment. On the other hand, the public often expects that scientists and engineers should have absolute knowledge about risks before making decisions that affect them. The purposeful use of conservative safety factors at each stage to protect human health serves to inflate resulting risk numbers, and the public will often express more concern with the possible magnitude of the worst-case scenario, either not understanding or electing to ignore how unlikely if not incredible it may be. Merely learning that the assessment is filled with uncertainty may make the public uneasy about the estimates, believing that the true risk may be even greater than indicated.

It is important to note that almost all of the resources applied in risk assessment are devoted to the first two items of Fischoff's list. The local citizens are just as influenced, perhaps more so, by the other three. Concentrating resources on the first two items is thought by decision makers to portray their decisions as purely objective. On the contrary, assessing risk and selecting from remedial alternatives is fraught with value judgments. Acknowledging the existence of values is far simpler than identifying and explaining the set of values underlying the decisions.

It is critical that decision makers sufficiently address the human element. This probably represents the most important avenue for closing the perception gap between the decision maker and the public. To not do so is a failure to deal with risk in its broadest sense: "Thinking about risks may be more productive than calculating them."[22] Finally, rarely do decision makers revisit the issue long after the decisions have been made to determine if the decision was a good one and whether their decision making process needs adjustment. Were the right questions asked? Did we solve the real problems?

Risk Communication

It is perhaps now understandable that given the five complexities listed by Fischoff, "The public often sees proponents of risk assessment as trying to convince people to accept risks that the proponents do not face rather than acting to remove them."[23] Not surprisingly, the remedy desired by the public is to remove the hazardous waste to "somewhere else" even though in scientific terms this may pose the largest risk.

While it is possible to provide reasonable quantitative estimates of risks, it is clear from the previous discussion that it is very difficult to explain these risks to the affected public. This derives only in part from the highly technical nature of the risk assessment process. Decision-making at hazardous waste sites is a political as well as a technical process. At many sites, the selection of an alternative is driven by the demands of the citizens living in the vicinity of the site. It is therefore essential that consideration be given to communicating risks as well as calculating them.

Santos and Edwards[24] suggest that to achieve effective risk communication three questions must be answered:

• Is the communicator listening and acknowledging the concerns of the audience?
• How capable is the spokesperson?
• Can the objectives of the presentation be met while still meeting the information needs of the public?

14-7 ECOLOGICAL RISK ASSESSMENT[†]

The previous sections have addressed the assessment of risk to human health. The primary mission of hazardous waste management also encompasses the protection of the environment, and risk assessment has an equal role with this objective. The two applications of risk assessment share many concepts and approaches. However, there are also some stark differences.

The introduction of chemicals into an ecosystem can provide a direct hazard to individual organisms, or it may affect their ability to reproduce. These perturbations can also affect the overall structure and function of an ecosystem or its components. The framework of this response is illustrated in Fig. 14-3.

This section introduces the current approaches used to estimate risks to the environment, broadly included under the category of **ecological risk assessment.** Ecological risk assessments are designed to evaluate the nature, magnitude, and transience or permanence of potential ecological effects. Somewhat similar with the four stage process for human health risk assessment, the ecological risk assessment consists of four interrelated tasks:

• Characterization of the baseline ecology and identification of potential receptor populations;
• Ecological toxicity assessment;

[†]Sec. 14-7 was prepared by Bruce Molholt and David Stout.

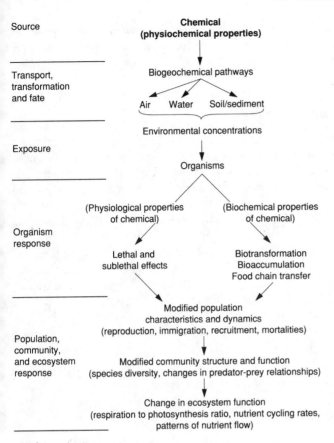

FIGURE 14-3
The ecotoxicological response to a chemical.[25]

- Evaluation of potential exposures; and
- Risk characterization.

The following subsections provide a description of each task.

Characterization of Baseline Ecology

This task begins with a review of previous site assessments and, if appropriate, a survey and evaluation of ecosystems and their components. This step serves as a synopsis of the prevailing ecology to provide an overview of the system diversity, as well as to identify potential sensitive subpopulations and any threatened or endangered species. As with human health risk assessment, it is often advantageous to select indicator species. Indicator species are selected for compelling reasons such as that they represent important commercial and/or recreational populations, sensitive species or endangered species present near the affected area. The concept of indicator species may also be important for reducing uncertainty in the analysis; it may not be pertinent to evaluate a sensitive species when more tolerant species actually dominate the site.

Ecological Toxicity Assessment

An ecotoxicity assessment both qualitatively and quantitatively evaluates the potential for adverse effects from the possible exposure to the chemicals of concern. The qualitative portion of the assessment provides a summary of ecotoxicity data for chemicals in the context of their presence in a sensitive ecosystem. This qualitative portion would draw on the information in the published literature. The quantitative portion entails identifying relevant indices of toxicity against which potential exposure concentrations can be compared in the risk characterization. Consideration is given to a number of endpoints (e.g., lethality, reproductive effects) and factors that may mitigate potential effects (e.g., temperature, water hardness).

Table 14-12 compares mammalian and aquatic/avian toxicology, indicating some of the concerns that must be addressed in ecotoxicology. It is particularly important to note that ecotoxicology is not nearly as well developed as human or mammalian toxicology, and is all the more difficult because of the diversity of species that must be protected. Ecotoxicology is discussed in Sec. 5-6.

Ecological Exposure Assessment

As with human health risk assessment, an exposure assessment is the process of measuring or estimating the intensity, frequency, and duration of exposure to an agent. More precisely, it evaluates the potential for exposure to each chemical of concern by the potential receptor populations (or indicator species) representing biotic components of any ecosystems in the site environs which may be affected. Two important components of an exposure assessment are an exposure pathway analysis and the estimation of exposure point concentrations.

Exposures are evaluated through the integration of baseline ecological data and information about the potential receptor populations. This information is then used to develop exposure scenarios. The exposure scenarios would consist of events that could occur in systems within or adjacent to the site. The routes of exposure considered for these scenarios would include ingestion and direct contact of contaminated surface water and sediments.

To evaluate these potential exposures, concentrations of the chemicals of concern in the exposure media (e.g., surface water and sediments) are required. These exposure point concentrations usually reflect collected environmental monitoring data. For example, a dissolved concentration in surface water may by used to represent the concentration of a chemical that is bioavailable to free-swimming organisms (i.e., nekton) inhabiting the water column.

TABLE 14-12

Comparison of mammalian toxicology to ecotoxicology[26]

Mammalian toxicology	Ecotoxicology (aquatic/avian toxicology)
Objective: to protect humans	Objective: to protect populations of many diverse species
Must almost always rely on animal models since experimentation with humans is not feasible	Can experiment directly on species of concern
Species of interest (humans) is known; thus, degree of extrapolation is certain	Not able to identify and test all species of concern; thus, degree of extrapolation is uncertain
Test organisms are warm-blooded (body temperature is relatively uniform and nearly independent of environmental temperature); thus, toxicity is predictable	Test organisms (aquatic) live in a variable environment and are often cold-blooded; thus, toxicity may not be sufficiently predictable
The dose of a test chemical usually can be measured directly and accurately and administered by a number of routes	The "dose" is known only in terms of the chemical's concentration in water (for aquatic organisms) and the length of exposure to it; the actual "absorbed dose" is sometimes determined experimentally using bioconcentration and metabolism studies
Extensive basic research has been conducted; emphasis has been on understanding mechanisms of toxic action	Little basic research has been conducted; emphasis has been on measuring toxic effects and generating numbers, with an eye toward regulatory needs
Test methods are well developed, their usefulness and limits are well understood	Test methods are either relatively new or generally not formalized (standardized); their usefulness in many cases is uncertain

Ecological Risk Characterization

The risk characterization task integrates information from all three other tasks to produce both a qualitative and quantitative estimate of risk. The risks to potential receptor populations that inhabit and/or use the affected ecosystems would be presented in a quantitative manner where quantitative indices are available, and a qualitative discussion would be presented regarding the possibility for adverse effects. Quantitative techniques involve the comparison for each compound of interest of the predicted and/or measured exposure point concentrations (from the exposure assessment) with the appropriate chemical-specific indices of toxicity (from the ecotoxicity assessment) to estimate the possibility of adverse effects. These chemical-specific comparisons represent ecological hazard indices and provide a means of assessing the potential significance of contamination.

Qualitative methods of characterization may not provide a definitive answer regarding potential risks, but they should provide a basis for professional judgment concerning the likelihood of adverse effects and yield important insight necessary for further evaluation. Finally, the limitations and uncertainties inherent to ecological risk assessment would be discussed to provide a proper perspective for use of the risk characterization results in decision making.

14-8   MONTE CARLO METHODS

Stochastic vs. Deterministic Calculations

Up to this point, the calculations in this chapter have been deterministic (i.e., all of the variables are treated as known constants). Many of these constants are estimates of either average or conservative if not worst-case conditions. For example, the carcinogen slope factor (SF) typically used is the 95th percentile value, so that 19 times out of twenty, the tabulated value is expected to exceed the actual potency.

Many of the values used in risk assessment are statistical point estimates (e.g., mean values, 95th percentile values) based on collected data. However, a point estimate provides only a portion of the information available in the data. This is illustrated in Fig. 14-4 showing two different probability distributions with the same point estimate for the 95th percentile. Clearly the underlying data sets are different.

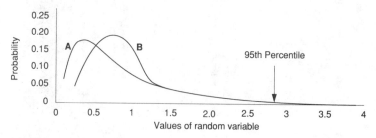

FIGURE 14-4
Example probability distributions.

Burmaster et al.[27] suggest three problems with the current practice in risk assessment to select point estimates so that the resulting risk is conservative.

- There is no way to know the degree of conservatism in the estimated risk (selecting 95th percentile values does not result in a 95th percentile of risk because the risk is calculated by numerous multiplications, divisions, summations, etc.).
- The resulting risk assessment will consider scenarios which will rarely if ever happen (e.g., every value at or near its respective maximum).
- Performing sensitivity analyses by varying input point estimates by $\pm 10\%$ or $\pm 25\%$ to determine the variability of the estimated risk is not meaningful because many of the inputs are at or near their maximum values.

These deficiencies can be addressed by using probability distributions in place of point estimates as input to the calculations. This is termed **stochastic modeling** and it provides a mechanism to utilize the full range of information. Stochastic models treat some variables as random variables drawn from known (or estimated) probability distributions.

Monte Carlo Methods

The **Monte Carlo** method is a well-established simulation procedure that replaces point estimates with random variables drawn from probability density functions. A complete description of the method is beyond the scope of this text and the reader is referred to standard texts on simulation.[28,29,30] The application of the method to risk assessment has been greatly simplified through the availability of commercial software that allows probability distributions to be input directly into spreadsheets.[31,32]

Once a probability distribution is incorporated into a spreadsheet cell, each time the spreadsheet is recalculated a new value of the random variable is selected from the distribution and used for calculations. The key to the method is to run the entire simulation hundreds or thousands of times. Each time new values of the random variables are selected and a new estimate of risk determined. The result of all of the calculations are summarized in a single histogram of risk values. This provides decision makers far more information than a single estimate of risk.

The process is summarized as follows:

1. Typical spreadsheet calculations are established for all chemicals and pathways to be modeled, following the methods used for deterministic calculations. For each of the random variables, discrete or continuous probability density functions are placed in the appropriate cells.
2. Any correlations among the random variables must be identified. For example, body weight and skin surface area would be positively correlated whereas age and ingestion of surface soils would be negatively correlated. It is important to identify these correlations so that individual simulations avoid selecting values of two different random variables that simply do not make sense. Commercially

available simulation software allows such dependencies to be easily input into the modeling process, but the risk analyst must identify the variables that are dependent. All other random variables are assumed to be independent.

3. A large number of simulations are run and the resulting risk values summarized. The summaries can include statistical tables and histograms of resulting risks and intermediate calculations.

A major difficulty in applying this method to quantitative risk assessment is estimating the probability distribution underlying many of the variables used. Table 14-13 provides some useful information.

Burmaster and von Stackelberg[33] suggest that SF data may vary by a factor of ten above or below the mean. From this they infer that a lognormal distribution may be appropriate. By setting two standard deviations of the log SF at ten and scaling the probability distribution to place the tabulated value of SF at the 95th percentile, they derive an estimate of the distribution as:

$$X1 \sim (SF/6.645) \times \exp[1.1513 \times N(0, 1)] \qquad (14\text{-}9)$$

where
$X1$ = distribution of slope factor values
SF = tabulated point estimate from App. B
$N(0, 1)$ = normal distribution with zero mean and unit standard deviation

Similarly, the reference doses (RfDs) for non-carcinogens are typically several orders of magnitude below levels that produce adverse effects in animals (NOAEL). Burmaster and von Stackelberg use the assumption that four standard deviations below

TABLE 14-13
Probability distributions for risk assessments[34]

Variable	Units	Point estimate	Distribution*	Point estimate location
Scenario specific data:				
Average body weight	kg	47	Normal (47,8.3)	mean
Time soil stays on skin	hr	8	Normal (6,1)	95th percentile
Average body surface area	m²	1.4	Normal (1.4,0.17)	mean
Fraction of skin area exposed	fraction	0.2	Lognormal (−2.15, 0.5)	85th percentile
Skin soil loading	mg/cm²	1	Uniform (0.75, 1.25)	mean
Soil ingestion rate	mg/d	50	Lognormal (3.44,0.80)	75th percentile
Soil bulk density	kg/m³	1600	Normal (1600,80)	mean
Cancer slope factors:				
SF, benzene	kg·d/mg	0.029	Lognormal (−4.33, 0.67)	88th percentile
SF, BaP	kg·d/mg	11.5	Lognormal (−0.79, 2.39)	91st percentile

*Note: For a normal distribution, the mean and standard deviation are used to describe the distribution. For a lognormal distribution, the mean and standard deviation of the underlying normal are used to describe the distribution. For a uniform distribution, the low and high are used to describe the distribution.

the logarithm of susceptibility equals a factor of five and two standard deviations above the mean of the logarithm of susceptibility is equivalent to the factor of ten used in developing the tabulated values of RfD. From these assumptions the following probability distribution is derived:

$$X2 \sim (2.236 \times \text{RfD})/\exp[0.402 \times N(0, 1)] \tag{14-10}$$

where $X2 =$ distribution of reference dose values
RfD $=$ tabulated point estimate from App. B

The method is illustrated in the following section.

An Example

As a simple example of the method, the carcinogenic risk from the incidental ingestion of soil containing benzo(a)pyrene (B(a)P) is examined. In Sec. 14-5, carcinogenic risk was determined using a constant called the carcinogenic slope factor (SF), with units (kg·day/mg). As shown in Sec. 5-5, this constant was determined by selecting the 95th percentile value of the slope of a plot of tumor incidence (risk) versus dose (mg chemical per kg body weight per day).

Data for cancer potency tend to be distributed lognormally. Burmaster, et al.[35] suggest that the SF for B(a)P is distributed lognormal (−0.79, 2.39) where the mean (−0.79) and the standard deviation (2.39) are from the underlying normal distribution. A graph of this probability density function is illustrated in Fig. 14-5.

The value of the point estimate of SF for benzo(a)pyrene is 11.5 kg·day/mg. This value occurs at the 91st percentile of the distribution shown in Fig. 14-5. This simply indicates that there is a 9% chance that a value randomly selected from the density function of Fig. 14-5 will exceed 11.5 kg·day/mg.

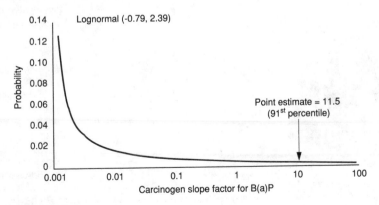

FIGURE 14-5
Probability density function of SF for benzo(a)pyrene.

Similarly, all of the variables in equations determining intake of contaminants, such as 14-3 and 14-4, are point estimates of random variables. In calculating the average daily intake, I (mg of chemical per kg body weight per day), conservative values of each of these variables were selected. In many instances the distribution of the data is available. For example, the concentration of a chemical in soil is estimated from many samples at a site. A histogram of this data provides the best estimate of the probability distribution underlying the data. Typical data for a site are shown in Fig. 14-6.

The data in Fig. 14-6 can be approximated by a distribution that is lognormal (0.165, 0.47). The 95th percentile of this distribution is approximately 2.5 mg/kg.

Suppose that the average daily intake at a specific site is related to the soil concentration by:

$$I = K \times C \tag{14-11}$$

where I = average daily intake (mg/kg·day)
        K = factor including variables such as body weight, frequency of exposure, etc., (i.e., all of the variables in Example 14-3) (day^{-1})
        C = concentration in soil (mg/kg)

Additionally assume that the variables in Example 14-3 have been statistically analyzed and K has been determined to be distributed normally with a mean of 4.5×10^{-5} and a standard deviation of 1.2×10^{-5}. The 95th percentile value of this distribution is 6.5×10^{-5}. From Eq. 14-6:

$$\text{Risk} = I \times \text{SF} = K \times C \times \text{SF} \tag{14-12}$$

As illustrated in Table 14-14, if mean values are selected for the three variables in Eq. 14-12 the resulting estimate of carcinogenic risk is 4.67×10^{-4} and if the point estimates are used the risk is 1.87×10^{-3}. However if random values are selected from

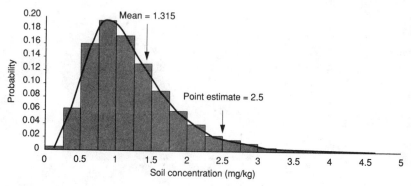

FIGURE 14-6
Histogram of benzo(a)pyrene concentrations in soil.

TABLE 14-14
Risk simulation

	SF (kg·day/mg)	Soil conc (mg/kg)	K	Risk (1/day)
Mean value:	7.89	1.315	4.5E-05	4.67E-04
Point estimate:	11.5	2.5	6.5E-05	1.87E-03
Run				
1	3.771	1.822	2.71E-05	1.86E-04
2	0.243	1.336	4.47E-05	1.45E-05
3	0.016	1.467	3.96E-05	9.17E-07
4	0.098	0.763	4.98E-05	3.71E-06
5	2.945	0.642	3.58E-05	6.77E-05

the probability distributions, the resulting risk is a random variable. Five iterations are shown in Table 14-14.

The second column shows 5 values randomly selected for SF using the distribution shown in Fig. 14-5, and the last column shows the resulting risks for that iteration. In order to obtain a true picture of the distribution of the risks, hundreds of iterations are performed and a histogram of risks produced. This is illustrated in Fig. 14-7.

Figure 14-7 is the result of 500 iterations of this calculation of risk from B(a)P in soil. The mean value of these 500 iterations was 4.675×10^{-4}, essentially the same as the mean value from Table 14-14. The median value of risk was 2.45×10^{-5} and for 90% of the iterations, the calculated risk was less than 5×10^{-4}. Clearly the histogram provides a clearer picture of estimated risks than a single point estimate.

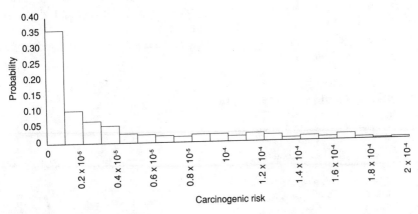

FIGURE 14-7
Histogram of carcinogenic risk simulation results.

14-9 CASE STUDY

At the Lackawanna Refuse Site in Old Forge, PA, some waste transporters had dumped drums of hazardous waste illegally in a landfill licensed to receive only municipal waste. The EPA placed the site on the National Priorities List and completed the Remedial Investigation[36]/Feasibility Study[37] (RI/FS) necessary under its Superfund Program. Most of the drums had been crushed when placed in the landfill; however some remained intact. The existence of intact drums of unknown contents made the safety of normal drilling questionable and special procedures were required to investigate subsurface conditions.[38]

After the RI/FS, the EPA issued in March 1985 a Record of Decision (ROD) which indicated that

> During ... hearings operators of the trucking firm testified that they brought drums of hazardous waste to the site. ... Estimates ranged between 10,000 and 20,000 drums. ... Other allegations [suggested] that bulk liquid wastes were disposed of.[39]

The ROD evaluated various remedial alternatives ranging from "No Action" to complete excavation and incineration. The EPA selected a remedy specifying the following:

> Removal of all drums and highly contaminated municipal refuse ... for off-site disposal ... Construction of a clay cap ... [and] installation of surface water drainage.

Subsequent to the issuance of the ROD, a Quantitative Risk Assessment[40] was performed in 1987. The results of this risk assessment are summarized in the following sections.

Hazard Identification

Indicator chemicals were selected based on an analysis of the data developed during the RI/FS on soils, ground water, surface water and leachate. All compounds detected were summed to one of the indicator compounds on the basis of having a similar effect. Of the more than 80 chemicals detected at the site, 13 final indicator compounds were selected as shown in Table 14-15.

TABLE 14-15
Indicator chemicals at the Lackawanna Refuse Site

Arsenic	Acetone
Barium	Ethylbenzene
Cresol	2-Hexanone
Trichloroethylene	4-Methyl-2-pentanone
1,2-Dichloroethane	Tetrachloroethylene
Methyl ethyl ketone	Xylene
Toluene	

Exposure Assessment

The exposure assessment modeled the migration of each of the indicator chemicals to the nearest receptors. This required developing a conceptual model of migration and exposure for the site using the geological cross section shown in Fig. 14-8. In all cases highly conservative assumptions were made to simplify the modeling efforts. Some of the assumptions included:

- All contamination observed off-site is from the site being studied
- Neither attenuation nor degradation of contaminants occurs

To develop the worst case exposure scenario, it was assumed that the mine pool under the site would be used as a future municipal water supply even though it had been heavily contaminated by acid mine drainage. The migration of contaminants to exposure points was estimated for each of the following remedial alternatives as developed in the ROD:

A. No action (baseline on which to compare other actions)

B. Capping and collecting and treating leachate

C. Alternative "B" plus excavation and removal of drums

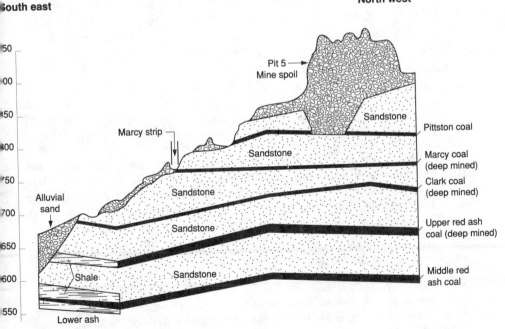

FIGURE 14-8
Geological cross section of the Lackawanna refuse site.[41]

D. Alternative "C" plus excavation and removal of all refuse that has been contaminated with hazardous waste.

The end product of the exposure assessment was the calculation of exposure level and dose incurred by the maximum exposed individual for each remedial alternative.

Toxicity Assessment

For each of the indicator chemicals, carcinogen slope factors (SFs) for carcinogens and reference doses (RfDs) for non-carcinogens were identified. Where standard references did not provide values for SFs or RfDs, estimates were made by toxicologists using the latest information available in the toxicology literature. In addition to the toxicity constants, appropriate environmental standards were identified to include EPA's proposed Maximum Concentration Limits (MCLs) and proposed Maximum Concentration Limit Guidelines (MCLGs) which had been established to protect drinking water. Other standards were the EPA Water Quality Criteria for Fish and Drinking Water, and Protection of Aquatic Life.

Risk Characterization

The final stage in the risk assessment brought together the results of the exposure assessment and toxicity assessment into a quantitative estimate of risk. For this site, the risk was determined for each remediation alternative. The results of the calculations for carcinogenic risk are presented as follows:

Alternative	Carcinogenic risk
A. No action	0.55×10^{-6}
B. Cap & leachate treatment	0.40×10^{-6}
C. B & drum removal	0.23×10^{-6}
D. C & removal of contaminated refuse	0.14×10^{-6}

As can be seen all of the alternatives, including the no action alternative, met EPA's acceptable level of carcinogenic risk (10^{-4} to 10^{-7} at that time). This would indicate that perhaps this site was not of sufficient risk to warrant being placed on the National Priorities List. As noted in Sec. 2-4, seven other criteria in addition to cost and risk are used to determine appropriate cleanup strategies at Superfund sites.

DISCUSSION TOPICS AND PROBLEMS

14-1. Determine the average daily intake and maximum hypothetical intake for all compounds listed in Table 14-3 from dermal contact of soils by on-site workers. Note that some chemicals may require evaluation as both carcinogens and non-carcinogens.

14-2. Make a complete list of all of the pathways for which risk calculations would be required for a site identified by your instructor.

14-3. Determine the total carcinogenic risk from the results of problem 14-1.

14-4. Determine the total non-carcinogenic risk from the results of problem 14-1.

14-5. Determine the average daily intake and maximum hypothetical intake for all compounds listed in Table 14-3 from incidental ingestion of soils by on-site workers.

14-6. Determine the total carcinogenic risk from the results of problem 14-5.

14-7. Determine the total non-carcinogenic risk from the results of problem 14-5.

14-8. Prepare a two-page summary of the article by Brown. [42]

14-9. Explain the difference between risk assessment and risk management.

14-10. A risk assessment is to be performed for a proposed hazardous waste incinerator. Briefly describe the specific factors that should be considered in each of the four steps.

14-11. A risk assessment is to be performed for a proposed hazardous waste land disposal facility. Briefly describe the specific factors that should be considered in each of the four steps.

14-12. Explain, in terms an ordinary citizen could understand, what is meant by a one in a million risk.

14-13. Explain the difference between total risk and incremental risk.

14-14. Briefly describe some of the problems encountered in communicating risk from hazardous waste activities to the public.

14-15. Explain, in terms an ordinary citizen could understand, what is meant by acceptable risk.

14-16. Describe how risk assessment might be used at a hazardous waste site to assist in establishing a specific clean up level for removal of trichloroethylene in ground water. What other methods might be employed to determine this clean up level?

14-17. Explain the difference between a hazard index and cancer risk.

14-18. Explain, in terms an ordinary citizen could understand, some of the weaknesses of the risk assessment process, and what steps may be taken to reduce the likelihood that unsafe decisions will be made based on incorrect risk calculations.

14-19. Describe the type of data that would be required to perform a quantitative risk assessment for a Superfund site consisting of a former unlined municipal landfill where drummed and bulk liquid hazardous wastes were disposed. Assume the wastes were known to contain benzene, toluene, ethylbenzene and xylene.

14-20. Describe the hazard identification stage of risk assessment as it might apply to a Superfund site.

14-21. What factors might be employed to reduce the number of chemicals for which fate and transport modeling will be required? For each factor explain the rationale for its use.

14-22. Review the remedial investigation report for a Superfund site, selected by your instructor, and identify what appear to be the major hazards; the most exposed population(s); the most significant pathways.

14-23. Review the article "An Environmental Risk Assessment of a Pesticide" by G.M. Rand,[25] and provide a two-page summary of the requirements for ecological risk assessments under the Federal FIFRA law.

14-24. Provide a preliminary estimate of the appropriate pathways for a site provided by your instructor using the flow chart provided in Fig. 14-3.

14-25. Perform a qualitative risk assessment on a laboratory identified by your instructor. Specifically identify:

(*a*) Hazards (e.g., several chemicals of concern)

(b) Potential receptors

(c) Pathways

(d) Toxicity of the chemicals listed in a.

14-26. Determine the potentially exposed population for a hypothetical hazardous waste site on the periphery of your campus. Specifically identify the most exposed individual(s) and sub-populations requiring special consideration.

14-27. Suppose that your campus water supply was contaminated with trichloroethylene at a concentration of 10 mg/L. Determine the total intake over a four-year undergraduate program using Eq. 14-3.

14-28. Repeat problem 14-27, assuming the contamination was in beer rather than water.

14-29. Discuss some of the special problems in assessing risk to carcinogens as opposed to non-carcinogens.

14-30. Comment on the validity of the assumptions underlying the exposure assessment described in Sec. 14-9 (Lackawanna Refuse Site).

14-31. Review the article by Cox and Blayney[44] and briefly describe some of the difficulties in obtaining public input in risk decisions.

14-32. Review the article by S. Brett, et al.[45] and discuss the special risks associated with excavation of wastes at a hazardous waste site.

14-33. Review the articles by Reinert[46] and Stephanatos.[47] Discuss how risk assessment may be used to help establish clean up levels at hazardous waste sites.

14-34. Using simulation software provided by your instructor, create a histogram showing the distribution of carcinogenic risk from ingestion of benzene in ground water. Assume the following site-specific information: Ingestion of water (L/day) is distributed normal (1.34, 0.4) and benzene concentration in water (mg/L) is distributed lognormal (.007, .76). Select distributions for body weight and slope factor from Table 14-13. Specifically calculate the probability of exceeding the 10^{-6} risk. Use EF = 365 days, ED = 170 yr, AT = 25,550 days.

REFERENCES

1. *Risk Assessment in the Federal Government: Managing the Process*, National Research Council, Comm. on Institutional Means for Assessment of Risks to Public Health, National Academy Press, Washington, D.C., p. 18, 1983.

2. Wilson, W. R., and E. A. C. Crouch: "Risk Assessment and Comparisons: An Introduction," *Science*, vol. 236, pp. 267–270, April 1987.

3. *Risk Assessment in the Federal Government*, op. cit., pp. 18–19.

4. Brown, H. S.: "A Critical Review of Current Approaches to Determining 'How Clean is Clean' at Hazardous Waste Sites," *Hazardous Waste & Hazardous Materials*, vol. 3, no. 3, Mary Ann Liebert, Inc. Publishers, pp. 233–260, 1986.

5. Huggins, A., and M. D. LaGrega: "Methods to Set Cleanup Goals to Protect Surface and Ground Water Quality at Hazardous Waste Sites," *Water Science & Technology*, vol. 23, pp. 405–412, 1991.

6. U.S. EPA: *Handbook for Conducting Endangerment Assessments*, U.S. EPA, Research Triangle Park, NC, 1987.

7. U.S. EPA: *Superfund Public Health Evaluation Manual (SPHEM)*, EPA 540/1-86/060, OSWER Directive 9285.4-1, U.S. EPA, October 1986.

8. U.S. EPA: *Risk Assessment Guidance for Superfund*, Vol. I. Human Health Evaluation Manual (Interim Final 9/29/89), OSWER Directive 9285.7-01a, September 1989.

9. U.S. EPA: *Superfund Exposure Assessment Manual*, OERR, EPA/540/1-88/001, OSWER Directive 9285.5-1, 1988.

10. U.S. EPA: *Exposure Factors Handbook*, Office of Health and Environmental Assessment, EPA/600/8-89/043, 1989.
11. U.S. Environmental Protection Agency: *Risk Assessment Guidance for Superfund*, 1989.
12. U.S. EPA: *Risk Assessment Guidance for Superfund*, 1989.
13. U.S. EPA: *Risk Assessment Guidance for Superfund*, 1989.
14. Conway, R. A., ed.: *Introduction to Environmental Risk Analysis*, Van Nostrand, p. 19, 1982.
15. U.S. EPA: *Superfund Exposure Assessment Manual*, EPA/540/1-88/001, OERR, OSWER Directive 9285.5-1, 1988.
16. U.S. EPA: *Exposure Assessment Methods Handbook*, Draft, Office of Health and Environmental Assessment, 1989.
17. U.S. EPA: *Risk Assessment Guidance for Superfund*, 1989.
18. U.S. EPA: *Superfund Exposure Assessment Manual*, 1988.
19. U.S. EPA: *Risk Assessment Guidance for Superfund*, 1989.
20. Wilson, R.: "Analyzing the Daily Risks of Life," *Technology Review* (M.I.T.), pp. 41-46, February 1979.
21. Fischoff, B., S. Lichtenstein, P. Slovic, S. L. Derby, and R. L. Keeney: *Acceptable Risk*, Cambridge University Press, 1981.
22. Hirschhorn, J. S., K. U. Oldenburg, and D. Dorau: "Using Risk Concepts in Superfund," *Superfund '87*, Hazardous Materials Control Research Inst., Washington, D.C., 1987.
23. Fischoff, B., S. Lichtenstein, P. Slovic, S. L. Derby, and R. L. Keeney: *Acceptable Risk*.
24. Santos, S. L., and S. Edwards: "Risk Communication: A Critical Part of the Public Participation Process," *Superfund '87*, Hazardous Materials Control Research Inst., Washington, D.C., pp. 254-259, 1987.
25. Rand, G. M.: "An Environmental Risk Assessment of a Pesticide," *The Risk Assessment of Environmental Hazards: A Textbook of Case Studies*, D.L. Paustenbach, ed., Wiley-Interscience, p. 903, 1989.
26. Rand, G. M.: "An Environmental Risk Assessment of a Pesticide," p. 902.
27. Burmaster, D. E., et al.: "Monte Carlo Techniques for Quantitative Uncertainty Analysis in Public Health Risk Assessments," *Superfund '90*, Hazardous Materials Control Research Inst., Washington, D.C., pp. 215-221, 1990.
28. Rubinstein, R. Y.: *Simulation and the Monte Carlo Method*, Wiley, 1981.
29. Bratley, P.: *A Guide to Simulation*, Springer-Verlag, 1983.
30. Kalos, M. H., and P. A. Whitlock: *Monte Carlo Methods, Vol. I: Basics*, Wiley, 1986.
31. Salmento, J. S., E. S. Rubin, and A. M. Finkel: "A Review of @RISK$^{\circledR}$," *Risk Analysis*, vol. 9, no. 2, pp. 255-257, 1989.
32. Burmaster, D. E., and E. C. Udell: "A Software Review of Crystal Ball$^{\circledR}$," *Risk Analysis*, vol. 10, no. 2, pp. 343-345, 1990.
33. Burmaster, D. E., and K. E. von Stackelberg: "Quantitative Uncertainty Analysis in Exposure and Dose Response Assessments in Public Health Risk Assessments Using Monte Carlo Techniques," *Superfund '89*, Hazardous Materials Control Research Inst., Washington, D.C., pp. 82-85, 1989.
35. Burmaster, D. E., et al.: "Monte Carlo Techniques for Quantitative Uncertainty Analysis."
36. United States Environmental Protection Agency: *Remedial Investigation Report, Lackawanna Refuse Site*, U.S. EPA, August 1984.
37. United States Environmental Protection Agency: *Feasibility Study of Alternatives, Lackawanna Refuse Site*, U.S. EPA, February 1985.
38. O'Hara, P. F., K. J. Bird, and W. A. Baughman: "Exploratory Drilling into a Buried Uncontrolled Drum Disposal Pit," *Superfund '86—Proceedings of the Ninth National Conference*, Hazardous Materials Control Research Inst., Washington, D.C., pp. 126-131, 1986.
39. United States Environmental Protection Agency: "Record of Decision—Lackawanna Refuse Site, Old Forge, PA," March 22, 1985.
40. Environmental Resources Management, Inc.: "Risk Assessment for the Lackawanna Refuse Site," April 1987.
41. Blasko, M. J., et al.: "Design of Remedial Measures and Waste Removal Program—Lackawanna

Refuse Superfund Site," *Superfund '87—Proceedings of the Ninth National Conference*, Hazardous Materials Control Research Inst., Washington D.C., pp. 367–370, 1987.

42. Brown, H. S.: "A Critical Review of Current Approaches to Determining 'How Clean is Clean,'" op. cit.

25. Rand, G. M.: "An Environmental Risk Assessment of a Pesticide," op. cit.

44. Cox, M., and E. Blayney: "Public Involvement in Acceptable Risk Decisions," *Mid-Atlantic Industrial Waste Conference*, G. D. Boardman, ed., pp. 454–461, Technomic Publishers, Lancaster, PA, 1986.

45. Brett, S. M., et al.: "Assessment of the Public Health Risks Associated with the Proposed Excavation of a Hazardous Waste Site," *The Risk Assessment of Environmental Hazards: A Textbook of Case Studies*, D. L. Paustenbach, ed., Wiley-Interscience, pp. 427–460, 1989.

46. Reinert, K. H.: "Risk-Based Cleanup Levels for Soils," *Superfund '90*, Hazardous Materials Control Research Inst., Washington, D.C., pp. 185–188, 1990.

47. Stephanatos, B. N.: "How Clean is Clean? The Importance of Using Site-Specific Factors in Developing Cleanup Levels at Hazardous Waste Sites," *Superfund '90*, Hazardous Materials Control Research Inst., Washington, D.C., pp. 612–617, 1990.

ADDITIONAL READING

Andelman, J. B., and D. W. Underhill: *Health Effects from Hazardous Waste Sites*, Lewis Publishers, Inc., Chelsea, MI, 1987.

Anon.: *Risk Assessment, Management, Communication: A Guide to Selected Sources*, NTIS No.: PB87-185500/GAR., NTIS, Springfield, VI, 1987.

Budd, W. W.: "A Comparison of Three Risk Assessment Techniques for Evaluating a Hazardous Waste Landfill," *Hazardous Waste and Hazardous Materials*, vol. 3, no. 3, pp. 309-320, 1986.

Clayson, D. B., et al.: *Toxicological Risk Assessment*, vols. 1 and 2, CRC Press, Boca Raton, FL, 1985.

Cohrssen, J. J., and V. T. Covello: "Risk Analysis: A Guide to Principles and Methods for Analyzing Health and Environmental Risks," NTIS PB89-137772, Council on Environ. Qual., Washington, D.C., 1989.

Covello, V. T., et al.: "Risk Evaluation and Management," *Contemporary Issues in Risk Analysis*, vol. 1, Plenum Press, New York, NY, 1986.

Federal Register: "Guidelines for Carcinogen Risk Assessment," vol. 51, no. 185, pp. 33992–34003, 1986.

Federal Register: "Guidelines for Development Toxicity Risk Assessment," vol. 51, no. 185, pp. 34028–34040, 1986.

Federal Register: "Guidelines for Exposure Assessment," vol. 51, no. 185, pp. 34041–34054, 1986.

Federal Register: "Guidelines for the Health Assessment of Chemical Mixtures," vol. 51, no. 185, pp. 34014–34025, 1986.

Fiksel, J.: "Quantitative Risk Analysis for Toxic Chemicals in the Environment," *Journal of Hazardous Materials*, vol. 10, no. 2–3, pp. 227–240, 1985.

Glickman, T. S., and M. Gough, eds.: *Readings in Risk,* Resources for the Future, 1990.

Goldman, B. A.: "The Use of Risk Assessment During Selection of Off-Site Response Actions," *Hazardous Waste and Hazardous Materials*, vol. 3, no. 2, 1987.

Hallenbeck, W. H., and K. M. Cunningham: *Quantitative Risk Assessment for Environmental and Occupational Health*, Lewis Publishing, 1986.

Layton, D. W., B. J. Mallon, D. H. Rosenblatt, and M. J. Small: "Deriving Allowable Daily Intakes for Systemic Toxicants Lacking Chronic Toxicity Data," *Regulatory Toxicology and Pharmacology*, vol. 7, no. 1, pp. 914-112, 1987.

Lowrance, W.: *Modern Science and Human Values*, Oxford University Press, London and New York, 1984.

Morgan, M. G., et al, "Communicating Risk to the Public," *Environ. Sci. & Tech.*, Vol. 26, No. 11, Nov. 1992, pp. 2048–2055.

Paustenbach, D. J., ed.: *The Risk Assessment of Environmental Hazards: A Textbook of Case Studies*, Wiley, 1989.

Scott, M. P.: "Applications of Risk Assessment Techniques to Hazardous Waste Management," *Waste Management Research*, vol. 5, no. 2, pp. 173–181, 1987.

Silbergeld, E. K.: "Five Types of Ambiguity: Scientific Uncertainty in Risk Assessment," *Hazardous Waste and Hazardous Materials*, vol. 4, no. 2, pp. 139–150, 1987.

Solomon, K. A., and P. Ricci, eds.: "Special Issue. Regulatory and Legal Aspects of Hazardous Materials Risks," *Journal of Hazardous Materials*, vol. 15, no. 1–2, 1987.

Tardiff, R. G., and J. V. Rodricks, eds.: *Toxic Substances and Human Risk—Principles of Data Interpretation*, Plenum Press, 1987.

Travis, C. C., et al.: "Cancer Risk Management: A Review of 132 Federal Regulatory Decisions," *Environ. Sci. Tech*, vol. 21, pp. 415–420, May 1987.

U.S. Environmental Protection Agency: *Health Effects Assessment Summary Tables (HEAST)*, U.S. EPA, 1991 (revised annually).

Whipple, C.: *De Minimis Risk*, Macmillan, New York, NY, 1988.

Wolf, K: "Hazardous Substances and Cancer Incidence: Introduction to the Special Issue on Risk Assessment and Risk Management," *Journal of Hazardous Materials*, vol. 10, no. 2-3, pp. 167–178, 1985.

ACKNOWLEDGMENTS

The authors wish to acknowledge the following individuals who assisted with the preparation of this chapter: Sec. 14-7 was prepared by Bruce Molholt, Ph.D., *ERM Inc.* and David Stout, Blasland, Bouck and Lee, Inc. Review assistance was provided by: Andrew Huggins, Ph.D., Basilis Stephanatos, Ph.D., and Robin Streeter, *ERM, Inc.* The principal reviewer for the ERM Group was B. C. Robison, D.V.M., Ph.D., D.A.B.T., *ERM-Southwest*.

CHAPTER
15

SITE AND
SUBSURFACE
CHARACTERIZATION

Bowed by the weight of century he leans
Upon his hoe and gazes on the ground
The emptiness of ages in his face
and on his back the burden of the world
    The Man with the Hoe, Edwin Markham

Whether designing a new facility or investigating potential contaminant migration at an existing site, proper characterization of the subsurface conditions and their interaction with surface features is critical to the process. For any hazardous waste facility, it is absolutely essential to know what lies on and beneath the site. In the context of this book, **site and subsurface characterization** may be defined as the qualitative and quantitative description of the conditions on and beneath the site which are pertinent to the hazardous waste management application.

A cross-section developed using site characterization tools and methodologies described in this chapter is shown in Fig. 15-1. This section illustrates the type of information that is required to define the site and subsurface conditions and answer the questions: What is there? Where is it going?

Site and subsurface characterization integrates a wide range of disciplines including geology, hydrology, hydrogeology, chemistry, biology and geotechnical engineering. Inadequate or, worse still, incorrect site and subsurface characterization can lead to ineffective remediation of contaminated sites and to inadequacies in new hazardous waste management facilities. For example, a new facility would be unwisely sited if a geologic fault were to go undetected. An undetected zone of ground water flow beneath an existing site can result in the remedial action functioning improp-

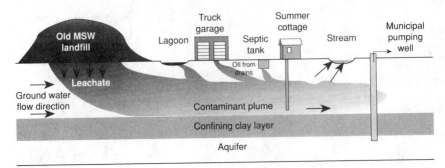

FIGURE 15-1
Site characterization.

erly. Subsurface investigations must generate a clear definition of the site geologic and hydrogeologic conditions; yet these investigations must reveal more than the hydrogeology. The investigations must delineate the influence of site and subsurface conditions upon the performance of the hazardous waste management system. The investigations must therefore reveal information on such things as prior site history, ground water flow direction, water quality and quantity, potential contaminant migration and constructability of remedial systems. In the case of site remediation, it is also necessary to estimate the total contaminant quantities.

This chapter defines and describes an approach to site and subsurface characterization and identifies the tools available. The approach and techniques are suitable for siting studies, design of new facilities, and for investigations for remediation of existing contaminated sites. It is the aim of this chapter to familiarize the student and hazardous waste manager with the available techniques and their applicability. Of the many hundreds of tools and techniques available for site and subsurface investigations (see Fig. 15-2 for one of the earliest examples), the most useful (and thus widely employed) are described in additional detail. This chapter places greater emphasis on subsurface characterization than on characterizing of site surface conditions. Air monitoring, surface soil sampling, water sampling of streams, springs and seeps, and aquatic surveys are examples of site surficial characterization studies that need to be conducted but which will not be described in this chapter.

Site and subsurface investigations are commonly implemented for CERCLA remedial investigations and RCRA corrective actions. The **remedial investigation (RI)** consists of the data collection and site characterization phase of a CERCLA remedial action and is begun in advance of the **feasibility study (FS)**. In the FS, alternative remedial measures are developed and evaluated. The feasibility study phase of the CERCLA action, therefore, cannot precede the remedial investigation. However, the RI and FS must be interdependent, but generic FS alternatives can be considered with the RI, especially to ascertain that key data are obtained during the RI. Without an adequate site assessment, the selected remedial alternative resulting from the feasibility study may be inadequate. Further, as remedial alternatives are tentatively chosen, additional site characterizations may be needed to confirm the suitability. In addition to

FIGURE 15-2
Early geophysicist (or modern diviner).

site characterizations conducted to assess contaminant migration and remedial action, characterizations are also conducted as part of the health and safety protection of site workers.

The techniques described in this chapter must be selected and employed with specific consideration given to the geologic and manmade setting of the site. For example, the necessity of avoiding buried underground utilities such as power and gas lines in an urban setting may dictate the use of different tools and techniques than those chosen in a rural setting, even with similar geologic conditions. The type and extent of the investigation will vary depending upon known or expected conditions. The chosen techniques must reflect the site specific nature of site and subsurface, must be flexible to adapt as conditions are revealed and must not evolve into a cookbook approach that is carried without additional thought from site to site.

15-1   METHODOLOGY

As with any comprehensive and complex environmental project, it is best to approach site and subsurface characterization in phases. It is recommended that these investigations be conceptualized in three phases as follows[1]:

1. Identify the problem and develop the scope of investigations (investigation planning),
2. Define the nature and extent of the site and subsurface conditions on a qualitative and quantitative basis, and
3. Complete a detailed site and subsurface characterization.

During the initial phase, it is necessary to collect and evaluate existing site and subsurface information. As much as possible with the available data, this phase defines specific site problems which exist at present, or potentially exist as a result of the past, present and/or proposed site activities. For example, site specific problems influencing the investigation of site and subsurface conditions at an old chemical manufacturing facility could include former waste disposal lagoons (now filled, graded and supporting a lawn), areas of buried underground storage tanks, or rail/truck loading and unloading areas. The initial phase also identifies site and subsurface features which may contribute to or reduce the severity of any problems or potential problems. For example, a limestone aquifer beneath the site may be overlain by fractured shale. In this case, the frequency and continuity of the shale fractures, as well as the hydraulic conductivity of the shale, govern the extent to which the shale can reduce the rate of contaminant transport to the underlying limestone aquifer. These conditions draw attention to the need, in this example, to define the continuity and properties of the shale.

In the context of site characterization, how do we define a "problem"? As described in general terms above, the problem definition guides the nature and extent of subsurface characterizations. The problem can best be defined on a risk basis. The assessment of risk is preliminary in the early stages of site characterization and more detailed in the later stages of site characterization. Detailed discussion of risk assessment is included in Chap. 14.

In the second phase, the collection of site specific data through sampling and field studies begins. During this phase, initial site characterization is completed. Site characterization is the representation or depiction of the site and subsurface into discrete units having generalized physical and engineering properties and is expressed in both qualitative and quantitative terms (see Example 15-1). During this phase, exposure pathways are identified and data for the screening of remedial alternatives or for the site specific design of new facilities are gathered.

Example 15-1 Initial site characterization. What items provide an example of the critical information developed for an initial site characterization of a former solid waste site in western Michigan?

Solution

1. The site was formally a sand borrow pit which excavated sands to the top of a clay layer.
2. At the site perimeter and underlying 6 to 12 cm of organic topsoil is a layer of fluvial "beach" sands averaging 10m in thickness.
3. The sand has a relatively high hydraulic conductivity, perhaps on the order of 10^{-2} cm/s.

4. The sand is underlain by a 15-m thick deposit of glacial lacustrine clays.

5. The hydraulic conductivity of the clay is estimated to be quite low, perhaps on the order of 10^{-7} cm/s.

6. The landfill accepted a wide range of municipal and industrial waters from the surrounding communities and light industry.

7. The water table is encountered at a depth of 3 m below the ground surface.

8. The nearest residences are within 500 m and rely upon ground water for their potable water.

The above summary developed from available information, can be used in the planning and design of the detailed site characterizations.

During the final phase, detailed site characterization, it is necessary to collect additional data required for the detailed analysis of the hazardous waste management alternatives. For example, the final phase of a remedial investigation could include tasks such as bench scale treatability studies employing proposed treatment processes and site specific wastes. The final phase of detailed site characterization would include analysis of the ground water chemistry in the site vicinity as well as at the nearby residences.

Although it is human nature to desire more and more information, it is impossible to define the site and subsurface conditions without some degree of uncertainty. Regardless of the number of samples taken and the number of in situ (in place) tests run, extrapolation is always required. To acquire "total knowledge" requires, in theory, an infinite number of monitoring wells, soil samples, ground water samples, and chemical analyses. It is necessary, and proper, to render informed scientific and engineering judgments regarding the site and subsurface conditions. It is cautioned that the inclination to define fully the subsurface conditions may be counter-productive, particularly in remedial investigations. In fact, additional data gathering and detailed site assessments have been employed to avoid or delay decisions regarding what remedy to implement and when to do it. The student is cautioned that a balance between additional study and moving forward with the appropriate action must be achieved. As of the time of writing this book, it is not unusual for a complete site and subsurface investigations in the United States to cost over a million dollars and require two or more years to complete. In the meantime, contaminants continue to migrate. Diligence is required in the design of investigations to reduce the time required, minimize the overall project costs, and yield the information that is **necessary and sufficient**.

15-2   PLANNING

It is safe to say that there is a wealth of information available regarding many sites under consideration. The information is available from a wide variety of sources including those listed on Table 15-1. Much can be learned from these sources of information without ever leaving the comfort of your office. In the following portion of this chapter, selected information sources identified in Table 15-1 will be discussed in detail.

TABLE 15-1

Sources of characterization information

Geological
    Local universities
    State agencies (e.g., Pennsylvania Geological Survey)
    Federal agencies (e.g., U.S. Geological Survey)
    Agricultural agencies (e.g., Soil Conservation Service)
    Satellite imagery
Geotechnical
    Prior site studies
    Regional/local building permits
    Consultants' reports
    Construction records
Operational
    Air photos
    Plant records
    Personnel interviews
    Operational and disposal permits

Although the site and subsurface investigations are conceptualized as phases, the investigation efforts are normally conducted along a time continuum. There must be an ongoing evaluation of data as they become available and a reevaluation of the investigation plan in light of the increasing knowledge gained as the investigation progresses. A work plan developed for site and subsurface investigations must be flexible enough to permit necessary changes which will inevitably result from data evaluation during the early stages of the investigation program.

While defining site specific data to gather during the field studies, it is essential to visit the site. The site inspection reveals information regarding topography, site access for investigation approaches, present drainage ways, seeps, existing landforms, and the potential for other interferences such as building foundations and site fill. This field survey may also identify the potential hazardous waste sources, migration pathways, receptors and any obvious impacts such as distressed vegetation. Further, during the initial phase of investigations of a prior hazardous waste site, the need for immediate remedial actions is identified. This initial remediation includes such action as immediate removal of leaking drums resulting in acute health and safety situations or planned removals where significant cost savings or minimizing risks of fire explosions could be accomplished. In addition, initial remedial measures often include fences and alternative water supplies to protect the public health and the environment.

Investigation Plan

Why have a plan? Administratively, the objectives of an **investigation plan** are to:

1. Provide specific guidance (details and protocols) for all work;
2. Provide mechanisms for planning and approving site activities;

3. Provide a basis for schedule and cost estimates; and

4. Insure that all the activities conducted are necessary and sufficient for planned activities.

For remedial investigations conducted under CERCLA, documents and reports required are shown in Table 15-2. Although it may not be necessary to include all of these documents and reports for investigations not conducted under CERCLA, Table 15-2 does provide a comprehensive check list for the types of documentation and reports that should be considered.

The major elements of an investigation and sampling plan include:

1. Scope and objectives of the investigation;

2. Summary of site background information;

3. Summary assessment of existing data;

4. Contaminants of interest;

TABLE 15-2
Work plan documents

Sampling plan
Quality assurance plan
Quality control plan
Health and safety plan
Operation plan
Operation schedule
Data management plan
Project management plan
Chain-of-custody procedures
Organizational chart
Field/laboratory documentation
    Project/field log books
    Sample tags
    Sample data sheets and logs
    Chain-of-custody records, seals
    Receipt of sample forms
    Laboratory log books
    Laboratory data, calculations, graphs
RI task reports
    Site description
    Contamination assessment
    Environmental assessment
    Public health assessment
    Endangerment assessment
    Draft/final RI report
Technical progress and financial reports
    Monthly technical progress report
    Monthly financial progress report
    Cumulative project cost report

5. Sample types;

6. Sampling locations and frequency;

7. Sampling and testing procedures;

8. Operation plan and schedule;

9. Cost estimate; and

10. Other supporting documents such as:

   a. Quality assurance and quality control plans,

   b. Health and safety plans, and

   c. Data management plans.

It is not uncommon for the organization responsible for the investigations (e.g. consulting firm, government agency, or industry) to have some of the above items as standard operating procedures. However, based upon the findings during the first phase of investigation, it is necessary to decide on the types and locations of samples required for the detailed site investigation.

The remainder of this section details the elements of the investigation and sampling plan. The remainder of the chapter then presents the methods and technologies appropriate for hazardous waste management site investigations.

Investigation and Information Objectives

What are the specific purposes of the proposed site investigations? Are they to define the distribution and migration of contaminants? Are they to assess the risk to the public health and the environment? Are the investigations being planned to develop remedial alternatives? Are they to design selected remediation techniques? Are the investigations to address all of the foregoing? What is the objective of the specific information which will result from the investigation? Will concentration of contaminants in a given ground water sample be used in a risk assessment or in a treatment stream design? The success of hazardous waste site investigations begins with a well thought out definition of the objectives of the investigation and the uses of the information gathered in that investigation.

Site Background

Considerable information is typically available for both the operational history and geology of the site. It is absolutely essential that this background information be assimilated prior to initiation of detailed site and subsurface investigations. Sources of background information regarding the operational history include interviews of personnel and reviews of property records.

The regional geology information for most sites is readily available as listed in Table 15-1. Geological information can come from a variety of sources including the federal and state geological surveys where regional geology reports describe the predominant geologic materials and their characteristics. This regional geological information can be supplemented with local geologic information from such sources as the geology department at the nearby universities or the county soil conservation office.

Air photos are another particularly useful source of information. Air photos are available for many areas of the United States dating back some 50 years. Interpretation of available air photos may show much about the site landforms, the engineering site history, past waste disposal operations, prior drainageways, potential for contaminant migration, topography, and vegetation. For example, shown in Fig. 15-3 are the interpreted site conditions from an air photo for a former waste disposal site. By 1966, the date of the air photo presented in Fig. 15-3, the site has been heavily developed as a landfill. The air photos show that the drainage which was generally north to south has been intercepted along the northern boundary and diverted along the eastern edge of the site. Lagoons and storage areas are revealed. Earlier photos reveal that the area began as predominantly grassy farmland with drainage proceeding from the northeast corner initially southwesterly and then southerly traversing the site. As time progressed, the early air photos reveal that, by 1958, the site is being developed as a landfill. By 1978, air photos show that the site has been essentially regraded with little evidence of prior landfill activity.

How are the results of air photo interpretation used in this case? First they help in identifying likely contaminant transport pathways. Specifically, the offsite drainage ways (old swales) would be more likely to contain contaminants transported from the landfill area than a randomly selected sampling location. Only from interpretation of air photos could this significant finding be developed. As a result, rather than **random** sampling, **focused** sampling can be undertaken to investigate contaminant levels in prior drainage ways. In addition, several lagoons are identified in the air photos. These lagoons constitute a likely source of contaminants within the landfill area. Interviews of plant personnel confirmed the lagoons revealed on the air photos were used for

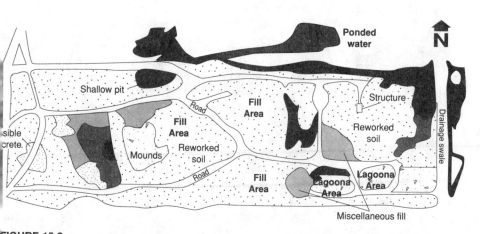

FIGURE 15-3
Air photo interpretation.

waste disposal. As a result, the lagoons could be targeted as potential "hot spots" during the on-site investigations.

Assessment of Existing Data

Assessment of available data includes an analysis of the data validity, sufficiency and sensitivity. Data validity should first be considered. For example, records such as well logs prepared by drillers are frequently available for wells at the site. What is the validity of water quality data from one of these pre-existing wells? A driller's log for a well installed many years ago may (or may not) include information about the depth of the casing, depth of the well, diameter of the borehole, depth and nature of different lithologies. Close examination of the driller's log commonly does not adequately characterize the geological conditions. (Why should they? That was often not their original purpose.) Thus a complete geologic or geotechnical description of all of the materials encountered would not normally be expected for a well installed in the distant past. The screened or monitored interval is often not fully defined, and as a result, water sampled from pre-existing wells may come from more than one aquifer or from some unknown depth in the subsurface. Water levels measured during drilling are often obtained before the water level has stabilized. Consequently, the usefulness of such data is limited. This brings to the forefront a basic tenet in hazardous waste management: *it is better to have no data than incorrect data.*

Data sufficiency can be assessed in terms of anticipated analyses. For example, ground water flow and contaminant transport analyses cannot be done without hydraulic conductivity and gradient information. The assessment of data helps shape the investigations.

Finally, it is necessary to look at data sensitivity. Will more data make any difference in the final outcome? Since generation of data takes time and costs money, evaluating the sensitivity of data is essential to a cost-effective investigation. Data sensitivity is illustrated by Example 15-2.

Example 15-2 Data sensitivity. A review of existing data for the former solid waste site in western Michigan described in example 15-1 included over one hundred grain size distribution analyses on the sand formally mined. In the samples the sand was found to be 30% ± 2% medium sand, 70% ± 2% fine sand and less than 1% silt and clay fines. How many more grain size analyses are needed?

Solution. It would appear that the site is well characterized with respect to grain size distribution, yet the answer to how many more tests are needed is not zero. A few confirmatory tests (perhaps three to five) are recommended to insure that the existing data is valid and free from systematic errors. However, once the previous laboratory results are confirmed, and provided field descriptions remain consistent, significant additional grain size testing would divert time, energy and money from other perhaps more important concerns. A statistical analysis of the data might be employed to document the engineering judgment that only a few additional tests are required.

Contaminants of Interest

The work plan for site and subsurface investigations must identify contaminants of interest. Identifying the contaminants of interest in a work plan being developed to guide an investigation to determine the contaminants at the site may appear as a "Catch 22." In one sense it is, since the investigation must detect the unexpected. On the other hand, the analysis of existing data should reveal suspected contaminants and contaminant transport pathways. This preliminary contaminant identification provides a basis for a preliminary risk assessment assessing the toxicological information and potential migration pathways. The toxicological and transport information is used to identify contaminants of particular interest for a site and subsurface investigation. With this information the types of samples, sampling locations and frequency, and sampling and analytical procedures may be better planned.

Sample Types

Samples of air, water, soil, and waste are typically obtained as part of the site and subsurface investigations. Water samples may include both surface and ground water. Soil samples include those for chemical assays and for measurement of physical and engineering properties (strength, permeability and compressibility). Sample types may be alternatively classified by whether or not they are samples of source (e.g. from an old waste disposal lagoon) or along transport pathways (i.e. ground water at a downgradient location).

Types of samples may also be indicated by the results of modeling studies which help identify those transport media which are most important for the particular site. As a result, samples could be of flora and fauna or of air quality as well as water quality. In some cases sampling of the human population may be included (not the whole human, just parts).

Sampling means more than physical samples; it is meant to include other data as well. For example the sampling plan might include monitoring of meteorological conditions including temperature, pressure, precipitation, wind velocity and humidity.

Sampling Locations and Frequency

The work plan must define the location and frequency of samples. By planning, in advance, sample locations and frequency, the probability of getting the proper number and types of samples at the proper locations is increased. It is advisable to consider the statistical validity of the data at this point in the planning process. Regardless of the number of samples obtained, the importance of the miniscule look at the subsurface must be recognized (see Example 15-3).

Example 15-3   Looking beneath the surface.   The work plan for site characterization of the relatively homogeneous clay beneath a former solid waste site in western Michigan calls for 20 undisturbed clay samples to be obtained for permeability testing. As described earlier, the sand is underlain by a 15-m thick deposit of glacial lacustrine clays. Each sample is 74 mm in diameter and 144 mm long. The site is 900 m by 600 m. Calculate, on a volumetric basis, the percentage of soil actually sampled.

Solution. Calculate the total volume of sample:

$$V_s = 20 \times \left(\frac{3.14 \times 74^2}{4} \right) \times 144/1000^3 = .012 \text{ m}^3$$

Calculate the total volume of clay layer:

$$V_t = 900 \times 600 \times 15 = 8.1 \times 10^6 \text{ m}^3$$

Calculate the ratio, in percent:

$$R = .012/8.1 \times 10^6 \times 100 = \underline{0.0000001\%}$$

Just think of the importance of the decisions based upon viewing so little of the subsurface! In fact, some would say that 20 tests are too many for a uniform clay deposit as described for this problem.

Although we actually sample and test a very small portion of the subsurface, caution needs to be exercised, however, to avoid obtaining and analyzing too many samples such that the information is unnecessary. For remedial investigations of sites with existing contamination, is it necessary to "define the plume?" It may not be necessary to know the location and concentration of all contaminants; a remedial system can often be developed based upon well-developed knowledge of site and subsurface conditions without a full definition of the plume location. Balanced against the need to control the quantity and quality of sampling information is the recognition that it is both expensive and time consuming to return to the site to gather additional samples should these be needed. Thus, it is recommended that the sampling frequency exceeds the minimum necessary in order to provide flexibility at future stages of the investigations.

There is much debate over random sample locations versus sampling in accordance to a uniform grid. As mentioned earlier, sampling locations should be based upon existing knowledge of the site conditions. Thus where possible, sampling locations should be neither random nor based upon a uniform grid; they should be purposely selected based upon site history and reconnaissance information, and most important, the geologic variability/uniformity.

Sampling and Testing Procedures

The work plan must include field sampling and testing procedures. The sampling procedures insure that the proper type and quality sample is obtained for subsequent laboratory analysis. Field testing procedures insure that in situ information is appropriately gathered. Since field explorations are costly, available opportunities for field testing must be properly anticipated and utilized. A list of procedures used in site investigations for field sampling and testing is presented in Table 15-3.

For samples obtained during the field investigations, protocols for chemical analysis are presented in such sources as Standard Methods.[2] Procedures for physical

TABLE 15-3
Sampling and field testing procedures

ASTM number	Test method
D653	Terminology relating to soil, rock, and contained fluids
D1452	Practice for soil investigation and sampling by auger borings
D1586	Method for penetration test and split-barrel sampling of soils
D1587	Practice for thin-walled sampling of soils
D2113	Practice for diamond core drilling of soils for site investigations
D2487	Test method for classification of soils for engineering purposes
D2488	Practice for description and identification of soils (visual-manual procedure)
D2573	Test method for field vane shear test in cohesive soil
D2607	Classification of peats, mosses, humus, and related products
D2937	Test method for density of soil in-place by the drive-cylinder method
D3385	Test method for infiltration rate of soils in field using double-ring infiltrometers
D3441	Method for deep, quasi-static cone and friction-cone penetration tests of soils
D4220	Practices for preserving and transporting soil samples

and engineering tests are also readily available from ASTM[3]. Where defined procedures are not available for specific contaminants or sample types, it may be necessary to develop specific analytical procedures for the project. However, where possible, sampling and analytical procedures are specified in the work plan by reference to standardized procedures. A listing of commonly employed physical and engineering test methods is presented on Table 15-4.

It is important to note that the bulk of the sampling and testing methods for site characterization were originally developed and used as geotechnical engineering tests. As these methods are employed in hazardous waste management studies, we must be conscious of the ever present need to obtain samples without altering their chemical composition and without cross-contamination between locations. It is recommended that the analytical laboratory responsible for chemical testing be consulted during this

TABLE 15-4
Laboratory test procedures

ASTM number	Test method
D422	Method for particle size analysis of soils
D854	Test method for specific gravity of soils
D2166	Test method for unconfined compressive strength of cohesive soil
D2216	Method for laboratory determination of water (moisture) content of soil, rock and aggregate mixtures
D2434	Test method for permeability of granular soils
D2113	Test method for one-dimensional consolidation properties of soils
D2664	Test method for triaxial compressive strength of undrained rock core specimens without pore pressure measurements
D3080	Method for direct shear test of soils under consolidated drained conditions
D4318	Test method for liquid limit, plastic limit and plasticity index of soils
EPA 9090	Hydraulic Conductivity

planning stage to ensure all sample handling and storage procedures are suitable for the types of chemical analyses required.

Sampling and test procedures must also carefully describe acceptable methods for cleaning equipment between samples and between borehole locations. Since the investigation will undoubtedly encounter both "hot spots" and areas deemed relatively clean, it is essential that contaminants not be carried between locations with the drilling and sampling equipment. Such cross-contamination is inevitable if equipment is not suitably cleaned between samples.

Operation Plan and Schedule

This element of the work plan defines the tasks to be completed, the team members who will perform the work and their role, the equipment to be used, the sampling order and the decontamination procedures. For example, who decides whether a boring should be terminated? The driller? The field inspector? The project manager? The owner's project manager? The regulatory agency? The operation plan will define the types of decisions expected and assigns authority for these decisions. The substitute for a well developed operation plan is confusion and delay at best and anarchy at worst.

The schedule realistically defines the start time, the duration of activities and the completion date. Contingencies, in terms of unexpected delays, must be accommodated.

Health and Safety

The protection of the personnel conducting site and subsurface investigations is too important to completely omit from this chapter, yet too extensive to cover in detail. The potential health and safety problems for workers at hazardous waste sites include:

- Exposure to toxic chemicals
- Electrical, fire, explosion and other safety hazards
- Exposure to extremely high or extremely low temperatures
- Exposure to high noise levels

In many cases, the use of personal protective equipment (PPE) while providing protection against site contaminants, posed additional safety hazards. PPE is usually cumbersome and restricts peripheral vision, increasing the possibility of injury from slips, trips and falls.

Addressing these and other hazards are the responsibility of a health and safety officer who will develop a site specific health and safety plan. This plan should include:[4]

- A listing of key safety and project personnel, emergency phone numbers and location of nearest emergency facilities such as fire and police departments, ambulance, and hospitals.

- A description of the risks associated with proposed site operations, including the appropriate personal protective equipment (see Table 15-5) and mitigation measures to be taken.
- A description of required training and medical monitoring (see below).
- Monitoring requirements (personnel and environmental).
- Decontamination procedures for people and equipment.
- Site maps, including hospital route, site location, and site feature maps.
- Required safe work practices while performing assigned/scheduled tasks.

Any worker at a superfund site must, by regulation (29 CFR 1910.120), undergo 40 hours of health and safety training. The training must explain the potential hazards of this type of work and provide the knowledge and skills required to use personal protective equipment. For example, the suits and face masks shown in Fig. 15-4 are cumbersome and uncomfortable. Special training and practice is required to work safely in such equipment. It is possible to dress someone up in a way that reduces

TABLE 15-5
Personal protective equipment at hazardous waste sites

Level	When Required	Description
A	High degree of skin contact and respiratory hazard or chemical substances at levels requiring fully encapsulated suits or when contaminants are unknown.	Level B plus fully encapsulated suits. Level A is rarely used at hazardous waste sites.
B	Minimum level when hazards are generally characterized such as during an initial site visit with limited background information. Required whenever toxic concentrations in air are a possibility, such as in enclosed spaces where volatile chemicals may be found.	Level C plus self-contained breathing apparatus (SCBA) such as typically worn by fire fighters or other supplied-air system.
C	Where airborne hazards have been measured and it is known and within conditions such that canister air purifiers are safe and where there is little likelihood of skin contact.	Level D plus air purifying (canister respirator and Tyvek) suit with a hood. Rubber boots are worn over work shoes and taped to Tyvek suit to provide a seal. Respirators must have been individually fitted to assure an adequate face-to-mask seal.
D	Minimum level for any site.	Steel toed work shoes. Hard hat with face mask if required. Work clothes with long sleeve shirt.

FIGURE 15-4
Workers in Level C protective equipment sampling a tank truck

the risk of chemical exposure but significantly increases the risk of physical injury. Perhaps the most important part of the training is to explain the limitation of safety equipment.

Special training is required for the use of self-contained breathing apparatus (SCBA) where the worker breathes only bottled air that is carried on a back pack. This training would include the proper assembly and inspection of the air pack, adjustments to the equipment, fitting of the face mask, and practice of performing manual labor wearing an SCBA.

Another important aspect of a health and safety program is a medical surveillance program. This would include pre-employment and annual medical examination, emergency and non-emergency treatment and maintenance of records documenting the program. The annual physical should include a pulmonary function test to demonstrate that the employee can safely use an air-purifying respirator while performing assigned duties. It is a regulatory requirement to obtain medical approval for all personnel who are required to wear respiratory protection during the course of their job responsibilities.

Health and safety programs are an integral part of any hazardous waste activity. Health and safety plans are scrutinized very closely by government agencies involved in the project to assure effectiveness and completeness. It is generally recommended that those programs are overseen by highly qualified health and safety professionals such as certified Industrial Hygienists and Safety Professionals.

Cost Estimate

Based upon the foregoing elements, the cost estimate can be prepared for the project. This is necessary to establish the required level of funding and assure that a proper and sufficiently thorough investigation can be conducted. It is necessary, however, to provide some flexibility in the plan as well as in the cost estimate to accommodate unexpected or troublesome site and subsurface conditions. It is common that, whatever the cost estimate, the actual costs are higher. As a result the cost estimate should always include a contingency on the order of 15%. A checklist showing the types of items included in a cost estimate is presented in Table 15-6.

Other Documents

In addition to the foregoing elements in the work plan, it is necessary to provide **Quality Assurance (QA)** and **Quality Control (QC)** documents, a data management plan, and a health and safety plan. The QA and QC procedures help ensure data validity in the analysis of the results. Specifically, a QC plan describes the specific procedures by which the implementation of sampling and analytical procedures designed to result in reliable data are documented. The QA plan describes the procedure by which the QC implementation is audited to ensure that the work and documentation is being conducted in accordance with established QC procedures.

Data management plans provide details as to how the data will be utilized and what engineering analyses can and will be conducted for the site with the acquired data. For example, the data management plan will describe the procedure to be implemented so that all engineers are using the most recent spreadsheet summarizing the hydraulic conditions and not an outdated or incorrect version of the spreadsheet. Finally, health and safety plans ensure that the proper personnel training is provided and that the appropriate level of protection is prescribed.

15-3   SITE CHARACTERIZATION

After the scope of the hazardous waste management problem has been defined and the preliminary investigations completed, the problem quantification can proceed. In this phase, the detailed site and subsurface characteristics are developed and quantified. It is the detailed information developed during this phase of study which permits the development of the facility design or the evaluation of the feasibility of the remedial alternatives. This phase of study has several tasks associated with it as described in the following sections of this chapter.

TABLE 15-6
Investigation cost estimate check-list

Field Costs
    Mobilization/demobilization
    Drilling
    Sampling—undisturbed
    Sampling—disturbed
    Field inspection (labor and expenses)
Laboratory costs
    Geotechnical tests
    Chemical tests
Engineering and scientific staff
    Geology
    Hydrogeology
    Chemistry
    Biology
    Civil engineering
        Geotechnology
        Hydrology
        Site development
    Computer modeling
Project management
    Cost control
    Client interface
    Accounts
    Meetings and presentations
    Senior project review
    Clerical/support staff graphics

Upon completion, of all the "paperwork" described in the problem identification and scoping phase, the collection of site specific data can begin. Site specific data are the result of sampling and field studies to characterize the site, to define exposure pathways, and to provide data for screening of remedial alternatives or the new facility design. Within these site specific studies, both indirect and direct methods are used. In employing **indirect methods**, we measure some other parameter which can then be related to the parameter of interest. In contrast, there are **direct methods**, such as boring, sampling, and testing which more directly measure the parameters of interest. Both methods are generally used for subsurface investigations.

The advantage of indirect methods is that they can be used for planning for direct methods, interpolation between points of hard data and optimization of drilling and sampling costs. This is in contrast with direct methods which, as noted, require boring, sampling and testing and provide direct information. The direct methods are usually somewhat limited due to the associated time and cost. With coupled and direct indirect methods, interpolations between test borings and monitoring wells can be made with considerably more confidence and the combination of both direct and indirect methods results in the most cost effective and technically sound investigation program.

The methods discussed below are frequently presented in isolation from the geologic setting in which they are carried out. Recognize that a few pages on geophysics will not make one a geophysicist or a couple of paragraphs on well drill techniques will not qualify anyone as a well driller.

However, the information will form a basis for chemical engineers to discuss alternatives with geophysicists, for biologists to consider optional methods with a driller and the generalist to interact with the specialist. Given the entirely multidisciplinary nature of hazardous waste management, such dialogue between specialists is essential to the successful outcome of the project.

15-4   GEOPHYSICS

Indirect methods are often called **geophysical** methods or methods of **remote sensing**. These techniques are useful for the reasons described above. Geophysical techniques provide information about the subsurface conditions through the measurement of certain properties (e.g. arrival time of a seismic wave) and the deduction of parameters of interest (e.g. depth to bedrock). Geophysical methods do *not* normally directly measure the parameter of interest. Before employing any geophysical investigations, it is necessary to review the site history and geology, identify areas of cultural interference and determine a potential target's response to a particular geophysical technique. The following sections describe in detail the most commonly employed surface geophysical methods utilized in site and subsurface investigations for hazardous waste management.

For most geophysical techniques, verification of interpretations is typically required by comparing with **ground truth** information. That is, confirmatory drilling, sampling and/or testing are completed at selected locations to verify the interpretations. As noted earlier, geophysical techniques are often used to strategically locate borings and, as such, the resulting direct investigation serves to verify interpretations.

Resistivity

Resistivity is defined as the resistance to current flow as a result of an applied electrical potential. Resistance is measured in ohms and resistivity is resistance times length (typically ohm m or ohm ft). In geophysical surveys resistivity is determined as the electrical resistance per length of a unit cross section area ($ohms/l/l^2 = ohms \times l$). Resistivity in the subsurface is a function of material type, porosity, water content and concentration of dissolved contaminants in the pore water. In general, dry soils and rock have very high resistance to electrical flow whereas inorganically contaminated saturated soils have a relatively low resistance. Soils contaminated with organic materials may have a relatively high resistance. Typical soil resistivity information for uncontaminated soils is shown in Table 15-7. Resistivity surveys are conducted to provide relatively detailed subsurface information to outline aquifer boundaries, depth to the water table and/or bedrock, and change in soil type or levels of contamination.

TABLE 15-7
Typical values of soil resistivity

Soils	Resistivity range (ohm-m)
Clays	1–150
Alluvium and sand	100–1,500
Fractured bedrock	Low 1,000s
Massive bedrock	High 1,000s

 Resistivity surveys are conducted by applying an electrical current across two electrodes into the earth, and measuring the change in voltage (electrical potential) across two receiving electrodes. In order to conduct a resistivity survey, electrodes (metal stakes) are driven about 8 inches into the ground, typically along a straight line. The current is supplied to two electrodes by means of a battery or small generator and then the voltage between the other electrodes is measured. A schematic of the principle is shown on Fig. 15-5.

 The soil resistivity, R, for a resistivity survey using the Wenner spacing (see Fig. 15-5) is found to be:

$$R = 2\pi s V / I \qquad (15\text{-}1)$$

where R = soil resistivity (ohm-meters)
   s = electrode spacing (meters)
   V = measured voltage (volts)
   I = applied current (amperes)

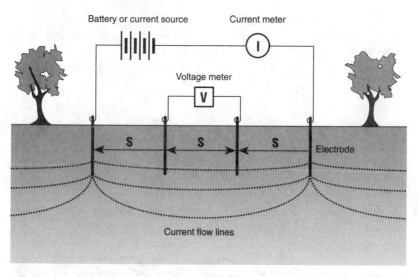

FIGURE 15-5
Schematic of resistivity surveys.

There are a variety of spacings (arrays) that can be utilized including the Wenner and Schlumberger arrays. By selecting different spacings the electrical resistivity survey is able to pinpoint conditions at specific soil depths. Note that conceptually the contours of equal resistivity are analogous to those for ground water flow. That is because the flow of electrical current is analogous to the flow of water through soil. For the flow of electricity, the voltage is the potential. For the flow of water, the total hydraulic head is the potential. Interpretation of resistivity data is best left to those specifically educated for this task, geophysicists.

Example 15-4 Soil resistivity. Using Eq. 15-1, calculate resistivity for three areas of a site under investigation and speculate as to soil type. In each case a current of 0.010 milliamps was applied and the spacing was 5 meters. The measured voltage at each location is given below.

Location	Measured voltage (volts)
1	0.467
2	0.034
3	0.492

Solution

$R = 2\pi s V / I$

Location 1: $R = 2 \times \pi \times 5 \times 0.467/0.01 = 1467$ Sand

Location 2: $R = 2 \times \pi \times 5 \times 0.034/0.01 = 107$ Clay

Location 3: $R = 2 \times \pi \times 5 \times 0.492/0.01 = 1546$ Sand

Seismic

In seismic surveys, a **seismic wave** is introduced into the subsurface as an energy impulse. The seismic wave velocity will vary depending upon material properties, mostly density. The seismic wave may be introduced by either a small explosive charge or by hitting a steel plate or ball into the ground with a sledge hammer. The detector of the seismic wave is termed a **geophone**. The seismic wave velocities generated using a sledge hammer are generally limited to survey depths of 50 feet or less. The seismic velocity for various materials is shown in Table 15-8.

Figure 15-6a presents a schematic of the technique. As shown the seismic wave propagates radially in the subsurface from the source. By placing a series of geophones at varying distances from the source, the arrival time at each geophone can be measured.

For the stratigraphy shown in Figure 15-6a the arrival time at the geophones near the source depends only on the velocity of the overburden. In this case the arrival time is linearly related to the distance. However, at increasing distances from the geophone the seismic wave will arrive faster along a path down through the

TABLE 15-8
Typical soil seismic velocities[5]

Material	Velocity (ft/s)	(m/s)
Weathered surface material	1,000–2,000	305–610
Gravel, rubble, or sand (dry)	1,500–3,000	468–915
Sand (wet)	2,000–6,000	610–1,830
Clay	3,000–9,000	915–2,750
Water (depending on temperature and salt content)	4,700–5,500	1,430–1,680
Sea water	4,800–5,000	1,460–1,530
Sandstone	6,000–31,000	1,830–3,970
Shale	9,000–41,000	2,750–4,270
Chalk	6,000–31,000	1,830–3,970
Limestone	7,000–20,000	2,140–6,100
Salt	14,000–17,000	4,270–5,190
Granite	15,000–19,000	4,580–5,800
Metamorphic rocks	10,000–23,000	3,050–7,020
Ice	12,050	

overburden horizontally through the bedrock and back up through the overburden. This represents a change in the linear relationship already established. Plotting the relationship between the distance of the geophones from the source and the arrival time for each geophone results in two straight lines of different slope as shown in Figure 15-6b. The inverse of the slope is velocity in length per unit time. The depth to bedrock (or any layer with a greater seismic velocity) is the thickness of the uppermost layer, t, and is calculated as:

$$t = (D/2)\sqrt{((v_2 - v_1)/(v_2 + v_1))} \qquad (15\text{-}2)$$

where $\quad D$ = distance to point of intersection
$\quad\quad\;\; v_1, v_2$ = velocity of layer 1, layer 2

Example 15-5 Seismic survey. Using the graph in Figure 15-5b, calculate the depth to bedrock and render an opinion as to the overburden and bedrock condition/type.

Solution
$v_1 = 73$ m/0.045 sec = 122 m/sec   Stiff clay
$v_2 = (140 - 73)/(.065 - 0.045) = 3350$ m/sec   Unweathered limestone
    Substituting $D = 74$ m, $v_1 = 122$ m/sec and $v_2 = 3350$ m/sec into Eq. 15-2, results in an overburden layer thickness of 35.7 m. Thus, the depth to bedrock is 35.7 m.

Electromagnetic Conductivity

Transmitting antennas can be used to induce an electromagnetic current into the ground which is transmitted through the soil similar to the transmission of electrical current

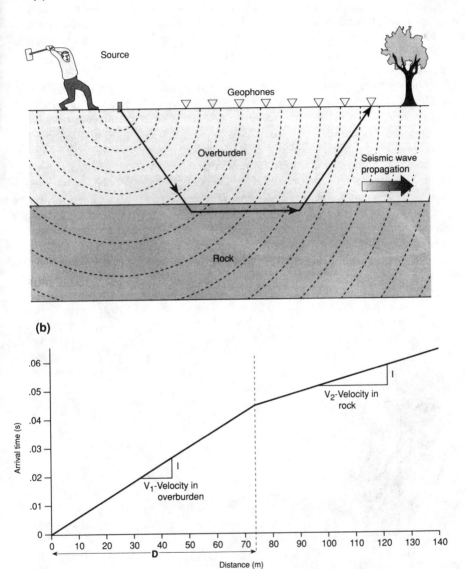

FIGURE 15-6
Schematic of seismic surveys.

directly induced in the ground by resistivity technique.[6] By indirectly inducing the current into the ground, a secondary magnetic field is generated and picked up by a receiving antenna. The terrain conductivity value is therefore an average conductivity over the effective depth of the survey. The effective survey depth is largely a function of the distance between the sending and receiving coils. The technique is much faster than the ground coupled resistivity technique. Terrain conductivity is influenced by material type, discontinuities (such as fractures), ground water chemistry, intercoil spacings and coil orientation. Since ground water chemistry has a significant influence on the measured conductivity, electromagnetics offer a rapid, cost-effective method of mapping contaminant plumes.

Shown in the Fig. 15-7 are terrain conductivity contours. The survey was conducted in a stratigraphy consisting of saturated sand in order to aid in the design of a preliminary monitoring system to assess a potential source of contamination. It was suspected that the site may have been used for illegal waste disposal although there was no remaining surficial evidence of such disposal. What can we see in Fig. 15-7? Readings of terrain conductivity (in millimhos per meter) were first obtained and plotted. From these data contours of equal conductivity were plotted. In this case, a conductivity anomaly (a high) is discovered. The center of the high can be interpreted to be the area of highest subsurface contamination. Next, observe the shape of the contours. In general the contours are elongated towards the south east. This elongation can be interpreted to show the direction of ground water flow and thus contaminant transport. Of course, drilling and sampling were required to verify the interpretations but the direct investigation costs were minimized and the information maximized.

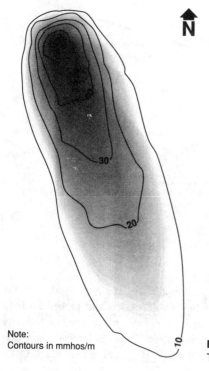

Note:
Contours in mmhos/m

FIGURE 15-7
Terrain conductivity contours.

Ground Penetrating Radar

In addition to seismic waves and electrical currents, it has been demonstrated that radio frequency microwaves can penetrate the soil and be transmitted through soil layers. This technology may be utilized to identify buried materials at hazardous waste sites.[7]Upon entering the soil and progressing downward through successively stiffer soil or rock layers, the waves reflect off of objects or layers possessing contrasting electron properties. Reflective waves or echoes are received and the resulting graph appears much like a sonar or sound profiling. An example of ground penetrating radar (GPR) is shown in Fig. 15-8 where reflections from ground penetrating radar indicate the presence of a buried gasoline tank.

The GPR techniques are therefore useful in detecting voids, underground trenches, buried drums of both plastic and metal, and floating hydrocarbon products in the subsurface. GPR is limited by rapid attenuation in conductive soils such as clays.

Gravimetric

Differences in the earth's gravimetric field can be measured with highly sensitive instrumentation. These technologies are typically used to detect features such as sink holes, fill zones, faults or buried valleys in the subsurface. The technique has found

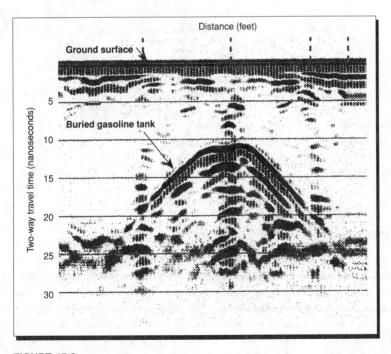

FIGURE 15-8
Ground penetrating radar.[7]

only limited use for detailed definition of subsurface conditions for hazardous waste management and will not be discussed further herein.

Magnetometer

A **magnetometer** picks up magnetic anomalies in the subsurface. A proton precession magnetometer uses spinning protons (the nuclei of hydrogen atoms) in a hydrocarbon fluid. The spinning protons act as magnetic dipoles and, as such, can be oriented (aligned, polarized) by application of a magnetic field. The magnetometer has a coil which is first used to align the spinning protons by applying a current. After aligning the spinning protons, the magnetic field in the subsurface causes the protons to precess, aligning themselves with respect to the subsurface magnetic field. This generates a current in the same coil used to originally align the protons. The frequency of the generated signal is proportional to the strength of the magnetic field (the total magnetic field intensity). The unit of measure is the gamma with the earth's field equal to about 50,000 gammas in South Florida. The value depends on location. In the U.S. it ranges from 45,000 to 60,000 gammas. But in South America, the value is as low as 25,000 gammas. Local disturbances in the earth's magnetic field, such as by buried drums, give rise to magnetic anomalies. Thus this technique is especially useful for the identification of buried drums or other ferromagnetic (iron and steel) foreign materials. It is noted however there are numerous sources of interference, including power lines, buried metal pipelines and fences (and geologists' big belt buckles). Magnetometer surveys are particularly useful for determining the location of underground storage tanks.

The results of a magnetometer survey to investigate the location of buried wastes at a manufacturing facility are shown in Fig. 15-9. In Fig. 15-9a the results as measured are shown. However, as noted above, known interferences (known as cultural anomalies) must be accounted for. For this particular project the engineered surface features included a building, an on-site decontamination trailer and a paved road. The data can be evaluated in light of these known interference, and represented as shown in Figure 15-9b to delineate anomalous areas. Test trenches verified the present of waste pits at each of the anomalies and the absence of waste pits in the areas without anomalies.[8]

Borehole Geophysics

Borehole geophysical methods are categorized as indirect methods in this text as the techniques generally measure one parameter (e.g. natural radiation) to deduce another (e.g. stratigraphy). They do, however, require penetration into the subsurface via the borehole making them more direct than the surface geophysical techniques described above. Borehole geophysical studies provide additional data to characterize the site and subsurface conditions.[9] The techniques available, the parameter actually measured, and the principle application of the measurement are shown in Table 15-9. Detailed discussion of these methods are beyond the scope of this text.

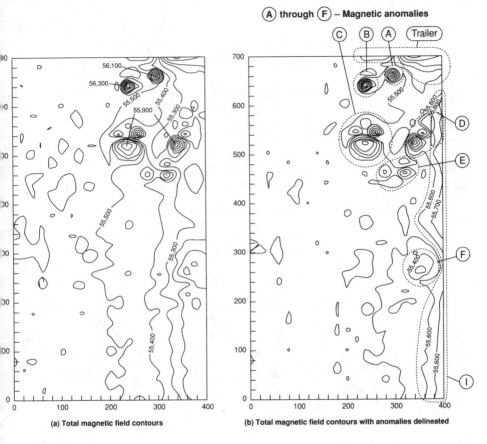

FIGURE 15-9
Magnetometer survey results.

15-5 BORING AND SAMPLING

Direct field methods of subsurface exploration generally consist of boring, sampling and testing. The most common of these techniques is the widely accepted and utilized installation of ground water monitoring wells. However, the diversity of subsurface investigation techniques available along with the specific information which they provide requires additional discussion. Direct subsurface investigation methods are necessary to provide in situ measurements of soil and rock properties, supply samples for laboratory characterizations of material properties, and to serve as the "ground-truth" for geophysical interpretations. The following sections of this chapter describe direct methods of field exploration and provide guidance for their use.

It is first necessary to classify direct exploration methods by their distinct individual components. First, there are a variety of drilling methods that are classified by the method by which the test boring is advanced into the subsurface. Next, there are distinct methods of soil and ground water sampling. These sampling techniques

TABLE 15-9
Borehole geophysics

Test technique	Measurement/application
Electrical techniques	
Spontaneous potential	Electrical potential/lithology
Resistance	Electrical resistivity/lithology
Single point resistance	Electrical resistivity/fracture
	Induction conductivity/lithology
Radiological techniques	
Gamma	Natural gamma radiation/lithology
Gamma gamma	Electron density/porosity, bulk density, moisture, lithology
Neutron	Hydrogen content/moisture, porosity
Miscellaneous techniques	
Acoustic velocity	Compression wave velocity/porosity, lithology, modulus
Caliper	Diameter/lithology, fractures
Temperature	Temperature/flow

may or may not overlap with the drilling methods. In addition, there are a number of in situ tests to measure specific soil properties. An in situ test may also result in the acquisition of a soil sample. Finally, any subsurface investigation must include laboratory analysis of representative samples of soil and ground water. Analysis can include geotechnical and chemical analysis. This chapter includes a discussion of the presentation of test results (i.e. data). The method of data presentation often reveals specific and informative aspects of the subsurface conditions.

Drilling Methods

A test boring may be advanced into the subsurface utilizing any one of a number of techniques. The selection of the technique depends on the subsurface conditions, the types of samples required, the type of in situ testing, and the method of monitoring well construction. The following paragraphs describe the most widely employed techniques used in hazardous waste management site and subsurface investigations. Conceptually, each drilling technique has two fundamental system requirements: 1) a method or tool to advance the borehole, and 2) a method or tool to carry the cuttings to the surface. As each of the drilling techniques are discussed below, bear in mind the operational function of the tools and techniques with respect to these two fundamental system requirements.

The selected drilling method must also reflect the objectives of the test borings. In many cases, boreholes may be used for in situ testing and sampling as well as for the installation of monitoring wells. In some cases, particularly for monitoring wells having multiple monitoring levels, boreholes should be drilled for the sole purpose of monitoring well installation. In either case, the influence of the drilling technique upon the borehole objectives must be considered.

Auger Drilling Methods

One of the most common methods of advancing of test borings though unconsolidated sediment (soil to the geotechnical engineer) is through the use of **continuous flight augers** as shown in Fig. 15-10. Auger borings are drilled by the rotation of the augers and cutting bit through the use of a drilling rig. The cutting bit loosens the soil and the auger flights carry the cuttings to the surface. Augers can vary in size from small (2″ diameter) augers that may be rotated by hand or with the help of a small engine (powered one or two person post hole digger) to large (up to 10′ diameter or larger) requiring specially equipped cranes. Most commonly augers used in site investigations are in the range of 4 to 12 inches in outside diameter used with truck mounted drilling rigs.

Continuous flight augers may be constructed as **solid stem augers** or **hollow stem augers**. The solid stem auger is constructed by wrapping the auger flight on a solid pipe or steel rod as shown in Fig. 15-11. The solid stem auger method permits the acquisition of a disturbed sample for examination and testing which employs remolded samples. With auger samples, the precise sampling location in the subsurface cannot be determined. Small and medium diameter continuous flight solid stem augers are typically used for shallow borings. Very large diameter holes are occasionally used to permit direct access of the engineer or geologist into the subsurface.

An improvement to solid stem auger technology appeared in the 1950's with the development of **hollow stem augers** as shown on Fig. 15-11.

In this case, the auger flight is connected to a pipe with an inside diameter varying from 2 to 12 inches permitting the auger to penetrate into the subsurface advancing the test boring while permitting the engineer/driller to sample the subsurface materials through the center of the hollow stem auger. In this way the boring can be advanced to a desired and precise sampling or testing depth. The testing and/or sampling tool can be lowered into the hole through the center of the augers and the in situ test or sampling can be conducted. In addition, hollow stem augers are useful for the installation of ground water monitoring wells through the inside of the auger. After well and seal installation, the auger is withdrawn and the monitoring well remains in place.

Rotary Drilling Methods

Rotary drilling techniques, among the earliest developed, are widely used to advance test borings in unconsolidated sediment. In this technique a cutter head is rotated in the presence of drilling fluid. The drilling fluid (either water or drilling mud) assists in loosening the soil and carries the cuttings to the surface. Borehole stability during rotary drilling may be maintained through the use of a steel casing inserted in the bore hole or through the use of the drilling mud. Drilling mud functions by means of the formation of a thin impervious membrane along the walls of the borehole coupled with the unit weight of the fluid to maintain the borehole stability. A schematic of mud rotary drilling is shown in Fig. 15-12.

The drilling of rock formations requires techniques and equipment different than that required for soil drilling. Rotary drilling is employed to advance a borehole

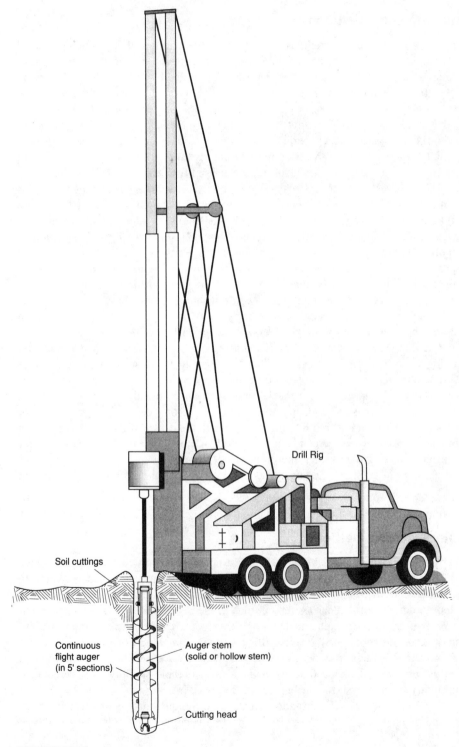

Drill Rig

Soil cuttings

Continuous
flight auger
(in 5' sections)

Auger stem
(solid or hollow stem)

Cutting head

FIGURE 15-10
Continuous flight augers.

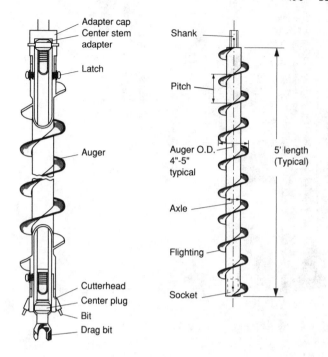

a.) Hollow stem auger   b.) Solid stem auger

FIGURE 15-11
Solid and hollow stem augers (after Acker[10]).

through rock. Air rotary methods are relatively rapid and inexpensive. In this method air is the drilling fluid employed to transport the cuttings to the surface. Rotary wash techniques use water to carry the cuttings to the surface. Rotary core drilling techniques both advance the borehole and allow for the recovery of a sample (a rock core). In addition to these techniques, a variety of drilling techniques including reverse dual tube rotary, cable tool percussion and reverse circulation become topics for detailed study.

The significance of knowledge of alternative drilling techniques lies in the ability to choose the most appropriate technique for the given site and geologic conditions. It is necessary to accommodate both the expertise of the driller with regard to specific drilling techniques and the expertise of the scientists and engineers with regard to the geology, site conditions and hazardous waste management activities planned. The selection of drilling procedures should be mutually agreeable between the driller and the scientists and engineers.

As emphasized later in this chapter in the discussion of monitoring well installations, it is important that underlying formations not be contaminated during the drilling process. Auger and rotary drilling techniques all quite obviously begin at the ground surface. Drilling deep through shallow contaminated formations and into underlying

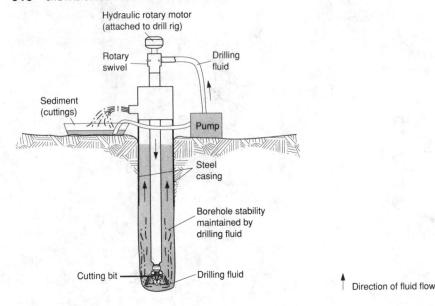

FIGURE 15-12
Mud rotary drilling.

formations may result in the transfer of contaminants from upper contaminated regions to deeper uncontaminated formations unless special precautions are taken to case
off the overlying zones of contamination. Drilling through contaminated formations
into underlying potentially uncontaminated formations requires the two-step process
described later.

In Situ Tests

Drilling and sampling to obtain subsurface samples for laboratory testing and analysis
provide one means of site characterization. No less important is testing **in situ**, that
is, in place within the formation. Sampling and testing in the laboratory has the disadvantage that the original in place properties of the material will be altered during
sampling, transport and preparing for laboratory tests. However, laboratory testing
may permit more sophisticated tests to be conducted under controlled, reproducible
laboratory conditions. In situ tests measure material properties as they exist in the
subsurface; there are no sampling, transport or handling costs. The cost lies in the
time and difficulty in testing, the uncertainty of the actual conditions at the test interval and the often semiempirical methods of analysis associated with in situ test data.
Nonetheless, in situ testing must be an integral part of any thorough site characterization. The tests described below are among those most commonly and successfully
employed.

Standard Penetration Test (SPT)

The most commonly employed in situ test is the **Standard Penetration Test** (SPT). This test is conducted using a standard split barrel sampler, driven into the ground with a 140 pound hammer falling 30 inches as shown in Fig. 15-13, in accordance with ASTM requirements.[11] The SPT value or N-value is obtained by summing the total number of hammer blows for the second and third six-inches of sampler penetration in an 18 inch drive. The advantages of this in situ test are that it incorporates information regarding the site in situ density/strength of the materials and provides a sample for classification and chemical and/or physical property testing. The sample is considered a disturbed sample in the sense that the sample should not be used for engineering property tests of strength, permeability or compressibility.

Although the SPT is the most widely employed method of in situ testing and sampling, the potential variability of test results is not widely considered. Studies have shown that the SPT values depend upon the energy imparted to the sampler and a number of factors can influence the energy.[12, 13] These factors include an inaccurate fall height, excessive rope friction caused by excessive turns of rope on the spinning cathead, and incomplete cleaning of the borehole prior to sampling and testing. Additional well-established factors[14] are shown in Table 15-10.

As a result of many years of standard penetration testing, a number of correlations to physical and engineering properties of soils have been developed. The SPT value may be correlated to relative density of sands,[15] angle of internal friction of

Drill rig boom

140 lb hammer used
in performing SPT

Drill coupling to advance augers

Drill rods with sampling support
attached at the bottom*
Hollow stem auger flight

Reeming rod

* Note: markings on rods spaced at 6
inch intervals for SPT measurement

FIGURE 15-13
Standard penetration test.

TABLE 15-10

Factors affecting the results of the SPT

Test detail	Effect on N-value
Inadequate cleaning of disturbed material in the borehole	Decreases
Failure to maintain sufficient hydrostatic head in the borehole	Decreases
Variations from the exact 30 inch drop	Either
Extreme length of drill rods P($>$ 175$'$)	Increases
Any interference with free fall	Increases
Using deformed sample spoon	Increases
Excessive driving of sample spoon before the blow count	Decreases
Failure of driller to completely release the tension on the rope	Increases
Driving sample spoon above the casing bottom	Increases
Use of wire line rather than manila rope	Increases
Carelessness in recording blow count	Either
Insufficient lubrication of the sheave	Increases
Larger size of borehole	Decreases

sands,[16] undrained shear strength of clays, settlement of foundations in sands,[17] and liquefaction potential.[18] As a minimum, granular soils can be classified by their relative density and cohesive soils by their consistency employing the standard penetration test N-value. Presented in Table 15-11 are correlations between the N-values and the soil relative density of a granular soil and the consistency of a cohesive soil.

TABLE 15-11

Correlations to the N-value from the SPT

Sands

N-value	Relative density	Description
0–4	0–15	Very loose
4–10	15–35	Loose
10–30	35–65	Medium
30–50	65–85	Dense
$>$ 50	85–100	Very dense

Clays

N-value	Unconfined compressive strength (kN/m^2)	Description
$<$ 2	$<$ 0.25	Very soft
2–4	0.25–0.50	Soft
4–8	0.5–1.0	Medium (firm)
8–15	1.0–2.0	Stiff
15–30	2.0–4.0	Very stiff
$>$ 30	$>$ 4.0	Hard

Cone Penetrometer Test (CPT)

Alternately, in situ shear strength of soil can be determined using a **cone penetrometer test** as shown in Fig. 15-14. The cone may be equipped with a piezometer to enable measurement of pore water pressures including rate of generation and dissipation which allows estimation of in situ permeability. The cone may also be modified to allow for pore water sampling. The electronic cone penetrometer provides an additional advantage over the standard penetration test in that a continuous vertical profile is obtained as opposed to sampling at fixed intervals. The primary disadvantage is that no sample is obtained.

Shown in Fig. 15-14a is a cross-sectional schematic showing the cone penetrometer test equipment. The cone in this case includes a friction sleeve and a pore water pressure transducer along with the conical end (often termed a piezocone). Also shown is a cone modified to acquire pore water samples. As the cone is pushed with the cone rig the tip resistance, friction sleeve resistance and porewater pressure is measured. Shown in Fig. 15-14b is a plot of the results. Since the cone produces a continuous profile, thin lenses of differing materials can be sensed. By calculating the ratio of the friction sleeve resistance to the cone tip resistance, a soil classification can be estimated. Finally, the pore water pressure profile can be determined from the pore pressure measurements.

Vane Shear Test

For soft clays the shear strength can be measured with **vane shear test** as shown in Fig. 15-16. In this test a vane is first pushed into the soil (always a clay rich soil, never a sand or gravel). The vane is then rotated slowly, measuring the torque, T_{max}, required to shear the soil. The dimensions of the vane (height H by diameter D) and thus the surface area of failure are coupled with the applied torque to compute the shear strength of the clay. The vane shear test provides a reliable estimation of the in situ range shear strength of clays. The shearing resistance, S_u, is calculated as:

$$S_u = (3T_{max})/(2\pi D^3) \text{ for } H/D = 1 \tag{15-3}$$

or

$$S_u = (6T_{max})/(7\pi D^3) \text{ for } H/fD = 2 \tag{15-4}$$

Extensive studies have shown that the results from vane shear testing need to be corrected for variations in clay type. The clay type can best be characterized for this correction by the plasticity index. The correction to the computed value is shown in Fig. 15-16. The field vane shear strength is multiplied by the correction factor to arrive at the actual field shear strength.

Dilatometer Test (DMT)

When it is necessary to understand lateral stresses in the subsurface, it may be desirable to run a dilatometer test. The dilatometer, shown in Fig. 15-17, is pushed into the ground with a conventional drilling rig to the desired test depth. The flexible membrane disk is then inflated using compressed air from the surface. As the pressure is

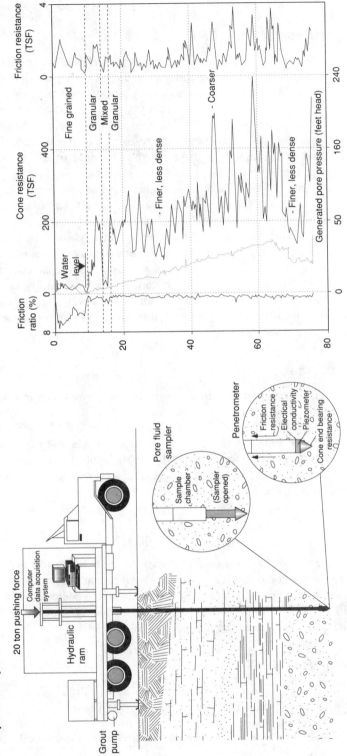

a.) Penetrometer subsurface exploration system

b.) CPT sounding log

FIGURE 15-14
Cone penetration testing.[19]

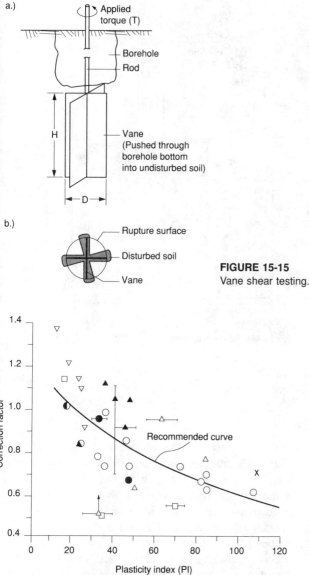

FIGURE 15-15
Vane shear testing.

S_u (field) = S_u (vane) · Correction factor

FIGURE 15-16
Correction to vane shear
test results (after Holtz and
Kovacs, 1981).

increased to the horizontal earth pressure pushing against the disk, the disk lifts off its
starting position. The pressure to cause this movement is considered the in situ horizontal earth pressure. The membrane is further expanded to cause movement through
a fixed distance and the ratio of pressure to displacement is taken as a soil modulus
(soil stiffness). Thus, the DMT allows for an estimation of lateral earth pressure and
elastic modulus.[20]

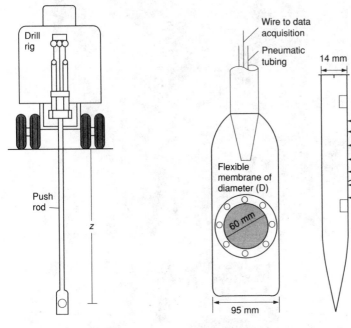

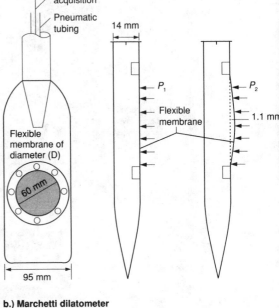

a.) The dilatometer pushed to depth z for test

b.) Marchetti dilatometer

FIGURE 15-17
Dilatometer testing.

Pressuremeter Test (PMT)

When it is necessary to understand lateral stresses in the subsurface, it may be desirable to run a pressure meter test. The PMT provides a measure of both shear strength and lateral pressures. In this test, shown schematically in Fig. 15-18, a cylinder is expanded in a borehole until the material has "failed." The data obtained by measuring the expansion pressures and the accompanying volume increases are analyzed using solutions derived from cavity expansion theory. Since the pressuremeter laterally loads the soil from its initial stress condition (after drilling the borehole) to failure, the device yields an estimate of the in situ horizontal earth pressure, the shear modulus and the shear strength. Self-boring pressuremeters have been developed to minimize drilling disturbance. The details of the test procedure and analysis of results are available elsewhere[21] and are beyond the scope of this book.

In Situ Water Level Measurement

The need to measure the depth to water in a monitoring well is universal. A **water level indicator** such as that shown in Fig. 15-19a is widely employed. This battery operated device is essentially a continuity meter employing a weighted, two wire cable, calibrated in length, with an open circuit at the tip. The weighted tip is lowered

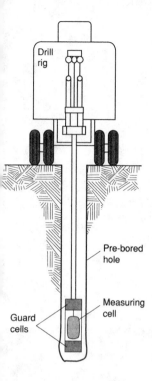

FIGURE 15-18
Pressuremeter testing.

into a monitoring well or piezometer (described in detail later in this chapter). When the tip encounters water, the circuit is completed and the current is registered on the amp meter. A measurement can be taken of the precise depth at which the circuit is completed and taken as the depth to water. The wire is spooled for ease of reeling out and in.

An economical alternative device (and, in the view of the authors, a technically superior method) is the **plopper**. The plopper, shown in Fig. 15-19b, consists of a measuring tape (preferably fiberglass) to which a weighted acoustical generator is added (the plopper). The acoustical generator is constructed of a steel pipe cap and threaded eye bolt available from the hardware store. The plopper is lowered into the well upon hitting the water surface emits a highly audible "plop." Precise depth to water is then made using the measuring tape. The readily audible nature of the plop is seemingly amplified by the long well casing.

In Situ Hydraulic Conductivity

In situ tests are commonly applied for determination of ground water properties as well. These tests can include single well permeability test such as infiltration or exfiltration (a slug test). Pumping tests, if the boundary conditions permit, provide insight into the macro behavior of the formation. Additional information regarding water level measurements, pump tests and hydraulic property tests is presented in Chap. 4.

a)

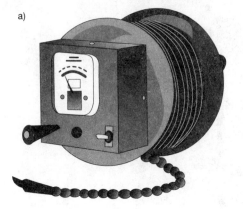

Battery powered water level indicator

b)

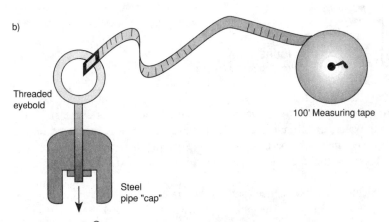

Threaded
eyebold

100' Measuring tape

Steel
pipe "cap"

Acoustical generator® for water level measurement (plopper)

FIGURE 15-19
Water level measurement.

In Situ Water Quality

In situ tests for concentrations of specific contaminants in water are currently in the development stage. For example, a remote sensing ground water monitor employing fiber-optics is under development.[22] This device would measure in situ both the presence and concentration of contaminants such as mercury, chromium and phenol. The development of a reliable in situ test device to measure contaminant concentrations will result in more reliable and cost effective site and subsurface characterizations.

An indication of organic contamination in ground water can be obtained using field organic vapor analyzers (OVA's). A field OVA can determine the total concen-

FIGURE 15-20
An organic vapor analyzer.

tration of organics in air. By measuring the air quality in the "head space" above the ground water in a monitoring well, an indication of organic contaminant concentration ground water is provided. Shown in Fig. 15-20 is the use of an OVA to determine concentration of organic vapors coming from an abandoned barrel.

Soil Sampling Methods

Soil samples can be obtained from the cuttings of the drilling methods described above. However, cuttings provide a 1) disturbed sample, with 2) difficulty in establishing exactly where in the formation the samples were obtained. Samples are more frequently obtained using specifically designed samplers. The resulting samples are classified as either disturbed or undisturbed depending upon how much alteration of the in situ state of stress and soil structure occurs. Disturbed samples are typically obtained through split barrel samplers during the standard penetration testing. Alternatively, undisturbed samplers have been developed to maintain soil structure.

Split-Barrel Samples

The SPT and the accompanying split-barrel sampler was described above. It is relevant to note that the split barrel sampler appears in a number of forms including those with one piece liners, segmented liners and differing dimensions. A larger split barrel, often known as a California sampler, may be driven with the 340 pound casing hammer. In all cases, split barrel samplers yield disturbed samples for use in physical or chemical property testing.

Thin Wall (Shelby Tube) Samplers

When samples of cohesive materials for test of strength, permeability or compressibility are required, it is necessary to use a thin wall sampler to obtain a "undisturbed" sample. Of course, the sample is never totally undisturbed. The very act of sampling changes the state of stress on the sample, alters the sample temperature and may impart some physical disturbance to the soil. There are a number of thin wall samplers in use including the Shelby tube, piston, Osterberg and Swedish foil samplers (see Fig. 15-21 for two examples).

The principle of these samplers is to:

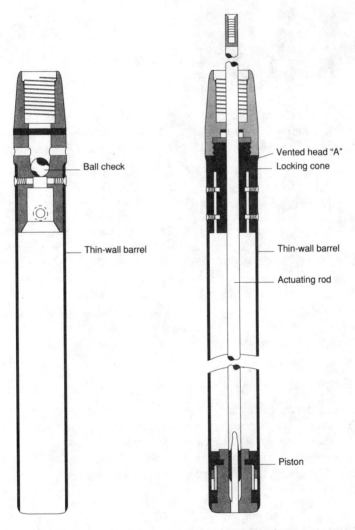

a.) Thin wall tube sampler   b.) Stationary piston sampler

FIGURE 15-21
Thin wall samplers.

1. Provide the least area of sampler as possible (thus the thin walls),
2. Provide a means to retain the sample in the tube (thus the check valves and pistons), and
3. Minimize the side wall friction as the sample enters the tube (thus the beveled cutting edge).

Rock Sampling Methods

Rock in the subsurface is normally sampled by coring. Rotary rock coring employs a **core barrel**, made up of a tube and cutting bit. As the core barrel is rotated, a cylindrical hole is formed and the core of rock is recovered from within the tube. The cutting tip is usually made of industrial grade diamonds. There are a number of core barrel designs including both single and double tube. Schematics of typical bits and core barrels are shown in Fig. 15-22. During coring, wash water is pumped down the hole to cool the cutting bit and to carry away the cuttings.

What good is a rock core for site characterization? First and foremost the core provides a sample for identification and laboratory testing. Much can be gleaned regarding the type of rock, the presence/absence of voids, bedding, fractures, and the overall rock quality. Quantitatively, the rock quality is measured by the **rock quality designator** (RQD). The RQD expressed as a percent is calculated as:

$$RQD = 100 \times L/R$$

where $L =$ cumulative length of all pieces over 100 mm (4 in)

$R =$ total recovery

Example 15-6 Rock quality designator. A 1.5 m core run was made from elevation 96 m to 94.5 m and 1.4 m was recovered. The recovered core was found to be broken into 8 pieces which were measured as follows:

Piece No.	Length (mm)
1	220
2	170
3	150
3	200
4	80
4	60
5	70
6	90
7	200
8	160

What is the RQD? What else can you say about rock quality?

Solution. The RQD is calculated from Eq. 15-3:
RQD $= 100 \times (220 + 170 + 150 + 200 + 200 + 160)/1400 = 78.6$
The zone of poorer rock quality occurs approximately between elevation 95.26 m (96 − (220 + 170 + 150 + 200)) and 94.86m (94.5 + 200 + 160). The precise depth range of the poorer quality rock cannot be known due to the loss (no recovery) of 0.1 m of rock. For the above estimate it is reasonable to assume that this 0.1 m core is missing from the zone of poorer quality.

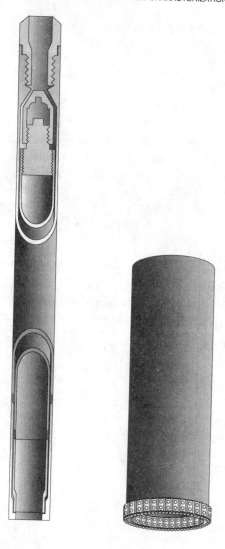

Double tube
core barrel

Bottom
discharge bit

FIGURE 15-22
Rock coring equipment.

15-6 MONITORING WELLS

The measurement of pressure head in the subsurface is accomplished via **piezometers**. The standpipe piezometer is a tube or pipe open at the top to the atmosphere and open at the bottom to limit portion of the formation (usually less than 2 feet). Piezometers can take the form of devices employing pneumatic systems, pressure transducers, and other electronic components. A **monitoring well** is used for both measurement of pressure (as a piezometer) and for sampling of ground water below the water table.

The piezometer (monitoring well) shown in Fig. 15-23 allows the measurement of average pressure head along the screened interval (and nowhere else). With the establishment of a datum, the elevation head at the screened interval can be determined and the total hydraulic head can be calculated. The total hydraulic head at a point (the location monitored by the well or piezometer) is found by first determining the height (with respect to the screened interval) of the water in the standpipe. This is the pressure head. The elevation head is the distance that the point of measurement (i.e., the screened interval) is above (or below) the datum. The total hydraulic head is then the sum of the pressure and elevation head.

Monitoring wells often serve an additional and equally important purpose beyond measuring the piezometric pressure. A properly installed monitoring well can be used to collect representative samples of the ground water for chemical analysis in the laboratory, or to measure **in situ** specific chemical properties of the ground water such as pH and conductivity. Monitoring wells can also be used in the measurement of hydraulic conductivity properties of the formation via single well permeability (slug) tests. Monitoring wells can also be used in the field measurement of hydraulic conductivity properties by using a number of monitoring wells and pumping wells in a pumping test.

Well Construction

Shown in Fig. 15-23 is a typical **monitoring well** installed in an unconfined aquifer. Each of the components labeled on the monitoring well schematic serve a specific

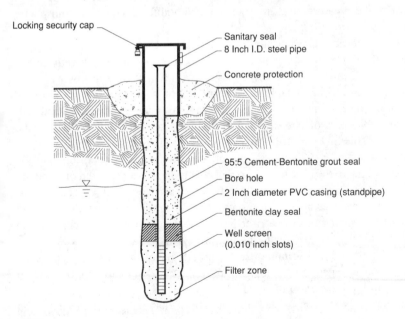

FIGURE 15-23
Typical monitoring well.

purpose in assuring that the monitoring well reflects the piezometric (potentiometric) pressure of the formation. The well is constructed of a standpipe (riser pipe), sealed along its length and connected to a short screen allowing formation water in and out of the riser pipe. The **screened interval** is that portion of the well which permits ground water to flow into or out of the monitoring well. The screened interval is normally encased in a **filter zone** often known as a **sand pack** or **gravel pack**. The purpose of the filter zone is to permit the free flow of water from the formation into and out of the screened interval without the movement of fines from the formation. It is necessary to provide a filter between the screen openings and the formation to reduce the potential that the fine fraction of the formation (the silty and clayey portion of the formation) does not get into the well. If fines were to enter the well, physical plugging could occur. Since the plugging is caused by materials with a relatively fine grain size, interactions between this fine material and the formation water could alter the well water chemistry. Chemical and biological activity within the well could also alter the water chemistry. The **casing** (riser pipe) connected to the screened interval is used to isolate the monitoring system upwards through the remainder of the formation to the surface.

Since the inside of the monitoring well is opened to the atmosphere, the pressure head in the formation forces the water level in the casing to rise to an equilibrium level. That is, the height of the column of water in the well is equal to the pressure head in the formation at the screened interval. This is the essence of a piezometer. The water level in a piezometer completed in an unconfined aquifer defines one point on the **water table**. The water table is an underground surface (three-dimensional locus of points) where the pressure head in an unconfined aquifer is equal to atmospheric pressure.

To preclude vertical leakage along the annular space between the riser pipe and the side wall of the boring, it is necessary to seal this space. Directly above the screened interval, this sealing is accomplished with a clay zone of low hydraulic conductivity. This clay zone is typically constructed of **bentonite**, a high swelling clay with **sodium montmorillonite** as its primary clay mineral. The clay seal will typically have a thickness of one to two feet and is located immediately above the screened interval. This low-permeability, high-swelling clay affords a tight barrier to the transport of water or contaminants from other portions of the bore hole into the screened interval. Above the clay seal is a **grout** seal typically constructed of a cement and bentonite mixture. This grout provides a further barrier preventing surface water or ground water from elsewhere in the formation from migrating into the screened interval. Finally a vented **cap** on the riser pipe is normally provided to prevent rain water or other surface water from entering into the well.

To provide security and to minimize the possibility of damage, a steel casing, embedded in a concrete pad, may also be employed. A locking cap may be used on the steel casing to limit access to the monitoring well.

Monitoring wells are constructed of a variety of materials including polyvinyl chloride (PVC), stainless steel, black steel and Teflon.[R] The selection of the material type for monitoring wells should consider well depth, types of contaminants expected, design life of the well, and geologic, biologic and geochemical nature of the formation. For a well whose primary function is the determination of piezometric conditions, polyvinyl chloride (PVC) is typically the material of choice. In a subsurface environment contaminated with organic contaminants, the organic nature of the PVC may confound the analytical chemistry test results. Further, the chemical resistance of the PVC to chlorinated solvents may raise questions regarding the casing durability. In these cases Teflon or stainless steel is often the material of choice. Black steel is typically not utilized in hazardous waste management monitoring well installations because of its relatively low corrosion resistance. In terms of costs, stainless steel and Teflon impart considerably greater materials costs to the project than PVC. When drilling, long term sampling and monitoring, and analytical chemistry costs are included in the economic considerations, the cost differential between wells of Teflon, stainless steel and PVC may become inconsequential. It is recommended that the material selection be based upon considerations of the geologic nature of the formation, the purposes of the monitoring well, and the nature of the expected ground water chemistry.

Aquifer Monitoring

At times the water table may be localized and discontinuous (related to geologic heterogeneity) as shown on Fig. 15-24. This is termed a **perched water table**. It is important to ascertain whether a given water level measurement in a monitoring well represents the water table in an unconfined aquifer, a perched water table or the hydraulic head in a confined aquifer. Measurements of hydraulic head in a confined aquifer can be contoured to develop a three dimensional locus of points termed the **piezometric** or **potentiometric surface**. The actual water surface in a confined aquifer must be at the top of the aquifer, that is, the bottom of the upper confining layer. In a confined aquifer, the water level in a piezometer open to the atmosphere will rise to a level which indicates the pressure in the formation at the screened interval.

As described, the primary purpose of a piezometer is to measure hydraulic head in the formation. A piezometer nest is a group of piezometers which act to measure the head at various depths at the same plan location. For example, the nest of piezometers shown in Fig. 15-25 are, for all practical purposes, located at the same location in plan view in an unconfined aquifer.

Note that 3 different water level elevations are observed in the piezometers shown on Fig. 15-25. The screened interval for each piezometer is at a different depth (and perhaps, but not necessarily, in a different geologic formation). These observations indicate that flow in this particular region is upward. The piezometer with the highest water level elevation is at a depth below the other two piezometers in the formation. There is a gradient or a change in head upward through the formation. In fact, the vertical gradient can be calculated as shown in Fig. 15-25.

Shown in Fig. 15-26 are 3 piezometers located at different locations in a confined aquifer. Assuming that the flow in the confined aquifer is only horizontal, flow as indicated by the varying water levels in the piezometers, is in a direction from left to

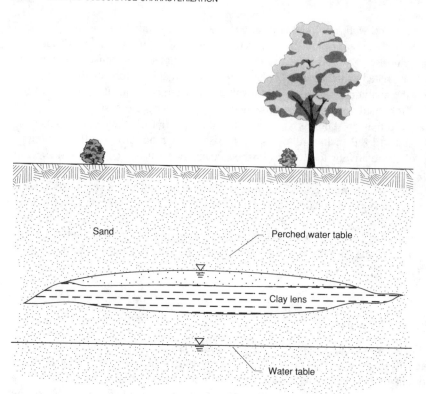

FIGURE 15-24
Perched water table.

right. As above, the hydraulic gradient and specific discharge can be calculated from the piezometer spacing and the water level elevations in the piezometers.

As in the example above, it is often necessary to measure the piezometric pressure or to collect water samples at various depths within the formation at the same coordinate location. This is done with multiple wells or nested wells as shown in Fig. 15-27. As shown, this is accomplished employing individual wells installed in separate but adjacent boreholes drilled to different depths in the formation. These are each physically close-by with a spacing on the order of about 10 feet on center. Alternatively, a multiple well piezometer installation can be constructed in a single larger diameter borehole. Multi-level wells in a single bore hole may offer advantages of lower cost. Because of the potential for cross contamination as a result of an imperfect seal between screened intervals, the well nest with multiple bore holes is preferred over the single bore hole installation.

As with unconsolidated formations described above, it is often necessary to monitor ground water levels and obtain ground water samples from rock formations as shown in Fig. 15-28. In the rock wells where the rock formation has adequate strength to maintain bore hole stability over the sampling interval, casing may not be

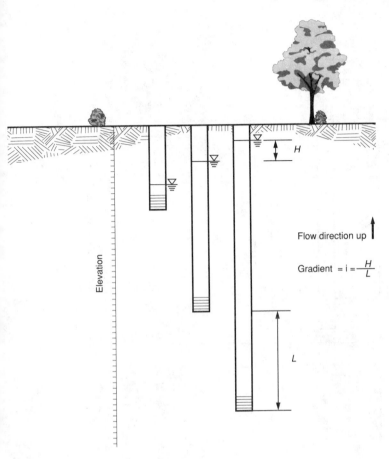

Flow direction up ↑

Gradient $= i = \dfrac{H}{L}$

FIGURE 15-25
Piezometer nest in unconfined aquifer.

needed. (Of course if regulations require casing independent of formation conditions, then one could say that casing is required.) Thus this rock formation may be monitored without the use of well screens. For stable rock formations, the bore hole is advanced from the surface to the desired sampling interval (analogous to the screened interval in unconsolidated formations). At this point a steel casing is installed in the bore hole and the annulus between the bore hole and the steel casing is sealed with grout to the surface. The casing serves as a riser pipe and lines the borehole from above the ground surface down to the top of the sampling interval. With the annulus sealed, the completed installation serves as the monitoring well. The stable, uncased hole in the sampling interval allows ground water to flow to the monitoring well. The installation allows ground water levels to be monitored and ground water to be withdrawn, sampled and analyzed for ground water quality. Uncased sampling intervals are commonly employed in rock, although the standard designs employing well screens are also used for rock formations.

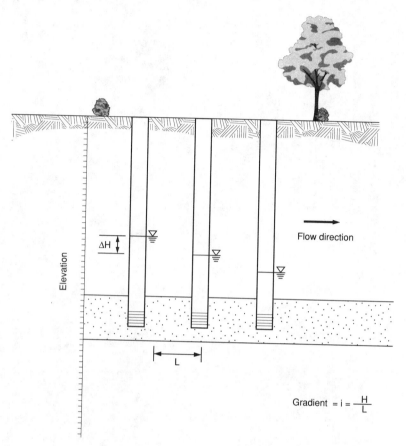

FIGURE 15-26
Piezometers in confined aquifer.

Well Installation on Contaminated Sites

It is not unusual in studies of hazardous waste management sites to encounter a stratigraphy which includes an unconfined aquifer separated from an underlying confined aquifer by a confining layer as shown in Fig. 15-29. In these cases it is often necessary to determine the ground water conditions, both piezometric and ground water quality, in both the overlying unconfined aquifer and the underlying confined aquifer. To accomplish these objectives requires drilling the bore hole through the potentially contaminated zone of the upper unconfined aquifer and into the confined aquifer. It is essential that precautions be taken to minimize the possibility of contaminating the underlying confined aquifer as a result of the investigation practices. Where possible, monitoring wells should be located beyond the limits of contamination. Many times this is not possible however. Consider the following question. How deep has the contaminant plume migrated beneath the site? Can this question be answered by drilling only beyond the limits of the known contamination?

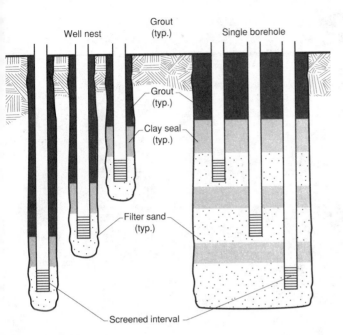

FIGURE 15-27
Multiple and nested wells.

Installation of a monitoring well into the underlying confined aquifer without introducing contamination from above requires special steps as shown in Fig. 15-30. This techinique is termed the **telescoping method of well installation**. In drilling of the bore hole for this monitoring well installation, it is recommended that the bore hole be advanced to the top of and slightly into the aquitard separating the unconfined and confined aquifers. A temporary steel casing is then sealed in place by grouting. The inside of the casing is then drilled, flushed, cleaned and washed to remove contamination within the temporary steel casing. After cleaning, drilling proceeds inside the temporary steel casing and into the underlying confined aquifer. This eliminates the potential for the drilling tools and fluids to carry contaminants from the overlying unconfined aquifer to the underlying confined aquifer. The monitoring well installation is then completed as previously described. Installing the well in this manner insures that the piezometric conditions and the ground water quality in the installed well are representative of the underlying confined aquifer. This well installation process minimizes any potential for either 1) bringing contaminants downward as a result from the upper zone to the lower zone during the process of drilling and, 2) allowing contaminants to migrate downward after the monitoring well is in place.

In the conduct of investigations at hazardous waste management sites it is crucial that contaminants not be transported from one location to another as a result of the drilling operations. It has been shown above that special precautions must be taken when advancing a bore hole through a contaminated zone into an underlying

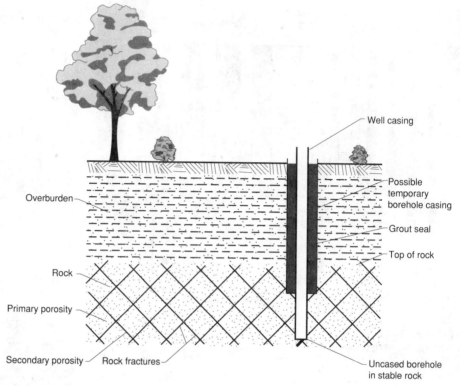

FIGURE 15-28
Monitoring rock formations.

and potentially uncontaminated zone. Likewise, special precautions must be taken to decontaminate equipment as it is moved from one location to another on the site. Decontamination is accomplished by washing with detergent and steam cleaning the drilling tools at each point in the drilling operation where there is concern for transport of chemical contaminants by the drilling tools. At a minimum, this would be done between boring locations as the drill equipment moves from one location to another on the site. Cleaning of augers, wrenches, cutting bits, sampling tools, and drill rods is necessary to prevent contaminant transport by way of the drilling equipment. In addition, the use of lubricating oils by the drilling contractor must be carefully controlled (minimized or eliminated). As a result of analyzing ground water quality parameters to parts per billion concentrations, a few drops of lubricating oil entering the formation at the time of drilling may lead the investigators to erroneous conclusions regarding the concentration/presence of organic contaminants. For additional information, an excellent treatise on ground water monitoring is provided by Driscoll.[23]

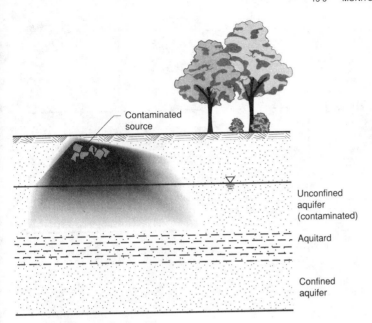

FIGURE 15-29
Contaminated uppermost aquifer.

Test Pits

An excellent but sometimes overlooked technique for investigating the subsurface condition is through the use of test pits. Test pits are typically advanced with a small backhoe and they provide opportunity for continuous visual inspection and to obtain large bulk samples. They are however, limited in depth due to the safety of the personnel entering the trench and the excavating depth of backhoe excavators. The implementation of robotics for test pit surveys can alleviate certain of these concerns.[24]

Soil Gas Surveys

The nature and extent of gases in the unsaturated (vadose) zone is often overlooked in efforts to obtain quality ground water samples. However, on many sites the contaminants in the unsaturated zone may be the focus of the remediation. Among the newer developments for the investigation of the distribution and transport of contaminants in the subsurface are soil gas surveys.[25,26] Soil gas surveys are conducted to provide the quantitative information needed to assess the risks associated with the unsaturated zone. Soil gas surveys have also been used as a low cost alternative to ground water sampling and testing to assess the nature of the ground water plume.

Soil gas surveys establish a vapor phase plume in the void space of the soil in the vadose zone. This information can then be interpreted in the context of source detection and plume delineation.[27] But how does it work? Soil gas surveys rely upon the contaminant property of vapor pressure (see Sec. 3-2) for success. A contaminant

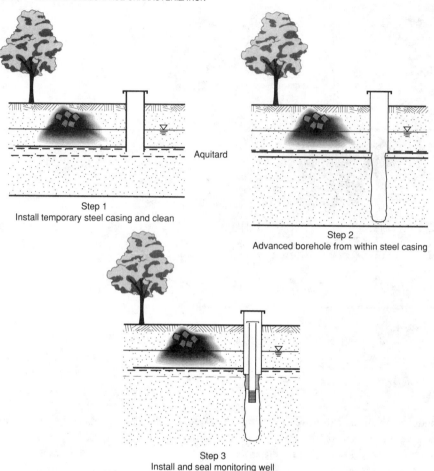

Step 1
Install temporary steel casing and clean

Aquitard

Step 2
Advanced borehole from within steel casing

Step 3
Install and seal monitoring well

FIGURE 15-30
Drilling through contaminated aquifer.

with a high vapor pressure and a low water solubility will volatilize from the ground water into the vadose zone as the plume migrates as shown in Fig. 15-31. Detailed study has shown that the soil-gas sample results correlate well with ground water sample results if the soil-gas sample is obtained within 5 feet of the water table for clayey soils and 10 feet of the water table for sandy soils.[28] The reliability of the technique is enhanced if the contaminant does not degrade rapidly into degradation products.

How are the contaminant concentrations measured in the soil voids? Although the details vary, soil gas surveys require the intrusion into the soil to either directly withdraw the soil gas, withdraw a soil sample from which the gas is obtained or leave a sampler in the soil from which gas is obtained. A sampler left in place[29] would seem appropriate where subsequent soil gas data are required to measure the time history

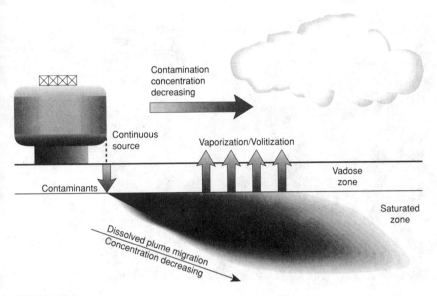

FIGURE 15-31
Principle of soil gas surveys.

of soil gas contaminant concentrations. After sampling, the gas is analyzed employing an organic vapor analyzer, a photoionization detector or gas chromatography. The analytical end of the soil gas survey may be done with field instruments, in a field laboratory or in a remote analytical laboratory. Shown on Figs. 15-32 and 15-33 is a field laboratory using a GC/MS for the analysis of soil gas. This Field Analytical Services Technology (FAST®) allows the site assessment to be adjusted as it progresses based upon the data generated.

Vadose Zone Monitoring

Monitoring in the vadose zone may require information regarding soil moisture potential, soil moisture content, soil salinity, temperature and soil pore water sampling. As described in Sec. 4-2, the pressure head is negative in the vadose zone and is dependent upon the soil moisture content. Thus the measurement device must be designed and installed so as not to alter these properties by the very act of measuring them. A thorough discussion of these topics constitute a separate textbook; the interested reader is referred to an excellent reference by Morrison.[30]

Ground Water Sampling

Obtaining a representative ground water sample can be a difficult task at best. However, securing representative and reliable ground water samples is perhaps the most critical element of a subsurface site characterization as ground water samples reveal what chemical constituents are present and at what concentrations and define the lat-

FIGURE 15-32
Mobile laboratory used for soil analysis

FIGURE 15-33
Interior of mobile laboratory

eral and vertical extent of contamination. In an on-going monitoring program, ground water samples reveal whether additional contamination is occurring, whether existing subsurface contamination is migrating or whether it is being contained or abated. Further, chemical concentrations obtained from ground water samples permit the performance of a risk assessment to determine the magnitude of any adverse impact on human health and the environment. Finally, chemical concentrations obtained from ground water samples are needed to guide the evaluation and selection of appropriate remedial actions. The entire RI/FS process depends upon representative and reliable ground water samples.

The decade of the eighties saw explosive growth in the field of ground water sampling and monitoring.[31] Extensive and detailed protocols were established[32] to ensure, as much as is possible, that proper sampling methods are employed. New and revised methodologies were developed to meet the increasing demand for measurement of contaminant concentrations in ground water at lower and lower concentrations. One reason for this increasing sophistication is the seemingly ubiquitous presence of organic solvents in ground water that are of concern at concentrations measured in parts per billion (ppb).

The most common method of ground water sampling is from monitoring wells described earlier in this chapter. Numerous factors, both engineered and natural, will affect the reliability and representativeness of the ground water sample. A number, but certainly not all, of the factors are the following:

1. Natural ground water chemistry and geochemical interactions which occur upon sample removal from the in situ temperature and stress conditions
2. Well installation methods
3. Extent of purging of monitoring well prior to sampling
4. Type of sampling device used
5. Decontamination of sampling devices between samples versus use of dedicated sampling devices for each location
6. Methods of sample preservation
7. Methods of sample shipping and storage
8. Duration of sample storage prior to testing

Presentation and Interpretation of Data and Results

The findings and conclusions developed as a result of the site and subsurface investigations must be clearly, concisely and accurately reported. Without such reporting, the value of investigations is considerably reduced. Those interested in the site and subsurface characterizations include owners, regulators, citizens and other scientists and engineers. The preparation of reports is an on-going effort and should not be left in its entirety to the end of the project. Report reviews include several types, each with different objectives. Editorial reviews are necessary to ensure the readability and grammatical accuracy of the product. Technical reviews are necessary to ensure the technical adequacy of the report. Details regarding methods of data preparation are incorporated into the case history at the end of this chapter.

15-7 GEOGRAPHIC INFORMATION SYSTEM[†]

GIS Overview

During the past several years extensive developments in the field of computer graphics have produced commercially available software systems that are applicable to the problems associated with characterizing hazardous waste sites. The use of digitized drawings for Computer-Aided Drafting and Design (CADD) has become common in the field of construction and civil engineering design. Likewise, automated mapping (AM) using digitized representations of map features has largely replaced conventional cartography, and facility/infrastructure management (FM) systems that associate facility drawings to text and numerical attributes have gained increasing application in both government and industrial organizations.

Each of these applications employs digitized drawings linked in varying degrees to data sets. The term **Geographic Information System (GIS)** is defined by Antenucci, et al. as

a computer system that stores and links non-graphic attributes or geographically referenced data with graphic map features to allow a wide range of information processing and display operations, as well as map production, analysis, and modeling.[33]

A thorough discussion of GIS and its applications is provided in this reference. The linking of data to maps or drawings provides a powerful tool for increased visibility and insight into the areal and subsurface extent of contamination in soils and ground water. GIS for site modeling is one particular application of the technology.

An industrial facility or any hazardous waste site where characterization and remediation are being conducted can be described by cartographic or map layers called planimetrics (e.g., roads, buildings, fences, property lines) and by data sets (e.g., sampling locations, drilling logs, groundwater elevations, topography samples, chemical concentrations). The cartographic layers in some systems may be composed of individual objects (e.g., buildings, sampling locations, rooms, waste sites), which have logical identity and spatial position. In other systems, it is more difficult or impossible to isolate the objects within the context of the layer or drawing phase in which they are stored.

[†]Section 15-7 was prepared by William Douglas, Ph.D., ERM, Inc.

Drawings or maps having objects linked to data are sometimes called *intelligent* because the objects are associated with information that is not necessarily graphical. The contamination level (mg/L) at a specific sampling location for a hazardous waste material, or for several chemicals, are examples of information that can be geographically referenced and related to spatial objects. Using GIS functions, the data sets can be accessed graphically to display the information associated with an object.

The technical data that are used to characterize the site (chemical, geological, geotechnical, geophysical, etc.) can be stored in a database that is accessible through conventional database management system functions. The linking of a database management system functionality with digitized cartography is another way of describing a GIS. Hence, a database of ground water elevations, chemical concentrations in soil and water, and drilling logs can be related to the sampling or drilling locations (objects) in the digitized drawing for a hazardous waste site. The database management system can be used to extract the data from the database and the automated mapping function can be used to display the data elements on a site drawing near the objects to which they relate. Data displays of this type (data posting) are commonly used in site investigations. GIS provides an efficient and flexible means for creating information products of this type. Computer-aided drafting functions provide a means for editing and modifying the display to make it more attractive or easier to understand.

Geographic data modeling is another element that has applicability to site investigation. Contamination plumes are simulated by modeling chemical concentrations in soil and ground water based on analytical laboratory results from actual samples. This approach is routinely used by scientists and engineers. A variety of contouring models are available that use different algorithms to estimate values at regular grid intervals from samples taken at irregular locations. These models develop lines of constant value (isopleths) from the gridded data through interpolation techniques. Linked to a GIS, they provide a means for defining a plume extent from database contents rapidly, and for displaying the isopleths as objects or as a layer to be used in conjunction with a site drawing. Topography and ground water elevation contours can likewise be developed and stored for display along with other layers or objects. The set of coordinates and values (x, y, elevation, or x, y, contamination) are often called *surfaces*. The surfaces may be physical (topography) or logical (contamination). Data modeling can also be used to produce a cross-section through a surface to display topography and subsurface layers or objects from the GIS. The integration of data modeling with the database management system and CADD functions in a GIS provides the site investigator with an effective means for analyzing, understanding, and conveying to others the extent and severity of site conditions, and for establishing a basis for defining remediation alternatives.

Applications

Two examples of applying GIS to site investigation are illustrated below. The first application relates to a site near the Washington D.C. National Airport contaminated with PCBs and heavy metals. A database management system named ORACLE®[35] was used to store the laboratory analytical results, which were transferred from EXCEL® spreadsheets into a technical database used for site investigations. A base map illustrating the property extent, fence line, nearby roads, buildings, north arrow, and sampling locations was digitized in AutoCAD.® The files were transferred to the Graphic Data System (GDS)®,[36] which is an integrated GIS/CADD software system. The Site Modeler module of GDS® uses object-based intelligent drawings to describe and characterize hazardous waste sites.

The objective of the investigation was to illustrate the contamination plume using depth-dependent sample results and varying action levels for each chemical parameter, the intersection of the groundwater table with the plume, and the relationship of the plume to the property fence line. Figure 15-34 illustrates the base map for the property, and Fig. 15-35 and 15-36 represent the contamination volume for two of eight scenarios that were investigated. The numerical values of the contaminated volume and the volume of excavation are also computed using GDS.® Hard copy drawings of the results as well as slide photographs of the screen images are used to support the presentations to the decision-making group.

A second application of GIS relates to the Washington Metro rail line where construction design alternatives were affected by a municipal incinerator ash pile situated in the right-of-way. Aerial surveys were conducted in 1976, before ash dumping was initiated, and in 1990. These surveys provided the planimetric features and topography around the site. The site investigation database included surface samples, soil borings, drilling logs, and soil and water sample results. A variety of information products were developed to support decisions relating to the hazardous classification of the ash pile, volume computations, cross-sections, and evaluations of alternative alignments through and around the waste pile. Figure 15-37 illustrates a site map with three alignments and a set of transverse cross-sections through the right-of way showing the ash fill above the natural topography. Figure 15-38 is a 3-dimensional view of the site topography, ash, and the three alignments. The darker shading in Fig. 15-38 represents areas covered by ash whereas the lighter shading represents the original topography. This information was used to assess the alternatives from a regulatory standpoint, and to evaluate tradeoffs between construction costs and remediation costs for the alternatives.

15-8 CASE HISTORY

An existing operating solid waste landfill site is located in the southeast of the U.S. The owner wished to consider expansion of the site. Site and subsurface investigations were undertaken to provide the needed technical information to adequately evaluate the suitability of expansion and to define design and operational constraints. As discussed earlier in this chapter, the effort first identified available preexisting information. For example, an investigation of the geological information showed the site to be located within the western limit of the coastal plain physiographic province. As a result the site could be expected to be underlain by Tertiary-Quaternary age sand and gravel deposits with an underlying silty clay deposit of Miocene age. The geologic studies

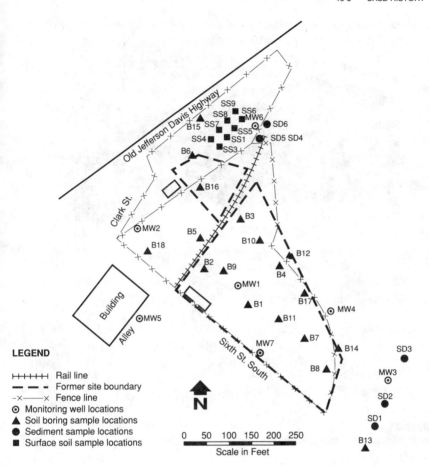

FIGURE 15-34
Site map.

further indicated that the silty clay is composed of quartz and montmorillonite. An examination of the site history showed that the site was formerly used to mine sand and gravel down to the Miocene clay. A lined landfill was then built on the northern portion and the owner was interested in expanding this landfill to the southern portion of the site. An industrial facility with the potential for ground water contamination was located to the east of the site.

Using the available subsurface and geologic information, subsurface investigations were conducted which included 25 test borings at 18 locations with monitoring wells installed in 24 test borings at 17 locations. In order to minimize formation disturbance, rotary drilling techniques were used employing Attapulgite clay drilling fluid to maintain borehole stability. Where the borings/monitoring wells penetrated the Miocene clay, special precautions were used as discussed earlier in the chapter so

Soils where contamination exceeds any of the values:
Lead 500 ppm, PCB 250 ppm, or TPH 1000 ppm

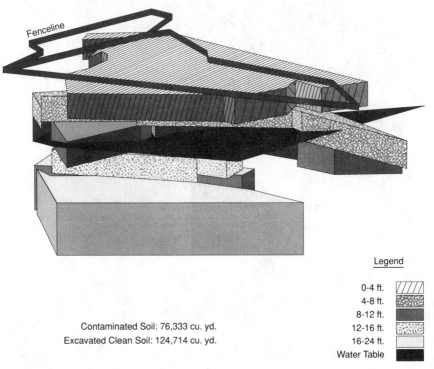

Fenceline

Contaminated Soil: 76,333 cu. yd.
Excavated Clean Soil: 124,714 cu. yd.

Legend

0-4 ft.
4-8 ft.
8-12 ft.
12-16 ft.
16-24 ft.
Water Table

FIGURE 15-35
GIS representation of contaminated soil (Scenario II)

as to not carry any potential contaminants down from the overlying sand and gravel to the materials beneath the clay. In situ testing and sampling of all unconsolidated sediment formations were by means of the standard penetration test. Thin-walled sampling of the Miocene clay was also carried out. Laboratory testing included physical property and classification tests of natural water content, liquid limit, plastic limit, unit weight, specific gravity, and grain size distribution by sieve and hydrometer (see Table 15-4 for the test methods). Engineering property tests included characterization of the undrained shear strength and the triaxial hydraulic conductivity of the Miocene clay.

How do we present all these data in a form that is understandable, usable and most importantly, revealing of the key site and subsurface features? For this project a number of presentation methods were used as presented in the following paragraphs. Normally, and for this project, a regional location plan and a boring and monitoring well site location plan are prepared. Each test boring is logged in the field and based upon the field log and the subsequent laboratory test data, a report form

Soils where contamination exceeds any of the values:
Lead 220 ppm, PCB 25 ppm, or TPH 100 ppm

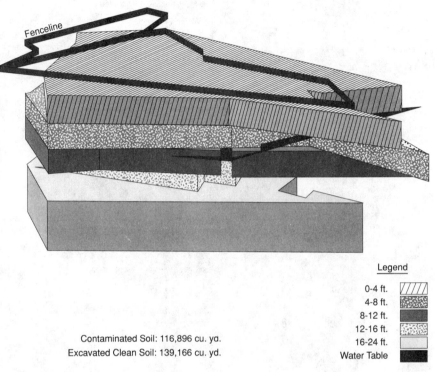

Legend

0-4 ft.	
4-8 ft.	
8-12 ft.	
12-16 ft.	
16-24 ft.	
Water Table	

Contaminated Soil: 116,896 cu. ya.
Excavated Clean Soil: 139,166 cu. yd.

FIGURE 15-36
GIS representation of contaminated soil (Scenario V)

boring log is prepared as shown in Fig. 15-39. The log of this particular test boring shows about 15 feet of fine grained alluvium overlying about 23 feet of dense sand and gravel. The underlying Miocene clay was encountered at a depth of about 37 feet at this location. The boring log also presents the results of SPT testing and the natural water content data obtained in the laboratory. The log also notes that grain size data were obtained on two samples and is presented elsewhere in the complete report.

Although much is revealed in a boring log, the information relates to only one point in plan at the site. A subsurface geological profile can be constructed by combining the information from several borings along a line as shown in Fig. 15-40. This section reveals the undisturbed natural section as well as the influence of mining operations upon the subsurface conditions. Additional insight into the subsurface conditions results from constructing and studying a subsurface profile behind the study of individual boring logs.

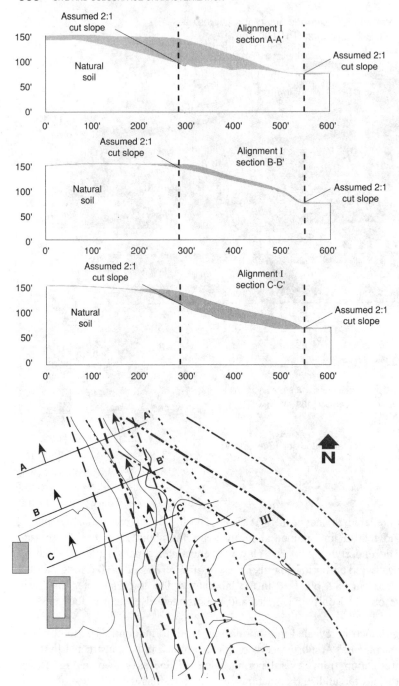

FIGURE 15-37
Contour map

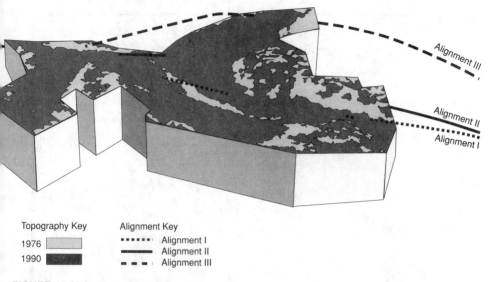

Topography Key

1976 ☐
1990 ■

Alignment Key

······ Alignment I
——— Alignment II
— — — Alignment III

FIGURE 15-38
Three-dimensional representation of site topography

As informative as subsurface profiles are in helping one visualize site and sub-surface conditions, we can do more. Structural contour maps can be prepared to map individual geologic units. In this case study a structural contour of the top of the Miocene clay is particularly revealing. Shown in Fig. 15-41 is a map prepared by taking the elevation of the interface between the sand and gravel and the Miocene from individual test borings and contouring these data. What can we see from this structural contour? Located at the southern end of the site is a buried valley in the Miocene clay, filled with sand and gravel and sloping towards the east northeast. We will look in a moment at how this information is used in the interpretation of ground water contours. The structural contours also reveal a northeast trending ridge in the clay.

Employing data from test boring regarding the elevation of both the top and bottom of any geologic unit, **isopac** maps can be prepared. Isopac maps present lines representing equal thickness of a formation material. Shown on Fig. 15-42 is an isopac map of the Miocene clay formation. This map is quite revealing with respect to the site characterization. We find that the clay is over 40 feet thick everywhere beneath the site and is as much as 50 feet in some locations. This thickness information coupled with the geologic information regarding the clay mineralogy (montmoril-lonite) and the hydraulic conductivity (varying from 3×10^{-7} to 2×10^{-8}, aver-aging 6×10^{-8} cm/sec) indicate the suitability of this layer as a natural confining layer.

Our case study must concern itself with ground water information as well as geologic information. To this end ground water monitoring wells were installed and the well installation details included on a monitoring well report (Fig. 15-43). Of

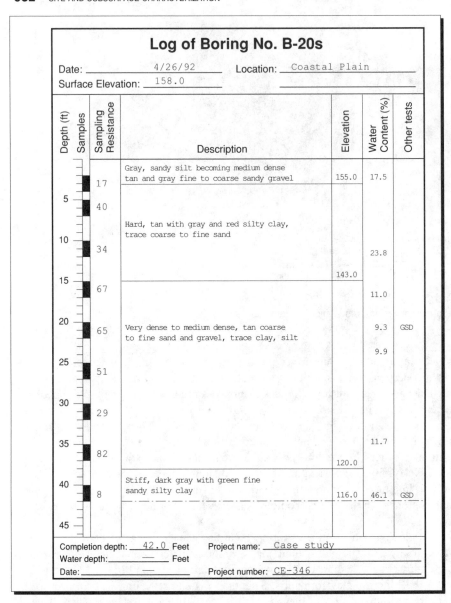

FIGURE 15-39
Typical boring log.

utmost importance is the clear definition of the **screened interval,** the zone from which water samples are obtained and piezometric levels are measured. A comparison of the elevations of the screened interval shown in Fig. 15-43 with the information shown in Fig. 15-39 reveals that the well is sampling the bottom 10 feet of the sand and gravel layer overlying the Miocene clay.

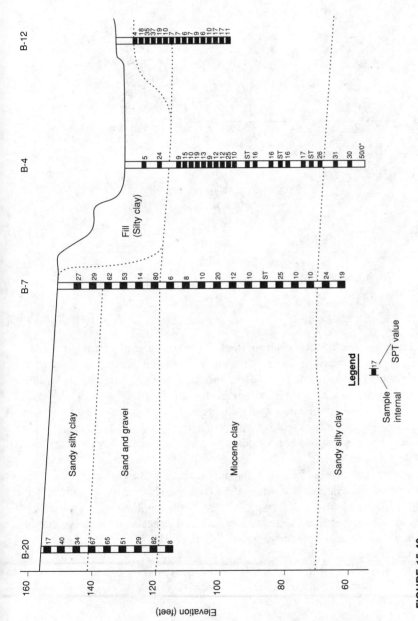

FIGURE 15-40
Geologic profile.

953

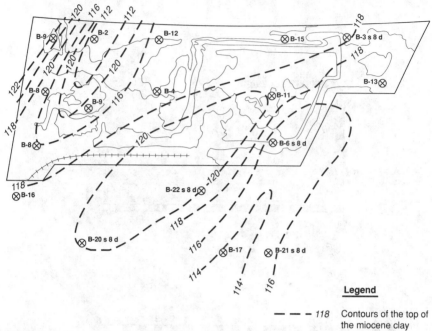

FIGURE 15-41
Structural contour.

Ground water contours for each aquifer can be constructed by taking water level readings from each of the wells using the methods described above. The water levels should be recorded "at the same time." Does this mean we need to be at all locations at once or use water level recorders to obtain this information? Sometimes yes and sometimes no. If water levels are rapidly changing such as during a pump test, we may need to go to such efforts. However, if water levels are steady state and do not change rapidly (over the course of hours) it is usually sufficient to record all the levels the same day. Shown in Fig. 15-44 are the ground water contours for the sand and gravel aquifer for our case study. These contours reveal that the buried valley exhibits control over the subsurface flow regime. Note the ground water flow is towards the buried valley from all portions of the site. Next note that the flow is from the adjacent industrial site. Thus, if ground water is contaminated beneath the adjacent site, the contaminated water would flow through our case study site.

Within the context of this case study, it is impossible to present all the data and interpretations available for this project. However, the case discussion above demonstrates the need to employ multiple means of data presentation to best understand the site and subsurface conditions. The case study also demonstrates that the selection of the means of site characterization must be appropriate to the information need for the project. So as to not leave the reader wondering what happened, it is noted that the case study site was eventually developed as a solid waste land disposal facility with geomembrane liner systems. In addition, a soil-bentonite slurry trench (chap. 16) was

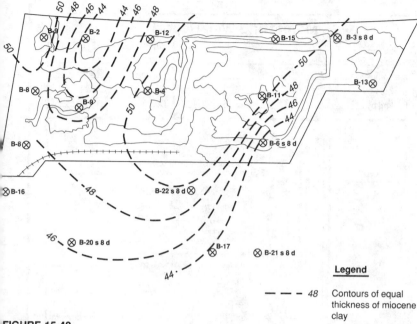

FIGURE 15-42
Isopac map.

constructed 1) to control the influx of ground water into the site to allow construction of a liner system (chap. 13), 2) to eliminate/minimize the influx of potentially contaminated water from the adjacent industrial site, and 3) to provide a secondary containment for added confidence regarding the long-term security of the leachate containment system.

DISCUSSION TOPICS AND PROBLEMS

15-1. Describe what is meant by necessary and sufficient.

15-2. An investigation of subsurface conditions at a 200 m by 200 m site employs an average of one test boring for each 250 m^2 of site (but not on uniform 50 m centers). Split-barrel samples are taken each 1.5 m from the ground surface down to the depth of investigation of 15 m. By volume, what percent of the subsurface materials is available for examination and testing. Why not use a uniform grid of 50 m on-center?

15-3. Visit the geology department at your school or your local soil conservation service office and study the air photos covering the nearest waste disposal facility. How do the existing drainage conditions influence the possible pathways for contaminant migration? Identify sites/areas you would target for purposive sampling and testing in the event of a spill at the facility.

15-4. In reference to ground water, define the term plume.

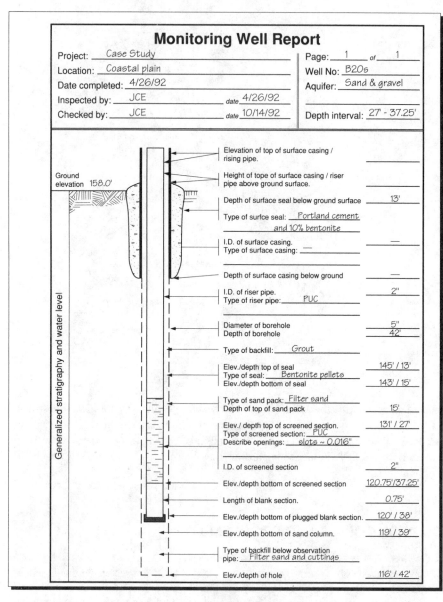

Monitoring Well Report

Project: _Case Study_   Page: _1_ of _1_

Location: _Coastal plain_   Well No: _B20s_

Date completed: _4/26/92_   Aquifer: _Sand & gravel_

Inspected by: _JCE_   date _4/26/92_

Checked by: _JCE_   date _10/14/92_   Depth interval: _27 - 37.25'_

Ground elevation 158.0'

Generalized stratigraphy and water level

Elevation of top of surface casing / rising pipe.	
Height of tope of surface casing / riser pipe above ground surface.	
Depth of surface seal below ground surface	13'
Type of surfce seal: _Portland cement and 10% bentonite_	
I.D. of surface casing.	—
Type of surface casing: _—_	
Depth of surface casing below ground	—
I.D. of riser pipe.	2"
Type of riser pipe: _PUC_	
Diameter of borehole	5"
Depth of borehole	42'
Type of backfill: _Grout_	
Elev./depth top of seal	145' / 13'
Type of seal: _Bentonite pellets_	
Elev./depth bottom of seal	143' / 15'
Type of sand pack: _Filter sand_	
Depth of top of sand pack	15'
Elev./ depth top of screened section.	131' / 27'
Type of screened section: _PUC_	
Describe openings: _slots ~ 0.016"_	
I.D. of screened section	2"
Elev./depth bottom of screened section	120.75'/37.25'
Length of blank section.	0.75'
Elev./depth bottom of plugged blank section.	120' / 38'
Elev./depth bottom of sand column.	119' / 39'
Type of backfill below observation pipe: _Filter sand and cuttings_	
Elev./depth of hole	116' / 42'

FIGURE 15-43
Monitoring well report.

15-5. Explain why direct and indirect methods should be used together and how this might be done for a specific site.

15-6. Seismic profiling is being considered for a site where dense material overlies soft material. Is this feasible? Why or why not?

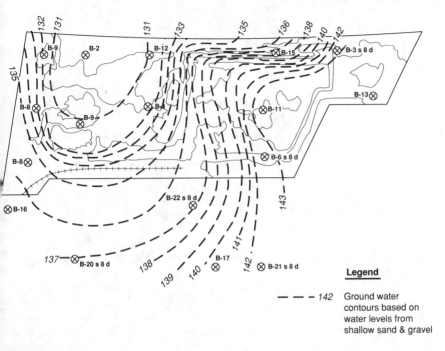

Legend

— — — *142* Ground water
contours based on
water levels from
shallow sand & gravel

FIGURE 15-44
Ground water contours.

15-7. Choose a geophysical technique and describe how the technique can be used to measure different subsurface properties in different situations.

15-8. For the following methods of subsurface characterization, state the measured parameter and the deduced parameter/information:

(*a*) Resistivity
(*b*) Seismic
(*c*) Magnetometry
(*d*) Electromagnetics

15-9. List the elements of a site and subsurface investigation plan.

15-10. Compare and contrast indirect and direct investigation methods.

15-11. Choose a direct investigation technique and describe how the technique can be used to measure different subsurface properties in different situations.

15-12. Describe three direct site investigation methods and the information each might provide pertaining to the site conditions.

15-13. An abandoned railroad car maintenance and refinishing facility is suspected of affecting the water supply of adjacent residential areas, but the specific impacts are uncertain. The facility operated from 1928 to 1983, and petroleum solvents and varied paints were both stored and directly used at the site. The location of those activities is poorly established. The site is situated on an unconsolidated sand unit containing thin clay laminae; the unit is up to 50 feet thick, and overlies a sequence of interbedded sandstone and shale, these rocks being penetrated by fracture systems dipping southeast. The residential areas are

northwest of the site. What steps would you recommend to:
(a) Locate buried metal (and other) containers on the site,
(b) Determine the location of the most heavily contaminated sands, if any, and the contaminants with them, and
(c) Establish the presence and (if within this unit) flow direction of the contaminants. Are they confined to the overlying unconsolidated sands, or flowing within the fracture system of the bedrock?

15-14. Why can't seismic profiling be conducted where hard or dense soil conditions overlie loose or soft soil conditions?

15-15. How does in situ testing differ from laboratory testing? What are the advantages and disadvantages of each approach?

15-16. An existing pesticide manufacturing plant is suspected of affecting the waters of a nearby stream. The site is set upon a sequence of sandstone and arkose, with these units lying horizontally and being penetrated by a weakly developed system of perpendicular vertical fractures. What investigative methods should be undertaken, and what should the Remedial Investigation specifically aim at to determine the nature and extent of contaminants in the area?

15-17. You've unexpectedly been transported to a site (as if by helicopter) which has a long history of industrial activity, and is thought to harbor a variety of wastes in a greater variety of containers. In other words, a little is known, and a quite a bit more is unknown, about the place. You're constrained by time and budgetary factors, and must limit yourself to two techniques to delineate the site conditions. What would you suggest be done? Why? Give pros and cons for each method you suggest. Based on whatever methods you have chosen, what would you not be able to tell about the site conditions? Which other sources of information, not field activities, would expand your knowledge of the site?

15-18. Identify and define the two Latin root words used to form isopac.

15-19. It has come to the attention of the Dean that the environmental science laboratory previously dumped its wastes in the field next to a campus building. The range of wastes would be expected to include acids, bases, organic solvents, detergents, and biological wastes. It is rumored that drums also may have been buried. The practice was stopped in 1982.

  You represent Country Club Consultants and have been asked to prepare a proposal to investigate this potential problem. The scope of work provided in the Request for Proposal is described below. Your assignment is to prepare the technical proposal for Country Club Consultants.

Scope of Work: Prepare a plan for Remedial Investigations for the aforementioned site. The plan should include you specific rationale and recommendations for the following:

- Remote sensing (geophysical methods)
- Borings: number, spacing, depth, type, locations, criteria, drilling method
- Monitoring wells
- Soil sampling
- Water sampling
- In situ tests
- Lab tests on soil and water samples
- Site characterization report

Basis of evaluation: The proposals will be evaluated on the basis of appropriateness of the technical scope of work, thoroughness, creativity, and proper selection of available techniques. Any available information that is developed as part of the proposal preparation (i.e. geology, etc.) should be included to strengthen the proposal. Listing of information for proposed work activities is suitable although such lists should only include relevant information. The proposal is limited to 5 pages plus tables and figures, if any.

REFERENCES

1. U.S. EPA: "Guidance on Remedial Investigations Under CERCLA," EPA/540/G-85/002.
2. *Standard Methods for the Examination of Waterways*, 17th ed., American Public Health Association, 1989.
3. American Society for Testing and Materials: Annual Book of Standards, Vol. 04.08, Soil and Rock; Building Stones; Geotextiles, Philadelphia, PA.
4. *Occupational Safety and Health Guidance Manual for Hazardous Waste Site Activities, National Institute of Occupational Safety and Health, Washington, D.C., October 1985.*
5. Jakosky, J. J.: *Exploration Physics*, Trija Publishing Co., Newport Beach, California, p. 660.
6. McNeill, J. D.: Electromagnetic Terrain Conductivity Measurement at Low Induction Numbers, Technical Note TN-6, Geonics Limited, Ontario, Canada, 15 pp., 1980.
7. Merin, I. S., "Identification of Previously Unrecognized Waste Pits Using Ground Penetrating Radar and Historical Aerial Photography," *Proceedings of the 11th National Superfund Conference*, HMCRI, Washington, D.C., Nov. 1990, pp. 314–319.
8. Struttman, T., and Anderson, T.: "A Comparison of Shallow Electromagnetic (EM-31) and Proton Precession Magnetometer Surface Geophysical Techniques to Effectively Delineate Pre-RCRA Buried Wastes at a Manufacturing Facility," Proceedings of the 10th National Conference on the Management of Uncontrolled Hazardous Waste Sites, Washington, D.C., pp. 27–34. November 1989.
9. Richardson, W. K., Kirkpatrick, G. L., and Cline, S. P.: "Integration of Borehole Geophysics and Aquifer Testing to Define a Fractured Bedrock Hydrogeologic System," Proceedings of the 10th National Conference on the Management of Uncontrolled Hazardous Waste Sites, Washington, D.C., pp. 277–281, November 1989.
10. Acker, W. L. "Basic Procedures for Soil Sampling and Core Drilling," Acker Drill Co., Scranton, PA, 1974, 246 pp.
11. ASTM: "Standard Test Method for Penetration Test and Split-Barrel Sampling of Soils," D 1586, American Society for Testing and Materials, Philadelphia, PA.
12. Kovacs, W. D., Griffith, A. H.: and Evans, J. C., "An Alternative to the Cathead and Rope for the Standard Penetration Test," *Geotechnical Testing Journal*, ASTM, vol. 1, no. 2, pp. 71–89, June 1978.
13. Kovacs, W. D., Evans, J. C., and Griffith, A. H.: "Towards a More Standardized SPT," Proceedings of the Ninth International Conference on Soil Mechanics and Foundation Engineering, Tokyo, Japan, vol. 2, pp. 269–276, 1977.
14. Fletcher, G. F. A.: "Standard Penetration Test: Its Uses and Abuses," *Journal of the Soil Mechanics and Foundations Division*, ASCE, vol. 91, no. SM4, pp. 67–75, July, 1965.
15. Gibbs, H.J. and Holtz, W.H., "Research on determining the Density of the Sands by Spoon Penetration Testing," *4th International Conference on Soil Mechanics and Foundation Engineering*, London, U.K., vol. I, p. 35.
16. Peck, R. B., Hanson W. E., and Thornburn, T. H.: *Foundation Engineering*, 2nd ed., J. Wiley & Sons, New York, NY, 464 pp., 1974.
17. Terzaghi, K. and Peck, R. B.: *Soil Mechanics in Engineering Practice*, 2nd ed., J. Wiley & Sons, New York, NY, 1967.
18. Seed, H. B., and Idriss, I. M.: "Simplified Procedure for Evaluating Soil Liquefaction Potential," *Journal of the Soil Mechanics and Foundations Division*, ASCE, vol. 97, no. SM9, pp. 1249–1273, Sept. 1971.

19. Strutynsky, A. I., and Sainey, T. J.: "Use of Piezometric Cone Penetration Testing and Penetrometer Groundwater Sampling for Volatile Organic Contaminant Plume Detection," Proc. 1990 Conf. on Petroleum Hydrocarbons and Organic Petroleum Inst., Houston, TX, Nov. 1990.

20. Marchetti, S.: "In situ Tests by Flat Dilatometer," *Journal of the Geotechnical Engineering Division*, ASCE, vol. 106, no. GT3, pp. 299–321, March, 1980.

21. Wroth, C. P.: "Interpretation of In Situ Soil Tests," *Geotechnique*, vol. 34, pp. 449–489, 1984.

22. HMCRI: "Two Technologies under Development for Access to Subsurface Contaminants," *Focus*, October 1989.

23. Driscoll: *Groundwater and Wells*, 2nd ed., 1986.

24. Pierdinock, M. J. et al.: "Remote Control Test Pit Excavation in a Shock Sensitive Drum Zone at a Superfund Site," presented at Haztech International 90, sponsored by the Institute for International Research, Inc., Pittsburgh, PA, October 2–4, 1990.

25. Kerfoot, H. B., and Barrows, L. J.: "Soil-Gas Measurement for the Detection of Subsurface Organic Contamination," EPA/600/S2-87/027, June, 1987.

26. Burns, R. B.: "Field Screening of Soil Vapors at a Subsurface Gasoline Contaminated Site in West Liberty, Morgan County, Kentucky," presented at Great Lakes 90, sponsored by the Hazardous Materials Control Research Institute, Cleveland, OH, September 26–28, 1990.

27. Godoy, F. E., and Naleid, D. S.: "Optimizing the Use of Soil Gas Surveys," *Hazardous Materials Control*, vol. 3, no. 5, pp. 23–29, Sept.–Oct., 1990.

28. Marks, B. J., and Singh, M.: "Comparison of Soil-Gas, Soil, and Groundwater Contaminant Levels of Benzene and Toluene," *Hazardous Materials Control*, vol. 3, no. 6, pp. 40–45, Nov.–Dec., 1990.

29. Lucero, D. P.: "A Soil Gas Sampler Implant for Monitoring Dump Site Subsurface Hazardous Fluids," *Hazardous Materials Control*, vol. 3, no. 5, pp. 36–44, Sept.–Oct., 1990.

30. Morrison, R. D.: Ground Water Monitoring Technology: Procedures, Equipment and Applications, Timco Mfg., Inc., Prairie du Sac, WI, 1983, 111 pp.

31. Nielsen, D. M., ed., *Practical Handbook of Ground-Water Monitoring*, Lewis Publishers, Inc., Chelsea, MI, 1991.

32. U.S. EPA: "A Compendium of Superfund Field Operations Methods," EPA/540/P-87/001, 1987.

33. Autennucci, J.C., et al., Geographic Information Systems, Van Nostrand Reinhold, New York, New York, 1991.

35. Dimmick, Shelly, Oracle RDBMS Database Administrators Guide, Oracle Corp., 1990.

36. GDS Operating Manual, Electronic Data Systems Corp., Maryland Heights, Missouri, 1992.

ADDITIONAL READING

Clark, Keith C.: *Analytical and Computer Cartography*, Prentice Hall, Englewood Cliffs, New Jersey, 1990.

Falotico, R. J. et al.: "The Necessity of Hydrogeologic Analysis for Successful In Situ Bioremediation," presented at Haztech International 90, sponsored by the Institute for International Research, Inc., Pittsburgh, PA, October 2–4, 1990.

Findley, L. A.: "Supplied Air Versus SCBA Respiratory Protection at Hazardous Waste Sites," presented at Great Lakes 90, sponsored by the Hazardous Materials Control Research Institute, Cleveland, OH, September 26–28, 1990.

Jensen, John R.: *Introductory Digital Image Processing—A Remove Sensing Perspective*, Prentice Hall, Englewood Cliffs, NJ, 1986.

Keith, L. H., ed.: *Principles of Environmental Sampling*, ACS Professional Reference Book, Washington, D.C., 1988.

Kerfoot, H. B.: "Soil-Gas Sampling for Detection and Delineation of Subsurface Organic Contamination," presented at Great Lakes 90, sponsored by the Hazardous Materials Control Research Institute, Cleveland, OH, September 26–28, 1990.

Medine, A. and Anderson, eds.: *Environmental Engineering—Proceedings of the ASCE Specialty Conference, Boulder, CO*, ASCE, New York, NY, 1983.

Pye, V. I., Patrick, R., and Quarles, J.: *Groundwater Contamination in the United States*, University of Pennsylvania Press, Philadelphia, PA, 1983.

Wilburn, D. J. et al.: "Applications of Geotechnical Software to Remedial Investigations and Feasibility Studies," presented at Haztech International 90, sponsored by the Institute for International Research, Inc., Pittsburgh, PA, October 2–4, 1990.

ACKNOWLEDGEMENTS

The authors wish to acknowledge the following individuals who assisted with the preparation of this chapter: Steve Fulton, Scott McQuown, Mike Moskow, John Harsh, William Douglas, *ERM, Inc.*; Gary Rabik, *Sun Refining and Marketing Co.;* John Link, Ben Myers, Jon Terry, *Bucknell University*; Don Fries, Jim Nash, *Villanova University*; and Ngayan Warren, *Pennsylvania State University—Harrisburg.* The principal reviewer for the ERM Group was Ronald Landon, *ERM, Inc.*

CHAPTER
16

CONTAINMENT

Πέμπτον ἐπέταξεν αὐτῷ ἆθλον τῶν Αὐγείου
βοσκημάτων ἐν ἡμέρᾳ μιᾷ μόνον ἐκφορῆσαι τὴν
ὄνθον...
Ἡρακλῆς... τῆς τε αὐγῆς τὸν θεμέλιον διεῖλε
καὶ τὸν Ἀλφειὸν καὶ τὸν Πηνειὸν σύνεγγυς
ῥέοντας παροχετεύσας ἐπήγαγεν,
ἔκρουν δὶ ἄλλης ἐξόδου ποιήσας.

The fifth labor [Eurystheus] laid on [Hercules] was to carry out the dung
of the cattle of Augeas in a single day by himself. ...
Hercules made a breach in the foundations of the cattle yard and then,
after diverting the courses of the Alpheus and Peneus rivers,
which flowed near one another, he turned them [into the yard],
having [first] made an outlet [for the water] through another opening.

The Library of Apollodorus, c. 100 B.C.

16-1 INTRODUCTION

Past industrial and waste management activities have contaminated soil and ground water at tens of thousands of sites in both the United States and abroad. A nationwide effort to clean up the "sins of the past" began in earnest as a result of the Comprehensive Environmental Response, Compensation and Liability Act (CERCLA)[1] commonly known as Superfund and described in Sec. 2-4. In the United States, more than 40,000 sites have been identified that are potential threats to the public health and/or the environment. Many of these sites include some risk of contaminant migration in the subsurface. Additionally, over 1200 sites have been cited for cleanup by their addition to the National Priorities List under Superfund. In the United Kingdom,[2] a registry of sites with contaminated land uses is expected to generate a list of up to 100,000 sites by the time it is completed. It is clear that a significant portion of our hazardous waste management resources are and will continue to be devoted to the assessment and cleanup of existing hazardous waste sites.

The cleanup of an existing hazardous waste site requires thorough site characterization as described in Chap. 15 and an assessment of risk as described in Chap. 14. The analysis of remedial alternatives is described in Chap. 17. But first, a knowledge of the remedial alternatives that are available is required. In some cases, removal of the source itself may be enough to eliminate contaminant transport into the environment. However, more frequently, it is necessary to minimize the rate of off site contaminant migration employing containment technologies to minimize risk to public health and the environment. Containment technologies may be associated with other technologies (such as in situ remedial alternatives) to implement a long-term cleanup strategy for the site.

This chapter describes containment technologies that may be appropriate for the cleanup of existing hazardous waste sites. Special emphasis is given attention to technologies that are frequently employed in conjunction with other techniques. The following chapter first describes passive controls such as slurry trench cutoff walls and covers. It then presents active ground water controls such as pump and treat and ground water collection systems. Removal technologies are also presented. Additional cleanup technologies which do not fit into the above categories are also discussed.

16-2 OBJECTIVES

The cleanup objectives vary from site to site and from organization to organization. In the broadest sense the objective of any cleanup program is to render the site safe for its intended future use. To do this the remediation scheme must minimize or eliminate the hazard to human health and the environment. How then do we minimize or eliminate this hazard? Is it necessary to remove the contaminants from the soil, water, and waste to effect a cleanup? If removal is selected as the cleanup approach, does this mean that every molecule of contaminant must be removed? Approaches to site cleanup are many and varied. In some cases removal is the most cost-effective and environmentally protective solution. In others, on-site treatment may be selected. In still others, containment may suffice. In all cases, the question of risk must be addressed and the selection of remedial alternatives will be influenced by their ability to reduce risk.

The cleanup or remediation program must be necessary and sufficient to accomplish the objective, that is, render the site and surrounding environment safe for the intended use. This objective will guide the selection of the appropriate remedial action. In turn, each component of the remedial program will have goals to meet. For some, complete removal of the contaminants is the only suitable remedial objective. For others, control of the contaminant migration pathways is adequate (and cost-effective).[3] In reality, no site will be completely cleaned of each molecule of contamination. Equally impractical is the digging up of all sites and hauling the materials to a landfill because this would merely transfer the contamination to another location. For many sites, mere containment is also inadequate. At present, the cleanup objectives depend upon a myriad of social, technical and political factors. This chapter focuses on the technical aspects of site remediation.

How much data are needed from the site characterization studies to select, design, and implement a remedial action program? Suffice it to say that all the desired information is never available; the site is never fully characterized. In scientific terms, we must recognize a certain degree of **uncertainty**. How can uncertainty be dealt with? One approach, tried and proven in the geotechnical engineering area, is the **observational approach.**[4] Application of this approach to hazardous waste site remediation requires the recognition of uncertainty and as such suggests:[5] 1) remedial design based upon the most probable site conditions, 2) identification of possible deviations from these conditions, 3) identification of parameters to monitor and confirm conditions and performance of the remedial design, and 4) contingency plans for potential deviation. Recognizing uncertainty, and application of the observational approach, offers the best opportunity to achieve the desired environmental protection at the lowest cost with the least risk.

Containment is likely a component in a remedial system because more than one technology may be used for site remediation. As the principles of each technology are described, their place in the universe of available methods is identified. To systematically discuss remediation in this text, the technologies are classified as either active or passive remedial system components.

- Active system components require considerable effort and on-going energy input to operate (for example, pumping wells).

- Passive system components work without much attention, except maintenance (such as a cover).

This framework enables us to examine systems by function and differentiate between systems performing the same function.

16-3 PASSIVE CONTAMINANT CONTROL SYSTEMS

The function of passive contaminant control systems in site remediation is to minimize contaminant transport rates. To examine this function, consider the potential pathways for contaminant migration at an uncontrolled hazardous waste site (see Fig. 16-1). Precipitation at the site either runs on or off, infiltrates or returns into the atmosphere through evapotranspiration. Precipitation runoff that encounters waste or contaminated soil may transport contaminants or contaminated sediment into the surrounding environment. Precipitation may also infiltrate into and through the waste generating leachate. Migration of leachate may introduce contaminants into the surface water and/or subsurface water environment. Contaminants present in leachate migrating to the ground water are then transported through the ground water environment and may

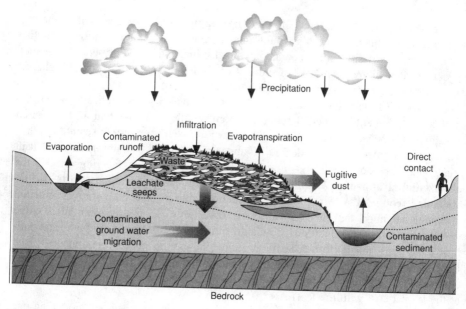

FIGURE 16-1
Pathways for contamination.

be discharged to surface water if the ground water influences surface water. Passive contaminant control technologies focus on controlling hydrologic pathways for contaminant migration. This approach is often termed **containment**. Due to uncertainties associated with design, construction, and long-term reliability of containment facilities, containment as a sole means of site remediation is not recommended for sites with significant levels of contamination. Containment is used at such sites in conjunction with source removal containment and institutional controls.

The technologies for the control of the contaminant transport pathways are discussed with respect to their function within the remedial program. For example, it is necessary to control the impact of precipitation to prevent it from becoming contaminated and to control the hydraulic transport of contaminants. We will thus look at surface water control technologies and then ground water control technologies.

The selection, design, and construction of containment technologies must recognize each of the exposure and transport processes. Ground water and contaminant transport processes, described in Chap. 4, include advection, dispersion (including both molecular diffusion and mechanical mixing), absorption, retardation and biological and chemical transformation. The selected containment technology must consider these transport processes to assess its effectiveness in controlling contaminant migration to the surrounding environment.

16-4   SURFACE WATER CONTROL TECHNOLOGIES

Figure 16-1 indicates the need to prevent precipitation from transporting contaminants off site by surface water runoff or by leachate generation. Covers/caps, surface water diversions, and sedimentation and erosion control systems are employed for this purpose. Systems that prevent/mitigate off-site transport of contamination are also incorporated into the design of new land disposal and storage facilities.

Although surface water control technologies are just one component of the overall site remediation system, site conditions may occasionally allow surface water control technologies to be the sole means of site remediation, particularly in countries other than the United States where the remedial technology for a contaminated site is selected consistent with the intended site usage. For example, consider a site, underlain by deep chalk deposits, having concentrations of residual metals in the soils from prior site activities. Site redevelopment for continued industrial usage could involve covering the site to preclude direct contact, eliminate contaminated runoff, and minimize/eliminate infiltration and leaching to reduce the risk and provide the appropriate level of protection for public health and the environment.

How are surface water control technologies integrated into the overall site remediation scheme? The answer is that it depends upon the other components of the remedial system. For example, the remedial program at Love Canal incorporates surface covers in conjunction with leachate/ground water collection and treatment systems. The cover reduces infiltration and eliminates the problems of direct contact previously evident.

Because the surface water control technologies have already been presented and discussed in the context of land disposal facilities, refer to Sec. 13-4 for additional detail regarding surface water control technologies.

16-5   GROUND WATER CONTROL

Subsurface vertical barriers are employed to contain contaminants and to redirect ground water flow. These barriers to horizontal contaminant transport and ground water flow used in hazardous waste management have developed from conventional engineering applications such as dewatering of excavations and control of ground water flow beneath dams and levees. A number of vertical barrier techniques are available including slurry trench cutoff walls (made of soil-bentonite, cement-bentonite, or plastic concrete), grout curtains (made of cement-based or chemical grouts), and steel sheet piling.

It is first necessary to establish the function of a vertical barrier in the framework of the site remediation system. One function of a vertical barrier wall is to provide containment to prevent migration of contaminants in the ground water. By employing a vertical barrier wall around the site, the rate of contaminant migration is slowed, particularly in anisotropic aquifers and formations having a horizontal hydraulic conductivity much higher than the vertical hydraulic conductivity. Passive containment of contaminants has not been widely employed as a method of hazardous waste site remediation in the United States. However, it is evident that after reduction in contaminant levels at sites undergoing cleanup by "pump and treat" methods, the vertical barrier wall left in place will, in the long term, serve as a passive barrier to reduce the rate at which any remaining contaminants may migrate into the subsurface environment.

The most common function of a vertical barrier in a remediation system is not containment but rather to inhibit the flow of clean ground water into the site. Performing this function, vertical barriers are commonly used in conjunction with ground water extraction and treatment systems. In this application, the ground water is extracted from within the cutoff wall and sent to treatment as shown in Fig. 16-2. Without the installation of vertical barrier, clean ground water would flow beneath the site, become contaminated, and increase the quantity of water for the pump and treat component of the remedial system. With the installation of a vertical barrier, clean regional ground water is prevented from entering the pump-and-treat system.

Vertical barriers also provide ground water control during construction where excavation into the subsurface is required for direct waste treatment, waste removal, or construction of liner systems. The vertical barrier initially serves to expedite construction.

Vertical barriers to control horizontal ground water flow and contaminant migration can be configured in several ways. The most common configuration is a circumferential cutoff wall that completely surrounds the waste or the waste site as shown in Fig. 16-2. Vertical barrier walls may also be located upgradient or downgradient of the site, depending upon the intended function of the barrier.

The principal purpose of an upgradient vertical barrier wall (Fig. 16-3) is to provide a cutoff controlling the influx of clean ground water flow from upgradient regions. When used in conjunction with a pumping system located on the downgradient side (and no vertical barrier on the downgradient side of the site), ground water withdrawal may recover contaminants that have migrated downgradient of the site. In this way, contaminants that have held site boundaries in total disregard may be captured and returned to the site contaminant treatment program.

The principal purpose of a downgradient cut-off wall, as shown in Fig. 16-4, is to utilize site ground water flowing beneath the site to actively flush contaminants from beneath the site. A downgradient wall is effective in aiding the pump-and-treat system to speed removal of contaminants from the subsurface. This method of vertical barrier wall deployment was selected for the remediation of the Rocky Mountain Arsenal Site.

Vertical barrier walls are typically embedded, or **keyed**, into a material of low permeability beneath the site (Fig. 16-5), although this is not always necessary. In the case of product that is lighter than water (for example, from a leaking underground fuel oil storage tank) the vertical barrier does not need to fully penetrate the aquifer into the underlying aquitard (see Fig. 16-5b).

An evaluation of the permissible rate of contaminant transport through or across the barrier is essential to the considerations of barrier technology. At present, it is

Plan view

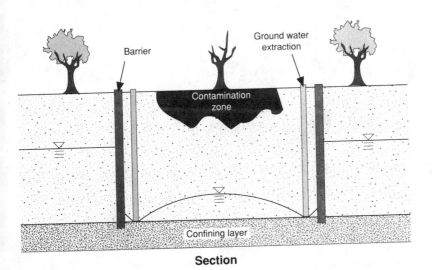

Section

FIGURE 16-2
Circumferential cutoff wall.

typical to simply specify a barrier as having an hydraulic conductivity of 1×10^{-7} centimeters per second (cm/sec) or less. This conductivity is chosen because, with careful design and construction, it is the lowest that can be readily and economically achieved. However, calculations of site-specific flow rates through the barrier may demonstrate for many projects that a barrier having an hydraulic conductivity 1×10^{-6} or 1×10^{-5} cm/sec may be suitable. This change in hydraulic conductivity requirement widens the choices available and potentially reduces the overall cost of the wall. It is usually necessary to employ ground water modeling to determine the anticipated impact of the barrier. For example, in one study of a proposed slurry trench cutoff wall, ground water modeling of the alternatives demonstrated that the presence of the vertical barrier was ineffective in reducing off-site contaminant loadings because

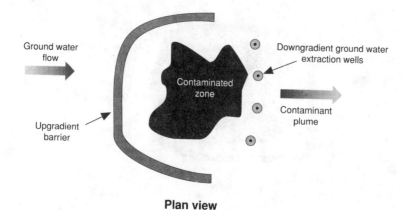

Ground water flow

Upgradient barrier

Contaminated zone

Downgradient ground water extraction wells

Contaminant plume

Plan view

Barrier

Ground water extraction

Ground water flow

Contaminated zone

Captured contaminant plume

Confining layer

Section

FIGURE 16-3
Upgradient vertical barrier.

modeling revealed the contaminant transport was virtually entirely in the fractured rock underlying the overburden. This led the designers to conclude that alternative remedial technologies were more appropriate for the specific site and subsurface conditions.

Example 16-1   Vertical barrier hydraulic conductivity. A 3-foot-thick circumferential barrier is planned to control the influx of clean ground water to extraction wells. The cost effective treatment plant capacity is 5 gpm. The perimeter is 3000 feet in length and the aquifer is 30 feet thick. An inward gradient across the barrier wall will be maintained by lowering the water level within the containment by 5 feet. What must the hydraulic conductivity of the wall be to control the influx through the wall to 5 gpm?

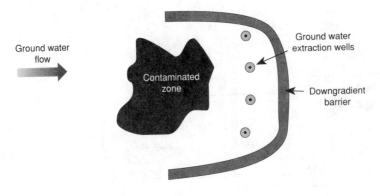

Plan view

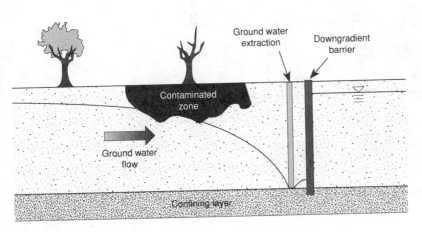

Section

FIGURE 16-4
Downgradient vertical barrier.

Solution. Assume one-dimensional flow reasonably models the flow conditions and apply Darcy's law (eq. 4-3) (Attention nonbelievers—construct a flow net to remove any doubt you may have in the reasonableness of this assumption).

$$q = ki\,A$$

where $q = 5$ gpm
      $k =$ hydraulic conductivity (unknown)
      $i = \Delta h \Delta L = 5/3 = 1.67$
      $A = 30 \times 3000 = 90{,}000$ ft^2

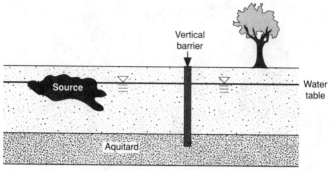

(a) Keyed vertical barrier wall

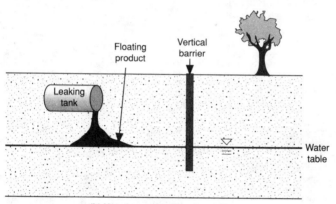

(b) Hanging vertical barrier wall

FIGURE 16-5
Embedment of vertical barrier walls.

substituting:

$k = 5/(1.67 \times 90{,}000 \times 7.48) = 4.5 \times 10^{-6}$ ft/min

$k = 4.5 \times 10^{-6}$ ft/min $\times$ 30.48 cm/ft $\times$ 1/60 min/sec

$k = 2.3 \times 10^{-6}$ cm/sec

At present, considerations of contaminant transport across barriers are usually limited to advective flow. As described in Sec. 4-2, a thorough analysis should include advective and diffusive flow and account for the soil-contaminant interactions utilizing diffusion and retardation coefficients. In one study the breakthrough time for carbon tetrachloride at an interior concentration of 1 mg/L was found to be about two years with a conventional soil-bentonite wall.[6,7] The study found that the breakthrough time increased to 30 years when the wall was designed using a high-carbon fly ash composite. These findings point to the importance of the consideration of contaminant transport when designing a cut-off wall.

The vertical barrier wall function and configuration are coupled with site and subsurface conditions to select the suitable type(s) of vertical barrier. For example, grout curtains are appropriate for vertical barriers in fractured rock whereas slurry trench cut-off wall techniques typically require softer materials that lend themselves to soil excavation technologies. It is noted, however, that the use of soil-bentonite slurry trench cut-off walls in Superfund site remediation appears to be the most common.[8] Guidance as to the most appropriate type of vertical barrier for specific subsurface conditions will be provided later in the chapter as each of the following techniques is discussed in more detail:

- Soil-bentonite slurry trench cut-off walls
- Cement-bentonite slurry trench cut-off walls
- Plastic concrete cut-off walls
- Diaphragm (structural) cut-off walls
- Vibrating beam cut-off walls
- Deep soil-mixed walls
- Composite cut-off walls
- Steel sheet pile cut-off walls
- Grout curtains

Soil-Bentonite Slurry Trench Cut-off Walls

A soil-bentonite slurry trench cut-off wall, often known simply as a **slurry wall**, takes its name from the method of construction. To construct this vertical barrier in the subsurface, a trench is first excavated below the existing grade utilizing a liquid slurry of bentonite and water to maintain the trench stability (Fig. 16-6). The slurry is comprised of approximately 95% water and 4–6% bentonite by weight. Bentonite is a sodium montmorillonitic clay (the structure is shown in Fig. 11-6). The resulting bentonite-water slurry is a viscous fluid with a unit weight of approximately 64–70 pounds per cubic foot (pcf).

Trench collapse is controlled by the resulting hydrostatic force system where the slurry counteracts the active soil pressures. Because a positive fluid pressure is maintained within the trench, the slurry exfiltrates through the walls of the trench, a filter cake is formed, and the filtrate enters into the formation. The filter cake is a very thin layer of fully hydrated bentonite, forming an impermeable boundary. The fluid pressure of the slurry opposes the active earth pressure to maintain the trench stability. Figure 16-7 is a photograph of an excavation showing the construction of a slurry trench. The excavation is made in the slurry-filled trench, and the excavation side walls are essentially vertical. The slurry maintains the trench stability and permits vertical excavations to depths of more than 100 feet.

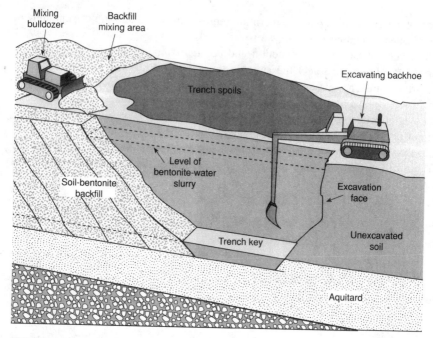

FIGURE 16-6
Soil-bentonite slurry trench cutoff construction wall.

FIGURE 16-7
Slurry trench excavation (photo courtesy of GEOCON Inc.).

Example 16-2   Slurry unit weight.   Calculate the expected unit weight of the bentonite-water slurry for a batch using one bag of bentonite (100 lbs) per 200 gallons of water. The unit weight of water is 62.4 lb/cu.ft and the specific gravity (G_s) of the bentonite is 2.77. What is the bentonite content? Why is the problem presented is such mixed units of pounds and gallons?

Solution

Bentonite content:

$$\% \text{ Bentonite} = 100/((200 \text{ gal}/7.48 \text{ gal/ft}^3) \times 62.4 \text{ lb/ft}^3) = 6.0\%$$

Determination of fluid unit weight:

$$G_s = W_s/V_s\gamma_w \text{ (definition of specific gravity)}$$

where

W_s = weight of solids

V_s = volume of solids

γ_w = specific weight of water

Substituting:

$$V_s = W_s/G_s\gamma_w = 100 \text{ lb}/(2.77 \times 62.4) = 0.58 \text{ ft}^3$$

Volume of water

$$V_w = 200 \text{ gal}/7.48 \text{ gal/ft}^3 = 26.74 \text{ ft}^3$$

Weight of water

$$W_w = V_w/\gamma_w = 26.74 \text{ ft}^3 \times 62.4 \text{ lb/ft}^3$$
$$W_w = 1668 \text{ lb.}$$

Fluid unit weight

$$\gamma_f = W_T/V_T = (1668 + 100)/(26.74 + 0.58)$$
$$\gamma_f = \underline{64.7\text{lb/ft}^3}$$

The units are representative of mix proportions used by contractors in the field.

Straightforward and well-documented methods of calculating trench stability are available.[9,10,11] Instability is modeled as an active soil wedge prone to slide into the trench. Resisting this instability is the strength of the soil adjacent to the excavation and the fluid pressure exerted by the bentonite-water slurry against the sidewalls of the trench. The filter cake is the vertical surface along which these forces must be balanced. Equations of equilibrium can be written for the driving forces (the active soil wedge) and the resisting forces (the soil strength and the bentonite fluid pressure).

Defining the factor of safety as the ratio of the resisting forces to the driving forces yields:

For clays:

$$F = 4S_u/H(\gamma - \gamma_f) \qquad (16\text{-}1)$$

For sands:

$$F = 2\sqrt{(\gamma \times \gamma_f)} \times \tan(\phi)/(\gamma - \gamma_f) \qquad (16\text{-}2)$$

where F = factor of safety

S_u = undrained shear strength (cohesion)

H = depth of the cut-off wall

γ = total unit weight of the soil

γ_f = unit weight of the bentonite-water slurry

ϕ = angle of internal friction

Example 16-3   Factor of safety. Calculate the factor of safety for a trench 30 feet deep excavated in sand having an angle of internal friction of 30. The unit weight of the fluid is 64.7 lb/ft^3 as calculated above. The sand has a total unit weight of 130 lb/ft^3.

Solution

$$F = 2\sqrt{(\gamma \times \gamma_f)} \times \tan(\phi)/(\gamma - \gamma_f)$$

where $\gamma = 130$
   $\gamma_f = 64.7$
   $\phi = 30$
   $F = 2\sqrt{(130 \times 64.7)} \times \tan(30)/(130 - 64.7)$
   $F = \underline{1.62}$

Note that the factor of safety is independent of trench depth.

The trench excavation is typically completed using a backhoe with a modified boom to depths of up to about 90 feet. Below this depth, it is necessary to use equipment such as drag lines and clam shells to excavate the trench. Table 16-1 summarizes the excavation equipment used for the construction of slurry cutoff walls.

Placing soil-bentonite backfill is the second part of the process of constructing a soil-bentonite slurry trench cutoff wall (see Fig. 16-6). After the trench excavation is completed to the entire depth, the bentonite-water slurry is then displaced with the soil-bentonite backfill. The soil-bentonite backfill at a consistency of high slump concrete is placed from the surface and moves (oozes, if you prefer) forward (in the direction the excavation is proceeding). The soil-bentonite backfill is a low permeability material composed of soil, water, and bentonite. The bentonite content of the backfill may be as low as 1% when there is an adequate presence of natural fines or may be as high as 4–5%.

The hydraulic conductivity of soil-bentonite backfill of field mixed samples typically varies in laboratory tests from a high of 1×10^{-5} cm/sec to a low of 1×10^{-9} cm/sec. In general, however, the hydraulic conductivity of soil-bentonite backfill ranges from 1×10^{-7} cm/sec to 1×10^{-8} cm/sec as shown in Fig. 16-8.

TABLE 16.1
Slurry trench excavation techniques

Type	Trench (m) width	Trench (m) depth	Comments
Standard backhoe	0.5–2.0	10	Rapid, least costly
Backhoe with extended boom	0.5–2.0	16	Rapid and low cost (extended excavation boom on unmodified backhoe)
Modified backhoe with extended boom	0.5–2.0	20	Extended excavation boom; modified power and counterweight
Clamshell	0.3–2.0	75	Attached to a crane
Dragline	1.0–3.0	50	Slow, for wide, deep excavations
Rotary	0.5–2.0	75	Slow, for boulders and rock
Specialty	Varies	Varies	e.g., Hydrofraize (from Soletanch

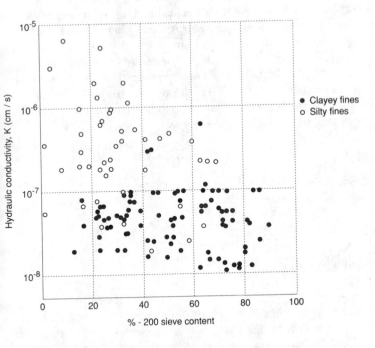

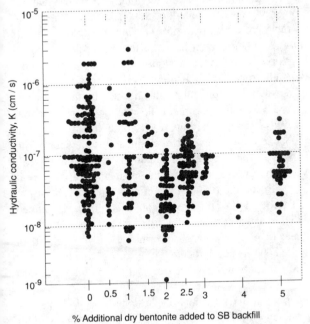

FIGURE 16-8
Soil-bentonite backfill hydraulic conductivity (from Ryan, 1987).

Note that the hydraulic conductivity is not strongly related, if related at all, to the bentonite content. This is because soils of a well-graded nature (coarse to fine sands and gravels with silt and clay) tend to be quite impermeable when homogenized and mixed with bentonite-water slurry. In contrast, clean sands may require the addition of considerable dry bentonite to the mix to achieve similar hydraulic conductivities. The hydraulic conductivity is strongly related to the presence and type of naturally occurring fine-grained materials as shown in Fig. 16-8.[12] A base soil with 20% or more natural clay fines should be able to produce a soil-bentonite backfill with a hydraulic conductivity less than 1×10^{-7} cm/sec.

Backfill Design

In order of importance for environmental application, characteristics to be designed into the soil-bentonite backfill include:

1. Chemical compatibility
2. Low permeability
3. Low compressibility
4. Moderate strength

In achieving these engineering characteristics, at the time of construction the backfill must flow freely into the trench and displace the bentonite-water slurry used to maintain trench stability. The most important soil parameters that affect these characteristics are grain size distribution and the water content of the backfill. Soil-bentonite backfills have been successfully fabricated from materials varying from clean sand to highly plastic clay. However, for the containment of hazardous waste, backfill requirements are necessarily stricter than those used for conventional dewatering applications. The difference in requirements result from two fundamental differences in expected performance. First, conventional dewatering applications may not require the degree of "perfection" because these systems are temporary. Small leaks may be inconsequential from a dewatering standpoint whereas from a hazardous waste containment standpoint they may be significant. Second, because these dewatering applications are short term, considerations of long-term changes in the hydraulic conductivity in the cutoff wall are not necessary. Specifically, for hazardous waste containment applications compatibility of the hazardous waste with the cut-off wall material must be evaluated to determine alteration of the permeability due to prolonged contact. Activity is defined as the plasticity index normalized with respect to the clay fraction; high activity clays, such as bentonite, are thus subject to greater ranges of shrinking and swelling than low-activity clays such as kaolinite. The results indicate while soil fabric changes are important in controlling hydraulic conductivity, at increasing activity volume changes become the controlling factor.

For environmental applications the backfill must be designed to minimize any potential changes in hydraulic conductivity. Generally, a lower hydraulic conductivity is required for hazardous waste containment than for conventional dewatering applications. These goals of long-term stability with low hydraulic conductivity can be

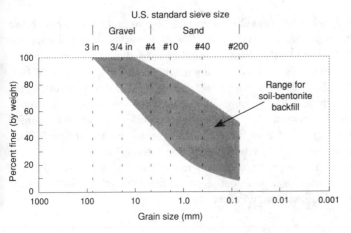

FIGURE 16-9
Grain size distribution for soil-bentonite backfill.

best achieved by fabricating the backfill from a well-graded soil blended with soil-bentonite slurry. This well-graded soil must contain all particle size ranges including coarse, medium, and fine gravel, coarse, medium and fine sand; silt; and clay. A recommended particle size range is shown in Fig. 16-9. Because the coarser granular materials in a well-graded backfill material are in point to point contact, a relatively low compressibility results. Further, with the well-graded nature of the material, the pore sizes (which become progressively finer) are filled with progressively finer material (silts and clays) resulting in a material with a relatively low hydraulic conductivity. Finally, by using a well-graded material, the potential for long-term changes in hydraulic conductivity due to soil-waste interactions is minimized. The components of a well graded soil are relatively stable (sand, silt, and low-activity clay). Clay of low activity (such as kaolinite) is not subject to dramatic permeability increases[13] as those of high activity (such as bentonite).[14] This approach differs from simply adding more and more bentonite to a poorly graded soil to achieve a low permeable backfill. The swelling of the bentonite is potentially reversible in the presence of contaminants (resulting in permeability increases over the long term).

As previously described, the use of a well-graded backfill coupled with the limited addition of bentonite to further reduce the hydraulic conductivity of the backfill is recommended. The reversible nature of the swelling of the bentonite may result in hydraulic conductivity increases in the presence of contaminants. As a worst case for hazardous waste containment applications, the hydraulic conductivity of the backfill may increase to the hydraulic conductivity of the soil without the bentonite. By selecting a well-graded material, a relatively low hydraulic conductivity is found in the base soil, similar to the relatively low hydraulic conductivity of naturally occurring well-graded glacial tills. The addition of bentonite-water slurry contributes to the low permeability of the material. Further, the slurry provides moisture and enhances the fluidity of the backfill to allow it to behave as high-slump concrete necessary to displace the bentonite-water slurry used in the trench for trench stability.

Some designers consider filter cake as a significant factor in the low permeability aspect of the completed soil-bentonite slurry cut-off wall. However, for environmental applications, the added benefit of the low permeability cake should be considered as an additional safety factor. This conservative approach results from several factors. First, the integrity of the filter cake cannot be ensured after backfill placement. As the backfill slides along, the backfill likely scrapes all or part of the filter cake off the trench wall. Second, the filter cake consists of high-swelling bentonite. It is well documented that bentonite swelling is reversible in the presence of many contaminants.[15] Thus, in the hazardous waste containment environment, the filter cake permeability may increase dramatically due to the presence of contaminants in the subsurface. Thus, it is recommended that any filter cake contribution to the low permeability nature of the wall be considered an additional safety factor.

Soil-Bentonite Cut-off Wall Construction

As shown in Fig. 16-6, the excavation is first completed using bentonite-water slurry to maintain the trench stability. Thus, the bentonite-water slurry is simply a construction expedient and is the basis for the name "slurry wall" commonly used in the industry. The backfill itself, however, constitutes the completed cutoff wall. Backfill is typically mixed along the side of the trench using materials excavated from the trench and then blended with bentonite-water slurry. For additional control over mixing, a remote batch process is utilized. The mixed backfill is transported by truck to the trench for placement. For waste containment applications, this is the recommended method of mixing because the backfill mixing can be better controlled using a remote mix area than by mixing along the trench.

It is noted that, on occasion, backfill has been mixed at a high enough water content to make the material pumpable. However, at the higher water content required to make the material pumpable,[8] the material is considerably more compressible. For hazardous waste containment applications, pumping of backfill is not recommended.

It has been shown that hydraulic conductivity testing can be accomplished relatively rapidly in on-site laboratories using a rigid wall consolidometer/permeameter to achieve results comparable to triaxial tests.[16] These data are shown in Fig. 16-10. The determination of hydraulic conductivity must be at the same gradient and consolidation pressure as expected in the field because soil-bentonite backfill hydraulic conductivity is stress dependent.[17]

Performance of Soil-Bentonite Cutoff Walls

As previously discussed, the management of hazardous waste requires an assessment of risk. This is also true with the construction of containment facilities. One must recognize the potential for construction defects and long-term property changes. The resulting design and construction procedure must minimize these risks. For purposes of presentation, the failure of containment systems such as slurry trench cut-off walls can be classified as construction defects or post-construction property changes.

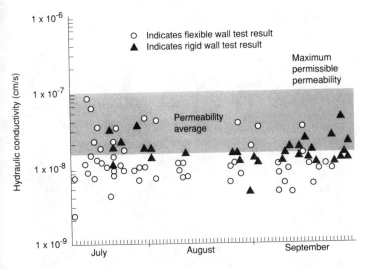

FIGURE 16-10
Soil-bentonite backfill permeability test results (from G. K. N. Hayward-Baker).[16]

Construction defects are those anomalies in the barrier wall that can result in localized greater rates of contaminant migration. Figure 16-11 depicts the potential construction defects for soil-bentonite slurry trench cutoff walls. Nonhomogeneous backfill results from improper mixing and includes lumps of unmixed soil and pockets of free slurry not fully blended with the soil. Also, during backfill placement slurry may become entrapped as a result of dropping backfill into the trench directly through the slurry. Slurry can also become entrapped as the backfill moves forward,[18] and the potential for this slurry entrapment increases with increasing wall depth.[19] If the trench is allowed to remain open for too long, sediment may accumulate and become trapped as part of the backfill. The trench sediment is coarse-grained with higher permeability than the designed backfill. Trench sediment may accumulate during periods of no activity such as overnight or over weekends and holidays. In addition, there may be a cave-in of the trench side or spalling from the formation resulting in materials falling from the side walls that may not be visible from the surface. Any of these contributors to nonhomogeneous backfill (see Fig. 16-11) results in a pocket or **window** within the wall.

Figure 16-11 also indicates the potential for inadequate excavation of the key. With regular soundings and inspection of the excavated soil this defect can normally be avoided. Keying into rock is more difficult, and, unless measures are taken (such as pre-blasting along the alignment), reduced cut-off wall performance may result due to flow beneath the wall. The Sylvester site (also known as Gilson Road) is an example of this.[20]

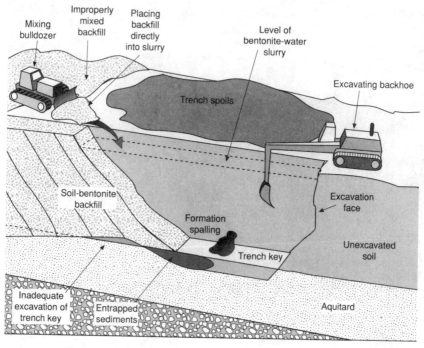

FIGURE 16-11
Potential defects in soil-bentonite slurry trench cutoff walls.

In addition to construction defects, long-term property changes and the backfill must be examined. Property changes can result from:

- Cycles of freezing and thawing
- Cycles of wetting and drying
- Desiccation of the backfill
- Chemical incompatibility

Cycles of freezing and thawing increase the hydraulic conductivity of the cut-off wall. This can be prevented with a properly designed and constructed final cover (see Sec. 13-4). Cycles of wetting and drying are inevitable with a fluctuating water table. Drying can cause the bentonite to shrink and increases in hydraulic conductivity will result. Portions of the cutoff wall that are semi-permanently above the water table will desiccate, resulting in increases in the hydraulic conductivity. Recent research has documented these permeability increases and shown that, even when resaturated, the permeability remains higher than the materials that have remained saturated since placement.[21] For hazardous waste management applications, special attention must be given to chemical incompatibility. Significant increases in permeability have been seen in the laboratory when soil-bentonite backfill is permeated with concentrated organic permeants.[22] These tests simulate a barrier wall in the presence of dense non-aqueous phase liquids (DNAPL). In these cases, designing the backfill using well-graded materials as previously described is recommended along with site-specific compatibility testing.

Cement-Bentonite Slurry Trench Cut-off Walls

Cement-bentonite slurry trench cutoff walls are excavated under a head of slurry composed of water, cement, and bentonite and have been widely employed for ground water control.[23] Unlike soil-bentonite slurry cut-off walls, however, the slurry is left to harden in a trench. As the cement hydrates, the combination of the water, cement, bentonite, and sand in suspension from excavating results in a hardened material with a consistency of stiff clay. Thus, the cement-bentonite wall is a one-step process. Because the excavated soils are not used in a backfill, the technique generates excess excavation materials for disposal. Although slurry trench cut-off walls of soil-bentonite can be consistently prepared having an hydraulic conductivity 1×10^{-7} cm/sec or less, cement-bentonite cut-off walls are generally limited to a hydraulic conductivity of approximately 1×10^{-5} to 1×10^{-6} cm/sec (Fig. 16-12). As shown in Fig. 16-12, the cement/water ratio is typically between 0.15 and 0.25 resulting in a cut-off wall that has a very low solids content; the wall is mostly water. Thus, cement-bentonite may be more prone to increases in hydraulic conductivity in response to the contaminated fluids, although limited data in this regard are available. Generally

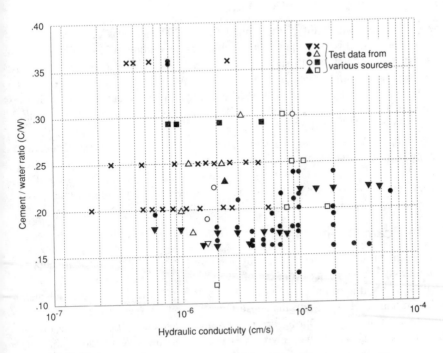

FIGURE 16-12
Cement-bentonite permeability test results.

speaking, for hazardous waste containment applications, cement-bentonite cutoff walls are rarely used.

Plastic Concrete Cut-off Walls

Another technique that may be utilized for containment systems is the plastic concrete slurry trench cut-off wall. How does a plastic concrete trench differ from a cement-bentonite trench? Plastic concrete is designed as a mixture of cement, bentonite, water and aggregate, unlike cement-bentonite which does not include aggregate and has a much higher water content. A second difference is that the plastic concrete trench is usually excavated in panels (see the following Diaphragm Walls section), and thus the plastic concrete is placed in the slurry-filled trench to replace the bentonite-water slurry used for excavation. For example, one project used a formulation of cement, pulverized fly ash, bentonite, and water to form the plastic concrete and employed the panel method of excavation to construct the cut-off to a depth of 160 feet below grade.[24] Studies reveal that plastic concrete has relatively low hydraulic conductivity, less than 1×10^{-7} cm/sec. Further, the shear strength is significantly greater than soil-bentonite or cement-bentonite. Finally, based upon the limited data available to date, plastic concrete may be more contaminant resistant than alternative techniques.[25] Although plastic concrete is stronger and less permeable than soil bentonite, it is more costly and has not found widespread use for hazardous waste applications to date.

Diaphragm (Structural) Cut-off Wall

The deepest cut-off wall built in the United States, 200 feet, was constructed as a diaphragm wall.[26] In this technique the trench is excavated under a head of bentonite-water slurry to maintain trench stability. However, it is excavated in panels, each some 20 feet long, using a clamshell or similar excavating tool. Steel reinforcement is then set into the bentonite-water slurry. The slurry is then displaced by high-slump concrete poured using the tremie method of concrete placement. A schematic of the method is shown in Fig. 16-13.

The diaphragm technique is widely used in applications requiring significant wall strength, such as where a cut-off wall is to be part of a structural earth retention system.[27] The hydraulic conductivity is very low, 1×10^{-8} cm/sec or lower. Due to its high cost the diaphragm wall has not, as yet, been used in a hazardous waste management applications.

Vibrating Beam Cut-off Walls

Vibrating beam cut-off walls have been employed for the horizontal control of contaminant migration. The vibrating beam cut-off wall constructed as shown in Fig. 16-14, is a grouting method appropriate for soils. A vibratory pile driver is used to advance a specially modified H-beam into the subsurface. The pile is specially equipped with injection nozzles at the tip. During driving, grout is injected to lubricate the driven pile. As the beam is withdrawn, a void is left equal to the size of the beam. The void is filled with grout pumped in through the injection nozzles as the beam is withdrawn. Subsequent beam penetrations are overlapped, resulting in a continuous barrier

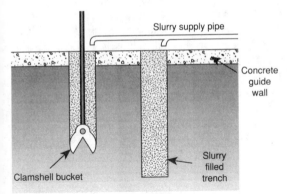

Excavate soil and replace with bentonite-water slurry

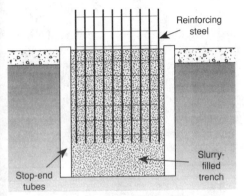

Place stop-end tubes and reinforcing steel
into fully excavated panel

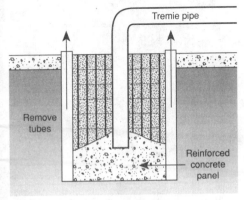

Pour tremie concrete to displace slurry,
remove stop-end tubes

FIGURE 16-13
Diaphragm slurry trench cutoff wall.

with a typical thickness of 2–3 inches. The slurry may be cement-bentonite (the most commonly employed) or a special bituminous grout. The principal advantage of this system is the elimination of the need to excavate potentially contaminated materials. Excavation of contaminated soils as in the slurry trench methods described earlier results in associated health and safety and disposal issues and costs.

The vibrating beam method does not inspire confidence in the integrity of the completed cut-off wall. First, the cut-off wall is relatively thin, only a few inches. Second, the location of the beam tip is uncertain, particularly with deep penetration. Although it has been demonstrated to be successful in sands of shallow depths,[28] it is difficult to penetrate dense soils for any significant depth. Further, even sands that may be initially loose are densified by the vibratory action of the pile driver resulting in difficulty in subsequent penetrations. This may lead to the need to pre-auger to achieve beam penetration.[29]

Deep Soil-Mixed Walls

Deep soil mixing has been employed to construct vertical barriers to horizontal ground water flows.[12] This technique was originally developed in Japan and has been recently employed in the United States. A special auger mixing shaft was developed that is rotated into the ground while simultaneously permitting the injection of bentonite and water or cement, bentonite and water slurry. The installation sequence, shown

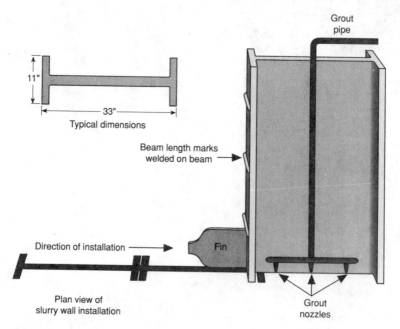

FIGURE 16-14
Vibrating beam-cutoff wall.

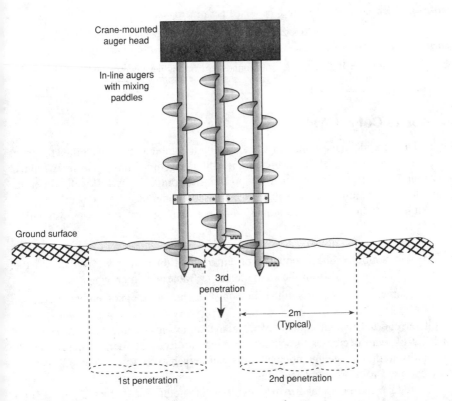

Crane-mounted
auger head

In-line augers
with mixing
paddles

Ground surface

3rd
penetration

2m
(Typical)

1st penetration   2nd penetration

FIGURE 16-15
Deep soil mixed cutoff wall

in Fig. 16-15, results in a column of treated soil when multiple mixing shafts are employed. Reinforcing can be added to the treated soil columns if additional strength is required.[30] A continuous wall, typically 20 to 36 inches wide, is obtained by over-lapping penetrations. Because the bentonite is added as bentonite-water slurry during mixing, the quantity of bentonite is typically limited to about 1%.

The advantages of the deep soil mixing technique are similar to those in the vibrating beam method. Health and safety risks are minimized by elimination of the need to excavate materials as required for conventional slurry trench cutoff wall techniques. The deep soil mixed wall can be constructed with added confidence because the wall is considerably wider than the vibrating beam wall. The hydraulic conductivity approaches 10^{-7} cm/sec.

Example 16-4   Deep soil mixed wall.   You are the field inspector for a deep soil mixed wall constructed to a depth of 30 feet and a width of 2.5 feet. The final cement content is to be 5% by dry weight. The dry unit weight of the soil in situ averages 112 lb/ft^3. The augers are configured to mix 7.5 feet of wall per penetration. How much cement needs to be added in each penetration?

Solution. Calculate weight of soil to be mixed per penetration:

$$W_t = 30 \times 2.5 \times 7.5 \times 112 = 63,000 \text{ pounds}$$

Cement weight at 5%:

$$W_c = W_t \times 0.05 = 3150 \text{ pounds} = \underline{\underline{1.6 \text{ tons per penetration}}}$$

Composite Cut-off Walls

Vertical barrier walls have been constructed using a combination of materials to result in a composite barrier. The concept of including a geomembrane barrier within a slurry trench cut-off wall is one such composite. The composite cut-off wall properties include lower hydraulic conductivity and increased contaminant resistance.

Although construction techniques vary between contractors, two installation methods are:

1. The geomembrane sheet mounted on an installation frame is lowered into the trench, disconnected from the frame, and the frame is withdrawn.
2. The geomembrane sheet is pulled in using weights on the geomembrane bottom.

It is reported that placement of a geomembrane in a cement-bentonite slurry trench cut-off wall decreases the overall permeability by two orders of magnitude. If high-quality joining of geomembrane sheets can be achieved, this decrease in permeability can be 4 to 5 orders of magnitude.[31]

To reduce the potential for the migration of methane gas from a landfill into adjacent land planned for residential construction, a slurry trench cut-off wall was built of soil-bentonite having a hydraulic conductivity of less than 1×10^{-6} cm/sec and incorporating a 40 mil HDPE geomembrane liner added to the upper portion of the wall to further reduce gas migration.[32] The geomembrane was considered necessary above the water table where the soil-bentonite is subject to wetting and drying cycles as well as potential desiccation drying and cracking (see Fig. 16-16).

Grout Curtains

Slurry trench cut-off wall technology is limited to soils that can be readily excavated. When it is necessary to install a vertical barrier to horizontal ground water flow in material that cannot be readily excavated (e.g., rock), then alternative technologies must be examined. Grout curtains are the most appropriate and commonly used alternative for these conditions. A grout curtain is installed in rock through the pressure injection of a pumpable material into the pores and fractures within the rock as shown in Fig. 16-17. The grout then gels or sets in place resulting in a material of lower permeability filling the voids and fractures. Grouts may be cement, pozzolanic, or chemical. Grouting technology utilized in hazardous waste containment systems has been developed for conventional geotechnical engineering applications such as beneath dams and levees to reduce seepage.[33] Grouting has been used both successfully

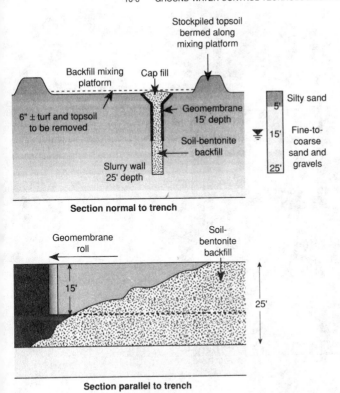

FIGURE 16-16
Composite cutoff wall.

and unsuccessfully[20] in hazardous waste management applications. Detailed studies have been conducted to show that the resistance of grout to attack depends upon the type of grout and the nature of the contaminant and their compatibility.[34]

16-6 GROUND WATER CONTROL TECHNOLOGIES: HORIZONTAL BARRIERS

Passive contaminant control systems include horizontal barriers as well as vertical barriers. Like vertical barriers, horizontal barriers in the subsurface are employed to contain contaminants and to redirect ground water flow. The subsurface horizontal barriers to vertical contaminant transport and ground water flow used in hazardous waste management can be categorized by their installation requirements. There are those that can and cannot be constructed in situ. "Constructed in situ" means that the barrier may be constructed beneath an existing waste site or contaminant plume. Most horizontal barriers must be constructed prior to the placement of overlying waste or contaminants. These horizontal barriers or liner systems, were developed

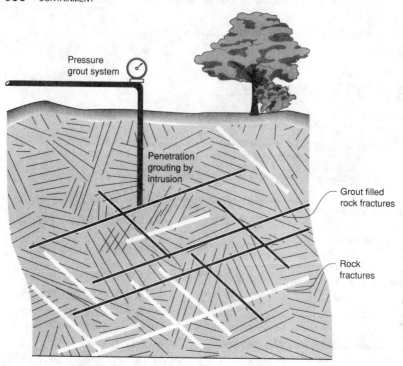

FIGURE 16-17
Grouting of rock fractures.

from conventional ground water control applications such as canals, dams, and water storage impoundments. The number of horizontal barrier techniques available is quite limited, particularly for in situ application.

The first step is to establish the function of a horizontal barrier in the framework of the site remediation system. One function of a horizontal barrier wall is to provide containment for contaminants. By employing a horizontal barrier constructed in situ beneath the site, the rate of contaminant migration is reduced. This is particularly important in sites having dense, non-aqueous phase liquids (DNAPL). Ground water pumping systems may not adequately control the downward migration of DNAPL in response to gravitational forces. Passive containment of contaminants using horizontal barriers (i.e., landfill liners) is widely employed as a method of hazardous waste site remediation in the United States—site remediation may be achieved by excavating wastes and contaminated soil and disposing of them in a landfill. In cases where the quantities requiring disposal are large, the landfill is often newly designed and constructed on site for the sole purpose of containing the site's contaminants.

Sec. 13-3, "Liner and Leachate Collection Systems," provides guidance to the construction of horizontal barriers for the disposal of contaminants on site. The range of liner systems and the design and construction considerations are the same whether the liner is for a new facility to accept off-site waste or a new facility to accept only on-site materials from the remediation program. Horizontal barriers constructed in situ are discussed below.

Grouted Liner Systems

As of the writing of this text, the use of grouted liner systems as a component of remediation is conceptual. The technology of grouting is well established for the control of water in conventional civil engineering applications such as seepage beneath dams and other water retaining structures. The grouting technology for liners is the same as was previously discussed for grout curtains. The difference relates to the design of the grout holes. For a grout curtain, the grout is injected along a linear alignment. For a grouted liner, the grout is injected uniformly in plan view to the desired depth beneath the site. Grouting for horizontal barriers has been successfully used to control the influx of ground water in excavations.[35] These applications make it possible to verify the effectiveness of the method. It is expected that grouted horizontal barriers will soon be used for waste containment applications. Where finer grained soils underlie the site, the naturally low hydraulic conductivity is utilized as the bottom containment barrier.

Block Displacement Technique

Another conceptual technique for the placement of a horizontal barrier is the block displacement technique. This method employs the principles of hydraulic fracturing commonly used to enhance oil recovery. Hydraulic fracturing of the subsurface materials occurs when the fluid pressure exceeds the total stress. The application of hydraulic fracturing to the installation of a horizontal barrier is shown in Fig. 16-18. The block displacement technique is actually a special form of grouting—high pressures are used to fracture the soil, and grout is then pumped into the fracture to create the horizontal barrier. As of this writing, the technique has been demonstrated in pilot studies but has not been employed in actual site remediation.[36]

Lagoon-Sealing Techniques

In some cases the leakage from the bottom of lagoons can be reduced by the installation of a horizontal barrier at the lagoon bottom constructed as the lagoon is still in service. The material used to construct this in situ horizontal barrier is bentonite, (see Sec. 11-3). Bentonite, a high swelling clay, can be processed to be in a granular form. Each granule is similar to a particle of sand in grain size but composed of thousands of colloidal-sized bentonite clay particles. When spread over the surface of a lagoon, the granules sink and form a blanket over the lagoon bottom. In time, and in the correct aqueous environment, the bentonite particles hydrate resulting in significant swelling. The in-place swelling process yields a liner of low permeability.

The principal constraint on this system is the chemistry of the lagoon liquids. Liquids that are acidic, organic, or with high electrolyte concentrations inhibit the hydration of the bentonite and impede the formation of a low-permeability barrier.

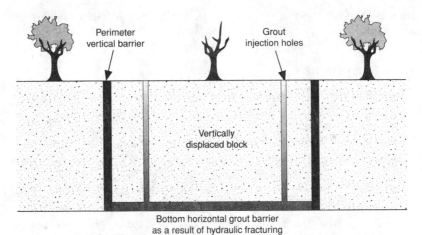

FIGURE 16-18
Block displacement technique.

In any case the completed barrier is not protected from erosion should the lagoon experience significant turbulence. In addition the ability to inspect the completed liner and ensure its performance is limited. Consideration of these constraints usually limits the use of lagoon sealing techniques to non-hazardous applications.

16-7 ACTIVE SYSTEMS

In addition to the various passive contaminant control systems (which are installed and then function without further ongoing energy input) active systems control and contain contaminant migration. These systems require ongoing energy input and include such systems as pump and treat, electrokinetics, in situ biotreatment and soil washing. As noted earlier, active systems such as those described below are frequently used in conjunction with passive containment systems. The following sections of this chapter describe remedial systems that actively "clean-up" the site.

Pump and Treat

The remediation of hazardous waste management sites requiring the cleanup of contaminated ground water often requires **pump and treat**. What is pump and treat? In this versatile approach, ground water is first extracted using any one of a number of recovery methods including wells, well points, and drain tile collection systems. Then the extracted water is treated using any one of a number of water treatment methods, described in Chaps. 9 and 10, such as air stripping, carbon adsorption, or biological treatment for organics and physical-chemical methods for inorganics. Thus, pump and treat schemes are developed on a site-specific basis for a range of site conditions and a range of contaminant types and concentrations. The technique is not without its limitations.[37] Monitoring of system performance often shows an initial decrease in the contaminant concentrations in the extracted water followed by a decreasing rate of decreasing contaminant concentrations. In the worst case the decline in contaminant concentrations is so small that the systems must operate for decades to clean up the ground water to drinking water standards. For example, the pump and treat system at Love Canal has been operating for more than 10 years with no end in sight. This uncertainty regarding the duration of operation makes the ultimate cost of pump and treat methods quite high. On the other hand, provided the annual cost is manageable, this technique protects the public health and the environment adequately for the present to allow for development of improved technologies in the future.

Drain Tile Collection Systems

A drain tile collection system, also known as a French drain, is constructed by excavating a continuous trench in the subsurface and installating of collection piping and filter media. By far Love Canal is the best known use of a drain tile collection system as a component in site remediation.[38] A cross-section through the system is shown in Figure 16-19.

The advantages of a drain tile collection system over wells or well points are several. The continuous nature of the drain tile collections systems intercepts a variety of subsurface features that otherwise allow water to circumnavigate its way around a pumping well including sand seams, root holes, buried conduits, and prior surface drainageways that have been filled in. Thus drain tile collection systems are more effective in the presence of subsurface heterogeneity and anisotropy. In the case of a floating contaminant, the drain tile collection system may require only a shallow interception of the ground water level. Alternatively, if a very deep trench is required, the need for excavation bracing and concerns for worker safety may limit the use of

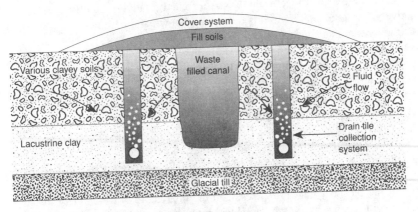

FIGURE 16-19
Section of containment system at Love Canal.

this technique. (The biopolymer technique discussed below is better suited for deep excavations.)

A drain tile collection system may be installed in the subsurface for several purposes, such as the Love Canal System.

- The system was used as an interceptor trench. Contaminants flowing with the ground water were intercepted by the trench precluding further migration.
- The system is used as a hydraulic barrier.
- The system serves as a means of ground water and contaminant extraction (pump and treat).
- The system serves to capture a plume that may have already migrated beyond the collection system alignment by reversing the gradient and "recalling" contaminants that may have left the site.

These last two reasons are frequently the basis for selection of a hydraulic barrier over a barrier of low hydraulic conductivity.

Biopolymer Extraction/Interception Trenches

The collection of ground water for treatment using wells or well points is limited in two ways: 1) wells are costly and time consuming to install, operate, and maintain; 2) ground water may move past the barrier in the region between the wells, even with relatively close well spacings. Alternatively, drain tile collection systems provide a continuum from which water can be withdrawn, providing a more positive cutoff. However, drain tile collections systems are costly and dangerous because they must be installed with workers entering the excavation. The excavation requires temporary shoring, ground water control, disposal of large volumes of soil, and risk to workers in the trench in a chemically and physically hazardous environment. The bio-polymer slurry trench (also known as trenching with vegetables) is a construction technique whereby a continuous subsurface drain can be installed relatively rapidly, safely, and cost-effectively.[39] The technique builds upon long-known and employed slurry trench methods to excavate a subsurface drain.[40, 41]

Using the slurry trench method of excavation, a trench is excavated under a head of slurry to maintain trench stability. In the conventional slurry trenches previously described, the slurry is formed by a mixture of bentonite and water. In the bio-polymer drain technique the slurry is formed by a mixture of bio-degradable materials (principally ground guar beans), additives (to extend the life of the biodegradable slurry), and water. In the conventional slurry trench method the slurry-filled trench is backfilled with a soil-bentonite backfill of low hydraulic conductivity. In the bio-polymer technique, the trench is backfilled with porous drainage media. Additives to the trench (to lower the pH to initiate enzyme action and to more rapidly enhance the breakdown of the bio-polymer slurry) result in the degradation of the slurry to water and natural carbohydrates. A sketch showing similarities and differences between the two techniques is shown in Fig. 16-20.

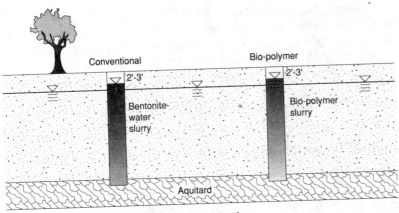

During construction

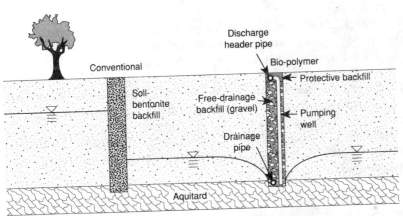

After construction and operation

FIGURE 16-20
Bio-polymer slurry trench.

The ground water is typically extracted from the trench through the use of well casings lowered vertically in the trench. Gravel is tremied around the well to maintain position and verticality while the remainder of the trench gravel backfill is pushed into the trench. Horizontal drain pipes can be installed along the bottom of the trench by using flexible pipe and a pipe-lying machine that follows along after the excavation. A typical completed installation is shown in Fig. 16-20.

16-8   OTHER REMEDIAL TECHNOLOGIES

All of the active and passive contaminant control technologies all have a common effect—they control the rate of contaminant migration. In that sense they are not "cleanup" technologies, but containment technologies. As noted, they may be and often are used in conjunction with other technologies. Even with pump and treat, where the treatment component may clean the ground water, very long time periods are required for operation resulting in this being used primarily as hydraulic containment. The remedial technologies described in the remainder of this chapter effect an actual clean-up of the site—they remove contaminants, rendering the site less contaminated. Additional treatment technologies are described in Chaps. 9 through 12.

Excavation and Disposal

The simplest, and oft applied, cleanup technology is to excavate the site and dispose of the material at another location. In times and localities where land disposal costs are low (usually due to inadequate environmental protection measures taken at the land disposal site) this is often the most economical alternative. Other treatment technologies may be selected including thermal (Chap. 12), biological (Chap. 10), and physico-chemical (Chap. 9).

How is the excavation and disposal remedial alternative implemented? Excavation and disposal employs conventional earth-moving equipment to excavate and transport the wastes and contaminated soils for disposal or treatment. Excavation equipment can include draglines, backhoes, clamshells, bulldozers, and loaders. Transportation is usually in containerized vehicles with adequate protection against spillage and leakage. In all cases, worker health and safety is a principal consideration. For contaminated sediment, hydraulic dredging equipment may be employed. Excavation of stream and river sediments typically requires handling a material that is about 90% water and 10% solids. Disposal and/or treatment may require dewatering of the sediment and treatment of the supernatant.

Soil Washing

What can be done about the thousands of tons of soil contaminated with a wide range of inorganic and organic contaminants? It does little good to clean ground water as described earlier in this chapter without remediating the source, such as contaminated soil. One approach is the application of soil-washing techniques. Washing may be with water, aqueous extractive agents, solvents, or even air.[42] Washing may be done on the entire soil matrix or on selected portions containing the contaminants that are separated from the clean portion by fractionation.[43] In a soil-washing demonstration project, 82% of the input sediment was classified as clean sand output, with the contaminants concentrated in the residuals.[44]

Electrokinetic Soil Decontamination

Electro-osmosis as a means to control ground water movement is a phenomena used and understood by civil engineers for more than 40 years.[45] In electro-osmosis, pore fluid moves in response to the application of a electrical potential, that is, a voltage difference (a constant, low, direct current). Chemical constituents also move in response to chemical potentials (chemical concentration gradients). Water and thus the dissolved chemical constituents also move in response to hydraulic gradients. Electrokinetics is the coupling of electrical, chemical and hydraulic gradients to effect contaminant removal. This technique has been demonstrated in the laboratory as effective in the removal of contaminants.[46,47] In one study, 75% to 95% of the adsorbed lead was removed by the process.[48]

Low-Temperature Soil Treatment

Soil contaminated with organics may be remediated by low-temperature thermal desorption. Soil is heated to 300°F to 400°F, well below incineration temperatures. Air is blown through the soil as a carrier for the desorbed organics, then captured and treated, typically by condensation.

In one such low-temperature soil treatment unit, the system removed the volatile organic compounds to below the detection limit.[49] A removal efficiency of 88% was found for a less volatile organic compound.

Permeable Treatment Beds

The use of permeable treatment beds for the in situ remediation of contaminated ground water is considered a conceptual technology and as such has not been used. The concept is to direct contaminated water through the permeable bed where it is subjected to treatment that could be chemical or biological. Chemical treatment could include activated carbon or ion exchange resins. Plugging and contact time are among the more important considerations.

The principal reason for the inclusion of a conceptual technology in this chapter is to alert the readers to the fact that new technologies are regularly under development in this rapidly changing field of hazardous waste management.

DISCUSSION TOPICS AND PROBLEMS

16-1. The site characterization at an old industrial facility on Long Island revealed a sandy site with shallow ground water contaminated with low levels of volatile organics. The contaminant plume was found to migrate a short distance beyond the property line but had not yet reached residences relying upon ground water for their drinking water supply. What might be the objective for cleanup of this site?

16-2. Cite three advantages and three disadvantages of containment as a means of hazardous waste site remediation.

16-3. Suppose that a section of the circumferential barrier wall described in Example 16-1 in the chapter text as constructed without adequately keying into an underlying layer of low hydraulic conductivity. A section of the original material 100 feet long and 3 feet deep remains with the original aquifer hydraulic conductivity of 1×10^{-2} cm/s (a shallow buried valley). What would the leakage through the wall be from this zone? Would this jeopardize the success of the associated pump and treat system?

16-4. A 1 m thick upgradient barrier is planned to control the influx of clean ground water to extraction wells. The downgradient ground water extraction wells and treatment plant are designed to operated at 2 m^3/s. The upgradient barrier is planned to be 500 m long, and the aquifer is 10 m thick. It is estimated that the average drawdown across the wall will be 4 m. By calculating the flow through the wall at various values of hydraulic conductivity, recommend a design value for the barrier.

16-5. Calculate the volume of water to be mixed with a 100 pound bag of bentonite for a 5% bentonite-water slurry. What will be the unit weight of this slurry?

16-6. What is the factor of safety for a trench 10 m deep excavated in sand having an angle of internal friction of 30°. The unit weight of the fluid is as calculated above. The sand has a total bulk density of 2.1 g/cm^3. How does the factor of safety change for a trench 20 m deep?

16-7. What is the factor of safety for a trench 10 m deep excavated in clay having a shear strength of 1 kg/cm^2. The unit weight of the fluid as calculated above. The clay has a total unit weight of 2.2 g/cm^3. How does the factor of safety change for a trench 20 m deep?

16-8. Describe the steps to be taken prior to determining a containment alternative at a hazardous waste site.

16-9. Describe the hydrologic pathways for contaminant migration at a typical waste disposal site.

16-10. List and briefly describe the types of vertical barriers available as passive containment alternatives.

16-11. What are active containment systems, and list some alternative techniques.

16-12. Construct the flow net for the conditions defined in Example 16-1.

REFERENCES

1. The Comprehensive Environmental Responsibility, Compensation and Liability Act)Superfund), 42 USC 9601 et seq., 1980.
2. Denner, J.: "Contaminated Land: Policy Development in the UK," *Proceedings of the 3rd Annual Conference on Contaminated Land: Policy, Regulation, Technology*, IBC Tech. Services, Ltd., London, England, Feb. 1992.
3. Loxham, M.: "The Control of Contaminant Migration Pathways as the Key to Cost Effective Remedial Measures for Contaminated Land," *Proceedings of the 2nd Conference on Contaminated Land: Policy, Regulation, and TechnologyK, IBC Technical Services, London, England, February 1991*.
4. Peck, R. B.: "Advantages and Limitations of the Observational Method in Applied Soil Mechanics," *Geotechnique*, vol. 19, pp. 171–187, 1969.
5. Brown, S. M., D. R. Lincoln, and W. A. Wallace: "Application of the Observational Approach to Remediation of Hazardous Waste Sites," CH2M-Hill, Bellevue, WA, April 1969.
6. Mott, H. V., and W. J. Weber: "Solute Migration in Soil-Bentonite Containment Barriers," *Proceedings of the 10th National Conference on the Management of Uncontrolled Hazardous Waste Sites*, HMCRI, Washington, D.C., pp. 526–533, November 1989.
7. Mott, H. V., and W. J. Weber: "Diffusion of Organic Contaminants Through Soil-Bentonite Cut-Off Barriers," *Research Journal WPCF*, vol. 63, no. 2, pp. 166-176, March/April 1991.
8. Ward, Linda M.: Close-up on Cleanup at Lipari Waste Site, *Hazardous Materials & Waste Management Magazine*, pp. 40–42, May-June 1984.
9. Xanthakos, P. P.: *Slurry Walls*, McGraw-Hill Book Company, New York, NY, 1974.
10. Spooner, Philip, et al.: *Slurry Trench Construction for Pollution Migration Control*, EPA-540/2-84-001, U.S. Environmental Protection Agency, Cincinnati, OH, February 1984.

11. Nash, K. L. "Stability of Trenches Filled with Fluids," ASCE, *J. Const. Div.*, vol. 100, no 4, pp. 533–542, 1974.

12. Ryan, C. R.: "Vertical Barriers in Soil for Pollution Containment," *Geotechnical Practice for Waste Disposal '87*, ASCE Geotechnical Special Publication No. 13, pp. 182–204, June 1987.

13. Acar, Y. B., A. Hamidon, S. Field, and L. Scott: "The Effects of Organic Fluids on Hydraulic Conductivity of Compacted Kaolinite," *Hydraulic Barriers in Soil and Rock*, ASTM 874, A. I. Johnson, R. K. Frobel, and N. J. Cavalli, eds., American Society for Testing and Materials, Philadelphia, PA, pp. 171–187, 1985.

14. Evans, J. C., H. Y. Fang, and I. J. Kugelman: "Containment of Hazardous Materials with Soil-Bentonite Slurry Walls," *Proceedings of the 6th National Conference on the Management of Uncontrolled Hazardous Waste Sites*, Washington, D.C., pp. 249–252, November 1985.

15. Alther, G. R., J. C. Evans, K. A. Witmer, and Hsai-Yang Fang: "Inorganic Permeant Effects Upon Bentonite," ASTM, *Hydraulic Barriers in Soil and Rock*, STP No. 874, Philadelphia, PA, pp. 64–77, July 1985.

16. GKN Hayward-Baker: *Case History—Mountain View Slurry Wall*, "Company Literature," Odenton, MD, 1989.

17. McCandless, R. M., and A. Bodocsi: "Hydraulic Characteristics of Model Soil-Bentonite Slurry Cutoff Walls," *Proceedings of the 5th National Conference on Hazardous Wastes and Hazardous Materials*, Las Vegas, NV, pp. 198–201, April 1988.

18. Evans, J. C., G. P. Lennon, and K. A. Witmer: "Analysis of Soil-Bentonite Backfill Placement in Slurry Walls," *Proceedings of the 6th National Conference on the Management of Uncontrolled Hazardous Waste Sites*, Washington, D.C., pp. 357–361, November 1985.

19. McCandless, R. M., and A. Bodocsi: "Investigation of Slurry Cutoff Wall Design and Construction Methods for Containing Hazardous Wastes," EPA 600-S2-87/063, U.S. Environmental Protection Agency, Cincinnati, OH, November 1987.

20. Trezek, G. J.: "Grout and Slurry Walls for Hazwaste Containment: The Down Side," *Hazardous Waste and Hazardous Materials*, vol. 3, no. 3, pp. 281–292, 1986.

21. Cooley, B.: "Behavior of Soil-Bentonite Slurry Trench Cutoff Walls," Honor's Thesis, Bucknell University, Dec. 1991.

22. Evans, J. C., Hsia-Yang Fang, and I. J. Kugelman: "Containment of Hazardous Materials with Soil-Bentonite Slurry Walls," *Proceedings of the 6th National Conference on the Management of Uncontrolled Hazardous Waste Sites*, Washington, D.C., pp. 249–252, November 1985.

23. Jefferis, S. A.: "Bentonite-Cement Slurries for Hydraulic Cut Offs," *Proceedings of the Tenth International Conference on Soil Mechanics and Foundation Engineering*, Stockholm, June 15–19, 1981b, A. A. Balkema, Rotterdam, pp. 435–440.

24. Reina, P.: "Slurry Trench Cutoff Surrounds Plant Site," *Engineering News Record*, p. 44, October 1, 1987.

25. Evans, J. C., E. D. Stahl, and E. Droof: "Plastic Concrete Cutoff Walls," *Geotechnical Practice for Waste Disposal '87*, ASCE Geotechnical Special Publication No. 13, pp. 462–472, June 1987.

26. *Engineering News Record*: "Cutoff Wall Scores a Record," *ENR*, pp. 17–18, Dec 1990.

27. Boyes, R. G. H.: *Structural and Cut-Off Diaphragm Walls*, Applied Science Publishers, Ltd., London, England, 1975.

28. Leonards, G. A., F. Schmednecht, J. L. Chameau, and S. Diamond: "Thin Slurry Cutoff Walls Installed by the Vibrated Beam Method," *Hydraulic Barriers in Soil and Rock*", ASTM STP 874, A. I. Johnson, R. K. Frobel, N. J. Cavalli, and C. B. Peterson, eds., American Society for Testing and Materials, Philadelphia, PA, pp. 34–44, 1985.

29. McLay, D.: "Installation of a Cement-Bentonite Slurry Wall Using the Vibrating Beam Method—A Case History," *Proceeding of the Nineteenth Mid-Atlantic Industrial Waste Conference*, Bucknell University, Lewisburg, PA, pp. 272–282, June 1987.

30. ASCE: "Deep Soil Mixing Builds Cut-off Wall," *Civil Engineering*, ASCE, p. 27, May 1990.

31. Jessberger, H. L. ed.: *Geotechnics of Landfills and Contaminated Land, Technical Recommendations*, European Technical Committtee, ETC 8, Ernst & Sohn, 1991.

32. GKN Hayward-Baker: *Case History—Oxnard Slurry Wall*, Company Literature, 1989.

33. Houlsby, A. C.: "Cement Grouting for Dams," *Proceedings of the 10th National Conference of the Conference on Grouting in Geotechnical Engineering*, W. H. Baker, ed., ASCE, New Orleans, LA, pp. 1–34, February 1982.

34. Bodocsi, A. and M. T. Bowers: "Permeability of Acrylate, Urethane, and Silicate Grouted Sands with Chemicals," *Journal of the Geotechnical Engineering Division*, ASCE, vol. 117, no. GT8, pp. 1227–1244, August 1991.

35. Raabe, E. W. and S. Toth: "Producing Slurry Walls and Sealing Slabs by Applying Soilcrete Process," *Proceedings of the Deep Foundations Conference, United Kingdom, pp. 111–118, 1989.*

36. Brunsing, T. P. and W. E. Grube, Jr.: "A Block Displacement Technique to Isolate Uncontrolled Hazardous Waste Sites," *Proceedings of the National Conference on Management of Uncontrolled Hazardous Waste Sites, HMCRI, pp. 249–253, November 1982.*

37. Mackay, D. M. and J. A. Cherry: "Groundwater Contamination: Pump-and-Treat Remediation," *Env. Sci. Technol.,* vol. 23, no. 6, pp. 630–636, 1989.

38. Schweitzer, G.: "Assessing Soil Contamination at Love Canal," *Proceedings of the National Conference on Management of Uncontrolled Hazardous Waste Sites, November 1982, pp. 399–405.*

39. Ralston, M.: "Trenching with Vegetables," *Civil Engineering*, ASCE.

40. Day, S.: "Extraction/Interception Trenches by the Bio-Polymer Slurry Drainage Trench Technique," *Hazardous Materials Control*, vol. 4, no. 5, pp. 27–31, Sept./Oct. 1991.

41. Day, S. R. and C. R. Ryan:"The State-of-the-Art in Biopolymer Drain Construction," *Slurry Walls: Design, Construction and Quality Control*, ASTM STP 1129, David B. Paul, Richard R. Davidson, and Nicholas J. Cavalli, eds., American Society of Testing Materials, Philadelphia, PA, 1992.

42. Raghavan, R., E. Coles, and D. Dietz: *Cleaning Excavated Soil Using Extraction Agents: A State-of-the-Art Review*, EPA/600/S@-89/034, January 1990.

43. Bardos, P. et. al.: "Contaminated Land Treatment Concepts under Development at Warren Spring Laboratory," *Euroenviron Contaminated Land Workshop,* Sept., 1990.

44. U.S. Environmental Protection Agency, "Soil Sediment Washing System," EPA/540/MR-921075, U.S. EPA, Oct. 1992.

45. Casagrande, L.: "The Application of Electro-osmosis to Practical Problems in Foundations and Earthwork," Technical Paper No. 30, Building Research Establishment, London, England.

46. Pamukcu, S., K. I.Lutful, and Hsia-Yang Fang: "Zinc Detoxification of Soils by Electro-Osmosis," *Transportation Research Record 1288*, Transportation Research Board, National Research Council, pp. 41–46, 1990.

47. Segall, B. A., O'Bannon, J. A. Mathia: "Electro-Osmosis Chemistry and Water Quality," *Journal of the Geotechnical Engineering Division*, ASCE, vol. 106, no. GT10, pp. 1148–1152, 1980.

48. Hamid, J., Y. B. Acar, and R. J. Gale: "PB(II) Removal from Kaolinite by Electrokinetics," *Journal of the Geotechnical Engineering Division*, ASCE, vol. 117, no. GT2, pp. 241–271, Febuary 1991.

49. U.S. Environmental Protection Agency, Low Temperature Therma Treatment System, EPA/540/MR-92-019, U.S. EPA, September 1992.

ADDITIONAL READING

Daniel, D.E. ed., *Geotechnical Practice for Waste Disposal*, Chapman and Hall, New York, NY, 1993, 683 pp.

ACKNOWLEDGMENTS

The authors wish to acknowledge the following individuals who assisted with the preparation of this chapter: Amuradha Ramana, ERM, Inc.; Rick Wolf, *The Pennsylvania State University—Harrisburg*; Jeff Burell, Mike Lucas, Jim Szykman, Bill Rinker, Mike Malusis, Ryan Bucceri, Dave Garg *Bucknell University*; and M. F. Weishaar, *Monsanto*. The principal reviewer for the ERM Group was Dayton Carpenter, *ERM-Midwest*.

CHAPTER
17

REMEDIAL
ALTERNATIVES
ANALYSIS

Who shall decide, when doctors disagree,
And soundest casuists doubt, like you and me?
                                                    Pope

After investigating a site and defining the hazards that it poses, two basic questions remain:

1. To what level should the site be remediated to protect human health and the environment (i.e., "how clean is clean")?
2. How should this level of protection be achieved (i.e., what is the "best" remedy)?

Deciding these two questions can be referred to collectively as "selection of remedy." Rarely is the decision self-evident. Several important factors compel the use of an open-ended, decision-making process that promotes full participation by interested parties, yet progresses to a conclusion such that the fundamental purpose is realized: the site is cleaned up. This chapter covers this process.

17-1 REMEDY SELECTION

The parties having a stake in the selection of a remedy come from diverse interests representing the local community, industry, environmentalists, the scientific and engineering professions, and government. Their concerns differ as do their values. They inevitably will disagree about what is the best remedy and even as to what level of

1001

protection is needed. In fact, positions can easily become inflexible with perhaps one party denying the existence of a problem and another insisting that the site poses grave risks. Adversarial posturing can occur, making communication difficult. A lack of trust can even enter the process, much as discussed in Secs. 1-4 and 8-5.

Such a climate invariably entangles and impedes the decision making process. However, even if decision making were confined to the technical community, engineers and scientists with experience in the same field would probably offer differing opinions regarding the needed level of protection and the best remedy. Four technical reasons explain the difficulty in selecting a remedy.

1. Site conditions can be very complicated.
2. Remediation is not straightforward but takes place usually over a long period of time and involves a number of interdependent steps.
3. A great number of remedial alternatives can exist, each having its own technical tradeoffs.
4. Experience does not exist yet to show clearly how best to proceed.

Because of these complicating factors, the selection of a remedy involves a great deal of judgment. The process of how this is done is important to ensure that relevant information and opinions are collected and considered. Selecting a remedy is equivalent to "problem solving," and the literature is replete with models usually embodying a classic process as follows:

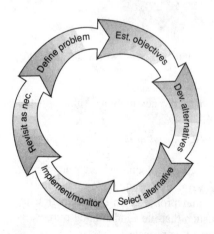

Figure 17-1 shows how this applies to the selection of remedies. It is a stepped process, featuring increasingly intensive analysis of fewer, yet ever more detailed alternatives.

Within the United States there are a number of separate and mostly independent remediation programs, with separate guidance for each. The same fundamental components of the classic "problem solving" process are found in each program even though specific regulatory guidance may prescribe some variants. This chapter avoids

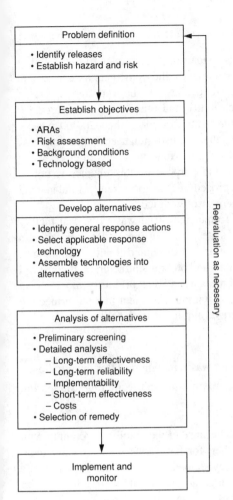

FIGURE 17-1
Process for selecting remedy.

using the specific guidance of any particular program. Regulatory guidance seems to change continuously. Yet, the *fundamental* process of selecting a remedy (i.e. the development and analysis of information) remains constant as it has in other planning endeavors. This chapter illustrates such a process. It probably does not comply with the exact requirements of any regulatory program, and a first step for practitioners in these endeavors is to obtain an understanding of the precise regulatory requirements and guidance applicable to the particular site.

The process covered in this chapter assumes that the site conditions are complicated—the site has complex geological characteristics, multiple sources of contamination, many different contaminants, or complicated pathways between sources and receptors. The selection of a remedy for sites lacking such factors should use a streamlined approach (e.g., minimizing, if not eliminating, screening efforts and limiting the number of detailed alternatives). Nevertheless, the decision-maker still should follow the same logical progression outlined in the remedy selection process of this chapter.

17-2   A GENERIC SITE

To illustrate the remedy selection process, this chapter takes examples from feasibility studies for the remediation of different Superfund sites and constructs a generic site. The documents prepared in studies of Superfund sites can be very lengthy (e.g., the site characterization, risk assessment, feasibility and other studies prepared for the Royal Hardage site near Criner, Oklahoma[1] number more than 20,000 pages). Therefore, some liberty had to be taken in the generic illustration used in this textbook. Simplifications of the procedural guidance of the Superfund programs in effect when the sites were studied can be noted because the purpose of this chapter is not to contrast or even to define precise regulatory procedure, but to present a fundamental process for selecting remedies. Some data were also modified.

The generic site for this textbook is a former industrial waste disposal facility that operated on a commercial basis under a state permit in the 1970s. It received a variety of industrial wastes including bulk liquids and drummed liquids and sludges. The wastes came from refineries, chemical plants, and other manufacturers. It consisted mostly of organic contaminants; however, some wastes were primarily inorganic in nature. Containerized sludges were placed intact without treatment in a pit termed the Drum Pit. Management of bulk and containerized liquids featured three steps:

1. Partial evaporation in shallow ponds.
2. In situ mixing of soil with the concentrated waste remaining after evaporation.
3. Excavation and disposal of the soil-waste mixture in a pit, called the Sludge Pit, excavated to bedrock.

The generic site was closed around 1980 because of the inability to comply with the more stringent regulations promulgated then for hazardous waste management. Closure of the site included scraping the visually stained soils from the mixing ponds and placing the soil in the Sludge Pit. The sludge and drum pits were covered with a soil cap, and they were referred to collectively as the "main source area." The site also contained a stormwater pond in the channel flanking the eastern boundary of the site, and it had received runoff from waste management areas. This and other surface features, including a nearby creek, are shown in Fig. 17-2.

The surface geology consists primarily of fine-grained sandstones and mudstones plus some unconsolidated alluvial sediments along the creek. The sandstones and mudstones, to a depth of 220 feet, can be divided into five distinctive strata as shown in Fig. 17-3. The upper three strata outcrop on the site. The uppermost stratum (Stratum I) yields submarginal quantities of ground water with the water table 15 to 30 feet below the surface. Stratum II is an aquitard. Stratum III resembles Stratum I, except that its yield is somewhat greater, serving as a marginal source of water supply to domestic wells in the region, although not near the site itself. Stratum IV yields only a few gallons per year of fresh to brackish water. Stratum V produces saline water only and has negligible hydraulic connection with overlying strata. In contrast with the bedrock strata, the alluvial deposits yield large quantities of ground water and are the source of domestic water supply for some of the residents living in the region.

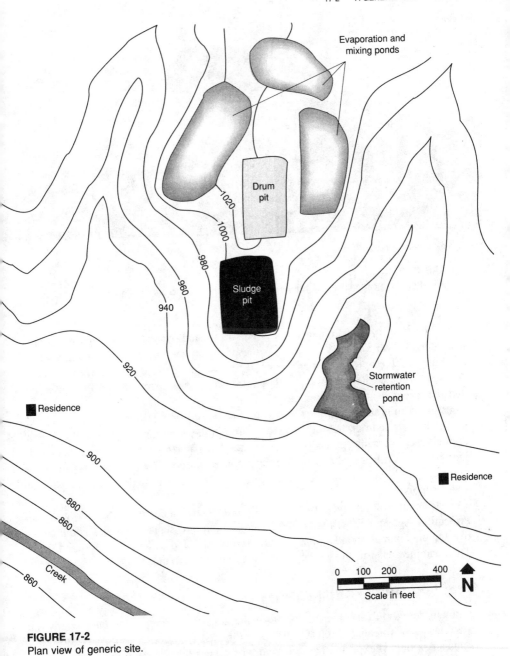

FIGURE 17-2
Plan view of generic site.

The historic migration of constituents from the main source area occurred primarily via stormwater runoff. Several areas of surficial soils and some sediments within site boundaries show significant levels of organic and inorganic constituents, including some pesticides and PCBs.

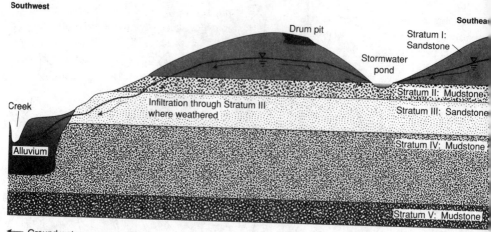

Southwest

Southeast

Drum pit

Stratum I: Sandstone

Stormwater pond

Stratum II: Mudstone

Creek

Infiltration through Stratum III where weathered

Stratum III: Sandstone

Alluvium

Stratum IV: Mudstone

Stratum V: Mudstone

←— Ground water movement

FIGURE 17-3
Simplified cross-section illustrating geological strata and ground water migration pathway.

Leaching of contaminants has affected the shallow ground water (the Stratum I aquifer) in two areas. A relatively widespread area exists east of the main source area, and affected ground water moves to the southeast toward the stormwater pond. A smaller area exists southwest of the main source area, extending into the alluvium, and the affected ground water moves to the southwest and south toward the creek. Constituent migration pathways via ground water are limited to very slow movement in the shallow bedrock but move rapidly through the alluvium, ultimately discharging to the creek. Organic constituents have not been detected in the creek, nor have inorganic constituents above background concentrations. The residences served by domestic wells in the alluvium near the site have been provided with an alternate water supply.

Remedial alternatives for the site were analyzed under the U.S. Environmental Protection Agency (EPA) Superfund program with two separate studies conducted—one for the sources (ponds, pits, soil and sediments) and the other for ground water. The examples used in this chapter pertain to just the ground water.

17-3 PROBLEM DEFINITION

Problem definition for a hazardous waste site consists of identifying the hazards (i.e. the releases or threatened releases of contaminants) (see Sec. 14-3). A hazard represents a risk only if exposure has occurred, or the possibility for future exposure exists. Relating hazards with current and possible future exposure is thus necessary to define the problems posed by a site. This is accomplished by defining a problem in terms of its **sources, pathways,** and **receptors**.

An example can be made for the case of the leaching of chemicals from contaminated soil to ground water. If the ground water is naturally brackish and, as a result, unusable by humans, there are no drinking water receptors; hence, no risks via this pathway. An analogous situation occurs at the generic site where natural geologic constraints prevent obtaining drinkable quantities from the affected shallow ground water in bedrock. However, the shallow ground water eventually discharges to the alluvium of the creek. Although no residents currently use the affected portion of the alluvium for drinking water, the alluvium does discharge into the surface water of the creek itself. Likewise, constituents in the affected ground water would discharge to the creek if not attenuated. Cattle may drink from the creek, and the human ingestion of milk and beef from such cattle does pose risk, however remote. Contaminants may be taken up by crops irrigated with surface water or shallow ground water and then ingested by humans directly or through animal products as described previously. The possibility also exists for dermal contact or incidental ingestion of the water in the creek.

At the site, migration of contaminants via the ground water to surface water was considered a problem. Some details defining this problem, particularly the receptors, are shown in Table 17-1.

17-4 REMEDIATION OBJECTIVES

Any alternative must satisfy the fundamental goal that it be protective of human health and the environment. To direct the study towards this goal requires setting specific objectives that, if satisfied by the ultimately selected remedy, will result in meeting this fundamental goal. Having defined the problems in the context of pathways, the remediation objectives for addressing these problems should relate explicitly to the pathways. It is ideal to express the objectives in measurable terms (e.g., contaminant concentrations at specific points of exposure). Only in that way can the effectiveness of a remedy be definitively ascertained. It is important to understand that the objectives may have to be revisited as more becomes known about the site, the receptors and the possible remedies.

TABLE 17-1
Problem definition example for the generic site: ground water pathway

Release mechanism	Leaching from source area to ground water
Environmental transport	Movement of constituents in ground water
Potential receptor points	Surface waters of the creek and pond; potential future drinking water wells into alluvium
Routes of human exposure	Incidental dermal absorption and incidental ingestion of surface water of the creek; ingestion of beef and milk from cattle drinking from the creek; ingestion of drinking water from potential future wells into alluvium
Potentially exposed human populations	Landowners of property on both sides of the creek downstream from the site and population served by dairy
Route of environmental exposure and exposed biota	Aquatic biota of the creek

There are at least four methods for establishing objectives:

Method	Example
Regulatory requirements and guidelines	Reduce trichloroethylene in all soil to not greater than the hypothetical regulatory limit of 10 mg/kg
Site-specific risk assessment	Reduce carcinogenic risk posed by all contaminants to not greater than 1×10^{-6}
Background conditions	Reduce all chemical constituents in soil and ground water to naturally occurring levels
Technology-based solutions	Excavate and incinerate all soil contaminated with greater than 10 mg/kg trichloroethylene

Objectives Based on Regulatory Requirements

In some cases there may exist under federal or state environmental law a "... standard, requirement, criterion or limitation" that is "... legally applicable or relevant and appropriate ..." under the circumstances of the release or threatened release of hazardous substances at a site.[2] In recognition of this as discussed in Chap. 2, these requirements represent remediation objectives for the actions taken to remedy the site. That is, the selected remedy must attain a level or standard of control equal to them.

The Superfund program in the United States has given rise to a new acronym, ARAR, or **A**pplicable or **R**elevant and **A**ppropriate **R**equirement (see Sec. 2-4). ARARs can be divided into two categories based on how specifically they were promulgated for hazardous substances encountered in site remediation. Requirements specifically addressing hazardous waste remediation are termed "applicable" requirements. They would be legally enforceable for virtually any remediation program. An example of an applicable standard would be one expressing the maximum allowable concentrations of a specific contaminant in soil. Few such explicit standards exist. These standards would apply to all sites regardless of location. The difficulty in developing national or even statewide standards is the enormous variation in geology, climate, population, environmental setting, and other site factors that determine risk. A standard may not provide adequate protection for one set of conditions and may be overly protective in another case.

Although few applicable requirements exist, there are potentially many "relevant" requirements. Environmental law is replete with requirements in the form of standards, criteria, and guidelines that have been developed as part of programs directed at issues other than hazardous waste (e.g., air quality). A requirement may address a constituent (e.g., lead), an action (e.g., incineration), or a location (e.g., wetlands). If a requirement generally pertains to the related conditions (i.e., constituent, action, or location) of the site or its remedy, the requirement is considered relevant. Many relevant requirements, however, simply are not appropriate for application to Superfund sites. Most simply "were developed to regulate very different types of problems than those involved in the Superfund program."[3] Some requirements, such as zero contaminant concentration goals, are technically infeasible. Many are arbitrary and bear little resemblance to site conditions. Their application to Superfund sites requires extensive judgment and this often creates disagreement as to whether the interpretation is correct.

Example 17-1 Relevant and appropriate requirements. Name the factors to be considered in determining whether a relevant requirement is appropriate.

Solution. The factors include:[4]

1. The purpose of the requirement.
2. The media regulated or affected.
3. The substances regulated.
4. The entities or interests affected or protected.
5. The actions or activities regulated.
6. Any variances, waivers, or exemptions to the requirements.
7. The type of place regulated.
8. The type and size of structure or facility regulated.
9. The use or potential use of affected resources.

At the generic site, a number of ARARs were identified. The key requirements are water quality standards for the surface water bodies that receive discharges of affected ground water, air quality regulations for fugitive dust control, and ambient air quality standards.

Objectives Based on Risk Assessment

The development of a remediation objective should take into account the hazards posed by the site, the potential for exposure, and the consequences if exposure were to occur. All of these are the elements embodied in the systematic framework of a quantitative risk assessment (Chap. 14). The concept requires establishing a target level of risk that becomes the remediation objective. An acceptable level of lifetime carcinogenic risk is generally considered to be in the range of 1×10^{-4} to 1×10^{-6} with 1×10^{-6} recognized as the point of departure for the purposes of conducting a comparative analysis. Similar guidelines exist for non-carcinogenic risk. The risks posed by a site for all source-pathway-receptor combinations under current and future exposures are calculated by receptor and compared with the objective—the target level of risk. If the risks posed by the site exceed the objective, remediation is required to reduce the risk to the target level.

The target risk level thus drives the selection of remedy. A remedy can reduce risk by reducing the hazard (e.g., removing highly contaminated soil) or reducing the potential for exposure (e.g., providing alternate drinking water supplies if the ground water were contaminated). In fact, by defining an acceptable level of risk, the risk assessment process can be used in reverse to calculate the acceptable concentration or mass of chemical contaminants that could remain on site.

To use risk assessment requires making some far-reaching decisions about future exposure scenarios. Although most sites have not resulted in significant exposure of off-site receptors, the potential is there. The potential exposure depends in part on the future uses made of the site and surrounding areas. To calculate risk, hypothetical scenarios must be developed about future uses of the site, the type of development that could occur on nearby properties, and the future use made of natural resources, particularly ground water. Starkly different amounts of residual soil or ground water contamination could be allowed to remain, dependent upon assumptions such as whether ground water will be consumed or whether houses will be built on or adjacent to the site. Risk assessment is an excellent tool for examining such scenarios.

The ground water migration pathway at the generic site provides an example of using risk assessment as a basis for setting remediation objectives. As stated above, the calculation of risks requires developing exposure scenarios. The key features of exposure scenarios assumed for the generic site are as follows:

- **Exposure Scenario I—Current site conditions.** This scenario appears to present no unacceptable risks due to the fencing of the site, the absence of detectable target contaminants in the creek, and the provision of an alternative water supply to prevent the domestic use of affected alluvial ground water.
- **Exposure Scenario II—Most probable land and water use over the next seventy years.** This scenario assumes more intensive use patterns and, hence, more potential exposure to hazardous substances than current conditions. Some contaminants are predicted to migrate, and exposure could occur from recreational use of the creek and consumption of beef and dairy products from cattle drinking from the creek.
- **Exposure Scenario III—Hypothetical "worst-case" scenario.** Future domestic use of affected alluvial ground water is assumed prior to it discharging to the creek. This scenario is characterized as unrealistic in that no credible future circumstances could be identified wherein this particular ground water would be used as a drinking water supply. For example, it would involve the adjacent residences knowingly consuming water from their affected wells rather than using the alternate water supply.

Based on these scenarios, the carcinogenic risks were calculated as shown in Table 17-2.

It should be noted that Exposure Scenarios II and III do not represent expected risks but are based on conservative assumptions and safety factors and are useful only in selecting a remedy. Even so, Exposure Scenario II does not pose unacceptable risk. Only when assuming future domestic use of the alluvial ground water in Scenario III do the estimated risks enter the debatable range as to what level of risk is acceptable.

Establishment of a risk-derived objective involves deciding what level of risk is acceptable and what is the appropriate exposure scenario. This is a policy decision that should consider "likely future land and resource use, the proximity of residents, the use of surrounding properties, the level of contamination of surrounding properties, and specific environmental issues."[5] To do this is to hypothesize, and it adds a

TABLE 17-2
Summary of carcinogenic risks posed by ground water pathway at the generic site

	Assuming upper bound predicted 70-year average concentrations	Assuming best estimate 70-year average concentration
Exposure scenario I	Risk $< 10^{-12}$	Risk $< 10^{-12}$
Exposure scenario II		
Child 0–6 years	8E-12	3E-12
Child 6–12 years	1E-10	4E-11
Adult	7E-11	2E-11
Exposure scenario III		
Child 0–6 years	1E-05	7E-06
Child 6–12 years	8E-06	3E-06
Adult	1E-04	8E-05

degree of uncertainty that has hindered public acceptance of risk-derived objectives. Nevertheless, risk assessment remains a primary tool for setting objectives. Without it, hypotheses must still be made but on an implicit, intuitive basis labeled as "professional judgment."

Other Methods for Setting Remediation Objectives

One simple method for establishing objectives is to have a single objective that the site be restored to the **background** conditions that existed before the site was contaminated. For the generic site, this would require elimination of all contamination other than that from cattle ranching. Restoration to background conditions is a flawed, if not infeasible, approach at many sites. An example is a site surrounded by heavy industrialization where risk to receptors may not diminish measurably unless the surrounding properties also restore their sites to background conditions. Further, the surrounding industrial use may conflict with the possible recreational and residential uses implied by returning an industrial site to background conditions.

Using **technology-based standards** is another method for developing objectives. Actually, the method would use a subset of the regulatory requirements discussed early in this section, and it has been applied in the United States in the past (e.g., the regulatory program for controlling water pollution). Under this concept, contamination would be reduced to a level consistent with what could be achieved by a specified available technology. For example, if the specified technology were incineration, the concentration of organic contaminants would have to be reduced by 99.99%. The decision would remain open whether this was achieved by incineration or an equivalent alternative. This required level of reduction would apply even if it were overly protective considering specific site conditions. If it were deemed underprotective, additional measures would be necessary. A variant of this concept is to develop a set of minimal remedial requirements based on general site conditions and potentially exposed populations.

Example of Remediation Objectives

A number of remedial objectives for the ground water pathway at the generic site could be considered.

1. Prevent spread of affected ground water into unaffected areas.
2. Protect drinking water supply at affected residences.
3. Control affected ground water migration to the alluvium and surface water bodies.
4. Reduce ground water constituent levels sufficiently to satisfy appropriate regulatory requirements.

Style and content are both important when writing objectives because all subsequent steps focus on meeting those objectives. It is important to note that a remediation objective should express implicitly if not explicitly: exposure route, receptors, contaminants of concern, and a measurable standard of performance or control. Ideally, the standard of performance could be expressed as an acceptable contaminant level in quantitative terms (e.g., concentration or amount). Objective No. 4 above is an example of such an approach in that regulatory standards as they exist provide the quantitative terms. It is not always possible or desirable to use or imply quantitative terms. In such a case, the standard of performance still should be definitive.

Example 17-2 Implication of remediation objective. What are the implications of the following three remediation objectives for a hypothetical site where leachate from contaminated soil is migrating via ground water to potential future locations for drinking water wells?

1. Reduce contamination of soil sufficiently such that levels of contaminants at wells represent a carcinogenic risk no greater than 1×10^{-6}.
2. Prevent spread of contamination to wells.
3. Provide safe drinking water to users of wells.

Solution. No. 1 is possibly too specific because it necessitates source control, either treatment or removal of contaminated soil. No. 2 is fully protective of human health but may be accomplished via various measures including containment, ground water treatment, as well as source control. By not using quantifiable terms, No. 2 is absolute and would imply to some parties that "zero" contamination is required. No. 3 could be accomplished via institutional controls and an alternate water supply, without treating, removing, or containing any contamination.

A remediation objective should not specify a technology, or even a general class of technologies. That comes later. Specifying a technology may preempt a more effective one, or it could even direct the study toward an infeasible one. For example, at the generic site, a sole objective stating "pump and treat bedrock ground water" is not technically feasible because even aggressive pumping would yield only negligible quantities of ground water due to hydrogeologic constraints and physiochemical limitations posed by the bedrock formation. This illustrates the need to keep an eye toward what is achievable to prevent establishing impossible objectives.

An exception, of course, to not specifying a technology in the objective is the case when ARARs clearly mandate a technology. As described earlier, however, this may create problems because even if a specific technology is deemed "appropriate" by some regulation, it may not be feasible at a given site.

Although it is not practical to pump and treat the shallow ground water at the generic site, it is appropriate to limit the amount of affected ground water discharging to surface water. Because possible future domestic use of the alluvial ground water poses what could be considered an unacceptable risk, it is also appropriate to prevent this from happening. Thus, the following remediation objectives were established for the ground water pathway:

1. Monitor/control affected ground water discharge to the creek to ensure that surface water quality standards continue to be met.
2. Prevent domestic and agricultural use of contaminated ground water.
3. Control the spread of the plume of contaminated ground water.

Remediation objectives represent steadfast requirements that should be considered mandatory for any analysis of alternatives—all alternatives developed for evaluation should, as a minimum, meet the objectives. Upon finding an alternative is unable to meet the objectives, it should be excluded from further consideration.

17-5   DEVELOPMENT OF ALTERNATIVES

Having established remediation objectives, there are five steps in developing alternatives:

1. Identify general response actions for each objective.
2. Characterize media to be remediated.
3. Identify potential technologies.
4. Screen the potential technologies.
5. Assemble the screened technologies into alternatives.

General Response Actions

The task of developing technical alternatives begins by identifying **general response actions** for each remediation objective. A general response action is a tersely cogent description of the remedial actions that will satisfy the objective. It could be thought of as a broad category of technologies. For a site with an objective to control leaching of constituents from contaminated soils, an example of a general response action could be "excavation and treatment of contaminated soil." It stops short of defining how the soil would be excavated or what treatment technology would be used. Other examples for the same objective are "control infiltration" and "contain migration of constituents." General response actions provide direction and focus to the subsequent tasks.

Characterization of Media to be Remediated

The general response actions, by originating from a source-pathway-receptor analysis, are already grouped by medium (e.g., ground water, soil). The step following the identification of general response actions is to quantify the extent to that the various media must undergo remediation. This need be done only in approximate terms, consistent with the general response action and sufficiently accurate to allow the general sizing of technologies. Examples of this characterization for the medium of contaminated soil include the following:

1. If the general response action was excavation and treatment of contaminated soil, it is necessary to estimate the volume of soil with contaminants exceeding the levels identified in a remediation objective.

2. If the general response action was controlling infiltration, it is necessary to estimate the areal content of contaminated soil.

3. If the general response action was containment, it is necessary to estimate the lateral and areal extent of contaminated soil.

Example 17-3 Excavation of contaminated soil. Soil samples were collected at a depth of 2 feet, 4 feet, and 6 feet from the property of a former battery-recycling plant. The property measures 1,000 feet by 300 feet and is suspected to be contaminated with lead. The sample locations are in an uniform grid as shown in Fig. 17-4. Table 17-3 lists the soil depth and the lead concentrations for each soil sample.

The nature of such contamination is that the concentration of lead decreases with depth. One response action is to excavate for off-site disposal all soil containing greater than 1000 mg/kg of lead. What is the approximate volume of affected soil requiring excavation under this alternative?

Solution. At each sample location, the depth to a concentration of 1000 mg/kg is approximated by interpolation. As an example, such depths for Sample Locations No. 6 and 14 are 4 feet

Plan View

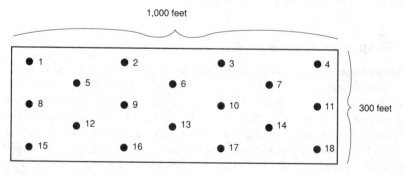

FIGURE 17-4
Contaminated soil sample location.

TABLE 17-3
Soil depth and lead concentrations

Soil sample location	Sample depth (ft)	Lead concentration (mg/kg)	Soil sample location	Sample depth (ft)	Lead concentration (mg/kg)
1	2	20	10	2	2000
2	2	150	10	4	1200
3	2	25	10	6	1000
4	2	20	11	2	200
5	2	300	12	2	2000
6	2	1200	12	4	250
6	6	800	13	2	800
7	4	1100	14	2	1400
7	6	800	14	4	1100
8	2	200	14	6	900
9	2	1800	15	2	100
9	4	1250	16	2	50
9	6	850	17	2	50
			18	2	100

and 5 feet, respectively. Next, around each sample location exceeding the 1000 mg/kg target, a polygon is delineated extending midway between any two sample stations. Because in this case the samples were located in an uniform grid, each represents an area of 17,000 ft². With this information, the volume of soil to be excavated can be estimated as follows:

Soil sample location	Approximate depth of soil exceeding target level (ft)	Approximate volume of soil exceeding target level (yd³)
6	4	2,500
7	5	3,100
9	5	3,100
10	6	3,700
12	3	1,900
13	1	600
14	5	3,100
		Total Volume = 18,000 yd³

Identification and Screening of Technologies

After characterizing the media, potentially applicable response technologies are identified. Many technologies are covered in Part Three of this book as well as in Chap. 16. Examples of two general response actions applicable to ground water at the generic site, and some types of potential technologies that conform to each general response action, are shown in Table 17-4.

The next step is to evaluate the identified technologies against a set of screening criteria. Important criteria are **suitability** (general technical ability to address contam-

TABLE 17-4
Potential ground water technologies at the generic site

General response action	Types of potential technologies
In situ treatment	Solvent flushing
	Bioreclamation
	Permeable treatment beds
Containment system	Cutoff walls or barriers
	Recovery wells
	Recovery trenches
	Horizontal barriers
	Vacuum extraction wells

inants at the site), **implementability** (primarily the ability to construct and operate the technology), **performance** (effectiveness and reliability), **environmental and health concerns, institutional concerns** (permitting), and **cost** (in relative, qualitative terms).

Table 17-5 shows examples of some of the technologies identified for ground water containment and in situ treatment and the results of the screening evaluation at the generic site. At this point in the process, the evaluation is very qualitative and must be based largely on judgment and experience. Thus, each technology is simply rated as being favorable (+), neutral (0), or unfavorable (−) with regard to the criteria. The usual reasons for eliminating technologies include unfavorable performance, difficulty in implementation, or excessive cost. For the generic site, more than 80 technologies were evaluated with about half retained for incorporation into alternatives. Table 17-5 provides summary remarks explaining the reason for eliminating a technology.

Assemblage of Technologies Into Alternatives

Technologies not eliminated are retained for further consideration. The retained technologies are assembled and configured into a set of alternatives covering the full range of actions that could be undertaken to remediate the site. The goal is to develop a sufficiently wide range of alternatives, each of which accomplishes the remediation objectives, to support a comparative analysis of tradeoffs.

The remediation of a site typically takes place over a long period of time and features a number of technologies that are implemented in an interdependent manner. An alternative is thus comprised of various components. The technologies represent building blocks in developing an alternative; some are **key components** while others are **ancillary**. A key component is one that establishes by itself a general approach for remediating one or more pathways. Ancillary components simply contribute to the overall performance of a key component or its approach. An alternative directed at ground water control for a hypothetical site may feature a cutoff wall (as the key component). A cutoff wall would require recovery of ground water that could be accomplished by either recovery wells or a recovery trench as ancillary components.

TABLE 17-5

Selected results of technology screening, generic site

Technologies	Suitability	Implement-ability	Performance	Environ. and health concerns	Cost	Retained	Remarks
Ground water containment							
Slurry wall	+	+	+	+	0	Yes	Can be applied to impede downgradient ground water movement or to reduce the recovery and treatment of clean ground water.
Cement diaphragm wall	+	+	+	+	−	Yes	Only where a slurry wall is inappropriate. Higher construction cost than a standard slurry wall.
Vibrating beam	0	+	−	+	0	No	Difficult to achieve a uniform low permeability barrier. Installation equipment can easily be deflected by buried debris.
Grout curtain	0	0	−	+	−	No	Difficult to achieve a uniform low permeability barrier. Prohibitive installation cost.
Sheet piling cutoff wall	0	0	0	+	0	No	Limited to cohesive soil applications. Difficult to achieve an effective low permeability barrier.
Recovery well system	+	+	0	0	+	Yes	Primarily used for control and recovery of affected ground water and light non-aqueous phase liquids. Can be used as an ancillary component to in situ treatment systems.
Recovery trench system	+	+	+	0	0	Yes	Implemented to control and recover affected shallow ground water and light non-aqueous phase liquids.
Recharge well system	0	0	0	0	0	Yes	Adjunct system to dispense in situ treatment fluids.
Well vacuum system	+	+	0	0	0	Yes	Enhances fluid recovery efficiency of a standard recovery well system. Application limited to fine-grained soil or rock formations and most confined aquifers.
In situ treatment							
Permeable treatment beds	0	0	−	+	−	No	Unproven technology and implementation costs are prohibitive.
In situ bioreclamation	0	0	0	−	0	Yes	Suitable for treating low concentrations of dissolved or organic compounds. Probably only applicable to low permeabilities of bedrock strata.
Soil ventilation	−	−	0	0	0	No	Low permeability alluvium due to site conditions. Low permeability formations at site would inhibit successful implementation.
Soil flushing	−	−	0	0	−	No	Not suitable to site conditions. Low permeability formations at site would inhibit successful implementation.

1017

To simplify the effort, ancillary components are tied together with key components to form larger building blocks. Some technologies are a necessary component of all alternatives, because no remedy would be protective without them. Typical examples include monitoring, removal of highly contaminated sources (i.e., "hot spots"), and control of surface water. They are usually grouped and referred to as "primary controls" for inclusion in all alternatives.

One alternative may feature treatment and another containment. If a site has several discrete sources needing remediation, alternatives can be structured with each using the same technical approach but providing a different spatial coverage. An example is the use of cutoff walls; in one alternative, a cutoff wall may flank more discrete sources, or in another it may follow a different configuration. Such represent different alternatives because the level of control differs substantially.

Two or more types of general treatment concepts may be retained as separate alternatives (e.g., incineration and soil vapor extraction). However, the range of alternatives should not be so narrow as to consist entirely of comparing one unit operation with another, both representing essentially the same concept (e.g., two types of incineration). This comes later as part of the design of the selected remedy.

Up to this point, the approach to remediation has been piecemeal rather than integrated. A typical case is that the contaminated media probably have been considered separately, as was done at the generic site. When assembling alternatives it is important to examine comprehensively the entire site and its remediation. For example, the approach to remediating a contaminated plume migrating with the ground water needs to be integrated with the steps taken to address the source of that contamination because the time frame for remediating ground water, if not the minimum concentration levels attainable, depends in large measure on such source control actions.

The alternatives developed for ground water at the generic site were assembled from the following components:

Primary Controls

Common to all alternatives
- Ground water monitoring
- Surface water monitoring
- Security fence

Common to all alternatives except "no action"
- Excavation of affected soils outside the main source area
- Surface cover over main source area
- Surface water controls
- Alternate water supply to residences

Optional key components

- Alluvial ground water recovery (by wells or trench) and treatment
- Alluvial in situ bioremediation
- Bedrock ground water recovery (by wells or trench) and treatment in the following areas:
  - Southeast area
  - Southwest area
- Cutoff wall (with recovery wells or trench) in the following areas:
  - Around main source area (drum pit and sludge pit)
  - Along southwest boundary
  - Along southeast boundary
- Deep recovery trench around main source area (drum pit and sludge pit)

These components were combined and assembled to yield 21 separate alternatives for the generic site, representing varied combinations of technical approaches, spatial coverage, and volumes of media remediated. In summary, the process of developing alternatives started with problem definition and the establishment of remediation objectives, and it has proceeded through the screening of technologies to assembling them into comprehensive alternatives. This is shown as Fig. 17-5 using examples from the generic site.

17-6 ANALYSIS OF ALTERNATIVES

Preliminary Screening

This assemblage of the technologies can easily yield a large number of viable, distinct alternatives. A large number, such as the 21 alternatives in the case of the generic site, needs to be screened against basic criteria to limit the total to a manageable level of the more promising alternatives for the final, detailed analysis. The screening criteria include effectiveness, implementability, and cost.

Information needs to be developed on important factors (e.g., the need for off-site disposal, required permits, time frame for realizing beneficial results) necessary to apply the screening criteria. For example, each alternative should be defined as to size and configuration:

- Treatment—flow rates and process sequence
- Containment—location and general dimensions

At this point, cost estimates can be of a preliminary nature but must include both capital cost and the cost of operation and maintenance. The cost estimates may derive from prior experience, cost curves, and vendor information. Nevertheless, it is still important to ensure that all estimates share the same degree of accuracy.

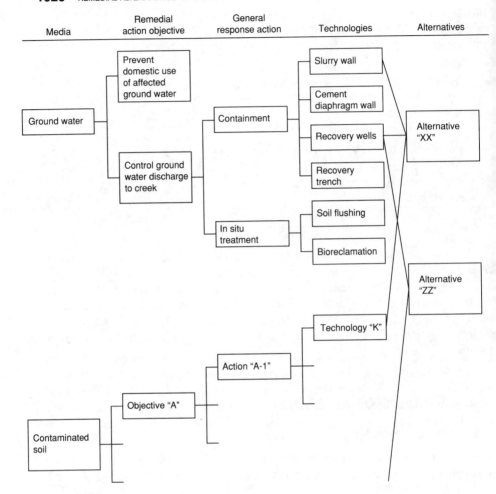

FIGURE 17-5
Development of alternatives.

The preliminary screening should yield the final set of discrete alternatives covering a wide range of remedial actions that could be implemented at the site. The number of final alternatives depend primarily on the degree of complexity presented by the site; whether compelling information shows that certain alternatives are likely to be chosen, and the possible need to evaluate certain alternatives because of regulatory requirements; preferences of regulatory agencies, persuasive public opinion, or institutional constraints. Usually, 4–6 alternatives are selected, and certainly not more than 10. One of the alternatives should be a "no action" alternative as a baseline for measuring the effectiveness of the others. Based on the screening of the 21 alternatives developed for the generic site, six were selected for further analysis as shown in Table 17-6. It should be noted that the six alternatives do not address source control.

TABLE 17-6
Summary of alternatives (generic site)

| Alternative | Primary controls | Ground water recovery and treatment | | | Cutoff wall/trench |
		Alluvium	Southwest area	Southeast area	
A	Partial				None
B	X				None
C	X	X			None
E	X		X	X	None
E1	X		X	X	Cutoff wall along southwest boundary.
L	X	X			Main source area surrounded by either cutoff wall or deep recovery trench.

Source control, because of regulatory procedures, was the subject of a separate study. Instead, each of the six address only ground water, and they had to assume a certain level of source control.

Technical Data Needed for Final Alternatives

Upon selecting the final set of alternatives, each should be reexamined and refined so that a thorough comparative analysis can be made. The level of detail should be consistent with that of a concept design but short of what would be required for a detailed design to implement an alternative. The particular technical details needed for each alternative are essentially as follows:

- Preparation of a basic component diagram and conceptual layout
- Major equipment needs and utility requirements
- Special engineering considerations for implementation of the alternative
- Potential environmental impacts and some methods for mitigating adverse impacts
- Operation, maintenance, and long-term monitoring requirements
- Implementation requirements including safety considerations, regulatory and permit requirements, temporary storage, off-site disposal needs and transportation
- Phasing and segmenting options
- Preliminary implementation schedule

Example 17-4 Concept design vs. engineering design. The types of information developed as part of an engineering design are unnecessarily detailed for the comparative analysis required to select an alternative. Instead, the level of detail should conform with that of a concept design. What are the differences between the two (use various categories of information as examples)?

Solution

Category of information	Concept design	Engineering design	
		Preliminary design	Final plans & specs
Layout	Component diagram	Piping and instrumentation diagrams	Piping drawings
Cost estimate	±50%	±30%	±10%
Required equipment	List of major equipment	Preliminary specifications for major equipment	Purchase or technical specifications for all equipment and material

Schedule requirements for completing the study can constrain the level of information obtained for each alternative. As an example, treatability tests may be necessary to confirm the applicability of a unit process or to define the parameters that have a major influence on costs. However, timeframes may not allow such studies, suggesting that promising technologies should be identified and tested as early as possible during the field investigation. If not, the study may have to proceed with literature data, and that may prove inconclusive and thus inhibit the selection of that particular remedy.

Criteria

The selection of the preferred alternative is based on a comparative analysis using a set of criteria. The specific criteria will vary depending upon the regulatory program and the latest guidance from the responsible agency. These criteria tend to fall in the following broad categories:

1. Long-term effectiveness
2. Long-term reliability
3. Implementability
4. Short-term effectiveness
5. Cost

Example 17-5 Comparison of criteria. What are the differences between the previously listed five criteria and the nine required under the EPA's National Contingency Plan required under SARA[6] (see Sec. 2-4)?

Solution. The EPA criteria are divided into three groups. The five EPA criteria referred to as the "Primary Balancing Criteria" group are essentially the same as the five previously listed. In fact, three read the same. A fourth is close; instead of "long-term effectiveness and permanence," it deletes the term "permanence." The fifth criterion in EPA's list is "reduction of toxicity, mobility, or volume through treatment" and is replaced herein by "long-term reliability." The reason for this replacement is that treatment is simply a means for achieving long-term reliability, with permanence the ultimate achievement. As expressed by Clean Sites, "It is not the use of treatment, but the result of using treatment, that is important."[7]

EPA's criteria include two other groups that are not addressed by the five criteria presented in this textbook. EPA's first group consists of two criteria referred to as "threshold criteria" (i.e., they must be satisfied for an alternative remedy to be eligible for selection). These two criteria are redundant because the process presented herein develops only alternatives that do indeed meet these requirements. The last two EPA criteria, "modifying criteria," are not true criteria; instead, they ensure that the selected remedy reflects the concerns of the state and local community, something that a proper public participation process would achieve.

Long-Term Effectiveness

Long-term effectiveness can be analyzed in two ways. The simple way is a largely qualitative assessment of how well an alternative meets the remedial action objective over the long term. On the other hand, a complete analysis would calculate the residual risk; this is the quantitative risk represented by untreated contaminants or treated residuals remaining at the site at the conclusion of the remedy's construction and operation period.

Many regulatory agencies require calculation of residual risk. This may seem unnecessary because a remedy that achieves the remedial objectives for a site is, by definition, resulting in acceptable residual risks. Is it meaningful, then, to calculate how acceptable is the residual risk? These calculations have value when examining tradeoffs among alternatives, provided the differences warrant the expense of such calculations. An example of such calculations was provided in Sec. 14-9 for the Lackawanna Refuse Site.

"Long-term" corresponds to a period of 70 years, (i.e., the life expectancy for calculating lifetime exposure to adults). Because residual risk depends on the toxicity, mobility, and mass of contamination, an obviously effective solution is one that treats the contaminants such as to reduce any or all of these partial components of risk. However, if a containment-type alternative provides barriers and controls sufficiently suitable in regard to the specific site conditions to meet the remedial action objective, the alternative is considered to be effective, even though it may not reduce the toxicity, mobility, or mass of contamination. The question remains whether a containment alternative provides long-term reliability, and that depends on the adequacy of its controls and is the subject of the next criterion.

Long-Term Reliability

Long-term reliability is only an issue with the alternatives that leave untreated contaminants or treatment residuals at the site at the conclusion of the implementation period. It is obvious that the alternatives providing the greatest degree of reliability are those providing "permanence." Stated in other words, if long-term reliability were measured on a continuum, permanence would represent the absolute maximum.

"Permanence" has received much debate among practitioners and is even listed as a statutory preference in the Superfund law, which unfortunately does not define the term. A definition proposed by a recent study seems to capture the spirit of the term "permanence":

> Permanent solutions are those that will endure indefinitely without posing the threat of any future release that would increase the risk above levels established for the site.[8]

Although such a definition in many cases would not consider alternatives that feature containment as being permanent, it would allow some level of residual risk to remain at the site following implementation.

Example 17-6 Definition of long-term. For what period should "long-term reliability" be evaluated?

Solution. As with "long-term effectiveness," it is at least 70 years, the life expectancy of humans. Whether to analyze for a longer period depends upon the expected life of the technological components of the remedy. For example, a slurry wall built to contain contaminated ground water without remediation of the contamination should be expected to provide effective containment for at least 70 years; however, it inevitably will leak at some time beyond that period due to natural factors. Therefore, the period of analysis for such remedies should extend beyond their expected life to allow an examination of replacement provisions or other contingencies.

When the selection of alternatives is done on a comparative basis, it encounters tradeoffs. **One tradeoff that requires careful consideration at most sites is whether to treat or to contain.** Treatment typically costs significantly more than containment but can diminish the hazard posed by the waste. Containment literally means that waste, whether untreated or the residuals of treatment, will be left on site. Containment requires technical controls, and sometimes controls of an institutional nature. An assessment of the long-term adequacy and reliability of these controls is crucial, and must address the following questions:

Containment/reduction of contamination

- How consistent over time is the alternative in effectively controlling migration of contaminants?
- How long will residual ground water constituent concentration levels persist?
- What is the amount of reduction with time?

Operation and maintenance

- What operation and maintenance functions must be performed?
- What difficulties and uncertainties may be associated with long-term operation and maintenance?

Management and monitoring

- What type and degree of long-term management is required?
- What are the requirements for long-term monitoring?
- What risk of exposure exists should monitoring be insufficient to detect failure?

Safety factors

- Do controls adequately handle potential problems?
- What is the relative safety factor (or amount of redundancy) incorporated into the alternative?
- What is the possibility that future remedial actions, not otherwise included in the remedy, may be required?
- How difficult would it be to implement the additional remedial actions, if required?
- What is the magnitude of the risk should the remedial action fail without being replaced?

Some alternatives feature treatment and leaving the residuals of this treatment in place without containment. Many of the questions previously listed also apply to such alternatives as to how the residuals are managed. A critical question that must be addressed for treatment alternatives is, "To what degree will the treatment be irreversible?"

Example 17-7 Long-term reliability. How should a remedy involving stabilization of organics be evaluated for "long-term reliability"?

Solution. Stabilization of organics can be reversed if the stabilized material deteriorates such as from repeated freezing and thawing. Specific durability testing can estimate the impact of wet/dry or freeze/thaw cycles (see Sec. 11-4). Treatment, as shown by this example, does not necessarily result in "permanence."

Implementability

Implementability is a function of three factors: (1) the history of the demonstrated performance of a technology, (2) the ability to construct and operate it given the existing conditions at the particular site, and (3) the ability to obtain the necessary permits from regulatory agencies. The types of questions raised by this issue include:

Demonstrated performance

- Are methods and equipment used in implementing the alternative common to commercial operations?
- Are technologies under consideration generally available and sufficiently demonstrated for the specific application?
- Will technologies require further development before they can be applied full-scale to the type of waste at the site?

- When should the technology be available for full-scale use?
- Are the qualified contractors limited in number?

Constructability

- How difficult is it to construct the alternative at this site?
- What uncertainties are related to construction?
- Are specialty contractors required?
- What is the likelihood that problems will develop and lead to schedule delays?

Operability

- What operating problems are anticipated and what are their impacts?
- Are the necessary equipment supplies and specialists available?
- For alternatives proposing off-site treatment and disposal, are such services available in adequate capacity?
- What difficulties and uncertainties may be associated with maintenance?

Regulatory feasibility

- What steps are required to coordinate with other agencies during the construction and operational period?
- What steps are required to set up long-term coordination among agencies?
- Can permits for on-site and off-site activities be obtained if and when required?

Example 17-8 Implication of stressing implementability. What implication does stressing ease of implementation have for the type of remedy selected?

Solution. It implies a preference for proven technologies, thus inhibiting the selection of innovative technologies. The EPA's Superfund program recognizes this as a potential obstacle to the development of new, improved technologies and requires the evaluation of innovative as well as proven technologies. However, overemphasizing innovation can turn cleanup efforts into research projects, thus delaying the original goal to remedy the site.

As stated earlier, a general policy in remedy selection is to examine the risk posed in the future under a hypothetical exposure scenario. This raises the issue of what investment should be made in the present to prevent future, hypothetical problems. The tradeoff boils down to whether to implement preventive measures now to eliminate the possibility of a future, hypothetical problem, or implement them at a future date if and when monitoring shows that without those measures the hypothetical problem would become real. This tradeoff could be examined on a stochastically derived present value basis, that typically would show that it is far less costly to implement them on a contingent basis—the problem may never materialize.

This decision, however, is not a strictly economic issue. The saga of Superfund shows many examples of why it is far better to prevent a problem now than to try to correct it in the future. Therefore, the first question to be asked is: will monitoring provide ample warning in the future to implement preventive measures? Also, will far more expensive remedial measures be necessary? Another question is: will funds be available in the future for the contingent responses? These questions and a number of others (listed under long-term effectiveness, long-term reliability, and implementability) require careful examination to decide this issue.

Short-Term Effectiveness

Short-term effectiveness deals primarily with the effects on human health and the environment of the remediation itself during its implementation phase. This would include the construction period and its operation period, up until the remedial action objectives are achieved. Even though the expression "short-term" is used, there is no limit to the timeframe for evaluating short-term effectiveness. As an example for a remedy involving excavation of soil, the risk to nearby residents from volatilization, dust, and other releases would continue for however long the excavation takes place, even if it occurs for 15 years.

Short-term effectiveness encompasses two issues other than health and environmental effects. It also examines worker safety and the time frame required to achieve the remedial action objectives. The questions for these issues include:

Health and environmental risks

- What emissions and discharges are released by the construction and operation of the alternative?
- What measures must be used during implementation and operation to mitigate health and environmental risk?
- How effective are the mitigation measures?
- Do migrational or exposure pathways exist that cannot be monitored adequately?

Worker safety

- Will the alternative threaten the safety of workers during implementation?
- Can these hazards be monitored adequately?

Implementation time

- What is the likelihood that treatment technologies will meet required process efficiencies and performance specifications?
- How long will it take to complete the construction after site work is initiated?
- How long until the remedial action objectives are achieved?

Cost

The weight given to cost when evaluating alternatives depends upon the particular guidance of the agency involved in regulating the remediation effort. Some practitioners have contended that the way SARA skirts the issue of cost means that cost does not play a major role in the selection of alternatives under the Superfund program. Others contend that cost is indeed an important factor and should be considered directly in the selection of alternatives.

Engineers recognize that practically any risk level can be achieved if cost is not a factor. Thus, cost must be a factor and should be considered as a last step in remedy selection. A comparison of costs inevitably ends up balancing non-cost criteria, such as long-term reliability and effectiveness, with cost. One way to illustrate this is to compare costs with the degree with which each alternative reduces risk posed by contamination at a site. Such a comparison is illustrated in Fig. 17-6 (Lackawanna Site) depicting a curve showing the risk reduction achieved by various alternatives versus their cost. The risk calculations for this particular example were discussed in Sec. 14-9. Risk calculations almost always assume that the alternative meets its objective. A highly sophisticated risk assessment would factor into its risk calculations the probability of failure represented by the long-term reliability of an alternative.

Costs consist of capital costs (the cost to construct the remedy) and annual operating and maintenance (O & M) costs (post-construction expenditures). Table 17-7 lists the types of costs included in these two categories.

Capital costs consist of direct and indirect costs. Direct costs consist primarily of expenditures for the materials, labor, and equipment necessary to construct the remedies. Direct costs also include the cost to transport off site and dispose of any waste (e.g., commercial incineration of excavated soil). Indirect costs include engineering design services and legal/financial services related to the installation of the remedy. Experience has shown the former amounts to be about 15% to 20% of the total direct costs while the latter can be as much as 10%. A contingency of at least 25% should be added to the total of direct and indirect costs.

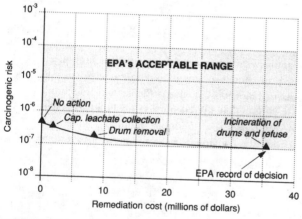

FIGURE 17-6
Carcinogenic risk versus cost for various remedial options.

TABLE 17-7
Types of costs

Capital costs	Annual operating and maintenance (O & M) costs
Construction costs	Operating labor costs
Equipment costs	Maintenance materials and labor costs
Land and site development costs	Auxiliary materials and energy
Building and services costs	Treatment residues disposal
Contractors' insurance and bonds	Environmental sampling and analysis
Relocation expenses	Purchased services
Surveying costs	Administrative costs
Disposal costs	Insurance, taxes, and licensing costs
Environmental sampling and analysis	Maintenance reserve and contingency funds
Engineering expenses	Rehabilitation costs
Construction material testing	Cost of periodic site reviews
License or permit costs	
Construction administration and reporting	
Construction interest	
Start up and shakedown costs	
Agency oversight	
Contingency allowances	

The comparison of the costs of different alternatives should be done on a present worth analysis. For alternatives involving long-term operating and monitoring costs, the selection of a period of performance is necessary to develop comparative costs. In such cases, a period of performance is usually considered to be 30 years. Most remedies featuring treatment, management, and monitoring can be accomplished in 10 to 15 years, usually far less. However, the measurement of the effectiveness of some remedies can require management and monitoring of a site for a substantial period of time following implementation of the remedy.

A present worth analysis, simply stated, defines the amount of money needed at the beginning (base year) to (1) fund the construction of the remedy and (2) provide an annual payment for operation and maintenance over the project life. The analysis involves discounting future costs to a common base year, usually the present. This requires the selection of a discount rate. The standard approach is to discount future costs at a rate equal to the interest paid on long-term bonds. Another approach is to assume that long-term inflation, taxes, and insurance will offset interest such that the effective discount rate would equal zero.

Table 17-8 tabulates the capital and O&M costs estimated for selected alternatives for the generic site as listed in Table 17-6. It also shows their values discounted at 0% and 10% for 30 years. The present worth factor for 10% and 30 years is 9.427 (see Sec. 7-2). In the case of alternative C, the annual O & M cost is $1,027,000. At an effective discount rate of zero, the present worth = $1.027 \times 30 = \$30.8$ million. At a rate of 10%, the present worth = $1.027 \times 9.427 = \$9.68$ million.

TABLE 17-8

Summary of cost of selected alternatives (generic site)

Alternative	Description	Capital cost*	Annual O & M cost*	O & M cost $I = 0\%$	O & M cost $I = 10\%$	Total c $I = 0\%$
A	No action	0.0	.067	2.0	0.6	2.0
B	Primary controls	3.9	.103	3.1	1.0	7.0
C	Primary controls and alluvial recovery	5.2	1.03	30.8	9.7	36.0
E	"B" plus southwest and southeast area recovery	6.1	1.019	32.6	10.2	38.7
E1	"E" plus southwest boundary cutoff wall	11.3	1.07	32.3	10.1	43.6
L	Primary controls, alluvial recovery, plus main source area cutoff wall	26.6	.31	9.3	2.9	35.9

* All costs in million dollars.

Example 17-9 Determining present values. The following cost estimate is provided for an alternative remedy:

Cost element	Cost
Initial capital cost	$3,000,000
Replacement capital cost, year 20	3,000,000
Annual O&M cost	600,000

The following are given:

- The useful life for the capital items is 20 years.
- Capital items have a salvage value at the end of their useful life of 10% of their initial cost.
- The depreciated values of the capital items are a linear function of time, but the entire depreciation is taken on the first day of the year.
- Engineering costs are 15% of the direct capital costs.
- Contingencies are 25% of the direct cost.
- The effective discount rate is 10%.
- Inflation is neglected.

Determine the present value cost of this alternative using a 30-year planning period.

Solution. The solution begins by calculating the present worth of individual elements.

The factor for determining the present worth of a single expenditure in 20 years may be determined from Eq. 7-3:

$$\text{Cost Factor} = \frac{1}{(1+i)^n} = \frac{1}{(1+0.1)^{20}} = 0.1486$$

The present worth factor (PWF) to determine the present worth of 10 equal annual payments may be determined from Eq. 7-4:

$$\text{PWF} = \frac{(1+i)^n - 1}{i(1+i)^n} = \frac{(1+0.1)^{10} - 1}{(0.1)(1+0.1)^{10}} = 6.1446$$

Dealing with depreciation is a bit more complicated. In this example, the capital equipment is replaced at the end of 20 years, but the planning period is 30 years. At the end of year 30 (the beginning of year 31), the capital equipment would still have approximately half of its useful life. Assume the value at this point is $1.35 million or 9/20 of the original value. To calculate the present worth we again use Eq. 7-3:

$$\text{PW} = 1.35 \left[\frac{1}{(1+0.1)^{30}} \right] = 1.35(0.0573) = \$0.077 \text{ million}$$

The remaining calculations are shown in the following table:

Item	Cost	Cost factor	Present value
Capital Costs:			
Initial capital cost	$3,000,000	1.0000	$3,000,000
Replacement cost	3,000,000	0.1486	445,800
Salvage value	(300,000)	0.1486	(44,600)
Undepreciated value	(1,350,000)	0.0573	(77,000)
Subtotal			3,324,200
Engineering (15%)			498,600
Contingencies (25%)			831,100
Subtotal, capital costs			4,653,900
O & M Costs:			
Annual O&M (year 1–10)	600,000	6.1446	3,686,800
Annual O&M (year 11–20)	600,000	6.1446	
	3,686,800	0.3855	1,421,300*
Annual O&M (year 21–30)	600,000	6.1446	
	3,686,800	0.1486	547,700**
Subtotal			5,655,800
Contingencies (25%)			1,414,000
Subtotal, O&M			7,069,800
Total estimated present value cost			11,700,000

* Present value at Year 10 discounted to Year 0
** Present value at Year 20 discounted to Year 0

Selection of Preferred Alternative

The selection of the preferred alternative involves a thorough evaluation of all alternatives on the basis of the criteria discussed earlier. The evaluation process is not structured. One reason is that the criteria actually are considerations rather than true criteria. A consideration is a specific characteristic or issue deemed important in evaluating alternatives. A wide range of quantitative data and qualitive information can be developed for each consideration. A criterion is a standard for making a decision with regard to a particular consideration; it would define a threshold between acceptable and unacceptable.

This distinction illustrates the nature of the process of evaluating the final alternatives. Thus, there should be no thresholds in the selection criteria; standards are reflected in the objectives and each alternative meets the objectives and is supposedly protective of human health and the environment, with the possible exception of the "no action" or any intentionally limited alternatives.

The process differs from a structured approach in other key ways. Alternatives are not ranked; hence, rigid methods such as a weighted summation model are avoided in that it would be difficult to establish the validity of weights and ratings. Finally, there should be no prescriptive approach on how to apply criteria. A prescriptive approach does not allow the full use of common sense.

Instead of a structured approach, the evaluation of alternatives should be done on a flexible, comparative basis featuring the following steps:

- Interactions (e.g., conflicts and synergies) among criteria are considered.
- Advantages and disadvantages of alternatives relative to one another become clear.
- Tradeoffs among alternatives are examined and balanced.
- This information converges to compelling reasons upon which a consensus can build for selecting one remedy.

If the analysis is made by knowledgeable persons under the scrutiny of interested parties, this approach should yield the most appropriate and defensible decisions.

17-7 REGULATORY GUIDANCE

A governmental agency usually provides guidance specifying the process and criteria to be used in conducting an analysis of alternatives.[9,10] The guidance varies from agency to agency, and the guidance from the same agency usually changes over time as policies change and more experience is obtained. The guidance can have a significant impact on the results of the analysis, ranging from making the study complicated and tedious to influencing the range of alternatives considered, possibly even changing the ultimate outcome.

An example is the EPA Superfund process as specified in its guidance document.[11] A critical study, drawing on the experience on more than 100 professionals familiar with the issues of site remediation, reported that the EPA process tended to work in a reverse order from that of the classic planning model.[12] The study found that instead of explicitly defining the required level of cleanup at the start of the process, the EPA process frequently identified this "... very late in the process and then only implicitly by what the selected remedy would achieve." In effect, alternatives were evaluated before clear objectives had been established.

The EPA approach also promotes developing and retaining a wide range of alternatives. The range extends from those failing to meet remediation objectives to some greatly exceeding them. In that way the final detailed analysis compares diverse courses of actions:

- "No action";
- Alternatives that fail, meet, or exceed ARARs;
- An alternative that involves off-site treatment or disposal;
- An alternative that features treatment;
- An alternative that eliminates, to the extent feasible, the need for any long-term management, including monitoring (i.e., the "permanence" alternative that requires virtually complete elimination or destruction of hazardous substances at the site);
- An alternative that features containment; and
- An alternative that features innovative technology.

Retaining this wide range serves to emphasize stark differences in the degree of effectiveness and the cost of the alternative remedies. Such is useful for fomenting participation by interested parties. However, it also requires expending effort in refining and evaluating specific alternatives that will not be selected either because they are not protective or they are overly protective at an unrealistically high cost.

The EPA approach does provide a special capability. There is a great deal of uncertainty at Superfund sites:

- The existing site conditions may not be fully defined.
- It is not evident what future uses of the site and its resources need to be protected.
- The effectiveness and performance of many remedial technologies are not clear.

Under such constraints and incomplete knowledge, the selection of a remedy must be subjective in large part. It must be dependent upon a large measure of common sense and have the flexibility to evaluate technologies on what they truly can achieve. Therefore, the EPA concept of delaying the setting of objectives, particularly those implicit in the hypothetical assumptions of future conditions, avoids pre-empting at too early of a stage some potentially viable alternatives.

DISCUSSION TOPICS AND PROBLEMS

17-1. What general response actions could be implemented to protect human health from contaminated ground water other than treating the ground water?

17-2. What is the difference between "long-term effectiveness" and "long-term reliability?"

17-3. Which would require a greater cost contingency: excavation of soil determined in the field to have contamination above a certain concentration or incineration of the sludge from a lagoon?

17-4. When should treatability tests be conducted? Why?

17-5. What are five technologies for achieving the general response action "treat soil contaminated with chlorinated solvents."

17-6. Based on the limited knowledge provided in this chapter, rank each of the final six alternatives selected for the generic site (see Table 17-6) with regard to long-term effectiveness? Long-term reliability?

17-7. For Problem 17-6, how would in situ biotreatment rank if the contaminants were degradable organics?

17-8. Which of the alternatives in Table 17-6 would you consider to be the preferred alternative based on information available in this chapter?

17-9. What additional information would you feel necessary for selecting a remedy from the six alternatives in Table 17-6.

17-10. What would this additional information need to show to convince you to select an alternative other than the one selected in Problem 17-8?

17-11. Assuming that the alluvial ground water at the generic site could be restored such that residences would not require an alternative water supply, up to what cost would you consider this the preferred alternative?

17-12. Describe the following data for a hypothetical or actual contaminated site: simplified plan view and one page of text and tables listing contaminant sources and concentrations and quantitative risks of exposure.

The remaining problems pertain to the site that you have described in Problem No. 17-12.

17-13. What are some possible remediation objectives for this site?

17-14. What is an example of a remediation objective that would be inappropriate?

17-15. What are some possible general response actions that could be used to achieve these objectives?

17-16. For one of the general response actions, which technologies are possibly applicable?

17-17. Which technologies should be retained for further analysis based on your preliminary screening of their effectiveness and reliability?

17-18. How could some of the technologies be combined to create different alternatives?

REFERENCES

1. "Feasibility Study Report, Management of Migration Second Operable Unit, Royal Hardage Site, Criner, Oklahoma," report prepared for the Hardage Steering Committee and the U.S. Environmental Protection Agency, by ERM-Southwest, Inc., Houston, TX, 30 May 1989.
2. *Federal Register*: "National Oil and Hazardous Substances Pollution Contingency Plan; Final Rule," 40 CFR Part 300.430, vol. 55, no. 46, March 1990.
3. Clean Sites, Inc.: "Improving Remedy Selection: An Explicit and Interactive Process for the Superfund Program," Alexandria, VA, October 1990.
4. *Federal Register*: "National Oil and Hazardous Substances Pollution Contingency Plan."
5. Clean Sites, Inc., "Improving Remedy Selection."
6. *Federal Register*: "National Oil and Hazardous Substances Pollution Contingency Plan."
7. Clean Sites, Inc., "Improving Remedy Selection," op. cit.
8. Ibid.
9. U.S. Environmental Protection Agency (EPA): "Guidance for Conducting Remedial Investigations and Feasibility Studies under CERCLA," EPA 540/G-89/004, Washington, D.C., October 1988.
10. U.S. Air Force Occupational & Environmental Health Laboratory: "Handbook to Support the Installation Restoration Program (IRP) Statements of Work for Remedial Investigation/Feasibility Studies," Brooks Air Base, TX, April 1988.

11. U.S. Environmental Protection Agency (EPA): "Guidance for Conducting Remedial Investigations and Feasibility Studies under CERCLA."
12. Clean Sites, Inc., "Improving Remedy Selection," op. cit.

ADDITIONAL READING

Bennett, G. F., F. S. Feates, and I. Wilder: *Hazardous Materials Spills Handbook,* McGraw-Hill, Inc., New York, NY, 1982.

Boutwell, S. H.: *Selection of Models for Remedial Action Assessment,* U.S. EPA, Washington, D.C., 1984.

Cohen, R. M., and W. J. Miller III: "Use of Analytical Models for Evaluating Corrective Actions at Hazardous Waste Sites," *Proceedings of the 3rd National Symposium of Aquifer Restoration and Groundwater Monitoring,* National Water Well Association, Dublin, OH, May 1983.

Cole, C. R., R. W. Bond, S. M. Brown, and G. W. Dawson: *Demonstration/Application of Groundwater Technology for Evaluation of Remedial Action Alternatives,* U.S. EPA, Washington, D.C., 1983.

Dawson, G. W., and B. W. Mercer: *Hazardous Waste Management,* John Wiley & Sons, New York, NY, 1986.

Ford, P. J., P. J. Turina, and D. E. Seely, *Characterization of Hazardous Waste Sites—A Methods Manual, Volume II, Available Sampling Methods,* EPA 600/4-83-040, NTIS no. PB-83-126920, U.S. EPA, Springfield, VA, 1983.

Hallstedt, P. A., M. A. Puskar, and S. P. Levine: "Application of the Hazard Ranking System to the Prioritization of Organic Compounds Identified at Hazardous Waste Remediation Action Sites," *Hazardous Waste and Hazardous Materials,* vol. 3, no. 2, 1986.

O'Brien & Gere Engineers, Inc.: *Hazardous Waste Site Remediation,* Van Nostrand-Reinhold, 1988.

U.S. Congress: *Superfund Strategy,* Office of Technology Assessment, Washington, D.C., April 1985.

U.S. Environmental Protection Agency: "Guidance for Conducting Remedial Investigation and Feasibility Studies under CERCLA, Interim Final," EPA/540/6-89/004, OSWER Directive 9355.3-01, October 1988.

U.S. Environmental Protection Agency: "Report on the Usage of Computer Models on Hazardous Waste/ Superfund Programs: Phase II, Final Report," U.S. EPA, OSWER, December 1990.

U.S. Environmental Protection Agency: "The Cost Digest: Cost Summaries for Selected Environmental Control Technologies," Office of Env. Engineering and Technology, EPA-600-8-84-010, October 1984.

U.S. Environmental Protection Agency: *Case Studies 1-23: Remedial Response at Hazardous Waste Sites,* U.S. EPA, Washington, D.C., March 1984.

U.S. Environmental Protection Agency: *EPA Guide for Identifying Cleanup Alternatives at Hazardous Waste Sites and Spills: Biological Treatment,* EPA-600/3-83-063, U.S. EPA, Springfield, VA, 1983.

U.S. Environmental Protection Agency: *Handbook for Evaluating Remedial Action Technology Plans,* EPA 600/2-83-076, U.S. EPA, Cincinnati, OH, 1983.

U.S. Environmental Protection Agency: *Handbook: Remedial Action at Waste Disposal Sites (Revised),* EPA/625/6-85/006, U.S. EPA, Cincinnati, OH, October 1985.

U.S. Environmental Protection Agency: *Modeling Remedial Actions at Uncontrolled Hazardous Waste Sites,* EPA 540/2-85-001, U.S. EPA, Springfield, VA, 1985.

U.S. Environmental Protection Agency: *Remedial Response at Hazardous Waste Sites, Summary Report,* EPA 540/2-82-002a, *Case Studies,* EPA 540/2-84-002b, U.S. EPA, Cincinnati, OH, 1984.

Wagner, K., et al.: *Remedial Action Technology for Waste Disposal Sites,* Noyes, Park Ridge, NJ, 1986.

ACKNOWLEDGMENTS

The authors wish to acknowledge the following individuals who assisted with the preparation of this chapter: Doyan Main and Gary Donnan, *ERM-Southwest*; Alan F. Hassett and Mark Schmittle, *ERM, Inc.*; Howie Wiseman, *ERM- Northeast*; and Chris Edgar, *Bucknell University.* The principal reviewer for the **ERM Group** was John Iannone, *ERM-Northeast.*

APPENDIX
A

CONTAMINANT
PROPERTIES

It is a capital mistake to theorize before one has data.
[Sherlock Holmes]   Sir Arthur Conan Doyle

The following tables were developed by starting with Table C-1 of the **Superfund Public Health Evaluation Manual** (EPA 540/1-86/060), October, 1986. This material was obtained from sources A through N (see sources of data below). Additional data were added from the references listed, and the source is denoted by the column labeled S.

Compounds are identified by their commonly used chemical names, chemical formulae, and the Chemical Abstract Service Registry Numbers (CASRN). Many chemicals are listed as toxic substances for specific regulatory purposes. The four columns labeled P, C, N, and H designate:

P Priority Pollutants, (designated under the Clean Water Act)
C Hazardous Substance List (designated under CERCLA)
N Appendix IX Chemicals (designated under RCRA)
H Hazardous Air Pollutant (designated under the 1991 Clear Air Act Amendment)

These designations are continuously updated and the latest *Federal Register* listing should be consulted.

Definitions and examples of the uses of the properties tabulated in this appendix may be found in Chap. 3. Use of the Freundlich parameters is discussed in Sec. 9.3.

In general the data included in this appendix are measured values. Where calculated values are utilized, the source is designated by the symbol "◇". Where solubility in water is reported as "infinitely soluble" the designation "Total" is shown in the table. The organic carbon partition coefficient, K_{oc}, is estimated from water solubility by the following equation:

$$K_{oc} = (-0.55 \log S) + 3.64$$
$$S = \text{solubility (mg/L)}$$

(A-1)

For parameters such as water solubility and vapor pressure, an additional column has been added to indicate the temperature at which the parameter was measured. The Henry's law constant is provided as a function of temperature:

$$H = \exp\{A - B/T\}$$

$H = $ Henry's law Constant (atm · m^3/mol)

$T = $ Temperature (Kelvins)

$A, B = $ Tabulated constants

(A-2)

While these tables were prepared with care, no warranty is intended or implied, and the user is advised to independently verify information prior to use. Most of these parameters were measured in pure water and may not be appropriate for environmental conditions different from those used in the measurement. The reader is strongly urged to consult the original reference before applying this information.

SOURCES OF DATA

A. U.S. Environmental Protection Agency: Health Effects Assessment for [Specific Chemical]. [Note: 58 individual documents available for specific chemicals or chemical groups], U.S. EPA, ECAO, 1985.

B. Jaber, et al.: *Data Acquisition for Environmental Transport and Fate Screening*, EPA 600/6-84-009, Office of Health and Environmental Assessment, U.S. EPA, Washington, D.C., 1984.

C. Mabey, et al.: *Aquatic Fate Process Data for Organic Priority Pollutants*, prepared by SRI International, EPA Contract Nos. 68-01-3867 and 68-03-2981, prepared for Monitoring and Data Support Division, Office of Water Regulations and Standards, Washington, D.C., 1982.

D. Callahan, et al.: *Water-Related Environmental Fate of 129 Priority Pollutants*, Volumes I and II, Office of Water Planning and Standards, Office of Water and Waste Management, U.S. EPA, EPA Contract Nos. 68-01-3852 and 68-01-3867, 1979.

E. U.S. Environmental Protection Agency: *Treatability Manual*, Volume I, EPA 600/2-82-001a, ORD, 1981.

F. Dawson, et al.: *Physical/Chemical Properties of Hazardous Waste Constituents*, prepared by Southeast Environmental Research Laboratory for U.S. EPA, 1980.

G. Lyman, et al.: *Handbook of Chemical Property Estimation Methods*, McGraw-Hill, New York, NY, 1982.

H. U.S. Environmental Protection Agency: Ambient Water Quality Criteria Documents for [Specific Chemical], OWRS, 1980.

I. Weast, et. al.: *CRC Handbook of Chemistry and Physics*, 1979.

J. Verschueren, Karel: *Handbook of Environmental Data for Organic Chemicals*, 2nd ed., Van Nostrand Reinhold Co., New York, NY, 1983.

L5. Perry, R. H., and Chilton: *Chemical Engineers' Handbook*, 5th ed., McGraw-Hill, New York, NY, 1973.

L6. Perry, R. H., and D. Green: *Chemical Engineers' Handbook*, 6th ed., McGraw-Hill, New York, NY, 1984.

M. U.S. Environmental Protection Agency: *Exposure Profiles for RCRA Risk-Cost Analysis Model*, prepared by Environ Corporation, 1984.

N. U.S. Environmental Protection Agency: *Characterization of Constituents from Selected Waste Streams Listed in 40 CFR Section 261*, prepared by Environ Corporation, 1984.

O. Dobbs, Richard A., and Jesse M. Cohen: *Carbon Adsorption Isotherms for Toxic Organics*, EPA-600/8-80-023, Wastewater Research Division, Cincinnati, OH.

P. Sax, N. Irving, and Richard J. Lewis, Sr.: *Dangerous Properties of Industrial Materials*, 7th ed., Van Nostrand Reinhold, New York, NY, 1989.

R. The Merck Index: Merck & Co., Inc., Rahway, New Jersey, NJ, 1989.

T. Yaws, Carl L., et al.: "Organic Chemicals: Water Solubility Data." *Chemical Engineering*, July 1990.

U. Schultz, H. Lee, et al.: *Superfund Exposure Assess Manual*, U.S. EPA, EPA/540/1-88/001, OSWER Directive 9285.5-1, April 1988.

V. U.S. Coast Guard: *Chemical Hazards Response Information System (CHRIS Manual)*, Hazardous Chemical Data, M.16465.12A, November 1984.

W. Haarhoff, J., and J. L. Cleasby: "Evaluation of Air Stripping for the Removal of Organic Drinking Water Contaminants," *Water South Africa*, January 1990.

X. Yaws, Carl L., et al.: "Hydrocarbons: Water Solubility Data," *Chemical Engineering*, April 1990.

Y. Howard, P. H.: *Handbook of Environmental Fate and Exposure Data for Organic Chemicals*, Lewis Publishers, 1989, 1990.

Z. Howe, G. B., M. E. Mullins, and T. N. Rogers: "Evaluation and Production of Henry's Law Constants and Aqueous Solubilities for Solvents and Hydrocarbon Fuel Components - Volume I: Technical Discussion," Final Report, Engineering and Services Lab, Tyndall Air Force Base, Report No. ESL-86-66, 1987.

AA. Gossett, J. M.: "Measurement of Henry's Law Constants for C1 and C2 Chlorinated Hydrocarbons," *Environmental Science and Technology*, vol. 21, pp. 202–208, 1987.

AB. Patterson, J. W.: *Industrial Wastewater Treatment Technology*, Butterworth Publishers, Stoneham, MA, pp. 331–334, 1985.

AC. Shen, T.: "Estimating Hazardous Air Emissions from Disposal Sites," *Pollution Engineering*, vol. 13, no. 8, pp. 31–34, 1981

AD. Lighton, David T., and Joseph M. Calo: "Distribution Coefficients of Chlorinated Hydrocarbons in Dilute Air-Water Systems for Ground Water Contamination Applications," *Journal of Chemical Engineering Data*, 1981.

ADDITIONAL REFERENCES

Aquatic Fate Process Data for Organic Priority Pollutants, prepared by SRI, International, EPA 440/4-81-09014 [NTIS: PB 87-169 090], December 1982.

American Mutual Insurance Alliance, *Handbook of Organic Industrial Solvents*, AMIA, Chicago, IL, 1988.

Kenaga, E., "Correlation of Bioconcentration Factors of Chemicals in Aquatic and Terrestrial Organisms with their Physical and Chemical Properties," *Ecotoxicol. Environ. Saf.*, vol. 4, pp. 26–38, 1980.

Patniak, Pradyot, *A Comprehensive Guide to the Hazardous Properties of Chemical Substances*, Van Nostrand-Rheinhold, New York, NY, 1992.

Chemical name	Chemical formula	P	C	N	H	CASRN	Molecular weight (g/mole)	Water solubility (mg/L)	(°C)
Acenaphthene	$C_{12}H_{10}$	P	C	N		83-32-9	154.21	3.42E+00	25
Acenaphthylene	$C_{12}H_8$	P	C	N		208-96-8	152.2	3.93E+00	20
Acetone	C_3H_6O		C	N		67-64-1	58.09	total	20
Acetonitrile	C_2H_3N	P		N	H	75-05-8	41.06	total	20
2-(Acetylamino) Fluorene	$C_{15}H_{13}NO$			N	H	53-96-3	223.29	6.50E+03	
Acrylic acid	$C_3H_4O_2$				H	79-10-7	72.07	total	20
Acrylonitrile	C_3H3N				H	107-13-1	53.07	7.35E+04	20
Aldicarb	$C_7H_{14}N_2O_2S$					116-06-3	190.29		
Aldrin	$C_{12}H_8Cl_6$	P	C	N		309-00-2	364.9	1.80E-01	25
Allyl Alcohol	C_3H_6O			N		107-18-6	58.09	total	
4-Aminobiphenyl	$C_{12}H_{11}N$			N	H	92-67-1	169.24	8.42E+02	
Amitrole	$C_2H_4N_4$					61-82-5	84.1	2.80E+02	25
Ammonia	H_3N					7664-41-7	17.04	5.30E+05	20
Anthracene	$C_{14}H_{10}$	P	C	N		120-12-7	178.24	.4.50E-02	25
Antimony	Sb				H	7440-36-0	121.75		
Arsenic	As				H	7440-38-2	74.92		
Auramine	$C_{17}H_{21}N_3 \cdot ClH \cdot H_2O$					2465-27-2	321.89	2.10E+00	
Azaserine	$C_5H_7N_3O_4$					115-02-6	173.15	1.36E+05	
Aziridine	C_2H_5N					151-56-4	43.08	total	20
Barium	Ba					7440-39-3	137.36	10mm	
Barium Cyanide	C_2BaN_2					542-62-1	189.38	6.86E+05	14
Benefin	$C_{13}H_{16}F_3N_3O_4$					1861-40-1	335	7.00E+01	25
Benz(c)acridine	$C_{17}H_{11}N$					225-51-4	229.29	1.40E+01	
Benz(a)anthracene	$C_{18}H_{12}$	P	C	N		56-55-3	228.3	5.70E-03	20
Benzene	C_6H_6	P	C	N	H	71-43-2	78.12	1.78E+03	25
Benzidine	$C_{12}H_{12}N_2$	P		N	H	92-87-5	184.26	4.00+02	12
Benzo(b)fluoranthene	$C_{20}H_{12}$					205-99-2	252.32	1.40E-02	20
Benzo(k)fluoranthene	$C_{20}H_{12}$					207-08-9	252.32	4.30E-03	25
Benzo(ghi)perylene	$C_{22}H_{12}$					191-24-2	276.34	2.60E-04	25
Benzo(a)pyrene	$C_{20}H_{12}$					50-32-8	252.32	3.80E-03	25
Benzotrichloride	$C_7H_5Cl_3$				H	98-07-7	195.47		
Benzyl Chloride	$C_6H_5CH_2Cl$				H	100-44-7	126.59	2.57E+01	25
Beryllium	Be				H	7440-41-7	9.01		
1,1′-Biphenyl	$C_{12}H_{10}$				H	92-52-4	154.22	7.50E+00	25
Bis(2-chloroethyl)ether	$C_4H_8Cl_2O$	P	C	N	H	111-44-4	143.02	1.20E+04	25
Bis(2-chloroisopropyl)ether	$C_6H_{12}Cl_2O$	P	C	N		108-60-1	171.08	1.70E+03	25
Bis(chloromethyl)ether	$C_2H_4Cl_2O$				H	542-88-1	114.96	2.20E+04	25
Bis(2-ethylhexyl)phthalate	$C_{24}H_{38}O_4$	P	C	N	H	117-81-7	390.62	4.00E-01	25
Bromomethane	CH_3Br					74-83-9	94.95	9.00E+02	20
1,3-Butadiene	C_4H_6				H	106-99-0	54.1	7.35+02	23
n-Butanol	$C_4H_{10}O$					71-36-3	74.14	6.69E+04	20
Butyl Phthalyl Butyl Glycolate	$C_{18}H_{24}O_6$					85-70-1	336.42	5.92E+05	20
Cacodylic Acid	$C_2H_7AsO_2$					75-60-5	138.01	5.88E+05	20

S = Source of data; P = Priority Poll; C = Haz. Substance List; N = App. IX; H = Haz. Air Poll.

apor pressure (Hg)	(°C)	S	Diffusion coefficient in air (cm²/s)	S	Henry's Constant H=exp(A-B/T) H(atm·m³/mol); T (K) A	B	S	Koc ml/g	S	Log Kow	S	Freundlich parameters pH	K (mg/kg)	1/n	S	Fish BCF (L/kg)	S
E-03	25	C						4.50E+03	C	4.00E+00	C	5.3	190	0.36	O	242	H
E-02	20	C						2.50E+03	C	3.70E+00	C	5.3	115	0.37	O		
E+02	30	J	0.1093	U				2.20E+00	◊	-2.40E-01	J						
E+01	20	J						2.20E+00	◊	-3.40E-01	F						
								1.60E+03	◊	3.28E+00		7.1	318	0.12	O		
E+00	20	J						1.61E-01	Y							0	F
E+02	23	C						8.50E-01	C	2.50E-01	C	5.3	1	0.51	AB	48	G
E-04	25	J															
E-06	25	C						9.60E+04	C	5.30E+00	C	5.3	651	0.92	O	28	H
E+01	20	J						1.70E-01	J	-2.20E-01	◊						
E+01	157	I						1.07E+02	◊	2.78E+00	B	7.2	200	0.26	O		
								4.40E+00	◊	-2.08E+00							
E+03	26	J						3.10E+00	◊	0.00E+00	F					0	F
E-05	25	C						1.40E+04	C	4.45E+00	A	5.3	376	0.70	O		
E+00	886	P														1	H
E+00	372	P														44	H
								2.90E+03	◊	4.16E+00	B						
								6.60E+00	◊	-1.08E+00	B						
E+02	20	J						1.30E+00	◊	-1.01E+00							
E+01	1049	P															
E-07	25	J															
								1.00E+03	◊	4.56E+00	B						
E-08	20	C						1.38E+06	G	5.60E+00	C						
E+01	25	C	0.09234	U	5.53E+00	3.19E+03	Z	8.30E+01	G	2.12E+00	A	7	1	2.90	O	5.2	H
E+01	176	I						1.05E+01	C	1.30E+00	C	7-9	220	0.35	O	87.5	H
E-07	20	C						5.50E+05	C	6.06E+00	A						
E-07	20	C						5.50E+05	C	6.06E+00	G	7.1	181	0.57	O		
E-10	25	C						1.60E+06	C	6.51E+00	A	7	11	0.37	O		
E-09	25	C						5.50E+06	C	6.06E+00	C	7.1	34	0.44	O		
E-01	20	Y								2.92E+00	Y						
E+00	22	J						5.00E+01	◊	2.63E+00	F						
E+01	1860	I															
E+01	117	I															
E-01	20	J						1.39E+01	C	1.50E+00	C	5.3	0	1.84	AB	6.9	H
E-01	20	J						6.10E+01	C	2.10E+00	C	5.4	24	0.57	O	0	D
E+01	22	Y						1.20E+00	C	3.80E-01	C					0.63	H
												5.3	11300	1.50	O		
E+03	37	J	0.1083	U													
E+00	20	J						1.20E+02	◊	1.99E+00	F						
E+00	21	V								8.80E-01	Y						
								2.40E+00	◊	0.00E+00	F						

alculated value.

Chemical name	Chemical formula	P	C	N	H	CASRN	Molecular weight (g/mole)	Water solubility (mg/L)	(°C)
Cadmium	Cd				H	7740-43-9	112.4		
Captan	$C_9H_8Cl_3NO_2S$				H	133-06-2	300.59	5.00E-01	20
Carbaryl	$C_{12}H_{11}NO_2$				H	63-25-2	201.24	4.00E+01	30
Carbon Disulfide	CS_2		C	N	H	75-15-0	76.13	2.30E+03	22
Carbon Tetrachloride	$CCl4$	P	C	N	H	56-23-5	153.81	8.00E+02	20
Chlordane	$C_{10}H_6Cl_8$	P	C	N	H	57-74-9	409.76	5.60E-01	25
Chlorobenzene	C_6H_5Cl	P	C	N	H	108-90-7	112.56	5.00E+02	20
Chlorobenzylate	$C_{16}H_{14}Cl_2O3$			N	H	510-15-6	325.2	2.19E+01	
Chlorodibromomethane	$CHBr_2Cl$	P	C	N		124-48-1	208.29	4.75E+03	20
Chloroform	$CHCl_3$	P	C	N	H	67-66-3	119.37	8.00E+03	20
Chloromethane	CH_3Cl					74-87-3	50.49		
Chloromethyl Methyl Ether	C_2H_5ClO				H	107-30-2	80.52		
Chromium	Cr				H	7440-47-3	52		
Chrysene	$C_{18}H_{12}$	P	C	N		218-01-9	228.3	1.80E-03	25
Copper	Cu					7440-50-8	63.55		
Cresol	C_7H_8O				H	1319-77-3	108.15	2.15E+04	20
Crotonaldehyde	C_4H_6O					4170-30-3	70.09	1.33E+05	20
Cyanogen	C_2N_2					460-19-5	52.04	9.49E+03	20
Cyanogen Chloride	CClN					506-77-4	61.47	1.00E+03	20
Cyclohexane	C_6H_{12}					110-82-7	84-18		
Cyclophosphamide	$C_7H_{15}Cl_2N_2O_2P$					50-18-0	261.1	1.31E+09	
Dalapon	$C_3H_4Cl_2O_2$					75-99-0	142.97		
4,4'-DDD	$C_{14}H_{10}Cl_4$	P	C	N		72-54-8	320.04	.02 - .09	25
4,4'-DDE	$C_{14}H_8Cl_4$	P	C	N	H	72-55-9	318.02	4.00E-02	20
4,4'-DDT	$C_{14}H_9Cl_5$	P	C	N		50-29-3	354.48	5.50E-03	25
Decalin	$C_{10}H_{18}$					91-17-8	138.28		
Diallate	$C_{10}H_{17}Cl_2NOS$					2303-16-4	270.24	1.40E+01	
2,4-Diaminotoluene	$C_7H_{10}N_2$					95-80-7	122.19	4.77E+04	
1,2,7,8-Dibenzopyrene	$C_{24}H_{14}$					189-55-9	302.38	1.10E-01	
Dibenz(a,h)Anthracene	$C_{22}H_{14}$	P	C	N		53-70-3	278.36	5.00E-04	25
Dibromochloromethane	$CHBr_2Cl$					124-48-1	208.29		
1,2-Dibromo-3- Chloropropane	$C_3H_5Br_2Cl$			N	H	96-12-8	236.36	1.00E+03	20
Di-n-Butyl Phthalate	$C_{16}H_{22}O_4$				H	84-74-2	278.38	1.30E+01	25
1,2-Dichlorobenzene	$C_6H_4Cl_2$	P	C	N		95-50-1	147	1.00E+02	20
1,3-Dichlorobenzene	$C_6H_4Cl_2$	P	C	N		541-73-1	147	1.23E+02	25
1,4-Dichlorobenzene	C_6H_4Cl2	P	C	N	H	106-46-7	147	7.90E+01	25
3,3'-Dichlorobenzidene	$C_{12}H_{10}Cl_2N_2$	P	C	N	H	91-94-1	253.14	3.99E+00	22
Dichlorodifluoromethane	CCl_2F_2			N		75-71-8	120.91	2.80E+02	25
1,1-Dichloroethane	$C_2H_4Cl_2$	P	C	N		75-34-3	98.96	5.50E+03	20
1,2-Dichloroethane	$C_2H_4Cl_2$	P	C	N		107-06-2	98.96	8.69E+03	20
1,1-Dichloroethylene	$C_2H_2Cl_2$	P	C	N		75-35-4	96.94	2.25E+03	20
1,1-Dichloroethene	$C_2H_2Cl_2$					75-35-4	96.94	2.25E+03	20
trans-1,2-Dichloroethene	$C_2H_2Cl_2$					156-60-5	96.94	6.00E+02	20

S = Source of data; P = Priority Poll; C = Haz. Substance List; N = App. IX; H = Haz. Air Poll.

por pressure Hg	(°C)	S	Diffusion coefficient in air (cm²/s)	S	A	B	S	Koc ml/g	S	Log Kow	S	pH	K (mg/kg)	1/n	S	Fish BCF (L/kg)	S
E+00	394	P														81	H
E-05		E						6.40E+03	◇	2.35E+00	F						
E-03	26	J								2.36E+00	F						
E+02	20	J						5.40E+01	◇	2.00E+00	F					0	
E+01	20	J	0.08451	U	1.13E+01	4.41E+03	Z	1.10E+02	◇	2.64E+00	A	7	40	0.84	O	19	H
E-05	25	C						1.40E+05	C	3.32E+00	A	5.3	245	0.38	O	14000	H
E+01	20	W	0.07627	U	3.47E+00	2.69E+03	Z	3.30E+02	C	2.84E+00	A	7.4	91	0.99	O	10	H
E-06		B						8.00E+02	◇	4.51E+00							
E+01	20	W			5.97E+00	3.12E+03	AA			2.09E+00	D	7	63	0.93	O		
E+02	20	J	0.09404	U	9.84E+00	4.61E+03	AA	3.10E+01	◇	1.97E+00	G	7	11	0.84	O	3.75	H
					9.36E+00	4.21E+03	AA										
E+02	21	V								0.00E+00	F						
E+01	1840	I														16	
E-09	25	C						2.00E+05	C	5.61E+00	A		716	0.46	26		
E+00	1628	P														200	
E+00	53	J						5.00E+02	G	1.97E+00	F					0	F
E+01	20	J															
E+03	21	V															
E+03	20	J								6.40E-01	J						
					9.14E+00	3.24E+03	Z										
								4.20E-02	◇	-3.22E+00	B						
E+00	71	V															
E-06	30	C						7.70E+05	C	6.20E+00	C						
E-06	20	C						4.40E+06	C	7.00E+00	C	5.3	232	0.37	O	51000	G
E-07	25	C						2.43E+05	G	6.19E+00	J	5.3	322	0.50	O	54000	H
					1.18E+01	4.13E+03	Z										
E-03		B						1.00E+03	◇	7.30E-01	B						
E-05		B						1.20E+01	◇	3.50E-01	B						
								1.20E+03	◇	6.62E+00	B						
E-10	20	C						3.30E+06	C	6.80E+00	C	7.1	69	0.75	O		
					1.46E+01	6.37E+03	Z										
E+00	20	W						9.80E+01	◇	2.29E+00	B						
E-05	25	C						1.70E+05	C	5.60E+00	C	3	220	0.45	AB		
E+00	20	C						1.70E+03	C	3.60E+00	C	5.5	129	0.43	O	56	H
E+00	25	C			2.88E+00	2.56E+03	Z	1.70E+03	C	3.60E+00	C	5.1	118	0.45	O	56	H
E+00	25	C			3.37E+00	2.72E+03	Z	1.70E+03	C	3.60E+00	C	5.1	121	0.47	O	56	H
E-05	22	C						1.55E+03	C	3.50E+00	C	7.2-9.1	300	0.20	O	312	H
E+00	25	C						5.80E+01	C	2.16E+00	D						
E+02	20	J	0.09643	U	8.64E+00	4.13E+03	AA	3.00E+01	C	1.79E+00	A	5.3	2	0.53	O		
E+01	20	J			-1.37E+00	1.52E+03	Z	1.40E+01	C	1.48E+00	A	5.3	4	0.83	O	1.2	H
E+02	20	W	0.08386	U	8.84E+00	3.73E+03	AA	6.50E+01	C	1.84E+00	A	5.3	5	0.54	O	5.6	H
E+02	20	W	0.08386	U	6.12E+00	2.91E+03	Z	6.50E+01	C	1.84E+00		5.3	4	0.83	O	5.6	
E+02	20	C			9.34E+00	4.18E+03	AA	5.90E+01	C	4.80E-01		6.7	3	0.51	O	1.6	

alculated value.

Chemical name	Chemical formula	P	C	N	H	CASRN	Molecular weight (g/mole)	Water solubility (mg/L)	(°C)
cis-Dichloroethylene	$C_2H_2Cl_2$					156-59-2	96.94		
cis,1,2-Dichloroethylene	$C_2H_2Cl_2$					540-59-0	96.94	8.00E+02	20
cis-1,2-Dichloroethene	$C_2H_2Cl_2$					540-59-0	96.94	8.00E+02	20
Dichloromethane	CH_2Cl_2	P	C	N		75-09-2	84.93	2.00E+04	20
2,4-Dichlorophenol	$C_6H_4Cl_2O$	P	C	N		120-83-2	163	4.50E+03	25
2,4-Dichlorophenoxyacetic Acid	$C_8H_6Cl_2O_3$			N		94-75-7	221.04	8.90E+02	25
Dichlorophenylarsine	$C_6H_5AsCl_2$					696-28-6	221.93		
1,2-Dichloropropane	$C_3H_6Cl_2$	P	C	N		78-87-5	112.99	2.70E+03	20
1,3-Dichloropropene	$C_3H_4Cl_2$				H	542-75-6	110.97	2.80E+03	25
Dieldrin	$C_{12}H_8Cl_6O$	P	C	N		60-57-1	380.9	1.95E-01	25
Diethyl Arsine	$C_4H_{11}As$					692-42-2	134.05	4.17E+02	
1,2-Diethylhydrazine	$C_4H_{12}N_2$					1615-80-1	88.18	2.88E+07	
Diethylnitrosamine	$C_4H_{10}N_2O$					55-18-5	102.16		
Diethyl-o-Phthalate	$C_{12}H_{14}O_4$	P	C	N		84-66-2	222.26	8.96E+02	25
Diethylstilbesterol	$C_{18}H_{20}O_2$					56-53-1	268.38	9.60E-03	
Dihydrosafrole	$C_{10}H_{12}O_2$					94-58-6	164.22	1.50E+03	
Dimethoate	$C_5H_{12}NO_3PS_2$					60-51-5	229.27	2.50E+04	23
Dimethylamine	C_2H_7N					124-40-3	45.1	6.20E+05	20
Dimethyl Phthalate	$C_{10}H_{10}O_4$				H	131-11-3	194.2		
Dimethyl Sulfate	$C_2H_6O_4S$				H	77-78-1	126.13	3.24E+05	
Dimethyl Terephthalate	$C_{10}H_{10}O_4$					120-61-6	194.2		
Dimethylaminoazobenzene	$C_{14}H_{15}N_3$					60-11-7	225.32	1.36E+01	
7,12-Dimethylbenz(a) anthracene	$C_{20}H_{16}$					57-97-6	256.36	4.40E-03	
Dimethylcarbamoyl Chloride	C_3H_6ClNO				H	79-44-7	107.55	1.44E+07	
1,1-Dimethylhydrazine	$C_2H_8N_2$				H	57-14-7	60.12	1.24E+08	
1,2-Dimethyl Hydrazine	$C_2H_8N_2$					540-73-8	60.12		
Dimethylnitrosamine	$C_2H_6N_2O$					62-75-9	74.1	1.00E+06	
1,3-Dinitrobenzene	$C_6H_4N_2O4$					99-65-0	168.12	8.50E+00	20
4,6-Dinitro-o-cresol	$C_7H_6N_2O_5$	P	C	N	H	534-52-1	198.15	2.90E+02	25
2,4-Dinitrophenol	$C_6H_4N_2O_5$	P	C	N	H	51-28-5	184.12	5.60E+03	18
2,3-Dinitrotoluene	$C_7H_6N_2O_4$					602-01-7	182.15	3.10E+03	
2,4-Dinitrotoluene	$C_7H_6N_2O_4$	P	C	N	H	121-14-2	182.15	2.70E+02	22
2,5-Dinitrotoluene	$C_7H_6N_2O_4$					619-15-8	182.15	1.80E+02	20
2,6-Dinitrotoluene	$C_7H_6N_2O_4$	P	C	N		606-20-2	182.15	1.80E+02	20
3,4-Dinitrotoluene	$C_7H_6N_2O_4$					610-39-9	182.15	1.08E+03	
Dinoseb	$C_{10}H_{12}N_2O_5$					88-85-7	240.24	5.00E+01	25
Dioxane-1,4	$C_4H_8O_2$			N	H	123-91-1	88.11	4.31E+05	
N,N-Diphenylamine	$C_{12}H_{11}N$			N		122-39-4	169.24	3.00E+02	25
1,2-Diphenylhydrazine	$C_{12}H_{12}N_2$	P		N	H	122-66-7	184.26	1.84E+03	20
Dipropylnitrosoamine	$C_6H_{14}N_2O$	P	C	N		621-64-7	130.22	9.90E+03	25
Disulfoton	$C_8H_{19}O_2PS_3$			N		298-04-4	274.42	1.72E+01	20
Endosulfan	$C_9H_6Cl_6O_3S$					115-29-7	406.91	5.30E-01	25

S = Source of data; P = Priority Poll; C = Haz. Substance List; N = App. IX; H = Haz. Air Poll.

por pressure (Hg)	(°C)	S	Diffusion coefficient in air (cm²/s)	S	Henry's Constant $H=\exp(A-B/T)$ $H(atm\cdot m^3/mol)$; T (K) A	B	S	Koc ml/g	S	Log Kow	S	Freundlich parameters pH	K	1/n	(mg/kg)	S	Fish BCF (L/kg)	S
					5.16E+00	3.14E+03	Z											
E+02	20	W			8.48E+00	4.19E+03	AA	4.90E+01	◊	7.00E-01	A						1.6	H
E+02	20	W			8.48E+00	4.19E+03	Y	4.90E+01	C	7.00E-01							1.6	
E+02	20	J			6.65E+00	3.82E+03	AA	8.80E+00	C	1.30E+00	C						5	H
E-02	20	C						3.80E+02	C	2.90E+00	C	5.3	157	0.15		O	41	H
E-01		F						2.00E+01	G	2.81E+00	F							
E-02	20	P																
E+01	20	J			9.84E+00	4.71E+03	Z	5.10E+01	C	2.00E+00	C	5.3	6	0.60		O		
E+01	25	J						4.80E+01	C	2.00E+00	C	5.3	8	0.46		O	1.9	H
E-07	20	C						1.70E+03	C	3.50E+00	C	5.3	606	0.51		O	4760	H
E+01		B						1.60E+02	◊	2.97E+00	B							
								3.00E-01	◊	-1.68E+00	B							
E+00		F								4.80E-01	F							
E-03	25	C						1.42E+02	C	2.50E+00	C	5.4	110	0.27		O	117	G
								2.80E+01	◊	5.46E+00	B							
								7.80E+01	◊	2.56E+00	B							
OE-02	25	J								2.71E+00	J							
2E+03	10	F	0.12577	U				2.20E+00	◊	-3.80E-01	F						0	F
OE-01	20	B						4.10E+00	◊	-1.24E+00	B							
OE+01	100	J																
OE-07		B						1.00E+03	◊	3.72E+00	B							
								4.76E+05	G	6.94E+00	B							
5E+00		B						5.00E-01	◊	-1.32E+00	B							
7E+02	25	J						2.00E-01	◊	-2.42E+00	B							
7E+02	25	J																
OE+00	25	C						1.00E-01	C	-6.80E-01	C						0	D
								1.50E+02	◊	1.62E+00	F							
OE-02	20	C						2.40E+02	C	2.70E+00	C	3.5	237			O	0	E
9E-05	18	C						1.66E+01	C	1.50E+00	C	7	33	0.61		O	0	D
								5.30E+01	◊	2.29E+00	B						3.8	H
OE-03	20	C						4.50E+01	C	2.00E+00	C	5.4	146	0.31		O	3.8	H
OE-05	25	Y						8.40E+01	◊	2.28E+00	B						3.8	H
OE-02	20	C						9.20E+01	C	2.00E+00	C	5.4	145	0.32		O	3.8	H
								9.40E+01	◊	2.29	B						3.8	H
OE+00	151																	
OE+01	25	B						3.50E+00	◊	0.01	B							
OE+00	108	P						4.70E+02	◊	3.60	B						30	G
OE-05	25	C						4.18E+02	C	2.90	C	5.3	16000	2.00		O	25	H
OE-01	25	C						1.50E+01	C	1.50	C							
OE-05	25	C										5.3	194	0.50		O		

Calculated value.

Chemical name	Chemical formula	P	C	N	H	CASRN	Molecular weight (g/mole)	Water solubility (mg/L)	(°C)
1-Chloro-2,3-Epoxypropane	C_3H_5ClO				H	106-89-8	92.53	6.00E+04	20
Ethanol	C_2H_6O					64-17-5	46.08	6.52E+04	20
2-Ethoxyethanol	$C_4H_{10}O_2$					110-80-5	90.14	1.00E+06	
Ethyl Acetate	$C_4H_8O_2$					141-78-6	88.12	7.40E+04	35
Ethyl Benzene	C_8H_{10}	P	C	N	H	100-41-4	106.18	1.52E+02	20
Ethyl Carbitol	$C_6H_{14}O_3$					111-90-0	134.2		
Ethylene Dibromide	$C_2H_4Br_2$				H	106-93-4	187.88	4.30E+03	20
Ethylene Glycol Monobutyl Ether	$C_6H_{14}O_2$					111-76-2	118.2	1.00E+06	
Ethylene Oxide	C_2H_4O				H	75-21-8	44.06	3.80E+08	20
Ethylene Thiourea	$C_3H_6N_2S$				H	96-45-7	102.17	2.00E+03	
Ethyl Methanesulfonate	$C_3H_8O_3S$					62-50-0	124.17	3.69E+05	
1-Ethyl-1-Nitrosourea	$C_3H_7N_3O_2$					759-73-9	117.13	3.31E+08	
Fluoranthene	$C_{16}H_{10}$	P	C	N		206-44-0	202.26	2.06E-01	25
Fluorene	$C_{13}H_{10}$					86-73-7	166.22	1.69E+00	25
Formaldehyde	CH_2O					50-00-0	30.03	4.00E+05	
Formic Acid	CH_2O_2					64-18-6	46.03	1.00E+06	
Furan	C_4H_4O					110-00-9	68.08	9.90E+03	20
Glycidaldehyde	$C_3H_4O_2$					765-34-4	72.1	1.70E+08	
Heptachlor	$C_{10}H_5Cl_7$	P	C	N	H	76-44-8	373.3	1.80E-01	25
Heptachlor Epoxide	$C_{10}H_5Cl_7O$					1024-57-3	389.3	3.50E-01	25
Hexachlorobenzene	C_6Cl_6	P	C	N	H	118-74-1	284.76	6.00E-03	25
Hexachlorbutadiene	C_4Cl_6	P	C	N	H	87-68-3	260.74	2.00E+00	20
Hexachlorocyclopentadiene	C_5Cl_6	P	C	N	H	77-47-4	272.75	1.80E+00	25
α-Hexachlorocyclohexane	$C_6H_6Cl_6$					319-84-6	290.82	1.63E+00	25
β-Hexachlorocyclohexane	$C_6H_6Cl_6$					319-85-7	290.82	2.40E-01	25
γ-Hexachlorocyclohexane	$C_6H_6Cl_6$					58-89-9	290.82	7.80E+00	25
δ-Hexachlorocyclohexane	$C_6H_6Cl_6$					319-86-8	290.82	3.14E+01	25
Hexachloroethane	C_2Cl_6	P	C	N	H	67-72-1	236.72	4.20E+04	20
Hexachlorophene	$C_{13}H_6Cl_6O_2$			N		70-30-4	406.89	4.00E-03	
Hexane	C_6H_{14}				H	110-54-3	86.17		
Hydrazine	H_4N_2				H	302-01-2	32.06	3.41E+08	
Hydrogen Cyanide	CHN					74-90-8	27.03	1.00E+06	
Hydrogen Sulfide	H_2S				H	7783-06-4	34.08	4.13E+03	
Indeno(1,2,3-cd)Pyrene	$C_{22}H_{12}$	P	C	N		193-39-5	276.34	5.30E-04	25
Iodomethane	CH_3I			N		74-88-4	141.94	1.40E+04	20
Iron	Fe					15438-31-0	55.85		
Isobutanol	$C_4H_{10}O$					78-83-1	74.14	7.29E+04	20
Isophorone	$C_9H_{14}O$	P	C		H	78-59-1	138.23	1.04E+04	20
Isoprene	C_5H_8					78-79-5	68.11		
Isosafrole	$C_{10}H_{10}O_2$			N		120-58-1	162.2	1.09E+03	
Kepone	$C_{10}Cl_{10}O$			N		143-50-0	490.6	7.60E+00	24
Lasiocarpine	$C_{21}H_{33}NO_7$					303-34-4	411.55	1.60E+03	
Lead	Pb				H	7439-92-1	207.19		
Malathion	$C_{10}H_{19}O_6PS_2$					121-75-5	330.38	1.45E+02	20

S = Source of data; P = Priority Poll; C = Haz. Substance List; N = App. IX; H = Haz. Air Poll.

oor pressure Hg)	(°C)	S	Diffusion coefficient in air (cm²/s)	S	Henry's Constant H=exp(A-B/T) H(atm·m³/mol); T (K) A	B	S	Koc ml/g	S	Log Kow	S	pH	K (mg/kg)	1/n	S	Fish BCF (L/kg)	S
E+01	20	J						1.00E+01	◊	0.15	B						
E+01	30	J	0.1273	U				2.20E+00	◊	-0.32	J	5.3					
E+00	20	J								0.00	F						
E+02	30	J	0.09005	U						0.73	y						
E+00	20	J	0.0707	U	6.10E+00	3.24E+03	Z	1.10E+03	C	3.15	A	7.3	53	0.79	O	37.5	H
E-02	16	V															
E+01	20	J			5.70E+00	3.88E+03	Z	4.40E+01	G	1.76	B						
E-01	20	J								0.00	F						
E+03	20	J						2.20E+00	◊	-0.22	B						
								6.70E+01	◊	-0.66	J					0	F
E-01		B						3.80E+00	◊	0.21	B						
								1.00E-01	◊								
E-06	25	C						3.80E+04	C	4.90	A	5.3	664	0.61	O	1150	H
E-04	20	C						7.30E+03			C	5.3	330	0.28	O	1300	G
E+03	25	Y						3.60E+00	◊	0.35	Y					0	F
E+01	20	J								-0.54	F					0	F
E+02	25	Y								1.34	Y						
E+01		B						1.00E-01	◊	-1.55	B						
E-04	25	C						1.20E+04	C	4.40	C	5.3	1220	0.95	O	15700	H
E-04	25	C						2.20E+02	G	2.70	C	5.3	1038	0.70	O	14400	G
E-05	20	C						3.90E+03	C	5.23	A	5.3	450	0.60	O	8690	H
E-01	20	C						2.90E+04	C	4.78	A	7	360	0.63	O	2.8	H
E-02	25	J						4.80E+03	C	5.04	A	5.3	370	0.17	O	4.3	H
E-05	20	C						3.80E+03	C	3.90	C					130	H
E-07	20	C						3.80E+03	C	3.90	C					130	H
E+01	20	C						1.08E+03	G	3.90	C					130	H
E-05	20	C						6.60E+03	C	4.10	C					130	H
E-01	20	J	3.74E+00		2.55E+03		Z	2.00E+04	C	4.60	C	5.3	97	0.38	O	87	H
								9.10E+04	◊	7.54	F						
			2.53E+01		7.53E+03		Z										
E+01	20	J						1.00E-01	◊	-3.08	B						
E+02	20	J								-2.50E-01	F					0	F
E+04	26	P															
E-10	20	C						1.60E+06	C	6.50	C						
E+02	25	J						2.30E+01	◊	1.69	J						
E+01	2040	I															
E+01	25	Y								0.76	Y						
E-01	20	C								2.22	Y	5.5	32	0.39	O		
E+02	20	J															
E+00	66	V						9.30E+01	◊	2.66	B						
								5.50E+04	◊	2.00	B						
								7.60E+01	◊	0.99	B					8400	G
E+01	1160	I														49	H
E-05	20	J								2.89	J					0	F

Calculated value.

Chemical name	Chemical formula	P	C	N	H	CASRN	Molecular weight (g/mole)	Water solubility (mg/L)	(°C)
Manganese	Mn				H	7439-96-5	54.94		
Mercury	Hg					7439-97-6	200.59		
Methanol	CH_4O				H	67-56-1	32.05	total	20
2-Methoxyethanol	$C_3H_8O_2$					109-86-4	76.11	1.00E+06	
2-Methyl Aziridine	C_3H_7N				H	75-55-8	57.11	9.44E+05	
Methyl Chloride	CH_3Cl					74-87-3	50.49	5.15E+03	20
Methylene chloride	CH_2Cl_2				H	75-09-2	84.93	2.00E+04	20
4,4′-Methylenebis (2-chloro-aniline)	$C_{13}H_{12}Cl_2N_2$			N	H	101-14-4	267.17		
Methyl Ethyl Ketone	C_4H_8O		C	N	H	78-93-3	72.12	3.50E+05	10
Methyl Isobutyl Ketone	$C_6H_{12}O$				H	108-10-1	100.18	1.72E+04	20
Methyl Methacrylate	$C_5H_8O_2$			N	H	80-62-6	100.13	1.28E+04	20
Methylnitrosourea	$C_2H_5N_3O_2$					684-93-5	103.1	6.89E+08	
Methyl Parathion	$C_8H_{10}NO_5PS$			N		298-00-0	263.23	1.72E+01	20
Methylvinylnitrosamine	$C_3H_6N_2O$					4549-40-0	86.11	7.60E+05	
Mustard Gas	$C_4H_8Cl_2S$					505-60-2	159.08	8.00E+02	
1-Naphthylamine	$C_{10}H_9N$			N		134-32-7	143.2	1.46E+03	20
2-Naphthylamine	$C_{10}H_9N$			N		91-59-8	143.2	5.86E+02	
Nickel	Ni					7440-02-0	58.71		
Nickel Cyanide, (solid)	C_2N_2Ni					557-19-7	110.75	5.15E+01	18
Nitric Oxide	NO					10102-43-9	30.01		
Nitrobenzene	$C_6H_5NO_2$	P	C	N	H	98-95-3	123.12	1.90E+03	20
Nitrogen Dioxide	NO_2					10102-44-0	46.01		
N-Nitrosopiperidine	$C_5H_{10}N_2O$			N		100-75-4	114.17	1.90E+06	
N-Nitrosopyrrolidine	$C_4H_8N_{20}$			N		930-55-2	100.14	7.00E+06	
Osmium Tetroxide	O_4Os					20816-12-0	254.2		
Pentachlorobenzene	C_6HCl_5			N		608-93-5	250.32	2.40E-01	22
Pentachloronitro Benzene	$C_6Cl_5NO_2$			N	H	82-68-8	295.32	7.11E-02	
Pentachlorophenol	C_6HCl_5O	P	C	N	H	87-86-5	266.32	1.40E+01	20
Phenanthrene	$C_{14}H_{10}$	P	C	N		85-01-8	178.24	1.00E+00	15
Phenobarbital	$C_{12}H_{12}N_2O_3$					50-06-6	232.26	1.00E+03	25
Phenol	$C_6H_6O_6$				H	108-95-2	94.12	8.20E+04	20
m-Phenylenediamine	$C_6H_8N_2$					108-45-2	108.16	3.51E+05	25
Phenyl Mercuric Acetate	$C_8H_8HgO_2$					62-38-4	336.75	1.67E+03	
Polychlorinated Biphenyls	NA				H	1336-36-3	NA	3.10E-02	25
Propyl Benzene	C_9H_{12}					103-65-1	120.21		
Pyrene	$C_{16}H_{10}$	P	C	N		129-00-0	202.26	1.30E-01	25
Pyridine	C_5H_5N			N		110-86-1	79.11	1.00E+06	
Safrole	$C_{10}H_{10}O_2$			N		94-59-7	162.2	1.50E+03	
Selenious Acid	H_2O_3Se					7783-00-8	128.98		
Selenium	Se					7782-49-2	78.96		
Silver	Ag					7440-22-4	107.87		
Strychnine	$C_{12}H_{22}N_2O_2$					57-24-9	334.45	1.43E+02	
Styrene	C_8H_8		C	N	H	100-42-5	104.16	3.00E+02	20

S = Source of data; P = Priority Poll; C = Haz. Substance List; N = App. IX; H = Haz. Air Poll.

vapor pressure (Hg)	(°C)	S	Diffusion coefficient in air (cm²/s)	S	A	B	S	Koc ml/g	S	Log Kow	S	pH	K (mg/kg)	1/n	S	Fish BCF (L/kg)	S
E+00	1292	P															
E-03	25	P														5500	H
E+02	21	P	0.16686	U						-0.77	Y						
E+00	20	J															
E+02	20	V						2.30E+00	◊	-0.48	B						
E+03	20	J	0.1083	U	5.56E+00	3.18E+03	Z	3.50E+01	◊	0.95	B	5.8	1	1.16	O		
E+02	20	J			8.48E+00	4.27E+03	Z	8.80E+00		1.30		5.8	1	1.16	AB	5	
												7.5	190	0.64	O		
E+01	20	K	0.09485	U	-4.84E+01	2.31E+03	Z	4.50E+00	◊	0.26	A					0	F
E+01	21	V								1.19	Y						
E+01	20	J						8.40E+02	◊	0.79	F						
								1.00E-01	◊	-3.81	B						
E-06		E						4.60E+02	◊	1.91	F					45	F
E+01		B						2.50E+00	◊	-0.23	B						
E-01		B						1.10E+02	◊	1.37	B						
E+00	104	P						6.10E+01		2.07	B						
E+00	108	P						1.30E+02		2.07	B						
E+00	1810	P														47	H
E+01	-178	I															
E-01	20	J						3.60E+01	C	1.85	D				O		
E+02	80	P															
E-01		B						1.50E+00	◊	-0.49	B						
E-01		B						8.00E-01	◊	-1.06	B						
E+01	26	P															
								1.30E+04	◊	5.19	F					2125	H
E-02	25	P						1.90E+04	◊	5.45	B						
E-04	20	J						5.30E+04	C	5.00	C	7	150	0.42	O	770	G
E-04	25	C						1.40E+04	C	4.46	A	5.3	215	0.44	O	2630	G
								9.80E+01	◊	-0.19	B						
E-01	20	J	0.08924	U	1.13E+01	4.66E+03	Z	1.42E+01	C	1.46	A	3.0-9	21	0.54	O	1.4	H
E+00	100	P															
												*	270	0.44	O		
E-05	25	C	0.05571	U				5.30E+05	C	6.04	C					1E+05	G
					7.84E+00	3.68E+03	Z										
E-06	25	C						3.80E+04	C	4.88	A		389	0.39	26		
E+01	25	J								0.66	F						
E+00	64	P						7.80E+01	◊	2.53	B						
E+00	15	P															
E+00	356	P														16	H
E+01	1540	I														3080	D
E+00	270	V															
E+00	20	V	0.0746	U						2.95	Y		120	0.56	O		

alculated value.

Chemical name	Chemical formula	P	C	N	H	CASRN	Molecular weight (g/mole)	Water solubility (mg/L)	(°C)
1,2,4,5-Tetrachlorobenzene	$C_6H_2Cl_4$			N		95-94-3	215.88	3.00E-01	22
2,3,7,8-Tetrachlorodi-									
benzo-p-dioxin	$C_{12}H_4Cl_4O_2$	P		N		1746-01-6	321.96	2.00E-04	25
1,1,1,2-Tetrachloroethane	$C_2H_2Cl_4$			N		630-20-6	167.84	2.90E+03	20
Tetrachloroethene	C_2Cl_4					127-18-4	165.82	1.50E+02	25
2,4,5,6-Tetrachlorophenol	$C_6H_2Cl_4O$					58-90-2	231.88	1.00E+03	
Tetraethyl Lead	$C_8H_{20}Pb$					78-00-2	323.47	8.00E-01	20
Thallium	Tl					7440-28-0	204.37		
Thallium Chloride	TlCl					7791-12-0	239.82	2.90E+03	
Thallium Sulfate	$Tl_2 \cdot O_4S$					7446-18-6	504.8	4.88E+04	21
Thioacetamide	C_2H_5NS					62-55-5	75.14		
Thiourea	Ch_4N_2S					62-56-6	76.13	9.18E+05	13
Tolidine	$C_{14}H_{16}N_2$					119-93-7	212.32	7.35E+01	
Toluene	C_7H_8	P	C	N	H	108-88-3	92.15	5.15E+02	20
o-Toluidine Hydrochloride	$C_7H_9N \cdot ClH$					636-21-5	143.63	1.50E+04	
Toxaphene	$C_{10}H_{10}Cl_8$ (approx)	P	C	N	H	8001-35-2	413.8	5.00E-01	25
Tribromomethane	$CHBr_3$	P	C	N		75-25-2	252.75	3.19E+03	30
1,2,4-Trichlorobenzene	$C_6H_3Cl_3$	P	C	N	H	120-82-1	181.44	3.00E+01	25
1,1,1-Trichloroethane	$C_2H_3Cl_3$	P	C	N		71-55-6	133.4	4.40E+03	20
1,1,2-Trichloroethane	$C_2H_3Cl_3$	P	C	N	H	79-00-5	133.4	4.50E+03	20
Trichloroethylene	C_2HCl_3					79-01-6	131..38	1.10E+00	25
Trichlorfon	$C_4H_8Cl_3O_4P$					52-68-6	257.44	1.54E+05	20
Trichlorofluoromethane	CCl_3F			N		75-69-4	137.36	1.10E+03	20
2,4,5-Trichlorophenol	$C_6H_3Cl_3O$		C	N	H	95-95-4	197.44	1.19E+03	25
2,4,6-Trichlorophenol	$C_6H_3Cl_3O$	P	C	N	H	88-06-2	197.44	8.00E+02	25
2,4,5-Trichlorophenoxy-									
acetic Acid	$C_8H_5Cl_3O_3$					93-76-5	255.48	2.06E+02	20
1,2,3-Trichloropropane	$C_3H_5Cl_3$			N		96-18-4	147.43	1.90E+03	20
1,1,2-Trichloro-1,2,2,-									
trifluoroethane	$C_2Cl_3F_3$					76-13-1	187.37	1.70E+01	20
1,3,5-Trimethylbenzene	C_9H_{12}					108-67-8	120.21		
syn-Trinitrotoluene (TNT)	$C_7H_5N_3O_6$					118-96-7	227.15	2.00E+02	15
Tris(2,3-dibromopropyl)									
phosphate	$C_9H_{15}Br_6O_4P$			N		126-72-7	697.67	8.00E+00	24
Uracil Mustard	$C_8H_{11}Cl_2N_3O_2$					66-75-1	252.12	6.41E+02	
Urethane	$C_3H_7NO_2$					51-79-6	89.11		
Vinyl Chloride	C_2H_3Cl	P	C	N	H	75-01-4	62.5	4.27E+03	20
m-Xylene	C_8H_{10}				H	95-47-6	106.18	1.75E+02	20
o-Xylene	C_8H_{10}				H	108-38-3	106.18	1.30E+02	20
p-Xylene	C_8H_{10}				H	106-42-3	106.18	1.98E+02	20
Xylene	C_8H_{10}				H	1330-20-7	106.18	1.98E+02	
Zinc	Zn					7440-66-6	65.37		

S = Source of data; P = Priority Poll; C = Haz. Substance List; N = App. IX; H = Haz. Air Poll.

Vapor pressure (Hg)	(°C)	S	Diffusion coefficient in air (cm²/s)	S	$H=\exp(A-B/T)$ $H(\text{atm·m}^3/\text{mol})$; T (K) A	B	S	Koc ml/g	S	Log Kow	S	Freundlich parameters pH	K (mg/kg)	1/n	S	Fish BCF (L/kg)	S
E-01	25	P						1.60E+03		4.67	F					1125	H
E-06	25	C						3.30E+06		6.72	A					5000	H
E+00	20	J	0.07729	U				5.40E+01		3.03	Y						
E+01	190	J	0.07729	U	1.24E+01	4.92E+03	AA	3.64E+02		2.60	A	5.3	51	0.56	O	31	H
								9.80E+01		4.10	F					240	H
E-01	20	J						4.90E+03	◊								
E+00	825	P															
E+01	517	P															
E+00		E															
										-0.46	J						
								1.60E+00	◊	-1.14	G						
								4.10E+02	◊	2.88	B						
E+01	20	J	0.08301	U	5.13E+00	3.02E+03	Z	3.00E+02	C	2.73	A	5.6	26	0.44	O	10.7	H
E-01		J						2.20E+01	◊	1.29	J						
E-01	25	C						9.64E+02	C	3.30	C					13100	H
E+00	25	J						1.16E+02	C	2.40	C						
E-01	25	C			7.36E+00	4.03E+03	Z	9.20E+03	C	4.30	C	5.3	157	0.31	O	2800	G
E+02	20	J			9.78E+00	4.13E+03	AA	1.52E+02	C	2.50	C	5.3	2	0.34	O	5.6	H
E+01	30	J						5.60E+01	C	2.47	A	5.3	6	0.60	O	5	H
E+01	20	J			1.14E+01	4.78E+03	AA	1.26E+02	C	2.38	A					10.6	H
E-06	20	J						6.10E+00	◊	2.29	A						
E+02	25	C	0.08329	U	9.48E+00	3.51E+03	Z	1.59E+02	C	2.53	D	5.3	5.6	0.24	AB		
E+02	25	J						8.90E+01	◊	3.72	A					110	H
E-02	25	C						2.00E+03	C	3.87	A	3	219	0.29	O	150	H
E+00	20	J															
E+02	20	J			9.65E+00	3.24E+03	Z			2.00	F						
					7.24E+00	3.63E+03	Z										
								3.10E+02	◊	4.12	B					2.7	G
								1.20E+02	◊	-1.09	B	*	11	0.63	O		
•E+01	78	P															
•E+03	20	W	0.11375	U	7.39E+00	3.29E+03	AA	5.70E+01	◊	1.38	A					1.17	H
•E+01	20	W			6.28E+00	3.34E+03	Z			2.95	F						
•E+01	20	W			5.54E+00	3.22E+03	Z			3.26	F						
•E+01	20	W			6.93E+00	3.52E+03	Z			3.15	F	7.3	85	0.19	O		
•E+01	20	W	0.07597	U	7.04E-03			2.40E+02	◊	3.26	F						
•E+00		D														47	H

alculated value.

APPENDIX
B

TOXICOLOGICAL
DATA

"If you drink much from a bottle marked 'poison,'
it is almost certain to disagree with you sooner or later."
Lewis Carroll

The following tables contain toxicological information taken from EPA's IRIS Database in July, 1993. Chemicals are identified by both their chemical names and their Chemical Abstract Service Registry Numbers (CASRN).

While these tables were prepared with care, no warranty is intended or implied and the user is advised to obtain current information directly for the IRIS Database.

The data in these tables consist of reference doses (RfD) for non-carcinogenic toxicity and slope factors (carcinogen potency factors)(CPF) for carcinogens. The derivation of these values is discussed in Secs. 5-4 and 5-5 and their application to risk assessment is discussed in Sec. 14-5.

Chemical name	CASRN	Oral RfD mg/kg-day	Date	Inhalation RfD mg/kg-day	Date	Oral CPF 1/mg/kg-day	Date	Inhalation CPF* 1/mg/kg-day	Date	Carc. Class.
Acenaphthene	83-32-9	6.00E-02	11/1/90	no data		no data		no data		D
Acenaphthylene	208-96-8	pending		no data		no data		no data		C
Acephate	30560-19-1	4.00E-03	2/1/90	no data		8.70E-03	7/1/89	no data (NA)		B2
Acetaldehyde	75-07-0	no data		2.57E-03	10/1/91	no data		7.70E-03 [I]	1/1/91	B2
Acetamide	60-35-5	no data		pending	10/1/92	no data		no data		
Acetone	67-64-1	1.00E-01	3/1/88	no data		no data		no data		D
Acetonitrile	75-05-8	6.00E-03 [I]	8/1/89	pending		no data		no data (none)		
Acetophenone	98-86-2	1.00E-01	1/1/89	pending		no data (none)		no data (none)		D
Acetyl chloride	75-36-5	no data		no data		no data (none)		no data (none)		D
Acifluorfen, sodium	62476-59-9	1.30E-02	12/1/88	no data		no data		no data		
Acrolein	107-02-8	no data		5.72E-06	10/1/91	no data (NA)		no data (NA)		C
Acrylamide	79-06-1	2.00E-04	3/1/91	message		4.50E-00	1/1/91	4.50E+00 [O]	1/1/91	B2
Acrylic Acid	79-10-7	8.00E-02	3/1/88	8.58E-05	10/1/90	no data		no data		
Acrylonitrile	107-13-1	no data		5.72E-04	11/1/91	5.40E-01	1/1/91	2.40E-01 [I]	1/1/91	B1
Adiponitrile	111-69-3	no data		no data		no data (none)		no data (none)		D
Alachlor	15972-60-8	1.00E-02	2/1/88	no data		pending		pending		
Alar	1596-84-5	1.50E-01	7/1/91	no data		pending		pending		
Aldicarb	116-06-3	2.00E-04	7/1/91	no data		no data (NA)		no data (NA)		
Aldicarb sulfone	1646-88-4	withdrawn	5/1/91	no data		no data		no data		
Aldrin	309-00-2	3.00E-05	12/1/88	no data		1.70E+01	1/1/91	1.70E+01 [O]	1/1/91	D
Alpha emitters	—	no data		no data		pending	3/1/93	pending	3/1/93	
Ally	74223-64-6	2.50E-01	6/30/88	no data		no data		no data		
Allyl alcohol	107-18-6	5.00E-03	8/1/89	no data		no data		no data		B2

The information in this table is based on information provided by the U.S. EPA's Integrated Risk Information System (IRIS) in July 1993.

*[I-%] - study based on inhalation study; percent absorption used (if none is listed 100% is assumed)

[O] - study based on oral study

NA - not analyzed

Chemical name	CASRN	Oral RfD mg/kg-day	Date	Inhalation RfD mg/kg-day	Date	Oral CPF 1/mg/kg-day	Date	Inhalation CPF* 1/mg/kg-day	Date	Carc. Class.
Allyl chloride	107-05-1	no data		2.86E-04	12/1/91	no data (NA)		no data (NA)		C
Aluminum phosphide	20859-73-8	4.00E-04	3/1/88	no data		no data		no data		
Amdro	67485-29-4	3.00E-04	9/30/87	no data		pending		pending		
Ametryn	834-12-8	9.00E-03	11/1/89	no data		no data		no data		D
4-Aminopyridine	504-24-5	no data		no data		no data (none)		no data (none)		
Amitraz	33089-61-1	2.50E-03	12/1/88	no data		no data		no data		
Ammonia	7664-41-1	no data		2.86E-02	5/1/91	no data		no data		
Ammonium acetate	631-61-8	no data		no data		no data (none)		no data (none)		D
Ammonium methacrylate	16325-47-6	no data		no data		no data (none)		no data (none)		D
Ammonium sulfamate	7773-06-0	2.00E-01	3/1/91	no data		no data		no data		
Aniline	62-53-3	no data		2.86E-04	11/1/90	5.70E-03	6/1/89	no data (NA)		B2
ortho-Anisidine	90-04-0	no data		message	10/1/91	no data		no data		
Anthracene	120-12-7	3.00E-01	7/1/91	no data		no data (none)		no data (none)		D
Antimony	7440-36-0	4.00E-04	2/1/91	no data		no data		no data		
Antimony trioxide	1309-64-4	no data		pending	3/1/93	no data		no data		
Apollo	74115-24-5	1.30E-02	11/1/89	no data		no data (NA)		no data (NA)		C
Aramite	140-57-8	pending		no data		2.50E-02	6/1/91	0.025 [O]	6/1/91	B2
Aroclor 1016	12674-11-2	7.00E-05	3/1/93	no data		no data (also see PCB)		no data		
Aroclor 1248	12672-29-6	pending	9/1/92	no data		no data (also see PCB)		no data		
Arsenic, inorganic	7440-38-2	3.00E-04	10/1/91	no data		1.75E+00	2/1/91	5.00E+01 [I-30%]	2/1/91	A
Arsine	7784-42-1	no data		pending	3/1/93	no data (NA)		no data		
Asbestos	1332-21-4	no data		no data		no data (none)		2.3E-1/fibers/mL [I]5/1/89		A
Assure	76578-14-8	9.00E-03	9/26/88	no data		no data (none)		no data (none)		D
Asulam	3337-71-1	5.00E-02	6/30/88	no data		no data		no data		
Atrazine	1912-24-9	5.00E-03	1/1/91	no data		pending		pending		

Chemical name	CASRN	Oral RfD mg/kg·day	Date	Inhalation RfD mg/kg·day	Date	Oral CPF 1/mg/kg·day	Date	Inhalation CPF* 1/mg/kg·day	Date	Carc. Class.
Avermectin B1	65195-55-3	4.00E-04	7/1/89	no data		no data		no data		
Azobenzene	103-33-3	no data		no data		1.10E-01	1/1/91	1.10E-01 [O]	1/1/91	B2
Barium	7440-39-3	7.00E-02	8/1/90	pending	12/1/91	no data		no data		
Barium cyanide	542-62-1	withdrawn	3/1/91	no data		no data		no data		
Baygon	114-26-1	4.00E-03	3/31/87	no data		pending		pending		
Bayleton	43121-43-3	3.00E-02	3/1/88	no data		no data		no data		
Baythroid	68359-37-5	2.50E-02	3/1/88	no data		no data		no data		
Benefin	1861-40-1	3.00E-01	3/1/88	no data		no data		no data		
Benomyl	17804-35-2	5.00E-02	3/1/89	no data		pending		pending		
Bentazon	25057-89-0	2.50E-03	5/1/91	no data		pending		pending		
Benz(a)anthracene	56-55-3	no data		no data		no data (NA)		no data (NA)		B2
Benzaldehyde	100-52-7	1.00E-01	12/1/88	no data		no data		no data		
Benzene	71-43-2	pending		pending		2.90E-02	1/1/91	2.90E-02 [I]	1/1/91	A
Benzidine	92-87-5	3.00E-03	1/1/90	message	7/1/91	2.30E+02	1/1/90	2.30E+02 [I]	11/1/90	A
Benzo(a)pyrene (BaP)	50-32-8	no data		no data		7.30E+00	7/1/92	no data		B2
Benzo(e)pyrene	192-97-2	no data		no data		pending	9/1/92	pending	9/1/92	B2
Benzo(b)fluoranthene	205-99-2	no data		no data		no data (NA)		no data (NA)		D
Benzo(g,h,i)perylene	191-24-2	no data		no data		no data (none)		no data (none)		
Benzo(j)fluoranthene	205-82-3	no data		no data		pending	1/1/93	pending	1/1/93	B2
Benzo(k)fluoranthene	207-08-9	no data		no data		no data (NA)		no data (NA)		D
Benzoic acid	65-85-0	4.00E+00	6/1/91	no data		no data (NA)		no data (NA)		B2
Benzotrichloride	98-07-7	no data		no data		1.30E+01 (for drinking water)	7/1/90	no data (NA)		B2
Benzyl chloride	100-44-7	no data		message	1/1/92	1.70E-01	1/1/90	no data (NA)		B2
Beryllium	7440-41-7	5.00E-03	9/1/90	no data		4.30E+00	1/1/91	8.40E+00 [I]	1/1/91	B2

Chemical name	CASRN	Oral RfD mg/kg·day	Date	Inhalation RfD mg/kg·day	Date	Oral CPF 1/mg/kg·day	Date	Inhalation CPF* 1/mg/kg·day	Date	Carc. Class.
Beryllium sulfate	13510-49-1	no data		pending	1/1/92	no data		no data		
Beta-photon emitters	–	no data		no data		pending	3/1/93	pending	3/1/93	D
Bidrin	141-66-2	1.00E-04	11/1/89	no data		no data		no data		D
Biphenthrin	82657-04-3	1.50E-02	8/22/88	no data		no data		no data		
1,1-Biphenyl	92-52-4	5.00E-02	8/1/89	message	9/20/90	no data (none)		no data (none)		
Bis(2-chloroethoxy)methane	111-91-1	no data		no data		no data (none)		no data (none)		
Bis(2-chloroisopropyl)ether	39638-32-9	4.00E-02	8/1/90	no data		no data		no data		
Bis(chloroethyl)ether (BCEE)	111-44-4	no data		message	10/1/91	1.10E+00	1/1/91	1.10E+00 [O]	1/1/91	B2
Bis(chloromethyl)ether (BCME)	542-88-1	no data		message	7/1/91	2.20E+02	1/1/91	2.20E+02 [I]	1/1/91	A
Bisphenol A	80-05-7	5.00E-02	9/26/88	no data		no data		no data		
Boron (Boron & Borates only)	7440-42-8	9.00E-02	10/1/89	no data		pending	1/1/93	pending	1/1/93	D
Brominated dibenzofurans	no CASRN	no data		no data		no data (none)		no data (none)		D
Bromochloromethane	74-97-5	no data		no data		no data (none)		no data (none)		
Bromodichloromethane	75-27-4	2.00E-02	3/1/91	no data		6.20E-02	2/1/93	no data (NA)		B2
p-Bromodiphenyl ether	101-55-3	no data		no data		no data (none)		no data (none)		D
Bromoethane	74-96-4	no data		no data		no data		no data		
Bromoform	75-25-2	2.00E-02	3/1/91	pending	3/1/93	pending	5/1/92	pending	5/1/92	B2
Bromomethane	74-83-9	1.40E-03	8/1/90	1.43E-03	4/1/92	7.90E-03	9/1/90	7.7E-07 [O-50%]	1/1/91	D
Bromotrichloromethane	75-62-7	no data		no data		no data		no data (none)		D
Bromoxynil	1689-84-5	2.00E-02	6/30/88	no data		no data		no data		
Bromoxynil octanoate	1689-99-2	2.00E-02	9/7/88	no data		no data (none)		1.80E+00 [I]	2/1/91	B2
1,3-Butadiene	106-99-0	no data		no data		no data (none)		no data (none)		
n-Butanol	71-36-3	1.00E-01	9/1/90	no data		no data		no data (NA)		D
Butyl benzyl phthalate	85-68-1	2.00E-01	8/1/91	no data		no data (none)		no data (none)		C
Butylate	2008-41-5	5.00E-02	6/1/89	no data		no data (NA)		no data (NA)		

Chemical name	CASRN	Oral RfD mg/kg·day	Date	Inhalation RfD mg/kg·day	Date	Oral CPF 1/mg/kg·day	Date	Inhalation CPF* 1/mg/kg·day	Date	Carc. Class.
t-Butylchloride	507-20-0	no data		no data		no data (none)		no data (none)		D
Butylphtalyl butylglycolate (BPBG)	85-70-1	1.00E+00	3/1/88	no data		no data		no data		
Cacodylic acid	75-60-5	pending	8/1/91	no data		no data (NA)	11/1/92	no data (NA)	11/1/92	D
Cadmium	7440-43-9	5.00E-04 (water)	10/1/89	pending		no data (NA)		no data (NA)		D
		1.00E-03 (food)	10/1/89					6.10E+00 [I]	3/1/91	B1
Calcium cyanide	592-01-8	4.00E-02	1/1/90	no data		no data		no data		
Caprolactam	105-60-2	5.00E-01	9/7/88	no data		no data		no data		
Captafol	2425-06-1	2.00E-03	9/30/87	no data		pending		pending		
Captan	133-06-2	1.30E-01	3/1/89	no data		pending		pending		
Carbaryl	63-25-2	1.00E-01	3/1/88	message	11/1/91	no data		no data		
Carbofuran	1563-66-2	5.00E-03	9/30/87	no data		no data		no data		
Carbon disulfide	75-15-0	1.00E-01 [I]	9/1/90	pending		no data		no data		
Carbon tetrachloride	56-23-5	7.00E-04	12/1/89	no data		1.30E-01	1/1/91	1.30E-01 [O-40%]	1/1/91	B2
Carbonyl sulfide	463-58-1	no data		message	11/1/91	no data		no data		
Carbosulfan	55285-14-8	1.00E-02	12/1/88	no data		no data		no data		
Carboxin	5234-68-4	1.00E-01	7/1/89	no data		no data		no data		
Catechol	120-80-9	no data		pending	10/1/92	no data		no data		
Chloral	75-87-6	2.00E-03	8/22/88	no data		no data		no data		
Chloramben	133-90-4	1.50E-02	3/1/88	no data		pending		pending		
Chlordane	57-74-9	6.00E-05	7/1/89	pending		1.30E+00	1/1/91	1.30E+00 [O]	1/1/91	B2
Chlorimuron-ethyl	90982-32-4	2.00E-02	11/1/89	no data		no data		no data		
Chlorine	7782-50-5	no data		pending	8/1/91	pending	1/1/93	pending	1/1/93	

1058

Chemical name	CASRN	Oral RfD mg/kg-day	Date	Inhalation RfD mg/kg-day	Date	Oral CPF 1/mg/kg-day	Date	Inhalation CPF* 1/mg/kg-day	Date	Carc. Class.
Chlorine cyanide	506-77-4	5.00E-02	3/1/88	no data		no data		no data		
Chlorine dioxide	10049-04-4	no data		5.72E-05	11/1/90	no data		no data		
Chlorite	14998-27-7	pending	9/1/92	no data		no data		no data		
p-Chloroaniline	106-47-8	4.00E-03	8/22/88	no data		no data		no data		D
2-Chloroacetophenone	532-27-4	no data		8.58E-06	10/1/91	no data		no data		
Chlorobenzene	108-90-7	2.00E-02	11/1/90	pending		no data		no data		D
Chlorobenzilate	510-15-6	2.00E-02	12/1/89	no data		no data		no data		D
1-Chlorobutane	109-69-3	no data		no data		no data		no data		D
2-Chlorobutane	78-68-4	no data		no data		no data		no data		
Chlorocyclopentadiene	41851-50-7	no data		pending	10/1/92	no data		no data		
1-Chloro-1,1-difluoroethane	75-68-3	no data		pending	10/1/92	no data		no data		
Chlorodifluoromethane	75-45-6	1.00E-02	6/30/88	pending		no data		no data		
Chloroform	67-66-3	pending		pending		6.10E-03	1/1/91	8.10E-02 [O]	1/1/91	B2
Chloromethane	74-87-3	no data		pending		pending	5/1/92	pending	5/1/92	
Chloromethyl methyl ether (CMME)	107-30-2	no data		pending		no data		no data		A
beta-Chloronaphthalene	91-58-7	8.00E-02	11/1/90	no data		no data		no data		
2-Chlorophenol	95-57-8	5.00E-03	8/22/88	no data		no data		no data		
p-Chlorophenyl methyl sulfide	123-09-1	pending	8/1/92	no data		no data		no data		
p-Chlorophenyl methyl sulfone	98-57-7	pending	8/1/92	no data		no data		no data		
p-Chlorophenyl methyl sulfoxide	934-73-6	pending	8/1/92	no data		no data		no data		
2-Chloro-1,1,1,2-tetra-fluoroethane	2837-89-0	no data		pending	3/1/93	no data		no data		
Chlorothalonil	1897-45-6	1.50E-02	3/1/88	no data		pending		pending		D
o-Chlorotoluene	95-49-8	2.00E-02	2/1/90	no data		no data		no data		D
Chlorpropham	101-21-3	2.00E-01	6/30/88	no data		no data		no data		D

Chemical name	CASRN	Oral RfD (mg/kg·day)	Date	Inhalation RfD (mg/kg·day)	Date	Oral CPF (1/mg/kg·day)	Date	Inhalation CPF* (1/mg/kg·day)	Date	Carc. Class.
Chlorpyrifos	2921-88-2	3.00E-03	3/1/88	no data		no data		no data		
Chlorsulfuron	64902-72-3	5.00E-02	1/1/90	no data		no data		no data		
Chromium (III)	16065-83-1	1.00E+00	3/1/88	pending		pending	5/1/92	pending		
Chromium (VI)	7440-47-3	5.00E-03	3/1/88	pending		no data (NA)		4.10E+01 [I]	5/1/92	A
Chrysene	218-01-9	no data		no data		no data (NA)		no data (NA)	1/1/91	B2
Cobalt	7440-48-4	pending	5/1/92	no data		no data		no data		
Coke oven emissions		no data		no data		no data		2.17E+00 [I]	5/1/89	A
Copper	7440-50-8	no data		no data		no data		no data		D
Copper cyanide	544-92-3	5.00E-03	9/7/88	no data		no data		no data		
Creosote	8001-58-9	no data		no data		no data		no data		B1
Crotonaldehyde	123-73-9	pending		pending		no data (NA)		no data (NA)		C
Cumene	98-82-8	4.00E-02	9/7/88	no data		no data		no data		D
Cyanazine	21725-46-2	withdrawn	7/1/92	pending		no data		no data		
Cyanide, free	57-12-5	2.00E-02	3/1/88	no data		no data		no data		
Cyanogen	460-19-5	4.00E-02	12/1/89	no data		no data		no data		
Cyanogen bromide	506-68-3	9.00E-02	9/26/88	no data		no data		no data		
Cyclohexanone	108-94-1	5.00E+00	9/30/87	no data		no data		no data		
Cyclohexylamine	108-91-8	2.00E-01	9/7/88	no data		no data		no data		
Cyhalothrin/Karate	68085-85-8	5.00E-03	6/3/88	no data		no data		no data		
Cypermethrin	52315-07-8	1.00E-02	1/1/90	no data		no data		no data		
Cyromazine	66215-27-8	7.50E-03	9/30/87	no data		no data		no data		
Dacthal	1861-32-1	5.00E-01	1/1/90	no data		no data		no data		
Dalapon, sodium salt	75-99-0	3.00E-02	6/1/89	no data		no data		no data		
Danitol	39515-41-8	withdrawn	5/1/93	no data		no data		no data		
Decabromdiphenyl ether (DBDPE)	1163-19-5	1.00E-02	12/1/89	no data		no data (NA)		no data (NA)		
DEGDN	693-21-0	no data		no data		pending	7/1/92	pending	7/1/92	C

Chemical name	CASRN	Oral RfD mg/kg·day	Date	Inhalation RfD mg/kg·day	Date	Oral CPF 1/mg/kg·day	Date	Inhalation CPF* 1/mg/kg·day	Date	Carc. Class.
Demeton	8065-48-3	4.00E-05	3/1/88	no data		no data		no data		
2,4-Diaminotoluene	95-80-7	no data		message	3/1/91	no data		no data		B2
Diazomethane	334-88-3	no data		message	11/1/91	no data		no data		
Dibenz(a,h)anthracene	53-70-3	no data		no data		no data (NA)		no data (NA)	5/1/93	D
Dibenzo(a,e)fluoranthene	5385-75-1	no data		no data		pending	5/1/93	pending		
Dibenzofuran	132-64-9	no data		pending	5/1/92	no data		no data		
1,2-Dibromo-3-chloro-propane (DBCP)	96-12-8	no data		5.72E-05	10/1/91	pending		pending		
1,4-Dibromobenzene	106-37-6	1.00E-02	3/1/88	no data		no data		no data		C
Dibromochloromethane	124-48-1	2.00E-02	3/1/91	no data		8.40E-02	11/1/90	no data (NA)		D
Dibromodichloromethane	594-18-3	no data		no data		no data		no data (NA)		D
p,p'-Dibromodiphenyl ether	2050-47-7	no data		no data		no data (NA)		no data (NA)	1/1/91	B2
1,2-Dibromoethane (EDB)	106-93-4	withdrawn	4/1/93	pending	12/1/92	8.50E+01	1/1/91	7.70E-01 [I]		
Dibromomethane	74-95-3	1.00E-01	8/1/90	pending	10/1/90	no data		no data		D
Dibutyl phthalate	84-74-2	1.00E-01	8/22/88	message		no data		no data		
Dicamba	1918-00-9	3.00E-02	10/1/92	no data		pending	3/1/93	pending	3/1/93	
Dichloroacetic acid	79-43-6	pending		no data		no data		no data		D
1,2-Dichlorobenzene	95-50-1	9.00E-02	6/1/90	no data		no data		no data		D
1,3-Dichlorobenzene	541-73-1	no data		no data		4.50E-01	8/1/90	no data		B2
3,3'-Dichlorobenzidine	91-94-1	no data		message	11/1/91	no data		no data		
Dichlorodifluoromethane	75-71-8	2.00E-01	8/1/90	no data		no data		no data		
p,p'-Dichlorodiphenyl-dichloroethane (DDD)	72-54-8	no data		no data		2.40E-01	6/1/89	no data		B2
p,p-Dichlorodiphenyl-dichloroethylene (DDE)	72-55-9	no data		no data		3.40E-01	8/22/88	no data		B2
p,p'-Dichlorodiphenyl-trichloroethane (DDT)	50-29-3	5.00E-04	9/30/87	no data		3.40E-01	1/1/91	3.40E-01 [O]	1/1/91	B2

Chemical name	CASRN	Oral RfD mg/kg·day	Date	Inhalation RfD mg/kg·day	Date	Oral CPF 1/mg/kg·day	Date	Inhalation CPF* 1/mg/kg·day	Date	Carc. Class.
1,1-Dichloroethane	75-34-3	pending		pending		no data		no data		C
1,2-Dichloroethane	107-06-2	no data		no data		9.10E-02	1/1/91	9.10E-02 [O]	1/1/91	B2
1,1-Dichloroethylene	75-35-4	9.00E-03	4/1/89	pending		6.00E-01	1/1/91	1.80E-01 [I]	1/1/91	C
cis-1,2-Dichloroethylene	156-59-2	pending		no data		no data		no data		
trans-1,2-Dichloroethylene	156-60-5	2.00E-02	1/1/89	no data		no data		no data		D
1,1-Dichloro-1-fluoroethane	1717-00-6	no data		pending	10/1/92	no data		no data		
Dichloromethane (methylene chloride)	75-09-2	6.00E-02	3/1/88	pending		7.50E-03	1/1/91	1.65E-03 [I]	1/1/91	B2
2,4-Dichlorophenol	120-83-2	3.00E-03	3/1/88	no data		no data		no data		
2,4-Dichlorophenoxy-acetic Acid (2,4-D)	94-75-7	1.00E-02	5/5/88	no data		no data		no data		
4-(2,4-Dichlorophenoxy) butiric acid (2,4-DB)	94-82-6	8.00E-03	3/1/88	no data		no data		no data		
1,2-Dichloropropane	78-87-5	no data		1.14E-03	12/1/91	no data		no data		
2,3-Dichloropropanol	616-23-9	3.00E-03	11/1/90	no data		no data		no data		
1,3-Dichloropropene	542-75-6	3.00E-04	10/1/90	5.72E-03	1/1/91	no data (NA)		no data (NA)		B2
1,1-Dichloro-2,2,2-trifluoroethane	306-82-2	no data		pending	10/1/92	no data (NA)		no data (NA)		
Dichlorvos	62-73-7	withdrawn	7/1/92	pending	12/1/92	2.90E-01	10/1/89	no data		B2
Dicofol	115-32-2	no data		no data		withdrawn	4/1/92	withdrawn	4/1/92	C
Dieldrin	60-57-1	5.00E-05	9/1/90	no data		1.60E+01	1/1/91	1.60E+01 [O]	1/1/91	B2
Diesel engine emissions	No CASRN	no data		pending	7/1/92	no data		no data		
Di(2-ethylhexyl)adipate	103-23-1	withdrawn	8/1/91	no data		1.20E-03	8/1/91	no data		C
Di(2-ethylhexyl)phthalate (DEHP)	117-81-7	2.00E-02	5/1/91	no data		1.40E-02	8/1/91	no data		B2
Diethyl phthalate	84-66-2	8.00E-01	8/1/91	no data		no data (NA)		no data (NA)		D
Diethyl-p-nitrophenylphosphate	311-45-5	no data		no data		no data (NA)		no data (NA)		D
Diethyl sulfate	64-67-5	no data		message	11/1/91	no data		no data (NA)	10/1/92	D

Chemical name	CASRN	Oral RfD mg/kg-day	Date	Inhalation RfD mg/kg-day	Date	Oral CPF 1/mg/kg-day	Date	Inhalation CPF* 1/mg/kg-day	Date	Carc. Class.
Diethylene glycol dinitrate (DEGDN)	693-21-0	no data		no data		no data		no data		D
Diethylene glycol monobutyl ether	112-34-5	no data		pending	5/1/92	no data		no data		
Diethylene glycol monobutyl ether acetate	124-17-4	no data		pending	5/1/92	no data		no data		
Difenzoquat	43222-48-6	8.00E-02	8/22/88	no data		no data		no data		
Diflubenzuron	35367-38-5	2.00E-02	9/30/87	pending	10/1/92	no data		no data		
1,1-Difluoroethane	75-37-6			no data		no data		no data		D
Diisopropyl methylphosphonate (DIMP)	1445-75-6	8.00E-	30	no data		no data		no data		C
Dimethipin	55290-64-7	2.00E-02	5/1/90	no data		pending		pending		
Dimethoate	60-15-5	2.00E-04	8/22/88	no data		no data		no data		
Dimethylamine	124-40-3	pending		withdrawn	1/1/91	no data		no data		
Dimethyl aminoazobenzene	60-11-7	no data		pending	10/1/92	no data		no data		
n-n-Dimethylaniline	121-69-7	2.00E-03	3/1/88	no data		no data		no data		
3,3-Dimethylbenzidine	119-93-7	no data		message	7/1/91	no data		no data		
Dimethylcarbamoyl chloride	79-44-7	no data		pending	5/1/93	no data		no data		
n,n-Dimethylformamide	68-12-2	no data		8.58E-03	10/1/90	no data		no data		
2,4-Dimethylphenol	105-67-9	2.00E-02	11/1/90	no data		no data		no data		
2,6-Dimethylphenol	576-26-1	6.00E-04	9/7/88	no data		no data		no data		
3,4-Dimethylphenol	95-65-8	1.00E-03	1/1/89	no data		no data		no data		
Dimethyl phthalate	131-11-3	pending		message	1/1/92	no data (NA)		no data (NA)		D
Dimethyl sulfate	77-78-1	no data		pending		no data (NA)		no data (NA)		B2
Dimethylterephthalate (DMT)	120-61-6	1.00E-01	3/1/88	no data		no data		no data		
m-Dinitrobenzene	99-65-0	1.00E-04	8/22/88	no data		no data		no data		D
o-Dinitrobenzene	528-29-0	no data		no data		pending	5/1/92	pending	5/1/92	D

Chemical name	CASRN	Oral RfD mg/kg·day	Date	Inhalation RfD mg/kg·day	Date	Oral CPF 1/mg/kg·day	Date	Inhalation CPF* 1/mg/kg·day	Date	Carc. Class.
4,6-Dinitro-o-cresol	534-52-1	no data		pending	3/1/93	no data		no data		
4,6-Dinitro-o-cyclohexylphenol	131-89-5	2.00E-03	1/1/89	no data		no data		no data		
2,4-Dinitrophenol	51-28-5	2.00E-03	7/1/91	message	10/1/91	no data		no data		
2,4-Dinitrotoluene	121-14-2	2.00E-03	6/1/92	message	3/1/91	no data		no data		
Dinitrotoluene mixture, 2,4-/2,6-	No CASRN	pending	8/1/91	no data		6.80E-01	9/1/90	no data (NA)		B2
Dinoseb	88-85-7	1.00E-03	8/1/89	no data		no data		no data		D
1,4-Dioxane	123-91-1	no data		no data		1.10E-02	9/1/90	no data (NA)		B2
Diphenamid	957-51-7	3.00E-02	9/30/87	no data		no data		no data		
Diphenylamine	122-39-4	2.50E-02	3/1/88	no data		no data		no data		
1,2-Diphenylhydrazine	122-66-7	no data		message	11/1/91	8.00E-01	1/1/91	8.00E-01 [O]	1/1/91	B2
Diquat	85-00-7	2.20E-03	3/1/88	no data		no data		no data		
Disulfoton	298-04-4	4.00E-05	3/1/88	no data		no data		no data		
1,4-Dithiane	505-29-3	pending	8/1/92	no data		no data		no data		
Diuron	330-54-1	2.00E-03	8/22/88	no data		no data (NA)		no data (NA)		D
Dodine	2439-10-3	4.00E-03	9/30/87	no data		no data		no data		
Endosulfan	115-29-7	5.00E-05	3/1/88	no data		no data		no data		
Endothall	145-73-3	2.00E-02	3/1/91	no data		no data		no data		
Endrin	72-20-8	3.00E-04	9/7/88	no data		pending		pending		D
Epichlorohydrin	106-89-8	withdrawn	4/1/92	2.86E-04	4/1/92	no data		no data		B2
1,2-Epoxybutane (EBU)	106-88-7	no data		1.37E+00	5/1/92	9.90E-03	1/1/91	4.20E-03 [I]	1/1/91	
Ethephon	16672-87-0	5.00E-03	8/22/88	no data		no data		no data		
Ethion	563-12-2	5.00E-04	9/1/89	no data		no data		no data		
2-Ethoxyethanol	110-80-5	no data		5.72E-02	5/1/91	no data		no data		
Ethylacetate	141-78-6	9.00E-01	3/1/88	no data		no data		no data		
Ethylbenzene	100-41-4	1.00E-01	3/1/88	2.86E-01	3/1/91	no data (NA)		no data (NA)		
Ethyl carbamate	51-79-6	no data		message	4/1/92	no data		no data		D

Chemical name	CASRN	Oral RfD mg/kg·day	Date	Inhalation RfD mg/kg·day	Date	Oral CPF 1/mg/kg·day	Date	Inhalation CPF* 1/mg/kg·day	Date	Carc. Class.
Ethyl chloride	75-00-3	no data		2.86E+00	4/1/91	no data		no data		
s-Ethyldipropylthiocarbamate	759-94-4	2.50E-02	3/1/88	no data		no data		no data		
Ethylene diamine	107-15-3	no data		message	5/1/91	pending		pending		
Ethylene thiourea (ETU)	96-45-7	8.00E-05	5/1/91	no data		pending		pending		
Ethyl ether	60-29-7	2.00E-01	7/1/91	no data		no data		no data		
Ethylene Glycol	107-21-1	2.00E+00	9/30/87	no data		no data		no data		
Ethyleneimine	151-56-4	no data		message	4/1/92	no data		no data		
Ethylphthalyl Ethylglycolate (EPEG)	84-72-0	3.00E+00	3/1/88	no data		no data		no data		
Ethyl p-nitrophenyl-phosphorothioate (EPN)	2104-64-5	1.00E-05	3/1/91	no data		no data		no data		
Express	101200-48-0	8.00E-03	4/1/90	no data		pending		pending		
Fenamiphos	22224-92-6	2.50E-04	9/1/90	no data		no data		no data		
Fluometuron	2164-17-2	1.30E-02	9/1/90	no data		no data		no data		
Fluoranthene	206-44-0	4.00E-02	7/1/91	no data		no data		no data		
Fluorene	86-73-7	4.00E-02	11/1/90	no data		no data		no data		
Fluorine (soluble fluoride)	7782-41-4	6.00E-02	6/1/89	no data		no data		no data		
Fluridone	59756-60-4	8.00E-02	8/22/88	no data		pending		pending		
Flurprimidol	56425-91-3	2.00E-02	4/1/90	no data		no data		no data		
Flutolanil	66332-96-5	6.00E-02	5/1/90	no data		no data		no data		
Fluvalinate	69409-94-5	1.00E-02	6/30/88	no data		no data		no data		
Folpet	133-07-3	1.00E-01	3/1/91	no data		3.50E-03	8/22/88	no data (NA)		B2
Fomesafen	72178-02-0	no data		no data		1.90E-01	1/1/89	no data (NA)		C
Fonofos	944-22-9	2.00E-03	3/1/91	no data		no data		no data		
Formaldehyde	50-00-0	2.00E-01	9/1/90	no data		no data		4.50E-02 [I]	1/1/91	B1

1065

Chemical name	CASRN	Oral RfD mg/kg-day	Date	Inhalation RfD mg/kg-day	Date	Oral CPF 1/mg/kg-day	Date	Inhalation CPF* 1/mg/kg-day	Date	Carc. Class.
Formic Acid	64-18-6	withdrawn	12/1/90	no data		pending	10/1/91	pending	10/1/91	
Foseytl-al	39148-24-8	3.00E+00	3/1/91	no data		no data (NA)		no data (NA)		C
Furan	110-00-9	1.00E-03	12/1/89	no data		no data		no data		
Furfural	98-01-1	3.00E-03	1/1/90	pending	8/1/91	no data		no data		
Furmecyclox	60568-05-0	no data		no data		3.00E-02	12/1/89	no data (NA)		B2
Glufosinate-ammonium	77182-82-2	4.00E-04	3/1/88	no data		no data		no data		
Glycidyaldehide	765-34-4	4.00E-04 [I]	4/1/91	no data		no data		no data		B2
Glyphosate	1071-83-6	1.00E-01	9/1/90	no data		no data (NA)		no data (NA)		D
Haloxyfop-methyl	69806-40-2	5.00E-05	1/1/91	no data		pending		pending		
Harmony	79277-27-3	1.30E-02	9/26/88	no data		no data		no data		
Heptachlor	76-44-8	5.00E-04	9/30/87	no data		4.50E+00	1/1/91	4.50E+00 [O]	1/1/91	B2
Heptachlor epoxide	1024-57-3	1.30E-05	3/1/88	no data		9.10E+00	1/1/91	9.10E+00 [O]	1/1/91	B2
n-Heptane	142-82-5	no data		no data		no data	1/1/93	no data	1/1/93	D
Hexabromobenzene	87-82-1	2.00E-03	3/1/88	no data		no data		no data		
Hexabromodiphenyl ether	36483-60-0	no data		no data		no data		no data		
Hexachlorobenzene	118-74-1	8.00E-04	4/1/91	message	3/1/91	1.60E+00	3/1/91	1.60E+00 [O]	3/1/91	D
Hexachlorobutadiene	87-68-3	withdrawn	5/1/93	no data		7.80E-02	4/1/91	7.80E-02 [O]	4/1/91	B2
alpha-Hexachlorocyclohexane (alpha-HCH)	319-84-6	no data		no data		6.30E+00	1/1/91	6.30E+00 [O]	1/1/91	C
beta-Hexachlorocyclohexane (beta-HCH)	319-85-7	no data		no data		no data		no data		B2
delta-Hexachlorocyclohexane (delta-HCH)	319-86-8	no data		no data		1.80E+00	1/1/91	1.80E+00 [O]	1/1/91	C
epsilon-Hexachlorocyclohexane	6108-10-7	no data		no data		no data		no data		D
gamma-Hexachlorocyclohexane (gamma-HCH); (Lindane)	58-89-9	3.00E-04	3/1/88	pending	7/1/92	pending		pending		D

Chemical name	CASRN	Oral RfD mg/kg·day	Date	Inhalation RfD mg/kg·day	Date	Oral CPF 1/mg·kg·day	Date	Inhalation CPF* 1/mg·kg·day	Date	Carc. Class.
technical-Hexachlorocyclo-hexane (t-HCH)	* 608-73-1	no data		no data		1.80E+00	1/1/91	1.80E+00 [O]	1/1/91	B2
Hexachlorocyclo-pentadiene (HCCPD)	77-47-4	7.00E-03	9/1/90	no data		no data		no data		D
Hexachlorodibenzo-p-dioxin,mixture (HxCDD)	19408-74-3	no data		no data		6.20E+03	1/1/91	6.20E+03 [O]	1/1/91	B2
Hexachloroethane	67-72-1	1.00E-03	4/1/91	pending	12/1/92	1.40E-02	1/1/91	1.40E-02 [O]	1/1/91	C
Hexachlorophene	70-30-4	3.00E-04	8/22/88	no data		no data		no data		
Hexahydro-1,3,5-trinitro--1,3,5-triazine (RDX)	121-82-4	3.00E-03	12/1/90	no data		1.10E-01	8/1/90	no data (NA)		C
1,6-Hexamethylene diisocynate	822-06-0	no data		pending	5/1/92	no data		no data		
Hexamethylphosphoramide	680-31-9	no data		pending	3/1/93	no data		no data		
n-Hexane	110-54-3	no data		5.72E-02	2/1/91	pending	9/1/91	pending	9/1/91	
Hexazinone	51235-04-2	3.30E-02	3/1/88	no data		no data		no data		
Hydrazine/Hydrazine Sulfate	302-01-2	no data		no data		3.00E+00	1/1/91	1.71E+01 [I]	1/1/91	B2
Hydrogen chloride	7647-01-0	no data		2.00E-03	1/1/91	no data		no data		
Hydrogen cyanide	74-90-8	2.00E-02	3/1/88	no data		no data		no data		
Hydrogen sulfide	7783-06-4	3.00E-03	3/1/88	2.57E-04	10/1/90	no data		no data		
Hydroquinone	123-31-9	no data		message		no data		no data		
Imazalil	35554-44-0	1.30E-02	3/1/88	no data		no data		no data		
Imazaquin	81335-37-7	2.50E-01	3/1/88	no data		no data		no data		
Indeno(1,2,3-cd)pyrene	193-39-5	no data		no data		no data (NA)		no data (NA)		B2
Ionizing radiation	—	no data		no data		pending	5/1/93	pending	5/1/93	
Iprodione	36734-19-7	4.00E-02	2/1/89	no data		no data		no data		
Isobutyl alcohol	78-83-1	3.00E-01	7/7/89	no data		no data		no data		

Chemical name	CASRN	Oral RfD mg/kg·day	Oral RfD Date	Inhalation RfD mg/kg·day	Inhalation RfD Date	Oral CPF 1/mg/kg·day	Oral CPF Date	Inhalation CPF* 1/mg/kg·day	Inhalation CPF* Date	Carc. Class.
Isophorone	78-59-1	2.00E-01	1/1/91	message		4.10E-03	3/1/91	no data		C
Isopropalin	33820-53-0	1.50E-02	6/1/89	no data		no data		no data		
Isopropyl methyl phosphonic acid (IMPA)	1832-54-8	1.00E-01	6/1/92	no data		no data		no data		
Isoxaben	82558-50-7	5.00E-02	4/1/91	no data		no data (NA)		no data (NA)		D
Lactofen	77501-63-4	2.00E-03	6/30/88	no data		no data		no data		C
Lead & compounds (inorganic)	7439-92-1	message	2/1/90	no data		not appropriate		no data (NA)		B2
Linuron	330-55-2	2.00E-03	8/1/90	no data		no data		no data		C
Londax	83055-99-6	2.00E-01	9/7/88	no data		no data		no data		
Malathion	121-75-5	2.00E-02	3/1/88	no data		pending		pending		
Maleic anhydride	108-31-6	1.00E-01	9/26/88	pending		no data		no data		
Maleic hydrazide	123-33-1	5.00E-01	3/1/88	no data		no data		no data		
Maneb	12427-38-2	5.00E-03	9/7/88	no data		no data (NA)		no data		
Manganese	7439-96-5	1.00E-02	8/1/90	1.14E-04	12/6/90	no data		no data (NA)		
Mepiquat chloride	24307-26-4	3.00E-02	8/22/88	no data		no data		no data		D
Mercury, inorganic	7439-97-6	pending		pending		pending		no data		
Merphos	150-50-5	3.00E-05	1/1/89	no data		no data		no data		D
Merphos oxide	78-48-8	3.00E-05	9/7/88	no data		no data		pending		
Metalaxyl	57837-19-1	6.00E-02	3/1/88	no data		pending		no data		
Methacrylontrile	126-98-7	1.00E-04	9/7/88	no data		no data		no data		
Methamidophos	10265-92-6	5.00E-05	3/1/88	no data		no data		no data		
Methanol	67-56-1	5.00E-01	9/7/88	no data		no data		pending		
Methidathion	950-37-8	1.00E-03	8/22/88	no data		pending		no data		
Methomyl	16752-77-5	2.50E-02	3/1/91	no data		no data		no data		C
Methoxychlor	72-43-5	5.00E-03	8/1/91	message	4/1/92	no data (NA)		no data (NA)		D

Chemical name	CASRN	Oral RfD mg/kg·day	Date	Inhalation RfD mg/kg·day	Date	Oral CPF 1/mg/kg·day	Date	Inhalation CPF* 1/mg/kg·day	Date	Carc. Class.
2-Methoxyethanol	109-86-4	pending	4/1/92	5.72E-03	5/1/91	no data		no data		D
Methyl acrylate	96-33-3	no data		no data		no data		no data		
Methyl chlorocarbonate	79-22-1	withdrawn	5/1/89	no data		pending		pending		
4-(2-Methyl-4-chlorophenoxy)-butric acid	94-81-5	1.00E-02	9/26/88	no data		no data		no data		
2-(2-Methyl-4-chlorophenoxy)-propionic acid	93-65-2	1.00E-03	1/1/89	no data		no data		no data		
2-Methyl-4-chlorophenoxy-acetic acid	94-76-6	5.00E-04	9/26/88	no data		no data		no data		
4,4'-Methylene bis (2-chloro-aniline)	101-14-4	no data		pending	3/1/93	no data		no data		
4,4'-Methylene bis (N,N'-di-methyl) aniline	101-61-1	no data		no data		4.60E-02	8/1/90	no data		B2
Methylene diphenyl isocya-nate (MDI)	101-68-8	no data		pending	3/1/93	no data		no data		
Methyl ethyl ketone (MEK)	78-93-3	withdrawn	8/1/91	2.86E-01	7/1/92	no data		no data		D
Methyl iodide	77-88-4	no data		pending	9/1/92	no data		no data		
Methyl isobutyl ketone (MIBK)	108-10-1	withdrawn	3/1/91	pending		no data		no data		
Methyl isocyanate	624-83-9	no data		message		no data		no data		
Methyl mercury	22967-92-6	3.00E-04	2/1/89	no data		no data		no data		
Methyl parathion	298-00-0	2.50E-04	3/1/91	no data		no data		no data		
2-Methylphenol (o-Cresol)	95-48-7	5.00E-02	9/1/90	message	4/1/92	no data		no data		C
3-Methylphenol (m-Cresol)	108-39-4	5.00E-02	9/1/90	message	4/1/92	no data		no data		C
4-Methylphenol (p-Cresol)	106-44-5	withdrawn	8/1/91	message	4/1/92	no data		no data		C
Methyl tert-butyl ether	1634-04-4	pending	3/1/93	1.43E-01	12/1/91	no data		no data		

Chemical name	CASRN	Oral RfD mg/kg-day	Date	Inhalation RfD mg/kg-day	Date	Oral CPF 1/mg/kg-day	Date	Inhalation CPF* 1/mg/kg-day	Date	Carc. Class.
Metolachlor	51218-45-2	1.50E-01	2/1/91	no data		no data (NA)		no data (NA)		C
Metribuzin	21087-64-9	2.5CE-02	3/1/91	no data		pending	3/1/93	pending	3/1/93	
Mirex	2385-85-5	2.00E-06	3/1/88	no data		pending	6/24/92	pending	6/24/92	
Molinate	2212-67-1	2.00E-03	9/26/88	no data		no data		no data		
Molybdenum	7439-98-7	5.00E-03	11/9/92	no data		no data		no data		
Monochloroamine	10599-90-3	1.00E-01	11/1/92	no data		pending	1/1/93	pending	1/1/93	
Naled	300-76-5	2.00E-03	6/1/89	no data		no data		no data		
Naphthalene	91-20-3	pending		no data		no data		no data		D
Napropamide	15299-99-7	1.00E-01	5/1/90	no data		no data		no data		
Nickel carbonyl	13463-39-3	no data		no data		no data		no data		B2
Nickel refinery dust	none	no data		no data		no data (NA)		8.40E-01 [I]	1/1/91	A
Nickel, soluble salts	7440-02-0	2.00E-02	12/1/91	pending		message		message		A
Nickel subsulfide	12035-72-2	no data		no data		no data		1.68E+00 [I]	1/1/91	A
Nitrapyrin	1929-82-4	withdrawn	7/1/92	no data		no data		no data		
Nitrate	14797-55-8	1.60E+00	5/1/91	no data		no data		no data		
Nitric oxide	10102-43-9	1.00E-01	8/1/91	no data		no data		no data		
Nitrite	14797-65-0	1.00E-01	8/1/92	no data		pending	8/1/92	pending	8/1/92	
2-Nitroaniline	88-74-4	pending	8/1/92	no data		no data		no data		
Nitrobenzene	98-95-3	5.00E-04 [I]	8/1/89	pending		no data		no data		D
4-Nitrobiphenyl	92-93-3	no data		no data		pending	9/1/92	pending	9/1/92	
Nitrogen dioxide	10102-44-0	1.00E+00	8/1/91	see NAAQS	2/1/93	pending	10/1/91	pending	10/1/91	
Nitroguanidine	556-88-7	1.00E-01	9/1/89	no data		no data		no data		D
p-Nitrophenol	100-02-7	pending		message	10/1/91	no data		no data		
2, Nitropropane	79-46-9	no data		5.72E-03	3/1/91	no data		no data		D
n-Nitroso-di-n-butylamine	924-16-3	no data		no data		5.40E+00	1/1/91	5.40E+00 [O]	1/1/91	B2

Chemical name	CASRN	Oral RfD mg/kg·day	Date	Inhalation RfD mg/kg·day	Date	Oral CPF 1/mg·kg·day	Date	Inhalation CPF* 1/mg·kg·day	Date	Carc. Class.
n-Nitroso-n-methylethylamine	10595-95-6	no data		no data		2.20E+01	12/1/89	no data		B2
n-Nitrosodi-n-propylamine	621-64-7	no data		no data		7.00E+00	3/1/88	no data		B2
n-Nitrosodiethanolamine	1116-54-7	no data		no data		2.80E+00	3/1/88	no data		B2
n-Nitrosodiethylamine	55-18-5	no data		no data		1.50E+02	1/1/91	1.50E+02 [O]	1/1/91	B2
n-Nitrosodimethylamine	62-75-9	no data		no data		5.10E+01	1/1/91	5.10E+01 [O]	1/1/91	B2
n-Nitrosodiphenylamine	86-30-6	no data		no data		4.90E-03	3/1/88	no data		B2
n-Nitrosopyrrolidine	930-55-2	no data		no data		2.10E+00	1/1/91	2.10E+00 [O]	1/1/91	B2
Nonabromodiphenyl ether	63936-56-1	no data		no data		no data		no data		D
Norflurazon	27314-13-2	4.00E-02	3/1/88	no data		pending		pending		
NuStar	85509-19-9	7.00E-04	9/26/88	no data		no data		no data		D
Octabromodiphenyl ether	32536-52-0	3.00E-03	8/1/90	no data		no data		no data		D
Octahydro-1,3,5,7-tetranitro -1,3,5,7-tetrazocine (HMX)	2691-41-0	5.00E-02	12/1/90	no data		no data (NA)		no data (NA)		
Oryzalin	19044-88-3	5.00E-02	7/1/89	no data		no data (NA)		no data (NA)		C
Oxadiazon	19666-30-9	5.00E-03	12/1/88	no data		pending		pending		
Oxamyl	23135-22-0	2.50E-02	3/1/88	no data		no data		no data		
Oxyfluorfen	42874-03-3	3.00E-03	12/1/88	no data		pending		pending		
Palcobutrazol	76738-62-0	1.30E-02	3/1/88	no data		no data		no data		
Parachlorophenyl methyl sulfide	123-09-1	pending	8/1/92	no data		pending	1/1/92	pending	1/1/92	
Parachlorophenyl methyl sulfone	98-57-7	pending	8/1/92	no data		pending	1/1/92	pending	1/1/92	
Parachlorophenyl methyl sulfoxide	934-73-6	pending	8/1/92	no data		pending	1/1/92	pending	1/1/92	
Paraquat	1910-42-5	4.50E-03	2/1/91	no data		no data		no data (NA)		C
Parathion	56-38-2	pending		no data		no data (NA)		no data (NA)		C
Pendimethalin	40487-42-1	4.00E-02	6/30/88	no data		no data		no data		

Chemical name	CASRN	Oral RfD mg/kg-day	Date	Inhalation RfD mg/kg-day	Date	Oral CPF 1/mg/kg-day	Date	Inhalation CPF* 1/mg/kg-day	Date	Carc. Class.
Pentabromodiphenyl ether	32534-81-9	2.00E-03	8/1/90	no data		no data		no data		D
Pentachlorobenzene	608-93-5	8.00E-04	3/1/88	no data		no data (NA)	11/9/92	no data (NA)	11/9/92	D
Pentachlorocyclopentadiene	25329-35-5	no data		no data		no data		no data		D
Pentachloroethane	76-01-7	pending	8/1/92	no data		no data		no data		
Pentachloronitrobenzene	82-68-8	3.00E-03	9/30/87	no data		pending		pending		B2
Pentachlorophenol	87-86-5	3.00E-02	4/1/90	pending		1.20E-01	3/1/91	no data (NA)		D
Phenanthrene	85-01-8	no data		no data		no data		no data		
Phenmedipham	13684-63-4	2.50E-01	6/1/90	no data		no data		no data		D
Phenol	108-95-2	6.00E-01	2/1/90	message		no data		no data		
m-Phenylenediamine	108-45-2	6.00E-03	3/31/87	no data		no data		no data		
Phenylmercuric acetate	62-38-4	8.00E-05	3/1/88	no data		no data		no data		
Phosalone	2310-17-0	withdrawn	12/1/88	no data		no data		no data		
Phosgene	75-44-5	no data		message	10/1/90	no data		no data		
Phosmet	732-11-6	2.00E-02	3/1/88	no data		no data		no data		
Phosphine	7803-51-2	3.00E-04	3/1/88	pending	3/1/93	no data		no data		
Phthalic anhydride	85-44-9	2.00E+00	9/7/88	no data		pending	5/1/92	pending	5/1/92	
Picloram	1918-02-1	7.00E-02	9/30/87	no data		pending		pending		
Pirimiphos-methyl	29232-93-7	1.00E-02	3/1/88	no data		no data		no data		
Polychlorinated biphenyls (PCBs)	1336-36-3	no data		no data		7.70E+00	5/1/89	no data		B2
Potassium bromate	7758-01-2	no data		no data		pending	11/9/92	no data	11/9/92	
Potassium cyanide	151-50-88	5.00E-02	3/1/88	no data		no data		no data		
Potassium silver cyanide	506-61-6	2.00E-01	1/1/90	no data		no data		no data		
Prochloraz	67747-09-5	9.00E-03	1/1/89	no data		1.50E-01	10/1/89	no data		C
Prometon	1610-18-0	1.50E-02	3/1/88	no data		no data		no data		
Prometryn	7287-19-6	4.00E-03	9/30/87	no data		no data		no data		
Pronamide	23950-58-5	7.50E-02	3/1/91	no data		pending		pending		

Chemical name	CASRN	Oral RfD mg/kg·day	Date	Inhalation RfD mg/kg·day	Date	Oral CPF 1/mg/kg·day	Date	Inhalation CPF* 1/mg/kg·day	Date	Carc. Class.
Propachlor	1918-16-7	1.30E-02	3/1/88	no data		no data		no data		
Propanil	709-98-8	5.00E-03	3/1/88	no data		no data		no data		
1,3-Propane sultone	1120-71-4	no data		pending	9/1/92	no data		no data		
Propargyl alcohol	107-19-7	2.00E-03	11/1/90	no data		no data		no data		
Propargite	2312-35-8	2.00E-02	5/1/90	no data		pending		pending		
Propazine	139-40-2	2.00E-02	10/1/90	no data		pending		no data		
Propham	122-42-9	2.00E-02	9/30/87	no data		no data		pending		
Propiconazole	60207-90-1	1.30E-02	8/22/88	message	7/1/91	no data		no data		
beta-Propiolactone	57-57-8	no data		pending	3/1/93	no data		no data		
Proprionaldehyde	123-38-6	no data		5.72E-01	6/1/91	no data		no data		
Propylene glycol	57-55-6	no data		5.72E-01	7/1/91	no data		no data		
Propylene glycol monoethyl ether	52125-53-8	no data		no data		no data		no data		
Propylene glycol monomethyl ether (PGME)	107-98-2	no dat		5.72E-01	7/1/91	no data		no data		
Propylene oxide	75-56-9	no data		8.58E-03	11/1/90	2.40E-01	1/1/91	2.60E-02 [1-50%]	1/1/91	B2
Propyleneimine	75-55-8	no data		message	1/1/92	no data		no data		
Pursuit	81335-77-5	2.50E-01	1/1/90	no data		no data		no data		
Pydrin	51630-58-1	2.50E-02	3/1/88	no data		no data		no data		
Pyrene	129-00-0	3.00E-02	7/1/91	no data		no data		no data		D
Pyridine	110-86-1	1.00E-03	6/1/89	no data		no data		no data		
Quinalphos	13593-03-8	5.00E-04	3/1/88	no data		no data		no data		
Quinclorac	84087-01-4	pending	5/1/92	message	10/1/90	no data		no data		
Quinone	106-51-4	no data		no data		withdrawn	1/1/89	withdrawn	1/1/89	
Radium 226, 228	7440-14-4	no data		no data		pending	3/1/93	pending	3/1/93	
Radon (inert gas only)	—	no data		no data		withdrawn	7/1/89	withdrawn	7/1/89	
Radon 222	14859-67-7	no data		no data						

Chemical name	CASRN	Oral RfD mg/kg·day	Date	Inhalation RfD mg/kg·day	Date	Oral CPF 1/mg/kg·day	Date	Inhalation CPF* 1/mg/kg·day	Date	Carc. Class.
Refractory ceramic fiber	NO CASRN	no data		no data		no data		no data		B2
Resemethrin	10453-86-8	3.00E-02	9/26/88	no data		no data		no data		
Rotenone	83-79-4	4.00E-03	9/7/88	no data		pending		pending		
Savey	78587-05-0	2.50E-02	9/26/88	no data		pending		pending		
Selenious acid	7783-00-8	5.00E-03	7/1/91	no data		no data		no data		D
Selenium & compounds	7782-49-2	5.00E-03	9/1/91	no data		no data		no data		D
Selenium sulfide	7446-34-6	no data		no data		no data (NA)		no data (NA)		B2
Selenourea	630-10-4	withdrawn	5/1/91	no data		no data		no data		
Sethoxydim	74051-80-2	9.00E-02	11/1/89	no data		no data		no data		
Silver	7440-22-4	5.00E-03	12/1/91	no data		no data (NA)		no data (NA)		D
Silver cyanide	506-64-9	1.00E-01	1/1/90	no data		no data		no data		
Simazine	122-34-9	withdrawn	9/1/91	no data		pending		pending		
Sodium acifluorfen	62476-59-9	1.30E-02	12/1/88	no data		no data		no data		
Sodium azide	26628-22-8	4.00E-03	3/1/88	no data		no data		no data		
Sodium cyanide	143-33-9	4.00E-02	3/31/87	no data		no data		no data		
Sodium diethyldithio-carbamate (Dithiocarb)	148-18-5	3.00E-02	3/1/88	no data		no data		no data		
Sodium fluoroacetate	62-74-8	2.00E-05	5/1/91	no data		no data		no data		
Strontium	7440-24-6	6.00E-01	10/1/92	no data		no data		no data		
Strychnine	57-24-9	3.00E-04	3/1/88	no data		no data		no data		
Styrene	100-42-5	2.00E-01	9/1/90	2.86E-01	11/9/92	pending		pending		
Systhane	88671-89-0	2.50E-02	9/26/88	no data		no data		no data		
Tebuthiuron	34014-18-1	7.00E-02	8/22/88	no data		no data		no data		
Terbacil	5902-51-2	1.30E-02	9/1/89	no data		no data		no data		
Terbutryn	886-50-0	1.00E-03	9/26/88	no data		no data		no data		
Tetrabromodiphenyl ether	40088-47-9	no data		no data		no data		no data		D

Chemical name	CASRN	Oral RfD mg/kg·day	Date	Inhalation RfD mg/kg·day	Date	Oral CPF 1/mg/kg·day	Date	Inhalation CPF* 1/mg/kg·day	Date	Carc. Class.
1,2,4,5- Tetrachlorobenzene	95-94-3	3.00E-04	3/1/91	no data		no data		no data		D
Tetrachlorocyclopentadiene	695-77-2	no data		no data		no data		no data		C
1,1,1,2- Tetrachloroethane	630-20-6	3.00E-02	8/1/90	no data		2.60E-02	1/1/91	2.60E-02 [O]	1/1/91	C
1,1,2,2-Tetrachloroethane	79-34-5	pending		no data		2.00E-01	1/1/91	2.00E-01 [O]	1/1/91	
Tetrachloroethylene	127-18-4	1.00E-02	3/1/88	no data		pending		pending		
2,3,4,6-Tetrachlorophenol	58-90-2	3.00E-02	3/1/88	no data		pending		pending		
Tetrachlorvinphos	961-11-5	3.00E-02	3/1/88	no data		no data		no data		
Tetraethyl lead	78-00-2	1.00E-07	3/1/88	no data		no data		no data		
Tetraethyldithiopyrophosphate	3689-24-5	5.00E-04	9/7/88	pending	10/1/92	no data		no data		D
1,1,1,2-Tetrafluoroethane	811-97-2	no data		no data		no data		no data		D
Thallic oxide	1314-32-5	withdrawn	5/1/89	no data		no data		no data		D
Thallium acetate	563-68-8	9.00E-05	9/1/90	no data		no data		no data		D
Thallium carbonate	6533-73-9	8.00E-05	9/1/90	no data		no data		no data		D
Thallium chloride	7791-12-0	8.00E-05	9/1/90	no data		no data		no data		D
Thallium nitrate	10102-45-1	9.00E-05	9/1/90	no data		no data		no data		D
Thallium selenite	12039-52-0	9.00E-05	9/1/90	no data		no data		no data		
Thallium (I) sulfate	7446-18-6	8.00E-05	9/1/90	no data		no data		no data		
Thiobencarb	28249-77-6	1.00E-02	9/30/87	no data		pending	5/1/92	pending	5/1/92	
Thiofanox	39196-18-4	pending		no data		no data		no data		
Thiophanate-methyl	23564-05-8	8.00E-02	3/1/88	no data		pending	9/1/91	pending	9/1/91	
Thiram	137-26-8	5.00E-03	2/1/89	no data		no data		pending		D
Toluene	108-88-3	2.00E-01	8/1/90	1.40E+00	9/1/92	no data		1.10E+00 [O]	1/1/91	B2
Toxaphene	8001-35-2	no data		no data		1.10E+00	1/1/91	no data		
Tralomethrin	66841-25-6	7.50E-03	8/1/90	no data		no data		no data		
Triallate	2303-17-5	1.30E-02	8/1/90	no data		no data		no data		
Triasulfuron	82097-50-5	1.00E-02	1/1/91	no data		no data		no data		

Chemical name	CASRN	Oral RfD		Inhalation RfD		Oral CPF		Inhalation CPF*		Carc. Class.
		mg/kg·day	Date	mg/kg·day	Date	1/mg/kg·day	Date	1/mg/kg·day	Date	
1,2,4-Tribromobenzene	615-54-3	5.00E-03	3/1/88	no data		no data		no data		D
Tribromochloromethane	594-15-0	no data		no data		no data		no data		D
Tribromodiphenyl ether	49690-94-0	no data		no data		no data		no data		
Tributylin oxide	56-35-9	3.00E-05	8/22/88	no data		no data		no data		
Trichloroacetic acid	76-03-9	pending	10/1/92	no data		pending	3/1/93	pending	3/1/93	
Tricresol	1319-77-3	pending	3/31/87	message	4/1/92	pending		pending		
1,1,2-Trichloro-1,2,2-trifluoroethane (CFC-113)	76-13-1	3.00E+01	1/1/90	pending		no data		no data		D
1,2,4-Trichlorobenzene	120-82-1	1.00E-02	5/1/92	pending	3/1/93	no data		no data		D
Trichlorocyclopentadiene	77323-84-3	no data		no data		no data (NA)		no data (NA)		D
1,1,1-Trichloroethane	71-55-6	withdrawn	8/1/91	pending		no data		no data		
1,1,2-Trichloroethane	79-00-5	4.00E-03	8/1/90	pending		5.70E-02	1/1/91	5.70E-02 [O]	1/1/91	C
Trichloroethylene	79-01-6	pending		no data		withdrawn	7/1/89	withdrawn	7/1/89	
Trichlorofluoromethane	75-69-4	3.00E-01	3/1/88	no data		no data		no data		
2,4,5-Trichlorophenol	95-95-4	1.00E-01	3/1/88	message	8/1/91	pending		pending		
2,4,6-Trichlorophenol	88-06-2	no data		message	8/1/91	1.10E-02	1/1/91	1.10E-02 [O]	1/1/91	B2
2(2,4,5-Trichlorophenoxy)-propionic acid	93-72-1	8.00E-03	9/7/08	no data		no data (none)		no data (none)		D
2,4,5-Trichlorophenoxy-acetic acid	93-76-5	1.00E-02	8/1/89	no data		no data		no data		
1,1,2-Trichloropropane	598-77-6	5.00E-03	9/26/88	no data		no data		no data		
1,2,3-Trichloropropane	96-18-4	6.00E-03	8/1/90	no data		pending		pending		
Tridiphane	58138-08-2	3.00E-03	3/1/88	no data		pending		pending		
Triethylamine	121-44-8	no data		2.00E-03	4/1/91	no data		no data		
Triethylene glycol monobutyl ether	143-22-6	no data		pending	5/1/92	no data		no data		

Chemical name	CASRN	Oral RfD mg/kg·day	Date	Inhalation RfD mg/kg·day	Date	Oral CPF 1/mg/kg·day	Date	Inhalation CPF* 1/mg/kg·day	Date	Carc. Class.
Triethylene glycol monoethyl ether	112-50-5	no data		pending	5/1/92	no data		no data		
Triethylene glycol monomethyl ether	112-35-6	no data		pending	5/1/92	no data		no data		
Trifluralin	1582-09-8	7.50E-03	7/1/89	no data		7.70E-03	8/22/88	no data		C
2,2,4-Trimethylpentane	540-84-1	no data		message	11/1/91	no data		no data		
1,3,5-Trinitrobenzene	99-35-4	5.00E-05	9/7/88	no data		pending		pending		
2,4,6-Trinitrotoluene (TNT)	118-96-7	5.00E-04	8/1/90	no data		3.00E-02	6/1/89	no data (NA)		C
Uranium, natural	7440-61-1	no data		no data		withdrawn	1/1/89	withdrawn	1/1/89	
Uranium (soluble salts)	–	3.00E-03	10/1/89	no data		no data		no data		
Vanadium pentoxide	1314-62-1	9.00E-03	6/30/88	no data		message		message		
Vernam	1929-77-7	1.00E-03	3/1/88	no data		no data		no data		
Vinclozolin	50471-44-8	2.50E-02	3/1/88	no data		no data		no data		
Vinyl acetate	108-05-4	no data		5.72E-02	4/1/91	pending		pending		
Vinyl bromide	593-60-2	no data		8.58E-04	5/1/93	no data		no data		
Vinyl chloride	75-01-4	no data		no data		pending	3/1/93	pending	3/1/93	
Warfin	81-81-2	3.00E-04	3/1/88	no data		no data		no data		
White phosphorus	7723-14-0	2.00E-05	8/1/90	no data		no data		no data		D
Xylenes	1330-20-7	2.00E+00	9/30/87	pending		no data		no data		
Zinc and compounds	7440-66-6	pending		no data		no data		no data		D
Zinc cyanide	557-21-1	5.00E-02	3/1/88	no data		no data		no data		
Zinc phosphide	1314-84-7	3.00E-04	1/1/90	no data		no data		no data		D
Zineb	12122-67-7	5.00E-02	3/1/88	no data		pending		pending		

APPENDIX
C

THERMODYNAMIC PROPERTIES

The Laws of Thermodynamics
1st law: *You can't win, you can only break even.*
2nd law: *You can only break even at absolute zero.*
3rd law: *You can't reach absolute zero.*

TABLES IN APPENDIX C

C-1   Values of Gas Constant, R
C-2   Superheated Steam SI Units
C-3   Temperature-dependent Heat Capacities (constant-pressure values)
C-4   Latent Heat of Fusion and Vaporization at 1 atm
C-5   Heat of Combustion for Common Waste Elements
C-6   Heats of Formation and Free Energies of Formation

This appendix presents some of the thermodynamic properties needed to work the problems in Chaps. 3 and 12. The source for the data is given at the end of each table. For further thermodynamic data the following sources are also suggested:

Dean, J. A., ed.: *Lange's Handbook of Chemistry*, 13th ed., McGraw-Hill, Inc., New York, NY, 1985.

Felder, R. M., and R. W. Rousseau: *Elementary Principles of Chemical Processes*, 2nd ed., John Wiley and Sons, New York, NY, 1986.

Perry, R. H., and D. W. Green, eds.: *Perry's Chemical Engineer's Handbook*, 6th ed., McGraw-Hill, Inc., New York, NY, 1984.

Reid, R. C., J. M. Prausnitz, and B. E. Poling: *The Properties of Gases and Liquids*, 4th ed., McGraw-Hill, Inc., New York, NY, 1987.

Sandler, S. I.: *Chemical and Engineering Thermodynamics*, 2nd ed., John Wiley and Sons, New York, NY, 1989.

Smith, J. M., and H. C. Van Ness: *Introduction to Chemical Engineering Thermodynamics*, 4th ed., McGraw-Hill, Inc., New York, NY, 1987.

Weast, R. C., ed.: *CRC Handbook of Chemistry and Physics*, CRC Press, Boca Raton, FL.

TABLE C-1

Values of gas constant, R	
0.730	$\dfrac{atm \cdot ft^3}{lb\text{-}mol^{\circ}R}$
10.73	$\dfrac{psia \cdot ft^3}{lb\text{-}mol^{\circ}R}$
297	$\dfrac{in\ H_2O \cdot ft^3}{lb\text{-}mol^{\circ}R}$
82.05	$\dfrac{atm \cdot cm^3}{gm\text{-}mol\ K}$
0.08205	$\dfrac{atm \cdot liter}{gm\text{-}mol\ K}$
8.314	$\dfrac{Pa \cdot m^3}{gm\text{-}mol\ K}$
62.4	$\dfrac{mm\ Hg \cdot liter}{gm\text{-}mol\ K}$

Source: Weast, Robert C., ed.: *Handbook of Chemistry and Physics*, 64th ed., CRC Press, Boca Raton, FL, p. F-197, 1983-84.

$$
\begin{aligned}
Pa &= Pascal = N/m^2 \\
J &= Joule = N \cdot m = Pa \cdot m^3 \\
1\ lb &= 453.592\ grams \\
1\ atm &= 760\ mm\ Hg\ @\ 0^{\circ}C \\
&= 101{,}325\ Pa \\
&= 14.696\ psia\ (lb/in^2\ absolute) \\
1\ psia &= 0.068046\ atm \\
&= 6895\ Pa
\end{aligned}
$$

TABLE C-2
Superheated steam SI units

V′ = specific volume (cm³/g), U′ = specific internal energy (kJ/kg), H′ = specific enthalpy (kJ/kg), S′ = specific entropy (kJ/(kg·K))

Temperature, °C (temperature, K)

Abs. press. kPa (sat. temp. °C)		Sat. water	Sat. steam	100 (373.15)	125 (398.15)	150 (423.15)	175 (448.15)	200 (473.15)	225 (498.15)	250 (523.15)	300 (573.15)	350 (623.15)	400 (673.15)	450 (723.15)	500 (773.15)	550 (823.15)	600 (873.15)	650 (923.15)
100 (99.63)	V′	1.043	1693.7	1695.5	1816.7	1936.3	2054.7	2172.3	2289.4	2406.1	2638.7	2870.4	3102.5	3334.0	3565.3	3796.5	4027.7	4258.8
	U′	417.406	2506.1	2506.6	2544.8	2582.7	2620.4	2658.1	2695.2	2733.9	2810.6	2888.6	2968.0	3049.0	3131.6	3216.0	3302.0	3389.8
	H′	417.511	2675.4	2676.2	2726.5	2776.3	2825.9	2875.4	2924.9	2974.5	3074.5	3175.6	3278.2	3382.4	3488.1	3595.6	3704.8	3815.7
	S′	1.3027	7.3598	7.3618	7.4923	7.6137	7.7275	7.8349	7.9369	8.0342	8.2166	8.3858	8.5442	8.6934	8.8348	8.9695	9.0982	9.2217
101.325 (100.00)	V′	1.044	1673.0	1673.0	1792.7	1910.7	2027.7	2143.8	2259.3	2374.5	2604.2	2833.2	3061.9	3290.3	3518.7	3746.9	3975.0	4203.1
	U′	418.959	2506.5	2506.5	2544.7	2582.6	2620.4	2658.1	2695.3	2733.9	2810.6	2888.5	2968.0	3048.9	3131.6	3215.9	3302.0	3389.8
	H′	419.064	2676.0	2676.0	2726.4	2776.2	2825.8	2875.3	2924.8	2974.5	3074.4	3175.6	3278.2	3382.3	3488.1	3595.6	3704.8	3815.7
	S′	1.3069	7.3554	7.3554	7.4860	7.6075	7.7213	7.8288	7.9308	8.0280	8.2105	8.3797	8.5381	8.6873	8.8287	8.9634	9.0922	9.2156
125 (105.99)	V′	1.049	1374.6	–	1449.1	1545.6	1641.0	1735.6	1829.6	1923.2	2109.7	2295.6	2481.5	2666.5	2851.7	3036.8	3221.8	3406.7
	U′	444.224	2513.4	–	2542.9	2581.2	2619.3	2657.2	2695.2	2733.3	2810.2	2888.2	2967.7	3048.7	3131.4	3215.8	3301.9	3389.7
	H′	444.356	2685.2	–	2724.0	2774.4	2824.4	2874.2	2923.9	2973.7	3073.9	3175.2	3277.8	3382.0	3487.9	3595.4	3704.6	3815.5
	S′	1.3740	7.2847	–	7.3844	7.5072	7.6219	7.7300	7.8324	7.9300	8.1129	8.2823	8.4408	8.5901	8.7316	8.8663	8.9951	9.1186
150 (111.37)	V′	1.053	1159.0	–	1204.0	1285.6	1365.2	1444.4	1523.0	1601.3	1757.0	1912.2	2066.2	2221.5	2375.9	2530.2	2684.5	2838.6
	U′	466.968	2519.5	–	2540.9	2579.7	2618.1	2656.3	2694.4	2732.7	2809.7	2887.9	2967.4	3048.5	3131.2	3215.6	3301.7	3389.5
	H′	467.126	2693.4	–	2721.5	2772.5	2822.9	2872.9	2922.9	2972.9	3073.3	3174.7	3277.5	3381.7	3487.6	3595.1	3704.4	3815.3
	S′	1.4336	7.2234	–	7.2953	7.4194	7.5352	7.6439	7.7468	7.8447	8.0280	8.1976	8.3562	8.5056	8.6472	8.7819	8.9108	9.0343
175 (116.06)	V′	1.057	1003.34	–	1028.8	1099.1	1168.2	1236.4	1304.1	1371.3	1505.1	1638.3	1771.1	1903.7	2036.1	2168.4	2300.7	2432.9
	U′	486.815	2524.7	–	2538.9	2578.2	2616.9	2655.3	2693.7	2732.1	2809.3	2887.5	2967.1	3048.3	3131.0	3215.4	3301.6	3389.4
	H′	487.000	2700.3	–	2719.0	2770.5	2821.3	2871.7	2921.9	2972.0	3072.7	3174.2	3277.1	3381.4	3487.3	3594.9	3704.2	3815.1
	S′	1.4849	7.1716	–	7.2191	7.3447	7.4614	7.5708	7.6741	7.7724	7.9561	8.1259	8.2847	8.4341	8.5758	8.7106	8.8394	8.9630

TABLE C-2 (continued)

V' = specific volume (cm³/g), U' = specific internal energy (kJ/kg), H' = specific enthalpy (kJ/kg), S' = specific entropy (kJ/(kg·K))

Temperature, °C (temperature, K)

Abs. press. kPa (sat. temp. °C)		Sat. water	Sat. steam	100 (373.15)	125 (398.15)	150 (423.15)	175 (448.15)	200 (473.15)	225 (498.15)	250 (523.15)	300 (573.15)	350 (623.15)	400 (673.15)	450 (723.15)	500 (773.15)	550 (823.15)	600 (873.15)	650 (923.15)
200 (120.23)	V'	1.061	885.44	—	897.47	959.54	1020.4	1080.4	1139.8	1198.9	1316.2	1432.8	1549.2	1665.3	1781.2	1897.1	2012.9	2128.6
	U'	504.489	2529.2	—	2536.9	2576.6	2615.7	2654.4	2692.9	2731.4	2808.8	2887.2	2966.9	3049.0	3130.8	3215.3	3301.4	3389.2
	H'	504.701	2706.3	—	2716.4	2768.5	2819.8	2870.5	2920.9	2971.2	3072.1	3173.8	3276.7	3381.1	3487.0	3594.7	3704.0	3815.0
	S'	1.5301	7.1268	—	7.1523	7.2794	7.3971	7.5072	7.6110	7.7096	7.8937	8.0638	8.2226	8.3722	8.5139	8.6487	8.7776	8.9012
225 (123.99)	V'	1.064	792.97	—	795.25	850.97	905.44	959.06	1012.1	1064.7	1169.2	1273.1	1376.6	1479.9	1583.0	1686.0	1789.0	1891.9
	U'	520.465	2533.2	—	2534.6	2575.1	2614.5	2653.5	2692.2	2730.8	2808.4	2886.9	2966.6	3047.8	3130.6	3215.1	3301.2	3389.1
	H'	520.705	2711.6	—	2713.8	2766.5	2818.2	2869.3	2919.9	2970.4	3071.5	3173.3	3276.3	3380.8	3486.8	3594.4	3703.8	3814.8
	S'	1.5705	7.0873	—	7.0928	7.2213	7.3400	7.4508	7.5551	7.6540	7.8385	8.0088	8.1679	8.3175	8.4593	8.5942	8.7231	8.8467
250 (127.43)	V'	1.068	718.44	—	—	764.09	813.47	861.98	909.91	957.41	1051.6	1145.2	1238.5	1331.5	1424.4	1517.2	1609.9	1702.5
	U'	535.077	2536.8	—	—	2573.5	2613.3	2652.5	2691.4	2730.2	2808.0	2886.5	2966.3	3047.6	3130.4	3214.9	3301.1	3389.0
	H'	535.343	2716.4	—	—	2764.5	2816.7	2868.0	2918.9	2969.6	3070.9	3172.8	3275.9	3380.4	3486.5	3594.2	3703.6	3814.6
	S'	1.6071	7.0520	—	—	7.1689	7.2886	7.4001	7.5050	7.6042	7.7891	7.9597	8.1188	8.2686	8.4104	8.5453	8.6743	8.7980
275 (130.60)	V'	1.071	657.04	—	—	693.00	738.21	782.55	826.29	869.61	955.45	1040.7	1125.5	1210.2	1294.7	1379.0	1463.3	1547.6
	U'	548.564	2540.0	—	—	2571.9	2612.1	2651.6	2690.7	2729.6	2807.5	2886.2	2966.0	3047.3	3130.2	3214.7	3300.9	3388.8
	H'	548.858	2720.7	—	—	2762.5	2815.1	2866.8	2917.9	2968.7	3070.3	3172.4	3275.5	3380.1	3486.2	3594.0	3703.4	3814.4
	S'	1.6407	7.0201	—	—	7.1211	7.2419	7.3541	7.4594	7.5590	7.7444	7.9151	8.0744	8.2243	8.3661	8.5011	8.6301	8.7538
300 (133.54)	V'	1.073	605.56	—	—	633.74	675.49	716.35	765.60	796.44	875.29	953.52	1031.4	1109.0	1186.5	1263.9	1341.2	1418.5
	U'	561.107	2543.0	—	—	2570.3	2610.8	2650.6	2689.9	2729.0	2807.1	2885.8	2965.8	3047.1	3130.0	3214.5	3300.8	3388.7
	H'	561.429	2724.7	—	—	2760.4	2813.5	2865.5	2916.9	2967.9	3069.7	3171.9	3275.2	3379.8	3486.0	3593.7	3703.2	3814.2
	S'	1.6716	6.9909	—	—	7.0771	7.1990	7.3119	7.4177	7.5176	7.7034	7.8744	8.0338	8.1838	8.3257	8.4608	8.5898	8.7135

TABLE C-2 (continued)

V' = specific volume (cm³/g), U' = specific internal energy (kJ/kg), H' = specific enthalpy (kJ/kg), S' = specific entropy (kJ/(kg·K))

Temperature, °C (temperature, K)

Abs. press. kPa (sat. temp. °C)		Sat. water	Sat. steam	150 (423.15)	175 (448.15)	200 (473.15)	220 (493.15)	240 (513.15)	260 (533.15)	280 (553.15)	300 (573.15)	325 (598.15)	350 (623.15)	400 (673.15)	450 (723.15)	500 (773.15)	550 (823.15)	600 (873.15)	650 (923.15)
325 (136.29)	V'	1.076	561.75	583.58	622.41	660.33	690.22	719.81	749.18	778.39	807.47	843.68	879.78	951.73	1023.5	1095.0	1166.5	1237.9	1309.2
	U'	572.847	2545.7	2568.7	2609.6	2649.6	2681.2	2712.7	2744.0	2775.3	2806.6	2845.9	2885.5	2965.5	3046.9	3129.6	3214.4	3300.6	3388.6
	H'	573.197	2728.3	2758.4	2811.9	2864.2	2905.6	2946.6	2987.5	3028.2	3069.0	3120.1	3171.4	3274.8	3379.5	3485.7	3593.6	3702.9	3814.1
	S'	1.7004	6.9640	7.0363	7.1592	7.2729	7.3585	7.4400	7.5181	7.5933	7.6657	7.7530	7.8369	7.9965	8.1465	8.2885	8.4236	8.5527	8.6764
350 (138.87)	V'	1.079	524.00	540.58	576.90	612.31	640.18	667.75	695.09	722.27	749.33	783.01	816.57	883.45	950.11	1016.6	1083.0	1149.3	1215.6
	U'	583.892	2548.2	2567.1	2608.3	2648.6	2680.4	2712.0	2743.4	2774.8	2806.2	2845.6	2885.1	2965.2	3046.2	3129.2	3214.2	3300.5	3388.4
	H'	584.270	2731.6	2756.5	2810.3	2863.0	2904.5	2945.7	2986.7	3027.6	3068.4	3119.6	3170.9	3274.4	3379.4	3485.4	3593.3	3702.7	3813.9
	S'	1.7273	6.9392	6.9982	7.1222	7.2366	7.3226	7.4045	7.4828	7.5581	7.6307	7.7181	7.8022	7.9619	8.1120	8.2540	8.3892	8.5183	8.6421
375 (141.31)	V'	1.081	491.13	503.29	537.46	570.69	596.81	622.62	648.22	673.64	698.94	730.42	761.79	824.28	886.54	948.66	1010.7	1072.6	1134.5
	U'	594.332	2550.6	2565.4	2607.1	2647.7	2679.6	2711.3	2742.8	2774.3	2805.7	2845.2	2884.8	2964.9	3046.4	3129.4	3214.0	3300.3	3388.3
	H'	594.737	2734.7	2754.1	2808.6	2861.7	2903.4	2944.8	2985.9	3026.9	3067.8	3119.1	3170.5	3274.0	3378.8	3485.1	3593.0	3702.6	3813.7
	S'	1.7526	6.9160	6.9624	7.0875	7.2027	7.2891	7.3713	7.4499	7.5254	7.5981	7.6856	7.7698	7.9296	8.0798	8.2219	8.3571	8.4863	8.6101
400 (143.62)	V'	1.084	462.22	470.66	502.93	534.26	558.85	583.14	607.20	631.09	654.85	684.41	713.85	772.50	830.92	889.19	947.35	1005.4	1063.4
	U'	604.237	2552.7	2563.7	2605.8	2646.7	2678.8	2710.6	2742.2	2773.7	2805.3	2844.8	2884.5	2964.6	3046.2	3129.2	3213.8	3300.2	3388.2
	H'	604.670	2737.6	2752.0	2807.0	2860.4	2902.3	2943.9	2985.1	3026.2	3067.2	3118.5	3170.0	3273.6	3378.5	3484.9	3592.8	3702.3	3813.5
	S'	1.7764	6.8943	6.9285	7.0548	7.1708	7.2576	7.3402	7.4190	7.4947	7.5675	7.6552	7.7395	7.8994	8.0497	8.1919	8.3271	8.4563	8.5802
425 (145.82)	V'	1.086	436.61	441.85	472.47	502.12	525.36	548.30	571.01	593.54	615.95	643.81	671.56	726.81	781.84	836.72	891.49	946.17	1000.8
	U'	613.667	2554.8	2562.0	2604.5	2645.7	2678.0	2709.9	2741.6	2773.2	2804.8	2844.4	2884.1	2964.4	3045.9	3129.0	3213.7	3300.0	3388.0
	H'	614.128	2740.3	2749.8	2805.3	2859.1	2901.2	2942.9	2984.3	3025.5	3066.6	3118.0	3169.5	3273.3	3378.2	3484.6	3592.5	3702.1	3813.4
	S'	1.7990	6.8739	6.8965	7.0239	7.1407	7.2280	7.3108	7.3899	7.4657	7.5388	7.6265	7.7109	7.8710	8.0214	8.1636	8.2989	8.4282	8.5520

TABLE C-2 (continued)

V′ = specific volume (cm³/g), U′ = specific internal energy (kJ/kg), H′ = specific enthalpy (kJ/kg), S′ = specific entropy (kJ/(kg· K))

Abs. press. kPa (sat. temp. °C)		Sat. water	Sat. steam	Temperature, °C (temperature, K)															
				150 (423.15)	175 (448.15)	200 (473.15)	220 (493.15)	240 (513.15)	260 (533.15)	280 (553.15)	300 (573.15)	325 (598.15)	350 (623.15)	400 (673.15)	450 (723.15)	500 (773.15)	550 (823.15)	600 (873.15)	650 (923.15)
450 (147.92)	V′	1.088	413.75	416.24	445.38	473.55	495.59	517.33	538.83	560.17	581.37	607.73	633.97	686.20	738.21	790.07	841.83	893.50	945.10
	U′	622.672	2556.7	2560.3	2603.2	2644.7	2677.1	2709.2	2741.0	2772.7	2804.4	2844.0	2883.8	2964.1	3045.7	3128.8	3213.5	3299.8	3387.9
	H′	623.162	2742.9	2747.7	2803.7	2857.8	2900.2	2942.0	2983.5	3024.8	3066.0	3117.5	3169.1	3272.9	3377.9	3484.3	3592.3	3701.9	3813.2
	S′	1.8204	6.8547	6.8660	6.9946	7.1121	7.1999	7.2831	7.3624	7.4384	7.5116	7.5995	7.6840	7.8442	7.9947	8.1370	8.2723	8.4016	8.5255
475 (149.92)	V′	1.091	393.22	393.31	421.14	447.97	468.95	489.62	510.05	530.30	550.43	575.44	600.33	649.87	699.18	748.34	797.40	846.37	895.27
	U′	631.294	2558.5	2558.6	2601.9	2643.7	2676.3	2708.5	2740.4	2772.2	2803.9	2843.6	2883.4	2963.8	3045.4	3128.6	3213.3	3299.7	3387.7
	H′	631.812	2745.3	2745.5	2802.0	2856.5	2899.1	2941.1	2982.7	3024.1	3065.4	3116.9	3168.6	3272.5	3377.6	3484.0	3592.1	3701.7	3813.0
	S′	1.8408	6.8365	6.8369	6.9667	7.0850	7.1732	7.2567	7.3363	7.4125	7.4858	7.5739	7.6585	7.8189	7.9694	8.1118	8.2472	8.3765	8.5004
500 (151.84)	V′	1.093	374.68	–	399.31	424.96	444.97	464.67	484.14	503.43	522.58	546.38	570.05	617.16	664.05	710.78	757.41	803.95	850.42
	U′	639.569	2560.2	–	2600.6	2642.7	2675.5	2707.8	2739.8	2771.7	2803.5	2843.2	2883.1	2963.5	3045.2	3128.4	3213.1	3299.5	3387.6
	H′	640.116	2747.5	–	2800.3	2855.1	2898.0	2940.1	2981.9	3023.4	3064.8	3116.4	3168.1	3272.1	3377.2	3483.8	3591.9	3701.5	3812.8
	S′	1.8604	6.8192	–	6.9400	7.0592	7.1478	7.2317	7.3115	7.3879	7.4614	7.5496	7.6343	7.7948	7.9454	8.0879	8.2233	8.3526	8.4766

Abs. press. kPa (sat. temp. °C)		Sat. water	Sat. steam	Temperature, °C (temperature, K)															
				225 (498.15)	250 (523.15)	275 (548.15)	300 (573.15)	325 (598.15)	350 (623.15)	375 (648.15)	400 (673.15)	425 (698.15)	450 (723.15)	475 (748.15)	500 (773.15)	525 (798.15)	550 (823.15)	600 (873.15)	650 (923.15)
2400 (221.78)	V′	1.193	83.199	84.149	91.075	97.411	103.36	109.05	114.55	119.93	125.22	130.44	135.61	140.73	145.82	150.88	155.91	165.92	175.86
	U′	949.066	2600.7	2608.6	2665.6	2717.3	2765.4	2811.1	2855.4	2898.8	2941.7	2984.5	3027.1	3069.9	3112.9	3156.1	3199.6	3287.7	3377.2
	H′	951.929	2800.4	2810.6	2884.2	2951.1	3013.4	3072.8	3130.4	3186.7	3242.3	3297.5	3352.6	3407.7	3462.9	3518.2	3573.8	3685.9	3799.3
	S′	2.5343	6.2690	6.2894	6.4338	6.5586	6.6699	6.7714	6.8656	6.9542	7.0384	7.1189	7.1964	7.2713	7.3439	7.4144	7.4830	7.6152	7.7414
2500 (223.94)	V′	1.197	79.905	80.210	86.985	93.154	98.925	104.43	109.75	114.94	120.04	125.07	130.04	134.97	139.87	144.74	149.58	159.21	168.76
	U′	958.969	2601.2	2603.8	2662.0	2714.5	2763.1	2809.3	2853.9	2897.5	2940.6	2983.4	3026.2	3069.0	3112.1	3155.4	3198.9	3287.1	3376.7
	H′	961.962	2800.9	2804.3	2879.5	2947.4	3010.4	3070.4	3128.2	3184.8	3240.7	3296.1	3351.3	3406.5	3461.7	3517.2	3572.9	3685.1	3798.6
	S′	2.5543	6.2536	6.2604	6.4077	6.5345	6.6470	6.7494	6.8442	6.9333	7.0178	7.0986	7.1763	7.2513	7.3240	7.3946	7.4633	7.5956	7.7220

TABLE C-2 (continued)

V′ = specific volume (cm³/g), U′ = specific internal energy (kJ/kg), H′ = specific enthalpy (kJ/kg), S′ = specific entropy (kJ/(kg·K))

Abs. press. kPa (sat. temp. °C)		Sat. water	Sat. steam	225 (498.15)	250 (523.15)	275 (548.15)	300 (573.15)	325 (598.15)	350 (623.15)	375 (648.15)	400 (673.15)	425 (698.15)	450 (723.15)	475 (748.15)	500 (773.15)	525 (798.15)	550 (823.15)	600 (873.15)	650 (923.15)
2600 (226.04)	V′	1.201	76.856	—	83.205	89.220	94.830	100.17	105.32	110.33	115.26	120.11	124.91	129.66	134.38	139.07	143.74	153.01	162.21
	U′	968.597	2601.5	—	2658.4	2711.7	2760.3	2807.4	2852.3	2896.1	2939.4	2982.3	3025.2	3068.1	3111.2	3154.6	3198.2	3286.5	3376.1
	H′	971.720	2801.4	—	2874.7	2943.6	3007.4	3067.9	3126.1	3183.0	3239.0	3294.6	3349.9	3405.3	3460.6	3516.3	3571.9	3684.3	3797.9
	S′	2.5736	6.2387	—	6.3823	6.5110	6.6249	6.7281	6.8236	6.9131	6.9979	7.0789	7.1568	7.2320	7.3048	7.3755	7.4443	7.5768	7.7033
2700 (228.07)	V′	1.205	74.025	—	79.698	85.575	91.036	96.218	101.21	106.07	110.83	115.52	120.15	124.74	129.30	133.82	138.33	147.27	156.14
	U′	977.968	2601.8	—	2654.7	2708.8	2758.6	2805.6	2850.7	2894.8	2938.2	2981.2	3024.2	3067.2	3110.4	3153.8	3197.5	3285.8	3375.6
	H′	981.222	2801.7	—	2869.9	2939.8	3004.4	3065.4	3124.0	3181.2	3237.4	3293.1	3348.6	3404.0	3459.5	3515.2	3571.0	3683.5	3797.1
	S′	2.5924	6.2244	—	6.3575	6.4882	6.6034	6.7075	6.8036	6.8935	6.9787	7.0600	7.1381	7.2134	7.2863	7.3571	7.4260	7.5587	7.6853
2800 (230.05)	V′	1.209	71.389	—	76.437	82.187	87.510	92.550	97.395	102.10	106.71	111.25	115.74	120.17	124.58	128.95	133.30	141.94	150.50
	U′	987.100	2602.1	—	2650.9	2705.9	2756.9	2803.7	2849.2	2893.4	2937.0	2980.2	3023.2	3066.3	3109.6	3153.1	3196.8	3285.2	3375.0
	H′	990.485	2802.0	—	2864.9	2936.0	3001.3	3062.8	3121.9	3179.3	3235.8	3291.7	3347.3	3402.8	3458.4	3514.1	3570.0	3682.6	3796.4
	S′	2.6106	6.2104	—	6.3331	6.4659	6.5824	6.6875	6.7842	6.8746	6.9601	7.0416	7.1199	7.1954	7.2685	7.3394	7.4084	7.5412	7.6679
2900 (231.97)	V′	1.213	68.928	—	73.395	79.029	84.226	89.133	93.843	98.414	102.88	107.28	111.62	115.92	120.18	124.42	128.62	136.97	145.26
	U′	996.008	2602.3	—	2647.1	2702.9	2754.0	2801.8	2847.6	2892.0	2935.8	2979.1	3022.3	3065.5	3108.8	3152.3	3196.1	3284.6	3374.5
	H′	999.524	2802.2	—	2859.9	2932.1	2998.2	3060.3	3119.7	3177.4	3234.1	3290.2	3346.0	3401.6	3457.3	3513.1	3569.1	3681.8	3795.7
	S′	2.6283	6.1969	—	6.3092	6.4441	6.5621	6.6681	6.7654	6.8563	6.9421	7.0239	7.1024	7.1780	7.2512	7.3222	7.3913	7.5243	7.6511
3000 (233.84)	V′	1.216	66.626	—	70.551	76.078	81.159	85.943	90.526	94.969	99.310	103.58	107.79	111.95	116.08	120.18	124.26	132.34	140.36
	U′	1004.7	2602.4	—	2643.2	2700.0	2751.6	2799.9	2846.0	2890.7	2934.6	2978.0	3021.3	3064.6	3107.9	3151.5	3195.4	3284.0	3373.9
	H′	1008.4	2802.3	—	2854.8	2928.2	2995.1	3057.7	3117.5	3175.6	3232.5	3288.7	3344.6	3400.4	3456.2	3512.1	3568.1	3681.0	3795.0
	S′	2.6455	6.1837	—	6.2857	6.4228	6.5422	6.6491	6.7471	6.8385	6.9246	7.0067	7.0854	7.1612	7.2345	7.3056	7.3748	7.5079	7.6349

Smith, J.M., and H.C. Van Ness: *Introduction to Chemical Engineering Thermodynamics*, 4th ed., McGraw-Hill, New York, NY, 1987.

TABLE C-3
Temperature-dependent heat capacities (constant-pressure values)

Form C_p (J/mol·°C) or (J/mol·K) = $a + bT + cT^2 + dT^3$

Compound	Formula	State*	Temp. unit	a	$b \times 10^2$	$c \times 10^5$	$d \times 10^9$	Range (units of T)
Acetone	CH_3COCH_3	l	°C	123.0	18.6			-30–60
Air		g	°C	28.94	0.4147	0.3191	-1.965	0–1500
Benzene	C_6H_6	l	K	62.55	23.4			279–350
		g	°C	74.06	32.95	-25.20	77.57	0–1200
Carbon tetrachloride	CCl_4	l	K	93.39	12.98			273–343
Cumene	C_9H_{12}	g	°C	139.2	53.76	-39.79	120.5	0–1200
Formaldehyde	CH_2O	g	°C	34.28	4.268	0.0000	-8.694	0–1200
n-Hexane	C_6H_{14}	l	°C	216.3				20–100
Hydrogen sulfide	H_2S	g	°C	33.51	1.547	0.3012	-3.292	0–1500
Methane	CH_4	g	°C	34.31	5.469	0.3661	-11.00	0–1200
		g	K	19.87	5.021	1.268	-11.00	273–1500
Methyl alcohol	CH_3OH	l	°C	75.86				0
				82.59				40
Methyl cyclopentane	C_6H_{12}	g	°C	98.83	45.857	-30.44	83.81	0–1200
Toluene	C_7H_8	l	°C	148.8				0
		l	°C	181.2				100
Water	H_2O	l	°C	75.4				0–100

Note: Formulas for gases are strictly applicable at pressures low enough for the ideal gas laws to apply.

*l - liquid; g - gas

Adapted from Felder, R. M., and R. W. Rousseau: *Elementary Principles of Chemical Processes*, 2nd ed., John Wiley & Sons, New York, NY, 1986.

Latent heat of fusion and vaporization at 1 atm

Compound	Formula	Melting point (°C)	Heat of fusion (cal/g)	Temperature (°C)	Heat of vaporization (cal/g)
Acetone	C_3H_6O	-95.5	23.42	56	124.41
Benzene	C_6H_6	5.533	30.100	25	103.57
				80.10	94.14
Benzoic acid	$C_7H_8O_2$	122.45	33.90	-	-
Carbon tetrachloride	CCl_4	-	-	0	52.06
				76.75	46.42
				200	32.73
Chlorobenzene	C_6H_5Cl	-	-	130.6	77.59
Cresol (p-)	C_7H_8O	34.6	26.28	202	100.58
Cyclohexane	C_6H_{12}	6.67	7.569	25	93.81
				80.74	85.6
Dibromobenzene (o-)	$C_6H_4Br_2$	1.8	12.78	-	-
Diphenyl amine	$C_{12}H_{11}N$	52.98	25.23	-	-
Ethyl amine	C_2H_7N	-	-	15	145.97
Ethylbenzene	C_8H_{10}	-94.950	20.629	25	95.11
				136.19	81.0
Methane	CH_4	-182.48	14.03	-161.6	121.87
Methanol	CH_4O	-97.8	23.7	0	284.29
				64.7	262.79
				100	241.29
Naphthalene	$C_{10}H_8$	80.0	36.0	218	75.49
Toluene	C_7H_8	-94.991	17.171	25	98.55
				110.62	86.8
Urethane	$C_3H_7NO_2$	48.7	40.85	-	-

Source: Perry, R. H., and D. W. Green, eds.: *Perry's Chemical Engineer's Handbook*, 6th ed., McGraw-Hill, New York, NY, pp. 3-123 to 3-126, 1984.

1 cal/g × 0.00419 = KJ/g

To convert to a molar basis multiply by mol. wt. e.g. for acetone, mol. wt. = 58.08 g/g mol

Heat of vaporization @ 56° C = 124.41 cal/g = 124.41 cal/g×58.08 g/g mol×0.00419 1×J/g = 30.2 1×J/g mol

TABLE C-5

Heat of combustion for common waste elements

Hazardous constituent	Heat of combustion kcal/gram	Hazardous constituent	Heat of combustion kcal/gram
Trichloromonofluoromethane	0.11	Octachlorobiphenyl	2.72
Tribromomethane	0.13	Acetyl chloride	2.77
Dichlorodifluoromethane	0.22	Trichloropropane, N.O.S.	2.81
Tetrachloromethane	0.24	1,2,3-Trichloropropane	2.81
Tetranitromethane	0.41	Dichloropropanol, N.O.S.	2.84
Hexachloroethane	0.46	Dimethyl sulfate	2.86
Dibromomethane	0.50	2,4,5-T	2.87
Pentachloroethane	0.53	2,4,5-Trichlorophenol	2.88
Hexachloropropene	0.70	2,4,6-Trichlorophenol	2.88
Chloroform	0.75	N-Nitroso-N-methylurea	2.89
Chloral (trichloroacetaldehyde)	0.80	Heptachlorobiphenyl	2.98
Cyanogen bromide	0.81	1,1-Dichloroethane	3.00
Trichloromethanetiol	0.84	1,2-Dichloroethane	3.00
Hexachlorocyclohexane	1.12	trans-1,2-Dichloroethane	3.00
Tetrachloroethene		Phenyl dichloroarsine	3.12
(Tetrachloroethylene)	1.19	N-Nitrosoarcosine	3.19
Cyanogen chloride	1.29	Azaserine	3.21
Formic acid	1.32	2-Fluoroacetamide	3.24
Iodomethane	1.34	Chloromethane	3.25
Tetrachloroethane, N.O.S.	1.39	Hexachlorobiphenyl	3.28
1,1,1,2-Tetrachloroethane	1.39	Bis (2-chloroethyl) ether	3.38
1,1,2,2-Tetrachloroethane	1.39	1,2,3,4,10,10-Hexachloro-	
1,2-Dibromomethane	1.43	1,4,4a,5,7,8a-hexahydro-	
1,2-Dibromo-3-chloropropane	1.48	1,4:5,8-endo, endo-	
Pentachloronitrobenzene	1.62	dimethanonaphthalene	3.38
Bromomethane	1.70	Benzenearsonic acid	3.40
Dichloromethane	1.70	Maleic anhydride	3.40
Trichloroethene		1,2,4-Trichlorobenzene	3.40
(Trichloroethylene)	1.74	TCDD	3.43
Hexachlorobenzene	1.79	Dichloropropene, N.O.S.	3.44
Bis (chloromethyl) ether	1.97	1,3-Dichloropropene	3.44
1,1,1-Trichloroethane	1.99	Endrin	3.46
1,1,2-Trichloroethane	1.99	Chloromethyl methyl ether	3.48
Pentachlorobenzene	2.05	2,4-Dinitrophenol	3.52
Pentachlorophenol	2.09	Nitrogen mustard N-oxide	
Hexachlorocyclopentadiene	2.10	and hydrochloride salt	3.56
Hexachlorobutadiene	2.12	Parathion	3.61
Kepone	2.15	2,4-D	3.62
2,3,4,6-Tetrachlorophenol	2.23	Pentachlorobiphenyl	3.66
Dichlorophenylaraine	2.31	1,3-Propane sultone	3.67
Decachlorobiphenyl	2.31	Methyl methanosulfonate	3.74
Endosulfan	2.33	Aldrin	3.75
Nonachlorobiphenyl	2.50	Nitroglycerine	3.79
Toxaphene	2.50	2,4-Dichlorophenol	3.81
1,2,4,5-Tetrachlorobenzene	2.61	2,6-Dichlorophenol	3.81
Bromoacetone	2.66	Hexachlorophene	3.82
Dichloroethylene, N.O.S.	2.70	Trypan blue	3.84
1,1-Dichloroethylene	2.70	Benzotrichloride	3.90
Chlordane	2.71	Cycasin	3.92
Heptachlor epoxide	2.71	N-Nitroso-N-ethylurea	3.92
Phenylmercury acetate	2.71	Cyclophosphamide	3.97

TABLE C-5 (continued)

Hazardous constituent	Heat of combustion kcal/gram	Hazardous constituent	Heat of combustion kcal/gram
Dichloropropane, N.O.S.	3.99	2-Methyl-2-(methylthio) propionaldehyde-o-(methylcarbonyl) oxime	5.34
1,2-Dichloropropane	3.99		
Methylparathion	4.00		
Uracil mustard	4.00	2-sec-Butyl-4,6 dinitrophenol (DNBP)	5.46
Amitrole	4.01		
Dimethoate	4.02	p-Nitroaniline	5.50
Tetraethyl lead	4.04	Chlorobenzilate	5.50
4,6-Dinitro-o-cresol and salts	4.06	Dieldrin	5.56
		2,4,5-TP	5.58
N-Methyl-N-nitro-N-nitrosoguanidine	4.06	Methoxychlor	5.59
		4-Nitroquinoline-1-oxide	5.59
Mustard gas	4.06	Diallate	5.62
Maleic hydrazide	4.10	Daunomycin	5.70
Dinitrobenzene, N.O.S.	4.15	Ethylenebisdithiocarbamate	5.70
N-Nitroso-N-methylurethane	4.18	3,3'-Dichlorobenzidine	5.72
1,4-Dichloro-2-butene	4.27	Pronamide	5.72
Nitrogen mustard and hydrochloride salt	4.28	Aflatoxins	5.73
		Disulfoton	5.73
Tetrachlorobiphenyl	4.29	Diepoxybutane	5.74
Hydrazine	4.44	Dimethyl phthalate	5.74
Vinyl chloride	4.45	Glycidylaldehyde	5.74
Formaldehyde	4.47	Acrylamide	5.75
Saccharin	4.49	3,3-Dimethyl-1-(methylthio)-2-butanone-0-(methylamino) carbonyl oxime	5.82
3-Chloropropionitrile	4.50		
DDT	4.51		
Thiourea	4.51	4-Bromophenyl phenyl ether	5.84
1-Acetyl-2-thiourea	4.55	Thiuram	5.85
Thiosemicarbazide	4.55	Methanethiol	5.91
Dichlorobenzene, N.O.S.	4.57	Tolulene diisocyanate	5.92
Ethyl cyanide	4.57	Chlorambucil	5.93
Bis (2-chloroethoxy) methane	4.60	Thioacetamide	5.95
2,4-Dinitrotoluene	4.68	Ethylenethiourea	5.98
Isocyanic acid, methyl ester	4.69	Malononitrile	5.98
7-Oxabicyclo (2.2.1)-heptane-2,3-dicarboxylic acid	4.70	5-Nitro-o-toluidine	5.98
		Nitrobenzene	6.01
Ethyl carbamate	4.73	3,4-Dihydroxy-alpha-(methylamino) methyl benzyl alcohol	6.05
5-(Aminomethyl)-3-isoxazolol	4.78		
Methylthiouracil	4.79		
4,4'-Methylene-bis-(2-chloroaniline)	4.84	Benzoquinone	6.07
		N-Nitrosomethylethylamine	6.13
Bis (2-chloroisopropyl) ether	4.93	p-Chloroaniline	6.14
4-Nitrophenol	4.95	Benzyl chloride	6.18
DDE	5.05	Resorcinol	6.19
Dimethylcarbamoyl chloride	5.08	Propylthiouracil	6.28
p-Chloro-m-cresol	5.08	Paraldehyde	6.30
Dichloromethylbenzene	5.09	Dichlorobiphenyl	6.36
Trichlorobiphenyl	5.10	Diethyl phthalate	6.39
DDD	5.14	Dioxane	6.41
Dimethylnitrosoamine	5.14	2-Methylacetonitrile	6.43
N-Nitrosodimethylamine	5.14	N-Nitrosopyrrolidone	6.43
Diethylarsine	5.25	Methyl methacrylate	6.52
Phthalic anhydride	5.29	Chlorobenzene	6.60
1-(o-chlorophenyl) thiourea	5.30	o-Toluidine hydrochloride	6.63

TABLE C-5 (continued)

Hazardous constituent	Heat of combustion kcal/gram	Hazardous constituent	Heat of combustion kcal/gram
N,N-Bis (2-chloroethyl)-2-naphthylamine	6.64	Methyl ethyl ketone (MEK)	8.07
2,6-Dinitrotoluene		Cresylic acid	8.09
di-n-octyl phthalate	6.67	Cresol	8.18
Reserpine	6.70	Toluene diamine	8.24
Methyl hydrazine	6.78	Acetophenone	8.26
Cyanogen	6.79	Butyl benzyl phthalate	8.29
Ethylene oxide	6.86	Ethyl cyanide	8.32
N-Nitrosodiethylamine	6.86	Bis (2-ethylhexyl) phthalate	8.42
2-Chlorophenol	6.89	Benzenethiol	8.43
N-Phenylthiourea	6.93	N-Nitrosodi-N-butylamine	8.46
Acrolein	6.96	2,4-Dimethylphenol	8.51
2-Butanone peroxide	6.96	Indenol (1,2,3-c,d) pyrene	8.52
p-Dimethylaminoazobenzene	6.97	Diethylstilbestrol	8.54
1,4-Naphthoquinone	6.97	1-Naphthylamine	8.54
3-(alpha-Acetonylbenzyl)-4-hydroxy-coumarin and salts (Warfarin)	7.00	2-Naphthylamine	8.54
		Methacrylonitrile	8.55
		Isobutyl alcohol	8.62
N-Nitrosodiethanolamine	7.02	1,2-Diethylhydrazine	8.68
N-Nitrosopiperidine	7.04	2-Picoline	8.72
N-Nitrosonornicotine	7.07	Aniline	8.73
Phenacetin	7.17	1,2-Diphenylhydrazine	8.73
Ethyl methacrylate	7.27	3,3'-Dimethoxybenzidine	8.81
Di-n-butyl phthalate	7.34	7N-Dibenzo (c,g) carbazole	8.90
3,3'-Dimethoxybenzidine	7.36	Benz (c) acridine	8.92
Acetonitrile	7.37	Nicotine and salts	8.92
4-Aminopyridine	7.37	4-Amino biphenyl	9.00
2-Chloronaphthalene	7.37	Diphenylamine	9.09
2 Propyn-1-o1	7.43	2-Methylaziridine	9.09
1-Naphthyl-2-thiourea	7.50	Benzidine	9.18
Isosafrole	7.62	Benzo (b) fluoranthene	9.25
Dihydrosafrole	7.66	Benzo (j) fluoranthene	9.25
Safrole	7.68	Benzo (a) pyrene	9.25
Auramine	7.69	Dibenzo (a,e) pyrene	9.33
Crotonaldehyde	7.73	Dibenzo (a,h) pyrene	9.33
Allyl alcohol	7.75	Dibenzo (a,i) pyrene	9.33
Monochlorobiphenyl	7.75	Fluoranthene	9.35
Phenol	7.78	Benz (a) anthracene	9.39
Phenylenediamine	7.81	Dibenz (a,h) anthracene (Dibenzo (a,h) anthracene)	9.40
Di-n-propylnitrosoamine	7.83	Dibenz (a,h) acridine	9.53
Pyridine	7.83	Dibenz (a,j) acirdine	9.53
Ethyleneimine	7.86	alpha, alpha-Dimethylphenethylamine	9.54
1,1-Dimethylhydrazine	7.87	3-Methylcholanthrene	9.57
1,2-Dimethylhydrazine	7.87	n-Propylamine	9.58
N-Nitrosomethylvinylamine	7.91	7,12-Dimethylbenz(a)anthracene	9.61
2-Acetylaminofluorine	7.82	Naphthalene	9.62
Acrylonitrile	7.93	Benzene	10.03
Methapyrilene	7.93	Toluene	10.14
Strychnine and salts	8.03		

Source: U.S. EPA: *Guidance Manual for Incinerator Permit Writers*, 1982.

TABLE C-6

Heats of formation and free energies of formation

Compound	State	Heat of formation at 25°C (kcal/mole)	Free energy of formation at 25°C (kcal/mole)
Cyclohexane	gas	−29.43	7.59
	liquid	−37.34	6.39
Benzene	gas	19.820	30.989
Ethylene	gas	12.496	16.282
Acetone	gas	−51.79	−36.45
	liquid	−59.32	−37.16
Isobutanol	gas	−69.05	−38.25
	liquid	−81.06	−39.36
SO	gas	19.02	12.75
SO_2	gas	−70.94	−71.68
SO_3	gas	−94.39	−88.59
	liquid	−103.03	−88.28
CO	gas	−26.416	−32.808
CO_2	gas	−94.052	−94.260
NO	gas	21.600	20.719
NO_2	gas	7.96	12.26
N_2O	gas	19.55	24.82

Source: Perry, R. H., and D. W. Green, eds.: *Perry's Chemical Engineer's Handbook*, 6th ed., McGraw-Hill, New York, NY, table 3-206, 1984.

APPENDIX
D

CONVERSION
FACTORS

Bring out number, weight, and measure in a year of dearth.

Blake

TABLES IN APPENDIX D

TABLE D-1

Metric conversion factors (U.S. customary units to SI units)

Multiply U.S. customary unit		By	To obtain the SI unit	
Name	Symbol		Symbol	Name
Acceleration				
feet per second squared	ft/s^2	0.3048^a	m/s^2	meters per second squared
inches per second squared	in/s^2	0.0254^a	m/s^2	meters per second squared
Area				
acre	acre	0.4047	ha	hectare
acre	acre	4.0469×10^{-3}	km^2	square kilometer
square foot	ft^2	9.2903×10^{-2}	m^2	square meter
square inch	in^2	6.4516^a	cm^2	square centimeter
square mile	mi^2	2.5900	km^2	square kilometer
square yard	yd^2	0.8361	m^2	square meter
Concentration/density				
slugs per cubic foot	slugs/ft^3	0.51541	g/cm^3	grams per cubic centimeter
pounds per cubic foot	lb/ft^3	16.0185	kg/m^3	kilograms per cubic meter
pounds per acre-foot	lb/acre-ft	3.6773×10^{-1}	g/m^3	grams per cubic meter
grains per cubic foot	grain/ft^3	2.2883	g/m^3	grams per cubic meter
pounds per million gallons	lb/Mgal	0.119826	mg/L	milligrams per liter
Diffusion coefficient				
square inches per second	in^2/s	6.4516	cm^2/s	square centimeter per second
square feet per second	ft^2/s	5.57418	m^2/min	square meter per minute
Energy				
British thermal unit	Btu	1.0551	kJ	kilojoule
foot-pound (force)	ft · lb$_f$	1.3558	J	joule (newton-meter)
horsepower-hour	hp · h	2.6845	MJ	megajoule
kilowatt-hour	kW · h	3,600^a	kJ	kilojoule
kilowatt-hour	kW · h	3.600×10^{6a}	J	joule
watt-hour	W · h	3.600^a	kJ	kilojoule
watt-second	W · s	1.000^a	J	joule
Force				
pound force	lb$_f$	4.4482	N	newton (kg · m/s^2)
Flowrate				
cubic feet per second	ft^3/s	2.8317×10^{-2}	m^3/s	cubic meters per second
gallons per day	gal/d	4.3813×10^{-5}	L/s	liters per second
gallons per day	gal/d	3.7854×10^{-3}	m^3/d	cubic meters per day
gallons per minute	gal/min	6.3090×10^{-5}	m^3/s	cubic meters per second
gallons per minute	gal/min	6.3090×10^{-2}	L/s	liters per second
million gallons per day	Mgal/d	43.8126	L/s	liters per second
million gallons per day	Mgal/d	3.7854×10^3	m^3/d	cubic meters per day
million gallons per day	Mgal/d	4.3813×10^{-2}	m^3/s	cubic meters per second
Length				
foot	ft	0.3048^a	m	meter
inch	in	2.54^a	cm	centimeter
inch	in	0.0254^a	m	meter
inch	in	25.4^a	mm	millimeter
mile	mi	1.6093	km	kilometer
yard	yd	0.9144^a	m	meter

Tables D-1 through D-3 adapted from Metcalf & Eddy, *Wastewater Engineering*, 3d. ed., McGraw-Hill, NY, 1991, by permission
^{a}Indicates exact conversion

TABLE D-1 (continued)

Multiply U.S. customary unit		By	To obtain the SI unit	
Name	Symbol		Symbol	Name
Loading rate				
pounds per square feet per second	$lb_f/ft^2 \cdot sec$	4.88	$kg/m^2 \cdot s$	kilograms per square meter per second
Mass				
ounce	oz	28.3495	g	gram
pound	lb	4.5359×10^2	g	gram
pound	lb	0.4536	kg	kilogram
ton (short: 2,000 lb)	ton	0.9072	Mg	megagram (10^3 kilogram) (metric ton)
ton (long: 2,240 lb)	ton	1.0160	Mg	megagram (10^3 kilogram) (metric ton)
Power				
British thermal units per second	Btu/s	1.0551	kW	kilowatt
foot-pounds (force) per second	$ft \cdot lb_f/s$	1.3558	W	watt
horsepower	hp	0.7457	kW	kilowatt
Pressure (force/area)				
atmosphere (standard)	atm	1.0133×10^2	kPa (kN/m^2)	kilopascal (kilonewtons per square meter)
inches of mercury (60°F)	inHg (60°F)	3.3768×10^3	Pa (N/m^2)	pascal (newtons per square meter)
inches of water (60°F)	inH_2O (60°F)	2.4884×10^2	Pa (N/m^2)	pascal (newtons per square meter)
pounds (force) per square foot	lb_f/ft^2	47.8803	Pa (N/m^2)	pascal (newtons per square meter)
pounds (force) per square inch	lb_f/in^2	6.8948×10^3	Pa (N/m^2)	pascal (newtons per square meter)
pounds (force) per square inch	lb_f/in^2	6.8948	kPa (kN/m^2)	kilopascal (kilonewtons per square meter)
Temperature				
degrees Fahrenheit	°F	0.555 (°F - 32)	°C	degrees Celsius (centigrade)
degrees Fahrenheit	°F	0.555 (°F + 459.67)	K	kelvin
Velocity				
feet per second	ft/s	0.3048^a	m/s	meters per second
miles per hour	mi/h	4.4704×10^{-1a}	m/s	meters per second
Absolute viscosity				
pounds second per square foot	$lb \cdot s/ft^2$	47.88	$kg/m \cdot s$	kilograms per meter per second
pounds (mass) per inches per second	$lb_m/in \cdot sec$	17.86	$kg/m \cdot s$	kilograms per meter per second
pounds per feet per second	$lb/ft \cdot s$	14.88	poise ($g/cm \cdot s$)	poise
pounds second per square inch	$lb_f \cdot s/in^2$	6895	$kg/m \cdot s$	kilograms per meter per second

[a] Indicates exact conversion

TABLE D-1 (continued)

Multiply U.S. customary unit		By	To obtain the SI unit	
Name	Symbol		Symbol	Name
Kinematic viscosity				
square feet per second	ft^2/s	929	stoke (cm^2/s)	stoke
square feet per second	ft^2/s	0.0929	m^2/s	square meters per second
square inches per second	in^2/s	6.4516	cm^2/s	square centimeter per second
Volume				
acre-foot	acre-ft	1.2335×10^3	m^3	cubic meter
cubic foot	ft^3	28.3168	L	liter
cubic foot	ft^3	2.8317×10^{-2}	m^3	cubic meter
cubic inch	in^3	16.3871	cm^3	cubic centimeter
cubic yard	yd^3	0.7646	m^3	cubic meter
gallon	gal	3.7854×10^{-3}	m^3	cubic meter
gallon	gal	3.7854	L	liter
ounce (U.S. fluid)	oz (U.S. fluid)	2.9573×10^{-2}	L	liter

[a] Indicates exact conversion

TABLE D-2

Metric conversion factors (SI units to U.S. customary units)

Multiply the SI unit		By	To obtain the U.S. customary unit	
Name	Symbol		Symbol	Name
Acceleration				
meters per second squared	m/s^2	3.2808	ft/s^2	feet per second squared
meters per second squared	m/s^2	39.3701	in/s^2	inches per second squared
Area				
hectare (10,000 m^2)	ha	2.4711	acre	acre
square centimeter	cm^2	0.1550	in^2	square inch
square kilometer	km^2	0.3861	mi^2	square mile
square kilometer	km^2	247.1054	acre	acre
square meter	m^2	10.7639	ft^2	square foot
square meter	m^2	1.1960	yd^2	square yard
Concentration/density				
grams per cubic centimeter	g/cm^3	1.9422	$slugs/ft^3$	slugs per cubic foot
kilograms per cubic meter	kg/m^3	0.06243	lb/ft^3	pounds per cubic foot
grams per cubic meter	g/m^3	2.7192	lb/acre-ft	pounds per acre-foot
grams per cubic meter	g/m^3	0.4370	$grain/ft^3$	grains per cubic foot
grams per liter	g/L	8.3454×10^{-3}	lb/gal	pounds per gallon
Diffusion coefficient				
square centimeters per second	cm^2/s	0.155	in^2/s	square inches per second
square meters per minute	m^2/min	0.1794	ft^2/s	square feet per second
Energy				
kilojoule	kJ	0.9478	Btu	British thermal unit
joule (newton-meter)	J	2.7778×10^{-7}	kW · h	kilowatt-hour
joule	J	0.7376	ft · lb_f	foot-pound (force)
joule	J	1.0000	W · s	watt-second
joule	J	0.2388	cal	calorie
kilojoule	kJ	2.7778×10^{-4}	kW · h	kilowatt-hour
kilojoule	kJ	0.2778	W · h	watt-hour
megajoule	MJ	0.3725	hp · h	horsepower-hour
Force				
newton	N	0.2248	lb_f	pound force
Flowrate				
cubic meters per day	m^3/d	264.1720	gal/d	gallons per day
cubic meters per day	m^3/d	2.6417×10^{-4}	Mgal/d	million gallons per day
cubic meters per second	m^3/s	35.3147	ft^3/s	cubic feet per second
cubic meters per second	m^3/s	22.8245	Mgal/d	million gallons per day
cubic meters per second	m^3/s	15,850.3	gal/min	gallons per minute
liters per second	L/s	22,824.5	gal/d	gallons per day
liters per second	L/s	0.0228	Mgal/d	million gallons per day
liters per second	L/s	15.8508	gal/min	gallons per minute
Length				
centimeter	cm	0.3937	in	inch
kilometer	km	0.6214	mi	mile
meter	m	39.3701	in	inch
meter	m	3.2808	ft	foot
meter	m	1.0936	yd	yard
millimeter	mm	0.03937	in	inch

TABLE D-2 (continued)

Multiply the SI unit		By	To obtain the U.S. customary unit	
Name	Symbol		Symbol	Name
Loading rate				
kilograms per square meter per second	kg/m^2 · s	0.2048	lb$_f$/ft^2 · sec	pounds per square feet per second
Mass				
gram	g	0.0353	oz	ounce
gram	g	0.0022	lb	pound
kilogram	kg	2.2046	lb	pound
kilogram	kg	0.068521	slug	slug
megagram (10^3 kilogram)	Mg	1.1023	ton	ton (short: 2,000 lb)
megagram (10^3 kilogram)	Mg	0.9842	ton	ton (long: 2,240 lb)
Power				
kilowatt	kW	0.9478	Btu/s	British thermal units per second
kilowatt	kW	1.3410	hp	horsepower
watt	W	0.7376	ft · lb$_f$/s	foot-pounds (force) per second
Pressure (force/area)				
pascal (newtons per square meter)	Pa (N/m^2)	1.4504 × 10^{-4}	lb$_f$/in^2	pounds (force) per square inch
pascal (newtons per square meter)	Pa (N/m^2)	2.0885 × 10^{-2}	lb$_f$/ft^2	pounds (force) per square foot
pascal (newtons per square meter)	Pa (N/m^2)	2.9613 × 10^{-4}	inHg	inches of mercury (60°F)
pascal (newtons per square meter)	Pa (N/m^2)	4.0187 × 10^{-3}	inH$_2$O	inches of water (60°F)
kilopascal (kilonewtons per square meter)	kPa (kN/m^2)	0.1450	lb$_f$/in^2	pounds (force) per square inch
kilopascal (kilonewtons per square meter)	kPa (kN/m^2)	0.0099	atm	atmosphere (standard)
Temperature				
degrees Celsius (centigrade)	°C	1.8 (°C) + 32	°F	degrees Fahrenheit
kelvin	K	1.8 (K) - 459.67	°F	degrees Fahrenheit
Velocity				
meters per second	m/s	2.2369	mi/h	miles per hour
meters per second	m/s	3.2808	ft/s	feet per second
Absolute viscosity				
kilograms per meter per second	kg/m · s	0.02089	lb · s/ft^2	pounds second per square feet
kilograms per meter per second	kg/m · s	0.05599	lb$_m$/in · sec	pounds (mass) per inches per second
poise	poise (g/cm · s)	0.0672	lb/ft · s	pounds per feet per second
kilograms per meter per second	kg/m · s	1.4503 × 10^{-4}	lb · s/in^2	pounds second per square inch
Kinematic viscosity				
stoke	stoke (cm^2/s)	1.076 × 10^{-3}	ft^2/s	square feet per second
square meters per second	m^2/s	10.764	ft^2/s	square feet per second
square centimeter per second	cm^2/s	0.155	in^2/s	square inches per second

TABLE D-2 (continued)

Multiply the SI unit		By	To obtain the U.S. customary unit	
Name	Symbol		Symbol	Name
Volume				
cubic centimeter	cm^3	0.0610	in^3	cubic inch
cubic meter	m^3	35.3147	ft^3	cubic foot
cubic meter	m^3	1.3079	yd^3	cubic yard
cubic meter	m^3	264.1720	gal	gallon
cubic meter	m^3	8.1071×10^{-4}	acre-ft	acre-foot
liter	L	0.2642	gal	gallon
liter	L	0.0353	ft^3	cubic foot
liter	L	33.8150	oz	ounce (U.S. fluid)

TABLE D-3

Common environmental unit conversion

To convert, multiply in direction shown by arrows

U.S. units	$\rightarrow$	$\leftarrow$	SI units
acre/(Mgal/d)	0.1069	9.3536	$ha/(10^3 m^3/d)$
Btu	1.0551	0.9478	kJ
Btu/lb	2.3241	0.4303	kJ/kg
$Btu/ft^2 \cdot °F \cdot h$	5.6735	0.1763	$W/m^2 \cdot °C$
$bu/acre \cdot yr$	2.4711	0.4047	$bu/ha \cdot yr$
ft/h	0.3048	3.2808	m/h
ft/min	18.2880	0.0547	m/h
$ft^2/capita$	0.0929	10.7639	$m^2/capita$
$ft^3/capita$	0.0283	35.3147	$m^3/capita$
ft^3/gal	7.4805	0.1337	m^3/m^3
$ft^3/ft \cdot min$	0.0929	10.7639	$m^3/m \cdot min$
ft^3/lb	0.0624	16.0185	m^3/kg
$ft^3/Mgal$	7.04805×10^{-3}	133.6805	$m^3/10^3 m^3$
$ft^3/Mgal \cdot d$	407.4611	0.0025	$m^2/10^3 m^3 \cdot d$
$ft^3/ft^2 \cdot h$	0.3048	3.2808	$m^3/m^2 \cdot h$
$ft^3/10^3 gal \cdot min$	7.04805×10^{-3}	133.6805	$m^3/m^3 \cdot min$
ft^3/min	1.6990	0.5886	m^3/h
ft^3/s	2.8317×10^{-2}	35.3145	m^3/s
$ft^3/10^3 ft^3 \cdot min$	0.001	1,000.0	$m^3/m^3 \cdot min$
gal	3.7854	0.2642	L
$gal/acre \cdot d$	0.0094	106.9064	$m^3/ha \cdot d$
$gal/ft \cdot d$	0.0124	80.5196	$m^3/m \cdot d$
$gal/ft^2 \cdot d$	0.0407	24.5424	$m^3/m^2 \cdot d$
$gal/ft^2 \cdot d$	0.0017	589.0173	$m^3/m^2 \cdot h$
$gal/ft^2 \cdot d$	0.0283	35.3420	$L/m^2 \cdot min$
$gal/ft^2 \cdot d$	40.7458	2.4542×10^{-2}	$L/m^2 \cdot d$
$gal/ft^2 \cdot min$	2.4448	0.4090	m/h
$gal/ft^2 \cdot min$	40.7458	0.0245	$L/m^2 \cdot min$
$gal/ft^2 \cdot min$	58.6740	0.0170	$m^3/m^2 \cdot d$
$gal/min \cdot ft$	12.4193	8.052×10^{-2}	$L/min \cdot m$
$hp/10^3 gal$	0.1970	5.0763	kW/m^3
$hp/10^3 ft^3$	26.3342	0.0380	$kW/10^3 m^3$
in	25.4	3.9370×10^{-2}	mm
inHg (60°F)	3.3768	0.2961	kPa Hg (60°F)
lb	0.4536	2.2046	kg
lb/acre	1.1209	0.8922	kg/ha
$lb/10^3 gal$	0.1198	8.3452	kg/m^3
$lb/hp \cdot h$	0.6083	1.6440	$kg/kW \cdot h$
lb/Mgal	0.1198	8.3454	g/m^3
lb/Mgal	1.1983×10^{-4}	8345.4	kg/m^3
lb/ft^2	4.8824	0.2048	kg/m^2
lb_f/in^2 (gage)	6.8948	0.1450	kPa (gage)
$lb/ft^3 \cdot h$	16.0185	0.0624	$kg/m^3 \cdot h$
$lb/10^3 ft^3 \cdot d$	0.0160	62.4280	$kg/m^3 \cdot d$
lb/ton	0.5000	2.0000	kg/tonne
$Mgal/acre \cdot d$	0.9354	1.0691	$m^3/m^2 \cdot d$
Mgal/d	3.7854×10^3	0.264×10^{-3}	m^3/d
Mgal/d	4.3813×10^{-2}	22.8245	m^3/s
min/in	3.9370	0.2540	$min/10^2 mm$
tons/acre	2.2417	0.4461	Mg/ha
yd^3	0.7646	1.3079	m^3

TABLE D-4
Equivalent values

Quantity	Equivalent values

Mass	1 kg = 1,000 g = 0.001 metric ton = 2.20462 lb_m = 35.27392 oz
	 1 lb_m = 16 oz = 5 $\times$ 10^{-4} ton = 453.593 g = 0.453593 kg
	 1 ton = 2,000 lb
	 1 m = 100 cm = 1,000 mm = 10^6 microns (μ) = 10^{10} angstroms (Å)
	     = 39.37 in = 3.2808 ft = 1.0936 yards = 0.0006214 mile
	 1 ft = 12 in = 1/3 yd = 0.3048 m = 30.48 cm
	 1 mile = 5,280 ft
	 1 m^3 = 1,000 liters = 10^6 cm^3 = 10^6 mL
	     = 35.3145 ft^3 = 220.83 imperial gallons = 264.17 gallons
	     = 1056.68 quarts
	 1 ft^3 = 1,728 in^3 = 7.4805 gallons = 0.028317 m^3
	     = 28.317 liters = 28,317 cm^3
	 1 acre-ft = 43,560 ft^3 = 1613.3 yd^3
	     = 0.325851 Mgal = 1.876 $\times$ 10^{-2} $mi^2 \cdot$ in

Force	1 N = 1 kg $\cdot$ m/s^2 = 10^5 dynes = 10^5 g $\cdot$ cm/s^2 = 0.22481 lb_f 1 lb_f = 32.174 $lb_m \cdot ft/s^2$ = 4.4482 N = 4.4482 $\times$ 10^5 dynes

Pressure	1 atm = 1.01325 $\times$ 10^5 N/m^2 (Pa) = 1.01325 bars
	     = 1.01325 $\times$ 10^6 $dynes/cm^2$
	     = 760 mm Hg @ 0°C (torr) = 10.333 m H_2O @ 4°C
	     = 14.696 lb_f/in^2 (psi) = 33.9 ft H_2O @ 4°C
	     = 29.921 inHg @ 0°C

Energy	1 J = 1 N $\cdot$ m = 10^7 ergs = 10^7 dyne $\cdot$ cm
	     = 2.778 $\times$ 10^{-7} kW $\cdot$ hr = 0.23901 cal
	     = 0.7376 ft-lb_f = 9.486 $\times$ 10^{-4} Btu

Power	1 W = 1 J/s = 0.23901 cal/s = 0.7376 ft $\cdot$ lb_f/s = 9.486 $\times$ 10^{-4} Btu/s = 1.341 $\times$ 10^{-3} hp
Concentrations	1 mg/L = $10^3 \mu$ g/L = 10^{-3} g/L = 10^{-6} kg/L = 1 g/m^3 ppm = parts per million = mg/L* ppb = parts per billion = 10^{-3} ppm = μg/L* lb_f/ft^3 = 7,000 grains/ft^3
Thermal conductivity	cm^2/s = 10^{-4} m^2/s
Emission rate	lb/1,000 $ft^3 \cdot$ day = 43.56 lb/acre-ft $\cdot$ day = 133.7 lb/Mgal $\cdot$ day lb_f/ft^3 = 7,000 grains/ft^3

*Converting dimensionless ppm or ppb to mass/volume assumes water is the solvent.

Table D-4 adapted from *Elementary Principles of Chemical Processes*, John Wiley & Sons, 1978, by permission.

INDEX